Ecology, Evolution and Behaviour of Wild Cattle

Implications for Conservation

Covering all thirteen species of wild cattle, the *Ecology, Evolution and Behaviour of Wild Cattle* brings together the contributions of international leading experts on the biology, evolution, conservation status and management of the tribe bovini, providing:

- a comprehensive review of current knowledge on systematic, anatomy and ecology of all wild cattle species (Chapters 1 to 8);
- a clear understanding of the conservation status of each species and the gaps in our current knowledge (Chapters 9 to 20);
- and presenting a number of case studies on conservation activities and investigating some of the most threatened and poorly understood species (Chapters 21 to 27).

An invaluable resource for students, researchers and professionals in behavioural ecology, evolutionary biology and conservation biology, this beautifully illustrated reference work reveals the extraordinary link between wild cattle and humans, the benefits some of these species have brought us, and their key roles in their natural ecosystems.

Mario Melletti is an independent researcher. For more than a decade he has been studying the ecology and behaviour of the forest buffalo, spending more than two years in a remote area in the Dzanga-Ndoki National Park, in the Central African Republic. He has obtained a PhD in animal ecology from the University of Rome in collaboration with the Department of Conservation Biology of Seville. He has collaborated in several projects and surveys on both mammals and birds. He is also a reviewer of many peer reviewed international journals.

James Burton is Chair of the IUCN Species Survival Commission's Asian Wild Cattle Specialist Group, leading a group of volunteer experts to monitor the status of, and facilitate conservation planning and action for, the nine wild cattle and buffalo species in Asia. Since 2000 he has worked on the conservation of the Anoa (*Bubalus* species) in Indonesia, including a PhD from the University of Edinburgh. He has conducted distribution surveys, ecological studies and population genetic research on a range of mammal species. He is currently Senior Research Manager (Oman) at the Earthwatch Institute.

Ecology, Evolution and Behaviour of Wild Cattle

Implications for Conservation

Edited by

Mario Melletti

Independent researcher

James Burton

Senior Research Manager (Oman),

Earthwatch Institute; Chair, IUCN-SSC Asian Wild Cattle Specialist Group

CAMBRIDGE
UNIVERSITY PRESS

CAMBRIDGE
UNIVERSITY PRESS

University Printing House, Cambridge CB2 8BS, United Kingdom

Cambridge University Press is part of the University of Cambridge.

It furthers the University's mission by disseminating knowledge in the pursuit of education, learning and research at the highest international levels of excellence.

www.cambridge.org
Information on this title: www.cambridge.org/9781107036642

First published 2014

A catalogue record for this publication is available from the British Library

Library of Congress Cataloguing in Publication data
Ecology, evolution, and behaviour of wild cattle : implications for conservation / edited by Mario Melletti, independent researcher, James Burton.
p. ; cm.
Includes bibliographical references and index.
ISBN 978-1-107-03664-2 (Hardback)
1. Bos–Conservation. 2. Bos–Evolution. 3. Cattle–Conservation. 4. Cattle–Evolution. 5. Bison–Conservation. 6. Bison–Evolution. I. Melletti, Mario, editor of compilation. II. Burton, James, 1973– editor of compilation.
[DNLM: 1. Animals, Wild–genetics. 2. Cattle–genetics. 3. Cattle Diseases–genetics. 4. Ecological and Environmental Phenomena. SF 756.65]
QL737.U53E255 2014
599.64'22–dc23 2014000263

ISBN 978-1-107-03664-2 Hardback

Additional resources for this publication at www.cambridge.org/9781107036642

To my father Guido Melletti and to my grandfather Amadio Romanelli whose support, encouragement and inspiration throughout my life have allowed me to pursue my dreams and have made this book possible.
Mario Melletti

Contents

List of contributors ix
Foreword by Simon Stuart xiii
Acknowledgements xiv

Introduction 1
Mario Melletti and James Burton

Part I – Systematic, ecology and domestication

1 **Systematic and evolution of Bovini** 7
Alexandre Hassanin

2 **Bovini as keystone species and landscape architects** 21
Herbert H. T. Prins and Herman van Oeveren

3 **Domestic cattle and buffaloes** 30
Johannes A. Lenstra, Marleen Felius and Bert Theunissen

Part II – Evolution, anatomy and function

4 **The evolution and skeletal anatomy of wild cattle (Bovini)** 39
Kris Kovarovic and Robert S. Scott

5 **Bovine mammary anatomy and function** 51
R. Michael Akers, Anthony V. Capuco and Stephen C. Nickerson

6 **The digestive system of ruminants, and peculiarities of (wild) cattle** 57
Marcus Clauss and Reinhold R. Hofmann

7 **Anatomy and evolution of teeth** 63
James Heywood

8 **Evolution, development and functional role of horns in cattle** 72
Edward Byrd Davis, Katherine A. Brakora and Kelsey Tull Stilson

Part III – Species accounts

9 **American bison *Bison bison* (Linnaeus, 1758)** 83
Glenn E. Plumb, P. J. White and Keith Aune

10 **European bison *Bison bonasus* (Linnaeus, 1758)** 115
Małgorzata Krasińska, Zbigniew A. Krasiński, Kajetan Perzanowski and Wanda Olech

11 **Gaur *Bos gaurus* C. H. Smith, 1827** 174
Farshid S. Ahrestani and K. Ullas Karanth

12 **Wild yak *Bos mutus* (Przewalski, 1883)** 194
Jianlin Han

13 **Banteng *Bos javanicus* d'Alton, 1823** 216
Penny C. Gardner, Satyawan Pudyatmoko, Naris Bhumpakphan, Marnoch Yindee, Datuk Laurentius N. Ambu and Benoit Goossens

14 **Kouprey *Bos sauveli* A. Urbain, 1937** 231
Mario Melletti, Alexandre Hassanin and Marzia Mirabile

15 **Aurochs *Bos primigenius* Bojanus, 1827** 240
T. van Vuure

16 **Wild water buffalo *Bubalus arnee* (Kerr, 1792)** 255
Anwaruddin Choudhury

17 **Anoas *Bubalus depressicornis* (C. H. Smith, 1827); *Bubalus quarlesi* (Ouwens, 1910)** 302
Philip M. Wheeler, Abdul Haris Mustari and James Burton

18 **Tamaraw *Bubalus mindorensis* Heude, 1888** 310
Merben R. Cebrian, Rodel M. Boyles, Josefina L. de Leon and James Burton

19 **Saola *Pseudoryx nghetinhensis* Dung, Giao, Chinh, Touc, Arctander & Mackinnon, 1993** 318
William Robichaud and Barney Long

20 **African buffalo *Syncerus caffer* (Sparrman, 1779)** 326
Daniel Cornélis, Mario Melletti, Lisa Korte, Sadie J. Ryan, Marzia Mirabile, Thomas Prin and Herbert H. T. Prins

Part IV – Conservation and management

21 **What is a wild bison? A case study of plains bison conservation in Canada** 373
C. Cormack Gates

22 **The case study: the restitution of the wisent (*Bison bonasus*) to the Carpathians** 385
Kajetan Perzanowski and Wanda Olech

23 **Capture and translocation of gaur (*Bos gaurus*) in India** 393
Parag Nigam, Sankar Kalyansundaram, Dave Cooper, Les Carlisle and Harbhajan Singh Pabla

24 **Status and management of the endangered wild water buffalo (*Bubalus arnee*) in the Koshi Tappu Wildlife Reserve, Nepal** 403
Ram Chandra Kandel, J. Stuart F. Barker and Mario Melletti

25 **Genetic structure of the African buffalo (*Syncerus caffer*) at continental and population scales: an evolutionary and conservation approach** 410
Nathalie Smitz, Daniel Cornélis, Philippe Chardonnet, Ettore Randi and Johan Michaux

26 **Livestock and buffalo (*Syncerus caffer*) interfaces in Africa: ecology of disease transmission and implications for conservation and development** 431
Richard Kock, Michael Kock, Michel de Garine-Wichatitsky, Philippe Chardonnet and Alexandre Caron

27 **Ex situ conservation of wild cattle: roles, status, management successes and challenges** 446
Daniel C. de Man

Index 459

Colour plates are to be found between pp. 272 and 273.

Contributors

Farshid S. Ahrestani, PhD,
Department of Ecology, Evolution & Environmental Biology, Columbia University, 10th Floor, Schermerhorn Ext., 1200 Amsterdam Avenue, New York 10027, USA.

R. Michael Akers,
Professor, Department Head Virginia Tech Department of Dairy Science, 2470 Litton Reaves Hall, Blacksburg VA 24061, USA.

Datuk Laurentius N. Ambu, PhD,
Director, Sabah Wildlife Department, 5th Floor, Block B, Wisma Muis 88100 Kota Kinabalu, Sabah, Borneo, Malaysia.

Keith Aune, PhD,
Wildlife Conservation Society, Senior Conservation Scientist, 301 N. Wilson, Bozeman MT 59715, USA.

J. Stuart F. Barker,
Professor, School of Environmental and Rural Science, University of New England, Armidale NSW 2351, Australia.

Naris Bhumpakphan, PhD,
Associate Professor, Department of Forest Biology, Faculty of Forestry, Kasetsart University, 50 Ngamwongvan Rd, Jatuchak, Bangkok 19000, Thailand.

Rodel M. Boyles,
Coordinator, Tamaraw Conservation Program, Department of Environment and Natural Resources, regional office IVB (MIMAROPA) Airport Road, San Roque I, San Jose, Occidental Mindoro 5100, Philippines.

Katherine A. Brakora, PhD,
Postdoctoral scholar, Department of Orthopedic Surgery, School of Medicine, University of California, San Francisco, CA 94110, USA.

James Burton, PhD,
Chair, IUCN-SSC Asian Wild Cattle Specialist Group. Earthwatch Institute, 256 Banbury Road, Oxford OX2 7D, UK.

Anthony V. Capuco, PhD,
Bovine Functional Genomics Lab Bldg 200, Rm 14, Powder Mill Rd, BARC-East Beltsville MD 20705, USA.

Les Carlisle,
Group Conservation Manager, & BEYOND, Post Box 966, White River 1240 Johannesburg, South Africa.

Alexandre Caron, PhD,
Researcher, CIRAD, Department of Environment and Societies, UPR AGIRs, 34398 Montpellier, France. Mammal Research Institute, University of Pretoria, South Africa. Based at the Research Platform Production and Conservation in Partnership, University of Zimbabwe.

Merben R. Cebrian,
USDOI Bureau of Land Management, 1150 University Avenue, Fairbanks, Alaska 99709, USA.

Philippe Chardonnet, PhD,
Wildlife Veterinarian, Director Fondation Internationale pour la Gestion de la Faune (Fondation IGF) 58 rue Beaubourg 75003 Paris, France.

Anwaruddin Choudhury, PhD, DSc,
IUCN-AWCSG, the Rhino Foundation for Nature in North East India Assam, House No. 7, Islampur Road, Guwahati 781007 (Assam), India.

Marcus Clauss,
Professor, Dr, MED, VET, Clinic for Zoo Animals, Exotic Pets and Wildlife, Vetsuisse Faculty, University of Zurich, Winterthurerstr. 260 CH-8057 Zurich, Switzerland.

Dave Cooper,
Senior Wildlife Veterinarian, Game Capture Unit, Ezemvelo KwaZulu-Natal Wildlife, Private Bag X01 St Lucia 3935, South Africa.

Daniel Cornélis, PhD,
Senior Researcher, CIRAD, Department of Environment and Societies, UPR AGIRs, 34398 Montpellier, France. Based at the Research Platform Production and Conservation in Partnership, University of Zimbabwe, Harare, Zimbabwe.

Edward Byrd Davis, PhD,
Fossil Collections Manager, Museum of Natural and Cultural History and Department of Geology, University of Oregon, Eugene OR 97403, USA.

Michel de Garine-Wichatitsky, PhD, DVM,
Senior Researcher, CIRAD, UPR AGIRs, TAC-EE/E, 34398 Cedex 5 Montpellier, France. Based at the Research Platform 'Production and Conservation in Partnership', University of Zimbabwe, Harare, Zimbabwe.

Josefina L. de Leon,
Chief, Wildlife Resources Division, Biodiversity Management Bureau, Department of Environment and Natural Resources. Quezon Avenue, Diliman, Quezon City, Philippines.

Daniel C. de Man,
Manager, Collection Coordination and Conservation, European Association of Zoos and Aquaria (EAZA), Amsterdam, the Netherlands, African buffalo European studbook (ESB) keeper, Banham Zoo, United Kingdom.

Marleen Felius,
Mauritsstraat 167, 3012 CH Rotterdam, the Netherlands.

Penny C. Gardner,
Bornean Banteng Programme, Cardiff University, United Kingdom; and Danau Girang Field Centre, Sabah Wildlife Department, Borneo, Malaysia.

C. Cormack Gates, PhD,
Faculty of Environmental Design, University of Calgary, 2500 University Drive NW, Calgary, Alberta T2N 1N4, Canada.

Benoit Goossens,
Danau Girang Field Centre, c/o Sabah Wildlife Department, Wisma Muis, Block B, 5th Floor, Sabah Borneo, Malaysia.

Jianlin Han, PhD,
CAAS-ILRI Joint Laboratory on Livestock and Forage Genetic Resources (JLLFGR), Institute of Animal Science (IAS), Chinese Academy of Agricultural Sciences (CAAS) No. 2, Yuan Ming Yuan Xi Lu, Haidian District Beijing 100193, P.R. China.

Alexandre Hassanin, PhD,
Muséum national d'Histoire naturelle (MNHN), Service de Systématique Moléculaire, Case postale N° 2643, rue Cuvier 75005 Paris, France.

James Heywood, PhD,
Department Paediatrics, Box 116, University of Cambridge, Addenbrooke's Hospital, Hills Road, Cambridge CB2 0QQ, UK.

Reinhold R. Hofmann,
Professor, Dr, MED, VET, Founding Director of the Leibniz-Institute of Zoo and Wildlife Research (IZW) Berlin, Germany.

Luong Viet Hung,
WWF Vietnam, 150 Truong Gia Mo Street Hue, Vietnam.

Sankar Kalyansundaram, PhD,
Wildlife Institute of India, PO Box 18, Chandrabani, Dehra Dun 248 001 Uttarakhand, India.

Ram Chandra Kandel,
Department of National Parks and Wildlife Conservation, PO Box 860 Kathmandu, Nepal.

K. Ullas Karanth, PhD,
Wildlife Conservation Society-India and Centre for Wildlife Studies, 26-2 Aga Abbas Ali Road Bengaluru (Bangalore), Karnataka 560042, India.

Michael Kock,
Senior Lecturer in Wildlife, Department of Production Animal Studies, Faculty of Veterinary Science, University of Pretoria, Onderstepoort, Gauteng, South Africa.

Richard Kock,
Professor of Wildlife Health and Emerging Diseases, Department of Pathology and Infectious Diseases, Royal Veterinary College, Hawkshead Lane North Mymms, Hertfordshire AL9 7TA, UK.

Lisa Korte, PhD,
Gabon Biodiversity Program Director, Centre for Conservation Education and Sustainability, Smithsonian Conservation Biology Institute, National Zoological Park MRC 0705, Washington DC 20013-7012, USA.

Kris Kovarovic, PhD,
Department of Anthropology, Durham University, South Road, Durham DH1 3LE, UK.

Małgorzata Krasińska,
Professor, Mammal Research Institute PAS, 17–230 Białowieża, Waszkiewicza str. 1, Poland.

Zbigniew A. Krasiński,
Professor, 17–230 Białowieza, Browska str. 1 B, Poland.

Johannes A. Lenstra, PhD,
Associate Professor, Faculty of Veterinary Medicine, Utrecht University Yalelaan 104, 3584 CM Utrecht University, the Netherlands.

Barney Long, PhD,
World Wildlife Fund, 1250 24th Street, NW, Washington DC 20037-1193, USA.

Mario Melletti, PhD,
Independent Researcher Via di Villa Chigi, 00199 Rome, Italy.

Johan Michaux, PhD,
Centre de Biologie et de Gestion des Populations (CBGP), UMR 1062, Campus International de Baillarguet, CS 30016, F-34988 Montferrier-le-Lez, France.

Marzia Mirabile, MSc,
Via Gian Pietro Talamini 00128 Rome, Italy.

Abdul Haris Mustari,
Department of Forest Resources Conservation, Faculty of Forestry, Bogor Agricultural University, Kampus IPB Darmaga Bogor 16680, West Java, Indonesia.

Stephen C. Nickerson,
Professor of Dairy Science, Animal and Dairy Science Department, Edgar L. Rhodes Centre for Animal and Dairy Science, University of Georgia, Athens GA 30602-2771, USA.

Parag Nigam,
Wildlife Institute of India, PO Box 18, Chandrabani, Dehra Dun – 248 001 Uttarakhand, India.

Wanda Olech,
Professor, Department of Genetics and Animal Breeding, Warsaw University of Life Sciences, 02-786 Warsaw, Ciszewskiego str. 8, Poland.

Harbhajan Singh Pabla,
Former Principal Chief Conservator of Forest (WL) & Chief Wildlife Warden, Madhya Pradesh, India.

Kajetan Perzanowski,
Carpathian Wildlife Research Station MIZ PAS, 38-700 Ustrzyki Dolne, Ogrodowa str. 10; Chair of Applied Ecology, Catholic University of Lublin, 20-708 Lublin Konstantynów str. 1H, Poland.

Glenn E. Plumb, PhD,
Chief Wildlife Biologist, National Park Service Natural Resource Stewardship and Science Directorate, Biological Resource Management Division, Wildlife Conservation Branch, 1201 Oakridge Drive, Fort Collins CO 80525, USA.

Thomas Prin, PhD,
Researcher, CIRAD, IGF Foundation, UMR CNRS 5558 – LBBE Université Claude Bernard Lyon 1, based at CIRAD – UPR AGIRs Department of Environment and Societies 34398 Montpellier Cedex 5, France.

Herbert H. T. Prins,
Professor, Resource Ecology Group, Wageningen University, Droevendaalsesteeg 3a, 6708 PB Wageningen, the Netherlands.

Satyawan Pudyatmoko,
Fakultas Kehutanan Universitas Gadjah Mada, Bulaksumur 55281,Yogyakarta, Indonesia.

Ettore Randi,
Conservation Genetics, Institute for Environmental Protection and Research (ISPRA), Via Cà Fornacetta 9, 40064 Ozzano dell'Emilia, Italy. Department 18/Section of Environmental Engineering. Aalborg University, Aalborg, Denmark.

William Robichaud,
IUCN/SSC Asian Wild Cattle Specialist Group; Coordinator Saola Working Group, c/o Global Wildlife Conservation PO Box 129, Austin TX 78767, USA.

Sadie J. Ryan, PhD,
134 Illick Hall, SUNY College of Environmental Science and Forestry (SUNY ESF), 1 Forestry Drive, Syracuse NY 13210, USA.

Robert S. Scott, PhD,
Department of Anthropology, School of Arts and Sciences, Rutgers, State University of New Jersey, 3rd floor RAB, 131 George Street, New Brunswick NJ 08901, USA.

Nathalie Smitz, PhD,
Conservation Genetics, University of Liège, Boulevard du Rectorat 26, 4000 Liège, Belgium.

Kelsey Tull Stilson,
c/o Edward Davis, Department of Geological Sciences 1272, University of Oregon Eugene, OR 97403, USA.

Bert Theunissen,
Professor, Descartes Centre for the History and Philosophy of the Sciences and the Humanities, Utrecht University, PO Box 80.010, 3508 TA Utrecht, the Netherlands.

Van Ngoc Thinh,
WWF-Vietnam, D13 Thang Long International Village, Cau Giay, Hanoi, Vietnam.

Le Ngoc Tuan,
Thua Thien – Hue Saola Nature Reserve, Son Thuy commune, A Luoi distict, Thua Thien Hue, Vietnam.

Herman van Oeveren,
Resource Ecology Group, Wageningen University Droevendaalsesteeg 3a, 6708 PB Wageningen, the Netherlands.

T. van Vuure, PhD,
Nude 45, 6702 DK Wageningen, the Netherlands.

Philip M. Wheeler, PhD,
Centre for Environmental and Marine Sciences, University of Hull Scarborough Campus, Filey Road, Scarborough YO11 3AZ, UK.

P. J. White, PhD,
Yellowstone National Park, PO Box 168, Mammoth WY 82190, USA.

Marnoch Yindee, DVM, PhD,
Department of Clinical Science and Public Health, Faculty of Veterinary Science, Mahidol University, Salaya Campus, 999 Phutthamonthon Sai 4 Road Salaya, Phutthamonthon Nakhonpathom 73170, Thailand.

Illustrators

• **Francesco Rinaldi,**
Via A. Panerai 36, 52037 Sansepolcro (Arezzo), Italy (12 colour illustrations).

• **Luciano Toma, PhD,**
Entomologist, Department of Infectious, Parasitic and Immuno-mediated Diseases – Section of Vector Borne Diseases and International Health – Istituto Superiore di Sanità, Viale Regina Elena 299, 00161 Rome, Italy (12 black/white illustrations).

• **Enrico Chiarelli,**
Via Rosa Govona 2, 00152 Rome, Italy (two greyscale illustrations).

Distribution Maps

• **Piero Visconti, PhD,**
Global Mammal Assessment programme, Department of Biology and Biotechnologies, University of Rome 'La Sapienza' Viale dell'Università 32, 00185 Rome, Italy.

Foreword

As the most common type of large domesticated ungulate, cattle play an integral role in supporting the world's human population, whether it is through food (meat, dairy products), the production of leather, dung for fuel or manure or as draft animals for agricultural use (mainly in poorer and less-developed countries). The Asian water buffalo is probably the most commonly used of all draft cattle, with as many as 165 million domesticated for the production of rice throughout tropical Asia.

Wild cattle are often overlooked, despite there being a number of incredibly striking species and despite the fact that most of the modern domesticated species of *Bos* are thought to have evolved from a single ancestor – the aurochs (*Bos primigenius*). This is possibly a result of the domestic cow somewhat overshadowing their existence due to its economic value; also, indeed visually, many domesticated breeds may be hard to distinguish from some of their wild cattle relatives which share many of the same morphological characteristics.

There are 13 species of wild cattle currently recognised. Some of these species are listed as Critically Endangered on the IUCN Red List of Threatened Species; one may now possibly be Extinct (the kouprey, *Bos sauveli*, has not been seen since the 1980s); and the last aurochs was recorded in Poland in 1627, having been hunted to extinction. The major threats impacting on the survival of all wild cattle are largely attributable to human activities, ranging from direct exploitation (hunting and trade), to the increased pressures of expansion, development and encroachment.

Undoubtedly, the effects of unregulated hunting have caused a rapid (and possibly irreversible) decline in the numbers of wild cattle now remaining. The kouprey is one of the most seriously threatened mammals in the world, with the number of living individuals not known – as noted above, it is thought possibly to now be extinct. As a desirable and extremely valuable target for trophy hunters, the kouprey has fallen victim to a trade fuelled by the demand for trophy horns and traditional medicines.

Conservation must be built on sound knowledge and scientific research. The IUCN Species Survival Commission (SSC) is a science-based network of more than 8,000 volunteer experts from almost every country of the world. SSC members are deployed in over 130 Specialist Groups, and it is one of these groups – the IUCN SSC Asian Wild Cattle Specialist Group – which has been instrumental in raising the profile of Asian wild cattle species and is committed to prioritising their long-term conservation to ensure their continued survival.

Over the years, I have been fortunate enough to see some of the remaining extremely impressive wild cattle species in their native environment. However, as these years have passed, I have also been unfortunate to witness the severe impacts human activities are having on a number of these species, particularly throughout Asia, where they now cling on in small, fragmented and isolated populations which are becoming increasingly non-viable.

I welcome this publication as a much-needed resource which, for the first time, will provide fundamental information for all 13 species of wild cattle and help to guide targeted conservation action. I also feel greatly encouraged that this book also represents the determination, skills and dedication of international experts who have contributed their invaluable knowledge and research to enable it to be written.

As with so many of the world's species faced with threats to their long-term existence, the future of the surviving 12 wild cattle species will depend on reconciling the needs of people and nature: to find pragmatic solutions that balance human population growth and expansion with the needs of wildlife and ecosystems this may conflict with. I hope that through this book we can raise the profile of wild cattle and help people to better understand their value and the urgency to protect this group of charismatic and impressive species.

Dr Simon Stuart
Chair, IUCN SSC

Acknowledgements

First of all, this book would not have been possible without the huge amount of collaboration and dedication of many people.

MM is immensely grateful to his friends Francesco Rinaldi, Luciano Toma and Enrico Chiarelli, who have worked for a year and a half and have made this book beautifully illustrated.

MM is deeply in debt to Vincenzo Penteriani, Maria Del Mar Delgado and Marzia Mirabile, who continually helped and advised him during the years of this work.

A special thanks goes to Piero Visconti of the Global Mammal Assessment Programme (Department of Biology and Biotechnologies, University 'La Sapienza' Rome), who modified the distribution maps on the basis of IUCN data.

We are extremely grateful to Professor Luigi Boitani (Department of Biology and Biotechnologies of University 'La Sapienza' Rome) for the useful comments on the first stage of this book.

We are particularly grateful to Hannah Henshaw and James McConnell for their proofreading of several chapters.

A special thanks goes to Dr Craig Hilton-Taylor, Head Red List Unit and to Katherine Rewinkel El-Darwish of the Global Communications Unit, IUCN, for their kind help to get the permissions for the distribution maps.

The editors are grateful to Dr Simon Stuart, Chair of IUCN SSC for writing the Foreword and to Rachel Roberts, SSC Network Coordination Officer, for the arrangement of the Foreword.

We are extremely grateful to all Cambridge staff; Martin Griffiths (Commissioning Editor) for his help during the approval process; Lynette Talbot (Assistant Editor) and Edward Bailey (Publishing Assistant) for the initial guidelines provided; Renee Duncan-Mestel (Publishing Assistant) and Ilaria Tassistro (Assistant Editor) for their crucial help during the final stages of the book. Also, a big thank you to Gary Smith for copyediting this manuscript.

For organising workshops where information, especially for mapping, was collated, we would like to thank: Miguel Pedrono (Southeast Asia maps), Christy Williams WWF-Nepal (South Asia), Tony Suhartono, Kishor Rithie, Vivek Menon, Rahul Kaul (Central India wild buffalo), Ikeu Sri Rejeki and IUCN AWCSG members for contributing information as follows: Farshid Ahrestani, Irzal Azharand, Tim Bean, Naris Bhumpakphan, Rodel M. Boyles, Le Xuan Canh, Philippe Chardonnet, Anwaruddin Choudhury, Vu Chi Cuong, Josefina de Leon, Thomas Gray, Bach Thanh Hai, Nguyen Thi Hanh, Simon Hedges, Joel Heinen, Saw Htun, Han Jianlin, Budsabong Kanchanasaka, Sun Kong, Barney Long, François Monicat, Abdul Haris Mustari, Hannah O'Kelly, Bouaphanh Phanthavong, Satyawan Pudyatmoko, William Robichaud, Ronilo Salac, Chairul Saleh, Sompoad Srikosamatara, Mattana Srikrachang, Rob Steinmetz, Herry Susilo, Dennis Ten, Robert Timmins, Tran Van Thanh, Nguyen Van Thanh, Le Thi Thuy, Frédéric Vallejo, Rosie Woodroffe and Choon Yung.

We would also like to thank Carmi G. Penny and Randy G. Rieches of the Husbandry Science/Curators of Mammals, San Diego Zoo; Lisa Smith, Director of Animal Programs Great Plains Zoo & Delbridge Museum of Natural History; and Pauline Buffle, IUCN Network Support Officer on Climate Change, for their availability and help.

MM is also grateful to Claudio Chiarelli and Bert Keightley for the permission to collect buffalo samples for genetic analyses and for the beautiful days spent together in the Zambezi Valley.

We would like to thank for their help during different stages of the book and during forest buffalo research in CAR: Will Duckworth, Andrea Turkalo, Francesco, Maria Isabella, Marta and Simone Verardi, Jacopo, Chiara, Alessandro, Kuma, India, Gaia, Red, Marco, Lucia and Luca Melletti, Maria Diletta and Alberto Melis, Andrea Rossi, Maria Teresa Paliotta, Igor Krugersberg, Gianfranco Tortellini, Luca Policastri, Giovanni Colicci, Alfredo Guillet, Jonathan S. Kingdon, Joe Walston, Sabrina Curran, Shelly Masi, Chloè Cipolletta, David Greer, Angelique Todd, Fiona Maisels, Lisa Steel, C. Oertle, D. Renner, G. Schwarzer, Noemi Spagnoletti, Any, Benzino, Bokaya, David, Mokoko, Arianna Manciocco, Riccardo Scalera, Anthony R. E. Sinclair, Daniel G. Bradley, John Williams and Paul Boettcher.

Thanks are also due to Luca Innocenzi for the editing of digital drawings, to Daniele and Andrea Guerrini for the digital scans of drawings, to Giuseppe Corte for editing some of the figures and to Giovanna Pentassuglia and Alessandro

Riccio of Historia Naturae (www.historianaturae.com) for useful advice.

Thanks go to Carlo Rondinini and Luigi Maiorano of the Global Mammal Assessment Programme (Department of Biology and Biotechnologies, University 'La Sapienza' Rome) for the useful discussion on forecast scenarios of wild cattle loss.

Last, but not least, MM would like to express his gratitude to his family, in particular his parents Guido Melletti and Carla Romanelli and his grandparents Mario Melletti, Mafalda Sperandini, Amadio Romanelli and Giovanna Gambini, whose support and enthusiasm have been crucial throughout his life and during the years of this project.

Introduction

Mario Melletti and James Burton

This is the first comprehensive work combining the state of knowledge of every known wild cattle species. It covers the biology, taxonomy, evolution, conservation status and management of all 13 species. This book is addressed to researchers, conservationists and anyone with an interest in wild cattle. We hope the work will inform and inspire greater interest and passion for this group of species to help in our understanding of them and in their successful conservation.

Wild cattle belong to an amazing group of animals that have fascinated humans since ancient times. There are species that are very well known and charismatic, such as the American bison (*Bison bison*), but also species that are very little known to humans, living in remote forest areas, such as the saola (*Pseudoryx nghetinhensis*) in Vietnam and Lao PDR and the anoas species (*Bubalus depressicornis* and *Bubalus quarlesi*) on the island of Sulawesi, Indonesia (see also Table 1).

The link between wild cattle and humans has existed for thousands of years. Five species of wild cattle have been domesticated in the last 7000 years, including the wild water buffalo and the yak. This domestication has provided many benefits to humans, from meat and milk to draught animals (see Chapter 3 for more details). They are therefore valuable today as a genetic pool for ensuring that domestic breeds can adapt to changing pressures, such as climate and disease. Also, some species may have key roles in their natural ecosystems. They are not only an important prey species for predators, but also play an interesting role in maintaining ecosystems by grazing and wallowing. For instance, wild yak at high densities may change grassland community (Qiao *et al.* 2009). It is also known that some other species, such as American bison and African buffalo (*Syncerus caffer*), similarly to yak, may increase grass species diversity in the prairies and savannas, respectively. Thus, these species play an important role in ecosystem management. So wild cattle deserve further investigation in the future and greater effort to preserve them and the ecosystems they are an integral part of. Conversely, some introduced species need management, such as the feral water buffalo in Australia, which has completely altered the coastal ecosystem and so needs proper management (Werner 2013).

The future survival of many of these species is of great concern, considering that almost one-fifth of extant vertebrate species are classified as Threatened (Hoffmann *et al.* 2010). Furthermore, the distribution ranges of many large mammals have recently contracted substantially in tropical Asia alongside further reductions in diversity (Schipper *et al.* 2008). The region has little tradition of effectively managed protected areas. Consequently, many Southeast Asian species will become extinct in the near future if current trends continue (Duckworth *et al.* 2012). The declining conservation status of many wild cattle species shows the pressures they face and highlights the urgent need for greater effective conservation action. Twelve of the 13 species of the tribe Bovini are threatened to some degree; one is already extinct (aurochs); and another is possibly extinct (kouprey; see Table 1 for the IUCN categories). Asia is the most important and the richest region in terms of wild cattle species. Nine of the 12 species in the world are found in this continent, where there is a need for proper conservation and management measures (see Figure 1 and Table 1 for the distribution by country of all wild cattle species).

Conserving these species is often a challenge, especially in certain regions. This is the case due to our poor knowledge on account of their behaviour, being shy, solitary, nocturnal, wide-ranging or occurring at low densities (Gray 2012). For example, the saola was only discovered by the scientific community in 1992 and was one of the most spectacular and surprising zoological discoveries of the twentieth century (see also Chapter 19). It is now restricted to a small geographic area between Lao PDR and Vietnam, and is highly threatened by illegal hunting. The saola is facing immense pressure in all locations where it is found and without scaling up conservation initiatives this species is likely to be driven to extinction in the coming years, as has probably happened to the kouprey (*Bos sauveli*) in Southeast Asia.

Many steps can be taken to prevent species and habitat loss. For example, efforts should concentrate on increasing effectiveness of protected-areas management where remaining populations of threatened species still persist. For example,

Ecology, Evolution and Behaviour of Wild Cattle: Implications for Conservation, ed. M. Melletti and J. Burton. Published by Cambridge University Press.

Table 1 Worldwide occurrences by country and conservation status of wild cattle according to the 2013 IUCN Red List of Threatened Species (for more details on the species distribution, see the related chapters in this book)

Species	Status (IUCN)	Occurrence	
American bison *Bison bison*[1]	Near Threatened	North America	Canada (Alberta, British Columbia, Manitoba, Northwest Territories, Saskatchewan, Yukon Territory); USA (Alaska, Arizona, California, Colorado, Iowa, Idaho, Illinois, Kansas, Kentucky, Minnesota, Missouri, Montana, Nebraska, North Dakota, Oklahoma, South Dakota, Texas, Utah, Wisconsin, Wyoming)
		Central America	Mexico
European bison *Bison bonasus*[2]	Vulnerable	Europe	Belarus, France, Germany, Lithuania, Poland, Romania, Russia, Slovakia, Ukraine
Gaur *Bos gaurus*	Vulnerable	Asia	Bangladesh, Bhutan, Cambodia, China, India, Lao PDR, Malaysia, Myanmar, Nepal, Thailand, Vietnam
Wild yak *Bos mutus*	Vulnerable	Asia	China, India
Banteng *Bos javanicus*	Endangered	Asia	Borneo, Cambodia, China?, Java, Lao PDR, Malaysia, Myanmar, Thailand, Vietnam
Kouprey *Bos sauveli*	Critically Endangered (possibly Extinct)	Asia	Cambodia, Lao PDR, Thailand, Vietnam
Wild water buffalo *Bubalus arnee*	Endangered	Asia	Bangladesh, Bhutan, Cambodia, India, Myanmar, Nepal, Thailand
Anoa *Bubalus depressicornis/quarlesi*	Endangered	Indonesia	Sulawesi and Buton Islands
Tamaraw *Bubalus mindorensis*	Critically Endangered	Philippines	Mindoro Island (Philippines)
Saola *Pseudoryx nghetinhensis*	Critically Endangered	Asia	Lao PDR, Vietnam
African buffalo *Syncerus caffer*	Least Concern	Africa	Angola, Benin, Botswana, Burkina Faso, Burundi, Cameroon, Central African Republic, Chad, Democratic Republic of Congo, Equatorial Guinea, Ethiopia, Gabon, Ghana, Guinea, Guinea-Bissau?, Kenya, Ivory Coast, Lesotho, Liberia?, Malawi, Mali?, Mozambique, Namibia, Nigeria, Niger, Republic of Congo, Rwanda, Senegal, Sierra Leone?, Somalia, South Africa, Sudan, Swaziland, Tanzania, Togo, Uganda, Zambia, Zimbabwe

[1] All bison herds managed for conservation purpose in North America are included.
[2] Free-ranging and semi-free populations included.
?: Probably Extinct.

the creation and the effective protection of many reserves and national parks along the Carpathian Mountains has allowed restoration of several populations of European bison from 1963 to today, with a trend that is still growing. This may secure the future viability of these populations (for more details, see Chapter 22). The involvement of local people is another crucial aspect that occurs when they are actively involved in conservation activities and understand the potential benefits of the species and the habitats they use for communities' sustainable resource use. An example comes from Vietnam, where a new approach was piloted using hired 'forest guards' coming primarily from communities surrounding the Thua-Thien Hue Saola Nature Reserve (see Box 19.1). The objective of this model is to provide intensive protection to the saola population of the reserve through the removal of direct threats.

One more aim of this book is to identify gaps in our knowledge, so that conservation practitioners can address those species and populations of highest priority for conservation efforts to prevent further declines. Greater global focus is beginning to occur on halting this decline in species, such as the Convention on Biological Diversity's Aichi target 12 – preventing extinctions – in its ten-year strategic plan (www.cbd.int/sp/targets).

Our intention for the book to inform those interested in wild cattle has three elements: (1) the book provides a comprehensive review of knowledge of all wild cattle species; (2) it gives a clear understanding of the conservation status of each

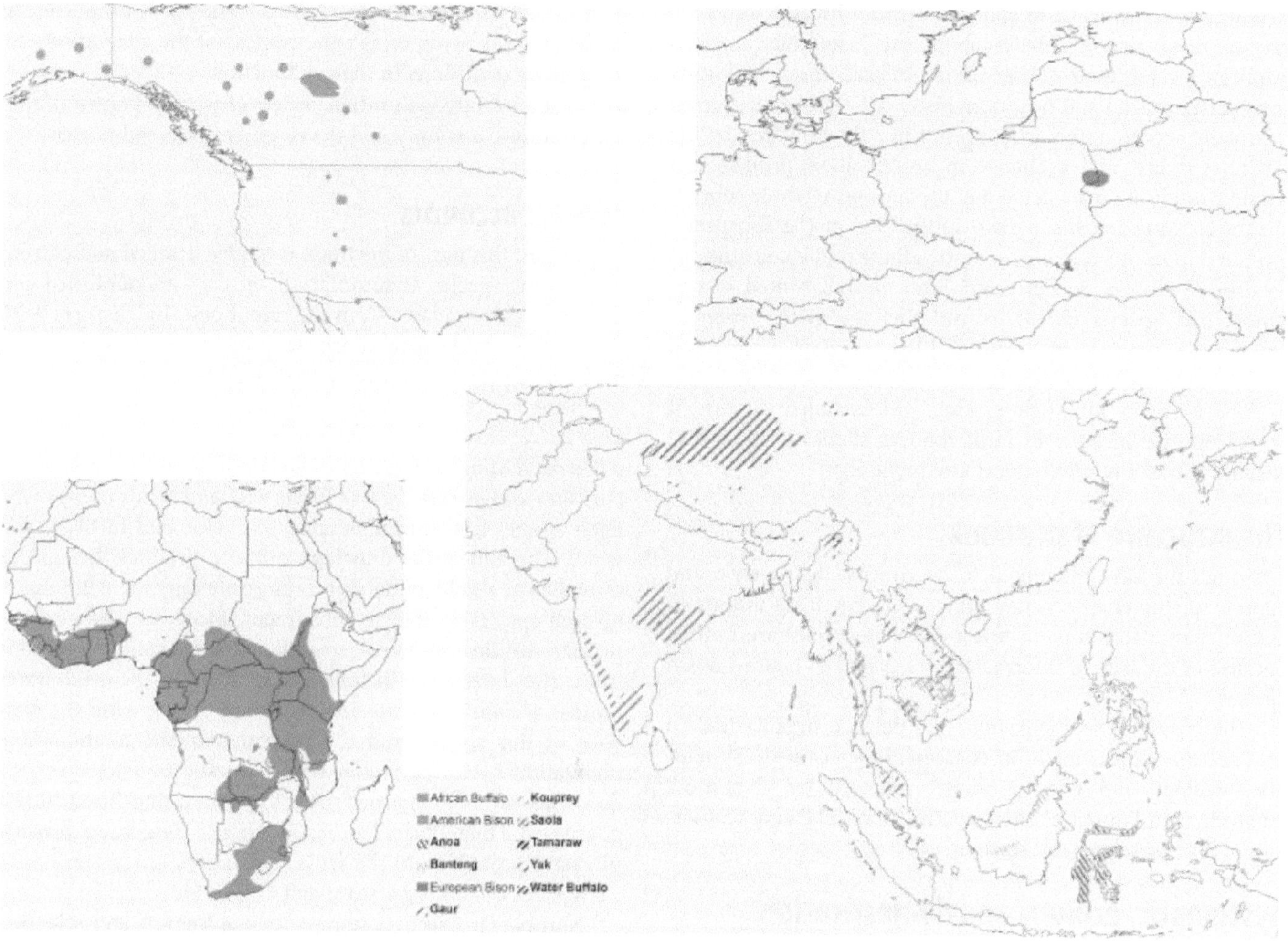

Figure 1 Worldwide distribution map of wild cattle according to the 2008 IUCN Red List of Threatened Species.

species and gaps in current knowledge; (3) it includes a number of case studies of conservation activities, including some of the most threatened and poorly understood species. These three aspects will hopefully contribute towards the conservation and management of wild cattle and their habitats.

General background

The initial stages of the work were begun in 2009, when one of the editors (Mario Melletti) set out to put together a basic idea and a collection of all the major papers and books on wild cattle. This huge amount of information gave an indication that there were so many important gaps in our knowledge of the tribe Bovini that at the beginning the book seemed almost impossible. However, after a year of collecting information, it started to look like the project could get under way, although the work would have to be considerable. The second step was to involve a second editor (James Burton), and work together to build up a suitable group of leading world experts. The third step was to form a team of artists: Francesco Rinaldi, who drew the colour plates of each species; and Luciano Toma and Enrico Chiarelli, who drew the black and white illustrations. This was a long process that occupied the artists for 18 months in an attempt to draw the most realistic animals. The fourth step was to involve a GIS expert, Piero Visconti of the Global Mammal Assessment Programme (University of Rome 'La Sapienza') to build up the distribution maps for all species on the basis of IUCN Red List data. From the beginning, the success of the book has depended on all the various contributors, located in many different countries, affiliated with many institutions including, among others, IUCN SSC, IUCN/SSC Asian Wild Cattle Specialist Groups (AWCSG) and AfBIG (African Buffalo Initiative Group ASG/SSC IUCN. This three-year study involved 70 experts from 21 countries.

It should be stated that this book is not a definitive taxonomic reference. We follow the most traditionally accepted version, such as the work of Wilson & Reeder (2005), *Mammal Species of the World*. For some species, for example the African buffalo, the recent work of Groves & Grubb (2011) separates them into more species – four in the case of the African buffalo. However, this taxonomic complexity remains to be resolved with further molecular studies; currently the editors, the authors and the available literature suggest that there is no strong genetic evidence to split the African buffalo into four species. Yet, current debates about the bison raise serious implications for their conservation. In fact, some biologists consider European and American bison to be the same species. If broadly accepted, this would result in a down-listing by the IUCN Red List and a change in conservation priorities; it would complicate management of the European (three subspecies) and American bison (two subspecies) at the subspecies level. Again in this case we prefer to adhere to the conventions used by Wilson & Reeder (2005) and the IUCN Red List of Threatened Species (2013), considering the genus *Bison* as comprising two species: European bison (*Bison bonasus*) and American bison (*Bison bison*).

The aurochs, although extinct, was included because it represents the ancestor of most modern domestic cattle and because it was a relatively recent extinction.

The structure of the book

Although we have attempted to be consistent, the style and content of the chapters may vary somewhat from author to author. Some accounts on forest-dwelling species are limited because of the lack of information compared to better-known species.

The layout of the book follows a detailed organisation in four sections consisting of 27 chapters. It begins with 'Systematic, Ecology and Domestication', followed by 'Evolution, Anatomy and Function', then 'Species Accounts' and 'Conservation and Management' sections.

Systematic, ecology and domestication

In this part of the book, the following are covered:

- Chapter 1: the ancient and recent hypotheses of interspecific relationships and the systematic position of the tribe Bovini are presented.
- Chapter 2: the role of Bovini as keystone species and landscape architects are presented.
- Chapter 3: the ecology and cultural aspects of domestication through millennia and across different regions of the world are described.

Evolution, anatomy and function

In this part of the book the main characteristics of some anatomical aspects of Bovini are discussed, such as mammary glands, digestive system, horns and teeth in the context of evolution and function. The chapters included within this part are as follows:

- Chapter 4 gives a review of what is known on the skeletal evolution of the entire Bovini tribe.
- Chapter 5 provides an overview of mammary development, anatomy and function.
- Chapter 6 is focused mainly on the general anatomy and morphophysiological adaptation of the digestive system in cattle.
- Chapter 7 describes a generalised bovine post-canine dental anatomy, placing this in the context of the adaptation of grazing dentitions to dietary function.
- Chapter 8 offers a unified review of what is known of the evolution, anatomy and development of bovid horns.

Species accounts

The aim of this part of the book is to give detailed and current data on the species. It summarises most of the published and grey literature available from the recent past. In Chapters 9–20 the 13 wild cattle species are described in detailed accounts prepared by one or more specialists. The chapters treat the following topics: *names*; *taxonomy*; *subspecies and distribution*; *descriptive notes*; *habitat*; *movements and home range*; *activity patterns*; *feeding ecology*; *reproduction and growth*; *behaviour*; *parasites and diseases*; *status in the wild and status in captivity*. Each species has been illustrated in colour and in black and white. The aim of the drawings is not to produce the age/sex classes as in a field guide, but to illustrate only the adult males of each species in their environment. Moreover, we also use pictures to illustrate most species and their habitats. In addition, two boxes are included: Box 16.1, in the wild water buffalo account, presents an interesting overview on the genetics of this species and the potential threats to the whole population by introgression with domestic buffalo. Box 19.1 describes the new approach for Vietnam using hired 'forest guards' in Thua-Thien Hue Saola Nature Reserve to provide intensive protection to the saola population.

Species accounts are arranged as follows:

Names: This section starts with the English and scientific names. Names are also given in French, German, Spanish and Italian. Where available, names in other languages are also provided. Common and scientific names generally follow the work of Wilson & Reeder's *Mammals Species of the World* (2005). Montgelard *et al.* (1997) proposed placing all species of Artiodactyla and Cetacea into the same order, called Cetartiodactyla, to render the classification compatible with the molecular phylogeny. We follow this classification, which is also adopted by the IUCN Red List of Threatened Species (2013).

Taxonomy: Taxonomic issues are discussed and, where genetic information is available, species relationships are assessed on the basis of the molecular and genetics studies.

Subspecies and distribution: In this section the global distributions of the species are described and changes from the past ranges are discussed. Where available, detailed information for each country is provided. The subspecies universally recognised are described and, in some cases, such as the African buffalo and banteng, the most recognised are treated in detail. For the distribution maps we used a combination of IUCN (2013) maps with the addition of data from the IUCN-SSC/AWCSG (2010) report. The maps cover the world distribution of the species. The information provided in the maps should always be taken in conjunction with this section, which often gives details on the distribution by countries or geographical features that may help to determine the location of the species.

Descriptive notes: The main morphological characteristics are summarised, including body, horns and hair characteristics (e.g. geographical variation in body/horns size, coat colour) and lifespan (e.g. maximum age recorded in the wild and in captivity, if known). The dental formula is provided as I = incisors, C = canines, P = premolars, M = molars. The numbers refer to the upper and lower side of the half of the jaw as described in this example: I 0/3 C 0/1 P 3/3 M 3/3 (×2) = 32.

Habitat: This section provides a description of the broad ecological requirements of the species and the factors affecting its distribution and the habitat used are discussed.

Movements and home range: In this section movement and home range are described and, when data are available in relation to age/sex classes, food resources, season and habitat type, these are provided.

Activity patterns: The general activity patterns and variation according to age/sex classes, food resources, period of day, seasons and habitat type are discussed.

Feeding ecology: This section describes the main feeding requirements of the species and, when available, the list of plant species eaten are given.

Reproduction and growth: Some aspects of reproduction and growth are analysed, at least for the most known species.

Behaviour: This section focuses on certain behavioural traits of the species, such as herd and individual behaviour, agonistic behaviour, sexual behaviour and sexual segregation.

Parasites and diseases: The most common diseases are described, including foot and mouth disease (FMD) and bovine tuberculosis (BTB). Where information is available, the most common parasite infections are briefly discussed.

Status in the wild: The status and conservation of the species worldwide is one of the main themes of this book, thus considerable efforts have been made to provide informative and up-to-date information. The current status of species and subspecies are described. This section also provides examples of the conservation measures, *in situ* and *ex situ*, that can be adopted to secure the survival of the species. In addition, information on legislation and classification within the IUCN Red List is given.

Status in captivity: This section discusses the status of the species that are kept in zoos, their breeding success and the size of the captive populations. Most information is obtained from the World Zoo and Aquarium Conservation Strategy (WZACS) and members of the European Association of Zoos and Aquaria (EAZA).

Reference: Each chapter aims to include most of the important literature in the form of papers, books, monographs, published and unpublished reports, grey literature, doctoral and master's theses. One of the aims of the reference section is to provide the reader with an idea of how much information is currently available on each species and give an overview of the state of our knowledge.

Conservation and management

In this section we have arranged a set of accounts, including some case studies, giving an overview on conservation, management, genetics and disease ecology.

In Chapter 21 the status of 'wild' plains bison in Canada is addressed in response to the questions: what is a wild plains bison, and are there any wild plains bison in Canada?

In Chapter 22 the amazing history of European bison restoration in the Carpathian Mountains is described.

In Chapter 23 the first project to capture and translocate gaur, conducted in India, is presented.

In Chapter 24 a conservation and management strategy of the only remaining small population of possibly pure wild water buffalo in Koshi Tappu Wildlife Reserve, Nepal, is presented.

Chapter 25 describes the genetic structure of the African buffalo on a continental scale, giving an interesting scenario on the complex relationships between different subspecies.

In Chapter 26 the importance of sanitation at buffalo–cattle interfaces is described for Africa in the face of intense agricultural and human development.

Finally, Chapter 27 describes the various roles that ex situ populations of wild cattle play in conservation of these species. An overview of wild cattle held in zoological collections and background on the status of these populations is discussed.

References

Duckworth, J. W., Batters, G., Belant, J. L., *et. al.* (2012). Why South-East Asia should be the world's priority for averting imminent species extinctions, and a call to join a developing cross-institutional programme to tackle this urgent issue. *SAPIENS* 5(2): 77–95.

Gray, T. N. E. (2012). Studying large mammals with imperfect detection: status and habitat preferences of wild cattle and large carnivores in eastern Cambodia. *Biotropica* **44**: 531–536.

Groves, C. & Grubb, P. (2011). *Ungulate Taxonomy.* Baltimore, MD: John Hopkins University Press.

Hoffmann, M., Hilton-Taylor, C., Angulo, A., *et al.* (2010). The impact of conservation on the status of the world's vertebrates. *Science* **330**: 1503.

IUCN (2013). *IUCN Red List of Threatened Species.* Version 2013.2. www.iucnredlist.org, accessed 29 May 2014.

IUCN-SSC Asian Wild Cattle Specialist Group (2010). *IUCN Regional Conservation Strategy for Wild Cattle and Buffaloes in South-east Asia, 2011 to 2020.*

Montgelard, C., Catzeflis, F. M. & Douzery, E. (1997). Phylogenetic relationships of artiodactyls and cetaceans as deduced from the comparison of cytochrome b and 12S rRNA mitochondrial sequences. *Molecular Biology and Evolution* **14**: 550–559.

Qiao, Y., Wang, Q. & Wang, W. (2009). Yak (*Bos grunniens* L.) grazing effects on vegetation of alpine meadow with *Potentilla fruticosa* L. (Rosaceae) shrub in

Qinghai-Tibet Plateau. *Polish Journal of Ecology* **57**: 769–777.

Schipper, J., Chanson, J. S., Chiozza, F., *et al.* (2008). The status of the world's land and marine mammals: diversity, threat, and knowledge. *Science* **322**: 225.

Werner, P. A. (2013). The rise and fall of the Asian water buffalo in the monsoonal tropics of Australia. In *Invasion Ecology and Ecosystem Theory: Insights from a Continent in Transformation.* Eds. Prins, H. H. T. & Gordon, I. J. Cambridge: Cambridge University Press.

Wilson, D. E. & Reeder, D. M. (2005). *Mammal Species of the World.* Baltimore, MD: John Hopkins University Press.

Part I Systematic, ecology and domestication

Chapter 1

Systematic and evolution of Bovini

Alexandre Hassanin

Introduction

According to the IUCN (2013), 13 species are currently recognized in the taxonomic group to which wild cattle belong (tribe Bovini): aurochs (now only represented by domestic breeds), banteng (in Southeast Asia), gaur (in India and mainland Southeast Asia), kouprey (in northern and eastern Cambodia and adjacent countries), yak (throughout the Tibetan Plateau), American bison (in the western half of North America), European bison (now confined to Eastern Europe), African buffalo (in sub-Saharan Africa), Asian water buffalo (only a few small populations in India and mainland Southeast Asia), lowland anoa and mountain anoa (both found only on Sulawesi and Buton Island), tamaraw (endemic to the Philippine island of Mindoro) and saola (in the Annamite Mountains of Vietnam and Laos).

The most fascinating aspect in the evolution of Bovini is that at least five species have been domesticated during the Holocene: aurochs, banteng, gaur, yak, Asian buffalo, and maybe the kouprey. As a result, their recent evolutionary history has been severely disrupted by the migration and explosive growth of human populations throughout the world. Before the Neolithic, populations of wild cattle were very successful and widely distributed across Europe, Africa, Asia and North America. Today, most wild species are restricted to small and isolated populations in a few countries, whereas domestic forms are present on all arable land on Earth. The aurochs, which was the ancestor of most breeds of domestic cattle, became extinct in 1627 (Tikhonov 2008). Field scientists have reported no living specimen of kouprey in the open deciduous dipterocarp forests of Southeast Asia since the 1980s, suggesting that it is possibly extinct (Timmins *et al.* 2008). All other wild species but the American bison are listed as Critically Endangered (CR), Endangered (EN) or Vulnerable (VN) in the IUCN Red List of Threatened Species (IUCN 2013).

In this chapter, I describe the systematic position of the tribe Bovini, and I review ancient and recent hypotheses of interspecific relationships. Then, I propose a biogeographic scenario to explain their past and current geographic distributions.

Systematic position of the tribe Bovini

Wild cattle belong to the order Cetartiodactyla (Figure 1.1), which is the second most diversified order of large mammals after Order Primates (IUCN 2013). Members of this taxonomic group were originally divided into two different orders: Artiodactyla and Cetacea. Artiodactyls are even-toed ungulates, including ruminants, pigs, hippos and camels. They were originally present on all continents, except the Australian region and Antarctica, and most domestic livestock come from this group, including cattle, sheep, goat, pig and camel. They are characterized by two main limb features: a paraxonic foot, which means the axis of the limb support passes between the third and fourth digits; and in the ankle, the astragalus is 'double-pulleyed', i.e. with a trochlea for the tibia and an opposing trochlea for the navicular, which enhances hindlimb flexion and extension, and allows very limited lateral rotation of the foot. Cetaceans include whales and dolphins. All are marine animals except for a few species of freshwater dolphins. The common ancestor of Cetacea acquired many adaptations to an aquatic life, such as a fusiform body, forelimbs modified into flippers, no hindlimbs or rudiments and a tail fin (fluke) used for propulsion. Molecular studies have recovered a sister-group relationship between cetaceans and hippos, indicating that Artiodactyla is paraphyletic (Irwin & Arnason 1994; Gatesy *et al.* 1996; Montgelard *et al.* 1997). To render the classification compatible with the molecular phylogeny, Montgelard *et al.* (1997) proposed placing all species of Artiodactyla and Cetacea into the same order, called Cetartiodactyla. After several years of controversy between molecular biologists and morphologists, a paraxonic foot and a double-pulley astragalus were found in Eocene whales (Gingerich *et al.* 2001), confirming that cetaceans evolved from terrestrial cetartiodactyls.

Wild cattle belong to the suborder Ruminantia (Figure 1.1), which is the most diversified group of Cetartiodactyla, with 215 species related to goat, sheep, deer, pronghorn, giraffe and chevrotains. All ruminants are able to digest cellulose through the action of various microorganisms (bacteria and eukaryotes, such as ciliates and fungi) that are contained in the rumen, the

Ecology, Evolution and Behaviour of Wild Cattle: Implications for Conservation, ed. M. Melletti and J. Burton. Published by Cambridge University Press.

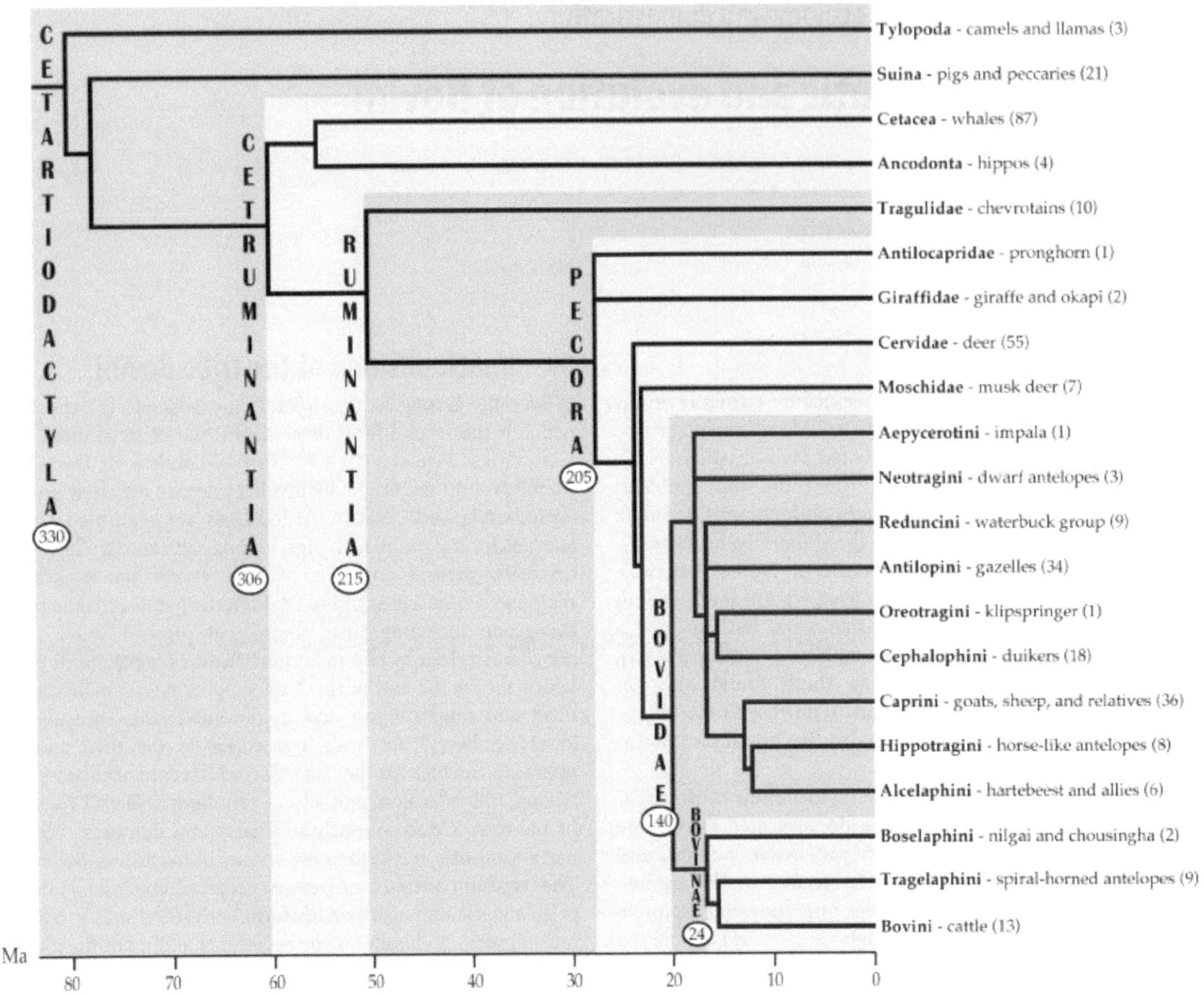

Figure 1.1 Systematic position of the tribe Bovini. The tree is a chronogram, which means that branch lengths are proportional to divergence times. A time scale is provided at the bottom. The ages were estimated using a DNA alignment of complete mitochondrial genomes (Hassanin *et al.* 2012). The number of species is indicated on the nodes for higher taxa, and between round brackets for terminal taxa (data extracted from the website of the IUCN).

most developed of the three or four compartments in their stomach. The compartmentalization of the ruminant stomach is probably the key innovation that explains their evolutionary success. In addition, ruminants have well-adapted teeth characterized by the presence of an incisiform lower canine, which is adjacent to lower incisors, and a horny pad that replaces the upper incisors. All modern and fossil representatives of the group are diagnosed by an osteological autapomorphy, which corresponds to the fusion of the cuboid and navicular bones in the tarsus. Molecular studies have confirmed the monophyly of Ruminantia, as well as the major division between Tragulina (composed of the family Tragulidae – African and Asian chevrotains) and Pecora (including all other ruminant families: Bovidae, Cervidae, Moschidae, Giraffidae and Antilocapridae) (Hassanin & Douzery 2003; Matthee *et al.* 2007; Hassanin *et al.* 2012). In addition, they have shown that Ruminantia, Cetacea and Hippopotamidae form a clade, named the Cetruminantia by Waddell *et al.* (1999) (e.g. Shimamura *et al.* 1997; Gatesy *et al.* 1999; Hassanin *et al.* 2012).

Wild cattle belong to the family Bovidae (Figure 1.1), which is the most successful family of the suborder Ruminantia (140 species). Bovids are characterized by the structure of the horns. Present in all males and sometimes females, these consist of a permanent bone core covered by a non-branched and non-deciduous sheath of keratin (Nowak 1999). The family is found on all continents except Australia, Antarctica and South America. Most ancient classifications recognize between five and eight subfamilies within the Bovidae (e.g. Simpson 1945; McKenna & Bell 1997; Grubb 2005), but molecular studies have concluded the existence of a major division within the family Bovidae, separating the subfamily Bovinae from all other species of Bovidae (Hassanin & Douzery 1999a, 1999b; Matthee & Robinson 1999; Matthee & Davis 2001; Ropiquet *et al.* 2009; Hassanin *et al.* 2012). This result is consistent with the taxonomic view of Kingdon (1997, 1982), who defined only two bovid subfamilies on the basis of morphology and behaviour: the subfamily Bovinae includes the three tribes Bovini (cattle, bison, yak, buffalo and saola), Boselaphini (nilgai and four-horned antelope) and Tragelaphini (bongo, eland, bushbuck, kudu, nyala and sitatunga); the subfamily Antilopinae includes all other bovid tribes (Antilopini, Aepycerotini, Alcelaphini, Caprini, Cephalophini, Hippotragini, Neotragini, Oreotragini and Reduncini; Figure 1.1).

Phylogenetic relationships among species of Bovini

Classifications of the tribe Bovini based on morphology

In the popular classification of Simpson (1945), all species of wild cattle were arranged into a new tribe, named Bovini, which included six different genera at that time: *Bos* (aurochs, domestic cattle and yak), *Anoa* (anoa and tamaraw), *Bibos* (gaur and banteng), *Bison* (European bison and American bison), *Bubalus* (Asian buffalo) and *Syncerus* (African buffalo; Figure 1.2a). This view was largely followed by Bohlken (1958), with the exception of two taxa: the tamaraw was considered a dwarf buffalo and therefore relegated to a subspecific rank within *Bubalus arnee*, and, similarly, the European bison was treated as a subspecies of *Bison bison* (Figure 1.2b). Simpson (1945) did not mention the kouprey in his classification, whereas Bohlken (1958) interpreted the kouprey as a hybrid between banteng and domestic cattle. Three years later, Bohlken (1961) recognized, however, that the kouprey belongs to a distinct species closely related to the banteng, but he reduced *Bibos* to a subgeneric rank under *Bos*. In the morphological study of Groves (1981), the kouprey was grouped with the aurochs and the yak with the bison (Figure 1.2c). As a consequence, the genera *Bibos* and *Bison* were synonymized with *Bos*. The phylogenetic results of Geraads (1992) were very similar (Figure 1.2d), but both yak and bison were found to be the sister-group of the aurochs, banteng, gaur and kouprey. Consequently, two genera were retained by Geraads (1992) within the subtribe Bovina: *Bison* for the two first species, and *Bos* for the four other species. Two other subtribes were proposed by Geraads (1992): Bubalina, comprising the Asian buffalo and anoas; and Syncerina, including only the African buffalo.

In 1992 a new bovid species, the saola, was discovered in the jungles of the Annamite Range of Vietnam and Laos: *Pseudoryx nghetinhensis* (Dung *et al.* 1993). This enigmatic and very rare species was first referred to the tribe Boselaphini (nilgai and four-horned antelope) on the basis of a preliminary DNA analysis and some morphological characters, such as the presence of pre-orbital glands and white markings of the pelage (Dung *et al.* 1993). Based on the morphology of the skull and dentition, Thomas (1994) concluded a close relationship of the saola with serow and goral of Asia (*Capricornis* and *Naemorhedus*, tribe Caprini). Schaller and Rabinowitz (1995) noted, however, resemblances to Bovini and Tragelaphini. These contradicting conclusions probably explain why the saola was not included in Bovini in the classification of McKenna & Bell (1997; Figure 1.2e). Thereafter, two independent molecular studies provided robust evidence for the placement of the saola within Bovini (Hassanin & Douzery 1999a; Gatesy & Arctander 2000). In more recent classifications, such as that of Grubb (2005), *Pseudoryx* is therefore classified in the tribe Bovini (Figure 1.2f).

According to the IUCN (2013), the tribe Bovini includes the following thirteen species:

(1) *Bos gaurus* C. H. Smith, 1827 – gaur
(2) *Bos javanicus* d'Alton, 1823 – banteng
(3) *Bos mutus* (Przewalski, 1883) – yak
(4) *Bos primigenius* Bojanus, 1827– aurochs
(5) *Bos sauveli* Urbain, 1937 – kouprey
(6) *Bison bison* (Linnaeus, 1758) – American bison
(7) *Bison bonasus* (Linnaeus, 1758) – European bison
(8) *Bubalus arnee* (Kerr, 1792) – Asian buffalo
(9) *Bubalus depressicornis* (C. H. Smith, 1827) – lowland anoa
(10) *Bubalus mindorensis* Heude, 1888 – tamaraw
(11) *Bubalus quarlesi* (Ouwens, 1910) – mountain anoa
(12) *Pseudoryx nghetinhensis* Dung, Giao, Chinh, Touc, Arctander & MacKinnon, 1993 – saola
(13) *Syncerus caffer* (Sparrman, 1779) – African buffalo.

Three taxa have sometimes been treated as subspecies rather than full species. For example, Bohlken (1958) proposed uniting *Bison bonasus* with *Bison bison*, *Bubalus mindorensis* with *Bubalus arnee* and *Bubalus quarlesi* with *Bubalus depressicornis* (Figure 1.2b). In addition, the two species of anoa are put into either the genus *Anoa* (Simpson 1945; Bohlken 1958; Geraads 1992; McKenna & Bell 1997) or the genus *Bubalus* (Groves 1981; Grubb 2005; IUCN 2013).

In the next section, domestic forms are treated as subspecies of the wild species name, in accordance with the stated aims of taxonomic freedom in the Code of Zoological Nomenclature (Gentry *et al.* 2004). Therefore, I use *Bos primigenius taurus* for the common cattle, *Bos primigenius indicus* for the zebu (or humped cattle), *Bos gaurus frontalis* for the mithan or gayal, *Bos mutus grunniens* for the domestic yak, *Bubalus arnee bubalis* for the river buffalo and *Bubalus arnee carabanesis* for the swamp buffalo.

Molecular phylogenies

Evidence for three divergent lineages within the tribe Bovini

Most phylogenetic studies based on DNA sequences have concluded the existence of three divergent lineages within the tribe Bovini, corresponding to the subtribes Bovina, Bubalina

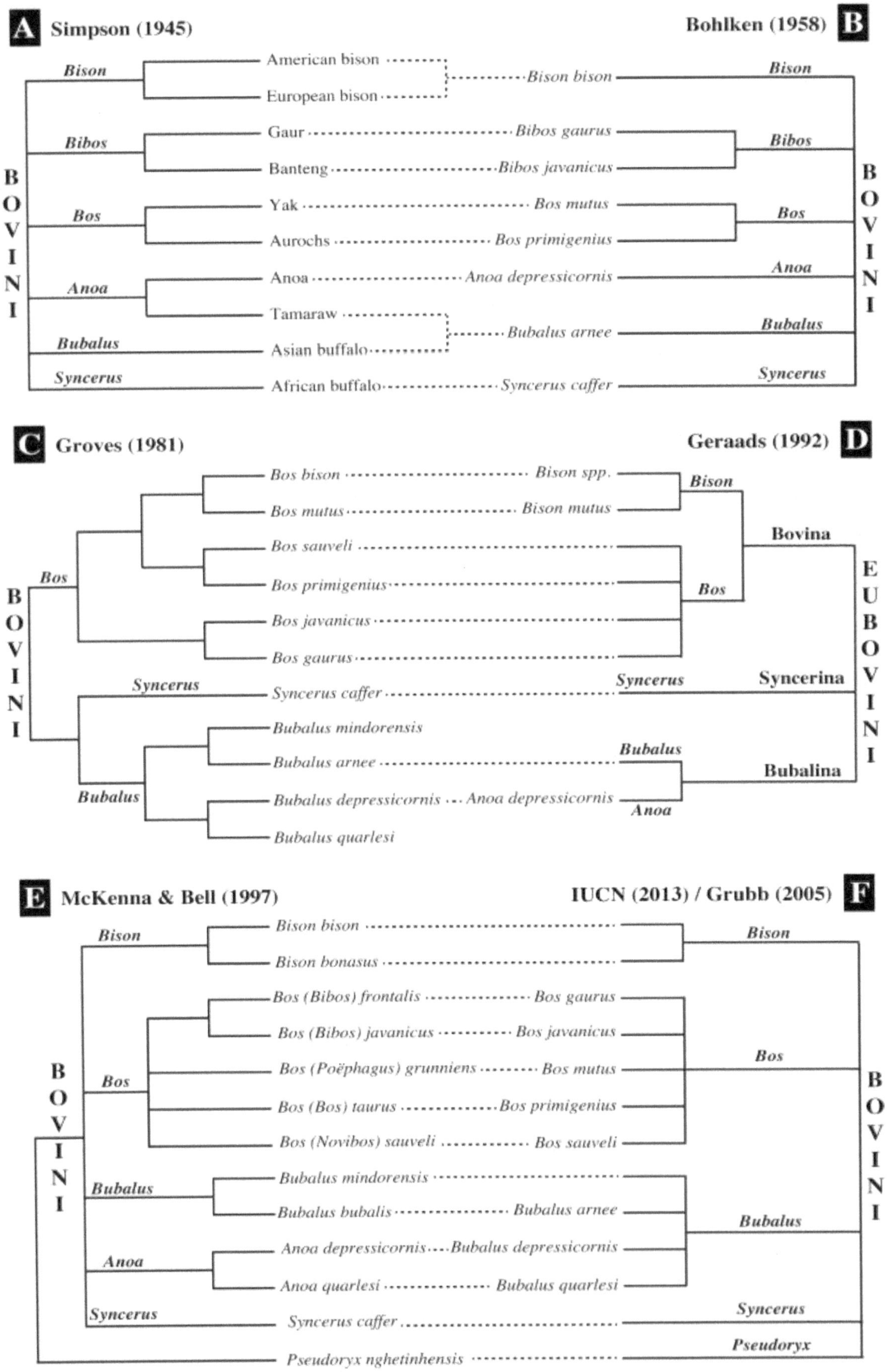

Figure 1.2 Classifications of Bovini based on the morphology.

and Pseudorygina (Hassanin & Douzery 1999a; Gatesy & Arctander 2000; Hassanin & Ropiquet 2004; Hassanin *et al.* 2012). The association of the saola with Bovini is robust and reliable, as it is supported by both mitochondrial and nuclear genomes. By contrast, the identification of the sister-group of the saola is problematic. By analysing complete mtDNA genomes, Hassanin *et al.* (2012) concluded that the most recent common ancestor of crown Bovini experienced rapid diversification into three lineages during the late Middle Miocene. Analyses of 18 autosomal markers have confirmed that the three subtribes Bovina, Bubalina and Pseudorygina diverged from each other over a short period of time (less than 0.5 Ma). However, Bovina and Bubalina were grouped together, a node supported by 12 molecular signatures (detected in several independent markers, including three diagnostic deletions; Hassanin *et al.* 2013).

Interspecific relationships within the subtribe Bovina

The question of interspecific relationships among species of Bovina remains highly debated among molecular biologists, as recent different data sets have produced conflicting phylogenetic results. I discuss below that there are only two cases of robust incongruence between mitochondrial and nuclear phylogenies of Bovini: one concerning the monophyly of banteng, and the other concerning the monophyly of bison. Moreover, I show that most conflicts between nuclear studies can be explained by a lack of robust phylogenetic signal.

The relationships among species of Bovina were first studied with DNA sequences of the mitochondrial genome, such as the subunit II of the cytochrome c oxidase gene, the cytochrome b gene, the small and large subunits ribosomal RNA genes (12S and 16S rRNAs), the control region (also named D-loop) or the complete mtDNA genome (Miyamoto *et al.* 1989; Janecek *et al.* 1996; Hassanin & Douzery 1999a, 1999b; Shreiber *et al.* 1999; Hassanin & Ropiquet 2004; Verkaar *et al.* 2004; Hassanin *et al.* 2012). The phylogenetic relationships supported by the mitochondrial data are summarized in Figure 1.3a. Surprisingly, the genus *Bison* and the species *Bos javanicus* were not found to be monophyletic with mtDNA data. Since these results are in strong disagreement with the analyses of morphological characters and nuclear markers, it is now obvious that the phylogenetic signal of the mtDNA genome can be misleading for inferring relationships among species of Bovina (Hassanin & Ropiquet 2004; Verkaar *et al.* 2004; Hassanin & Ropiquet 2007; Nijman *et al.* 2008).

The monophyly of *Bison* has been confirmed by analysing three nuclear markers from the Y-chromosome (Verkaar *et al.* 2004; Nijman *et al.* 2008) and 18 autosomal genes (Hassanin *et al.* 2013). Although discordant with mtDNA data, these results are in good agreement with the morphology, and the fact that hybrids between American and European bison are fertile in both sexes (Van Gelder 1977). Verkaar *et al.* (2004) proposed two hypotheses for explaining the anomalous divergence of the mtDNAs from the two bison species: the first hypothesis is lineage sorting, which implies that two distinct mitochondrial lineages coexisted until the recent divergence of American and European bison; alternatively, the second hypothesis is that the European bison has emerged by species hybridization initiated by introgression of bison bulls in another ancestral species. A recent study has shown that American and European bison have very similar autosomal sequences (Hassanin *et al.* 2013). Their nucleotide variation is similar to that found for intraspecific variation in *Bos gaurus*, *Bos javanicus* and *Bos primigenius*. There is no trace of hybrid origin in the nuclear markers of the European bison. Therefore, the data suggest that the mitochondrial genome of European bison was acquired by introgression after one or several past events of interspecific hybridization between a male of European bison and a female of an extinct species, which was related to *Bos primigenius*. The mtDNA introgression probably arose somewhere in Europe after the divergence between American and European bison, i.e. during the Middle Pleistocene according to the fossil record (Scott 2010). From the taxonomic point of view, these molecular results suggest, first, synonymy of the genus *Bison* with *Bos*, and second, treatment of the American and European bison as subspecies of *Bos bison*.

The sister-group relationship between Cambodian and Java banteng, i.e. the monophyly of *Bos javanicus*, has been confirmed by analysing DNA sequences from the *SRY* gene and different autosomal genes (Hassanin & Ropiquet 2007; Hassanin *et al.* 2013). Although slightly different, the Cambodian and Java banteng share many morphological characteristics that are not observed in other species of Bovina, including a large white patch on the hindquarters, a white band around the muzzle, the typical reddish-brown colour of females and young males, the shape and size of the horns in females, and the horny shield that connects the bases of the horns on the forehead of adult males. As in the case of bison, mtDNA introgression has been invoked to explain mtDNA sequence similarity between Cambodian banteng (subspecies *Bos javanicus birmanicus*) and kouprey (*Bos sauveli*) (Hassanin & Ropiquet 2007). According to this hypothesis, the mitochondrial genome of kouprey was transferred into the ancestor of Cambodian banteng by natural hybridization. Molecular dating estimates have suggested that the hybridization occurred during the Pleistocene epoch, at 1.34 ± 0.45 Ma (Hassanin & Ropiquet 2007). The mitochondrial introgression hypothesis assumes that at least one kouprey female, which was probably young in order to overcome interspecific ethological barriers, was adopted into a herd of banteng. The event may have happened in open, dry, deciduous forests of Northern Cambodia, where several field biologists have reported the existence of temporary mixed herds between banteng and kouprey (Edmond-Blanc 1947; Wharton 1957; Pfeffer 1969).

These two detected cases of ancient mtDNA introgression suggest that interspecific hybridization may have been a relatively common process during the evolutionary history of Bovina. Several arguments can be advanced to support this idea. First of all, several hybrids have been described between

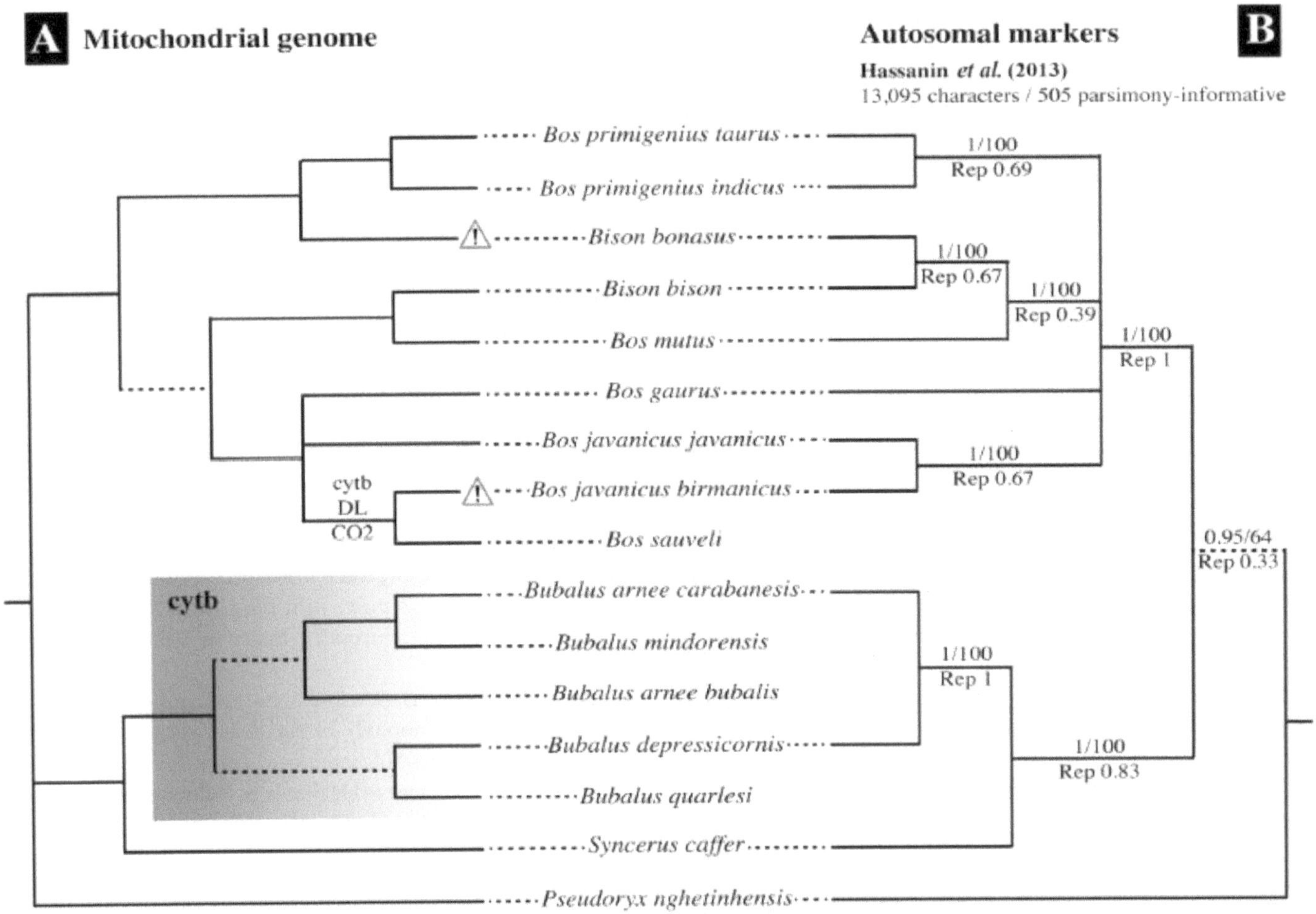

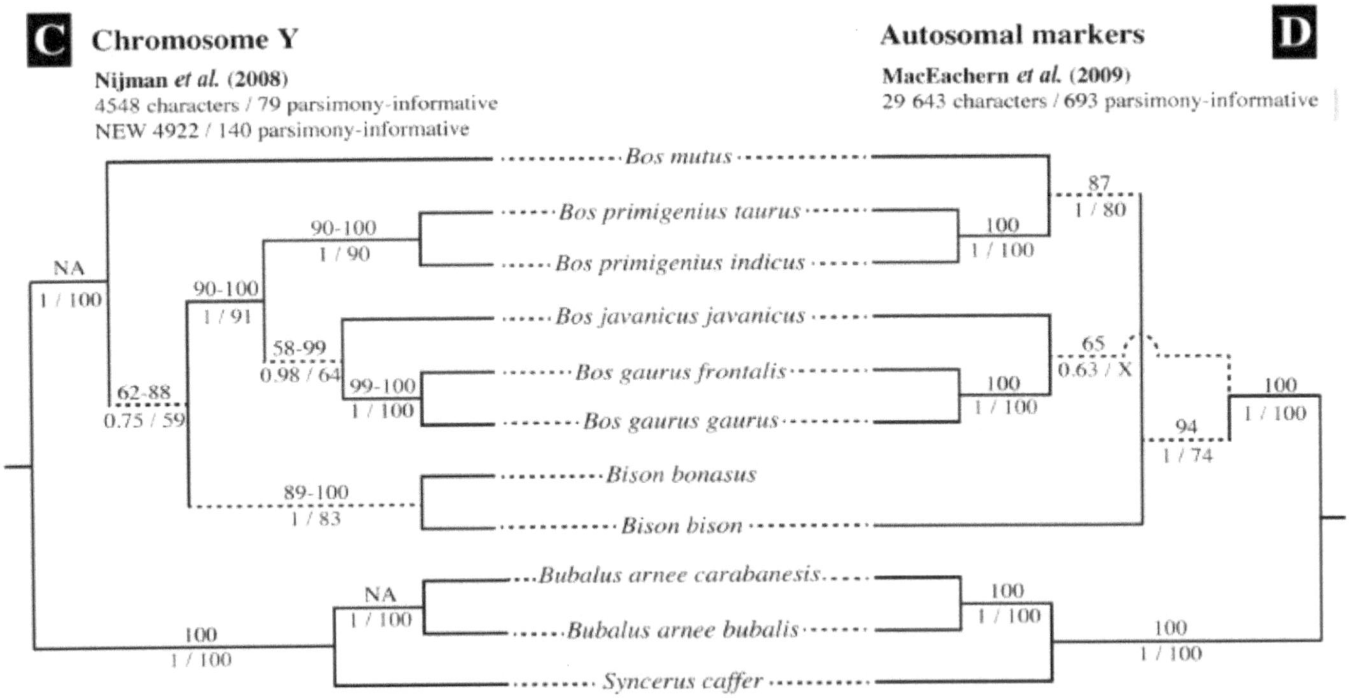

Figure 1.3 Molecular phylogenies of the tribe Bovini. (a) The tree summarizes the phylogeny of Bovini as reconstructed from a DNA alignment of complete mitochondrial genomes (Hassanin *et al.* 2012). A terminal branch with a danger sign highlights a taxon concerned by an ancient mitochondrial

wild and domestic species of Bovina, such as the yakow, which is a hybrid between yak and domestic cow, the selembu, which is a hybrid between gaur and zebu, and the beefalo, which is a hybrid between domestic bull and bison (Mamat-Hamidi & Hilmi 2009). Such hybrids can be easily obtained because most species of Bovina share very similar karyotypes with the same diploid number of chromosomes (2n = 60): *Bos primigenius, Bos javanicus, Bos mutus, Bison bison, Bison bonasus.* Only *Bos gaurus* has slightly different karyotypes (2n = 58 or 56), which can be, however, easily explained by one or two Robertsonian fusions: (1;29) and (2;28) (Ropiquet *et al.* 2008). Similar Robertsonian fusions (1;29) and (2;28) have also been described in the subspecies *Bos javanicus birmanicus*, as well as in several individuals of *Bos primigenius taurus* and *Bos primigenius indicus* (Ropiquet et al. 2008). All these cytogenetic data explain why viable and fertile hybrids have been produced in captivity among various species of Bovina (Van Gelder 1977). In general, the males of the first hybrid generation are sterile, whereas the females are not. The sterility of hybrids of the heterogametic sex (males XY) is commonly encountered in mammals (Haldane's rule; Coyne & Orr 2004), and this characteristic explains why introgression occurs more often on the maternally inherited mtDNA rather than on the paternally inherited markers of the Y-chromosome (Ballard & Whitlock 2004).

By analysing 18 autosomal markers, Hassanin *et al.* (2013) have concluded that bison and yak are sister-groups. This hypothesis is robust, as the node was recovered in the separate analyses of seven independent nuclear markers. In addition, it confirms the morphological studies of Groves (1981) and Geraads (1992), as well as the palaeogeographic scenario of Tibetan mammals proposed by Deng *et al.* (2011). The analyses of amplified fragment length polymorphism (AFLP) fingerprinting published by Buntjer *et al.* (2002) have also provided a signal in favour of the clade uniting bison and yak. By contrast, other molecular studies have concluded different relationships, including a basal position of *Bison* (microsatellites; Ritz *et al.* 2000), a basal position of *Bos mutus* (Y-chromosomal genes; Nijman *et al.* 2008) or an association of *Bos mutus* with *Bos primigenius* (autosomal genes; MacEachern *et al.* 2009). However, my reanalyses have shown that the datasets used by Nijman *et al.* (2008) and MacEachern *et al.* (2009) do not contain a strong signal for the position of both bison and yak ($BP_{ML} < 80$; Figure 1.3c,d). For instance, the early divergence of *Bos mutus* is only supported by the fact that all other species of Bovina share a G nucleotide in position 3186 of the Y-chromosomal alignment. I consider therefore that the hypothesis of a sister-group relationship between yak and bison cannot be rejected by these previous molecular studies.

To conclude, only a few nodes can be considered as being robust and reliable within the subtribe Bovina. They include the association of American and European bison, their sister-group relationship with the yak and the monophyly of the species *Bos javanicus, Bos gaurus* and *Bos primigenius.* In addition, it has been suggested that the species *Bos javanicus* and *Bos gaurus* share close phylogenetic affinities using three independent data sets (mtDNA genome – Hassanin *et al.* 2012; Y-chromosomal genes – Nijman *et al.* 2008; autosomal genes – MacEachern *et al.* 2009). All other relationships are unstable and need further testing with additional molecular data.

Interspecies relationships within the subtribe Bubalina

In agreement with morphological data, all molecular studies indicate an ancient divergence between *Bubalus* and *Syncerus* (Tanaka *et al.* 1996; Hassanin & Ropiquet 2004; Nijman *et al.* 2008; MacEachern *et al.* 2009; Hassanin *et al.* 2012). By contrast, phylogenetic relationships between Asian species of Bubalina remain poorly understood, because only a few molecular studies have focused on this question. To date, most studies dealing with systematic relationships among species of *Bubalus* were based on the analyses of the mitochondrial cytochrome b (*cytb*) gene only (Tanaka *et al.* 1996; Kikkawa *et al.* 1997; Schreiber *et al.* 1999). These analyses have revealed close affinities between the tamaraw (*Bubalus mindorensis*) and swamp buffalo (*Bubalus arnee carabanensis*) (Tanaka *et al.* 1996; Schreiber *et al.* 1999), with the river buffalo (*Bubalus arnee bubalis*) being more distant. Despite its inclusive position within the clade of domestic buffaloes, the tamaraw was not included in all subsequent phylogenetic analyses (e.g. Kierstein *et al.* 2004; Kumar *et al.* 2007; Lei *et al.* 2011) and no data other than the *cytb* gene have been generated for this crucial taxon. Interestingly, the two types of domestic buffalo can be distinguished by their chromosome numbers, with 2n = 48 in swamp buffalo and 2n = 50 in river buffalo. The

Caption for Figure 1.3 (*cont.*) introgression (see text for details). Note that the complete mitochondrial genome was not available for *Bos sauveli, Bubalus mindorensis* and *Bubalus quarlesi*. The phylogenetic position of *Bos sauveli* was determined using only three mitochondrial markers, corresponding to the cytochrome b gene (*cytb*), subunit II of the cytochrome c oxidase (*CO2*), and D-loop (*DL*; Hassanin & Ropiquet 2007). Similarly, interspecific relationships within the genus *Bubalus* were interpreted from the analyses of the cytochrome b gene published by Tanaka *et al.* (1996), Kikkawa *et al.* (1997) and Schreiber *et al.* (1999). Dashed branches are not highly supported by the data. (b) The tree was reconstructed from the analyses of 18 autosomal genes (Hassanin *et al.* 2013). The values above the branches indicate Bayesian posterior probabilities (PP, at the left of the slash) and bootstrap percentages obtained with maximum likelihood method (BP_{ML}, at the right of the slash). The value below the branches indicates the index of reproducibility (Rep), which is the ratio between the number of markers supporting the node of interest to the total number of markers (i.e. 18). (c) The tree was reconstructed from the Y-chromosomal sequences published by Nijman *et al.* (2008). I also performed phylogenetic analyses using an alignment with two additional taxa: one outgroup species, *Capra hircus*, to root the Bovini tree, and *Bubalus arnee carabanensis*. (d) The tree was reconstructed using the nuclear data published by MacEachern *et al.* (2009). For the C and D trees, the values above the branches indicate BP values published by the authors. The results of my analyses are indicated below the branches, with PP at the left of the slash, and BP_{ML} at the right of the slash. The X indicates that the node was not recovered with the maximum likelihood method. Dashed branches indicate nodes that are not highly supported by the data ($BP_{ML} < 90$).

wild water buffalo has also 2n = 50 chromosomes, whereas the tamaraw has a distinct karyotype with 2n = 46 (Gallagher *et al.* 1999; Tanaka *et al.* 2000; Nguyen *et al.* 2008). The *cytb* sequences available in the nucleotide sequence databases for lowland anoa (*Bubalus depressicornis*) and mountain anoas (*Bubalus quarlesi*) can be grouped into four, five or six divergent mtDNA haplogroups (data not shown). These results suggest that several new species of anoa could be described. This mtDNA hypothesis is also supported by the fact that several different karyotypes were found in anoas, with 2n = 36, 38, 42, 44, 45, 46, 47 and 48 chromosomes (Schreiber *et al.* 1999; Burton *et al.* 2005).

Given that genetic introgression has been reported between swamp, river and wild buffaloes (Zhang *et al.* 2011), interspecific relationships within *Bubalus* should be reassessed using nuclear data. Microsatellites and Y-chromosomal data have confirmed that swamp and river buffaloes have been domesticated independently (Yindee *et al.* 2010; Zhang *et al.* 2011). A nuclear phylogeny including all wild species currently recognized within *Bubalus*, as well as the divergent mtDNA lineages identified for anoas, needs to be reconstructed in order to better understand the taxonomy of this genus.

Biogeographic interpretations

Where did Bovini first diversify?

Bibi (2007 2009), following Pilgrim (1939), argued that fossils from the Late Miocene of South Asia attributed to *Selenoportax* are stem representatives of Bovini. However, Bibi did not consider the saola (*Pseudoryx*), and so meant only the clade defined by Bovina and Bubalina (Bibi, personal communication 2012). The earliest fossils of *Selenoportax* are from the Siwaliks of Pakistan at 10.2 Ma (Badgley *et al.* 2008). *Selenoportax* has also been reported from Myanmar (Pilgrim 1939). Records from China (Qiu & Qiu 1995) and Greece (Solounias 1981) are doubtful (Bibi 2007 2009). The biogeographic analyses of Hassanin & Ropiquet (2004) have supported a scenario with an Asian origin for Bovini. Since all three subtribes are currently more diversified in Southeast Asia, this region is likely to constitute the centre of origin of crown Bovini. A basal position of Pseudorygina, as supported by recent nuclear analyses (Hassanin *et al.* 2013), gives even more credence to this hypothesis. Molecular dating estimates have suggested that the tribe Bovini rapidly diversified into the three subtribes Bovina, Bubalina and Pseudorygina during the late Middle Miocene, between 13.7 and 13.0 ± 2.0 Ma (Hassanin & Ropiquet 2004; Hassanin *et al.* 2012, 2013). The diversification of Bovini during the late Middle Miocene of Southeast Asia, specifically the evolution of Bubalina and Bovina, may have been driven by changes in vegetation distribution, as palaeoecological data have documented oscillations between tropical woodlands and grasslands in northern Thailand between 13.3 and 13.1 Ma (Sepulchre *et al.* 2010). This scenario implies that there is a long undocumented gap in the fossil record of Southeast Asia, since the oldest Bovini found in this region were described in the Late Miocene of Myanmar at around 6 Ma (*Selenoportax*; Pilgrim 1939; Zin-Maung-Maung-Thein *et al.* 2011). However, *Selenoportax* was recently found in the Late Miocene of the Irrawaddy Formation (Myanmar), a site dated between 10.4 and 8.8 Ma (Bibi, personal communication 2012). The morphology of the earliest Bovini must have been very different from that of extant members of the subtribes Bovina and Bubalina. Their general appearance was probably more similar to that of *Pseudoryx*, a genus that apparently does not share any derived dental or osteological character with other Bovini. Indeed, Thomas (1994) wrote that *'plus d'une dizaine de caractères dérivés qui définisssent la tribu des Bovini* (cf. Geraads 1992) *ne se retrouvent pas chez Pseuroryx* [More than a dozen of features that characterize the tribe Bovini (cf. Geraads 1992) are not found in *Pseudoryx*]'. As a consequence, Thomas (1994) concluded that *Pseudoryx* share more affinities with Caprini than Bovini. As mentioned above, this morphological view has been refuted by all molecular studies. The difficulty in interpreting the evolution of morphological characters suggests that bovid fossils discovered in Middle Miocene deposits might not be easily assigned to the tribe Bovini, even if they belong to the crown group.

An Asian origin for Bubalina?

Today, members of the subtribe Bubalina remain confined to tropical regions of the Old World. There are two divergent biogeographic lineages: the first is represented by the genus *Bubalus*, which contains species living near abundant water sources in the Indian subcontinent and Southeast Asia; the second includes only *Syncerus caffer*, which is found in various habitats, including semi-arid bushland in sub-Saharan Africa (IUCN 2013). Molecular estimations have indicated that these two groups diverged from each other in the Late Miocene, between 8.2 ± 1.7 Ma and 9.1 ± 1.6 Ma (Hassanin & Ropiquet 2004; Hassanin *et al.* 2012). The palaeontological data indicate a more recent origin, since these two groups appeared only during the Pliocene. The lineage leading to *Syncerus* is represented by *Ugandax* in eastern and southern Africa between 5.7 and 2 Ma, followed by Plio-Pleistocene species of *Syncerus*, such as *S. acoelotus* in East Africa and *S. antiquus* in East, North and South Africa (Bibi 2009; Gentry 2010). There is also an undescribed Bovini from Toros-Menalla in Chad, dated between 7 and 6 Ma (Vignaud *et al.* 2002), which could be a species of Bubalina. The lineage leading to *Bubalus* is represented by *Proamphibos* in the Pliocene of the Siwaliks between 3.5 and 2.2 Ma, followed by *Hemibos*, which is also recorded during Early Pleistocene times in Europe and China (Qui *et al.* 2004; Khan & Akhtar 2011; Martínez-Navarro *et al.* 2011), and by *Bubalus*, which is well diversified in the Middle and Late Pleistocene of the Indian subcontinent, Europe, China and Southeast Asia (van den Bergh 1999; Tong 2006; Louys *et al.* 2007; Chauhan 2008; Martínez-Navarro *et al.* 2011). Therefore, a 3 Ma hiatus exists in the fossil record of crown

Bubalina. However, a simple biogeographic scenario can be proposed if we accept the idea that several Miocene fossils of *Pachyportax* can be related to crown Bubalina. Indeed, *Pachyportax* is known from the Late Miocene of eastern Africa, Arabia, South Asia and Southeast Asia (Pilgrim 1939; Bibi 2007; Bibi 2009; Khan *et al.* 2009). The age and distribution of these fossils suggest that a dispersal event occurred during the Late Miocene, from South Asia or Southeast Asia into Africa. In Arabia, tropical savanna and grasslands have replaced tropical woodlands during the Tortonian, i.e. between 11.6 and 7.2 Ma (Pound *et al.* 2012). Thanks to these favourable conditions, the ancestor of the *Syncerus* lineage may have migrated from Asia to Africa at around 8–9 Ma.

Pleistocene migration of *Bubalus* to Southeast Asian islands?

Today, three of the four wild species of *Bubalus* are present on the islands of Southeast Asia, with *B. depressicornis* and *B. quarlesi* on Sulawesi and Buton Island, and *B. mindorensis* on the Philippine island of Mindoro. Although *B. arnee* is the only mainland species, many fossils of *Bubalus* were retrieved from Pleistocene deposits of the Asian mainland, in India, China and Southeast Asia (Tong 2006; Louys *et al.* 2007; Chauhan 2008). In addition, *Proamphibos* from the Siwaliks is frequently designated as the Pliocene ancestor for both *Bubalus* and *Hemibos* (e.g. Pilgrim 1939; Martínez-Navarro *et al.* 2007; Bibi 2009). The genus *Bubalus* has also been reported in the Middle and Late Pleistocene of Europe (Breda *et al.* 2010; Masini *et al.* 2012). The fossil record suggests therefore that insular species of *Bubalus* originated from mainland species during the Pleistocene epoch. This scenario is supported by a series of arguments. First, all molecular studies indicate that extant species of *Bubalus* diverged during the Pleistocene epoch. Indeed, the divergence time between *B. arnee* and *B. depressicornis* was estimated at 2 Ma (Tanaka *et al.* 1996), 2.2 ± 0.5 Ma (Hassanin & Ropiquet 2004), or 1.3 ± 0.5 Ma in two recent studies based on either mitochondrial or nuclear data sets (Hassanin *et al.* 2012, 2013). In addition, Tanaka *et al.* (1996) estimated that the tamaraw and swamp buffalo diverged at around 1.5 Ma. Second, all fossil remains attributed to *Bubalus* on Indonesian and Philippine islands were discovered in Pleistocene and Holocene deposits (e.g. Custodio *et al.* 1996; van den Bergh 1999; Croft *et al.* 2006; Bouteaux 2008). Third, the island rule, which implies that resource limitation and the absence of large predators are selecting for dwarf bovids on islands (Lomolino 2005), can explain why tamaraw and anoas are smaller than the wild water buffalo (<150 kg *versus* 700–1200 kg; Nowak 1999; Burton *et al.* 2005). In agreement with the theory, larger Pleistocene buffaloes (>500 kg) have been found on Borneo, which is the largest island of Southeast Asia (Bouteaux 2008), whereas a small Pleistocene buffalo (150–165 kg) has been described from the small island of Cebu (Philippines; Croft *et al.* 2006). Fourth, van den Bergh *et al.* (2001) stated that the modern mammalian fauna of Sulawesi and the Philippines have little in common, indicating different migration routes. They proposed therefore that the species of anoas (*B. depressicornis* and *B. quarlesi*) colonized Sulawesi during the Middle or Late Pleistocene from an Asian mainland ancestor, probably closely related to *Hemibos*, whereas the Philippine dwarf buffalo (*B. mindorensis*) is a more recent derivative of the wild water buffalo (*B. arnee*).

Rapid radiation and dispersal of Bovina during the middle Pliocene

Several extinct genera, which are possibly related to Bovina, first appeared during the Pliocene: *Simatherium* and *Pelorovis* in Africa, *Leptobos* in Africa and Europe. During the Early Pleistocene, *Pelorovis* spread in the Middle East, *Leptobos* was present in northeastern India and China and *Epileptobos* was found on Java (Figure 1.4; Duvernois 1992; Vislobokova 2005; Martínez-Navarro *et al.* 2007; Dong 2008; Bibi 2009). The phylogenetic relationships between extinct and extant species of Bovina remain problematic (Geraads 1992; Bibi 2009), but the morphometric analyses of Martínez-Navarro *et al.* (2007) have evidenced two major groups: *Pelorovis* and *Bos primigenius*, and *Leptobos* and *Bison*. Unfortunately, most extant species of Bovina were not included in the analyses (i.e. *B. gaurus*, *B. javanicus*, *B. mutus* and *B. sauveli*). However, the authors proposed to synonymize the genus *Pelorovis* with *Bos*, and suggested that *Bos* originated in the Late Pliocene of eastern Africa, and dispersed into the Middle East at around 1.4 Ma, and finally into Europe at around 0.6–0.5 Ma. My interpretation is that this biogeographic scenario could hold only for the lineage leading to *Bos primigenius*. Molecular and cytogenetic studies have shown that the genus *Bison* should be treated as a synonym of *Bos*. If the extinct genera *Epileptobos*, *Leptobos* and *Pelorovis* are confirmed to be related to living species of *Bos* (and *Bison*), then this would imply their synonymy with *Bos* as well. This would result in the genus *Bos* being already represented by different morphological and biogeographic lineages by the Pliocene of Africa, Europe and northeastern South Asia. Molecular dating analyses support this early age, suggesting that the last common ancestor of extant Bovina underwent a rapid radiation during the Early to Middle Pliocene (3.3–4.8 Ma in Hassanin & Douzery 1999b; 4.7 ± 0.8 Ma in Hassanin & Ropiquet 2004; 3.3 ± 0.9 Ma in Hassanin *et al.* 2012; 3.7 ± 1.1 Ma in Hassanin *et al.* 2013). At that time, most of the modern arid and semi-arid climate zones in Africa and Arabia were covered with temperate and tropical xerophytic shrublands and grasslands (Salzmann *et al.* 2011). Simulations have also indicated that there was a northward expansion of temperate forests and grasslands in Eurasia, with vast tracts of grassland in Siberia (Haywood *et al.* 2009). I suggest that these palaeoenvironmental conditions triggered the radiation and dispersal of Bovina in a huge area covering northern and eastern Africa, southern Europe, western Asia, central Asia and Siberia.

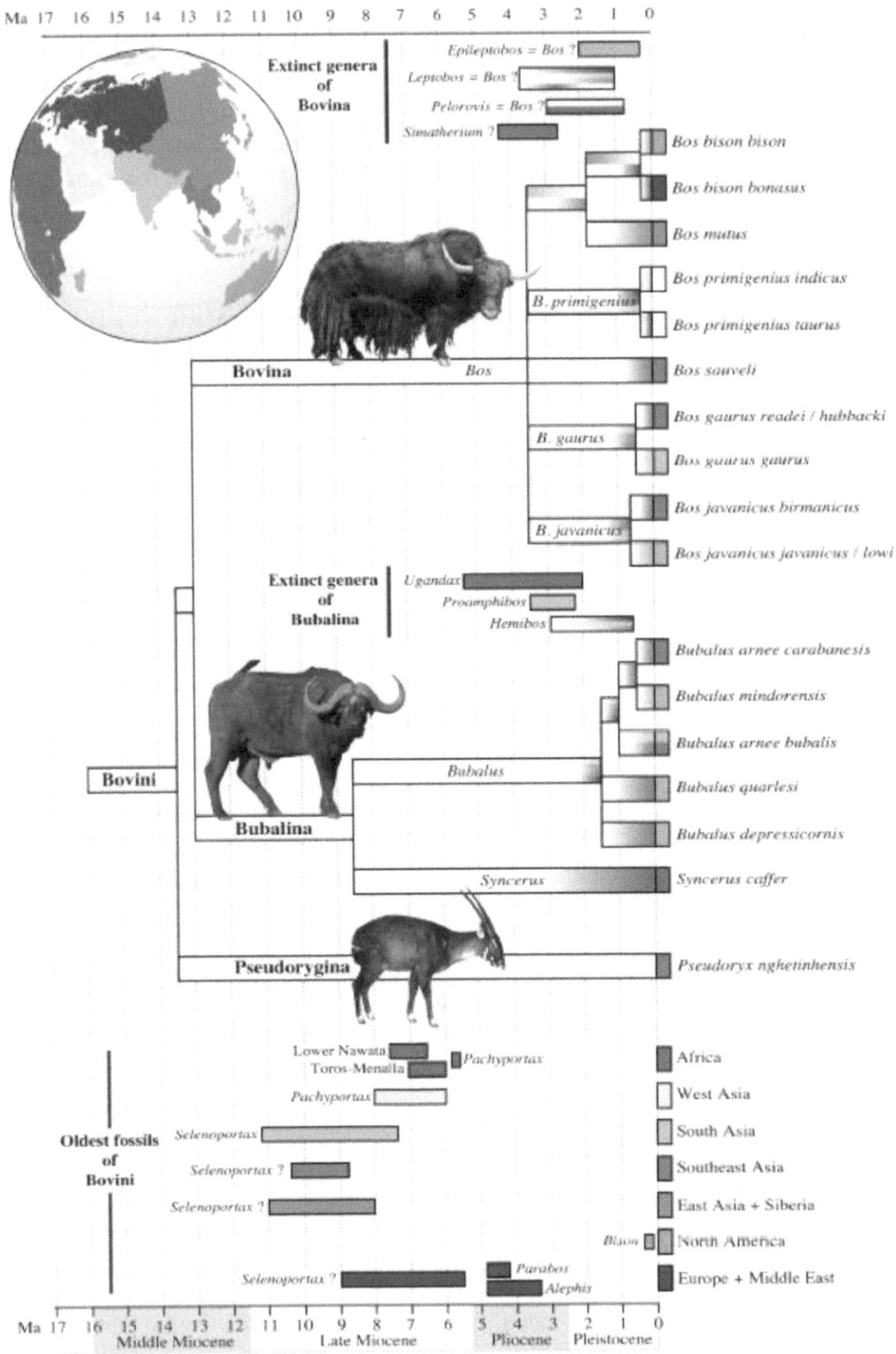

Figure 1.4 Timetree of Bovini. The chronogram summarizes the divergence dates recently estimated from DNA sequences using a Bayesian relaxed molecular clock approach (Hassanin *et al.* 2012; 2013). The time scale comes from the Geological Society of America (2009, available at www.geosociety.org/science/timescale/timescl.pdf). The coloured rectangles at the left of species names indicate the geographic distribution of extant taxa. Internal branches are coloured according to the fossil record. Most palaeontological data were interpreted from Bibi (2009), the Paleobiology database (www.paleodb.org) and references cited in the text. The oldest fossils assigned to the tribe Bovini are indicated at the bottom for each geographic region (i.e. Africa; West Asia; South Asia; Southeast Asia; East Asia and Siberia; North America; and Europe and the Middle East; see the globe for the geographic regions). The age and geographic distribution of extinct genera assigned to Bovina and Bubalina are also detailed. (Photos of yak and African buffalo by Alexandre Hassanin. Photo of Saola by William G. Robichaud.) A black and white version of this figure will appear in some formats. For the colour version, please refer to the plate section.

Evolution of the yak–bison lineage in the Northern Hemisphere

As discussed previously, nuclear data strongly support the association of American and European bison, and their sister-group relationship with the yak. Interestingly, yak and bison share at least two osteological synapomorphies, which may be useful for identifying their ancestry in the fossil record: unlike other species of Bovina, the premaxillae does not touch the nasal (Olsen 1990); also they have 14 thoracic vertebrae and five lumbars, whereas other Bovina have 13 thoracic vertebrae and six lumbars (Groves 1981). Unfortunately, these characters are only rarely preserved in the fossil remains. To date, fossils related to the yak have been found in the Late Pleistocene of Siberia, Tibet and Nepal (Leslie & Schaller 2009), whereas fossils ascribed to *Bison* are much older, with *B. sivalensis* from the Late Pliocene of the Upper Siwaliks (dated between 3.3 and 2.6 Ma; Khan *et al.* 2010). Bison appeared in Eastern Europe at 1.77 Ma, and were present in the Middle East at around 1.6–1.2 Ma, and thereafter in

Western Europe at around 1.5 Ma (Martínez-Navarro *et al.* 2012). Bison entered North America at around 240–220 ka, and then they rapidly spread out across the continent, where they diversified into different species (Scott 2010). The palaeontological data suggest therefore that the common ancestor of yak and bison emerged somewhere in Asia, and possibly during the Pliocene. Based on the discovery of a Himalayan woolly rhino in the Pliocene (dated to 3.7 Ma), Deng *et al.* (2011) have proposed that some Ice Age megaherbivores, such as the woolly rhino and yak, first evolved in Tibet before the beginning of the Ice Age. Such a scenario is compatible with our molecular estimates, since the yak–bison lineage separated from other Bovina between 3.7 ± 1.1 Ma and 2 ± 0.5 Ma (Hassanin *et al.* 2013). Subsequently, yak remained in East Asia, while bison spread in the Siwaliks, Middle East and Europe, and crossed into North America through Beringia in the Middle Pleistocene.

Acknowledgement

I am grateful to Faysal Bibi for offering critical and useful comments on the first draft of this chapter. I thank Faysal Bibi and Olivier Chavasseau for the bibliography and information on bovid fossils, William G. Robichaud for the photo of Saola and Mario Melletti for discussion on the taxonomic status of *Syncerus* species.

References

Badgley, C., Barry, J. C., Morgan, M. E., Nelson, S. V., Behrensmeyer, A. K., Cerling, T. E. & Pilbeam, D. (2008). Ecological changes in Miocene mammalian record show impact of prolonged climatic forcing. *Proceedings of the National Academy of Sciences of the United States of America* **105**: 12145–12149.

Ballard, J. W. & Whitlock, M. C. (2004). The incomplete natural history of mitochondria. *Molecular Ecology* **13**: 729–744.

Bibi, F. (2007). Origin, paleoecology, and paleobiogeography of early Bovini. *Palaeogeography Palaeoclimatology Palaeoecology* **248**: 60–72.

Bibi, F. (2009). Evolution, systematics, and paleoecology of Bovinae (Mammalia: Artiodactyla) from the Late Miocene to the Recent. PhD Thesis, Yale University.

Bohlken, H. (1958). Vergleichende untersuchungen an wildrindern (Tribus Bovini Simpson, 1945). *Zoologische Jahrbücher* **68**: 113–202.

Bohlken, H. (1961). Der Kouprey, *Bos (Bibos) sauveli* Urbain 1937. *Zeitschrift für Saügetierkunde* **26**: 193–254.

Bouteaux, A. (2008). Étude taphonomique d'assemblages fauniques de sites à *Homo erectus* du dôme de Sangiran (Pléistocène moyen, Java central, Indonésie). *Annales de Paléontologie* **94**: 229–243.

Breda, M., Collinge, S. E., Parfitt, S. A. & Lister, A. M. (2010). Metric analysis of ungulate mammals in the early Middle Pleistocene of Britain, in relation to taxonomy and biostratigraphy: I – Rhinocerotidae and Bovidae. *Quaternary International* **228**: 136–156.

Buntjer, J. B., Otsen, M., Nijman, I. J., Kuiper, M. T. & Lenstra, J. A. (2002). Phylogeny of bovine species based on AFLP fingerprinting. *Heredity (Edinb)* **88**: 46–51.

Burton, J. A., Hedges, S. & Mustari, A. H. (2005). The taxonomic status, distribution and conservation of the lowland anoa *Bubalus depressicornis* and mountain anoa *Bubalus quarlesi*. *Mammal Review* **35**: 25–50.

Chauhan, P. R. (2008). Large mammal fossil occurrences and associated archaeological evidence in Pleistocene contexts of peninsular India and Sri Lanka. *Quaternary International* **192**: 20–42.

Coyne, J. A. & Orr, H. A. (2004). *Speciation.* Sunderland, MA: Sinauer Associates.

Croft, D. A., Heaney, L. R., Flynn, J. J. & Bautista, A. P. (2006). Fossil remains of a new, diminutive *Bubalus* (Artiodactyla: Bovidae: Bovini) from Cebu Island, Philippines. *Journal of Mammalogy* **87**: 1037–1051.

Custodio, C. C., Lepiten, M. V. & Heaney, L. R. (1996). *Bubalus mindorensis. Mammalian Species* **520**: 1–5.

Deng, T., Wang, X., Fortelius, M., Li, Q., Wang, Y., Tseng, Z. J., Takeuchi, G. T., Saylor, J. E., Säilä, L. K. & Xie, G. (2011). Out of Tibet: Pliocene woolly rhino suggests high-plateau origin of Ice Age megaherbivores. *Science* **333**: 1285–1288.

Dong, W. (2008). Nouveau matériel de *Leptobos (Smertiobos) crassus* (Artiodactyla, Mammalia) du Pléistocène inférieur à Renzidong (Chine de l'Est). *Geobios* **41**: 355–364.

Dung, V. V., Giao, P. M., Chinh, N. N., Tuoc, D. & MacKinnon, J. (1993). A new species of living bovid from Vietnam. *Nature* **363**: 443–445.

Duvernois, M. -P. (1992). Mise au point sur le genre *Leptobos* (Mammalia, Artiodactyla, Bovidae): implications biostratigraphiques et phylogénétiques. *Geobios* **25**: 155–166.

Edmond-Blanc, F. (1947). A contribution of the knowledge of the Cambodian wild ox or kouproh. *Journal of Mammalogy* **28**: 245–248.

Gallagher, D. S. Jr., Davis, S. K., De Donato, M., Burzlaff, J. D., Womack, J. E., Taylor, J. F. & Kumamoto, A. T. (1999). A molecular cytogenetic analysis of the tribe Bovini (Artiodactyla: Bovidae: Bovinae) with an emphasis on sex chromosome morphology and NOR distribution. *Chromosome Research* **7**: 481–492.

Gatesy, J. & Arctander, P. (2000). Hidden morphological support for the phylogenetic placement of *Pseudoryx nghetinhensis* with bovine bovids: a combined analysis of gross anatomical evidence and DNA sequences from five genes. *Systematic Biology* **49**: 515–538.

Gatesy, J., Hayashi, C., Cronin, M. A. & Arctander, P. (1996). Evidence from milk casein genes that cetaceans are close relatives of hippopotamid artiodactyls. *Molecular Biology and Evolution* **13**: 954-963.

Gatesy, J., O'Grady, P. & Baker, R. H. (1999). Corroboration among data sets in simultaneous analysis: hidden support for phylogenetic relationships among higher level artiodactyl taxa. *Cladistics* **15**: 271–313.

Gentry, A. W. (2010). Bovidae. In *Cenozoic Mammals of Africa*, pp. 741–796. Eds. Werdelin, L. & Sanders, W. J.. Berkeley, CA: University of California Press.

Gentry, A., Clutton-Brock, J. & Groves, C. P. (2004). The naming of wild animal species and their domestic derivatives. *Journal of Archaeological Science* **31**: 645–651.

Geraads, D. (1992). Phylogenetic analysis of the tribe Bovini (Mammalia, Artiodactyla). *Zoological Journal of the Linnean Society* **104**: 193–207.

Gingerich, P. D., ul Haq, M., Zalmout, I. S., Khan, I. H. & Malkani, M. S. (2001). Origin of whales from early artiodactyls: hands and feet of Eocene Protocetidae from Pakistan. *Science* **293**: 2239–2242.

Groves, C. P. (1981). Systematic relationships in the Bovini (Artiodactyla, Bovidae). *Zeitschrift für Zoologische Systematik und Evolutionsforschung* **19**: 264–278.

Grubb, P. (2005). Order Artiodactyla. In *Mammal Species of the World: A Taxonomic and Geographic Reference* (3rd edn) 2 vols, pp. 637–722. Eds. Wilson, D. E., & Reeder, D. M. Baltimore, MD: Johns Hopkins University Press.

Hassanin, A. & Douzery, E. J. (1999a). Evolutionary affinities of the enigmatic saola (*Pseudoryx nghetinhensis*) in the context of the molecular phylogeny of Bovidae. *Proceedings of the Royal Society B: Biological Sciences* **266**: 893–900.

Hassanin, A. & Douzery, E. J. (1999b). The tribal radiation of the family Bovidae (Artiodactyla) and the evolution of the mitochondrial cytochrome b gene. *Molecular Phylogenetics and Evolution* **13**: 227–243.

Hassanin, A. & Douzery, E. J. (2003). Molecular and morphological phylogenies of Ruminantia and the alternative position of the Moschidae. *Systematic Biology* **52**: 206–228.

Hassanin, A. & Ropiquet, A. (2004). Molecular phylogeny of the tribe Bovini (Bovidae, Bovinae) and the taxonomic status of the Kouprey, *Bos sauveli* Urbain 1937. *Molecular Phylogenetics and Evolution* **33**: 896–907.

Hassanin, A. & Ropiquet, A. (2007). Resolving a zoological mystery: the kouprey is a real species. *Proceedings of the Royal Society B: Biological Sciences* **274**: 2849–2855.

Hassanin, A., Delsuc, F., Ropiquet, A., *et al.* (2012). Pattern and timing of diversification of Cetartiodactyla (Mammalia, Laurasiatheria), as revealed by a comprehensive analysis of mitochondrial genomes. *Comptes Rendus Biologies* **335**: 32–50.

Hassanin, A., An, J., Ropiquet, A., Nguyen, T. T. & Couloux, A. (2013). Combining multiple autosomal introns for phylogeny and taxonomy of Laurasiatherian mammals: application to the tribe Bovini (Cetartiodactyla, Bovidae). *Molecular Phylogenetics and Evolution* **66**: 766–775.

Haywood, A. M., Chandler, M. A., Valdes, P. J., Salzmann, U., Lunt, D. J. & Dowsett, H. J. (2009). Comparison of mid-Pliocene climate predictions produced by the HadAM3 and GCMAM3 General Circulation Models. *Global and Planetary Change* **66**: 208–224.

Irwin, D. & Arnason, U. (1994). Cytochrome b gene of marine mammals: phylogeny and evolution. *Journal of Mammalian Evolution* **2**: 37–55.

IUCN (2013). IUCN Red List of Threatened Species. Version 2011.2. www.iucnredlist.org. Downloaded 22 May 2014.

Janecek, L. L., Honeycutt, R. L., Adkins, R. M. & Davis, S. K. (1996). Mitochondrial gene sequences and the molecular systematics of the artiodactyl subfamily bovinae. *Molecular Phylogenetics and Evolution* **6**: 107–119.

Khan, M. A. & Akhtar, M. (2011). *Proamphibos kashmiricus* (Bovini: Bovidae: Mammalia) lower case from the Pinjor Formation of Pakistan. *Pakistan Journal of Zoology* **43**: 615–621.

Khan, M. A., Iliopoulos, G. & Akhtar, M. (2009). Boselaphines (Artiodactyla, Ruminantia, Bovidae) from the Middle Siwaliks of Hasnot, Pakistan. *Geobios* **42**: 739–753.

Khan, M. A., Kostopoulos, D. S., Akhtar, M. & Nazir, M. (2010). Bison remains from the Upper Siwaliks of Pakistan. *Neues Jahrbuch für Geologie und Paläontologie-Abhandlungen* **258**: 121–128.

Kierstein, G., Vallinoto, M., Silva, A., Schneider, M. P., Iannuzzi, L. & Brenig, B. (2004). Analysis of mitochondrial D-loop region casts new light on domestic water buffalo (*Bubalus bubalis*) phylogeny. *Molecular Phylogenetics and Evolution* **30**: 308–324.

Kikkawa, Y., Yonekawa, H., Suzuki, H. & Amano, T. (1997). Analysis of genetic diversity of domestic water buffaloes and anoas based on variations in the mitochondrial gene for cytochrome b. *Animal Genetics* **28**: 195–201.

Kingdon, J. (1982). *East African Mammals: An Atlas of Evolution in Africa*, Vol. IIIC. London: Academic Press.

Kingdon, J. (1997). *The Kingdon Field Guide to African Mammals.* London: Natural World Academic Press.

Kumar, S., Nagarajan, M., Sandhu, J. S., Kumar, N., Behl, V. & Nishanth, G. (2007). Mitochondrial DNA analyses of Indian water buffalo support a distinct genetic origin of river and swamp buffalo. *Animal Genetics* **38**: 227–232.

Lei, C. Z., Zhang, C. M., Weining, S., Campana, M. G., Bower, M. A., Zhang, X. M., Liu, L., Lan, X. Y. & Chen, H. (2011). Genetic diversity of mitochondrial cytochrome b gene in Chinese native buffalo. *Animal Genetics* **42**: 432–436.

Leslie, D. M., Jr & Schaller, G. B. (2009). *Bos grunniens* and *Bos mutus* (Artiodactyla: Bovidae). *Mammalian Species* **836**: 1–17.

Lomolino, M. V. (2005). Body size evolution in insular vertebrates: generality of the island rule. *Journal of Biogeography* **32**: 1683–1699.

Louys, J., Curnoe, D. & Tong, H. (2007). Characteristics of Pleistocene megafauna extinctions in Southeast Asia. *Palaeogeography, Palaeoclimatology, Palaeoecology* **243**: 152–173.

MacEachern, S., McEwan, J. & Goddard, M. (2009). Phylogenetic reconstruction and the identification of ancient polymorphism in the Bovini tribe (Bovidae, Bovinae). *BMC Genomics* **10**: 177.

Mamat-Hamidi, I. I & Hilmi, M. (2009). Karyotype of Malayan gaur (*Bos gaurus hubbacki*), Sahiwal-Friesian cattle and Gaur × cattle hybrid backcrosses. *Pakistan Journal of Biological Sciences* **12**: 896–901.

Martínez-Navarro, B., Perez-Claros, J. A., Palombo, M. R., Lorenzo, R. & Palmqvist, P. (2007). The olduvai buffalo *Pelorovis* and the origin of *Bos*. *Quaternary Research* **68**: 220–226.

Martínez-Navarro, B., Ros-Montoya, S., Patrocinio, E. M. & Palmqvist, P. (2011). Presence of the Asian origin Bovini, *Hemibos sp* aff. *Hemibos gracilis* and *Bison sp*., at the early Pleistocene site of Venta Micena (Orce, Spain). *Quaternary International* **243**: 54–60.

Martínez-Navarro, B., Belmaker, M. & Bar-Yosef, O. (2012). The Bovid assemblage (Bovidae, Mammalia) from the early pleistocene site of 'Ubeidiya, Israel: Biochronological and paleoecological implications for the fossil and lithic bearing strata. *Quaternary International* **267**: 78–97.

Masini, F., Palombo, M. R. & Rozzi, R. (2012). A reappraisal of the Early to Middle Pleistocene Italian Bovidae. *Quaternary International* **288**: 45–62.

Matthee, C. A. & Davis, S. K. (2001). Molecular insights into the evolution of the family Bovidae: a nuclear DNA perspective. *Molecular Biology and Evolution* **18**: 1220–1230.

Matthee, C. A. & Robinson, T. J. (1999). Cytochrome b phylogeny of the family bovidae: resolution within the alcelaphini, antilopini, neotragini, and tragelaphini. *Molecular Phylogenetics and Evolution* **12**: 31–46.

Matthee, C. A., Eick, G., Willows-Munro, S., Montgelard, C., Pardini, A. T. & Robinson, T. J. (2007). Indel evolution of mammalian introns and the utility of non-coding nuclear markers in eutherian phylogenetics. *Molecular Phylogenetics and Evolution* **42**: 827–837.

McKenna, M. C. & Bell, S. K. (1997). *Classification of Mammals Above the Species Level.* New York: Columbia University Press.

Miyamoto, M. M., Tanhauser, S. M. & Laipis, P. J. (1989). Systematic relationships in the artiodactyl tribe Bovini (family Bovidae), as determined from mitochondrial-DNA sequences. *Systematic Zoology* **38**: 342–349.

Montgelard, C., Catzeflis, F. M. & Douzery, E. (1997). Phylogenetic relationships of artiodactyls and cetaceans as deduced from the comparison of cytochrome b and 12S rRNA mitochondrial sequences. *Molecular Biology and Evolution* **14**: 550–559.

Nguyen, T. T., Aniskin, V. M., Gerbault-Seureau, M., Planton, H., Renard, J. P., Nguyen, B. X., Hassanin, A. & Volobouev, V. T. (2008). Phylogenetic position of the saola (*Pseudoryx nghetinhensis*) inferred from cytogenetic analysis of eleven species of Bovidae. *Cytogenetic and Genome Research* **122**: 41–54.

Nijman, I. J., van Boxtel, D. C. J., van Cann, L. M., Marnoch, Y., Cuppen, E. & Lenstra, J. A. (2008). Phylogeny of Y chromosomes from bovine species. *Cladistics* **24**: 723–726.

Nowak, R. M. (1999). *Walker's Mammals of the World.* Vol. 1. Baltimore, MD and London: Johns Hopkins University Press.

Olsen, S. J. (1990). Fossil ancestry of the yak, its cultural significance, and domestication in Tibet. *Proceedings of the Academy of Natural Sciences of Philadelphia* **142**: 73–100.

Pfeffer, P. (1969). Considérations sur l'écologie des forêts claires du Cambodge oriental. *Terre et Vie* **1**: 3–24.

Pilgrim, G. E. (1939). The fossil Bovidae of India. *Palaeontologia Indica NS* **26**: 1–356.

Pound, M. J., Haywood, A. M., Salzmann, U. & Riding, J. B. (2012). Global vegetation dynamics and latitudinal temperature gradients during the Mid to Late Miocene (15.97–5.33 Ma). *Earth-Science Reviews* **112**: 1–22.

Qiu, Z., & Qiu, Z. (1995). Chronological sequence and subdivision of Chinese Neogene mammalian faunas. *Palaeogeography Palaeoclimatology Palaeoecology* **116**: 41–70.

Qui, Z., Deng, T. & Wang, B. (2004). Early Pleistocene mammalian fauna from Longdan, Dongxiang, Gansu, China. *Palaeontologia Sinica* **191**: 1–198.

Ritz, L. R., Glowatzki-Mullis, M. L., MacHugh, D. E. & Gaillard, C. (2000). Phylogenetic analysis of the tribe Bovini using microsatellites. *Animal Genetics* **31**: 178–185.

Ropiquet, A., Gerbault-Seureau, M., Deuve, J. L., Gilbert, C., Pagacova, E., Chai, N., Rubes, J. & Hassanin, A. (2008). Chromosome evolution in the subtribe Bovina (Mammalia, Bovidae): the karyotype of the Cambodian banteng (*Bos javanicus birmanicus*) suggests that Robertsonian translocations are related to interspecific hybridization. *Chromosome Research* **16**: 1107–1118.

Ropiquet, A., Li, B. & Hassanin, A. (2009). SuperTRI: a new approach based on branch support analyses of multiple independent data sets for assessing reliability of phylogenetic inferences. *Comptes Rendus Biologies* **332**: 832–847.

Salzmann, U., Williams, M., Haywood, A. M., Johnson, A. L. A., Kender, S. & Zalasiewicz, J. (2011). Climate and environment of a Pliocene warm world. *Palaeogeography, Palaeoclimatology, Palaeoecology* **309**: 1–8.

Schaller, G. B. & Rabinowitz, A. (1995). The saola or spindle-horn bovid *Pseudoryx nghetinhensis* in Laos. *Oryx* **29**: 107–114.

Schreiber, A., Seibold, I., Nötzold, G. & Wink, M. (1999). Cytochrome b gene haplotypes characterize chromosomal lineages of anoa, the Sulawesi dwarf buffalo (Bovidae: *Bubalus sp.*). *Journal of Heredity* **90**: 165–176.

Scott, E. (2010). Extinctions, scenarios, and assumptions: changes in latest Pleistocene large herbivore abundance and distribution in western North America. *Quaternary International* **217**: 225–239.

Sepulchre, P., Jolly, D., Ducrocq, S., Chaimanee, Y., Jaeger, J. J. & Raillard, A. (2010). Mid-Tertiary paleoenvironments in Thailand: pollen evidence. *Climate of the Past* **6**: 461–473.

Shimamura, M., Yasue, H., Ohshima, K., Abe, H., Kato, H., Kishiro, T., Goto, M., Munechika, I. & Okada, N. (1997). Molecular evidence from retroposons that whales form a clade within even-toed ungulates. *Nature* **388**: 666–670.

Simpson, G. G. (1945). The principles of classification and a classification of mammals. *Bulletin of the American Museum of Natural History* **85**: 1–350.

Solounias, N. (1981). The Turolian fauna from the Island of Samos, Greece; with special emphasis on the hyaenids and the bovids. *Contributions to Vertebrate Evolution* **6**: 1–248.

Tanaka, K., Solis, C. D., Masangkay, J. S., Maeda, K., Kawamoto, Y. & Namikawa, T. (1996). Phylogenetic relationship among all living species of the genus *Bubalus* based on DNA sequences of the cytochrome b gene. *Biochemical Genetics* **34**: 443–452.

Tanaka, K., Matsuda, Y., Masangkay, J. S., Solis, C. D., Anunciado, R. V., Kuro-o, M. & Namikawa, T. (2000). Cytogenetic analysis of the tamaraw (*Bubalus mindorensis*): a comparison of R-banded karyotype and chromosomal distribution of centromeric satellite DNAs, telomeric sequence, and 18S-28S rRNA genes with domestic water buffaloes. *Journal of Heredity* **91**: 117–121.

Thomas, H. (1994). Anatomie crânienne et relations phylogénétiques du nouveau bovidé (*Pseudoryx nghetinhensis*) découvert dans la cordillère annamitique au Vietnam. *Mammalia* **58**: 453–481.

Tikhonov, A. (2008). *Bos primigenius.* In IUCN 2013. *IUCN Red List of Threatened Species.* Version 2013.2. www.iucnredlist.org.

Timmins, R. J., Hedges, S. & Duckworth, J. W. (2008). *Bos sauveli.* In *IUCN 2013. IUCN Red List of Threatened Species.* Version 2013.2. www.iucnredlist.org.

Tong, H. (2006). Mammalian faunal differentiations between North and South China during the Quaternary Period. *Anthropologie* **110**: 870–887.

van den Bergh, G. D. (1999). The Late Neogene elephantoid-bearing faunas of Indonesia and their palaeozoogeographic implications: a study of the terrestrial faunal succession of Sulawesi, Flores and Java, including evidence for early hominid dispersal east of Wallace's Line. *Scripta Geologica* **117**: 1–419.

van den Bergh, G. D., de Vos, J. & Sondaar, P. Y. (2001). The Late Quaternary palaeogeography of mammal evolution in the Indonesian Archipelago. *Palaeogeography Palaeoclimatology Palaeoecology* **171**: 385–408.

Van Gelder, R. G. (1977). Mammalian hybrids and generic limits. *American Museum Novitates* **2635**: 1–25.

Verkaar, E. L., Nijman, I. J., Beeke, M., Hanekamp, E. & Lenstra, J. A. (2004). Maternal and paternal lineages in cross-breeding bovine species: has wisent a hybrid origin? *Molecular Biology and Evolution* **21**: 1165–1170.

Vignaud, P., Duringer, P., Mackaye, H. T., Likius, A., Blondel, C., Boisserie, J. R., de Bonis, L., Eisenmann, V., Etienne, M. E., Geraads, D., Guy, F., Lehmann, T., Lihoreau, F., Lopez-Martinez, N., Mourer-Chauviré, C., Otero, O., Rage, J. C., Schuster, M., Viriot, L., Zazzo, A. & Brunet, M. (2002). Geology and palaeontology of the Upper Miocene Toros-Menalla hominid locality, Chad. *Nature* **418**: 152–155.

Vislobokova, I. (2005). On Pliocene faunas with Proboscideans in the territory of the former Soviet Union. *Quaternary International* **126–128**: 93–105.

Waddell, P. J., Okada, N. & Hasegawa, M. (1999). Towards resolving the interordinal relationships of placental mammals. *Systematic Biology* **48**: 1–5.

Wharton, C. H. (1957). An ecological study of the kouprey, *Novibos sauveli* (Urbain). *Monographs of the Institute of Science and Technology, Manila, Philippines* **5**: 1–107.

Yindee, M., Vlamings, B. H., Wajjwalku, W., Techakumphu, M., Lohachit, C., Sirivaidyapong, S., Thitaram, C., Amarasinghe, A. A., Alexander, P. A., Colenbrander, B. & Lenstra, J. A. (2010). Y-chromosomal variation confirms independent domestications of swamp and river buffalo. *Animal Genetics* **41**: 433–435.

Zhang, Y., Vankan, D., Zhang, Y. & Barker, J. S. (2011). Genetic differentiation of water buffalo (*Bubalus bubalis*) populations in China, Nepal and south-east Asia: inferences on the region of domestication of the swamp buffalo. *Animal Genetics* **42**: 366–377.

Zin-Maung-Maung-Thein, Takai, M., Uno, H., Wynn, J. G., Egi, N., Tsubamoto, T., Thaung-Htike, Aung-Naing-Soe, Maung-Maung, Nishimura, T. & Yoneda, M. (2011). Stable isotope analysis of the tooth enamel of Chaingzauk mammalian fauna (late Neogene, Myanmar) and its implication to paleoenvironment and paleogeography. *Palaeogeography Palaeoclimatology Palaeoecology* **300**: 11–22.

Chapter 2

Bovini as keystone species and landscape architects

Herbert H. T. Prins and Herman van Oeveren

Introduction

In this chapter we ask what role Bovini play in ecosystems. We concentrate on some of the more unexpected impacts bovines have, and not on the fact that they serve as food for large predators, or that if they are not killed by these and have not been found by vultures, that they are devoured by maggots. We also do not concentrate on their role as nutrient recyclers. In that respect, the study of Vinod & Sabu (2007) illustrates well the cascading effects of the (local) extinction of large herbivores on the community structure of, in their case, dung beetles. In the wild, the dung of wild water buffalo, *Bubalus arnee*, is decomposed by fungi; flies and beetles and insects can be found in it, thus providing food for the red junglefowl, *Gallus gallus*, and green peafowl, *Pavo muticus* (Chaiyarat *et al.* 2004; also see Dean & MacDonald 1981; Middleton 2013). The dung of domestic yaks, *Bos grunniens*, is similarly used (Wu & Sun 2010). A hypothesis we ignore here suggests that large herbivores have a negative effect on reptile biomass (Janzen 1976). The reason to investigate would be that our experience is the opposite of one of Janzen's suppositions, namely that large herbivores promote fire, and his other supposition that trampling would be bad for reptiles and amphibians. However, Friend & Taylor (1984) find both positive and negative effects of large herbivores. Other functions one normally does not think about include provision by water buffalo of blood meals for ticks, which may be infected with parasites, after which the buffalo can act as a reservoir for some diseases (Miranpuri 1988). Much literature on this subject is available (see, for example, Jongejan & Uilenberg 2004; Fyumagwa *et al.* 2007; Anderson *et al.* 2013 for African buffalo, *Syncerus caffer*, and ticks). It is also noteworthy that water buffalo were found to be near-free of ticks (Tulloch 1968).

In this review we will not pay attention to the effect of American bison, *Bison bison*, on prairie grass species diversity, simply because that is well established: their effects are far ranging and profound and can be studied so well because in many of the North American reserves bison occur in high enough densities to have measurable effects (see Knapp *et al.* 1999; Truett *et al.* 2001). Also, yaks change the grassland community, at least in captivity (Li *et al.* 2007; Qiao *et al.* 2009) but again, this is as expected. We also purposefully pay little attention to the African buffalo, simply because, as in the case of the American bison, so much is known about this species (see Chapter 20). Moreover, because this species is always part of a much larger assemblage of ungulates, its effects are difficult to disentangle from those of the other species. Finally, because aurochs, *Bos primigenius*, is extinct we only pay passing attention to the effects of the feral descendants of *Bos taurus*, since it is unclear whether their effects are identical to their wild forebears. We have never come across references to the functioning of feral zebu, and nature restoration projects with feral zebu in Pakistan or India are unknown to us. While the search term 'feral cattle' yielded 1290 returns using Google Scholar (28 January 2013), the term 'feral zebu' yielded only six, none of which was about nature restoration.

The roles of wild cattle: the near-unknowns – kouprey, gaur, banteng, wisent and yak

It is important to realize that the different Bovini are individual species and that they all have their own adaptations and followed their unique evolutionary pathways. It is thus certain that they function differently from cattle (e.g. Allred *et al.* 2011). This can be neatly illustrated by the nearly extinct yak. In the early 1990s there were an estimated 15 000 (Miller *et al.* 1994); how many there were before nomads started building up the competing herds of small stock and domestic yaks is unknown. An order of magnitude can be estimated, though, because approximately 14 million domestic yaks live at present in Tibet and some insight can be derived from these (see, for instance, Han *et al.* 2002; Long *et al.* 2008). As expected for animals adapted to living at high altitudes, they have larger hearts and lungs, higher concentrations of red blood cells and haemoglobin than other cattle species (Cai & Wiener 1995), their energy expenditure does not increase with altitude (but, of course, with lower temperatures: Han *et al.* 2002), their basal metabolic rate is lower than for comparable animals

Ecology, Evolution and Behaviour of Wild Cattle: Implications for Conservation, ed. M. Melletti and J. Burton. Published by Cambridge University Press.

and they are well adapted to low-quality food (Han *et al.* 2002 and references therein). The 2.5 million km^2 plateau is characterized by mismanagement and overgrazing (Long *et al.* 2008; Wang & Ba 2008), so it is unclear what is to be learnt from present-day domestic yak systems about the wild yak's natural functioning. However, some unexpected observations are made, such as that yaks do not use their tongue to graze (possibly as adaptation to eating frozen feed), the papillae on their tongue are better adapted to grinding fibrous food than the tongues of cattle, and the structure of the gustatory papillae suggest that they have a lesser sense of taste than cattle, which is interpreted as an adaptation to cope with antifeedants (Shao *et al.* 2010). They retain food for more time in their digestive tract than cattle and thus absorb nitrogen better (Schaefer *et al.* 1978). Indeed, yaks are not 'just a different cow'! Similar to the yak, the banteng, *Bos javanicus*, has also been domesticated, and its domestic descendants, the Bali cattle, number about 2–3 million. Studies are under way concerning the adaptations of the Bali cattle (Purwantara *et al.* 2012; see also Mohamad *et al.* 2012) but it is still too early to draw conclusions concerning its specific adaptations. Nevertheless, since the water retention capacity of banteng is much higher than of *Bos taurus* cattle (Siebert & MacFarlane 1969), it is very likely that Bali cattle are unlike ordinary cattle. Modern agricultural studies can teach a lot about how the banteng is adapted to its environment, and thus what its ecological functioning is. Genetic studies may reveal particular adaptations to hot environments or tick resistance, such as that found between taurine cattle and zebu (Chan *et al.* 2010), but it is unlikely that these studies will yield insight into the functional roles these bovines play in the landscape.

As with many of the Asian wild cattle, the situation of the gaur, *Bos gaurus*, is also precarious. Their total number is about 30 000 (Choudhury 2002), and as Choudhury states, 'surprisingly, this majestic animal has not been well studied'. Of course that is not so surprising given that administrative and safety issues abound in the range states, making research nearly impossible, especially for non-local scientists. It is less of a floodplain species than the water buffalo and moves out of these areas during the wet season. Apparently the habitats suiting its needs best are open areas in low-lying river valleys with fringing forests (Conry 1989; Choudhury 2002), although camera-trapping data suggest that more may occur in dense forests than based on visual estimates (Gray 2012). It is most likely the species did not occur in the drier northwest of the Indian subcontinent (Choudhury 2002).

Nguyen (2009) points out that both banteng and gaur, *Bos gaurus*, are severely endangered in Vietnam, mostly because of illegal hunting and habitat loss. In his words, 'only a small fraction of the gaur's former population remains in some isolated areas'. Nguyen (2009) even reports trophy hunting of these threatened and protected species in the national parks, but logging and fishing are also taking place. The numbers of banteng and gaur are low, and the kouprey, *Bos sauveli*, is extinct. In other words, the chance that information concerning the functioning of these Bovini can be deduced at the moment from Vietnam, or even Laos, Malaysia, Cambodia, Thailand or Myanmar, is nil. Indeed, as Gray (2012) points out 'monitoring charismatic large mammal species in tropical forests is a conservation challenge. Many such species are shy, solitary, nocturnal, wide-ranging, or occur at low densities.' One should question, however, the validity of this defence for not having much knowledge on the ecology of banteng or gaur; indeed, gaur in Bandipur NP (India) or banteng in Baluran NP (Indonesia) are neither shy nor solitary (personal observation), and we know already quite a lot about their basic ecology (e.g. Ahrestani *et al.* 2011 2012). It is likely that the incessant persecution, harassment, competition with livestock, marginalization, etc., have made these animals rare and shy. A simple lack of effective conservation (see Heinen & Srikosamatara 1996) precludes the opportunity to study these animals in the natural state with natural densities so as to learn what their function is, or could have been. Not only a 'conservation tragedy' (Pedrono *et al.* 2009) is taking place in Asia, but also a 'knowledge tragedy', and this conservation disaster is not 'impending', as Sodhi *et al.* (2004) write. It is not only in Southeast Asia, but has taken place and is taking place in West , South, East and Central Asia under the very eyes of governments that sign bills and treaties concerning biodiversity but do not stem the conservation crisis. Yet, when conservation is taken seriously and becomes effective, population recovery can be fast, even for Asian bovines (Khan *et al.* 1996; Madhusudan 2004; Steinmetz *et al.* 2010).

Currently 15 000 wild yaks still live in areas where there are nearly 1000 times more domestic yaks (see above). Where there are now 2–3 million Bali cattle, there are only about 5000 banteng in the wild in Indonesia, again pointing at a 1000-times reduction of this species. The water buffalo is a critically endangered species, with only a few hundred or thousand individuals in the wild (Chaiyarat *et al.* 2004), but 160–180 million live in a domestic form (Miranpuri 1988). The likely reduction is in the order of about 10 000 times. Given the small proportion of (sub-)tropical Asia under effective conservation, it would not be an exaggeration to assume that presently less than 0.1% of the potential number of gaurs still survive in India and Southeast Asia (about 20 000 and 10 000, respectively), which yields a rough estimate of potentially some 20 million and 10 million gaurs 'missing' from these lands. From Europe through the Middle East to India, the aurochs, *Bos primigenius*, is extinct, and from Siberia to Europe the European bison, *Bison bonasus*, nearly so. Given these enormous reductions in numbers and the paucity of in-depth ecological studies into the functioning of wild cattle, can anything meaningfully still be deduced about their functioning? It is only because the American bison is rebounding so well in terms of numbers that it becomes increasingly clear that the role of Bovini in ecosystems is not merely epiphenomenal!

Effects on forest structure

Wild cattle are not merely cattle that look a bit different. They are species with their own adaptations and thus do not function just like other cows. Gray's (2012) remark that 'By providing graze, browse, and fruits at a single location the mixed deciduous and semi-evergreen forest may, seasonally, offer a richer array of food types than other habitats' is interesting because it mentions frugivory by a bovine. Also, Purwantara *et al.* (2012) mention fruit eating by banteng. These observations point in the direction of seed dispersal by large Bovini in tropical forests (but see Dinerstein 1989). Cattle have been shown to be able to disperse seeds (Middleton & Mason 1992; Traba *et al.* 2003; Nogales *et al.* 2005; for reviews see Corlett 1998 and references therein; Traveset 1998), while water buffalo are suspected (Khan *et al.* 2005). We think the role of bovines in seed distribution is an under-researched area, especially in tropical rainforests. Elsewhere, the European bison is an important seed disperser (Pykälä 2001; Jaroszewicz *et al.* 2008; Jaroszewicz & Pirożnikow 2011), as is the North American bison (Rosas *et al.* 2008) and the yak (Yu *et al.* 2012). Even Aristotle (~300 BCE) noted that cattle *Bos taurus* were fruit eaters (Aristotle 1984, p. 931). It is thus clear that, in general, Bovini may have an important function in structuring vegetation communities through seed dispersal and creating situations that are good for seed germination.

Because Bovini are effective seed dispersers, they affect forest structure, but their effects are exerted in other life stages of trees too. In an exclosure experiment with European bison and other ungulates in a temperate natural forest, they did not affect seedling density but they had a strong effect on saplings and thus affected forest vegetation structure (Kuijper *et al.* 2010a 2010b). In another study in the same area it was found that oak, *Quercus robur*, regeneration was possible in the presence of large herbivores, including European bison (Bobiec *et al.* 2011), in contrast to studies of cattle by, for example, Olff *et al.* (1999) or Bakker *et al.* (2004). This may either point towards a different functioning of *Bison bonasus* as compared to *Bos taurus*, or it may have been a scale-effect too, since the cattle studies were conducted in areas that were fenced. As a result of these insights, there is an increasing effort to reintroduce European bison in European protected areas (e.g. Perzanowski *et al.* 2004; Belousova *et al.* 2005; Heurich *et al.* 2011), but also in the Netherlands, Sweden and France.

Much insight could be gleaned from two completely unforeseen 'experiments' with large bovines: in the mid 1800s both water buffalo and banteng were introduced into northern Australia, and this led to major insights into the functioning of these bovines.

Even though banteng populations became quite large in northern Australia (some 70 per km^2) their effect was not very noticeable (Bowman & Panton 1991). Apparently little research has been done since that time on this species, which increased from around 20 individuals in 1849 to 10 000 or more in 2005 (deKoninck 2005) an increase (~9% per year) that must have been close to its intrinsic growth rate.

In contrast, however, the build-up of the water buffalo population in Australia (see next section) resulted in an enormous impact on woodland structure. Analysis of aerial photographs show that woody cover declined during the period when buffalo numbers rapidly increased, and increased again after the first buffalo were removed (Petty *et al.* 2007; Petty & Werner 2010), an effect which was confirmed by experimental removals of water buffalo (Skeat *et al.* 1996). Experimental studies showed that the direct effect of buffalo on woodlands was to reduce herbaceous vegetation (mainly grasses), which in turn enhanced the growth and survival of established trees of all sizes (Petty *et al.* 2007). Competition for water among grasses, juvenile plants and canopy trees is very strong in the top soil during the early dry season in this part of Australia (Chen *et al.* 2004), as we have shown too in southern Africa (van der Waal *et al.* 2009; 2011). An important insight gained from the build-up of the water buffalo population and its subsequent removal was that the ecological cascades were not identical, thus showing excellent proof of ecosystem hysteresis (Petty *et al.* 2007). In other words, the functions of these Bovini is much more significant than can be deduced from simple observational studies on marginal surviving populations in pauperized ecosystems in Asia.

Very large effects of water buffalo

Because of the rarity of the wild water buffalo, we cannot hope to learn much from observations in the wild about how the species functions. Like the water buffalo, the related tamaraw, *Bubalus mindorensis*, make mud wallows (Lydekker 1898; Talbot & Talbot 1966), but otherwise little is known of this bovine and it suffers from poor protection (Kuehn 1986; Ishihara & Kanai 2010). Of the other two dwarf buffalo species, lowland anoa, *Bubalus depressicornis*, and mountain anoa, *Bubalus quarlesi*, essentially nothing is known about their ecology, but their protection status is as deficient as the other Asian buffalo species (Burton *et al.* 2005). To make any meaningful statement about the functioning of these buffalo species in the forests or floodplains of tropical Asia is impossible: these species occur in unnatural low densities and are nowadays so rare that they cannot be well studied. The functions that can be deduced from wild Bovini pale in comparison to those of domesticated forms – for instance, in 1988 at least 7000 cattle were in the 'protected' Kosi Tappu Wildlife Reserve of Nepal, together with at least 130 resident domestic buffalo and 97 wild Asian water buffalo (Heinen 1993). This is akin to the situation in Hemis National Park in India, where about 8000 domestic sheep and goats co-occurred with fewer than 100 wild sheep, but where the decline of these wild sheep was ascribed to wild predators in 2007 (personal communication with the local protection agency). However, a little can still be learned from these wild Asian studies. A critical point that emerges from the Chaiyarat *et al.* (2004) study, for instance, is the high number of wallows in the area where water buffalo are still left in Thailand: this sheds important light on their functioning (see below).

Much is known about water buffalo because of their long history of domestication and their present use in agriculture (see Granum *et al.* 2007; Wanapat *et al.* 2009). Their digestive efficiency is as well understood as that of cattle (see the many references in Wanapat *et al.* 2009). Water buffalo are perhaps not as successful as the descendants of aurochs, yet they also number in the millions (see above). Although now rare in the wild, mainly due to lack of effective conservation, faulty reserve design, disease transfer from domestic stock and competition with livestock in reserves (Heinen 1993; Khan *et al.* 1996; Heinen & Singh 2001; Chaiyarat *et al.* 2004), water buffalo can show a sustained growth rate when not hunted or outcompeted. For instance, in a new pristine environment they increased from around 65 animals in 1849 to 120 000 in *c.*1960 in northern Australia (Tulloch 1969), which represents an annual rate of increase of ~7%. In the early 1980s their number had further increased to about 280 000 individuals (Petty *et al.* 2007). Tulloch (1969) described their behaviour, saying:

> All animals had well-defined tracks along which they travelled. They had, in their respective areas, manure heaps which they regularly visited. Each animal, in the wet season, would go to a particular pool of water, smell it, and lie in it . . . they moved over a very limited area; here they lived and carried out their daily activities. Within this area (or home range) were certain fixed points or places visited regularly. These included three toilet areas, a drinking point, feeding grounds, a pool in which the animals could lie, a camping area, and certain trees on which they rubbed their bodies. There were tracks leading from one fixed point to another.

This encapsulates their functioning in a tropical ecosystem without many natural enemies, because in the period that the water buffalo increased, the salt water crocodile of northern Australia, *Crocodylus porosus*, was nearly eradicated: concentrated grazing and wallowing were their main activities. This led to competition with feral cattle and a complete altering of coastal ecosystems (Werner 2014). The functioning of the dung heaps that Tulloch (1969) describes was not investigated (Werner, personal communication) but it could be functioning the same way (abandoned) cattle kraals do in Africa (e.g. van der Waal *et al.* 2009; 2011). These water buffalo dung heaps were typically 6 × 4 m and 40 cm thick (Tulloch 1979), which makes them very comparable to these kraals in Africa.

The functioning of water buffalo in this exotic landscape brought to the fore the huge impact wallowing can have. Indeed, the eradication programme that was set by the Australian government was aimed at the total removal of water buffalo from the landscape as their wallowing altered pools of clear water with abundant water-lilies and other plants on which magpie geese, *Anseranas semipalmata*, but also Aboriginal people, depended, changing them to murky waters with much lower biodiversity (e.g. Corbett *et al.* 1996; Petty *et al.* 2007; Werner 2014). Apart from the swamp structure, the buffalo had much impact on the tidal creeks and channels. By removing the vegetation, the erosion rate increased and tidal channels became much deeper (Petty *et al.* 2007). A similar effect is seen on European saltmarshes that are grazed by cattle (personal observation in the Dutch Waddenzee saltmarshes), as is the disappearance of *Phragmites* from floodplains grazed by free-ranging feral water buffalo (Petty *et al.* 2007). It is worth citing the series of impacts that the buffalo had in Kakadoo:

> buffalo removed most of the surface vegetation, starting with freshwater perennial grasses, including *Hymenachne* and *Phragmites*, then turning to the less palatable but more abundant *Paspalum distichum*, a salt-tolerant grass which stabilizes tidal levees and is key in preventing the formation of saltwater bearing channels. Channel formation rates were correlated with signs of buffalo impact . . . The denuded floodplain surface was scoured by freshwater draining off the floodplains at the end of the wet season. This created channels that connected freshwater floodplains directly to tide-bearing channels. These channels directly and substantially impacted floodplain ecology by permitting the influx of tidal water to regions of salt-intolerant vegetation, and permitting early drainage of impounded freshwater, thus leading to early drying of floodplains in the dry season, and increased subsoil salinity. High sediment loads carried by incoming tides also filled in permanent freshwater lakes as has occurred on at least two occasions.
>
> (Petty *et al.* 2007: 451)

As in any good cascade, we may expect that water buffalo had thus far-reaching impacts on other species. Indeed, magpie geese were reported to have declined when buffalo increased (Werner 2014; but see Corbett *et al.* 1996), but also dusky rat, *Rattus colletti*, numbers and other small vertebrates declined (Werner 2014). When the buffalo were removed, feral wild pigs increased three-fold (Petty *et al.* 2007), thus exposing an unexpected causal link between these two species. Buffalo trampling in wet terrains has far-reaching consequences for other species, like the barn owl, *Tyto alba*, lapwing, *Vanella miles*, or insects and amphibians (Friend & Taylor 1984).

In this context it is worth noting that the Australian reaction – kill and remove water buffalo – is quite different from new management insights developed in Germany in the Project Bubalus, where water buffalo are used to manage wetlands. One of the first things the water buffalo did was digging wallows; the project has not been under way long enough to assess its impact on biodiversity, but the first signs are positive (Wiegleb & Krawczynski 2010).

Bovini as keystone species

Water buffalo have all the hallmarks of a keystone species and landscape architect. In Australia this was undesired, because they are aliens and there are many reasons why the Australian reserve managers and local people opted for a landscape without water buffalo (Russell-Smith & Bowman 1992; Werner 2014) but (not at the same location) with banteng (deKoninck 2005). Do other bovines have these very large effects on landscapes too? The answer is 'yes'. The effect of American bison on water quality, silt load and invertebrates seems to be much smaller than that of cattle in prairie streams (Fritz *et al.* 1999), but cattle do have very large impacts in areas of flowing water. In contrast to the Australian situation, their landscape impact is viewed positively in some other ecosystems. For instance, in the Dutch, German and Danish Waddenzee area there are extensive saltmarshes that are protected for biodiversity purposes, aesthetics and foreshore protection. Ungrazed saltmarshes are dominated by *Elytrigia atherica*, which is basically unpalatable to livestock but also to other animals like hares (Kuijper & Bakker 2008). These saltmarshes can accommodate large numbers of barnacle geese, *Branta leucopsis*, which forage on *Festuca rubra* (Ydenberg & Prins 1981), a grass species that is replaced by *Elytrigia*. Heavy cattle grazing reverts the succession (Bakker 1985) back to *Festuca rubra* and *Puccinellia maritima*, thus enabling goose grazing. More examples of the positive effects of cattle grazing on other bird species can be found in van Wieren (1998). The impact of water buffalo in northern Australia was evaluated negatively, but the point here is that the impact of this Bovini on the tidal and riverine landscape was as great as what was experienced in the Waddenzee area.

American bison, in contrast to water buffalo and African buffalo, use dry wallows instead of wet ones. There may have once been some 100 million bison wallows in the North American prairies (McMillan *et al.* 2011), often still visible a century after their local eradication (personal observation), but their effect is very different from the wallowing behaviour in areas where flowing water gets access to the substrate through the wallowing behaviour. Yet even under these circumstances, their wallowing behaviour influences hydrology and run-off patterns (Butler 2006). Because of their strong impact on the landscape, Fox *et al.* (2012) even argue that American bison should be reintroduced in a particular area for the benefit of restoring an ecosystem and its endangered plant species. In this respect cattle are different because they wallow much less than American bison and rarely construct their own wallows (Hartnett *et al.* 1997): it is their trampling that mainly changes the landscape in wet terrains (Mosley *et al.* 1997; Clark 1998; Belsky *et al.* 1999). Gaur and banteng apparently do not wallow (Hubback 1937; Edmond-Blanc 1947) but perhaps kouprey did (Edmond-Blanc 1947). European bison do wallow, but the explanation for their behaviour appears simpler than for American bison, where seven explanations have been given for the behaviour (McMillan *et al.* 2000, 2011).

Finally, yak also wallow, like American bison, and their breaking of the turf cover of *Kobresia* vegetation may have (had) far-reaching consequences for the stability of the Alpine Steppe of the Tibetan Plateau (Miehe *et al.* 2011; Unteregelsbacher *et al.* 2012). We do not know what yak numbers were before they were persecuted, hunted and outcompeted, but it is likely that yaks had a keystone function (cf. Linder *et al.* 2012), like American bison.

Conclusion

The forest-living bovines interact with the forest mainly as seed dispersers, but they also change the forest structure by affecting the interspecific differential mortality rates of the different woody species, as was shown by the studies of European bison and Asian water buffalo. It is likely that the other Asian forest species (gaur, banteng and gayal) and extinct Eurasian aurochs did the same. The sorry state of survival of the Asian forest species, however, makes it difficult to prove. The most important functioning of Bovini is their ability to modify landscapes at quite an unprecedented scale, mainly through their interaction with rivulets and creeks, but possibly with frost and permafrost too. Because of their vast impact on the landscape, at least three of these bovines are keystone species, namely American bison, yak and water buffalo. This vast impact of bovines has been recognized in the rangeland management in the United States and Canada, where stream protection against cattle is a very important issue. The underlying cause of these cascading effects on the landscape is the behaviour of the Bovini – wallowing and seeking cooling in streams. From studies on water turnover in the 1960s, it may be expected that the effect of these species can be arranged in increasing order as banteng < zebu (*Bos indicus*) < shorthorn (*Bos taurus*) < water buffalo (Siebert & MacFarlane 1969). African buffalo, like water buffalo, are known for their wallowing in pools and high water turnover (Prins 1996). On the basis of their physiology we expect a very large impact of buffalo on riparian and estuarine systems, and less so by bovines that are less water demanding. This clearly is an avenue for further research into the functioning of these magnificent animals. With the realization that Bovini are keystone species, the call for their protection and reintroduction gains in power: only when their densities reach those that they had in the past can their keystone role be played out again.

Acknowledgements

The study for this chapter was possible through a Fulbright Grant for the first author and a research stint at Princeton University. The long-term research of Jan P. Bakker and his penetrating insights into saltmarsh ecology and the effects of cattle there have shaped our insights profoundly, for which we are grateful.

References

Ahrestani, F. S., Heitkönig, I. M. A. & Prins, H. H. T. (2011). Herbaceous production in South India: limiting factors and implications for large herbivores. *Plant and Soil* **349**: 319–330.

Ahrestani, F. S., Heitkönig, I. M. A. & Prins, H. H. T. (2012). Diet and habitat-niche relationships within an assemblage of large herbivores in a seasonal tropical forest. *Journal of Tropical Ecology* **28**: 385–394.

Allred, B. W., Fuhlendorf, S. D. & Hamilton, R. G. (2011). The role of herbivores in Great Plains conservation: comparative ecology of bison and cattle. *Ecosphere* **2(26)**: 1–17.

Anderson, K., Ezenwa, V. & Jolles, A. (2013). Tick infestation patterns in free ranging African buffalo (*Syncerus caffer*): effects of host innate immunity and niche segregation among tick species. *International Journal for Parasitology: Parasites and Wildlife* **2**: 1–9.

Aristotle (1984). History of Animals, Book VIII, 7. In *The Complete Works of Aristotle, the Revised Oxford Translation. Bollinger Series LXXI.2*, p. 931. Ed. Barnes, J. Princeton, NJ: Princeton University Press.

Bakker, E. S., Olff, H., Vandenberghe, C., De Maeyer, K., Smit, R., Gleichman, J. M. & Vera, F. W. M. (2004). Ecological anachronisms in the recruitment of temperate light demanding tree species in wooded pastures. *Journal of Applied Ecology* **41**: 571–582.

Bakker, J. P. (1985). The impact of grazing on plant communities, plant populations and soil conditions on salt marshes. *Plant Ecology* **62**: 391–398.

Belousova, I. P., Smirnov, K. A., Kazmin, V. D. & Kudrjavtsev, I. V. (2005). Reintroduction of the European bison into the forest ecosystem of the Orlovskoe Poles'e National Park. *Russian Journal of Ecology* **36**: 115–119.

Belsky, A. J., Matzke, A. & Uselman, S. (1999). Survey of livestock influences on stream and riparian ecosystems in the western United States. *Journal of Soil and Water Conservation* **54**: 413–431.

Bobiec, A., Kuijper, D. P., Niklasson, M., Romankiewicz, A. & Solecka, K. (2011). Oak (*Quercus robur* L.) regeneration in early successional woodlands grazed by wild ungulates in the absence of livestock. *Forest Ecology and Management* **262**: 780–790.

Bowman, D. M. J. S. & Panton, W. J. (1991). Sign and habitat impact of banteng (*Bos javanicus*) and pig (*Sus scrofa*), Cobourg Peninsula, northern Australia. *Australian Journal of Ecology* **16**: 15–17.

Burton, J. A., Hedges, S. & Mustari, A. H. (2005). The taxonomic status, distribution and conservation of the lowland anoa *Bubalus depressicornis* and mountain anoa *Bubalus quarlesi*. *Mammal Review* **35**: 25–50.

Butler, D. R. (2006). Human-induced changes in animal populations and distributions, and the subsequent effects on fluvial systems. *Geomorphology* **79**: 448–459.

Cai, L. & Wiener, G. (1995). *The Yak*. Bangkok: FAO Regional Office for Asia and the Pacific.

Chaiyarat, R., Lauhacindra, V., Kutintara, U., Bhumpakphan, N. & Prayurasiddhi, T. (2004). Population of wild water buffalo (*Bubalus bubalis*) in Huai Kha Khaeng Wildlife Sanctuary, Thailand. *Natural History Bulletin of the Siam Society* **52**: 151–162.

Chan, E. K., Nagaraj, S. H. & Reverter, A. (2010). The evolution of tropical adaptation: comparing taurine and zebu cattle. *Animal Genetics* **41**: 467–477.

Chen, X., Eamus, D. & Hutley, L. B. (2004). Seasonal patterns of fine-root productivity and turnover in a tropical savanna of northern Australia. *Journal of Tropical Ecology* **20**: 221–224.

Choudhury, A. (2002). Distribution and conservation of the Gaur *Bos gaurus* in the Indian Subcontinent. *Mammal Review* **32**: 199–226.

Clark, E. A. (1998). Landscape variables affecting livestock impacts on water quality in the humid temperature zone. *Canadian Journal of Plant Science* **78**: 181–190.

Conry, P. J. (1989). Gaur *Bos gaurus* and development in Malaysia. *Biological Conservation* **49**: 47–65.

Corbett, L., Hertog, A. L. & Muller, W. J. (1996). An experimental study of the impact of feral swamp buffalo *Bubalus bubalis* on the breeding habitat and nesting success of magpie geese *Anseranas semipalmata* in Kakadu National Park. *Biological Conservation* **76**: 277–287.

Corlett, R. T. (1998). Frugivory and seed dispersal by vertebrates in the Oriental (Indomalayan) Region. *Biological Reviews of the Cambridge Philosophical Society* **73**: 413–448.

Dean, W. R. J. & MacDonald, I. A. W. (1981). A review of African birds feeding in association with mammals. *Ostrich* **52**: 135–155.

deKoninck, V. (2005). Joint management of banteng (*Bos javanicus*) in a contested cultural landscape: observations and implications. *Human Dimensions of Wildlife* **10**: 123–135.

Dinerstein, E. (1989). The foliage-as-fruit hypothesis and the feeding behavior of South Asian ungulates. *Biotropica* **21**: 214–218.

Edmond-Blanc, F. (1947). A contribution to the knowledge of the Cambodian wild ox or kouproh. *Journal of Mammalogy* **28**: 245–248.

Fox, T. A., Hugenholtz, C. H., Bender, D. & Gates, C. C. (2012). Can bison play a role in conserving habitat for endangered sandhills species in Canada? *Biodiversity and Conservation* **21**: 1441–1455

Friend, G. R. & Taylor, J. A. (1984). Ground surface features attributable to feral buffalo, *Bubalus bubalis*: II. Their relationship to the abundance of small animal species. *Australian Wildlife Resource* **11**: 311–323.

Fritz, K. M., Dodds, W. K. & Pontius, J. (1999). The effects of bison crossings on the macroinvertebrate community in a tallgrass prairie stream. *The American Midland Naturalist* **141**: 253–265.

Fyumagwa, R. D., Runyoro, V. Horak, I. G. & Hoare, R. (2007). Ecology and control of ticks as disease vectors in wildlife of the Ngorongoro Crater, Tanzania. *South African Journal of Wildlife Research* **37**: 79–90.

Granum, G., Wanapat, M., Pakdee, P., Wachirapakorn, C. & Toburan, W. (2007). A comparative study on the effect of cassava hay supplementation in swamp buffaloes (*Bubalus bubalis*) and cattle (*Bos indicus*). *Asian-Australian Journal of Animal Science* **20**: 1389–1396.

Gray, T. N. E. (2012). Studying large mammals with imperfect detection: status and habitat preferences of wild cattle and large carnivores in eastern Cambodia. *Biotropica* **44**: 531–536.

Han, X. T., Xie, A. Y., Bi, X. C., Liu, S. J. & Hu, L. H. (2002). Effects of high altitude and season on fasting heat production in the yak *Bos grunniens* or *Poephagus grunniens*. *British Journal of Nutrition* **88**: 189–197.

Hartnett, D. C., Steuter, A. A. & Hickman, K. R. (1997). Comparative ecology of native and introduced ungulates. In *Ecology and Conservation of Great Plains Vertebrates*, pp. 72–101. Eds. Knopf, F. & Samsons, F. New York: Springer

Heinen, J. T. (1993). Population viability and management recommendations for wild water buffalo *Bubalus bubalis* in Kosi Tappu Wildlife Reserve, Nepal. *Biological Conservation* **65**: 29–34.

Heinen, J. T. & Singh, G. R. (2001). A census and some management implications for wild buffalo (*Bubalus bubalis*) in Nepal. *Biological Conservation* **101**: 391–394.

Heinen, J. T. & Srikosamatara, S. (1996). Status and protection of Asian wild cattle and buffalo. *Conservation Biology* **10**: 931–934.

Heurich, M., Baierl, F., Günther, S. & Sinner, K. F. (2011). Management and conservation of large mammals in the Bavarian Forest National Park. *Silva Gabreta* **17**: 1–18.

Hubback, T. R. (1937). The Malayan gaur or seladang. *Journal of Mammalogy* **18**(3): 267–279.

Ishihara, S. & Kanai, Y. (2010). Population dynamics, herd behavior, and genetic analysis of the critically endangered tamaraw (*Bubalus mindorensis*): a mini review. *Journal of Developments in Sustainable Agriculture* **5**: 136–140.

Janzen, D. H. (1976). The depression of reptile biomass by large herbivores. *American Naturalist* **110**: 371–400.

Jaroszewicz, B. J. B. & Pirożnikow E. P. E. (2011). Dung longevity influences the fate of endozoochorically dispersed seeds in forest ecosystems. *Botany* **89**: 779–785.

Jaroszewicz, B. J. B., Pirożnikow, E. P. E. & Sagehorn, R. S. R. (2008). The European bison as seed dispersers: the effect on the species composition of a disturbed pine forest community. *Botany* **86**: 475–484.

Jongejan, F. & Uilenberg, G. (2004). The global importance of ticks. *Parasitology* **129**: S3–S14.

Khan, J. A., Chellam, R., Rodgers, W. A. & Johnsingh, A. J. T. (1996). Ungulate densities and biomass in the tropical dry deciduous forests of Gir, Gujarat, India. *Journal of Tropical Ecology* **12**: 149–162.

Khan, M. L., Bhuyan, P. & Tripathi, R. S. (2005). Effects of forest disturbance on fruit set, seed dispersal and predation of Rudraksh (*Elaeocarpus ganitrus* Roxb.) in northeast India. *Current Science* **88**: 133–142.

Knapp, A. K., Blair, J. M., Briggs, J. M., Collins, S. L., Hartnett, D. C., Johnson, L. C. & Towne, E. G. (1999). The keystone role of bison in North American tallgrass prairie. *BioScience* **49**: 39–50.

Kuehn, D. W. (1986). Population and social characteristics of the tamarao (*Bubalus mindorensis*). *Biotropica* **18**: 263–266.

Kuijper, D. P. J. & Bakker, J. P. (2008). Unpreferred plants affect patch choice and spatial distribution of European brown hares. *Acta Oecologica* **34**: 339–344.

Kuijper, D. P., Cromsigt, J. P., Jędrzejewska, B., Miścicki, S., Churski, M., Jędrzejewski, W. & Kweczlich, I. (2010a). Bottom-up versus top-down control of tree regeneration in the Białowieża Primeval Forest, Poland. *Journal of Ecology* **98**: 888–899.

Kuijper, D. P., Jędrzejewska, B., Brzeziecki, B., Churski, M., Jędrzejewski, W. & Żybura, H. (2010b). Fluctuating ungulate density shapes tree recruitment in natural stands of the Białowieża Primeval Forest, Poland. *Journal of Vegetation Science* **21**: 1082–1098.

Li, G., Hasegawa, N., Song, R., Idota, S., Wang, Y. & Feng, S. (2007). Effects of rotational grazing of yak (*Bos grunniens*) in warm and cold seasons on floral diversity and biomass of *Potentilla fruticosa* alpine rangeland in northern Qinghai-Tibetan Plateau. *Animal Behaviour and Management* **43**: 1–8.

Linder, H. P., Bykova, O., Dyke, J., Etienne, R. S., Hickler, T., Kühn, I., Marion, G., Ohlemüller, R., Schymanski, S. J. & Singer, A. (2012). Biotic modifiers, environmental modulation and species distribution models. *Journal of Biogeography* **39**: 2179–2190.

Long, R. J., Ding, L. M., Shang, Z. H. & Guo, X. H. (2008). The yak grazing system on the Qinghai-Tibetan plateau and its status. *The Rangeland Journal* **30**: 241–246.

Lydekker, R. (1898). *Wild Oxen, Sheep and Goats of All Lands, Living and Extinct*. London: Rowland Wards.

Madhusudan, M. D. (2004). Recovery of wild large herbivores following livestock decline in a tropical Indian wildlife reserve. *Journal of Applied Ecology* **41**: 858–869.

McMillan, B. R., Cottam, M. R. & Kaufman, D. W. (2000). Wallowing behavior of American bison (*Bos bison*) in tallgrass prairie: an examination of alternate explanations. *The American Midland Naturalist* **144**: 159–167.

McMillan, B. R., Pfeiffer, K. A. & Kaufman, D. W. (2011). Vegetation responses to an animal-generated disturbance (bison wallows) in tallgrass prairie. *The American Midland Naturalist* **165**: 60–73.

Middleton, B. A. (2013). Rediscovering traditional vegetation management in preserves: trading experiences between cultures and continents. *Biological Conservation* **158**: 271–279.

Middleton, B. A. & Mason, D. H. (1992). Seed herbivory by nilgai, feral cattle, and wild boar in the Keoladeo National Park, India. *Biotropica* **24**: 538–543.

Miehe, G., Miehe, S., Bach, K., Nölling, J., Hanspach, J., Reudenbach, C., Kaiser, K., Wesche, K., Mosbrugger, V., Yang, Y. P. & Ma, Y. M. (2011). Plant communities of central Tibetan pastures in the Alpine Steppe/*Kobresia pygmaea* ecotone. *Journal of Arid Environments* **75**: 711–723.

Miller, D., Harris, R. & Cai, G. (1994). Wild yaks and their conservation on the Tibetan Plateau. In *Proceedings of the 1st International Congress on Yak*, pp. 24–34. Lanzhou: Gansu Agricultural University.

Miranpuri, G. S. (1988). Ticks parasitising the Indian buffalo (*Bubalus bubalis*) and their possible role in disease transmission. *Veterinary Parasitology* **27**: 357–362.

Mohamad, K., Olsson, M., Andersson, G., Purwantara, B., Van Tol, H. T. A., Rodriguez-Martinez, H., Colenbrander, B. & Lenstra, J. A. (2012). The origin of Indonesian cattle and conservation genetics of the Bali cattle breed. *Reproduction in Domestic Animals* **47**(1): 18–20.

Mosley, J. C., Cook, P. S., Griffis, A. J. & O'Laughlin, J. (1997). Guidelines for managing cattle grazing in riparian areas to protect water quality: review of research and best management practices policy. Idaho Forest, Wildlife and Range Policy Analysis Group Report 15. oregonstate.edu/dept/range/sites/default/files/RNG455-555PDFLinks/Mosley_grazing_strategies.pdf

Nguyen, M. H. (2009). The status of vulnerable gaur *Bos gaurus* and endangered banteng *Bos javanicus* in Ea So Nature Reserve and Yok Don and Cat Tien National Parks, Vietnam. *Oryx* **43**: 129–135.

Nogales, M., Nieves, C., Illera, J. C., Padilla, D. P. & Traveset, A. (2005). Effect of native and alien vertebrate frugivores on seed viability and germination patterns of *Rubia fruticosa* (Rubiaceae) in the eastern Canary Islands. *Functional Ecology* **19**: 429–436.

Olff, H., Vera, F. W. M., Bokdam, J., Bakker, E. S., Gleichman, J. M., De Maeyer, K. & Smit, R. (1999). Shifting mosaics in grazed woodlands driven by the alteration of plant facilitation and competition. *Plant Biology* **1**: 127–137.

Pedrono, M., Tuan, H. M., Chouteau, P. & Vallejo, F. (2009). Status and distribution of the endangered banteng *Bos javanicus birmanicus* in Vietnam: a conservation tragedy. *Oryx* **43**: 618–625.

Perzanowski, K., Olech, W. & Kozak, I. (2004). Constraints for re-establishing a meta-population of the European bison in Ukraine. *Biological Conservation* **120**: 345–353.

Petty, A. M. & Werner, P. A. (2010). How many buffalo does it take to change a savanna? A response to Bowman et al. (2008). *Journal of Biogeography* **37**: 193–199.

Petty, A. M., Werner, P. A., Lehmann, C. E. R., Riley, J. E., Banfai, D. S. & Elliott, L. P. (2007) Savanna responses to feral buffalo in Kakadu National Park, Australia. *Ecological Monographs* **77**: 441–463.

Prins, H. H. T. (1996). *Behaviour and Ecology of the African Buffalo: Social Inequality and Decision Making*. London: Chapman & Hall.

Purwantara, B., Noor, R. R., Andersson, G. & Rodriguez-Martinez, H. (2012). Banteng and Bali cattle in Indonesia: status and forecasts. *Reproduction in Domestic Animals* **47(1)**: 2–6.

Pykälä, J. (2001). Mitigating human effects on European biodiversity through traditional animal husbandry. *Conservation Biology* **14**: 705–712.

Qiao, Y., Wang, Q. & Wang, W. (2009). Yak (*Bos grunniens* L.) grazing effects on vegetation of alpine meadow with *Potentilla fruticosa* L. (Rosaceae) shrub in Qinghai-Tibet Plateau. *Polish Journal of Ecology* **57**: 769–777.

Rosas, C. A., Engle, D. M., Shaw, J. H. & Palmer, M. W. (2008). Seed dispersal by *Bison bison* in a tallgrass prairie. *Journal of Vegetation Science* **19**: 769–778.

Russell-Smith, J. & Bowman, D. M. J. S. (1992). Conservation of monsoon rainforest isolates in the Northern Territory, Australia. *Biological Conservation* **59**: 51–63.

Schaefer, A. L., Young, B. A. & Chimwano, A. M. (1978). Ration digestion and retention times of digesta in domestic cattle (*Bos taurus*), American bison (*Bison bison*), and Tibetan yak (*Bos grunniens*). *Canadian Journal of Zoology* **56**: 2355–2358.

Shao, B., Long, R., Ding, Y., Wang, J., Ding, L. & Wang, H. (2010). Morphological adaptations of yak (*Bos grunniens*) tongue to the foraging environment of the Qinghai-Tibetan Plateau. *Journal of Animal Science* **88**: 2594–2603.

Siebert, B. D. & MacFarlane, W. V. (1969). Body water content and water turnover of tropical *Bos taurus, Bos indicus, Bibos banteng*, and *Bos bubalus bubalis*. *Australian Journal of Agricultural Research* **20**: 613–622.

Skeat, A. J., East, T. J. & Corbett, L. K. (1996). Impact of feral water buffalo. In *Landscape and Vegetation Ecology of the Kakadu Region, Northern Australia*, pp. 155–177. Eds. Finlayson, C. M. & Von Oertzen, I. Dordrecht: Kluwer Academic.

Sodhi, N. S., Koh, L. P., Brook, B. W. & Ng, P. K. L. (2004) Southeast Asian biodiversity: an impending disaster. *Trends in Ecology & Evolution* **19**: 654–660.

Steinmetz, R., Chutipong, W., Seuaturien, N., Chirngsaard, E. & Khaengkhetkarn, M. (2010). Population recovery patterns of Southeast Asian ungulates after poaching. *Biological Conservation* **143**: 42–51.

Talbot, L. M. & Talbot, M. H. (1966). The tamarau (*Bubalus mindorensis* Heude): observations and recommendations. *Mammalia* **30**: 1–12.

Traba, J., Levassor, C. & Peco, B. (2003). Restoration of species richness in abandoned Mediterranean grasslands: seeds in cattle dung. *Restoration Ecology* **11**: 378–384.

Traveset, A. (1998). Effect of seed passage through vertebrate frugivores' guts on germination: a review. *Perspectives in Plant Ecology, Evolution and Systematics* **1**: 151–190.

Truett, J. C., Phillips, M., Kunkel, K. & Miller, R. (2001). Managing bison to restore biodiversity. *Great Plains Research: A Journal of Natural and Social Sciences* **541**: 123–144.

Tulloch, D. G. (1968). Incidence of calving and birth weights of domesticated buffalo in the Northern Territory. *Proceedings, Australian Society for Animal Production* **7**: 144–147.

Tulloch, D. G. (1969). Home range in feral water buffalo, *Bubalus bubalis* Lydekker. *Australian Journal of Zoology* **17**: 143–152.

Tulloch, D. G. (1979). The water buffalo, *Bubalus bubalis*, in Australia: reproductive and parent–offspring behaviour. *Australian Wildlife Research* **6**: 265–287.

Unteregelsbacher, S., Hafner, S., Guggenberger, G., Miehe, G., Xu, X., Liu, J. & Kuzyakov, Y. (2012). Response of long-, medium- and short-term processes of the carbon budget to overgrazing-induced crusts in the Tibetan Plateau. *Biogeochemistry* **111**: 187–201.

van der Waal, C., de Kroon, H., de Boer, W. F., Heitkönig, I. M. A., Skidmore, A. K., de Knegt, H. J., van Langevelde, F., van Wieren, S. E., Grant, R. C., Page, B. R., Slotow, R., Kohi, E., Mwakiwa, E. & Prins, H. H. T. (2009). Water and nutrients alter herbaceous competitive effects on tree seedlings in a semi-arid savanna. *Journal of Ecology* **97**: 430–439.

van der Waal, C., Kool, A., Meijer, S. S., Kohi, E., Heitkonig, I. M. A., de Boer, W. F., van Langevelde, F., Grant, R. C., Peel, M. J. S., Slotow, R., de Knegt, H. J., Prins, H. H. T. & de Kroon, H. (2011). Large herbivores may alter vegetation structure of semi-arid savannas through soil nutrient mediation. *Oecologia* **165**: 1095–1107.

van Wieren, S. E. (1998). Effects of large herbivores upon the animal community: grazing and conservation management. In *Grazing and Conservation Management*. pp. 185–214. Eds. Wallis DeVries, M. F., Bakker J. P. & Van Wieren, S. E. Boston: Kluwer.

Vinod, K. V. & Sabu, T. K. (2007). Species composition and community structure of dung beetles attracted to dung of gaur and elephant in the moist forests of South Western Ghats. *Journal of Insect Science* **7**: 1–14.

Wanapat, M., Pilajun, R. & Kongmun, P. (2009). Ruminal ecology of swamp buffalo as influenced by dietary sources. *Animal Feed Science and Technology* **151**: 205–214.

Wang, D. & Ba, L. (2008). Ecology of meadow steppe in northeast China. *The Rangeland Journal* **30**: 247–254.

Werner, P. A. (2014). The rise and fall of the Asian water buffalo in the monsoonal tropics of Australia. In *Invasion Ecology and Ecosystem Theory: Insights From a Continent in Transformation*, pp. 452–496. Eds. Prins, H. H. T. & Gordon, I. J. Cambridge: Cambridge University Press.

Wiegleb, G. & Krawczynski, R. (2010). Biodiversity management by water buffalos in restored wetlands. *Waldökologie, Landschaftsforschung und Naturschutz* **10**: 17–22.

Wu, X. & Sun, S. (2010). The roles of beetles and flies in yak dung removal in an alpine meadow of eastern Qinghai-Tibetan Plateau. *Ecoscience* **17**: 146–155.

Ydenberg, R. C. & Prins, H. H. T. (1981). Spring grazing and the manipulation of food quality by barnacle geese. *Journal of Applied Ecology* **18**: 443–453.

Yu, X., Xu, C., Wang, F., Shang, Z. & Long, R. (2012). Recovery and germinability of seeds ingested by yaks and Tibetan sheep could have important effects on the population dynamics of alpine meadow plants on the Qinghai-Tibetan Plateau. *The Rangeland Journal* **34**: 249–255.

Chapter 3

Domestic cattle and buffaloes

Johannes A. Lenstra, Marleen Felius and Bert Theunissen

Domestic bovine species

By the process of domestication, several bovine species have adapted to an agricultural habitat. As shown in Table 3.1, taurine and indicine cattle are numerically the most important, followed by river and swamp buffalo. These species have spread to several continents, while the domestic forms of the yak, banteng and gaur are concentrated near the distribution areas of their wild ancestors. Domestication of the extinct kouprey (*Bos sauveli*) has been proposed after finding kouprey mtDNA in a museum specimen of a Cambodian bull (Hassanin *et al.* 2006).

In line with molecular phylogenies (Buntjer *et al.* 2002; Verkaar *et al.* 2004; Nijman *et al.* 2008; Decker *et al.* 2009; MacEachern *et al.* 2009), taurine and zebu cattle can be crossed with other bovines, except buffaloes (Lenstra & Bradley 1999). Interspecific breeding may occur spontaneously or is carried out for terminal crossing or upgrading of breeds (Table 3.2). Hybrid taurine–zebu offspring are fertile, but crossing of zebu or taurine cattle to other species results in fertile cows and sterile bulls.

Because of their complete cross-fertility, taurine and zebu cattle should both be considered as subspecies of the wild ancestor *Bos primigenius*. However, because they resulted from different domestications, here they are described separately. The same applies to the swamp and river types of water buffalo, the cross-fertile subspecies of the wild *Bubalus arnee* (Groeneveld *et al.* 2010; Yindee *et al.* 2010).

Domestication, ecology and cultural aspects

For all domestic species the wild ancestor has been identified by DNA comparison with wild species. For narrowing down the region and period of domestication, we rely mainly on palaeontological findings. Skeletal remains are identified as belonging to domestic animals on the basis of their relatively small size, distorted age and gender distributions, reduced sexual dimorphism and morphological changes associated with the domestic status (Hall 1994; Zeder *et al.* 2006).

Since textual or pictorial documentation is scarce, our understanding of the early history of domestic bovines is based mainly on analysis of mitochondrial, Y-chromosomal and autosomal DNA (Groeneveld *et al.* 2010). Diversity of the *MHC* regions indicates that backcrossing of the earliest domestic animals with their wild ancestors contributed to the genetic diversity of livestock (Vilá *et al.* 2005).

In addition to the skeletal changes, domestication attenuated behaviour and increased the intramuscular fat deposition. Other changes relative to the wild ancestor also increased the within-species diversity. First, the dispersal of domestic taurine cattle and, to a lesser degree, zebu and river buffalo led to the development of several ecotypes, which differ in adaptation to their local environment. For instance, African N'Dama cattle have developed a trypanotolerance and Siberian Yakut and Scottish Galloway and Highland cattle adapted their hair coats to the cold.

Second, human selection generated many different 'agrotypes', which preceded the formation of breeds and differ in coat colour, horn development, docility (with aggression promoted in fighting cattle), other morphological traits or productivity. For *c.*200 years, diversity has been accentuated by systematic selection within genetically isolated uniform populations – the breeds. This initiated for taurine cattle, as well as for other livestock species, a new and consequential phase in their evolution (Ajmone-Marsan *et al.* 2010) and was accompanied by a further growth of the domestic population. With the increase of dairy production (Barker 1985), many cattle have acquired a wedge shape and oversized udders at the expense of the animal charismatic appeal.

In spite of the variety of species, ecotypes, agrotypes and habitats, a specific ecological niche of cattle can be defined. Similar to the smaller ruminants, sheep and goat, cattle utilize the energy stored in cellulose from grass and foliage, which is not directly accessible to humans. Depending on the species and breed, this energy is converted to milk, meat, hide, dung and draught power.

From the beginning, most bovine species have also participated in religious rituals, festivals, races and fighting games. As

Ecology, Evolution and Behaviour of Wild Cattle: Implications for Conservation, ed. M. Melletti and J. Burton. Published by Cambridge University Press.

Table 3.1 Cattle species

Species	Ancestor species	Approximate domestication time (years BP)	Distribution range	Estimated population size (million, 2007)	Main distinctive traits	Agricultural purpose
Taurine cattle (*Bos primigenius taurus*)	Aurochs (*Bos primigenius*)	8000–10 000	Worldwide	600	High dairy and beef productivity	Dairy, beef and hide production, draught power
Zebu (*Bos primigenius indicus*)	Aurochs (*Bos primigenius namadicus*)	8000	Tropical zones of Asia, Africa and America	721	Adaptation to hot and dry climate	Dairy and beef production, hides, draught power
Bali cattle (*Bos javanicus*)	Banteng (*Bos javanicus*)	5000	Bali, East Java, southeast Sulawesi, Kalimantan, South Sumatra, Lampung, North Australia (feral)	4	High fertility, adaptation to low-quality fodder and extensive management, meat quality	Draught power, beef production
Gayal or mithun (*Bos gaurus frontalis*)	Gaur (*Bos gaurus*)	Not known	India–Myanmar border region, easternmost India near Himalaya, Bhutan	0.2	Size, adaptation to steep hills, friendliness, semi-feral	Beef production; terminal crossing with zebu
Yak (*Bos mutus grunniens*)	Wild yak (*Bos mutus*)	4500	Central Asia above 2000 m, Tibetan plateau	13	Adaptation to high altitude	Dairy, beef, hide, wool and dung production, draught power
River buffalo (*Bubalus arnee bubalis*)	Wild water buffalo (*Bubalus arnee*)	5000	Brazil, Italy, Balkan, Egypt, Southwest Asia, South Asia west of Indochina, North Australia (feral)	122	Fat content of milk, dairy productivity, extensive management, compulsive wallowing	Dairy production, draught power
Swamp buffalo (*Bubalus arnee carabanesis*)	Wild water buffalo (*Bubalus arnee*)	5000	Brazil, China, Indochina, Indonesia, North Australia (feral)	38	Extensive management, strength, compulsive wallowing	Draught power for rice cultivation

Population sites have been estimated on the basis of the data of Wint and Robinson (2007)

in all domestic species, breeding has been governed by both subjective and objective criteria (e.g. see Theunissen 2008). Even in modern agriculture, breed traditions are kept alive via agricultural shows, the interest of which goes beyond the rational requirements of agricultural production.

Domestication and history

Taurine cattle

Around 8500 BC, after the domestications of sheep and goat, taurine cattle (*Bos primigenius taurus*) were domesticated in the Southwest Asian Fertile Crescent by taming the wild and now extinct aurochs (*Bos primigenius*; Bollongino *et al.* 2012). These aurochs carried a high diversity of mitochondrial DNA (mtDNA) in Southwest Asian cattle with haplogroups T, T1, T2, T3 and T5. Independent taurine domestications were postulated to have occurred in East Asia and Africa. However, this has been refuted by a phylogeny of complete mtDNA sequences, which shows short links of the African T1 and East Asian T4 with the Southwest Asian T and T3 mtDNA haplotypes, respectively (Achilli *et al.* 2009; Ajmone-Marsan *et al.* 2010; Groeneveld *et al.* 2010).

From Southwest Asia, taurine cattle spread to a large variety of habitats (Figure 3.1a), ranging from tropical Africa to the more temperate Mediterranean and Central European climates and the harsh winters of North Siberia (Ajmone-Marsan *et al.* 2010). Male introgression by Asian, African or European aurochs may very well have contributed to the local adaptation and changed the nuclear, but not the mitochondrial genome. Remains of Neolithic farms in Europe revealed that cattle and other livestock migrated along two routes, the Mediterranean coasts and the Danube river, respectively, arriving at the North Sea coast around 3000 BC (Barker 1985). North Asia

Table 3.2 Hybrid cattle (Felius 1995)

Name	Paternal origin/ sire	Maternal origin/ dam	Geographic range	Breeding status	Agricultural purpose
Taurindicine	Zebu	Taurine	Worldwide near tropical zones	Established breeds	As zebu
Sanga	Taurine, zebu	Taurine	Africa	Established breeds	As zebu
Canchim, Ibagé, etc.	Taurine (beef type)	Zebu (Nelore)	Brazil	Established breeds	Tropical beef
Girolando[1]	Taurine (Holstein)	Zebu (Gir)	Brazil	Established breed	Tropical dairy
Thai, Indonesian cattle	Zebu	Banteng, Bali cattle	Indochina, Indonesia	Established breeds	As zebu
Selembu, jatsa/jatsun	Zebu/gayal	Gayal/zebu	Malaysia	Terminal crossing	Dairy, work
Dulong[2]	Gayal	Zebu, taurine	Yunnan, China	Semi-feral	As gayal
Yakow, khainag[3]	Zebu, taurine/yak	Yak/zebu, taurine	Near yak at 1500–2000 m altitude	Terminal crossing	As yak
Beefalo	Bison/taurine	Taurine/bison	USA	Breed (3/8 bison)	Low-fat beef
American breed	Bison, taurine	Taurine, zebu	USA	Breed (1/8 bison)	Low-fat beef
Zubron	Wisent/ taurine	Taurine/ wisent	Poland	Experimental herd	Extensive management

[1] Also used for other Brazilian taurine–zebu crossbreeds.
[2] Gou *et al.* (2010)
[3] Tumennasan *et al.* (1997)

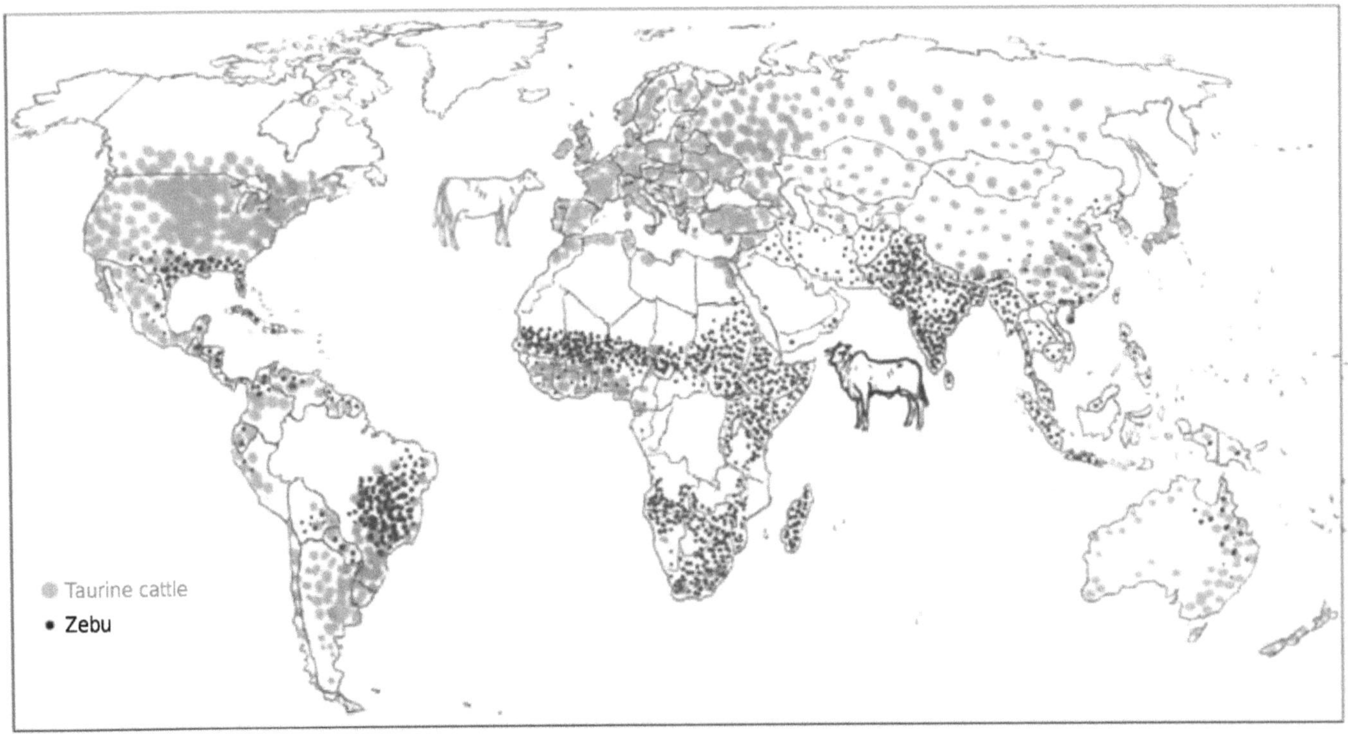

Figure 3.1 Global distributions of (a) taurine and zebu cattle; (b) domestic banteng, gayal and yak; and (c) swamp, river and Mediterranean river buffalo.

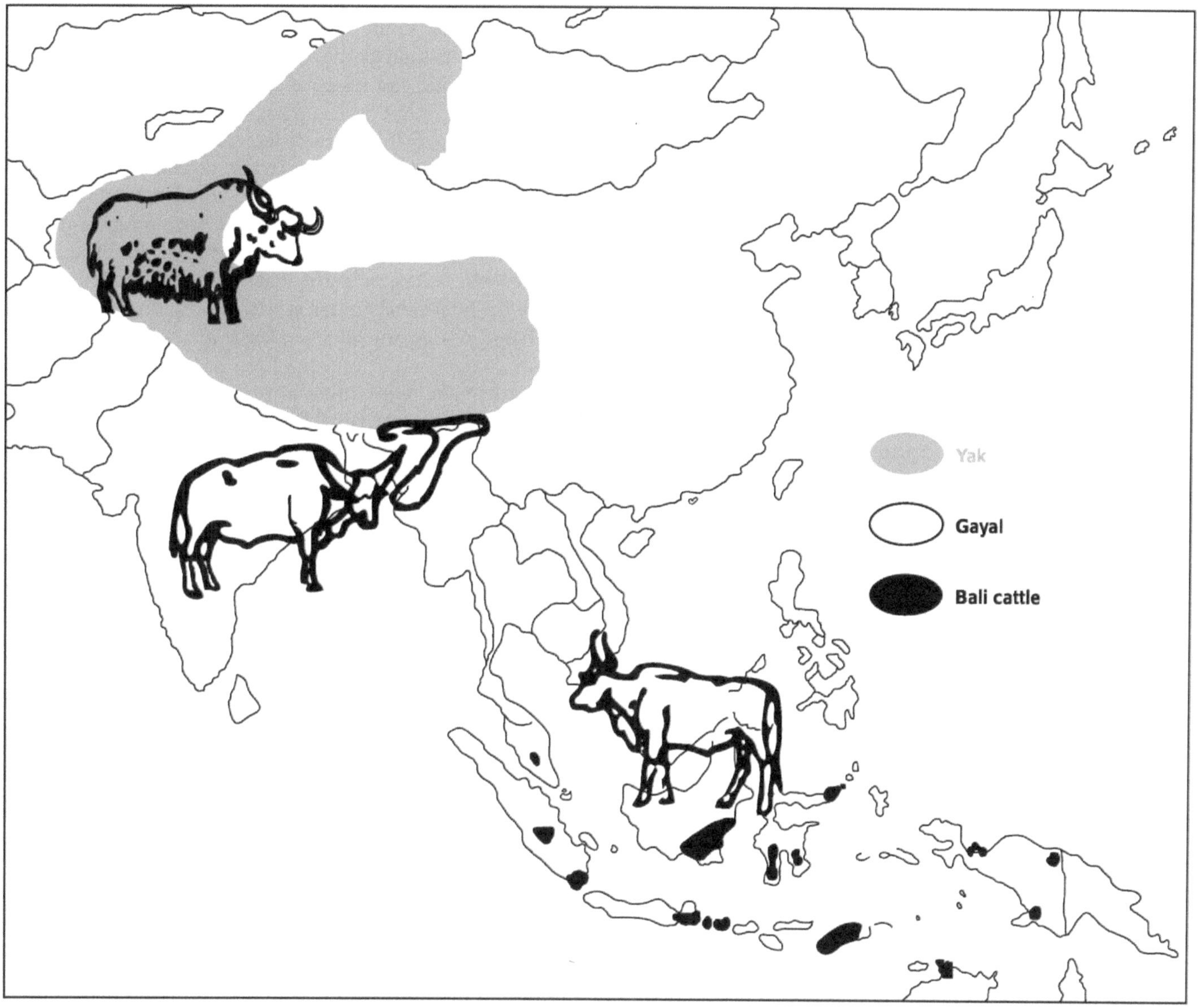

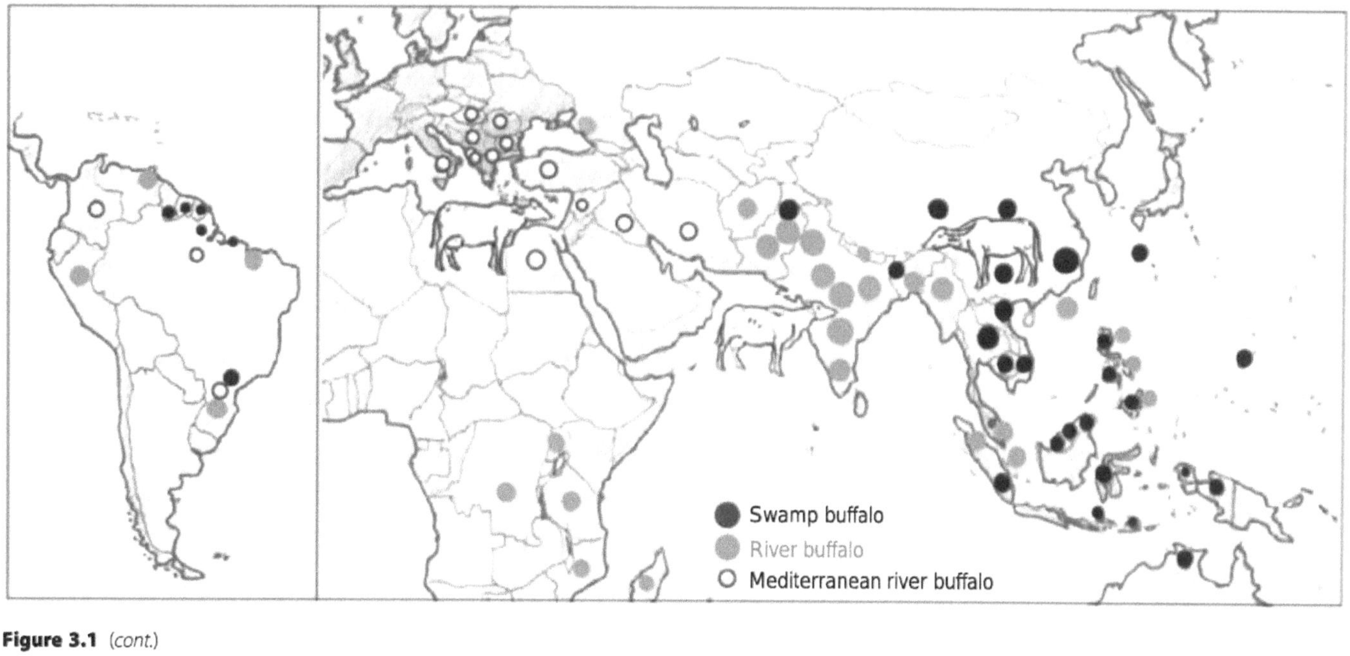

Figure 3.1 *(cont.)*

may have been populated via the Caucasus or via Europe. From 4000–5000 BC, cattle migrated into Africa (Hodges & Payne 1997).

The first domestic cattle were long-horned, a phenotype that is still common in several British, French, Mediterranean and African breeds. The first short-horned cattle appeared in Mesopotamia around 3000 BC. This phenotype fitted better in the domestic habitat and replaced, either by a second wave of migrations or by independent mutations (Epstein & Mason 1984), most long-horned cattle in Asia and neighbouring continents, reaching Britain in 1000–2000 BC and predominant in Europe from *c.*1000 BC. In Scotland and the Nordic regions the necessity to house cattle during the winter period even favoured the polled phenotype (Felius 1995).

MtDNA haplotypes have revealed several introgressions and transcontinental migrations (Ajmone-Marsan *et al.* 2010; Groeneveld *et al.* 2010). Presumably, by serial bottlenecks during the migrations, T1 and T3 became the major haplotypes in Africa and Europe, respectively. However, the finding of T3 mtDNA haplotypes in pre-Neolithic Italian aurochs (Ajmone-Marsan *et al.* 2010; Lari *et al.* 2011) may suggest an alternative scenario, in which a pre-domestic population bottleneck during one of the glacial periods made T3 dominant in the ancestors of European cattle in South European or Anatolian refugia.

Central and North European aurochs carried mtDNA P haplotypes, which have a low frequency (*c.*1/1000) in European domestic cattle (Ajmone-Marsan *et al.* 2010). This, as well as the sporadic finding of R haplotypes (Bonfiglio *et al.* 2010), indicates a rare recruitment of cows from the European aurochs population. A relatively high frequency of T1 in Iberian cattle reflects prehistoric or medieval immigration of cattle from Africa. Because Central and South American cattle were imported from Portugal, Spain and Africa, T1 is also found frequently in Criollo cattle (Ajmone-Marsan *et al.* 2010; Ginja *et al.* 2010; Groeneveld *et al.* 2010).

Domestication of taurine cattle resulted in a decrease of size, which continued until the Middle Ages. A selective disadvantage of large cattle may have been imposed by (1) castration of the strongest bulls for use as work animals; (2) slaughtering of the largest animals just before the winter; and (3) food shortage during the winter period in the temperate zones, likely to become acute during the frequent periods of unrest. In post-medieval society, improvements in the husbandry practices and a better infrastructure allowed larger cattle to be kept. The decrease in size was less pronounced in long-horned Italian beef and draught cattle. Palaeontological findings suggest that during the Roman era this large cattle coexisted with smaller local short-horned cattle in the peripheral regions of the empire (Bökönyi 1974).

Other large-scale migrations of cattle accompanied the migrations of Germanic people during the collapse of the Western Roman Empire. In addition, regional cattle populations often became decimated by famines, epidemics and plundering, which necessitated imports from neighbouring regions.

The eighteenth century saw the first breed formation, which, especially in the Western world, had a profound influence on the exterior, productivity and genetic constitution of cattle. Systematic and selective breeding and genetic isolation eventually led to the development of hundreds of specialized breeds, which over time became integral parts of local tradition (Felius 1995). Europe has the largest diversity of cattle with hundreds of well-separated breeds, while differentiation of indicine and taurine breeds in Asia and Africa is more gradual. However, crossbreeding was also common in Europe and genetic surveys demonstrated that breeds still contain a large part of the total variety of the species (European Cattle Genetic Diversity Consortium 2006; Decker *et al.* 2009).

For comparison of breeds within continents, nuclear genetic markers are more informative than mtDNA (Groeneveld *et al.* 2010). Y-chromosomal haplotypes as markers of paternal lineages reveal a sharp genetic boundary that divides both France and Germany in northern and southern parts. This may reflect the contrast of specialized dairy cattle from the northwestern continental lowland dairy with the Central European beef or dual-purpose cattle, but also coincides with traditional cultural differences between Northern and Southern Europe (Edwards *et al.* 2011).

Since the second half of the last century a growing demand for food led to a focus on breeds that are the most productive in a given environment. For instance, the Holstein Friesian is the first choice for industrial dairy production, but the Swiss Brown mountain cattle is often used for dairy production under more extensive conditions. Large-sized continental beef breeds such as Limousin, Charolais and Piemontese are used for intensive fattening, while British Angus and Hereford are kept for traditional pasture fattening and the hardy Galloway, Highland and Salers for extensive grazing. Spotted cattle (*Fleckvieh*) such as Simmental and related breeds serve as dual-purpose cattle in Central Europe, where the landscape is less suitable for large-scale dairy farming than in Northwestern Europe.

Although this has led to the disappearance of several local breeds, hundreds of breeds still exist. Analysis of microsatellites and single nucleotide polymorphisms showed, in agreement with the Y-chromosomal study, a broad division of European cattle in northern and central European, Mediterranean and Balkan-Anatolian cattle. Within these groups, several clusters of related breeds were identified. These clusters correspond to groups of breeds from the same region (British, Nordic, Iberian and South of France breeds, respectively) or to the expansion of popular types of cattle (black- or red-pied dairy, Baltic red, Nordic Ayrshire breeds, Alpine spotted and Alpine Brown mountain cattle, long-horned Podolic draught breeds). This molecular-genetic classification broadly agrees with an earlier biochemical and an integrated geographic-morphological classification (Felius *et al.* 2011).

Export of Iberian cattle to America and limited introgression of zebu gave rise to the Criollo cattle (Ginja *et al.* 2010). Several North European, Central European and Italian breeds became cosmopolitan by export to North America and Australia. For at least three breeds the American breeding regime led to allopatric development and differences with the European ancestors (Felius *et al.* 2011): systematic selection turned the Dutch Friesian into the highly productive Holstein Friesian; the brown dual-purpose Swiss Brown became the lightly coloured dairy Brown Swiss; and after crossing to Angus the dual-purpose red-pied Simmental changed for a large part into black and polled beef cattle. In addition, American as well as Australian breeders created several taurine or taurindicine synthetic breeds as original contributions to the diversity and productivity of cattle (Felius 1995; Hodges & Payne 1997).

Zebu

About 6000 BC, domestication of the related *Bos primigenius namadicus* in the Indus Valley resulted in the zebu (*Bos primigenius indicus*), which acquired its characteristic hump only after domestication (Magee *et al.* 2007; Chen *et al.* 2010). Several traits contribute to their heat adaptation: a low metabolic rate, many large sweat glands, a large skin surface, intramuscular instead of subcutaneous fat, a short, smooth coat, a low susceptibility to insects, ticks and protozoa and good utilization of low-quality fodder (Turton 1991).

From the Indus Valley zebu moved to the tropical zones of all continents (Payne & Hodges 1997; Ajmone-Marsan *et al.* 2010). Migrating eastward, zebu reached China, Indochina and Indonesia, which led to crossing with taurine cattle from North China or with banteng and gayal cattle (Payne & Hodges 1997).

A westward movement from 2000 BC brought zebu to the Southwest Asian region of taurine domestication and to Africa. Pictures of zebu in Egypt date from ~1800 BC, but large-scale introduction of zebu in Africa started only about 2000 years ago (Epstein and Mason 1984). Immigration of zebu in Africa was stimulated by the Islamic invasions after AD 700 (Hodges & Payne 1997; Ajmone-Marsan *et al.* 2010). The exclusive finding in African zebu breeds of taurine mtDNA demonstrated that zebu expanded via incrossing of zebu bulls (Bradley 2006). Hybridization of zebu and indigenous taurine cattle resulted in the sanga, which reached South Africa around AD 500 and were, around AD 1500, the dominant type of cattle in East and Central Africa. Because of a higher resistance to rinderpest, zebu largely replaced sanga in East Africa after the epidemic of 1887–1897.

From the nineteenth century, male-mediated introgression of zebu in taurine cattle resulted in the American Brahman zebu breed and the large zebu population in Brazil (Ajmone-Marsan *et al.* 2010). A minority of the Brazilian Nelore and Gyr cattle that descends from Indian cows may be the only zebus outside Asia with the original indicine mtDNA (Meirelles *et al.* 1999). Zebu now occupies the tropical zones of all continents and may numerically surpass taurine cattle (Table 3.1).

Banteng

Banteng may have been domesticated in Southeast Asia as early as 5000 BC (Felius 1995; Higham 2002). Although Indochinese cattle are now almost exclusively zebu, Thai cattle are said to have resembled Indonesian Bali cattle (M. Felius, personal communication), which is essentially pure Indonesian banteng. Continental bantengs (*tsine*) are fawn-coloured, while in Indonesia the banteng or Bali cows are slightly darker and the bulls dark brown. Bali bulls are smaller than wild bulls (Felius 1995).

As evidenced by the finding of banteng mtDNA in most Indonesian zebu breeds, banteng-derived cattle used to be widespread. Replacement by zebu started 1500 years ago and was accelerated by imports of Ongole cattle since the end of the nineteenth century. This zebu was required as a draught animal on paved roads, for which the banteng hooves are too soft (Mohamad *et al.* 2009).

Bali cattle have been kept pure since 1913 by a ban on crossbreeding on the Isle of Bali. They are used for ploughing rice paddy fields. Advantages of Bali cattle relative to zebu are their disease resistance, high fertility (McCool 1992), ability to adapt to low-quality fodder and stressful climatic conditions (Martojo 2012; Purwantara *et al.* 2012) and the tenderness of the meat from young animals – traits that may very well be exploited in other countries and continents (Mohamad *et al.* 2009). However, susceptibility to malignant catarrh precludes rearing together with sheep, while their deer-like temperament makes banteng most suitable for small-scale farming (Martojo 2012).

Bali cattle populations are kept on several Indonesian islands (Figure 3.1c) and have been exported to Papua New Guinea, Australia, Malaysia and the Philippines. A Malaysian population was clearly introgressed with zebu (Nijman *et al.* 2003). The viability of a feral population in the Australian Coburg peninsula (Bradshaw *et al.* 2007) shows that Bali cattle, like goat, cat and horse, have retained the fitness of their wild ancestors (Bradshaw *et al.* 2007).

Gayal

In Assam and Myanmar (Figure 3.1b) the semi-feral gayal or mithun (*Bos gaurus frontalis*) was derived from the large wild gaur (*Bos gaurus*). Gayal and gaur have the same mtDNA, Y-chromosomal DNA (Verkaar *et al.* 2004; Nijman *et al.* 2008; Tanaka *et al.* 2011) and karyotype. Introgression from wild gaur bulls is supposed to take place frequently, although gayal have different horns and a more friendly character. Only a minority of the gayal in Myanmar and Bhutan carry taurine or zebu mtDNA (Tanaka *et al.* 2011), but Chinese Yunnan gayal, also named Dulong cattle, have a zebu maternal origin (Gou *et al.* 2010). Gayal confer social status to their owners within their tribal society. The primary use of gayal is for a ceremonial sacrifice, after which their meat is consumed (Mason 1984).

Of more economical importance is the use of gayal for terminal crossing with zebu for producing, in Malaysia, the selembu and, in India and Bhutan, the dairy Jatsun cows and Jatsa bulls, which are strong draught animals in a cold and dry climate unsuitable for the parental gayal. Jatsun cows are backcrossed to Siri bulls to rejuvenate the Bhutanese taurindicine cattle (Felius 1995; Hickman & Tenzin 1982).

Yak

About 2500 BC a domestic form (*Bos mutus grunniens*) of the wild yak (*Bos mutus*) was developed and now occupies a large area on the Qinghai-Tibetan Plateau and adjacent areas above 3000 m (Mason 1984; Wiener *et al.* 2003; Figure 3.1b). The current gene pool is supposed to have resulted from reuniting pre-domestic populations that during Pleistocene glacial periods were in several refugia. This would explain the observation of two major deeply divergent mtDNA haplogroups in a single gene pool (Qi *et al.* 2008; Wang *et al.* 2010). In China, 12 breeds are recognized, which form three clusters (Wiener *et al.* 2003; Zhang *et al.* 2008). Differentiation of breeds is weak, but clear geographic trends were observed (Qi *et al.* 2008; Zhang *et al.* 2008). Maternal introgression of taurine cattle was observed in less than 2% of the animals, while taurine-specific microsatellite alleles indicate a taurine introgression of up to 7% in the northern and eastern extremes of the yak distribution range (Qi *et al.* 2010).

The yak has long been intricately involved in the pastoral Himalayan economy, culture and religion. It produces milk, meat, dung and fibres, serves as draught and pack animals in impassable terrain, and participates in cultural and ritual activities. In Tibet and Nepal at altitudes of 1500–2000 m, terminal crossing of taurine or zebu bulls with yak cows is carried out to breed hybrid yakows, large animals and excellent dairy producers (Felius 1995; Wiener *et al.* 2003).

Swamp buffalo

The swamp buffalo (*Bubalus arnee carabanesis*) or carabao emerged by domestication of the wild water buffalo (*Bubalus arnee*). Analysis of microsatellites and mtDNA diversity (Groeneveld *et al.* 2010; Yindee *et al.* 2010; Zhang *et al.* 2011) combined with archaeological evidence (Higham 2002) indicates domestication in South China and/or Indochina around 2000 BC. The domestication of the swamp buffalo coincides with the start of wet rice cultivation, which requires strong draught power for ploughing the rice paddies.

Subsequent migrations brought the swamp buffalo to central China, the Malaysian peninsula and Indonesia. Swamp buffaloes are supposed to have been brought to the Philippines by Malaysian immigrants around 200–300 BC (Mason 1984), whence they were taken to Guam at the end of the seventeenth century. In both countries water buffaloes are now a national symbol. Imports to Australia from 1826 led to the establishment of a feral population. This grew to 350 000 animals by 1980, but was then culled and reduced in size. Swamp buffaloes brought to Brazil from 1896 appeared to be less suitable for extensive management than river buffaloes due to a tendency to become feral, but small populations still exist (Wilkins 1991).

The swamp buffalo may still be considered as the living tractor of Southeast Asia and China. It has adapted to hot and humid environments by compulsive wallowing, and surpasses zebu and taurine cattle in strength, ability to thrive in marshy areas, utilization of coarse fibrous feed and disease resistance (Turton 1991). Fertility is lower than in cattle, which is compensated by a longer life span.

MtDNA with two well-separated lineages A and B, as well as Y-chromosomal sequences, indicate a higher nucleotide diversity than in the river buffalo (Groeneveld *et al.* 2010; Yindee 2010; Yindee *et al.* 2010). There are no recognized breeds, but microsatellite (Zhang *et al.* 2011) and especially Y-chromosomal sequences in Thailand (Yindee *et al.* 2010) indicate spatial trends.

River buffalo

River buffaloes (*Bubalus arnee bubalis*) and swamp buffaloes are cross-fertile, but are at least as divergent as taurine and zebu cattle (Yindee *et al.* 2010) and even have different karyotypes. The river buffalo has generally the same characteristics as the swamp buffalo, but has smaller horns and prefers clear water for wallowing (Cockrill 1981). It has been domesticated since 2500 BC in the Indus Valley (Mason 1984; Kumar *et al.* 2007). Water buffaloes were not known in the Roman Empire, which indicates that river buffaloes migrated westwards well after domestication. Available evidence suggests that river buffaloes were in Palestine in AD 723, entered Egypt with the Arabs or later and were numerous in the ninth century in Anatolia, in 1154 in the Campania and in the thirteenth century in Thrace and Macedonia (Mason 1984).

The generally accepted notion that returning crusaders brought water buffaloes to Italy is not supported by contemporary documentation. Alternative routes of import are from Sicily by its Norman rulers, from the Balkans by invading Goths or Lombards or from Byzantine Greece across the Ionian Sea. However, it is not clear yet if water buffaloes were present in the postulated regions of origin at the time of the supposed import.

In contrast to swamp buffaloes, most river buffaloes are primarily kept for milk. Because of its high fat content, buffalo milk can be converted easily to butterfat, which can be stored without cooling. The most well-known buffalo product is the Italian mozzarella cheese, the basis of the authentic Neapolitan pizza.

Although 70% of the Indian river buffaloes are nondescript, 12 breeds are recognized, which differ in appearance and dairy production. Mediterranean river buffaloes have larger horns than Indian river buffaloes, but are not kept as separate breeds. From the beginning of the twentieth century, river buffaloes were exported to several South American and African countries (Figure 3.1c). Brazil has now 1.5 million animals, most of which are Mediterranean or Indian Jafarabadi and Murrah animals. Breeding for beef on Trinidad resulted in the Buffalypso type, which in turn has been exported as a dairy/beef breed to other countries and even continents (Cockrill 1981; Mason 1984).

Although several local studies have been carried out (Groeneveld *et al.* 2010), there is no global-level picture of breed relationships and patterns of diversity. As already demonstrated for other livestock species, modern genomic investigations of a comprehensive sample would complement our present knowledge about the history and genetic constitution of both buffalo species.

Future perspectives

The future of cattle unfolds along different lines. For the most industrial breeds, selective breeding is going to be intensified by genetic selection of desired traits and genomic selection for a high breeding value. This is to be based on genome-level research that mainly targets taurine cattle with a growing emphasis on Holstein Friesian dairy cattle, but other dairy and beef taurine and indicine breeds are being investigated as well. Expectedly, an increase in production and a narrowing of the genetic basis will have to be balanced by outbreeding and crossbreeding in order to restore disease resistance and fertility.

Artificial insemination and multiple-ovulation-embryo-transfer (MOET) are commonly practised for taurine cattle. However, cloning is not yet practical and genetic manipulation is not even on the agenda.

A separate development is a renewed interest in local breeds. This not only maintains the diversity of the genetic resources, which are not only suitable for more extensive management than required for the highly productive breeds, but also preserves a potential adaptation to other environments and changes in the climate.

Acknowledgements

We would like to thank Marleen Felius for drawing Figure 3.1.

References

Achilli, A., Bonfiglio, S., Olivieri, A., *et al.* (2009). The multifaceted origin of taurine cattle reflected by the mitochondrial genome. *PLoS One* **4**: e5753.

Ajmone-Marsan, P., Garcia, J. F. & Lenstra, J. A. (2010). On the origin of cattle: how aurochs became cattle and colonized the world. *Evolutionary Anthropology* **19**: 148–157.

Barker, G. (1985). *Prehistoric Farming in Europe*. Cambridge: Cambridge University Press.

Bökönyi, S. (1974). *History of Domestic Mammals in Central and Eastern Europe*. Budapest: Akadémiai Kiadó.

Bollongino, R., Burger, J., Powell, A., Mashkour, M., Vigne, J. D. & Thomas, M. G. (2012). Modern taurine cattle descended from small number of Near-Eastern founders. *Molecular Biology and Evolution* **29**(9): 2101–2104.

Bonfiglio, S., Achilli, A., Olivieri, A., *et al.* (2010). The enigmatic origin of bovine mtDNA haplogroup R: sporadic interbreeding or an independent event of *Bos primigenius* domestication in Italy? *PLoS One* **5**: e15760.

Bradley, D. G. (2006). Genetics and origins of domestic cattle. In *Documenting Domestication: New Genetic and Archaeological Paradigms*, pp. 317–328. Eds. Zeder, M. A., Bradley, D. G., Emshwiller, E. & Smith. B. D. Berkeley, CA: University of California Press.

Bradshaw, C. J., Isagi, Y., Kaneko, S., Brook, B. W., Bowman, D. M. & Frankham, R. (2007). Low genetic diversity in the bottlenecked population of endangered non-native banteng in northern Australia. *Molecular Ecology* **16**: 2998–3008.

Buntjer, J. B., Otsen, M., Nijman, I. J., Kuiper, M. T. & Lenstra, J. A. (2002). Phylogeny of bovine species based on AFLP fingerprinting. *Heredity* **88**: 46–51.

Chen, S., Lin, B. Z., Baig, M., *et al.* (2010). Zebu cattle are an exclusive legacy of the South Asia neolithic. *Molecular Biology and Evolution* **27**: 1–6.

Cockrill, W. R. (1981). The water buffalo: a review. *British Veterinary Journal* **137**: 8–17.

Decker, J. E., Pires, J. C., Conant, G. C., *et al.* (2009). Resolving the evolution of extant and extinct ruminants with high-throughput phylogenomics. *Proceedings of the National Academy of Sciences of the United States of America* **106**: 18644–18649.

Edwards, C. J., Ginja, C., Kantanen, J., *et al.* (2011). Dual origins of dairy cattle farming: evidence from a comprehensive survey of European Y-chromosomal variation. *PLoS One* **6**: e15922.

Epstein, H. & Mason, I. L. (1984). Cattle. In *Evolution of Domesticated* Animals. Ed. Mason, I. L. Harlow: Longman Group.

European Cattle Genetic Diversity Consortium (2006). Marker-assisted conservation of European cattle breeds: an evaluation. *Animal Genetics* **37**: 475–481.

Felius, M. (1995). *Cattle Breeds, an Encyclopedia*. Doetinchem: Misset Uitgeverij.

Felius, M., Koolmees, P. A., Theunissen, B., European Cattle Genetic Diversity Consortium & Lenstra, J. A. (2011). On the breeds of cattle: historic and current classifications. *Diversity* **3**: 660–692.

Ginja, C., Penedo, M. C., Melucci, L., *et al.* (2010). Origins and genetic diversity of New World Creole cattle: inferences from mitochondrial and Y chromosome polymorphisms. *Animal Genetics* **41**: 128–141.

Gou, X., Wang, Y., Yang, S., Deng, W. & Mao, H. (2010). Genetic diversity and origin of Gayal and cattle in Yunnan revealed by mtDNA control region and SRY gene sequence variation. *Journal of Animal Breeding and Genetics* **127**: 154–160.

Groeneveld, L. F., Lenstra, J. A., Eding, H., *et al.* (2010). Genetic diversity in farm animals: a review. *Animal Genetics* **41**: 6–31.

Hall, S. J. G. (1994). *Livestock Biodiversity*. Oxford: Blackwell Publishing.

Hassanin, A., Ropiquet, A., Cornette, R., *et al.* (2006). Has the kouprey (*Bos sauveli* Urbain, 1937) been domesticated in Cambodia? *Comptes Rendus Biologies* **329**: 124–135.

Hickman, C. G. & Tenzin, D. (1982). The classical crossbreeding systems in Bhutan. *Bhutan Journal of Animal Husbandry* **5**: 19–22.

Higham, C. (2002). *Early Cultures of Mainland Southeast Asia*. Bangkok: River Books.

Hodges, J. & Payne, W. J. A. (1997). *Tropical Cattle: Origin, Breeding and Breeding Policies*. Oxford: Blackwell Science.

Kumar, S., Nagarajan, M., Sandhu, J. S., Kumar, N. & Behl, V. (2007). Phylogeography and domestication of Indian river buffalo. *BMC Evolutionary Biology* 7: 186.

Lari, M., Rizzi, E., Mona, S., *et al.* (2011). The complete mitochondrial genome of an 11,450-year-old aurochsen (*Bos primigenius*) from Central Italy. *BMC Evolutionary Biology* **11**: 32.

Lenstra, J. A. & Bradley, D. G. (1999). Systematics and phylogeny of cattle. In *The Genetics of Cattle*, pp. 1–14. Eds. Fries R. & Ruvinsky,A. Oxon and New York: CABI Publishing.

MacEachern, S., McEwan, J. & Goddard, M. (2009). Phylogenetic reconstruction and the identification of ancient polymorphism in the *Bovini* tribe (*Bovidae, Bovinae*). *BMC Genomics* **10**: 177.

Magee, D. A., Mannen, H. & Bradley, D. G. (2007). Duality in *Bos indicus* mtDNA diversity: support for geographical complexity in zebu domestication. In *The Evolution and History of Human Populations in South Asia*, pp. 385–391. Eds. Petraglia, M. D. & Allchin,B. New York: Springer.

Martojo, H. (2012). Indigenous Bali cattle is most suitable for sustainable small farming in Indonesia. *Reproduction in Domestic Animals* 47 (Suppl 1): 10–14.

Mason, I. L. (1984). *Evolution of Domesticated Animals*. London and New York: Longman.

McCool, C. (1992). Buffalo and Bali cattle: exploiting their reproductive behaviour and physiology. *Tropical Animal Health and Production* **24**: 165–172.

Meirelles, F. V., Rosa, A. J. M., Lôbo, R. B., Garcia, J. M., Smith, L. C. & Duarte, F. A. M. (1999). Is the American zebu really *Bos indicus?*. *Genetics and Molecular Biology* **22**: 543–546.

Mohamad, K., Olsson, M., van Tol, H. T., *et al.* (2009). On the origin of Indonesian cattle. *PLoS One* **4**: e5490.

Nijman, I. J., Otsen, M., Verkaar, E. L., *et al.* (2003). Hybridization of banteng (*Bos javanicus*) and zebu (*Bos indicus*) revealed by mitochondrial DNA, satellite DNA, AFLP and microsatellites. *Heredity* **90**: 10–16.

Nijman, I. J., Van Boxtel, D. J., Van Cann, L. M., Cuppen, E. & Lenstra, J. A. (2008). Phylogeny of Y-chromosomes from interbreeding bovine species. *Cladistics* **24**: 723–726.

Payne, W. J. A. & Hodges, J. (1997). Classification systems. In *Tropical Cattle: Origins, Breeds and Breeding Policies*, pp. 89–94. Eds. Hodges, J. & Payne, W. J. A. Oxford: Blackwell Science.

Purwantara, B., Noor, R. R., Andersson, G. & Rodriguez-Martinez, H. (2012). Banteng and Bali cattle in Indonesia: status and forecasts. *Reproduction in Domestic Animals* **47** (Suppl 1): 2–6.

Qi, X., Jianlin, H., Blench, R., Rege, J. E. O. & Hanotte, O. (2008). Understanding yak pastoralism in central Asian highlands: mitochondrial DNA evidence for origin, domestication and dispersal of domestic yak. In *Past Human Migrations in East Asia: Matching Archaeology, Linguistics and Genetics*, pp. 427–442. Eds. Sanchez-Mazas, A., Blench, R., Ross, M. D., Peiros, I. & Lin, M. Oxon: Routledge.

Qi, X. B., Jianlin, H., Wang, G., Rege, J. E. & Hanotte, O. (2010). Assessment of cattle genetic introgression into domestic yak populations using mitochondrial and microsatellite DNA markers. *Animal Genetics* **41**: 242–252.

Tanaka, K., Takizawa, T., Murakoshi, H., *et al.* (2011). Molecular phylogeny and diversity of Myanmar and Bhutan mithun based on mtDNA sequences. *Animal Science Journal* **82**: 52–56.

Theunissen, B. (2008). Breeding without Mendelism: theory and practice of dairy cattle breeding in the Netherlands 1900–1950. *Journal of the History of Biology* **41**: 637–676.

Tumennasan, K., Tuya, T., Hotta, Y., Takase, H., Speed, R. M. & Chandley, A. C. (1997). Fertility investigations in the F1 hybrid and backcross progeny of cattle (*Bos taurus*) and yak (*B. grunniens*) in Mongolia. *Cytogenetics and Cell Genetics* **78**: 69–73.

Turton, J. D. (1991). Modern needs for different genetic types. In *Cattle Genetic Resources*, pp. 21–49. Ed. Hickman, C. G. Amsterdam: Elsevier.

Verkaar, E. L., Nijman, I. J., Beeke, M., Hanekamp, E. & Lenstra, J. A. (2004). Maternal and paternal lineages in cross-breeding bovine species: has wisent a hybrid origin? *Molecular Biology and Evolution* **21**: 1165–1170.

Vilá, C., Seddon, J. & Ellegren, H. (2005). Genes of domestic mammals augmented by backcrossing with wild ancestors. *Trends in Genetics* **21**: 214–218.

Wang, Z. F., Shen, X., Liu, B., *et al.* (2010). Phylogeographical analyses of domestic and wild yaks based on mitochondrial DNA: new data and reappraisal. *Journal of Biogeography* **37**: 2332–2344.

Wiener, G., Jianlin, H. & Ruijun, L. (2003). *The Yak*. Bangkok: FAO.

Wilkins, J. V. (1991). Special regional problems of breeding resources: Latin American and Caribbean examples. In *Cattle Genetic Resources*, pp. 21–49. Ed. Hickman, C. G. Amsterdam: Elsevier.

Wint, W. & Robinson, T. (2007). *Gridded Livestock of the World 2007*. Rome: FAO.

Yindee, M. (2010). Reproduction and genetic diversity of the swamp buffalo. PhD thesis. Utrecht University.

Yindee, M., Vlamings, B. H., Wajjwalku, W., *et al.* (2010). Y-chromosomal variation confirms independent domestications of swamp and river buffalo. *Animal Genetics* **41**: 433–435.

Zeder, M. A., Emshwiller, E., Smith, B. D. & Bradley, D. G. (2006). Documenting domestication: the intersection of genetics and archaeology. *Trends in Genetics* **22**: 139–155.

Zhang, G., Chen, W., Xue, M., *et al.* (2008). Analysis of genetic diversity and population structure of Chinese yak breeds (*Bos grunniens*) using microsatellite markers. *Journal of Genetics and Genomics* **35**: 233–238.

Zhang, Y., Vankan, D., Zhang, Y. & Barker, J. S. (2011). Genetic differentiation of water buffalo (*Bubalus bubalis*) populations in China, Nepal and south-east Asia: inferences on the region of domestication of the swamp buffalo. *Animal Genetics* **42**: 366–377.

Chapter 4

The evolution and skeletal anatomy of wild cattle (Bovini)

Kris Kovarovic and Robert S. Scott

Introduction

The primacy of domestic cattle as a food resource in many cultures in recent human history might perhaps lead one to erroneously assume that we have a long-term ancestral association with members of the cattle tribe, Bovini. However, the domestication of wild cattle is a process which began relatively recently from an evolutionary perspective. Bovini had only just evolved and arrived on the African continent at the time that our hominin ancestors also became established there, at the end of the Miocene. Near the Pliocene–Pleistocene transition, the fossil record shows that hominins learnt to exploit animal resources and that the family Bovidae was certainly an important aspect of this dietary adaptation. Although some fossil Bovini bear evidence of early hominin consumption, they were not favoured over other species and it is likely that their large body size made them difficult prey in the initial stages of active hunting (e.g. Blumenschine & Pobiner 2007).

There were multiple bovine species once present in Africa, where humanity's roots are firmly planted, but the tribe is now restricted on that continent to only one wild species in the genus *Syncerus*. On a global scale, however, the tribe is moderately diverse. The aims of this chapter are three-fold: (1) review what is known of the evolution of the entire Bovini tribe from the deep evolutionary past, with a focus on the skeletal morphologies that define each taxonomic group; (2) describe the basic skeletal anatomy of the extant wild cattle and what can be inferred about their behaviour from these morphologies; and (3) briefly describe some apparent evolutionary changes in the limbs of the Bovini skeleton, given what is known of the phylogeny of the group.

Evolutionary history and morphology

Cetartiodactyla is one of several mammalian orders of ungulates, including even-toed hoofed mammals, whales and dolphins. Distinguishing features found within this order include even-numbered toes and paraxonic feet, the axis of which runs through the third and fourth digits, reduced parietals and enlarged frontals, generally absent clavicle, reduced ulna and fibula, simple premolars, diastema between the canine or incisors and the premolars and a talus with an inferior pulley surface and a superior rolling articular surface, among other characteristics (Romer 1945; Koopman 1967; Gentry 2000). Members of this group evolved 55 million years ago (Ma) at the beginning of the Eocene, when a drop in global sea level and a widespread mammalian radiation occurred concurrently (Rose 1996; Gentry 2000). They were relatively abundant in North America, but scarcer in Europe and Asia. Throughout the Eocene, they were outnumbered by Perissodactyla, their odd-toed ungulate cousins, until the Eocene–Oligocene transition, when the feeding strategies of the specialized folivorous forms were better suited to increasingly more seasonal habitats (Janis 1989; 1993).

Wild cattle are members of this suborder, Ruminantia, or ruminants, named for their unique three- or four-chambered stomachs in which vegetation is continuously digested. Living and fossil ruminants, which emerge in the middle to late Eocene, are united by the fusion of the navicular and cuboid tarsals and, although not restricted entirely to this group (it is also seen in horses), the third and fourth metapodials have fused into a single long bone, an adaptation for cursoriality (Romer 1945; Gentry 2000).

The fossil record presents the first evidence for the evolution of Bovidae in the Miocene, with the small-bodied genus *Eotragus* appearing approximately 18 Ma in the Kamlial Formation, Pakistan, if not somewhat earlier in the Bugti Hills (Solounias *et al.* 1995; Ginsburg *et al.* 2001). However, molecular evidence suggests that the emergence of the family was approximately 20–24 Ma (Miyamoto *et al.* 1993; Hassanin & Douzery 1999; Hernández Fernández & Vrba 2005). Extant bovids lack true lateral digits, but possess a reduced or absent upper canine and a generally more hypsodont dentition (Romer 1945). The two features said to define the family exclusively are permanent sheaths of keratin over a bony horn core in males and sometimes females, and the presence of relatively large foramina ovales at the base of the skull (Janis & Scott 1987). An in-depth description of Bovidae morphological characters can be found in Solounias (2007) and Gentry (1992, 2010).

Ecology, Evolution and Behaviour of Wild Cattle: Implications for Conservation, ed. M. Melletti and J. Burton. Published by Cambridge University Press.

Bovids diversified at a rapid pace during the Miocene and homoplasy was common, making their phylogenetic relationships difficult to establish (Vrba 1985; Gentry 1992). However, all manner of morphological and molecular studies have shown that not long after the family's evolution, it differentiated into two main sub-familial lineages, the Bovinae (bovines) and Antilopinae (antelopes, goats and sheep), both of which radiated throughout the middle Miocene (e.g. Kingdon 1997; Hassanin & Douzery 1999; Vrba & Schaller 2000). The Bovinae is the more primitive of the two lineages; bovines are relatively larger than other bovids and possess smooth and often keeled horns (Gentry 1978; Groves & Grubb 2011).

Bovini: the wild cattle

Early fossil Bovini are known from both Asia and Africa at approximately 6–7 million years ago and, although it is not clear where the tribe evolved, many researchers suggest a Eurasian origin (Gentry 1992; Harris 2003; Hassanin & Ropiquet 2004; Bibi 2007b). Fossil evidence indicates that the tribe becomes increasingly diverse at the expense of the more primitive boselaphines, from which they likely evolved, on both continents between the late Miocene and Pleistocene (Gentry 2000, 2010; Kostopoulos 2006). They may owe their early success to some species' ability to broaden their trophic niche to exploit drier and more seasonal habitats (Bibi 2007a, 2007b). However, even the earliest fossil evidence for Bovini (Gentry 1992; Harris 2003; Hassanin & Ropiquet 2004; Bibi 2007b) antedates considerably the apparent origin of the tribe at around 16.9 Ma (Hernández Fernández & Vrba 2005) and provides no direct evidence of the evolution of their postcranial skeleton.

Bovini are large and increasingly evolved to be more sexually dimorphic (Vrba & Schaller 2000). They have generally shortened braincases, low skulls with posteriorly elongated frontals, wide occipital regions, short faces, well-developed internal sinuses in the horn cores, short necks, robust legs and vertically positioned tibiae (Kingdon 1982; Gentry 1992, 2010; Solounias 2007). Females have horns (Gentry 2010). Morphoclines in some taxa, including the African buffalo (*Syncerus caffer*), the gaur (*Bos gaurus*) and Asian buffaloes in the genus *Bubalus*, have been identified on the basis of successively more derived aspects of their horns and crania (Grubb 2000).

Within the Bovini tribe, two clades have been identified often by molecular studies: cattle (genera *Bos* and *Bison*, which some advocate collapsing into a single genus (i.e. Hernández Fernández & Vrba 2005) and buffalo (genera *Bubalus* and *Syncerus*) (Hassanin & Douzery 1999 and references therein; Hassanin & Ropiquet 2004). Some morphological features also link these genera, including a large development above the scapula in *Bison* and some *Bos* (Solounias 2007). Some differences include the shorter metapodials of the African *Syncerus* compared to *Bos* or *Bubalus* (Gentry 2010).

Ecomorphology of wild cattle

Many skeletal traits of wild cattle can be interpreted as adaptations to habitat-specific behaviours. Such postcranial 'ecomorphologies' pertain to the exploitation of the spatial niche (i.e. locomotion and movement) and craniodental ecomorphologies relate to the exploitation of the trophic niche (i.e. dietary behaviour). For instance, wild cattle are largely grazing or mixed grazing–browsing species and display craniodental adaptations for this dietary regime, including fairly high-crowned molars, a square premaxilla and a long incisor row, which facilitate the efficient intake and mastication of grass (Gordon & Illius 1988; Solounias 2007; see also Chapter 1).

Postcranially, wild cattle limb dimensions and morphology may be interpreted in regards to both locomotor repertoire and graviportality. Their large size might appear to preclude them from closed environments, but in fact *Bison bison* is the only species which inhabits a consistently open habitat. The remaining living species prefer vegetation cover to some degree and, within the context of the family, they are not as well adapted for cursorial locomotion as observed in open-country bovids such as gazelles. Bovini limbs are relatively short and possess features typical of other wooded or moist habitat-dwelling taxa, which allow the animal to manoeuvre around objects and change the direction of movement in habitats where space is limited (Köhler 1993). Stout limbs and the vertical positioning of the tibia may also provide support for their massive bodies. Their short limbs, in addition to certain behaviours such as a dependence on water, have also been said to suggest a forest origin for the entire tribe (Kingdon 1982).

Analysis of limb bones

Given the paucity of evidence from the postcranial skeleton relevant to most nodes internal to the phylogenetic consensus for the Bovini, very little can be said about the evolution of the skeleton of Bovini species. However, given the large body size, sexual dimorphism and aforementioned ecomorphological observations, the rate and direction of changes in the postcrania, especially as they relate to overall body form, are of some interest. Thus, we offer a preliminary and exploratory reconstruction of distal limb bone morphology and lengths at nodes within a consensus phylogeny of the Bovini (Hernández Fernández & Vrba 2005). These reconstructions provide a starting point for a consideration of Bovini fossils to be uncovered in the future.

Sample and methods

A small sample of measurements for extant Bovini and Boselaphini as an outgroup was extracted from databases comprising linear postcranial metrics compiled by both authors. Available measurements were the maximum lengths of the metatarsal and metacarpal bones, as well as the mediolateral and anteroposterior diameters of these elements taken at the midshaft. The sample and measurements are shown in Table 4.1.

Table 4.1 Measurements used.

Species	Museum	Specimen	Sex	Metatarsal length	Metatarsal midshaft mediolateral diameter	Metatarsal midshaft anteroposterior diameter	Metacarpal length	Metacarpal midshaft mediolateral diameter	Metacarpal midshaft anteroposterior diameter
Bison bison bison	AMNH	AMNH130211	Female	240.7	30.89	30.59	197.2	37.91	26.63
Bison bison athabascae	AMNH	AMNH98957	Female	262.75	29.55	30.03	216.65	36.24	26.63
Bison bonasus	BERLIN	BERLIN33838	Female	247.47	30.7	30.32	201.6	34.11	26.63
Bubalus depressicornis	LEIDEN	LEIDEN43541 (cat a)	Female	148.71	21.27	19.33	130.31	24.14	26.63
Bubalus mindorensis	LEIDEN	LEIDENcat b	Female	170.2	29.5	23.48	144.02	35.73	26.63
Bos javanicus	NMNH	NMNH154385	Female	244.05	30.75	27.36	214	32.97	26.63
Syncerus caffer caffer	NMNH	NMNH163311	Female	206.5	35.58	31.14	170.7	43.62	26.63
Bison bison athabascae	NMNH	NMNH197705	Female	247.35	33.29	30.8	201.9	40.17	26.63
Syncerus caffer caffer	STOCK	NRM 592667	Female	210.6	35.77	28.54	177.69	44.14	26.63
Syncerus caffer nanus	RMCA	RMCA25492-M	Female	194.25	26.38	26.13	170.56	32.13	26.63
Syncerus caffer nanus	RMCA	RMCA37469-M	Female	212.4	33.31	30.95	185.52	38.8	26.63
Bos mutus	AMNH	AMNH22909	Female	173.5	23.21	21.03	143.5	29.59	16.31
Syncerus caffer nanus	AMNH	AMNH53579	Female	205	33	26.63	175	38.19	24.05
Tetracerus quadricornis	AMNH	AMNH54981	Female	153	11.77	11.89	142	12.12	9.67
Bison bison athabascae	AMNH	AMNH98954	Female	259	32.38	30.07	211.5	39.92	25.9
Bison bison athabascae	AMNH	AMNH98957	Female	263	30	28.7	218	36.69	24.76
Bubalus quarlesi	MCZ	MCZ27999	Female	120	15.67	14.7	103	18.57	12.92
Bubalus arnee	MCZ	MCZ5002	Female	197	23.4	23.45	165.5	27.73	19.14

Table 4.1 *(cont.)*

Species	Museum	Specimen	Sex	Metatarsal length	Metatarsal midshaft mediolateral diameter	Metatarsal midshaft anteroposterior diameter	Metacarpal length	Metacarpal midshaft mediolateral diameter	Metacarpal midshaft anteroposterior diameter
Bos sauveli	NMNH	NMNH361392	Female	250	26.55	28.79	221	30.5	22.4
Bison bison bison	NMNH	NMNHA22377	Female	266	36.88	36.67	215	47.03	32.29
Bos javanicus	AMNH	AMNH54551	Male	267.6	37.11	35.82	239.9	43.08	26.63
Syncerus caffer caffer	BUDA	BUDA60.147.1	Male	232	41.17	36.83	192.6	51.89	26.63
Bubalus mindorensis	FMNH	FMNH18817	Male	157.3	31.53	23.12	133.78	38.36	26.63
Syncerus caffer caffer	NMNH	NMNH164768	Male	228.95	43.7	36.31	193.25	53.39	26.63
Bos javanicus	NMNH	NMNH198317	Male	244.82	34.03	31.8	218.6	37.3	26.63
Bison bison bison	NMNH	NMNHA22375	Male	254.7	31.91	33.71	204.45	41.88	26.63
Bison bonasus	STOCK	NRM 587698	Male	259.94	43.49	37.17	212.24	52.79	26.63
Bos javanicus	STOCK	NRM 602553	Male	276.98	41.87	36.13	236.2	39.75	26.63
Syncerus caffer caffer	RMCA	RMCA25430-M	Male	218.2	34.73	29.67	182.96	41.56	26.63
Syncerus caffer nanus	RMCA	RMCA25527-M	Male	189.27	33.67	27.17	155.86	40.72	26.63
Syncerus caffer nanus	RMCA	RMCA25536-M	Male	195.31	30.08	26.5	168.96	38.07	26.63
Bos gaurus	AMNH	AMNH113746	Male	275	40.37	39.77	243.5	47.71	33.34
Bubalus depressicornis	AMNH	AMNH150050	Male	137	19.15	17.71	120.5	22.04	15.79
Bubalus mindorensis	AMNH	AMNH31660	Male	162.5	28.33	22.41	138.5	35.22	20.14
Tetracerus quadricornis	AMNH	AMNH35306	Male	165.5	10.56	10.78	153	11.15	9.13
Bubalus depressicornis	AMNH	AMNH35380	Male	130.5	19.09	16.87	111.5	21.4	14.6
Boselaphus tragocamelus	AMNH	AMNH35390	Male	251.5	25.21	26.16	246	26.44	20.67

Syncerus caffer nanus	AMNH	AMNH53576	Male	210.5	34.19	30.14	182.5	42.57	26.71
Syncerus caffer nanus	AMNH	AMNH53587	Male	199	33.28	26.88	171	38.53	23.72
Bos gaurus	AMNH	AMNH54469	Male	275	34.39	35.52	244	38.9	26.51
Bos gaurus	AMNH	AMNH54470	Male	267	37.34	35.36	240.5	45.79	32.38
Tetracerus quadricornis	AMNH	AMNH54980	Male	140.5	11.13	11.08	129	11.49	9.74
Syncerus caffer caffer	AMNH	AMNH82004	Male	235	44.74	36.43	193	55.22	31.08
Syncerus caffer caffer	AMNH	AMNH82005	Male	226	42.33	34.66	192.5	55.47	31.92
Bos mutus	AMNH	AMNH90186	Male	191	26.75	24.41	151.5	33.88	19.48
Bison bison athabascae	AMNH	AMNH98953	Male	277.5	37.28	37.34	231.5	48.31	32.61
Bison bison bison	MCZ	MCZ10	Male	264	37.25	35.95	215	48.16	30.61
Bison bison bison	MCZ	MCZ11	Male	251	35.77	34.28	202	43.99	29.07
Bison bison bison	MCZ	MCZ1770	Male	263.5	21.74	36.57	212	42.28	29.37
Boselaphus tragocamelus	NMNH	NMNH269127	Male	246	23.36	22.84	242	25.07	19.89
Syncerus caffer nanus	PC	PC CONGO256	Male	191	31.23	25.53	162	38.55	22.97
Syncerus caffer nanus	PC	PC FC56	Male	204	28.57	26.72	182	33.46	23.91
Bison bonasus	STOCK	NRM 587692	Unknown	256.48	31.17	27.91	204	36.32	26.63
Bos mutus	AMNH	AMNH10263	Unknown	189.5	25.41	22.29	156	31.94	18.25
Bos gaurus	AMNH	AMNH112979	Unknown	280	39.92	38.37	251.5	44.84	32.44
Bos gaurus	AMNH	AMNH113747	Unknown	272.5	39.46	38.02	243	46.22	33.44
Bubalus depressicornis	AMNH	AMNH214382	Unknown	132	17.05	15.65	117	20.1	13.29
Bos mutus	AMNH	AMNH35258	Unknown	184.5	29.47	25.3	150	38.27	21.22
Boselaphus tragocamelus	AMNH	AMNH35620	Unknown	258	24.8	25.23	253	25.64	19.74

Table 4.1 *(cont.)*

Species	Museum	Specimen	Sex	Metatarsal length	Metatarsal midshaft mediolateral diameter	Metatarsal midshaft anteroposterior diameter	Metacarpal length	Metacarpal midshaft mediolateral diameter	Metacarpal midshaft anteroposterior diameter
Boselaphus tragocamelus	AMNH	AMNH35629	Unknown	249	23.34	23.79	241	24.87	18.94
Tetracerus quadricornis	AMNH	AMNH54941	Unknown	150	11.92	12.63			
Bison bison athabascae	AMNH	AMNH73615	Unknown	281	36.86	37.83	231.5	50.3	31.98
Bubalus depressicornis	AMNH	AMNH80257	Unknown	141.5	17.83	16.14	125	21.69	14.53
Syncerus caffer caffer	AMNH	AMNH82006	Unknown	229	39.87	36.08	196	51.63	32.61
Bison bison athabascae	AMNH	AMNH98228	Unknown	256	33.89	34.17	205.5	42.95	29.41
Bison bison bison	MCZ	MCZ1771	Unknown	260	32.61	32.59	211.5	42.85	29.96
Boselaphus tragocamelus	MCZ	MCZ1936	Unknown	277.5	30	27.7	276.5	33.19	24.22
Bubalus mindorensis	NMNH	NMNH219049	Unknown	156	30.87	23.39	130.5	37.91	22.2
Bubalus depressicornis	NMNH	NMNH219297	Unknown	129	19.53	16.86	109	23.75	14.59
Bos sauveli	NMNH	NMNH399379	Unknown	247	26	29.75	218	28.88	33.74
Bison bison bison	TMM	TMM M-6116	Unknown	245	36.77	33.18			

The specimens used and their respective collections are shown with the raw measurements for each specimen.

Species means for each measurement were analysed using the ace function in R to generate reconstructed measurements or phylogenetically weighted means at the nodes of the Bovini phylogeny using Felsenstein's (1985) phylogenetic independent contrasts (PICs). The results are estimates at each node based on the node descendants that assume a Brownian motion model of evolution. Accordingly, changes along a given branch are independent of previous changes and are drawn from a distribution with a mean change of zero and a variance proportional to branch lengths. Thus, change typically increases with branch length and branch length or the time since a given divergence is a critical variable. The key point is that using PIC, characters at nodes within a phylogeny are reconstructed: (1) based on descendant taxa, the structure of the phylogeny and branch lengths; and (2) that the model of evolution is Brownian.

The consensus phylogeny and branch lengths of Hernández Fernández & Vrba (2005) were used with three modifications. First, the divergence of *Bubalus depressicornis* and *Bubalus quarlesi* was set at 1.25 Ma following Schreiber *et al.* (1999). Second, the subspecies *Bison bison bison* and *Bison bison athabascae* were included separately in the analysis with a divergence time of 5000 BP (see Gates *et al.* 2001, 2010). Third, the subspecies *Syncerus caffer caffer* and *Syncerus caffer nanus* were also included separately in the analysis with a divergence date of 155 000 BP (estimate after Van Hooft *et al.* 2002).

After ancestral state reconstruction, the tip data (extant species means) and the reconstructed phylogenetically weighted means at each node were used to generate a size variable equal to the geometric mean of all four metapodial diameters (anteroposterior and mediolateral diameters of both metacarpals and metatarsals). All four diameters and metapodial lengths were then plotted against size, which allowed an assessment of apparent evolutionary trends in size, distal limb elongation or shortening and relative metapodial robusticity.

Results

In terms of overall size as estimated by the proxy of metapodial midshaft diameters, the common ancestor of the Bovini is reconstructed as similar in size to *Bos sauveli*, the kouprey or *Syncerus caffer nanus*, the forest buffalo. This is shown graphically in Figure 4.1 by the position along the horizontal size axis of the reconstructed common ancestor (CA) of the Bovini.

The branch leading to the *Bos–Bison* node implies an increase in size, while size is reconstructed as basically unchanged from the ancestral Bovini state for the *Syncerus–Bubalus* node. *Bubalus* would appear to further decrease in size while *Syncerus* appears to increase in size. The common ancestor of *Bubalus quarlesi* and *Bubalus depressicornis* is reconstructed as even smaller in size. The *Bos–Bison* clade is reconstructed as remaining in general larger, with *Bos mutus* and *Bos sauveli* both likely to have experienced a trajectory leading to their decreased size compared to the rest of the *Bos–Bison* clade.

Trends in metatarsal length and midshaft diameters are relative to both reconstructed ancestral nodes and the overall size trendline for the Bovini. Thus, changes from an ancestral node that parallel the size trend for the Bovini indicate only morphological change consistent with scaling. However, trajectories at an angle with the Bovini size trendline indicate morphological change from an ancestral state that may not be explained by changes in size alone. The trajectory of apparent changes is shown by arrows in Figure 4.1.

Overall the apparent evolutionary trends in distal limb elongation or shortening are similar for the metatarsal (see Figure 4.1a) and metacarpal. Even relative to a likely decrease in body size there appears to be distal limb shortening leading from the common ancestor of the Bovini to the *Syncerus–Bubalus* node. The reverse trend, relative elongation, is apparent for the branch leading to the *Bos–Bison* node.

Other trends in terms of distal limb length are:

(1) elongation for *Bubalus arnee* and shortening for *Bubalus mindorensis*;
(2) elongation for *Bos sauveli* and shortening for *Bos mutus*.

Comparing reconstructed (for nodes) and measured (for tips) mediolateral diameters of the metatarsal with the size proxy variable shows little in terms of variation in the relative metatarsal mediolateral dimension. The only exception is *Bos sauveli*, with a more mediolaterally restricted metatarsal than seems typical for the rest of the Bovini.

In contrast, evolutionary change appears more likely to have influenced the anteroposterior dimension of the metatarsal (see Figure 4.1b). Thus, compared to the two Boselaphini included in the analysis, the Bovini as a whole appear to have a reduced anteroposterior metatarsal diameter relative to expectations based on size. Within the Bovini, trends largely mirror those already noted for relative metatarsal length. Thus, concomitant with metatarsal shortening for *Syncerus–Bubalus* there appears to be a possibly correlated reduction in relative anteroposterior metatarsal diameter. The reverse trend appears evident, leading to the *Bos–Bison* clade. Other apparent trends include:

(1) increased relative anteroposterior diameter for *Bubalus arnee* and decreased relative anteroposterior diameter for *Bubalus mindorensis*, paralleling the metatarsal length trends for these two species;
(2) increased relative anteroposterior diameter for *Bos sauveli*, also paralleling its metatarsal elongation.

The anteroposterior and mediolateral diameters of the metacarpal again appear to track the overall size proxy, with two exceptions. *Bos mutus* has a mediolaterally expanded and

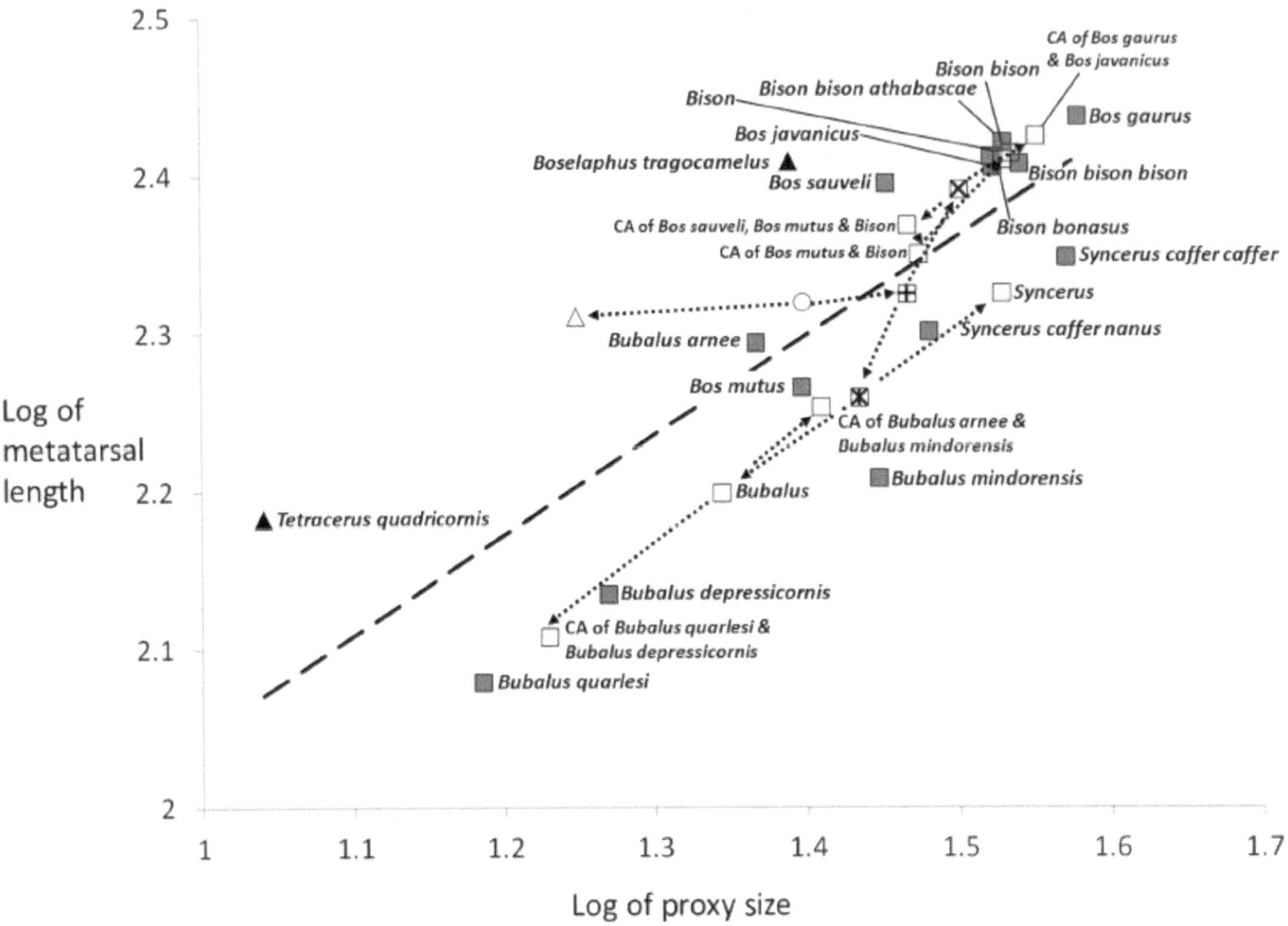

Figure 4.1 Bivariate plots of (a) log metatarsal length and (b) log anteroposterior diameter at metatarsal midshaft versus log proxy size (geometric mean of the four midshaft diameters). Means of extant species in the data set (tip taxa) are plotted using shaded symbols (triangles for Boselaphini and squares for Bovini) and are labelled by species name. Phylogenetically weighted means for the nodes (reconstructed CAs) are plotted using open symbols. The open circle denotes the basal node and the open triangle the CA of Boselaphini. Open squares denote the nodes within the Bovini phylogeny: a plus symbol marks the CA of the Bovini, a star marks the CA of *Syncerus* and *Bubalus*, and an X marks the CA of *Bos* and *Bison*. The remaining reconstructed CAs within the Bovini marked by open squares are labelled. A dashed line is an ordinary least squares regression for all the plotted data points (all tips and all nodes) and is shown for reference. Arrows connect the nodes from ancestor to descendant and show apparent evolutionary trajectories within the phylogeny. An arrow parallel or sub-parallel with the dashed line suggests mainly changes in size. Arrows that angle away from the dashed line suggest changes in morphology: shortening or elongation of the metatarsal (a) and changes in anteroposterior diameter at midshaft of the metatarsal (b).

anteroposteriorly restricted metacarpal. *Bos sauveli* evinces the reverse trend: mediolateral restriction and anteroposterior expansion of the metacarpal.

Discussion

The analysis and ancestral state reconstructions conducted here is at the outset simplistic and primarily descriptive. What the results describe are evolutionary trajectories implied by the limited data set of extant wild cattle. There is an apparent divide whereby a lineage leading to a *Bos–Bison* ancestor increases in body size, and elongates and deepens (anteroposterior expansion) its metapodials. In contrast, a lineage leading to *Syncerus–Bubalus* shortens and broadens its metapodials without a reduction in body size.

In general, these trends suggest possible adaptation to more open and/or dry habitats in a *Bos–Bison* common ancestor, while a *Syncerus–Bubalus* ancestor may have adapted to more moist, wooded habitats. This interpretation flows from the suggestion of Kingdon (1982) regarding shorter limbs and more moist, forested habitats as well as the perspective of Köhler (1993) on bovid metapodial adaptation (see also Scott 2004).

However, since the reconstructions here are simply phylogenetically weighted means of extant species and do not

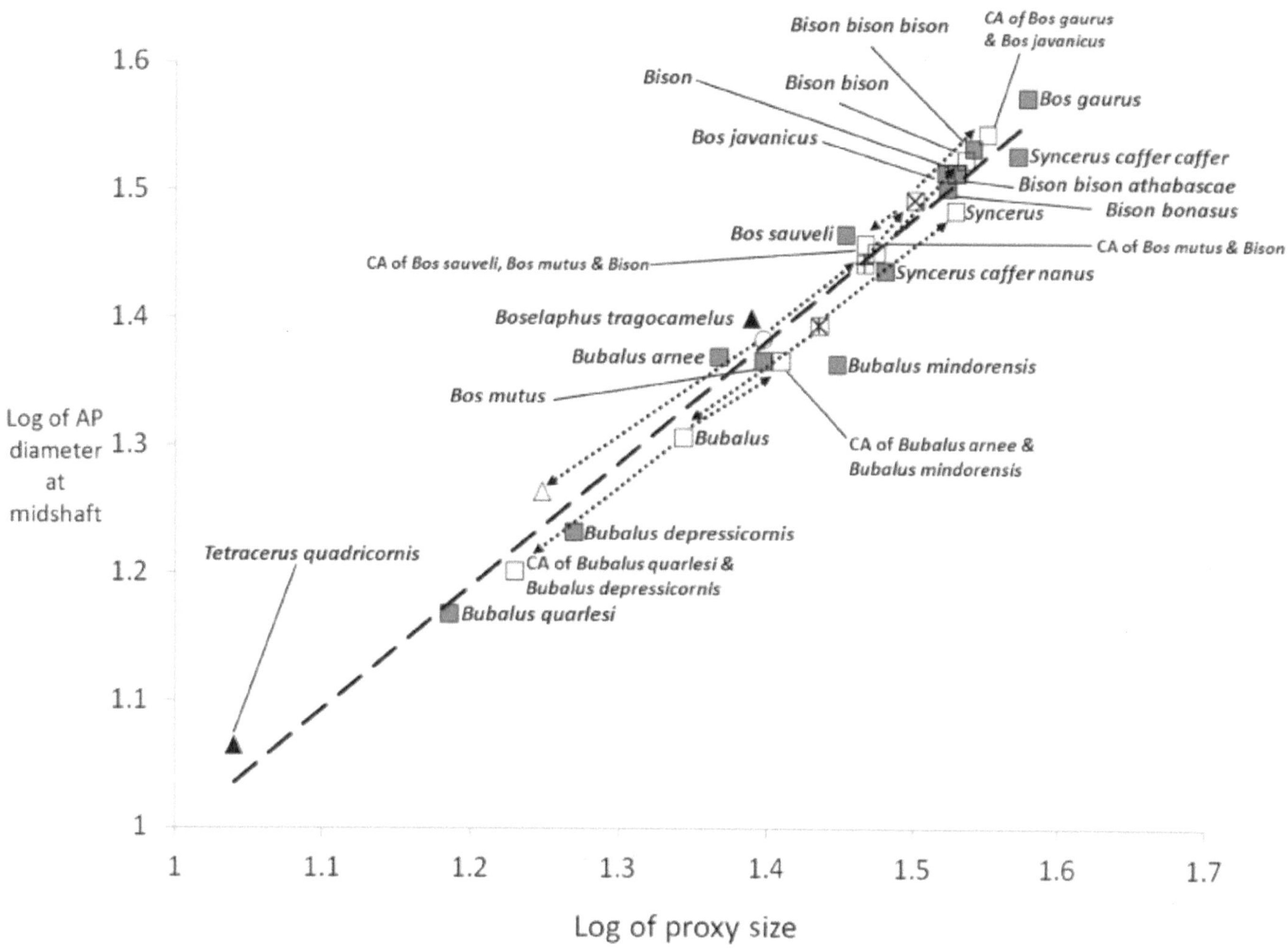

Figure 4.1 *(cont.)*

incorporate data from fossils of relevant ages, the actual evolutionary trend might emphasize change along either lineage to a greater of lesser extent and a different common ancestor of the Bovini would be implied. Thus, the results presented here are best viewed as hypotheses to be tested by fossils of the right age.

The reconstructions here imply changes fairly deep within the Bovini. Testing these requires fossils identifiable as members of the Bovini tribe with ages greater than seven million years ago to predate the likely common ancestor of *Bos* and *Bison*, or older than 11 million years ago to predate the *Syncerus–Bubalus* clade. The other apparent evolutionary trends would occur later in time, but most importantly are associated with specific extant species. Thus, morphologically distinct trends evident for *Bos mutus*, *Bubalus arnee* and *Bubalus mindorensis* could be relatively recent innovations. Again, cases such as the contrasting apparent trajectories of *Bubalus arnee* and *Bubalus mindorensis* are not resolvable without relevant fossil remains.

Overall, the reconstructed morphological trends shown in Figure 4.1 and the reconstructed states in Table 4.2 represent starting hypotheses regarding one aspect of skeletal evolution in the Bovini. Future work will require the application of fossil evidence and an investigation of other aspects of the postcranial skeleton.

Acknowledgements

Many thanks go to Mario Melletti for the kind invitation to contribute to this book. Our gratitude also goes to Emily Field-Lucas (Durham University BSc, 2012) for her patient and thorough research assistance in the early stages of the development of this chapter. We are grateful to Dr Dimitris S. Kostopoulus for the critical and useful comments on the first draft.

Table 4.2 Species means and reconstructed measurements

Species mean or reconstructed mean for node	Metatarsal length	Metatarsal midshaft mediolateral diameter	Metatarsal midshaft anteroposterior diameter	Metacarpal length	Metacarpal midshaft mediolateral diameter	Metacarpal midshaft anteroposterior diameter	Proxy size
Bison bison athabascae	263.8	33.3	32.7	216.7	33.3	28.3	42.1
Bison bison bison	255.6	33.0	34.2	208.2	33.0	29.2	43.4
Bison bonasus	254.6	35.1	31.8	205.9	35.1	26.6	41.1
Bos gaurus	273.9	38.3	37.4	244.5	38.3	31.6	44.7
Bos javanicus	258.4	35.9	32.8	227.2	35.9	26.6	38.3
Bos mutus	184.6	26.2	23.3	150.3	26.2	18.8	33.4
Bos sauveli	248.5	26.3	29.3	219.5	26.3	28.1	29.7
Bubalus arnee	197.0	23.4	23.5	165.5	23.4	19.1	27.7
Bubalus depressicornis	136.5	19.0	17.1	118.9	19.0	16.6	22.2
Bubalus mindorensis	161.5	30.1	23.1	136.7	30.1	23.9	36.8
Bubalus quarlesi	120.0	15.7	14.7	103.0	15.7	12.9	18.6
Syncerus caffer caffer	223.3	39.7	33.7	187.3	39.7	28.6	49.6
Syncerus caffer nanus	200.1	31.5	27.4	172.6	31.5	25.3	37.9
Boselaphus tragocamelus	256.4	25.3	25.1	251.7	25.3	20.7	27.0
Tetracerus quadricornis	152.3	11.3	11.6	141.3	11.3	9.5	11.6
CA of all included	208.7	25.7	24.2	186.8	25.7	20.9	29.6
CA of *Tetracerus* and *Boselpahus*	204.3	18.3	18.4	196.5	18.3	15.1	19.3
CA of Bovini	211.4	30.1	27.7	181.0	30.1	24.4	35.8
CA of *Bos* and *Bison*	246.3	32.0	31.2	212.1	32.0	26.8	37.3
CA of *Bos sauveli*, *Bos mutus* and *Bison*	234.0	28.8	28.7	197.4	28.8	25.4	34.7
CA of *Bos mutus* and *Bison*	224.1	30.5	28.4	182.3	30.5	23.6	38.0
Bison	257.2	34.1	32.6	209.2	34.1	27.7	41.9
Bison bison	259.7	33.1	33.4	212.4	33.1	28.7	42.8
CA of *Bos gaurus* and *Bos javanicus*	266.1	37.1	35.1	235.8	37.1	29.1	41.5
CA of *Syncerus* and *Bubalus*	181.8	28.5	24.8	154.6	28.5	22.4	34.6

Table 4.2 (cont.)

Species mean or reconstructed mean for node	Metatarsal length	Metatarsal midshaft mediolateral diameter	Metatarsal midshaft anteroposterior diameter	Metacarpal length	Metacarpal midshaft mediolateral diameter	Metacarpal midshaft anteroposterior diameter	Proxy size
Bubalus	158.2	22.9	20.2	134.5	22.9	18.7	27.4
CA of *Bubalus arnee* and *Bubalus mindorensis*	179.3	26.7	23.3	151.1	26.7	21.5	32.3
CA of *Bubalus depressicornis* and *Bubalus quarlesi*	128.2	17.3	15.9	110.9	17.3	14.7	20.4
Syncerus caffer	211.7	35.6	30.6	180.0	35.6	27.0	43.8

Means for extant species based on the data in Table 4.1 are shown along with the phylogenetically weighted means for reconstructed CAs. These data are used to generate Figure 4.1. Proxy size is the geometric mean of the four midshaft diameters

References

Bibi, F. (2007a). Dietary niche partitioning among fossil bovids in late Miocene C_3 habitats: consilience of functional morphology and stable isotope analysis. *Palaeogeography, Palaeoclimatology, Palaeoecology* **253**: 529–538.

Bibi, F. (2007b). Origin, paleoecology, and paleobiology of early Bovini. *Palaeogeography, Palaeoclimatology, Palaeoecology* **248**: 60–72.

Blumenschine, R. J. & Pobiner, B. L. (2007). Zooarchaeology and the ecology of Oldowan hominin carnivory. In *Evolution of the Human Diet: The Known, the Unknown and the Unknowable*, pp. 167–190. Ed. Ungar, P. S. New York: Oxford University Press.

Felsenstein, J. (1985). Phylogenies and the comparative method. *American Naturalist* **125**: 1–15.

Gates, C. C., Ellison, K., Freese, C. H., Aune, K. & Boyd, D. P. (2010). Numerical and geographic status. In *American Bison Status Survey and Conservation Guidelines 2010*, pp. 55–62. Eds. Gates, C. C., Freese, C. H., Gogan, J. P. & Kotzman, M. Gland: IUCN.

Gates, C. C., Stephenson, R. O., Reynolds, H. W., *et al.* (2001). *National Recovery Plan for the Wood Bison (*Bison bison athabascae*)*. Ottawa: National Recovery Plan.

Gentry, A. W. (1978). Bovidae. In *Evolution of African Mammals*, pp. 540–572. Eds. Maglio, V. J. & Cooke, H. B. S. Cambridge, MA: Harvard University Press.

Gentry, A. W. (1992). The subfamilies and tribes of the family Bovidae. *Mammal Review* **22**: 1–32.

Gentry, A. W. (2000). The ruminant radiation. In *Antelopes, Deer and Relatives: Fossil Record, Behavioral Ecology, Systematics and Conservation*, pp. 11–20. Eds. Vrba, E. S. & Schaller, G. B. New Haven, CT: Yale University Press.

Gentry, A. W. (2010). Bovidae. In *Cenozoic Mammals of Africa*, pp. 741–796. Eds. Werdelin, L. & Sanders, W. J. Berkeley, CA: University of California Press.

Ginsburg, L., Morales, J. & Soria, D. (2001). Les ruminantia (Artiodactyla, Mammalia) du Miocène des Bugti (Balouchistan, Pakistan). *Estudios Geológicos* **57**: 155–170.

Gordon, I. J. & Illius, A. W. (1988). Incisor arcade structure and diet selection in ruminants. *Functional Ecology* **2**: 15–22.

Groves, C. & Grubb, P. (2011). *Ungulate Taxonomy*. Baltimore, MD: Johns Hopkins University Press.

Grubb, P. (2000). Morphoclinal evolution in ungulates. In *Antelopes, Deer and Relatives: Fossil Record, Behavioral Ecology, Systematics and Conservation*, pp. 156–170. Eds. Vrba, E. S. & Schaller, G. B. New Haven, CT: Yale University Press.

Harris, J. M. (2003). Bovidae from the Lothagam succession. In *Lothagam: The Dawn of Humanity in Eastern Africa*, pp. 531–579. Eds. Leakey, M. G. & Harris, J. M. New York: Columbia University Press.

Hassanin, A. & Douzery, E. J. P. (1999). The tribal radiation of the family Bovidae (Artiodactyla) and the evolution of the mitochondrial cytochrome b gene. *Molecular Phylogenetics and Evolution* **13**: 227–243.

Hassanin, A. & Ropiquet, A. (2004). Molecular phylogeny of the tribe Bovini (Bovidae, Bovinae) and the taxonomic status of the kouprey, *Bos sauveli* Urbain 1937. *Molecular Phylogenetics and Evolution* **33**: 896–907.

Hernández Fernández, M. & Vrba, E. S. (2005). A complete estimate of the phylogenetic relationships in Ruminantia: a dated species-level supertree of the extant ruminants. *Biological Reviews* **80**: 269–302.

Janis, C. M. (1989). A climatic explanation for patterns of evolutionary diversity in ungulate mammals. *Palaeontology* **32**: 463–481.

Janis, C. M. (1993). Tertiary mammal evolution in the context of changing climates, vegetation, and tectonic events. *Annual Review of Ecological Systematics* **24**: 467–500.

Janis, C. M. & Scott, K. M. (1987). The interrelationships of higher ruminant families, with special emphasis on the members of Cervoidea. *American Museum Novitates* **2893**: 1–85.

Kingdon, J. (1982). *East African Mammals*, Vol. **IIIC,D**. Chicago, IL: University of Chicago Press.

Kingdon, J. (1997). *The Kingdon Field Guide to African Mammals*. San Diego, CA: Academic Press.

Köhler, M. (1993). *Skeleton and Habitat of Recent and Fossil Ruminants*. München: F. Pfeil.

Koopman, K. F. (1967). Artiodactyls. In *Recent Mammals of the World: A Synopsis of Families*, pp. 385–406. Eds. Andersen, S. & Knox, Jones J. Jr. New York: The Ronald Press.

Kostopoulos, D. S. (2006). Greek bovids through time. *Hellenic Journal of Geosciences* **41**: 141–152.

Miyamoto, M. M, Kraus, F., Laipis, P. J., Tanhauser, S. M. & Webb, S. D. (1993). Mitochondrial DNA phylogenies within Artiodactyla. In *Mammal Phylogeny: Placentals*, pp. 268–281. Eds. Szalay, F. S., Novacek, M. J. & McKenna. M. C. New York: Springer-Verlag.

Romer, A. S. (1945). *Vertebrate Paleontology*. Chicago, IL: University of Chicago Press.

Rose, K. D. (1996). On the origin of the order Artiodactyla. *Proceedings of the National Academy of Sciences* **93**: 1705–1709.

Schreiber, A., Seibold, I., Notzold, G. & Wink, M. (1999). Cytochrome b gene haplotypes characterize chromosomal lineages of anoa, the Sulawesi dwarf buffalo (Bovidae: *Bubalus* sp.). *Journal of Heredity* **90**: 165–176.

Scott, R. S. (2004). The comparative paleoecology of late Miocene Eurasian hominoids. PhD thesis. University of Texas at Austin.

Solounias, N. (2007). Bovidae. In *The Evolution of Artiodactyls*, pp. 278–291. Eds. Prothero, D. R. & Foss, S. E. Baltimore, MD: Johns Hopkins University Press.

Solounias, N., Barry, J. C., Bernor, R. L., Lindsay, E. H. & Raza, S. M. (1995). The oldest bovid from the Siwaliks, Pakistan. *Journal of Vertebrate Paleontology* **15**: 806–814.

Van Hooft, W. F., Groen, A. F. & Prins, H. H. T. (2002). Phylogeography of the African buffalo based on mitochondrial and Y-chromosomal loci: Pleistocene origin and population expansion of the Cape buffalo subspecies. *Molecular Ecology* **11**: 267–279.

Vrba, E. S. (1985). African Bovidae: evolutionary events since the Miocene. *South African Journal of Science* **81**: 263–266.

Vrba, E. S. & Schaller, G. B. (2000). Phylogeny of Bovidae based on behavior, glands, skulls, and postcrania. In *Antelopes, Deer and Relatives: Fossil Record, Behavioral Ecology, Systematics and Conservation*, pp. 203–222. Eds. Vrba, E. S. & Schaller, G. B. New Haven, CT: Yale University Press.

Chapter 5

Bovine mammary anatomy and function

R. Michael Akers, Anthony V. Capuco and Stephen C. Nickerson

Evolution of lactation

Every biology student quickly learns that the presence of functioning mammary glands is a hallmark of mammals. As outlined in his recent review, Oftedal (2012) provides support for the idea that primitive lactation began as a reproductive advantage in synapsids (the direct ancestors of mammals), likely during the Pennsylvania period. The capacity of 'modern' mammary glands to provide secretions with both vital nutrients (proteins, fats and carbohydrates) and health benefits (immunoglobulins, lysozyme, etc.) likely evolved from apocrine-like glands that were associated with hair follicles. It is believed that secretions from these glands provided moisture and antimicrobial agents for the parchment-like shelled eggs laid by these ancestors. Fossil evidence suggests that some therapsids and mammalian forms present during the Triassic period produced milk-like secretions. Evolutionary pressure likely promoted the incorporation of molecules such as lysozyme or iron-binding lactoferrin into these secretions. Certainly the capacity to prevent desiccation of the eggs and protection from microbial attack would have been highly beneficial.

The disaccharide lactose appears in all milks, with the exception of some marine mammals. Its synthesis depends on the enzyme lactose synthetase, which is active as a complex of β1-4-galactosyl transferase and the hormonally regulated milk protein α-lactalbumin. It is known that α-lactalbumin evolved from lysozyme prior to the evolution of sauropsids (leading to birds and reptiles). Thus, the ability to produce lactose appeared before it was utilized as a milk component. It is likely, then, that many of these early milk-like secretions contained antimicrobial oligosaccharides but no lactose. Interestingly, the cellular biosynthesis of lactose in the modern mammary gland also acts to promote hydration of suckling young because of the capacity of lactose contained within secretory vesicles to attract water via osmosis, thus effectively diluting mammary secretions. It is also likely that the presence of lactose improves milk let-down and the supply of milk to the suckling young because of the production of less viscous secretions. Indeed, experiments in transgenic mice have demonstrated that when the α-lactalbumin gene is silenced, the pups fail to thrive because of a failure to productively nurse (Stinnakre *et al.* 1994).

Fossil evidence also suggests it is likely that caseins were present during the Triassic period. The extensive bone and tooth development that occurred in neonates before the onset of independent feeding would necessitate the consumption of considerable quantities of calcium in the milk-like secretions. This suggests the presence of caseins in these secretions, which could provide for the delivery of calcium, phosphorus and protein to the newborns. With the synthesis of lactose, these modified secretions would have been even more nutritious and energy-rich. All of the primary milk constituents are believed to have origins that appeared before the divergence of synapsids from sauropsids. The subsequent evolution of placental-based reproduction in eutherian mammals removed the necessity for milk as a source of water, antimicrobials and nutrients for the egg, and substituted the need for a complex milk to support the neonate throughout an extended lactation.

With the domestication of plants and animals, the Neolithic Revolution began with dramatic changes in society and human culture. Dairy husbandry provided important sources of food, meat and fiber throughout the world. As Oftedal (2012: 355–368) concludes, 'the modern dairy industry is built upon a very old foundation, the cornerstones of which were laid before dinosaurs ruled the earth in the Jurassic and Cretaceous'. Regardless of evolutionary details, milk is indeed central to all mammals as it provides for both the immediate and long-term needs of the offspring (Capuco & Akers 2009). It is rather ironic that while we have long appreciated the nutrient value of mammary secretions, we are seemingly only now beginning to appreciate the rich diversity of the bioactive agents in milk and milk products. Certainly, there are likely beneficial effects for the suckling offspring and other consumers of milk and milk products (Clare & Swaisgood 2000; D'Alessandro *et al.* 2010; Mills *et al.* 2011).

The goal of this chapter is to provide an overview of mammary development, anatomy and function specific to the bovine. By necessity, our model is the modern domestic

Ecology, Evolution and Behaviour of Wild Cattle: Implications for Conservation, ed. M. Melletti and J. Burton. Published by Cambridge University Press.

dairy cow and nearly exclusively those of the Holstein breed. In North America, Holsteins account for about 90% of the dairy cattle population.

Mammary growth and development

It should come as no surprise that in bovine the vast majority of mammary-related studies have focused on dairy cattle, including the impacts of genetics, management and disease on mammary development, milk production and milk composition (Akers 2000, 2006; Davis Rincker *et al.* 2011; De Vries *et al.* 2011; Wall & McFadden 2012). There are few direct comparisons between even beef and dairy cows specific to lactation (Keys *et al.* 1989; Akers *et al.* 2006). To our knowledge, virtually nothing specific to mammary anatomy that directly compares wild cattle and domestic animals exists in the literature. Nonetheless, a review of udder development in dairy cows provides a basis for understanding lactation and mammary development in all cattle.

Early development in the bovine

In the 30-day-old bovine embryo, a slight thickening of the ventrolateral ectoderm (the mammary band) occurs at the time of limb bud appearance. These condensed ectodermal cells appear on either side of the midline and extend from the upper limb to the lower limb. A further condensation of these cells produces the mammary streak. By about five weeks, the mammary streak becomes the mammary line, a thin ridge of ectodermal cells that adjoins a deeper layer of compact mesenchymal cells. The mammary line shortens and the ectodermal cells proliferate and penetrate the lower mesenchymal cell layer, creating the mammary crest. A distinct basement membrane separates the ectodermal cells from the underlying mesenchymal cells. When shortening of the mammary crest is complete, the structure remains prominent in those areas where the mammary glands will finally form. For bovines this area is the inguinal region.

The dome-like cluster of the ectodermal cells that appears at about embryonic day 40 is called the mammary hillock. Increased rounding of hillock cells by about day 43 of gestation produces the hemispherical mammary bud. In the bovine, there are four mammary buds in the inguinal region, two on each side of the former mammary band. These buds give rise to the fore and rear quarters (mammary glands) of the udder. The mammary buds appear when the crown–rump length of the embryo is about 25 mm, and their formation is a critical stage in mammary development. Soon after the bud appears, the ectodermal layers sink into the mesenchyme, forming a depression (mammary pit) on the embryo's surface. The mammary bud stage marks the beginning of development patterns which distinguish mammary structures among various species. Differences between patterns of development in males and females also become apparent after this time. By day 80, the teat and primary sprout form. The primary sprout gives rise to the teat and gland cisterns. Secondary sprouts, which arise from the primary sprouts, appear by day 90, and by day 100 the primary and secondary sprouts are canalized. The canalized secondary sprouts give rise to ducts that will drain into the gland cistern. By the time of calving, the teat, teat cistern and gland cistern are formed (Anderson 1979).

Soon after birth, the gland is relatively quiescent, but beginning at about 50 days postpartum the parenchymal tissue (the epithelial portion of the gland that will give rise to the major ducts, lobules and alveoli) begins an allometric growth phase that continues as puberty approaches. During this early prepubertal period, the epithelial portion of the gland associated with the gland cistern is very dense and compact. Growth advances as epithelial structures are formed from sprouts associated with the gland cistern, a process referred to as budding. These closely packed ducts radiate from the gland cistern in a pattern that is reminiscent of broccoli florets; however, the parenchymal ducts fill only a fraction of the mammary fat pad (Ellis *et al.* 2012). This is quite different from the pattern of development in rodents (Akers 2002; Hovey *et al.* 1999). Continuing ductal growth into the surrounding mammary fat pad yields the major ducts that drain groups of alveoli (lobules) in the mature gland. Development of the lobules and alveoli is usually restricted to the last third of pregnancy. The histological development associated with postnatal stages of mammary development and structure is illustrated in Figure 5.1. Figure 5.2 provides a diagrammatic view of the udder in a mature animal, while Figure 5.3 illustrates the arrangement of lobules, major mammary ducts and the gland and teat cisterns.

Regulators of mammary growth and development

The endocrine system (Capuco & Akers 2011a, 2011b) is critical to all aspects of mammary development (mammogenesis), onset of lactation (lactogenesis) and maintenance of milk secretion (galactopoiesis). Lactogenesis is usually described as a two-stage process. Stage I consists of limited structural and functional differentiation of the secretory epithelium during the last third of pregnancy. Stage II allows for completion of cellular differentiation during the immediate periparturient period coinciding with the onset of copious milk synthesis and secretion. The degree to which the alveolar cells successfully differentiate near the time of calving and the total number of alveolar cells explains much of the reduced milk production in beef cows compared with dairy cows (Keys *et al.* 1989; Akers *et al.* 2006).

Responses to various anterior pituitary preparations exogenously administered in early studies made it clear that mammogenic and lactogenic substances were present, but the efforts of C. H. Li and colleagues in the 1940s to purify prolactin (Prl) and growth hormone (GH) were critical to our understanding of the fundamental roles of Prl and GH in mammary development and lactation. Specific roles for these hormones in controlling mammogenesis in rodents were

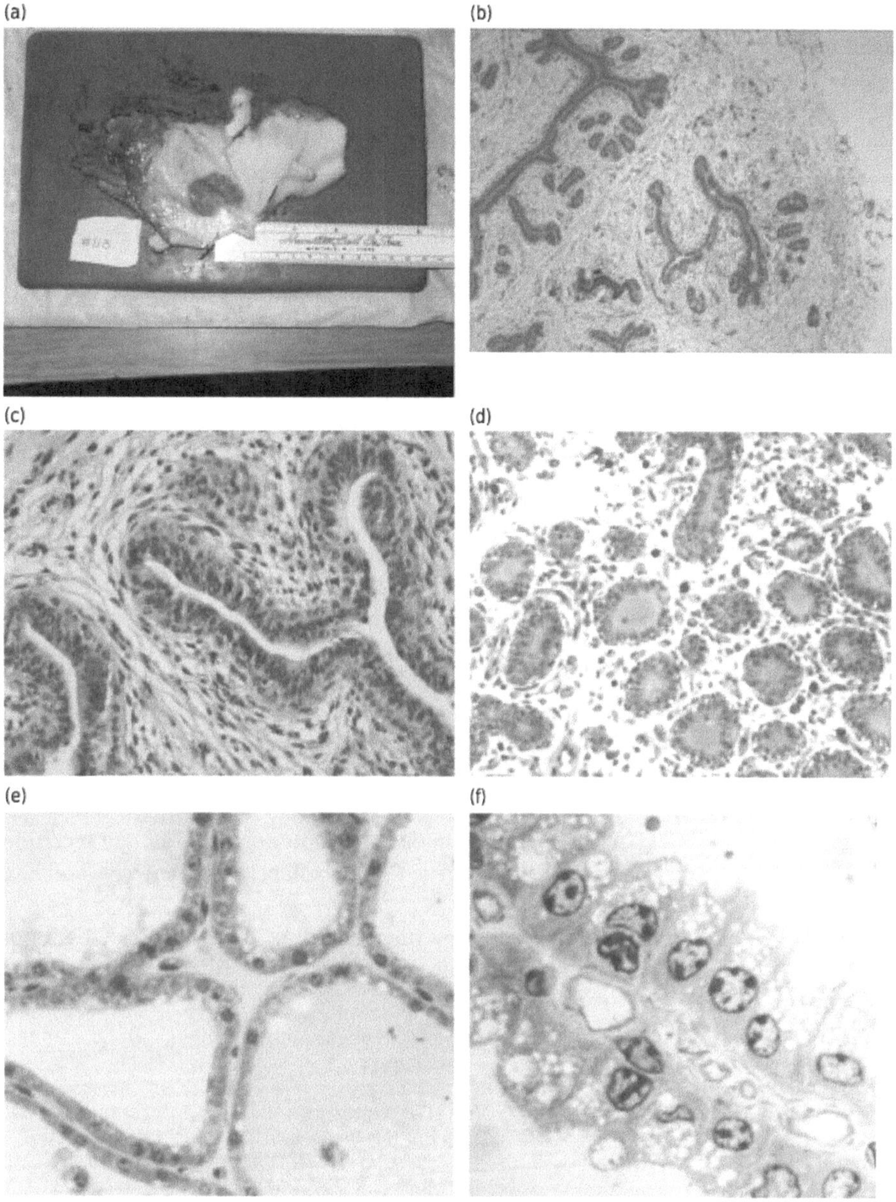

Figure 5.1 (a) A sagittal section of the mammary gland of a 75-day-old Holstein heifer. Note the very dense, compact mass of mammary parenchymal tissue surrounded by the mammary fat pad. (b) Developing ducts within the parenchyma of a four-month-old Holstein heifer. Note profiles of ducts surrounded by stromal cells and extracellular matrix but few apparent adipocytes. (c) The multilayered epithelium of mammary ducts from a four-month-old Holstein heifer. (d) The early development of mammary alveoli from a Holstein heifer at 160 days of gestation. (e) A cross-section of the closely aligned mammary alveoli from a lactating Holstein cow. (f) A profile of two adjacent alveoli and the marked cytological differentiation of the alveolar epithelial cells from a lactating Holstein cow.

delineated in classic hormone ablation and replacement experiments (Lyons *et al.* 1958; Nandi 1958). Triply operated (adrenalectomized, ovariectomized and hypophysectomized) rats and mice were treated with various combinations of hormones to determine if normal mammary development could be restored. Injections of estrogen and GH together caused proliferation of mammary ducts. But combinations of estrogen, progesterone, Prl and GH were needed for lobulo-alveolar development. The maximum ductular and lobulo-alveolar development, although still less than that observed during pregnancy, was obtained in animals given glucocorticoids in addition to the aforementioned hormones.

British researchers working to improve and maintain milk supplies during World War II, initiated multiple endocrine studies on mammary development and function in dairy animals (Cowie *et al.* 1980). For example, the effects of estrogen and progesterone on mammogenesis were extensively evaluated in attempts to induce lactation in non-pregnant animals. Because surgical complexities and experimental costs limit use of ablation replacement experiments to study mammary development and function in cattle, Cowie (1969) used hypophysectomized–ovariectomized goats and showed that mammary development comparable to that observed in mid-gestation goats occurred in surgically prepared goats given a combination of estrogen, progesterone, Prl, GH and ACTH. Such experiments confirmed that the fundamental endocrine regulators of mammary development elucidated in rodents played similar roles in dairy animals. Indeed, changes in concentrations of hormones in circulation, expression of receptors and direct effects of these hormones on mammary cells or tissue explants in culture (Akers 2006) all supported the significance of these hormones on mammary gland development in virtually all species. However, it is also important to appreciate that while these classic mammogenic and lactogenic hormones and their signaling cascades explain the fundamentals of mammary growth and function, molecular biology, proteomics and genomics have generated a plethora of growth factors and regulatory proteins that appear to coordinate, mediate and perhaps modify actions of these classic mammary acting hormones (Akers *et al.* 2000). Table 5.1 provides a summary of the impacts of 'classic' hormones and selected growth factors on mammary development and function (Tucker 2000; Akers 2006).

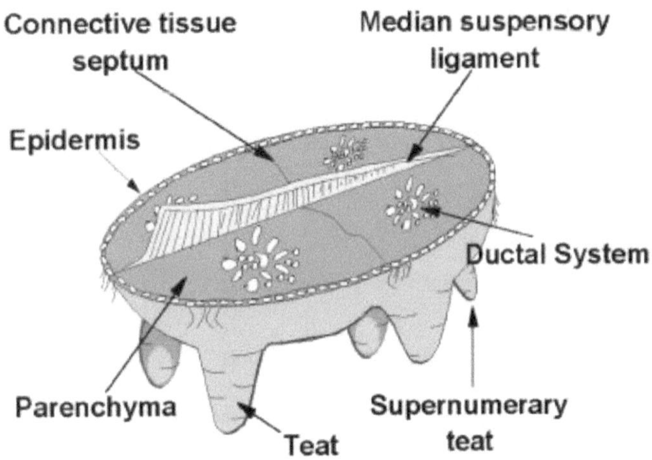

Figure 5.2 The gross anatomy of the mature mammary gland of a lactating Holstein dairy cow.

Hormonal and growth factor regulation of lactogenesis is closely reflected in the coordinated changes in serum concentrations of hormones around the time of parturition (Tucker 1981). For example, in pregnant heifers elevated blood concentrations of progesterone (P4) act to inhibit the onset of milk secretion. During the week before calving, P4 concentrations decline dramatically. Simultaneously, concentrations of estradiol (E2), glucocorticoids and, perhaps most critically, Prl increase at calving. These changes (release from the negative impact of P4) and positive stimulation of alveolar cell differentiation, primarily by Prl and glucocorticoids, act to convert the alveolar epithelial cells from a quiescent state to a biochemically and structurally active state of milk component synthesis (stage II lactogenesis). Along with changes in hormone concentrations, increased expression of Prl and glucocorticoid receptors and decreased P4 receptor expression further drive milk synthesis and secretion (Akers 1985). Furthermore, decreased secretion of P4 when coupled with increased secretion of E2 stimulates expression of oxytocin receptors on the myoepithelial cells surrounding the alveoli. Thus, the stage is set not only for milk synthesis and secretion into the alveolar luminal spaces, but an enhanced milk ejection response to oxytocin, which is released in response to suckling or milking. That Prl is a key to lactogenesis was demonstrated by chemical ablation and replacement experiments in dairy cows (Akers *et al.* 1981a, 1981b), as well as by a study of the effects of Prl addition on α-lactalbumin synthesis and secretion in cultured mammary tissue explants from cows (Goodman *et al.* 1983).

Undoubtedly, an understanding of the classic mammogenic and lactogenic hormones, their receptors and related

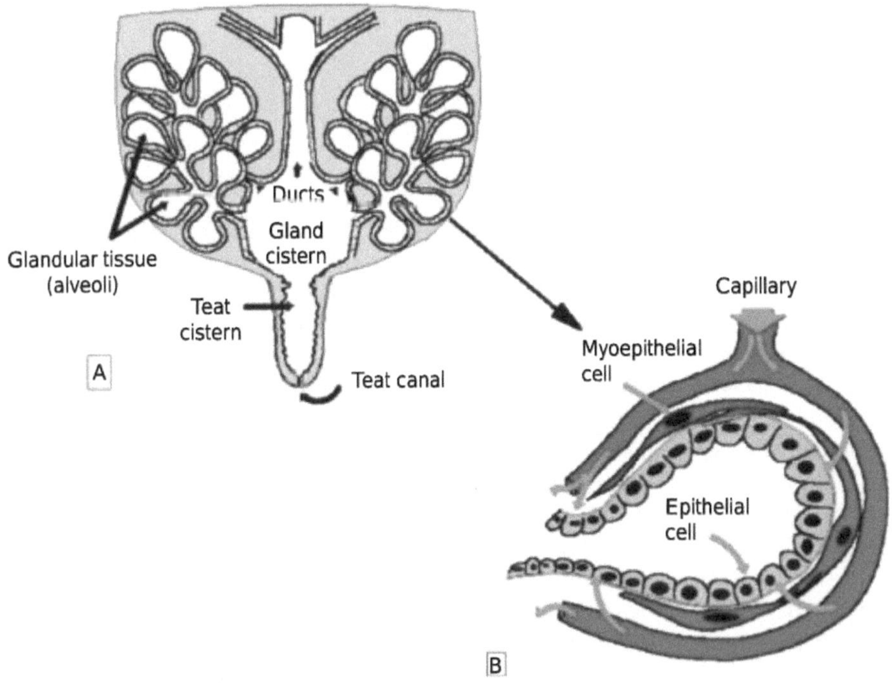

Figure 5.3 A stylized view (a and b) of the arrangement of lobules of alveoli with major mammary ducts, the gland and teat cistern, as well as the arrangement cells within an alveolus and blood supply (with permission from Nickerson and Akers 2011). A black and white version of this figure will appear in some formats. For the colour version, please refer to the plate section.

Table 5.1 Hormones and growth factors (GFs) in mammary development and function

Hormone or GF[5]	Source	Major action
Estrogen	Ovary/placenta/local[1] (?)	Ductal cell proliferation and morphogenesis
Progesterone	Ovary/placental	Ductal cell proliferation and morphogenesis
Prolactin	Anterior pituitary/local[1] (?)	Alveolar cell differentiation
Growth hormone	Anterior pituitary	Alveolar cell function and metabolism
Glucocorticoids	Adrenal cortex	Alveolar cell differentiation
Insulin	Endocrine pancreas	Metabolism protein synthesis
Oxytocin	Posterior pituitary	Milk ejection
IGF-I	Liver/local[1]	Ductal cell proliferation/metabolic support
TGF-β family[2]	Various tissues/local[1]	Stromal tissue remodeling, ductal development
EGF family[3]	Various tissues/local[1]	Proliferation parenchyma and remodeling
FGF family[4]	Various tissues/local[1]	Proliferation and ductal development
Leptin	Adipose tissue/local[1]	Cell proliferation, metabolic effects

[1] Local refers to the mammary tissue (parenchyma or stroma) production of the hormone or growth factor.
[2] TGF-β family refers to a group of at least five proteins and related receptors involved in extracellular matrix remodeling and inhibition of epithelial cell proliferation.
[3] EFG family refers to at least ten proteins including EGF, TGF-α, amphiregulin, etc. and companion ERB receptors (ErbB-1, ErbB-2 also called Her2 or Neu). The proteins and receptors are involved in human breast carcinomas and are known to stimulate the growth of bovine mammary cells.
[4] FGF family refers to a group of 20 proteins that have shared structures and high affinity for heparin and other extracellular matrix proteins. They were first identified from bovine tissue extracts and are believed to be stroma-derived mitogens in the mammary gland that act in a paracrine manner to influence epithelial cell proliferation and ductal morphogenesis.
[5] Related references can be found in Akers (2006).

signaling pathways help to explain much of mammary development, at least in broad strokes. However, as suggested earlier and represented in Table 5.1, how the plethora of growth factors and regulators interact to modify mammary development and function is poorly understood. This is especially true in economically important dairy ruminants.

References

Akers, R. M. (1985). Lactogenic hormones: binding sites, mammary growth, secretory cells differentiation and milk biosynthesis in ruminants. *Journal of Dairy Science* **68**: 413–437.

Akers, R. M. (2000). Selection for milk production from a lactation biology viewpoint. *Journal of Dairy Science* **83**: 1151–1158.

Akers, R. M. (2002). *Lactation and the Mammary Gland.* Ames, IA: Iowa State Press.

Akers, R. M. (2006). Major advances associated with hormone and growth factor regulation of mammary growth and lactation in dairy cows. *Journal of Dairy Science* **89**: 1222–1234.

Akers, R. M., Bauman, D. E., Capuco, A. V., Goodman, G. T. & Tucker, H. A. (1981a). Prolactin regulation of milk secretion and biochemical differentiation of mammary epithelial cells in periparturient cows. *Endocrinology* **109**: 23–30.

Akers, R. M., Bauman, D. E., Goodman, G. T., Capuco, A. V. & Tucker, H. A. (1981b). Prolactin regulation of cytological differentiation of mammary epithelial cells in periparturient cows. *Endocrinology* **109**: 31–40.

Akers, R. M., McFadden, T. B., Purup, S., Vestergaard, M., Sejrsen, K. & Capuco, A. V. (2000). Local IGF-I axis in peripubertal ruminant mammary development. *Journal of Mammary Gland Biology and Neoplasia* **5**: 43–51.

Akers, R. M., Capuco, A. V. & Keys, J. E. (2006). Cellular differentiation in mammary tissue of beef and dairy heifers. *Livestock Science* **105**: 44–49.

Anderson, R. R. (1979). Embryonic and fetal development of the mammary apparatus. In *Lactation: A Comprehensive Treatise*, vol. 4, pp. 3–40. Ed. Larson, B. L. New York: Academic Press.

Capuco, A. V. & Akers, R. M. (2009). The origin and evolution of lactation. *Journal of Biology* **8**: 1–4.

Capuco, A. V. & Akers, R. M. (2011a). Galactopoiesis, effect of hormones and growth factors. In *Encyclopedia of Dairy Science*, 2nd edition, vol. 2, pp. 26–31. New York: Academic Press.

Capuco, A. V. & Akers, R. M. (2011b). Galactopoiesis, effect of treatment with bovine somatotropin. In *Encyclopedia of Dairy Science*, 2nd edition, vol. 2, pp. 32–37. New York: Academic Press.

Clare, D. A. & Swaisgood, H. E. (2000). Bioactive peptides: a prospectus. *Journal of Dairy Science* **83**: 1187–1195.

Cowie, A. T. (1969). Lactogenesis. In *The Initiation of Milk Secretion at Parturition*, pp. 157–169. Eds. Reynolds, M. & Folley, S. J. Philadelphia, PA: University of Pennsylvania Press.

Cowie, A. T., Forsyth, I. A. & Hart, I. C. (1980). *Hormonal Control of Lactation.* Berlin, Heidelberg: Springer-Verlag.

D'Alessandro, A., Zolla, L. & Scaloni, A. (2010). The bovine milk proteome: cherishing, nourishing and fostering molecular complexity. An interactomics and functional overview. *Molecular Biosystems* 7: 579–597.

Davis Rincker, L. E., VanderHaar, M. J., Wolf, C. A., Liesman, J. S., Chapin, L. T. & Weber-Nielsen, M. S. (2011). Effect of intensified feeding of heifer calves on growth, pubertal age, calving age, milk yield and economics. *Journal of Dairy Science* **94**: 3554–3567.

De Vries, L. D., Casey, T., Dover, H., VanderHaar, M. J. & Plaut, K. (2011). Effects of transforming growth factor-β on mammary remodeling during the dry period of dairy cows. *Journal of Dairy Science* **94**: 6036–6046.

Ellis, S., Akers, R. M., Capuco, A. V. & Safayi, S. (2012). Bovine mammary epithelial cell lineages and parenchymal development. *Journal of Animal Science* **90**: 1666–1673.

Goodman, G. T., Akers, R. M., Friderici, K. H. & Tucker, H. A. (1983). Hormonal regulation of alpha-lactalbumin secretion from bovine mammary tissue cultured in vitro. *Endocrinology* **112**: 1324–1330.

Hovey, R. C., McFadden, T. B. & Akers, R. M. (1999). Regulation of mammary gland growth and morphogenesis by mammary fat pad: a species comparison. *Journal of Mammary Gland Biology and Neoplasia* **4**: 53–68.

Keys, J. E., Capuco, A. V., Akers, R. M. & Djiane, J. (1989). Comparative study of mammary gland development and differentiation between beef and dairy heifers. *Domestic Animal Endocrinology* **6**: 311–319.

Lyons, W. R., Li, C. H. & Johnson, R. E. (1958). The hormonal control of mammary growth. *Recent Progress in Hormone Research* **14**: 219–254.

Mills, S., Ross, R. P., Hill, C., Fitzgerald, G. F. & Stanton, C. (2011). Milk intelligence: mining milk for bioactive substances associated with human health. *International Dairy Journal* **21**: 377–401.

Nandi, S. (1958). Endocrine control of mammary gland development and function in the C3H/He Crgl mouse. *Journal of the National Cancer Institute* **21**: 1039–1360.

Nickerson, S. C. & Akers, R. M. (2011). Mammary gland anatomy. In *Encyclopedia of Dairy Science*, 2nd edition, vol. 2, pp. 328–337. New York: Academic Press.

Oftedal, O. (2012). The evolution of milk secretion and its ancient origins. *Animal* **6**: 1–14.

Stinnakre, M. G., Vilotte, J. L., Soulier, S. & Mercier, J. C. (1994). Creation and phenotypic analysis of alpha-lactalbumin-deficient mice. *Proceedings of the National Academy Sciences of the United States of America* **91**: 6544–6548.

Tucker, H. A. (1981). Physiological control of mammary growth, lactogenesis, and lactation. *Journal of Dairy Science* **64**: 1403–1421.

Tucker, H. A. (2000). Hormones, mammary growth, and lactation: a 41-year perspective. *Journal of Dairy Science* **83**: 874–884.

Wall, E. H. & McFadden, T. B. (2012). A local affair: how the mammary gland adapts to changes in milking frequency. *Journal of Animal Science* **90**: 1695–1707.

Chapter 6

The digestive system of ruminants, and peculiarities of (wild) cattle

Marcus Clauss and Reinhold R. Hofmann

Introduction

Cattle are members of the Ruminantia, which represent the most successful group of extant large herbivores in terms of species diversity. Ruminants have a multi-chambered forestomach, similar to other foregut fermenters like kangaroo, hippos, peccaries or sloths (Langer 1988); this means that plant material is partly digested by symbiotic microbes before the whole digesta – which includes the partly digested diet *and* the microbes – is passed on to the lower digestive tract where the herbivore's own enzymes further digest this mixture. Because microbial protein is a major component of this mixture, foregut fermenters produce a set of specific enzymes in their glandular stomach and small intestine that help break down microbial cells, so that their protein can be used (Pacheco *et al.* 2007). Functional ruminants – the phylogenetic ruminants as well as the camelids – combine simple foregut fermentation with peculiar sorting mechanisms that assure that larger digesta particles are regurgitated and re-masticated (ruminated). This process of rumination is an obligatory physiological feature, facilitates a more efficient particle size reduction (Fritz *et al.* 2009), higher digestive efficiencies (Foose 1982) and potentially also higher food intake levels than observed in non-ruminant foregut fermenters (Clauss *et al.* 2010a).

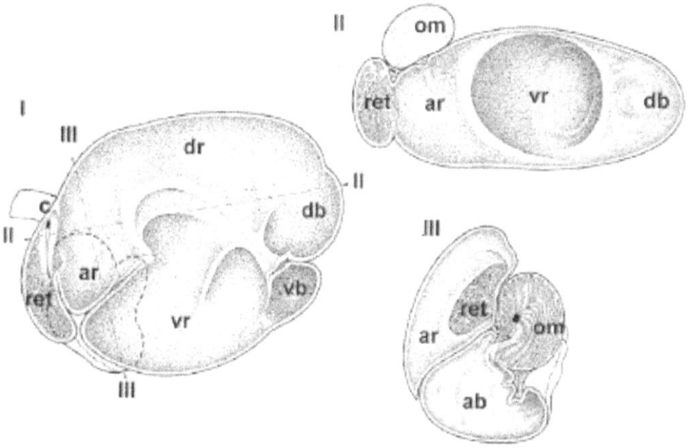

Figure 6.1 Schematic presentation of the ruminant forestomach modified from Hofmann (1973). (I) view from the left side into the reticulorumen with the dorsal rumen (dr), ventral rumen (vr), the dorsal and ventral blindsacs (db, vb), the atrium ruminis (ar) and the reticulum (ret). Note the remnants of the 'ventricular groove', through which, in suckling ruminants, milk is diverted directly from the cardia (c) into the ostium reticulo-omasale; this groove is structurally maintained in all adult ruminants. It leads via the omasal canal into the abomasum. Food is digested by symbiotic microbes in the rumen compartments and sorted, according to density, in the reticulum. The transects that lead to views II and III are indicated. (II) view from the top of the reticulorumen, through the intraruminal orifice from the dorsal to the ventral rumen; this orifice is framed by muscular ridges called 'rumen pillars'; note the omasum (om) which is not opened in this view. (III) view from within the forestomach towards the front, displaying the ruminoreticular orifice that links the atrium ruminis and the reticulum, and the inside of the omasum with its laminae, and the abomasum (ab). Digesta is passed from the reticulum either back into the atrium ruminis, or on to the omasum, where moisture is removed, and from there to the abomasum, where digestion by the ruminant's own enzymes is initiated.

Ruminant digestive anatomy and physiology

The ruminant stomach consists of four compartments – three representing the forestomach complex, and the last representing the glandular stomach ('abomasum'), the equivalent of the stomach of monogastric animals (Hofmann & Schnorr 1982). The three forestomach compartments are, in the sequence of the digestive process, the rumen, the reticulum and the omasum (Figure 6.1). From the outside, the rumen and the reticulum form a unit – a large fermentation chamber with several sub-compartments, including the dorsal and the ventral rumen, the dorsal and ventral rumen blindsacs, the atrium ruminis and the reticulum. The whole complex is often referred to as the reticulorumen (RR). The reticulum is the most cranial part of the RR. On the right side of the RR, the omasum is a distinct structure. In contrast to the RR, which has a consistency of the digesta it contains, the omasum is more solid to the touch, and ball- or bean-shaped. The omasum leads to the abomasum, which in turn leads to the small and then the large intestine.

Due to the economic relevance of domestic ruminants, an enormous amount of research has been published on detailed functions of the ruminant gastrointestinal tract (GIT) in relation to digestion and absorption. Without doing this body of research justice, the function of the different parts of the ruminant digestive tract can be crudely summarized as follows.

Ecology, Evolution and Behaviour of Wild Cattle: Implications for Conservation, ed. M. Melletti and J. Burton. Published by Cambridge University Press.

(1) Plant material is fermented by microbes in the rumen, a process in which fermentation products originate that are absorbed through the rumen wall and used by the host. The rumen's internal mucosa is papillated to increase its absorptive surface. Because microbial digestion of plant material takes time, the rumen itself, by its sheer volume, plays an important role in delaying digesta passage (Demment 1982). Strong muscular ridges (the rumen 'pillars') help mix the rumen contents.

The diet itself can have an additional effect on the delay of particle escape from the rumen. Forage particles that are ingested by ruminants may form a 'fibre mat' or 'raft' in the rumen, a more or less coherent matrix of entangled particles, through which fluid may pass, but in which small particles are entrapped. This so-called 'filter-bed effect' appears to be more pronounced on grass diets as opposed to browse diets (Clauss *et al.* 2011a; Lauper *et al.* 2013), which has interesting implications for anatomical and evolutionary diversification within ruminants (see below).

(2) Digesta is sorted in the reticulum by a separation mechanism based on particle density – larger, lighter and not yet completely digested particles float in the moist reticulum contents and are rejected back into the rumen, whereas small, dense and mostly digested particles are passed on together with fluid into the omasum (Baumont & Deswysen 1991; Lechner-Doll *et al.* 1991). The mucosa of the reticulum has a typical 'honeycomb' pattern of muscular ridges that help in contracting the organ (Hofmann & Schnorr 1982).

That larger, not-yet-digested particles are lighter than smaller, digested ones appears to be a common feature of plant material subjected to microbial digestion (Sutherland 1988; Clauss *et al.* 2009a, 2009b). Sorting according to these properties requires two things: a liquid environment where particles can float and sediment, and a constant position of the anatomical structures in relation to gravity. The reticulum contents are always very moist, usually more moist than the overall rumen contents (Clauss *et al.* 2009a, 2009b). As the reticulum is positioned underneath the cardia – the opening of the oesophagus into the RR – saliva 'drops' into this compartment first, assuring the high moisture content. Sorting in a moist compartment is the precondition for particle separation and hence rumination; rumination, i.e. mastication, in turn increases salivary inflow and hence guarantees this precondition. With respect to the position in relation to gravity, the typical ruminant resting and sleeping position, sternal recumbency (as opposed to lying on the side), guarantees this second prerequisite (Clauss 2004). As a side effect, the sorting mechanism (and not foregut fermentation as such) appears to be linked to an increased methane production when compared to non-ruminant herbivores (Franz *et al.* 2010, 2011); the reasons for this remain to be understood.

The high moisture content in the reticulum represents another challenge: when passing digesta on to the lower digestive tract, the small particles move together with the fluid – it is physically impossible to retain the fluid. Hence, highly diluted digesta would reach the sites of acid and enzyme secretion in the glandular stomach, which would require high secretory activity to compensate for the dilution. Therefore, it would be favourable if this side effect of the sorting mechanism could be counteracted:

(3) In the omasum, fluid is re-absorbed from the digesta passed out of the reticulum. The omasum has an internal structure of different-sized 'leaves' that enlarge its internal absorptive surface (Hofmann & Schnorr 1982). Digesta in the omasum is invariably much drier than in the reticulum (Clauss *et al.* 2009a, 2009b) than in the subsequent glandular stomach where acid and enzyme secretions increase the moisture content again.
(4) In the glandular stomach, microbes are killed by gastric acid, and enzymatic digestion of both microbes and diet remnants is initiated.
(5) In the small intestine, enzymatic digestion continues and nutrients are absorbed.
(6) In the first part of the large intestine, secondary fermentation of hitherto refractory plant material takes place in the caecum and proximal colon.
(7) In the second part of the large intestine, fluid is re-absorbed from the digesta to reduce faecal water losses. This part of the GIT is particularly developed in desert-adapted ruminants (Woodall & Skinner 1993), which therefore have drier faeces (Clauss *et al.* 2004).

Note that points 4–7 are similar to the digestive functions found in monogastric animals.

Peculiarities of cattle digestive anatomy and physiology

A large anatomic variability in basically all sections of the RR and GIT has been demonstrated in ruminants (Hofmann 1969, 1973, 1985, 1988, 1989, 2000; Figure 6.2). The most impressive characteristics of cattle documented in these studies as compared to other ruminants are the enormous absolute RR capacity, the disproportionately large omasum, very deep reticular honeycomb cells, very pronounced rumen pillars, a distinct gradient in the papillation pattern in the rumen with an often completely non-papillated dorsal area, and a short large intestine. In terms of RR physiology, cattle have a very distinct stratification of rumen contents, with a gas dome on top, a fibre mat filling the majority of the rumen and a more liquid phase underneath with sediment (Clauss *et al.* 2009b; Hummel *et al.* 2009). They have a distinct separation of the passage of fluids and particles through the RR, which is not only due to very pronounced particle retention, but also due to a very high fluid throughput (Clauss *et al.* 2006b). Cattle achieve this with a very high saliva production – in domestic cattle, amounts of up to 240 litres of saliva per day have been measured

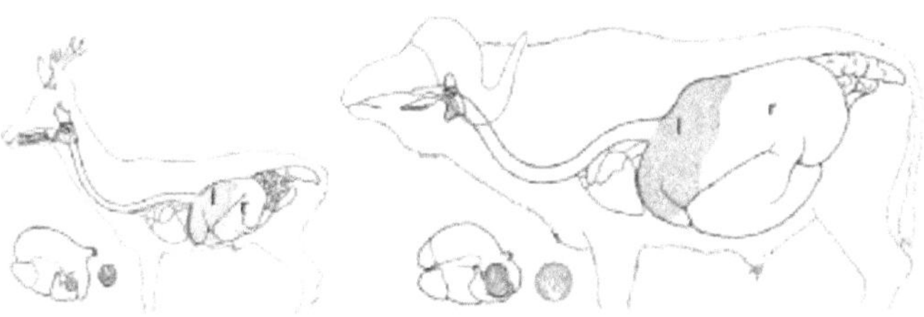

Figure 6.2 Morphological variation in the ruminant digestive system, exemplified in roe deer (*Capreolus capreolus*) on the left and African buffalo (*Syncerus caffer*) on the right, as seen on their left body side. Note the comparatively larger salivary glands (s), smaller rumen (r), larger liver (shaded area, l), and larger intestinal tract (i) in roe deer; when viewing the forestomach from its right side, differences in the size of the omasum are evident (left inlays), which are also visible in schematic transects of this organ (right inlays). Modified from Hofmann (1985).

(Maekawa *et al.* 2002). Cattle are the only group of wild ruminants in which a certain type of rumen protozoa, the Isotrichidae ('holotrichs'), occur in high proportions (Clauss *et al.* 2011b).

Some of these observations may be linked to each other. In the overall *bauplan* of the body cavity, the exceptionally large rumen of cattle matches the reduction of other organs for which less space may be available – for example, with respect to the lungs, which may be reduced in size so that a higher breathing frequency is needed to compensate (Mortolaa & Lanthier 2005), with respect to the liver, which appears to be relatively smaller in large cattle compared to other ruminants (Hofmann 1989) and with respect to the large intestine, which can no longer absorb enough water for faecal pellet formation – hence cattle typically defecate in 'pies' and not in 'pellets', as most other ruminants (Clauss *et al.* 2003a). With respect to the forestomach, a high fluid throughput and hence moisture content in the RR may have allowed cattle to evolve higher reticular honeycomb ridges than other ruminants, which close the reticulum during contractions – because the fluid freely available in the rumen can immediately re-fill the reticulum after a contraction (Clauss *et al.* 2010b). A high fluid throughput through the RR also matches a very large omasum that re-absorbs this fluid (Clauss *et al.* 2006a). The distinct papillation gradient matches both the separation of particle and fluid passage through the RR (Clauss *et al.* 2009c) and the stratification of rumen contents (Codron & Clauss 2010).

Hypotheses on function: what is the adaptive relevance of the morphophysiological characteristics of cattle?

The observations on the digestive tract of cattle beg for functional interpretations. Ideally, such interpretations should be backed by evidence, either from controlled experiments, from functional models or from comparative analyses testing for convergence. In the absence of such evidence, interpretations must remain speculative, and their main yardstick is their inherent logic. Whereas the adaptive value of morphological variations of the prehensile and processing units (craniodentals, masticatory apparatus) follows evident mechanical principles and has been rarely questioned, the functional interpretation of the variation observed in the digestive tract itself has been less unanimous (Clauss *et al.* 2008). Comparative analyses of morphology and physiology are our best approach for understanding the peculiarities of adaptations in cattle.

From the very beginning of the documentation of variation in ruminant digestive physiology onwards, Hofmann (1968, 1989; Hofmann & Stewart 1972) recognized a relationship between digestive tract morphology and feeding type (along the browser–intermediate feeder–grazer continuum). This has been corroborated in a series of statistical evaluations that tested the variability in morphophysiological parameters either between feeding types, or along a diet gradient (usually percentage of grass in the natural diet; Clauss *et al.* 2003b, 2006a, 2009c, 2010b, 2011b; Pérez-Barberìa *et al.* 2004; Hofmann *et al.* 2008), so that the existence of the relationship should be beyond doubt. The task to interpret this relationship, however, remains. To start with, one should not take a direct causal relationship between the diet and the morphophysiology for granted, but should also be open to the possibility that the diet is just a proxy for another, really causative factor. Additionally, one should probably not approach any interpretation with a concept of exclusiveness (in the sense that one particular adaptation is only suitable for one type of diet, for example), but probably rather in a concept of comparative/competitive advantages. In particular, the 'cattle-type' digestive physiology has been suggested to accommodate a large spectrum of diets, from exclusive grass diets to mixed diets with a high proportion of browse, but not browse-only diets (Codron & Clauss 2010). This corresponds to the fact that among wild cattle, species or subspecies that are not exclusive grazers but tolerate mixed diets or have seasonally high intakes of browse in the wild have evolved – among *Bison*, the European bison (*Bison bonasus*) (Gebczynska *et al.* 1991) and the wood bison (*Bison bison athabascae*) (Larter & Gates 1991) potentially ingest higher proportions of browse than the plains bison (*Bison bison*); within *Syncerus*, the forest buffalo (*Syncerus caffer nanus*) potentially ingests higher proportions of browse (Bekhuis *et al.* 2008) than the Cape buffalo (*Syncerus caffer caffer*), but also the Cape buffalo was demonstrated to ingest mixed diets (Landman & Kerley 2001). Among Asian wild cattle, relevant proportions of browse have been reported or are suspected in the gaur (*Bos gaurus*; Chetri 2006), water buffalo (*Bubalus arnee*), banteng (*Bos javanicus*; Bowman *et al.* 2010) or the anoas (*Anoa* spp.; Pujaningsih *et al.* 2009); for fossil feral cattle, evidence for higher proportions of browse in their diet when compared to typically grazing extant representatives has been produced (Noe-Nygaard *et al.* 2005; Rivals *et al.* 2007; Schulz & Kaiser 2007); domestic cattle ranging corresponding pastures can ingest high proportions of browse material (Connor *et al.* 1963; Thilenius & Hungerford 1967; Holechek *et al.* 1982; Bayer 1990); and in traditional agriculture, browse has been used as a supplement in domestic cattle husbandry (Machatschek 2002). Although most of these reports indicate that grass is the preferred forage of cattle, they also indicate flexibility in feeding habits. In contrast, the 'moose-type' digestive physiology – characterized by a low relative fluid throughput and an absence or low degree of RR contents stratification (Clauss *et al.* 2010a) – appears to be linked to browse-only diets in the wild (Codron & Clauss 2010), corresponding to the statement that 'browsers are non-grazers' (Van Wieren 1996). In practice, it may also be possible for such animals to ingest higher-than-predicted proportions of grass, such as in captive moose (*Alces alces*) kept on grass silage (Lechner *et al.* 2010) or free-ranging roe deer (*Capreolus capreolus*) in agriculturally shaped landscapes (Abbas *et al.* 2013). In the wild this may be limited to situations where grass availability is not limiting and these animals are not faced with competition from 'cattle-type' species that exploit this niche more efficiently, such as the reindeer (*Rangifer tarandus*), the major outlier in Codron & Clauss (2010).

The interpretation of the enormous relative RR capacity of cattle is straightforward. A large RR facilitates long digesta retention in the organ without compromising food intake. As suggested by Hofmann (1989), this can be interpreted as an adaptation to grass diets: grass is peculiar among forages because it does not only have high fibre content, but an exceptionally high proportion of this fibre is digestible (Hummel *et al.* 2006). Thus, the absolute digestibility of grasses is usually higher than that of browse, but because the digestible material is fibre, it takes a longer time to reach this maximum digestibility. Therefore, animals adapted to grazing should benefit from long retention times. As mentioned above, grass material itself enhances its own retention in the rumen by inducing a fibre mat and the 'filter-bed effect'. It needs to be noted that this does not mean that animals with a capacious rumen cannot digest browse diets efficiently. A comparatively small rumen might represent a conceptual challenge for animals that have to compete on grass diets with animals that have comparatively larger rumens. The converse does not hold: a comparatively large rumen does not pose problems in the processing of browse material.

A more challenging question to answer is about the adaptive value of a high fluid throughput, as evidenced by actual measurements and the various morphological features conceptually compatible with it. On the one hand, it has been suggested that a higher throughput of low-viscosity fluid might enhance rumen contents stratification by creating an environment suitable for flotation/sedimentation to occur not only in the reticulum, but also in the rumen itself, hence enhancing the 'filter-bed effect' and consequentially fibre digestion (Clauss *et al.* 2006b). However, given that indications for a 'filter-bed effect' were also noticed in moose on grass diets (Clauss *et al.* 2011a; Lauper *et al.* 2013), this interpretation may be questionable, and the diet itself, rather than animal characteristics, may be the major factor determining this process. Another consequence of a high fluid throughput in wild ruminants was, to our knowledge, first proposed by Hummel *et al.* (2008). These authors referred to literature on domestic ruminants that shows that a high fluid throughput enhances the harvest of microbes from the forestomach and leads to a high inflow of microbes into the lower digestive tract. Additional evidence for this effect from experimental studies in which fluid throughput through the forestomach of domestic ruminants was artificially enhanced support this interpretation (reviewed by Clauss *et al.* 2010a). If such an increase of the efficiency of microbial growth in, and harvest from, the forestomach was the major driver in the evolution of the observed morphophysiological features, then they should be comparatively independent from the natural diet, as suggested in Codron & Clauss (2010) and by the examples of mixed diets observed in cattle species mentioned above. Some (but not necessarily all) ruminant species that specialized on browse-only diets might be excluded from this evolutionary scenario, because they depend on certain salivary proteins to defend themselves against negative effects of the secondary plant compounds in browse. These species are characterized by large salivary glands possibly necessary to maintain high levels of these defensive proteins in the saliva; they might be more limited in the absolute amount of saliva they can produce (Hofmann *et al.* 2008) and cannot achieve the levels of fluid throughput, and the morphological adaptations to it, that are observed in other ruminants, and especially in the cattle group. These concepts, evidently, need further corroboration from comparative and experimental studies.

When comparing the anatomy and physiology of ruminant species, cattle appear to be one extreme – with very pronounced features linked to the consumption of high-fibre diets and a high fluid throughput through their forestomach. Potentially, the evolution of grasslands fostered the evolution of the cattle digestive adaptations, liberating the ruminant system from potential constraints set by browse forage, without limiting them to exclusive grass diets.

References

Venturing into the vast realm of scientific publications on ruminant digestive physiology is a rewarding adventure. Doing justice to the researchers that have worked in and advanced this field is impossible in the space available here. The publications listed should not be understood as the most important ones, but as gates into this research area.

Abbas, F., Picot, D., Merlet, J., *et al.* (2013). A typical browser, the roe deer, may consume substantial quantities of grasses in open landscapes. *European Journal of Wildlife Research* **59**: 69–75.

Baumont, R. & Deswysen, A. G. (1991). Mélange et propulsion du contenu du réticulo-rumen. *Reproduction Nutrition Development* **31**: 335–359.

Bayer, W. (1990). Use of native browse by Fulani cattle in central Nigeria. *Agroforestry Systems* **12**: 217–228.

Bekhuis, P. D. B. M., De Jong, C. B. & Prins, H. H. T. (2008). Diet selection and density estimates of forest buffalo in Campo-Ma'an National Park, Cameroon. *African Journal of Ecology* **46**: 668–675.

Bowman, D. M. J. S., Murphy, B. P. & McMahon, C. R. (2010). Using carbon isotope analysis of the diet of two introduced Australian megaherbivores to understand Pleistocene megafaunal extinctions. *Journal of Biogeography* **37**: 499–505.

Chetri, M. (2006). Diet analysis of gaur (*Bos gaurus*) by micro-histological analysis of fecal samples in Parsa Wildlife Reserve, Nepal. *Our Nature* **4**: 20–28.

Clauss, M. (2004). The potential interplay of posture, digestive anatomy, ingesta density and gravity in mammalian herbivores, or why sloths do not rest hanging upside down. *Mammal Review* **34**: 241–245.

Clauss, M., Frey, R., Kiefer, B., *et al.* (2003a). The maximum attainable body size of herbivorous mammals: morphophysiological constraints on foregut, and adaptations of hindgut fermenters. *Oecologia* **136**: 14–27.

Clauss, M., Lechner-Doll, M. & Streich, W. J. (2003b). Ruminant diversification as an adaptation to the physicomechanical characteristics of forage: a reevaluation of an old debate and a new hypothesis. *Oikos* **102**: 253–262.

Clauss, M., Lechner-Doll, M. & Streich, W. J. (2004). Differences in the range of faecal dry matter content between feeding types of captive wild ruminants. *Acta Theriologica* **49**: 259–267.

Clauss, M., Hofmann, R. R., Hummel, J., *et al.* (2006a). The macroscopic anatomy of the omasum of free-ranging moose (*Alces alces*) and muskoxen (*Ovibos moschatus*) and a comparison of the omasal laminal surface area in 34 ruminant species. *Journal of Zoology* **270**: 346–358.

Clauss, M., Hummel, J. & Streich, W. J. (2006b). The dissociation of the fluid and particle phase in the forestomach as a physiological characteristic of large grazing ruminants: an evaluation of available, comparable ruminant passage data. *European Journal of Wildlife Research* **52**: 88–98.

Clauss, M., Kaiser, T. & Hummel, J. (2008) The morphophysiological adaptations of browsing and grazing mammals. In The Ecology of Browsing and Grazing, pp. 47–88. Eds. Gordon, I. J. & Prins, H. H. T. Heidelberg: Springer.

Clauss, M., Fritz, J., Bayer, D., *et al.* (2009a). Physical characteristics of rumen contents in two small ruminants of different feeding type, the mouflon (*Ovis ammon musimon*) and the roe deer (*Capreolus capreolus*). *Zoology* **112**: 195–205.

Clauss, M., Fritz, J., Bayer, D., *et al.* (2009b). Physical characteristics of rumen contents in four large ruminants of different feeding type, the addax (*Addax nasomaculatus*), bison (*Bison bison*), red deer (*Cervus elaphus*) and moose (*Alces alces*). *Comparative Biochemistry and Physiology A* **152**: 398–406.

Clauss, M., Hofmann, R. R., Fickel, J., Streich, W. J. & Hummel, J. (2009c). The intraruminal papillation gradient in wild ruminants of different feeding types: implications for rumen physiology. *Journal of Morphology* **270**: 929–942.

Clauss, M., Hume, I. D. & Hummel, J. (2010a). Evolutionary adaptations of ruminants and their potential relevance for modern production systems. *Animal* **4**: 979–992.

Clauss, M., Hofmann, R. R., Streich, W. J., Fickel, J. & Hummel, J. (2010b). Convergence in the macroscopic anatomy of the reticulum in wild ruminant species of different feeding types and a new resulting hypothesis on reticular function. *Journal of Zoology* **281**: 26–38.

Clauss, M., Lechner, I., Barboza, P., *et al.* (2011a). The effect of size and density on the mean retention time of particles in the reticulorumen of cattle (*Bos primigenius* f. *taurus*), muskoxen (*Ovibos moschatus*) and moose (*Alces alces*). *British Journal of Nutrition* **105**: 634–644.

Clauss, M., Müller, K., Fickel, J., Streich, W. J., Hatt, J. M. & Südekum, K.-H. (2011b). Macroecology of the host determines microecology of endobionts: protozoal faunas vary with wild ruminant feeding type and body mass. *Journal of Zoology* **283**: 169–185.

Codron, D. & Clauss, M. (2010). Rumen physiology constrains diet niche: linking digestive physiology and food selection across wild ruminant species. *Canadian Journal of Zoology* **88**: 1129–1138.

Connor, J. M., Bohman, V. R., Lesperance, A. L. & Kinsinger, F. E. (1963). Nutritive evaluation of summer range forage with cattle. *Journal of Animal Science* **22**: 961–969.

Demment, M. W. (1982). The scaling of ruminoreticulum size with body weight in East African ungulates. *African Journal of Ecology* **20**: 43–47.

Foose, T. J. (1982) Trophic strategies of ruminant versus nonruminant ungulates. PhD thesis. University of Chicago.

Franz, R., Soliva, C. R., Kreuzer, M., Steuer, P., Hummel, J. & Clauss, M. (2010). Methane production and body mass in ruminants and equids. *Evolutionary Ecology Research* **12**: 727–738.

Franz, R., Soliva, C. R., Kreuzer, M., Hummel, J. & Clauss, M. (2011). Methane in rabbits (*Oryctolagus cuniculus*) and guinea pigs (*Cavia porcellus*) on a hay-only diet: implications for the scaling of methane procution with body mass in nonruminant mammalian herbivores. *Comparative Biochemistry and Physiology A* **158**: 177–181.

Fritz, J., Hummel, J., Kienzle, E., Arnold, C., Nunn, C. & Clauss, M. (2009). Comparative chewing efficiency in mammalian herbivores. *Oikos* **118**: 1623–1632.

Gebczynska, Z., Gebczynski, M. & Martynowicz, E. (1991). Food eaten by the free-living European bison in Bialowieza Forest. *Acta Theriologica* **36**: 307–313.

Hofmann, R. R. (1968). Comparison of the rumen and omasum structure in East African game ruminants in relation to their feeding habits. *Symposium of the Zoological Society of London* **21**: 179–194.

Hofmann, R. R. (1969). Zur Topographie und Morphologie des Wiederkäuermagens im Hinblick auf seine Funktion (nach vergleichenden Untersuchungen an Material ostafrikanischer Wildarten). *Zentralblatt für Veterinärmedizin* **10** (Suppl): 1–180.

Hofmann, R. R. (1973). *The Ruminant Stomach*. Nairobi: East African Literature Bureau.

Hofmann, R. R. (1985). Digestive physiology of deer: their morphophysiological specialisation and adaptation. *Royal Society of New Zealand Bulletin* **22**: 393–407.

Hofmann, R. R. (1988) Morphophysiological evolutionary adaptations of the ruminant digestive system. In *Aspects of Digestive Physiology in Ruminants*, pp. 1–20. Eds. Dobson, A. & Dobson, M. J. Ithaca, NY: Cornell University Press.

Hofmann, R. R. (1989). Evolutionary steps of ecophysiological adaptation and diversification of ruminants: a comparative view of their digestive system. *Oecologia* **78**: 443–457.

Hofmann, R. R. (2000). Functional and comparative digestive system anatomy of Arctic ungulates. *Rangifer* **20**: 71–81.

Hofmann, R. R. & Schnorr, B. (1982). *Die funktionelle Morphologie des Wiederkäuer-Magens*. Stuttgart: Ferdinand Enke.

Hofmann, R. R. & Stewart, D. R. M. (1972). Grazer or browser: a classification based on the stomach-structure and feeding habit of East African ruminants. *Mammalia* **36**: 226–240.

Hofmann, R. R., Streich, W. J., Fickel, J., Hummel, J. & Clauss, M. (2008). Convergent evolution in feeding types: salivary gland mass differences in wild ruminant species. *Journal of Morphology* **269**: 240–257.

Holechek, J. L., Vavra, M., Skovlin, J. & Krueger, W. C. (1982). Cattle diets in the Blue Mountains of Oregon II: forests. *Journal of Range Management* **35**: 239–242.

Hummel, J., Südekum, K.-H., Streich, W. J. & Clauss, M. (2006). Forage fermentation patterns and their implications for herbivore ingesta retention times. *Functional Ecology* **20**: 989–1002.

Hummel, J., Steuer, P., Südekum, K.-H., *et al.* (2008). Fluid and particle retention in the digestive tract of the addax antelope (*Addax nasomaculatus*): adaptations of a grazing desert ruminant. *Comparative Biochemistry and Physiology A* **149**: 142–149.

Hummel, J., Südekum, K.-H., Bayer, D., *et al.* (2009). Physical characteristics of reticuloruminal contents of cattle in relation to forage type and time after feeding. *Journal of Animal Physiology and Animal Nutrition* **93**: 209–220.

Landman, M. & Kerley, G. I. H. (2001). Dietary shifts: do grazers become browsers in the Thicket Biome? *Koedoe* **44**: 31–36.

Langer, P. (1988). *The Mammalian Herbivore Stomach*. Stuttgart and New York: Gustav Fischer Verlag.

Larter, N. C. & Gates, C. C. (1991). Diet and habitat selection of wood bison in relation to seasonal changes in forage quantity and quality. *Canadian Journal of Zoology* **69**: 2677–2685.

Lauper, M., Lechner, I., Barboza, P., *et al.* (2013). Rumination of different-sized particles in muskoxen (*Ovibos moschatus*) and moose (*Alces alces*) on grass and browse diets, and implications for rumination in different ruminant feeding types. *Mammalian Biology* **78**: 142–152.

Lechner, I., Barboza, P., Collins, W., *et al.* (2010). Differential passage of fluids and different-sized particles in fistulated oxen (*Bos primigenius* f. *taurus*), muskoxen (*Ovibos moschatus*), reindeer (*Rangifer tarandus*) and moose (*Alces alces*): rumen particle size discrimination is independent from contents stratification. *Comparative Biochemistry and Physiology A* **155**: 211–222.

Lechner-Doll, M., Kaske, M. & Engelhardt, W. V. (1991) Factors affecting the mean retention time of particles in the forestomach of ruminants and camelids. In *Physiological Aspects of Digestion and Metabolism in Ruminants*, pp. 455–482. Eds. Tsuda, T., Sasaki, Y. & Kawashima, R. San Diego, CA: Academic Press.

Machatschek, M. (2002). *Laubgeschichten: Gebrauchswissen einer alten Baumwirtschaft, Speise- und Futterlaubkultur*. Vienna: Böhlau.

Maekawa, M., Beauchemin, K. A. & Christensen, D. A. (2002). Effect of concentrate level and feeding management on chewing activities, saliva production, and ruminal pH of lactating dairy cows. *Journal of Dairy Science* **85**: 1165–1175.

Mortolaa, J. P. & Lanthier, C. (2005). Breathing frequency in ruminants: a comparative analysis with non-ruminant mammals. *Respiratory Physiology & Neurobiology* **145**: 265–277

Noe-Nygaard, N., Price, T. D. & Hede, S. U. (2005). Diet of aurochs and early cattle in southern Scandinavia: evidence from ^{15}N and ^{13}C stable isotopes. *Journal of Archaeological Science* **32**: 855–871.

Pacheco, M. A., Concepción, J. L., Rangel, J. D., Ruiz, M. C., Michelangeli, F. & Domínguez-Bello, M. G. (2007). Stomach lysozymes of the three-toed sloth (*Bradypus variegatus*), an arboreal folivore from the Neotropics. *Comparative Biochemistry and Physiology A* **147**: 808–819.

Pérez-Barberìa, F. J., Elston, D. A., Gordon, I. J. & Illius, A. W. (2004). The evolution of phylogenetic differences in the efficiency of digestion in ruminants. *Proceedings of the Royal Society B* **271**: 1081–1090.

Pujaningsih, R. I., Sutrisno, C. I., Supriondho, Y., *et al.* (2009). Diet composition of anoa (*Bubalus* spp.) studied using direct observation and dung analysis method in their habitat. *Journal of the Indonesian Tropical Animal Agriculture* **34**: 223–228.

Rivals, F., Solounias, N. & Mihlbachler, M. C. (2007). Evidence for geographic variation in the diets of late Pleistocene and early Holocene *Bison* in North America, and differences from the diets of recent *Bison*. *Quaternary Research* **68**: 338–346.

Schulz, E. & Kaiser, T. M. (2007). Feeding strategy of the Urus *Bos primigenius* from the Holocene of Denmark. *Courier Forschungsinstitut Senckenberg* **259**: 155–164.

Sutherland, T. M. (1988) Particle separation in the forestomach of sheep. In *Aspects of Digestive Physiology in Ruminants*, pp. 43–73. Eds. Dobson, A. & Dobson, M. J. Ithaca, NY: Cornell University Press.

Thilenius, J. F. & Hungerford, K. E. (1967). Browse use by cattle and deer in northern Idaho. *Journal of Wildlife Management* **31**: 141–145.

Van Wieren, S. E. (1996) Digestive strategies in ruminants and nonruminants, Thesis Landbouw, University of Wageningen.

Woodall, P. F. & Skinner, J. D. (1993). Dimensions of the intestine, diet and faecal water loss in some African antelope. *Journal of Zoology* **229**: 457–471.

Chapter 7

Anatomy and evolution of teeth

James Heywood

Introduction

The Bovini is a diverse tribe of the ruminant family Bovidae, this diversity extending biogeographically, morphologically and ecologically. They are part of a larger subfamily – the Bovinae – that includes the now exclusively African Tragelaphini (bushbuck, kudus, etc.) and the South Asian Boselaphini (nilgai and cousingha). The Bovini – hereafter bovins, *sensu* Bibi & Vrba (2010) – inhabit a range of ecological realms unparalleled by other bovid tribes, occupying, among others, montane, edaphic lowland and seasonally arid grassland habitats across four continents. They are perhaps particularly well known for the acquisition of some of the largest body sizes in ruminant history. As is the case with most mammalian groups, much of our knowledge of bovin evolution and associated palaeobiological interpretation is focused around the dentition of these animals. This chapter describes a generalised bovin postcanine dental anatomy, placing this in the context of the adaptation of grazing dentitions to dietary function, and makes comparisons with other bovid grazers. Interspecific dental form will also be considered within the Bovini, though these comparisons should be regarded as little more than a pilot study. Finally, an attempt is made to tie in the observed variation in dental form between bovid grazers and between bovin species with their known evolutionary histories.

Dental anatomy

While some functional morphological studies do consider the precanine dentition (e.g. Janis & Ehrhardt 1988), this chapter will focus entirely on the postcanine dentition – particularly the molars – as is the case in a host of other studies that use the dentition for either phylogenetic or functional-morphological use (e.g. Archer & Sanson 2002; Heywood 2010; Kaiser *et al.* 2010).

Figure 7.1 provides a comparison of the upper and lower molar dentition of a generalised bovin with a boselaphine. The modern boselaphine (*Boselaphus tragocamelus*) can be thought of for these purposes as representing something akin to a more primitive dental form for the Bovinae.

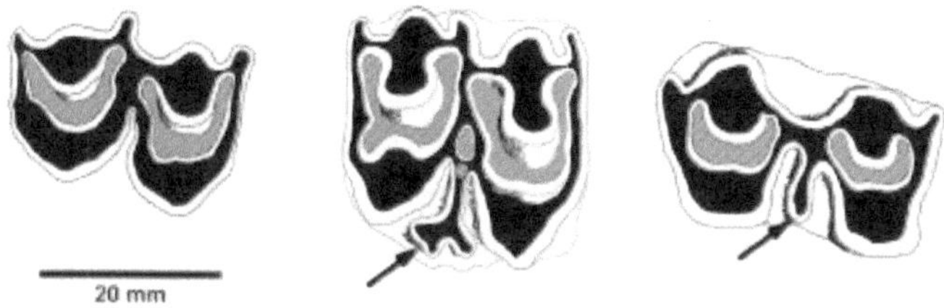

Figure 7.1 Occlusal view of generalised bovin upper and lower molars (*Bos gaurus* – centre, *Bison bison* – right) and upper molar of a Boselaphine (*Boselaphus tragocamelus* – left). Arrows point to entostyles (upper) and entostylids (lower).

A number of features present in the bovin dentition are noteworthy. The teeth are altogether larger with higher crowns, and the occlusal form is more complex and robust when compared with the boselaphine. In the molars, the entostyles/entostylids (arrows in Figure 7.1) are not only present but are often highly folded, particularly when viewed at more advanced wear stages. The labial styles of the upper molars project strongly and the central cavities (formed by the continuity of the preparaconule and preprotoconule, and postparaconule and postprotoconule in the case of the anterior central cavity of the upper molars) are often highly folded. There are usually accessory cavities (small islands of enamel) located between the two central cavities and there may often be two or three present. In all areas, the exposed enamel ridges are very thick and the teeth are heavily coated with thick cementum. The premolars are also rather more complex in their occlusal outline, compared with the somewhat more 'globular' (Bibi 2007a) outline of those found in boselaphines, both modern and extinct.

Dental function

The features described above are part of a broad suite of (presumably) adaptive characteristics that very much place the bovin dentition among those highly specialised to a grazing habit. Archer & Sanson (2002) found that African grazing ruminants possess thicker enamel with more evenly spaced dentine basins, plus the presence of more complex accessory

Ecology, Evolution and Behaviour of Wild Cattle: Implications for Conservation, ed. M. Melletti and J. Burton. Published by Cambridge University Press.

enamel structures (accessory cavities/central pillars and entostyles/basal pillars) than did intermediate feeders or browsers. Kaiser *et al.* (2010), in an analysis of the enamel alignment of 37 species of ruminants, found that grazers possessed more enamel aligned parallel to the (transverse) chewing stroke.

Heywood (2010), combined the approach of the above two studies in a more taxonomically diverse group of bovids and found very similar results. There it was shown in particular that grazers possess both thicker and greater quantities of enamel that exhibit a more bimodal orientation of the internal enamel structures compared with other feeding groups. This trio of more recent studies confirms earlier findings in horses (Rensberger *et al.* 1984). Furthermore, a broader ruminant-wide analysis (Heywood 2009) found similar, though less well-defined patterns in cervids. An overall conclusion can be reached; that these occlusal features have appeared in a multitude of herbivorous mammal groups by convergent evolution.

What is less clear is quite how these anatomical traits relate to the functional demands of a grazing habit. A range of speculative suggestions have come from the aforementioned studies. Kaiser *et al.* (2010) present three well-considered options:

(1) Lower angles (parallel to chewing stroke) facilitate re-aligning the linear grass fibres more optimally for shearing.
(2) The generally flatter crowns of grazers may introduce more propalinal mandibular movements and with it more capacity for shearing when jaw movements depart from the transverse.
(3) Lower angles (particularly in the form of the 'O'-shaped paracones and metacones in upper molars) help to better distribute the necessarily greater chewing forces implicit in consuming a generally tougher diet of grass.

Heywood (2010) further considers the possibility that the observations are more of a corollary of space-filling constraints, the aim being simply to increase the amount of enamel in the tooth crown under the constraints of dental development and the four-cusp pattern seen in many ungulates. Certainly, bovids that consume more grass possess more internal enamel (accessory cavities and central ridges). In any event, such conclusions can rarely be anything other than speculative, not least in the absence of complementary data from corresponding lower molars *and* the fact that much of the data described above account for tooth form in only two dimensions. While it is certainly true that grazers have generally flatter dental surfaces compared with browsers (Fortelius & Solounias 2000), there is still a significant three-dimensional aspect to dental occlusion that needs to be explored.

One possible insight into the problem of identifying the exact nature of dental function in herbivorous mammals more generally, but grazers in particular, may reside in the work of Codron *et al.* (2008b). In a comparison of the craniodental morphology of ten grazing African ruminants and one equid they found that molar crown height, mandibular angle, premolar/molar ratio and palatal width varied consistently with the dietary nitrogen isotope composition. This was equated to dietary quality – higher proportions of $\delta^{15}N$ corresponding to increased consumption of more metabolically active proteins and more generally higher-quality diets.

A comparison of bovin dental form with other bovid grazers

That the Bovini, Hippotragini (minus the *Oryx*, *Oryx gazella*) and Reduncini clustered together to the exclusion of the Alcelaphini and *Oryx* in the above study is of little surprise. Bibi (2007b) remarks that there are many convergent dental features between these three tribes. Below I present a new analysis that extends the work of Codron *et al.* (2008b) and is consistent with the observations of Bibi (2007b) and others. The analysis follows naturally from the observations of the former study and aims to assess whether the same taxonomic and morphological complexes are borne out using occlusal form as they were for craniodental morphology.

Methods

The data presented below are modified from a ruminant-wide analysis on upper second molar occlusal morphology (Heywood 2009). Full details of the data collection methods are provided by Heywood (2010) in an analysis in the Bovidae. Twenty-two species were selected from the tribes Alcelaphini, Bovini, Hippotragini and Reduncini (Table 7.1). The modern representatives of these four tribes are exclusively grazers, the exception being the Bovini in the form of the anoa (*Bubalus depressicornis*), of which more shall be considered in later sections and which is excluded in this analysis on account of its less certain grazing habit (see Flores-Miyamoto *et al.* 2005 and references therein). All four tribes are also those represented in the analyses of Codron *et al.* (2008b).

The 'occlusal parameters' (enamel length and thicknesses, plus dentine basin width) are described in detail by Heywood (2010) and are modifications to those originally conceived by Archer & Sanson (2002). The variables quantify length and thickness of enamel formations in moderately worn molars. Two additional variables are included in this analysis, interior and exterior enamel, which comprise the 'enamel distribution ratio' (EDR), which is a measure of the degree to which enamel is distributed externally or internally. In effect, it quantifies the degree of structural investment in external structures (such as the entostyle) versus internal structures (e.g. the accessory cavities) and the degree of enamel folding. Enamel thickness is pooled into a single measure, both Archer & Sanson (2002) and Heywood (2010) showing that enamel thickness varies more between bovids of differing feeding types, rather than within them.

'Occlusal profiles' from Heywood (2010) are also employed in much the same way. Enamel alignment relative to the transverse direction of the chewing stroke is measured at 200 μm intervals and the proportions of enamel intervals

Table 7.1 Species used in all analyses of bovid/bovin dental occlusal morphology

Tribe	Species	N
Bovini	*Bison bison*	5
	Bos gaurus	8
	Bubalus arnee	4
	*Bubalus depressicornis**	4
	Syncerus caffer	13
Alcelaphini	*Alcelaphus buselaphus*	25
	Alcelaphus sigmoceros	4
	Beatragus hunteri	10
	Connochaetes taurinus	10
	Damaliscus dorcas	6
	Damaliscus lunatus	13
Hippotragini	*Addax nasomaculatus*	9
	Hippotragus equinus	10
	Hippotragus niger	8
	Oryx gazella	10
Reduncini	*Kobus ellipsiprymnus*	8
	Kobus kob	14
	Kobus leche	5
	Kobus megaceros	3
	Kobus vardonii	10
	Redunca arundinum	10
	Redunca fulvorufula	8
	Redunca redunca	10

* Used only in comparison of bovin species.

falling into 10° bins between zero and 90° are used as shape variables. In this study, these proportions are recorded separately across the entire tooth surface and in the central ridges alone (the latter as per Heywood 2010). These variables are again modified from those initially devised by Rensberger *et al.* (1984) and employed more recently by Kaiser (2002) and Kaiser *et al.* (2010). All variables used in the analyses in the current section and the section 'A comparison of dental form within the Bovini: a pilot study' are summarised in Table 7.2. All parameters are corrected for size by dividing by tooth length, except EDR which is a dimensionless trait. Occlusal profiles are self-corrected as they are proportions of the total enamel distribution. Species means from samples collected from museum collections in the United Kingdom are used for these analyses.

A number of multivariate statistical methods are employed to ascertain the similarity of members of the four tribes. Hierarchical cluster analysis (HCA) employing the within groups average linkage based on squared Euclidean distances is used to identify natural groupings among the 22 species. A principal component analysis (PCA) using an oblique rotation method on the multivariate correlation matrix is then employed to identify the pattern of variation between members of the groups, with univariate analyses of variance (ANOVA) used to further identify the strength of variation between tribes. All statistical analyses were conducted using SPSS v18.0.

Table 7.2 Summary of variables used in the multivariate analyses of bovid grazing tribes and bovin species

Variable category	Abbreviation	Description	Used in analysis of bovid grazers	Used in analysis of bovin species
Parameter	IEL	Relative internal enamel length		✓
	EEL	Relative external enamel length		✓
	TEL	Relative total enamel length	✓	✓
	EDR	Enamel distribution ratio	✓	
	DA	Relative paracone width	✓	✓
	DC	Relative protocone width	✓	
	W	Relative tooth width	✓	
	R1	Relative width of the labial paracone enamel ridge		✓
	R2	Relative width of the lingual paracone enamel ridge		✓
	R3	Relative width of the labial protocone enamel ridge		✓
	R4	Relative width of the lingual protocone enamel ridge		✓
	R*	Mean enamel thickness at above four loci	✓	
Profiles	Tx	Proportion of all enamel between x–10° and x° measured between 0° and 90° relative to the transverse chewing direction	✓	
	Cx	Proportion of *central ridge* enamel between x–10° and x° measured between 0° and 90° relative to the transverse chewing direction		✓*

* Only the ranges 0°–10°, 30°–40°, 40°–50° and 80°–90°

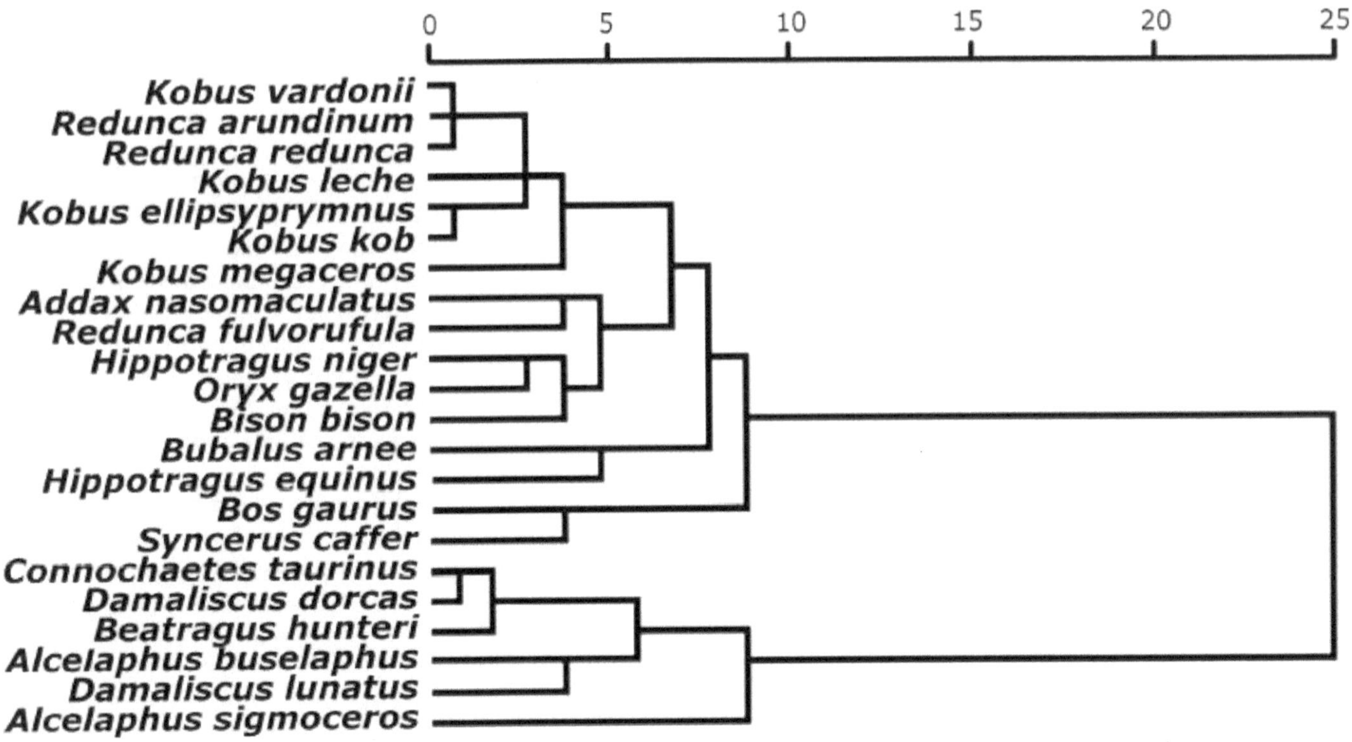

Figure 7.2 Tree diagram representing cluster analysis using the squared Euclidean distance matrix of means of occlusal parameters and profiles from 22 species of bovid grazer.

Results

Figure 7.2 represents the results of the HCA and shows two distinct clusters that find members of the Alcelaphini rendered distinct from members of the other three tribes. PCA reveals a similar pattern of separation. Two principal components representing 80% (PC1 = 67%, PC2 = 13%) of the total variation were extracted. Plots of the factor scores for each of the 22 species are presented in Figure 7.3.

PC1 correlates positively with mean enamel thickness, the proportion of enamel distributed externally (EDR) as well as total enamel length, paracone, protocone and therefore tooth width, and with enamel ridges aligned closer to parallel to the transverse chewing direction. Negative correlations are found with the proportions of enamel distributed more perpendicular to the chewing direction.

PC1 scores agree broadly with the separation of Alcelaphini from the Hippotragini/Bovini/Reduncini complex. Figures 7.4 and 7.5 demonstrate how enamel distribution and alignment, respectively, differ between the four tribes. Table 7.3 shows the results of univariate ANOVAs for all variables used in the analysis, altogether highlighting the differences between the Alcelaphini and the remaining tribes.

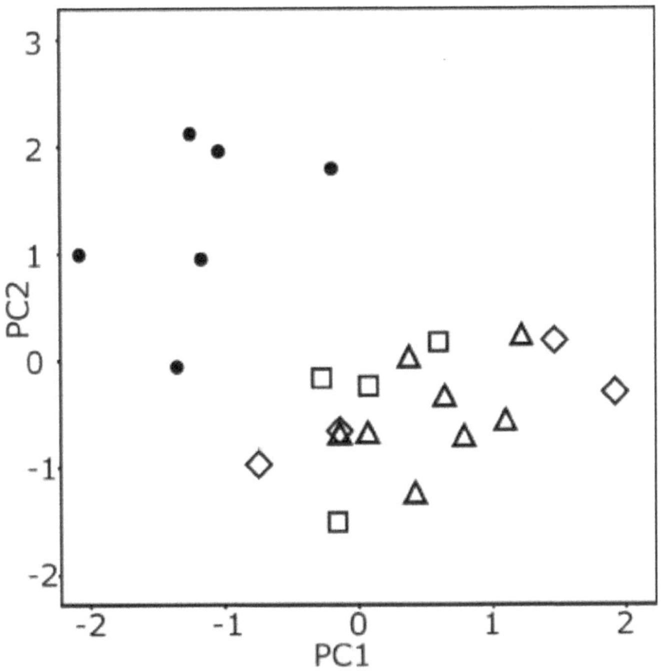

Figure 7.3 Plots of the first two extracted principal components using species means of dental occlusal variables in bovid grazing tribes. Symbols: ● = Alcelaphini; ◇ = Bovini; □ = Hippotragini; Δ = Reduncini.

Discussion

It has been previously shown that bovid grazers have a very characteristic dental occlusal morphology when compared with other feeding types. Heywood (2010) showed that grazers possess much thicker enamel of greater length and aligned in a much more bimodal distribution compared with browsers, mixed feeders and frugivores. However, the above analyses demonstrate what is clear from simple observation and has

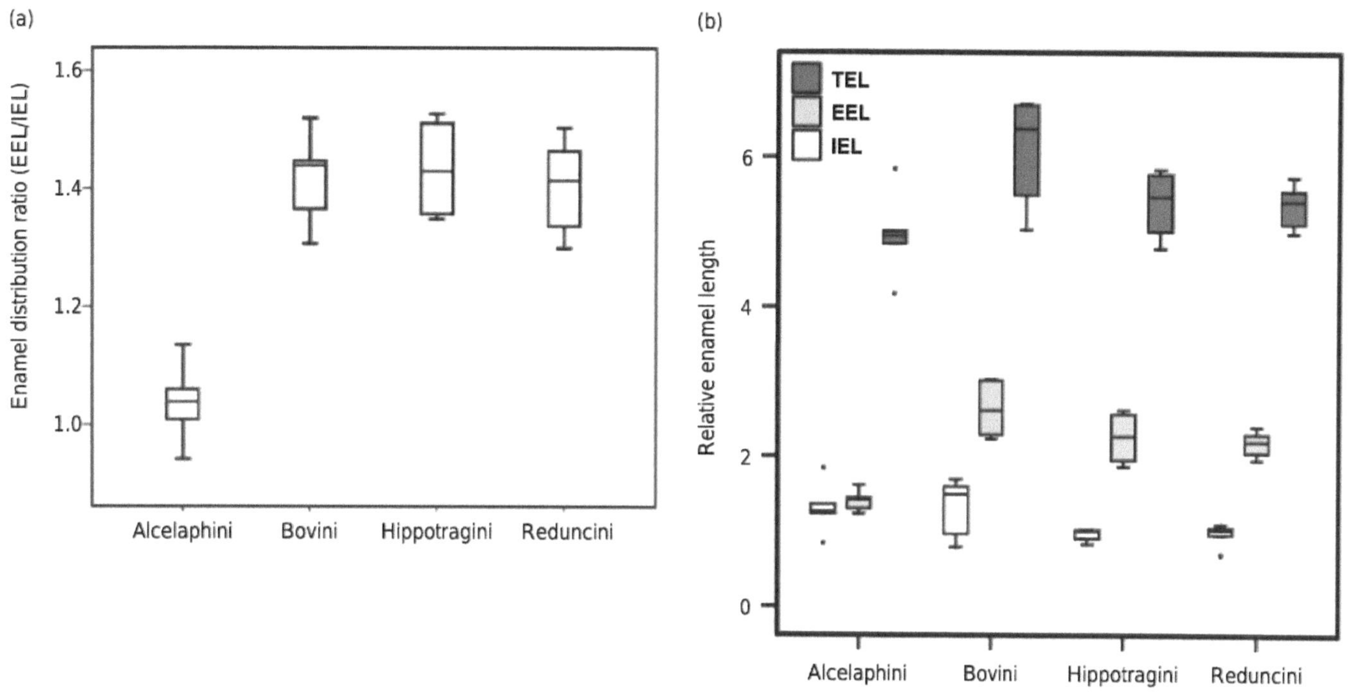

Figure 7.4 Boxplots of enamel distribution ratios (a) and relative enamel lengths (b) grouped by bovid tribe.

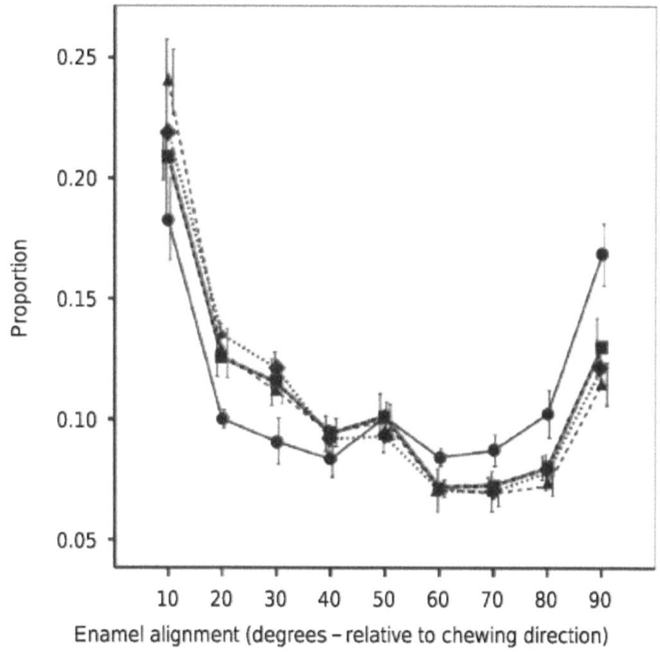

Figure 7.5 Enamel alignment distribution of four bovid grazing tribes. Points = tribe means per enamel alignment bin, error bars = one standard deviation. Symbols: ● = Alcelaphini; ♦ = Bovini; ■ = Hippotragini; ▲ = Reduncini.

been previously suggested informally (e.g. Bibi 2007b); among bovid grazers, the Bovini have converged on a very similar dental occlusal form to the Hippotragini and Reduncini. This form can be characterised by the possession of relatively wider teeth, with a particularly wide paracone, with more enamel distributed on the exterior of the tooth and with this enamel being aligned more parallel to the chewing direction. Such a tooth form is characterised by the presence of an (often complex) entostyle, a U-shaped central ridge and very pronounced styles (Figure 7.6). By contrast, the alcelaphine molar form is characterised by more complex internal enamel formations, particularly the central ridge, which might be characterised as butterfly-shaped.

These observations are consistent with the general findings of Codron *et al.* (2008b), though the *Oryx* (*Oryx gazella*) remained true to its hippotragine affiliation in clustering in the Bovini/Hippotragini/Reduncini complex. Together this pair of studies very much suggests that among bovids, two distinct methods of grazing have evolved and at least one of those has done so convergently – the Hippotragini, Bovini and Reduncini all having independent phylogenetic origins within the Bovidae (e.g. see Chapter 1; Marcot 2007; see Figure 7.7).

The narrower tooth rows of alcelaphines are consistent with a shorter chewing stroke to cope with the presumably shorter grass blades of short-/medium-height vegetation. By contrast, the wider teeth with a greater amount of enamel in the remaining three tribes are consistent with the need to cope with taller, tougher grasses higher in nitrogen, certainly facilitating a longer chewing stroke with an increased surface area of enamel.

What is still not yet understood, however, is whether the occlusal shape of such dentitions reflects the specific laminar structure of grass with parallel fibres, or more general structural properties such as toughness. It is tempting to attribute the overall higher proportion of enamel nearer the perpendicular in alcelaphines, especially in the central ridge, as the result of selection towards increasing the number of shearing

Table 7.3 Results of univariate analyses of variance of species means of occlusal form variables grouping by the bovid tribe

Variable	$F_{3,22}$	*p*	Posthoc comparisons
R*	9.94	**<0.001**	A < H = B < R
IEL	1.83	0.17	
EEL	13.49	**<0.001**	A < H = B = R
TEL	3.38	**0.037**	A < H = B = R
EDR	41.61	**<0.001**	A < H = B = R
DA	13.29	**<0.001**	A < H = B = R
DC	2.36	0.09	
W	5.72	**0.005**	A < H = B = R
T10	7.76	**0.001**	A < H = B = R
T20	15.40	**<0.001**	A < H = B = R
T30	13.62	**<0.001**	A < H = B = R
T40	2.78	0.065	
T50	1.84	0.167	
T60	6.87	**0.002**	A > H = B = R
T70	7.55	**0.001**	A > H = B = R
T80	15.21	**<0.001**	A > H = B = R
T90	20.38	**<0.001**	A > H = B = R

p values of comparisons significant to the 5% level are highlighted in bold.

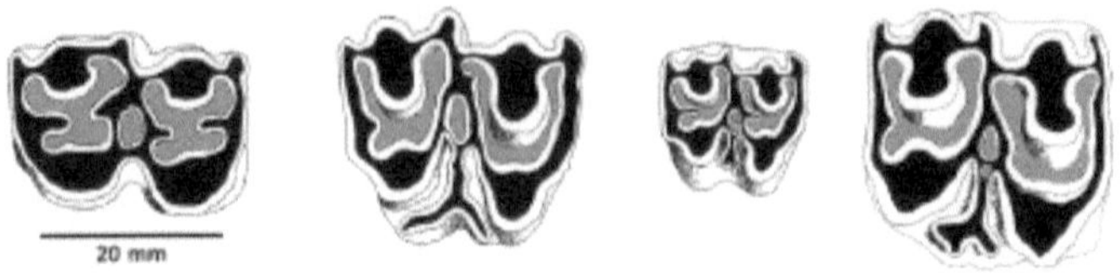

Figure 7.6 Representative upper second molars of the four studied bovid grazing tribes. From left to right: *Connochaetes taurinus* (Alcelaphini), *Hippotragus equinus* (Hippotragini), *Kobus vardonii* (Reduncini), *Bos gaurus* (Bovini).

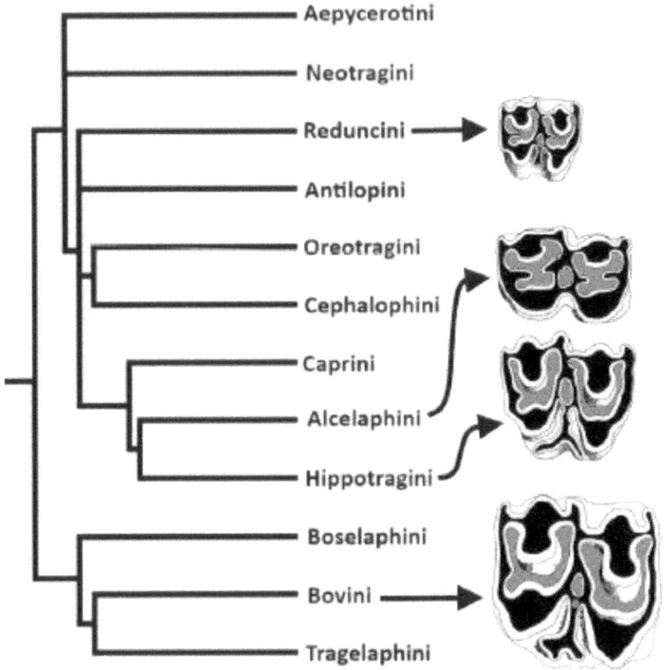

Figure 7.7 Representative bovid grazing tribe tooth morphologies mapped onto a molecular phylogeny of the Bovidae (Hassanin, 2014).

contacts to cope with the need for shearing a laminar structure with parallel fibres, as suggested by Kaiser (2002).

However, would such a morphology not be just as likely to evolve in other grazers, the hippotragines, reduncines and bovins in particular? Perhaps it is grass toughness, rather than its particular laminar form, that has a greater role in explaining variation in the dental occlusal form of grazing ruminants. A higher quantity of thicker enamel arranged in a columnar framework would resist the higher stresses imparted on the dental battery when comminuting the altogether tougher plant parts that the Hippotragini, Reduncini and Bovini have been suggested to consume (Codron *et al.* 2008b). This interpretation aligns itself most closely with the third interpretation provided by Kaiser *et al.* (2010). That duikers, often referred to as a 'frugivorous' group (Bodmer 1990) also display relatively wide teeth with thickened enamel coincident with a diet high in fibre (Dierenfeld *et al.* 2002), and thus potentially tough, lends support to this notion.

A comparison of dental form within the Bovini: a pilot study

The above discussion implies a degree of homogeneity among the Bovini in terms of dental form and function, with the corollary being that bovin dietary ecology is similarly lacking in variation. However, as alluded to previously, the Bovini are one of the more diverse bovid tribes, morphologically, biogeographically and ecologically. Some of this diversity may be pinned to the presence of a number of endemic island forms from Southeast Asia, in particular the anoa (*Bubalus depressicornis/quarlesi*) and the tamaraw (*B. mindorensis*). It should not be forgotten that aside from these forms, much ecological diversity exists: the largely forest-dwelling Southeast Asian *Bos* species *B. gaurus*, *B. javanicus* and *B. sauveli*, and European forest-dwelling *Bison bonasus*, the montane *B. mutus*, the plains-dwelling *Bison bison*; swamp-inhabiting *Bubalus arnee* and savanna and forest subspecies of *Syncerus caffer*. Does this diversity in habitat translate into any appreciable variation in dental form?

A similar set of analyses to those described above are employed on five bovin species as listed in Table 7.1. Though by no means an exhaustive array, it does cover a number of the ecological groups mentioned above, including importantly the anoa, the one bovin species thought to occupy a more mixed diet compared with its larger continental relatives (see Flores-Miyamoto *et al.* 2005 and references therein). The variables

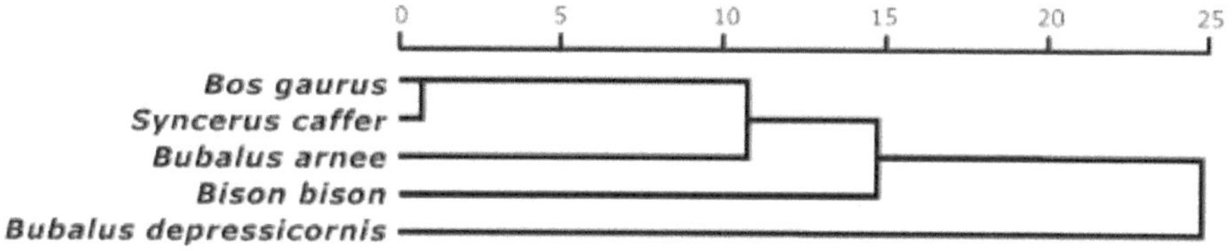

Figure 7.8 Tree diagram representing cluster analysis using the squared Euclidean distance matrix of means of occlusal parameters and enamel distribution profiles for five bovin species.

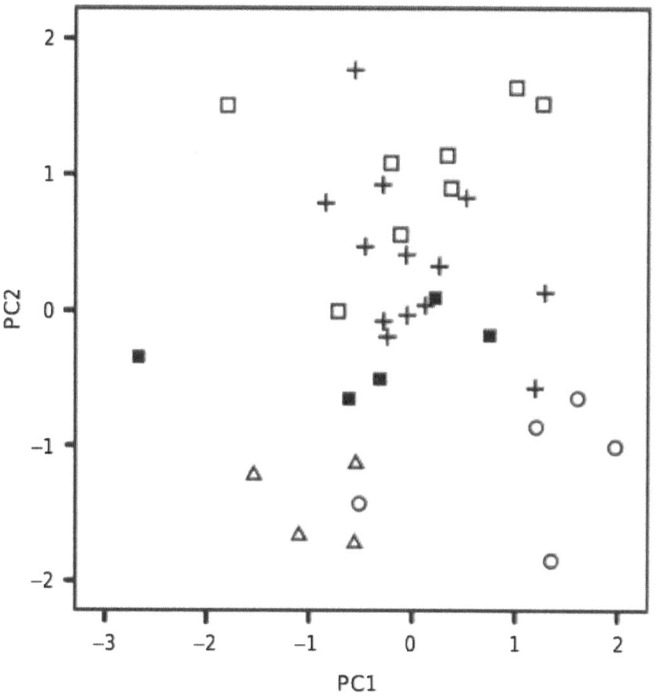

Figure 7.9 Plots of the first two extracted principal components using specimen data of dental occlusal variables from five bovin species. Symbols: ○ = *Bison bison*; □ = *Bos gaurus*; ■ = *Bubalus arnee*; Δ = *Bubalus depressicornis*; + = *Syncerus caffer*.

used are indicated in Table 7.2 and differ slightly to those employed in the previous analysis. Central ridge profiles are used instead of those across the whole tooth; and the mean enamel thickness is replaced by thickness at all four loci (R1–R4), these being used as separate variables in light of the findings of Archer & Sanson (2002) and Heywood (2010). 'EDR' is dispensed in favour of its constituent 'EEL' and 'IEL'.

Results

Figure 7.8 shows the resulting dendrogram from a hierarchical cluster analysis of species means. The anoa is clustered separately to the four larger bovin species, with the bison as the next most dissimilar group. To further validate this clustering pattern, PCA was run on individual specimens to identify whether the same grouping patterns resulted. An orthogonal method of rotation was employed.

Figure 7.9 presents plots of the first two extracted components, which together explain 51.7% of the variance (PC1 = 39.8%, PC2 = 11.9%). PC1 is associated with central ridge profiles, positive correlations existing between PC1 higher proportions of enamel parallel and perpendicular to the chewing direction and negatively with those aligned around 40°. PC2 is associated more with enamel thickness – especially the lingual protocone ridge (R3) and length, high values of PC2 associated with both longer and thicker enamel.

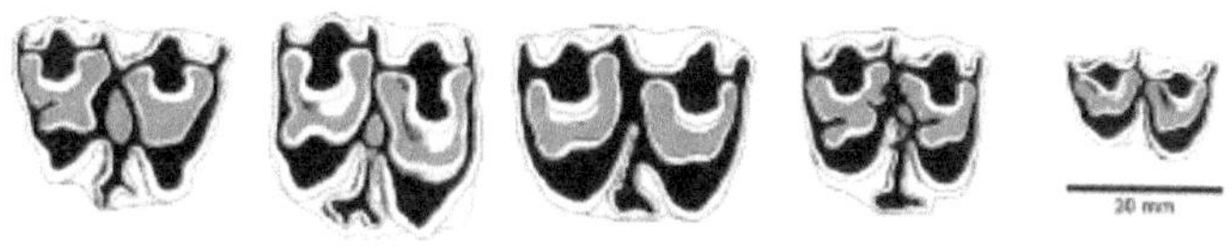

Figure 7.10 Representative upper second molars of five bovin species (from left to right): *Bubalus arnee, Bos gaurus, Bison bison, Syncerus caffer, Bubalus depressicornis*.

Discussion

The results of multivariate analyses of occlusal morphology of five species of bovin are consistent with the known dietary ecology of these animals. The anoa (*Bubalus depressicornis*), traditionally classified as a mixed feeder in ungulate-wide analyses of craniodental morphology (e.g. Mendoza *et al.* 2002) was often found to be categorised as a grazer when compared with all other members of the Bovidae (Heywood 2010). This is unsurprising as it still displays relatively thick enamel, a trait commonly found to be much reduced in other mixed feeders and browsers. However, when compared to other bovins, the anoa shows a somewhat reduced R3, and relatively less enamel, both internally and externally – a result of a general absence of accessory cavities and entostyles (Figure 7.10). Finally, its central ridge is aligned more to the diagonal relative to the chewing direction, again a shape profile more akin to that found in mixed feeders and browsers in other bovids.

In addition to this pattern, it is also evident that the plains bison, *Bison bison*, shows a somewhat discrete morphology compared with the remaining bovin species. The bison is characterised by a relatively reduced quantity of enamel, though still with more in external formations than seen in the anoa ($T = 2.754$, $p = 0.014$), but with a more characteristic pattern of enamel alignment both perpendicular and parallel to the chewing direction as seen in all grazers.

Two interpretations of this pattern are possible, not withstanding of course the very distinct possibility that poor sampling, particularly in the genera *Bubalus* and *Bison*, may be influencing the results.

The first is that the bison have adopted a more 'alcelaphine'-like grazing niche of low-level bulk feeding and lesser selectivity. That bison should also possess the largest hypsodonty index of any bovin (Janis 1988) and should be regarded as ground-level feeders (Spencer 1995) is consistent with this observation.

Another intriguing possibility is that bison have over the course of their history inhabited C_3 grasslands in the Northern Hemisphere. Hoppe *et al.* (2006) report that bison do not preferentially select C_3 or C_4 grasses across their range; instead the grass type consumed reflects directly the available resources. Feranec (2004) reports a similar pattern from the Pleistocene of Florida. C_4 grasses are widely regarded as being more physically and metabolically demanding for herbivores compared with C_3 grasses (see Codron *et al.* 2008a and references therein). The relatively lengthy exposure of bison to C_3 grasses in their evolutionary history, though with the consumption of C_4 as and when geographical and/or historical changes in grass communities demanded, raises the possibility that bison invest less in their occlusal structures compared with lower latitude dwelling bovins, which should be explored further.

As a brief addendum, the author has not encountered the dentition of the enigmatic saola, *Pseudoryx nghetinhensis*, and so does not feel it appropriate to make any functional or phylogenetic inferences. However, from the image presented by Thomas (1994) it would seem that the saola displays a very primitive and generalised dental arrangement for a member of the Bovinae, resembling something akin to a tragelaphine.

Bovin dental evolution

A thorough review of the early evolution of the Bovini and the associated dental characteristics is presented by Bibi and colleagues (Bibi 2007b; Bibi *et al.* 2009). It is apparent that the earliest bovins originated in the Indian subcontinent around nine million years ago (though the poor African fossil record at this time potentially confounds this interpretation). Notwithstanding this, however, this location and period are coincident with the increasing prevalence of C_4 grasses (Morgan *et al.* 1994). The first known occurrence of reduncines is from the Siwaliks at a similar time and hippotragines and alcelaphines, while known initially in Africa, are also known to have dispersed into South Asia.

The precision of first occurrences of all these tribes is still very much in question. One thing that is clear is that the convergently evolved and highly specialised dental forms, of which two major complexes appear to be evident, evolved at a time when grass, and more specifically the more physically and metabolically demanding C_4 grasses, were becoming more and more prevalent.

Such specialism has been retained across the diversity of the modern Bovini, with exceptions present in the form of the more fixed feeding anoa and possibly the bison as a mixed $C_{3/4}$ grazer, though more work following from the preliminary observations above needs to be conducted. What the precise driving force was behind the evolution of such a specialised dentition remains unanswered, though an attempt has been made above with reference to the variation observed among grazing bovids.

Summary and conclusion

The Bovini are renowned for their large body sizes and primarily grazing habit. As such, their dental anatomy reflects the demands of a grass diet, possessing as it does large quantities of often highly folded enamel and with higher tooth crowns compared with browsing forms. However, what is also clear when compared with other bovid grazers is that the Bovini possess a particular type of grazing dentition and one that has convergently evolved in at least two other bovid tribes, the Hippotragini and Reduncini. This type is quite distinct from the form found in the Alcelaphini and is coincident with other craniodental differences described previously.

Nevertheless, some variation within the tribe is evident. The anoa – an island endemic – when compared to other bovins, possesses many traits that are found in mixed feeding and browsing bovids. There is also an intriguing suggestion that North American bison have evolved a less complex, though still unmistakably grazing, dentition compared with the other large-bodied bovins. Whether this difference represents nothing more than a statistical deficiency of the analysis, a more alcelaphine mode of grazing or the lengthy exposure over evolutionary time scales to C_3 grasses remains to be seen.

The analyses presented here, while admittedly tentative in their scope on account of the sample sizes obtained and fairly limited taxon sampling, do hint at the notion that large grazing ruminants are functionally more diverse than perhaps previously supposed. That the Bovini should apparently have originated at a time of increase in C_4 grasses further highlights the fact that large grazers face a range of physical and metabolic challenges, each perhaps with their own solution. Clearly, further research to elucidate such patterns is needed, with greater synergy between zoologists and botanists surely being a primary path to such a goal.

Acknowledgements

Data collection was funded by Domestic Research Studentship and Pembroke College, both University of Cambridge. The author would like to thank the museum curation staff at the University of Cambridge Museum of Zoology, the Natural History Museum (London), Quex Park (Birchington-on-Sea) and the University of Oxford Natural History Museum for access to their collections, and Edward Davis and Kelsey Stilton for their insightful comments and significant improvements to this chapter. I am extremely grateful to Mario Melletti for the opportunity to contribute to this volume.

References

Archer, D. & Sanson, G. (2002). Form and function of the selenodont molar in southern African ruminants in relation to their feeding habits. *Journal of Zoology* **257**: 13–26.

Bibi, F. (2007a). Dietary niche partitioning among fossil bovids in late Miocene C3 habitats: consilience of functional morphology and stable isotope analysis. *Paleogeography, Paleoclimatology, Paleoecology* **253**: 529–538.

Bibi, F. (2007b). Origin, paleoecology, and paleobiogeography of early Bovini. *Paleogeography, Paleoclimatology, Paleoecology* **248**: 60–72.

Bibi, F. & Vrba, E. S. (2010). Unravelling bovin phylogeny: accomplishments and challenges. *BMC Biology* **8**: 50.

Bibi, F., Bukhsianidze, M., Gentry, A. W., Geraads, D., Kostopoulos, D. S. & Vrba., E. S. (2009). The fossil record and evolution of Bovidae: state of the field. *Palaeontologica Electronica* **12**(3): 10A.

Bodmer, R. E. (1990) Ungulate frugivores and the browser-grazer continuum. *Oikos* **57**: 319–325.

Codron, D., Brink, J. S., Rossouw, L. & Clauss, M. (2008a) The evolution of ecological specialization in southern African ungulates: competition or physical environmental turnover? *Oikos* **117**: 344–353.

Codron, D. L., Brink, J. S., Rossouw, L. & Clauss, M. (2008b). Functional differentiation of African grazing ruminants: an example of specialized adaptations to very small changes in diet. *Biological Journal of the Linnean Society* **94**: 755–764.

Dierenfeld, E. S., Mueller, P. J. & Hall, M. B. (2002). Duikers: native food composition, micronutrient assessment and implications for improving captive diets. *Zoo Biology* **21**: 185–196.

Feranec, R. S. (2004). Geographic variation in the diet of hypsodont herbivores from the Rancholabrean of Florida. *Paleogeography, Paleoclimatology, Paleoecology* **207**: 359–369.

Flores-Miyamoto, K., Clauss, M., Ortmann, N. S. & Sainsbury, A. W. (2005). Nutrition of captive lowland anoa (*Bubalus depressicornis*): a study on ingesta passage, intake, digestibility and a diet survey. *Zoo Biology* **24**: 125–134.

Fortelius, M. & Solounias, N. (2000). Functional characterization of ungulate molars using the abrasion-attrition wear gradient: a new method for reconstructing paleodiets. *American Museum Novitates* **3301**: 1–36.

Hassanin, A. (2014). Systematic and evolution of Bovini. In *Ecology, Evolution and Behaviour of Wild Cattle: Implications for Conservation*. Eds. Melletti, M. & Burton, J. Cambridge: Cambridge University Press.

Heywood, J. J. N. (2009). Ruminant palaeodietary reconstruction using occlusal morphology of upper molars. PhD thesis. University of Cambridge.

Heywood, J. J. N. (2010). Functional anatomy of bovid upper molar occlusal surfaces with respect to diet. *Journal of Zoology, London* **281**(1): 1–11.

Hoppe, K. A., Paytan, A. & Chamberlain, P. (2006). Reconstructing grassland vegetation and paleotemperatures using carbon isotope ratios of bison tooth enamel. *Geology* **34**(8): 649–652.

Janis, C. M. (1988) An estimation of tooth volume and hypsodonty indices in ungulate mammals, and the correlation of these factors with dietary preferences. In *Teeth Revisited*, pp. 367–387. Eds. Russell, D. E., Santoro, J. P. & Sigogneau-Russell, D. Paris: Editions du Museum.

Janis, C. M. & Ehrhardt, D. (1988). Correlation of relative muzzle width and relative incisor width with dietary preference in ungulates. *Zoological Journal of the Linnean Society* **92**: 267–284.

Kaiser, T. M. (2002). Functional significance of ontogenetic gradients in the enamel ridge pattern of the upper cheek dentition of the Miocene hipparionine horse *Cormohipparion occidentale* (Equidae, Perissodactyla). *Senckenbergiana lethaea* **82**: 167–180.

Kaiser, T. M., Fickel, J. Streich, W. J., Hummel, J. & Clauss, M. (2010). Enamel ridge alignment in upper molars of ruminants in relation to their natural diet. *Journal of Zoology, London* **281**(1): 12–25.

Marcot, J. D. (2007). Molecular phylogeny of terrestrial artiodactyls: conflicts and resolution. In *The Evolution of Artiodactyls*. Eds. Prothero, D. R. & Foss, S. E. Baltimore, MD: Johns Hopkins University Press.

Mendoza, M., Janis, C. M. & Palmqvist, P. (2002). Characterizing complex craniodental patterns related to feeding behaviour in ungulates: a multivariate approach. *Journal of Zoology* **258**: 223–246.

Morgan, M. E., Kingston, J. D. & Marino, B. D. (1994). Carbon isotopic evidence for the emergence of C4 plants in the Neogene from Pakistan and Kenya. *Nature* **367**: 162–165.

Rensberger, J. M., Forsten, A. & Fortelius, M. (1984). Functional evolution of the cheek tooth pattern and chewing direction in tertiary horses. *Paleobiology*, **10**: 439–452.

Spencer, L. M. (1995). Morphological correlates of dietary resource partitioning in the African-Bovidae. *Journal of Mammalogy* **76**: 448–471.

Thomas, H. (1994). Cranial anatomy and phylogenetic relationships of a new bovine (*Pseudoryx nghetinhensis*) discovered in the Vietnamese Annamite mountain range. *Mammalia* **58**(3): 453–481.

Chapter 8

Evolution, development and functional role of horns in cattle

Edward Byrd Davis, Katherine A. Brakora and Kelsey Tull Stilson

Introduction

Horns are one of the most distinctive physical characters of cattle and of bovids in general. In fact, the presence of cattle horns has been the signature character of a number of critical religious and artistic figures through human history, including the seminal Ba'al of the ancient Middle East, the Minotaur of Greek legend, Apis of Egyptian mythology and the Ushi-Oni of Japan, among others (Figure 8.1). While there are many other, more subtle skeletal and soft tissue characters that distinguish bovids from other ruminants, the presence of paired frontal horns (and not antlers, ossicones or pronghorns: Davis *et al.* 2011) is an obvious character that clearly diagnoses the evolutionary group (Bibi *et al.* 2009). Among bovids, cattle horns are notable for the presence of horns in both sexes (before domestication) and their lack of surface elaboration (e.g. keels and rings). Consequently, the presence of horns in at least one sex is plesiomorphic within cattle, so their presence tells us little of the evolutionary relationships within the group. Rather, the distinctive shape of cattle horns, with broad, straight beams and a lack of surface elaboration, aids in distinguishing their fossil relatives from the closely related tragelaphine antelopes.

Besides their iconic appearance and role in evolutionary studies, horns affect the economics of raising and managing cattle. Horned animals can injure herd mates, harm stock handlers and interfere with squeeze chutes, feeders and other devices (Hopkins *et al.* 2004; Anderson 2009; AVMA 2012). Horned beef cattle also fetch lower prices, and bruising caused by horns reduces the market value of herds (Hopkins *et al.* 2004; Anderson 2009). Options for controlling horn growth include developing genetically hornless (polled) strains (Prayaga 2007), disbudding calves (removing or destroying the developing horn tissue) or manually dehorning animals. In the latter case, horns should be removed within the first 3–4 months of life, before the horn cores are invaded by the frontal sinuses (Frandson *et al.* 2009). Better knowledge of the developmental controls of normal horn growth could provide easier, more economical or more humane alternatives for polling cattle until genetically polled strains of the desired breeds are developed.

Figure 8.1 Illustrations of cultural figures signified by their possession of the horns of cattle. First row, from left to right: **Ushi-Oni**: literally the 'cow demon' of Japan, a common symbol to ward away evil spirits. **Nandi**: the mount of the Hindu god Shiva, the destroyer or transformer. **The Minotaur**: mythical offspring of human and bull, the Minotaur was trapped in a labyrinth on his home island of Crete and given seven youths and seven maidens as a sacrifice each year. **Apis**: one of the earliest Egyptian gods, closely linked with fertility and strength, whose symbol (a bull with a sun above it) became the ankh. **Ba'al**: originally a fertility god, Ba'al has most famously been known as a false idol, first as the Golden Calf of Exodus and subsequently as Ba'al Zebub (i.e. Beelzebub or 'The Lord of the Flies'). **Viking**: the iconic horned Viking helmet was actually introduced in the repopularisation of Norse mythology in the nineteenth century, with the horns 'legitimising' the Vikings by connecting them with Classical European myths such as the Minotaur. Second row, left to right: **Chauvet-Pont-d'Arc Cave**: paintings of cattle by humans have been found from up to 30 000 years ago in southern France. **Constellation of Taurus**: Latin for 'bull' and one of the 12 members of the zodiac, this constellation dates to the Bronze Age, when its stars aligned with the sun during the spring equinox.

Without a detailed knowledge of horn development in living animals, it is difficult to identify the series of steps by which cattle horns evolved their current form. Our goal is to present a unified review of what is known of the evolution and development of bovid horns, with emphasis where appropriate for wild cattle. We will begin with an overview of the anatomy

Ecology, Evolution and Behaviour of Wild Cattle: Implications for Conservation, ed. M. Melletti and J. Burton. Published by Cambridge University Press.

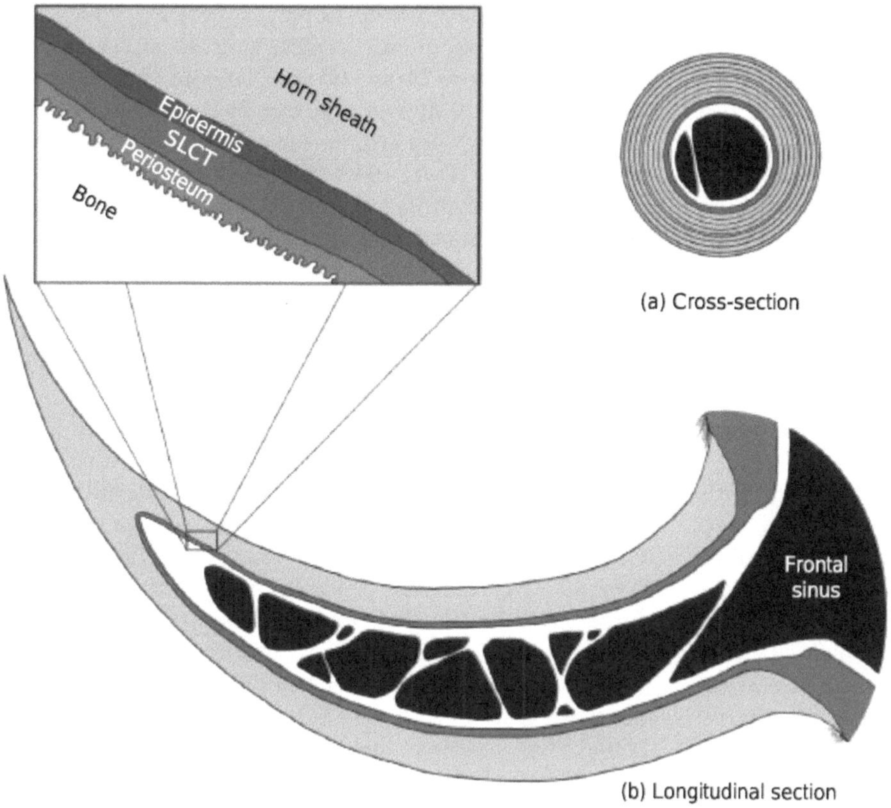

Figure 8.2 Sections of adult horn of a domestic cow, showing the bony horncore composed of compact bone invaded by the frontal sinus. (a) Cross-section, showing the annular growth rings within the keratin sheath. (b) Longitudinal section, with a magnified view of the layers of bone, connective tissue and external keratinous horn sheath. Although the periosteum is responsible for growth of the bony horncore, development of the horn organ is initiated and controlled by the connective tissue above the periosteum.

of the horns, followed by a discussion of the development of those tissues. We will then briefly review the known genetics of horn expression, in terms of polling, scurs (incompletely developed horns) and intersexuality as it applies to horn expression. Our final section will review current hypotheses regarding the adaptive significance of different shapes of horns. Given the dearth of empirical knowledge of horn evolution and development, hopefully our review will spur new research into this rich area of study.

Anatomy

While the headgear of other animals may be colloquially referred to as 'horns' (e.g. horned toads, rhinoceros 'horns' and giraffe 'horns'), we define horns phylogenetically, so that only bovid ruminants may possess true horns. Even so, terminological ambiguity has complicated the study of bovid horns from the beginning: the term 'horn' has been used to refer to the overall organ, parts of it or to a specific tissue – the keratinised epidermis that forms the hard outer surface (and other hard structures such as hooves and claws). Additionally, in palaeontological work, where the non-skeletal tissues are not preserved, the inner bone structure is often mistakenly called the 'horn'. For the sake of simplicity, we will refer to the whole organ as the 'horn', the inner bone as the 'horncore' and the keratinous covering as the 'horn sheath', following Davis *et al.* (2011).

Horns are positioned in the posterolateral region of the dorsal frontal bone in the majority of bovids. In cattle, the apparently very lateral position of the horns owes more to the modified proportions of the frontal bone than to the relative position of the horns. Macroscopically, the horncore looks like an outgrowth of the frontal bone. Within the frontal bone, a frontal sinus develops on each side; left and right frontal sinuses are separated by a septum. Usually in month three or four in cattle, an extension of the frontal sinus (the cornual diverticulum) penetrates the bone of the growing horn (Frandson *et al.* 2009).

Bovid horns consist of a series of five distinct tissues (Figure 8.2); from internal to external, these include bone, periosteum, subcutaneous loose connective tissue (SLCT), modified dermis often called 'corium' (Colville & Bassert 2009) and the epidermis; the horn sheath is the dead, outermost layer of the epidermis and is highly keratinised. The distinctive shapes of bovid horns are primarily determined by differential growth of the horn sheath. To date, no research has directly investigated the developmental processes by which this differential growth is controlled.

The live tissues of the horns are supplied by the standard blood vessels of the head. The cornual artery and frontal rami diverge from the superficial temporal artery, supplying the anterior and posterior aspects of the horn, respectively. The cornual vein drains the horn and joins the superficial temporal vein.

Innervation of the horns is predominantly achieved by the cornual branch of the zygomaticotemporal nerve in cattle (Popesko & Getty 1979; Dugdale 2010; Budras *et al.* 2011) and buffalo (Fouad *et al.* 1979), although the branching pattern is variable, at least in cattle (Madekurozwa *et al.* 2000). Three additional nerves have been inconsistently implicated in the nerve supply of the horn. In a study of four calves and two heifers, Madekurozwa *et al.* (2000) reported that all animals' horns were innervated by a cornual branch of the infratrochlear nerve, which branches off the nasociliary nerve, from the opthalmic nerve (Budras *et al.* 2004). The infratrochlear nerve is also reported to supply the horns in buffalo (Fouad *et al.* 1979) and goats (Dugdale 2010). However, Budras *et al.* (2011) indicate that, at least in cattle, innervation of the horn by the infratrochlear nerve may not be universal and, when it occurs, results from abnormally long branches of these nerves. The frontal nerve, which terminates in the skin of the frontal region as the supraorbital nerve (Budras *et al.* 2004), is also reported to innervate the horn in some, but not all, cattle (Madekurozwa *et al.* 2000; Budras *et al.* 2011). Whether cervical nerves supply the caudal regions of the horn base is the least clear; Popesko & Getty (1979) show branches of the first cervical nerve present in the superficial tissues just caudal to the horn, whereas Dugdale (2010) indicate that cutaneous branches of the second cervical nerve may need to be desensitised when dehorning adult cattle with well-developed horns. By contrast, Madekurozwa *et al.* (2000) found no involvement by cervical nerves in two calves and two heifers, and Budras *et al.* (2011) state that there are no indications for cervical nerve involvement. This disagreement about the nerve supply to the horn could suggest that horn innervation is more variable in domestic cattle than in other species. Alternatively, variation in other species may be under-reported because these species are less intensively studied. Nevertheless, the main factors responsible for the variation in domestic cattle (species, breed, sex, age or environmental factors) are as yet unknown.

Development

Our discussion of horn development draws extensively from the supplementary literature review we contributed as part of Davis *et al.* (2011), and most knowledge of horn development today traces to Dove's (1936) tissue transplantation work on cattle and goats.

Clinically, horns begin as palpable, slightly mobile 'buds' under the skin, allowing dehorning techniques that scrape off the developing horn tissues. Around two months of age, the buds become attached to the skull, necessitating dehorning techniques that cut through the base of the horncore (Anderson 2009; AVMA 2012). However, visible evidence of this transition is exceedingly rare, giving the distinct impression that the horncore is an outgrowth of the frontal bone – an assertion that still persists in many veterinary textbooks (e.g. Samuelson 2007; Bacha and Bacha 2012) even though separate (dermal) ossification centres have been regularly discovered and reported in the literature since 1829 (reviewed in Dove 1936).

Early in bovid horn development there are three structures that deserve discussion (Davis *et al.* 2011): the soft tissue anlage, the *os cornu* and the horncore bud. Above the presumptive horn sites on the frontal bone, the epidermal and connective tissues acquire their potentials and irreversibly differentiate (Dove 1936). Within the connective tissues, the anlage (German: layout, structure) differentiates before birth, retaining its connective tissue character (Dove 1936). The anlage appears to use as-yet unknown mechanisms to induce the subsequent development of the bony horncore (Davis *et al.* 2011). It is unknown how long the anlage remains structurally and functionally distinct from the other connective tissues.

The first potentially hard-tissue growth appears to be the *os cornu* (not to be confused with the horncore, or cornual process, in adults), a separate, palpable nodule in the soft tissue immediately above the periosteum. The *os cornu* has been reported as dermis/SLCT (Durst 1902; Brandt 1928; Dove 1936) or cartilage (Gadow 1902; Atzenkern 1923). If the *os cornu* directly abuts the periosteum or frontal bone, it will fuse to the presumptive horn site; however, the presence of SLCT between the structure and the frontal bone causes it to ossify separately, forming a scur (or 'loose horn') (Dove 1936). The suggestion that the *os cornu* is a cartilaginous precursor of the horncore is rejected by several workers, who argue for intramembranous ossification in all stages of horn development (Durst 1902; Brandt 1928; Janis & Scott 1987). Transplantation of horn-site connective tissues, but not the periosteum or the *os cornu*, to a non-horn site causes ectopic horn growth, and excision of the *os cornu* does not by itself prevent future horn development (which proceeds without re-growing an *os cornu*) (Dove 1936). Damage to a developing horn can prevent proper fusion to the frontal bone (Kysely 2010). These results indicate that it is the connective tissues, not the periosteum or the *os cornu*, that are responsible for early horn development (including the horncore).

Importantly, the few studies of horn growth that have been performed have not controlled for sex and genotype of the several genes and multiple loci that are now known to determine the presence of horns and scurs in cattle (Mariasegaram *et al.* 2010; Capitan *et al.* 2011; Medugorac *et al.* 2012; Seichter *et al.* 2012). Consequently, the role of the *os cornu* in horn development is controversial, with many researchers convinced that it is only present in abnormal individuals (Sandifort 1829; Dove 1936; Bubenik 1983; Kysely 2010). As the genetic basis of horns is not yet fully understood, it is possible that there are several stages or contributing pathways in horn development which are variously affected by sex and genotype, producing a range of slightly different developmental histories, even when adult horn phenotypes are considered equivalent.

While there is debate on whether the horn is induced by the anlage or by fusion of the *os cornu* to the frontal bone, the next step in horn development is uncontroversial: the bony horncore begins to ossify in earnest and the overlying epidermis begins to produce cornified horn tissue (Durst 1902). Early layers of juvenile horn tissue are typically softer and more fibrous than those produced later and may slough off (George 1956; O'Gara & Matson 1975). The horncore subsequently elongates and widens with appositional bone growth, adding new bone material at the tip and external surface (Durst 1902, 1926; Janis & Scott 1987). As a sign of the poor state of knowledge of horn development, there is controversy over whether this growth is intramembranous or endochondral; that is, whether the bone of the horncore develops directly from the periosteum, as with other bones of the skull, or develops from a cartilaginous precursor, as in the growth plates of long bones (Saint-Hilaire 1837; Durst 1902; Fambach 1909; Dove 1936; Kosasih 1959; Goss 1983; Janis & Scott 1987; Bubenik 1990). Whether or not the growth is intramembranous, the pattern of horncore bone growth is similar to the way cervids develop their antlers (appositionally at the tip and surface) and importantly distinct from the way giraffids enlarge their ossicones (adding bone in the non-cartilaginous connective tissue at the base) (Davis *et al.* 2011).

As with other living skeletal elements, horncores are actively remodelled throughout life in response to developmental, physiological and mechanical demands (Albarella 1995). In cattle, the horncore is usually invaded by the frontal sinus by four months of age (AVMA 2012). The extent of sinus invasion can leave the horncore extremely thin-walled in some bovid species (Farke 2010). As the animal approaches adult size, horn growth slows and the horncore increases in density, remodelling to produce more compact bone in a wave that progresses from external to internal and proximal to distal, beginning at the exterior of the base and ending at the tip of the horncore. Bone remodelling likely obscures most histological traces of early horncore development.

The horn sheath grows from the epidermis, beginning very early in postnatal ontogeny. It is clear that the epidermis has the potential to make horn tissue from early in development, as indicated by transplant experiments; however, the presence of the *os cornu* appears necessary for normal sheath development (Dove 1936). The horn sheath grows continually from the epithelium surrounding the entire horncore, pushing out in all directions. Thus, the structure of the mature horn sheath is similar to a stack of ever larger cones (Figure 8.2), with horn sheaths in some temperate species showing dramatic layering as a consequence of seasonal variation in growth rates (Goss 1983; Santiago-Moreno *et al.* 2005). At least in some bovids, keratin production rates are not uniform over the horn, with the tip of the horn sheath often thicker than the walls (Goss 1983); in fact, variation in the rate of horn tissue production in different areas of the horn is thought to contribute to, if not determine, the many shapes and surface ornamentation of bovid horns (Kingdon 1982).

Genetics

Research into breeding hornless ('polled') strains has resulted in better understanding of the genetics of horn expression than of the processes of horn development (e.g. White & Ibsen 1936; Cole & Johannson 1948; Long & Gregory 1978). The presence of horns is a recessive trait, controlled by the interaction of four genes that also control the presence of scurs (Long & Gregory 1978; Asai *et al.* 2004; Capitan *et al.* 2009; Mariasegaram *et al.* 2010; Medugorac *et al.* 2012). These genes are designated *Poll* for polledness (with polledness dominant), *Horn* for horn expression, *African horn* for the 'African horn gene' expressed most often in zebu cattle in Africa (White & Ibsen 1936) and *Scurs* for the presence of scurs (Long & Gregory 1978). *Poll* is epistatic to (suppresses) *Horn*, which is, in turn, epistatic to *Scurs*. *African horn* is at a low frequency in most domestic breeds, but common in wild cattle, and it is epistatic to *Poll*, causing the expression of horns in animals that should otherwise be hornless or scurred. Some workers had suggested that *Poll*, *Horn* and *African horn* may be alleles of the same locus (Georges *et al.* 1993) instead of epistatically linked separate loci, but recent genetic work rejects this hypothesis. The expression of *Scurs* has been suggested to be controlled by the sex of the animal, with expression in both heterozygous and homozygous males, but with only homozygous females expressing loose horns (Long & Gregory 1978). More recent research has shown no direct sex-chromosome linkage for *Scurs* (Asai *et al.* 2004), suggesting a hormonal linkage in the relative expression of scurs in males and females. The complex interaction between these genes can make predicting the horned/polled phenotype of a calf difficult without adequate knowledge of the genotypes of the parents.

Recent genetic work has pursued the location of these genes within the cattle genome, with positive results that may lead to genotyping in aid of artificial selection for polledness (Prayaga 2007). The *Poll* locus has been mapped to the centromeric region of bovine chromosome 1 (Georges *et al.* 1993; Brenneman *et al.* 1996; Harlizius *et al.* 1997; Brockmann *et al.* 2000; Drogemuller *et al.* 2005), with the recent work of Medugorac *et al.* (2012) isolating the locus to a ~381 kb interval and providing clear evidence that polledness is expressed by one or both of two distinct mutations in this region. Attempts to map the *Scur* locus have suggested the possibility of multiple, independent loci producing scurred phenotypes, with some workers mapping the locus to bovine chromosome 19 (Asai *et al.* 2004; but see Capitan *et al.* 2009), and others mapping it to bovine chromosome 4 (Capitan *et al.* 2011). Both of these results confirm the hypothesised lack of linkage with the *Poll* locus suggested by breeding experiments, but suggest the inheritance of scurring is more complex than initially hypothesised. To date, the *Horn* and *African horn* loci have not been mapped to bovine chromosomes (Prayaga 2007; Mariasegaram *et al.* 2010; Nicholas 2012), but work with sheep has suggested that the locus might be in bovine chromosome 12, analogous to a locus for *Horn* on sheep chromosome 10

(Montgomery *et al.* 1996), which is currently being sequenced (Pickering *et al.* 2009).

Recent work has moved beyond simply mapping the loci of these genes towards a developmental process-based understanding of gene expression in horns. Mariasegaram *et al.* (2010) used transcript profiling to compare the expression of known bovine genes in polled, horned and scurred Brahman cattle, revealing the networks of mostly soft-tissue-related genes involved in the development of horns. Such studies, combined with the growing genomic knowledge of *Bos taurus* (e.g. Ihara *et al.* 2004), should lead to a practical knowledge of the developmental processes involved in initiating and maintaining horn growth, so that future workers may either artificially select for polled cattle or artificially induce polling in cattle through directed intervention in these developmental pathways.

The expression of horns in goats is autosomally linked to sex determination; the polled phenotype is highly correlated with intersexuality (Pailhoux *et al.* 2001a, 2001b). Polling in goats is linked to XX individuals expressing hermaphrodite or male phenotypes, called the Polled Intersex Syndrome (*PIS*: Soller & Angel 1964). The locus responsible for this phenotype has been isolated in the distal region of goat chromosome 1 (Pailhoux *et al.* 2001a), but to date no research has connected this *PIS* locus to the *Poll* locus for cattle. Although *PIS* appears to be unique to goats, the relationship is conserved enough to provide insight on a human condition affecting both eyelid development and ovarian function (Schibler *et al.* 2000), suggesting an ancient relationship between the development of select integumentary structures and sex expression or function. Consequently, identification and investigation of the homologous region in the cattle genome could provide insights on the relationship between sex and horn expression in bovines, but to date no research has investigated this connection.

Adaptive significance of horns

The presence of horns is apomorphic for the whole bovid family (Bibi *et al.* 2009; Davis *et al.* 2011), making horns plesiomorphic among bovines and all other bovids (Figure 8.3). Previous research on the adaptive significance of bovid horns has therefore focused on the many horn shapes observed among bovids, and among African antelope in particular. Likewise, the question of adaptive significance of cattle horns must centre on other features of horns in wild cattle, such as their shape or lack of dramatic sexual dimorphism. Unfortunately, the lack of resolution among the bovines hampers phylogenetic inference, allowing the clear inference only of the presence of a lateral orientation of the horns at the split between Bubalina and Bovina and a single origin of the distinct water buffalo horn shape in *Syncerus* (Figure 8.3). A hypothesised function for horn traits must be evaluated in the context of all similar animals with headgear, including non-bovids, such as deer, rhinoceroses and even elephants (with their tusks) to identify the selective pressure(s) and the phylogenetic locus or loci of morphological change (vs maintenance). In many cases, morphological adaptations originate long before extant lineages evolved, invalidating claims of novel adaptation in extant species.

In his seminal work on horn evolution, Geist (1966) suggested that the plesiomorphic state for horns was as short, stabbing weapons intended for broadside engagement between males, and that other forms were modified to reflect both the mating system and ecology of the descendant lineages. The evolution of broader horns (or tusks, as in elephants), was suggested to be a mechanism for decreasing combat mortality by allowing males to enter into more ritualised head-to-head (literally) combat. Geist (1966) suggested that many horns and analogous structures were selected to maximise the effectiveness of male–male combat in testing individual strength while keeping mortality to a minimum, particularly in ecosystems with a high concentration and diversity of predators (Geist 1974). In Geist's (1966) system, the horns of cattle (as well as those of bison and the tusks of elephants) represent the first step away from simple straight horns, providing a stout skull for withstanding impact, and forward/upward curvature of the tips to contain the opponent's skull, preventing a flank attack. Subsequent work has, unsurprisingly, shown that the true evolutionary picture is much more complex, rejecting Geist's (1966) idealised linear progression of horn type (Lundrigan 1996; Brashares *et al.* 2000). However, the association of horn type with mode of fighting has withstood additional tests and refinements, and it is clear that the broad-based, widely splayed horns seen in bovids like cattle are strongly associated with head-to-head ramming, even when phylogenetic interdependence is taken into account (Lundrigan 1996; Caro *et al.* 2003).

Research has also suggested an interplay between body mass and habitat in controlling the combination of horn morphology, male–male combat method, antipredator strategy and sexual dimorphism. The large body mass of wild cattle co-evolved with their non-selective feeding ecology, which in turn is connected to their large herd size and preferred predator defence strategy of communal attack (Brashares *et al.* 2000; Caro *et al.* 2004). In addition, evidence suggests that the movement into open habitats fostered polygyny and subsequent body size dimorphism in wild cattle (Perez-Barbaria *et al.* 2002), which has had an influence on their mating system (Bro-Jorgensen 2007), and, consequently, their horn morphology (Caro *et al.* 2003). All of these characteristics cause them to fall into social group E of Jarman (1974), the category of animals living in large herds that contain many males of breeding age, who determine their mating rights through social dominance determined by previous contests. In this way, they are using their horns not to establish physical territories, but to carve a position in the social hierarchy (Jarman 1983).

The presence of female horns is not plesiomorphic within bovids (Figure 8.4), so the common presence of horned females among unrelated groups of large-bodied bovids, like wild cattle, has elicited several competing hypotheses in the

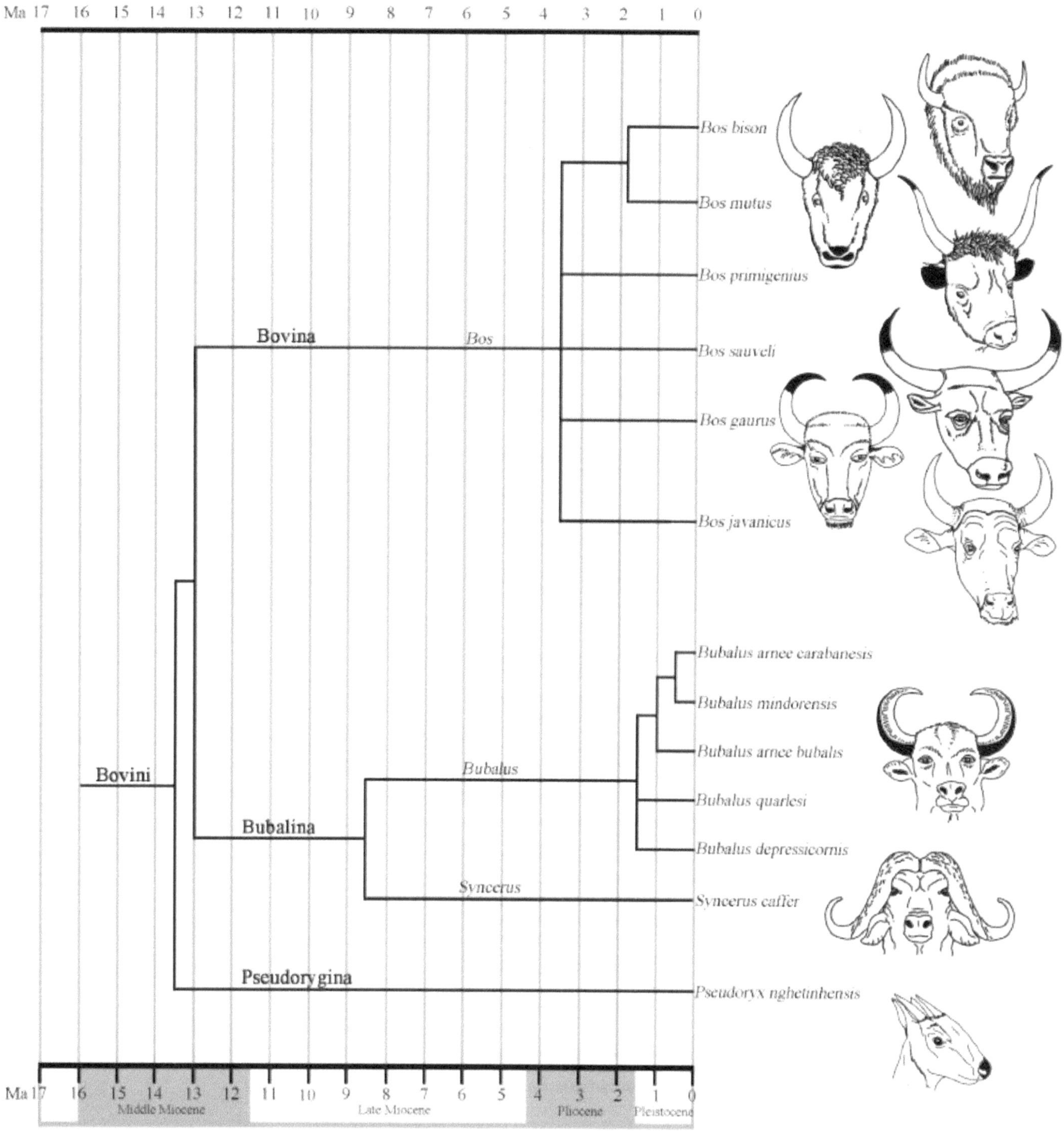

Figure 8.3 Horn shape mapped onto the phylogeny of wild cattle (see Chapter 1). The illustrations display the general dominant phenotypic expression of the lowest clade with discernible differences. Laterally extending horns seem to have evolved early within the Bovini and display a limited amount of phenotypic variability, with the notable expression of the water buffalo (*Syncerus caffer*).

literature. Geist (1974) suggested that gregarious herbivores competing within a herd in an open landscape would experience a strong selection against sexual dimorphism, resulting in females that are equal in size and armament to males, ensuring their success in competition with males for resources. A modified version of this idea, proposed by Roberts (1996), suggests that female horns are an adaptation for fighting for resources among females, arising in ecological and social settings that combine large groups of females and scarce resources. Packer (1983) championed the hypothesis that

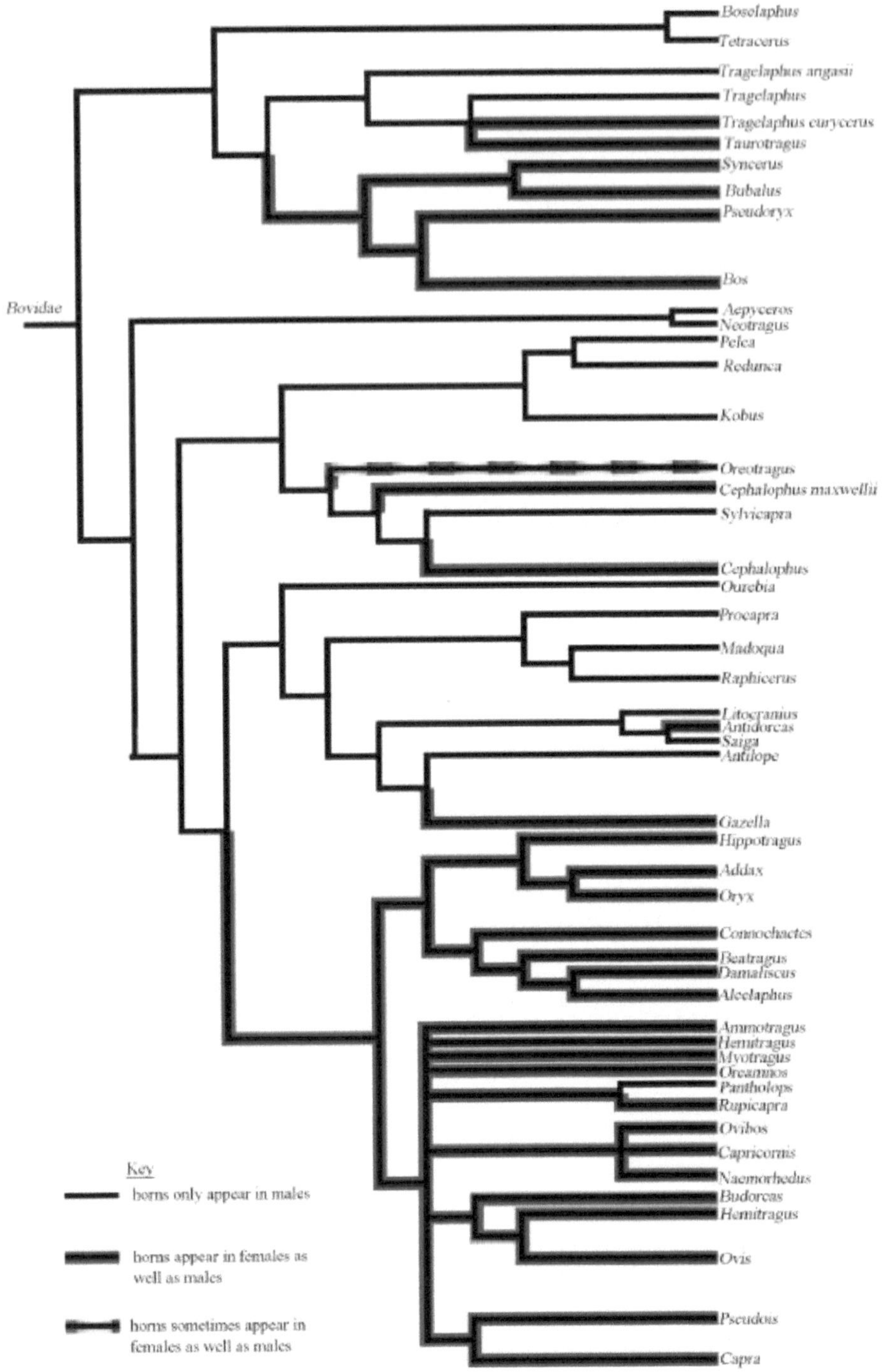

Figure 8.4 Female horn expression mapped onto the phylogeny of the Bovidae. A plain black line indicates the presence of horns in only males; black lines with a grey highlight indicate the presence of horns in both females and males; and black lines with a dashed grey highlight indicate the sporadic occurrence of horns in females. We mapped the characters onto the phylogeny of Marcot (2007) using parsimony with the DELTRAN assumption when necessary.

female horns were primarily weapons for defence against predators, arguing that the horns of females seem to be better adapted as stabbing weapons, while male horns tend to be structured to control attacks and reduce the rate of injuries in male–male combat (Packer 1983; but see Estes 1991). Estes (1991) has suggested that female horns could be an adaptation to protect juvenile male offspring from aggression by adult males: horned females would provide social camouflage to hide developing males. In contrast, Kiltie (1985) proposed a null hypothesis, suggesting that female horns were evolved and maintained because of a non-adaptive mutation linking the expression of horns to reproductive virility (as seen in *PIS* for goats). Subsequent phylogenetically informed empirical work has rejected this null hypothesis (Roberts 1996; Stankowich & Caro 2009), with the most recent results suggesting that the expression of female horns in large-bodied bovids like wild cattle is controlled by a combination of the need for antipredator defence and the competition between females over territory (Stankowich & Caro 2009). Intriguingly, new research on African buffalo (*Syncerus caffer*) has suggested that both female and male horns play a significant role as advertisement of parasite load, with healthier individuals having larger horns (Ezenwa & Jolles 2008). This idea deserves further study.

In the end, plotting expression of female horns on the phylogeny of the Bovidae (Marcot 2007) shows the independent expression of horned females in multiple subfamilies and a range of body sizes (Figure 8.4). Large, sexually dimorphic bovines such as *Synercus* and *Bos* display female horn expression, while the smaller bovine species do not. However, within the generally non-sexually dimorphic, cursorial species of the Antilopinae, *Gazella* and *Antidorcas* females exhibit horns, while others do not. This expression of female horns across a range of body sizes suggests that presence of female horns is more closely related to ecology or mating system than physiology. Female horn expression may be a consequence of a number of natural and sexual selection regimes, rather than one dominant driver.

Conclusions

Horns are an essential part of the evolutionary, ecological and economic story of wild cattle. More than a century of research into the genetics, development and evolution of bovid horns has produced many reasonable hypotheses about the governing processes. Nevertheless, our knowledge is surprisingly scant given the long-standing cultural and economic value of these animals. The most progress has been made on understanding the genetic basis of horn expression; conversely, horn development is the least studied aspect of horn biology, with no systematic investigations for more than 75 years.

Horns develop through a complex interaction between soft tissues and the underlying bone, thought to be governed mostly by the connective tissues above the presumptive horn site. Signalling for the onset of horn formation is controlled by three or more genes, one of which has been located on bovine chromosome 1; signalling for complete fusion is controlled by a gene on bovine chromosome 12 with at least two causative mutations. Horn expression is linked in some way to the sex of the animal, but not through a simple locus on the sex chromosome. Current results from evolutionary studies suggest that horns in wild cattle have evolved in concert with their body sizes, antipredator strategies and mating system, producing structures that reduce mortality in male-to-male tests of strength and facilitate antipredator defence in both males and females.

Pressing research questions on the origin and function of horns include: Where are the other loci that control horn expression? How does sex affect the expression of these genes? How does genotype and allelic identity at one locus affect the expression of genes elsewhere? What are the signalling pathways that determine horn location, induce tissue differentiation and control subsequent horn growth? What have been the roles of health advertisement, inter- and intrasexual competition and inter- and intraspecific defence (particularly of juveniles), and other factors, which determined the selective regime affecting cattle horn phenotypes, for both extant and extinct taxa? What events (perhaps unique to one or more cattle lineages), such as genomic rearrangements, cross-lineage hybridisation, population bottlenecks or periods of geographic isolation, have affected the evolution of cattle and their horns?

Much exciting research lies ahead before we attain a satisfying grasp of the underlying biology of bovid horns. Better knowledge of genetics and development in particular will greatly improve our ability to wisely manage cattle of all kinds, whose history has been entwined with ours for tens of thousands of years (Figure 8.1).

Acknowledgements

We would first like to thank A. Lee, who contributed to our 2011 review of ruminant headgear and provided important insight into this review of bovid horn evolution, development and function. We would also like to thank the Hopkins Lab at the UO and the Padian Lab at UCB for enduring long discussions of the importance of bovid horns and the intricacies of their scientific history. M. Hanneson of UO Campus Operations graciously allowed use of the bandsaw to prepare the horncore that formed the basis for Figure 8.2.

References

Albarella, U. (1995). Depressions on sheep horncores. *Journal of Archaeological Science* **22**(5): 699–704.

Anderson, N. (2009). Dehorning of calves. *IMAFRA Factsheet* **09-003**: 1–12.

Asai, M., Berryere, T. G. & Schmutz, S. M. (2004). The scurs locus in cattle maps to bovine chromosome 19. *Animal Genetics* **35**(1): 34–39.

Atzenkern, J. (1923). Zur entwicklung der os cornu der cavicornier. *Anatomischer Anzeiger* **57**: 125–130.

AVMA (American Veterinary Medical Association). (2012). Welfare implications of the dehorning and disbudding of cattle. American Veterinary Medical Association informational pamphlet.

Bacha, W. J. and Bacha L. M. (2012). *Color Atlas of Veterinary Histology*. Chichester: John Wiley & Sons.

Bibi, F. M., Bukhsianidze, M., Gentry, A. W., Geraads, D., Kostopoulos, D. S. & Vrba, E. S. (2009). The fossil record and evolution of Bovidae: state of the field. *Palaeontologia Electronica* **12**(3): 1–11.

Brandt, K. (1928). Die Entwicklung des Hornes beim Rinde bis zum Beginn der Pneumatisation des Hornzapfens. *Gegenbaurs Morphologisches Jahrbuch* **60**: 428–468.

Brashares, J. S., Garland Jr., T. & Arcese, P. (2000). Phylogenetic analysis of coadaptation in behavior, diet, and body size in the African antelope. *Behavioral Ecology* **11**(4): 452–463.

Brenneman, R. A., Davis, S. K., Sanders, J. O., *et al.* (1996). The polled locus maps to BTA1 in a *Bos indicus* x *Bos taurus* cross. *Journal of Heredity* **87**(2): 156–161.

Brockmann, G. A., Martin, J., Teuscher, F. & Schwerin, M. (2000). Marker controlled inheritance of the polled locus in Simmental cattle. *Archiv Fur Tierzucht-Archives of Animal Breeding* **43**(3): 207–212.

Bro-Jorgensen, J. (2007). The intensity of sexual selection predicts weapon size in male bovids. *Evolution* **61**(6): 1316–1326.

Bubenik, G. A. (1983). Shift of seasonal cycle in white tailed deer by oral administration of melatonin. *Journal of Experimental Zoology* **225**: 155–156.

Bubenik, G. A. (1990). Neuroendocrine regulation of the antler cycle. In *Horns, Pronghorns and Antlers*, pp. 261–293. Eds. Bubenik, G. A. & Bubenik, A. B. New York: Springer-Verlag.

Bubenik, G. A. & Bubenick, A. B. (1990). *Horns, Pronghorns, and Antlers: Evolution, Morphology, Physiology, and Social Significance.* New York: Springer-Verlag.

Budras, K.-D., Habel, R. E., Wunsche, A. & Buda, S. (2004). *Bovine Anatomy: An Illustrated Text.* Hannover: Schluetersche.

Budras, K.-D., Habel, R. E., Mulling, C. K. W. & Greenough, P. R. (2011). *Bovine Anatomy: An Illustrated Text.* 2nd edition. Hannover: Schluetersche.

Capitan, A., Grohs, C., Gautier, M. & Eggen, A. (2009). The scurs inheritance: new insights from the French Charolais breed. *BMC Genetics* **10**(1): 33.

Capitan, A., Grohs, C., Weiss, B., Rossignol, M. N., Reversé, P. & Eggen, A. (2011). A newly described bovine Type 2 Scurs Syndrome segregates with a frame-shift mutation in *TWIST1*. *PLoS ONE* **6**(7): e22242.

Caro, T. M., Graham, C. M., Stoner, C. J. & Flores, M. M. (2003). Correlates of horn and antler shape in bovids and cervids. *Behavioral Ecology and Sociobiology* **55** (1): 32–41.

Caro, T. M., Graham, C. M., Stoner, C. J. & Vargas, J. K. (2004). Adaptive significance of antipredator behaviour in artiodactyls. *Animal Behaviour* **67**: 205–228.

Cole, L. J. & Johansson, I. (1948). Inheritance in crosses of Jersey and Holstein-Friesian with Aberdeen-angus cattle. *American Naturalist* **82**(804): 145–170.

Colville, T. P. & Bassert, J. M. (2009). *Clinical Anatomy and Physiology Laboratory Manual for Veterinary Technicians.* London: Mosby.

Davis, E. B., Brakora, K. A. & Lee, A. H. (2011). Evolution of ruminant headgear: a review. *Proceedings of the Royal Society B: Biological Sciences* **278**: 2857–2865.

Dove, W. F. (1936). Artificial production of the fabulous unicorn. *Scientific Monthly* **42**(5): 431–436.

Drogemuller, C., Wohlke, A., Momke, S. & Distl, O. (2005). Fine mapping of the polled locus to a 1-Mb region on bovine chromosome 1q12. *Mammalian Genome* **16**(8): 613–620.

Dugdale, A. (2010). *Veterinary Anaesthesia: Principles to Practice.* Ames, IA: Wiley-Blackwell.

Dürst, J. U. (1902). Sur le developpement des cornes chez les cavicorne. *Bulletin du Museum d'Histoire Naturelle Paris* **3**: 197–203.

Dürst, J. U. (1926). Das Horn der Cavicornia. Seine Entstchungsursache, seine Entwicklung, Gestaltung auf den Schadel der horntragenden Wiederkauer. *Denkschriften des Schweizerischen Naturforschenden Gesellschaft* **68**: 1–180.

Estes, R. D. (1974). Social organization of the African Bovidae. In *The Behaviour of Ungulates and its Relationship to Management*, pp. 166–205. Eds. Geist, V. & Walther, F. Morges: IUCN Publications.

Estes, R. D. (1991). *Behavior Guide to African Mammals: Including Hoofed Mammals, Carnivores, Primates.* Berkeley and Los Angeles, CA: University of California Press.

Ezenwa, V. O. & Jolles, A. E. (2008). Horns honestly advertise parasite infection in male and female African buffalo. *Animal Behaviour* **75**: 2013–2021.

Fambach, O. (1909). Geweih und Gehörn: Ein kritisches Referat. *Zeitschrift für Naturwissenschaften* **81**: 225–264.

Farke, A. A. (2010). Evolution and functional morphology of the frontal sinuses in Bovidae (Mammalia: Artiodactyla), and implications for the evolution of cranial pneumaticity. *Zoological Journal of the Linnean Society* **159**: 988–1014.

Fouad, K., Shokry, M. & Fahmy, L. (1979). Anaesthesia of the horn in buffaloes. *Zentralblatt für Veterinärmedizin Reihe A* **26**: 78–82.

Frandson, R. D., Wilke, W. L. & Fails, A. D. (2009). *Anatomy and Physiology of Farm Animals.* 7th editon. Ames, IA: Wiley-Blackwell.

Gadow, H. (1902). The evolution of horns and antlers. *Proceedings of the Zoological Society of London* **1**: 206–222.

Geist, V. (1966). Evolution of horn-like organs. *Behaviour* **27**: 175–214.

Geist, V. (1974). Relationship of social evolution and ecology in ungulates. *American Zoologist* **14**(1): 205–220.

George, A. N. (1956). The post-natal development of the horn tubules and fibres (intertubular horn) in the horns of sheep. *British Veterinary Journal* **112**(1): 30–34.

Georges, M., Drinkwater, R., King, T. *et al.* (1993). Microsatellite mapping of a gene affecting horn development in *Bos taurus*. *Nature Genetics* **4**(2): 206–210.

Goss, R. J. (1983). *Deer Antlers: Regeneration, Function, and Evolution.* New York: Academic Press.

Harlizius, B., Tammen, I., Eichler, K., Eggen, A. & Hetzel, D. J. S. (1997). New markers on bovine chromosome 1 are closely linked to the polled gene in Simmental and Pinzgauer cattle. *Mammalian Genome* **8**(4): 255–257.

Hopkins, F. M., Neel, J. B. & Kirkpatrick, F. D. (2004). *Dehorning calves.* Agricultural Extension Service, the University of Tennessee PB 1684.

Ihara, N., Takasuga, A., Mizoshita, K., *et al.* (2004). A comprehensive genetic map of the cattle genome based on 3802 microsatellites. *Genome Research* **14** (10A): 1987–1998.

Janis, C. M. & Scott, K. M. (1987). The interrelationships of higher ruminant families with special emphasis on the members of the Cervoidea. *American Museum Novitates* **2893**: 1–85.

Jarman, P. (1974). The social organisation of antelope in relation to their ecology. *Behaviour* **48**: 215–267.

Jarman, P. (1983). Mating system and sexual dimorphism in large, terrestrial mammalian herbivores. *Biological Reviews of the Cambridge Philosophical Society* **58**(4): 485–520.

Kiltie, R. A. (1985). Evolution and function of horns and hornlike organs in female ungulates. *Biological Journal of the Linnean Society* **24**: 299–320.

Kingdon, J. (1982). *East African Mammals: An Atlas of Evolution in Africa.* Vol. **IIIC**. Chicago, IL: University of Chicago Press.

Kosasih, R. (1959). Observations on the structure of 'loose horns' in a zebu cow. *Communicationes Veterinariae* 3(2): 107–119.

Kysely, R. (2010). Breed character or pathology? Cattle with loose horns from the Eneolithic site of Hostivice-Litovice (Czech Republic). *Journal of Archaeological Science* 37(6): 1241–1246.

Long, C. & Gregory, K. (1978). Inheritance of the horned, scurred and polled condition in cattle. *Journal of Heredity* **69**: 395–400.

Lundrigan, B. (1996). Morphology of horns and fighting behavior in the family Bovidae. *Journal of Mammalogy* 77(2): 462–475.

Madekurozwa, M. C., Purton, M. & Boyd, J. S. (2000). Re-examination of horn innervation in cattle using Fluorogold. *Zimbabwe Veterinary Journal* **31**: 65–71.

Marcot J. D. (2007). Molecular phylogeny of terrestrial artiodactyls: conflicts and resolution. In *The Evolution of Artiodactyls*, pp. 4–18. Eds. Prothero, D. R. & Foss, S. E. Baltimore, MD: Johns Hopkins University Press.

Mariasegaram, M., Reverter, A., Barris, W., Lehnert, S. A., Dalrymple, B. & Prayaga, K. (2010). Transcription profiling provides insights into gene pathways involved in horn and scurs development in cattle. *BMC Genomics* **11**: 370.

Medugorac, I., Seichter, D., Graf, A., *et al.* (2012). Bovine polledness: an autosomal dominant trait with allelic heterogeneity. *PLoS ONE* 7(6): e39477.

Montgomery, G. W., Henry, H. M., Dodds, K. G., Beattie, A. E., Wuliji, T. & Crawford, A. M. (1996). Mapping the *Horns (Ho)* locus in sheep: a further locus controlling horn development in domestic animals. *Journal of Heredity* **87**: 358–363.

Nicholas, F. W. (2012). Medelian inheritance in cattle. In *Bovine Genomics*, p. 284. Ed. Womack, J. Ames, IA: Wiley-Blackwell.

O'Gara, B. W. & Matson, G. (1975). Growth and casting of horns by pronghorns and exfoliation of horns by bovids. *Journal of Mammalogy* **56**(4): 829–846.

Packer, C. (1983). Sexual dimorphism: the horns of African antelopes. *Science* **221**: 1191–1193.

Pailhoux, E., Vigier, B., Chaffaux, S., *et al.* (2001a). A 11.7-kb deletion triggers intersexuality and polledness in goats. *Nature Genetics* **29**(4): 453–458.

Pailhoux, E., Vigier, B., Vaiman, D., *et al.* (2001b). Contribution of domestic animals to the identification of new genes involved in sex determination. *Journal of Experimental Zoology* **290**(7): 700–708.

Perez-Barberia, F. J., Robles, L. & Nores, C. (1996). Horn growth pattern in Cantabrian chamois *Rupicapra pyrenaica parva*: influence of sex, location and phaenology. *Acta Theriologica* **41**(1): 83–92.

Perez-Barberia, F. J., Gordon, I. J. & Pagel, M. (2002). The origins of sexual dimorphism in body size in ungulates. *Evolution* **56**: 1276–1285.

Pickering, N. K., Johnson, P. L., Auvray, B., Dodds, K. G. & McEwan, J. C. (2009). Mapping the horn locus in sheep. *Proceedings of the Association for the Advancement of Animal Breeding and Genetics* **18**: 88–91.

Popesko, P. & Getty, R. (1979). *Atlas of Topographical Anatomy of the Domestic Animals.* Vols **I–III**. Philadelphia, PA: Saunders.

Prayaga, K. C. (2007). Genetic options to replace dehorning in beef cattle: a review. *Australian Journal of Agricultural Research* **58**(1): 1–8.

Roberts, S. C. (1996). The evolution of hornedness in female ruminants. *Behaviour* **133**: 399–442.

Saint-Hilaire, M. G. (1837). Sur le nouveau genre Sivatherium, trouvé fossile au bas du versant méridional de l'Himalaya, dans la vallée du Markanda; animal gigantesque de l'ancien honde, que je propose de rapporter au genre *Camelopardalis*. *C. R. Séances hebdomadaires de l'Académie des Sciences* **1**: 53.

Samuelson, D. A. (2007). *Textbook of Veterinary Histology.* St Louis, MO: Elsevier Health.

Sandifort, G. (1829). Over de Vorming en Ontwikkeling der Horens van zogende dieren in het algmeen en van die der Hertenbeesten in het bijzonder. *Nieuwe Verhandelingen der Eerste Klasse Koninklijk-Nederlandsche Instituut van Wetenschappen* **2**: 67–106.

Santiago-Moreno, J., Gomez-Brunet, J., Toledano-Diaz, A., Gonzalez-Bulnes, A., Picazo, R. A. & Lopez-Sebastian, A. (2005). Influence of age on the relationship between annual changes in horn growth rate and prolactin secretion in the European mouflon (*Ovis gmelini musimon*). *Animal Reproduction Science* **85**(3): 251–261.

Schibler, L., Cribiu, E. P., Oustry-Vaiman, A., Furet, J.-P. & Vaiman, D. (2000). Fine mapping suggests that the goat polled intersex syndrome and the human Blepharophimosis Ptosis Epicanthus Syndrome map to a 100-kb homologous region. *Genome Research* **10**: 311–318.

Seichter, D., Russ, I., Rothammer, S., Eder, J., Forster, M. & Medugorac, I. (2012). SNP-based association mapping of the polled gene in divergent cattle breeds. *Animal Genetics* **43**(5): 595–598.

Soller, M. & Angel, H. (1964). Polledness and abnormal sex ratios in saanen goats. *Journal of Heredity* **55**: 139–142.

Stankowich, T. & Caro, T. (2009). Evolution of weaponry in female bovids. *Proceedings of the Royal Society B: Biological Sciences* **276**: 4329–4334.

White, W. T. & Ibsen, H. L. (1936). Horn inheritance in Galloway–Holstein cattle crosses. *Journal of Genetics* **32**(1): 33–49.

Chapter 9

American bison *Bison bison* (Linnaeus, 1758)

Glenn E. Plumb, P. J. White and Keith Aune

Names

Genus: *Bison* H. Smith, 1827

Species: American bison *Bison bison* (Linnaeus, 1758)

Subspecies: plains bison *Bison bison bison* (Linnaeus, 1758); wood bison *Bison bison athabascae* (Rhoads, 1898)

Names in other languages: **French**: Bison d'Amérique; **German**: Amerikanischer Bison; **Spanish**: Bisonte Americano; **Italian**: Bisonte Americano; **Lakota**: Tatanka

Other common names: plains bison, buffalo

Ecology, Evolution and Behaviour of Wild Cattle: Implications for Conservation, ed. M. Melletti and J. Burton. Published by Cambridge University Press.

Previous reviews (Shaw & Meagher 2000; Reynolds *et al.* 2003; Gates *et al.* 2010) have commented on the common use of 'buffalo' to describe the American bison (*Bison bison*), as compared to the true African buffalo (*Syncerus caffer*) or Asian water buffalo (*Bubalus arnee*). This interchange freely occurs in documents written by federal, state and local government, tribes and private profit and non-profit sectors (e.g. United States Department of the Interior Bison Conservation Initiative, Inter-Tribal Buffalo Council, National Bison Association, Buffalo Field Campaign), and even in science publications such as Halbert *et al.* (2007). Historically, derivatives such as bisonte, buffes, buffelo and buffles were encountered (Reynolds *et al.* 1982; Dary 1989; Danz 1997) and tribal names may vary considerably, with the Sioux 'tatanka' having received widespread reference and usage through popular media. Notwithstanding the taxonomic basis for bison described below, the common name is now interchanged freely between buffalo and bison without apparent conflict in meaning or understanding towards effective stewardship. It appears that the prevailing free and synonymous interchange has become an entrenched societal preference without negative consequences for conservation, and as such there seems little widespread interest to adopt either as a single common name.

Taxonomy

Bison are mammals from the order Cetartiodactyla and family Bovidae, subfamily Bovinae and tribe Bovini. This ox-like tribe includes four genera: *Bubalus* (Asian water buffalo); *Syncerus* (African buffalo); *Bos* (domestic cattle and their wild relatives) and *Bison* (bison) (Sinclair 1977; Wall *et al.* 1992, Wilson & Reeder 2005). The genus *Bison* includes two extant species, including *Bison bonasus*, the European bison or wisent, and *Bison bison*, the bison of North America.

The Bovidae family has been known since the early Miocene (Pilgrim 1947). The genus *Bison* first appeared on the Asian subcontinent around the Pliocene–Pleistocene border (Geist 1971; McDonald 1981). Most members of the Bovinae subfamily evolved and remain in tropical/subtropical environments, but both bison and Yak radiated to colder climates (Geist 1971; Schaller 1976). McDonald (1981) suggested that the genus bison occupied three major habitats during the late quaternary period: forests and woodlands; savannas and steppes; and open grasslands. Field studies demonstrate that bison occupying different ecotypes around the world clearly vary in behavior and morphology (Geist & Karsten 1977). The taxonomy of bison has been somewhat contentious and there are several historic and ongoing taxonomic debates that are not fully resolved.

Bos versus *Bison*

When Linnaeus first classified the bison in 1758 for his tenth edition of *Systema Naturae*, he assigned the animal to *Bos* (Wilson & Reeder 2005). During the nineteenth century, taxonomists determined that there was adequate anatomical distinctiveness to warrant assigning the bison to its own genus (Shaw & Meagher 2000). Therefore, in 1827, C. Hamilton Smith assigned the generic name *Bison* to the American bison and the European bison (Skinner & Kaisen 1947). In 1849 Knight elevated the subgenus *Bison* to the level of genus (Skinner & Kaisen 1947). Since then, taxonomists have debated the validity of the genus, some arguing that bison are not sufficiently distinct from cattle, gaur, yak and oxen to warrant a distinct genus.

Over the past four decades, there have been suggestions to combine *Bison* and *Bos* into one genus (Stormont *et al.* 1961; Van Gelder 1977; Gentry 1978; Groves 1981; Baccus *et al.* 1983; Miyamoto *et al.* 1989; Modi *et al.* 1996). Studies of DNA, blood types and chromosomal, immunological and protein sequences demonstrate that *Bison* and *Bos* are genetically similar (Stormont *et al.* 1961; Bhambhani & Kuspira 1969; Dayhoff 1972; Wilson *et al.* 1985; Beintema *et al.* 1986; Kleinschmidt & Sgouros 1987). Divergences among mitochondrial DNA sequences of *Bison bison*, *Bos grunniens*, and *Bos taurus* were comparable to those calculated among other sets of congeneric species (Miyamoto *et al.* 1989). Reproductive information also support arguments for a close phylogenetic relationship between *Bos* and *Bison* and some members of *Bos* can hybridize and produce partially fertile female offspring (Van Gelder 1977; Miyamoto *et al.* 1989; Wall *et al.* 1992; Ward 2000). On the other hand, species divergence and reproductive incompatibility are evident by the low fertility of first generation (F1) bison × cattle offspring (Boyd 1908; Steklenev & Yasinetskaya 1982) and the difficulty producing viable male offspring (Boyd 1908; Goodnight 1914; Steklenev & Yasinetskaya 1982; Steklenev *et al.* 1986). Behavioral incompatibility is also evident (Gates *et al.* 2010). Although mating of bison and cattle can readily be achieved in a controlled environment, they preferentially associate and mate with individuals of their own species under open range conditions (Jones 1907; Boyd 1908, 1914; Goodnight 1914). Paleontological evidence indicates that the genera *Bos* and *Bison* shared a common ancestor 1 000 000 to 1 400 000 years ago (Loftus *et al.* 1994).

A change of *Bison* to *Bos* may reflect inferred evolutionary relationships and genetic similarities and provide continuity and stability to the scientific reference for bison, which currently has two species names in use (*B. bonasus* and *B. bison*) (Gates *et al.* 2010). However, changing the genus from *Bison* back to *Bos* would complicate management of European (three subspecies) and American bison (two subspecies) at the subspecies level and disrupt an established history of public policy and scientific community identification with the genus *Bison* (Gates *et al.* 2010). In consideration of the uncertainties explained above, and in keeping with the naming conventions for mammals used for the 1996 Red List and the 2008 Red List (Wilson & Reeder 1993, 2005), the IUCN American Bison Specialist Group adheres to the genus *Bison* with two species, European bison (*B. bonasus*) and American bison (*B. bison*), in its conservation status survey and action plan (Gates *et al.* 2010).

Plains and wood bison subspecies designations

A controversial aspect of American bison taxonomy is the legitimacy of subspecies designations for plains bison (*B. bison bison*) and wood bison (*B. bison athabascae*) (Gates *et al.* 2010). The two subspecies were first distinguished in 1897, when Rhoads formally recognized the wood bison subspecies as *B. bison athabascae* based on morphologic descriptions of the animal (Rhoads 1897). The fossil record, morphological studies and observations of variability among living bison suggest that the species exhibited considerable geographic variation (Wilson 1975; McDonald 1981; van Zyll de Jong 1993). This variation led to claims identifying a northern wood bison and a southern plains bison, which differed in pelage and conformation (Rhoads 1897; Guthrie 1980; van Zyll de Jong 1993). Recent studies of DNA microsatellites indicate that the genetic distances between plains bison and wood bison are greater than those within either of the two subspecies (Wilson & Strobeck 1999; Wilson 2001). However, Douglas *et al.* (2011) found no significant difference in whole mitochondrial DNA sequences between these two subspecies.

Although the two variants visibly differ in morphological characteristics, some scientists have argued that these differences alone do not adequately substantiate subspecies designation (Geist 1991). The issue is further complicated by the human-induced hybridization between plains bison and wood bison in Wood Buffalo National Park during the 1920s (Gates *et al.* 2010). Despite these questions and complications, Canada's National Wood Bison Recovery Team concluded: (1) historically, multiple morphological and genetic characteristics distinguished wood bison from the plains bison; (2) wood bison and plains bison continue to be morphologically and genetically distinct, despite hybridization; and (3) wood bison constitute a subspecies of bison, and therefore should be managed separately from plains bison (Gates *et al.* 2001). Debating subspecies designation in a taxonomic system does not absolve humans of the responsibility to recognize and maintain intraspecific diversity as the raw material of evolution (Gates *et al.* 2010).

Subspecies and distribution

As described above, there are currently two recognized subspecies of American bison. These two subspecies are separated in distribution by a rather fuzzy line stretching across central Canada (van Zyll de Jong 1986; Reynolds *et al.* 2003; Gates *et al.* 2010). The most recent range map for these two species is extracted from Gates *et al.* (2010) and draws on van Zyll de Jong (1986), Sanderson *et al.* (2008) and Stephenson *et al.* (2001). Unfortunately, several plains bison populations were introduced into historic wood bison habitat and wood bison have been reintroduced south of their historic range into plains bison habitat, thereby confusing the current subspecies distributions (Freese *et al.* 2007; Sanderson *et al.* 2008; Gates *et al.* 2010).

The extent of the original range of all American bison was estimated at 9.4 million km^2, encompassing 22 major biomes (Sanderson *et al.* 2008). Wood bison historically ranged to the northwest of plains bison, with 79% of their original range found in Canada and the remainder in Alaska (Reynolds *et al.* 2003). Plains bison historically ranged from northern Mexico to northern Canada and were mostly found west of the Mississippi River. Plains bison occupied habitats in eastern United States, from Florida to the Great Lakes, but were extirpated by the late 1700s as Euro-Americans colonized this area (Shoemaker 1915; Cahalane 1947; Sanderson *et al.* 2008) (see Figures 9.2–9.3). Figure 9.1 shows the historic range (see Plate 1 for the present range).

Descriptive notes

Body measurements: **Shoulder height**: 150–200 cm; **Head and body length**: 210–350 cm; **Tail length**: 30–60 cm; **Body mass**: 600–1000 kg. Dental formula: I 0/3, C 0/1, P 3/3, M 3/3 (×2) = 32 (Nowak 1991; Wilson & Mittermeier 2011)

The American bison is the largest terrestrial mammal in the Western Hemisphere (Shaw & Meagher 2000; Reynolds *et al.* 2003). The plains bison is generally smaller than the wood bison (Reynolds *et al.* 1982). Sexual dimorphism is apparent among adults, yet females generally resemble males in color, body configuration and presence of permanent horns (Reynolds *et al.* 1982), and males are, on average, larger (Nowak 1991; Shaw & Meagher 2000; Reynolds *et al.* 2003). Kelsall *et al.* (1978) found males ~9% heavier than females of equal chest girth, with an apparently linear relationship of weight and chest girth, while Berger & Peacock (1988) found that chest girth accounted for only 64% and 33% of size–weight variation in males and females, respectively. Males reach adult size at about six years, while females reach adult size by about four years (Banfield 1974). The forehead is short and broad, with narrowed muzzle and pointed nasal bones. The male skull length is 49–60 cm and width is ~24–35 cm; female skull length is 42.5–50 cm and width is ~23.5–27.5 cm (Allen 1876; Skinner & Kaisen 1947; Hall & Kelson 1959; Shakelton *et al.* 1975). The muzzle is narrow with long, pointed nasal bones that do not reach the premaxillae; the nasal opening is composed of premaxillae, maxillae and nasals; and the orbits are tubular, composed of frontals, lacrimals and jugals without preorbital vacuities in the skull (Allen 1876). Pelage of the head, neck and shoulders and forelegs is long, shaggy and brownish-black, and the remainder of the body is covered with shorter hairs of brown that lighten through sun bleaching as summer progresses (Nowak 1991; Shaw & Meagher 2000), with albino or gray hair color rare (Meagher 1973). There is a tufted tail, long, coarse guard hairs with thick wooly undercoat, with chin hair that usually resembles a goatee-type beard (Banfield 1974). Geist & Karsten (1977) found wood bison to have shorter, less dense, darker head and neck hair, a shorter and more pointed beard, beard-to-chest mane shorter or absent, and leg 'chaps' long hair absent compared to plains

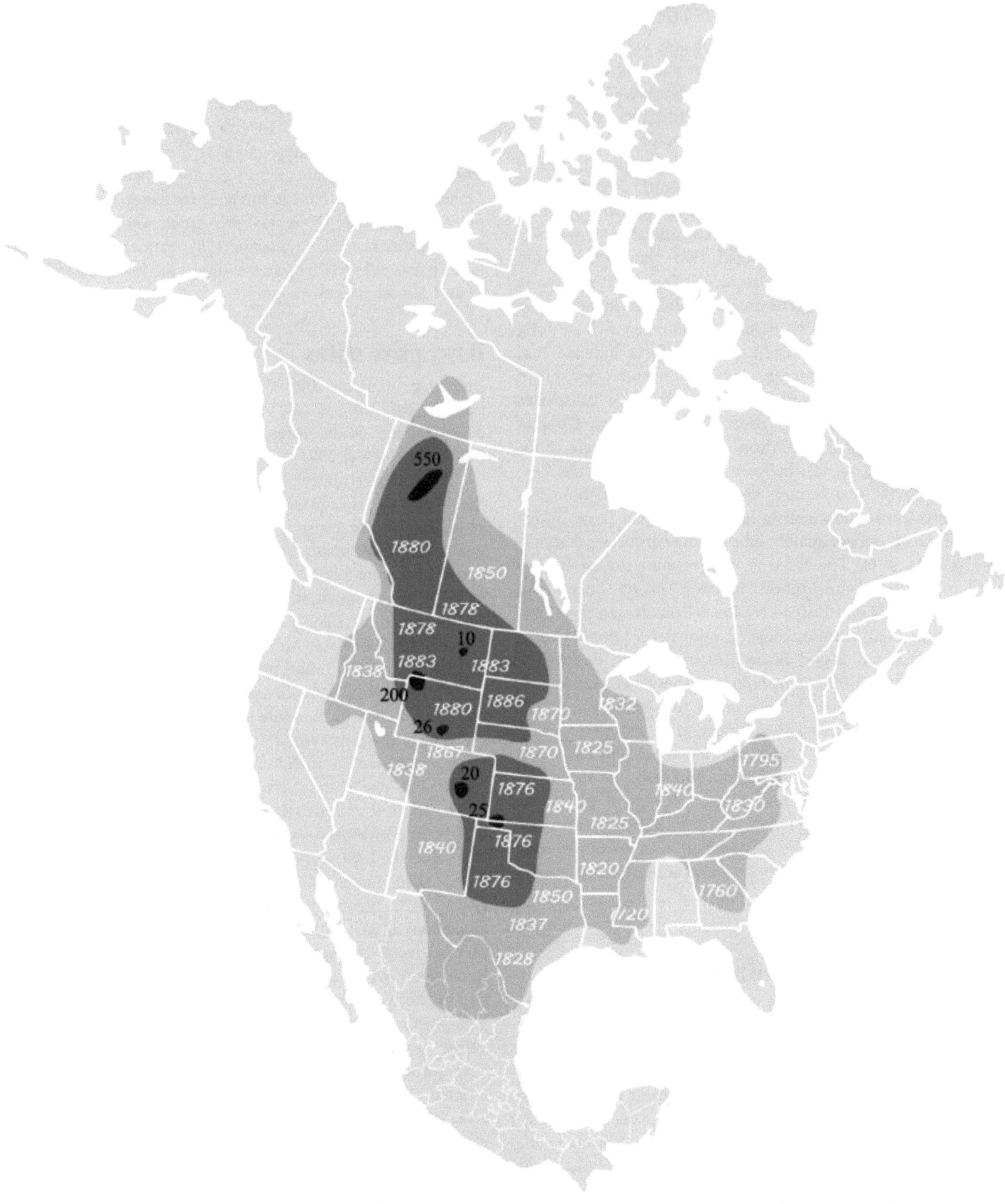

Figure 9.1 Original home range, light brown, and dates of estimated local exterminations of plains bison and wood bison. Dark number indicate number of bison as January 1st 1889. Light numbers give the date of local extermination. Light brown, range as of 1870; dark brown, range as of 1889, with modern state and province borders. Recreated from Hornaday (1889). Courtesy of Wildlife Conservation Society. A black and white version of this figure will appear in some formats. For the colour version, please refer to the plate section.

Figure 9.2 Soldiers of the First United States Cavalry with bison heads confiscated in 1894 inside Yellowstone National Park from a notorious illegal hunter. This incident generated national outrage that helped launch new US federal legislation to protect wildlife and usher in a new era of conservation. Courtesy of the United States National Park Service.

Figure 9.3 Late nineteenth-century mound of plains bison skulls stacked in a rough rectangular pyramid (~30 m side, ~15 m high), yielding a minimum of 15 000 skulls; indicative of the magnitude of slaughter at near industrial pace. Courtesy of the Wildlife Conservation Society Archives.

bison. Young calves are an orange-brown to reddish-brown 'buff' color that gradually darkens to adult coloration by four months (Nowak 1991; Shaw & Meagher 2000). The head is heavy, neck short, and there is a shoulder hump of variable distinction, suggesting forequarters out of proportion to smaller-appearing hindquarters, with short legs and large, rounded hooves and tracks similar to domestic cattle (Reynolds *et al.* 1982). Horns are short and sharp from the side of the head, and laterally upcurve over the head, with female horns more slender and showing a greater tendency to curve inward toward the tips (Reynolds *et al.* 1982; Nowak 1991; Shaw & Meagher 2000; Reynolds *et al.* 2003). Bison produce a variety of sounds, including a male 'bellow' heard most frequently during the breeding season, a 'snort' and 'cough' associated with agnostic behavior. Cows searching for calves will exhibit a series of snorts, and calves can exhibit a

Figure 9.4 Plains bison bull at Yellowstone National Park. Courtesy of Wildlife Conservation Society.

bawling (Fuller 1960). Bison possess hypsodont cheek teeth and no upper incisors, with upper canines reduced or absent, and selenodont molars that have median style or enamel fold between anterior and posterior lobes that disappear with age and wear (Skinner & Kaisen 1947). Selenodont molars section and grind vegetation, with a dental formula as all other bovids: 0/3, 0/1, 3/3, 3/3 = 32 (Vaughn 1972). Christianson *et al.* (2005) found subtle but detectable sex differences in tooth wear patterns at Yellowstone, with Moffit (1998) confirming that bison do produce regulate cementum annuli in the roots of teeth that correspond accurately with age.

Habitat

American bison historically ranged from the arid grasslands of northern Mexico, through the grasslands of the Great Plains, to the floodplain meadows and northern forests of interior Alaska and Canada (Gates & Gogan 2010). Radiocarbon dating of skeletal remains has confirmed that eighteenth- and nineteenth-century bison periodically occupied Rocky Mountain elevations approaching 4000 m (Cannon 2007). Thus, they are adapted to use a variety of habitats through the year. Bison are grazers during all seasons and feed primarily on grasses and sedges in open grassland and meadow communities in all regions (Meagher 1973; Reynolds *et al.* 1978; Larter & Gates 1991).

Significant portions of the original ranges of bison are no longer available for recovery due to agricultural and urban development and political and social concerns about allowing these massive wild animals outside existing reserves, including human safety and property damage, conflicts with private landowners, depredation of agricultural crops, competition with livestock grazing, lack of local public support and lack of funds for state or provincial management (Gates *et al.* 2001; Boyd 2003; Plumb *et al.* 2009). Substantial recovery of wild, free-ranging bison populations in the greater Yellowstone area is constrained by the availability of low-elevation winter habitat where forage is relatively accessible (Plumb *et al.* 2009). Substantial recovery of bison in the Northern Territories of Canada is constrained by the availability of summer forage and the limited extent of graminoid wetlands that provide winter foraging habitat (Strong & Gates 2009). Also, habitat management may be necessary in areas where the natural fire regime has been altered and trees and shrubs have encroached into meadows (Gates *et al.* 2001).

Movements and home range

Bison are highly mobile and can traverse large expanses of land in relatively brief spans of time. For example, Yellowstone bison can travel more than 30 km in a single day (Meagher 1989). Bison in the greater Wood Buffalo area travel up to 50 km annually from their center of activity and range over areas of approximately 180–1440 km^2 (Larter & Gates 1990; Chen & Morley 2005). Factors influencing movement within a home range include seasonal vegetation changes, interspersion and size or quality of foraging sites, the rut, presence of biting insects, snow accumulation, the need to conserve or gain energy and the availability of water (McHugh 1958; Meagher 1973, 1986). The sizes of bison home ranges tend to increase with bison density and in areas of poorer habitat quality or lower forage availability (Gates & Broberg 2011). Also, population substructure has been observed in the greater Yellowstone and Wood Buffalo areas, with different segments of the population using different portions of the landscape due to traditional use patterns (Joly & Messier 2001; Olexa & Gogan 2007).

Most free-ranging bison are seasonally migratory, with documented directional and altitudinal movements of as much

Figure 9.5 A herd of plains bison in mixed grassland forest habitat at ~2400 m elevation at Yellowstone National Park. Courtesy of the United States National Park Service.

as 150 km in mountainous habitat and 240 km in boreal habitat (Soper 1941; McHugh 1958; Meagher 1973, 1986; Geremia *et al.* 2011). For example, most Yellowstone bison spend summer at high elevations in the mountains and move to lower elevations along river corridors and in valleys during winter (Meagher 1973; Geremia *et al.* 2011). Migration allows bison to access greater food supplies through the year, efficiently use available resources and release portions of their range from intensive use for a portion of the year (Gates *et al.* 2005; Plumb *et al.* 2009). Factors influencing the timing and magnitude of seasonal migrations include changes in the quality and quantity of forage, weather conditions and events (e.g. snow pack and storms) that alter access to forage, bison density, learned behaviors and interactions with other ungulates and predators (Soper 1941; Meagher 1973; Bruggeman *et al.* 2009c; Gates & Broberg 2011; Geremia *et al.* 2011).

Travel between summer and winter ranges is usually along well-defined routes in mountainous areas, but more fan-like in boreal forests (Soper 1941; Meagher 1973, 1986). For example, Yellowstone bison tend to travel along rivers and streams during winter, with the spatial travel network largely being defined by streams connecting foraging areas (Bruggeman *et al.* 2009a, 2009b). The probability of bison movements and the spatial distribution of travel corridors are affected by topographic and habitat attributes including slope, landscape roughness, vegetation type and distances to streams, foraging areas and forests (Bruggeman *et al.* 2009b). Temporal trends in travel are driven by factors affecting resource availability, including bison density, landscape and snow pack heterogeneity, and vegetation phenology (Bruggeman *et al.* 2009a). Population substructure may contribute to different segments of the population (e.g. breeding herds) migrating to different seasonal use areas due to learning and traditional use patterns, as has been observed in Yellowstone bison (Meagher 1989; Geremia *et al.* 2011).

Adult male bison may travel considerable distances and colonize new habitat in previously unoccupied areas, especially when bison densities get relatively high (Meager 1989; Gates & Larter 1990; Gates *et al.* 2005). These males then seasonally return to join females during the rut and, subsequently, mixed-sex groups may move into the expanded peripheral range (Gates & Larter 1990; Gates *et al.* 2005). For example, Yellowstone bison gradually expanded their range as density increased above 0.41 bison per square kilometer and bison began to experience nutritional shortages that were exacerbated by winters with deep snow pack (Meagher 1989; Taper *et al.* 2000; Bruggeman *et al.* 2009c; Plumb *et al.* 2009; Geremia *et al.* 2011). In other words, increases in bison abundance were followed by range expansion (Gates & Broberg 2011). In contrast, Mackenzie wood bison in the Northwest Territories of Canada showed pulsed population growth and range expansion. A slowing in the growth rate of the population of Mackenzie bison was detected as density approached approximately 0.55 bison per square kilometer, followed by dispersal to previously unoccupied habitat and a subsequent increase in the population growth rate as density decreased (Gates & Larter 1990; Larter *et al.* 2000). Thus, range expansion may delay responses to food limitation such as diminished survival and fecundity, or allow increased population growth until new areas can no longer be colonized to provide additional forage (Plumb *et al.* 2009; Geremia *et al.* 2011).

Activity patterns

Most ruminants exhibit regular polyphasic activity patterns, alternating between foraging and resting/ruminating activities

Figure 9.6 Bison group size declines amidst peak summer green biomass at Yellowstone National Park. Courtesy of the United States National Park Service.

(Gates & Hudson 1983; Hudson & Watkins 1986; Mysterud 1998). This basic pattern is preempted by walking between bedding and feeding sites, moving to alternate feeding sites, social interactions and predator avoidance (Renecker & Hudson 1993). Wild ruminants typically alter their daily activity pattern in response to intrinsic energy demand and environmental constraints (Trudell & White 1981; Hudson & White 1985; Hudson & Frank 1987).

Specific studies of bison activity generally focus on describing their social structure, behavior and foraging strategies (Soper 1941; McHugh 1958; Meagher 1973; Van Vuren 1984; Plumb & Dodd 1993). Like other large ruminants, bison spend a significant portion of their activity foraging, with smaller amounts of time invested in resting and traveling (Plumb & Dodd; 1993, Bergman *et al.* 2001; Fortin *et al.* 2002; Bruggeman *et al.* 2006). However, bison are social animals and also invest time toward reinforcing social structures, caring for young and reproducing (Fuller 1960; Lott & Minta 1983; Rutberg 1983; Green 1992; Shaw 2012). The literature documents many other social activities, including play (Lumia 1972), aggression (Fuller 1960), defense (McHugh 1972) and grooming (Soper 1941).

Bison activity patterns are typically described in the literature at two, somewhat interrelated, temporal scales – seasonal and daily. These seasonal and daily activity patterns are modified by a wide range of environmental and biological factors such as forage quantity and quality, weather conditions, reproductive status and interspecific interactions like predation, competition, human disturbance and insect harassment (Mysterud 1998). Like other large ruminants, the typical daily activity pattern of American bison is polyphasic, with intense feeding bouts and interspersed rests to ruminate and digest forage (Rutley & Hudson 2001). Most authors report a diurnal pattern, but some levels of nocturnal movement and foraging activity are reported (McHugh 1958; Hein & Preston 1998). Nocturnal grazing is more frequent during periods of warmer, dryer weather; especially in late summer and early fall (Hein & Preston 1998).

Daily activity budgets are modified by specific environmental factors that operate at that temporal scale. Bison practice behavioral thermoregulation and adjust daily activity patterns according to changes in temperature and precipitation (McHugh 1958; Belovsky & Slade 1986; Hein & Preston 1998). Severe snow storms can temporarily limit foraging activity and increase resting time to conserve energy (Rutley & Hudson 2001). In some locales, bison activity patterns are greatly influenced by insect harassment. At high levels of black fly infestation, wood bison decrease their daily foraging time by up to 15% (Melton *et al.* 1989). Bison may respond to human disturbance by fleeing the area, reducing use near the disturbance or changing the timing of daily activities such as feeding, traveling and resting (McHugh 1958; Aune 1981; Bjornlie & Garrott 2001; Hardy 2001, Fortin & Andruskiw 2003). Predation can influence daily activity patterns of bison due to the necessity for vigilance and periodic escape behaviors (Carbyn & Trottier 1987; Laundre *et al.* 2001). These necessary adjustments in daily activity patterns and increased daily movement radii can have energy costs and affect survival (Hudson & White 1985; Picton 1999; Fortin *et al.* 2004).

Bison activity patterns are broadly adjusted to meet seasonal changes in the environment, like forage quality and quantity, water availability and weather patterns, and express seasonal behaviors like mating, calving and migration. Rutley & Hudson (2001) found that during winter bison displayed two main daytime foraging bouts and significant nighttime foraging. During summer, phasic activity is less defined due to an increased number of foraging cycles. Total foraging time declines from summer to winter, while resting time increases. Increased snow depth and density as winter progresses affects the foraging behavior and distribution of bison (Fortin *et al.* 2003; Bruggeman *et al.* 2006). Seasonal patterns of Yellowstone bison traveling during winter are similar across years, but decrease through the middle of winter to reduce energy costs (Bruggeman *et al.* 2009b). Reproductive status influences activity patterns of American bison, especially during rutting and calving seasons (McHugh 1958; Haugen & Shult 1968; Lott 1981). Mate searching and travel distances of males increase during the rut (Berger & Cunningham 1994) and foraging time decreases during the breeding season for both males and females (Melton *et al.* 1989; Plumb & Dodd 1993). Lactating mothers must allocate more time to nursing and less time to resting (Green 1986, 1990). The seasonal distribution and abundance of water can impact bison movements and increase daytime activity around water sources (Van Vuren 1984; Kohl 2012). Long-distance migrations are rare in bison populations today, but in migratory bison herds daily activity patterns change while bison travel between seasonal ranges (Plumb *et al.* 2009; Gates & Broberg 2011).

Feeding ecology

The bison is a ruminant with a four-chambered stomach and complex ruminant physiology that provides the capacity to digest fibrous forage (Plumb & Dodd 1993). Towne *et al.* (1988) showed that bison average greater ruminal ammonia nitrogen concentrations, more total ciliate protozoal counts and higher ciliate cell volume than cattle, while the percentages of protozoans and cellulolytic bacteria in the bison rumen are different from those of cattle (Varel & Dehority 1989). Bison exhibit a Type B ruminal protozoa population, whereas cattle exhibit a Type A or mixed A–B protozoal population (Towne *et al.* 1988). Type B is commonly found in other wild herbivores such as water buffalo (*Bubalus arnee*) and musk oxen (*Ovibus moschatus*), while domestic ruminants usually exhibit a Type A or mixed A–B (Towne *et al.* 1988). It appears that the difference in protozoan populations is partially responsible for higher ciliate cell volume in bison (Towne *et al.* 1988), which in turn increases degradation rates of plant fragments (Clarke *et al.* 1982). Peden *et al.* (1974) hypothesized that bison may be capable of more efficient digestion of low-quality forages than cattle due to urea recycling from body pools. Keith *et al.* (1981) substantiated that bison urea metabolism responds to dietary nitrogen levels. Bison may also avoid shifts in diet to facilitate maintaining a consistent microbial rumen flora (Fortin *et al.* 2002).

Across all habitat types of their contemporary range, bison consume primarily herbaceous graminoids (grasses and sedges), supplemented in order by herbaceous non-graminoid forbs, half-shrubs and occasionally woody vegetation when higher-order foods are limited (Soper 1941; Meagher 1973; Peden *et al.* 1974; Peden 1976; Schwartz & Nagy 1976; Reynolds *et al.* 1978; Telfer & Cairns 1979; Cairns & Telfer 1980; McCullough 1980; Campbell & Hinkes 1983; Waggoner & Hinkes 1986; Shaw & Carter 1990; Larter & Gates 1991; Matthews 1991; Plumb & Dodd 1993; Vinton *et al.* 1993; Singer & Norland 1994; Pfieffer & Hartnett 1995; Steuter *et al.* 1995; Coppedge & Shaw 1998; Coppedge *et al.* 1998a; Post *et al.* 2001; Fortin *et al.* 2002; Olenicki & Irby 2004). When habitats include mixed herbaceous and non-herbaceous vegetation communities, bison exhibit variable levels of local and seasonal forage preference (Shaw & Carter 1990; Plumb & Dodd 1993; Vinton *et al.* 1993; Steuter *et al.* 1995; Hartnett *et al.* 1996). Grasses comprise more than 60% of the diet through the year in grassland and prairie habitats (Gogan *et al.* 2010). Warm season grasses predominate in the diets of bison on shortgrass prairie, while cool season grasses and sedges predominate on mixed prairie (Peden *et al.* 1974; Popp 1980; Meagher 1986). Grasses dominate diets in shrub–steppe habitats, while sedges comprise more than 40% of the diet through the year in mountains and northern forests where sedge–grass meadows are a major habitat component (Meagher 1973, 1986; Reynolds *et al.* 1978; Telfer & Cairns 1979; Singer & Norland 1994; Gogan *et al.* 2010).

Bison respond to differences and changes in the distribution, quality and quantity of forages by moving within habitats to patches of highly nutritious forages and moving between habitats or seasonal ranges (Gogan *et al.* 2010). As a result, bison use a broad variety of habitats to exploit short- and long-term heterogeneity in forages from patch to landscape scales (Gross *et al.* 2010). Like most ungulates of western North America, bison in northern areas vacate their higher elevation summer ranges as winter snow pack accumulates. These bison may substantially change their diets among seasons, preferentially foraging in lowland meadows or wet meadows interspersed among forests with relatively high biomass of sedges and grasses during winter and eating a more diverse diet of grasses, sedges and riparian and shrub foliage during spring, summer and autumn (Reynolds *et al.* 1978; Cairns & Telfer 1980; Larter & Gates 1991; Gates *et al.* 2001; Barmore 2003; Gogan *et al.* 2010). Relatively high densities of bison may deplete forage in high-quality patches during winter, resulting in increased use of poorer quality habitat and a lower diet quality (Gates & Broberg 2011). Plains bison use of forested areas is limited except for shade or shelter from storms, escape from insects and other disturbances, and to traverse between foraging areas or seasonal ranges (Fuller 1962; Meagher 1973, 1986).

Peden *et al.* (1974) suggested that bison foraging was influenced by variations in crude protein and fiber among spatially variable plant groups or species, as well as the proportion of plant groups in respective diets. Forb and browse forages generally have greater net energy content than graminoids. However, since more graminoids can generally be ingested per unit of time than forbs, it may be possible to achieve similar dietary energy intake levels with different diets (Belovsky 1986). To achieve essential nutrient and energy intake, bison should select strongly in favor of high-quality graminoids or increase daily intake so that essential requirements are derived from a greater volume of food. Competition with sympatric herbivores could modify foraging strategy such as with moose (*Alces alces*) in the Northwest Territories, where both moose and bison diets have a high browse component (Fischer & Gates 2005). However, in general, studies of bison foraging have found that plains bison actively select more nutritious forages, and forage in a highly efficient manner that satisfies their nutritional needs and complements diet selection by sympatric herbivores (Coppock *et al.* 1983; Hudson & Frank 1987; Singer & Norland 1994; Wallace *et al.* 1995). Indeed, bison appear to behave as time minimizers rather than energy maximizers; that is, they minimize daily foraging time (Bergman *et al.* 2001; Fortin *et al.* 2002, 2003). These authors speculate that bison may have evolved in an environment where foraging time competes directly with other activities, affecting total fitness. Feeding patterns may also be attributed to dedicating time to avoid insect harassment, scanning for predators, maintaining thermal balance or social interactions (Bergman *et al.* 2001).

At the landscape level, through foraging and associated non-foraging activities, contemporary captive, semi-free or free-ranging bison are capable of serving as key species and ecosystem engineers, both responding to and affecting landscape heterogeneity (Collins & Uno 1983; Polley & Wallace 1986; Knapp *et al.* 1999); edaphic variables such as soil temperature, light availability and soil moisture (Fahnestock & Knapp 1993, 1994); vegetation community composition and quality through consumption of primary productivity (Coppock *et al.* 1983; Coppock & Detling 1986; Frank & McNaughton 1992; Fahnestock & Knapp 1993; Frank & Evans 1997; Detling 1998; Frank 1998); interactions with fire (Edwards 1978; Vinton & Hartnett 1992; Campbell *et al.* 1994; Pfeiffer & Steuter 1994; Pfieffer & Hartnett 1995; Chowns *et al.* 1998; Coppedge *et al.* 1998b; Biondini *et al.* 1999; Crockett & Engle 1999; Johnson & Matchett 2001; Bakker *et al.* 2003; Trager *et al.* 2004; Collins & Smith 2006); interactions with prairie dogs and pocket gophers (Coppock *et al.* 1983; Coppock & Detling 1986; Krueger 1986; Cid *et al.* 1991; Steuter *et al.* 1995; Knowles *et al.* 2002; Virchow & Hygnstrom 2002); direct physical impacts such as rubbing or wallowing (Meagher 1973; Coppedge & Shaw 1997; Coppedge *et al.* 1999; Polley & Collins 1984; Towne 2000); nutrient recycling and seed dispersal (Berthoud 1892; McHugh 1958; Detling *et al.* 1980; Day & Detling 1990a, 1990b; Steinauer & Collins 2001; Gokbulak 2002; Rosas *et al.* 2008); surface hydrology and runoff (Butler 2006); habitat values for sympatric species (Hornaday 1889; Allen 1967; England & DeVos 1969; Lovaas & Bromley 1972; Schwartz & Nagy 1976; Cairns & Telfer 1980; McCullough 1980; Coppock *et al.* 1983; Marlow *et al.* 1984; Wydeven & Dahlgren 1985; Krueger 1986; Waggoner & Hinkes 1986; Whicker & Detling 1988; Fahnestock & Knapp 1993; Singer & Norland 1994; Damhoureyeh & Hartnett 1997; Matlack *et al.* 2001; Miller 2002; Virchow & Hygnstrom 2002; Barmore 2003; Yoakum 2004); trophic relations (Van Camp 1989; Carbyn *et al.* 1993; Larter *et al.* 1994; Fritz *et al.* 1999; Goguen & Mathews 1999; Gerlanc & Kaufman 2003; Joern 2005; Webster 2005; Jonas & Joern 2007); and habitat values for rare and sensitive species (Friedman 1929; Bragg 1940; Corn & Peterson 1996; Knopf 1996; Fahnestock & Detling 2002; Powell 2006; Fuhlendorf *et al.* 2009).

The future of a wild-type bison is dependent on the restoration of wide-ranging herds on large-scale, ecologically intact landscapes where the full suite of natural behaviors can be expressed (Freese *et al.* 2007; Redford *et al.* 2009). Confinement and increasing domestication of American bison may have long-term consequences for behavior and activity patterns (Gates *et al.* 2010). Free-ranging bison may allocate considerably more time to movement, social interaction and migration between seasonal ranges than confined bison. Free-ranging bison that migrate seasonally may allocate as much as 11% of their time to travel (Bruggeman *et al.* 2006). Confined bison typically spend less than 6% of their time on minor activities (including traveling) and allocate large portions of their activity budget to foraging and resting (Rutley & Hudson 2001). Unnatural selection for a constrained suite of behavior traits could result in a maladapted bison not fitted to the natural environment (Lott 1998; Gates *et al.* 2010).

Reproduction and growth

Mature male bison join female groups for the mating season, which occurs from late June through September (McHugh 1958; Haugen 1974; Lott 1981). Many bulls are solitary or in small bachelor groups before the rut, but join mixed herds when females come into estrous (Meagher 1973). Dominance among bulls is established through threat displays and sparring matches that can be quite violent (McHugh 1958; Lott 1981). There is often seasonal synchrony to breeding, with most copulations occurring during a two- to four-week period (Haugen 1974; Lott 1981; Meagher 1986). Breeding may occur later at more northern latitudes (Fuller 1962).

Bison reach sexual maturity between two to four years old and females typically give birth to their first calf at three years of age (McHugh 1958; Fuller 1962; Halloran 1968; Meagher 1973; Gogan *et al.* 2010). Males often do not breed until they are at least five or six years old (McHugh 1958; Lott 1981). Younger bulls are capable of breeding, but are prevented from doing so by older, larger and more experienced males (Halloran 1968). The estrous cycle lasts about three weeks (19–26 days), with females being receptive for 1–2 days and having one or two ovulations during a single breeding season (Haugen 1974; Rutberg 1986a; Kirkpatrick *et al.* 1991).

Female bison generally produce young every 1–3 years depending on their age and physical condition (Kirkpatrick *et al.* 1993). Females continue to breed until approximately 16 years of age (Green 1990). The gestation period lasts about 285 days, or about 9.5 months (Haugen 1974; Rutberg 1986a).

Fetal sex ratios are often biased toward more males (Fuller 1960; Meagher 1973; Haugen 1974; Rutberg 1986a). Bison typically produce a single calf at birth that weighs 15–30 kg (McHugh 1958; Fuller 1960; Halloran 1961; Meagher 1986).

The majority of births take place in spring (April/May/June) and there appears to be some degree of local synchrony in birthing, with 80% of births occurring during a 1–2-month period (Haugen 1974; Rutberg 1984; Jones *et al.* 2010). Birth synchrony is especially noticeable in northern populations of bison where wolf predation on calves is high (Gates & Larter 1990). Birthing may occur earlier in herds or years that experience an earlier onset of spring plant growth (Gogan *et al.* 2005). Also, bison in poor nutritional condition or with debilitating diseases may calve later and/or show low synchrony in birthing (Berger 1992; Green & Rothstein 1993a; Berger & Cain 1999). Females born early in the calving season may, in turn, be more fecund than those born later (Green & Rothstein 1993b).

Females often segregate themselves from the herd or group prior to calving, but sometimes give birth within the herd (McHugh 1958). Birthing normally occurs while the female is lying down, and newborn calves have reddish-tan pelage and can stand and nurse within 30 minutes of birth (Meagher 1986). Typically the mother frees the calf through consumption of placental membranes, and then licks amniotic fluid from the calf's fur and begins suckling up to ten minutes after birth (Lott 2002; McHugh 1958). Calves begin grazing and drinking water within a week of birth, but continue to nurse for seven months or longer. Their pelage begins to darken after 2.5 months and gradually becomes dark chocolate brown (McHugh 1958). Males may be larger than females by four months of age and calves usually weigh between 135 and 180 kg by 8–9 months of age (Meagher 1973, 1986; Gogan *et al.* 2010). Calves are weaned at 7–12 months of age (McHugh 1958; Halloran 1968; Meagher 1986).

Behavior

Seasonally dynamic aggregation is well documented for the American bison, with reliable historic references to immense herds on the Great Plains (Hornaday 1889; Roe 1970; Dary 1989; Isenberg 2000). Contemporary plains and wood bison under semi-free ranging and wild free-ranging conditions continue to exhibit seasonally dynamic aggregation. Considerable variability in seasonal group size is thought to be dynamically associated with forage conditions (Jarman 1974; Lott 1991; Schuler *et al.* 2006), landscape position (Berger & Cunningham 1994), seasonal behavior (Meagher 1973; Melton *et al.* 1989; Komers *et al.* 1993; Berger & Cunningham 1994), environmental conditions (Hess 2002; White *et al.* 2011) and population size (Schuler *et al.* 2006). In contemporary situations it is during the breeding season (rut) that largest aggregations occur when mature bulls, that otherwise exhibit sexual segregation, join the mixed-sex and -age groups (Komers *et al.* 1992). For both plains and wood bison, group size builds during the summer breeding season, and then rapidly diminishes during autumn (Hornaday 1889; Komers *et al.* 1992; Berger & Cunningham 1994; Schuler *et al.* 2006).

Groups of more than 1000 bison have been observed during the rut in contemporary Oklahoma (Schuler *et al.* 2006) and Yellowstone National Park, Wyoming (White *et al.* 2011). Among free-ranging Yellowstone bison, group size declined from winter (~250) to spring (~15) as the area occupied by bison increased (Hess 2002). Lott (1991) indicated that highly variable group fidelity by individuals was primarily an adaptive response to seasonal patchiness and cycles of forage abundance and distribution. In a fenced herd at Badlands National Park, South Dakota, group size varied from a mean of 79 to 157, depending on terrain position (Berger & Cunningham 1994), whereas in an unfenced herd at Yellowstone mean maximum group size increased from 140 during May to more than 250 in September (Hess 2002), regardless of terrain. Bowyer *et al.* (2007) found that, among Delta Junction, Alaska bison, groups with larger males were larger than groups with smaller adult males, and that while most groups were mixed sex, large males only occurred in one-half of groups.

Herd structure and substructure characterization have been observed in plains and wood bison wherein adult females, calves and immature males form unstable mixed-sex and -age groups, and large bulls form separate, smaller groups throughout much of the year (Komers *et al.* 1992; Carbyn *et al.* 1998; Joly & Messier 2001, 2004a; Chen & Morley 2005; Gogan *et al.* 2005; Olexa & Gogan 2007). Seasonal fidelity to traditional landscape use patterns is thought to also contribute to plains and wood bison population substructure across large landscape scales, with stronger cohesion and coordinated movements during summer than in winter (Joly & Messier 2001; Chen & Morley 2005; Olexa & Gogan 2007).

Patterns of hierarchical social dominance are well documented in plains and wood bison. Reinhardt (1985) found that dominance status increased with increasing age, with males dominant over females and aggression more likely against an individual of similar social rank. Lott (1979) found that a linear dominance hierarchy accounted for 78% of observed outcomes of aggressive interactions among mature males during the breeding season, with 88% of interactions won by the aggressive initiator, although neither age nor weight appeared to explain high social standing. Rutberg (1983) found that dominance status was related to age in female bison, and older bulls initiated a higher proportion of dominance displays of aggressive interactions/fights than younger bulls among free-ranging bison during the peak of rut (Komers *et al.* 1994a, 1994b, 1994c). During early to mid rut mature bull aggressiveness increases, and by late rut older bulls decrease aggressiveness while younger bulls take advantage of mating opportunities by adapting to the prevailing social environment. Komers *et al.* (1994b) suggested that willingness to risk injury in fights increases with age and represents increased reproductive effort that is unaffected by social environment.

Figure 9.7 Aggression between adult male plains bison peaks during the summer rut at Yellowstone National Park. Although uncommon, serious injury would subsequently predispose that animal to selective wolf or bear predation. Courtesy of United States National Park Service.

Lott (1979) found that bulls with higher social standing had higher breeding rates.

Rutberg (1986b) observed that aggressive fights between adult females are rare and that female dominance status periodically reverses with no evidence of dominance inhibition of female fitness. The frequency of aggressive dominance displays among females was highest in areas with sparse forage where dominance likely had the most effect on short-term feeding efficiency. Coppedge *et al.* (1997) observed that orphan calves were recipients of more aggression, especially by yearlings, than any age class, and concluded that dominance rank of bison is linear and established by older cohorts over each subsequent year's cohort. Similarly, Rothstein & Griswold (1991) suggested that behavioral interactions between young bison serve as training for later life stages, rather than competition for present resources. Komers *et al.* (1994b) found that agonistic behavior among young bison resulted in no effect on growth rate.

While social dominance is well documented, the flexibility of group cohesion suggests that the social system and ecological circumstances of bison probably produced an outbred species that, combined with a polygynous breeding system, suggests relatively high effective breeding populations (Lott 1991). In a mixed herd on Catalina Island, California, Lott & Minta (1983) observed that individual associations were essentially random, except for mother–calf bonds, and hypothesized that random association confers a higher probability of matching individual physiological needs at any moment with needs of another potential co-group member. Vervaecke *et al.* (2005) examined whether high dominance results in increased bison fitness and found dominance rank correlated with three indices of body condition – weight, speed of hair loss and fatness – with only fatness reaching significance. Although higher dominant females or their daughters did not show higher fecundity, weight at weaning correlated significantly with maternal rank (Vervaecke *et al.* 2005). Variation in birth date was related to individual juvenile and adult quality, with earlier-born individuals socially dominant to later-born peers (Green & Rothstein 1993a). These findings suggest that benefits of early birth persist well into adulthood, with positive fitness consequences, while later parturition imposed immediate costs on mothers, including subsequent reproductive failure (Green & Rothstein 1993b).

Adult male bison move away from mixed-sex and -age groups at the end of the rut and form temporary groups that exhibit a linear dominance hierarchy, with older heavier animals occupying the highest social positions and dominant over younger, smaller males (Berger & Cunningham 1994; Komers *et al.* 1994b; Lott 2002; Roden *et al.* 2005). At Badlands National Park, South Dakota, groups of adult or subadult males rarely exceed ten individuals (Berger & Cunningham 1994). Male wood bison become more solitary with increasing age, are more frequently aggressive and test females for estrous more frequently than do younger bulls (Komers *et al.* 1992). Mooring *et al.* (2005) found that sexually segregated bull groups used areas with greater per capita forage of lower quality compared with females in mixed groups. This finding suggests that segregated bull groups must employ a forage-maximizing strategy compared to female groups that employ an activity tradeoff foraging strategy.

Reproductive success among prime-age males is highly differentiated, with single males accounting for 70–100% of calves (Roden *et al.* 2003). Maher & Byers (1987) found distinct patterns of male reproductive effort: participation in rut beginning at 5–6 years of age; increasing reproductive effort associated with reduced foraging time; aggression and risk of injury at 6–12 years of age; and sharply declining reproductive effort after 12 years of age that more closely resemble younger bulls. Komers *et al.* (1992) concluded that males periodically leave female breeding groups to recover, not for the purpose of searching for new female groups, and that group fluidity suggests that males can encounter new females even when remaining in a female mixed group. Breeding in bison is strongly seasonal and promiscuous, with copulation in a brief one–one relationship, wherein females stimulate bulls to compete for the opportunity to tend, which leads to cooperative copulations (Lott 1981). While Green & Rothstein (1993a) observed asynchronous parturition at Wind Cave National Park, with births over extended periods across years, Berger & Cain (1999) found strong birth synchrony of free-ranging bison at Grand Teton National Park. Berger (1992) found that females use olfactory clues to explore the estrous status of other females prior to their own estrous but not after. Females in good body condition that mated after the seasonal peak shortened their gestation to synchronize with other females, but no synchronizing occurred in females in poor condition.

Komers *et al.* (1994c) observed that higher male copulation rates were associated with higher social standing, and Lott (1981) found that in late rut 20% of cows had copulated more than once. Dominant bulls form so-called 'tending bonds' with individual females just prior to, or during, estrous and typically attempt to keep other bulls away and to keep the female near the edge of a mixed-sex and -age group until she accepts copulation (McHugh 1958; Fuller 1960; Meagher 1973; Berger & Cunningham 1994; Lott 2002). Socially dominant bulls had higher reproductive effort during peak rut and tended ovulating females regardless of reproductive status. Maintenance behaviors decreased when more females were synchronously in estrous (Roden *et al.* 2011), while females in groups with larger males were more likely to copulate than females in groups with smaller adult males, with large males seeking adult females rather than vice versa (Bowyer *et al.* 2007). Lactating females are more likely to ovulate than non-lactating females, but among ovulating females there is no difference in parturition among lactating versus non-lactating females. These findings suggest that reproductive status affects mostly ovulation, not conception (Roden *et al.* 2011).

During the rut, mature males join mixed-sex groups to compete for mating opportunities and temporarily leave these groups to recover from high-cost breeding activities (Komers *et al.* 1992). Mating is highly synchronous during mid-August, with male scent marking associated most often with sexual activity, and group size increasing with increasing distance from forest edge as a likely response to predation risk. Roden *et al.* (2011) observed that dominant bulls spent less time foraging than younger bulls, all males decreased foraging during peak rut, resting decreased with age and that rut phase and number of females synchronously in estrous best explained reproductive effort.

Older females nurse their calves longer each bout than do young females (Green 1986), and there is little evidence of sex-based parental investment or that mothers in better condition bear more sons (Green & Rothstein 1991). Minor trade-offs between growth and reproduction did not reflect substantial costs as measured by long-term reproductive success (Green & Rothstein 1991). Green *et al.* (1993) found that weaning occurs by 7–8 months of age, gradually through maturational changes rather than through parent–offspring conflict. Post-weaning mother–daughter associations continue due to maternal contributions to maintenance or proximity, with daughters benefiting through more time in the center of groups and less displacement when their mother is in the group (Green *et al.* 1989). The longest mother–calf associations are with female offspring, which may remain with their mother through three summers. Male offspring may remain with their mother through two summers (Green *et al.* 1989; Shaw & Carter 1989).

Komers *et al.* (1993) observed higher calving in mixed groups where young were present, suggesting fecund mothers associated with other fecund mothers. Females express higher bite-step rates and feeding selectivity than males, resulting in temporal but not spatial segregation. Thus, female-led nursery groups can also serve to protect calves through dilution of predation (Komers *et al.* 1993). Bison females use quick charges or steady advances to defend a calf against threats (Hornaday 1889; Garretson 1938; McHugh 1958). Specific examples have been documented in which an isolated plains bison female vigorously defended her calf from a grizzly bear (*Ursus arctos*), even though the bear was ultimately successful in killing the calf (Varley & Gunther 2002). Also, a mixed-sex and -age group of adult plains bison responded to an approaching grizzly bear by facing the bear in a compact group, with the calves running behind the adults (Gunther 1991). Adult females and other members of mixed-sex and -age groups may cooperatively protect calves from predators, and Carbyn & Trottier (1987, 1988) described how, during wolf attacks, calves moved to the center of the bison group (Figure 9.8).

Bison exhibit a wide array of individual and group behaviors. Perhaps the best known individual bison behaviors are horning, wherein all age and sex classes (McHugh 1958) rub against an object (typically a shrub or small tree) with their head, horns, neck or shoulders; and wallowing, which involves rolling on dry ground while tearing at the earth with horns and hooves (Coppedge & Shaw 1997). Catlin (in Hornaday 1889) described bison pelage matted with mud by wallowing in areas of shallow water table. Insect avoidance has been suggested as a primary outcome of wallowing and rubbing (Soper 1941; McHugh 1958; Meagher 1973; Edwards 1978; Coppedge & Shaw 1997; Mooring & Samuel 1998; McMillan *et al.* 2000), but all age classes engage in wallowing behavior year-round (Reynolds *et al.* 2003). Also, sexually mature males wallow more frequently during the rut, urinating in the wallow before pawing and rolling (McHugh 1958) – perhaps to advertise their physical condition to other males (Lott 2002) and/or stimulate estrous in females (Bowyer *et al.* 1998).

Figure 9.8 Despite defensive group behavior by plains and wood bison, wolf predation of bison can occur, especially when local terrain conditions favor wolf aggression. In this example from Yellowstone National Park, wolves were unable to gain sufficient advantage, resulting only in harassment of these bison for a short period of time. Courtesy of the United States National Park Service.

Mooring & Samuel (1998) documented higher rates of wallowing and rubbing during October, when tick larvae were blood feeding, but little grooming when nymphal and adult ticks predominated, suggesting that such grooming removes most larval ticks before they can attach, and then the dense hair coat serves as a physical barrier resulting in overall low tick infestation rates compared to moose, elk (*Cervus elaphus*) and deer (*Odocoileus* spp.).

Parasites and diseases

Three major bacterial diseases (brucellosis, tuberculosis and anthrax) are prevalent in some populations of wild bison. When these diseases occur in livestock they may restrict trade or pose a risk to human health, and as such are designated in Canada and the United States as 'reportable' or 'notifiable' under federal and provincial/state legislation (Aune *et al.* 2010). This regulatory framework for livestock disease can have direct consequences for both wild and captive bison. For example, bison in the greater Yellowstone area were infected with the non-native disease brucellosis by domestic cattle some time before 1917 (Meagher & Meyer 1994). Also, brucellosis and tuberculosis were introduced into the greater Wood Buffalo area when infected plains bison were relocated there during 1925–1928. Bison were likely infected with tuberculosis by domestic cattle (Tessaro 1992). In addition, outbreaks of anthrax have affected several populations of bison in the greater Yellowstone area and the Northwest Territories (Dragon & Elkin 2001; Aune *et al.* 2010). Malignant catarrhal fever can occur as acute and chronic cases in bison, most often associated with mixing with a carrier such as domestic sheep (Schultheiss *et al.* 1998, 2001).

Bovine brucellosis is caused by the bacterium *Brucella abortus*, which lives as a facultative intracellular parasite (Thorne 2001). Bacteria can be transmitted between individual bison and also among bison, elk and domestic cattle through the ingestion of live bacteria from infected birth tissues (Cheville *et al.* 1998). Brucellosis causes abortions and still births, and can cause crippling arthritis in infected joints (Williams *et al.* 1997; Thorne 2001; Geremia *et al.* 2009). Only pregnant females become infectious with a high probability of shedding *B. abortus*, and several studies have failed to document sexual transmission from males to females in either domestic cattle or bison (e.g. Robison 1994). Another mode of transmission is through milk as actively infected females nurse their calves, though this mode appears less important than transmission between unrelated bison (Cheville *et al.* 1998; Rhyan *et al.* 2009). While brucellosis does not naturally affect adult survival in bison, it diminishes reproduction through lower pregnancy rates and can reduce population growth rates by approximately 15% (Geremia *et al.* 2009). However, the disease has not prevented rapid population increases (Meagher 1973; Fuller *et al.* 2007a, 2007b).

Bovine tuberculosis is caused by the bacterium *Mycobacterium bovis* and is primarily transmitted by ingestion or inhalation, though it can also pass from mother to offspring through the placental connection or contaminated milk (Tessaro *et al.* 1990). The disease may cause debilitating effects to the respiratory, digestive, urinary, nervous, skeletal and reproductive systems (Aune *et al.* 2010). Tuberculosis reduces reproduction in bison through fetal losses and decreased pregnancy rates due to poorer condition (Choquette *et al.* 1961; Joly & Messier 2005). This disease can also increase the vulnerability of bison (especially older animals) to predation by causing debilitation or poorer condition due to ineffective foraging (Tessaro *et al.* 1990). In combination with brucellosis, tuberculosis may have contributed to substantially decreased bison numbers in Wood Buffalo National Park – though the specific contribution of tuberculosis to this trend remains unclear (Joly & Messier 2005).

Anthrax is caused by the bacterium *Bacillus anthracis* and is transmitted by inhalation or ingestion of endospores that are often found in alkaline, calcium-rich soils (Aune *et al.* 2010). Endospores are non-reproductive forms of the bacterium that can remain viable but dormant for decades before reactivating when environmental conditions become favorable (Dragon & Rennie 1995). Spores are dispersed into the air by activities that break up soil (e.g. wallowing) or carcasses and infect bison that inhale or ingest them. The bacterium then replicates in the bloodstream and releases toxins that cause septicemia (infection) and death (Dragon & Rennie 1995). Behavioral and climatic factors such as summer months, high temperatures, drought, intense mating activity and high bison density have been associated with anthrax outbreaks (Aune *et al.* 2010). Anthrax appears to infect mature male bison more than females or other age classes (Dragon *et al.* 1999). The intermittent occurrence of outbreaks and tendency for male deaths suggests this disease has a minor influence on bison population dynamics unless operating in conjunction with other limiting factors (Aune *et al.* 2010).

The chronic exposure of bison in and around the greater Yellowstone and Wood Buffalo areas to diseases that can be transmitted to livestock and humans is an important factor affecting the original range and potential for further recovery of bison outside existing reserve boundaries (Gates *et al.* 2001; Plumb *et al.* 2009). The management of diseases such as brucellosis and tuberculosis is warranted based on their exotic origins and the threat they pose to livestock and humans (Aune *et al.* 2010). However, these contagious diseases present a challenge to management agencies that are attempting to conserve or restore bison and their ecological processes (e.g. migration, dispersal) while preventing transmission to domestic cattle near reserve boundaries (White *et al.* 2011). The processes for long-term conservation of free-ranging bison operate on large landscapes over long periods of time, while the effectiveness of maintaining livestock health can be observed annually. Managers can quickly react to prevent disease transmission from bison to livestock, but some of the effects to bison associated with these actions may not be detectable for decades and, as a result, unintended consequences may occur. For example, culls to prevent Yellowstone bison chronically exposed to brucellosis from mingling with cattle differentially affected breeding herds and altered gender structure, created reduced female cohorts, and dampened productivity (White *et al.* 2011). Thus, best management practices for preventing disease transmission should be conservative and applied within a risk-management framework to avoid undermining long-term conservation efforts where impacts are more subtle and occur over a longer time period (Nishi 2010).

Additional mortality factors

Survival rates for bison approach 90% in fenced protected areas or those areas that are only lighted hunted in the absence of predators and diseases (Gates *et al.* 2010). The survival rate for calves in the first six months at Wood Buffalo National

Figure 9.9 In the face of extreme weather, plains bison will often reduce movement and thus energy expenditure. In this case, this female waited out a three-day winter snow blizzard on a nature preserve in South Dakota. Courtesy of Glenn Plumb.

Park was 47% (Bradley & Wilmshurst 2005). Bison infected with both brucellosis and bovine tuberculosis at Wood Buffalo National Park experienced lower survival rates than did those infected with only one of the two diseases, or not infected at all (Joly & Messier 2001, 2004a, 2004b, 2005; Bradley & Wilmshurst 2005). In increasing populations well below carrying capacity, survival rates for both sexes averaged 75% for wood bison at the Mackenzie Bison Sanctuary, and 95% for the Jackson plains bison herd (Larter *et al.* 2000; USFWS-NPS 2007). Episodic droughts that reduce late growing season forage quality increase the probability of wildland fires that reduce the amount of winter forage available (Frank & McNaughton 1992). Droughts and wildfire followed by severe winters, alone or in combination, have led to episodic over-winter mortality in the absence of wolf predation (Green *et al.* 1997). Over-winter survival of Yellowstone bison is most strongly influenced by winter severity and bison density (Turner *et al.* 1994; Wallace *et al.* 2004; Geremia *et al.* 2009). Drowning of bison by falling through thin ice in spring and fall has been reported (Roe 1970). Once bison break through lake or river ice, they are generally unable to haul themselves out and become trapped (Carbyn *et al.* 1993), such as when 177 animals drowned in the Mackenzie Bison Sanctuary after breaking through the spring ice of Falaise Lake (Gates *et al.* 1991). Likewise, abnormally high temperatures (+12 °C) in January 2009 affected ice conditions, which likely caused the drowning of up to 13 animals of the Nahanni wood bison population (N.C. Larter, unpublished data cited in Gates *et al.* 2010). Still, bison are known to swim across short and long distances. Larter *et al.* (2003) observed bison swimming across a 1.7-km wide section of Liard River, taking 27 minutes with downstream movement to swim a total of 3.6 km.

Status in the wild

IUCN Red List category: Near Threatened

International and global status

The Convention on International Trade in Endangered Species of Wild Fauna and Flora (CITES) is a multinational agreement among nations to regulate the exploitation of endangered species. Species listed in Appendix I are those threatened with extinction, while species under Appendix II might soon be if trade is not controlled. Wood bison were transferred from CITES Appendix I to Appendix II in 1997 based on Canada's ability to satisfy the 'precautionary measures' of Resolution Conf. 9.24 (Annex 4, paragraphs B.2.b.i and ii). Although bison are in demand for trade, they are managed according to the requirements of Article IV. It was determined that Canada maintains appropriate enforcement controls to prevent the unauthorized taking of wild bison for commercial farming, and that the transfer to Appendix II was consistent with the goals of the government's recovery plan, and would not hamper progress toward the recovery of wood bison in the wild within their historic range. Import and export of wood bison is regulated under permit by CITES authorities within member nations. Plains bison are not listed under CITES (www.cites.org).

Bison were recently listed as 'Near Threatened' in the IUCN/World Conservation Union's Red book (Gates & Aune 2008). A taxon is Near Threatened when it has been evaluated against the criteria but does not qualify for Critically Endangered, Endangered or Vulnerable now, but is close to qualifying for or is likely to qualify for a threatened category in the near future. No distinction is made between wood and plains bison in the IUCN Red Book. NatureServe is a non-profit conservation organization and international network of biological inventories known as natural heritage programs or conservation data centers operating in the United States, Canada, Latin America and the Caribbean. It assigned an overall conservation status rank to American bison of G4 (Apparently Secure), meaning they are globally common (>100 occurrences) and generally widespread, but may be rare in parts of their range, and although they are secure in their global range, there may be a concern for their security in the long term (NatureServe 2006). The wood bison is ranked by NatureServe as G4T2Q, where T refers to it being an intraspecific taxon (trinomial), 2 means imperiled and Q refers to questionable taxonomy. The plains bison is ranked as G4TU, where U means currently unrankable due to a lack of information or substantially conflicting information about status or trends.

Legal status in North America

The wood bison was designated by Canada as 'Endangered' in 1978. Owing to progress made toward recovery it was down-listed to 'Threatened' in 1988. This designation was re-evaluated and affirmed in May 2000. The wood bison is protected under the Canadian Species at Risk Act (COSEWIC 2003), but hunting is allowed in Alberta, the Northwest Territories and the Yukon, subject to conservation strategies and management regulation. In June 1970, the wood bison was listed under the US Endangered Species Act (ESA) as 'Endangered in Canada' to reflect its status in Canada at that time. Canada and the United States are undertaking efforts to harmonize the national listings of this subspecies (Gates *et al.* 2001).

Although plains bison are currently not listed in the United States or Canada under species at risk of extinction legislation, there have been efforts undertaken to consider a listing status (COSEWIC 2004). In 2004 COSEWIC recommended designating plains bison as Threatened under the Species at Risk Act in Canada (Wilson & Zittlau 2004). The proposed change was listed for comment on the public registry in 2005. Criticism ensued from commercial bison producers concerned with the impact on their industry and international trade, and there was a lack of support by Agriculture and Agri-Food Canada and the provincial governments. In July 2006 the Federal Minister of the Environment recommended that the plains bison not be listed because of potential economic implications for the Canadian bison industry (http://canadagazette.gc.ca/partII/2005/20050727/html/si72-e.html).

Under Mexican law, wildlife species belong to the nation and are managed by the federal government. However, Mexico has only recently developed a wildlife conservation and management system that entitles the landowner to be registered into the program (*Unidades de Manejo y Aprovechamiento*) to receive the benefits of harvest and commercial use of the wildlife. This program has doubled the landscape available for wildlife protection in Mexico. In 1995 the federal government established a bureau managed by the Secretary of the Environment. Within this organization is a department for the administration of wildlife conservation programs. In 2007 the conservation of threatened species became the responsibility of the National System of Protected Areas. Conservation and protection of the only free-ranging bison herd in Mexico, the Janos-Hildago bison herd, is an example of rising conservation interest in Mexico. Bison in this specific population are protected by endangered species status under federal law. All other bison in Mexico are privately owned and maintained on fenced private property.

Ten states of the United States, four provinces in Canada and one state in Mexico classify bison as wildlife (Gates *et al.* 2010). All other states and provinces within their historic range designate bison solely as domestic livestock. Plains bison are designated and managed as wildlife in Alaska, Arizona, Utah, Montana, Wyoming, British Columbia, Alberta, Saskatchewan and Chihuahua. Four states consider bison as wildlife but do not have free-ranging populations to manage: Idaho (extreme rarity); Missouri (extirpated); New Mexico (no longer occurring); and Texas (extirpated). Plains bison are listed and managed as wildlife but are considered extirpated in Alberta and Manitoba. Wild bison are preserved, as a public trust resource, managed to protect natural selection processes, and hunted as free-roaming wildlife in Alaska, Arizona, Utah, Montana, Wyoming, British Columbia, Alberta and Saskatchewan. Wood bison are designated and managed as wildlife under provincial statutes in Manitoba, Alberta, British Columbia, Yukon and the Northwest Territories. Wood bison enjoy protected status in all of these provinces. There are legal restrictions on hunting and other activities such as capture and harassment. Subsistence hunting by Aboriginal peoples is allowed under strict regulation in the Northwest Territories and Yukon. Treaty harvests are conducted by some American Indian tribes with recognized hunting rights on unoccupied federal lands in Montana near Yellowstone National Park.

Status in the wild

Although their decline and subsequent recovery is frequently recounted, the ecological recovery of several wild bison populations was never really considered and, consequently, their restoration has never been fully accomplished (Sanderson *et al.* 2008). Most bison in North America are found on farms and ranches (~400 000), while relatively few (~30 000) are located on provincial/state, federal and non-profit conservation reserves (Gates *et al.* 2010). The IUCN recently completed a status review and identified 62 conservation herds which hosted 20 504 plains bison and 10 871 wood bison (Gates *et al.* 2010). Most bison populations are small (75% are <400) and isolated. Few populations are distributed broadly on native prairie or woodland landscapes throughout suitable habitat and most do not enjoy equal legal or policy status when compared to other important wildlife species such as caribou (*Rangifer tarandus*), elk, deer, or pronghorn (*Antilocapra americana*). In addition, many populations are not subject to the forces of natural selection, where ecological processes like predation and competition for food or mates are allowed to operate. Bison currently occupy less than 1% of their historic range (Gates *et al.* 2010). Many scientists are concerned that the American bison is ecologically extinct (Freese *et al.* 2007; Sanderson *et al.* 2008) (Table 9.1 and Figure 9.10).

Another threat to bison conservation and restoration is maintaining the genetic integrity of the species. Four major concerns identified at a workshop on bison conservation genetics (American Bison Society, Tulsa, Oklahoma, May 2011) and by the IUCN American Bison Specialist Group (Gates *et al.* 2010) include: (1) loss of genetic variation; (2) cattle gene introgression; (3) artificial selection; and (4) loss of genetic lineages (including hybridization between wood and plains bison). These four concerns are interrelated such that management decisions designed to address one concern are likely to affect others.

Despite isolation, small population size and a historic bottleneck, McClenaghan *et al.* (1990) found that plains bison have greater genetic variability than several other mammals. Furthermore, Wilson & Strobeck (1999), Halbert (2003) and Halbert & Derr (2008) found levels of DNA microsatellite variability in bison populations to be similar to other North American ungulates. However, some populations are demonstrating serious reductions in diversity and interventions are necessary (Halbert & Derr 2008). Hybridization between plains and wood bison is extensive and has caused irreversible changes to the genetic integrity of wood bison in Canada (Gates *et al.* 2010). The introgression of cattle genes is pervasive among most conservation, tribal and commercial herds (Ward *et al.* 1999; Ward 2000; Halbert 2003; Halbert *et al.* 2005a, 2005b; Halbert & Derr 2007), and strategies have been proposed to reduce the amount of cattle genes in conservation herds provided they do not reduce diversity (Hedrick 2009, 2010). One of the greatest long-term threats to wild bison genetics may be the gradual domestication of bison through husbandry and unnatural selection (Gates *et al.* 2010). White & Wallen (2012) suggest that while a full suite of natural genetics selection processes such as predation, environmental conditions, seasonal migration, etc., may be fully operational only at the few wild, free-ranging populations, conservation herds should promote conservation of wildness and natural selection to retain

Table 9.1 Herd (location) name, jurisdiction, authority and number of bison managed for conservation purposes in North America

State/ province	Herd (location) name	Jurisdiction	Primary management authority	Number		Trend
				Plains bison	Wood bison	
Alaska	Anchorage	Non-government organization	Alaskan Wildlife Conservation Center	82		Increasing
Alaska	Chitina	State	Alaska Department of Fish and Game	38		Stable
Alaska	Copper River	State	Alaska Department of Fish and Game	108		Stable or increasing
Alaska	Delta	State	Alaska Department of Fish and Game	360		Stable
Alaska	Farewell Lake	State	Alaska Department of Fish and Game	400		Increasing
Alaska	Popof Island	State	Alaska Department of Fish and Game	175		Stable
Alberta	Camp Wainwright	Federal-Canada	Department of National Defense; Canadian Forces	16		Stable
Alberta	Elk Island National Park	Federal-Canada	Parks Canada Agency	430	311	Stable
Alberta	Hay-Zama	Provincial	Government of Alberta, Fish and Wildlife Division		750	Decreasing
Alberta/ Northwest Territories	Wood Buffalo National Park	Federal Canada	Parks Canada Agency		5641	Increasing
Alberta/ Saskatchewan	Primrose Lake Air Weapons Range	Federal-Canada	Department of National Defense; SK Environment	100		Stable
Alberta	Waterton Lakes National Park	Federal-Canada	Parks Canada Agency	16		Decreasing
Arizona	House Rock State Wildlife Area	State	Arizona Fish and Game Department	217		Increasing
Arizona	Raymond State Wildlife Area	State	Arizona Fish and Game Department	72		Stable
British Columbia	Pink Mountain	Provincial	British Columbia Ministry of Water; Land and Air Protection	1000		Unknown
British Columbia	Etthithun Lake	Provincial	British Columbia Department of Water, Lands, and Air Protection		124	Increasing
British Columbia	Nordquist	Provincial	British Columbia Department of Water, Lands, and Air Protection		140	Stable to decreasing
California	Camp Pendleton	Military	U.S. Department of Defense; Marine Corp.	150		Decreasing
California	Santa Catalina Island	Non-government organization	Catalina Island Conservancy	170		Decreasing to stable
Colorado	Daniels Park	Municipal	Denver Parks and Recreation	26		Unknown
Colorado	Genesee Park	Municipal	Denver Parks and Recreation	26		Unknown
Colorado	Medano-Zapata Ranch	Non-government organization	The Nature Conservancy	2000		Increasing
Colorado	Rocky Mountain Arsenal	Federal-US	US Fish and Wildlife Service	28		Increasing
Idaho	Three Islands State Park	State	State of Idaho Parks & Recreation	5		Stable

Table 9.1 *(cont.)*

State/ province	Herd (location) name	Jurisdiction	Primary management authority	Number		Trend
				Plains bison	Wood bison	
Illinois	Fermi National Accelerator Laboratory	Federal-US	Department of Energy	18		Stable
Iowa	Broken Kettle Grasslands	Non-government organization	The Nature Conservancy	28		Increasing
Iowa	Neal Smith National Wildlife Refuge	Federal-US	US Fish and Wildlife Service	58		Increasing
Kansas	Konza Prairie Biological Station	State/non-government organization	Kansas State University; The Nature Conservancy	321		Increasing to stable
Kansas	Maxwell Wildlife Refuge	State	Kansas Department of Wildlife and Parks	210		Stable
Kansas	Sandsage Bison Range & Wildlife Area	State	Kansas Department of Wildlife and Parks	72		Decreasing
Kansas	Smoky Valley Ranch	Non-government organization	The Nature Conservancy	120		Stable
Kansas	Tallgrass Prairie National Preserve	Federal-US	US National Park Service	14		Increasing
Kentucky	Land Between the Lakes Natl. Rec. Area	Federal-US	USDA Forest Service	130		Decreasing
Kentucky	Big Bone Lick State Park	State	Kentucky	14		Unknown
Manitoba	Riding Mountain National Park	Federal-Canada	Parks Canada Agency	33		Unknown
Manitoba	Chitek Lake	Provincial	Government of Manitoba, Department of Natural Resources, Waterhen First Nation		225	Increasing
Mexico	Janos-Hidalgo	Federal Mexico, non-government organization	The Nature Conservancy	130		Unknown
Minnesota	Blue Mounds State Park	State	Minnesota Department of Natural Resources; Parks and Recreation	56		Stable
Missouri	Prairie State Park	State	Missouri Department of Natural Resources	76		Stable
Montana	American Prairie Refuge	Non-government organization	American Prairie Foundation	87		Increasing
Montana	National Bison Range	Federal-US	US Fish and Wildlife Service	340		Stable
Nebraska	Fort Niobrara National Wildlife Refuge	Federal-US	US Fish and Wildlife Service	300		Decreasing
Nebraska	Sully's Hill at Fort Niobrara	Federal-US	US Fish and Wildlife Service	48		Stable

Table 9.1 *(cont.)*

State/ province	Herd (location) name	Jurisdiction	Primary management authority	Number		Trend
				Plains bison	Wood bison	
Nebraska	Fort Robinson State Park	State	Nebraska Game and Parks	500		Stable
Nebraska	Niobrara Valley Preserve	Non-government organization	The Nature Conservancy	473		Stable
Nebraska	Wildcat Hills State Recreation Area	State	Nebraska Game and Parks	10		Stable
North Dakota	Cross Ranch Nature Preserve	Non-government organization	The Nature Conservancy	140		Stable
North Dakota	Sullys Hill National Game Preserve	Federal-US	US Fish and Wildlife Service	7		Stable
North Dakota	Theodore Roosevelt National Park	Federal-US	US National Parks Service	750		Stable to increasing
Northwest Territories	Mackenzie Bison Sanctuary	Territorial	Government of Northwest Territories, Resources, Wildlife, and Economic Development		1600	Stable
Northwest Territories	Nahanni	Territorial	Government of Northwest Territories, Resources, Wildlife, and Economic Development		400	Increasing
Northwest Territories	Slave River Lowlands	Territorial	Government of Northwest Territories, Resources, Wildlife, and Economic Development		500	Stable to decreasing
Oklahoma	Tallgrass Prairie Preserve	Non-government organization	The Nature Conservancy	2400		Stable
Oklahoma	Wichita Mountains National Wildlife Refuge	Federal-US	US Fish and Wildlife Service	650		Stable
Saskatchewan	Prince Albert National Park	Federal-Canada	Parks Canada Agency	400		Stable
Saskatchewan	Grasslands National Park	Federal-Canada	Parks Canada Agency	100		Stable
Saskatchewan	Buffalo Pound Provincial Park	Provincial	Saskatchewan Environment; Parks Branch	35		Stable
Saskatchewan	Old-Man-On-His-Back Conservation Area	ENGO	Nature Conservancy of Canada	100		Stable
South Dakota	Badlands National Park	Federal-US	US National Parks Service	750		Stable
South Dakota	Bear Butte State Park	State	South Dakota Game Fish and Parks Dept.	16		Stable
South Dakota	Custer State Park	State	South Dakota Game Fish and Parks Dept.	1100		Stable
South Dakota	Ordway Prairie Preserve	Non- government organization	The Nature Conservancy	255		Stable
South Dakota	Lame Johnny Creek Ranch	Non- government organization	The Nature Conservancy	24		Stable

Table 9.1 *(cont.)*

State/ province	Herd (location) name	Jurisdiction	Primary management authority	Number		Trend
				Plains bison	Wood bison	
South Dakota	Wind Cave National Park	Federal-US	US National Parks Service	400		Stable
Texas	Caprock Canyons State Park	State	Texas Parks and Wildlife Department	40		Stable
Texas	Clymer Meadow Preserve	Non-government organization/private	The Nature Conservancy; private rancher	320		Stable
Utah	Antelope Island State Park	State	Utah Department of Natural Resources; Parks and Recreation	600		Stable
Utah	Henry Mountains	State	Utah Division of Wildlife Resources	600		Stable
Utah	Book Cliffs	State/tribe	State of Utah; Ute Tribe	50		Stable to increasing
Wisconsin	Sandhill Wildlife Area	State	Wisconsin Department of Natural Resources	15		Unknown
Wyoming	Bear River State Park	State	Wyoming State Parks and Historic Sites	8		Unknown
Wyoming	Grand Teton National Park/Nat. Elk Refuge	Federal/State	US National Park Service; US Fish and Wildlife Service	922		Decreasing
Wyoming	Hot Springs State Park	State	Wyoming State Parks and Historic Sites	11		Stable
Wyoming/ Montana/ Idaho	Yellowstone National Park	Federal/State	US National Park Service; State of Montana	3800		Stable to increasing
Yukon Territory	Aishihik	Territorial	Government of Yukon		1100	Increasing
TOTAL				21 950	10 791	

The number of bison is as 2010, and vary periodically according to management programs within each herd.
Data adapted from Gates *et al.* 2010.

adaptive capabilities, rather than settling for genetic or population substructures that were likely created or exacerbated by human actions. The incremental loss of natural behaviors and limited natural selection through processes like predation and mate competition will significantly alter the evolutionary path of American bison.

Other threats to bison restoration and conservation include declining suitable habitat; competition with cattle and other livestock; lack of public tolerance for free-ranging bison; reportable diseases that transmit to livestock; and current patterns of land use. Important social challenges to conservation of American bison are gaining public recognition for bison as wildlife and building the cooperation for shared stewardship of bison ranging across multiple jurisdictions. Decades of managing bison as a special wildlife case requiring agriculture-based management practices has hindered the ecological recovery of this important keystone species (Redford *et al.* 2009)

Bison managed by Native American tribes

Many North American Native Peoples have strong cultural, spiritual and symbolic relationships with bison (Notzke 1994; Zontek 2007). Some tribes believe that because the animals once sustained their Indian way of life, they in turn must help the bison to sustain their place on the earth. The conservation of wild bison includes intangible values these tribes hold for bison. Values vary greatly between tribes and, in some cases, between members of the same tribe. Some tribal people believe that the status of the bison is an indicator of the treatment of North American Indians. Interest in preserving the cultural significance of bison and in restoring cultural connections to the species is important for native governments and communities who participate in bison conservation (Notzke 1994; Zontek 2007).

The InterTribal Buffalo Council (ITBC) was formed in 1990 with the mission to restore bison to Indian Nations in a

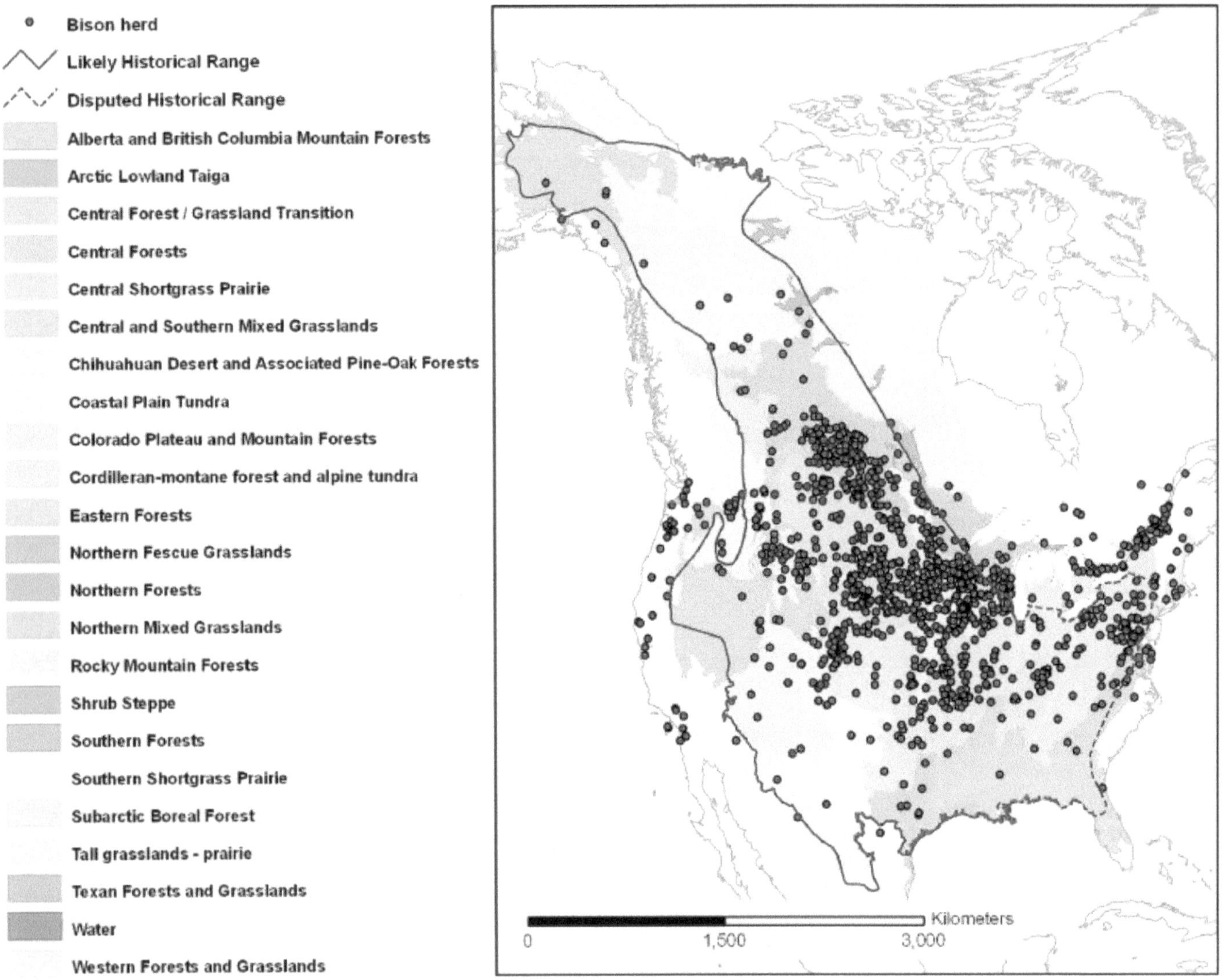

Figure 9.10 Distribution of a sample of existing American bison herds across the historical range (and beyond) in North America. The size of the dots overestimates the actual area occupied. Best estimates are that bison currently occupy < 1% of their *c.* 1500 historical range (modified from Hall & Kelson 1959), shown by the dark line. Major habitat types are indicated by grey areas in the background. Reprinted with permission from Sanderson *et al.* (2008).

manner that is compatible with their spiritual and cultural beliefs and practices (ITBC website: www.intertribalbison.org). In cooperation with the Native American Fish and Wildlife Society, the ITBC was able to secure US congressional support for bison restoration in 1991. In 1992, tribal representatives met and the ITBC became an officially recognized tribal organization in the United States. The ITBC was approved in 2010 as a federally chartered Indian organization by the Bureau of Indian Affairs under Section 17 of the Indian Reorganization Act and is governed by a Board of Directors comprising a tribal representative from each member tribe. Currently there are about 57 member tribes in 19 states that collectively manage more than 15 000 bison. The role of the ITBC is to act as a facilitator for education and training, developing market strategies, coordinating transfer of bison from federal ownership to tribal lands and providing technical assistance to tribal members to encourage sound management. The ITBC does not have a presence in Canada, nor is there an equivalent organization there.

Some tribal bison managers consider bison as wild animals regardless of the source of stock, genetic introgression from cattle or domestication history. This can be the basis for conflict with conservation biologists who apply biological criteria for evaluating conservation merit of a herd. Numerous Native tribes own or influence the management of a significant land base that could sustain large bison herds. However, there has not been a systematic survey of the number of herds or distribution of bison under Native management. The source of bison managed in Native American herds includes several federal herds and some commercial herds. There has been little genetic testing of these bison, but some degree of cattle gene introgression is inferred.

Privately owned bison for commerce

Over 96% of the bison in North America are privately owned and managed for commercial production (Boyd & Gates 2006). Bison can be kept as domestic livestock in all of the United States and portions of Canada. These bison are privately owned and typically managed for meat production or breeding. In Alberta, Saskatchewan and Manitoba individuals in the private sector may own bison where they are regulated as livestock. In British Columbia, bison may be produced commercially, but a game-farming license is required. Commercial herds owned by individuals, corporations or non-government organizations are managed independently, subject to market forces, and regulations governing animal health and trade. In the Yukon and Northwest Territories, existing policy prevents the establishment of plains bison ranches or their introduction into the wild. There is no unified conservation effort or regulatory framework that encourages or facilitates conservation of commercial bison as wildlife at national, or state and provincial levels.

In 2005 it was estimated that there were over 500 000 bison on farms and ranches in North America. The US Department of Agriculture's most recent Census of Agriculture, conducted in 2007, estimated there were approximately 200 000 on private lands. In Canada it is estimated that there are almost 2000 bison producers who own over 250 000 bison. Because of consumer demand and the initiatives of farmers and ranchers, the commercial captive population of this almost extinct animal continues to grow. In 2010 the National Bison Association and Canadian Bison Association issued a statement noting that even though sales of bison meat have doubled since 2005, the sector is still a small fraction of the meat marketplace. According to that statement, the 92 000 head of bison processed in the United States and Canada last year represented less than - one-day's processing of beef in the United States. Market prices in late 2011 clearly illustrate the burgeoning growth of the buffalo business. The average price paid by marketers for a young bull carcass in late 2011 was $3.98 per pound, which was 65% higher than the price paid only three years earlier.

The genomic condition of commercial bison in North America varies widely and is not coordinated under the current conservation strategy for American bison (Gates *et al.* 2010). Because production bison are privately owned, there is no means to assure the genetic integrity of this population. There are known subspecies hybrids, and cattle gene introgression has been reported within the commercial production herds of the United States and Canada (Newton 1993; Ward 2000; http://americanbeefalo.org). However, some bison producers are developing herds with reputable genetics and the National Bison Association recently drafted guidelines to improve conservation genetics within the industry. American bison are commercially raised for meat production on other continents around the world, including New Zealand, Australia, Europe and Asia. In Europe this creates a conservation challenge for efforts to recover the European bison (Pucek *et al.* 2004).

Status in captivity

American bison held at zoological institutions

There is a long and rich history of American bison in zoos, dating back to 1887 when famed naturalist William Hornaday was instrumental in obtaining a few animals for what later became the Smithsonian's National Zoo. Hornaday went on to become the first director of the Bronx Zoo and bison were first displayed there in 1899. Bison have continued to be important collection elements at both zoos. Today, bison are displayed in more than 80 zoos in the United States (Redford & Fearn 2007). Reports of 2007 and 2012 by the International Species Information System (ISIS) indicate there are 974 and 678 bison in North American zoos, respectively (see also Chapter 27). There are eight zoos in Canada hosting about 75 bison (CAZA Institutional Members: www.caza.ca). Most of the bison in Canadian zoological institutions are wood bison rather than plains bison.

Management of bison at zoological institutions is coordinated through a Taxon Advisory Group. The Bison, Buffalo, and Cattle Taxon Advisory Group of the America Zoo Association was formed in 1992 to address captive management issues for wild cattle species and to support both *in situ* and *ex situ* conservation efforts. Thirteen species of wild cattle are covered under this Taxon Advisory Group, whose goals are to:

- support programs for wild cattle;
- maintain genetic and demographic reservoirs in the captive populations that may eventually provide animals or gametes to bolster wild populations;
- increase knowledge of natural history, husbandry and behavior of wild cattle;
- serve as a resource for information on the status, management and conservation of wild cattle species in nature and in zoological institutions;
- promote research that will enhance the well-being of captive wild cattle;
- foster links and provide logistical and/or financial support for insitu efforts;
- coordinate Taxon Advisory Group efforts with other applicable conservation groups; and
- promote public education programs that highlight wild cattle species and habitat conservation. More than 143 million people visit zoos associated with the America Zoo Association annually.

Both plains and wood bison from North America are hosted at several wildlife reserves in Europe, New Zealand and Australia (World Association of Zoos and Aquariums: www.waza.org). The introduction and hosting of American bison on reserves in Europe is considered a potential threat to European bison recovery because these bison can readily hybridize (Pucek *et al.* 2004).

References

Allen, D. L. (1967). *The Life of Prairies and Plains*. New York: McGraw Hill Book Company.

Allen, J. A. (1876). *The American Bisons: Living and Extinct*. Cambridge, MA: University Press.

Aune, K. (1981). Impacts of winter recreationists on wildlife in a portion of Yellowstone National Park, Wyoming. Masters thesis. Montana State University.

Aune, K., Gates, C. C., Elkin, B. T., Hugh-Jones, M., Joly, D. O. & Nishi, J. (2010). Reportable or notifiable diseases. In *American Bison: Status Survey and Conservation Guidelines 2010*, pp. 27–37. Eds. Gates, C. C., Freese, C. H., Gogan, P. J. P. & Kotzman, M. Gland: International Union for the Conservation of Nature and Natural Resources.

Baccus, R., Ryman, N., Smith, M. H., Reuterwall, C. & Cameron, D. (1983). Genetic variability and differentiation of large grazing mammals. *Journal of Mammalogy*, **64**: 109–120.

Bakker, C., Blair, J. M. & Knapp, A. K. (2003). Does resource availability, resource heterogeneity or species turnover mediate changes in plant species richness in grazed grasslands? *Oecologia* **137**: 385–391.

Banfield, A. W. F. (1974). *The Mammals of Canada*. Toronto: University of Toronto Press.

Barmore, W. J. (2003). *Ecology of Ungulates and their Winter Range in Northern Yellowstone National Park: Research and Synthesis 1962–1970*. Mammoth Hot Springs, WY: National Park Service.

Beintema, J. J., Fitch, W. M. & Carsana, A. (1986). Molecular evolution of pancreatic-type ribonucleases. *Molecular Biology and Evolution*, **3**: 262–275.

Belovsky, G. E. (1986). Optimal foraging and community structure: implications for a guild of generalist grassland herbivores. *Oecologia* **70**: 35–52.

Belovsky, G. E. & Slade, J. B. (1986). Time budgets of grassland herbivores: body size similarities *Oecologia* **70**: 53–62.

Berger, J. (1992). Facilitation of reproductive synchrony by gestation adjustment in gregarious mammals: a new hypothesis. *Ecology* **73**: 323–329.

Berger, J. & Cain, S. L. (1999). Reproductive synchrony in brucellosis-exposed bison in the southern Greater Yellowstone Ecosystem and in noninfected populations. *Conservation Biology* **13**: 357–366.

Berger, J. & Cunningham, C. (1994). *Bison: Mating and Conservation in Small Populations: Methods and Cases in Conservation Science*. New York: Columbia University Press.

Berger, J. & Peacock, M. (1988). Variability in size-weight relationships of *Bison bison*. *Journal of Mammalogy* **69**: 618–624.

Bergman, C. M., Fryxell, J. M., Gates, C. C. & Fortin, D. (2001). Ungulate foraging strategies: energy maximizing or time minimizing. *Journal of Animal Ecology* **70**: 289–300.

Berthoud, E. L. (1892). A peculiar case of plant dissemination. *Botanical Gazette* **17**: 321–326.

Bhambhani, R. & Kuspira, J. (1969). The somatic karyotypes of American bison and domestic cattle. *Canadian Journal of Genetics and Cytology* **11**: 243–249.

Biondini, M. E., Steuter, A. A. & Hamilton, R. G. (1999). Bison use of fire-managed remnant prairies. *Journal of Range Management* **52**: 454–461.

Bjornlie, D. D. & Garrott, R. A. (2001). Effects of winter road grooming on bison in Yellowstone National Park. *Journal of Wildlife Management* **65**: 560–572.

Bowyer, R. T., Manteca, X. & Hoymork, A. (1998). Scent marking in American bison: morphological and spatial characteristics of wallows and rubbed trees. In *International Symposium of Bison Ecology and Management in North America*, pp. 81–91. Eds. Irby, L. & Knight, J. Bozeman, MT: Montana State University.

Bowyer, R. T., Bleich, V. C., Manteca, X., Whiting, J. C. & Stewart, K. M. (2007). Sociality, mate choice, and timing of mating in American bison (*Bison bison*): effects of large males. *Ethology* **113**: 1048–1060.

Boyd, D. P. (2003). Conservation of North American bison: status and recommendations. Master's thesis. University of Calgary.

Boyd, D. P. & Gates, C. C. (2006). A brief review of the status of plains bison in North America. *Journal of the West* **45**: 15–21.

Boyd, M. M. (1908). A short account of an experiment in crossing the American bison with domestic cattle. *Annual Report of the American Breeders' Association* **4**: 324–331.

Boyd, M. M. (1914). Crossing bison and cattle. *Journal of Heredity* **5**: 189–197.

Bradley, M. & Wilmshurst, J. (2005). The fall and rise of bison populations in Wood Buffalo National Park: 1971 to 2003. *Canadian Journal of Zoology* **83**: 1195–1205.

Bragg, A. N. (1940). Observations on the ecology and natural history of Anura: I. Habits, habitat, and breeding of *Bufo cognatus* Say. *American Naturalist* **74**: 322–349.

Bruggeman J. E., Garrott, R. A., Bjornlie, D. D., White, P. J., Watson, F. G. R. & Borkowski, J. (2006). Temporal variability in winter travel patterns of Yellowstone bison: the effects of road grooming. *Ecological Applications* **16**: 1539–1554.

Bruggeman, J. E., Garrott, R. A., White, P. J., Bjornlie, D. D., Watson, F. G. R. & Borkowski, J. J. (2009a). Bison winter road travel: facilitated by road grooming or a manifestation of natural trends. In *The Ecology of Large Mammals in Central Yellowstone: Sixteen Years of Integrated Field Studies*, pp. 603–621. Eds. Garrott, R. A., White, P. J. & Watson, F. G. R. San Diego, CA: Elsevier.

Bruggeman, J. E., Garrott, R. A., White, P. J., Watson, F. G. R. & Wallen, R. W. (2009b). Effects of snow and landscape attributes on bison winter travel patterns and habitat use. In *The Ecology of Large Mammals in Central Yellowstone: Sixteen Years of Integrated Field Studies*, pp. 623–647. Eds. Garrott, R. A., White, P. J. & Watson, F. G. R. San Diego, CA: Elsevier.

Bruggeman, J. E., White, P. J., Garrott, R. A. & Watson, F. G. R. (2009c). Partial migration in central Yellowstone bison. In *The Ecology of Large Mammals in Central Yellowstone: Sixteen Years of Integrated Field Studies*, pp. 217–235. Eds. Garrott, R. A., White, P. J. & Watson, F. G. R. San Diego, CA: Elsevier.

Butler, D. R. (2006). Human-induced changes in animal populations and distributions and the subsequent effects on fluvial systems. *Geomorphology* **79**: 448–459.

Cahalane, V. H. (1947). *Mammals of North America*. New York: Macmillan.

Cairns, A. L. & Telfer, E. S. (1980). Habitat use by four sympatric ungulates in boreal mixedwood forest. *Journal of Wildlife Management* **44**: 849–857.

Campbell, B. H. & Hinkes, M. (1983). Winter diets and habitat use of Alaska bison after wildfire. *Wildlife Society Bulletin* **11**: 16–21.

Campbell, C., Campbell, I. D., Blyth, C. B. & McAndrews, J. H. (1994). Bison extirpation may have caused aspen expansion in western Canada. *Ecography* **17**: 360–362.

Cannon, K. P. (2007). They went as high as they choose: what an isolated skull can tell us about the biogeography of high-altitude bison. *Arctic, Antarctic, and Alpine Research* **39**: 44–56.

Carbyn, L. N. & Trottier, T. (1987). Responses of bison on their calving grounds to predation by wolves in Wood Buffalo National Park. *Canadian Journal of Zoology* **65**: 2072–2078.

Carbyn, L. N. & Trottier, T. (1988). Description of wolf attacks on bison calves in Wood Buffalo National Park. *Arctic* **41**: 297–302.

Carbyn, L. N., Oosenbrug, S. M. & Anions, D. W. (1993). *Wolves, Bison and the Dynamics Related to the Peace-Athabasca Delta in Canada's Wood Buffalo National Park*. Edmonton: University of Alberta.

Carbyn, L. N., Lunn, N. & Timoney, K. (1998). Trends in the distribution and abundance of bison in Wood Buffalo National Park. *Wildlife Society Bulletin* **26**: 463–470.

Chen, S. & Morley, R. S. (2005). Observed herd size and animal association. *Ecological Modeling* **189**: 425–435.

Cheville, N. F., McCullough, D. R. & Paulson, L. R. (1998). *Brucellosis in the Greater Yellowstone Area*. Washington, DC: National Academy Press.

Choquette, L. P. E., Gallivan, J. F., Byrne, J. L. & Pilipavisus, J. (1961). Parasites and diseases of bison in Canada: I. Tuberculosis and some other pathological conditions in bison at Wood Buffalo and Elk Island National Parks in the fall and winter of 1959–60. *Canadian Veterinary Journal* **2**: 168–174.

Chowns, T., Gates, C. C. & Lepine, F. (1998). Large scale prescribed burning to improve wood bison habitat in northern Canada. In *Proceedings of the International Symposium on Bison Ecology and Management in North America, June 4–7, 1997*, pp. 205–210. Eds. Irby, L. & Knight, J. Bozeman, MT: Montana State University.

Christianson, D. A., Gogan, P. J. P., Podruzny, K.M & Olexa, E. M. (2005). Incisor wear and age in Yellowstone bison. *Wildlife Society Bulletin* **33**: 669–676.

Cid, M. S., Detling, J. K., Whicker, A. D. & Brizuela, M. A. (1991). Vegetational responses of a mixed-grass prairie site following exclusion of prairie dogs and bison. *Journal of Range Management* **44**: 100–105.

Clarke, R. T. J., Ulyatt, M. J. & John, A. (1982). Variation in numbers and mass of ciliate protozoa in the rumens of sheep fed chaffed alfalfa (*Medicago sativa*). *Applied and Environmental Microbiology* **43**: 1201–1204.

Collins, S. L. & Smith, M. D. (2006). Scale-dependent interaction of fire and grazing on community heterogeneity in tallgrass prairie. *Ecology* **87**: 2058–2067.

Collins, S. L. and Uno, G. E. (1983). The effect of early spring burning on vegetation in buffalo wallows. *Bulletin of the Torrey Botanical Club* **110**: 474–481.

Coppedge, B. R. & Shaw, J. H. (1997). Effects of horning and rubbing behavior by bison (*Bison bison*) on woody vegetation in a tallgrass prairie landscape. *American Midland Naturalist* **138**: 189–196.

Coppedge, B. R. & Shaw, J. H. (1998). Bison grazing patterns on seasonally burned tallgrass prairie. *Journal of Range Management* **51**: 258–264.

Coppedge, B. R., Carter, T. S., Shaw, J. H. & Hamilton, R. G. (1997). Agnostic behavior associated with orphan bison (*Bison bison* L.) calves released into a mixed resident population. *Applied Animal Behavior Science* **55**: 1–10.

Coppedge, B. R., Leslie, D. M. & Shaw, J. H. (1998a). Botanical composition of bison diets on tallgrass prairie in Oklahoma. *Journal of Range Management* **51**: 379–382.

Coppedge, B. R., Engle, D. M., Toepfer, C. S. & Shaw, J. H. (1998b). Effects of seasonal fire, bison grazing and climatic variation on tallgrass prairie vegetation. *Plant Ecology* **139**: 235–246.

Coppedge, B. R., Fuhlendorf, S. D., Engle, D. M., Carter, B. J. & Shaw, J. H. (1999). Grassland soil depressions: relict bison wallows or inherent landscape heterogeneity? *American Midland Naturalist* **142**: 382–392.

Coppock, D. L. & Detling, J. K. (1986). Alteration of bison and black-tailed prairie dog grazing interaction by prescribed burning. *Journal of Wildlife Management* **50**: 452–455.

Coppock, D. L., Ellis, J. E., Detling, J. K. & Dyer, M. I. (1983). Plant–herbivore interactions in a North American mixed-grass prairie: II. Responses of bison to modification of vegetation by prairie dogs. *Oecologia* **56**: 10–15.

Corn, S. P. & Peterson, C. R. (1996). Prairie legacies: amphibians and reptiles. *PrairieConservation*, pp. 125–134. Eds. Samson, F. B. & Knopf, F. L. Washington, DC: Island Press.

COSEWIC (2003). *Canadian Species at Risk*. Ottawa: Committee on the Status of Endangered Wildlife in Canada.

COSEWIC (2004). *Canadian Species at Risk*. Ottawa: Committee on the Status of Endangered Wildlife in Canada.

Crockett, J. S. & Engle, D. M. (1999). Combustion characteristics of bison (*Bison bison*) fecal pats ignited by grassland fires. *American Midland Naturalist* **141**: 12–18.

Damhoureyeh, S. A. & Hartnett, D. C. (1997). Effects of bison and cattle on growth, reproduction, and abundance of five tallgrass prairie forbs. *American Journal of Botany* **84**: 1719–1728.

Danz, H. P. (1997). *Of Bison and Man*. Niwot, CO: University Press of Colorado.

Dary, D. A. (1989). *The Buffalo Book: The Full Saga of the American Animal*. Chicago, IL: Swallow Press.

Day, T. A. & Detling, J. K. (1990a). Grassland patch dynamics and herbivore grazing preference following urine deposition. *Ecology* **71**: 180–188.

Day, T. A. & Detling, J. K. (1990b). Changes in grass leaf water relations following bison urine deposition. *American Midland Naturalist* **123**: 171–178.

Dayhoff, M. O. (1972). *Atlas of Protein Sequence and Structure*, Vol. **5**. Silver Spring, MD: National Biomedical Research Foundation.

Detling, J. K. (1998). Mammalian herbivores: ecosystem-level effects in two grassland national parks. *Wildlife Society Bulletin* **26**: 438–448.

Detling, J. K., Dyer, M. I., Procter-Gregg, C. & Winn, D. T. (1980). Plant–herbivore interactions: examination of potential effects of bison saliva on regrowth of

Bouteloua gracilis (H. B. K.) Lag. *Oecologia* **45**: 26–31.

Douglas, K. C., Halbert, N. D., Kolenda, C., Childers, C., Hunter, D. L. & Derr, J. N. (2011). Complete mitochondrial DNA sequence analysis of *Bison bison* and bison-cattle hybrids: function and phylogeny. *Mitochondrion* **11**: 166–175.

Dragon, D. C. & Elkin, B.T. (2001). An overview of early anthrax outbreaks in northern Canada: field reports of the Health of Animals Branch, Agriculture Canada, 1962–1971. *Arctic* **54**: 32–40.

Dragon, D. C. & Rennie, R. P. (1995). The ecology of anthrax spores: tough but not invincible. *Canadian Veterinary Journal* **36**: 295–301.

Dragon, D. C., Elkin, B. T., Nishi, J. S. & Ellsworth, T. R. (1999). A review of anthrax in Canada and implications for research on the disease in northern bison. *Journal of Applied Microbiology* **87**: 208–213.

Edwards, T. (1978). Buffalo and prairie ecology, In *Fifth Midwest Prairie Conference Proceedings*, pp. 110–112. Eds. Glenn-Lewin, D. C. & Landers, R. Q., Jr. Ames, IA: Iowa State University.

England, R. E. & DeVos, A. (1969). Influence of animals on pristine conditions of the Canadian grasslands. *Journal of Range Management* **22**: 87–94.

Fahnestock, J. T. & Detling, J. K. (2002). Bison–prairie dog–plant interactions in a North American mixed-grass prairie. *Oecologia* **132**: 86–95.

Fahnestock, J. T. & Knapp, A. K. (1993). Water relations and growth of tallgrass prairie forbs in response to selective grass herbivory by bison. *International Journal of Plant Science* **154**: 432–440.

Fahnestock, J. T. and Knapp, A. K. (1994). Plant responses to selective grazing by bison: interactions between light, herbivory and water stress. *Vegetatio* **115**: 123–131.

Fischer, L. & Gates, C. C. (2005). Competition potential between sympatric woodland caribou and wood bison. *Canadian Journal of Zoology* **83**: 1162–1173.

Fortin, D. & Andruskiw, M. (2003). Behavioral response of free-ranging bison to human disturbance. *Wildlife Society Bulletin* **31**: 804–813.

Fortin, D., Fryxell, J. M. & Pilote, R. (2002). The temporal scale of foraging decisions in bison. *Ecology* **83**: 970–982.

Fortin, D., Fryxell, J. M., O'Brodovich, L. & Frandsen, D. (2003). Foraging ecology of bison at the landscape and plant community levels: the applicability of energy maximization principles. *Oecologia* **134**: 219–227.

Fortin, D., Boyce, M. S. & Merrill, E. H. (2004). Multi-tasking by mammalian herbivores: overlapping processes during foraging. *Ecology* **85**: 2312–2322.

Frank, D. A. (1998). Ungulate regulation of ecosystem processes in Yellowstone National Park: direct and feedback effects. *Wildlife Society Bulletin* **26**: 410–418.

Frank, D. A. & Evans, R. D. (1997). Effects of native grazers on grassland N cycling in Yellowstone National Park. *Ecology* **78**: 2238–2248.

Frank, D. A. & McNaughton, S. J. (1992). The ecology of plants, large mammalian herbivores, and drought in Yellowstone National Park. *Ecology* **73**: 2043–2058.

Freese, C. H., Aune, Boyd, D. P., *et al.* (2007). Second chance for the plains bison. *Biological Conservation* **136**: 175–184.

Friedman, H. (1929). *The Cowbirds*. Springfield, IL: Charles C. Thomas.

Fritz, K. M., Dodds, W. K. & Pontius, J. (1999). The effects of bison crossings on the macroinvertebrate community in a tallgrass prairie stream. *American Midland Naturalist* **141**: 253–265.

Fuhlendorf, S. D., Engle, D. M., Kerby, J. & Hamilton, R. (2009). Pyric herbivory: rewilding landscapes through the recoupling of fire and grazing. *Conservation Biology* **23**: 588–598.

Fuller, J. A., Garrott, R. A., White, P. J., Aune, K. E., Roffe, T. J. & Rhyan, J. (2007a). Reproduction and survival of Yellowstone bison. *Journal of Wildlife Management* **71**: 2365–2372.

Fuller, J. A., Garrott, R. A., & White, P. J. (2007b). Emigration and density dependence in Yellowstone bison. *Journal of Wildlife Management* **71**: 1924–1933.

Fuller, W. A. (1960). Behavior and social organization of the wild bison of Wood Buffalo National Park, Canada. *Arctic* **13**: 2–19.

Fuller, W. A. (1962). The biology and management of the bison of Wood Buffalo National Park. *Canadian Wildlife Service Wildlife Management Bulletin Series* **16**: 1–52.

Garretson, M. S. (1938). *The American Bison: The Story of its Extermination as a Wild Species and Its Restoration Under Federal Protection*. New York: New York Zoological Society.

Gates, C. C. & Aune, K. (2008). *Bison bison*. 2008 IUCN Red List of Threatened Species, www.iucnredlist.org, accessed 25 January 2009.

Gates, C. C. & Broberg, L. (2011). *Yellowstone Bison: The Science and Management of a Migratory Wildlife Population*. Missoula, MT: University of Montana Press.

Gates, C. C. & Gogan, P. J. P. (2010). Introduction: the context. In *American Bison: Status Survey and Conservation Guidelines 2010*, pp. 1–4. Eds. Gates, C. C., Freese, C. H., Gogan, P. J. P. & Kotzman, M. Gland: International Union for the Conservation of Nature and Natural Resources.

Gates, C. C. & Hudson, R. J. (1983). Foraging behavior of wapiti in a boreal forest enclosure. *Naturaliste Canada* **110**: 197–206.

Gates, C. C. & Larter, N. C. (1990). Growth and dispersal of an erupting large herbivore population in northern Canada: the Mackenzie wood bison (*Bison bison athabascae*). *Arctic* **43**: 231–238.

Gates, C. C., Larter, N. C. & Komers, P. K. (1991). *Size and Composition of the Mackenzie Bison Population in 1989*. Yellowknife: Renewable Resources.

Gates, C. C., Stephenson, R. O., Reynolds, H. W., *et al.* (2001). *National Recovery Plan for the Wood Bison (*Bison bison athabascae*)*. Ottawa: Canadian Wildlife Service.

Gates, C. C., Stelfox, B., Muhly, T., Chowns, T. & Hudson, R. J. (2005). *The Ecology of Bison Movements and Distribution in and Beyond Yellowstone National Park: A Critical Review with Implications for Winter Use and Transboundary Population Management*. Alberta: University of Calgary.

Gates, C. C., Freese, C. H., Gogan, P. J.P & Kotzman, M. (eds) (2010). *American Bison: Status Survey and Conservation Guidelines 2010*. Gland: IUCN.

Geist, V. (1971). The relation of social evolution and dispersal in ungulates during the Pleistocene with emphasis on the Old World deer and the genus *Bison*. *Quaternary Research* **1**: 285–315.

Geist, V. (1991). Phantom subspecies: the wood bison *Bison bison 'athabascae'*

Rhoads 1897 is not a valid taxon, but an ecotype. *Arctic* **44**: 283–300.

Geist, V. & Karsten, P. (1977). The wood bison (*Bison bison athabascae* Rhoads) in relation to hypothesis on the origin of American bison (*Bison bison* Linnaeus). *Zeitschrift fur Saugeteirkunde* **42**: 119–127.

Gentry, A. W. (1978). Bovidae. In *Evolution of African Mammals*, pp. 540–572. Eds. Maglioand, V. J. & Cooke, H. B. S. Cambridge, MA: Harvard University Press.

Geremia, C., White, P. J., Garrott, R. A., *et al.* (2009). Demography of central Yellowstone bison: effects of climate, density and disease. In *The Ecology of Large Mammals in Central Yellowstone: Sixteen Years of Integrated Field Studies*, pp. 255–279. Eds. Garrott, R. A., White, P. J. & Watson, F. G. R. San Diego, CA: Elsevier.

Geremia, C., White, P. J., Wallen, R. L., *et al.* (2011). Predicting bison migration out of Yellowstone National Park using Bayesian models. *PLoS ONE* **6**: e16848.

Gerlanc, N. M. & Kaufman, G. A. (2003). Use of bison wallows by anurans on Konza Prairie. *American Midland Naturalist* **150**: 158–168.

Gogan, P. J. P., Podruzny, K. M., Olexa, E. M., Ihsle Pac, H. & Frey, K. L. (2005). Yellowstone bison fetal development and phenology of parturition. *Journal of Wildlife Management* **69**: 1716–1730.

Gogan, P. J. P., Larter, N. C., Shaw, J. H., Gross, J. E., Gates, C. C. & Truett, J. (2010). General biology, ecology and demographics. In *American Bison: Status Survey and Conservation Guidelines 2010*, pp. 39–54. Eds. Gates, C. C., Freese, C. H., Gogan, P. J. P. & Kotzman, M. Gland: International Union for the Conservation of Nature and Natural Resources.

Goguen, C. B. & Mathews, N. E. (1999). Review of the causes and implications of the association between cowbirds and livestock. *Studies in Avian Biology* **18**: 10–17.

Gokbulak, F. (2002). Effect of American bison (*Bison bison* L.) on the recovery and germinability of seeds of range forage species. *Grass and Forage Science* **57**: 395–400.

Goodnight, C. (1914). My experience with bison hybrids. *Journal of Heredity* **5**: 197–199.

Green, G. I., Mattson, D. J. and Peek, J. M. (1997). Spring feeding on ungulate carcasses by grizzly bears in Yellowstone National Park. *Journal of Wildlife Management* **61**: 1040–1055.

Green W. C. H. (1986). Age-related differences in nursing behavior among american bison cows (*Bison bison*). *Journal of Mammalogy* **67**: 739–741.

Green, W. C. H. (1990). Reproductive effort and associated costs in bison (*Bison bison*): do older mothers try harder? *Behavioral Ecology* **1**: 148–160.

Green, W. C. H. (1992). The development of independence in bison: pre-weaning spatial relations between mothers and calves. *Animal Behaviour* **43**: 759–773.

Green, W. C. H. & Rothstein, A. (1991). Trade-offs between growth and reproduction in female bison. *Oecologia* **86**: 521–527.

Green, W. C. H. & Rothstein, A. (1993a). Asynchronous parturition in bison: implications for the hider–follower dichotomy. *Journal of Mammalogy* **74**: 920–925.

Green, W. C. H. & Rothstein, A. (1993b). Persistent influences of birth date on dominance, growth, and reproductive success in bison. *Journal of Zoology* **230**: 177–186.

Green, W. C. H., Griswold, J. G. & Rothstein, A. (1989). Postweaning associations among bison mothers and daughters. *Animal Behavior* **38**: 847–858.

Green, W. C. H., Rothstein, A. & Griswold, J. G. (1993). Weaning and parent–offspring conflict: variation relative to interbirth interval in bison. *Ethology* **95**: 105–125.

Gross, J. E., Halbert, N. D., Derr, J. N., *et al.* (2010). Conservation guidelines for population, genetic, and disease management. In *American Bison: Status Survey and Conservation Guidelines 2010*, pp. 85–101. Eds. Gates, C. C., Freese, C. H., Gogan, P. J. P. & Kotzman, M. Gland: International Union for the Conservation of Nature and Natural Resources.

Groves, C. P. (1981). Systematic relationships in the Bovini (Artiodactyla, Bovidae). *Zeitschrift für Zoologischen Systematik und Evolutionsforschung* **19**: 264–278.

Gunther, K. A. (1991). Grizzly bear activity and human-induced modification in Pelican Valley, Yellowstone National Park. Master's thesis. Montana State University.

Guthrie, R. D. (1980). Bison and man in North America. *Canadian Journal of Anthropology* **1**: 55–73.

Halbert, N. D. (2003). The utilization of genetic markers to resolve modern management issues in historic bison populations: implications for species conservation. PhD dissertation. Texas A&M University.

Halbert, N. D. & Derr, J. (2007). A comprehensive evaluation of cattle introgression into U.S. federal bison herds. *Journal of Heredity* **98**: 1–12.

Halbert, N. D. & Derr, J. N. (2008). Patterns of genetic variation in the U.S. federal bison herds. *Molecular Ecology* **17**: 4963–4977.

Halbert, N. D., Grant, W. E. & Derr, J. N. (2005a). Genetic and demographic consequences of importing animals into a small population: a simulation model of the Texas State Bison Herd (USA). *Ecological Modelling* **181**: 263–276.

Halbert, N. D., Ward, T. J., Schnabel, R. D., Taylor, J. F. & Derr, J. N. (2005b). Conservation genomics: disequilibrium mapping of domestic cattle chromosomal segments in North American bison populations. *Molecular Ecology* **14**: 2343–2362.

Halbert, N. D., Gogan, P. J. P., Hiebert, R. & Derr, J. N. (2007). Where the buffalo roam: the role of history and genetics in the conservation of bison on U.S. federal lands. *Park Science* **24**: 22–29.

Hall, E. R. & Kelson, K. R. (1959). *The Mammals of North America*, Vol. **2**. New York: Ronald Press Company.

Halloran, A. F. (1961). American bison weights and measurements from the Wichita Mts. Refuge. *Proceedings Oklahoma Academy Science* **41**: 212–218.

Halloran, A. F. (1968). Bison (Bovidae) productivity on the Wichita Mountains Wildlife Refuge, Oklahoma. *Southwestern Naturalist* **13**: 23–26.

Hardy, A. R. (2001). Bison and elk responses to winter recreation in Yellowstone National Park. Masters thesis. Montana State University.

Hartnett, D. C., Hickman, K. R. & Fischer Walter, L. E. (1996). Effects of bison grazing, fire, and topography on floristic diversity in tallgrass prairie. *Journal of Range Management* **49**: 413–420.

Haugen, A. O. (1974). Reproduction in the plains bison. *Iowa State Journal of Research* **49**: 1–8.

Haugen, A. O. and Shult, M. J. (1968). Breeding biology and behavior of American bison. *Iowa State Journal of Research* **49**: 25–44.

Hedrick, P. W. (2009). Conservation genetics and North American bison (*Bison bison*). *Journal of Heredity* **100**: 411–420.

Hedrick, P. W. (2010). Cattle ancestry in bison: explanations for higher mtDNA than autosomal ancestry. *Molecular Ecology* **19**: 3328–3335.

Hein, F. J. & Preston, C. R. (1998). Summer nocturnal movements and habitat selection by *Bison bison* in Colorado. In *Proceedings of the International Symposium on Bison Ecology and Management in North America, June 4–7, 1997*, pp. 205–210. Eds. Irby, L. and Knight, J. Bozeman, MT: Montana State University.

Hess, S. C. (2002). Aerial survey methodology for bison population estimation in Yellowstone National Park. PhD dissertation. Montana State University.

Hornaday, W. T. (1889). *The Extermination of the American Bison, with a Sketch of its Discovery and Life History: Annual Report (1887)*. Washington, DC: Smithsonian Institution.

Hudson, R. J. and Frank, S. (1987). Foraging ecology of bison in aspen boreal habitat. *Journal of Range Management* **40**: 71–75.

Hudson, R. J. and Watkins, G. W. (1986). Foraging rates of wapiti on green and cured pastures. *Canadian Journal of Zoology* **64**: 1705–1708.

Hudson, R. J. and White, R. G. (1985). *Bioenergetics of Wild Herbivores*. Boca Raton, FL: CRC Press, Inc.

Isenberg, A. C. (2000). *The Destruction of the Bison: An Environmental History 1750–1920*. Cambridge: Cambridge University Press.

Jarman, P. J. (1974). The social organisation of antelope in relation to their ecology. *Behaviour* **48**: 215–267.

Joern, A. (2005). Disturbance by fire frequency and bison grazing modulate grasshopper assemblages in tallgrass prairie. *Ecology* **86**: 861–873.

Johnnson, L. C. & Matchett, J. R. (2001). Fire and grazing regulate belowground processes in tallgrass prairie. *Ecology* **82**: 3377–3389.

Joly, D. O. & Messier, F. (2001). *Limiting Effects of Bovine Brucellosis and Tuberculosis on Bison within Wood Buffalo National Park*. Saskatoon: University of Saskatchewan.

Joly, D. O. & Messier, F. (2004a). Testing hypotheses of bison population decline, 1970–1999, in Wood Buffalo National Park: synergism between exotic disease and predation. *Canadian Journal of Zoology* **82**: 1165–1176.

Joly, D. O. & Messier, F. (2004b). Factors affecting apparent prevalence of tuberculosis and brucellosis in wood bison. *Journal of Animal Ecology* **73**: 623–631.

Joly, D. O. & Messier, F. (2005). The effect of bovine tuberculosis and brucellosis on reproduction and survival of wood bison in Wood Buffalo National Park. *Journal of Animal Ecology* **74**: 543–551.

Jonas, J. L. & Joern, A. (2007). Grasshopper (Orthoptera: Acrididae) communities respond to fire, bison grazing and weather in North American tallgrass prairie: a long term study. *Oecologia* **153**: 699–711.

Jones, C. J. (1907). Breeding cattelo. *Annual Report of the American Breeders' Association* **3**: 161–165.

Jones, J. D., Treanor, J. J., Wallen, R. L. & White, P. J. (2010). Timing of parturition events in Yellowstone bison, *Bison bison*: implications for bison conservation and brucellosis transmission risk to cattle. *Wildlife Biology* **16**: 333–339.

Keith, E. O., Ellis, J. E., Phillips, R. W., Dyer, M. I. & Ward, G. M. (1981). Some aspects of urea metabolism in North American bison. *Acta Theriologica* **26**: 257–268.

Kelsall, J. P., Telfer, E. S. & Kingsley, M,. (1978). Relationship of bison weight to chest girth. *Journal of Wildlife Management* **42**: 659–661.

Kirkpatrick, J. F., Kincy, V., Bancroft, K., Shideler, S. E. & Lasley, B. L. (1991). Oestrous cycle of the North American bison (*Bison bison*) characterized by urinary pregnanediol-3-glucuronide. *Journal of Reproduction and Fertility* **93**: 541–547.

Kirkpatrick, J. F., Gudermuth, D. F., Flagan, R. L., McCarthy, J. C. & Lasley, B. L. (1993). Remote monitoring of ovulation and pregnancy of Yellowstone bison. *Journal of Wildlife Management* **57**: 407–412.

Kleinschmidt, T. & Sgouros, J. G. (1987). Hemoglobin sequences. *Biological Chemistry Hoppe-Seyler* **368**: 579–615.

Knapp, A. K., Blair, J. M., Briggs, J. M., *et al.* (1999). The keystone role of bison in North American tallgrass prairie. *BioScience* **49**: 39–50.

Knopf, F. L. (1996). Prairie legacies: birds. In *Prairie Conservation*, pp. 135–148. Eds. Samson, F.B & Knopf, F. L. Washington, DC: Island Press.

Knowles, C. J., Proctor, J. D. & Forrest, S. C. (2002). Black-tailed prairie dog abundance and distribution in the Great Plains based on historic and contemporary information. *Great Plains Research* **12**: 219–254.

Kohl, M. T. (2012). Bison conservation in the northern Great Plains. Masters thesis. University of Montana.

Komers, P. E., Messier, F. & Gates, C. C. (1992). Search or relax: the case of bachelor wood bison. *Behavioural Ecology and Sociobiology* **31**: 195–203.

Komers, P. E., Messier, F. & Gates, C. C. (1993). Group structure in wood bison: nutritional and reproductive determinants. *Canadian Journal of Zoology* **71**: 1367–1371.

Komers, P. E., Messier, F. & Gates, C. C. (1994a). Plasticity of reproductive behaviour in wood bison bulls: when subadults are given a chance. *Ethology, Ecology and Evolution* **6**: 313–330.

Komers, P. E., Messier, F. & Gates, C. C. (1994b). Plasticity of reproductive behaviour in wood bison bulls: on risks and opportunities. *Ethology, Ecology and Evolution* **6**: 481–495.

Komers, P. E., Messier, F., Flood, P. F. & Gates, C. C. (1994c). Reproductive behavior of male wood bison in relation to progesterone level in females. *Journal of Mammology* **75**: 757–765.

Krueger, K. (1986). Feeding relationships among bison, pronghorn, and prairie dogs: an experimental analysis. *Ecology* **67**: 760–770.

Larter, N. C. & Gates, C. C. (1990). Home ranges of wood bison in an expanding population. *Journal of Mammalogy* **71**: 604–607.

Larter, N. C. & Gates, C. C. (1991). Diet and habitat selection of wood bison in relation to seasonal changes in forage quantity and quality. *Canadian Journal of Zoology* **69**: 2677–2685.

Larter, N. C., Sinclair, A. R. E. & Gates, C. C. (1994). The response of predators to an erupting bison (*Bison bison athabascae*) population. *Canadian Field-Naturalist* **108**: 318–327.

Larter, N. C., Sinclair, A. R. E., Ellsworth, T., Nishi, J. & Gates, C. C. (2000). Dynamics of reintroduction in an indigenous large ungulate: the wood bison of Northern Canada. *Animal Conservation* **4**: 299–309.

Larter, N. C., Nishi, J. S., Ellsworth, T., Johnson, D., More, G. & Allaire, D. G. (2003). Observations of wood bison swimming across the Liard River, Northwest Territories. *Arctic* **56**: 408–412.

Laundre, J. W., Hernandez, L. & Altendorf, K. B. (2001). Wolves, elk, and bison: reestablishing the 'landscape of fear' in Yellowstone National Park, USA. *Canadian Journal of Zoology* **79**: 1401–1409.

Loftus, R. T., MacHugh, D. E., Bradley, D. G., Sharp, P. M. & Cunningham, P. (1994). Evidence for two independent domestications of cattle. *Proceedings of the National Academy of Sciences of the United States of America* **91**: 2757–2761.

Lott, D. F. (1979). Dominance relations and breeding rate in mature male American bison. *Zeitschrift fur Tierpsychology* **49**: 418–432.

Lott, D. F. (1981). Sexual behavior and intersexual strategies in American bison. *Zeitschrift fur Tierpsychology* **56**: 97–114.

Lott, D. F. (1991). American bison socioecology. *Applied Animal Behaviour Science* **29**: 135–145.

Lott, D. F. (1998). The impact of domestication on bison behavior. *Proceedings of the International Symposium on Bison Ecology and Management in North America, June 4–7, 1997*, pp. 205–210. Eds. Irby, L. & Knight, J.Bozeman, MT: Montana State University.

Lott, D. F. (2002). *American Bison: A Natural History*. Berkeley and Los Angeles, CA: University of California Press.

Lott, D. F. & Minta, S. (1983). Home ranges of American bison cows on Santa Catalina Island, California. *Journal of Mammalogy* **64**: 161–162.

Lovaas, A. L. & Bromley, P. T. (1972). Preliminary studies of pronghorn antelope–black-tail prairie dog relations in Wind Cave National Park. *Antelope States Workshop Proceedings* **5**: 115–156.

Lumia, A. R. (1972). The relationship between dominance and play behavior in the American buffalo, *Bison bison*. *Zeitschrift fuer Tierpsychologie* **30**: 416–419.

Maher, C. R. & Byers, J. A. (1987). Age-related changes in reproductive effort of male bison. *Behavioral Ecology and Sociobiology* **21**: 91–96.

Marlow, C. B., Irby, L. R. & Norland, J. E. (1984). *Optimum Carrying Capacity for Bison in Theodore Roosevelt National Park*. Denver, CO: National Park Service.

Matlack, R. S., Kaufman, D. W. & Kaufman, G. A. (2001). Influence of grazing by bison and cattle on deer mice in burned tallgrass prairie. *American Midland Naturalist* **146**: 361–368.

Matthews, S. B. (1991). An assessment of bison habitat in the Mills/Mink Lakes area, Northwest Territories, using LANDSAT thematic mapper data. *Arctic* **44**: 75–80.

McClenaghan, L. R., Berger, J. & Truesdale, H. D. (1990). Founding lineages and genetic variability in plains bison (*Bison Bison*) from Badlands National Park, South Dakota. *Conservation Biology* **4**: 285–289.

McCullough, Y. B. (1980). Niche separation of seven North American ungulates on the National Bison Range, Montana. PhD dissertation. University of Michigan.

McDonald, J. N. (1981). *North American Bison: Their Classification and Evolution*. Berkeley, CA: University of California Press.

McHugh, T. (1958). Social behavior of the American buffalo (*Bison bison bison*). *Zoologica* **43**: 1–40.

McHugh, T. (1972). *The Time of the Buffalo*. New York: Alfred A. Knopf, Inc.

McMillan, B. R., Cottam, M. R. & Kaufman, D. W. (2000). Wallowing behavior of American bison (*Bison bison*) in tallgrass prairie: an examination of alternate explanations. *American Midland Naturalist* **144**: 159–167.

Meagher, M. (1973). *The Bison of Yellowstone National Park*. Washington, DC: National Park Service.

Meagher, M. (1986). *Bison bison*. *Mammalian Species* **266**: 1–8.

Meagher, M. M. (1989). Range expansion by bison of Yellowstone National Park. *Journal of Mammalogy* **70**: 670–675.

Meagher, M. & Meyer, M. E. (1994). On the origin of brucellosis in bison of Yellowstone National Park: a review. *Conservation Biology* **8**: 645–653.

Melton, D. A., Larter, N. C., Gates, C. C. & Virgil, J. A. (1989). The influence of rut and environmental factors on the behaviour of wood bison. *Acta Theriologica* **34**: 179–193.

Miller, W. (2002). Elk interactions with other ungulates. In *North American Elk: Ecology and Management*, pp. 435–447. Eds. Toweill, D. E. & Thomas, J. W. Washington, DC:Smithsonian Institution Press.

Miyamoto, M. M., Tanhauser, S. M. & Laipis, P. J. (1989). Systematic relationships in the Artiodactyl tribe Bovini (family Bovidae), as determined from mitochondrial DNA sequences. *Systematic Zoology* **38**: 342–349.

Modi, W. S., Gallagher, D. S. & Womack, J. E. (1996). Evolutionary histories of highly repeated DNA families among the Artiodactyla (Mammalia). *Journal of Molecular Evolution* **42**: 337–349.

Moffitt, S. A. (1998). Aging bison by the incremental cementum growth layers in teeth. *Journal of Wildlife Management* **62**: 1276–1280.

Mooring, M. S. & Samuel, W. M. (1998). Tick defense strategies in bison: the role of grooming and hair coat. *Behaviour* **135**: 693–718.

Mooring, M. S., Reisig, D. D., Osborne, E. R., *et al.* (2005). Sexual segregation in bison: a test of multiple hypotheses. *Behaviour* **142**: 897–927.

Mysterud, A. (1998). The relative roles of body size and feeding type on activity time of temperate ruminants. *Oecologia* **113**: 442–446.

NatureServe. (2006). NatureServe explorer: an online encyclopedia of life. Web application. Version 6.1. www.natureserve.org/explorer, accessed 26 February 2007.

Newton, R. D. (1993). *Bison Hybridization and Bison-influenced Cattle in North America*. Waco, TX: Baylor University Press.

Nishi, J. S. (2010). A review of best practices and principles for bison disease issues: Greater Yellowstone and Wood Buffalo areas. *American Bison Society Working Paper Series* **3**: 1–101.

Notzke, C. (1994). *Aboriginal Peoples and Natural Resources in Canada*. North York, ON: Captus Press Inc.

Nowak, R. M. (1991). *Walker's Mammals of the World*. Baltimore, MD and London: Johns Hopkins University Press.

Olenicki, T. J. & Irby, L. R. (2004). *Determining Forage Availability and Use*

Patterns for Bison in the Hayden Valley of Yellowstone National Park. Bozeman, MT: USGS.

Olexa, E. M. & Gogan, P. J. P. (2007). Spatial population structure of Yellowstone bison. *Journal of Wildlife Management* **71**: 1531–1538.

Peden, D. G. (1976). Botanical composition of bison diets on shortgrass plains. *American Midland Naturalist* **96**: 225–229.

Peden, D. G., Van Dyne, G. M., Rice, R. W. & Hansen, R. M. (1974). The trophic ecology of *Bison bison* L. on shortgrass prairie. *Journal of Applied Ecology* **11**: 489–498.

Pfeiffer, K. E. & Hartnett, D. C. (1995). Bison selectivity and grazing response of little bluestem in tallgrass prairie. *Journal of Range Management* **48**: 26–31.

Pfeiffer, K. E. & Steuter, A. A. (1994). Preliminary response of sandhills prairie to fire and bison grazing. *Journal of Range Management* **47**: 395–397.

Picton, H. D. (1999). Energetic costs of wildlife displacement by winter recreationists. In *Effects of Winter Recreation on Wildlife of the Greater Yellowstone Area: Report to the Greater Yellowstone Coordinating Committee*, pp. 135–144. Eds. Olliff, T., Legg, K. & Kaeding, B. Yellowstone National Park, WY: National Park Service.

Pilgrim, G. E., (1947). The evolution of buffaloes, oxen, sheep and goats. *Journal of the Linnean Society of London* **41**: 272–286.

Plumb, G. E. & Dodd, J. L. (1993). Foraging ecology of bison and cattle on mixed prairie: implications for natural areas management. *Ecological Applications* **3**: 631–643.

Plumb, G. E., White, P. J., Coughenour, M. B. & Wallen, R. L. (2009). Carrying capacity, migration, and dispersal in Yellowstone bison. *Biological Conservation* **142**: 2377–2387.

Polley, E. W. & Collins, S. L. (1984). Relationships of vegetation and environment in buffalo wallows. *American Midland Naturalist* **112**: 178–186.

Polley, E. W. & Wallace, L. L. (1986). The relationship of plant species heterogeneity to soil variation in buffalo wallows. *Southwestern Naturalist* **31**: 493–501.

Popp, J. (1980). Range ecology of bison on mixed grass prairie at Wind Cave National Park. Masters thesis. Iowa State University.

Post, D. M., Armbrust, T. S., Horne, E. A. & Goheen, J. R. (2001). Sexual segregation results in differences in content of quality of bison (*Bos bison*) diets. *Journal of Mammalogy* **82**: 407–413.

Powell, F. L. A. (2006). Effects of prescribed burns and bison (*Bos bison*) grazing on breeding bird abundances in tallgrass prairie. *The Auk* **123**: 183–197.

Pucek, Z., Belousova, I. P., Krasińska, M., Krasiński, Z. A. & Olech, W. (2004). *European Bison. Status Survey and Conservation Action Plan: UCN/SSC Bison Specialist Group*. Gland: and Cambridge: IUCN.

Redford, K. H. & Fearn, E. (2007). *The Ecological Future of the North American Bison.* Bronx, NY: Wildlife Conservation Society.

Redford, K. H., Aune, K. & Fearn, E. (2009). The second recovery of bison: ecological restoration of North America's largest land mammal. *The Wildlife Professional* **3**: 46–50.

Reinhardt, V. (1985). Social behaviour in a confined bison herd. *Behaviour* **92**: 209–226.

Renecker, L. A. & Hudson, R. J. (1993). Morphology, bioenergetics and resource use: patterns and processes. In *Alberta's Hoofed Mammals: Their Ecology, Status and Management*, pp. 141–163. Ed. Stelfox, J. B. Edmonton: Lone Pine Press.

Reynolds, H. W., Hansen, R. M. & Peden, D. G. (1978). Diets of the Slave River Lowland bison herd, Northwest Territories, Canada. *Journal of Wildlife Management* **42**: 581–590.

Reynolds, H. W., Glaholt, R. D. & Hawley, A. W. L. (1982). Bison. In *Wild Mammals of North America: Biology, Management, and Economics*, pp. 972–1007. Eds. Chapman, J. A. and Feldhamer, G. A. Baltimore, MD and London: Johns Hopkins University Press.

Reynolds, H. W., Gates, C. C. & Glaholt, R. D. (2003). Bison (*Bison bison*). In *Wild Mammals of North America: Biology, Management and Conservation*, pp. 1009–1060. Eds. Chapman, J. A. and Feldhamer, G. A. Baltimore, MD and London: Johns Hopkins University Press.

Rhoads, S. N. (1897). Notes on living and extinct species of North American Bovidae. *Proceedings of the Academy of Natural Sciences* **49**: 483–502.

Rhyan, J. C., Aune, K., Roffe, T., *et al.* (2009). Pathogenesis and epidemiology of brucellosis in Yellowstone bison: serologic and culture results from adult females and their progeny. *Journal of Wildlife Diseases* **45**: 729–739.

Robison, C. D. (1994). *Conservation of Germ Plasm from Brucella Abortus-infected Bison (*Bison bison*) Using Natural Service, Embryo Transfer, and In Vitro Maturation/In Vitro Fertilization.* College Station, TX: Texas A & M University.

Roden C., Hilde, V., Guy, M. & Linda, V. E. (2003). Reproductive success of bison bulls (*Bison bison bison*) in seminatural conditions. *Animal Reproduction Science* **79**: 33–43.

Roden, C., Vervaecke, H. & Van Elsacker, L. (2005). Dominance, age and weight in American bison males (*Bison bison*) during nonrut in semi-natural conditions. *Applied Animal Behaviour Science* **92**: 169–177.

Roden, C., Stevens, J. M. G., Vervaecke, H. & Van Elsacker, L. (2011). Reproductive effort of bison bulls (*Bison bison*) in semi-natural conditions. *Journal of Ethology* **29**: 285–291.

Roe, F. G. (1970). *The North American Buffalo: A Critical Study of the Species in its Wild State.* 2nd edition. Toronto: University of Toronto Press.

Rosas, C. A., Engle, D. M., Shaw, J. H. & Palmer, M. W. (2008). Seed dispersal by *Bison bison* in a tallgrass prairie. *Journal of Vegetation Science* **19**: 769–778.

Rothstein, A. & Griswold, J. G. (1991). Age and sex preferences for social partners by juvenile bison bulls, *Bison bison. Animal Behaviour* **41**: 227–237.

Rutberg, A. T. (1983). Factors influencing dominance status in American bison. *Zeitshrift fur Tierpsychologie* **63**: 206–212.

Rutberg, A. T. (1984). Birth synchrony in American bison (*Bison bison*): response to predation or season? *Journal of Mammalogy* **65**: 418–423.

Rutberg, A. T. (1986a). Lactation and fetal sex ratios in American bison. *American Naturalist* **127**: 89–94.

Rutberg, A. T. (1986b). Dominance and its fitness consequences in American bison cows. *Behaviour* **96**: 62–91.

Rutley, B. D. & Hudson, R. J. (2001). Activity budgets and foraging behavior of bison on seeded pastures. *Journal of Range Management* **54**: 218–225.

Sanderson, E. W., Redford, K. H., Weber, B., *et al.* (2008). The ecological future of the North American Bison: conceiving long-term, large-scale conservation of wildlife. *Conservation Biology* **22**: 252–266.

Schaller, G. B. (1976). Aggressive behavior of the domestic yak. *Journal of the Bombay Natural History Society* **73**: 385–389.

Schuler, K. L., Leslie, D. M., Jr., Shaw, J. H. & Maichak, E. J. (2006). Temporal-spatial distribution of American bison (*Bison bison*) in a tallgrass prairie fire mosaic. *Journal of Mammalogy* **87**: 539–544.

Schultheiss, P. C., Collins, J. K., Austgen, L. E. & DeMartini, J. C. (1998). Malignant catarrhal fever in bison: acute and chronic cases. *Journal of Veterinary Diagnostic Investigation* **10**: 255–262.

Schultheiss, P. C., Collins, J. K., Miller, M.W., *et al.* (2001). Malignant catarrhal fever in bison. *Veterinary Pathology* **38**: 577.

Schwartz, C. C. & Nagy, J. G. (1976). Pronghorn diets relative to forage availability in northeastern Colorado. *Journal of Wildlife Management* **40**: 469–478.

Shakelton, D. M., Hills, L. V. & Hutton, D. A. (1975). Aspects of variation in cranial characters of plains bison (*Bison bison bison*) from Elk Island National Park, Alberta. *Journal of Mammalogy* **48**: 871–887.

Shaw, J. H. & Carter, T. S. (1989). Calving patterns among American bison. *Journal of Wildlife Management* **53**: 896–898.

Shaw, J. H. & Carter, T. S. (1990). Bison movements in relation to fire and seasonality. *Wildlife Society Bulletin* **18**: 426–430.

Shaw, J. & Meagher, M. (2000). Bison. In *Ecology and Management of Large Mammals in North America*, pp. 447–466. Eds. Desmarais, S. and Krausman, P. R. Upper Saddle River, NJ: Prentice Hall.

Shaw, R. A. (2012). Social organization and decision making in North American bison: implications for management. Graduate theses and dissertations, Paper 1204. Utah State University. http://digitalcommons.usu.edu/etd/1204.

Shoemaker, H. W. (1915). *A Pennsylvania Bison Hunt*. Lewisburg, PA: Wennawoods Publishing.

Sinclair, A. R. E. (1977). *The African Buffalo: A Study of Resource Limitations of Populations*. Chicago, IL: University of Chicago Press.

Singer, F. J. & Norland, J. E. (1994). Niche relationships within a guild of ungulate species in Yellowstone National Park, Wyoming, following release from artificial controls. *Canadian Journal of Zoology* **72**: 1383–1394.

Skinner, M. F. & Kaisen, O. C. (1947). The fossil bison of Alaska and preliminary revision of the genus. *Bulletin of the American Museum of Natural History* **89**: 123–256.

Soper, J. D. (1941). History, range, and home life of the northern bison. *Ecological Monographs* **11**: 348–412.

Steinauer E. M. & Collins, S. L. (2001). Feedback loops in ecological hierarchies following urine deposition in tallgrass prairie. *Ecology* **82**: 1319–1329.

Steklenev, E. P. & Yasinetskaya, N. I. (1982). Results of crossing of the bison (*Bison bison bison* L.), with the domestic cow (*Bos (Bos) taurus typicus*) and characteristics of the chromosome complexes of the hybrid progeny. *Tsitologiia I Genetika* **16**: 28–33.

Steklenev, E. P., Yasinetskaya, N. I. & Nechiporenko, V. K. (1986). Spontaneous variability and associative ability of chromosomes of hybrids of bison with domestic cattle. *Tsitologiia I Genetika (Kiev)* **20**: 284–287.

Stephenson, R. O., Gerlach, S. C., Guthrie, R. D., Harington, C. R., Mills, R. O. & Hare, G. (2001). Wood bison in late Holocene Alaska and adjacent Canada: paleontological, archaeological and historical records. In *People and Wildlife in Northern North America: Essays in Honor of R. Dale Guthrie*, pp. 124–158. Eds. Gerlach, S. C. & Murray, M. S. Oxford: British Archaeological Reports.

Steuter, A. A., Steinauer, E. M., Hill, G. L., Bowers, P. A. & Tieszen, L. L. (1995). Distribution and diet of bison and pocket gophers in a sandhills prairie. *Ecological Applications* **5**: 756–766.

Stormont, C., Miller, W. J. & Suzuki, Y. (1961). Blood groups and the taxonomic status of American buffalo and domestic cattle. *Evolution* **15**: 196–208.

Strong, W. L. & Gates, C. C. (2009). Wood bison population recovery and forage availability in northwestern Canada. *Journal of Environmental Management* **90**: 434–440.

Taper, M. L., Meagher, M. & Jerde, C. L. (2000). *The Phenology of Space: Spatial Aspects of Bison Density Dependence in Yellowstone National Park*. Bozeman, MT: USGS.

Telfer, E. S. & Cairns, A. L. (1979). Bison-wapiti interrelationships in Elk Island National Park, Alberta. In *North American Elk: Ecology, Behavior and Management*, pp 114–121. Eds. Boyce, M. S. & Hayden-Wing, L. D. Laramie, WY: University of Wyoming.

Tessaro, S. V. (1992). Bovine tuberculosis and brucellosis in animals, including man. In *Buffalo*, pp. 207–224. Eds. Foster, J., Harrison, D. & MacLaren, I. S. Edmonton: University of Alberta Press.

Tessaro, S. V., Forbes, L. B. & Turcotte, C. (1990). A survey of brucellosis and tuberculosis in bison in and around Wood Buffalo National Park, Canada. *Canadian Veterinary Journal* **31**: 174–180.

Thorne, E. T. (2001). Brucellosis. In *Infectious Diseases of Wild Mammals*, pp. 372–395. Eds. Williams, E. S. & Baker, I. K. Ames, IA: Iowa State University Press.

Towne, E. G. (2000). Prairie vegetation and soil nutrient response to ungulate carcasses. *Oecologia* **122**: 232–239.

Towne, G., Nagaraja, T. G., Cochran, R. C., Harmon, D. L., Owensby, C. E. & Kaufman, D. W. (1988). Comparison of ruminal fermentation characteristics and microbial population in bison and cattle. *Applied and Environmental Microbiology* **10**: 2510–2514.

Trager, M. D., Wilson, G. W. T. & Hartnett, D. C. (2004). Concurrent effects of fire regime, grazing and bison wallowing on tallgrass prairie vegetation. *American Midland Naturalist* **152**: 237–247.

Trudell, J. & White, R. G. (1981). The effect of forage structure and availability on food intake, biting rate and daily eating of reindeer. *Journal of Applied Ecology* **18**: 63–81.

Turner, M. G., Wu, Y., Wallace, L. L., Romme, W. H. & Brenkert, A. (1994). Simulating winter interactions among ungulates, vegetation, and fire in northern Yellowstone Park. *Ecological Applications* **4**: 472–496.

USFWS-NPS (US Fish and Wildlife Service and National Park Service). (2007). *Final*

Bison and Elk Management Plan and Environmental Impact Statement for the National Elk Refuge/Grand Teton National Park/John D. Rockefeller Jr., Memorial Parkway, Teton County, Wyoming Volume 1. Washington, DC: US Government Printing Office.

Van Camp, J. (1989). A surviving herd of endangered wood bison at Hook Lake, Northwest Territories. *Arctic* **42**: 314–322.

Van Gelder, R. G. (1977). Mammalian hybrids and generic limits. *American Museum Novitates* **2635**: 1–25.

Van Vuren, D. (1984). Summer diets of bison and cattle in southern Utah. *Journal of Range Management* **37**: 260–261.

van Zyll de Jong, C. G. (1986). *A Systematic Study of Recent Bison, with Particular Consideration of the Wood Bison (*Bison bison athabascae *Rhodes 1898).* Ottawa: National Museums of Canada.

van Zyll de Jong, C. G. (1993). Origin and geographic variation of recent North American bison. *Alberta: Studies in the Arts and Sciences* **3**: 21–35.

Varel, V. H. & Dehority, B. A. (1989). Ruminal cellulytic bacteria and protozoa from bison, cattle–bison hybrids, and cattle fed three alfalfa–corn diets. *Applied and Environmental Microbiology* **55**: 148–153.

Varley, N. & Gunther, K. A. (2002). Grizzly bear predation on a bison calf in Yellowstone National Park. *Ursus* **13**: 377–381.

Vaughn, T. A. (1972). *Mammalogy.* Philadelphia, PA: W. B. Saunders.

Vervaecke, H., Roden, C. & DeVries, H. (2005). Dominance, fatness and fitness in female American bison, *Bison bison*. *Animal Behavior* **70**: 763–770.

Vinton, M. A. & Hartnett, D. C. (1992). Effects of bison grazing on *Andropogon gerardii* and *Panicum virgatum* in burned and unburned tallgrass prairie. *Oecologia* **90**: 374–382.

Vinton, M. A., Hartnett, D. C., Finck, E. J. & Briggs, J. M. (1993). Interactive effects of fire, bison (*Bison bison*) grazing and plant community composition in tallgrass prairie. *American Midland Naturalist* **129**: 10–18.

Virchow, D. R. & Hygnstrom, S. E. (2002). Estimation of presettlement populations of the black-tailed prairie dog: a reply. *Great Plains Research* **12**: 255–260.

Waggoner, V. & Hinkes, M. (1986). Summer and fall browse utilization by an Alaskan bison herd. *Journal of Wildlife Management* **50**: 322–324.

Wall, D. A., Davis, S. K. & Read, B. M. (1992). Phylogenetic relationships in the subfamily Bovinae (Mammalia: Artiodactyla) based on ribosomal DNA. *Journal of Mammalogy* **73**(2): 262–275.

Wallace, L. L., Turner, M. G., Romme, W. H., O'Neil, R. V. and Wu, Y. (1995). Scale of heterogeneity of forage production and winter foraging by elk and bison. *Landscape Ecology* **10**: 75–83.

Wallace, L. L., Coughenour, M. B., Turner, M. G. & Romme, W. H. (2004). Ungulate survival in northern Yellowstone park: the results of two independent models. In *After the Fires: The Ecology of Change in Yellowstone National Park*, pp. 299–317. Ed. Wallace, L. L. New Haven, CT: Yale University Press.

Ward, T. J. (2000). *An Evaluation of the Outcome of Interspecific Hybridization Events Coincident With a Dramatic Demographic Decline in North American Bison.* PhD dissertation, College Station: Texas A&M University.

Ward, T. J., Bielawski, J. P., Davis, S. K., Templeton, J. W. & Derr, J. N. (1999). Identification of domestic cattle hybrids in wild cattle and bison species: a general approach using mtDNA markers and the parametric bootstrap. *Animal Conservation* **2**: 51–57.

Webster, E. M. (2005). Behavior of brown-headed cowbirds foraging with American bison and insect communities of a New Mexico shortgrass prairie. Masters thesis University of Wisconsin.

Whicker, A. D. & Detling, J. K. (1988). Ecological consequences of prairie dog disturbances. *BioScience* **38**: 778–785.

White, P. J. and Wallen, R. L. (2012). Yellowstone bison: should we preserve artificial population substructure or rely on ecological processes? *Journal of Heredity Advances.* Published online 23 August 2012.

White, P. J., Wallen, R. L., Geremia, C., Treanor, J. J. & Blanton, D. W. (2011). Management of Yellowstone bison and brucellosis transmission risk: implications for conservation and restoration. *Biological Conservation* **144**: 1322–1334.

Williams, E. S., Cain, S. L. & Davis, D. S. (1997). Brucellosis: the disease in bison. In *Brucellosis, Bison, Elk and Cattle in the Greater Yellowstone Area: Defining the Problem, Exploring the Solutions*, pp. 7–19. Eds. Thorne, E. T., Boyce, M. S., Nicoletti, P. & Kreeger, T. J. Cheyenne, WY: Wyoming Game and Fish.

Wilson, A. C., Cann, R. L., Carr, S. M., *et al.* (1985). Mitochondrial DNA and two perspectives on evolutionary genetics. *Biological Journal of the Linnaean Society* **26**: 375–400.

Wilson, D. E. & Mittermeier, R. A. (2011). *Handbook of the Mammals of the World*, Vol. **2**. Barcelona: Lynx Edicions.

Wilson, D. E. & Reeder, D. M. (1993). *Mammal Species of the World.* Washington, DC: Smithsonian Institution Press.

Wilson, D. E. & Reeder, D. M. (2005). *Mammal Species of the World: A Taxonomic and Geographic Reference.* 3rd edition. Baltimore, MD: Johns Hopkins University Press.

Wilson, G. A. (2001). Population genetic studies of wood and plains bison populations. PhD dissertation. University of Alberta.

Wilson, G. A. & Strobeck, C. M. (1999). Genetic variation within and relatedness among wood and plains bison populations. *Genome* **42**: 483–496.

Wilson, G. A. & Zittlau, K. A. (2004). *Management Strategies for Minimizing Loss of Genetic Diversity in Wood and Plains Bison Populations at Elk Island National Park.* Edmonton: Elk Island National Park, Parks Canada.

Wilson, M. (1975). Holocene fossil bison from Wyoming and adjacent areas. Masters thesis. University of Wyoming.

Wydeven, A. P. & Dahlgren, R. B. (1985). Ungulate habitat relationships in Wind Cave National Park. *Journal of Wildlife Management* **49**: 805–813.

Yoakum, J. D. (2004). Relationships with other herbivores. In *Pronghorn Ecology and Management*, pp. 501–538. Eds. O'Gara, B. W. & Yoakum, J. D. Boulder, CO: University Press of Colorado.

Zontek, K. (2007). *Buffalo Nation: American Indian Efforts to Restore the Bison.* Winnipeg: Bison Books.

Chapter 10

European bison *Bison bonasus* (Linnaeus, 1758)

Małgorzata Krasińska, Zbigniew A. Krasiński, Kajetan Perzanowski and Wanda Olech

Names

Genus: *Bison* H. Smith, 1827

Species: European bison *Bison bonasus* (Linnaeus, 1758)

Subspecies: Lowland (Białowieża) European bison, *Bison bonasus bonasus* (Linnaeus, 1758); Caucasian (mountain) European bison, *Bison bonasus caucasicus* (Turkin & Satunin, 1904); Carpathian (Transylvanian) European bison, *Bison bonasus hungarorum* (Kretzoi, 1946)

Names in other languages: **French:** Bison d'Europe; **German:** Wisent; **Spanish:** Bisonte Europeo; **Italian:** Bisonte Europeo; **Polish:** Żubr; **Russian:** Zubr; **Slovak:** Zubor; **Romanian:** Zimbru.

Other common names: Wisent

Ecology, Evolution and Behaviour of Wild Cattle: Implications for Conservation, ed. M. Melletti and J. Burton. Published by Cambridge University Press.

Taxonomy

The European bison belong to the sub-order of ruminants, whose representatives have a four-chambered stomach comprising the three pre-stomach compartments – the rumen, reticulum and omasum – and the true stomach – the abomasum. It is in the pre-stomach, especially the massive rumen (whose volume in adult bison may exceed 100 litres), that plant food is gathered for periodic regurgitation as 'the cud is chewed' (for more details see Chapter 6).

Horn structure qualifies European bison as a member of the Bovidae family, in which horns are hollow inside. This is in contrast with less closely related animals like red and roe deer or moose, whose bony antlers have no empty space inside. The horns of European bison are composed of a bony core (cornual process) and hard sheath of horny material.

Ruminants have 32 teeth, lacking as they do the upper incisors and canines, while their molars with their broad, folded crowns are adapted for chewing plant material.

The genus *Bison* includes large and massive herbivorous mammals now present in Europe and North America. According to Flerov (1979) the genus was widely spread in the temperate zones of Asia and Europe during late Pliocene and early Pleistocene (2–5 million years ago), but other authors (Martínez-Navarro *et al.* 2012) claim that *Bison* appeared in Eastern Europe at 1.77 million years ago, became present in the Middle East at around 1.6–1.2 million years ago and reached Western Europe at around 1.5 million years ago. Similarly, Shapiro *et al.*(2004) estimate that *Bison* first entered eastern Beringia from Asia during the middle Pleistocene between 300 000 and 130 000 years ago, and later spread southward into central North America between 130 000 and 75 000 years ago, while according to Scott (2010) this happened between 240 000 and 220 000 years ago. Final separation of bison living in North America and Asia occurred after the retreat of the last glaciation over 10 000 years ago, when the land connection (Beringia) between Asia and America was flooded.

This genus is represented now by two species – the European bison, *Bison bonasus* (L.), and North American bison, *Bison bison*. Within vast areas inhabited by those animals, further differentiation took place. In the conditions of treeless prairie, the prairie bison has developed – *Bison bison bison* (Linnaeus, 1758), while in forested areas of present Canada – appeared the wood bison *Bison bison athabascae* (Rhoads, 1897).

Similar was the case of the European bison, where besides the lowland, called also Białowieża, bison (*Bison bonasus bonasus* L.) – inhabiting Central and Eastern Europe – occurred also a subspecies of the Caucasian (or mountain) bison (*Bison bonasus caucasicus* Turkin & Satunin, 1904) in the Caucasus Mountains.

Some authors claim that the European bison may originate from the late Pleistocene re-emigrants from North America (McDonald 1981). Craniometrical studies proved a close similarity between late Pleistocene bison of Eurasia and North America (Zyll de Jong 1986), confirming earlier hypotheses stating that the American and European bison may belong to the same species (Skinner & Kaisen 1947; Bohlken 1967).

A majority of authors consider, however, the currently existing representatives of the genus *Bison* as two separate species, taking into account morphological differences (Figures 10.1–10. 2) and their spatial isolation, despite the fact that progeny of both species obtained through their hybridisation are fertile (Pucek 1986; Pucek *et al.* 2004; Krasińska & Krasiński 2007).

Figure 10.1 Comparison of silhouettes of European and American bison. Figure by T. Samojlik. Source: Krasiński, 2010, modified.

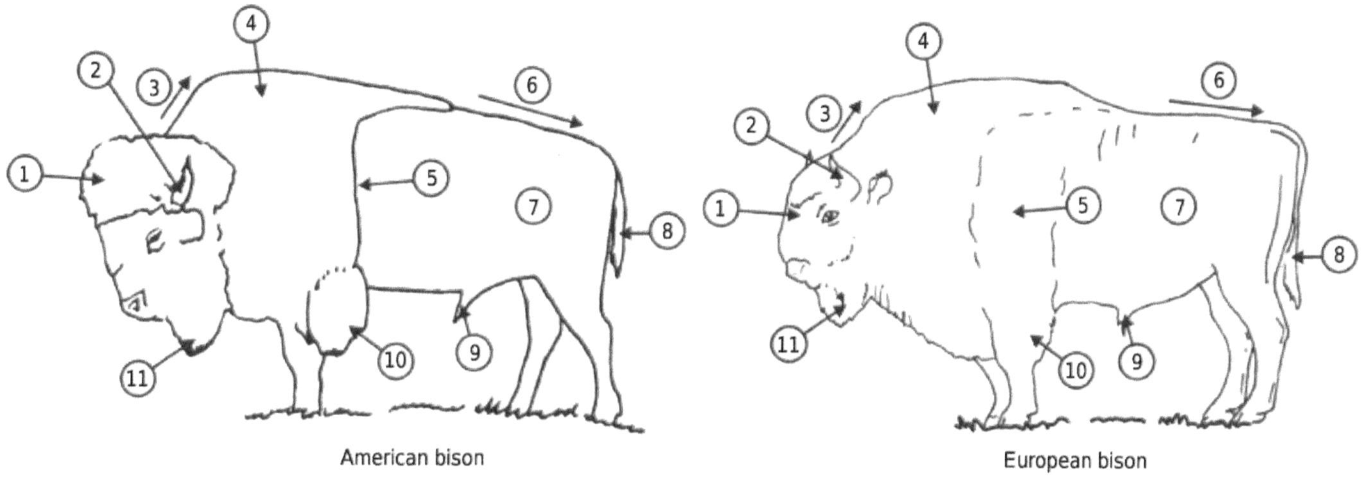

Figure 10.2 Comparison of body features of European and American bison. Silhouette of American bison, source: Reynolds *et al.* 1982. Figure by T. Samojlik. Source: Krasiński, 2010, modified.

American bison

(1) Hairs on the head longer than in European bison, a distinct fringe on a forehead.
(2) Short horns, frequently hidden in the hair, ends of horns directed upwards.
(3) Line of the back (dorsum) behind the head is sharply directed upwards.
(4) Forequarters are covered by longer, curly hair, coloured chocolate-brown.
(5) The border between coat of forequarters and hindquarters is well visible.
(6) A posterior line of the spine (dorsum) descends in a gentle arch.
(7) Hindquarters are covered with smooth and short hair, coloured copper-brown.
(8) Short tail covered with smooth hair, ended with a brush or a tuft of long hair.
(9) In males the tuft of hair at underbelly around prepuce is well visible.
(10) Longer hairs covering the upper part of forelegs end at the wrist, suggesting the bison is wearing 'puffed pants'.
(11) Distinct beard of triangular shape – 'goatee beard'.

European bison

(1) Well-formed head set higher than in American bison.
(2) Horns longer, more massive than in American bison, turning to the centre of the head in a gentle arch, clearly visible in females.
(3) Line of the back (dorsum) is gently ascending and then gradually lowers towards the hindquarters. Its shape is described as 'carp-like'.
(4) Hair covering forequarters is shorter than in American bison, coloured flaxen-brown with reddish tinge.
(5) The border between the coat covering forequarters and hindquarters it not so distinctly marked as in American bison.
(6) Line of back (dorsum) along hindquarters is not descending.
(7) Hair covering hindquarters is shorter than in forequarters.
(8) The tail reaches the ankle joint, and is covered with longer hair than in American bison.
(9) In adult males a little bulge is visible behind the prepuce.
(10) Hair covering the upper part of the forelegs is shorter than in American bison.
(11) The beard is shorter than in American bison, gently descending to the chest.

Genetic information

Bison have 60 chromosomes, of which 58 are acrocentric autosomes, and two the sex chromosomes (a large sub-metacentric X and a small sub-metacentric Y) (Melander 1959; Koulischer *et al.* 1967; Fedyk & Sysa 1971).

Bison bonasus is a very special species from the genetic point of view. Its gene pool is very narrow, which causes a deep concern about the future of the species. The reason for such low genetic variability is that the species has passed through a serious bottleneck more than once. In the nineteenth century, the species was reduced to its last two populations. One of them was *Bison bonasus bonasus* in Białowieża Primeval Forest and the second *B. b. caucasicus* in the Caucasus Mountains. Neither population was very large, the first counted at ~700 animals and the second being three times larger. Because of low population numbers it could be expected that the European bison lost some of its original variability due to genetic drift. Then, during and just after World War I, there occurred a major bottleneck. Both wild populations disappeared and only a few animals survived in captivity (Groeben 1932; Pucek *et al.* 2004).

The inventory carried out in 1924 showed only 54 (M29, F25) living European bison with proven genetic purity. Those individuals, as well as their known ancestors, were listed in the first edition of the *European Bison Pedigree Book* (EBPB), where every animal was assigned a name and number (Groeben 1932). The detailed pedigree analysis was done first by Slatis (1960). He found that all European bison (the whole population in 1954) are descendants of 17 ancestors, but since some of these were represented by only one progeny, the author identified 12 unique genotypes as founders of the current population. Since then, in all subsequent analyses, this set of 12 founders is used (Table 10.1) (Olech 1989, 1999; Belousova 1993; Krasiński 1994; Pucek *et al.* 2004). Now, genes of all of these 12 animals are represented in the present population of the species. Simple facts – that the lowest

Table 10.1 Founders of the contemporary population of the European bison

No.	Sex	Pedigree No.	Name	Year of birth/death	Place of birth	Main place of breeding
1	F	42	**Planta**	ca 1904–1931	Pszczyna	Pszczyna (von Pless)
2	M	45	**Plebejer**	1917–1937	Pszczyna	Pszczyna (von Pless)
3	M	87	**Bill**	1913–1929	Zoo Budapest	From 1916: Zoo Stockholm
4	F	89	**Bilma**	1913–1939	Białowieża	From 1916: Zoo Stockholm; From 1935: Białowieża
5	M	15	**Begründer**	1903–1919	Zoo Berlin	Zoo Berlin
6	F	16	**Plavia**	1906–1932	Pszczyna	From 1919: Zoo Berlin
7	M	147	**Bismarck**	1925–1934	Zoo Schönbrunn (Austria)	From 1927: Zoo Berlin. From 1928: Zoo München
8	M	100	Kaukasus	1907–1925	Caucasus	From 1908: Stellingen near/Hamburg. From 1922: Boitzenburg (von Arnim)
9	F	96	**Gatschina (Gatczyna)**	1911–1932	Białowieża	From 1916: Scharbow (von Beyme) From 1924: Zoo Poznań. From 1931: Białowieża
10	F	95	**Garde**	1907–1922	Białowieża	From 1909: Stellingen near/Hamburg. From 1922: Boitzenburg (von Arnim)
11	F	35	**Plewna**	1912–1922	Zoo Frankfurt	Zoo Frankfurt
12	F	46	**Placida**	1918–1926	Pszczyna	From 1921: Scharbow (von Beyme)

Table after: Groeben (1932); Slatis (1960); Krasiński (1994); Olech (1989).
The Lowland line originates from the first seven founders (No. 1–7), and the Lowland–Caucasian line from all 12 founders. M – male. F – female. Names of lowland bison are in bold.

population number of European bison was 54 individuals, and that the gene pool of contemporary European bison originates from only 12 founders – are very often mixed up, and seem to be confused even by some people working on species genetics.

A thorough analysis of the pedigree of 54 animals alive in 1924 has revealed that 28 animals could be recognised as ancestors of the contemporary population. Seventeen of them were identified and their contribution explained 30 years later by Slatis (1960), but genes of the remaining (11 founders) existed only at the beginning of the restitution of the population. The comparison of genetic structure of the European bison population in 1924 and 1954 showed that 11 potential founders had a considerable representation in 1924, equalling to 26.9% of total gene pool (Olech 2009). Some potential founders died 1–2 years after the first inventory, but there were four founders represented by their progeny up to the year 1951. All those animals were kept in Woburn Reserve by Herzog von Bedford, who was very much involved in endangered species conservation. Among others he has succeeded in breeding Père David's deer (*Elaphurus davidianus*), but probably did not hold the European bison in high regard (Bostock 2001). This was truly unfortunate, because for a species with such a narrow gene pool, the survival of this group could be very important for improving genetic variability.

Presently, the species is divided into two genetic lines: one-pure lowland subspecies named **Lowland (or Białowieża) line (LB)**, and the second with inclusion of genes of the male No. 100 Kaukasus, the only one representing the Caucasian subspecies. The name of the second line is **Lowland-Caucasian (LC)**. LB is derived from seven founders but LC from all 12 – seven common for both lines and five unique only for the LC line. Changes in time of founders' contribution within both lines are presented in (Figure 10.3).

When the contribution of the 12 founders is compared, it is interesting to notice considerable changes of their values in time. In 1924, animals No. 42 and 45 (EBPB) had together only 5.6% contribution to the total genetic pool, but later, because of the large number of their descendants, their representation increased up to 45.2% in 1954, and to almost 84% in the contemporary LB (Pucek *et al.* 2004). The opposite happened to the contribution of Kaukasus (No. 100), with an initial contribution larger than 10%, now decreased to just 6%.

It is important for species conservation that the long period of breeding only in captivity and the establishment of the Pedigree Book allow now for determination of the origins of particular animals using the pedigree analysis. In the case of free-ranging herds, the pedigree data allow determination of the structure of the reintroduced population. It is also possible to calculate various coefficients characterising the genetic value of single animals or populations using the pedigree list containing 10–12 generations of ancestors as a source of data. One coefficient calculated in this way is a degree of inbreeding.

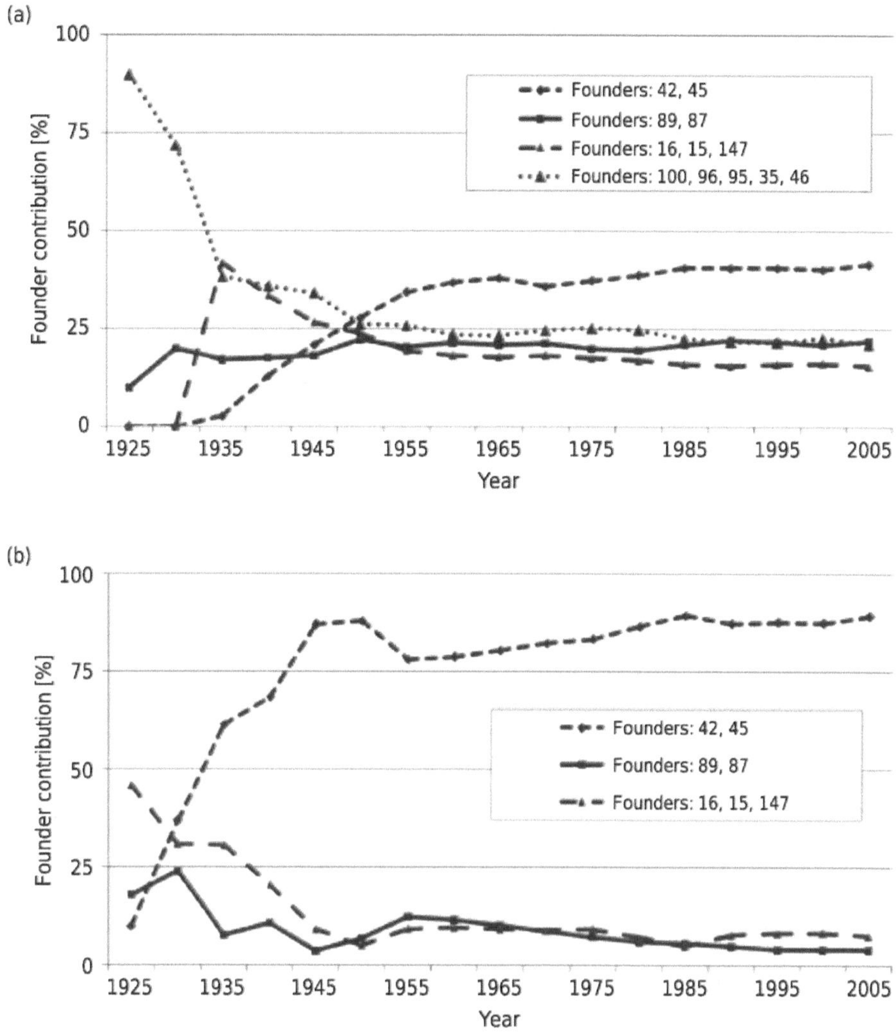

Figure 10.3 The changes of founders' contribution in time within the LB line (a) and the LC line (b).

The average level of inbreeding coefficient for animals born recently equals 48% for LB and 28.9% for LC (Olech 2003).

On the basis of pedigree it is possible to calculate not only the inbreeding coefficient, which reflects the homozygosity of an individual, but also the mean kinship (MK), which shows the relation of the individual to other members of a population, and the founder genome contribution/retention revealing the structure of the population. Presently in both lines there are over-represented genes of two founders (male No. 45, 'Plebejer' and female No. 42, 'Planta'); their genetic contribution in the gene pool of LB is higher than 84%. The same situation concerns the retention of founder genes, and only in the case of those two founders (No. 42 and 45) more than 50% of their genotype is saved in the contemporary LB. Of the other five founders' genes, only 17–34% is saved in the present population. Due to the bottleneck that occurred between 1940 and 1945, a majority of males from the present world population of LB, both in free-living herds and captive groups, have copies of the same Y-chromosome of founder No. 45, 'Plebejer' (Pucek *et al.* 2004).

The Y-chromosome of founder No. 100, 'Kaukasus', can be found in the free-living herd of Bieszczady and in some captive groups. The Y-chromosome of founders No. 15, 'Begrunder', and No. 147, 'Bismarck', were lost during the breeding process between 1945 to 1997. The last male descendant of founder No. 87, 'Bill', died childless in 1935 (Sipko *et al.* 1996). This means that three of five Y-chromosomes existing some 70 years ago are now lost.

However, there are no serious signs of inbreeding depression in this species, but there are still concerns about its influence upon the stability of development and demographic parameters of bison herds (Sipko *et al.* 1996). In fact, the only significant negative influence of inbreeding found up to now is the lowered viability of young animals. Recent studies show significant positive correlations between the level of inbreeding and the mortality of young animals belonging to the LC line (Slatis 1960; Olech 1987; Belousova 1993). For animals belonging to the LB line, though, with a much higher level of inbreeding, such a relationship between inbreeding and mortality has not been found (Olech 2003).

Some depressive effects of inbreeding were identified upon skeletal growth, more expressed in females and within LC animals (Kobryńczuk 1985). Increased inbreeding causes neurocranium shortening and elongation of the skull basis, which results in elongation and narrowing of the splanchnocranium.

Furthermore, the skeleton of the distal appendage elements is also elongated, while the scapula is shortened.

The lack of evident signs of inbreeding depression in this species could be explained by the difficulties of detecting inbreeding depression effects under natural conditions (Reed *et al.* 2003), and that it may be just a matter of time. But we believe that within the LB line lack of inbreeding depression can be a result of purging of the gene pool after the bottleneck (Olech 2003). According to Nei *et al.* (1975), fast recovery of the species after a severe bottleneck may minimise its impact by more effective purging of the genetic load. Also, we must remember that very low resistance to illnesses and parasites observed within the European bison population can also be a sign of inbreeding depression.

Many studies have proved low levels of genetic variability in European bison populations. For the population of Białowieża Forest (LB line) the average heterozygosity of proteins coded by 20 loci was estimated as 3.5% (Gębczyński & Tomaszewska-Guszkiewicz 1987). Other studies showed for 69 structural loci an average heterozygosity equal to 1.2%, and only 5.8% of loci were polymorphic (Hartl & Pucek 1994). In other studies the differentiation of proteins (22 loci; Sipko *et al.* 1996), blood group systems (nine systems, 57 antigens; Sipko *et al.* 1995), kappa-casein gene (Sipko *et al.* 1994; Burzyńska & Topczewski 1995) and the major histocompatibility complex class II DRB and DQB gene (Udina *et al.* 1994) were investigated. These studies generally show that genetic diversity within and among all analysed populations of European bison are much lower than for other species (Olech 2006).

Current studies with the use of DNA markers provide the same conclusion. The European bison shows about half the microsatellite heterozygosity of the closely related American bison (*B. bison*). The effective population size (N_e) for the highly genetically homogenous LB line in the Polish part of the Białowieża Forest is estimated at 23.5, far less than the actual population size of 450 (Tokarska *et al.* 2011). Microsatellites are used widely for checking the polymorphism level (Nowak & Olech 2008a). They allow also assessing the purity of the species *Bison bonasus*. That is important due to a possibility of its hybridisation with *Bison bison* and *Bos taurus* (Nowak & Olech 2008b).

It is obvious that genetic variation in the European bison has been seriously decreased by the historical population bottlenecks; however, the mentioned studies were only conducted on a part of the whole European bison population. Nevertheless, it is still important to find a tool, except pedigree analysis of captive individuals, for assessing the genetic diversity between individuals. The application of routine sets of cattle microsatellite markers in parentage and identity analysis of European bison was inefficient in comparison to the use of bovine SNP (single nucleotide polymorphism) markers (Tokarska *et al.* 2009). Moreover, SNP markers were effectively used to evaluate genetic variability between European and American bison (Pertoldi *et al.* 2010). Also, it was proved in a study by Kamiński *et al.* (2011) that SNP markers can be used for genetic analysis within *Bison bonasus* species.

European Bison Pedigree Book

During the International Congress for Nature Conservation in Paris, on 2 June 1923, the Polish representative, Jan Sztolcman (1854–1928), a traveller, scientist and the vice-director of the State Zoological Museum in Warsaw, delivered an 'Appeal on the necessity of saving the European bison'. The Congress approved the resolution, which stated an aim: 'To create as possibly as quickly an international society between countries in who's territories European bison occur.' An inaugural meeting of the Society for the Protection of European Bison took place on 25–26 August 1923 in Berlin. Elected as its chairman was Kurt Priemel, the director of the Zoological Garden at Frankfurt am Main. The first goal of the Society for the Protection of European Bison was to prepare a register of all still-living European bison. Such a register (*Das Zuchtbuch*) was published in 1932, under the editorship of Gerd von der Groeben in the fifth volume of the journal *Berichte der Internationalen Gesellschaft zur Erhaltung des Wisent*. This first register, by the end of 1923, demonstrated the existence of 55 (29, 26) European bison (Raczyński & Bołbot 2009). At No. 1 in the pedigree book was registered 'Planet', born in 1881 at the Pszczyna breeding centre; number two was 'Bellona', born in 1883 at Berlin Zoo. In the first volume all European bison that took part in the restitution of the species were registered (Slatis 1960; Olech 1989). The following two volumes of the pedigree book (1933 and 1937) were elaborated and edited in the same journal by Erna Mohr from Berlin. After World War II, the editorial office of the EBPB was transferred to Poland, and editorial work was conducted by Jan Żabiński (1897–1974), the director of the Zoological Garden of Warsaw. In cooperation with Erna Mohr, Jan Żabiński determined and verified pedigrees of European bison that survived the war, and published the first post-war volume of the EBPB. Since 1987 the editorial office of the EBPB has been situated at Białowieski National Park, and its publication is coordinated by Jan Raczyński. At the beginning, every volume covered two-year periods, but since 1992 the EBPB has been published annually. All European bison reported by breeders obtain pedigree numbers. Names are given by breeders according to the system approved by the editorial office, with the first two letters indicating the place of origin of an individual. The EBPB contains the World Register of European bison at the end of each year in captivity, and the list of free-ranging populations, separately for every country. Registration in the EBPB is a guarantee of origin and genetic purity, which is crucial information for breeders (Pucek 1984; Raczyński & Bołbot 2009).

Subspecies and distribution

Subspecies

Three subspecies are recognised within the European bison:

(1) **Lowland (Białowieża) European bison, *Bison bonasus bonasus*** (Linnaeus, 1758). In historic times (middle and

late Holocene) it was an inhabitant of mixed grassland/forest and forested areas of central and eastern Europe, as far east as the River Don (Heptner *et al.* 1966; Benecke 2005; Kuemmerle *et al.* 2012). Encroachment of settlements and agriculture, particularly in last 1000 years, caused habitat fragmentation and limited the area suitable for bison, which as a consequence reduced their numbers considerably (Kuemmerle *et al.* 2012). The gradual decrease of population, leading to the extinction of bison in the wild, initiated about the eleventh century, lasted until the nineteenth century. European bison survived in this region of present Lithuania, which at that time was a part of East Prussia, until the end of the seventeenth century. The last individual was killed there in 1775. Since the nineteenth century, lowland European bison have occurred only at Białowieska Forest (Sztolcman 1924). In the beginning of the nineteenth century, representatives of this subspecies lived exclusively in the Białowieża Forest. According to K. Wróblewski (1927) before World War I, in the Białowieża Forest there were still 727 bison. The last of them was killed in April 1919 (Sztolcman 1924; Pucek 2004; Pucek *et al.* 2004). From the Białowieża Forest, between the eighteenth and beginning of the twentieth centuries, European bison were caught and sent to zoological gardens and various private breeding centres in Europe (Krasińska & Krasiński 2007).

(2) **The Caucasian (mountain) European bison, *Bison bonasus caucasicus*** (Turkin & Satunin, 1904), used to live in the forested zone of a northern slope of the main ridge of a mountain massif of Caucasus and its foothills. At the southern slope of this ridge, Caucasian bison lived only within its western part, up to the border with Abkhazia. In the nineteenth century, this population was estimated at about 2000 animals. The Caucasian War and the expansion of settlements within the foothills, as well as the development of livestock breeding, exploitation of forests and hunting, caused a shift in the home range of this population into the area between the rivers Belaya and Laba (Heptner *et al.* 1966). At the end of the nineteenth century there were only 442 bison left there (Kulagin 1919). Since the beginning of the twentieth century, the disappearance of this subspecies became very rapid. In 1920 only 50 animals were still alive there. Even the creation of the Caucasian Nature Reserve in 1924 did not save it from extinction. The last Caucasian bison, according to Bashkirov (1940) was killed in 1927, while Heptner *et al.* (1966) placed this event in 1926.

(3) **The Carpathian (Transylvanian) European bison, *Bison bonasus hungarorum*** (Kretzoi, 1946), inhabited the Carpathians (including Transylvania – a historical land in central Romania). Its extinction is estimated at the second half of the eighteenth century (Szalay 1913; Nahlik 1991; Nakhlyk 1992).

Recently, Russian scientists put forward a controversial proposal for consideration of a new subspecies, the hybrids of European bison living in the Caucasus Biosphere Reserve (Rautian *et al.* 2000).

'Mountain bison' in the Caucasus Biosphere Reserve

In the central part of the main massif of the Caucasus Mountains, between the rivers Belaya and Laba, at the area of the Caucasus Biosphere Reserve, lives a population of so-called 'mountain bison'. They originate from five (one male and four females) hybrids of European and American bison, brought in 1940 to the Caucasus from the steppe reserve of Askania Nova (southern Ukraine). Those animals were kept in the enclosure. Since 1949 the share of European bison blood increased in this group due to supplementing it with males belonging to the LC line, i.e. 15 bulls from Poland. In 1954 those animals were released into the wild, and since then their numbers have grown, in 1965 reaching 449 individuals (Kalugin 1968). In 1984 this population counted about 1300 animals (Yazan & Nemcev 1985). However, in following years, due to poaching, the number of those hybrids dropped down to 550 in 1999 and to 400 in 2000 (Nemtsev *et al.* 2003; Pucek *et al.* 2004). Because of the admixture of the blood of American bison, those animals are not registered in the EBPB. Rautian *et al.* (2000) proposed the separation of those animals into a distinct subspecies of 'mountain bison', *Bison bonasus montanus*.

World distribution of European bison

Historical times

The European bison was quite widely distributed in Europe until the early medieval period. Because of unpropitious climatic conditions it did not occur in the south of the continent (Apennine and Iberian Peninsulas and in the southern part of the Balkan Peninsula). In the north, its range of occurrence was generally limited by the Baltic Sea, except the most southern part of Scandinavia. There are no data confirming the presence of this species in Great Britain. In the east, its limit of occurrence went south from the Finland Gulf, more or less as far as today's Moscow and Ryazan or River Don. In the south, its range extended as far as the Azov and Black Seas (Heptner *et al.* 1966; Pucek 2004; Pucek *et al.* 2004; Benecke 2005) (Figure 10.4; see also Figure 10.5). However, according to recent analysis of archaeological remnants of European bison from Europe, the western limit of the occurrence of this species in middle and late Holocene could not extend so far to the west as was thought earlier (Benecke 2005).

Population numbers and the range of distribution shrank quickly in following centuries, however, due to intensive hunting and the loss of habitat to agriculture. The process of gradual decline towards extinction had begun as early as the eleventh century, and continued without reverse into the nineteenth century (Figure 10.4). Already in the mid sixteenth century, in the west of Europe, the species was so rare that it was commonly confused with the aurochs.

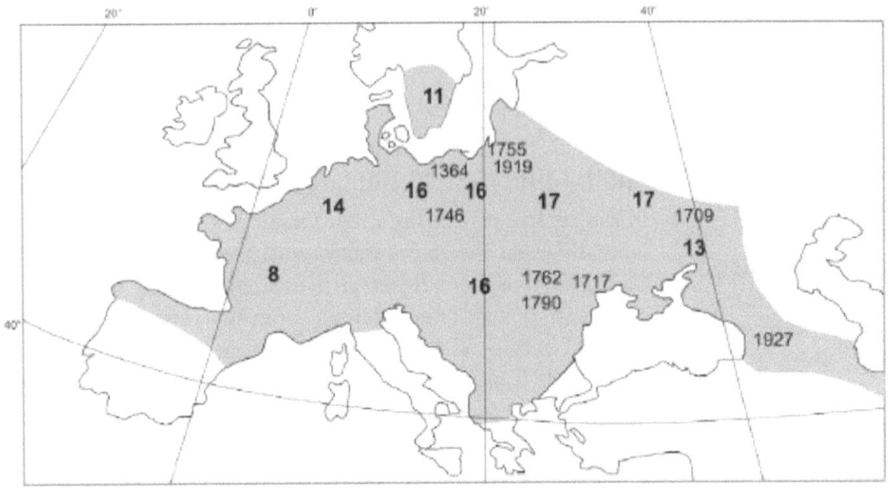

Figure 10.4 Reconstructed range of the European bison in Holocene and early historical times (grey area) and process of extinction of the species. The century (bold) and year of the last bison records in the given part of the range. Source: Pucek, 2004.

Figure 10.5 Distribution of free-ranging and semi-free populations of the European bison in 2010, and reconstructed range of the European bison in Holocene and early historical times based on bone findings of this species. Figure by Marcin Górny. Based on the European Bison Pedigree Book (2010) and Heptner *et al.* 1966; Benecke, 2005; Kuemmerle *et al.* 2011.

The last historic sites of the occurrence of the species in the wild were Białowieża Forest (until 1919) (Sztolcman 1924; Pucek 2004; Pucek *et al.* 2004) and Caucasus Mountains (until 1926 or 1927) (Bashkirov 1940; Heptner *et al.* 1966).

Present distribution

Bison bonasus is among few species that were completely extirpated in the wild and successfully restored to nature only after a long period of breeding in captivity.

The last animals living at the end of 1924 were mostly kept in estates of the European aristocracy – including seven at Pszczyna (Prince von Pless), ten at Woburn (Herzog von Bedford), two at Boitzenburg (Count Arnim) and four at Scharbow (von Beyme). In zoological gardens there were 28 animals (Austria, Holland, Germany, Sweden, England and the Soviet Union) and one individual at the reserve Askania Nova. Created later were breeding centres for European bison in Poland (Białowieża and Pszczyna); Germany (Springe and Hellabrun); the Soviet Union (Sierpuchov (later Priokskoterrasnyj)); and Sweden (Avesta); followed by Niepołomice and Smardzewice in Poland, Hardehausen and Damerower Werder in Germany and Okskij Reserve in the Soviet Union.

After more than 80 years of efforts towards the restitution of the species, its total number exceeded 4000 individuals (4663 in 2011) both in freedom and in captivity (see Table 10.3; see also plate 2). Since there were no European bison living in the wild after 1919 and 1927, present distribution of the species is a result of efforts to restore this species in a number of European countries. In many cases, free-ranging herds were established within sites formerly occupied by the species; however, there are some examples (especially in Russia) of experimental location of new herds far outside their original range.

Captive breeding

By the end of 2011 there were 220 captive centres for European bison, maintained in 29 European countries (breeding centres, zoos, etc.). The most numerous (83) are breeders in Germany, comprising reserves, breeding stations, zoos, etc., where more than 500 animals now live. The most important breeding centres there are Damerower Werder, Hardehausen and Springe, as well as two new centres, Döberitzer Heide and Neuburg. The distribution of European breeding centres in relation to the number of animals maintained in every centre is given in Figure 10.5 (see also Figure 10.33). The frequency of centres maintaining herds belonging to LB and LC genetic lines is presented in Table 10.2. Because of such domination of very small herds, a coordinated programme for the exchange of animals among herds is necessary, which is now the main task of the European Bison Conservation Centre.

The number of animals kept in captivity has grown steadily since the 1960s, from 562 in 1960 to 1497 by the end of 2011, including 440 LB animals and 1057 belonging to the LC line. Both genetic lines are separated in captivity and only 12 breeding stations keep animals belonging to both lines (together 76 animals: 32 LB and 44 LC). Outside of Europe, European bison are now kept in zoos of North and South America (Canada and Brazil).

Table 10.2 The numbers and frequency of European bison breeding centres maintaining herds belonging to LB and LC genetic lines

Size of the herd	LC	LB	Both	Σ	%
1–3	64	14	3	81	36.8
4–7	63	13	5	81	36.8
8–12	20	10	3	33	15.0
13+	15	9	1	25	11.4
Σ	162	46	12	220	100

Free-ranging and semi-free populations

The history of free-ranging herds begins in 1952 at Białowieża, Poland, followed by 1953 in Belarus (Krasiński 1978a, 1978b; Krasińska & Krasiński 2007; Kozlo & Bunevich 2009).

Free-ranging herds are now distributed east from the Odra River (a border between Poland and Germany) as far as central European Russia (Figure 10.5). The majority of free-ranging populations of European bison are distributed within the eastern part of the species' historical range. However, in Russia, outside of its original geographical range, several populations were established, even as far as the Altai Mountains near the border with Mongolia (Cherga population).

Herds consisting of animals belonging to the LB line occur now in Poland (except the Bieszczady Mountains), Belarus and Lithuania, while the LC line is represented in all Carpathian herds, Ukraine and Russia (except Cherga) (Figure 10.5 and 10.6).

The numbers of animals in free- and semi-free-ranging herds exceeded the captive part of world population in 1986, and reached 3111 animals in 2011 (Table 10.3). The free-living part of the world population consists now of 32 herds of various size dispersed across six countries (Belarus, Lithuania, Poland, Russia, Slovakia and Ukraine) and eight semi-free herds in six countries (Figure 10.5). By the end of 2011 (EBPB 2011), free-ranging populations counted 2892 animals. Five free-ranging herds existed at that time in Poland (1041 individuals), seven in Belarus (1061), 12 in Russia (484), one in Slovakia (nine animals), six in Ukraine (225) and one in Lithuania (72). The remaining 219 animals were living in semi-free herds, one in Belarus (14 animals), France (36), Germany (28), Ukraine (21) and two such herds in Romania (67 animals) and Russia (53) (Table 10.4). In 2012 the first free-ranging herd was established in Romania – Vanatori Neamt Nature Park, and in spring of 2013 the first semi-free herd in Germany at Rothaargebirge. The creation of a herd on the Island of Bornholm in the Baltic Sea has also been negotiated.

Regarding the size of free-ranging herds, the absolute majority (17) counts fewer than 50 animals and only seven

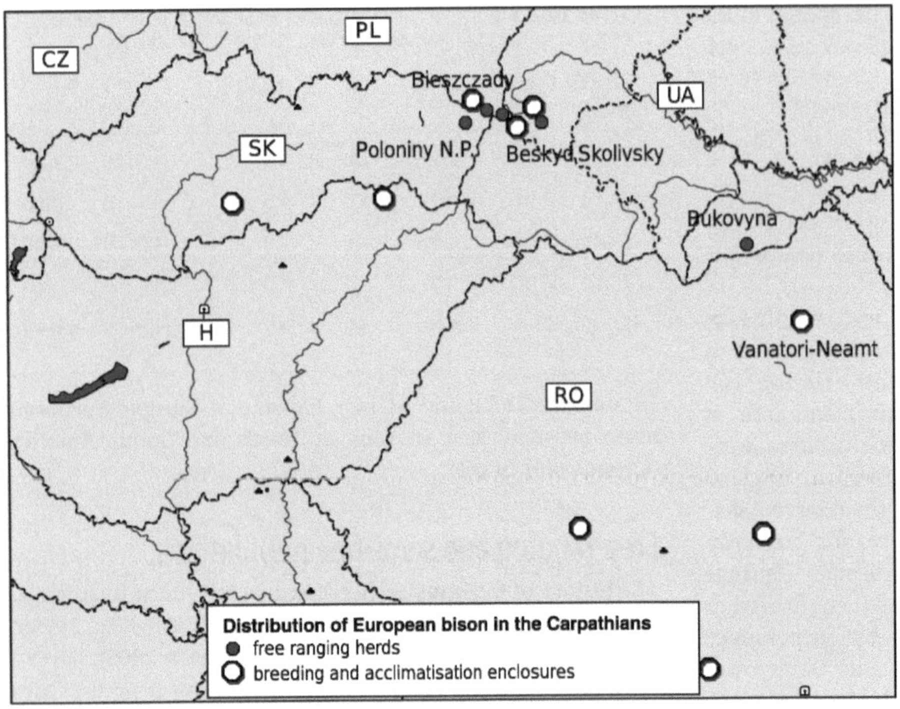

Figure 10.6 Distribution of the European bison in the Carpathian region.

(regarded as demographically safe) exceed 100 bison (Table 10.4). At such small sizes of isolated populations there is no guarantee for their proper functioning over a longer time scale without proper management and planned exchange of animals. The highest potential for further development of sizeable populations is in Russia because availability of vast areas that remain mostly uninhabited by people.

However, the creation and development of so many free-ranging populations should be regarded as a big success of the restitution programme, unthinkable at the beginning. Such wide dispersal of the species, although not desirable from the genetic and demographic point of view, may ensure its survival in the case of unpredictable events at the local scale. Demographic and genetic deficiencies are dealt with by planned exchange of animals.

European bison in the Carpathians

The situation of the species in the Carpathians is special, because this eco-region provides a unique chance to create a meta-population of European bison occupying a continuous home range over the area of several countries.

Apparently in the medieval period, the European bison was quite common and well known in the Carpathians, since from Romania through Ukraine to Poland there are a number of geographical names connected with this species. Its numbers were considerably reduced some 200–300 years ago, probably mostly due to overhunting and poaching. The most recent records on the presence of bison in the wild come from Ukraine (early eighteenth century) and Transylvania, from where the last captured European bison was transferred to the imperial menagerie at Schönbrunn, near Vienna (Nakhlyk 1992; Parnikoza *et al.* 2009).

All bison presently dwelling in the Carpathians belong to the mixed LC genetic line of the species (*Bison bonasus* L., 1758). Some authors recognise also a third subspecies native to the Carpathians (*Bison bonasus hungarorum*, Kretzoi, 1946), which apparently ceased to exist in the second half of the eighteenth century (Krasińska & Krasiński 2007). Therefore, the whole present Carpathian population of the species originates from planned introductions, performed since the 1960s.

The largest (and the oldest) bison population inhabits the eastern part of the Bieszczady Mountains, the range situated in the southeast tip of Poland, bordering Slovakia and Ukraine (see Plate 2 and Figure 10.6). This population has been initiated through introductions in 1963 and in 1980 to two forest districts of this region (Stuposiany and Komańcza). Founders of this population were animals of the LC line, selected from Polish breeding centres. Between 2002 and 2011 this population was supplemented several times with genetically selected animals (altogether 19 individuals) from various foreign breeding centres (Sweden, Denmark, Ireland, Germany and Czech Republic).

The whole population of Bieszczady is about 300 animals, the second largest free-ranging herd of European bison in Poland (after Białowieża), and the most numerous mountain herd of this species in the world (Perzanowski & Paszkiewicz 2000; Perzanowski 2001, 2004; Perzanowski *et al.* 2006b).

In Slovakia the reintroduction of European bison started in National Park Poloniny in 2004, with five genetically selected animals offered for the programme from Holland, Italy and Switzerland, supplemented with four more individuals from

Table 10.3 The world population of European bison from 1965 to 2010

Year	Captive breeding		Free-ranging populations		Total	Remarks
	Numbers	%	Numbers	%		
1965	562	72.0	218	28.0	780	
1966	605	70.4	254	29.6	859	
1967	593	63.0	349	37.0	942	
1968	628	61.6	391	38.4	**1019**	Numbers of bison exceeded 1000 individuals
1969	711	62.1	434	37.9	1145	
1970	700	58.3	500	41.7	1200	
1971	737	57.2	551	42.8	1288	
1972	833	57.0	629	43.0	1462	
1973	901	58.0	652	42.0	1553	
1974	984	59.1	680	40.9	1664	
1975	979	56.7	748	43.3	1727	
1976	1075	57.7	789	42.3	1864	
1977	1139	57.1	855	42.9	1994	
1978	1192	56.5	919	43.5	**2111**	Numbers of bison exceeded 2000 individuals
1979	1211	55.1	987	44.9	2198	
1980	1247	56.6	**1079**	46.4	2326	Numbers of bison in free-ranging populations exceeded 1000 individuals
1981	1242	51.7	1161	48.3	2403	
1982	1273	51.7	1190	48.3	2463	
1983	**1300**	50.0	**1300**	50.0	2600	Balanced numbers of both forms of breeding
1984	1329	49.1	1377	50.9	2706	
1985	1454	50.8	1408	49.2	2862	
1986	1441	49.1	1495	50.9	2936	
1987	1377	44.7	1705	55.3	**3082**	Numbers of bison exceeded 3000 individuals for the first time
1988	1412	44.9	1734	55.1	3146	
1989	1452	45.6	1735	54.4	3187	
1990	1477	42.3	1855	55.7	3332	
1991	1512	44.4	1895	55.6	**3407**	
1992	1403	41.4	**1990**	58.6	3393	
1993	1450	42.2	1831	55.8	3281	
1994	1481	45.5	1778	54.5	3259	
1995	1229	39.6	1870	60.4	3099	
1996	1162	38.7	1840	61.3	3002	
1997	1096	37.5	1829	62.5	2925	
1998	1133	39.9	1709	60.1	2842	
1999	1177	40.4	1738	59.6	2915	
2000	1154	40.2	1710	59.7	2864	
2001	1155	39.0	1809	61.0	2964	

Table 10.3 *(cont.)*

Year	Captive breeding		Free-ranging populations		Total	Remarks
	Numbers	%	Numbers	%		
2002	1186	38.5	1893	61.5	3079	
2003	1166	38.2	1887	61.8	3053	
2004	1200	38.0	1955	62.0	3155	
2005	1305	38.0	2134	62.0	3439	
2006	1376	39.0	2163	61.0	3539	
2007	1443	37.9	2367	62.1	3810	
2008	1392	34.5	2640	65.5	**4032**	Numbers of bison exceeded 4000 individuals for the first time
2009	1444	34.0	2787	66.0	4231	
2010	1475	33.0	**2956**	67.0	**4431**	
2011	1552	33.0	**3111**	67.0	**4663**	Maximal numbers of bison in free-ranging populations and of world population

Data from the *European Bison Pedigree Book* (1965–2011).
Source: Krasińska and Krasiński (2007), updated.

Slovak zoos in 2005 and 2006. Several times the herd was joined by bison migrating from Bieszczady. By the beginning of 2012 it counted nine animals, and its home range overlaps mostly with the area of the national park (Perzanowski *et al.* 2006a; Pčola & Gurecka 2008).

In southern Slovakia, in an enclosure near the village of Topol'cianki, a semi-free herd comprising 15 animals is maintained.

In the Ukrainian part of the Carpathians, European bison were introduced first in 1965, to the forestry of Maydan (Majdanska population) situated in the Skolyvsky Beskyd range at the area of Lvov Province. Later (1969) they were brought also to Bukovyna (Chernivtsi Province) and (1976) near Nadvirna (Ivano-Frankivsk Province) (Bukovynska and Nadvirnjanska populations). Individuals used for introduction originated from Belarus and Russia.

The herd at Maydan Forestry grew to 50 individuals in 1987 but, due to various factors, mostly poaching, this herd disappeared in winter 2008/2009.

In 2009/2010, ten individuals from German and Austrian breeding stations were brought to a new 3-ha acclimatisation enclosure in Skolyvsky Beskydy National Park. The free-ranging herd there counts 13 individuals (three born in Ukraine).

The herd at Navirna, which at its maximum did not exceed 30 individuals (1990–1991), was the only herd in the Carpathians in which a disease of prepuce (balanoposthitis) was recorded. The herd ceased to exist in 1999.

The home range of the population established in the Chernivtsi Province extends over the area of two forest-game enterprises (Beregometske i Storozhinecke), and partially the nature reserve Vyzhnytsja. For a number of years it was the most numerous bison population in the Carpathians, reaching almost 225 individuals in 1994. However, according to independent data from 2009, only 28 individuals remain (officially 46) (Parnikoza *et al.* 2009; Smagol *et al.* 2010; Khoyetskyy 2011).

The reintroduction of European bison to Romania was initiated in 1958, when a pair of animals from Białowieża and Niepołomice (Poland) were brought to a captive breeding station at Hateg-Slivut Reserve in Hunedoara Province. Another breeding reserve was established in 1969 at Dragoš Voda, in the northeastern part of the country, with the first three animals brought there from Hateg-Slivut. In 1981, in the southern part of Romania, the Neagra-Bucšani Bison Reserve was established on the basis of bison from Hateg-Slivut, Okski Zapovednik in Russia, Bulgaria and Białowieża. Now at Hateg-Slivut live four individuals; however, in the past there were up to 13 animals there. The herd in the Neagra-Bucšani reserve numbers 34. The Dragoš Voda Reserve, being a part of Vanatori-Neamt Natural Park, for a long time maintained only several animals. However, since 2004 animals from Romania (Neagra- Bucšani), Switzerland (Goldau, Berno) and Germany (Springe, Hardehausen, Karslruhe) have gradually been brought to the site. In 2011 this herd reached 24 animals living in the 176-ha enclosure. The first group of five animals was released to the wild in March 2012 (Perzanowski & Marszałek 2012).

The most recently established (2008) breeding centre for bison in Romania is Vama Buzalaui Bison Reserve at Brasov county. At present this herd has 11 individuals (Pascu 2009).

In the northern part of Hungary, close to its border with Slovakia at the village of Füzérkomlós, a breeding enclosure for European bison has been built. In 2009 five animals were brought from Poland.

Table 10.4 Categories of free-ranging and semi-free populations of European bison listed in the *European Bison Pedigree Book* 2011

Name or number* of populations	Country	Number of animals
Populations demographically safe, consisting of >100 animals		
Puszcza Białowieska	Poland	481
Belovezhskaja Pushha	Belarus	429
	Total for Białowieża Forest	910
Bieszczady	Poland	277
Orlovskoe Poles'e National'nijj Park	Russia	240
Osipovichskijj Leskhoz Mogilevskojj Oblasti	Belarus	186
Kolkhoz 'Ozery' Grodnenskojj Oblasti	Belarus	164
Puszcza Knyszyńska	Poland	103
Populations consisting of <100 animals, functioning at risk of the loss of genetic variability		
Puszcza Borecka	Poland	91
Uladivska	Ukraine	91
Stado Zachodniopomorskie	Poland	89
Poleskijj	Belarus	85
Pripjatskijj National Park	Belarus	83
Nalibokskijj Zakaznik	Belarus	79
Panavėžys-Pašilių Stumbrynas	Lithuania	72
Populations consisting of <50 animals, not ensuring the normal population functioning over a long period		
Ten populations	Russia	244
Five populations	Ukraine	134
Berezyna	Belarus	35
Narodny Park Poloniny	Slovakia	9
Semi-free populations		
Cherga	Russia	48
Neagra Buscani	Romania	43
Vanatori Neamt	Romania	24
Reserve Biologique Des Mont' D'azur	France	36
Dőberitzer Heide	Germany	28
Zalisska	Ukraine	21
Mogilev	Belarus	14
Plejjstocenovyjj Park 'Sakha Jakutija'	Russia	5
Total		**3111**

Names of breeding centres after the *European Bison Pedigree Book* (2011).
Source: Krasińska and Krasiński (2007), updated.

The project initiated in 1998 by the Large Herbivore Initiative and Polish Academy of Sciences continues now with the Bison Friends Society and the European Bison Conservation Centre. It aims at the creation of viable populations – one at the borders of Poland, Slovakia and Ukraine and the second at the border of Romania and Ukraine – and further dissemination of animals through natural dispersion along the Carpathian arch (Figure 10.6).

Descriptive notes

Body measurements: **Shoulder height:** 152–188 cm; **Head and body length:** 236–300 cm; **Tail length:** 30–60 cm; **Body mass:** 340–840 kg. **Dental formula:** I 0/3, C 0/1, P 3/3, M 3/3 (×2) = 32. The number of teeth in European bison is 32, with 12 in the upper jaw, and 20 in the lower jaw. Differences between sexes are described in body mass and body size sections.

The European bison is Europe's largest extant mammal. Its strong and compact outline gives the impression of the animal being thick-set, but a sense of majesty emanates from the beast, as well as of the force of primeval nature. Its powerful body is covered with a reddish-brown coat that blends remarkably well with the animal's surroundings. Even experienced bison watchers can, in a forest, mistake a bison for the fallen trunk of a tree (Krasiński 1993).

European bison display a distinct sexual dimorphism in their build. Bulls are markedly larger and more heavily built than cows, with a greater disproportion between the hind and forequarters of the body (Figure 10.7). The bull's particularly tall spinal processes of the thoracic vertebrae are well muscled (Krysiak & Świeżynski 1967), the effect being to form the hump conferring that very impressive appearance upon males. The humps of cows are much smaller (Figure 10.8; Krysiak 1951; Kobryńczuk 2000, 2008).

The trunk of the European bison is relatively short and narrow, which makes its movements through forest vegetation easier. The head is large and low-set, its profile slightly convex. The ears are small, quite wide and barely distinguishable from the fur. The fur of adult animals is brownish, but there is an individual variety of shades in their colouring (between tawny or dark brown), which was indicated by Wróblewski (1927). After birth, calves are reddish-brown, but after the first change of coat, which occurs about 3–4 month of life, they become tawny-brown. Late-born calves (end of autumn to beginning of winter) already have a colouring after the first moult like calves born in spring.

The sides of the head and lower parts of the legs are darker. The head, neck and front of the body are covered with longer hair, which along the lower part of the neck and the breast form a so-called 'beard'. The rear part of the body is covered with short fur and ends with a tail, with long hair reaching the ankle joint (see also Figure 10.2). The skin of the European bison is thick and elastic.

Adult European bison begin to moult at the end of winter/beginning of spring, most often at the beginning of March. The first moult is observed in bulls around the head and neck. Depending on individual condition, the moult may be

Figure 10.7 The fine bull 'Poranek' (pedigree number 980) was one of the most impressive individuals known. His body mass reached 920 kg – a record never beaten at Białowieża. Photo by Z. A. Krasiński.

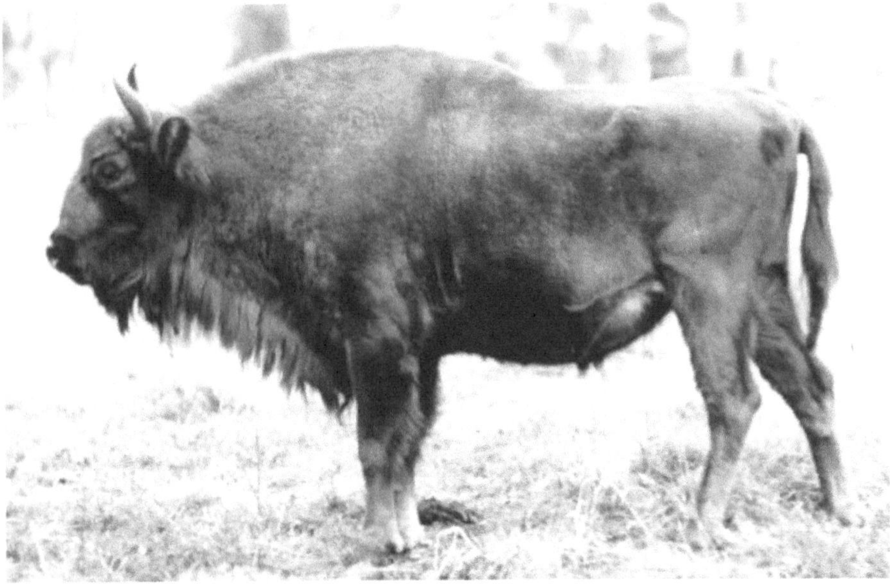

Figure 10.8 A typical cow of lowland European bison in Białowieża Breeding Centre. Photo from Z. A. Krasinski's archive.

short or continue even until summer. According to Kiseleva (1974), in the Okskij Breeding Centre (Russia), moulting of European bison begins in mid-February. In males, the moult lasts on average 128 days, in females 183, in yearlings 157 and in calves 163 days. During the moult, animals get rid of their winter coat, intensively rubbing themselves against wind-thrown trees, rough tree trunks and stumps. Such items, described by Wróblewski (1927) as 'bison combs', become well polished during the spring, and covered with hanging tufts of shed coat.

The brown eyes of European bison are relatively small and set in protruding eye sockets. Those animals do not have very good eyesight, and may not even notice a man wearing camouflage clothing, standing still among trees (Krasiński 1993).

The muzzle of the European bison is black, with its upper part surrounded with a silver-white ring of hair. The hooves are black, though sometimes they may have lighter, ivory-like colouring. The udder in a cow is small and highly set. The prepuce in males ends with a tuft of hair; the scrotum is small and situated close to the underbelly.

While wandering through the forest, European bison leave distinct tracks (Figure 10.9). Tracks of the front hooves are larger than those of the hind legs. Tracks of males are larger than those of females. Individuals of both sexes and of

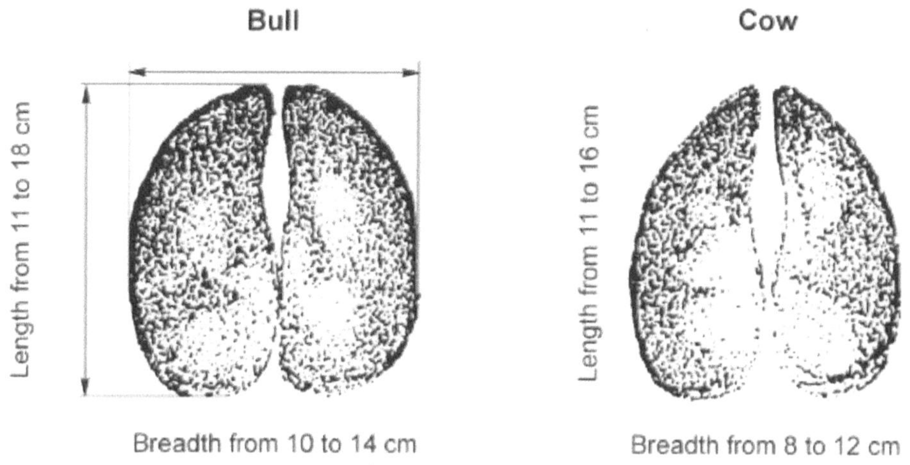

Figure 10.9 A sketch of the tracks of adult bison. Source: Krasińska & Krasiński, 2007; photo by M. Januszczak.

any age emit a musky smell that remains for some time in places frequented by these animals. Because the whole coat of the European bison is saturated with this smell, it may be assumed that the smell is produced by vesicular glands of skin (tallow), with ducts that lead to the area of hair follicles.

European bison never roar. The term used for the call of a bison is 'grunt'. Most frequently it is given by cows communicating with calves; bulls emit such a call mostly during the rut. Cows unmistakably identify the voices of their calves, even among a large group, and calves find their mothers by calling them with grunts for feeding. The voice of a bull is lower and hoarser than that of a cow.

The sense of smell is well developed in European bison. Males follow herds during the rut by smelling tracks of cows left on the ground. Members of a herd that are temporarily separated from the group act in a similar fashion.

European bison are able to run quite fast, but only for a short distance. Usually after running for some 100 m they stop and pant heavily.

Both sexes have horns, growing close to the rear edge of the head. Calves have nuclei of horns at birth. In two-year-old animals, horns begin to curve inwards. This curvature increases as the animal grows, particularly in females, which have horns that are much more strongly curved than those of males. The horns of bulls are wider set, longer and thicker than those of cows. In both sexes horns are usually completely black; however, in the free-ranging herd of Białowieża Forest some individuals have an ivory tint at the tips of their horns. Ends of horns in old males are often abraded as a result of many years of struggling with various obstacles like fallen trees, tree stumps or young spruces uprooted during the rutting season. Because of the sickle-like shape of the horns, the numerous contests among animals usually do not have serious consequences. A maximal distance between horn curves of adult bulls ranges across 63–84 cm, and the horn length is 25–70 cm; in adult cows those parameters are 47–70 cm and 41–52 cm, respectively (Pucek 1986; Krasińska & Krasiński 2002).

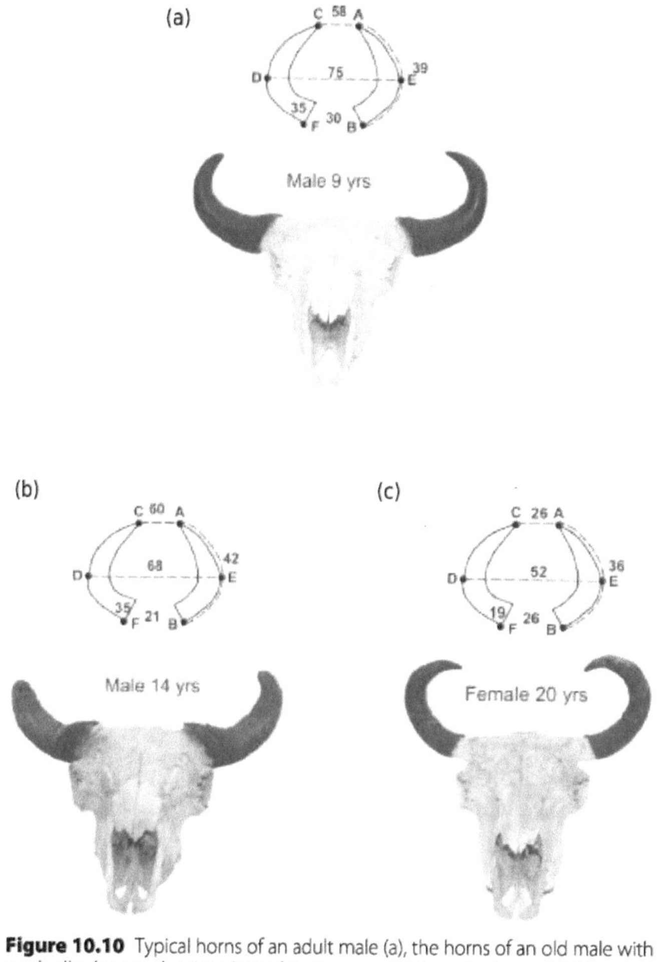

Figure 10.10 Typical horns of an adult male (a), the horns of an old male with markedly damaged points. Irregular rings are visible at the bases (b). The corneous sheaths of female horns are distinctly pointing upward, and the ends of the horns are sharp (c).
Source: Krasińska & Krasiński, 2007.

The skull of an adult bull is wide and massively built, with eye sockets to the sides. Protruding bony rims around eye sockets protect the eyes against damage by tree branches during movement through the forest. In males skulls are longer, wider and higher than in females; the largest differences are observed in the width at the eye sockets (Figure 10.10). Additionally, skulls of males have a greater cavity volume than those of females (Kobryńczuk *et al.* 2008; Krasińska *et al.* 2008). The powerful head is held up by a sinewy *ligamentum nuchae.* This ligament and the *ligamentum flavum* between the spinal processes of the thoracic vertebrae allow for holding in place and rising of the head while foraging at the forest floor.

The spinal column of European bison consists of seven cervical, 14 dorsal, five lumbar, five sacral and 13–16 caudal vertebrae (Roskosz 1962; Empel & Roskosz 1963). In European bison there are 14 pairs of ribs, while in domestic cattle there are only 13 pairs. This was noticed for the first time by an eminent zoologist and anatomist, Ludwik Henryk Bojanus (1827), from Vilnius University.

Body mass

Body mass of the European bison kept in captivity is higher than that of animals in the wild. This difference is noticeable already by the age of six months. In the free-ranging population of Białowieża Forest, the average body mass of adult European bison (six years or older) is 634 kg (436–840 kg) in males, while in females it is 424 kg (340–540 kg). The average body weight of animals from the breeding reserve is 747 kg (580–920 kg), and in cows 460 kg (320–640 kg) (Table 10.5). The maximal weight of a bull from the breeding reserve is 920 kg and a cow 640 kg, while for animals in the wild it is 840 kg and 540 kg, respectively (Figure 10.11). In the Belarusian population, the average body mass of a bull is 635 kg (maximally 850 kg), and the cow 435 kg (maximally 540 kg) (Sipko *et al.* 2002; Krasochko *et al.* 2004; Bunevich 2006; Kozlo & Bunevich 2009). Animals of the mixed LC line living in Okskij Breeding Centre (Russia) have body weights similar to those from the Polish population of Białowieża Forest.

We have access to only a few records on the weight of European bison in the past. At the obelisk raised at Białowieża (Poland) to commemorate the hunt of King August III the Saxon in 1752, there is an inscription stating that among those killed were 42 European bison, the largest weighed at 14 quintals and 50 pounds (about 720 kg). According to Bojanus (1827) the bison that was killed in 1612 weighed 1770 pounds (a little over 700 kg). Values of body mass of European bison from the last natural population of Białowieża Forest, given by Wróblewski (1927), were much lower than presently recorded.

Body size

Average dimensions of adult bulls at six years of age and older are: height at withers 172 cm (maximal 188 cm), body length 257 cm (maximal 300 cm), heart girth 247 cm (maximal 280 cm). The same dimensions in cows are respectively as follows: 152 cm (maximal 167 cm), 236 cm (maximal 270 cm) and 209 cm (maximal 246 cm) (Krasińska & Krasiński 2002) (Figure 10.11),

In the Belarusian population, dimensions of adult animals are similar or slightly larger, which may be a result of a different method used to take the measurements (Sipko *et al.* 2002; Kozlo & Bunevich 2009).

Comparing body mass and dimensions of contemporary animals and those living in a distant past, it is possible to say that the high level of inbreeding among lowland European bison (Olech 2003) is not reflected inauspiciously in the development, build and body size of those animals. At the point of reaching physical maturity, contemporary European bison are correctly developed and achieve body mass and dimensions even considerably larger that those given by Wróblewski (1927) for animals belonging to the last natural population of Białowieża Forest, before World War I.

Life span

Precise determination of life span in European bison can be achieved by reference to data contained in the *European Bison Pedigree Book.* The analysis of longevity among bison kept in captivity reveals that females have life spans of over 30 years in some cases, and males – even exceptionally – just up to 20 (Koch 1956; Krasińska & Krasiński 2007; EBPB 2010). In the free-ranging population the oldest marked female lived to be 24 – far beyond the maximum of 18 years noted for marked bulls. Overall, then, females can be said to pass the 25-year mark only rarely, while bulls rarely exceed 20 years of age. Bison in the wild have shorter lives than those in captivity, although the determination of age using teeth-wear pattern and horns is obviously less exact than reference to known dates of birth.

Habitat

By the end of 2011 the *European Bison Pedigree Book* registered 34 free-ranging populations in six countries. The detailed ecological studies on spatial distribution, social structure, habitat use or the natural diet of that species were conducted regularly only in a few populations (in both parts of Białowieża Forest and the Bieszczady Mountains). Other populations were studied only occasionally, e.g. Knyszyńska and Borecka Forests (Poland), or they are only at their beginning (other populations of Belarus, Russia and Slovakia). Data on feeding strategy, reproduction and various kinds of behaviour come mostly from the population living in the Polish part of Białowieża Forest or from captivity, which makes it impossible to assess the variability of those parameters in various habitat types.

Only recently did the assessment of habitat types used by European bison in ancient times become possible. Kuemmerle *et al.* (2012), using a time-calibrated niche model together with the model of vegetation dynamics, found that European bison existed during the Holocene either in semi-open areas or in broadleaved, mixed and coniferous forests. The areas inhabited by the European bison were characterised by high vegetation cover and a blend of woody vegetation and grass. During the last 8000 years, growth of the human population and expansion of agriculture reduced and fragmented European bison habitats. According to some authors (Heptner *et al.* 1966; Kaplan *et al.* 2009; Kuemmerle *et al.* 2011), the reason that

Table 10.5 Comparison of body masses (kg) of European bison living in the free-ranging population and at the captive breeding centre in Białowieża Forest

Age	Free-ranging population			Captive breeding		
	Mean	Min.	Max.	Mean	Min.	Max.
Males						
0	–	–	–	28	16	35
1 month	–	–	–	52	32	83
2 months	41	35	48	–	–	–
3 months	53	32	70	–	–	–
4 months	76	46	96	83	82	84
5 months	82	48	108			
6 months	99	90	160	143	131	160
7 months	101	78	140	150	–	–
8 months	103	70	180	170	130	180
9 months	109	80	140	–	–	–
10 months	–	–	–	–	–	–
11 months	–	–	–	187	155	220
12 months	124	80	170	175	130	220
2 years	220	130	300	331	220	453
3 years	303	240	390	387	310	502
4 years	366	260	430	502	380	609
5 years	467	345	660	591	572	610
≥6 years	634	436	840	747	580	920
Females						
0	–	–	–	24	15	33
1 month	–	–	–	30	–	–
2 months		–	–	–	–	–
3 months	47	32	70	–	–	–
4 months	66	43	81	66	65	67
5 months	72	48	100	–	–	–
6 months	91	52	120	111	105	118
7 months	97	62	148	129	120	148
8 months	111	70	190	159	148	170
9 months	124	96	146	–	–	–
10 months	–	–	–	–	–	–
11 months	–	–	–	200	–	–
12 months	119	96	140	190	170	220
2 years	213	170	280	228	170	306
3 years	252	150	310	227	220	240
4 years	341	260	458	400	378	432
5 years	407	340	460	439	420	458
≥6 years	424	340	540	460	320	640

After Krasińska & Krasiński (2002).

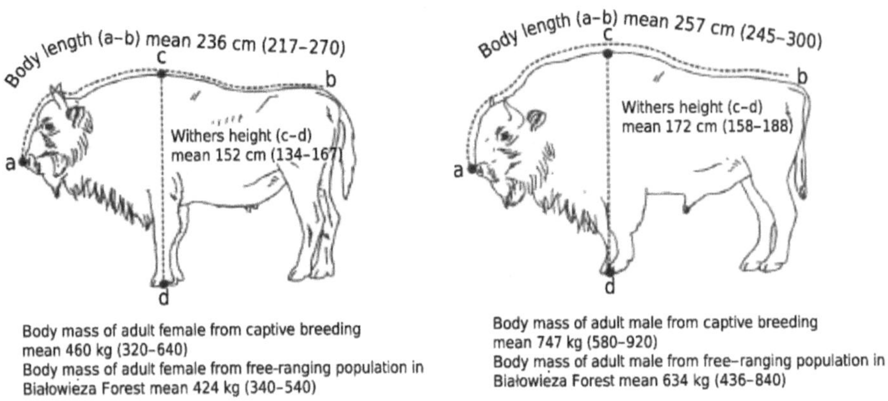

Figure 10.11 The silhouette of an adult male bison (a bull), and an adult female (a cow). Source: Krasińska & Krasiński, 2007.

European bison survived the longest in the north of Central and Eastern Europe was delayed expansion of agriculture, lower human density and the fact that suitable forest cover was retained in this region until the nineteenth century. The last natural population of lowland bison survived until 1919 in the Białowieża Primeval Forest. Wróblewski (1927), who studied this last population, found that the most willingly frequented habitats were mixed and deciduous forests (Tilio-carpinetum), alder woods and meadows in valleys of rivers flowing through the forest, used in various proportions in different seasons of the year.

During the species' restitution, European bison were introduced into various types of habitats belonging to continental or mountain regions (the Carpathians, Caucasus Mountains and Altai Mountains). The free-ranging population of the LB line (in Poland and Belarus) most frequently inhabits mixed/deciduous forests, and sometimes also forest complexes dominated by coniferous stands, including various proportions of mown meadows and forest glades. Animals of the LC line were introduced into forests of continental regions, and in the subzone of the southern taiga (one population in Vologda region, a northern region of the European part of Russia), as well as in mountain regions of Poland, Slovakia, Ukraine and Russia. Only one population of the LB line of European bison inhabits foothills of the Altai Mountains (east Russia). In Poland, four populations of the LB line live in the continental region, including two in mixed forest/field habitats and two in lowland forests. The only Polish population of the LC line has been introduced to the mountain region of Bieszczady Mountains, part of the Carpathians.

Białowieża Forest

Białowieża Forest is one of the best preserved lowland forests in Europe; it is located at the Polish–Belarusian border and its surroundings (52°30'–52°54' N, 23°26'–23°56' E), covering 1500 km^2. As a consequence of World War II, the Białowieża Forest became bisected by the state border, the western part remaining in Poland, while the eastern part found itself in the USSR (within the Belarusian SSR). The eastern (Belarusian) part is larger at 875 km^2, while the western part located in Poland occupies 625 km^2 (Sokołowski 2004). The forest is situated within the belt of lowlands running across Eastern Europe. The terrain there is almost entirely flat, and its average altitude is 170 m above sea level. This area is situated within the zone of temperate climate.

In the Polish part of Białowieża Forest, forests constitute 95% of the area, with the remainder covered by meadows (0.1%), fields, water, wastelands and roads, and forest clearings (0.2%). In this area the prevailing habitats are dominated by broadleaved species. Mixed, fresh and moist deciduous forests account there for 61.5% of the forested area, while alder woods cover 13.5%. The coniferous forest habitats (25% of area) are in turn dominated by fresh mixed coniferous stands (over 17% of the area) (Gątkiewicz & Zbrożek 2003; Sokołowski 2004).

In the Belarusian part, forest covers 89%, and all types of non-forest land make-up the remaining 11%. The most common tree stands are those with a predominance of coniferous species, which account for 69% of the forested area. Deciduous forests occupy only 12% of the forest area there and alder woods 19%. Open land accounts for just 2% of the area (comprising fields, forest glades and meadows), but is an important component providing a considerable part of the feeding base for ungulates (Kozul'ko & Zhukov 1999; Bunevich 2003).

Poland and Belarus established two separate free-ranging populations of European bison in their respective parts of the Białowieża Forest. These developing populations did not make effective contact with each other, remaining mostly isolated due to the species' typically sedentary lifestyle. Only very occasionally observed was the migration of single bulls across the state border. However, the construction of a fence along the border in 1981 terminated even this limited natural exchange of individuals.

Between 1976–1989 the density of European bison within its home range at Białowieża Forest was very high, ranging from 13 to 16 animals per 10 km^2 (Krasiński 1978a; Krasińska & Krasiński 1994). Since 1989, when the population home range started to extend, its density diminished to six individuals per 10 km^2, and to date remains at the same level.

At the Belarusian part of Białowieża Forest the density between 1970–1984 was at the level of 12 individuals

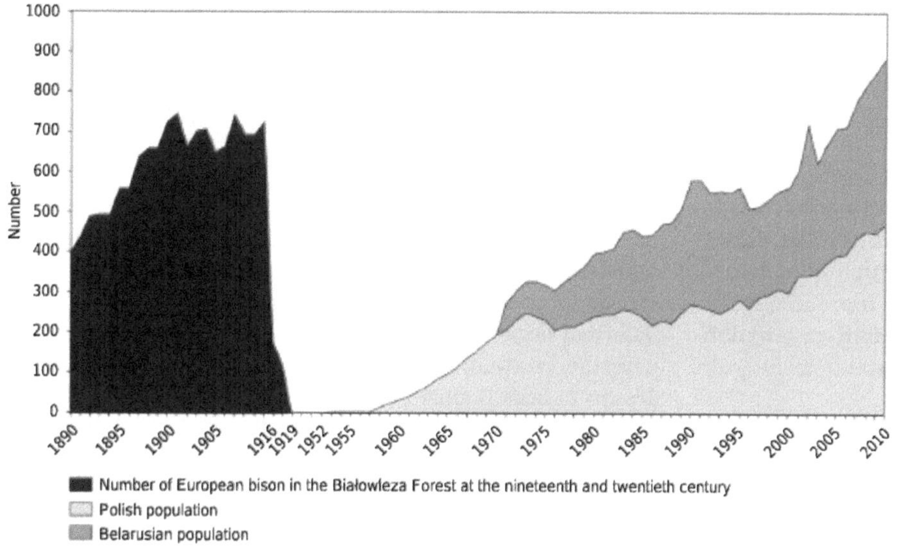

Figure 10.12 Numbers of European bison in the Białowieża Forest in 1890–2010. Data from Karcov, 1903; Rörig, 1918 (after Pucek, 1994); Sztolcman, 1924; Wróblewski, 1927; Krasiński 1978a; Krasińska & Krasiński 2007; Bunevich, 2003; Bunevich *et al.* 2006; Kozlo & Bunevich 2009; European Bison Pedigree Book 2008–2010.
Source: Krasińska & Krasiński, 2007, modified.

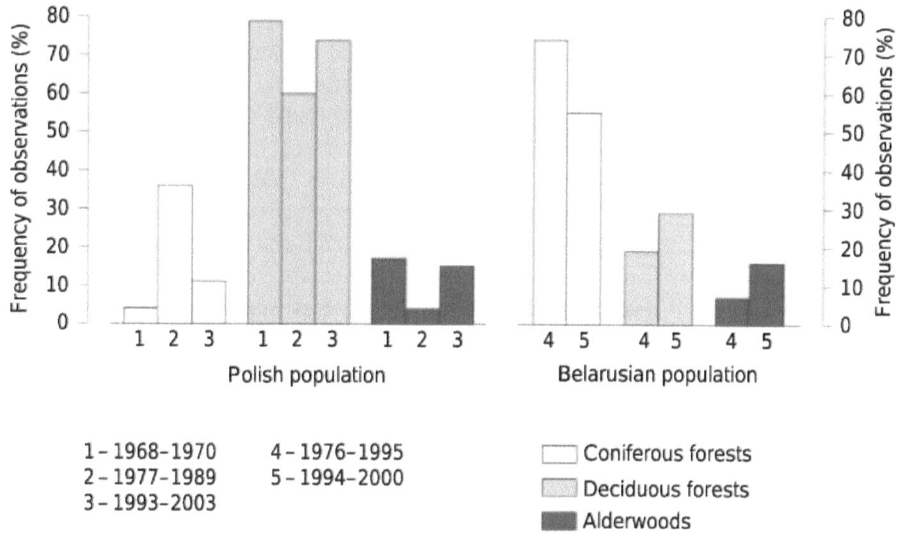

Figure 10.13 Utilisation of various forest types by European bison in the Polish and Belarusian parts of Białowieża Forest, in the vegetative seasons of 1976–2003.
Source: Krasińska & Krasiński, 2007.

per 10 km^2, and decreased to three individuals per 10 km^2 as the home range increased (Bunevich & Kochko 1988; Bunevich 2003).

By the end of 2011 the whole Białowieża Forest was inhabited by 910 bison (481 in the Polish part and 429 in the Belarusian). Those numbers exceed the size of the last wild population living there at the beginning of the twentieth century (Figure 10.12).

Habitat differences between both parts of Białowieża Forest cause different uses of the environment by European bison. In the Polish part the main habitats inhabited by bison are mixed, fresh and moist deciduous forests (54–79% of observations) (Borowski & Kossak 1972; Krasiński 1978a; Krasińska *et al.* 1987; Daleszczyk *et al.* 2007) (Figure 10.13). Utilisation of alder woods (wet forests with an admixture of ash) fluctuated from 4.5% to 17%. The use of alder woods in the last decade has increased to 15% of animals' observations, probably because of climatic changes, since in dry and warm years alder woods become more accessible to bison as water level drops. Bison rarely used fresh mixed coniferous forest, with the exception of the period between 1976 and 1995 (Krasiński 1978a; Krasińska *et al.* 1987; Daleszczyk *et al.* 2007).

The patterns of habitat utilisation by the bison from the Belarusian part of Białowieża Forest were quite different (Figure 10.13). Mostly fresh and wet coniferous forests were utilised there, as well as fresh mixed coniferous stands (about 70% of observations). The frequency of utilisation of deciduous forests and alder woods is distinctly lower though (Bunevich 1990). Nevertheless, European bison there have increased their use of deciduous forests and alder woods since the mid 1990s (Daleszczyk *et al.* 2007).

The main foods of European bison, being typical ruminants, are herbs and grasses covering the forest floor and grassland habitats. In Białowieża Forest, phenological differences in the development of vegetation in various forest habitats influence seasonal changes in bison environmental selectivity

(Falińska 1973). In deciduous and mixed forests feeding bison can be observed throughout the growing season. In these habitats plants of both short and long growing seasons occur, the latter lasting into late autumn. In spring (April–May) bison concentrate their foraging activity in deciduous forests (51% of observations). The fresh mixed coniferous forest becomes increasingly visited in June and July (71% of observations), when deciduous forests are less used (22%). Alder forest is most often visited in early spring, only rarely in summer (and then, during droughts), but once again more frequently in autumn. From August to late autumn penetration of deciduous forests by the bison increases (up to 58% of observations) (Krasińska *et al.* 1987).

In the Belarusian part of Białowieża Forest, the utilisation of coniferous forests increases in the first half of the vegetative season, and declines between August and October. During summer the utilisation of mixed and fresh deciduous forests and alder woods increases. In autumn bison tend to move to open areas (Krasiński *et al.* 1994a; Daleszczyk *et al.* 2007).

The utilisation of open areas (clear-cuts, young plantations, forest glades and meadows) within the area of Białowieża Forest depends on the way glades/meadows and the forest are managed. Forest clearings and mown meadows are used mostly during spring and from August to October. It is known from earlier studies that until the beginning of the 1980s bison, both in Polish and Belarusian parts of Białowieża Forest, frequented to the same degree forest and open habitats, as long as the latter remained managed (Korochkina 1973; Krasiński 1978a; Krasiński *et al.* 1994a). Termination of mowing of forest meadows or its limitation (in the Belarusian part) caused a decline of the use of those areas by bison in both parts of Białowieża Forest (Daleszczyk *et al.* 2007; Krasińska & Krasiński 2007). In the Polish part of Białowieża Forest, re-cultivation of glades and meadows and restoration of their regular mowing since 2006 caused an increase in bison use of open areas within and outside the forest (Krasińska *et al.* 2010). Radio-tracking data collected between 2006 and 2010 showed strong preference of European bison to open areas situated within the home range of the population; this includes both openings inside the forest and agricultural lands beyond its boundaries (Krasińska & Krasiński 2007; Hofman-Kamińska & Kowalczyk 2010; Kowalczyk 2010).

Winter feeding of ungulates has a long tradition in the area of Białowieża Forest. As early as the sixteenth and seventeenth centuries, hay collected at forest meadows was left in haystacks for winter to provide food for ungulates (Hedemann 1939). At present, in both parts of Białowieża Forest, bison are fed in winter with hay at permanent feeding points, which causes an aggregation of animals around such sites. On one hand, supplemental feeding contributes to mitigation of damage to tree stands in winter and decreases natural mortality, but on the other hand, such concentration of a large number of animals in winter may be harmful to their health and change their use of natural forages. During snowless winters European bison do not feed much upon the supplemental food, but consume browse and tree bark (Krasińska & Krasiński 2007). At present, about 14% of the population feeds in winter at agricultural areas outside of Białowieża Forest. Animals search under the snow for winter crops: cereals and rape-seed. The most preferred are winter rye and rape-seed (Hofman-Kamińska & Kowalczyk 2010)

In both parts of Białowieża Forest habitat selection by adult bulls and mixed herds is different. Adult bulls prefer fresh and moist deciduous forest and avoid all categories of coniferous forests, whereas mixed groups use more types of forest than bulls and negatively select only a category of coniferous forest with the exception of fresh mixed coniferous forest. Mixed groups in both populations positively select deciduous forests, whereas alder woods are used proportionally to their availability (Daleszczyk *et al.* 2007).

Other lowland populations

In other European bison populations in the northeast of Poland (e.g. Borecka Forest) between 1962 and 1991 the fresh deciduous forest was also the habitat mostly frequented by the bison (97% of observations in the forest). During the vegetative season, animals used forest habitat more often (63% of observations) than open areas (37%). Among categories of open areas where bison used to feed, most frequently used were meadows (75% of observations), clear-cuts and young plantations (15%) and cultivated fields (10%). Cultivated fields (rape-seed, winter crops) were used frequently in early spring and late autumn. In Borecka Forest supplemental winter feeding is applied routinely (Krasiński & Krasińska 1992). Between 2002 and 2006, bison there mostly were present in forest habitats, but less frequently compared to the earlier period (65% of observations), whereas in open areas (meadows and young plantations) in 35% of cases. The use of fresh deciduous forest decreased in this period (56% of observations). Between June and August they were also recorded in alder woods (4% of observations). Seasonally (in May), they used the mixed fresh deciduous forest. Between July and December bison fed outside of the forest at cultivated fields. However, a sudden (following a hurricane in 2002) increase in the proportion of open areas within the Borecka Forest (a belt about 2 km wide across the whole forest complex) caused a decrease of the home range of this population and reduction of summer movements outside of the forest (Żoch 2007).

The lowland population of European bison from Volozhinskaya living in Nalibokskaya Pushcha (western part of Belarus) demonstrates a different manner of habitat use. These animals forage during spring and midsummer at cultivated meadows situated close to the forest, while in the autumn and winter they remain within tree stands. During autumn, bison mostly forage in fresh and moist types of forest (birch, alder and mixed spruce–deciduous forests) (over 64% of observations). Nevertheless, their basic habitats are open areas – cultivated meadows of 4 km^2 and mown meadows along the rivers (Shakun 2011).

Lowland bison of the West-Pomeranian population (northwest Poland) occupy a mixed forest–field landscape in which forests occupy about 42% of the population home range, and cultivated fields over 50%. In that population, open areas – mostly cultivated fields – are the basic foraging ground for European bison. They prefer large-scale fields where they are not disturbed by farmers (Janicka *et al.* 2008; Karnecki 2008; Tracz *et al.* 2008). Similarly, in Knyszyńska Forest (Poland), where coniferous stands dominate, every year from late autumn up to the early spring, the whole herd leaves the forest and moves to cultivated fields. Between 2005 and 2007, together with an increase of population numbers, its home range increased up 250 km^2, out of which only 74 km^2 (i.e. 6% of Knyszyńska Forest area) are forests (A. Lickiewicz *et al.* unpublished; Krasińska & Krasiński 2007). Animals mostly use forest habitats in summer, while in autumn and winter they move towards the north into agricultural areas, mostly into fields sown with winter rye and rape-seed (Hofman-Kamińska & Kowalczyk 2010). In Lithuania the wild bison population occupies a mixed forest–field landscape, where the buffer zone comprising mixed deciduous forest covers some 50 km^2. Those bison feed intensively at fields and managed meadows (Balčiauskas 1999).

The Carpathians

The Carpathians are the mountain chain covering an area of about 210 000 km^2, extending over six countries from the borders of Austria, through the Czech Republic, Hungary, Slovakia, Poland, Ukraine and Romania, to the borders of Serbia. They form a continuous eco-region, connecting Central and Southeastern Europe, creating a corridor allowing for migrations of numerous wildlife species. The region is a mainstay of many large mammals, including brown bear, lynx, wolf, red and roe deer, chamois, wild boar and, for the last 50 years, bison (Webster *et al.* 2001).

In the ranges of the Carpathians inhabited by European bison, the dominating habitat is the mixed beech–fir forest, i.e. the Carpathian beechwood (*Fagetum carpaticum*). The analysis done for all free-ranging herds in the eco-region showed that bison there prefer forest-dominated habitat with a mosaic of grassland patches, situated in areas of low human disturbance. No clear preference for particular forest habitats was found, though bison preferred managed grasslands over fallow and abandoned fields. All other land cover types were avoided. Also, the presence of protected areas was important in determining a location's suitability for bison (Kuemmerle *et al.* 2010).

In the Bieszczady Mountains (Poland), where bison of the LC line were first reintroduced (1963) (Figure 10.6; see also Figure 10.32) and where so far the absolute majority of research on this species related to its life in mountain conditions was performed, the percentage of forest within the home ranges of bison is 85% on average. This region of the Carpathians belongs to the lowest parts of this mountain chain, with its highest peak reaching only 1346 m above sea level. Human population density there is extremely low for Europe, at 5–15 people per km^2 as a result of compulsory resettlements imposed there after World War II. Therefore, the proportion of cultivated fields is marginal in the region, and the majority of open areas there are composed of meadows or pastures in various stages of secondary succession (Augustyn & Kozak 1997; Winnicki & Zemanek 2009).

In winter 93.0% of bison records there were coming from the forested area, while 7.0% were from open areas outside (meadows, pastures, fallow land) or inside of forest complexes (glades, openings, etc.). In vegetative seasons, only 3.4% of records of bison presence were from open areas, and as much as 96.6% from the forest. In relation to availability of both forms of land cover, it has been found that bison there were slightly avoiding open areas (Perzanowski *et al.* 2011b).

Data from Poloniny National Park (Slovakia), where nine bison of the LC line were introduced to the wild in 2004, show that within the summer home range of this herd 72.6% constituted forested areas, and 27.4% open areas in various stages of secondary succession, while in winter this proportion was 51.1% and 48.9%, respectively. The site selected for introduction there is within the former area of several villages, which in the 1970s were compulsorily abandoned in order to create a protective zone for a dam reservoir, Starina (Pčola *et al.* 2006).

Within home ranges of Bieszczady bison herds (MCP) in the vegetative season the most intensively penetrated (percentage of bison records) tree stands, depending on a dominating tree species, were: beech (*Fagus sylvatica*) (61.5%), spruce (*Picea abies*) (10%) and mixed deciduous–coniferous stands (12%). In winter those were: beech (44%), scotch pine (*Pinus sylvestris*) (23%) and alder (*Alnus incana*) stands (20%).

According to the results of the analysis of stand preference (frequency of use/stand availability), the most preferred in winter were pine and ash–sycamore (*Fraxinus excelsior/Acer pseudoplatanus*) stands, while in summer those were spruce and mixed coniferous stands.

In the vegetative season, bison were most often present at stands with loose canopy (68% of records), while in winter they selected stands with closed canopy (78% of records) (Perzanowski *et al.* 2008).

At Poloniny National Park, among available forest types, bison frequented mostly associations belonging to *Fagetum typicum* (83.3% of records), and stands in the age of 21–40 years (41.7% of records). Out of available categories of open areas, most often bison were observed at abandoned meadows (43.6%) and abandoned orchards (21.8%). Over the whole year, the most frequently visited land cover type was abandoned meadows (26.3% of observations), followed by forest association *Dentario glandulosae–Fagetum* (20.9%) and abandoned orchards (13.2%) (Pčola *et al.* 2006).

Regarding the use of the terrain, bison of Bieszczady in summer seasons were most frequently recorded at elevations of 550–849 m above sea level (92.9%), while in winter they were found at 450–749 m above sea level (98.0%). No records of bison's presence were from elevations below 450 m above sea level in summer, or from above 949 m above sea level in winter. Those seasonal differences are mostly connected with presence and depth of snow cover, which at higher elevations (above 700–800 m) may last even from late October to early April, and exceed 1 m depth. In spring, in their vertical movements, bison follow the development of fresh vegetation, while in autumn they move towards lower elevations, usually synchronically with the appearance of first snow.

In Slovakia, a free-ranging herd was most frequently observed within elevations of 500–700 m above sea level, but occasionally they were recorded as high as 900 m and down in a valley at 360 m (Pčola *et al.* 2006).

Comparing bison's presence over three categories of terrain features (slopes, valleys and mountain ridges), it has been found that bison slightly preferred slopes in both seasons (Ivlev's index values: +0.048 in vegetative season and +0.050 in winter) while valleys and ridges were avoided (–0.93 and –0.88; –0.33 and –0.55, respectively) (Perzanowski *et al.* 2011a, 2012).

In both Carpathian populations, bison do not rely much on a supplemental winter feeding, only when a deep snow hampers their movements they more often use haystacks or hay-silage, which in this region are commonly used for winter feeding of deer.

So far in the Carpathians there have not been studies concerning the differences in habitat use by bulls and cows, as well as by animals from various age classes.

Concluding, it is possible to say that optimal habitats for lowland European bison in the continental region are areas dominated by mixed/deciduous forests, with some share of alder woods with an ash and fresh mixed coniferous forest. In the areas where coniferous stands dominate, the significance of open areas increases because they provide important supplemental forage for bison. Within forest complexes where a large percentage of the areas are open (meadows, glades, etc.) bison may seasonally use such habitats to a high degree (20–45% of observations). For proper functioning of contemporary European bison populations living mostly in the forest, the heterogeneity of the environment is indispensable, assuring them a possibility of finding adequate food in all seasons of the year (Krasiński 2010). In the case of populations inhabiting landscapes of forests mixed with fields, their main habitats are cultivated fields and meadows. Development of such populations depends on the social acceptance of the level of damage caused by bison to crops.

In mountains (e.g. the Carpathians) the pattern of the use of space and habitats by European bison of the LC line is different than in continental landscapes. In the Bieszczady Mountains, the main habitat for bison in the vegetative season is forest (96.6% of records of bison's presence), mainly *Fagetum typicum* association, while the use of open areas is very low there (only 3.4%). In relation to availability of both forms of land cover, it has been found that bison there were slightly avoiding open areas. For the bison from Poloniny National Park (Slovakia), introduced to the area of now depopulated former villages, the most frequently visited are open areas, such as abandoned meadows or orchards, followed by forest association *Dentario glandulosae-Fagetum*.

Movements and home range

Movements

In all contemporary, free-ranging populations of European bison, movements of males, sometimes for long distances (Figure 10.14), that do not have any connection with an abundance and availability of food resources are observed. Long-distance movements are usually of a one-way nature, and the majority of wandering males become either culled or poached. Such movements were observed in the first years after the creation of free-ranging populations for both lines of the species (Krasińska & Krasiński 2007). The most famous was a solitary bull that left the area of the Bieszczady Mountains (southeast Poland) and moved in a northerly direction over a distance of 300 km in 1966 and 700 km in 1981 (Figure 10.14). Departures of bulls in various directions from Borecka Forest (northeast Poland) over distances of 40–150 km are recorded relatively often (Krasińska & Krasiński 2007). Lone males from the other population living in the mixed forest–field lowland landscape of Poland (the West Pomeranian herd) wandered for distances of up to 100 km (Bereszyński 2005; Krasińska & Krasiński 2007).

In the Polish population of Białowieża Forest, the first movements outside of its boundaries took place in 1956. In 1969 an adult bull made a move to the Knyszyńska Forest – a distance of 30 km (Krasiński *et al.* 1994b). The longest movement of a bull from this population (34 km) was recorded in 2007. In the Belarusian part of the Białowieża Forest, long-distance movements of bulls have been recorded since the creation of the free-roaming population; bulls moved there on average over distances ranging from 10 km to 180 km. One of them even managed to reach Lithuania (350 km) (Korochkina 1973; Bunevich 2003).

A detailed analysis of the nature of bison movements beyond the forest complexes was done only in the Polish population of Białowieża Forest. During a period of 46 years (1964–2010) there were on average nine males each year (the annual range of 0–33 individuals) observed outside the forest. Agricultural lands were most frequently visited by lone bulls (50% of observations) and definitely less often by groups of 2–3 males (25%) (Figure 10.15). Such areas were most often visited by adult males of 5–18 years old (96% of males of known age). The average distance of bulls' movements from the forest edge was 1.8 ± 0.17 km (range: 0.1–33.9 km). After 1990 males were moving away from the forest for significantly longer distances (2.1 ± 0.18 km) compared to earlier years (0.9 ± 1.5 km). The annual number of males moving outside of forest boundaries towards agricultural land is significantly dependent on population numbers and the depth of

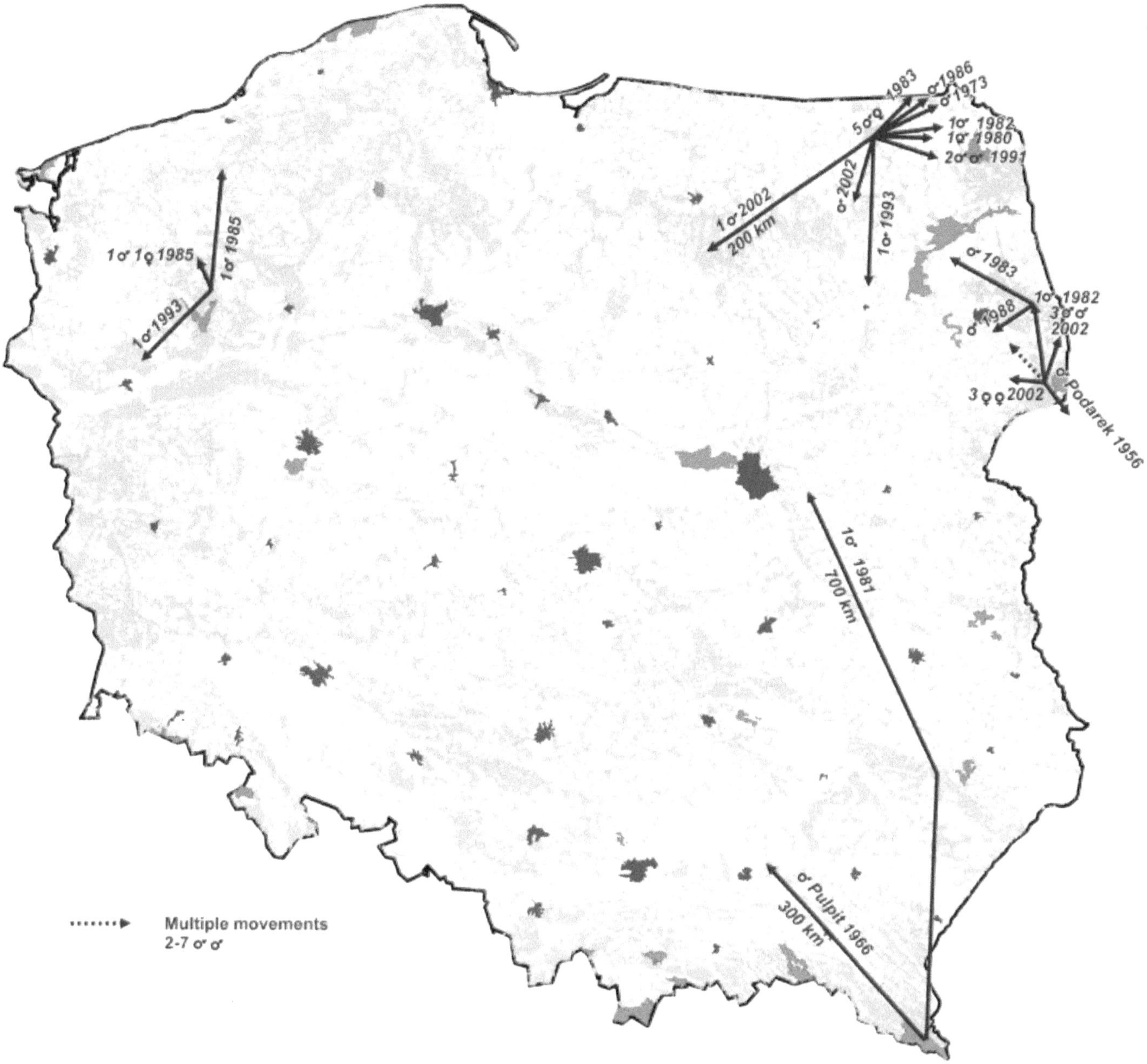

Figure 10.14 Long-distance movements of European bison recorded in Poland in 1956–2002. Source: Krasińska & Krasiński, 2007.

snow, but is not dependent on the proportion of adult males in the population or the ratio of adult males to cows (Kowalczyk *et al.* 2013).

Cows living in mixed groups are more numerous than bull groups, and demonstrate different spatial behaviour to males, related to the abundance and availability of the nutritional resources. Since the population of Białowieska Forest exceeded 260 individuals, the number of cows migrating to agricultural areas grew systematically from 26 in 1994 to 106 in 2010 (Kowalczyk *et al.* 2013). The number of bison in mixed groups out of the Białowieża Forest was correlated with population size (Kowalczyk *et al.* 2013). It may be supposed that at present, in the autumn–winter period there are not enough nutritional resources in the Białowieża Forest for a population numbering near 500 bison (Krasińska *et al.* 2010). The average distance of the movements of cows beyond the boundaries of the forest was 2.1 ± 0.18 km (range: 0.1–17.2 km) and did not differ significantly from average distances of relocation recorded for males.

The number of males moving beyond boundaries of the forest varied in particular years, while the number of such cases involving females grew systematically. Directions of movements were significantly different for males and females. Males wandered mostly towards the west and southwest while the cows moved towards the north. Moreover, the direction of movements was different to the expected one. The range of a

Figure 10.15 Adult bulls foraging on the meadow outside of Białowieża Forest, March 2011. Photo by T. Kamiński. A black and white version of this figure will appear in some formats. For the colour version, please refer to the plate section.

majority of translocations (94% of bulls and 93% of cows) was up to 5 km from the forest boundary. Movements of bison into the neighbourhood of Białowieska Forest were strictly seasonal. A majority of such observations (78% of males and 88% of females) comes from the period between November and April. An increase of penetration into areas situated outside of the forest may be stimulated by various factors, among which many can be dependent on range expansion and the preference of wisents for open areas. Clear seasonality of the movement pattern may indicate the beginning of partial seasonal migrations in the population of Białowieża (Kowalczyk *et al.* 2013).

Typical seasonal migrations of bison herds are also observed in other Polish populations of European bison living in the forest–field landscape. In the Knyszyńska Forest, bison spend the vegetative season in the forest habitats, while in autumn they migrate to agricultural areas and stay there until spring (Krasińska & Krasiński 2007; Hofman-Kamińska & Kowalczyk 2010).

The distances covered in the course of a 24-hour period tend not to be large, although the search for entirely new foraging grounds may extend up to 12 km per day (Korochkina 1972a; Krasińska *et al.* 1987). In periods with no snow, the distances covered by radio-collared bison averaged 7.2 ± 3.0 km per day (max. 16.3) (Rouys 2000; Rouys & Krasińska 2001). During winter, radio-tracked individuals spending time around supplementary feeding sites cover on average just 1.25 ± 0.55 km per day (max. 2.3 km). In contrast, individuals not making use of the hay during the winter covered daily distances on average of 4.74 km per day (max. 13.4) (Rouys & Krasińska 2001). Recently it has been found that radio-collared bison regularly fed with hay moved on average 560 m in 24 hours, while animals without supplemental feeding moved 860–900 m (Kowalczyk *et al.* 2010). Also, other factors, like the presence of snow cover and low temperatures, considerably reduce the mobility of bison (Krasińska & Krasiński 1994, 1995; Krasińska *et al.* 2000).

Information on movements of European bison belonging to the LC line, living in mountain regions, were collected for two populations living in the Bieszczady Mountains (Poland) and within Poloniny National Park (Slovakia). Bison inhabiting the Bieszczady Mountains undertake their seasonal movements in late autumn, heading northwards towards areas situated at lower elevations than those occupied in summer. They make their way back to higher elevations usually by the end of March/beginning of April (Paszkiewicz 2004; Perzanowski & Januszczak 2004).The distance measured between the acclimatisation enclosure, where founders of the herd were released to the wild in 1980, and the farthest locations of a bison presence recorded in 2008 was almost 36 km. A distance between farthest locations to the north of bison belonging to the western herd of Bieszczady, measured annually, varied between 0.7 km in 2005 and 5.33 km in 2003. During eight years of the study an average annual rate of home range extension (to the north) was 2.6 km (Perzanowski & Januszczak 2010).

A distance of individual seasonal movements was estimated on the basis of telemetric bearings of four adult animals (three bulls and one cow) obtained between 2002 and 2005. The two most distant locations of an animal, recorded in winter and in a following vegetative season, provided a measure of a maximal length of individual annual movements. Their values varied between 12.8 km and 18.5 km for cows, and between 5.06 km and 22.9 km for bulls (Perzanowski *et al.* 2012)

Movements of bison introduced to Poloniny National Park in Slovakia were measured in consecutive months between 2004 and 2006. In that period the herd changed its location between 0.53 km and 9.77 km during one month. Once, its monthly movement slightly exceeded 11 km (in June 2005) (Adamec *et al.* 2006).

Daily individual movements of Carpathian bison have not been an object of a separate study yet.

Population home range

The home range of the population of European bison belonging to the LB line, inhabiting lowland forests in the continental region (Białowieża Forest, Poland), did not increase systematically alongside population growth. The Polish population of Białowieża Forest increased its range successively until its size exceeded 100 individuals in 1965. During the next 20 years, in spite of the growth of population numbers, mixed groups and the majority of bulls occupied a fairly stable range of about 140 km^2 (22% of Białowieża Forest), situated in the centre of the Polish part of this forest complex (Krasiński 1978a; Krasińska & Krasiński 2007). At the beginning of the 1990s, the population dispersed spontaneously over the entire forest, and its home range gradually increased (Krasińska & Krasiński 2007). Together with the growth of population numbers, the movements of bison outside forest boundaries into agricultural areas (fields and meadows) became more frequent. In 2009 about 14% of the population penetrated those areas between autumn and spring, and the home range of this population grew to 780 km^2 (Kowalczyk 2010; Krasińska *et al.* 2010).

In the Belarusian part of Białowieża Forest, in the beginning of the 1970s, when the population of lowland bison was already formed and counted 50–60 individuals, it occupied a small home range of 90–100 km^2 (Korochkina 1973). Until the mid 1980s, despite over three-fold increase of population numbers, the home range of its major part (cows, yearlings, calves and some bulls) did not increase, and equalled 90–100 km^2 (one-sixth of the forest area) (Korochkina & Kochko 1983; Bunevich & Kochko 1988). In order to reduce the density of bison, between 1983 and 1986, in total 37 animals were captured and transferred to central and northeastern parts of the forest. In the mid 1990s the home range of the major fraction of the Belarusian population consisted of four parts. Since the end of the 1990s, two small groups have dwelled permanently outside the forest, and two others used to leave the forest periodically. The total home range of the Belarusian population after the year 2000 was 600 km^2, i.e. 70% of the eastern part of Białowieża Forest, which is similar to the home range of the Polish population in the same period (Bunevich 2003).

The population of the lowland bison at Borecka Forest in the mid 1980s occupied a home range of 8.7 km^2. In following years, despite a noticeable increase of population numbers, its home range grew up only slightly, and by the end of the 1990s it reached 10.7 km^2, i.e. almost 60% of this forest complex area (Krasińska & Krasiński 2007). In 2000–2006 the size of the home range of this population did not change, and during winter it dwindled to a mere 0.2 km^2 (200 ha) (Żoch 2007). European bison of this population are intensively fed in winter with hay and a mix of concentrated foods with some amount of tree bark, and therefore during this season they remain within a very small home range.

The Volozhinskaya population of the LB line of European bison, inhabiting lowland forest of Belarus (Nalibokskaya Pushcha) and counting 53 animals, in the vegetative season of 2004 occupied a home range of 350 km^2, while in winter this was only 65 km^2 (Kozlo & Bunevich 2009). However, in 2010, when the population grew to 79 animals, its home range in the vegetative season increased to 400 km^2 and covered almost the whole area of Nalibokskaya Pushcha. In winter, though, its range dwindled to 70 km^2 (Shakun 2011). The range of this population in the vegetative season is markedly larger than that of bison living in both parts of Białowieża Forest in the time when their numbers were at similar levels.

The herd of the LC line inhabiting the forested area of Vologda region (northern region of the European part of Russia) belongs to the sub-zone of the southern taiga. Results of preliminary studies showed that the estimated home range of this herd, counting 8–13 individuals, covers up to 10 km^2 (Tyapougin & Gusarov 2004; Gusarov 2011).

The home range of the other population of lowland European bison, inhabiting mixed forest–field landscape in the continental region, grew systematically together with an increase in the numbers of animals. In Knyszyńska Forest, in the first years after introduction, European bison occupied an area of 67 km^2, situated near the acclimatisation enclosure. In 2005–2007 their home range increased to 250 km^2, out of which 74 km^2, (i.e. 6% of the area of Knyszyńska Forest) consists of forests (A. Lickiewicz *et al.* unpublished data). In summer, bison frequent mostly forest habitats, and with the termination of the vegetative season for ground flora in the forest – from September and October – almost all animals move out into agricultural areas (meadows and cultivated fields), and remain there through the whole winter until the onset of spring (Krasińska & Krasiński 2007; Hofman-Kamińska & Kowalczyk 2010).

The home range of the population inhabiting mountains (Bieszczady) estimated as minimum convex polygon (MCP) during eight years (2001–2008) varied between 144 km^2 and 300 km^2. During that time, bison numbers there grew from 140 to 271 individuals. Population density, however, did not change much within that period (respectively, 0.97–0.90 individuals per km^2). The correlation between population numbers and the area of the home range was, however, not very high ($r^2 = 0.567$, $p < 0.5$). In comparison to the whole home range (MCP), the area of kernel 95% was 38% in winter, and 12% in summer, while the area of kernel 50% was 1.5% and 0.7%, respectively. These data indicate a very strong tendency of European bison for aggregation (Perzanowski & Januszczak 2004).

Home ranges

Information on home range sizes and locations of lowland European bison in the Białowieża Forest, was obtained in 1993–2002 through radio-tracking of animals living on both sides of the national border. In total, 34 individuals (24 males

Figure 10.16 Dr R. Kowalczyk from Mammal Research Institute PAS and Dr Z. A. Krasiński from Białowieża National Park are fitting the radio-collar to a bison cow, Białowieża Forest, winter 2005. Photo by T. Kamiński. A black and white version of this figure will appear in some formats. For the colour version, please refer to the plate section.

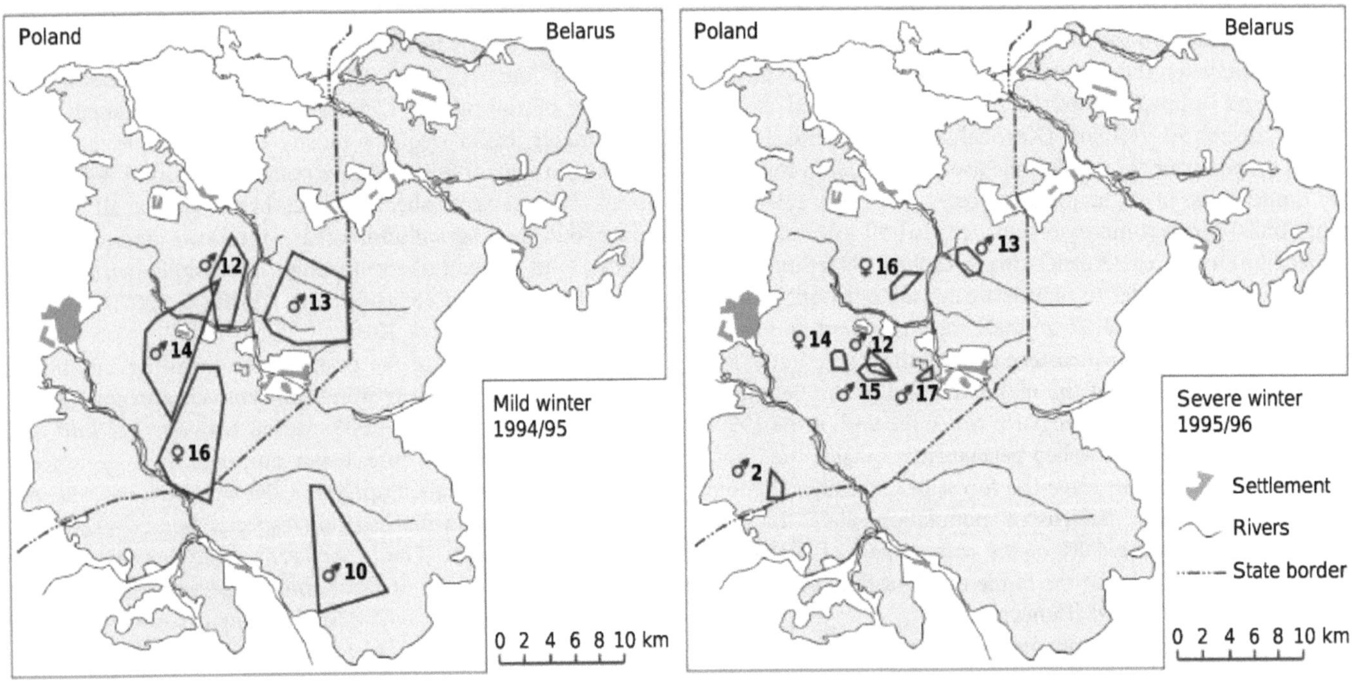

Figure 10.17 Sizes of winter home ranges of Białowieża Forest bison in relation to weather conditions. Source: Krasińska *et al.* 2000.

and 10 females) were fitted with radio-transmitters, 25 (20, 5) within the Polish part and 9 (4, 5) on the Belarusian side. The study was continued in 2005–2010 in the Polish part of Białowieża Forest, when 42 (22, 20) individuals were radio-collared (Figure 10.16).

The size of winter home ranges depends on the frequency of providing the animals with hay. Bison gathering around supplementary feeding sites in Białowieża Forest occupy very small areas. Home ranges of bulls covered 11 km^2 on average (range 0.8–32 km^2), and those of cows 8 km^2 on average (range: 0.4–34 km^2). The largest area of home ranges (32–34 km^2) coincided with the mild winter of 1994/1995 (Figure 10.17). (Krasińska *et al.* 2000). Studies conducted in 2007/2008 proved that the size of winter home ranges

Figure 10.18 A young bull imported from Sweden, fitted with radio-collar – after its release at Bieszczady Mountains. Photo by K. Perzanowski. A black and white version of this figure will appear in some formats. For the colour version, please refer to the plate section.

obtained for radio-collared bison regularly benefiting from supplemental food was even smaller, being on average just 3 km^2, while in the case of those being fed 1–2 times per week is was 16 km^2; animals not attending feeding sites at all covered 30 km^2 on average. It may indicate that together with a decrease of the frequency of supplemental feeding, the size of home ranges increases, which is connected with a necessity for more intensive searching for natural food in the forest (Kowalczyk *et al.* 2010).

During the vegetative period, home ranges of cows living in mixed groups averaged 69 km^2 (range: 45–97 km^2) and were similar to those of bulls 70 km^2 (29–152 km^2). Young adult bulls (4–6 years old) occupy an area of 44 km^2 on average, while those aged six and over (i.e. those participating in breeding) penetrate larger areas covering 84 km^2 on average. Those differences mainly resulted from the higher mobility of older bulls in the breeding season (Krasińska *et al.* 2000). In general, home ranges of bulls are more influenced by their sexual activity than by food-related factors. Home range sizes of cows and mixed groups are dependent on the abundance and availability of nutritional resources. Within the Belarusian part of Białowieża Forest in August–October, groups of 40–70 individuals fed at mown meadows and forest clearings. They occupied small home ranges (24 km^2 on average). At the same time, much smaller groups (7–14 animals) penetrated in the forest, covering areas almost twice as large (45 km^2 on average; Krasińska *et al.* 2000). No significant differences between the sizes of individuals' home ranges in the Polish and Belarusian parts of Białowieża Forest have been found, even though the latter are known to offer a less abundant feeding base. This may be the reason for bison living in the eastern part of Białowieża Forest to use more frequently mown forest clearings and meadows (Daleszczyk *et al.* 2007). European bison show a specific spatial fidelity, meaning annual home ranges of the same individual may for several years remain within the same part of the forest, and mutually overlap to a high degree (87–91%). Mixed groups appearing in spring, when aggregations typical for winter time are dispersing, occupy separate areas in the growing season (Krasińska *et al.* 2000). Such a situation, however, is not beneficial for the population. Data obtained from radio-tracking in the last five years indicate that about 90% of bulls mate with cows belonging to the same winter refuge. Only a small number of bulls penetrate the home ranges of cows originating from other winter aggregations during the rut (Kowalczyk *et al.* 2010). Central parts of home ranges of all radio-tracked individuals were penetrated most intensively out of the whole home range area. These centres of activity or 'core areas' (including 50% of a given individual's radio-tracked locations) were usually situated by sources of water, as well as in the vicinity of clearings or meadows (Krasińska *et al.* 2000). Activity centres, estimated as the area with 95% probability of an individual's occurrence, for cows fitted with satellite GPS collars, covered on average 40 km^2, i.e. some 20% of the whole home range. However, if the probability level of 50%, is assumed, the area of activity centre was only 1 km^2 (0.5% of the whole home range area) (Kowalczyk 2010).

Bison living in a mixed forest–field landscape (Knyszyńska Forest, Poland), where the forest is dominated by coniferous stands, occupy larger (130 km^2 on average) individual home ranges compared to those from Białowieża Forest. Activity centres of cows fitted with radio-collars equalled there on average 40 km^2 (20% of a home range area) (Kowalczyk 2010).

Individual home ranges estimated between 2002 and 2004 with ground telemetry for four individuals (three bulls and one cow) released at the Bieszczady Mountain (Figure 10.18) varied between 58.7 km^2 and 122.6 km^2 in winter (average 83.3 km^2), and 1.0–62.1 km^2 (average 26.8 km^2) in summer. Solitary bulls were mostly observed at outer parts of the herd's home range (Perzanowski & Januszczak 2004). These differences may reflect the Bieszczady bison habit of undertaking a late autumn migration to areas at lower elevations than are occupied in summer, as well as the fact that the animals here make only limited use of supplementary feeding in winter (Paszkiewicz 2004; Perzanowski & Januszczak 2004). Therefore it is possible to state that European bison living in natural conditions of mountainous regions occupy much larger ranges in winter than supplementary fed lowland bison in the continental region. That may be connected with the necessity for penetration of larger areas in a search for naturally available forage. In the Bieszczady population, sizes of home ranges during the growing season are much smaller than those of bison in Białowieża Forest. This may reflect the much more abundant food base present in the forests of the Bieszczady Mountains, as opposed to lowland forests. In this mountain population, meadow communities compose only some 20% of its home range (Perzanowski & Januszczak 2004).

Preliminary results of telemetric monitoring of three individuals (one bull and two cows) performed after the introduction of European bison to Poloniny National Park at Slovakia between 2004 and 2006 allowed estimation of the home range of the whole herd (nine animals) at 69 km^2 in the vegetative season and 22 km^2 in winter (78 km^2 during the whole year) (Pčola *et al.* 2006).

Activity patterns

The multi-phasic, daily activity rhythm of European bison is typical of ruminants. It consists of three phases: feeding, resting and movements. Feeding is an alternate activity to resting, which is spent mostly on rumination (Korochkina 1972a; Caboń-Raczyńska *et al.* 1983, 1987). Such an activity rhythm was also observed in bison living in large enclosures (Brink 1980). Also, in the condition of a zoological garden (e.g. Prague Zoo) all bison exhibited complete coincidence in their activity rhythm (Vajner 1980).

In the growing season, bison spend around 60% of their daily activity on feeding. This leaves 30% of the time available for resting, and the remaining 10% on movements not connected with feeding. Summer brings a high degree of synchronicity to the activity of all members of a group, thus confirming the consolidation of the herd as a structural unit of the population. In populations where supplemental food is provided for winter, an opposite proportion is observed between feeding on hay, which takes some 30% of a day, and resting, which takes 60% (Caboń-Raczyńska *et al.* 1983, 1987).

Animals living in the Belarusian part of Białowieża Forest spent more time on eating than those in the Polish part (80% on average), probably as a consequence of less nutritional food available in that part of the forest. The amount of time devoted to feeding obviously depends on the abundance and availability of food. More time is spent by the bison on feeding in spring and autumn, when the quantity of available food is limited (Korochkina 1972b).

Feeding activity of bison living in the mixed groups is, in the vegetative season, divided into feeding upon plants of the herb layer (95%), browsing (3%) and debarking (2%) (Caboń-Raczyńska *et al.* 1987). The intensity of debarking changes seasonally, being the most intensive (18% of activity) at the turn of winter and spring (in April in Białowieża Forest).

In winter it is possible to observe three peaks of feeding activity during a 24-hour period, although there is little synchrony in activity between individuals. Even when a majority of herd members are resting, there are always some individuals occupied by feeding. Winter herds display two periods of rest: in the evening between 18:00 and 21:00, and again between midnight and the early hours of the morning. In winter, bison ruminate for 10–22% of diurnal hours. Resting is in fact the most highly synchronised winter activity, along with walking to watering places (Caboń-Raczyńska *et al.* 1983). Bulls that did not join feeding sites are more active, and walk significantly more than animals staying around the feeding sites. The factor responsible for the change in the feeding behaviour of the bison at feeding sites is the abundance of easily available food (hay) there (Rouys 2000; Kowalczyk *et al.* 2011).

In the vegetative period, bison are active through the day. Observations of diurnal (dawn–dusk) activity among bison in the Polish part of Białowieża Forest has led to the identification of four feeding phases coinciding with sunrise/dawn, two periods during the day and a fourth period associated with the approach of sunset (Caboń-Raczyńska *et al.* 1987). Rouys (2000), studying the activity of radio-collared bison in summer, selected two peaks of activity: one in the early morning (2:00–7:00), and the second in the afternoon (12:00–18:00). Korochkina (1972a) reported that bison feed at night too, devoting just 3.0–3.5 hours to rest during the hours of darkness. Radio telemetry confirms that, while nighttime activity does take place, it is of less duration than that during the day. In particular, the period between 23:00 and 02:00 seems to be set aside for minimal activity or movements (Rouys 2000).

While the feeding periods before dawn and around dusk are of a routine character, the remaining sessions of feeding during the day are subject to variable timing, depending on the weather, harassment by insects, availability of food and the degree of human-related disturbance. In the last natural population, Wróblewski (1927) observed early morning feeding by bison as a regular phenomenon, as well as repeated bouts of feeding through the day until the dusk and the evening. During periods of hot weather, bison fed in the morning, in the evening and at night, spending much of the day motionless (Korochkina 1972a). The activity of European bison becomes restricted when the temperature exceeds 30 °C (Rouys 2000).

In both populations inhabiting Białowieża Forest, bison tend to come out of the forest into cultivated fields in the early morning or in the late evening. A single feeding bout ranges

Figure 10.19 Wallowing is the most often observed behaviour among adult bulls in the breeding season. Photo by M. Hławiczka. A black and white version of this figure will appear in some formats. For the colour version, please refer to the plate section.

from 15 minutes up to five hours and 15 minutes, while periods of rest – mostly associated with rumination – last between 15 minutes up to more than four hours (Caboń-Raczyńska *et al.* 1987). In Prague Zoo animals spend 40% of a 24-hour period ruminating (Vajner 1980).

Drinking during the growing season is not a regular activity in the daily rhythm of European bison. In the Białowieża Forest there are a number of natural watercourses, along with the stagnant water present in depressions, fens and bogs, as well as artificial watering holes. Bison living in groups generally drink from permanent bodies of water or watercourses. Lone males on the move tend to use puddles they meet on their way. In winter, bison make water accessible to themselves by breaking the ice along streams, or by trampling frozen ground in alder forests to stimulate inflow of water to small depressions created in this way. The consumption of snow, being observed at 1–4 times per day, is another way of assuring water intake in winter.

Wallowing is most often observed among adults, and only rarely among calves and young individuals. As many as 73% of all observed cases of wallowing among bulls occurs in the breeding season (Figure 10.19). Its frequency increases in the case of the dominant bull. When a bull is mating with a cow in oestrus, the subordinate bulls engage themselves in more frequent wallowing. The favourite places for wallowing are sandy hillocks, gravelly or sandy forest roads, roadsides and areas of exposed soil in clear-cut areas or gravel pits. Such sites are known as 'dust baths'. Before beginning to wallow, a bull may paw the ground with its fore legs, scattering sand or soil all over the place and itself. Sometimes bulls also urinate immediately before starting to wallow. During wallowing, a bison lies on its side and kicks out with straightened legs, but never makes the turn over to its spine. Thus, if bison want to wallow on the other side of their bodies, they will not roll over, but stand up and then lie down on the other side to wallow again. Wallowing also may take place in winter; however, most often it is performed then by adult individuals (in 76% of observations), while the young (2–3 years old) do it in just 16% of cases, and calves only exceptionally (8%) (Krasińska & Krasiński 2007).

Feeding ecology

The European bison is a ruminating herbivore. Morphology of its alimentary tract reflects the adaptation to feeding on grass and fibrous fodder, consumed in large amounts and with little selectivity (Hoffman 1978). Detailed studies concerning the composition of European bison diet and its seasonal changes were conducted mostly for populations living in the continental region of Białowieża Forest. There are few data on the diet of European bison living in mountains.

The fresh mass of food eaten over a 24-hour period by an adult bison ranges over 30–60 kg (Aleksandrov & Golgovskaya 1965; Kalugin 1968). Holodova & Belousova (1989) wrote that 2–3-year-old bison take on average 27 kg (25–30 kg) of fresh food per day, as compared with 48 kg (40–50 kg) in the case of 4–5-year-olds. Gębczyńska & Krasińska (1972) reported that five-year-old bison ate 23–32 kg of fresh food each day, but its diet consisted of 39% browse (tree leaves and twigs).

A study on food digestibility in European bison, conducted on a diet consisting of meadow hay (70% of the diet) and pearl barley (30%), proved that comparing to domestic cattle the bison is able to digest food more effectively, especially regarding the fibre (lignin) (Gębczyńska *et al.* 1974). High digestion of lignin is a specific adaptation to life in the forest, making possible the use of foods abundant there, which are rich in that component. The other aspect of the study was the digestibility of natural food consisting of plants that are easily accessible to the bison and willingly consumed by animals in various periods of the growing season.

European bison feed on ground flora vegetation: herbs, grasses and sedges of the forest floor, clearings and meadows

Figure 10.20 A young European bison feeds on natural forage after being released to the wild. Photo by K. Perzanowski. A black and white version of this figure will appear in some formats. For the colour version, please refer to the plate section.

(Figure 10.20), as well as upon woody material: browse and tree bark. Wróblewski (1927), studying in Białowieża the last free-ranging population of the early twentieth century, concluded on the basis of stomach-content analysis that ground vegetation makes up to 80% of a bison's diet, while the remaining 20% consists of woody species. In the Polish part of Białowieża Forest, Borowski & Kossak (1972) estimated that the diet of European bison then included 137 plant species, 27 of which were trees and shrubs, compared with 14 of grasses and sedges and 96 of herbs. Quantitatively, 33% of consumed food was of woody origin, while 67% comprised grasses, sedges and herbs. Tree species of significant importance for forestry made up 19% of total food intake. The same authors found that among the preferred tree and shrub species were: hornbeam (*Carpinus betulus*), sallow (*Salix caprea*), ash (*Fraxinus excelsior*) and raspberry (*Rubus idaeus*). From the group of grasses and sedges, the species eaten most frequently were: rough small-reed (*Calamagrostis arundinacea*), wood sedge (*Carex sylvatica*) and hairy sedge (*Carex hirta*). Among herbaceous plants the most often selected were: ground elder (*Aegopodium podagraria*), common nettle (*Urtica dioica*), woolly buttercup (*Ranunculus lanuginosus*) and cabbage thistle (*Cirsium oleraceum*). The most intensively debarked were: oak (*Quercus robur*), hornbeam (*Carpinus betulus*), ash (*Fraxinus excelsior*) and Norway spruce (*Picea abies*). Acorns are among the favourite natural food items of European bison, although they are only abundant in the masts occurring every 6–9 years. It has been found that the recruitment rate in bison population of Białowieża Forest was higher after mast years (Mysterud *et al.* 2007). Additionally, the occurrence of mast years corresponded to male-biased sex ratios in newborn calves (Hayward *et al.* 2011). Similarly, in the Bieszczady Mountains (Poland), bison willingly feast on beech nuts. In mast years this is a cause for bison to remain longer in the autumn within beech stands at higher elevations (Paszkiewicz 2004). Gębczyńska *et al.* (1991) found that during the vegetative season, the participation of herbaceous plants in the bison diet fluctuated from 88.2% in spring to 93.2% (percentage of rumen capacity) in autumn, while the proportion of woody material changed from 11.2% in spring to 6.7% in autumn. From spring to autumn the bison chose most frequently *Picea excelsa*, *Pinus silvestris*, *Quercus robur*, *Betula pubescens*, *Carpinus betulus*, *Populus tremula*. Browsing happens quite commonly and is the most intensive in June and July, mainly in deciduous or fresh mixed-coniferous forests. The proportion of browse in the European bison diet coincides with the period of full development of tree foliage, when its nutritional value is the highest (Drożdż *et al.* 1975).

There are more plant species in the diet of European bison living in the Belarusian part of Białowieża Forest than in the Polish part. According to Korochkina (1969, 1971, 1972b) the diet of bison there includes 331 species of grassy and herbaceous plants from 42 families, among which some 85 species can be considered the main food (being eaten willingly and present in the diet frequently). Moreover, bison feed there on 38 species of trees and shrubs (Korochkina 1969) – i.e. on more taxons than in the Polish part. Nevertheless, the most important remain ash, maple, aspen, rowan, sallow and elm.

Recently, Jaroszewicz & Pirożnikow (2008), using the method of seed identification in faeces, determined that the diet of bison from the Polish population of Białowieża Forest consists of 454 vascular plant species of which 409 were herbaceous and 45 woody. In studies conducted recently in the Polish part of Białowieża Forest, similar proportions of herbaceous plants, trees and shrubs in European bison diet were found; however, some changes were identified regarding the selection of certain plant species in the comparison to earlier analyses. A comparison of the most often consumed plants with their abundance showed a preference for red raspberry and American millet grass (*Milium efusum*), but the avoidance of wood anemone (*Anemone nemorosa*) and creeping buttercup (*Ranunculus repens*). The percentage of herbs in the bison diet was stable during the whole vegetative period (45–48%), the consumption of grasses decreased, while foraging on trees, shrubs and ferns increased in late summer. Also recorded were statistically significant differences in the diet of males and females. Cows consumed more grasses and sedges while males foraged more on trees, shrubs and ferns, while the proportion of intake of herbs was similar (Kamiński *et al.* 2010). Similarly, Hofman-Kamińska *et al.* (2011) determined that the diet of both sexes differed significantly across months, which may indicate the existence of different foraging strategies of bulls and cows.

Winter diet of bison living in lowland forests depends on weather conditions and the level of supplementary feeding. During mild winters, and when there is only a little snow, bison prefer natural food. In severe winters, though, bison rely mainly on hay (Korochkina 1974, 1978; Krasińska & Krasiński 2007). They are also strongly attracted by fallen trees. A long-practised custom in the Białowieża Primeval Forest is cutting down aspen and leaving it as food for ungulates. Gębczyńska *et al.* (1991) found that in the winter, when bison have access to supplemental food (hay), their main foods are grasses, sedges and herbs (90.4% of rumen capacity), while trees and shrubs are only a small fraction (9.5%). Preferred species of trees were then: *Picea excelsa*, *Betula pubescens*, *Carpinus betulus* and *Quercur robus*. In the winter diet of lowland bison living in Nalibokskaya Pushcha (western Belarus), browse constitutes 64% of consumed food; the most frequently eaten species are: pine, birch, buckthorn and spruce (Shakun 2011).

A degree of damage in agriculture may indicate which forages are preferred by European bison spending winters outside of Białowieża Forest and Knyszyńska Forest. Those animals penetrate fields searching for winter cereals and rape-seed hidden under the snow. At areas situated to the north of the Polish boundary of Białowieża Forest, 50% of all damage was to cereals, 13% to rape-seed and 35% to hay left in stacks for winter at meadows. Preferred are rape-seed and rye. Around Knyszyńska Forest, though, 76% of damage occurs in cereals (Hofman 2008; Hofman-Kamińska & Kowalczyk 2010).

Bison in Białowieża Forest that were intensively fed with hay during the winter also used to browse and debark trees, mostly those growing in the nearest neighbourhood of feeding points. Rubbing against tree trunks or chewing off the bark becomes more frequent in late winter and early spring, especially in April when there is not yet much ground vegetation in the forest. According to Korochkina (1969, 1974), bison debark trees of a diameter less than or equal to 10 cm. Bunevich (1994), while studying the natural foods of bison in winter, came to the conclusion that the species most affected by debarking are hornbeam, lime, hazel and spruce, primarily those with diameters of up to 15 cm, although in the case of ash and lime there were also trees of up to 30 cm. Dackiewicz (1999) studied the influence of European bison upon the forest stands at specially designated trial plots at two feeding points, where the two largest winter groups of bison in the Polish part of Białowieża Forest used to concentrate. He has found that the most intensively browsed there was oak (60% of trees belonging to this species were damaged), followed by ash (46%, despite only few trees of that species at studied plots), lime (42%) and hornbeam (32%). The least intensively browsed were spruce, birch and alder. Debarked were primarily the hornbeam and lime, while oak, ash and spruce were debarked only to a low degree and birch not at all. Cut-down aspen trees, lying on the ground, were browsed and debarked 100%. Intensive debarking of young ash trees around supplemental feeding points was also observed at Bieszczady Mountains (K. Perzanowski, personal communication).

Kowalczyk *et al.* (2011), while studying the behaviour of radio-collared bison in the Polish part of Białowieża Forest, found that the extent of use of supplementary food affects the composition of bison diet. The amount of woody material (trees and shrubs) consumed by bison increases with decreasing use of feeding sites – from 16% in the case of intensively fed animals to 65% in the case of animals not using the supplemental food during winter. On the other hand, the amount of herbs, grasses and sedges decreases from 82% when animals are intensively fed, down to 32% for bison not using the supplemental food. The most often consumed tree species there were *Carpinus/Coryllus*, followed by *Betula* sp., *Salix* sp., *Pinus silvestris* and *Quecus robur*. *Carpinus/Coryllus*, *Quercus* and *Alnus* were taken proportionally to their availability. The study proved a strong preference for *Salix* sp., *Pinus silvestris*, *Ulnus* sp., and *Betula* sp., and the avoidance of *Tilia cordata*, *Acer platanoides* and *Fraxinus excelsior*. The most browsed species by all studied groups of European bison were *Carpinus/Coryllus* (34–49% of woody material). Bison foraging at agricultural areas consumed mainly *Pinus*, *Salix* sp., *Betula* sp. and *Quercus* (84% of woody material in total). Those results may indicate that at the beginning of the twenty-first century some changes in the preference towards certain tree species in the Polish part of Białowieża Forest could occur, which may be connected with an increase of bison population numbers.

Concluding, it is possible to say that in the vegetative period, the main component of diet of lowland bison living in the continental region are herbaceous plants, while food derived from trees and shrubs is, at most, a supplementary item (Wróblewski 1927; Korochkina 1969, 1972b, 1974; Borowski & Kossak 1972; Dzięciołowski *et al.* 1975; Dzięciołowski 1991; Gębczyńska *et al.* 1991; Jaroszewicz & Pirożnikow 2008; Hofman 2008; Kamiński *et al.* 2010; Hofman-Kamińska *et al.* 2011). In winter, though, the proportion of natural forages in the diet of European bison depends on weather conditions and the extent of supplemental feeding.

In Vologda region, which belongs to the sub-zone of the southern taiga (northern region of the European part of Russia), bison of the LC line in the vegetative period consumed a variety of herbaceous plants. During summer the role of browse forage increased, and in winter it became the basic diet. Those bison were provided in winter with hay and crushed grain only occasionally and in minimal quantities (Tyapougin & Gusarov 2004; Gusarov 2011).

Bison of the LC line living in a mountain region of Poloniny National Park (northeastern Slovakia) during the whole year consumed herbs, grasses and leaves and twigs of bramble (*Rubus fructicosus*), raspberry (*Rubus idaeus*) and bilberry (*Vaccinium myrtillus*). They used also sprouts and bark of trees like aspen (*Populus tremula*), grey alder (*Alnus incana*), sallow (*Salix caprea*), wild cherry (*Cerasus avium*), mountain ash (*Fraxinus excelsior*), beech (*Fagus silvatica*), Scotch pine (*Pinus silvestris*) and spruce (*Picea abies*). During the autumn/winter period they also ate apples in abandoned orchards, dry grass and dry beech leaves. This herd was also provided with hay at feeding points (Pčola *et al.* 2006).

The most important species, especially for the winter survival of bison in the whole of the Carpathians, is the bramble (*Rubus* sp.), a standing crop that may reach even several hundred kilograms per hectare. Because of its height, it is relatively easily accessible even under the snow, and its winter-green leaves remain highly digestible and protein-rich through the whole winter (Aleksandrowicz *et al.* 2009; Mazurek 2010).

Bison from the free-ranging population of Cejskijj Zakaznik, inhabiting the central Caucasus Mountains (North Ossetia region) do not receive any supplementary food during winter. At the beginning of that season, when snow cover is still thin, the bison diet mainly consists of blackberry and herbaceous plants (59%), while bark accounts for 39%. However, in the second half of winter – when snow becomes deep – the share of bark increases to 79%, while bilberry and green plants account for just 6% of total food intake. The bark consumed there at that time comes from eight tree species, mainly lime, maple and elm (Kaz'min & Smirnov 1992).

A herd of European bison makes use of their home range in a rotational way. Foraging in a continuous, slow movement; bison make a loop, coming back to the same area after several days or a couple of weeks. Following their experienced herd leaders, they move around their ranges, never remaining in one place for a long time. Because they forage on the upper parts of ground vegetation, the passage of even a large group of bison does not leave visible signs of their presence. Such a manner of foraging allows for the effective regeneration of plant cover (Korochkina 1973; Krasińska *et al.* 1987; Krasińska & Krasiński 2007; Kowalczyk 2010).

Foraging on clearings (glades) and mown meadows was analysed within the area of Białowieża Forest, and it has been found that it is not regular. Mixed groups of bison go out there for grazing mostly during early morning hours and in the evening, and rarely visit the same place for several days in a row. Most often, even on the same day, they leave such a site and for several days feed in the forest. Afterwards they again visit a glade or a meadow, but not necessarily the same one. Solitary bulls or groups of males also visit open areas within the forest irregularly, though more frequently than cows, and they may graze on the same meadow for several consecutive days. Mixed herds and groups of males foraging at fields outside the forest visit such areas systematically during several winter months, changing the foraging ground every several days. They forage most often in the early morning and in the evening, hiding for the day in wooded patches within fields.

Reproduction and growth

Breeding season

Wróblewski (1927) wrote that breeding season in the last natural population of European bison living in Białowieża Forest occurred in August and September. In the present free-ranging population of Białowieża Forest rutting season lasts from August to October, although the highest intensity of activity definitely falls between August and September (Figure 10.21). Cows of the free-ranging population may rarely engage themselves in mating at the end of July, or in November–December, while breeding activity is still less frequent in January and February (Krasiński 1978b; Caboń-Raczyńska *et al.* 1983).

Receptivity of cows out of the true breeding season (i.e. in autumn or early winter) occurs much more frequently in captivity, where the high-protein diet available year-round, tends to disrupt the normal seasonality of reproductive activity. Thus, the breeding season at the Białowieża Breeding Centre begins in July and continues on until March, albeit with 70–80% of mating taking place in the August–October period (Krasiński & Raczyński 1967; Krasiński 1978b). A similar situation occurs in Russia's Okskjj Reserve, at which in the years 1960–1970 the mating season lasted between the first half of July until January, sometimes extending even to April (Kiseleva 1974).

Litter size

European bison cows give birth to a single calf. Twins happen only occasionally. Bison twins have been recorded only four times in captivity: at Pszczyna Breeding Centre (Poland), the Prioksko-Terrasnyjj Reserve near Moscow (Russia), Vienna Zoo (Austria) (EBPB 1947, 1961, 1970–1971) and recently at Damerower Werden Breeding Centre (Germany) (Kelterborn *et al.* 2009). The first time calving with twins in the free-ranging population was observed was twice in the herd of bison of the LC line, living in Vologda region (Russia) (Tyapougin & Gusarov 2004; Gusarov 2011).

Duration of gestation and parturition

In European bison the gestation period lasts on average for 264 (254–270) days (Krasiński & Raczyński 1967). Other authors affirm similar periods of 266 (256–279) (Jaczewski 1958) and 267 (259–279) days (Kiseleva 1974). However, since there are few visible signs of pregnancy in female bison (because of the build of cows and small size of the foetus) the precise determination of its advancement is difficult.

Figure 10.21 A bull mates a cow during the rutting season at the European Bison Breeding Centre in Pszczyna (Poland). Photo by M. Hławiczka. A black and white version of this figure will appear in some formats. For the colour version, please refer to the plate section.

Parturition does not last long. In four observed cases at the Białowieża Reserve (Poland) it took between one hour and two hours and 11 minutes (Daleszczyk & Krasiński 2001). In the Pszczyna Reserve (Poland) observed time of parturition ranged between 2.5 and 3.5 hours (Zięba 2007).

Age of first reproduction

In captivity male bison can reach sexual maturity at the age of 15–20 months, although such cases are rarely observed (Zablockij 1949; Mohr 1952; Jaczewski 1958). Histological studies of the testes of captive-bred males or individuals from the free-ranging population in Białowieża Forest make it clear that some males enter sexual maturity in their third year of life. However, all studied males had fully developed spermatogenesis by the age of four years, thus they were sexually mature (Czykier *et al.* 1999). According to Belarusian data, in that part of Białowieża Forest males are sexually mature at the age of three years (Korochkina 1971; Kozlo & Bunevich 2009).

In captivity, females become pregnant at the age of between 24 and 28 months (Jaczewski 1958). However, most often they become sexually mature at the age of three years, so the first calves are born in their fourth year of life. Such a pattern was observed for 54% of cows being bred in captivity (Krasiński & Raczyński 1967; Krasiński 1978b; Belousova 1986). According to Daleszczyk (2011), recently in the Białowieża Breeding Centre females calved for the first time at 3–5 years old (mean: 4.04 years). In the Belarusian part of Białowieża Forest cows were calving for the first time very rarely at the age of three years, mostly in the fourth and sometimes in the fifth or sixth years of life (Kozlo & Bunevich 2009). In Damerower Werden European Breeding Centre the age of first calving averages 50 months (4.2 years) (range: 31–68 months). The youngest bull able to reproduce was 26 months old (Kelterborn *et al.* 2009).

Length of the cycle of oestrus

Females remain in oestrus for one to three days (average 2.6), which is associated with a short-term relationship with a bull, during which each partner displays a characteristic behaviour (Krasiński & Raczyński 1967; Krasiński 1978b; Daleszczyk 2002). The phase of oestrus when the female allows being covered by the bull may last for 2–3 days. Bison females most often display repeated oestrus at intervals lasting for about 20 days. In the Breeding Centre Niepołomice (Poland), 80% of the oestrus repetitions last between 18–22 days (Krasiński & Raczyński 1967).

Calving intervals

European bison calves are most often born in late spring. Such timing is optimal for their subsequent development. According to the oldest captive-breeding data (for 1891–1946) the peak of calving (66% of cases) occurs in May, June and July. Only occasionally calves may be born in January and February (Koch 1956). At the Białowieża Breeding Reserve, the calving season extends from April to December, with a peak of births (80%) in May–July (Krasiński & Raczyński 1967). In Russian breeding reserves, bison births most often occur between May and September, although they are also known from November and December (Kiseleva 1969, 1974; Belousova 1986).

In the free-ranging population, the calving season is limited to the period from May to September, and is the most intensive between May and July, when over 80% of cows calve (Krasiński & Raczyński 1967).

In reserve conditions it is normal for females to be fertile until 20 years of age. Furthermore, 63–91% of females in the reserve capable of reproducing calve in a given year. On average, therefore, a cow there produces nine calves in a lifetime (Krasiński & Raczyński 1967; Krasiński 1978b). In Damerower Werden Centre females living in a semi-free herd have on average 7.8 calves per cow, while females from breeding herds average 12.5 calves per cow (Kelterborn *et al.* 2009). Cows are the most fecund between 5 and 13 years of life (Daleszczyk 2011).

In both free-ranging populations of Białowieża Forest bison cows calve every second year (Krasiński 1978b; Bunevich & Kochko 1988; Bunevich 1999; Krasińska & Krasiński 2007; Kozlo & Bunevich 2009).

Body weight of newborn calves

An average body mass of males at birth is 28 kg (16–35 kg) and is higher than that of females, which is 24 kg (15–33 kg) (Table 10.5). In the Belarusian population recorded body mass for both sexes at birth was lower, on average 20.4 kg (17.5–24 kg) (Krasochko *et al.* 2004; Bunevich 2006; Kozlo & Bunevich 2009).

Length of lactation

In captive conditions the duration of lactation depends on the sex and age of the calf. Young males suckle longer than females, but less frequently. During the first three months, a male calf suckles on average seven minutes each time; cf. 5–6 minutes in heifers. A single suckling may last from several seconds to 22 minutes (Daleszczyk 2002, 2004).

In the case of free-ranging bison, between May and July calves of several days up to three months old suck for short periods – four minutes on average (range: 2–6 minutes) – and frequently. Those aged three months and older feed longer (on average for seven minutes; range: 4–19) but less often (Caboń-Raczyńska *et al.* 1987). Calves may suck up to the age of 12 months or longer. Although the suckling most often ends in spring, when a cow is not pregnant again she may continue to feed the last year calf even until October.

Rate of growth

Body mass of both sexes in European bison is positively and significantly correlated with age ($p < 0.0005$). In males body mass increases proportionally to the age until they reach seven years, while in females this increase lasts until five years of age (Figure 10.22). Calves double their body mass at the age of three months. Among young up to two years, the body mass of males and females does not differ significantly. At three years, males both from captivity and free-ranging herds have on average higher body mass than females. This sex dimorphism increases gradually during the post-natal development. It becomes distinctive from the third year and continues for the whole life. The body mass of adult bulls in the sixth year and older is significantly higher than that of cows (Table 10.5). All body parameters of European bison are significantly correlated with age ($p < 0.0005$) (Figures 10.22–10.23). The highest increase rate takes place during the first 12 months of life. In older animals, the growth rate is slower and ends at the age of 5–6 years. The rate of increase in body dimensions becomes higher in males after the first year of life. Its fastest increment is observed until animals attain 200 kg of body weight. The maximal values of body measurements in European bison are reached at the age of 5–6 years (Krasińska & Krasiński 2002).

At birth, the trunk of a calf is short and tall. At the age of 8–10 months its proportions change, and the length and the height of the trunk become equal. During the whole body development, the trunk of females is shorter than that of males

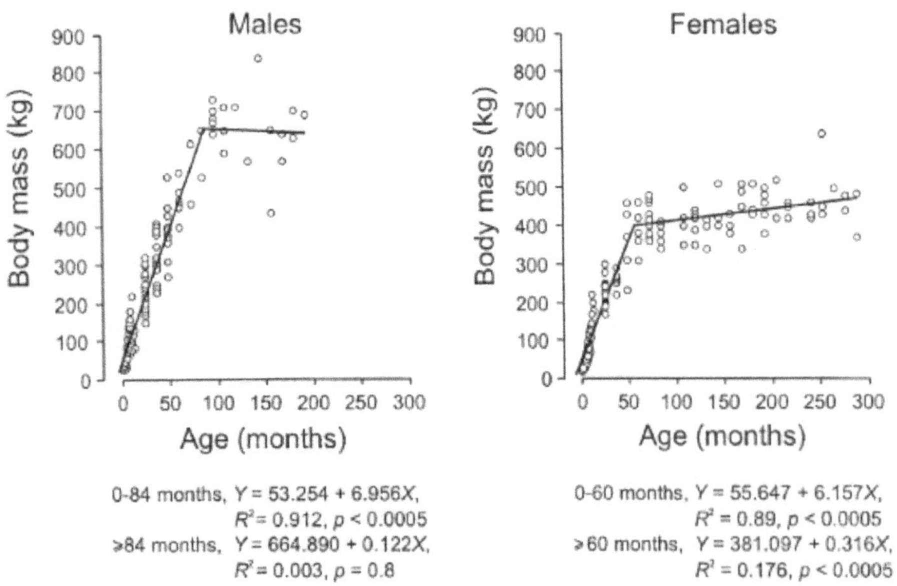

Figure 10.22 Body mass of European bison (average for 165 males and 210 females) in relation to age. Source: Krasińska & Krasiński, 2002.

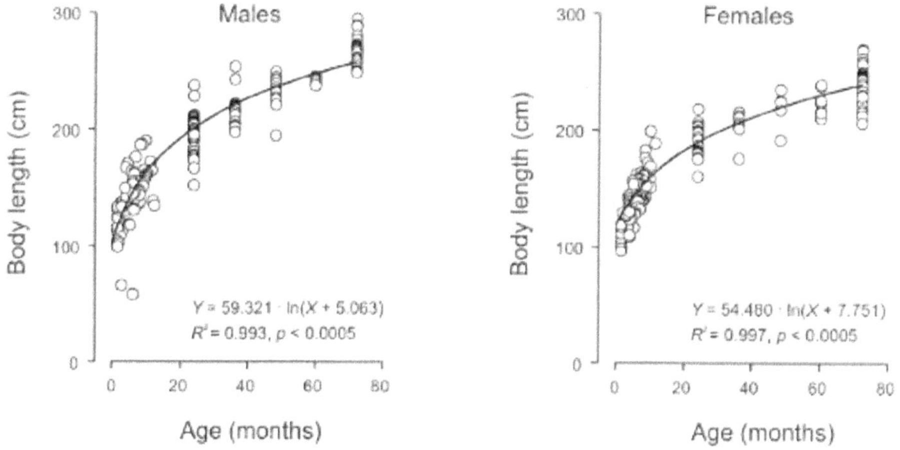

Figure 10.23 Body length of European bison (average for 165 males and 210 females) in relation to age. Source: Krasińska & Krasiński, 2002.

Figure 10.24 Several days old calf, rescued after a torrent at Bieszczady Mountains. Photo by K. Perzanowski. A black and white version of this figure will appear in some formats. For the colour version, please refer to the plate section.

(approximated to the height at the hump). Newborn bulls are slightly larger than heifers. The sex dimorphism visible from the age of three years remains for the whole life span (Krasińska & Krasiński 2002).

On the basis of changes in body mass and dimensions connected with age, it has been determined that the development in European bison is completed after the fifth year of life (Empel & Roskosz 1963; Krysiak & Świeżyński 1967). According to Jarocki (1830) the development in this species lasts until six years of age, while Brincken (1826) stated that the European bison grows until six or seven years of age. However, Belarusian scientists are of the opinion that the development in this species lasts until an old age, in males up to 16–18 years and in cows until 14–15 years (Kozlo & Bunevich 2009).

Determination of sex and age

Among traits allowing for the determination of the sex of a bison from a distance are: its silhouette, body size and size/shape of horns (see also the description notes given earlier in this chapter).

The age of a bison can be determined by the body size and shape of its horns and, during a close examination, by the teeth wear pattern or by the number of cement layers in the cross-section of a tooth.

Determination of age and sex by size and horn shape

Calves are born with horn buds that develop rapidly (Figure 10.24). Two-month-old animals already have horns about 2 cm long. Sex-related differences are visible much earlier in the shape and size of horns than in other parameters of the animal's body. During the first 12 months of life horns grow by around 5 cm every three months, and their circumference at the horn base extends by some 2–3 cm in the same period. Among bison of 12 months old or younger, it is not possible to distinguish both sexes on the basis of minimal differences in horn dimensions. At the age of six months, all bison still have straight horns. A slight inward curvature begins to develop by the age of 12 months. Two-year-old males already have markedly larger horns than females of the same age, and those differences continue throughout the whole life. At this age, European bison have markedly inward-curving horns, and the curvature is already more pronounced in females. Horns grow quickly during the first three years of life (Figure 10.25), and by the end of this period differences in the shape of horns between males and females become distinct. Horns of adult bulls are longer, thicker and set more widely apart than those of cows (see also Figure 10.10) (Krasińska & Krasiński 2007).

The outer surface of the horns of young bison is matt, but in older animals it becomes glossy. Horns attain their full development in 7–8-year-old individuals. Males older than ten years frequently have worn-down and rounded ends of horns as a result of rubbing them against trees (Krasińska & Krasiński 2007). Horns of such individuals become shorter for this reason. In contrast, even 20-year-old females do not have worn-down horns (see also Figure 10.10).

Figure 10.25 Determination of a bison's age by its horns. Left to right: horns of an animal about half a year old, two years old and three years old. Photo: Z. A. Krasiński, modified. Source: Krasińska & Krasiński, 2007.

Determination of age by reference to teeth pattern

The deciduous teeth in European bison appear in the first three months of life, although the full formation of their crowns continues until the sixth month (Table 10.6). The age at which particular molars erupt are listed in Table 10.7. The overall growth of the molar crown lasts 5–13 months (Węgrzyn & Serwatka 1984). The age at which the deciduous incisors are replaced by permanent ones is given in Table 10.8. From the moment the milk teeth are lost, the incisors and canines obtain a final shape of their crowns within some 3–6 months. The replacement of premolars and incisors occurs 2–5 months earlier in females than in males.

European bison at the age of 12 months have a set of eight deciduous teeth in the mandible. At the age of about two years a gradual replacement of milk teeth with permanent ones is initiated. This process begins with central incisors followed by further incisors. Around the age of three and a half years, the whole set of deciduous teeth is replaced, so at the age of four years, bison have a full set of permanent teeth in the mandible. Molars emerge at the same time in both male and female bison. The appearance of mandible molars and exchange of the premolars takes place earlier than the replacement of teeth in the maxilla. In general, the majority of teeth (other than the canines) erupt and are replaced in the same way in European bison as in late-maturing breeds of cattle (Węgrzyn & Serwatka 1984).

In old bison the crowns of the incisors become severely worn – at times barely protruding beyond the gum line. This is particularly visible in free-ranging bison and is obviously linked with chewing of tree bark. Wróblewski (1927) drew

Table 10.6 Age of eruption of deciduous teeth in European bison

Mandible		Maxilla	
Name of a tooth	**Age in months**	**Name of a tooth**	**Age in months**
Id_1	Before birth	Lack Id^1	–
Id_2	Before or after birth	Lack Id^2	–
Id_3	Before birth	Lack Id^3	–
Cd_1	Before or after birth	Lack Cd^1	–
Pd_2	1.0–1.5	Pd^2	Before birth
Pd_3	Before birth	Pd^3	Before birth
Pd_4 Cranial lobe Medial lobe Caudal lobe	Before birth Before birth 1.0–1.5	Pd^4 Cranial lobe Caudal lobe	Before birth 2–3
Deciduous premolars are almost fully formed in four-month-old European bison			

After: Węgrzyn & Serwatka (1984), modified.
Id_1, Id_2, Id_3 – incisors; Cd_1 – canine; Pd_2, Pd_3, Pd_4 – premolars in the mandible; Pd^2, Pd^3, Pd^4 – premolars in the maxilla.

Table 10.7 Age of eruption of molars in European bison

Mandible		Maxilla	
Name of permanent molar	**Mean age in years**	**Name of permanent molar**	**Mean age in years**
M_1		M^1	
Cranial lobe	0.5	Cranial lobe	1
Caudal lobe		Caudal lobe	
M_2		M^2	
Cranial lobe	1.5	Cranial lobe	1.5–2
Caudal lobe		Caudal lobe	2.0–2.5
M_3		M^3	
Cranial lobe	2.5	Cranial lobe	2.5–3.0
Medial lobe	3.0	Caudal lobe	2.5–3.0
Caudal lobe	3.5		

Molars in the European bison are permanent dentition

After: Węgrzyn & Serwatka (1984), modified.
M_1, M_2, M_3 – molars in the mandible; M^1, M^2, M^3 – molars in the maxilla.

Table 10.8 Age at which deciduous teeth are replaced by permanent teeth in European bison.

Mandible		Maxilla	
Name of tooth	**Mean age in years**	**Name of tooth**	**Mean age in years**
I_1	2.0	Lack I^1	–
I_2	2.5	Lack I^2	–
I_3	3.0	Lack I^3	–
C_1	3.5	Lack C^1	–
P_2	2.0	P^2	2.5
P_3	2.5	P^3	2.5
P_4	3.0	Pd^4	3.0

After: Węgrzyn & Serwatka (1984), modified.
Abbreviations as in Table 10.7; C – permanent canine.

attention to this, suggesting that the process begins in middle age and intensifies gradually. Some bulls not exceeding 15 years may even experience difficulties with chewing of beets. Considering that the ancestors of European bison evolved in steppe and tundra habitats, this may indicate that the level of adaptation of European bison to forest life in general, and feeding on tree bark in particular, is not yet complete. Furthermore, since bulls tend to debark trees more intensively than cows, the degree of wear of their incisors is higher.

Ageing on the basis of cement layers in teeth

This method is based on counting under magnification the annual rings of cement, visible in decalcified and microtome-sliced sections of a tooth. Attempts to determine European bison age in this way were described by Klevezal & Pucek (1987).

Behaviour

Social organisation

Among populations of lowland European bison inhabiting the continental region, multi-annual studies on population structure were conducted only at Białowieża Forest. Poland and Belarus have established two separate free-ranging populations in their respective parts of Białowieża Forest at the beginning of the 1950s. Natural mortality in both populations is very low (coefficient of mortality about 3%). In both populations animals in poor condition or that are sick are culled – at a level of 11% of the population in the Polish population and 4.5% in the Belarus. By the end of 2011 both parts of the forest were inhabited by 910 bison (481 in the Polish part and 429 in the Belarusian), i.e. more than the numbers of the last wild population in the forest at the beginning of the twentieth century (see also Figure 10.12).

The age/sex structure of the bison population of Białowieża Forest is described according to three categories comprising: adult bulls and cows (four years and older); 2-3-year-old adolescents; and calves up to one year old. Such categories, though, have a practical aspect, because they allow for age determination in field conditions. Over the period of 1960–2009, the Białowieża population on average consisted of 25% adult bulls, 37% adult cows, 2.5% adolescents and 15.5% calves. Long-term analysis of the population structure shows that over time there is a relative stability in the proportion of adult bulls and calves. What increased steadily in this period, however, was the contribution of adult cows. In 1981–2009 this value was higher than the average for the whole analysed period.

The European bison is a gregarious animal. In snow-free periods, mixed and bull groups are the basic units of the population. Mixed groups contain cows, the young (2–3 years old), calves and – temporarily – adult bulls. The mean size of mixed groups in the Polish population is 13 animals (maximum observed 92) (Krasińska & Krasiński 2007). In the Belarusian population the size of mixed groups is greater, at 21 animals on average (maximum 120) (Kozlo & Bunevich 2009). In free-ranging populations of the lowland European bison (Białowieża Forest, Borecka Forest) mixed groups of up to 20 animals are the most common, so it is assumed that this is the optimal group size for European bison in the forest (Figure 10.26) (Krasińska & Krasiński 2007; Żoch 2007). The average size of mixed groups is environment-dependent, which becomes visible in diverse habitats that include the forest and meadows. Within Białowieża Forest, large groups of European bison are observed most frequently between August and November at mown meadows and forest glades. At open areas bison form numerous groups, because it is

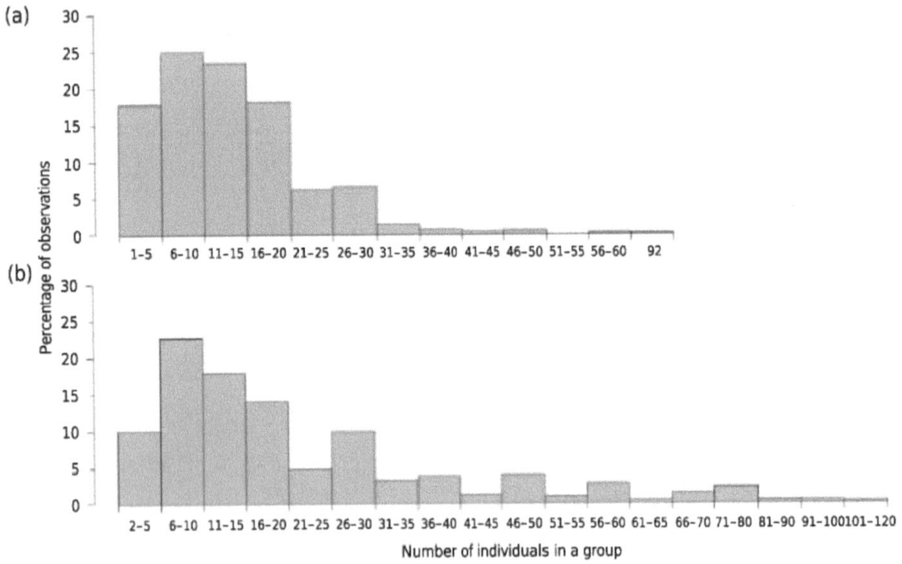

Figure 10.26 Frequency distribution of the size of mixed groups in the Polish (a) and Belarusian (b) populations of the European bison of the Białowieża Forest, within the vegetative period of 1976–1993. Source: Krasińska & Krasiński, 2007.

possible for a larger number of animals to fulfil their energy requirements at the same place and time than it is at other areas of the forest.

Young males (1–2 years old) stay with mixed groups. Bulls that are 3–5 years old appear in mixed groups as often as in groups of bulls. Sexually mature bulls (4–5 years old) form bachelor groups (Figure 10.27; see also Figure 10.15) consisting of 2–8 individuals. Adult bulls (six years old and above) and elder males move around in pairs or remain solitary. About 60% of males lead a solitary life (Figure 10.28; Krasińska & Krasiński 2007).

European bison living in both parts of Białowieża Forest during winter form several large groups (aggregations). Now, in the Polish part of Białowieża Forest large aggregations consisting of 50–100 (maximally 125) individuals are observed. Additionally, several smaller groups exist there, consisting of 10–40 bison. During winter in the Polish part of Białowieża Forest, only 35% of all adult males participate in mixed aggregations); the remaining males form separate groups of bulls counting up to 15 individuals, or lead a solitary life (Krasińska & Krasiński 2007). On the agricultural fields and meadows outside of the area of Białowieża Forest and Knyszyńska Forest, bison aggregate into large winter groups (up to 50 and over 70, respectively) (Hofman-Kamińska & Kowalczyk 2010).

In the Bieszczady Mountains the structure of the LC line population estimated for a four-year period (2006–2009) was: bulls 38% (35–39%, $n = 384$); cows 48% (46–51%, $n = 488$); and calves 13% (11–15%, $n = 136$). In this population, regardless of the season, the most frequent were groups counting 3-10 individuals (29.4–64.3% of all observations). Relatively often recorded were also groups composed of 11–20 individuals (15.6–32.1% of all observations). Thus, 51.5–96.4% of all observed groups counted up to 20 animals, similar to

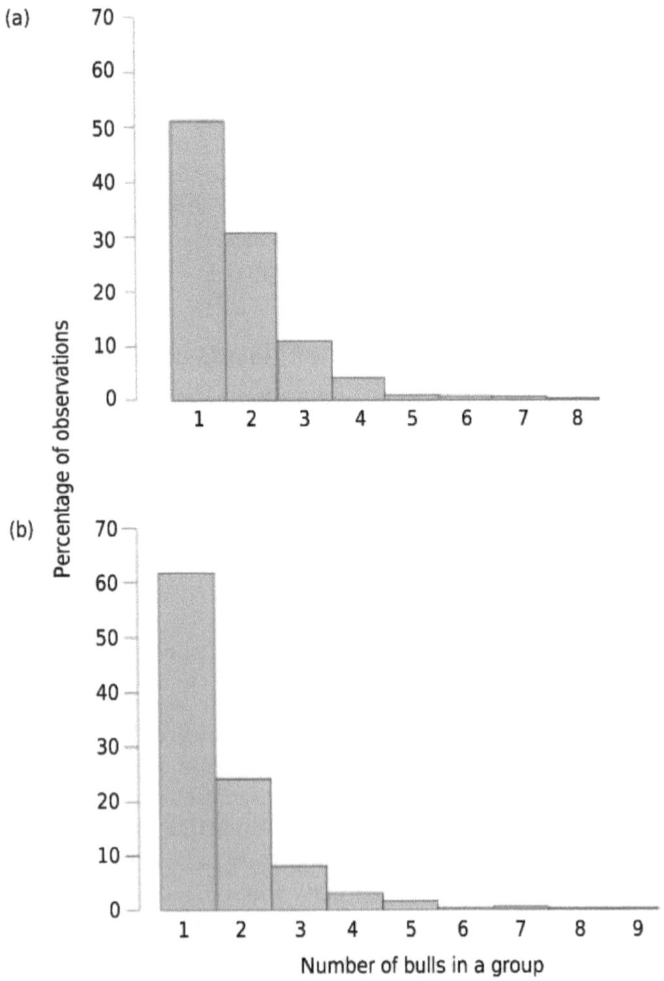

Figure 10.27 Group sizes of bulls in the Polish (a) and Belarusian (b) populations of the European bison of Białowieża Forest, within the vegetative period of 1976–1993. Source: Krasińska & Krasiński, 2007.

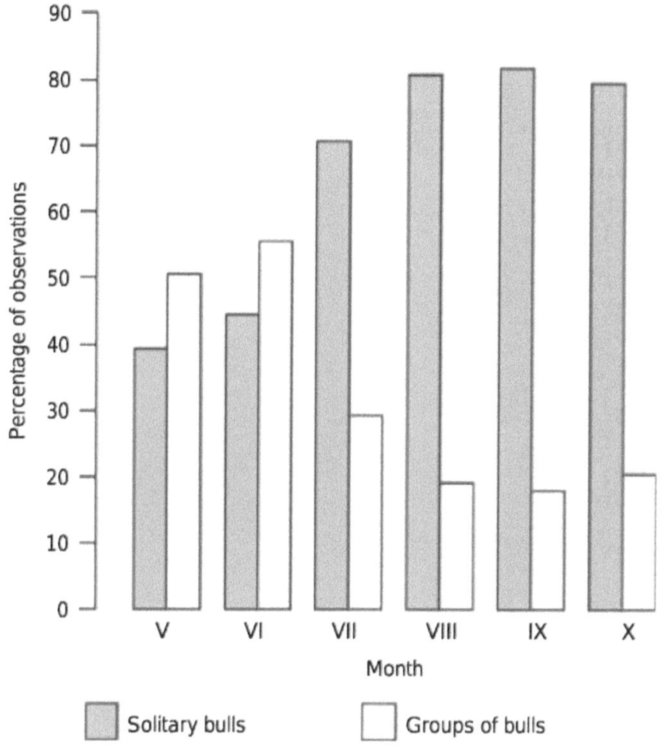

Figure 10.28 Seasonal changes in the frequency of observations of lone bulls and bulls in bachelor groups during the vegetative seasons of 1976–1990. Source: Krasińska & Krasiński, 2007.

the case of lowland bison. Solitary individuals (mostly bulls) were recorded in 3.6–24.5% of all observations. Large groups (over 30 individuals) were seen only rarely (0.7–2.4% of all observations), which indicates that in the mountains large groups are less frequent than in the lowland, continental region. Groups of various sizes were recorded with significantly different frequency, depending on a season. Within bison groups a relatively high rotation of animals is observed during the whole year. It is a result of movements of single individuals as an effect of establishing the hierarchy in the group, reaching sexual maturity by males and either division of herds into smaller groups or aggregation of animals into large herds, depending on changes in food availability, weather conditions or human-related disturbance. Results obtained at Bieszczady suggest that this species tends to create groups of a specific size, optimal for current conditions (Perzanowski *et al.* 2005).

Herd dynamics

Observations on lowland populations confirm that herds of European bison are not stable and longlasting social units. There is a constant rotation of individuals among particular groups. To some extent there is a separation of cows, which form mixed groups together with calves and some young bulls, from adolescent bulls which mostly stay within so-called bachelor groups, as well as elder bulls leading a solitary life. Nevertheless, in most cases, especially when animals are provided with supplemental food for winter, those groups finally merge together.

Behaviour of mixed groups is determined mostly by foraging activity. Herds move within the home range in search of food, led by an experienced cow. Such elder cows, usually having a calf, are leaders of mixed groups. Bulls do not lead mixed groups, because their presence in a group is only periodic (Krasińska & Krasiński 2007). During the pre-rut season (May–July) behaviour of cows in a herd is to a large degree a consequence of calving and nursing of the calf. In autumn (August–September) a sexual behaviour connected with the rutting period dominates.

Behaviour of males is also determined by foraging patterns as well as sexual activity. Groups of bulls are not stable; the relationships between males are feeble. In pre-rut season males are very active, occupying smaller home ranges than in the rutting season ($p < 0.01$). The number of solitary males increases in the rutting season when bulls tend to disperse, while searching for rutting females.

Mother–calf behaviour

In the free-ranging population a cow about to give birth leaves the herd to deliver the calf in a safe place. After several days the mother and calf rejoin the herd. Only once has it been possible to observe the arrival of a female with newborn (still wet) calf to a free-ranging group. When that herd moved off, the cow encouraged the calf to follow on by nudging it with her head.

Small calves generally lie close to nursing mothers. One-month-old calves are already very mobile and they follow their mothers closely when groups are on the move, while calves that are several months old take positions in the middle of the group (Krasińska & Krasiński 2007).

Observation of parturition and of the subsequent development of relationships between mothers and calves is only truly possible in captivity, where the identification of particular individuals is possible, as well as the constant observation of selected animals. Licking of the calf immediately follows the delivery (Figure 10.29) and the placenta is eaten within 4–5 hours. The calf begins to stand 20–45 minutes after birth, and starts suckling within the first hour of life (Daleszczyk & Krasiński 2001). The first week in the life of a calf is the period of the closest relationship with its mother (Daleszczyk 2002, 2004). The calf then remains very close to its mother, being no more than three metres away. At this time the female limits her own time for feeding (27% of observation time) to allow stronger involvement in maternal behaviour, like licking and feeding the calf, and the rest (8% and 59% of observation time, respectively). In the following weeks, cows extend the time spent on feeding and devote less time to the care of the calf and resting (Daleszczyk 2002). Contact between the mother and the offspring are most often initiated by the cow. The calf, during this period, learns to recognise the scent, voice and

Figure 10.29 The first steps are not very successful. The mother moves barely a single pace away from the calf, and is quite capable of mounting a fierce attack on any intruder coming within her zone of security. Photo by Z. A. Krasiński.

appearance of its mother. Calves after the birth start to move around with their mothers as soon as they are strong enough to follow them. European bison do not hide their young as is the case with some other ungulates (e.g. roe deer). The integration of the calf into a group begins in the second week and it begins through incorporation of the calf into a peer subgroup (Daleszczyk 2005). In free-ranging populations offspring remain in a mixed group together with their mothers until they are two years old.

Antagonistic behaviour

An active, aggressive behaviour is rare among cows, adolescents of either sex or calves. As it was already mentioned, groups of bison do not have permanent composition; frequently they merge, and subsequently break-up into smaller herds. Out of the rutting season, 99% of observed cases of new animals joining a herd take place without any visible reaction from members of the group. Most probably the hierarchical structure of groups is maintained in the course of the process and is recognised by newcomers.

In contrast, during the rutting period aggressive interactions between adult bulls are observed quite frequently as groups merge. If elder bulls (aged 6–12) are present in each group, they may assume the characteristic pose intended to determine who the dominant is. Such a bull may approach the rival, circling him slowly, coming to a halt, hoofing the ground, wallowing on it and damaging young spruce trees. If the other male is not displaying a submissive attitude, or tries to demonstrate symptoms of aggression, then a direct contest takes place.

Aggressive behaviour towards younger rivals is also displayed by dominant bulls covering females. Therefore, young males usually keep a safe distance between themselves and the mating pair, but among themselves they often start to push each other around, initiating trials of strength, etc., although these are only mock conflicts.

In the free-ranging populations fights between bulls during the rut are witnessed only rarely. Nevertheless, wounded bulls are observed from time to time, sometimes bearing traces of a blow received from another bull's horns. Males occasionally even die from the wounds they have sustained in a fight. Real fights are rare, though, because it is unlikely for males of the same size and strength to enter into direct contact. Prior to any actual contest, the rivals show just how strong they are, displaying characteristic behaviour, e.g. breaking and/or uprooting young trees (mostly shallow-rooted spruce in the case of Białowieża Forest), repeatedly hoofing the ground, urinating and then wallowing in such places. Each trial of strength among bulls is different; its course and duration depends on the character, temperament and force of both rivals.

Actual fights are seen only sporadically. In the Polish part of Białowieża Forest one such event was observed in detail in August 1984. A herd within which a large eight-year old bull was mating with a cow in oestrus was approached by another male of similar age and condition. The newcomer quickly made his way into the centre of the group, sniffed one of the cows, raised its head to assume the *flehmen* position and immediately set off in the direction of the mating bull. He circled the latter, and then the two of them stood alongside each other with their heads turned in opposite directions. Further preparations for the fight lasted some ten minutes.

Figure 10.30 Beginning of the contest, bulls with low-bowed heads are pushing at each other. Photo by Z. A. Krasiński. A black and white version of this image will appear in some formats. For the colour version, please refer to the plate section.

During this time both bulls came head to head with each other several times, butting and pushing. Every time, rising of the head by one of the bulls immediately encouraged the other to take a head-on position.

In the next phase the bulls stood alongside, turning their heads to face each other. During the next five minutes each rival hit the other bull several times in the back or head with its horns. Following, the main phase of the conflict lasted for 15 minutes. Both bulls stood head to head (Figure 10.30), regularly locking horns and heaving powerfully against each other. In constantly repeated attacks, the males struck each other with their horns or slammed horns into the head of their opponent. This behaviour was accompanied by powerful flicks of the tail. The wrestling bulls trampled down the ground vegetation until bare soil was exposed. Such a struggle required a huge effort, and the males gradually became more and more tired, snorting and steadily slowing the pace of their activity. The surroundings of the site were saturated with the characteristic, very strong musky scent of bison. Finally, the defeated bull turned away and slowly backed off. Other animals of the group paid no attention to the conflict, continuing to feed or rest. The whole contest lasted about 30 minutes, and was ultimately won by the bull which had first mated with the cow. Looking exhausted, he went to lie down in the bushes, only rejoining his herd after two hours of repose. Shortly thereafter he had to take another break and lie down. When another hour had passed, he rose, uprooted a young spruce by digging it out with his horns, and began to feed.

Sexual behaviour

Sexual behaviour displayed by bulls is not limited to the breeding period only, but is observed also at other times of the year. Young (2–4 years old) males attending mixed groups sniff the outer parts of the female genitals from time to time, adopting so-called *flehmen* posture, with raised head and neck and parted lips with the upper lip folded back. Flehmen allows the male to detect the hormone concentrations in the female's urine and thereby assess when she is sexually receptive. Sometimes they also try to mount a cow.

Lone bulls dwelling in the free-ranging population sniff out the tracks of other bison and then pursue mixed groups. Adult males, when aroused, demonstrate their strength, pawing impatiently at the ground, snapping young trees, bending down to the ground or even breaking taller spruces, and sometimes using their horns to uproot trees. During the breeding season males also quite frequently emit a croaking sound, known as a grunt.

The definite sign that a female is in oestrus is when she becomes covered by a bull (see also Figure 10.21). In a rut, a bull accompanies a cow uninterruptedly for 1–2 days, rarely even bothering to eat or rest. The bull is trying to isolate the cow from the herd, remaining with her on its perimeter and engaging in a series of ritual activities. The male frequently licks or sniffs the area around the vulva, mounts the cow fully or partially, adopts the aforementioned *flehmen* position repeatedly and frequently grunts and snorts, pawing repeatedly at the ground and sometimes scattering the soil around. From time to time the rutting male wallows, frequently urinating into the wallow. Mounting of a cow ending in actual copulation is a brief event (Krasiński & Raczyński 1967; Krasiński 1978b; Caboń-Raczyńska *et al.* 1987). Young, (4–5 years old) bulls present in the same group also become aroused at this point, brushing against one another with their horns, breaking branches, sometimes wallowing on the ground and trying to manoeuvre themselves into the immediate vicinity of a cow in heat, whenever the dominant male becomes even slightly separated from her. However, they are immediately repelled by the male intent on covering the female.

Vocal communication

It is rather rare for bison to communicate vocally. A mother uses the grunt to call a calf over to feed, sometimes a calf grunts to bring its mother. A mother also uses this means to summon her calf when she is feeling unsure of the situation. Bulls also grunt when they feel threatened. Their voice is lower and harsher than that of cows. During the rut the call of a bull may be described as a threat grunt when it serves to repel or discourage potential rivals. At this time, bulls also produce other sounds that are described as blowing and puffing.

Antipredator behaviour

Antipredator behaviour is difficult to observe in European conditions, since there are few places where large predators and European bison occur together. At the present time, European bison inhabiting lowland forests do not have any natural enemies. In Białowieża Forest, wolves, because of their low numbers, do not play any significant role in the regulation of the bison population. Nevertheless, recently, several cases of bear predation on bison were reported from the Caucasus, as well as two cases from free-ranging populations in Bukovynska in the eastern Carpathians (Ukraine) and two others in the Bieszczady Mountains (Poland) (Krasińska & Krasiński 2007; Paszkiewicz 2008). In the population of the Bieszczady Mountains, a situation was observed several times in which a group of bison was approached by a single wolf. Animals from the herd reacted to the appearance of the wolf by gathering together, close to each other, with calves inside the group and adult animals in the outer ring, facing the wolf until it gave up and left. At the beginning of February 2012 a calf about eight months old was killed and partially eaten by a small pack of wolves (Figure 10.31) (M. Januszczak, personal communication).

Sexual segregation

The size and composition of mixed groups is not stable, and changes seasonally. Groups meet frequently, combine and then quickly split, exchanging some individuals. The bonds between the young are the least permanent; young bulls exchange most frequently.

Young 1–2-year-old males spend most of their time in mixed groups. Bulls of 3–5 years are as likely to be found in a mixed group as in a bull-only group. Bull groups are not stable and frequently break down. Bulls of 6–12 years are generally only present in mixed groups during the breeding season. At other times they join small bachelor herds or remain alone. At the end of July, just prior to the breeding season, bulls make more and more frequent visits to mixed groups to check if any of the cows there are receptive.

Parasites and diseases

The health problems of European bison from the last free-ranging population living in Białowieża Forest at the beginning of the twentieth century were described by Wróblewski (1927). He recorded foot-and-mouth disease in bison three times during his stay in Białowieża Forest. He has mentioned that only few, among dead animals, were entirely free of flukes and/or lungworms, and identified a new species of trypanosome from bison blood. Contemporary populations of this species are also threatened by various disease factors (Kita & Anusz 2006). Less dangerous are traditional infectious diseases typical for both cattle and wild ungulates like foot-and-mouth disease (*Aphtae epizooticae*) or Bollinger disease,

Figure 10.31 A young calf killed and eaten by a pack of wolves at Bieszczady Mountain, February 2012. Photo by M. Januszczak.

but increasing is the threat of sicknesses with complicated aetiology, like that known as posthitis – a disease occurring in bison involving inflammation of the prepuce. In more advanced cases, when the penis becomes entirely inflamed, it is described as balanoposthitis. Bacteriological studies demonstrated that pathologically affected tissues support bacteria of the genera *Corynebacterium* and *Bacillus*, as well as *Staphylococcus koagulazo* negative, *Pseudomonas aeruginosa* and *E. coli* (Kita *et al.* 1990, 2003; Anusz *et al.* 2010). The anaerobic bacterium *Fusobacterium necrophorum* has also been found there (Jakob *et al.* 2000). Nevertheless, all studies done until now have identified neither the source of any infection, nor any single pathogenic agent that can be linked unambiguously with the disease. Contributing to its incidence may also be environmental factors like insects, bacteria and viruses. The disease is not contagious, because only some males become affected: on average 6.5% of males in the Polish population of Białowieża Forest. It occurs in a similar intensity in males in both parts of Białowieża Forest, but it appeared sporadically also in other Polish, Ukrainian and German populations (Krasińska & Krasiński 2007; Krasińska & Krasiński 2010; Bunevich 2010).

An example of a disease most probably transmitted from cattle to European bison is tuberculosis. It is caused by *Mycobacterium bovis*, and was detected in European bison for the first time in the population of the Bieszczady Mountains (Poland) (Żurawski & Lipiec 1997). At the area penetrated by bison, in 1992–1996 several cases of cattle being infected with tuberculosis were confirmed. The next cases of tuberculosis were recorded in a dead bison cow (2010) and culled bull (2011) (Brewczyński & Welz 2011). The epizootiologic analysis of the livestock and selected wildlife species in Bieszczady suggest the possibility of transmission of this infectious agent from livestock to wildlife (Sawla *et al.* 2011). *Mycobacterium* belongs to the most resistant pathogens; it may remain in the environment for several years, so prevention of this disease is particularly difficult. In European bison of Białowieża Forest, so far tuberculosis has not been detected.

A new threat for European bison is blue tongue disease. This illness is caused by a virus, and the sources of infection are sick animals that are bearers of the virus. The infection is transmitted by small, several millimetres long insects of genus *Culiciodes*. The disease originates from Africa and was brought to Europe via infected animals. In 2007 an outbreak occurred within Western Europe in sheep, causing 80% mortality. Also infected were cattle. In 2007 this disease caused the death of ten European bison at the European Bison Breeding Centre at Hardehausen (Rhine – Northern Westphalia). During the autopsies the following were confirmed: swollen and bluish tongue and ulcers inside of the mouth, hence the synonym of this disease: foot-and-mouth disease (Glunz 2002).

The European bison offers exceptional opportunities for parasitological studies, having been eradicated in the natural state, only to be reintroduced to the wild after some years, being in the meantime maintained in captivity in semi-natural conditions. So far, an incidence of some 38 helminth (flatworm and roundworm) species has been reported in European bison of Białowieża Forest (Table 10.9). An overwhelming majority of these species would also be typical for domesticated ruminants; only the trypanosome *T. wrublewski*, the biting-louse *Bisonicola sedecimdecembrii* and the mite *Demodex bisonianus* can be regarded as specific or typical of the European bison.

Two endoparasites widespread in bison are the liver fluke *Fasciola hepatica* (present in 44% of animals studied) and the lungworm *Dictyocaulus viviparus* (a roundworm (nematode) recorded in 58% of studied animals) (Demiaszkiewicz *et al.* 1999). Intense infestations of both of these parasites produce obvious clinical symptoms in bison, among them loss of body weight, poor development and diarrhoea. Occasionally, they may even lead to an animal's death. In the population of European bison inhabiting the Bieszczady Mountains, lung nematodes have not yet been detected, while the lungworm occurs sporadically.

Nematodes of the stomach and gastrointestinal tract are very common, with 100% of animals potentially being affected (Dróżdż *et al.* 1989, 1990, 1999, 2002a, 2002b). A typical helminthofauna of free-ranging bison has been augmented by 15 species initially characteristic for other ungulates (red and roe deer, and moose) (Table 10.9).

For the first time in Poland, in 1977 in the Bieszczady Mountains, a bloodsucking nematode *Ashworthius sidemi* has been found in European bison (Dróżdż *et al.* 1998). Originally characteristic for Asian deer species (mainly the sika), following this species' numerous introductions to Central Europe (Germany, Austria, Czech Republic), it dispersed into neighbouring countries like Poland, Ukraine, Slovakia and Belarus (Dróżdż *et al.* 2000, 2002b). The species was first detected in the Belarusian part of the forest in 1999 (Dróżdż *et al.* 2002a) and then in the bison on the Polish side just two years later. By 2003 as many as 100% of the Białowieża bison had already been infected by this nematode (Dróżdż *et al.* 2003; Kochko 2003).

Furthermore, it has been found that European bison are among numerous intermediate hosts of an intracellular protozoan parasite *Neospora caninum*, whose final hosts are the dog, the coyote and probably also other species of subfamily *Caninae*. In 1988–2009 the presence of antibodies against this parasite was found in 7.3–16.7% of tested animals from several bison populations (Cabaj *et al.* 2005, 2009, 2010). Fortunately, the presence of *Neospora caninum* in bison has not yet been associated with any distinct negative effects on its health.

The European bison is host to 14 species of ectoparasites (parasitic arthropods), 13 of these known to be present in the Białowieża population (Izdebska 2001a, 2001b, 2001c, 2005, 2006, 2011; Izdebska & Cydzik 2010) (Table 10.10).

Status in the wild

IUCN Red List Category: Vulnerable

Table 10.9 The helminth fauna of free-ranging European bison in the Białowieża Forest, Borecka Forest and Bieszczady Mountains.

No.	Species of parasite	Percentage of infected bison		
		Białowieża Forest	Borecka Forest	Bieszczady
1	*Fasciola hepatica*	84	20	12
2	***Parafasciolopsis fasciolaemorpha***	**5**	–	–
3	*Dicrocoelium dendricum*	16	–	–
4	*Paramphistomum cervi*	+	x	x
5	*Moniezia benedeni*	+	–	–
6	*Chabertia ovina*	33	x	x
7	*Oesophagostomum radiatum*	14	x	x
8	*Oe. Venulosum*	71	x	x
9	*Trichostrongulus axei*	25	60	46
10	***T. capricola***	**12**	**40**	**15**
11	*T. vitrinus*	–	–	8
12	*Ostertagia ostertagi*	100	100	69
13	*O. lyrata*	87	80	8
14	***O. leptospicularis***	**87**	**80**	**100**
15	***O. kolchida***	**87**	**80**	**61**
16	***O. antipini***	**12**	**–**	**–**
17	***Spiculopteragia boehmi***	**87**	**60**	**53**
18	***S. mathevossiani***	**19**	**–**	**–**
19	***S. asymetrica***	**2**	–	–
20	*Cooperia oncophora*	100	–	–
21	*C. surnabada*	62	–	8
22	***C. pectinata***	**12**	**–**	**46**
23	*C. punctata*	19	20	–
24	*Haemonchus placei*	37	–	–
25	***Ashworthius sidemi***	**100**	**–**	**100**
26	*Nematodirus helvetianus*	87	20	–
27	***N. roscidus***	**100**	**80**	**–**
28	***N. europeaus***	**25**	**60**	**–**
29	***Namatodirella alcidis***	**2**	–	–
30	*Dictyocaulus viviparous*	94	80	–
31	*Thelasia gulosa*	5	–	–
32	*Onchocerca lienalis*	70	x	75
33	*O. gutturosa*	47	x	x
34	*Setaria cervi*	3	–	–
35	*Trichiuris ovis*	43	x	x
36	*Aonchotheca bilobata*	75	–	100
37	***Mazamastrongylus dagestanicus***	**12**	–	–
38	***Bunostomum trigonocephalum***	**22**	–	–
	Number of species	**38**	**13**	**11**

x: not. tested; –: no determined presence of a parasite; +: confirmed presence of a parasite. Typical parasites of ungulates are in bold.
Data from: Dróżdż *et al.* (1989, 1999, 2002a, 2002b); Demiaszkiewicz *et al.* (2008, 2009a, 2009b, 2010).
Source: Krasińska & Krasińska (2007), modified by. A. Demiaszkiewicz.

Table 10.10 Ectoparasites of the European bison

No	Species	Place of occurrence			Remarks
		Puszcza Białowieska Free-ranging and captive population	Bieszczady Mountains	Other regions of Poland (Niepołomice, Smardzewice, zoos)	
1	*Bisonicola sedecimdecembrii* (Insecta, Phthiraptera, Trichodectidae)	+	+	+	Biting louse, specific parasite of European bison (prevalence above 50%); mallophagosis is manifested only rarely
2	*Lipoptena cervi* (Insecta, Diptera, Hippoboscidae)	+	–	+	Deer ked – a typical parasite of cervids though noted rather sporadically on bison (prevalence about 8%)
3	*Melophagus ovinus* (Insecta, Diptera, Hippoboscidae)	+	–	–	Sheep ked is a sheep-specific parasite (it was found on few bison in captivity)
4	*Chrysozona (Haematopota) pluvialis* (Insecta, Diptera, Tabanidae)	+	–	–	Common horse fly was found only in the Białowieża Forest in the summer; however, this is a common species in Poland
5	*Dermacentor reticulatus* (Acari, Ixodida, Ixodidae)	+	–	+	Marsh tick – in winter ticks were noted on above 90% of European bison from the Białowieża Forest (free-ranging population)
6	*Ixodes ricinus* (Acari, Ixodida, Ixodidae)	+	+	+	Common tick – frequent and numerous in Poland; it occurs commonly on European bison (in winter prevalence is almost 40%)
7	*Ixodes persulcatus* (Acari, Ixodida, Ixodidae)	+	–	–	Taiga tick – rare in Poland (only in the Białowieża Forest); occasionally on bison
8	*Ixodes hexagonus* (Acari, Ixodida, Ixodidae)	–	–	+	Hedgehog tick – only one individual found on the European bison (incidental parasite)
9	*Chorioptes bovis* (Acari, Astigmata, Psoroptidae)	+	+	+	Chorioptic mite – prevalence above 50%; occasional manifestations of chorioptosis
10	*Psoroptes equi* (Acari, Astigmata, Psoroptidae)	+	–	–	Psoroptic mite – noted occasionally on bison
11	*Sarcoptes scabiei* (Acari, Astigmata, Sarcoptidae)	+	–	–	Sarcoptic mite (itch mite) – occasionally on bison
12	*Demodex bisonianus* (Acari, Prostigmata, Demodecidae)	+	+	+	Demodecid mite – specific to European bison (prevalence about 50%)
13	*Demodex bovis* (Acari, Prostigmata, Demodecidae)	+	–	–	Demodecid mite – specific parasite of cattle, noted incidentally on European bison
14	*Demodex* sp. (Acari, Prostigmata, Demodecidae)	+	–	–	Unknown species of demodecid mites – found only on few individuals of European bison from Białowieża Forest

+: confirmed presence of a parasite. –: no evidence of presence of a parasite.
Data from: Izdebska (2001a, 2001b, 2001c, 2005, 2006, 2011); Izdebska & Cydzik (2010).
Source: Krasińska & Krasiński (2007), modified by J. N. Izdebska.

Worldwide status

The achievements in restitution of highly endangered species can be measured as the number of breeding individuals, even if the species is maintained only in captive conditions, but the true success is accomplished if the species becomes restored to the wild. In the case of the European bison, the first step in this process was the creation of the first free-ranging herd at Białowieża, Poland in 1952. Since 1986 the number of animals in the wild exceeded the number of individuals raised in captivity. Nevertheless, even with a total number of over 4400 individuals, the species still remains threatened with extinction and requires further attention and protective measures (see also Table 10.3).

The European bison had been listed as an endangered species (EN) in the *IUCN Red Data Book* since 1966. However, with new criteria introduced in 2001, the species is listed now as vulnerable (VU). It is mentioned in Annex III to the Bern Convention, and in the Annex II to the Habitat Directive (92/43/EEC changed into 92/62 EEC) as a priority species. However, in the Washington Convention, CITES – (i.e. the 1973 Convention on the International Trade in Endangered Species of Wild Flora and Fauna held in Washington on 3 March 1973) – the European bison is not listed in the Annex to the Regulation of the Minister of the Environment of 27 February 2001 (Krasińska & Krasiński 2007).

Main threats

Undoubtedly, the main challenge in the conservation of this species remains the mitigation of further loss of genetic diversity. Until recently, this parameter could be assessed only for animals in captivity, on the basis of pedigree records. Since the introduction of DNA analyses, it is possible to evaluate the level of homogeneity also in free-ranging herds. Results of the preliminary study on this subject in the Carpathians indicate that the loss of genetic variability in the wild happens faster than could be expected on the basis of known pedigrees of ancestors of the herd (Łopieńska *et al.* 2011). In the case of this species there are no available wild populations that could become a reservoir of genetically diverse animals, so the only way to deal with this issue is planned supplementation of free-ranging populations with individuals of known pedigree, bearing genes of underrepresented ancestors.

This problem is additionally complicated in this species because of the existence of two genetic lines which differ considerably regarding the contribution of particular ancestors. Moreover, the genetic pool of the species becomes increasingly dominated by genes of just one pair of founders of the species. Therefore, to minimise the further loss of underrepresented genes, the separation of both lines should be maintained as long as possible.

Low genetic diversity is connected with a weak resistance to diseases. In the case of the European bison, the most dangerous are foot and mouth disease, bovine tuberculosis, brucellosis and the recently spread within Europe of the blue tongue virus. Outbreaks of foot-and-mouth disease in European bison herds took place in the past (e.g. at the Pszczyna Breeding Centre) and resulted in eradication of the whole stock. Similarly, in Bieszczady Mountains between 1996 and 2001 the herd of 16 animals was eliminated because of an outbreak of bovine tuberculosis, which originated from cattle, but germs were found later in other wildlife species (badgers, deer). In 2007 the breeding centre at Hardehausen lost ten animals to blue tongue disease. A problem peculiar to European bison is an inflammation of the prepuce (Balanoposthitis), leading to infertility of males. This disease has been observed mostly in free-ranging populations of bison in both parts of Białowieża Forest and occasionally in a few other wild populations (Kyrgizya, Ukraine); however, the factor responsible for its incidence has not yet been identified (Krasińska & Krasiński 2007, 2010; Glunz 2002; Perzanowski 2010).

Locally, free-ranging herds of European bison may be threatened by poaching, which in countries like Poland, Slovakia or Lithuania is incidental, but in others (e.g. Ukraine) poses a serious problem. The largest population of this species in the Carpathians, in Ukraine or Bukovyna, that counted almost 250 individuals in the mid 1990s, decreased to fewer than 50 in 2010. In Ukraine poaching for European bison is either organised for wealthy people in pursuit of a trophy or by inhabitants from local settlements, for whom the main purpose is the meat (Parnikoza *et al.* 2009; Khoyetsky 2011).

In regions where fragmented forest–field habitats dominate within home ranges of European bison herds (e.g. Lithuania, northeastern Poland), a danger of traffic accidents appears, especially if there are highways intersecting the forest (Tracz & Tracz 2010, 2011).

Recently, a growing problem is the fragmentation of former natural or semi-natural habitats, e.g. due to the development of recreational infrastructure, which for a number of years has been a significant issue in the Carpathians. Through preventing the natural exchange of individuals, it may lead to permanent functional isolation of free-ranging herds, which would contribute to the further loss of their genetic variability (Bashta *et al.* 2010).

So far a marginal issue seems to be the impact of predation; however, in regions where sizeable populations of large predators (brown bears, wolves) occur (e.g. the Carpathians, Russia) some cases are observed. Together with an increase of European bison numbers and gradual habituation of predators to such prey it may be expected that the frequency of such incidents would increase (Perzanowski & Marszałek 2008).

Competition with cattle and conflicts with livestock breeders occur when there is, within European bison ranges, a dense network of settlements bordering directly with the forest or settlements surrounded by the forest (e.g. Białowieża, Lithuania). In such places bison are seen grazing at the pasture among cows or eating hay from hay stacks (Balciauskas 2000). An additional and serious problem in such areas appears a possibility of transmission of diseases and parasites between bison and livestock.

Within European bison ranges there have so far been no attempts to restore wild cattle. A threat of hybridisation could appear if there was any contact with the American bison, which is already farmed in Europe. There are some fears about the possibility of crossbreeding of European bison with domestic cattle, but this seems to be rather unlikely since the experimental mating of both species proved to be difficult and of very low efficiency.

Minimum area required

Estimation of the minimum area required (MAR) should be based on data on population density and home range size obtained for free-ranging, non-restricted populations.

In the lowland population of Białowieża, densities of European bison in the Polish part ranged between 13–16 animals per 1000 ha in years 1976–1989, but later on together with an increase of home range it declined to six animals per 1000 ha. On the Belarusian side of Białowieża Forest densities that were up to 12 animals per 1000 ha until 1984 decreased later to three bison per 1000 ha. In Białowieża the most frequently observed are groups of about 20 animals, while herds counting 40 or more individuals are seen only rarely, and when animals graze on open meadows (Krasińska & Krasiński 2007).

In free-ranging bison populations dwelling in the Carpathian Mountains within their home ranges densities were observed of between 6 and 14 animals per 1000 ha, depending on the year and season. Those estimates were done for whole subpopulations, the size of which changed over the last ten years from about 80 to 150 individuals. Those subpopulations consisted of some number of smaller herds and groups which were not monitored separately. In most cases small bison groups grow until they reach a number of 30–40 animals, and afterwards break into smaller units counting 10–20 animals, which remain fairly stable for some time (Perzanowski 2001–2011).

Therefore, assuming that a group of about 20 animals is the most typical and stable, it is possible to say that a minimal area required for such a unit ranges between some 1500–6500 ha, depending on the type of forest (its carrying capacity) and the season of the year.

The population viability analysis of Polish populations of European bison performed by Suchecka (2009) has demonstrated that due to a risk of extinction and loss of genetic diversity, isolated populations of this species should number at least 50 animals. Also important is the sex structure of the initial group, which should be slightly biased towards females (62–75%). Nevertheless, on the basis of experiences with free-ranging herds in Poland, populations that exceed 100 animals are regarded as demographically and genetically safe. So far there are only two such populations in Poland (Białowieża and Bieszczady) and in the whole of Europe only six such herds exist at the moment. Therefore, one direction in the conservation of this species should include efforts towards creation of larger numbers of numerous, viable herds.

Conservation measures *ex situ* and *in situ*

The main factors that should be considered in planning the conservation measures for European bison are: their extremely high level of inbreeding and a danger of further loss of genetic diversity. So far, detailed information on the genetic status of particular animals only allowed assessment on the basis of pedigree, so such an approach was possible only for animals in captivity. The European bison is in a different situation to other endangered species, because at the very beginning of the restitution programme there was a pedigree book for all animals with known origin, and those records have continued to the present. Such a tool allows for precise planning of mating and an exchange of individuals among breeders. Those actions should be coordinated at the European level and become included in EU legislation concerning conservation of this species.

An additional aspect of *ex situ* conservation is the possibility to reduce a risk of extinction due to diseases. Because of low heterogeneity in the species, these animals have only a weak resistance to various infections. Therefore, maintaining a large number of isolated groups allows for individually dealing with outbreaks of diseases with no risk of spreading them over a significant part of the world population.

Conservation *in situ* should be focused upon the restoration of the species to its natural habitats (see also Figure 10.32) and creation of self-sustainable populations. This requires consideration of several aspects.

One of these aspects is the identification of the most suitable introduction sites, regarding the quality of habitat, the size of an available area and human dimension/socio-economic facet (i.e. possibility of anthropogenic conflicts) (Perzanowski & Olech 2007).

The second issue, which depends mostly on the carrying capacity of the area selected for introduction, is the potential size of the established population. Preferably, the new population should be demographically and genetically stable, but if there are some restrictions regarding the home range size, creation of a smaller herd should be considered, with numbers and structure periodically monitored and regulated. A network of such small herds may be maintained and actively managed as a meta-population, with exchanges of individuals performed according to a prescribed pattern (Olech & Perzanowski 2010, 2011).

The third problem is connected with the necessity to prevent further loss of genetic variability. In the wild there is no possibility to control mating patterns, and pedigree is known only for the initial group of herd founders. Furthermore, recent analysis of genetic variability (microsatellites) in European bison herds established in the Bieszczady Mountains some 50 and 30 years ago gave much lower results than values expected on the basis of the pedigree of the ancestors of those herds (Nowak and Olech 2008a). This indicates the necessity of periodical monitoring of free-ranging herds and considering the supplementation of them with genetically selected, captive

Figure 10.32 Young European bison released to acclimatisation enclosure at Bieszczady National Park. Photo by K. Perzanowski. A black and white version of this figure will appear in some formats. For the colour version, please refer to the plate section.

animals. In wild herds there is also a threat of an outbreak of diseases; therefore, their health status needs systematic monitoring and preventive actions regarding, e.g. contact with livestock, should be undertaken (Olech *et al.* 2008).

Finally, other factors possibly threatening the reproduction rate, survival of young, longevity of adults, restricting movements, etc. – like all kinds of human-related disturbance, habitat fragmentation, conflicts with agriculture and forestry – should be assessed for both already existing herds and sites selected for future introductions and appropriate mitigation measures implemented.

Status in captivity

The status of captive animals, i.e. the number and distribution of breeding centres, zoos, etc. as well as the numbers of animals being maintained in particular countries, group size and representation of both genetic lines in captivity has been described in detail in earlier sections. Location of breeding centres in Europe is presented in Figure 10.33.

In Europe, European bison are bred in various conditions, ranging from relatively small pens (1000–3000 m^2), where animals are completely dependent on forage available in feeders (Figure 10.34) to quite large enclosures (a few hundred hectares) (Figure 10.35 and 10.36), where there is access to pasture and the possibility of browsing; and cows may even separate themselves from the herd during the calving season in these larger enclosures. The number of animals maintained in every centre in relation to herd size is illustrated in Figure 10.37.

The main problem regarding captive animals, apart from the maintenance of their genetic variability, is the lack of international standards for living conditions of those animals in captivity, ensuring their well-being. Recommendations for the protection of European bison in captive conditions have already been published by the IUCN as the Status Survey and Conservation Action Plan European bison (Pucek *et al.* 2004). Also available is a book summarising requirements regarding food, access to water and space, as well as describing optimal design of facilities for European bison in captivity (Olech *et al.* 2008). However, those recommendations and suggestions are not commonly followed, which should be improved considering the species' high protective status. The European organisation of zoos (EAZA) provides some indications for minimal size of enclosures for bison, which are 500 m^2 for a pair and 200 m^2 for each additional animal. This area is very small, so generally the European Bison Conservation Centre tries to avoid keeping European bison in small pens. Tendencies for improper feeding are also observed in captive conditions – animals are treated like cattle and often overfed. As a consequence, problems with diarrhoea and acidosis are likely to appear.

Inconsistent is also the status of the species among European countries regarding rules for its handling, transportation and international trade, since in some countries it is treated within the same category as common cattle. In countries like Poland or Slovakia, the European bison is a part of national fauna and is included in the National Red List and protected by national law. In countries where European bison are kept only in captive or semi-free conditions the status of the species is not clear; the species is not protected by national law, so in many situations is treated as cattle. According to the EU Habitat Directive, all EU countries are supposed to protect European bison, but locally

Figure 10.33 Present distribution of European bison breeding centres and semi-free herds of LB and LC lines. Source: European Bison Pedigree Book 2010.

Figure 10.34 Animals in Busolengo Zoo (Italy). (Photo by W. Olech.). A black and white version of this figure will appear in some formats. For the colour version, please refer to the plate section.

Figure 10.35 Herd in the European Bison Breeding Centre at Pszczyna (Poland). Photo by M. Hławiczka. A black and white version of this figure will appear in some formats. For the colour version, please refer to the plate section.

Figure 10.36 Group of European bison foraging at a forest clearing during the rut period; on the left is a radio-collared bull, September 2011. Photo by T. Kamiński. A black and white version of this figure will appear in some formats. For the colour version, please refer to the plate section.

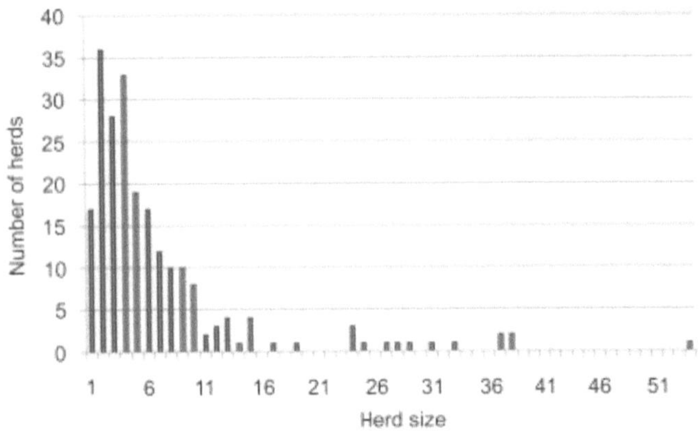

Figure 10.37 Numbers of European bison captive centres in relation to the herd size (number of animals maintained) in every centre.

the interpretation of this rule varies and, for example, does not include animals in captivity.

It is important to mention that 40% of the world population is kept in captivity, and because of the history of species restitution, this part of the world population has a better balanced gene pool and therefore may serve as a genetic resource for the whole species and a source of animals for reintroduction.

Acknowledgements

The authors would like to thank Mr Tomasz Kamiński from the Mammal Research Institute PAS in Białowieża for the preparation of drawings and photos.

References

Adamec, M., Perdiková, Z., Pčola, Š. & Adamcová, M. (2006). Priestorová activita zubrov hrivnatých Bison bonasus v NP Poloniny. [Spatial activity of European bison in Poloniny N. P.]. In *Perspektywy rozwoju populacji żubrów*, pp. 55–62. Ed. Olech, W. Goczałkowice Zdrój: Artisco. [In Slovak].

Aleksandrov, V. N. & Golgovskaya, K. J. (1965). Kormovyje ugodia zubrov Kavkazskogo Zapovednika. [Food abundance for European bison in Caucasus Reserve]. *Trudy Kavkazskogo Gosudarstnennogo Zapovednika* **8**: 129–154. [In Russian]

Aleksandrowicz, K., Perzanowski, K., Cătănoiu, S. & Deju, R. (2009). Browse and ground flora supply in selected tree stands of Vanatori Neamt Nature Park. *Studies and Research in Vanatori Neamt Nature Park* **2**: 40–46.

Anusz, K., Kita, J. & Zaleska, M. (2010). Threats to the health of the European bison. In *European Bison Conservation in the Białowieża Forest, Threats and Prospects of the Population Development*, pp. 51–62. Eds. Kowalczyk, R., Ławreszuk, R. &. Wójcik, J. M. Białowieża: Mammal Research Institute PAS. [In Polish with English summary].

Augustyn, M. & Kozak, I. (1997). The trends of anthropogenic pressure in Polish and Ukrainian Carpathians. In *Proceedings of the 2nd Annual Meeting of International Centre of Ecology, Polish Academy of Sciences*, pp. 15–22. Eds. Perzanowski, K. & Augustyn, M. Ustrzyki Dolne Poland: Carpathian Branch ICE PAS.

Balčiauskas, L. (1999). European bison (*Bison bonasus*) in Lithuania: status and possibilities of range extension. *Acta Zoologica Lithuanica, Biodiversity*, **9**: 3–18.

Balčiauskas, L. (2000). Restoration of European bison to Lithuania: achievements and problems. In *Proceedings International Symposium 'European Bison – Yesterday, Today and Tomorrow', 9–10.12.2000 Šiaulai, Lithuania*, pp. 8–15. Ed. Balčiauskas, L. Šiaulai: UAB, Šiaulai University.

Bashkirov, V. S. (1940). Kavkazskii zubr. Monograficheskii ocherk. [Caucasian European bison, Monograph]. In *Kavkazskij zubr*, pp. 7–22. Ed. Kulagin, N. M. Moskva: Glavnoe Upravelenije po zapovednikam, zooparkam i zoosadam. [In Russian].

Bashta, A., Kruhlov, I., Korzhyk, V., *et al.* (2010). Creation of ecological corridors in Ukraine. State Agency for Protected Areas of the Ministry of Environmental Protection of Ukraine, Interecocentre Kyiv, Altenburg & Wymenga Ecological Consultants.

Belousova, I. P. (1986). Prioksko-terrasnyj biosfernyi zapovednik, pos. Danki. [Prioksko-terrasnyj Biosphere Reserve]. In *Piervoe Vsesojuznoe Soveshchanie po problemam zookul'tury. Tezisy dokladov chast' 2*. pp. 11–13. Moskwa: Zookul'tura. [In Russian].

Belousova, I. P. (1993). About modern breeding lines' origin of European bison (*Bison bonasus* L.). In *K voprosu o vozmiozhnosti sokhraneniya zubra v Rossii*], pp. 6–30. Ed. Belousova, I. E. Pushchino, Russia: ONTI PNTS RAN. [In Russian with English summary].

Benecke, N. (2005). The Holocene distribution of European bison: the archeozoological record. *Munibe (Antropoliogie-Arkeologie)* **57**: 421–428.

Bereszyński, A. (2005). Restytucja żubra *Bison bonasus* w Polsce zachodniej koncepcja, stan aktualny i strategia rozwoju. [Restitution of the European bison to the western Poland, a concept, present status and the strategy of development]. In *Materiały konferencji naukowej Ochrona Żubrów Zachodniopomorskich, Ińsko, 15–16 wrzesień 2005*, pp. 5–11. Borzkowce: WFOŚ i GW Szczecin. [In Polish].

Bohlken, H. (1967). Beitreg zur Systematic der rezenten Formen der Gattung Bison H. Smith, 1827. *Zeitschrift für zoologische Systematic und Evolutionsforschung* **5**: 51–110.

Bojanus, L. H. (1827). De Uro nostrate eiusque sceleto commentatio. *Nova Acta Physico-Medica. Academiae Cesareae Leopoldino-Caroline Naturae Curiosum* **32**: 1–184.

Borowski, S. & Kossak, S. (1972). The natural food preferences of the European bison in seasons free of snow cover. *Acta Theriologica* **17**: 151–169.

Bostock, S. S. C. (2001). *Zoos and Animals Rights: The Ethics of Keeping Snimals.* New York: Routledge.

Brewczyński, P. & Welz, M. (2011). Threat of tuberculosis of European bison in the Bieszczady. *European Bison Conservation Newsletter* **4**: 63–70.

Brincken, J. (1826). *Memoire Descriptif sur la Foret Imperiale de Białowieża, en Lithuanie.* Varsovie N: Glücksberg.

Brink Van den, W. J. (1980). The behaviour of wisent and bison in larger enclosures. *Acta Theriologica* **25**: 115–130.

Bunevich, A. N. (1990). Teritorjal'noe razmeshchenie zubrov v Belovezhskoi Pushchi. [Spatial distribution of European bison in the Białowieża Forest]. In *Materialy nauchno-prakticheskoi konferencii, posvyashchennoi 50-letiu regulyarnych issledovanii v Belovezhskoi Pushche, 19–21 dekabrja 1989*, pp. 129–130. Minsk: Belosovprof [In Russian].

Bunevich, A. N. (1994). Povrezddaemost' zubrami drevostoev v mestach zimnej koncentracii. [Damages of trees by European bison at feeding sites]. In *Materialy konferencii: "Problemy izucheniya sokhraneniya i isopol'zovaniya biologicheskogo raznobraziya zhivotnogo mira"*, pp. 241–242. Minsk: PKP Junilend. [In Russian].

Bunevich, A. N. (1999). Bioritmy reprodukcii zubra (Bison bonasus bonasus) Belovezhskoj Pushchi. [Breeding rhythms of lowland European bison in Białowieża Forest]. In *Biologicheskie Ritmy. Materialy Mezhdunarodnoi nauchno- prakticheskoi konferencii. Brest 1999*, pp. 55–56. Brest: Brestskii Gosudarstvennyi Universitet. [In Russian].

Bunevich, A. N. (2003). Dynamics and structure of the home range of European bison population in Białowieża Forest. *Belovezhskaya Pushcha. Issledovaniya* **11**: 160–177. [In Russian with English summary].

Bunevich, A. N. (2006). Body mass and measurements of the European bison in the Belovezhskaya Pushcha. *Belovezhskaya Pushcha, Issedovaniya* **12**: 121–141. [In Russian with English summary].

Bunevich, A. N. (2010). Dynamics of the intensity of disease of uro-genital organs (balanoposthitis) in males of European bison in the Belarussian part of the Białowieża Forest. *Parki narodowe i Rezerwaty. Przyrody* **29**(3): 77–87. [In Polish with English summary].

Bunevich, A. N. & Kochko, F. P. (1988). Dinamika chislennosti i struktura populacii zubrov Belovezhskoi Pushchi. [Dynamics of population numbers and structure of the European bison in

Białowieża Forest]. In *Populacjonnye issledovanniya zhivotnykh v zapovednikakh. Sbornik nauchnykh trudov*, pp. 96–114. Moskva: Nauka. [In Russian].

Bunevich, A. N., Krasińska, M. & Daleszczyk, K. (2006). Creation and growth of the free-ranging population of lowland European bison, *Bison bonasus bonasus* (L.) in the Belarussian part of Białowieża Forest. *Parki narodowe i Rezerwaty. Przyrody* **25**(4): 101–118. [In Polish with English summary].

Burzyńska, B. & Topczewski, J. (1995). Genotyping of *Bison bonasus* β-casein gene following DNA sequence amplification. *Animal Genetics* **26**: 335–336.

Cabaj, W., Moskwa, B., Pastusiak, K. & Gill, J. (2005). Antibodies to *Neospora caninum* in the blood of European bison (*Bison bonasus bonasus* L.) living in Poland. *Veterinary Parasitology* **128**: 163–168.

Cabaj, W., Bień, J., Goździk, K. & Moskwa, B. (2009). *Neospora caninum* in European bison in Białowieża, Poland: the current research status. *European Bison Conservation Newsletter* **2**: 102–111.

Cabaj, W., Bień, J., Goździk, K. & Moskwa, B. (2010) *Neospora caninum* in European bison in Białowieża, Poland. *European Bison Conservation Newsletter* **3**: 63–68.

Caboń-Raczyńska, K., Krasińska, M. & Krasiński, Z. A. (1983). Behaviour and daily activity rhythm of European bison in winter. *Acta Theriologica* **28**: 273–299.

Caboń-Raczyńska, K., Krasińska, M., Krasiński, Z. A. & Wójcik, J. M. (1987). Rhythm of daily activity and behaviour of European bison in Białowieża Forest in the period without snow cover. *Acta Theriologica* **32**: 335–372.

Czykier, E., Sawicki, B. & Krasińska, M. (1999). Postnatal development of the European bison spermatogenesis. *Acta Theriologica* **44**: 77–90.

Dackiewicz, J. (1999). Oddziaływanie żubrów na las w miejscach ich zimowej koncentracji w Puszczy Bialowieskiej. [Influence of European bison upon the forest at winter feeding sites of Białowieża Forest]. MSc thesis. Faculty of Forestry, Warsaw University of Life Sciences-SGGW. [In Polish].

Daleszczyk, K. (2002). *Behawior rozrodczy Bison bonaus (L.)*. [Breeding behaviour of European bison]. PhD thesis. Wydział Bioinżynierii Zwierząt Uniwersytetu Warmińsko-Mazurskiego w Olsztynie. [In Polish].

Daleszczyk, K. (2004). Mother–calf relationships and maternal investment in European bison *Bison bonasus. Acta Theriologica* **49**: 555–566.

Daleszczyk, K. (2005). Activity pattern of calf and its integration into herd in European bison, *Bison bonasus. Folia Zoologica* **54**: 359–363.

Daleszczyk, K. (2011). Some factors influencing reproductive parameters of European bison cows. *European Bison Conservation Newsletter* **4**: 45–54.

Daleszczyk, K. & Krasiński, Z. A. (2001). Parturition behaviour of European bison living in reserves. *Folia Zoologia* **50**: 75–78.

Daleszczyk, K., Krasińska, M., Krasiński, Z. A. & Bunevich, A. N. (2007). Habitat structure, climatic factors and habitat use by European bison (*Bison bonasus*) in Polish and Belarusian parts of Białowieża Forest, Poland. *Canadian Journal of Zoology* **85**: 261–272.

Deju, R. (2006). Restoration program for free ranging population of the European bison (Bison bonasus L., 1759) in Romania. In *Proceedings of the Workshop 'European Bison and Local Communities in Carpathian Mountains Vanatori N.P 23–26.08.2006'*, pp. 5–10. Eds. Cătănoiu, S. & Deju, R. Piatra-neamt: Vanatori-Neamt N. P.

Demiaszkiewicz, A. W., Dróżdż, J. & Lachowicz, J. (1999). Occurence of lung nematodes in red deer of Białowieża Forest. *Medycyna Weterynaryjna* **55**: 519–520. [In Polish with English summary].

Demiaszkiewicz, A. W., Pyziel, A. M. & Lachowicz, J. (2008). Helminthological status of European bison in Białowieża Forest in the winter 2007/2008. *European Bison Conservation Newsletter* **1**: 42–53. [In Polish with English summary].

Demiaszkiewicz, A., Pyziel, A. M., Lachowicz, J. & Kuligowska, I. (2009a). Dictyocaulosis of European bison in Białowieża Forest. *European Bison Conservation Newsletter* **2**: 112–118.

Demiaszkiewicz, A. W., Pyziel, A. M., Lachowicz, J. & Kuligowska, I. (2009b). Robaczyca płucna żubrów w Puszczy Białowieskiej. [Dictyocaulosis in European bison of Białowieża Forest]. *European Bison Conservation Newsletter* **2**: 112–118. [In Polish with English summary].

Demaszkiewicz, A., Pyziel, A. M., Kuligowska, J. & Lachowicz, J. (2010). Parasitoses of the European bison caused by nematodes located in the large intestine. *European Bison Conservation Newsletter* **3**: 69–74. [In Polish with English summary].

Deodatus, F. & Protsenko, L. (eds) (2010). *Creation of Ecological Corridors in Ukraine*. Kyiv: State Agency for Protected Areas of the Ministry of Environmental Protection of Ukraine, Altenburg & Wymenga Ecological Consultants, Interecocentre.

Drożdż, A., Weiner, J., Gębczyńska, Z. & Krasińska, M. (1975). Some bioenergetic parameters of wild ruminants. *Polish Ecological Studies* **1**: 85–101.

Dróżdż, J., Demiaszkiewicz, A. W. & Lachowicz, J. (1989). The helminth fauna of free-ranging European bison, *Bison bonasus* (L.). *Acta Parasitologica Polonica* **34**: 117–124. [In Polish with English summary].

Dróżdż, J., Demiaszkiewicz, A. W. & Lachowicz, J. (1990). Nematodes in large intestine of European bison. *Wiadomości Parazytologiczne* **36**: 35–38. [In Polish with English summary].

Dróżdż, J., Demiaszkiewicz, A. W. & Lachowicz, J. (1998). *Ashworthius sidemi* (Nematoda, Trichostrongylidae) a new parasite of the European bison *Bison bonasus* (L.) and the question of independence of *A. gagarini. Acta Parasitologica* **43**: 75–80.

Dróżdż, J., Demiaszkiewicz, A. W. & Lachowicz, J. (1999). Forming of helminthofauna of free-living European bison in three different forest biotopes (Białowieża Forest, Borecka Forest and Bieszczady Mountains). In *Abstracts of International Scientific Conference 'Health Protection of Free-ranging* Bison bonasus *in Białowieża Forest', Warszawa, 26–27 listopada 1999*, pp. 25–27. Warszawa: Katedra Patologii i Epizootiologii Wydziału Weterynaryjnego SGGW.

Dróżdż, J., Demiaszkiewicz, A. W. & Lachowicz, J. (2000). Ashwortiosis: a new parasitic disease of wild ruminants. *Medycyna Weterynaryjna* **56**: 32–35. [In Polish with English summary].

Dróżdż, J., Demiaszkiewicz, A. W. & Lachowicz, J. (2002a). Forming of gastro-intestinal nematodes fauna of free ranging European bison in Białowieża Primeval Forest during last 17 years (1984–2001. *Wiadomości Parazytologiczne* **48**: 375–381. [In Polish with English summary].

Dróżdż, J., Demiaszkiewicz, A. W. & Lachowicz, J. (2002b). Exchange of gastrointestinal nematodes between roe and red deer (Cervidae) and European bison (Bovidae) in the Bieszczady Mountains (Carpathians, Poland). *Acta Parasitologica* **47**: 314–317.

Dróżdż, J., Demiaszkiewicz, A. W., Lachowicz, J., Przybysz, I. & Kochko, Yu. (2003). Rola jeleniowatych w rozprzestrzenieniu aswortiozy wśród wolno żyjących żubrów w Polsce. [The role of deer in dissemination of ashwortiosis among free-living wisents in Poland]. In *Streszczenia referatów konferencji 'Znaczenie badań naukowych dla ochrony żubrów w Polsce', Warszawa, 13 czerwca 2003*, pp. 16–17. Warszawa: Katedra Genetyki i Ogólnej Hodowli Zwierząt, Wydział Nauk o Zwierzętach SGGW. [In Polish].

Dzięciołowski, R. M. (1991). Ecological niches of five big ungulates in forest tracts. *Folia Forestalia Polonica, Series A: Forestry* **33**: 56–70.

Dzięciołowski, R., Kossak, S., Borowski, S. & Morow, K. (1975). Diets of herbivorous mammals. *Polish Ecological Studies* **1**: 33–50.

EBPB (European Bison Pedigree Book) (1947). Ed. Żabiński, J. Warszawa: International Society for the Protection of the European Bison.

EBPB (European Bison Pedigree Book) (1961). Ed. Żabiński, J. Warszawa: International Society for the Protection of the European bison.

EBPB (European Bison Pedigree Book) (1970–1971). Ed. Żabiński, J. Warszawa: Państwowe Wydawnictwo Naukowe.

EBPB (European Bison Pedigree Book) (2010). Ed. Raczyński, J. Białowieża: Białowieża National Park.

EBPB (European Bison Pedigree Book) (2011). Ed. Raczyński, J. Białowieża: Białowieża National Park.

Empel, W. & Roskosz, T. (1963). Das Skelett der Gliedmassen des Wisents *Bison bonasus* (Linnaeus, 1958). *Acta Theriologica*, **7**: 259–300.

Falińska, K. (1973). Seasonal dynamics of herb layer in forest communities of Białowieża National Park. *Phytocoenosis* **2**: 1–120.

Fedyk, S. & Sysa, P. (1971). Chromosomes of European bison, domestic cattle and their hybrids. *Acta Theriologica* **16**: 465–470.

Flerov, K. K. (1979). Sistematika i evoluciya. [Systematic and evolution]. In *Zubr. Morfologiya, sitematika, evoluciya, ekologia*, pp. 1–127. Ed. Sokolov, V. E. Moskva: Izdatel'stvo Nauka. [In Russian].

Gątkiewicz, T. & Zbrożek, P. (eds) (2003). *Leśny Kompleks Promocyjny Puszcza Białowieska*. [Forest Promotional Complex – Białowieża Forest]. Białystok, Bydgoszcz: Media Art. Anna Włodarczyk. [In Polish].

Gębczyńska, Z. & Krasińska, M. (1972). Food preferences and requirements of the European bison. *Acta Theriologica* **17**: 105–117.

Gębczyńska, Z., Kowalczyk, J., Krasińska, M. & Ziołecka, A. (1974). A comparison of the digestibility of nutrients by European bison and cattle. *Acta Theriologica* **19**: 283–289.

Gębczyńska, Z., Gębczyński, M. & Martynowicz, E. (1991). Food eaten by free-living European bison. *Acta Theriologica* **36**: 307–313.

Gębczyński, M. & Tomaszewska-Guszkiewicz, K. (1987). Genetic variability in the European bison. *Biochemical Systematics and Ecology* **15**: 285–288.

Glunz, R. (2002). History, present situation and future prospects of the European Bison Breeding Centre in Hardehausen. In *Abstracts 1st International European Bison Congress. Hardehausen 9-12.09.2008*, p. 34.

Groeben, G. von der, (ed.) (1932). Das Zuchtbuch 1932. *Berichte der International Gesellschaft zur Erhaltung des Wisents Berlin* **5**(1): 5–50.

Gusarov, I. (2011). Condition of population of European bison in northern region of Russia. *European Bison Conservation Newsletter* **4**: 117–120.

Hartl, G. B. & Pucek, Z. (1994). Genetic depletion in the European bison (*Bison bonasus*) and the significance of electrophoretic heterozygosity for conservation. *Conservation Biology* **8**: 67–174.

Hayward, M. W., Kowalczyk, R., Krasiński, Z. A., Krasińska, M., Dackiewicz, J. & Cornulier, T. (2011). Restoration and intensive management have no effect on evolutionary strategies. *Endangered Species Research* **15**: 53–61.

Hedemann, O. (1939.) *Dzieje Puszczy Białowieskiej w Polsce przedrozbiorowej (w okresie do 1798)*. [History of Białowieża Forest in Poland before the partition (until 1798)]. Warszawa: Instytut Badawczy Lasów Państwowych, Rozprawy i Sprawozdania Seria A, 41. [In Polish].

Heptner, V. G., Nasimovich, A. A. & Bannikov, A. G. (1966). *Die Sägetiere der Sowietunion: Part I Paarhufer und Unpaarhufer*. Jena: G. Fischer Verlag.

Hoffman, R. R. (1978). Die Stellung europeischän Wildwiederkäuer im System des Assungstypen. In *Wildbiologische information für den Jäger*, pp. 1–142. St. Gallen: Jagd+Hege Verlag

Hofman, E. (2008). Wykorzystanie przez żubry (Bison bonasus) terenów otwartych przyległych do Puszczy Białowieskiej i Puszczy Knyszyńskiej w latach 2000-2007. [Utilisation of open areas adjoining the Białowieża Forest and Knyszyńska Forest in 2000–2007]. MSc thesis. Catholic University of Lublin. [In Polish].

Hofman-Kamińska, E. & Kowalczyk, R. (2010). What are bison searching for in agricultural fields?. In *European Bison Conservation in the Białowieża Forest: Threats and Prospects of the Population Development*, pp. 136–146. Eds. Kowalczyk, R., Ławreszuk, D. &. Wójcik, J. M. Białowieża: Mammal Research Institute PAS. [In Polish with English summary].

Hofman-Kamińska, E., Kamiński, T. & Kowalczyk, R. (2011). Microhistological analysis of European bison diet in the Białowieża Forest. In *Abstracts of International Conference 'European Bison, Forests, and Lakes' 22–23.09.2011*, pp. 18–19. Malinówka: European Bison Friends Society. [In Polish with English summary].

Holodova, M. V. & Belousova, I. P. (1989). Potreblenie i usvoenie pitatel'nykh veshchestv i energii zubrami (*Bison bonasus*). [Digestibility of nutrients by European bison (*Bison bonasus*)]. *Zoologicheskij Zhurnal* **68**: 107–117. [In Russian].

Izdebska, J. N. (2001a). European bison arthropods from closed Polish breeding facilities. *Acta Parasitologica* **46**: 135–137.

Izdebska, J. N. (2001b). The occurrence of parasitic arthropods in two groups of

European bison in the Białowieża Primeval Forest. *Wiadomości Parazytologiczne* **47**: 801–804.

Izdebska, J. N. (2001c). Parasitic arthropods in bison from Bieszczady Moutains. *Scientific Messenger of Lviv State Academy of Veterinary Medicine named after S. Z. Gzhytskyj* 3(2): 208–211.

Izdebska, J. N. (2005). Porównanie infestacji stawonogami pasożytniczymi żubrów ze stad wolnych i hodowli w Polsce. [Comparison of infestations by arthropod ectoparasites of the European bison in Bieszczady Mts]. In *Ochrona żubrów zachodniopomorskich*, pp. 38–42. Ińsko: Wydawca ZODR Barzkowice. [In Polish].

Izdebska, J. N. (2006). Skin mites (Acari: Demodecidae, Psoroptidae, Sarcoptidae) in the European bison, *Bison bonasus*. *Biological Letters* **43**: 169–174.

Izdebska, J. N. (2011). *Bisonicola sedecimdecembrii* (Phthiraptera: trochodectodae] from European bison: redescription of adults and description of juvenile stages. *Entomologica Fennica* **22**: 69–77.

Izdebska, J. N. & Cydzik, K. (2010). Analysis of the reasons for differences in topical specificity among various species of tick (Acari, Ixodidae) infesting European bison. *European Bison Conservation Newsletter* **3**: 75–84.

Jaczewski, Z. (1958). Reproduction of the European bison, *Bison bonasus* (L.) in reserves. *Acta Theriologica* **1**: 333–376.

Jakob, W., Schröder, H. D., Rudolph, M., *et al.* (2000). Necrobacillosis in free-living males of European bison in Poland. *Journal of Wildlife Diseases* **36**: 248–256.

Janicka, D., Hołubczat, E., Karnecki, R., *et al.* (2008). Restitution of the European bison *Bison bonasus* (L.1758) as a component of active conservation within Natura 2000 sites of Western Pomerania. *European Bison Conservation Newsletter* **1**: 132–139. [In Polish with English summary].

Jarocki, F. P.(1830). *O Puszczy Białowieskiej i o celniejszych w niej zwierzętach.* [About Białowieża Forest and its more important animals]. Warszawa: Komisja Rządowa Wyznań Religijnych i Oświecenia Publicznego. [In Polish].

Jaroszewicz, B. & Pirożnikow, E. (2008). Diversity of plant species eaten and dispersed by the European bison *Bison bonasus* in Białowieża forest. *European Bison Conservation Newsletter* **1**: 14–29.

Kalugin, C. G. (1968). Vostanovleniye zubra na severo-zapadnom Kavkaze. [Restoration of European bison in north-western Caucasus.] *TrudyKavkazskogo Zapovednika* **10**: 3–94.[In Russian].

Kamiński, T., Hofman-Kamińska, E. & Kowalczyk, R. (2010). *European bison diet in Białowieża Forest.* In *Streszczenia referatów VIII Miedzynarodowej Konferencji 'Żubr w Puszczy Królewskiej', Niepołomice 9–10 września 2010*, pp. 26–27. Niepołomice: Stowarzyszenie Miłośników Żubrów i Nadleśnictwo Niepołomice. [In Polish with English summary].

Kamiński, S., Olech, W., Olenski, K., Nowak, Z. & Ruść, A. (2011). Single nucleotide polymorphisms between two lines of European bison (*Bison bonasus*) detected by the use of Illumina Bovine 50 K BeadChip. *Conservation Genetics Resources.* DOI: 10.1007/s12686-011-9535-z.

Kaplan, J. O., Krumhardt, K. M. & Zimmermann, N. (2009). The prehistoric and preindustrial deforestration of Europe. *Quaternary Science Reviews* **28**: 3016–3034.

Karnecki, R. (2008). Karta obserwacji gatunku *Bison bonasus* (Stado Zachodniopomorskie) dla obszaru Natura 2000 – Mirosławiec PLH320045. [Observation card for the species *Bison bonasus* (West Pomeranian herd) for Natura 2000 site – Mirosławiec PLH320045]. Instytut Ochrony Przyrody PAN. [In Polish].

Kaz'min, V. D. & Smirnov, K. A. (1992). Zimnee pitanie, kormovye resursy i troficheskoe vosdeistvie zubra na lesnye fitocenozy central'nogo Kavkaza. [Winter feeding, food abundance and European bison influence upon the forest habitats of central Caucasus]. *Biuletin Moskovskogo obshchestva ispitatelei prirody. Odl. Biol.* **97**(2): 26–35. [In Russian].

Kelterborn, T., Center, F. & Zacharias, K. (2009). 52 years of European bison breeding on the Wisent-Island in the heart of Meklemburg-Vorpommern. *European Bison Conservation Newsletter* **2**: 172–181.

Khoyetskyy, P. (2011). The history and current state of the *Bison bonasus* L. population in western Ukraine. *European Bison Conservation Newsletter* **4**: 21–30.

Kiseleva, E. G. (1969). Breeding of the Caucasus-Białowieża European bison in the Okski Reservation: Postępy Restytucji Żubra. In *Materiały III Konferencji Polsko-Radzieckiej. Białowieża-Kamieniuki, 18–21 kwietnia 1967 r.*, pp. 233–249. Warszawa: Państwowe Wydawnictwo Naukowe [In Polish and Russian with English summary].

Kiseleva, E. G. (1974). Razvedenie zubrov v pitomnike Okskogo Zapovednika. [Reproduction of European bison in Okski Reserve]. *Berezinskii Zapovednik, Issledovaniya* **3**: 103–138. [In Russian].

Kita, J. & Anusz, K. (2006). Infectious diseases in *Bison bonasus* in the years 1910–1988. In *Health Threats for the European Bison Particularly in Free-roaming Populations in Poland*, pp. 17–26. Eds. Kita, J. & Anusz, K. Warszawa: SGGW Publisher.

Kita, J., Dziąba, K., Piusiński, W., *et al.* (1990). A disease of genital organ of free-roaming male of the European bison in the Białowieża Primeval Forest, Poland. *Medycyna Weterynaryjna* **44**: 474–476. [In Polish with English summary].

Kita, J., Anusz, K., Zaleska, M., *et al.* (2003). Relationship among ecology, demography and diseases of European bison (*Bison bonasus*). *Journal of Veterinary Sciences* **6**(4): 261–206.

Klevezal, G. & Pucek, Z. (1987). Growth layers in tooth cement and dentine of European bison and its hybrids with domestic cattle. *Acta Theriologica* **32**: 115–128.

Kobryńczuk, F. (1985). The influence of inbreeding on the shape and size of the skeleton of the European bison. *Acta Theriologica* **30**: 379–422.

Kobryńczuk, F. (2000). Żubr w 'Pieśni' Mikołaja Hussowskiego i w polskich opracowaniach morfologicznych (cz. II). [European bison in Mikołaj Hussowski poem and in the Polish morphological publications (part II)]. *Agricola, Pismo SGGW* **47**: 16–19.

Kobryńczuk, F. (2008). Special features of the European bison' body. *Matecznik Białowieski Wydanie Specjalne*: 17–18. [In Polish with English summary].

Kobryńczuk, F., Krasińska, M. & Szara, T. (2008). Sexual dimorphism in skulls of the lowland European bison, *Bison bonasus bonasus. Annales Zoologici Fennici* **45**: 335–340.

Koch, W. (1956). Jahreszeitlische Schwankungen der Fruchtbarkeit bei Wildrindern. *Fortpflanzung,*

Zuchthygiene und Haustierbesamung **6**: 85–87.

Kochko, Yu. (2003). The results of the study on the fauna of helminths of Belovezhskaya Pushcha in XX century. *Belovezhskaya Pushcha, Issledovaniya* **11**: 205–223. [In Russian with English summary].

Korochkina, L. N. (1969). Drevesnaya rastitel'nost' v pitanii zubrov Belovezhskoi Pushchi. [Woody plants in feeding of the European bison in Białowieża Forest]. *Belovezhskaya Pushcha, Issledovaniya* **2**: 120–126. [In Russian].

Korochkina, L. N. (1971). Kormovaya baza i nekotorye itogi reaklimatyzacii zubrov v Belovezhskoi Pushchi. [Food resources and adaptation of European bison to the conditions of Białowieża Forest]. *Belovezhskaya Pushcha, Issledovaniya* **5**: 164–176. [In Russian].

Korochkina, L. N. (1972a). Sutochnyj ritm aktivnosti zubrov Belovezhskoi Pushchi" [Daily activity rhythm of the European bison in Białowieża Forest]. *Belovezhskaya Pushcha, Issledovaniya* **6**: 125–131. [In Russian].

Korochkina, L. N. (1972b). Travjanistaya rastitel'nost' v pitanii zubrov Belovezhskoi Pushchi. [Monocotyledonous plants in feeding of European bison in Białowieża Forest]. *Belovezhskaya Pushcha, Issledovaniya* **6**: 110–124. [In Russian].

Korochkina, L. N. (1973). Rajon obitaniya i stacional'noe razmieshchenie zubrov v Belovezhsko Pushchi. [Population home range and spatial distribution of European bison in Białowieża Forest]. *Belovezhskaya Pushcha. Issledovaniya* **7**: 148–165. [In Russian].

Korochkina, L. N. (1974). Voprosy reaklimatyzacii zubrov i emkost' kormovych ugodii Belovezhskoi Pushchi. [Adaptation of European bison and food abundance in Białowieża Forest]. *Belovezhskaya Pushcha, Issledovaniya* **8**: 118–133. [In Russian].

Korochkina, L. N. (1978). Vlijanie snezhnogo pokrova na osobennosti zimnego pitaniya olen'ich. [Influence of snow cover on the specificity of winter foraging of deer]. *Zapovedniki Belarusi, Issledovaniya* **2**: 100–110. [In Russian].

Korochkina, L. N. & Kochko, F. P. (1983). Dinamika chislennosti vol'nozhivushchich zubrov Belovezhskoi Pushchi. [Dynamics of number of free-ranging population of European bison in Białowieża Forest]. *Zapovedniki Belarusi, Issledovaniya* **7**: 52–59. [In Russian].

Koulischer, L., Tijskens, T. & Mortelmans, J. (1967). Mammalian cytogenetics: I. The chromosomes of three species of Bovidae: *Bos taurus, Bison bonasus, Cephalophus grimmi. Acta Zoologica Pathologica Antverpiensia* **43**: 135–141.

Kowalczyk, R. (2010). European bison: king of the forest or meadows and river valleys?. In *European Bison Conservation in the Białowieża Forest: Threats and Prospects of the Population Development*, pp. 123–134. Eds. Kowalczyk, R., Ławreszuk, D. & Wójcik, J. M. Białowieża: Mammal Research Institute PAS. [In Polish with English summary].

Kowalczyk, R., Kamiński, T. & Schneider, T. C. (2010). Influence of supplementary feeding in bison in Białowieża Primeval Forest. In *European Bison Conservation in the Białowieża Forest: Threats and Prospects of the Population Development*, pp. 147–159. Eds. Kowalczyk, R., Ławreszuk, D. & Wójcik, J. M. Białowieża: Mammal Research Institute PAS. [In Polish with English summary].

Kowalczyk, R., Taberlet, P., Coissac, E., *et al.* (2011). Influence of forest management practice on large herbivore diet: case of European bison in Białowieża Primeval Forest (Poland). *Forest Ecology and Management* **261**: 821–828.

Kowalczyk, R., Krasińska, M., Kamiński, T., *et al.* (2013). Movements of European bison (*Bison bonasus*) beyond the Bialowieza Forest (NE Poland): range expansion or partial migrations? *Acta Theriologica*. DOI: 10.1007/s13364-013-0136-y.

Kozlo, P. G. & Bunevich, A. N. (2009). *Zubr v Belarusi.* [The European bison in Belarus]. Minsk: Belaruskaya Navuka. [In Russian].

Kozul'ko, G. A. & Zhukov, V. P. (1999). Gosudarstvennyi nacional'nyi park 'Belovezhskaya Pushcha': stareishii zapovednik v Europie. [State National Park 'Belovezhskaya Pushcha': the oldest reserve of Europe]. In *Belovezhskaya Pushcha na rubiezhe tret'ego tysucholetiya. Materialy nauchno-prakticheskoi konferencii, posvashchonnoi 60-letii co dniya obrazovaniya zapovednika 'Belovezhskaya Pushcha' 22–24 dekabrya 1999 g. p.Kameniuki, Bretskaya obl. Minsk*, pp. 16–33. Kameniuki: State National Park Belovezhskaya Pushcha. [In Russian].

Krasińska, M. & Krasiński, Z. A. (1994). Spatial structure of the European bison population in the Polish part of the Białowieza Forest 1976–1993. *Parki Narodowe i Rezerwaty Przyrody* **13**(4): 69–87. [In Polish with English summary].

Krasińska, M. & Krasiński, Z. A. (1995). Composition, group size, and spatial distribution of European bison bulls in Białowieża Forest. *Acta Theriologica* **40**: 1–21.

Krasińska, M. & Krasiński, Z. A. (2002). Body mass and measurements of the European bison during postnatal development. *Acta Theriologica* **47**: 85–106.

Krasińska, M. & Krasiński, Z. A. (2007). *European Bison: The Nature Monograph.* Białowieża: Mammal Research Institute PAS.

Krasińska, M. & Krasiński, Z. A. (2010). Course and dispersal of the posthitis/balanoposthitis in males of the European bison in the Polish part of Białowieża Forest. *Parki narodowe Rezerwaty Przyrody* **29**(2): 197–128. [In Polish with English summary].

Krasińska, M., Caboń-Raczyńska, K. & Krasiński, Z. A. (1987). Strategy of habitat utilization by European bison in the Białowieża Forest. *Acta Theriologica* **32**: 147–202.

Krasińska, M., Krasiński, Z. A. & Bunevich, A. N. (2000). Factors affecting the variability in home range size and distribution in the Polish and Belarusian part of the Białowieża Forest. *Acta Theriologica* **45**: 321–334.

Krasińska, M., Szuma, E., Kobryńczuk, F. & Szara, T. (2008). Morphometric variation of the skull during postnatal development in the Lowland European bison (*Bison bonasus bonasus*). *Acta Theriologica* **53**: 193–216.

Krasińska, M., Krasiński, Z. A. & Ławreszuk, D. (2010). Dispersion of the population and use of open areas by European bison in Białowieża Forest. In *European Bison Conservation in the Białowieża Forest: Threats and Prospects of the Population Development*, pp. 111–122. Eds. Kowalczyk, R., Ławreszuk, D. & Wójcik, J. M. Białowieża: Mammal Research Institute PAS. [In Polish with English summary].

Krasiński, Z. A. (1978a). Dynamics and structure of the European bison population in the Białowieża Primeval Forest. *Acta Theriologica* **23**: 3–48.

Krasiński, Z. A. (1978b). Restytucja i ewolucja populacji żubrów w Polsce. [Restitution and evolution of European bison population in Poland]. In *Człowiek i Nauka*, pp. 207–229. Warszawa: Wiedza Powszechna. [In Polish].

Krasiński, Z. A. (1993). *Bison: A Relict of Ancient Times*. Białowieża: Wydawnictwo BPN.

Krasiński, Z. A. (1994). Restitution of the European bison in the Białowieża reserve in the years 1929–1952. *Parki Narodowe i Rezerwaty Przyrody* **13**(4): 3–23. [In Polish with English summary].

Krasiński, Z. A. (2010). European bison and his close relative from North America. In *European Bison Conservation in the Białowieża Forest: Threats and Prospects of the Population Development*, pp. 83–92. Eds. Kowalczyk, R., Ławreszuk, D. & Wójcik, J. M. Białowieża: Mammal Research Institute PAS. [In Polish with English summary].

Krasiński, Z. A. & Krasińska, M. (1992). Free ranging European bison in Borecka Forest. *Acta Theriologica* **37**: 301–317.

Krasiński, Z. & Raczyński, I. J. (1967). The reproduction biology of European bison living in reserves and freedom. *Acta Theriologica* **12**: 407–444.

Krasiński, Z. A., Bunevich, A. N. & Krasińska, M., (1994a). Characteristics of the European bison populations in the Polish and Belarusian part of the Białowieża Forest. *Parki Narodowe i Rezerwaty Przyrody* **13**(4): 25–67. [In Polish with English summary].

Krasiński, Z. A., Krasińska, M. & Leniec, H. (1994b). The European bison in the Knyszyńska Forest. *Parki Narodowe i Rezerwaty Przyrody* **13**(4): 107–114. [In Polish with English summary].

Krasochko, P. A, Kozlo, P. G., Krasochko, I. A., Yakubovskij, M. V., Bunevich, A. N. & Kochko, Y. P. (2004). *Ekologicheskie i veterinarnye aspekty zubrov v Belarusi*. [Ecological and veterinarian aspects of European bison in Belarus]. Minsk: Biznesofset. [In Russian].

Krysiak, K. (1951). Więzadło karkowe, *ligamentum nuchae* żubra (*Bison bonasus*). [*Ligamentum nuchae* of European bison, *Bison bonasus*]. *Folia Morphologica* **2**: 271–283. [In Polish].

Krysiak, K. & Świeżyński, K. (1967). The present state of research on the morphology of the European bison. *Acta Theriologica* **12**: 323–331.

Kuemmerle, T., Perzanowski, K., Chaskovsky, O., *et al.* (2010). European bison habitat in the Carpathian Mountains. *Biological Conservation* **143**: 908–916.

Kuemmerle, T., Radeloff, V. C., Perzanowski, K., *et al.* (2011). Predicting potential European bison habitat across its former range. *Ecological Applications* **21**(3): 830–843.

Kuemmerle, T., Hickler, T., Olofsson, J., Schurgers, G. & Radeloff, V. C. (2012). Reconstruction of range dynamics and range fragmentation of European bison for the last 8000 years. *Diversity and Distribution* **18**: 47–59

Kulagin, N. M. (1919). *Zubr Belovezhskoi Pushchi*. [European bison of Białowieża Forest]. Moskva: Izdatelstvo Moskovskogo Nauchnogo Instituta. [In Russian].

Łopieńska, M., Nowak, Z., Wojciechowska, M., Macheta, L., Charon, K. M. & Olech, W. (2011). A comparison of polymorphism of exon 2 DRB3 belonging to two genetic lines of European bison: Lowland and Lowland–Caucasian lines. In *Abstracts of International Conference 'European Bison, Forests, Lakes', 22–23.09.2011*, pp. 38. Malinówka: European Bison Friends Society. [In Polish with English summary].

Martínez-Navarro, B., Belmaker, M. & Bar-Yosef, O. (2012). The Bovid assemblage (Bovidae, Mammalia) from the Early Pleistocene site of 'Ubeidiya, Israel: biochronological and paleoecological implications for the fossil and lithic bearing strata. *Quaternary International* **267**: 78–97.

Mazurek, A. (2010). Ocena zasobności runa i biomasy ściółki w wybranych siedliskach Magurskiego Parku Narodowego. [The assessment of ground flora and litter standing crop in selected habitats of Magurski National Park]. Bachelor thesis, Catholic University of Lublin. [In Polish].

McDonald, J. N. (1981). *North American Bison: Their Classification and Evolution*. Berkeley and Los Angeles, CA, London: University of California Press.

Melander, Y. (1959). The mitotic chromosomes of some cavicorn mammals (*Bos taurus* L., *Bison bonasus* L. and *Ovis aries* L.). *Hereditas* **45**: 649–664.

Mohr, E. (1952). *Der Wisent*. Leiptzig: Die Neue Brehm-Bücherei 74.

Mysterud, A., Bartoń, K. A., Jędrzejewska, B., *et al.* (2007). Population ecology and conservation of endangered megafauna: the case of European bison in Białowieża Primeval Forest, Poland. *Animal Conservation* **10**: 77–87.

Nahlik, A. (1991). W sprawie transylwańskiego żubra. [About the Transylvanian European bison]. *Sylwan* **85**: 63–68.

Nakhlyk, A. (1992). Pro Transilvanskoho zubra. [Transylvanian European bison]. *Lisove hospodarstwo, lisowa, paperova i derevoobrona promyslovist* **23**: 22–27. [In Ukrainian].

Nei, M., Maruyama, T. & Chakraborty, R. (1975). The bottleneck effect and genetic variability in populations. *Evolution* **29**: 1–10.

Nemtsev, A. S., Rautian, G. S., Puzarenko, A. Ju., Sipko, T. P., Kalabushkin, B. A. & Mironenko, I. V. (2003). *Zubr na Kavkaze*. [European bison at Caucasus]. Moskva-Maikop: Kachestvo. [In Russian].

Nowak, Z. & Olech, W. (2008a). The microsatellite variability within European bison sex chromosomes. *European Bison Conservation Newsletter* **1**: 72–78. [In Polish with English summary].

Nowak, Z. & Olech, W. (2008b). Verification of phylogenetic hypothesis concerning the evolution of genus Bison. *Annals of Warsaw University of Life Sciences – SGGW, Animal Science* **45**: 60–65.

Olech, W. (1987). Analysis of inbreeding in European bison. *Acta Theriologica* **32**: 373–387.

Olech, W. (1989). The participation of ancestral genes in the existing population of European bison. *Acta Theriologica* **34** (29): 397–407.

Olech, W. (1999). The number of ancestors and their contribution to European bison (*Bison bonasus* L.) population. *Annals of WAU, Animal Science* **35**: 111–117.

Olech, W. (2003). *Influence of Individual Inbred and Mother's Inbred on Calves' Survival in European Bison (*Bison bonasus*)*. Warszawa: Rozprawy Naukowe i Monografie, Wydawnictwo SGGW. [In Polish with English summary].

Olech, W. (2006). The analysis of European bison genetic diversity using the pedigree data. In *Health Threats for the European Bison Particularly in Free-roaming Populations in Poland*, pp. 205–210. Eds. Kita, J. & Anusz, K. Warszawa: SGGW Publishers.

Olech, W. (2009). The changes of founders number and their contribution to the European bison population during 80 years of species restitution. *European Bison Conservation Newsletter* **2**: 54–60.

Olech, W. & Perzanowski, K. (2010). Conservation strategy for the European bison in Poland. In *Abstracts from the International Conference 'Conservation of the European Bison in the Białowieża Forest' 25-27 February 2010, Białowieża Poland*, pp. 47–48. Białowieża: Mammal Research Institute PAS.

Olech, W. & Perzanowski, K. (2011). Ochrona i modelowanie populacji żubra. [Conservation and modelling of European bison populations]. *Annals of Warsaw University of Life Sciences – SGGW* **50**: 57–65.

Olech, W., Bielecki, W., Bołbot, A., *et al.* (eds) (2008). *Hodowla żubrów, poradnik utrzymania w niewoli.* [Breeding of the European Bison, a Guide for Captivity]. Warszawa: Stowarzyszenie Miłośników Żubrów. ([In Polish].

Parnikoza, I., Boreiko, V., Sesin, V. & Kaliuzhna, M. (2009). History, current state and perspectives of conservation of European Bison in Ukraine. *European Bison Conservation Newsletter* **2**: 5–16.

Pascu, M. (2009). Valea Zimbrilor Bison Reserve: a stimulus of local sustainable development. *European Bison Conservation Newsletter* **2**: 182–186.

Paszkiewicz, R. (2004). The use of tree stands by the European bison population of Lowland–Caucasian line of western Bieszczady. *Parki Narodowe i Rezerwaty Przyrody* **23**(4): 647–656. [In Polish with English summary].

Paszkiewicz, R. (2008). Monitoring of European bison in Baligród Forest District. *European Bison Conservation Newsletter* **1**: 140–145.

Pčola, Š., & Gurecka, J. (2008). Najnovšie poznatky z reštitúcie zubra hrivnatého (*Bison bonasus*) v Národnom Parku Poloniny. [The most recent information about the restitution of European bison (*Bison bonasus*) to Poloniny National Park]. *European Bison Conservation Newsletter* **1**: 146–152. [in Slovak].

Pčola, Š., Adamec, M. & Pčola, Š. jun. (2006). Reintrodukcja żubra *Bison bonasus* w Parku Narodowym Połoniny. [Reintroduction of the European bison, *Bison bonasus* to Poloniny National Park]. In *Perspektywy rozwoju populacji żubrów [Perspectives of the development of European bison population]*, pp. 43–54. Ed. Olech, W. Goczałkowice Zdrój: Artisco. [In Polish].

Pertoldi, C., Wójcik, J. M., Tokarska, M., *et al.* (2010). Genome variability in European and American bison detected using the BovineSNP50 BeadChip. *Conservation Genetics* **11**: 627–634.

Perzanowski, K. (2001). The restitution of the European bison in Bieszczady as a component of re-naturalisation of mountain ecosystems. *Roczniki Bieszczadzkie* **9**: 151–156.

Perzanowski, K. (2004). European bison at Bieszczady mountains as a part of the Carpathian population of the species. In *Proceedings of the Conference 'European Bison Conservation' 30 September–2 October 2004, Białowieża Poland*. pp. 108–111. Eds. Krasińska, M. & Daleszczyk, K. Białowieża: Mammal Research Institute PAS.

Perzanowski, K. (2001–2011). *Stały monitoring żubrów na terenie nadleśnictw bieszczadzkich. Raporty roczne.* [Permanent Monitoring of European Bison Within the Area of Forest Districts of Bieszaczady Mts. Annual reports]. Krosno: RDLP. [In Polish].

Perzanowski, K. (2010). Kaszel na połoninach: pojawienie się gruźlicy w stadzie żubrów bieszczadzkich. ['The cough' in alpine meadows: the appearance of tuberculosis in European bison herd of Bieszczady Mts]. *Las Polski* **11**: 23. [In Polish].

Perzanowski, K. & Deju, R. (2006). Is there a future for European bison in Romania? In *Proceedings of the Workshop 'European Bison and Local Communities in Carpathian Mountains'. Vanatori N. P. 23-26. 08.2006*, pp. 11–16. Eds. Cătănoiu, S. & Deju, R. Piatra-Neamt: Editura Nona.

Perzanowski, K. & Januszczak, M. (2004). Preliminary assessment of the home range dynamics of the European bison *Bison bonasus* in Bieszczady Mountains. *Parki Narodowe i Rezerwaty Przyrody* **23**(4): 639–646. [In Polish with English summary].

Perzanowski, K. & Januszczak, M. (2010). A dispersal rate in a wisent population of Bieszczady Mountains. *European Bison Conservation Newsletter* **3**: 45–52.

Perzanowski, K. & Marszałek, E. (2008). *Żubr przywrócony górom.* [European Bison Restored to the Mountains]. Rzeszów: RS Druk. [In Polish].

Perzanowski, K. & Marszałek, E. (2012). *The Return of the Wisent to the Carpathians.* Krosno: RDLP w Krośnie.

Perzanowski, K. & Olech, W. (2007). A future for the European bison *Bison bonasus* in the Carpathian ecoregion. *Wildlife Biology* **13**(1): 108–112.

Perzanowski, K. & Paszkiewicz, R. (2000). Restytucja i współczesny stan populacji żubrów w Bieszczadach. [Restitution and present status of European bison population in Bieszczady]. In *Monografie bieszczadzkie: Kręgowce Bieszczadów Zachodnich*, vol.**9**, pp. 219–229. Ed. Głowaciński, Z. Ustrzyki Dolne: Ośrodek naukowo-Dydaktyczny Bieszczadzkiego Parku Narodowego. [In Polish].

Perzanowski, K., Wołoszyn, A. & Januszczak, M. (2005). Wielkość ugrupowań w nizinnych i górskich populacjach żubrów. [Size of groups in lowland and mountain populations of European bison]. In *Zmiany w populacjach ssaków jako pochodna dynamiki zmian środowiska*, pp. 3–10. Krakow: Akademia Rolnicza w Krakowie. [In Polish].

Perzanowski, K., Adamec, M. & Pčola, Š. (2006a). Status and restitution of a transboundary Polish–Slovak population of the European bison. In *Proceedings of the Conference 'Vyskum a ochrana cicavcov na Slovensku' VII, Zvolen 2005*, pp. 159–164. Eds. Adamec, M. & Urban, P. Banská Bystrica: Štátna ochrana prírody SR.

Perzanowski, K., Olech, W., Januszczak, M. & Wołoszyn-Gałęza, A. (2006b). Ocena efektów introdukcji żubra w Bieszczadach. [Estimation of effects of the introduction of European bison to Bieszczady]. In *Sposoby rozpoznawania, oceny i monitoringu wartości przyrodniczych polskich lasów*, vol. **4**, pp. 201–212. Ed. Anderwald, D. Rogów: Studia i Materiały CEPL. [In Polish].

Perzanowski, K., Wołoszyn-Gałęza, A. & Januszczak, M. (2008). Założenia do wyznaczenia ostoi żubra w Bieszczadach. [Assumptions for delineation of European bison refuges at Bieszczady]. *European Bison Conservation Newsletter* **1**: 79–86.

Perzanowski, K., Januszczak, M. & Wołoszyn-Gałęza, A. (2011a). Utilisation of the terrain by wisents in Bieszczady Mountains. *European Bison Conservation Newsletter* **4**: 5–12.

Perzanowski, K., Januszczak, M. & Wołoszyn-Gałęza, A. (2011b). Użytkowanie terenów otwartych i zalesionych przez żubry należące do zachodniej subpopulacji w Bieszczadach. [Utilisation of open and forested areas by European bison belonging to the western subpopulation of Bieszczady]. *Roczniki Bieszczadzkie* **19**: 191–206. [In Polish].

Perzanowski, K., Januszczak, M. & Wołoszyn-Gałęza, A. (2012). Seasonal movements of European bison (*Bison bonasus* L. 1758) in Bieszczady Mountains. *Biological Letters* **49**: 139–145.

Pucek, Z. (1984). European Bison Pedigree Book. *Przegląd Zoologiczny*, **28**: 426–435. [In Polish with English summary].

Pucek, Z. (1986). *Bison bonasus* (Linnaeus 1758): Wisent. In *Handbuch der Säugetiere Europas. Band 2/II (Paarhuf)*, pp. 278–315. Eds. Niethammer, P. J. & Krapp, F. Wiesbaden: AULA-Verlag GmbH.

Pucek, Z. (1994). Progress and threats to the European bison. *Kosmos* **43**: 147–169. [In Polish with English sumary].

Pucek, Z. (2004). European bison: history of a flag species. In *Essays on Mammals of Białowieża Forest*, pp. 25–34. Eds. Jedrzejewska, B. & Wójcik, J. M. Białowieża: Mammal Research Institute PAS.

Pucek, Z., Belousova, I. P., Krasińska, M., Krasiński, Z. A. & Olech,W. (eds) (2004). *European Bison: Status Survey and Conservation Action Plan.* Gland and Cambridge: UK IUCN/SSB Bison Specialist Group IUCN.

Raczyński, J. & Bołbot, M. (2009). *The European Bison Pedigree Book in History and Today*. Białowieża: Białowieża National Park.

Rautian, G. S., Kalabushkin, B. A. & Nemcev, A. S. (2000). A new species of the European bison, *Bison bonasus montanus* ssp. Nov. (Bovidae, Artiodactyla). *Doklady Biological Sciences* **375**: 636–640.

Reed, D. H., Lowe, E. H., Briscoe, D. A. & Frankham, R. (2003). Inbreeding and extinction: effects of rate of inbreeding. *Conservation Genetics* **4**: 405–410.

Reynolds, H. W., Glaholt, R. D. & Hawley, A. W. L. (1982). *Bison Bison bison*. In *Wild Mammals in North America*, pp. 972–1007. Eds. Chapman, I. A. & Feldhamer, G. A. Baltimore, MD: Johns Hopkins University Press.

Roskosz, T. (1962). Morphoplogie der Wirbelsäule des Wisents, *Bison bonasus* (Linnaeus 1758). *Acta Theriologica* **6**: 113–164.

Rouys, S. (2000). Activity and movements of European bison in the Białowieża Forest, Poland. PhD thesis. University of Greenwich.

Rouys, S. & Krasińska, M. (2001). Influence of supplementary winter feeding on the daily travel distance of European bison in the Białowieża Forest, Poland. In *Wildlife, Land and People: Priorities for the 21st Century*, pp. 315–316. Eds. Fields, R., Warren, R. J., Okarma, H. & Sievert, P. R. Bethesda, MD: Wildlife Society USA.

Sawla, A., Anusz, K., Welz, M., Wozikowski, R., Zaleska, M. & Kita, J. (2011). Analysis of epizootiological situation in livestock and free-living wildlife on the Bieszczady Mountains in relation to occurrence of tuberculosis in European bison (*Bison bonasus*). *European Bison Conservation Newsletter* **4**: 71–80.

Scott, E. (2010). Extinctions, scenarios, and assumptions: changes in latest Pleistocene large herbivore abundance and distribution in western North America. *Quaternary International* **217**: 225–239.

Shakun, V. (2011). Formation of the free-ranging population of European bison (*Bison bonasus*) Volozhinskaya. *European Bison Conservation Newsletter* **4**: 31–36.

Shapiro, B., Drummond, A. J., Rambaut, A., *et al.* (2004). Rise and fall of the Beringian Steppe bison. *Science* **306**: 1561–1565.

Sipko, T. P., Udina, I. G., Badagueva, Yu. N. & Suklimova G. E. (1994). Polymorphism of k-casein gene in the family Bovidae: comparative analysis. *Russian Journal of Genetics* **30**: 203–207.

Sipko, T. P., Rautian, G. S., Udina, I. G., Ukhanov, S. V. & Berendyaeva, Z. I. (1995). Investigation of blood group polymorphism in European bison (*Bison bonasus*). *Russian Journal of Genetics* **31**: 93–100.

Sipko, T. P., Rautian, G. S., Udina, I. G. & Rakitskaya, T. A. (1996). Polymorphism of biochemical markers in European bison (*Bison bonasus*). *Russian Journal of Genetics* **32**: 356–351.

Sipko, T. P., Puzachenko, A. Ju., Bunevich, A. N., Kiseleva, E. G. & Nemcev, A. S. (2002). Variability exterior tags of the European bison. In *'Voprosy sovremennogo okhotovedeniya' Materialy mezhdunarodnoi nauchno-prakticheskoi konferencii 5–6 dekabrya 2002 g.* pp. 367–385. Moskva: Centrokhotkontrol. [In Russian with English summary].

Skinner, M. F. & Kaisen, O. C. (1947). The fossil bison of Alaska and preliminary revision of the genus. *Bulletin of the American Museum of Natural History* **89**: 123–236.

Slatis, H. M. (1960). An analysis of inbreeding in the European bison. *Genetics* **45**: 275–287.

Smagol, V., Gavrys, G. & Sharapa, O. (2010). Analysis of numbers and origin of wisent herds in Ukraine. *European Bison Conservation Newsletter* **3**: 133–136.

Sokołowski, A. W. (2004). *Lasy Puszczy Białowieskiej*. [Forest types of Białowieża Forest]. Warszawa: Centrum Informacyjne Lasów Państwowych. [In Polish].

Suchecka, A. (2009). Evaluation of the influence of genetic and demographic factors on the success of European bison small herds' reintroduction. PhD dissertation. Warsaw University of Life Sciences-SGGW.

Szalay, B. (1913). Bölényadatok kritikája. *Vadászlap* **14**: 260. [In Hungarian].

Sztolcman, J. (1924). Matèriaux pour l'histoire naturelle et pour l'historique du Bison d'Europe (*Bison bonasus* Linn.). *Annales zoologici Musei Polonici Historiae Naturalis* **2**: 49–136.

Tokarska, M., Kawałko, A., Wójcik, J. M. & Pertoldi, C. (2009). Genetic variability in the European bison (*Bison bonasus*) population from Białowieża forest over 50 years. *Biological Journal of the Linnean Society* **97**: 801–809.

Tokarska, M., Pertoldi, C., Kowalczyk, R. & Perzanowski, K. (2011). Genetic status of the European bison *Bison bonasus* after extinction in the wild and subsequent recovery. *Mammal Review* **41**(2): 151–162.

Tracz, M. & Tracz, M. (2010). Protection of European bison in Westpomeranian voivodeship. *European Bison Conservation Newsletter* **3**: 119–124. [In Polish with English summary].

Tracz, M. & Tracz, M. (2011). Using the GPS telemetry in the management of European bison herd. *European Bison Conservation Newsletter* **4**: 10–134. [In Polish with English summary].

Tracz, M., Olech, W. & Wysocki, D. (2008). New herds of European bison in West-Pomeranian voivodeship. *European Bison Conservation Newsletter* **1**: 153–160. [In Polish with English summary].

Tyapougin, Y. A. & Gusarov, I. V. (2004). Creation of population of European bison in condition of northern region of Russia. In *Proceedings of the Conference European Bison Conservation 30 september–2 October 2004 Białowieża, Poland*, pp. 137–139. Eds. Krasińska, M. & Daleszczyk, K. Białowieża: Mammals Research Institute PAS.

Udina, I. G., Sokolova, S. S., Sipko, T. P. & Sulimova, G. E. (1994). Comparative analysis of polymorphism for DQB and DRB DNA loci of the major histocompatibility complex in representatives of the family *Bovidae*. *Russian Journal of Genetics* **30**: 313–317.

Vajner, L. (1980). Etologická studie stáda zubrů evropských, *Bison bonasus* (Linnaeus 1758) v pražkè zoo. [Ecological investigations of European bison herd in Prague Zoo]. *Gazela* **2**: 59–69. [In Czech].

Webster, R., Holt, S. & Avis, C. (eds) (2001). *The status of the Carpathians: WWF – The Carpathian Ecoregion Initiative Report*. Vienna: WWF.

Węgrzyn, M. & Serwatka, S. (1984). Teeth eruption in the European bison. *Acta Theriologica* **29**: 11–121.

Winnicki, T. & Zemanek, B. (2009). *Nature in the Bieszczady National Park*. Ustrzyki Dolne: Wyd. Bieszczadzkiego Parku Narodowego.

Wróblewski, K. (1927). *Żubr Puszczy Białowieskiej*. [European bison of Białowieża Forest]. Poznań, Wydawnictwo Polskie. [In Polish].

Yazan, Ju. & Nemcev, A. (1985). Kavkazkie chistoporodnye gornyje zubry. [Caucasian pure blood mountain bison]. *Okhota i Okhotnichye Khozyaistvo* **1**: 16–17. [In Russian].

Zablockij, M. A. (1949). Tridcat' piat' let razvedeniya zubrov i ikh gibridov v Zooparke Askania Nova (1902–1937). [Thirty five years of European bison breeding and their hybrids in Askania Nova reserve]. In *Kavkazskii zubr*, pp. 73–124. Ed. Kulagin, N. M. Moskva: Glav. Upravlen. po Zoopark. Zoosadam. [In Russian].

Zięba, K. (2007). Behaviour during parturition and breeding results of lowland European bison (*Bison bonasus bonasus*) in Pszczyna herd. In *The Importance of Ex Situ Breeding for the Restoration of European Bison*, pp 37–60. Ed. Olech, W. Gołuchów: European Bison Friends Society.

Zyll de Jong, C. G., van (1986). *A Systematic Study of Recent Bison, with Particular Consideration of the Wood Bison* (Bison bison athabascae *Rhoads, 1898)*. Ottawa: National Museum of Canada, Publication of Natural Sciences.

Żoch, K. (2007). Wpływ huraganu na letni i zimowy areał występowania populacji żubra Bison bonasus (Linnaeus, 1758) w Puszczy Boreckiej. [An influence of windstorm on summer and winter ranges of European bison in Borecka Forest]. MSc thesis. Faculty of Forestry, Warsaw University of Life Sciences-SGGW. [In Polish].

Żurawski, C. & Lipiec, M. (1997). Generalized tuberculosis in a European bison. *Medycyna Weterynaryjna* **53**: 90–92. [In Polish with English summary].

Chapter 11

Gaur *Bos gaurus* C. H. Smith, 1827

Farshid S. Ahrestani and K. Ullas Karanth

Names

Genus: *Bos* Linnaeus, 1758

Species: Gaur *Bos gaurus* C. H. Smith, 1827

Names in other languages: **French**: Gaur; **German**: Gaur; **Spanish**: Gaur; **Italian**: Gaur; **Adi**: Tadok; **Burmese**: Peeoug; **Kannada**: Kaati, Kaadu kona, Kaadu yemme; **Kannada-in Uttara Kannada District kulga**: Gameya; **Northern Udupi district, Kannada**: Duddu; **Lao**: Meuay; **Malay**: Seladang; **Mandarin**: Da-E-Ni, 大额牛; **Marathi**: Raangawa; **Malayalam**: Kattu pothu; **Nepali**: Gauri gai; **Tamil**: Kaatu maadu; **Vietnamese**: Bò Tót; **Telugu**: Adavi dunna; **Thai**: Krating กระทิง; **Gonds**: Moo.

Other common names: Indian bison

Note: although *Bos frontalis* (Lambert 1804), is often used as the scientific name for the gaur (see Grubb 2005), the International Commission of Zoological Nomenclature (2003) ruled that *Bos gaurus* be adopted as the name for the gaur despite *Bos gaurus* being antedated by *Bos frontalis*. *Bos frontalis* is used as the scientific name for the mithun, the domestic form of the gaur.

Ecology, Evolution and Behaviour of Wild Cattle: Implications for Conservation, ed. M. Melletti and J. Burton. Published by Cambridge University Press.

Taxonomy

The gaur is a member of the Bovini clade (tribe), one of three clades (the other two being Boselaphini and Tragelaphini) recognized to be part of the larger clade (subfamily) Bovinae; the Bovinae along with the Antilopinae comprise the family Bovidae (Bibi 2007). Members of the bovine lineage were present by 8.9 Ma, appear to have originated in Asia, south of the Himalayas, and are understood to have expanded their ancestral dietary niche to include roughage in more open environments by diversifying and extending their range in response to the vegetation changes in the late Miocene (Bibi 2007). Using 15 complete or partially sequenced autosomal genes, MacEachern *et al.* (2009) identified three lineages in the Bovini clade after it split from the Boselaphini and Tragelaphini clades: the buffalo clade (*Bubalus* and *Syncerus* species), the banteng *Bos javanicus* and gaur clade, and the cattle clade. These findings were consistent with what the following five other studies also found: (1) Hassanin & Ropiquet (2004) used phylogenetic analyses of 2065 nucleotide characters representing three different genetic markers and found that the subtribe Bovina radiated into three clades: the first included the domestic ox, zebu and European bison, the second the yak and American bison, and the third contained the kouprey *Bos sauveli*, banteng and gaur; (2) Buntjer *et al.* (2002) used amplified fragment length polymorphism (AFLP) fingerprinting and comparative morphology and found that the gaur was most closely related to the banteng, followed by zebu cattle, and then taurine cattle; (3) Hassanin & Douzery (1999) compared the nucleotide sequence of the complete mitochondrial cytochrome b gene between 51 species of the family Bovidae and found that the subfamily *Bovinae* included four tribes and subtribes: Boselaphini, Tragelaphini, cattle-Bovini (*Bos* and *Bison*), and buffalo-Bovini (*Bubalus* and *Syncerus*); (4) Janecek et al. (1996) analysed the nucleotide sequence evolution of the mitochondrial cytochrome c oxidase subunit II and found that the gaur was related most closely to the banteng, followed by *Bison bison*, then the cattle group of *Bos taurus* and *Bos grunniens*, then *Bison bonasus*, and finally the cluster of *Syncerus* and *Bubalus*; (5) Hartl *et al.* (1988) studied 15 genetic loci in ten species of Bovini and found that the gaur was most closely related to banteng, followed by the aurochs *Bos primigenius* and yak *Bos mutus*, then bison (*Bison bison* and *Bison bonasus*), and finally the African buffalo *Syncerus caffer* and anoa *Bubalus depressicornis*.

Subspecies and distribution

Subspecies

Based on morphological differences between 5–6 specimens, Lydekker (1907) classified three subspecies of the gaur: *Bos gaurus gaurus* (Bangladesh, India and Nepal), *Bos gaurus readei* (Burma, China) and *Bos gaurus hubbacki* (Malaysia). Lydekker classified *Bos gaurus hubbacki* based on three adult male skulls from Malaysia, while adding that the Malaysian gaur subspecies was different because Hubback had reported that adult male gaur in Malaysia do not have dewlaps. Hubback (1937), however, later clarified that adult male gaur in Malaysia do have dewlaps, just not all the ones he had seen. A detailed study of gaur and gayal skulls conducted in the last few years of the nineteenth century (Baker 1903) found that many gaur in India did not have a concave forehead, which debunked another Lydekker claim that 'the Indian and Burmese gaur races have a forwardly inclined and prominent ridge between the horns, which gives the forehead a concave outline and distinguishes them from the Malay race'. Considering that Lydekker classified the three gaur subspecies based on a small sample size of adult male gaur skulls, and much of what he used to justify his classification has subsequently been shown to be incorrect, his classification of three gaur subspecies is no longer recognized.

More recently, based on skull and horn measurements, two subspecies of the gaur have been proposed (Groves 2003; Groves & Grubb 2011): *Bos gaurus gaurus* in India and Nepal (and possibly Bangladesh), and *Bos gaurus laosiensis* in Myanmar (Burma), Lao PDR, Vietnam, Cambodia, Thailand and West Malaysia (and presumably southern China). This classification was based on the analysis of six cranial variables (greatest skull length, nasal length, postorbital breadth, breadth between horn bases and greatest and least occipital breadths), which showed a separation between the South and Southeast Asian samples ($n = 7$ and 10, respectively). However, again given the small sample size of skulls analysed, and the fact that no rigorous genetic analyses have corroborated this classification, the existence of gaur subspecies is inconclusive.

Mithun: the domestic form of the gaur

The mithun is considered the domestic form of the gaur, and is found in the northeast region in India and Bhutan; there are no reports of gaur being domesticated anywhere else in the world. The mithun looks very similar to the gaur, except for its disproportionately formed horns and visibly smaller size. There have been reports of mithun–gaur hybrids (Gee 1964).

Distribution

Globally, the gaur has ranged, both historically and in the present time, in South and Southeast Asia (Plate 3), which includes Bangladesh (where it's nearly extirpated), Bhutan, Cambodia, China (Yunnan and South Tibet), India, Lao PDR, Malaysia (peninsular Malaysia), Myanmar (Burma), Nepal, Sri Lanka (extinct), Thailand and South Vietnam. The distribution area of the gaur, however, has shrunk by over 80% in the last 100 years and the gaur is now mainly found in protected areas (Groves & Grubb 2011).

Figure 11.1 Adult male gaur (photo by Ramki Sreenivasan; location: Nagarahole, India). A black and white version of this figure will appear in some formats. For the colour version, please refer to the plate section.

Descriptive notes

Body measurements: **Shoulder height**: 145–197 cm; **Head and body length**: 249–330 cm; **Tail length**: 70–89 cm; **Body mass**: 440–941 kg. The dental formula for the gaur is: I 0/4, C 0/0, P 3/3, M 3/3 (×2) = 32 (A. Bourgeois, personal communication; F. S. Ahrestani, unpublished data; Wilson & Mittermeier 2011). Differences in body measurements between sexes are detailed below.

The gaur is a sexually dimorphic species and the conspicuous differences between the sexes begin to be noticeable beyond the age of three years (Figures 11.1, 11.2, 11.3, 11.6). These differences are in body size, horn shape and size, pelage and prominent physical features such as the dewlap and dorsal ridge that are found only in adult males (Figure 11.1 and Figure 11.3). A detailed description of the differences in morphological characteristics between the sexes in different age classes is provided in Table 11.1 and illustrated in Figure 11.6.

Based on nine measurements of adult males and 14 of adult females, Ahrestani & Prins (2011) reported a maximum shoulder height of 175 cm for males and 148 cm for females for captive gaur. The other records of gaur shoulder height are from gaur killed in hunts, and they are all from males: the record is Pillay's (1952) measurement of 197 cm; the largest gaur Inverarity (1889) measured was 176 cm; Cameron (1929) measured a gaur at 145 cm; and Forsyth (1889) at 178 cm. The length from nose to root of the tail of an adult male body has been measured to be 249 cm (Cameron 1929) and 284 cm (Dunbar-Brander 1923), and the length of the tail was measured to be 86 cm (Dunbar-Brander 1923) and 89 cm (Cameron 1929). Weights of adult males shot in the wild have been reported as 590 kg, 782 kg, 864 kg, 864 kg, 931 kg and 941 kg (Dunbar-Brander 1923; Meinertzhagen 1939; Morris 1947); and 440 kg (excluding blood; Schaller 1967) and 703 kg

Figure 11.2 Adult female gaur (photo by Ramki Sreenivasan; location: Nagarahole, India). A black and white version of this figure will appear in some formats. For the colour version, please refer to the plate section.

Figure 11.3 Adult male gaur with conspicuous dorsal ridge and dewlap – features that distinguish males from females (photo by Ramki Sreenivasan; location: Valparai, India). A black and white version of this figure will appear in some formats. For the colour version, please refer to the plate section.

(Meinertzhagen 1939) for adult females. Robinson (1941) reported the weight of a gaur skull to be ~21 kg. The length of a male gaur horn has been measured to be 61–96 cm (Forsyth 1889; Inverarity 1889; Ward 1922; Cameron 1929; Robinson 1941; Hundley 1952; Pillay 1952; Schaller 1967), and the (outer) width or spread of the male horns have been measured at 89–120 cm (Inverarity 1889; Cameron 1929; Robinson 1941; Hundley 1952; Pillay 1952; F. S. Ahrestani, unpublished data).

One of the distinguishing features of the gaur, its dorsal ridge, is 'formed by a row of single bones springing from the back bone immediately behind the junction of each pair of ribs, of which the gaur has 13 pairs' (Inverarity 1889) (Figure 11.3). The highest point of the dorsal ridge is above the fifth/sixth rib, after which the ridge reduces in height and ends abruptly at the last rib, which is the same as that of cattle (J. Sheppard, personal communication). The gaur secretes an oily chemical through its skin, first reported by Hubback (1937), which has been recently identified as 5-(1-hydroxynonyl)-2-tetrahydrofuranpentanoic acid (18-carbon bovidic acid) (Oliver *et al.* 2003; Ishii *et al.* 2004). Studies have shown that this chemical has insect repellent properties; it was tested to be a mosquito repellent in two studies (Oliver *et al.* 2003; Tran & Chauhan 2007). While the gaur does not share its oily secretion trait, it does share physical features such as white feet in both sexes and the dewlap and dorsal ridge in adult males with other Bovini species found within its range (kouprey and banteng).

In general, females live longer than males; the maximum recorded life span for captive female gaur has been ~24 years (Crandall 1964) and for captive males has been 23.6 years (Ahrestani *et al.* 2011). Based on field observations by multiple naturalists and field biologists (Inverarity 1889; Dunbar-Brander 1923; Hubback 1937; Schaller 1967; Krishnan 1972), it is generally accepted that the gaur depends less on its sense of sight (though it is capable of seeing well at night), and more on its sense of smell and hearing to detect danger.

Habitat

The availability of both forage (determined by both vegetation type and phenology) and water are likely to strongly influence habitat selection by the gaur. Currently, gaur distribution is mainly restricted to forested areas (dry and moist deciduous, semi-evergreen, evergreen) and hilly terrain (<2500 m) (Figures 11.4–11.5). Steinmetz *et al.* (2010) found that in Southeast Asia the gaur consistently selected deciduous over evergreen forest as its populations recovered after poaching; in Lao PDR, Vietnam and Cambodia gaur were found in evergreen forests, montane forests and in lowland open and closed forests, the species appearing to frequent grassy openings in closed canopy forests (Duckworth *et al.* 1999; Grieser Johns 2000; Timmins & Ou Rattanak 2001). Both naturalists (Forsyth 1889; Inverarity 1889; Dunbar-Brander 1923) and field scientists (Schaller 1967) noted that the gaur prefers hilly terrain: although the naturalists (Forsyth 1889; Dunbar-Brander 1923) suggested that the gaur seasonally moves to hilly terrain to 'escape insects', which was speculative at best; Schaller (1967), on the other hand, proposed that the gaur was found in hilly terrain mainly because of more intensive anthropogenic pressures, such as hunting and extensive forest clearance at lower elevations where human population densities are higher.

Even within a selected habitat type, the gaur appears to show some local seasonal migrations, but these are not long-distance seasonal migrations as is often the case with African bovids. In Nagarahole and Banidpur, South India, which support high-density gaur populations, the seasonal movements of

Table 11.1 Morphological characteristics of different age classes of male and female gaur (Bos gaurus)

Calves	**Females and males**	
0–2 months	Easily distinguishable by light orange-brown body coat and absence of 'white stockings'. At birth gaur stand ~70 cm at the shoulder and grow ~15 cm in height and 50% in body mass in the first two months. A white throat patch is visible for the first 2–3 weeks, horns appear only as bumps on the head and the insides of legs above the knees are white.	
3–5 months	Not much bigger than the previous age class, but colour of coat has changed to brown and 'white stockings' become visible. Shoulder height at five months is ~100 cm. Horns are visible, completely black and are ~10–15 cm long. Horns of males are longer than those of females, but it is not possible to discern between the sexes based on horn length. The presence of the scrotum is the only morphological difference between the sexes, but is difficult to detect in free-ranging gaur of this age.	
6–15 months	Both females and males grow equally and significantly in this age class and reach two-thirds adult female shoulder height. Horns of males in this age class are longer, but do not differ in shape from those of females. Besides the scrotum, which is difficult to observe in free-ranging gaur of this age class too, no other visible morphological difference can be used to distinguish between the sexes.	
Juveniles	**Females**	**Males**
15–24 months The sexes are similar in height and size, but are distinguishable by horn shape and size.	**Size** 90% of height at shoulder, and 80% of body size of adult females. **Horns** Reach 15–25 cm in length, remain completely black, curve less away from the head, making the horns closer together; their ends point inwards.	**Horns** Begin to whiten from the base up, are 25–30% white by the age of two years, ~30 cm in length, and curve spreading away from the head to taper into sharp ends pointing upwards.
2–3 years Males are bigger than females, but sexes are best differentiated based on horn differences.	**Size** By the age of three years they are 90–95% as tall at the shoulder and 80–90% as big as adult females. **Horns** Develop significantly in length and thickness, are 20–30% white, and have a semi-circular shape with ends pointing towards each other.	**Size** By the age of three years, males begin to exceed the size of adult females. **Horns** 50–60% white with ends that still point upwards. They are significantly longer, thicker and wider apart than those of females of the same age and males of the previous age class.
Adults	**Females**	**Males**
>3 years It is easy to differentiate between the two sexes.	**Size and colour** Not much growth in height, but they continue to increase in size till the age of 5–6 years. Females remain dark brown and do not turn black. **Horns** Continue to get thicker, longer and to curve inwards and closer to each other with increasing age. If the amount of white on a female's horns exceeds 80% it would be safe to assume that the female is older than ten years.	3–5 years **Size and colour** Bigger in size; dorsal ridge begins to get prominent. The anterior of the back and flanks begins to blacken. **Horns** Are 60–70% white by the age of five years, and are less pointed at the ends. Horns begin to curve inwards at the ends, an important distinguishing feature from the previous age class. >5 years **Size and colour** Beyond the age of five, males can be called 'black bulls'. With advancing age bulls continue to increase in height and mass, their coats continue to blacken, their dorsal ridge continues to grow in height, thickness and prominence, and their dewlaps continue to get bigger. **Horns** Males with horns that are over 85% white, worn at the ends and heavily corrugated closer to the head can be considered to be older than eight years old.

Measurements of shoulder height are precise while those of horns are best estimates.

the gaur appear to be linked to seasonal changes in plant phenology and forage availability rather than altitudinal differences.

In Malaysia, most naturalists have associated gaur habitat use with secondary vegetation and stressed the importance of jungle clearings, abandoned fields of shifting cultivators, forest fringe and openings along rivers (Hubback 1937; Foenander 1952; Stevens 1968; Weigum 1972; Conry 1989). In the only study of radio-collared gaur in Malaysia, Conry (1989) found that home ranges were primarily located in secondary forest

Figure 11.4 Montane evergreen grassland–shola habitat (photo by Rohit Rao; location: Kudremukh National Park, India). A black and white version of this figure will appear in some formats. For the colour version, please refer to the plate section.

Figure 11.5 Dry deciduous forest (photo by R. Killivalavan; location: Cauvery Wildlife Sanctuary, India).

and agricultural estates – no use was made of primary forest, although it was available within 1.5 km of their home ranges; the zone within 500 m of a river was used the most, and the greatest distance the gaur were found from a river was 10 km. This information, however, only applies to habitat selection by day, which is when Conry (1989) collected his data; the gaur is known to be active at night in many areas.

Foraging along stream banks in the summer by gaur has been reported by many naturalists in India (Forsyth 1889; Inverarity 1889; Dunbar-Brander 1923; Krishnan 1972). The gaur has rarely been seen wallowing, unlike the buffalo that wallows often. Multiple authors have reported the use and apparent importance of mineral licks for the gaur (Inverarity 1889; Hubback 1937; Foenander 1952; Ogilvie 1954; Schaller 1967; Krishnan 1972; Weigum 1972; Steinmetz 2004), and even today the gaur is often seen at mineral licks throughout its habitat.

Population densities

Gaur populations have not been rigorously monitored because of the effort required to monitor their low-density populations in dense habitats. Consequently, reliable data on their

Table 11.2 Population density estimates of gaur at different protected reserves in Karnataka, India measured using line transect sampling in 2011

Location	Year	Habitat	Sampling effort (km)	Sample size	Detection probability	Sampled area (km^2)	Density/km^2 (Mean ± SE)	Biomass (kg/km^2)
Anshi-Dandeli Tiger Reserve	2011	Semi-evergreen and moist-deciduous forests	1358.2	19	0.5	1820.49	0.41 ± 0.15	223.45
Bhadra Tiger Reserve	2011	Moist-deciduous and grassland–shola forests	913.19	44	0.45	734.11	1.01 ± 0.37	550.45
Nagarahole Tiger Reserve	2011	Moist-deciduous forest	956.05	67	0.44	798.77	5.14 ± 1.3	2801.3
Bandipur Tiger Reserve	2011	Dry-deciduous and moist-deciduous forests	984.63	113	0.37	612.27	5.25 ± 1.45	2861.25

Source: K. U. Karanth, unpublished data.
Sample size = number of visual detections of gaur clusters; detection probability = proportion of clusters in the surveyed area estimated as detected; density = number of individual gaur/km^2.

population densities and biomass are scarce. However, a recent study applying rigorous line transect sampling on foot (Karanth *et al.* unpublished data) has measured gaur population densities at a number of sites in Karnataka State, India (Table 11.2). The lower densities reported from some of these sites are possibly a consequence of hunting and other anthropogenic pressures, and the higher estimates are from well-protected sites such as Nagarahole and Bandipur.

Movements and home range

The primary driver of gaur movement, both daily and seasonally, is the quest to consume the nutrients required to maintain their large body mass.

Seasonal movement and home range

Often gaur, both solitary males as well as herds, can be seen in the same area for a few days or even a couple of weeks (Inverarity 1889). An adult female, recognized because of a horn deformity, remained within a 30 km^2 area for over a decade in Nagarahaole (K. U. Karanth, unpublished data), which demonstrated high site fidelity and restricted seasonal movement. However, multiple other studies have found the gaur to seasonally migrate, but across short distances: Guin & Pal (1982) found that the gaur is not found in all months within Garumara Wildlife Sanctuary (a moist deciduous sal *Shorea robusta* forest) in India – they were found between November and June, and their absence between July and October coincides with the rainy season; Imam (1985) reports that the gaur is not found during the dry season, mid March to July, in Megpal, Orissa; and in a recent study from the Bandipur–Mudumalai region, Ahrestani *et al.* (2012) proposed that the gaur inhabit different areas in different seasons within an area of 200 km^2.

A guar herd closely monitored using only visual observations and fresh spoor by Weigum (1972) in the late 1960s in Taman Negara National Park ranged over 13 km^2, mainly along the Tembeling River. In addition there have been three studies of radio-collared gaur that have estimated the gaur's movement patterns and home range:

(1) Conry (1989) found the home range size for a yearling male (November 1978–May 1979) to be 29.9 km^2, 52.1 km^2 for a yearling female (October 1977–December 1978) and 137.3 km^2 for a ~5-year-old adult male (January–November 1978). Average distance between consecutive daily locations was 1.6 km ($n = 10$), and the greatest distance across the home range measured 20.8 km for the adult male. Physiographical features such as rivers, mineral licks, agricultural fringe and forest clearings appeared to greatly influence home range size and shape. Disturbed habitats such as secondary forest and agricultural estates received the greatest use. Areas within 500 m of agricultural fringe, within 500 m of major rivers and below 61 m elevation were heavily used by gaur, often to a proportionately greater extent than their availability in home ranges. Human habitation did not appear to greatly affect use patterns; areas adjacent to settlements were not avoided but were used less than those at intermediate distances. Limited logging and agricultural development appeared to benefit gaur by creating early seral habitats. The availability of mineral licks seemed very important as all home ranges contained one or more licks.
(2) An adult gaur male and adult female were radio-collared and monitored for approximately a year in 1997–1998 in Pench Tiger Reserve, India. The summer home ranges of the male and female were 12.61 km^2 and 7.25 km^2, and the winter home ranges were 7.62 km^2 and 13.8 km^2 (Sankar *et al.* 2002).
(3) In 2011 19 gaur and in 2012 an additional 31 gaur were reintroduced to Bandhavgarh Tiger Reserve, India. Bandhavgarh had lost its gaur population by 1998 for

unknown reasons. The reintroduced gaur were equipped with radio-collars and their movements were monitored. The annual home ranges of the males and females were 254.4 km^2 and 200.4 km^2, respectively. The summer (April 2011–June 2011) home ranges for males and females were 231.5 km^2 and 161.2 km^2, respectively. In monsoon and post-monsoon periods (July 2011 to October 2011) the home ranges of males and females were 110.6 km^2 and 135.7 km^2, respectively. In winter (November 2011–February 2012) the home ranges of males and females were 98.5 km^2 and 152.5 km^2, respectively (K. Sankar, personal communication).

Activity patterns

The gaur forages both in the day and night, seeming to prefer nights, dawn and dusk, particularly in areas of warm climate. For most of the day, gaur often retreat to forested areas to chew their cud, generally lying down (Forsyth 1889; Krishnan 1972; F. S. Ahrestani, unpublished data); Schaller (1967) noted during his study that most gaur were in the shelter of trees by 07:00 and rarely seen in the open after 08:00. One explanation for retreating to forested areas would be that of seeking shade, but another could be concealment from predators as Forsyth (1889) noted how difficult it was to distinguish a gaur lying down in the flickering shadow of a tree. The gaur grooms itself, licks its hide, rubs its neck, rumps or sides against a tree trunk, or scratches itself with the tips of its horns in a manner similar to cattle. Gaur generally lie down on their belly with their legs folded under the body, but occasionally an animal will lie flat on its side with its legs stretched out. In Pench Tiger Reserve, India, a radio-telemetry study recorded the mean daily (daytime) movement of an adult male in summer to be 1.75 km (± 1.23), and 1.31 km (± 0.79) in the monsoon, and that of the adult female as 1.39 km (± 0.98) in the monsoon and 1.23 km (± 0.79) in the summer (Sankar *et al.* 2002).

Feeding ecology

The gaur, because of its large body size, is a bulk feeder, and a growing body of evidence suggests that it's primarily a grazer: Inverarity (1889) was the first to state that the gaur primarily consume grasses; Schaller (1967) found that the mean proportion of grasses in the rumen contents of the four gaur he autopsied was 85% (range: 66–100%); Chetri (2003, 2006) found that grasses made up 66%, browse 26% and herbs and others 6% of the gaur diet; Gad & Shyama (2011) found that graminoids (grasses, sedges and bamboo) accounted for >60% of epidermal fragments found in gaur faeces; and a recent study (Ahrestani *et al.* 2012) found the gaur to be a grazer throughout the year. Leaves and shoots of bamboo, the world's largest grass species, are understood to be preferred forage for the gaur throughout its range: all three dominant bamboo species in South India (*Bambusa arundinacea*, *Dendrocalamus strictus* and *Oxytenanthera monostigma*) are eaten (K. U. Karanth, unpublished data), and in Thailand bamboo shoots in the wet season and bamboo leaves in the dry season were preferentially eaten (Steinmetz 2004; Prayurasiddhi 1997). The gaur being primarily a grazer fits Hofmann & Stewart's (1972) classification of Bovini species as bulk feeders (grazers) – other closely related Bovini species, like the water buffalo, African buffalo and American bison are also all primarily grazers (Ahrestani *et al.* 2012).

Although the gaur eats primarily grasses, it is also a browser: Schaller (1967) recorded that the gaur ate at least six grass species, seven shrub species, three vine species, four forb species and leaves from at least 17 tree species and fruits from two tree species; Dunbar-Brander (1923) recorded gaur feeding on more than 40 plant species; Chetri (2003, 2006), using microhistologial analysis, found that the gaur consumed 49 plant species (23 grasses, 17 browse species, and nine herbs and others); and Gad & Shyama (2011) found gaur to feed on seven grass species, five herb species, eight shrub species and 12 tree species.

Among the different plant parts, gaur primarily consume leaves. In Gad & Shyama's (2011) study, leaves made up 87% of what gaur consumed; Krishnan (1972) and Schaller (1967) report that the gaur extends its tongue to curl around leaves and twigs, which it pulls into its mouth and bites off with a tug. The gaur also eats other plant parts, such as fruits, young shoots, bark and flowers. Studies have reported the gaur eating fruits of woody plants, and eating young shoots is often associated with foraging on bamboo (Table 11.3). Forsyth (1889) reported that gaur loved to eat bamboo; Schaller (1967) noted that gaur consumed bamboo; Krishnan (1972) reported that gaur consumed the leaves and culms of the common species of bamboo in South India (*Bambusa arundinacea* and *Dendrocalamus strictus*) and observed gaur feeding on bamboo branches that had been brought to the ground by elephants; and a recent study by Gad & Shyama (2011) and multiple field researchers and naturalists who have worked in South India have also reported that various parts of bamboo are consumed by the gaur.

It is likely that grasses, including bamboo, dominate a gaur's diet throughout the year, but the gaur seasonally changes the amount of browse that it consumes; Gad & Shyama (2011) found in their four-year study that the proportion of browse consumed by gaur increased from 15% in winter to 40% in summer. During the dry season in deciduous forests there's little to graze or browse (the leaves that are available on the forest floor are dry, too), which is when multiple studies have recorded the gaur eating the bark of trees, such as *Adina cordifolia* (Dunbar-Brander 1923), teak *Tectona grandis* (Pasha *et al.* 2002; Gad & Shyama 2011), and an exotic cultivate, the cashew (*Anacardium occidentale*) in India; and *Holarrhena antidysentrica* and *Wendlandia natoniana* in Malaysia (Ogilvie 1954). The eating of bark is understood to be in response to either a shortage in forage, mineral resources or both.

Table 11.3 Plant species eaten by gaur recorded by different studies

Location	Grass species eaten (B = bamboo species)		Browse species eaten (F = fruits eaten, B = bark eaten)		Forbs (F), vines (V) and shrubs (S) species eaten		References
India			*Aegle marmefos*	F			Dunbar-Brander, 1923
			Adina cordifolia	B			
			Randia dumetorum	F			
Malaysia	*Imperata*		*Atalantia monophylla*				Hubback, 1937
	arundinacea		*Eugenia zeylanica*				
	Paspalum		*Trema amboinensis*				
	conjugatum						
Kanha, India	*Coix lachrymajobi*		*Bauhinea racemosa*		*Asparagus racemosus*	V	Schaller, 1967
	Dendrocalamus		*Bridelia sqamosa*		*Bauhinea vahlii*	V	
	strictus	B	*Casseria graveolens*		*Blumea virens*	F	
	Imperata cylindrica		*Cassia fistula*	F	*Curcuma ferruginea*	F	
	Themeda triandra		*Cordia myxa*		*Embelia tseriamcottam*	S	
	Thysanolaena		*Ehretia laevis*		*Kydia calycina*	S	
	mzxima		*Emblica officinalis*	F	*Leucas mollissima*	F	
	Vetiveria zizanioides		*Gardenia latifolia*		*Moghania stricta*	S	
			Gmelina arborea		*Ougeinia oojenensis*	S	
			Grewia abutilifolia		*Phoenix humilis*	S	
			Miliusa velutina		*Schrebera swientenioides*	S	
			Mallotus phillippinensis		*Smilax zeylanica*	V	
			Odina wodier		*Sterculia foetida*	S	
			Shorea robusta		*Vernonia divergens*	F	
			Syzygium cumunii				
			Terminalia alata				
			Terminalia tomentosa				
			Xeromphis uliginosa				
			Zizyphus xylophorus				
Central India	*Bambusa*		*Aegle mormelos*	F	*Ardisia solanacea*	S	Krishnan, 1972
	arundinacea	B	*Butea monosperma*		*Costus speciosus*	S	
	Dendrocalamus		*Cassia fistula*	F	*Desmodium pulchellum*	S	
	strictus	B	*Cordia myxa*		*Grewia aspera*	S	
			Emblica spp.	F	*Grewia hirsute*	S	
			Gmelina arborea	F	*Hibiscus lamps*	S	
			Grewia tiliaefolia		*Helicteres isora*	S	
			Phyllanthus sp.	F	*Smilax zeylancia*	V	
			Terminalia bellerica	F			
			Zizyphus trinervia				
			Zizyphus xylopyrus				
Pench Wildlife	*Bambusa*				*Adina cordifolia*		Shukla & Khare, 1998
Sanctuary	*arundinacea*	B			*Egle marmelos*	F	
	Dendrocalamus				*Randia dumetorum*	F	
	strictus	B					
	Themeda triandra						
	Themeda						
	quadrivalis						
	Saccharum						
	spontaneum						
Parsa Wildlife	*Apluda* sp.		*Phaulopsis impbricata*		*Asparagus racemosus*	V	Chetri, 2003, 2006
Reserve, Nepal	*Cymbopogon*		*Phyllantus emblica*	F			
	microtheca		*Terminalia chebula*	F			
	Cymbopogon spp.		*Wendlandia exserta*				
	Imperata cylindrica						

Table 11.3 (cont.)

Location	Grass species eaten (B = bamboo species)	Browse species eaten (F = fruits eaten, B = bark eaten)	Forbs (F), vines (V) and shrubs (S) species eaten	References
	Paspalidium punctatum *Phragmites karka* *Saccharum spontaneum* *Themeda* sp. *Vetiveria zizanoides*			
Mollem region, Goa, India	*Bambusa arundanacea* B *Cynodon dactylon* *Cyperus rotundus* *Dendrocalamus strictus* B *Digitaria* sp. *Spermacoce* sp. *Vetiveria zizanioides*	*Anacardium occidentale* B *Dillenia pentagyna* *Tectona grandis*		Gad & Shyama, 2011
Ulu Lepar, Pahang, Malaysia	*Danim triganum* B *Dryobalanops aromatica* *Endospermum malaccense* *Imperata cylindrica* *Panicum maximum* *Paspalum conjugatum* *Paspalum vaginatum*	*Atalantia monophylla* *Dillenia ovate* *Erythrina variegata* *Intsia palembanica* *Macaranga gigantea* *Shorea acuminata* *Shorea parvifolia*	*Centrosema pubescens* V *Goniothalamus* sp. S *Macaranga tanarius* S *Melastoma malabathricum* S *Parkia javanica* S *Trema angustifolia* S *Trema angustifolia* S	Yusof, 2009

Reproduction and growth

Different annual mating and calving periods have been reported for gaur in different regions: Dunbar-Brander (1923) reported that the majority of mating in central India occurred in December and January, with calves being born in August and September; Stebbing (1911) and Sanderson (1912) also reported similar months, but noted that calves may be born in April, May and June; Hubback (1937) reported that young gaur in Malaysia were seen at all times except from October to December; Morris (1938) reported that the peak rutting period in southern India was from November to March; K. U. Karanth (unpublished data) found rutting activity to peak from June to September (the wet season) in Nagarahole. However, a recent study by Ahrestani *et al.* (2011) reported that gaur calves were found throughout the year with no discernable peak, which is similar to what Inverarity (1889) reported for gaur in India, Peacock (1933) reported for gaur in Burma and Schaller (1967) reported for Kanha, India. Schaller (1967), however, also reported that 'the frequency of rutting call and other aspects of sexual behaviour reached a peak in March and April'.

Hubback (1937) reported that the gestation period of the gaur was similar to the gestation period of domestic cattle, which is about nine months. Data from both Omaha Zoo, United States, and Mysore Zoo, India, indicate that the gestation period for gaur is 9–10 months, ~280 days (Ahrestani & Prins 2011). There is no evidence to contradict Schaller's (1967) statement that 'single births are the rule, there being no records of twins'; no birth of twins was recorded in: 319 gaur births (before 2011) in European zoos (A. Bourgeois, personal communication); ~150 gaur births (1968–2006) in Omaha Zoo; and the 41 gaur births (1994–2006) in Mysore Zoo, India (F. S. Ahrestani, unpublished data). Ahrestani *et al.* (2011) found the ratio of males to females at birth was 0.86 ($n = 41$) in Mysore Zoo and 1.36 ($n = 180$) in Omaha Zoo; Reed (1959) reported a ratio of 1.29 ($n = 16$) from the National Zoological Park, Calcutta, India. These reports are conflicting, and based on data of the other Bovini species, Ahrestani *et al.* (2011) concluded that it was reasonable to

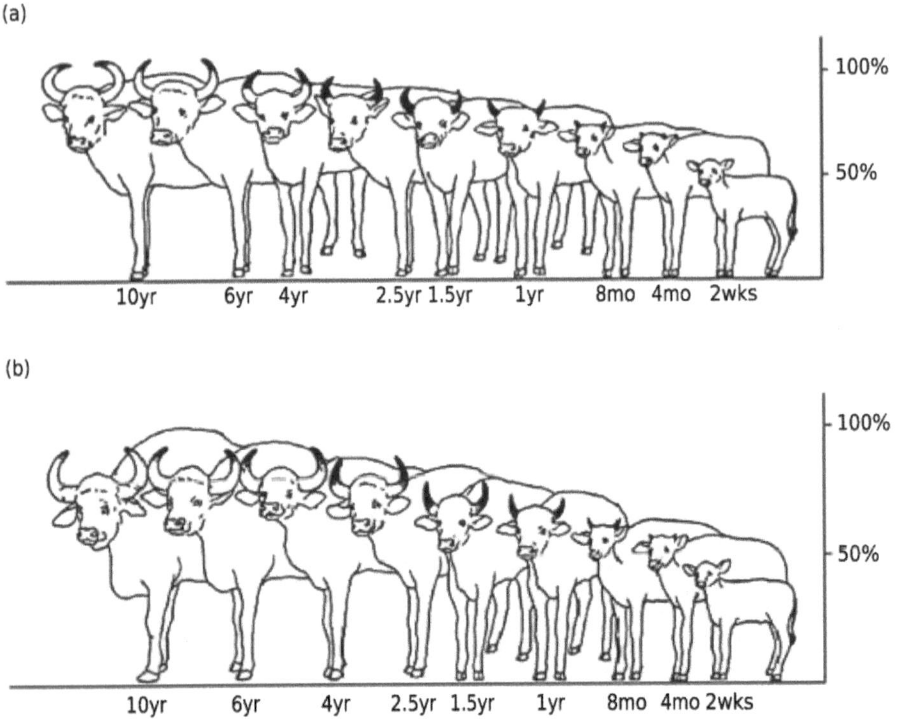

Figure 11.6 Age-specific relative growth of (a) female and (b) male gaur *Bos gaurus*. 100% = shoulder height of ten-year-old. (Reprinted from Ahrestani & Prins, 2011, DOI: 10.1515/MAMM.2010.078, by kind permission from Walter De Gruyter.)

assume that the sex ratio at birth for gaur was at parity. Keeping this in mind, and the fact that there is no discernible difference between the sexes in the growth of gaur in the first two years (Ahrestani & Prins 2011), it is not surprising that of the 109 medium-sized and large calves that Schaller (1967) sexed, 55 were male and 54 were female, and of the 282 gaur sexed below the age of one year and the 153 gaur sexed between the age of one and two years, Ahrestani *et al.* (2011) found 140 males and 142 females, and 76 males and 77 females, respectively.

While there are no specific data for gaur, the average length of the oestrous cycle in gayal was found to be 22 days and the average duration of heat was found to be 45 days (Giasuddin *et al.* 2003). Primiparity for captive gaur is three years, which is similar to other Bovini species (Ahrestani *et al.* 2011). Captive gaur females are known to give birth at the age of 18 years, and therefore it is assumed similarly aged free-ranging female gaur are capable of giving birth too (Ahrestani *et al.* 2011). Under captive conditions – given their 9–10 month gestation period – gaur often give birth every year (Ahrestani *et al.* 2011). There are no data regarding calving intervals in free-ranging gaur, but based on data of similar-sized free-ranging Bovini species, Ahrestani *et al.* (2011) suggest that female gaur that reach the age of 20 years may be expected to have produced 8–10 calves. Milk consumed by a captive gayal calf was found to be 304 ml/day and the length of lactation was 116 days (Giasuddin *et al.* 2003). Calves as old as 180–240 days have been seen suckling females both in captivity and the wild (F. S. Ahrestani, unpublished data).

Cattle have been seen within gaur herds, but there is no evidence to show that these interactions have led to interbreeding. The weight of newly born calves is 25–30 kg (mean weight of 17 male calves = 30.1 kg and mean weight of 17 female calves = 25.9 kg born in Omaha Zoo; J. Shepard, personal communication) and by the end of the first year captive gaur can be as heavy as 200 kg (A. Bourgeois, personal communication). Ahrestani & Prins (2011) found that both sexes grow in body size similarly until the age of 15 months, after which males grow more than females (Figures 11.6–11.7). Quoting Ahrestani & Prins (2011), 'Although females appear to attain 90–95% of their maximum shoulder height by three years, they continue to grow in body mass and probably attain their maximum body size by the age of six years. Males continue to grow for a greater number of years than females, attain their maximum shoulder height at 8–10 years, and probably attain maximum body size by 10–12 years.'

It is possible that there's a tradeoff between attaining a large body mass and longevity in male gaur, as every study of gaur demographics has found that there are more adult females (45–62%) than adult males (20–38%) (Table 11.4; Figure 11.8). In general, calves comprised 4–24% and the gender ratio until the age of two years was found to be parity in gaur populations (Ahrestani *et al.* 2011).

Behaviour

Social and community behaviour

The gaur is socially organized predominantly into herds (Figures 11.9 and 11.10). Females of all ages and males below the age of three years are nearly always found in herds, and adult males are found both within herds and alone (Inverarity 1889; Hubback 1937; Morris 1937; Schaller 1967). There exists no long-term study to reliably comment on how gaur herds are

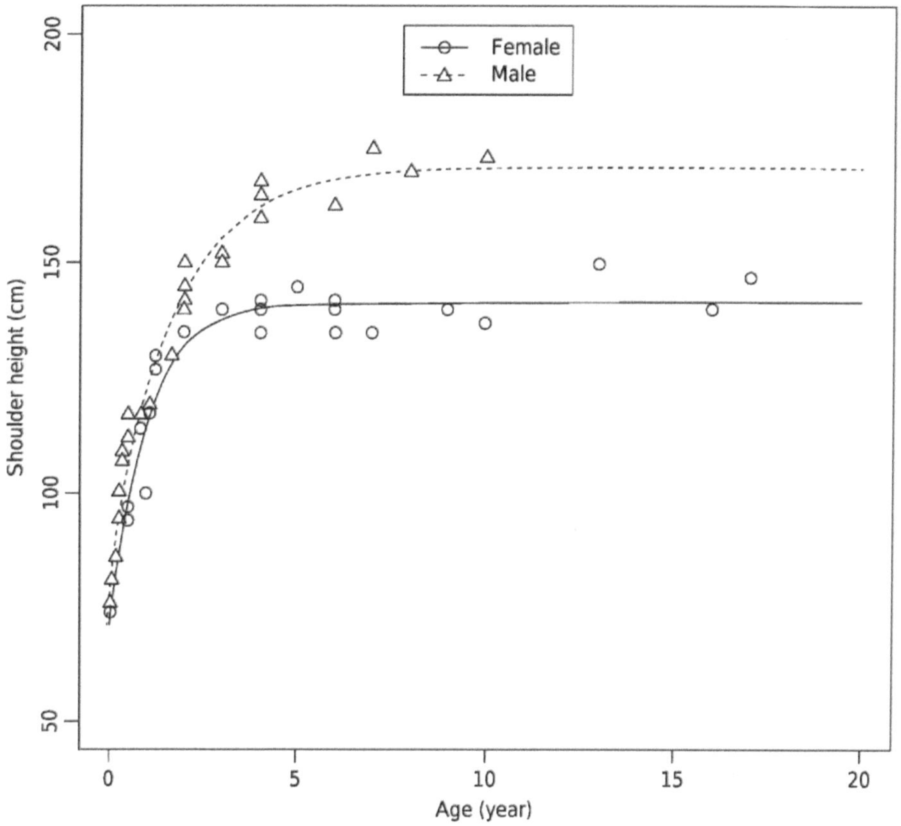

Figure 11.7 Growth of male and female gaur *Bos gaurus* measured at shoulder height. The shoulder height data (51 measurements) were fitted by plotting von Bertalanffy (1938) growth functions for both sexes: for females, $L_t = 142\,\{1 - \exp[-0.19\,(t + 0.69)]\}$ and for males, $L_t = 171\,\{1 - \exp[-0.57\,(t + 1.18)]\}$. Note: the 51 measurements included females at 18 different ages and males at 17 different ages, derived from 12 females and 10 males at Mysore Zoo, Mysore, India, and eight females and six males at Henry Doorly Zoo, Omaha, United States. Multiple individuals in Mysore Zoo were measured at multiple ages. (Reprinted from Ahrestani & Prins, 2011, DOI: 10.1515/MAMM.2010.078, by kind permission from Walter De Gruyter.)

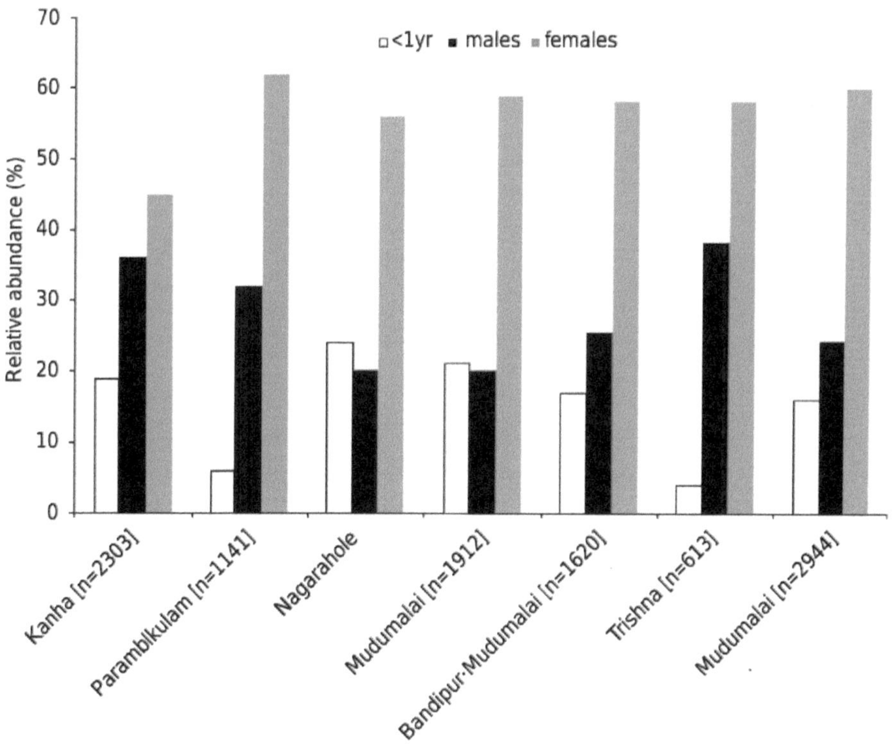

Figure 11.8 Relative abundance of calves (<1 year), males (>1 year) and females (>1 year) found in the field study in comparison with four other gaur studies in India. Literature sources are: Kanha National Park, Madhya Pradesh – Schaller (1967); Parambikulam Wildlife Sanctuary, Kerala – Vairavel (1998); Nagarahole National Park, Karnataka – Karanth & Sunquist (1995); Mudumalai National Park, Tamil Nadu – Ashokkumar *et al.* (2004); Mudumalai and Bandipur National Parks, South India – Ahrestani *et al.* (2011); Trishna National Park, Tripura – Dasgupta *et al.* (2008); Mudumalai National Park, Tamil Nadu – Ramesh *et al.* (2012).

formed, maintained, change and what hierarchies exist in their social organization, and therefore it is unclear when adult males join and leave gaur herds, how long they stay with a herd and how exactly this behaviour changes with age. It is, however, generally accepted that the solitary life of an adult male gaur increases with age – multiple old-looking gaur have been observed living solitary lives for months. Seeing a bachelor herd or an individual adult female or an individual calf are

Table 11.4 The range, mean and maximum size of gaur herds in different populations

Mean group size	Maximum group size	Adult males: adult females: calves	Location	Reference
	~75	33:100	Mudumalai–Bandipur Tiger Reserves, India	Ahrestani *et al.* 2011
9.8 (± 7.6 SD)	47	25:100:16	Mudumalai Tiger Reserve, India	Ashokkumar *et al.* 2010
7.5 (± 0.40 SE)	42	39:100:39	Mudumalai Tiger Reserve, India	Ramesh *et al.* 2012
5.5 (± 5.2 SD)			Vietnam	Nguyen 2009
4 (± 0.21 SE) 4.6 (± 0.26 SE) in winter 3.1 (± 0.51 SE) in summer]	14	58:100 56:100:27 in summer 61:100:9 in winter	Trishna Wildlife Sanctuary, India	Dasgupta *et al.* 2008
2.3			Bhadra Tiger Reserve, India	Jathanna *et al.* 2003
6	19	45:100:16	Parambikulam Tiger Reserve, India	Vairavel 1998
	70		Jaldapara, India	Bhattacharya *et al.* 1997
6.9	47	18:100:43	Nagarahole Tiger Reserve, India	Karanth & Sunquist 1992
	30		Erivakulam National Park, India	Rice 1986
6–7		50:100	Kanha Tiger Reserve, India	Belsare *et al.* 1984
	60		Mudumalai Tiger Reserve, India	Johnsingh 1983
5–7	27		Palamau Wildlife Sanctuary, India	Sahai 1977
8–11	40	80:100: 39–45 January–June 29 July–December	Kanha Tiger Reserve, India	Schaller 1967
	~100		Pidaung Wildlife Sanctuary, India	Mustill 1938
10–12			Malaysia	Hubback 1937
10–20			Burma	Peacock 1933
8–12			India	Dunbar-Brander 1923
~12	30–40, rarely 100		India	Sanderson 1912
10–20			India	Russell 1900
10–12			India	Forsyth 1889

all uncommon sights, though Ahrestani (unpublished data) once observed 16 adult male gaur together. Of the 200 herds tallied in Kanha, Schaller (1967) found that ~50% had an adult black bull, 52% of Johnsingh's (1983) observations of gaur were of individual adult males and 48% of male gaur that Ramesh *et al.* (2012) observed were found to be solitary. Often herds include more than one black bull; Schaller (1967) has seen herds commonly with 2–3 black bulls and herds with both six and eight black bulls in them.

From Table 11.4 it appears that the average herd size is between five and ten. There have been, however, observations of large (>50 individuals) aggregations of gaur (Sanderson 1912; Mustill 1938; Johnsingh 1983; F. S. Ahrestani, unpublished data). These are generally recognized as being temporary composite herds, i.e. large herds that emerge when multiple smaller herds come together. The majority of these large composite herds have been observed during the rainy season, and it is possible that the gaur come together to exploit a favourable forage resource or area.

There's little in-depth understanding of the dominance relationships between gaur males, but there exists a rich collection of observational information. Sparring is not a regular occurrence between adult (>5 years) males, but occurs fairly commonly between younger males. Ahrestani (unpublished data) observed sparring between young males aged 1.5–3 years on multiple occasions: coming face to face against each other, lowering their heads, locking horns and twisting their heads. Iverarity (1889) reports seeing two adult males sparring with their horns and head butting. Schaller (1967) was the first to describe the lateral display, which is when a bull gaur is seen

Figure 11.9 Adult females in a gaur herd (photo by Ramki Sreenivasan; location: Bandipur, India). A black and white version of this figure will appear in some formats. For the colour version, please refer to the plate section.

Figure 11.10 Gaur herd with adult females and calves (photo by M. N. Naveen; location: Bandipur, India). A black and white version of this figure will appear in some formats. For the colour version, please refer to the plate section.

standing still with its feet together and back held somewhat hunched in the presence of another bull. This has come to be interpreted as a non-physical method used by gaur to establish dominance. Krishnan (1972), based on his observations of the position of the bulls involved in these displays, was unsure, however, if lateral was the correct word, as it could be an oblique or an end-on display.

Threatening gestures between male gaur consist of walking directly at an opponent with the head slightly lowered, accompanied by lateral sweeps of the horns (Belsare *et al.* 1984). Krishnan (1972) has observed adult gaur tossing their heads side to side while holding their head high in what appeared to be a threat display towards a domestic elephant, and reports seeing males emit a low moaning and grumbling, varied with occasional expulsions of breath, while running or walking (often in circles) and swiping at undergrowth with lowered horns. Dunbar-Brander (1923) described the threatening behaviour of gaur as 'whistling snorts made through the nose as in the first instance, but are generally accompanied by fighting the undergrowth and by pawing up the ground'.

Multiple authors (Inverarity 1889; Schaller 1967; Krishnan 1972) have observed adult male gaur with wounds that seemed to be a result of fights with other gaur. Johnson (1986) once witnessed a couple of adult male gaur fighting next to a gaur herd. The gaur displayed at each other, thumping the ground and snorting repeatedly, and the younger of the two repeatedly threatened the older combatant with his horns held low; the two bulls made contact with their heads multiple times, pushing against each other and once twisting their entwined horns. The confrontation finally ended by the older bull managing to push the younger bull down a slope. The pushed bull was found limping a couple of days later. Dunbar-Brander (1923) reports once seeing two bison fighting, shoving each other about, and when disengaged, they swung their heads from side to side with a twisting motion, and in so doing gave each other terrific blows on the horns. Krishnan (1972) has observed adult gaur move other adult gaur of the same gender out of the way by lunging and thrusting their horns with their heads kept low.

In general, the gaur is considered to be tolerant of other herbivores. Krishnan (1972) reported that the gaur avoids people as much as possible, a behaviour that is often observed of gaur even today. However, occasional reports of humans being killed by gaur do occur: the majority of these are associated with incidents of gaur charging after being shot or wounded (Forsyth 1889), or when surprised (Inverarity 1889), including a recent incident when a person was killed while trying to photograph a gaur using flash photography in darkness (R. Kannan, personal communication); Morris (1952, 1953) reported being personally charged when encountering an adult male gaur at a distance of ~5 m; a hunter being charged by an adult male gaur; two beaters being killed by an adult male gaur during a hunt; and a person being killed by an adult male gaur. K. U. Karanth (unpublished data) was charged by an adult female gaur that had been badly injured by a tiger he was radio-tracking.

Defensive and antipredator behaviour

When a gaur becomes alert, it will normally have its head raised (sometimes with the muzzle raised as if the animal is sniffing the air), nostrils flared and facing the direction of the threat (Schaller 1967; Krishnan 1972; Belsare *et al.* 1984; Ahrestani, personal observation). Bulls and cows emit short, deep grunts and snorts when excited (Krishnan 1972), and often emit loud snorts that have been described as 'violent and rapid expulsions of air through the nostrils' (Hubback 1937), 'a trumpet-like blast of air through the nose, accompanied occasionally by a growling sound – a harsh, rolling bru-u-u-u' (Schaller 1967); and 'a call spelled pff-hong; the pff is the noise made by the rush of air past the lips before the note is struck' (Dunbar-Brander 1923). Usually, a snort of this nature often 'raises every head in the herd' (Dunbar-Brander 1923; F. S. Ahrestani, unpublished data).

Gaur often flee from danger, and Schaller (1967) has observed that 'when they do, they often give a series of rather stiff-legged bounds, some two to eight in number, with their forelegs brought down hard in unison to produce a series of distinct thumps. After thumping, it usually trots off without making further sounds other than the usual commotion of a large animal moving through the forest.'

Gaur calves fall prey to leopards, wild dog and tigers, but adult gaur are normally only predated upon by tigers (Ahrestani *et al.* 2011). Dunbar-Brander (1923) reported that tigers kill calves as well as adults, even solitary bulls; Schaller (1967) claimed that the primary cause of death for gaur in 1964 in Kanha was predation by tiger, principally affecting calves; Rice (1986) found the carcass of an adult female gaur killed by a tiger and the remains of gaur in three of four tiger scats collected in Erivakulam National Park, India. Johnsingh (1983) found that gaur (calves) made up 0.19% of dhole diet, and ~5% of tiger diet (analysed from scats) in Mudumalai Tiger Reserve, India. Karanth & Sunquist (1995, 2000) quantified predation by gaur in Nagarahole, South India, and found that adult female and adult male gaur comprised 23% and 15% of tiger *Panthera tigris* kills, respectively – this 1.5:1 mortality ratio, however, was reported with a 3:1 (56:20) adult female: adult male ratio, suggesting that a greater proportion of adult males than females were killed by tigers – and that 88% of the 33 gaur they recorded killed by tigers (including all adults) resulted from throat bites. Adult bulls have a thick neck and often a dewlap, which makes killing them by asphyxiation difficult, and given their strong back as well, one tactic used by tigers to bring down a bull gaur is by biting the bull's hock, thereby hamstringing the bull (Dunbar-Brander 1923).

Adult gaur bulls are capable of killing tigers that attack them: Blackburn (1934) reports finding a dead tigress and a bull gaur after what was evidently a fight to the death, and K. U. Karanth (unpublished data) reports that one of the radio-collared adult male tigers in his study apparently died during an attack on a gaur. If a tiger is detected by a herd that has calves, the adult female gaur are known to bunch together in a mux-ox fashion, i.e. backs to each other and facing outwards (Krishnan 1972), and have been observed leaping and kicking their hind legs, snorting and tossing their horns and holding their heads low and advancing, sometimes even rapidly, towards detected leopards and tigers (Krishnan 1972; Johnsingh 1983; Karanth 1984; Tyabji 1989; F. S. Ahrestani, unpublished data).

Reproductive and sexual behaviour

Females leave the herd to give birth alone, and can remain separated from a herd for as many as four days (Sanderson 1912). The mother licks a calf a lot after birth (Krishnan 1972). Young (under two months old) calves have been found alone, lying motionless in the undergrowth with their head and neck stretched out close to the ground, apparently trying to make themselves less visible (Inverarity 1889; Hubback 1937; Ahrestani *et al.* 2011). However, it is rare to see a calf travelling alone as calves are nearly always seen in the company of females, and the ones that are below the age of three months are often at the heel of a female, presumably their mother. It is understood that gaur mothers suckle calves till they are six months old (Schaller 1967; F. S. Ahrestani, unpublished data), and the longest continuous suckling by a calf has been observed to be approximately nine minutes (Schaller 1967). Gaur, mostly females, are known to moo, sounding just like cattle (Dunbar-Brander 1923; Krishnan 1972; F. S. Ahrestani, unpublished data), but it's unclear whether this is a mother communicating to its calf.

Licking in gaur is also conspicuous during courtship; Schaller (1967) reported that 'cows and bulls lick each other's necks, shoulders, and rump, occasionally for as long as ten minutes without interruption, and bulls also lick each other at times, in five out of seven instances the subordinate animal licked a gaur of higher rank'. Bulls exhibit flehmen – raising the muzzle and curling their lip after licking a female's vulva, and particularly after sniffing a female's urine or faeces – a habit that is common to many ungulates (Schaller 1967). Schaller also reported observing tending, a behaviour first described for American bison by McHugh (1958), in which males stand beside a female, wait while the females graze, and follow the female step by step.

The gaur's rutting call is so distinctive that it distinguishes adult male gaur as much as the dorsal ridge and dewlap do. The manner in which this call is emitted is best described by Krishnan (1972): the call begins by keeping the head low, and as the call progresses the head is raised slowly, stretching the neck till at the end of the call the muzzle with the lips partly open is pointing up, with the whites of the eyes showing with the effort. The call has been described as: 'the most absurd piping or whistling sound, more like the call of a bird than anything else, and absurd by reason of it emanating from so large and powerful an animal' (Dunbar-Brander 1923); 'a most peculiar sound, a cross between the bugling of a wapiti (elk) and the trumpeting of an elephant, but at the same time a melodious sound that carries a long way' (Hubback 1937); a song, because of its musical qualities (Schaller 1967) 'that is not low but can be heard a mile away; a long, drawn, resonant, low and high pitched at the same time, not necessarily loud if heard at close quarters, but has the ability to be heard half a mile away' (Krishnan 1972). The sound of the call was described in detail by Schaller (1967): 'a clear, resonant u-u-u-u-u about one to three seconds long, either constant in pitch or slightly rising and falling; this note may be followed by a second one somewhat lower in tone, by a third one still lower, and so forth, giving the impression of someone practicing musical scales; and as many as ten seconds sometimes elapse between the first and last note. When this call is made, it is common to hear similar calls from other bulls.'

Parasites and diseases

The cattle diseases of rinderpest and foot-and-mouth have been long associated with gaur populations in India. They were reported to affect gaur populations in the nineteenth century (Inverarity 1889; Baker 1890), and since then there have been periodic reports of the diseases, particularly rinderpest, devastating gaur populations around the country: Dunbar-Brander (1923) and Stewart (1928) reported gaur in Central India being impacted by rinderpest and foot-and-mouth; a rinderpest epidemic killed many gaur in the Kanha Park area in 1925–1926 (Schaller 1967); Morris (1949) reported shooting an adult male gaur that had foot-and-mouth; Ali (1953) reported a case of foot-and-mouth; and Anderson (1959) reported rinderpest in southern India. A rinderpest epidemic all but wiped out the gaur population in Mudumalai in 1968 (Krishnan 1972), when hundreds of gaur died; Imam (1985) reported rinderpest epidemics that killed several hundred gaur in Raimora, Assam, India, in 1967, in the Rairakhol region, Orissa, India, in 1972 and also in Berbera and Dhuanali in Orissa in 1973. A rinderpest epidemic killed hundreds of gaur in Bhadra in 1989 (D. V. Girish, personal communication); four gayal in Calcutta Zoo contracted foot-and-mouth in October 1990 and died within ten days (Choudhury *et al.* 1992); in March 2007 a gaur bull in Bandipur was found uncharacteristically sitting in full view at the side of a road – the gaur had multiple lesions on its mouth, typical of a foot-and-mouth attack (Ahrestani, personal observation). There has also been one report of gaur dying from anthrax (Peacock 1933).

Only Schaller (1967) has reported on parasites of gaur – a gaur cow that he autopsied contained light infestations of nematodes (*Oesophagostomum radiatum*) in the large intestine and trematodes (*Gastrothylax crumenifer*) in the rumen, and ticks (*Boophilus microplus*) were found on two adults that were checked for ectoparasites.

Status in the wild

The gaur is listed as Vulnerable according to the IUCN's Red List criteria. Populations of gaur in much of its Southeast Asian range are believed to have decreased by over 70% in the last 30 years, while in India and Nepal the decrease has been considerably less. Currently, gaur populations in Southeast Asia continue to decrease, while in parts of India, especially South India, gaur populations are increasing. The overall global gaur population (estimated to be 13 000–30 000) is likely to decrease by 30% over the next 30 years. Besides the inevitable threat of shrinking habitat, poaching for its meat and horns are serious threats for the long-term survival of gaur, especially in Southeast Asia.

Bangladesh: the gaur is probably extinct in Bangladesh; no records have been reported since the 1970s (Khan 1985). It is possible, though, that individuals from Mizoram and Tripura cross over into Bangladesh occasionally (Choudhury 2002).

Bhutan: the gaur is apparently found all along the southern foothill zone, notably in protected areas, and there are recent reports of gaur sightings from outside protected areas.

Cambodia: the gaur was widespread till the 1960s, after which the overall country population decreased by nearly

90%. The largest population can be found in East Cambodia (Mondulkiri Province), and recent protection measures may have stabilized this population (Timmins & Ou Rattanak 2001; Tordoff *et al.* 2005).

China: the gaur occurs in Yunnan and Southeast Tibet, though its status and populations in these regions are unknown. It would be surprising if the overall population in these regions exceeds 100–200; a report from nearly 20 years ago stated that the gaur had been wiped out from much of Yunnan Province (Xiang & Santiapillai 1993).

India: the gaur is found in three widely separated geographic areas in India: the Western Ghats, Central India, and Northeast India. The largest gaur population (3000–5000) in the world is found in the Nilgiri Biosphere, Western Ghats (a complex of five contiguous Tiger Reserves and additional Reserved Forests). Gaur populations in India are secure in the Western Ghats; multiple populations in the region are reputedly increasing. Populations in Central India are less secure, while those in Northeast India are the most vulnerable. The greatest threat to the gaur in India is shrinking habitat, although there are periodic reports of gaur being poached for their meat.

Lao DPR: the national population was estimated to be ~1000 in the 1990s (Byers *et al.* 1995). However, since then multiple populations in the country have been extirpated and it's estimated that the current national population is no more than 500.

Malaysia: it is possible that the country population estimate of ~500 in 1994 has been halved since that time. The outlook for gaur, except for maybe in Taman Negara (peninsular Malaysia's largest National Park), is grim (Lynam *et al.* 2007). The main reason for the decrease is poaching of animals for meat and horns.

Myanmar: the status of gaur in Myanmar is poorly understood, though results from a nationwide camera-trap survey from a decade ago painted a positive picture of its status (Lynam 2003).

Nepal: the national gaur population probably does not exceed 500, is considered to be currently stable and the majority of animals are confined to Chitwan National Park and Parsa Wildlife Reserve.

Thailand: the current outlook for gaur in northern Thailand is favourable. An effective anti-poaching campaign and reforestation programme over the last couple of decades has led to currently increasing multiple gaur populations, most notably in Huai Kha Khaeng and Thung Yai Naresuan Wildlife Sanctuaries and Khao Yai National Park. Currently, southern Thailand has small fragments of forests and therefore it can be assumed that the gaur has largely been lost in this region. However, it may be safe along the Malaysian border, where the human population is low and forest tracts are larger because of an insurgency. The national population could be in excess of 1000 individuals.

Vietnam: the current status of gaur is unknown, and it is thought that the populations that remain are maybe in serious decline. The majority of animals in the country are confined to Cat Tien National Park (Polet & Ling 2004; Nguyen 2009).

Conservation of the gaur

Besides India, Nepal, Thailand and maybe Myanmar, the current outlook on the long-term survival of the gaur in other countries within its geographic range looks bleak. Besides the omnipresent threat of habitat loss to satisfy human needs, poaching for its meat and its horns for trophies is a serious threat. Thailand has recently shown that it's possible to reverse the trend of declining gaur populations that is prevalent in all its Southeast Asian neighbours. Government institutions, and both international and local conservation organizations, continue to combat the threat of poaching in all the Southeast Asian countries using various *in situ* conservation programmes. With the additional threat of climate change also looming large, the future of the gaur, not unlike most of its sympatric ungulates, remains vulnerable.

Status in captivity

Although initial attempts to breed gaur in captivity failed (Forsyth 1889), gaur have been successfully bred in captivity since 1956 (Hintz & Foose 1982). There have been 319 gaur births in European zoos; an equal number have been born in US zoos; and there have been over 50 births in India. Currently, the largest captive population of gaur is located in Omaha Zoo (there is a total population of ~70 housed in two subpopulations); the total captive gaur population in India is ~70. There has been a lot of inbreeding within these gaur populations (for example, the 23 gaur that were present between 2005 and 2007 in Mysore Zoo, India, were all bred from only three founder gaur individuals captured in the wild 15 years ago), which has led to higher juvenile mortality when compared to non-inbred gaur – something that had previously been demonstrated to be higher in inbred versus non-inbred young in 15 out of 16 other species (not including the gaur) of ungulates in zoos (Hintz & Foose 1982). Inbred gaur populations were found to have a male-skewed sex ratio in gaur births (62:38) in contrast to a 46:54 male:female ratio born to gaur that were not inbred (Hintz & Foose 1982). The figures from EEP (EAZA), ISB (WAZA), SEAZA and CZAI breeding programmes are probably an underestimate of all gaur kept globally in captivity (see Chapter 27 for more details).

Acknowledgements

We would like to thank Joe Shepard of Omaha Zoo and Aude Bourgeois of Muséum National d'Histoire Naturelle, France, for their assistance in collecting data on captive gaur, and N. Samba Kumar for assistance in collating the density estimates presented in Table 11.2. We would also like to thank Dr Thopsie Gopal, Animal Health Consultant & Technical Adviser on Wildlife Health, Bangalore, India, for the information on gaur diseases.

References

Ahrestani, F. S. & Prins, H. H. T. (2011). Age and sex determination of gaur *Bos gaurus* (Bovidae). *Mammalia* **75**: 151–155.

Ahrestani, F. S., Iyer, S., Heitkonig, I. M. A. & Prins, H. H. T. (2011). Life-history traits of gaur *Bos gaurus*: a first analysis. *Mammal Review* **41**: 75–84.

Ahrestani, F. S., Heitkonig, I. M. A. & Prins, H. H. T. (2012). Diet and habitat-niche relationships within an assemblage of large herbivores in a seasonal tropical forest. *Journal of Tropical Ecology* **28**: 385–394.

Ali, S. (1953) Hyderabad state. In *The Preservation of Wild Life in India*. Ed. Burton, R. Bangalore: Bangalore Press.

Anderson, K. (1959) The big bull bison of Gedesal. In *The Black Panther of Sivanipalli and Other Adventures of the Indian Jungle*. London: Allen & Unwin.

Ashokkumar, M., Swaminathan, S. & Daniel, J. C. (2004) A study on ecology and conservation of gaur in Mudumalai Wildlife Sanctuary and National Park. Report submitted to Tamil Forest Department (Wildlife Wing), Tamil Nadu, India.

Ashokkumar, M., Nagarajan, R. & Desai, A. A. (2010). Group size and age-sex composition of Asian elephant and gaur in Mudumalai Tiger Reserve, southern India. *Gajah* **32**: 27–34.

Baker, E. C. S. (1903) The gaur and the gayal (*Bos gaurus* and *Bos frontalis*). *Journal of the Bombay Natural History Society* **15**: 227–248.

Baker, S. W. (1890) *Wild Beasts and Their Ways*. London: Macmillan.

Belsare, D. K., Chandiramani, S. C. & Dutta, J. J. (1984). Composition and behaviour of gaur herd in Kanha National Park. *Cheetal* **26**: 28–31.

Bhattacharya, S., Choudhury, A. & Biswas, G. G. (1997). *A Collaborative Study on the Gaur (Bos gaurus H. Smith) in North Bengal, West Bengal*. Calcutta: World Wide Fund for Nature.

Bibi, F. (2007). Origin, paleoecology, and paleobiogeography of early Bovini. *Palaeogeography, Palaeoclimatology, Palaeoecology* **248**: 60–72.

Blackburn, H. V. (1934) A bull gaur (*B. gaurus*) and a tigress fight to death. *Journal of the Bombay Natural History Society* **37**: 950–951.

Buntjer, J. B., Otsen, M., Nijman, I. J., Kuiper, M. T. R. & Lenstra, J. A. (2002). Phylogeny of bovine species based on AFLP fingerprinting. *Heredity* **88**: 46–51.

Byers, O., Hedges, S. & Seal, U. S. (1995). *Asian Wild Cattle Conservation Assessment and Management Plan Workshop*. Apple Valley, MN: IUCN/SSC Conservation Breeding Specialist Group.

Cameron, I. L. (1929). Body measurements of a gaur (*Bibos gaurus*). *Journal of the Bombay Natural History Society* **33**: 983–985.

Chetri, M. (2003). Food habits of gaur (*Bos gaurus*) and livestock (cows and buffaloes) in Parsa Wildlife Sanctuary, central Nepal. *Himalayan Journal of Sciences* **1**: 31–36.

Chetri, M. (2006). Diet analysis of gaur, *Bos gaurus gaurus* (Smith, 1827) by micro-histological analysis of fecal samples in Parsa Wildlife Reserve, Nepal. *Our Nature* **4**: 20–28.

Choudhury, A. (2002). Distribution and conservation of the Gaur *Bos gaurus* in the Indian subcontinent. *Mammal Review* **32**: 199–226.

Choudhury, B., Mazumder, R. & Bhattacharya, A. K. (1992) Foot and mouth disease in gayal (*Bos gaurus frontalis*) in Calcutta Zoo. *Scientific and Technical Review of the Office International des Epizooties* **11**: 797–798.

Conry, P. J. (1989). Gaur (*Bos gaurus*) and development in Malaysia. *Biological Conservation* **49**: 47–65.

Crandall, L. S. (1964). *The Management of Wild Animals in Captivity*. Chicago, IL: Chicago University Press.

Dasgupta, S., Sankar, K. & Gupta, A. K. (2008). Density, group size and sex ratios of gaur (*Bos gaurus* H. Smith) in a sub-tropical semi-evergreen forest of northeast India. *Indian Forester* **134**: 1281–1288.

Duckworth, J. W., Salter, R. E. & Khounbline, K. (1999). *Wildlife in Lao PDR: 1999 Status Report*. Vientiane, Laos: IUCN,.

Dunbar-Brander, A. A. (1923). *Wild Animals in Central India*. London: Edward Arnold.

Foenander, E. C. (1952). *Big Game of Malaya*. London: Batchworth Press.

Forsyth, J. (1889). *The Highlands of Central India*. London: Chapman and Hall.

Gad, S. D. & Shyama, S. K. (2011). Diet composition and quality in Indian bison (*Bos gaurus*) based on fecal analysis. *Zoological Science* **28**: 264–267.

Gee, E. P. (1964). *The Wildlife of India*. London: Fontana.

Gentry, A., Clutton-Brock, J. & Groves, C. P. (2004). The naming of wild animal species and their domestic derivatives. *Journal of Archaeological Science* **31**: 645–651.

Giasuddin, M., Huque, K. S. & Alam, J. (2003) Reproductive potentials of gayal (*Bos frontalis*) under semi-intensive management. *Asian-Australasian Journal of Animal Science* **16**: 331–334.

Grieser Johns, A. (2000). *Pu Mat: A Biodiversity Survey of a Vietnamese Protected Area*. Vinh, Vietnam: Social Forestry and Nature Conservation in Nghe An Province.

Groves, C. P. (2003). Taxonomy of ungulates of the Indian subcontinent. *Journal of the Bombay Natural History Society* **100**: 341–362.

Groves, C. P. & Grubb, P. (2011). *Ungulate Taxonomy*. Baltimore, MD: Johns Hopkins University Press.

Grubb, P. (2005). Artiodactyla. In *Mammal Species of the World: A Taxonomic and Geographic Reference*, pp. 637–722. Eds. Wilson, D. E. & Reeder, D. M. Baltimore, MD: Johns Hopkins University Press.

Guin, D. P. & Pal, B. C. (1982). Seasonal migration of gaur (*Bos gaurus* H.Smith) at Garumara Wildlife Sanctuary, West Bengal, India. *Cheetal* **23**: 6.

Hartl, G. B., Goltenboth, R., Grillitsch, M. & Willing, R. (1988). On the biochemical systematics of the Bovini. *Biochemical Systematics and Ecology* **16**: 576–579.

Hassanin, A. & Douzery, E. J. P. (1999). The tribal radiation of the family Bovidae (Artiodactyla) and the evolution of the mitochondrial cytochrome b gene. *Molecular Phylogenetics and Evolution* **13**: 227–243.

Hassanin, A. & Ropiquet, A. (2004). Molecular phylogeny of the tribe Bovini (Bovidae, Bovinae) and the taxonomic status of the Kouprey, *Bos sauveli* Urbain 1937. *Molecular Phylogenetics and Evolution* **33**: 896–907.

Hintz, R. L. & Foose, T. J. (1982). Inbreeding, mortality and sex ratio in gaur (*Bos gaurus*) under captivity. *The Journal of Heredity* **73**: 297–298.

Hofmann, R. R. & Stewart, D. R. M. (1972). Grazer or browser: a classification based on the stomach-structure and feeding habits of East African ruminants. *Mammalia* **36**: 226–240.

Hubback, T. R. (1937). The Malayan gaur or seladang. *Journal of Mammalogy* **18**: 267–279.

Hundley, H. G. (1952). Measurements of an Indian bison head (*Bibos gaurus*). *Journal of the Bombay Natural History Society* **50**: 933–934.

Imam, A. R. H. B. (1985). Seasonal migrations of gaur (*Bos gaurus*). *Cheetal* **26**: 45–48.

International Commission of Zoological Nomenclature. (2003). Usage of 17 specific names based on wild species which are pre-dated by or contemporary with those based on domestic animals (Lepidoptera, Osteichthyes, Mammalia): conserved. *Bulletin of Zoological Nomenclature* **60**: 81–84.

Inverarity, J. D. (1889). The Indian bison. *Journal of the Bombay Natural History Society* **3**: 294–311.

Ishii, H., Krane, S., Itagaki, Y., Berova, N., Nakanishi, K. & Weldon, P. J. (2004). Absolute configuration of a hydroxyfuranoid acid from the pelage of the genus *Bos*, 18-(6S,9R,10R)-bovidic acid. *Journal of Natural Products* **67**: 1426–1430.

Janecek, L. L., Honeycutt, R. L., Adkins, R. M. & Savis, S. K. (1996). Mitochondrial gene sequences and the molecular systematics of the Artiodactyl subfamily Bovinae. *Molecular Phylogenetics and Evolution* **6**: 107–119.

Jathanna, D., Karanth, K. U. & Johnsingh, A. J. T. (2003). Estimation of large herbivore densities in the tropical forests of southern India using distance sampling. *Journal of Zoology* **261**: 285–290.

Johnsingh, A. J. T. (1983). Large mammalian prey and predators in Bandipur, India. *Journal of the Bombay Natural History Society* **80**: 1–49.

Johnson, J. M. (1986) A fight between bull gaurs in Mudumalai. *Journal of the Bombay Natural History Society* **83**: 659–660.

Karanth, K. U. (1984). Interaction between gaur and tiger in Bhadra Wildlife Sanctuary. *Journal of the Bombay Natural History Society* **81**: 460.

Karanth, K. U. & Sunquist, M. E. (1992). Population structure, density and biomass of large herbivores in the tropical forests of Nagarahole, India. *Journal of Tropical Ecology* **8**: 21–35.

Karanth, K. U. & Sunquist, M. E. (1995). Prey selection by tiger, leopard and dhole in tropical forests. *Journal of Animal Ecology* **64**: 439–450.

Karanth, K. U. & Sunquist, M. E. (2000). Behavioural correlates of predation by tiger (*Panthera tigris*), leopard (*Panthera pardus*) and dhole (*Cuon alpinus*) in Nagarahole, India. *Journal of Zoology* **250**: 255–265.

Khan, M. A. R. (1985). *Future Conservation Priorities for Bangladesh: 25th Working Session of IUCN's Commission on National Parks and Protected Areas* Corbett National Park, India: IUCN.

Krishnan, M. (1972). An ecological survey of the larger mammals of peninsular India. *Journal of the Bombay Natural History Society* **69**: 322–349.

Lydekker, R. (1907). *The Game Animals of India, Burma, Malaya and Tibet.* London: Rowland Ward.

Lynam, A. J. (2003). *A National Tiger Action Plan for the Union of Myanmar.* Yangon, Myanmar: Myanmar Forest Department, Ministry of Forestry.

Lynam, A. J., Laidlaw, R., Noordin, W. S. W., Elagupillay, S. & Bennett, E. L. (2007). Assessing the conservation status of the tiger *Panthera tigris* at priority sites in peninsular Malaysia. *Oryx* **41**: 454–462.

MacEachern, S., McEwan, J. & Goddard, M. (2009). Phylogenetic reconstruction and the identification of ancient polymorphism in the Bovini tribe (Bovidae, Bovinae). *BMC Genomics* **10**: 1–17.

Madhusudan, M. D. (2004). Recovery of wild large herbivores following livestock decline in a tropical Indian wildlife reserve. *Journal of Applied Ecology* **41**: 858–869.

McHugh, T. C. (1958). Social behaviour of the American buffalo (*Bison bison bison*). *Zoologica* **43**: 1–40.

Meinertzhagen, R. (1939). Some weights and measurements of large mammals. *Proceedings of the Zoological Society, London* **108**: 433–440.

Morris, R. C. (1937) Solitary bull bison (*Bibos gaurus* H. Sm.). *Journal of the Bombay Natural History Society* **39**: 617.

Morris, R. C. (1938) On whistling of gaur. *Journal of the Bombay Natural History Society* **40**: 117.

Morris, R. C. (1947). Weight of bull bison. *Journal of the Bombay Natural History Society* **45**: 153.

Morris, R. C. (1949) A diseased gaur. *Journal of the Bombay Natural History Society* **48**: 578.

Morris, R. C. (1952). Cases of unwounded gaur or Indian bison (*Bibos gaurus*) charging. *Journal of the Bombay Natural History Society* **51**: 266.

Morris, R. C. (1953). Gaur attacking man. *Journal of the Bombay Natural History Society* **52**: 204–205.

Mustill, F. J. (1938). Behaviour of gaur or Indian bison (*Bibos gaurus*). *Journal of the Bombay Natural History Society* **40**: 731–733.

Nguyen, M. H. (2009). The status of Vulnerable gaur *Bos gaurus* and Endangered banteng *Bos javanicus* in Ea So Nature Reserve and Yok Don and Cat Tien National Parks, Vietnam. *Oryx* **43**: 129–135.

Ogilvie, C. S. (1954). The behaviour of seladang. *Malay Nature Journal* **9**: 1–10.

Oliver, J. E., Weldon, P. J., Peterson, K. S., Schmidt, W. F. & Debboun, M. (2003) Synthesis and absolute configuration reassignment of 5-(1-hydroxynonyl)-2-tetrahydrofuranpentanoic acid, a mosquito deterrent from gaur. Abstracts of Papers, 225th American Chemical Society National Meeting, New Orleans, Abstract #30.

Pasha, M. K. S., Areendran, G., Sankar, K. & Qureshi, Q. (2002). Debarking of teak *Tectona grandis* Linn. by gaur *Bos gaurus* H. Smith during summer in a tropical dry deciduous habitat of central India. *Journal of the Bombay Natural History Society* **99**: 238–244.

Peacock, E. H. (1933). *A Game-book for Burma and Adjoining Territories.* London: Witherby.

Pillay, B. S. (1952). The record spread of gaur horns (*Bibos gaurus*). *Journal of the Bombay Natural History Society* **50**: 935–936.

Polet, G. & Ling, S. (2004). Protecting mammal diversity: opportunities and constraints for pragmatic conservation management in Cat Tien National Park, Vietnam. *Oryx* **38**: 186.

Prayurasiddhi, T. (1997). The ecological separation between gaur (*Bos gaurus*) and banteng (*Bos javanicus*) in Huai Kha Khaeng Wildlife Sanctuary, Thailand. PhD thesis. University of Minnesota.

Ramesh, T., Sankar, K., Qureshi, Q. & Kalle, R. (2012). Group size and population structure of megaherbivores (gaur *Bos gaurus* and Asian elephant *Elephas maximus*) in a deciduous habitat of Western Ghats, India. *Mammal Study* **37**: 47–54.

Reed, T. (1959) Report on the National Zoological Park for the year ended June 30, 1958. *Smithsonian Institutional Report* **1958**: 140–179.

Rice, C. G. (1986). Observations on predators and prey at Eravikulam National Park, Kerala. *Journal of the Bombay Natural History Society* **83**: 283–305.

Robinson, S. C. H. (1941). Measurements of a Indian bison head (*Bibos gaurus*). *Journal of the Bombay Natural History Society* **43**: 100–101.

Russell, C. E. M. (1900). *Bullet and Shot in Indian Forest, Plain and Hill.* London: W. Thacker and Co.

Sahai, S. P. (1977). *Backs to the Walls: Saga of Wildlife in Bihar, India.* New Delhi: Affiliated East-West Press.

Sanderson, G. P. (1912). *Thirteen Years Among the Wild Beasts of India.* Edinburgh: John Grant.

Sankar, K., Qureshi, Q., Pasha, M. K. S. & Areendran, G. (2002). *Ecology of Gaur (Bos gaurus) in Pench Tiger Reserve, Madhya Pradesh.* Dehradun, India: Wildlife Institute of India.

Schaller, G. B. (1967). *The Deer and the Tiger: A Study of Wildlife in India.* Chicago, IL: University of Chicago Press.

Shukla, R. & Khare, P. K. (1998). Food habits of wild ungulates and their competition with livestock in Pench Wildlife Reserve, central India. *Journal of the Bombay Natural History Society* **95**: 418–421.

Stebbing, E. P. (1911). *Jungle By-ways in India.* London: John Lane.

Steinmetz, R. (2004). Gaur (*Bos gaurus*) and banteng (*B-javanicus*) in the lowland forest mosaic of Xe Pian Protected Area, Lao PDR: abundance, habitat use, and conservation. *Mammalia* **68**: 141–157.

Steinmetz, R., Chutipong, W., Seuaturien, N., Chirngsaard, E. & Khaengkhetkarn, M. (2010). Population recovery patterns of Southeast Asian ungulates after poaching. *Biological Conservation* **143**: 42–51.

Stevens, W. E. (1968). *The Conservation of Wildlife in West Malaysia.* Kuala Lumpur: Department of Wildlife and National Parks.

Stewart, A. E. (1928) *Tiger and Other Game.* London: Longman Green.

Timmins, R. J. & Ou Rattanak. (2001). *The Importance of Phnom Prich Wildlife Sanctuary and Adjacent Areas for the Conservation of Tigers and Other Key Species.* Hanoi, Vietnam and Phonm Penh, Cambodia: WWF.

Tordoff, A. W., Timmins, R. J., Maxwell, A., Keavuth, H., Vuthy, L. & Hourt, K. E. (2005). *Biological Assessment of the Lower Mekong Dry Forests Ecoregion.* Phnom Penh, Cambodia: WWF.

Tran, K. & Chauhan, K. R. (2007). Structural activity of Bovidic acid and related compounds as feeding deterrents against *Aedes aegypti. Biopesticides International* **3**: 53–57.

Tyabji, H. N. (1989). Reaction of a group of gaur (*Bos gaurus*) to the presence of an unidentified predator at Bandhavgarh National Park, Madhya Pradesh. *Journal of the Bombay Natural History Society* **87**: 287–288.

Vairavel, S. M. (1998). Ecology of gaur (*Bos gaurus* H. Smith) with special reference to habitat utilization in Parambikulum Wildlife Sanctuary, Kerala, India. PhD thesis, Forest Research Institute, Dehra Dun.

von Bertalanffy, L. (1938) A quantitative theory of organic growth. *Human Biology* **10**: 181–213.

Ward, R. (1922) *Records of Big Game.* London: Rowland Ward.

Weigum, L. E. (1972). The problems in the preservation of the seladang in the Malaysian National Park. MSc thesis, Michigan State University.

Wilson, D. E. & Mittermeier, R. A. (2011). *Handbook of the Mammals of the World*, Vol. **2**. Barcelona: Lynx Edicions

Xiang, Z. & Santiapillai, C. (1993). Asian Elephant Specialist Group report. *Species* **20**: 54–55.

Yusof, E. (2009). Food preferences of seledang (*Bos gaurus hubbackki*) in Ulu Lepar, Pahang, Peninsular Malaysia. *Journal of Sustainable Development* **2**: 201–207.

Part III

Species accounts

Chapter 12

Wild yak *Bos mutus* (Przewalski, 1883)

Jianlin Han

Names

Genus: *Bos* Linnaeus, 1758

Species: Wild yak *Bos mutus* (Przewalski, 1883)

Names in other languages: **French:** Yack; **German:** Wildyak; **Spanish:** Yak; **Italian:** Yak; **Chinese:** 牦牛; **Tibetan:** Yag

The wild yak *Bos mutus* (Przewalski, 1883) is a bovine species living in the wild of northern Qinghai-Tibetan Plateau (QTP) (Figure 12.1) with a small, vulnerable population, and in a few zoos with very limited numbers of individuals. It has been considered as the direct ancestor of domesticated yak (*Bos grunniens* Linnaeus, 1766), which was originally recognized as a 'grunting bull' due to its deep and short sounds, like a hog. However, since the first establishment of the genus *Bos* for bovids by Linnaeus in 1758 and the assignment of *Bos grunniens* for domestic yak, also by Linnaeus (1766, 1788), the Latin names and taxonomic classifications of both wild and domestic yak had undergone numerous changes under the genera of *Bos*, *Poephagus* or *Bison* based on very limited morphological or skeleton (e.g. skull) evidence, which has thus led to great confusion (see Olsen 1990; Wiener *et al.* 2003; Leslie & Schaller 2009; Han 2014).

Ecology, Evolution and Behaviour of Wild Cattle: Implications for Conservation, ed. M. Melletti and J. Burton. Published by Cambridge University Press.

Figure 12.1 A group of wild yak around Budongquan station (4611 metres above sea level). Photo was taken in Kekexili Nature Reserve, western Qinghai, China in late November 2009 (photo by Damin He). A black and white version of this figure will appear in some formats. For the colour version, please refer to the plate section.

Taxonomy

History of recognition and taxonomic assignment of wild and domestic yak

Griffith *et al.* (1827) introduced some basic morphological differences of yak from the American bison, which was referred to as *Bos americanus* by these authors at the same time, and directly cited the yak as *B. poëphagus* with a simple claim that 'this animal was originally noticed by Aelian under the above name, and since described by Pallas, who preferred as a specific designation *grunniens* or grunting; but it should rather be groaning, as its voice has no similarity with the grunt of a hog'. It is not clear if '*B.*' is an abbreviation of '*Bos*' or '*Bison*', although the yak was included in the 'Bisontine Group'. Soon after, yak was introduced as '*B. Poephagus*' clearly under the 'Sub-genus II – *Bison*', while '*Poëphagus*' of Aelian and '*B. Grunniens*' of Pallas were also referred to (Griffith 1827). Jardine (1836) then cited it as '*Bison poephagus* of Smith' as the taxonomic designation for yak. He also listed it as '*Poephagus* of Elian' and '*Bos Grunniens* of Pallas and Authors', with a simple description of domestic yak alone. The domestic yak (and its skin collections) from Nepal and Tibet, China was first referred to as '*Bos grunniens* Linn.' (Hodgson 1832, 1834) but later changed to *Bisonus* (or *Bison*) *poephagus* based on the skull and skeleton difference or similarity of yak from domestic cattle and Gayal in Nepal or with the American bison. The major difference noticed was that both cattle and Gayal have 13 pairs of ribs, while the yak shares 14 pairs of ribs with the American bison (Hodgson 1841a, 1841b, 1842, 1844).

In reference to *Bos grunniens*, Linn., *Bisonus poephagus* of Hodgson or *Bos poephagus* of Smith for yak, Gray (1843) first assigned *Poephagus grunniens* to the collection of yak in the British Museum. He then argued for this independent genus based on its short and triangular intermaxillaries not reaching to the nasal, in comparison to that of *Bos taurus* with elongated intermaxillaries reaching to the suture between the nasal and cheekbone (Gray 1846a). *Poëphagus grunniens* was later applied to the British Museum's collections with the notice that wild yak from Tibet were much larger than domestic yak and the withers of the wild yak bull were relatively high (Gray 1852, 1872). Gray (1846b, 1863) also used *Bison poephagus* (or *poëphagus*) for the skins of two wild yak and one domestic yak collected from Tibet. Gray (1852) recorded for the first time all the different names for yak, varying from *Bos grunniens*, Linn. to *Bos* (*Bison*) *poëphagus* of Smith, *Bison poëphagus* of Gray, *Poëphagus grunniens* of Gray, *Bisonius poëphagus* of Hodgson, *Bos poëphagus* of Pallas and *Poëphagus* of Aelian. However, Turner (1850), concerned that 'Gray's distinctions, in the extent of the intermaxillary bones upon the sides of the nasal aperture, will not always hold good', thus rejected the adoption of genus of *Poëphagus* but included yak under the genus of *Bison* as *B. grunniens*. Smith (1858) continued the usage of *Bison poephagus* for the yak as a species under the subgenus *Bison*. Jerdon (1874) referred to both wild and domestic yak as *Poephagus grunniens*, L. and provided some body measurements of wild yak which were shot by several British sportsmen at that time. Vasey (1857) introduced domestic yak and also a few early specimen of yak in the British Museum as being of *Bos grunniens*. Sterndale (1884) treated the domestic yak as *Poephagus grunniens* and compared its general description with the common ox. Blanford (1888) followed *Bos grunniens* to introduce both domestic and wild yak.

While retaining the Latin name as *Bos grunniens* for domestic yak at the very beginning (Flower & Lydekker 1891;

Lydekker 1898, 1900), Lydekker (1901) once introduced yak as *Bos (Bison) grunniens*. For the first time Lydekker (1898, 1900, 1907, 1913) introduced the detailed morphology of domestic yak and some basic data on wild yak, e.g. the body weight of a wild yak bull being around 1100–1200 lb. He indicated that

> the 'grunting' practice is strictly confined to the domesticated breeds, the wild yak uttering no such sound. It has, therefore, been proposed to regard the latter as a distinct species, under the name of *mutus*; but this seems a quite unnecessary refinement in nomenclature, and the most that would be justifiable in this direction would be to designate the wild race as *Bos grunniens mutus*. It is true that such a combination of names would involve a contradiction; but such little inconsistencies are not regarded as matters of any importance by modern naturalists
>
> (Lydekker 1900)

At the same time he clearly stated that 'the name *Bos grunniens* was doubtless given by Linnaeus to the domesticated breed of yak, and Prezewalzki has proposed to designate the wild race as *B. mutus*. If the latter term were used at all, it should be employed in a subspecific sense' (Lydekker 1898). However, he realized that it was impossible to distinguish wild yak from its domestic counterpart except by their body sizes, with domestic yak always being much smaller than wild yak as an outcome of the domestication process. Therefore it was unnecessary to define their subspecies status (Lydekker 1898).

Considering the intermediate characters of yak skulls between taurine oxen and true bison, and the presence of 14 pairs of ribs in yak as an essential bisontine character, Lydekker (1898, 1900, 1907) concluded that 'it seems preferable to regard the yak as an aberrant member of the bisontine group, rather than the representative of a group by itself'. This explained why he started in later days to consider the yak under the subgenus '*Poëphagus*' of *Bos* and introduced domestic yak as *Bos (Poëphagus) grunniens* and wild yak as *Bos grunniens mutus* (Lydekker 1907, 1913). Lydekker (1898, 1913) further realized various changes in the Latin names for yak and documented all the assignments of domestic and wild yak as different Latin names from *Bos grunniens*, Linn. (1758, 1766) to *Bos poëphagus* of Pallas (1811) and Smith (1827), *Bos (Bison) poëphagus* of Smith (1827), *Bison poëphagus* of Jardine (1836), *Poëphagus grunniens* of Gray (1843), *Poëphagus mutus* of Prezewalzki (1884, 1887), *Bos (Poëphagus) grunniens* of Huet (1891) and *Bos (Bison) grunniens* of Lydekker (1901).

Apparently, in translation of the original Russian edition of Przewalski's *Third Journey into Central Asia*, several names of the wild yak were proposed, such as *Poëphagus mutus* in the German version, which was thought to have been applied particularly to the wild yak of a specific type locality in its extreme northern limit of distribution in the alpine region of Nan Shan range, where Przewalski crossed (approximately 39° 20′ N, 95° E, located in today's Qilian Mountains in the northeast edge of the QTP), while *Poëphagus grunniens* was used in the English version, which merely referred to 'Tibet' as the type locality (Lydekker 1913), not extending the range farther north than the Kunlun Mountains. The latter was converted into *Poephagus grunniens mutus* (Harper 1940). Allen (1940) considered the domestic yak as a member of the genus *Poëphagus* of Gray, and used *Poëphagus grunniens* (Linnaeus) for wild yak with an argument that 'this does not seem sufficient ground for making the wild yak a subspecies, using Przewalski's name *P. mutus*, until it can be shown that this is different from domesticated yaks as a geographical race. The name *P. mutus* was given by Przewalski to Tibetan wild yaks.' He nevertheless provided details in description and measurements of two wild yak specimens collected from eastern Tibet, China and also expressed for the first time his concern on the issue that 'to what extent these wild individuals represent feral animals that have descended from domesticated stock is not possible to tell'.

Harper (1945) used *Poëphagus grunniens mutus* Przewalski to refer to wild yak which had been suffering from 'severely intensive hunting with modern rifles in a region where game laws are unknown, and its range as well as its numbers have become considerably reduced'. Pilgrim (1947) simply mentioned that 'The yak (*Bos grunniens*) of Tibet is closely related to the oxen, but differs from them in the possession of fourteen pairs of ribs. It has been made the type of a separate genus, *Poëphagus* [wrongly written as *Poëpgagus*]'. Although a subgenus name of yak as *Poephagus* Gray (1843) and a Latin name as *Bos grunniens* Linnaeus (1766) were indicated, Ellerman & Morrison-Scott (1966) first treated the wild and domestic yak as separate subspecies, with the former carrying the Latin name as *Bos grunniens mutus* Przewalski (1883) and the latter as *Bos grunniens grunniens* Linnaeus (1766). Groves (1971) seemed to have misunderstood Ellerman & Morrison-Scott (1966), who did not 'make the wild yak a sub-species of the domestic yak' but rather followed the reporting process with the domestic yak earlier than its wild ancestor. It was in fact Lydekker (1898, 1900) who first proposed this concept of classification for wild yak as a subspecies of domestic yak (see earlier). However, Groves (1971) first 'referred to as *Bos mutus* Przewalski, 1873, this being the earliest name applied to a wild yak – whereas the domestic yak has been known as *Bos grunniens* Linnaeus, 1758', which was nonetheless incorrectly cited from Linnaeus, 1766. Groves (1981) examined 30 characters of skulls of Bovini and Boselaphini preserved in the British Museum and Paris Museum, and proposed a phylogeny of living Bovini including yak (*Bos mutus*) and bison (*Bos bison*) together, based on their special kinship.

Feng *et al.* (1984) first applied '*Poëphagus mutus* Przewalski' to the wild yak in Tibet and later they suggested yak and bison as separate species to be included in the same genus in parallel to the genus *Bos* under the tribe Bovini (Feng *et al.* 1986). Olsen (1990) reviewed the historical changes in Latin names for yak and suggested a 'return' to use '*Phoephagus grunniens* (Linnaeus)' for yak based on his comprehensive comparison of yak skulls against other related bovids, including *Bos* and *Bison*. In fact, this name is not accurate because the genus '*Phoephagus*' was first assigned by Gray (1843) while the species '*grunniens*' was named by Linnaeus (1766, 1788) based on the limited specimen of domestic yak. Olsen (1990) nevertheless questioned whether the fragmented bovid fossils found in a few localities in Nepal, India, China and Russia may relate to the ancestry of yak. Leslie & Schaller (2009) recently provided a detailed documentation of the evolving taxonomy of yak in general and the domestic yak as well as wild yak in particular based on historical literature in English, French, German, Russian and Swedish, while the author of this chapter only has a capacity to retrieve all relevant literature in English.

Except for the issues in translation of common and scientific names of yak, either domestic or wild, between different languages, it is clear that the intermediate morphological and anatomical structures of yak between *Bos* and *Bison* species have been the considerations for proposing all different taxonomic placements over the past two-and-a-half centuries. In general, yak and *Bison* species share a long, hairy coat, 14 dorsal vertebrae, 14 pairs of ribs and also the same arrangement of premaxillaries, maxillaries and nasals (see figure 3 in Olsen 1990), but yak's dorsal margin of the maxilla has an intervening extension which prevents the nasal from reaching the premaxillae (see figure 4 in Olsen 1990). Compared to *Bos* species, yak has one more dorsal vertebra and one more pair of ribs, but one fewer lumbar and four to zero fewer coccygeal vertebrae. On the other hand, the nasals of *Bos* species are elongated to reach the premaxillae. These different arrangements in skull structures are also associated to relevant facial muscles (Olsen 1990, 1991; Wiener *et al.* 2003; Leslie & Schaller 2009). The historical hybridization of domestic yak with local cattle (*Bos taurus* or *Bos indicus*) practised since the Yin Dynasty in China (approximately 1100 BC) (Cai 1980, 1989; Zhang 1989, 2000; Qi *et al.* 2010) may further complicate this matter.

Molecular taxonomy and nomenclature of yak

Yak is placed in the order Cetartiodactyla, suborder Ruminantia, family Bovidae, subfamily Bovinae, tribe Bovini, genus *Bos*, in which five other living species of domestic gayal or mithan (*Bos frontalis* Lambert, 1804) and/or wild gaur (*B. gaurus* Smith, 1827), banteng (*Bos javanicus* d'Alton, 1823) and kouprey (*Bos sauveli* Urbain, 1937), as well as the extinct aurochs (*Bos primigenius* Bojanus, 1827) are included (Grubb 2005; IUCN 2013). However, a matrix of 57 characters (mainly cranial) and 32 taxa of fossil and living Bovini (Geraads 1992), biochemical polymorphisms at 15 blood protein loci (Hartl *et al.* 1988), partial mitochondrial DNA (mtDNA) sequences (Miyamoto *et al.* 1989; Kraus *et al.* 1992; Janecek *et al.* 1996; Ward *et al.* 1999; Jacoby & Fonseca 2000; Guo *et al.* 2006a; Q. Li *et al.* 2006a, 2006b, 2008; Finlay *et al.* 2007; Gu *et al.* 2007; Yang *et al.* 2009; Xie *et al.* 2010), Y-chromosome-specific PCR-RFLP markers (Verkaar *et al.* 2003), 15 endonuclease site mapping of nuclear ribosomal DNA (Wall *et al.* 1992), 20 autosomal microsatellite DNA markers (Ritz 1997; Ritz *et al.* 2000), AFLP fingerprinting for nuclear DNA (Buntjer 1997; Buntjer *et al.* 2002), sequences of single nuclear functional genes (Pitra *et al.* 1997; Fan *et al.* 2000; Ma *et al.* 2007, 2012; Ma 2008; Xie *et al.* 2010; Bai *et al.* 2011), as well as combined mtDNA, Y-chromosome-specific and nuclear DNA sequencing data (Hassanin & Douzery 1999; Hassanin & Ropiquet 2004, 2007; Verkaar *et al.* 2004) reached controversial conclusions, and therefore did not resolve the taxonomic position of yak (Lenstra & Bradley 1999; Han 2003; Li *et al.* 2006b).

Apart from the uncertainty of taxonomic placement of yak in general, there have also been specific cases of wrong usage of the scientific name given to wild yak for domestic yak, e.g. in Bohlken (1958), Hartl *et al.* (1988) and Pitra *et al.* (1997), or mixed names for domestic and wild yak, e.g. *Bos grunniens* in the early studies of Hassanin & Douzery (1999), Hassanin & Ropiquet (2004, 2007) and Hassanin *et al.* (2012), but *Bos mutus* throughout their latest publication (Hassanin *et al.* 2013). Schreiber *et al.* (1999) wrongly applied *Bos grunniens mutus* to the partial mtDNA cytochrome b gene sequence (GenBank accession no. Y16063) derived from a domestic yak collected from Berlin Zoo but *Bos* (*Poephagus*) *mutus* f. grunniens was used in their paper. Such poor scientific practice can certainly lead to significant confusion when molecular data of unknown genetic background/ancestry or unidentified sources of 'yak' samples are reported.

Based on the review for problems in scientific naming of domestic animals (Groves 1971, 1995) and the proposal to use *Poephagus* (currently *Bos*) *mutus* Przewalski, 1883 for wild yak from Gentry *et al.* (1996) to the International Commission on Zoological Nomenclature, a ruling was eventually published by the Commission with wild yak being assigned as *Bos mutus* (Przewalski, 1883) and domestic yak as *Bos grunniens* Linnaeus, 1766, both seem to have been treated as independent species under the genus *Bos* (International Commission on Zoological Nomenclature 2003; Gentry *et al.* 2004). Nevertheless, no deep explanation or scientific support was available to verify this ruling, though it intended to clarify issues on the nomenclature of wild species which have a domesticated counterpart. It is well known that female hybrids of Bovini species are all fertile, as are the male hybrids from the more recently diverged taurine cattle (*Bos taurus*) and zebu (*Bos indicus*), gaur (*Bos gaurus*) and gayal (*Bos frontalis*), and American bison (*Bison bison*) and European bison (*Bison bonasus*), respectively (Lenstra & Bradley 1999). The hybridization of yak with all other Bovini species produces fertile F1 females but sterile F1 males; therefore yak is certainly justified for a separate genus, and thus *Poephagus* genus may likely hold true. The following general key is considered to differentiate yak from other *Bos* species in colour, pattern and length of pelage, horn characteristics and morphology: long skirts of hair on the chest, flanks and rump; tail fully haired and horse-like; 14 dorsal and five lumbar vertebrae; and 14 pairs of ribs (Leslie & Schaller 2009).

In consideration of the shared mtDNA haplotypes (Schaller 1998), Grubb (2005) thought that 'it may still be valid for those who consider *B. grunniens* and *B. mutus* to be con-specific to employ the senior name for the name of the species', therefore he also provisionally treated the wild yak *B. grunniens mutus* as a subspecies of domestic yak *B. grunniens*. This was misunderstood by Zhang *et al.* (2009), who took *B. g. grunniens* for wild yak but *B. g. mutus* for domestic yak, though both were considered as subspecies of *B. grunniens*. Recently a Y-chromosome phylogeny based on 4685 nucleotides of *DBY*, *ZFY* and *SRY* genes clearly linked domestic yak to *Bison*, which challenged the commonly accepted monophyletic genus *Bos* and suggested a separate genus *Poephagus* as the designation of yak in general and *Poephagus grunniens* for domestic yak and *Poephagus mutus* for wild yak in particular (Nijman *et al.* 2008). MacEachern *et al.* (2009) sequenced 15 complete or partial autosomal genes around 30 000 bp with 1786 segregating sites and realized that the American bison and domestic yak formed one separate but ambiguous lineage that was attributed to the lineage sorting of ancient polymorphisms at 133 polymorphic sites, leading to ambiguous inheritance in Bovini tribe. Janecek *et al.* (1996) and Verkaar *et al.* (2004) also proposed the lineage sorting to explain the anomalous divergence for partial Bovini mtDNA sequences. Decker *et al.* (2009) applied the genome-wide single nucleotide polymorphism (SNP) chip developed initially for taurine cattle with 54 693 SNP loci to genotype 678 individuals of 61 Pecoran species, including domestic yak. Genotypes of 40 483 scored SNPs were used to reconstruct a completely bifurcating tree with highly supported nodes for all the species, and the only two yak individuals included in the study first joined the cluster including American bison (*Bison bison*) and European bison (*Bison bonasus*), but clearly separated from gaur, banteng, taurine and zebu cattle.

Hassanin *et al.* (2012) and Zeyland *et al.* (2012) analysed 14 902 bp and 15 419 bp coding sequences of complete mtDNA (excluding the non-coding D-loop region) of domestic yak, American and European bison, together with other bovine complete mtDNA reference sequences, respectively. The results revealed a first clustering of yak with American bison. It was believed that the lineage-sorting hypothesis cannot readily interpret the high number of shared European bison/American bison/yak mutations; therefore mitochondrial transfer, most likely from yak to bison following ancient hybridization between yak and American bison as well as subsequent backcrossing (Janecek *et al.* 1996; Hassanin & Ropiquet 2004, 2007; Verkaar *et al.* 2004), was postulated to have generated their close mtDNA relationship (Zeyland *et al.* 2012). Hassanin *et al.* (2012) further suggested that *Bos* and *Bison* genera were to be para- or polyphyletic, while the non-robust intertribal relationships within Bovinae were due to their rapid diversifications.

Hassanin *et al.* (2013) analysed 13 095 nucleotides of 18 autosomal loci, mostly located on different bovine chromosomes, for phylogenetic reconstruction of 17 taxa of the tribe Bovini, including domestic yak. A paraphyletic pattern was observed again for the genus *Bos* due to a highly associated sister-group relationship between yak and bison, but nine out of the 18 loci strongly supported the genus *Bison* to be monophyletic, similar to what was detected in AFLP fingerprinting for nuclear DNA (Buntjer 1997; Buntjer *et al.* 2002) and Y-chromosome sequences (Verkaar *et al.* 2004; Nijman *et al.* 2008). Re-analyses of the data sets of Y-chromosome sequences from Nijman *et al.* (2008) and other autosomal gene sequences from MacEachern *et al.* (2009) did not show a strong signal for the position of both bison and yak. In consideration of the failure in using mtDNA sequencing information for a clear taxonomic signal due to maternal introgression and higher levels of homoplasy among the Bovini species, as well as the agreement of the latest genetic data with morphological studies (Groves 1981; Olsen 1990; Geraads 1992), Hassanin *et al.* (2013) concluded that the genus *Bison* should be regarded as a synonym of *Bos*, with the European bison relegated to a subspecies rank within *Bos bison* as *B. bison bison*. If this proposal holds true, wild yak could only be justified as a subspecies in *Bos grunniens mutus* of domestic yak *Bos grunniens*.

It is interesting to note that the English word 'yak' was derived from Tibetan གཡག་, which is translated into 'yag' or 'gyag' following the Wylie scheme or 'yak' in phonetics for a domestic yak bull, while a wild yak bull is called འབྲོང་།(Drong in phonetics), a wild yak cow འབྲོང་མོ་(drong mo) and a wild yak calf འབྲོང་ཕྲུག་(drong truk) in Tibetan. There are also specific and rich Tibetan scripts for different sexes and age groups of domestic yak, as well as hybrids between domestic yak and taurine or zebu cattle, an indication of the unique and historical interaction and association of Tibetan people with yak and yak husbandry. It is clear that 'yak' is simply used for both sexes of all age groups in other languages which have borrowed this word. The historical changes in Chinese names and scripts since 3000 BC for both domestic and wild yak have been documented by Wiener *et al.* (2003). The wild yak was recognized along with its domestication several thousand years ago. The first description of wild yak was recorded in the famous 'Compendium of Materia Medica' written in 1578 (http://en.wikipedia.org/wiki/Bencao_Gangmu) by Li Shizhen (1596), as follows: 'In the south-western area around the yak country, Li Niu [犛牛 in Chinese script denoted wild yak at that time; Li 1986] lives in the high mountains. Its appearance, hair colour and tail are the same as those of the domestic yak, but its body is larger. The coat colour of wild yak is black and its horns are very long.'

Typical specimen and genetic research of wild yak

Apart from the partial specimen in skins and skulls of wild yak collected from Ladakh in India and Tibet in China, and kept in the British Museum (Lydekker 1913), the first two specimen of complete bodies of wild yak, both males of ages 3 and 4.5 years and body weights of 450 kg and 500 kg, respectively, were collected from Tuole Yematan in western Qilian Mountains in August 1962, and are now preserved in the State Zoological

Museum of China in Beijing and the Northwest Institute of Plateau Biology, Chinese Academy of Sciences in Xining, Qinghai Province, China (Wang & Wang 1963).

The research progress on molecular and cytogenetics of both domestic and wild yak made in the last quarter of the twentieth century was intensively reviewed by the author (Han 2003). The chromosomes of wild yak were investigated and compared with those of domestic yak. They shared the same karyotype with 29 pairs of acrocentric autosomes and one pair of sex chromosomes of XX/XY, with sub-metacentromeres, but no obvious structural difference was detected between them (Guo *et al.* 1991; Li *et al.* 1998). The basic biochemical and physiological profiles in blood of wild yak were well documented (Li *et al.* 1997; Xu & Xu 1998; Wang 2001). Higher total and specific activities of serum lactate dehydrogenase were detected in wild yak than domestic yak, yak–cattle hybrids and taurine cattle, implying a stronger adaptability of wild yak (Xu & Li 1991). A wider active boundary of pH value but one band less in serum α-amylase was observed in wild yak than in domestic yak (Xu *et al.* 1995).

Guo *et al.* (2006b) first sequenced 637 nucleotides in a partial mtDNA D-loop of 13 wild yak (GenBank accession no. DQ139202-DG139214 carrying a scientific name of *Bos grunniens*), of which eight were sampled from western Qinghai and five from Datong Yak Farm and the museum of Northwest Plateau Institute of Biology, the Chinese Academy of Sciences, which were captured or killed on the QTP but lack precise details of the locations of the captures or kills. Ma *et al.* (2009, 2010) reported the genetic variation in the same D-loop fragment of six wild yak (FJ548840-FJ548845, carrying a scientific name of *Bos grunniens mutus*) from Datong Yak Farm. Another 34 wild yak samples were collected from Kekexili in western Qinghai, of which 15 animals were derived from one location among 100 dead yak following a snow disaster, while the remaining 19 samples were from dead yak scattered in central Kekexili (Wang *et al.* 2010). The same partial D-loop fragment was sequenced from all 34 animals (GQ464116-GQ464149, carrying a scientific name of *Bos grunniens*) and 21 complete mtDNA sequences (GQ464246-GQ464266, carrying a scientific name of *Bos grunniens*) of wild yak were also generated by the same group for depth analyses, together with a large number of domestic yak (Wang *et al.* 2011). There are also another two complete D-loop sequences (AY722118 and AY749414) of wild yak carrying a scientific name of *Bos grunniens mutus* and sampled from Xinjiang, China. All these sequences were compared with homologous domestic yak sequences (Mipam *et al.* 2012). Great caution must be taken regarding the different scientific names of these four sets of wild yak samples, because the last two sequences have already been wrongly treated as domestic yak from Xinjiang, China by Lai *et al.* (2007) and Wang *et al.* (2010).

A joint analysis of these 55 sequences identifies two insertions/deletions (gaps) and 53 polymorphic sites, including 11 singletons and 42 parsimony informative sites, which define 32 haplotypes (including gaps). The haplotype diversity (Hd) is 0.961 ± 0.012, and nucleotide diversity (Pi) is 0.022 ± 0.001 (excluding gaps), among these 55 D-loop sequences in 637 nucleotides (Figure 12.2), which are higher than in most of the domestic yak populations (Guo *et al.* 2006b; Lai *et al.* 2007; Qi *et al.* 2008; Wang *et al.* 2010). There are only ten shared haplotypes with sequences ranging from two to six. A median joining network (Bandelt *et al.* 1999) is constructed to illustrate their phylogenetic relationship (Figure 12.3). Because 49 out of these 55 sequences were already included in the comparative analysis with domestic yak (Wang *et al.* 2010), the general pattern of this network overlays with all major haplogroups of domestic yak and around 60% of these haplotypes fall within domestic yak lineages. The sequences of six captive wild yak sampled from Datong Yak Farm (Ma *et al.* 2009, 2010) are located in all major haplotypes along with other wild yak. Most of these haplotypes are specific to wild yak, while there is no indication of population expansion or bottleneck events.

The levels of divergence among the haplotypes of wild yak are deeper and more complicated than that of domestic yak. Among the 21 complete mtDNA genome sequences of wild yak, there are 17 haplotypes with only two shared with domestic yak. There is also a higher genetic diversity in wild yak than domestic yak in the complete mtDNA genome (Wang *et al.* 2010). These data suggest that the extant wild yak populations harbour very rich genetic diversity, which is believed to have survived in the refugee wild yak populations over the last glacial period and have been subsequently recruited into today's domestic yak gene pool (Guo *et al.* 2006b; Qi *et al.* 2008; Wang *et al.* 2010).

It seems that the power of mtDNA sequences, either in specific fragments or complete genome, is relatively limited in resolving the Bovini phylogeny due not only to its tiny genome following maternal inheritance, but also the possibility of female introgression and incomplete lineage sorting of ancestral polymorphisms leading to unreliable phylogeny (Hassanin *et al.* 2013). Recently, large sequencing data sets of multiple nuclear genes have been applied to reconstruct the phylogeny of tribe Bovini (MacEachern *et al.* 2009; Hassanin *et al.* 2013); however, these studies only included one or two yak samples which were collected from the United States or Europe with unknown history or unreliable genetic ancestry. For example, Hassanin *et al.* (2013) argued about the genetic purity of a 'wild yak' as one of two yak samples included in the study of MacEachern *et al.* (2009). This sample was collected from Kansas City Zoo, which keeps several so-called Tibetan yak, and they are in fact domestic yak. Nevertheless, Ward *et al.* (1999) identified one of the three yak in this zoo as having domestic cattle mtDNA. Decker *et al.* (2009) demonstrated a high-throughput phylogenomics approach to resolve the phylogeny of ruminants using current Illumina BovineSNP50 v2 BeadChip; however, apart from the issue of ascertainment biases, only two yak sampled in the United States were included in their genotyping. As is shown in the comprehensive mtDNA analysis, both wild and domestic yak carry a rather complicated and deeply divergent genetic background.

Haplotype	Polymorphic sites / sequence	Frequency
	3455678811111111122222222222222222222222233333444444455	
	8445194601335669990011222345555556677778800334333448879	
	57288352340118068700345891302383601562189560314	
DQ139202	ATACGCAG-TATGAATCCAAGGGACGCATCACTGTCCACCGTTAATCTTTTGATC	1
DQ139203	..T.AT.A-..CAT...T...AA.T.......C.C.T.T........CC..A...	1
DQ139204, DQ139210 GQ464134, GQ464141 GQ464143, GQ464144	....AT.A-..CA....T...A..T.......C.C.T.T........CC..A...	6
DQ139205, GQ464136 FJ548840, FJ548842	..T.AT.A-..CAT...T...A..T.......C.C.T.T........CC..A...	4
DQ139206	........A.G.A.G.T.G..AA.TA..CT..CACT.....C.GG..CCC.AG..	1
DQ139207, GQ464117	........-.G.A.G.T.G.AAA.TAT.CT..CACTT.T..C.GG..CCC.AG..	2
DQ139208	........-.G.A.G...G..AA.TA..CT...ACTT....C..G..CCC.AG..	1
DQ139209, DQ139213 GQ464119, GQ464120 GQ464135, FJ548841	........A.G.A.G.T.GG.AA.TAT.CT..CACTT....C.GG..CCC.AG..	6
DQ139211, DQ139212	.C......A.G.A.G...GG.AA.TAT.CT..CA.TT....C.GG..CCC.AG..	2
DQ139214	........-.G.A.G.T.G.AAA.TA..CT..CACTT..T....GC.CCC.AG..	1
GQ464116	..T.AT.A-..CAT...T...A..T.........C.T.T........CC..A...	1
GQ464118	........-.G.A.G.T.G.AAA.TATGCT..CACTT.T..C.GG..CCC.AG..	1
GQ464121	..T.AT.A-..CAT...T...A..T.......C.C.T..........CC..A...	1
GQ464122, GQ464124	........-.G.A.G.T.G..AAGTAT.CT..CACTT.T..C.GG..CCC.AG..	2
GQ464123	........-.G.A.GCT.G.AAA.TA..CT..CACTT....C.GGC.CCC.AG..	1
GQ464125, GQ464126 GQ464128, FJ548843	.....T.A-..CA.G.....A..T.....G.C.C.T.T..C.....CC..A...	4
GQ464127, GQ464129	..T.AT.A-..CAT...T...A..T.......C.C.T.T.......TCC..A...	2
GQ464130, GQ464131 GQ464149	........-.G.A.G.T.G..AA.TA..CT..CACTTG...C.GG..CCC.AG..	3
GQ464132	........-.G.A.G.T.G..AA.TA..CT..CACTTG..AC.GG..CCC.AG..	1
GQ464133	........-...A...G	1
GQ464137	........-.G.A.G.T.G.AAA.TA..CT..CACTT....C.GGC.CCC.AG..	1
GQ464138	.......-.A.G.A.G.T.GG.AA.TAT.CT..CACTT....C.GG..CCC.AG..	1
GQ464139	.......-.-.G.A.G.T.G.AAA.TA..CT...ACTT....CCGGC.CCC.AG..	1
GQ464140	........-.G.A.G.T.G.AAA.TA..CT..CACTT....C.GG..CCC.AG..	1
GQ464142	...T....A.G.A.G.T.GG.AA.TAT.CT..CACTT....C.GG..CCC.AG..	1
GQ464145	........-.G.A.G.T.G.AAA.TA.GCT..CACTT....C..GC.CCC.AG..	1
GQ464146, FJ548845	........-.G.A.G.T.G..AA.TA..CT..CACT.....C.GG..CCC.AG..	2
GQ464147	........-.G.A.G.T.G..AA.TA..CT.TCACTT....C.GG..C.C.AG..	1
GQ464148	........-.GCA.G.T.GG.AA.TAT.CT..CACTT....C.GG..CCC.AG..	1
FJ548844	........A.G.A.G.T.GG.AA.TAT.CT.TCACTT....C.GG..CCC.AG..	1
AY722118	G...AT.A-A.CA....T...A..T.......C.C.T.T........CC.CA.C.	1
AY749414	G...AT.A-AGCA.G..T...A..T.......C.C.T.T.....G..CC.CA.C.	1

Figure 12.2 Thirty-two haplotypes and their frequencies (last column) defined by 55 polymorphic sites (top three lines) in 617 nucleotides of 55 partial wild yak mtDNA D-loop sequences. '.' represents identical nucleotides to DQ139202 while '-' stands for gaps.

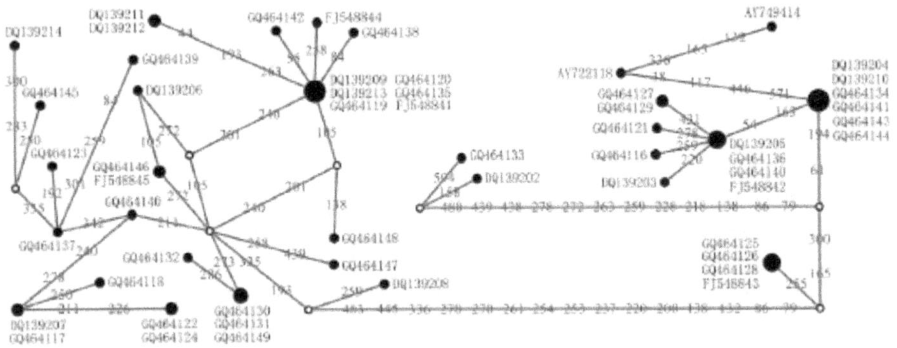

Figure 12.3 A median joining network of the 32 haplotypes identified from 55 wild yak mtDNA D-loop sequences. Black circles are 32 haplotypes with their sizes in proportion to the number of shared sequences. Empty circles represent the median vectors. Numbers linking every two haplotypes are specific polymorphic sites (including two gaps).

Only a population genomics study involving a good number of wild and domestic yak samples from its homeland and covering its geographic scope can eventually contribute to a deeper understanding of the complete Bovini phylogeny in general, and the evolutionary relationship between wild and domestic yak in particular. The draft genome assembly of a female domestic yak and some associated genomic DNA sequences of a wild yak (Qiu *et al.* 2012) are expected to pave the way in this direction.

Subspecies and distribution

There is only a single species of wild yak, which was assigned the scientific name of *Bos mutus* (Przewalski, 1883) by the International Commission on Zoological Nomenclature in 2003. Blanford (1888) and Lydekker (1898, 1900, 1913) first indicated that wild yak were restricted to the QTP, ranging from the eastern part of Ladakh, India as far as Gansu in northwest China, and extending northwards as far as the chain of the Kunlun Mountains. In summer they were found at elevations between 4200 and 6100 metres, and even in winter it was probable that, in Ladakh at least, they seldom, if ever, descended much below 3900 metres. Zhao & Gao (1991) once specified the distribution range of wild yak in Hindu Kush Himalayan region covering Nepal, Bhutan and India and on the QTP in China with a geographic scope between 78°36′ and 102° E and between 27°30′ and 40° N. Piao *et al.* (1999) reported a northward contraction of the range to 29° N. The wild yak is believed to have lived in much wider geographic areas and also at lower elevations, down to 3200 metres before

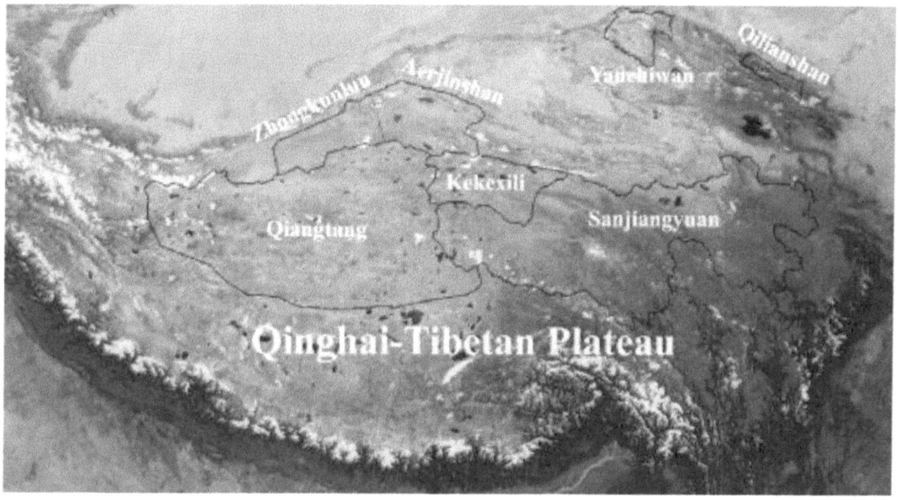

Figure 12.4 Geographic localities and relative sizes of the six major nature reserves where the majority of wild yak are distributed in China. A black and white version of this figure will appear in some formats. For the colour version, please refer to the plate section.

the rangelands were cultivated in the recent past (Feng 1990; Lu *et al.* 1993; Miller *et al.* 1994; Schaller & Liu 1996; Piao *et al.* 1999). For example, wild yak were once recorded in Shiqu in northwestern Sichuan, but had died out – probably by the 1980s (Piao *et al.* 1999; Feng 2001; Zhang *et al.* 2009). Under increasing pressure of ever-growing human activities and illegal hunting, in particular in the last century, the distribution range of wild yak has been shrinking by at least 50% and is now restricted to today's core areas at higher elevations (>4500 metres) in Qiangtang in northern Tibet, Kekexili in western Qinghai and inaccessible regions of Kunlun Mountains in southern Xinjiang (see figure 1 in Lu *et al.* 1993; figure 2 in Schaller & Liu 1996; figure 1 in Piao *et al.* 1999; figure 2 in Leslie & Schaller 2009; and figure 1 in Zhang *et al.* 2009). A few small and fragmented populations also survive in isolated areas in the Aerjin and Qilian Mountains adjoining northeastern QTP, as well as in western Tibet and southern Qinghai, China (Feng 1990; Lu *et al.* 1993; Schaller & Liu 1996; Piao *et al.* 1999; Liu & Schaller 2003; Wiener *et al.* 2003; Liu 2007a, 2007b; Leslie & Schaller 2009; Zhang *et al.* 2009). A very small number (<100) of wild yak can be found in Ladakh area in India and in northern Nepal, probably following seasonal movements across borders between China and India, as well as between China and Nepal (Fox *et al.* 1991, 1994; Miller *et al.* 1994; Schaller & Liu 1996; Liu 2007a, 2007b; Leslie & Schaller 2009). Nonetheless, Namgail (2009) made an estimate of 200 wild yak in Ladakh, India based on secondary information.

The estimated number of wild yak was around 15 000–20 000 in the early 1990s (Schaller & Liu 1996; Piao *et al.* 1999; Liu & Schaller 2003; Wiener *et al.* 2003; Liu 2007a, 2007b; Leslie & Schaller 2009), of which 7959 ± 802 were in Tibet, 4092 in Qinghai, 3031 in Xinjiang and 103 in Gansu in the late 1980s (Piao *et al.* 1999). A survey early this century reported an estimated 19 740 ± 3952 wild yak in China, of which 9050 ± 1840 were found in Tibet, 8400 ± 1406 in Qinghai, 1850 ± 410 in Qinghai, 350 ± 91 in Gansu and 90 ± 25 in Sichuan (Yang *et al.* 2006; Yao *et al.* 2006; Guo *et al.* 2007). Feng (1990) overestimated the number of wild yak in Kunlun Mountains in southern Xinjiang alone as 35 000 in the late 1980s. Lu *et al.* (1993) believed the number of wild yak in China could be between 20 000 and 40 000, while Zhang *et al.* (2009) recently reported a similar estimate between 24 000 and 30 000. It has been recognized that protection efforts in the last two to three decades have led to a positive response in population recovery of wild yak in some national or provincial nature reserves (Figure 12.4) responsible for and/or involved in the protection of wild yak from illegal hunting in China (Harris & Loggers 2004; Harris *et al.* 1999, 2005; Leslie & Schaller 2009; Zhang *et al.* 2009). The worldwide distribution range of wild yak is shown in Plate 4.

Qiangtang (Changtang) Nature Reserve (298 000 km^2; 79° 42′–92°59′ E, 32°10′–36°32′ N; established in 1993 in Tibet and upgraded in 2000 to the second largest national nature reserve in China; V in the IUCN categories, meaning a Protected Landscape/Seascape): located in north-central Tibet, between Kunlun and Kekexili Mountains in the north and Gangdisi and Nianqingtanggula Mountains in the south. Wild yak are mostly found at elevations between 5000 to 6000 metres in this reserve. There were around 7000–7500 wild yak in the 1990s in this reserve (Liu & Schaller 2003; Liu 2007a, 2007b). The latest estimation is slightly more than 10 000 wild yak. Some wild yak with a light golden-brown coat colour are found in this reserve (Lu *et al.* 1993; Liu & Schaller 2003; Yang *et al.* 2006; Liu 2007a, 2007b) and the current number of this mutant variety is estimated at close to 200 (Figure 12.5). Special attention should be given to this rare and unique genotype of wild yak.

Kekexili (Kokoxili or Koh Xil) Nature Reserve (45 000 km^2; 89°25′–94°05′ E, 34°19′ – 36°16′ N; a core area 25 500 km^2 in 89°30′–92°30′ E, 34°40′– 35°50′ N; established in 1995 in Qinghai and upgraded in 1997 to national-level status; V in the IUCN categories): located between Kunlun Mountains in the north and Wulanwula Mountains in the south in Yushu of southwestern Qinghai, bordering Tibet in the west and Xinjiang

Figure 12.5 A wild yak bull with an atypical light coat in Qiangtang Nature Reserve. Photo was taken in Ritu, Tibet, China in March 2010 (photo by Yong Wang). A black and white version of this figure will appear in some formats. For the colour version, please refer to the plate section.

in the north. The average elevation is round 4600 metres. There were 9416 ± 1832 wild yak in 2003 in Qinghai, most of them found in this reserve. This reserve is considered as a refuge for a substantial wild yak population (Schaller *et al.* 2007). Harris & Loggers (2004) and Harris *et al.* (2005) estimated an increase of wild yak from 1200–1300 in 1991–1997 (Harris & Miller 1995; Harris *et al.* 1996; Harris *et al.* 1999) to 1700 in 2002 in the famous Yeniugou valley (93°10′–94°40′ E, 35°45′–35°55′ N), which is located immediately northeast to this reserve. It is interesting to note that Yeniugou, meaning 'Wild Yak Valley' in Chinese, is within Kunlun Mountains in Haixi Mongolian Autonomous Prefecture in western Qinghai and is not included within any reserve (Miller & Bedunah 1994); therefore wildlife in the valley have only been protected by the national wildlife law for the last three decades. Zhang *et al.* (2009) proposed establishing a nature reserve to cover the Yeniugou area.

Sanjiangyuan Nature Reserve (152 300 km^2; 31°39′–36°12′ E, 89°45′–102°23′ N; newly established in 2000 in Qinghai and upgraded in 2003 to the largest national nature reserve in China; V in the IUCN categories): located in southern Qinghai, east of Qiangtang Reserve. There are 18 protection zones, each having core, buffer and experimental areas. The total core areas are 31 200 km^2, buffer areas 39 200 km^2 and experimental areas 81 900 km^2. Average elevations are around 4400 metres. Three zones are dedicated to wildlife protection, including Suojia-Qumahe in Zhiduo County, Jiangxi in Yushu County and Baiza in Nangqian County of Yushu Tibetan Autonomous Prefecture. The number of wild yak was estimated to have increased to around 2000 in Qumahe area alone.

Aerjinshan Nature Reserve (45 000 km^2; 87°10′–91°18′ E, 36°00′–37°49′ N; established in 1983 in Xinjiang and upgraded in 1985 to the first national nature reserve in China; once the largest of its kind in Asia; V in the IUCN categories): located between the Aerjin Mountains in the north and east section of Kunlun Mountains in the south in Ruoqiang County of southeastern Xinjiang. It borders Qiangtang Reserve in Tibet in the south and Kekexili Reserve in Qinghai in the southeast. It consists of core area of 33 124.5 km^2, buffer areas of 8675.5 km^2 and experiment areas of 3200 km^2. The average elevations are around 4600 metres. Zhou *et al.* (1985) and Gu (1987) recorded wild yak at elevations of 3800–5000 metres in herds from 3–10 animals to herds of up to 80 animals, and estimated 3031 wild yak in the Kumukule Basin in the eastern part of this reserve, where most of the large ungulates were distributed in the early 1980s. Zhang (1991) once reported an estimate of 20 000 wild yak in this basin in the late 1980s, which was clearly a miscalculation. A recent survey in this area only estimated 985 wild yak early this century (Z. C. Li *et al.* 2006). The total number of wild yak in this reserve were approximately 2500 in 1984, 3200 in 1995 and 5800 in 2006–2010. There are almost no domestic yak in this reserve; therefore the wild yak in this reserve are believed to have the best genetic purity.

Zhongkunlun (Middle Kunlun) Nature Reserve (32 000 km^2; established in 2001 in Xinjiang at provincial status; V in the IUCN categories): located on the northern slope of the QTP and middle part of the Kunlun Mountains in the southern mountainous area of Qiemo County of southeastern Xinjiang. This reserve borders the Qiangtang Reserve in Tibet in the south and the Aerjinshan Reserve in the east. The average elevation is around 4000 metres. It was created with a focus on the protection of ungulates, including wild yak, which was recorded in the late 1980s (Feng 1990).

Qilianshan Nature Reserve (expanded from 2300 km^2 in 1988 to 26 530 km^2 in 2012; 97°25′–103°46′ E, 36°43′–39°36′

N; established in 1987 in Gansu and upgraded in 1988 to national-level status; V in the IUCN categories): located in northwestern Gansu and borders Qinghai in the west. This reserve covers the range of Qilian Mountains and includes a core area of 8023 km^2. The average elevations range from 4000 to 5000 metres. Wild yak were recorded in recent years, but there is no estimation of population size (Xia *et al.* 2003; Zhou 2003; Yin 2005).

Yanchiwan Nature Reserve (13 600 km^2; 95°21′–97°10′ E, 38°26′–39°52′; established in 1982 in Gansu and updated in 2006 to national-level status; V in the IUCN categories): located in the western part of the Qilian Mountains in northwestern Gansu. Average elevation is around 4000 metres. A small number of wild yak were encountered in the late 1980s (Zheng *et al.* 1989), but there is no record in recent surveys (Huang *et al.* 2005; Zhao *et al.* 2011).

Gonggashan Nature Reserve (4091 km^2; 101°29′–102°12′ E, 29°01′–30°05′ N; established in 1996 in Sichuan and upgraded in 1997 to national-level status; V in the IUCN categories): located in the southwest QTP in western Sichuan. Close to 100 wild yak are estimated to be present in this reserve (Yang *et al.* 2006; Yao *et al.* 2006; Guo *et al.* 2007).

Feng (1990) once also reported a small number of wild yak in the western Kunlun Mountains. It is clear that although protection efforts of wild yak from illegal poaching have to be consistently made across its entire distribution range, three huge continuous reserves of Qiangtang, Kekexili and Aerjinshan will definitely serve as the most critical core areas and provide a long-term safeguard to the survival of wild yak.

Descriptive notes

Body measurements: **Shoulder height**: 160–205 cm; **Head and body length**: 240–380 cm; **Tail length**: up to 100 cm; **Body mass**: 500–1200 kg.; **Heart girth**: 210–270 cm. Measurements are combined for both Qilian and Kunlun types. Yak and taurine or zebu cattle share the same dental formula: dI 0/3, dC 0/1, dP 3/3 for deciduous teeth; I 0/3, C 0/1, P 3/3, M 3/3 (×2) = 32 for permanent teeth (I = incisors, C = canines, P = premolars, M = molars, on the upper/lower side of half the jaw).

As Leslie & Schaller (2009) indicated, the descriptions of wild yak remain unchanged since the first documentation and in comparison with domestic yak (Blanford 1888; Lydekker 1898, 1900, 1913). An adult wild yak has a large and compact body with short, strong legs. Its ears are small and hooves are large and rounded. It develops a long, hairy coat consisting of long, coarse hairs, and a dense woolly undercoat covering its chest, flanks and rump, almost reaching the ground in mature males. Its tail is also fully haired and horse-like. The coat colour is dark brown to black, except the recently discovered golden-brown genotype in Qiangtang. Compared to its domestic counterpart, the wild yak has a silver-grey colour line down the back of the body and behind the withers, as well as grey-white hairs around the muzzle. The large horns are smooth and round, slightly oval at the base in old animals, curving outward and upward at first, then forward, inward and upward, and slightly backward at the end (Figure 12.6). The lifespan of wild yak ranges from 20 to 25 years both in the wild and in captivity. Major general anatomical distinctiveness of yak in comparison with other bovine species is mentioned earlier. Further details and differences between the wild and domestic yak can be found in Wiener *et al.* (2003).

According to the variation and difference in body structure and size, skull and horn shape, behaviour and geographic distribution of wild yak, Lu *et al.* (1993) defined the *Qilian type* and the *Kunlun type* (Figure 12.7).

The Qilian type is distributed in the western part of the Qilian Mountains and eastern part of the Aerjin Mountains,

Figure 12.6 A wild yak bull in Kekexili Nature Reserve, western Qinghai, China in late November 2009 (photo by Damin He). A black and white version of this figure will appear in some formats. For the colour version, please refer to the plate section.

Figure 12.7 Illustration of the Qilian and Kunlun types of wild yak bulls at four years of age (adapted from figure 2 in Lu *et al.* 1993).

where the Qilian, Yanchiwan and eastern section of Aerjin Shan Reserves are located. The rangelands are alpine steppe and desert steppe. There is a conspicuous hump but no dewlap. Face and leg are relatively long. Muzzle and ears are relatively small. Coat colour in adults is generally brownish-black. Young and elder animals often look dark brown. Eye sockets and tip of muzzle are greyish. There is an obvious back line in grey-white colour. No hair is present in the facial part, while chest, shoulders, flanks, belly and legs are all covered with long hairs. The tail has loosing, brush-like long hairs. Males have much larger horns than females. The open distance between horns can reach 70 cm. Horns are round but slightly oval at their base. They first curve outward and upward, then slightly backward, inward and upward, and slightly backward at the tips. Females are much smaller than male (only one-third to one-half in size). These yak are generally peaceful and make no attempt to harass people or other animals.

The Kunlun type is found in northern Tibet and the range of Kunlun Mountains where the Qiangtang, Kekexili, Sanjiangyuan and west section of Aerjinshan Reserves are located, and also in small pockets around the headwaters of Yaluzangbu (Yarlung Tsangpo) River in western Tibet. The rangelands are alpine meadow, alpine steppe and alpine cold desert steppe. Adult males are extremely large. The skull or head is heavily developed with a wide but relatively short facial part. The hump is remarkable but there is no dewlap. The legs are relatively short but sturdy. Coat colour is black or blackish brown. Hairs around the muzzle and eye sockets are greyish. There is a clear grey-white or brown back line. No long hair is found in the facial part, but there are several clusters of long hair in the forehead. Chest, forelegs and flanks are covered with thick, long hairs, almost reaching to the ground in mature males. The horns are very thick, with a circumference up to 50 cm at their bases and often used by some herdsmen as milking vessels. The shape of the horns is relatively open and they curve remarkably inward and forward in their middle sections, but the tips are obviously backward (see also Figure 12.6). Both sexes of this type are aggressive and readily attack people and other animals. The cow has a strong defensive ability to protect her calf, while bulls who fail the fighting for mating would stay alone and are often violent.

Habitat

Wild yak are mainly distributed in northern and northeastern QTP, which ranges approximately from 78° to 102° E and from 29° to 40° N and covers 2.5 million km^2, of which 85% is above 3000 metres and 50% is above 4500 metres. There are a number of unique topographic regions characterized by different drainage patterns and parallel high mountains that confine and dissect the QTP. The headwaters of Yellow, Yangtze, Mekong and Yarlung Tsangpo all originate from Qinghai and Tibetan regions on the QTP. Moisture from the southwest monsoon moves up from east and south directions, and therefore rainfall and temperatures on the QTP vary significantly from place to place according to their latitudes and longitudes. The annual precipitation in most parts of the QTP is limited to less than 400 mm and rains are concentrated in the very short warm (summer) season (June to August), often in the form of snow or ice storms. The ecosystems on the QTP are unique and fragile in terms of their high elevations, low temperatures and thin oxygen contents. The wild yak are restricted to the extremely cold and treeless uplands where the annual average temperatures range between 0 °C and 6 °C, while the lowest one in winter can drop to −40 °C and below. From central QTP, starting from southern Qiangtang in Tibet towards Qilian Mountains in northeastern Qinghai, the weather falls into the second frigid zone of continental climate, while the northwestern part of the QTP has a semi-arid climate and the northern Qiangtang area up to Kunlun Mountains is characterized as a cold and arid climate (Feng *et al.* 1986; Miller *et al.* 1994; Miller & Schaller 1996; Schaller & Liu 1996; Peng & Ouyang 1999; Liu & Schaller 2003; Yao *et al.* 2006).

The cold alpine rangelands are the most common ecosystems on the QTP and there is very diversified and unique vegetation in high spatial heterogeneities which are largely determined by rainfall and elevations, often following different narrow and vertical distribution patterns in mountainous areas or wide and horizontal patterns in open plains. The main distribution range of wild yak within the nature reserves introduced above has three major vegetation types: *alpine meadow*, *alpine steppe* and *desert steppe*.

Alpine meadow is located in extensive areas with an annual rainfall around 350–400 mm, including the plains of eastern

Qinghai and in Tibet along the Lhasa–Golmud highway (railway) and west close to the Siling Co (lake). The precipitation is gradually reduced in north and west directions; therefore the alpine meadow is broken up into patches of different sizes, mainly found along streams, seepages and swamps. The dominant grasses are short and thick sedges of genus *Kobresia*, up to 10–40 cm tall, followed by rich forbs of genus *Bistorta*, *Gentiana* and *Pedicularis*. Other minor companying species include *Stipa* spp., *Artemisia* spp., *Poa* spp., *Elymus* spp. and *Carex* spp. Most of the southern Qiangtang area in Tibet is covered by this type of vegetation.

Alpine steppe vegetation is generally found at high elevations between 4300 and 5100 metres, represented by the central Qiangtang area in Tibet. The weather is very cold and windy; its soils consist of sand, gravel or silt with few grasses; and annual rainfall is reduced to 100–350 mm. Therefore, vegetation coverage is only around 30–50% and the sod layer may reach 15–20 cm. The dominant plants are species of genus *Stipa*, comprising 50–60% of total vegetation composition, while other species – including *Poa* spp., *Festuca* spp., *Kobresia* spp., *Artemisia* spp. and *Carex* spp. – are common but vary among different localities, with 15–20% of the composition. Common forbs are represented by species of genus *Potentilla*, *Oxytropis* and *Astragalus*, making up 10–15% of the composition. Shrubs are normally less than 30 cm tall and include species of genus *Ceratoides*, *Ajania* and *Myricaria*, among other minor species. Alpine steppe can extend up to elevations around 5200–5300 metres with a shift of dominant plants, e.g. from *Stipa* spp. into *Kobresia* spp.

Desert steppe vegetation is distributed in northern Qiangtang area in Tibet between 34°30′ and 35° N and further extended into southern Xinjiang where the annual rainfall is only 50–100 mm and the soils are sandy. Vegetation coverage is thus less than 30%, although the composition of plant species is similar to the vegetation of alpine steppe. Dominant plants are thus changed to sedge, *Carex moorcroftii* and *Ceratoides compacta*, which are often found in small patches carrying a minor proportion of *Stipa* spp. The forb species only make up 5% of the total composition.

Movements and home range

Wild yak are not typical migratory animals, but they make movements following water sources and vegetation phenological patterns over the summer season, in general from south to north and/or from low to high elevations along slopes for nutritious forage while reserve directions are followed from late autumn through winter. It is probable that the avoidance of attacks by flies and parasites, as well as interference of livestock grazing, is a reason to drive wild yak to higher elevations with limited vegetation even in summer.

However, the movements of wild yak are restricted in the deep mountains and valleys where there is very limited, if any, human interference. They do not have permanent residential sites but wander rather long distances – e.g. around 50 km – in search for grasses, the longer distance in winter due to scarcity of grasses compared to summer, in particular over the four vast continuous reserves of Qiangtang, Kekexili, Aerjinshan and Zhongkunlun (Zhao & Gao 1991; Yao *et al.* 2006). It is apparent that the increasing human activities, in particular the fast development of road and railway networks on the QTP in the past two decades, have definitely further contracted the movements and home range of the wild yak. Lian *et al.* (2012) recently observed that wild yak are more sensitive to road traffic than Tibetan wild ass, Tibetan antelope and Tibetan gazelle, and the avoidance distance from roads for wild yak was around 1000 metres for small herds but 3000–4000 metres for big herds. They nevertheless noted the positive sign that road traffic has been reducing following the operation of the Qinghai–Tibet railway which, it is claimed, Tibetan wildlife have adapted to (Yang & Xia 2008).

Activity patterns

Wild yak spend most of the daytime grazing, often in the morning and late afternoon. Around midday they rest, often in a southern slope of glacial hills or mountains, where there is less wind but sufficient sunshine so that they can comfortably regurgitate either standing or lying down. At night the herds cluster together for warmth and the calves are protected in the centre of the groups.

Feeding ecology

The wild yak favours grasses. *Stipa* grasses can represent 30–50% of the forage that an animal consumes, followed by *Carex* spp. at 10–20% of total consumption in June. In general, non-Gramineae grasses at higher elevations (up to 6000 metres) are relatively important sources of forage (up to 30%) for wild yak in the short summer season, which only lasts around three months from June to August, while grasses at lower elevations (down to 3800 metres) are used by wild yak mainly from late autumn through the long winter when the high mountainous areas are covered by heavy snow.

Wild yak live together with Tibetan antelope, Tibetan gazelle, Tibetan argali, blue sheep and Tibetan wild ass in most of the protected reserves on the QTP. These ungulates share a broadly overlapping habitat, although wild yak, blue sheep and argalis usually occupy the mountainous areas. However, the limited vegetation and relatively low biomass offer very limited choice to these species. They graze on a few plant species dominated by *Kobresia* spp., *Stipa* spp. and *Carex* spp. Forage in summer is relatively abundant and nutritious and animals spread to different areas and elevations. Wild yak move up to high mountains, so the competition among these species is minimal.

Harris & Miller (1995) found that wild yak in Yeniugou were divided into large, mostly nursery groups, and small bachelor groups. The nursery groups include only 4.4% adult males, and generally graze sedge meadows at high elevations and then rest for regurgitation and loafing in nearby glacial moraines. Small bachelor groups occupy relatively wide areas from high to mid and low elevations where the grasslands are dominated by *Stipa* spp. Harris and Miller identified 72.41% sedge (*Kobresia* spp.), 13.14% grass and 10.15% forb species in the sampled diets of wild yak during summer in 1991 and 1992. A grazing range shift of nursery groups took place between early August and late September from high elevations down to *Stipa* grasslands over the winter season. *Stipa* spp. made up 66.2% of the relative density of plant fragments in the faecal samples of wild yak in autumn, while *Kobresia* spp. were reduced down to 23.2%. This shift seemed to coincide with the maturation of *Kobresia* spp. in sedge meadows, suggesting a maximal use of the forage resources. Wild yak shared a high overlap index of summer diets with white-lipped deer (0.967), blue sheep (0.978), argali (0.920) and male Tibetan antelope (0.873), followed by domestic sheep (0.472), indicating that they consumed very similar diets of a high proportion of sedges. Although the forage competition among ungulates as well as between wild ungulates and domestic sheep was weak in summer, domestic sheep were expected to play a role in limiting the populations of all Tibetan wild ungulates in the area.

Cao *et al.* (2009) observed similar diet compositions based on the examination of faecal samples of wild yak in Kekexili Reserve. They identified plants of 14 genera, including *Stipa* spp., *Festuca* spp., *Poa* spp., *Littledalea racemosa, Kengyilia thoroldiana, Deyeuxia tibetica, Kobresia* spp., *Carex* spp., *Leontopodium pusilum, Astragalus* spp., *Oxytrops* spp., *Draba* spp., *Triglochin maritimum* and *Renumuria soongarica* in both summer and winter diets of wild yak, of which sedges were 67.44% and grasses 32.03% in July 2005, while the former was reduced to 56.96% but the latter increased to 36.78% in January 2005. The diets of domestic yak and sheep consisted of 67.27% and 37.76% sedges and 28.96% and 57.29% of grasses in summer, but 61.84% and 32.32% of sedges, 18.70% and 43.61% of grasses, and 14.49% and 16.70% of legumes in winter, respectively. Wild yak consumed more species in winter than in summer. Wild yak were found sharing almost identical diets with domestic yak (0.934), followed by domestic sheep, (0.685) in summer. In winter such share was reduced to 0.707 for wild and domestic yak, due probably to the movement of domestic yak herds back to their winter pastures at lower elevations, but increased a little to 0.722 for wild yak and domestic sheep. Wild yak, Tibetan antelope and Tibetan wild ass also had similar but moderately overlapped summer and winter diets.

The long winter from October to the following May, when leaves are dormant or dead, is a real challenge to the survival of Tibetan wild ungulates in general and newborns in particular. Therefore, the highest competition for identical forage resources between domestic and wild yak in summer is clearly a specific and critical issue to be addressed, while general competition between livestock and Tibetan wild ungulates must be considered as a major, common problem for the management of most of the protected areas on the QTP. Ma & Jiang (2006) also argued that environmental degradation following climate change impacts on the survival of wildlife on the QTP. They claimed that wild yak was one of the large mammals which had disappeared from the Qinghai Lake region.

Reproduction and growth

Male yak reach sexual maturation at three years of age and their peak sexual behaviour occurs at 4–6 years old. Females start to show the first oestrus at three to four years of age. The sexual ratio of mature males to females in regular herds is around 1:10. Oestrus and mating take place from July to December and parturition mostly happens from April to June. Cows at relatively low elevations come to heat earlier than those at high elevations. Cows without a calf come to oestrus in July to August, while a few cows with a calf born early in the year arrive to oestrus from September. The gestation length varies from 258 to 270 days, with a single calf per litter. The calving interval varies from every two to three years, depending largely on the weather and vegetation conditions. Under the challenge of harsh climate and unpredictable weather conditions on the QTP, the seasonal oestrus of wild yak cows only persists for a short period of 1–3 days and arises often silently and invisible, mostly once per year. The oestrus cycle is around 20 days. Wild yak can be reproductive for 8–10 years.

Similar to other bovines, the size of the horns increases as the wild yak ages; the older a yak, the larger its horns. In addition, the horns of wild yak often carry the marks of nutritional fluctuations over winter. Therefore both size and nutritional marks can be used to estimate the age of an animal, together with its tooth wear and eruption sequence, which is similar to that of other bovines. The general descriptions of wild yak were covered earlier, in particular those features associated with the two ecological types. Ma (2012) and Wang (2012) recently reported the body weights and measurements from birth to three years of wild yak reproduced by the parents captured from Kunlun Mountains in western Qinghai and also in northeastern Tibet (see later) and kept in 'captivity' on Datong Yak Farm (100°52′–101°26′ E, 37°11′–37°32′; elevations at 2900–4280 metres; located in the eastern section and western edge of the Qilian Mountains) (Table 12.1). These data show that, similar to domestic yak, males grow faster than females and cold winter has a more pronounced negative effect on the development of males than females, suggesting there is a need to provide supplement feeds to young males during the winter season.

Behaviour

The wild yak's thick, hairy coat is very well adapted to freezing temperatures. The enlarged lungs and heart allow the yak to breathe the thin air at high elevations. Such adaptability may imply that wild yak cannot thrive in warmer climates, possibly another reason that drives them to higher elevations in

Table 12.1 Body weights (average ± SD in kilograms) and measurements (average ± SD in centimetres) of young wild yak reproduced and kept on the Datong Yak Farm

Age	Sex	No.	Body weight (kg)	Height at withers (cm)	Body length (cm)	Heart girth (cm)	Circumference of cannon bone (cm)
Birth	M	7	26.00 ± 1.94	69.50 ± 2.26	61.33 ± 3.93	73.67 ± 2.66	9.80 ± 0.45
	F	11	23.55 ± 5.43	68.25 ± 3.75	60.75 ± 5.01	72.08 ± 5.01	9.75 ± 0.92
Six months	M	7	105.50 ± 46.26	96.50 ± 16.26	100.50 ± 20.51	120.50 ± 27.58	12.04 ± 0.45
	F	11	79.80 ± 40.53	87.56 ± 12.69	88.33 ± 12.86	108.70 ± 20.43	12.01 ± 1.31
One year	M	7	127.97 ± 70.95	100.07 ± 8.54	103.70 ± 10.61	125.40 ± 9.88	13.33 ± 1.15
	F	11	123.20 ± 27.95	101.10 ± 10.89	106.70 ± 12.60	126.00 ± 18.00	13.00 ± 3.71
Two years	M	7	227.94 ± 73.83	118.43 ± 8.30	124.10 ± 8.43	158.00 ± 7.02	16.00 ± 1.10
	F	11	196.80 ± 27.31	122.40 ± 13.88	121.90 ± 14.29	152.00 ± 28.32	15.20 ± 4.32
Three years	M	7	310.72 ± 114.81	134.83 ± 12.16	135.20 ± 13.17	177.30 ± 24.31	17.80 ± 2.16
	F	11	263.50 ± 130.11	127.00 ± 14.14	138.00 ± 16.97	162.00 ± 31.11	17.00 ± 1.10

summer. Wild yak prefer to stay in groups and often move in big herds comprising junior and adult animals of both sexes, except the mature bulls who lost the fight for mating. The size of a herd varies from around ten animals (commonly) to around 200 animals (occasionally); herds are often bigger in winter than in summer. The fighting for dominance takes place before every mating season and adult bulls are sensitive and battle each other seriously. In the mating season a single winning or dominant bull leads a herd of reproductive cows along with their newborns and yearling calves. In the non-mating season, nearly all mature bulls depart from the nursery herds.

Compositions of herds vary from place to place and are hardly recorded. It is expected that 10–15% of animals in herds are juniors. Once in August a specific herd of 48 animals consisted of 32 cows, seven yearling males and nine newborns (Liu & Schaller 2003). A higher proportion, up to 25%, of junior animals was also observed in Aerjinshan Reserve in late June (Gu 1987). However, juniors made up only 5.7% of 700 yak in and around the Aru Basin of Qiangtang Reserve, while 12.7% of a herd was young animals among 267 yak in other areas in the reserve in August 1990. In June 1994 only 2.5% of juniors were observed among 690 yak in eastern Qiangtang Reserve. The lower proportion of juniors in those wild yak populations was suspected to be associated with reproductive failures following disease challenges (Schaller & Liu 1996).

Parasites and diseases

Wild yak suffer the same diseases and external and internal parasites as domestic yak, which were reviewed in detail by Wiener *et al.* (2003). The overlapped grazing areas between wild yak and livestock including domestic yak, sheep, goats and Tibetan cattle increase the risk of infections, in particular over the long winter season when most of them share very limited habitats. Chang *et al.* (2013) recently investigated the carrier status of foot-and-mouth disease (FMD) in domestic yak following an experimental challenge with FMD virus O/Akesu/58 strain. The infected yak showed typical clinical signs with high levels of total antibody against the FMD virus and maintained the carrier status for at least eight months.

Diseases are believed to contribute to population reduction and reproductive failures of wild yak, e.g. abortion of pregnant wild yak cows and early death of calves which suffer from brucellosis, contagious bovine pleuropneumonia and bovine ephemeral fever, among others (Yao *et al.* 2006). However, there is almost no specific investigation of important diseases in wild yak, and only a few cases of particular diseases of wild yak have been reported.

Han (2001) diagnosed and treated six young wild yak which were freshly captured from western Qinghai (see later) and transported to Datong Yak Farm in August 1999. They were between six months to one year old and suffered from infections of *Clostridium welchii* due probably to the enclosed management and artificial feeding regime. They developed severe enteritis necroticans and two of them died while the other four recovered following treatment. Drölma *et al.* (2006) once rescued a wild yak cow who suffered dystocia. The cow was left behind by the herds in Qumalai County of western Qinghai, where relatively dense wild yak populations are distributed. It was her first birth and she seemed to have struggled for two days. The calf was dead and eventually taken out through a successful caesarean section. The cow recovered with care after 13 days and joined the herds 20 days after the operation. A case of bovine *Pasteurella multocida* infection of a wild yak in Lanzhou Wuquanshan Zoo has also been reported (Liu & Wang 2006).

Cao *et al.* (2006) surveyed the helminth eggs in faeces of 16 wild yak in Kekexili from December 2004 to January 2005. The wild yak had high infection rates ranging from 75% to 87.5%, but only carried a few species of genus *Cooperia* and *Trichostrongylus* at low density of faecal egg counts of 9.06 and 5.93, respectively, suggesting that the eggs of these helminth species had strong cold resistance in order to survive the extremely low temperatures and then to reinfect the animals during the next warm season, while wild yak had competent immunity over the winter season.

Status in the wild

Since the 1990s, intensive efforts have been made to help save the wild yak population. One of the major threats in the wild within the nature reserves is the hybridization with domestic yak when livestock are moved into the reserves for grazing on overlapped pastures, in particular in the summer season (Schaller & Liu 1996; Harris 2008; Leslie & Schaller 2009). It has been observed by local nomads that wild yak bulls occasionally run into domestic herds to mate yak cows in breeding season. Hybrids of the first generation, the 'semi-wild yak' as they are called, are born naturally and may stay in domestic herds. Such hybrids are favoured by some nomads because of their comparative performance, well-known as the 'hybrid vigour'. However, a few domestic yak cows can sometimes be swept away from the herds in breeding season by strong wild yak bulls, and they may subsequently join the wild yak populations (Feng 2001; Zhao 2006; Liu 2007b). This can lead to an erosion of wild yak genetics, e.g. non-black coat colours (except the recently discovered 'golden wild yak' genotype in Qiangtang) or spots (e.g. white-faced) are observed in some wild yak populations (Liu & Schaller 2003; Harris & Loggers 2004). Such genetic erosion can have a negative impact on the adaptability of wild yak, while also the spotted animals may more easily attract predators to the herds. Certainly, increased risk of disease transmission from domestic yak cows to wild yak populations is a hazard (Schaller & Liu 1996; Zhao 2006).

As recently indicated by Leslie & Schaller (2009) and Zhang *et al.* (2009), the growing competition between wild and domestic yak for forage on the summer pasture in particular and the increasing expansion of human and livestock activities in both number and spatial dominance in general on the QTP have been a rapidly increasing problem over the past two or three decades. Governmental policy changes and reforms in livestock and rangeland use from 'communal' to 'private' management systems on the QTP since the 1980s had first motivated a rapid increase in the number of livestock around the 1990s, followed by heterogeneous rangeland allocation regimes implemented from the mid 1990s according to either the number of livestock of a household in some areas or the number of people in the household in other areas. Most of the open rangelands are now divided and fenced into small parcels that seriously affect the movement and access of wild yak to grazing and water resources.

Wild yak are also vulnerable to all diseases of domestic yak and also cattle. It is claimed that diseases cause 15–30% of population loss, while insufficient nutrition due to snow disaster and drought among other natural factors account for a further 20% of deaths in wild yak (Yao *et al.* 2006). Other than humans, the possible predators of wild yak are Tibetan wolf (*Canis lupus chanco*), Tibetan brown bear (*Ursus arctos pruinosus*) and snow leopard (*Uncia uncia*); however, no specific data on the level of their predation of wild yak is available (Harris *et al.* 1996; Leslie & Schaller 2009). Xu *et al.* (2006, 2010) investigated the food habits of Tibetan brown bear in two summer seasons (June–August) of 2005 and 2009. A combined analysis of faecal samples of the bears showed that wild yak were counted by 18.7% in relative frequency and 30.2% in dry matter residuals, but no active hunting of wild yak by the Tibetan brown bear was observed.

The wild yak is certainly one of the most endangered species on the QTP and has been restricted to isolated refuge in the most remote, relatively human-free areas within a few nature reserves (Piao *et al.* 1999; Yang *et al.* 2006; Yao *et al.* 2006; Guo *et al.* 2007; Leslie & Schaller 2009; Zhang *et al.* 2009). It was once considered a rare and unique species in China with a limited population, and thus recommended to be included in the second group of protected species, with its hunting to be banned but allowing a restricted exploration with permission in 1962. Nonetheless, large-scale and organized hunting for wild yak was not uncommon and has led to a further serious population reduction in general and extinction of wild yak in some specific areas in Sichuan, Gansu, Xinjiang and Qinghai between the 1950s and 1980s. For example, a place in the Aerjin Mountains was named as 'Niutougou' in Chinese, meaning 'a valley full of wild yak heads' after the massive killing of wild yak in the 1960s (Gu 1987; Feng 1990; Piao *et al.* 1999; Zhang *et al.* 2009). After the adoption of the 'Law of the People's Republic of China on the Protection of Wildlife' in November 1988, the wild yak was added to the first group of Class I protected species which were endorsed by the State Council of the Peoples Republic of China in December 1988 and became effective since January 1989; thereafter hunting of wild yak has been illegal and can be punished. The protection actions, in particular within the nature reserves, have been enforced since the 1990s. However, poaching for meat, though minimal and often by illegal miners, is still the primary threat to the wild yak, in particular within the vast nature reserves where effective mobile patrol forces are largely lacking.

The IUCN once recognized wild yak as Endangered between 1986 and 1996, but reconsidered it as Vulnerable A2ac+3c+4c since 1996 (Harris & Leslie 2008), which is argued to have probably been underclassified (Leslie & Schaller 2009). Wild yak have been protected under CITES Appendix I since 1975 and are currently maintained in the same appendix, which was valid from 25 September 2012 (www.cites.org, last accessed April 2013).

Status in captivity

Beijing Zoo started keeping and showing wild yak in 1969, but the first arrivals died due to poor transportation procedures and adaptation. Two pairs of wild yak were obtained from Qumalai in Qinghai in February 1972, and one bull died from injury during transportation. Another bull and two cows eventually survived in the zoo and successfully reproduced in captivity in 1973 (Zhou & Zhang 1982). By 1980 six births were achieved, of which two were transferred to other zoos. A protocol for the management and monitoring of wild yak was summarized by Zhou (1989). A shift in oestrus was observed from July through December in their native home range to October through the following March in the zoo. Calving for a single calf with a gestation of 252–265 days took place in July to December. Oestrus persists for two to three days and its cycle was 25 days (21–28). Lanzhou Wuquanshan Zoo, which was established in 1992 in Gansu, and Xining Zoo, which was established in 1960 in Qinghai, have been keeping a very small number of wild yak. The origins of wild yak in other zoos, though only few in China, are often unclear or unreliable. Some zoos even wrongly claimed domestic yak to be wild yak. Therefore, the role of very limited numbers of wild yak in a few zoos for its protection is not clear. At the time of writing only 15 wild yak are kept in zoological institutions in Europe. For more details on wild and domestic yak kept in zoological collections, see Chapter 27.

The 'hybrid vigour' observed in the accidently born 'semi-wild' yak through the uncontrolled mating of wild yak bulls with domestic yak cows (see earlier) motivated long-term scientific efforts to capture and tame four wild yak bulls in the early 1980s on Datong Yak Farm in Qinghai, China. These animals were the only materials accessible for and subjected to extensive scientific research described earlier. Two male calves were obtained from Lanzhou Wuquanshan Zoo in Gansu in 1983. They were born in 1981 and 1983 by a pair of wild yak captured from Subei, an area of Qilian Mountains in Gansu, and kept at the zoo in Lanzhou. The other two male calves were directly captured, one also from Qilian Mountains in 1987 and the other from an area of Kunlun Mountains in southwestern Qinghai in 1987 (Li *et al.* 1997; Dai *et al.* 1997a, 1997c). Three additional wild yak, one male and two females, were captured from Qumalai in western Qinghai in 1989. These animals were considered as representatives of the Qilian and Kunlun types of wild yak (Lu *et al.* 1993). The males were successfully tamed and trained for semen collection and significantly contributed to the genetic improvement programmes of domestic yak (Lu *et al.* 1991; Yang *et al.* 1992; Dai *et al.* 1997a, 1997b, 1997c; Li 2004; Li *et al.* 2005; Lu 2005; Wang & Sun 2005; Yang 2008). To renew the genetic advantage of the wild yak, another 14 wild yak calves from six months to one year of age were captured from Yushu in southwestern Qinghai in August 1998, following official permission (Han 2001). The surviving animals of this group led to the establishment of a small, pure wild yak herd, now self-sustained on the farm (Wang 2012). The latest single wild yak calf was rescued from Kekexili area in western Qinghai in 2008. Currently there are more than 20 pure wild yak kept on the farm (Mr He Xiaolin, former head of the Datong Yak Farm, personal communication, April 2013).

The success story of achievements in exploration of the superior genetics of wild yak for the genetic improvement of domestic yak was documented in detail by Wiener *et al.* (2003). The first new yak breed, the 'Datong Yak', was developed based on continuous genetic selection and breeding for 20 years, since the mid 1980s. Testing and comparison of different combinations, each having more than 120 yak carrying different levels of wild yak genes (blood) for their performance and management were conducted. The data showed that the hybrids carrying 75% of wild yak blood had relatively high growth rates and body weights at different ages, but they were too 'wild' to be managed, while the animals carrying 25% of wild yak blood were easy to manage, but their performances were relatively low. Therefore the F1 generation hybrids (Figure 12.8) carrying 50% of wild yak blood were selected for further breeding towards the fixation of the superior genes of wild yak. A total of 1086 domestic yak cows were selected and used for artificial insemination with the semen collected from four well-trained wild yak bulls. The breeding goals were achieved at the F3 and F4 generations with all animals (around 2200) expressing a very uniform body form. The body weights of Datong Yak were 15–27% higher than the other domestic yak and the conception rate of Datong Yak reached to 70–80%,

Figure 12.8 Semi-wild yak bulls in Zhiduo County of western Qinghai, China in August 2012 (photo by Damin He). A black and white version of this figure will appear in some formats. For the colour version, please refer to the plate section.

which was 15–20% higher than other domestic yak at the same age on the farm. The bulls of Datong Yak can be employed for semen collection at 24–28 months of age, which was one year earlier than domestic yak. This new breed was finally endorsed and granted an official certificate by the Ministry of Agriculture, PR China in 2005. Since then around 2000 breeding bulls of the Datong Yak have been bred on the farm and distributed annually to yak farmers in Qinghai, Gansu, Xinjiang and Tibet for the genetic enhancement of domestic yak (Lu *et al.* 2005; Hu *et al.* 2005; He 2009; Xue 2012). This is probably the first case of large-scale successful exploration and appreciation by humans of the unique genetic resources of a wildlife species which was once domesticated several thousand years ago.

Acknowledgements

Special thanks is given to Lu Tsi and Pematso for their assistance in checking and translating the relevant Tibetan names, as well to He Daming and Wang Yong for their generous provision of photos of wild yak. The author also acknowledges the funding from the Chinese contribution to the ILRI for collaborative research at the CAAS-ILRI Joint Laboratory on Livestock and Forage Genetic Resources.

References

Allen, G. M. (1940). The mammals of China and Mongolia. *American Museum of Natural History*. **11**(2): 1258–1261.

Bai, W. L., Yin, R. H., Dou, Q. L., *et al.* (2011). Molecular characterization and phylogenetic analysis of a yak (*Bos grunniens*) κ-casein cDNA from lactating mammary gland. *Molecular Biology Reports* **38**(4): 2711–2718.

Bandelt, H. J., Forster, P. & Röhl, A. (1999). Median-joining networks for inferring intraspecific phylogenies. *Molecular Biology and Evolution* **16**(1): 37–48.

Blanford, W. T. (1888). *The Fauna of British India, Including Ceylon and Burma: Mammalia*. London: Taylor and Francis.

Bohlken, H. (1958). Vergleichende Untersuchungen an Wildrindern (Tribus Bovini Simpson, 1945). Zoologische Jahrbücher. *Abteilung für Allgemeine Zoologie und Physiologie der Tiere* **68**: 113–202.

Buntjer, J. B. (1997). DAN Repeats in the vertebrate genome as probes in phylogeny and species identification. Academic thesis. Utrecht University.

Buntjer, J. B., Otsen, M., Nijman, I. J., Kuiper, M. T. & Lenstra, J. A. (2002). Phylogeny of bovine species based on AFLP fingerprinting. *Heredity* **88**(1): 46–51.

Cai, L. (1980). The nomenclature for the hybrids of yak with cattle. *Journal of China Yak*, **2**: 20–23. (In Chinese)

Cai, L. (1989). *Sichuan Yak*. Chengdu: Sichuan Ethnic Press. (In Chinese)

Cao, Y. F., Su, J. P., Zhang, T. Z. & Lian, X. M. (2006). Investigation on helminth eggs in feces of Tibetan antelope, wild yak, Tibetan wild ass and Tibetan gazelle. *Sichuan Journal of Zoology* **25**(3): 611–613. (In Chinese with English abstract)

Cao, Y. F., Zhang, T. Z., Lian, Y. M., Cui, Q. H., Deng, D. D. & Su, J. P. (2009). Diet overlap among selected ungulates in Kekexili region, Qinghai Province. *Sichuan Journal of Zoology*, **28**(1): 49–54. (In Chinese with English abstract)

Chang, H. Y., Ma, Y. B., Lin, T., Cong, G. Z, Du, J. Z. & Ma, J. L. (2013). Foot-and-mouth disease virus carrier status in *Bos grunniens* yaks. *Virology Journal* **10**(1): 81.

Dai, H. Q., Li, C. Y., Yang, G. D., Yin, M. C., Yang, Q. S. & Zhang, C. B. (1997a). Feeding and management of wild yak bulls. *Chinese Qinghai Journal of Animal and Veterinary Sciences* **27**(1): 42–43. (In Chinese)

Dai, H. Q., Li, C. Y., Yin, M. C., Yang, Q. S. & Hu, S. H. (1997b). Semen volume and its influencing factors in wild yaks. *Chinese Qinghai Journal of Animal and Veterinary Sciences* **27**(1): 9–11. (In Chinese with English abstract)

Dai, H. Q., Yin, M. C., Li, C. Y., Yang, G. D. & Yang, Q. S. (1997c). Domestication of wild and semi-wild yak bulls. *Chinese Qinghai Journal of Animal and Veterinary Sciences* **27**(1): 7–8. (In Chinese with English abstract)

Decker, J. E., Pires, J. C., Conant, G. C., *et al.* (2009). Resolving the evolution of extant and extinct ruminants with high-throughput phylogenomics. *Proceedings of the National Academy of Sciences, of the United States of America* **106**(44): 18644–18649.

Drölma, Chuktso, Döndruptsomo & Yumkyap. (2006). A case of caesarean operation for wild yak on Qinghai-Tibetan Plateau. *Chinese Journal of Traditional Veterinary Science* **2**: 35–36. (In Chinese)

Ellerman, J. R. & Morrison-Scott, T. C. S. (1966). *Checklist of Palaearctic and Indian Mammals 1758 to 1946*. 2nd edition. Oxford: Alden Press.

Fan, B. L., Li, N. & Wu, C. X. (2000). Research on constructing phylogenetics trees of ruminants basing on the database of milk protein gene sequences. *Acta Genetica Sinica*, **27**(6): 485–497. (In Chinese with English abstract)

Feng, Z. J. (1990). On the status and conservation of the wild animal resources in the Karkorum–Kunlun Mountains region, China. *Journal of Natural Resources* **5**(4): 343–352. (In Chinese with English abstract)

Feng, Z. J. (2001). Unique animals on the Qinghai-Tibetan Plateau. *Man and the Biosphere* **2**: 12–14. (In Chinese)

Feng, Z. J., Cai, G. Q. & Zheng, C. L. (1984). A checklist of the mammals of Xizang (Tibet). *Acta Theriologica Sinina* **4**(4): 341–358. (In Chinese with English abstract)

Feng, Z. J., Cai, G. Q. & Zheng, C. L. (1986). *The Mammals of Xizang (Tibet)*. Beijing: Science Press. (In Chinese)

Finlay, E. K., Gaillard, C., Vahidi, S., *et al.* (2007). Bayesian inference of population expansions in domestic bovines. *Biology Letters* **3**(4): 449–452.

Flower, W. H. & Lydekker, R. (1891). *An Introduction to the Study of Mammals: Living and Extinct*. London: Adam and Charles Black.

Fox, J. L., Nurbu, C. & Chundawat, R. S. (1991). The mountain ungulates of Ladakh, India. *Biological Conservation* **58**(2): 167–190.

Fox, J. L., Nurbu, C., Bhatt, S. & Chandola, A. (1994). Wildlife conservation and land-use changes in the Transhimalayan region of Ladakh, India. *Mountain Research and Development* **14**(1): 39–60.

Gentry, A., Clutton-Brock, J. & Groves, C. P. (1996). Case 3010: proposed conservation of usage of 15 mammal specific names based on wild species which are antedated by or contemporary with those based on domestic animals. *Bulletin of Zoological Nomenclature* **53**: 28–37.

Gentry, A., Clutton-Brock, J. & Groves, C. P. (2004). The naming of wild animals and their domestic derivatives. *Journal of Archaeological Science* **31**(5): 645–651.

Geraads, D. (1992). Phylogenetic analysis of the tribe Bovini (Mammalia: Artiodactyl). *Zoological Journal of the Linnean Society* **104**: 193–207.

Gray, J. E. (1843). *List of the Specimens of Mammalia in the Collection of the British Museum*. London: George Woodfall and Son.

Gray, J. E. (1846a). On the arrangement of the hollow-horned ruminants (Bovidae). *Annals and Magazine of Natural History* **18**: 227–233.

Gray, J. E. (1846b) *Catalogue of the Specimens and Drawings of Mammals and Birds of Nepal and Tibet, Presented by B. H. Hodgson, Esq. to the British Museum*. London: E. Newman.

Gray, J. E. (1852). *Catalogue of the Specimens of Mammalia in the Collection of the British Museum, Part III. Ungulata Furcipeda*. London: Taylor & Francis.

Gray, J. E. (1863). *Catalogue of the Specimens and Drawings of Mammals, Birds, Reptiles, and Fishes of Nepal and Tibet, Presented by B.H. Hodgson, Esq., to the British Museum*. 2nd edition. London: Taylor & Francis.

Gray, J. E. (1872). *Catalogue of Ruminant Mammalia (Pecora, Linnaeus) in the British Museum*. London: Taylor & Francis.

Griffith, E. (1827). Synopsis of the species of the class Mammalia, as arranged with reference to their organization, by Cuvier, and other naturalists with specific characters, synopsis. In *The Animal Kingdom*, Vol. 5, pp. 373–375. Eds. Griffith, E. *et al.* London: G. B. Whittaker.

Griffith, E., Smith, C. H. & Pidgeon, E. (1827). The class Mammalia arranged by the Baron Cuvier with specific descriptions. In *The Animal Kingdom*, Vol. **4**, pp. 395–406. Eds. Griffith, E. *et al.* London: G. B. Whittaker.

Groves, C. P. (1971). Request for a declaration modifying Article 1 so as to exclude names proposed for domestic animals from zoological nomenclature. *Bulletin of Zoological Nomenclature* **27**: 269–272.

Groves, C. P. (1981). Systematic relationships in the Bovini (Artiodactyla, Bovidae). *Journal of Zoological Systematics and Evolutionary Research* **19**(4): 264–278.

Groves, C. P. (1995). On the nomenclature of domestic animals. *Bulletin of Zoological Nomenclature* **52**(2): 137–141.

Grubb, P. (2005). Order Artiodactyla. In *Mammal Species of the World: A Taxonomic and Geographic Reference*, pp. 637–722. 3rd edition. Eds. Wilson, D. E. & Reeder, D. M. Baltimore, MD: Johns Hopkins University Press.

Gu, J. H. (1987). Ungulates in eastern Kunlun-Arjin Shan region in Xinjiang. *Arid Zone Research* **3**: 56–68. (In Chinese)

Gu, Z. L., Zhao, X. B., Li, N. & Wu. C. X. (2007). Complete sequence of the yak (*Bos grunniens*) mitochondrial genome and its evolutionary relationship with other ruminants. *Molecular Phylogenetics and Evolution* **42**(1): 248–255.

Guo, G., Jia, Y. H., Yan, P., Lu, Z. L. & Li, K. L. (1991). Comparative karyotyping between wild and domestic yak. *Journal of China Yak* **2**: 46–48. (In Chinese)

Guo, S. C., Liu, J. Q., Qi, D. L., Yang, J. & Zhao, X. Q. (2006a). Taxonomic placement and origin of yaks: implications from analyses of mtDNA D-loop fragment sequences. *Acta Theriologica Sinica* **26**(4): 325–330. (In Chinese with English abstract)

Guo, S. C., Savolainen, P., Su, J. P., *et al.* (2006b). Origin of mitochondrial DNA diversity of domestic yaks. *BMC Evolutionary Biology* **6**: 73.

Guo, X., Yan, P., Chen, S. L., Yang, B. H., Zeng, Y. H. & Liang C. N. (2007). Conservation and utilization of genetic resources of wild yak in China. *Acta Ecologiae Animalis Domastici* **28**(5): 96–98. (In Chinese with English abstract)

Han, J. L. (2003). Molecular and cytogenetics in yak: a scientific basis for breeding and evidence for phylogeny. In *The Yak*, pp. 415–435. 2nd edition. Eds. Wiener, G., Han, J. L. & Long, R. J. Bangkok: The Regional Office for Asia and the Pacific, Food and Agriculture Organization of the United Nations.

Han, J. L. (2014). Domestication of yak. In *Encyclopedia of Global Archaeology*, pp. 7939–7941 Ed. Smith, C. New York: Springer Science + Business Media. DOI: 10.1007/978-1-4419-0465-2.

Han, Z. H. (2001). Diagnosis and treatment of enteritis necroticans due to infection of *Clostridium welchii* in wild yak. *Journal of Traditional Chinese Veterinary Medicine* **2**: 27–28. (In Chinese)

Harper, F. (1940). The nomenclature and type localities of certain Old World mammals. *Journal of Mammalogy* **21**(3): 322–332.

Harper, F. (1945). *Extinct and Vanishing Mammals of the Old World*. Baltimore, MD: The Lord Baltimore Press.

Harris, R. B. (2008). *Wildlife Conservation in China: Preserving the Habitat of China's Wild West*. Armonk, NY: M. E. Sharpe.

Harris, R. B. & Leslie, D. (2008). *Bos mutus*. In *Threatened Species: IUCN Red List 2012*. Version 2012.2. www.iucnredlist.org, accessed 11 April 2013.

Harris, R. B. & Loggers, C. O. (2004). Status of Tibetan plateau mammals in Yeniugou, China. *Wildlife Biology* **10**(2): 91–99.

Harris, R. B. & Miller, D. J. (1995). Overlap in summer habitats and diets of Tibetan Plateau ungulates. *Mammalia* **59**(2): 197–212.

Harris, R. B., Miller, D. J. & Cai, G. Q. (1996). Mammalian status and conservation in Yeniugou, Qinghai, China. *Acta Theriologica Sinica* **16**(2): 113–118. (In Chinese with English abstract)

Harris, R. B., Pletscher, D. H., Loggers, C. O. & Miller, D. J. (1999). Status and trends of Tibetan plateau mammalian fauna, Yeniugou, China. *Biological Conservation* **87**(1): 13–19.

Harris, R. B., Ali, A. & Loggers, C. (2005). Trend monitoring of large mammals: two case studies. *Acta Theriologica Sinina* **25** (4): 319–325. (In English with Chinese abstract)

Hartl, G. B., Göltenboth, R., Grilltsch, M. & Willing, R. (1988). On the biochemical systematics of the Bovini. *Biochemical Systematics and Ecology* **16**(6): 575–579.

Hassanin, A. & Douzery, E. J. P. (1999). Evolutionary affinities of the enigmatic saola (*Pseudoryx nghetinhensis*) in the context of the molecular phylogeny of Bovidae. *Proceedings of the Royal Society B: Biological Sciences* **266**(1422): 893–900.

Hassanin, A. & Ropiquet, A. (2004). Molecular phylogeny of the tribe Bovini (Bovidae, Bovinae) and the taxonomic status of the Kouprey, *Bos sauveli* Urbain 1937. *Molecular Phylogenetics and Evolution* **33**(3): 896–907.

Hassanin, A. & Ropiquet, A. (2007). Resolving a zoological mystery: the kouprey is a real species. *Proceedings of the Royal Society B: Biological Sciences* **274**(1627): 2849–2855.

Hassanin, A., Delsuc, F., Ropiquet, A., *et al.* (2012). Pattern and timing of diversification of Cetartiodactyla (Mammalia, Laurasiatheria), as revealed by a comprehensive analysis of mitochondrial genomes. *Comptes Rendus Biologies*, **335**: 32–50.

Hassanin, A., An, J., Ropiquet, A., Nguyen, T. T. and Couloux, A. (2013). Combining multiple autosomal introns for studying shallow phylogeny and taxonomy of Laurasiatherian mammals: application to the tribe Bovini (Cetartiodactyla, Bovidae). *Molecular Phylogenetics and Evolution* **66**(3): 766–775.

He, X. L. (2009). Breeding and extending experience of Datong Yak breed. *China Cattle Science* **35**(5): 27–31. (In Chinese)

Hodgson, B. H. (1832). On the mammalia of Nepal. *Journal of the Asiatic Society of Bengal* **1**: 335–349.

Hodgson, B. H. (1834). On the mammalia of Nepal. *Proceedings of the Zoological Society of London* **1834**: 95–104.

Hodgson, B. H. (1841a). Illustrations of the genera of the Bovinae: part 1. Skeletons of *Bos*, *Bibos*, and *Bison*; the individuals examined being the common bull of Nepal, the Gowri Gao of Nepal and the yak. *Journal of the Asiatic Society of Bengal* **10**: 449–470.

Hodgson, B. H. (1841b). Classified catalogue of mammals of Nepal (corrected to end of 1841, first presented 1832). *Journal of the Asiatic Society of Bengal* **10**: 907–916.

Hodgson, B. H. (1842). Notice of the mammals of Tibet, with description and plates of some new species. *Journal of the Asiatic Society of Bengal* **11**: 275–289.

Hodgson, B. H. (1844). Classified catalogue of mammals of Nepal (corrected to end of 1841, first presented 1832). *Calcutta Journal of Natural History and Miscellany of the Art and Sciences in India* **4**: 284–294.

Hu, J., Zhang, R. C. & Lu, Z. L. (2005). Biological feature and genetic diversity of wild yak, domestic yak and their hybrids. *China Herbivores*, Special Issue: 104–112. (In Chinese)

Huang, Z. H., Liu, N. F., Zhang, L. X., Long, J. & Liu, C. B. (2005). Change of ungulate species resource in Gansu Yanchiwan Natural Reserve. *Journal of Economic Animal* **9**(4): 246–248. (In Chinese with English abstract)

International Commission on Zoological Nomenclature. (2003). Opinion 2027 (Case 3010): usage of 17 specific names based on wild species which are predated by or contemporary with those based on domestic animals (Lepidoptera, Osteichthyes, Mammalia): conserved. *Bulletin of Zoological Nomenclature* **60**: 81–84.

IUCN (2013). *IUCN Red List of Threatened Species*. Version 2013.2. www.iucnredlist.org, accessed 29 May 2014.

Jacoby, C. P. & Fonseca, C. G. (2000). Linguistic analysis in phylogeny estimation: a case study of mtDNAs of Bovidae. *Journal of Biomolecular Structure and Dynamics* **17**: 1047–1055.

Janecek, L. L., Honeycutt, R. L., Adkins, R. M. & Davis, S. K. (1996). Mitochondrial gene sequences and the molecular systematics of the artiodactyl subfamily bovinae. *Molecular and Phylogenetic Evolution* **6**(1): 107–119.

Jardine, W. (1836). *The Naturalist's Library*. Vol. **22**. Mammalia. Ruminantia: Part II. London: Chatto & Windus.

Jerdon, T. C. (1874). *The Mammals of India: A Natural History of All the Animals Known to Inhabit Continental India*. London: John Wheldon.

Kraus, F., Jarecki, L., Miyamoto, M. M., Tanhauser, S. M. & Laipis, P. J. (1992). Mispairing and compensational changes during the evolution of mitochondrial ribosomal RNA. *Molecular and Biological Evolution* **9**(4): 770–774.

Lai, S. J., Chen, S. Y., Liu, Y. P. & Yao, Y. G. (2007). Mitochondrial DNA sequence diversity and origin of Chinese domestic yak. *Animal Genetics* **38**(1): 77–80.

Lenstra, J. A. & Bradley, D. G. (1999). Systematics and phylogeny of cattle. In *The Genetics of Cattle*, pp. 1–14. Eds. Fries, R. & Ruvinsky, A. Wallingford: CABI.

Leslie, D. M., Jr & Schaller, G. B. (2009). *Bos grunniens and Bos mutus* (Artiodactyla: Bovidae). *Mammalian Species* **836**: 1–17.

Li, J. T. (2004). Taming and utilization of wild yak. *Journal of Minorities Teachers College of Qinghai Teachers University* **15** (1): 61–64. (In Chinese)

Li, J. X., Qi, D. L. & Zhang, W. X. (1998). Study on karyotype of wild yak. *Gansu Animal and Veterinary Sciences* **3**: 8–10. (In Chinese)

Li, J. Y., Han, K. & Ma, Z. C. (1997). Physiological parameters in bloods of wild yak bull. *Chinese Qinghai Journal of Animal and Veterinary Sciences* **27**(1): 12–13. (In Chinese with English abstract)

Li, K. L., Lu, H. J., Guo, G., *et al.* (2005). Taming for semen collection and utilization of wild yak. *China Herbivores*, Special Issue: 17–18, 45. (In Chinese)

Li, Q. F., Li, Y. X., Zhao, X. B., *et al.* (2006a). Sequencing *cytochrome b* gene of mitochondrial DNA in yak and researching its origin and taxonomic status. *Acta Veterinaria et Zootechnica Sinica* **37**(11): 1118–1123. (In Chinese with English abstract)

Li, Q. F., Zhao, X. B., Liu, H. L., Li, N. & Xie, Z. (2006b). A review of the research on taxonomic status in yak (*Poephagus*). *Acta Zootaxonomica Sinica* **31**(3): 520–524. (In Chinese with English abstract)

Li, Q. F., Li, Y. X., Zhao, X. B., *et al.* (2008). Sequencing D-loop region of mitochondrial DNA in yak and study on its taxonomic status in Bovinae. *Acta Veterinaria et Zootechnica Sinica* **39**(1): 1–6. (In Chinese with English abstract)

Li, R. M. (1986). Records of bovine species in ancient Chinese history. In *Proceedings of Papers about History of Animal Husbandry in China*, pp. 146–149. Eds. Zhang, Z. & Zhu, X. Beijing: China Science Press. (In Chinese)

Li, Shih-Chen (Li Shizhen). (1596). *Compendium of Materia Medica, or The Great Herbal. (Pen Ts'ao Kang Mu, or Ben Cao Gang Mu)*. (In Chinese).

Li, Z. C., Yang, Q. S., Zhang, H. B., Liu, Z. H. & Yuan, G. (2006). Population number and distribution of big mammals in eastern Arjin Mountain Nature Reserve, Xinjiang. *Sichuan Journal of Zoology* **25** (1): 92–95. (In Chinese with English abstract)

Lian, X. M., Li, X. X. & Xu, T. (2012). Avoidance distances of four ungulates from roads in Kekexili and related protection suggestions. *Chinese Journal of Ecology* **31**(1): 81–86. (In Chinese with English abstract)

Linnaeus (Linnaei), C. (1758). *Systema Naturae per Regna Tria Naturae, secvndum Classes, Ordines, Genera, Species, cum Characteribus, Differentiis, Synonymis, Locis. Tomus I. Editio Decima, Reformata. Holmiae, Impensis Direct. Laurentii Salvii.*

Linnaeus (Linné), C. (1766). *Systema Naturae per Regna Tria Naturae, secvndum Classes, Ordines, Genera, Species cum Characteribus, Differentiis, Synonymis, Locis. Tomus I. Editio decima tertia, ad Etitionem duodecimam reformatam Holmienfem. Vindobonae, Typis Ioannis Thomae nob. de Trattnern, Cars. Reg. Aulae Typogr. et Bibliopolae.*

Linnaeus (Linné), C. (1788). *Systema Naturae per Regna Tria Naturae, secvndum Classes, Ordines, Genera, Species, cum Characteribus, Differentiis, Synonymis, Locis. Tomus I. Editio Decima, Tertia, Aucta, Reformata (Cora J. F. Gmelin). Lipsiae.*

Liu, W. L. (2007a). Yak. *Tibet Geographic* **1**: 100–107. (In Chinese)

Liu, W. L. (2007b). The origins and current situation of wild yaks. *Journal of Tibet University* **22**(1): 114–117. (In Chinese with English abstract)

Liu, W. L. & Schaller, G. B. (2003). Distribution and current status of wild yak. *Tibet's Science and Technology* **11**: 17–23. (In Chinese)

Liu, Z. Y. & Wang, H. T. (2006). A case of bovine *Pasteurella multocida* in wild yak in zoo. *Proceedings of Zoos in China* **2**: 198. (In Chinese)

Lu, Z. L. (2005). Reproductive characteristics of wild yak bulls. *China Herbivores*, Special Issue: 139–140, 61. (In Chinese)

Lu, Z. L., Li, K. L., Lu, H. J., Guo, G. & Wang, M. Q. (1991). Investigation of high reproductivity of wild yak bull. *Gansu Animal and Veterinary Sciences* **21**(5): 5–7. (In Chinese)

Lu, Z. L., Li, K. L. & Lu, H. L. (1993). Distribution, ecological types and utilization of wild yak in China. *Chinese Journal of Zoology* **28**(4): 41–45. (In Chinese with English abstract)

Lu, Z. L., He, X. L. & Yan, P. (2005). Introduction to the first new yak breed in the world: 'Datong yak'. *China Herbivores* **25**(5): 59–61. (In Chinese)

Lydekker, R. (1898). *Wild Oxen, Sheep, Goats of All Lands*. London: Rowland Ward.

Lydekker, R. (1900). *The Great and Small Game of India, Burma, & Tibet*. London: Rowland Ward.

Lydekker, R. (1901). *The Great and Small Game of Europe, Western and Northern Asia and America: Their Distribution, Habits, and Structure*. London: Rowland Ward.

Lydekker, R. (1907). *The Game of India, Burma, Malaya, and Tibet*. London: Rowland Ward.

Lydekker, R. (1913). *Catalogue of the Ungulate Mammals in the Britain Museum (Natural History), Vol. I, Artiodactyla, Family Bovidae, Subfamilies Bovinae to Ovibovinae*. London: William Clowes and Sons, Limited.

Ma, G. J. (2012). Comparison of birth weights of wild yak and Datong yak. *Heilongjiang Journal of Animal Reproduction*, **20**(5): 11–12. (In Chinese).

Ma, R. J. & Jiang, Z. G. (2006). Impacts of environmental degradation on wild vertebrates in the Qinghai Lake drainage, China. *Acta Ecologica Sinica*, **26**(9): 3066–3073. (In Chinese with English abstract)

Ma, Z. J. (2008). Sequence comparison of partial *GHRHR* gene among wild yak (*Bos grunniens mutus*) and other species (Bovidae). *Journal of Qinghai University (Natural Science)* **26**(5): 69–73. (In Chinese with English abstract)

Ma, Z. J., Zhong, J. C., Cheng, Z. H., Liu, L., Chang, H. P. & Luo, X. L. (2007). Sequence variation and molecular evolution of hormone-sensitive lipase genes in species of Bovidae. *Journal of Genetics and Genomics* **34**(1): 26–34.

Ma, Z. J., Zhong, J. C., Han, J. L., Xu, J. T., Dou, Q. L. & Chang, H. P. (2009). Genetic diversity in the mtDNA D-Loop region of wild yak (*Bos grunniens mutus*). *Acta Ecologica Sinina* **29**(9): 4798–4803. (In Chinese with English abstract)

Ma, Z. J., Zhong, J., Han, J. L., *et al.* (2010). Genetic diversity and demographic history of wild yak (*Bos grunniens mutus*) inferred from mtDNA D-loop sequences. *African Journal of Biotechnology* **9**(46): 7805–7810.

Ma, Z. J., Song, H., Zhong, J., Stanton, D., Wei, Y. & Sun, Y. (2012). Molecular characterisation of the wild yak (*Bos grunniens mutus*) melanocortin receptor-4 (*MC4R*) gene and a comparative analysis with that of other Bovinae species. *Journal of Applied Animal Research* **40**(2): 81–85.

MacEachern, S., McEwan, J. & Goddard, M. (2009). Phylogenetic reconstruction and the identification of ancient polymorphism in the Bovini tribe (Bovidae, Bovinae). *BMC Genomics* **10**: 177.

Miller, D. J. & Bedunah, D. J. (1994). Rangelands of the Kunlun Mountains in western China. *Rangelands* **16**(2): 71–76.

Miller, D. J. & Schaller, G. B. (1996). Rangelands of the Chang Tang Wildlife Reserve in Tibet. *Rangelands* **18**(3): 91–96.

Miller, D. J., Harris, R. D. & Cai, G. Q. (1994). Wild yaks and their conservation on the Tibetan Plateau. In *Proceedings of the First International Congress on Yak*, pp. 27–34. Eds. Zhang, R. C., Wu, J. P. & Han, J. L. Lanzhou, Gansu.

Mipam, T. D., Wen, Y. L., Fu, C. X., *et al.* (2012). Maternal phylogeny of a newly-found yak population in China. *International Journal of Molecular Sciences* **13**(9): 11455–11470.

Miyamoto, M. M., Tanhauser, S. M. & Laipis, P. J. (1989). Systematic relationship in the artiodactyls tribe Bovini (family Bovidae), as determined from mitochondrial DNA sequences. *Systematic Zoology* **38**(4): 342–349.

Namgail, T. (2009). Mountain ungulates of the Trans-Himalayan region of Ladakh, India. *International Journal of Wilderness* **15**(2): 35–40.

Nijman, I. J., van Boxtel, D. C. J., van Cann, L. M., Marnoch, Y., Cuppen, E. & Lenstra, J. A. (2008). Phylogeny of Y chromosomes from bovine species. *Cladistics* **24**: 723–726.

Olsen, S. J. (1990). Fossil ancestry of the yak, its cultural significance and domestication in Tibet. *Proceedings of the Academy of Natural Sciences of Philadelphia* **142**: 73–100.

Olsen, S. J. (1991). Confused yak taxonomy and evidence of domestication. *Illinois State Museum Scientific Papers* **23**: 387–393.

Peng, X. W. & Ouyang, X. (1999). Ecological adaptability of wild yak (*Bos mutus*). *Ecology of Domestic Animals* **20**(3): 20–23. (In Chinese)

Piao, R. Z., Ma, Y. Q. & Cui, H. S. (1999). Current situation of the yak population in China. *Life Science Research* **3**(2): 161–169. (In Chinese with English abstract)

Pilgrim, G. E. (1947). The evolution of the buffaloes, oxen, sheep and goats. *Journal of the Linnean Society* **41**: 272–286.

Pitra, C., Fürbass, R. & Seyfert, H. (1997). Molecular phylogeny of the tribe Bovini (Mammalia: Artiodactyla): alternative placement of the Anoa. *Journal of Evolutionary Biology* **10**(4): 589–600.

Przewalski, N. M. (1883). *From Zaisan through Khami to the Tibet and upper reaches of the Yellow River: Third journey in the Central Asia (1879–1880)*. St Petersburg: Imperial Russian Geographic Society. (In Russian)

Qi, X. B., Han, J. L., Blench, R., Rege, J. E. O. & Hanotte, O. (2008). Understanding the yak pastoralism in Central Asian Highlands: genetic evidence for origin, domestication and dispersion of domestic yak. In *Past Human Migrations in East Asia: Matching Archaeology, Linguistics and Genetics*, pp. 427–442. Eds. Sanchez-Mazas, A., Blench, R., Ross, M. D., Peiros, I. & Lin, M. London and New York: Routledge.

Qi, X. B., Han, J. L., Wang, G., Rege, J. E. O. & Hanotte, O. (2010). Assessment of cattle genetic introgression into domestic yak populations using mitochondrial and microsatellite DNA markers. *Animal Genetics* **41**(3): 242–252.

Qiu, Q., Zhang, G. J., Ma, T., *et al.* (2012). The yak genome and adaptation to life at high altitude. *Nature Genetics* **44**(8): 946–949.

Ritz, L. R. (1997). Genetic diversity in the Tribe Bovini. PhD Dissertation. University of Bern.

Ritz, L. R., Glowatzki-Mullis, M. L., MacHugh, D. E. & Gaillard, C. (2000). Phylogenetic analysis of the tribe Bovini using microsatellites. *Animal Genetics* **31** (3): 178–185.

Schaller, G. B. (1998). *Wildlife of the Tibetan Steppe*. Chicago, IL: University of Chicago Press.

Schaller, G. B. & Liu, W. (1996). Distribution, status, and conservation of wild yak *Bos grunniens*. *Biological Conservation* **76**(1): 1–8.

Schaller, G. B., Kang, A. L., Hashi, T. D. & Cai, P. (2007). A winter wildlife survey in the northern Qiangtang of Tibet Autonomous Region and Qinghai Province, China. *Acta Theriologica Sinica* **27**(4): 309–316. (In English with Chinese abstract)

Schreiber, A., Seibold, I., Nötzold, G. & Wink, M. (1999). Cytochrome *b* gene haplotypes characterize chromosomal lineages of anoa, the Sulawesi dwarf buffalo. *Journal of Heredity* **90**(1): 165–176.

Smith, C. H. (1858). Introduction to Mammalia. In *Naturalist's Library*, Vol. **15**, pp. 288–290. Ed. Jardine, W. London: Chatto & Windus.

Sterndale, R. A. (1884). *Natural History of the Mammalia of India and Ceylon*. London: W. Thacker and Co.

Turner, H. N. (1850). On the generic subdivision of the Bovidae, or hollow-horned ruminants. *Proceedings of the Zoological Society of London*, **18**: 177–178.

Vasey, G. (1857). *A Monograph of the Genus* Bos: *The Natural History of Bulls, Bison, and Buffaloes, Exhibiting all the Known Species, and the More Remarkable Varieties*. London: John Russell Smith.

Verkaar, E. L. C., Vervaecke, H., Roden, C., *et al.* (2003). Paternally inherited markers in bovine hybrid populations. *Heredity* **91** (6): 565–569.

Verkaar, E. L. C., Nijman, I. J., Beeke, M., Hanekamp, E. & Lenstra, J. A. (2004). Maternal and paternal lineages in cross-breeding bovine species. Has wisent a hybrid origin? *Molecular Biology and Evolution* **21**(7): 1165–1170.

Wall, D. A., Davis, S. K. & Read, B. M. (1992). Phylogenetic relationships in the subfamily Bovinae (Mammalia: Artiodactyla) based on ribosomal DNA. *Journal of Mammalogy* **73**(2): 262–275.

Wang, J. J. & Wang C. Z. (1963). The first specimen of wild yak *Bos grunniens mutus* Przewalski in China. *Chinese Journal of Zoology* **2**: 18. (In Chinese)

Wang, W. (2012). Research on growth and development in wild yak. *Chinese Qinghai Journal of Animal and Veterinary Sciences* **42**(2): 8–10. (In Chinese)

Wang, Y. (2001). Biochemical parameters in blood of wild yak bulls. *Chinese Journal of Veterinary Science and Technology* **31**(9): 24–25. (In Chinese)

Wang, Z. F., Shen, X., Liu, B., *et al.* (2010). Phylogeographical analyses of domestic and wild yaks based on mitochondrial DNA: new data and reappraisal. *Journal of Biogeography* **37**(12): 2332–2344.

Wang, Z. F., Yonezawa, T., Liu, B., *et al.* (2011). Domestication relaxed selective constraints on the yak mitochondrial genome. *Molecular Biology and Evolution* **28**(5): 1553–1556.

Wang, Z. J. & Sun, Y. S. (2005). Experience and suggestion on the management during taming of pure wild yak. *Chinese Qinghai Journal of Animal and Veterinary Sciences* **35**(1): 34–35. (In Chinese)

Ward, T. J., Bielawski, J. P., Davis, S. K., Templeton, J. W. & Derr, J. N. (1999). Identification of domestic cattle hybrids in wild cattle and bison species: a general approach using mtDNA markers and the parametric bootstrap. *Animal Conservation* **2**(1): 51–57.

Wiener, G., Han, J. L. & Long, R. J. (2003). *The Yak*. 2nd edition. Bangkok: The Regional Office for Asia and the Pacific, Food and Agriculture Organization of the United Nations.

Xia, L., Yang, Q. S., Xiang, Y. & Feng, Z. J. (2003). Distribution pattern and zoogeographical analysis of mammals in Qilian Mountain areas, Qinghai, China. *Acta Theriologica Sinica*, **23**(4): 295–303. (In Chinese with English abstract)

Xie, Y. L., Li, Y. X., Zhao, X. B., *et al.* (2010). Origins of the Chinese yak: evidence from maternal and paternal inheritance. Paper presented at the 4th International Conference on Bioinformatics and Biomedical Engineering, 18–20 June.

Xu, A. C., Jiang, Z. G., Li, C. W., Guo, J. X., Wu, G. S. &. Cai, P. (2006). Summer food habits of brown bears in Kekexili Nature Reserve, Qinghai-Tibetan Plateau, China. *Ursus* **17**(2): 132–137.

Xu, A. C., Jiang, Z. G., Li, C. W. & Cai, P. (2010). Food habits and hunting patterns of Tibetan brown bear during warm seasons in Kekexili region on Qinghai-Tibetan Plateau. *Zoological Research*, **31** (6): 670–674. (In Chinese with English abstract)

Xu, Y. D. & Li, K. L. (1991). Breeding utility value of wild yak by analyzing serum LDH. *Journal of Xiamen University (Natural Science)* **30**(4): 424–427. (In Chinese with English abstract)

Xu, Y. D. & Xu, L. (1998). Comparison of the contents of serum proteins and amino acids in wild yak and domestic yak in Qinghai. *Chinese Journal of Grass and Livestock* **3**: 9–10. (In Chinese)

Xu, Y. D., Ye, J. N., Lu, Z. L. & Li, K. L. (1995). Characteristics of serum α-amylase isozyme in wild yak under domestic conditions. *Journal of Xiamen University (Natural Science)* **34**(5): 786–790. (In Chinese with English abstract)

Xue, C. A. (2012). Notes on the extension and utilization of Datong Yak breed. *Chinese Qinghai Journal of Animal and Veterinary Sciences* **42**(2): 56.

Yang, B. H., Lang, X., Yan, P., Liang, C. N. & Sun, X. P. (2006). Construction for the system of genetic resource banks on wild yak in China. *Acta Ecologiae Animals Domastici*, **27**(1): 1-5. (In Chinese with English abstract)

Yang, G. D. (2008). Technical protocols for taming and management of wild yak. *China Cattle Science* **34**(6): 78–79. (In Chinese)

Yang, G. L., Meng, T. F., Li, K. L., Lu, Z. L. & Lu, Z. G. (1992). Relationship between enzyme activities in sperm of wild yak and conception rate. *Journal of China Yak*, **1**: 30–34. (In Chinese)

Yang, Q. S. & Xia, L. (2008). Tibetan wildlife is getting used to the railway. *Nature* **452** (7189): 810–811.

Yang W. Y., Chen, X. M., Zhong, J. C., *et al.* (2009). Sequence analysis and evolution relationship of Cyt *b* gene in Bovidae. *China Herbivores* **29**(3): 8–13. (In Chinese with English abstract)

Yao, J., Yang, B. H., Yan, P., Liang C. N., Guo, J. & Jiao, S. (2006). Analysis on habitat variance and behaviours of *Bos grunniens* in China. *Acta Prataculturae Sinica* **15**(2): 124–128. (In Chinese with English abstract)

Yin, C. L. (2005). Conservation of biodiversity and sustainable pest management of forest in the Natural Reserve of Qilian Mountains. *Journal of Lanzhou University (Natural Sciences)* **41**(5): 33–37. (In Chinese with English abstract)

Zeyland, J., Wolko, Ł., Lipiński, D., *et al.* (2012). Tracking of wisent–bison–yak mitochondrial evolution. *Journal of Applied Genetics* **53**(3): 317–322.

Zhang, B. P. (1991). Physical environment and animal resources in the Alking natural protection region (Kumukule basin). *Journal of Arid Land Resources and Environment* **5**(1): 87–95. (In Chinese with English abstract)

Zhang, R. C. (1989). *China Yak*. Lanzhou: Gansu Scientific & Technology Press. (In Chinese.)

Zhang, R. C. (2000). Interspecies hybridization between yak, Bos taurus and Bos indicus and reproduction of the hybrids. International Veterinary Information Service, Paper No. A1304.0900.

Zhang, Z. G., Xia, L. & Yang, Q. S. (2009). Distribution and conservation of yak (*Bos grunniens*). *Chinese Journal of Zoology* **44** (1): 148–150. (In Chinese with English abstract)

Zhao, J. N. & Gao, X. Y. (1991). The wild yak. *Arid Zone Research* **1**: 34. (In Chinese)

Zhao, Z. Z. (2006). Reasoning of the 'genetic erosion' of wild yak in Tibet. *Zhejianglinye* **1**: 40–41. (In Chinese)

Zhao, Z., He, Y., Yang, P. Y., *et al.* (2011). Investigation on wild animal resource in Sunan and Subei counties. *Acta Prataculturae Sinica* **20**(2): 67–75. (In Chinese with English abstract)

Zheng, S. W., Yu, Y. Q., Han, Y. P., Wu, J. Y. & Zuo, Q. Y. (1989). Studies on the ungulates community at Yanchiwan Nature Reserve. *Acta Theriologica Sinica* **9**(2): 130–136. (In Chinese with English abstract)

Zhou, C. Y. (1989). Feeding and reproduction of wild yak. *Chinese Journal of Wildlife* **5**: 28–29. (In Chinese)

Zhou, C. Y. & Zhang, G. S. (1982). Wild yak born in zoo. *China Nature* **3**: 58. (In Chinese)

Zhou, J. D., Li, S. H. & Gu, J. H. (1985). The preliminary observation on the mammals in Kunlun-Altun basin. *Acta Theriologica Sinica* **5**(2): 160. (In Chinese)

Zhou, J. Q. (2003). Current situation, challenge and strategy of management of National Qilian Shan Nature Reserve. *Journal of Gansu Forestry Science and Technology* **28**(3): 36–38. (In Chinese)

Chapter 13

Banteng *Bos javanicus* d'Alton, 1823

Penny C. Gardner, Satyawan Pudyatmoko, Naris Bhumpakphan, Marnoch Yindee, Datuk Laurentius N. Ambu and Benoit Goossens

Names

Genus: *Bos* Linnaeus, 1758

Species: Banteng *Bos javanicus* d'Alton, 1823

Subspecies: Mainland banteng *Bos javanicus birmanicus* Lydekker, 1898. Java banteng *Bos javanicus javanicus* d'Alton, 1823. Bornean banteng *Bos javanicus lowi* Lydekker, 1912

Names in other languages: **French**: Banteng; **German:** Banteng; **Spanish**: Banteng; **Italian**: Banteng.

Other names: **Malaysian** (Sabah): Tembadau, Banting, Sapi hutan; **Thai**: Wau daeng วัวป่า; **Laotian**: Ngua Pha; **Myanmar**: Tsiane

Ecology, Evolution and Behaviour of Wild Cattle: Implications for Conservation, ed. M. Melletti and J. Burton. Published by Cambridge University Press.

Taxonomy

There is some evidence that suggests that the banteng is not a monophyletic species (see Hassanin & Ropiquet 2006); however, phylogenetic studies of this species are few and continue to be confounded by hybridization, small sample sizes (Nijman *et al.* 2003) and the inability to obtain tissue samples from all subspecies. Grzimek (1987) did not accept the separation of banteng into three subspecies as valid. He argued that interbreeding of small populations of wild banteng with domestic or feral cattle still occurred continuously. This suggestion was supported by Corbert & Hill (1992), who also did not recognize the existence of banteng subspecies. Timmins *et al.* (2008) proposed that *Bos javanicus lowi* should be included into *Bos javanicus javanicus*, yet no molecular studies have investigated the phylogeny of *B. j. lowi* due to the difficulty of obtaining tissue samples from wild individuals. Timmins *et al.* (2008) also proposed *Bos javanicus javanicus* and *Bos javanicus birmanicus* should be tentatively accepted as different subspecies. It is essential to clarify the genetic description of all subspecies through morphometric and molecular analysis prior to a merger of subspecies. For instance, four haplotypes from six faecal samples of banteng in Thai forests similar to Cambodia's banteng have been reported (Manatchaiworakul *et al.* 2011). However, these first analyses require further investigations to better understand the relationships between different banteng populations.

Until the beginning of the Quaternary, Java Island was still isolated from the Asian mainland, indicated by animals which were common dwellers of the island. The fossils found from this period demonstrate that the oldest faunal composition in Java is estimated as from 2–1. Ma (van den Berg *et al.* 2001), while the oldest fossils of early banteng (*Bibos palaesondaicus*) were discovered in Trinil HK (from 0.9 Ma to late Pleistocene) in East Java. During this period, the main habitat type in Java was open woodland. In East Java, early Banteng fossils were found in the younger stages of the fauna unit in Kedung Brubus (0.8–0.7 Ma), and in Ngandong in the late Pleistocene in Java. During the period of Kedung Brubus and Ngandong fauna unit, the environment was still dominated by open woodland, but in the next stage of fauna unit succession, namely Punung fauna unit (125–60 ka), the environment changed into moister forest. Fossils of early banteng were not found in this period, which might be due to acidic soils that decomposed banteng bones. The extinction of species might be due to dramatic climatic change. This finding suggests that early banteng were more adapted to open woodland than to rainforest (van den Berg *et al.* 2001). The oldest fossils of modern banteng (*Bos javanicus*) were found in Holocene caves in Sampung, Gua Jimbe and possibly Wajak, in East Java. These fossils were dated to approximately 10–6 ka (Whitten *et al.* 1996; van den Berg *et al.* 2001).

Subspecies and distribution

The species (*Bos javanicus*) is generally recognized as three distinct subspecies, mainly based on pronounced differences in phenotype appearance and colour (Halder 1976; Byers *et al.* 1995): *Bos javanicus birmanicus* in Myanmar, Thailand, Laos, Vietnam and Cambodia; *B. j. javanicus* in Java and Bali; and *B. j. lowi* in Borneo (Tun Yin 1967; Lekagul & McNeely 1988; Hassanin & Ropiquet 2006; Timmins *et al.* 2008). In general, the body size of *B. j. javanicus* in Java is larger than *B. j. birmanicus* in mainland Southeast Asia; *B. j. lowi* in Borneo is the smallest (Hoogerwerf 1970).

In the past, mainland banteng were found in larger distribution ranges than the other subspecies. Nowadays, mainland banteng are known to have disappeared from Bangladesh, India and South Malaysia (Medway 1969; Francis 2008). The worldwide distribution of banteng is represented in Plate 5.

Present distribution (mainland)

In Thailand, banteng can be found in 21 reserves within ten forest complexes. There are only four populations present in Huai Kha Khaeng Wildlife Sanctuary and Khaeng Krachan National Park in the west, and in Thong Pha Phum National Park and Khao Ang Ruenai Wildlife Sanctuary in the southeast part of Thailand (Bhumpakphan & McShea 2011).

In Myanmar, banteng were found in Piduang Wildlife Sanctuary (Tun Yin 1967). The species still occurs in scattered reserves in Indang forest in the north of the country (dry deciduous dipterocarp forest).

In Vietnam only five disconnected populations (estimated population 74–103 individuals) still exist. Herds in Vietnam are found in Yok Don National Park (30–40 individuals) and the Ea So Nature Reserve (23–31 individuals; Pedrono *et al.* 2009).

In Laos banteng were reported to be close to extinction, but recently herds have been observed roaming at the Phu Foi Lom mountainous area along the border of Thailand, and in both Buntharik-Yod Mon Wildlife Sanctuary and Dong Khanthung Proposed National Biodiversity Conservation Area.

In Cambodia banteng still persist in the dry forest on the northern plain of northern Cambodia along the border with Thailand. Banteng have also been reported in the northeastern plain to the east in Yok Don National Park of Vietnam.

Historical distribution (Borneo)

The historic distribution of banteng included Sarawak and Sabah (Malaysian Borneo) and Kalimantan (Indonesian Borneo). It is widely accepted that banteng previously occurred within Brunei and this has been noted many times (see Payne *et al.* 1985; National Research Council 2002; Timmins *et al.* 2008); however, no firsthand accounts of banteng encounters were found in any published book, peer-reviewed journal article or using an online search engine. Furthermore, it was suggested by Pedrono *et al.* (2009) that the banteng in Brunei were in fact *B. j. birmanicus* and not *B. j. lowi*; there is, however, no evidence to support this theory. With the exception of Sabah, no information is available on the historic distribution of banteng within Sarawak or Kalimantan. The distribution of banteng in Sabah and their population sizes were first estimated in 1982 (Davies & Payne 1982) using community-based discussions. Populations of banteng were characterized according to seven categories – definitely present (post-1978); unknown (post-1978); pre-1978 records; present but rare or abundance unknown; present with intermediate abundance; present and common; and possible cattle hybrids – and mapped using a grid of 10^2 miles (Davies & Payne 1982). The total population size of banteng (not including possible hybrids) in Sabah was estimated to be between 300 and 550 in 1982 (Davies & Payne 1982). At the time of the survey the road infrastructure of Sabah was limited and remote forest access was severely restricted; this probably impeded data collection in more remote areas. After 1982 widespread timber production and logging activity increased the access to forested areas and their interior and, as a consequence, the conversion of secondary forest to vast monoculture plantations accelerated. This, together with human population expansion and encroachment of forest habitat, has probably exterminated some banteng populations identified by Davies & Payne (1982).

Present distribution (Borneo)

The present-day distribution of banteng includes Sabah, (possibly) Sarawak and Kalimantan. In eastern Sabah, surveys of banteng were conducted by P. Gardner (unpublished) in Tabin Wildlife Reserve and Malua Forest Reserve in 2011–2013 using camera traps and sign detection. From the photographic captures obtained from the camera traps, multiple herds were identified in both locations and were recognized to be breeding successfully. Supplementary information on banteng in other forest reserves was obtained from other researchers using camera traps to study a range of species since 2008, and from the Sabah Wildlife Department, the Sabah Forestry Department and Malua BioBank. Minimum population sizes were estimated from photographs of identifiable individuals, from presence of signs (tracks and dung) in areas where no photographs were obtained and from direct sightings. Minimum population size estimates were >52 individuals in Tabin Wildlife Reserve and >35 individuals in Malua Forest Reserve (Table 13.1). An estimated 100 individuals were thought to be present in Kulamba Forest Reserve (Timmins *et al.* 2008); however, it is unclear how this figure was obtained. The current presence of banteng was confirmed in a number of historic locations (also identified in 1982 by Davies & Payne (1982)), including Segaliud and Tangkulap Forest Reserves. However, current data are too deficient to estimate present-day minimum population sizes. The current distribution of banteng in Sabah (Figure 13.1) is largely confined to the east and central districts; surveys of remnant populations in the north, south and west are yet to be conducted. One historic population in the Dent Peninsular of the Lahad Datu district was confirmed as extinct in 2012 due to complete loss of habitat, which occurred in the past three decades. Confirmation of the status of the remaining historic populations is required in order to establish the complete current distribution of banteng in Sabah. On the Sarawak/Sabah border and the northeast Sarawak/north Kalimantan border small populations of banteng may still persist (Belding Giman, personal communication 2012). In Kalimantan six banteng populations were identified by Pudyatmoko *et al.* (2012) using information collected from various sources during 2003 to 2009; the collective minimum population size estimate was approximately 106 in 2010; however, this information was based on guesses or counts using unsystematic techniques. The present distribution of banteng in Kalimantan extends from East Kalimantan to Central Kalimantan and includes protected national parks and unprotected land, including plantations.

Descriptive notes

Body measurements: **Shoulder height**: 120–190 cm; **Head and body length**: 245–383 cm; **Tail length**: 60 cm; **Body mass**: 400–900 kg. Dental formula: I 0/3, C 0/1, P 3/3, M 3/3 (× 2) = 32 (Wilson & Mittermeier 2011).

The three subspecies share many morphological characteristics, including a large white patch on the buttocks and white stockings on their lower legs (Lydekker 1913). The whitish or pale brown spot above the eye and the white band around the muzzle are not present in Bornean banteng (P. Gardner, personal observation).

The National Research Council (2002) stated that banteng in Borneo are the smallest of all three subspecies, but no guideline weights of Bornean banteng were provided to substantiate that claim. The weight of 'average-sized bulls' from Java and Burma was 635 kg and 825 kg for large bulls, while cows weighed 400 kg (National Research Council 2002). The estimated weight of a Bali cattle (domesticated form of banteng) bullock that was approximately three years of age from Lok Kawi Wildlife Park (Kota Kinabalu, Sabah) was 400 kg, recorded in 2012 (Sen Nathan, personal communication). The shoulder height of the same Bali cattle bullock was measured at 114 cm. The shoulder height of banteng in Java was estimated to be 160–190 cm for bulls and 140 cm for cows (National Research Council 2002), but the ages of the individuals were unknown.

Banteng are sexually dimorphic in both pelage and body shape. Mainland bulls vary from chestnut to rusty brown, while females and juveniles are chestnut. In Cambodia, 20% of the bulls are blackish, but in the Southern Peninsula of Thailand most bulls are black. On average the head measures 60 cm (Hoogerwerf 1970). The horn length varies from 25 cm to 39 cm from tip to tip and the horn girth about 16 cm to 20 cm in Java banteng for old cows (Hoogerwerf 1970).

Male calves are born with a light pelage and white buttocks (Figures 13.2–13.3). During the first month the male Bornean banteng develops white stockings and the first signs of poles or horns (P. Gardner, personal observation). After two months of age their pelage shows the first signs of darkening, beginning with the undercarriage, along the spine, the tip of the tail and facial features (P. Gardner, personal observation). Pelage

Table 13.1 Collated information on the status, location and size of remnant herds of banteng in Sabah (Borneo) documented over the past five years (2009–2013) by various sources.

Location	Population status	Minimum population size estimate	Data type	Source of information and year
		Sabah (Malaysia), Borneo		
Tabin Wildlife Reserve	Present and breeding	>52	Camera trap photographs, signs and direct observations	Gardner (unpublished), 2013
			Camera trap photographs, signs and dung sample	IZW/BORA/SWD/SFD (unpublished), 2012
			Direct observations and signs	Sabah Wildlife Department (unpublished), 2011
			Camera trap photographs	Hearn & Ross (unpublished), 2009
Dent Peninsula (Lahad Datu district)	Extinct	Extinct	Forest converted to oil palm plantation	Gardner (unpublished), 2013
Malua Forest Reserve	Present and breeding	>35	Camera trap photographs and signs	Gardner (unpublished), 2013
			Direct observations and signs	Sabah Forestry Department/Malua BioBank (unpublished), 2011–2012
			Camera trap photographs	Hearn & Ross (unpublished), 2010
Tangkulap Forest Reserve	Present but breeding status unknown	Unknown	Camera trap photographs	Wilting (unpublished), 2009
Segaliud Forest Reserve	Present and breeding	Unknown	Camera trap photographs	Wilting (unpublished), 2010
Danum Valley Conservation Area	Present but breeding status unknown	Unknown	Personal communication	Sabah Forestry Department/Malua BioBank (unpublished), 2013
Kulamba Forest Reserve	Present but breeding status unknown	100	Unknown Direct observation	(Timmins *et al.* 2008) (Sabah Wildlife Department, personal communication)
Deramakot Forest Reserve	Present and breeding but possible hybrid	100	Unknown Camera trap photographs	(Timmins *et al.* 2008) (Samejima *et al.* 2012)
Kinabatangan Wildlife Sanctuary (Sukau)	Unknown	Unknown	Community surveys	(Davies & Payne 1982)
Tongod District	Unknown	Unknown	Community surveys	(Davies & Payne 1982)
Kudat District	Unknown	Unknown	Community surveys	(Davies & Payne 1982)
Kudat Marudu Bay Forest Reserve	Unknown	Unknown	Community surveys	(Davies & Payne 1982)
Pitas district	Unknown	Unknown	Community surveys	(Davies & Payne 1982)
Pitas district	Unknown	Unknown	Community surveys	(Davies & Payne 1982)
Sugut district	Unknown	Unknown	Community surveys	(Davies & Payne 1982)
Kpg. Mansud (Kota Klias District, Interior Division)	Unknown	Unknown	Community surveys	(Davies & Payne 1982)
Sigapon (Keningau district, Interior Divison)	Unknown	Unknown	Community surveys	(Davies & Payne 1982)
Sipitang Forest Reserve (Interior Division)	Unknown	Unknown	Community surveys	(Davies & Payne 1982)
Nabawan (Interior Division)	Unknown	Unknown	Community surveys	(Davies & Payne 1982)
Tenom (Interior Division)	Unknown	Unknown	Community surveys	(Davies & Payne 1982)

Table 13.1 *(cont.)*

Location	Population status	Minimum population size estimate	Data type	Source of information and year
Maliau Basin Conservation Area	Present and breeding	>10	Camera trap photographs Direct observation	Brodie (unpublished), 2010 Amos (personal communication), 2011 Gardner (unpublished), 2013
Semporna District	Unknown	Unknown	Community surveys	(Davies & Payne 1982)
Ulu Kalumpang Forest Reserve (Kunak District)	Unknown	Unknown	Community surveys	(Davies & Payne 1982)
Tawau Hills National Park	Unknown	Unknown	Community surveys	(Davies & Payne 1982)
Tinagat Forest Reserve (Tawau District)	Unknown	Unknown	Community surveys	(Davies & Payne 1982)
		Kalimantan (Indonesia), Borneo		
Lamandau Regency	Unknown	Unknown	Population distributed in some villages	(Pudyatmoko *et al.* 2012)
Kutai National Park	Unknown	1989: 48 individuals 1993: 40 individuals 2002: 34 individuals 2003: 34 individuals	Unknown	(Pudyatmoko *et al.* 2012)
KayanMentarang National Park	Unknown	2008: 72 individuals	Unknown	(Pudyatmoko *et al.* 2012)
Nunukan Regency	Unknown	No quantitative data	Unknown	(Pudyatmoko *et al.* 2012)
Malinau Regency	Unknown	No quantitative data	Unknown	(Pudyatmoko *et al.* 2012)
Berau Regency	Unknown	No quantitative data	Unknown	(Pudyatmoko *et al.* 2012)

progressively darkens, and poles rapidly develop after two months; growth (in length) is estimated (approximately) to be one inch per month until at least nine months of age (P. Gardner, personal observation). At approximately 9–10 months coat colour is dark brown and horns are approximately five inches long and noticeably wider in spread, with tips pointing outward (P. Gardner, personal observation), and at approximately two years of age males have changed coat colour to black and poles are 'crescent shaped' (National Research Council 2002). Over two years of age banteng bulls develop heavier forequarters and a small dewlap, the coat turns black, with a black tail and white stockings and white buttocks (Figure 13.4). The poles continue to expand in dimension until approximately ten years of age and change from all black to black with brown at the base (P. Gardner, personal observation). The shape of the poles for a mature bull can range from broad, symmetrical and slightly crescent-shaped to broad, symmetrical and tightly crescent-shaped (P. Gardner, personal observation). Pole spread of bulls was estimated to be approximately 60–75 cm (National Research Council 2002); however, it is not clear which subspecies this measurement relates to.

Like males, female calves are also born with a light tan coat and white buttocks, and are identical in appearance to male calves (P. Gardner, personal observation). They develop poles and change pelage colour at a slower rate than males; at approximately 2–2.5 months old female calves develop white stockings and the first sign of poles (P. Gardner, personal observation). After this time their coat colour progressively darkens to brown with a dark undercarriage, black tip of the tail and slightly darker facial features, and they may or may not develop a dark-brown/black streak along the spine. At six months of age the poles of female calves are approximately 1–2 inches long and, at one year of age, poles are approximately six inches long and straight (P. Gardner, personal observation). By two years of age the poles are noticeably crescent-shaped, and unique features such as light forehead patches, imperfections on the poles and scars on the neck resulting from fly bites become more apparent (P. Gardner, personal observation).

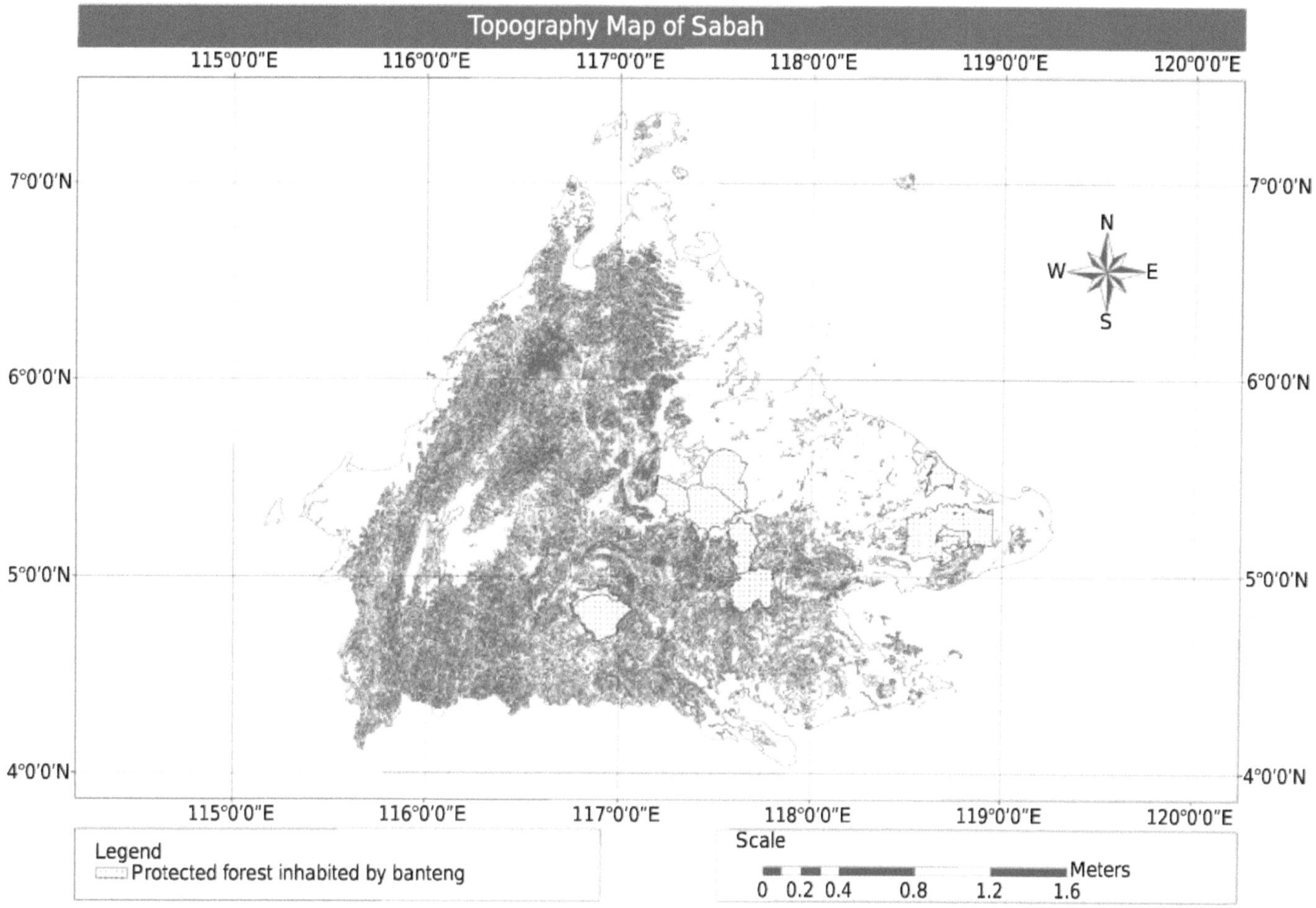

Figure 13.1 Map of Sabah (Malaysia, Borneo) depicting the protected forest areas inhabited by banteng during the past five years (2009–2013, Penny C. Gardner).

Figure 13.2 A mixed herd of banteng in Alas Purwo National Park (Java) consisting of bulls, cows, subadults and juveniles (photo by Pudyatmoko). A black and white version of this figure will appear in some formats. For the colour version, please refer to the plate section.

Figure 13.3 The banteng form herds with several clans (cows with calf and juvenile in Thailand); herds range from 2 to 30 individuals. Each herd is usually led by an old cow. A black and white version of this figure will appear in some formats. For the colour version, please refer to the plate section.

Figure 13.4 Mature dominant bull in Malua Forest Reserve/Malua BioBank (Borneo), captured on camera trap in 2011 (© Danau Girang Field Centre/Sabah Wildlife Department using Reconyx PC800 HyperFire Professional Semi-Covert IR camera trap). A black and white version of this figure will appear in some formats. For the colour version, please refer to the plate section.

The pelage colour of mature cows ranges from very light brown to dark brown with characteristic white rump and stockings (P. Gardner, personal observation). The poles of cows darken with age from brown to dark brown or black, with a shape that ranges from symmetrical and slightly crescent-shaped, to non-symmetrical, long and slightly inverted, and broken (P. Gardner, personal observation).

Habitat

Mainland banteng mostly inhabit dry forests in the lowland areas in Thailand (Nakhasathien *et al.* 1987; Prayurasithi 1987). The banteng population in Huai Kha Khaeng Wildlife Sanctuary, western Thailand, strictly uses the lowland valleys of 160–600 m elevation. The species at Xe Pian Biodiversity

Figure 13.5 Tropical lowland dipterocarp forest is the principal habitat type of banteng in Sabah, Borneo (photo by Penny C. Gardner). A black and white version of this figure will appear in some formats. For the colour version, please refer to the plate section.

Conservation Area in Laos use similar habitats (Steinmetz 2004). However, banteng may show some seasonal difference in habitat use.

Until the beginning of the nineteenth century banteng were still present in almost all areas of Java Island (Hoogerwerf 1970). The altitudinal range of this subspecies included coastal forest, lowland and highland forest up to 2000 m above sea level. The German botanist Junghuhn reported in the eighteenth century that banteng preferred to live in high forests in west Java. Interestingly, he had never seen banteng eastward of Mount Tengger and Mount Semeru. Hence, he never reported the occurrence of banteng in the current main habitats in Java, namely in Meru Betiri, Baluran and Alas Purwo National Park. Due to rapid human population growth and intensive development in Java, large forest areas were converted for agricultural purposes and buildings. Habitat loss and fragmentation were the main causes of population decrease and range shrinking of banteng in Java.

In Sabah (Borneo) the predominant habitat type inhabited by banteng is tropical lowland dipterocarp forest (Timmins *et al.* 2008; P. Gardner, unpublished), which occurs <300 m in elevation and comprises numerous species of *Dipterocarpacea*, including *Dipterocarpus*, *Shorea* and *Dryobalanops aromatica* (Whitemore 1990). Secondary tropical lowland dipterocarp forest that is regenerating is characterized by dense vegetation at ground level and includes species such as wild ginger, rattan and vines and lianas, which provide favourable refuge for banteng (Figure 13.5). Banteng are also found in mixed hill dipterocarp forest (P. Gardner, personal observation), seasonal freshwater swamp forest and beach forest (Davies & Payne 1982; Timmins *et al.* 2008). Banteng have not been observed in mangrove or Nipah forests.

Seasonal variation in habitat use was only observed in one instance in seasonal freshwater swamp forest in Eastern Sabah; local villagers often observed banteng foraging in a secluded open grassland swamp during the summer months/dry season (April–September) but never during the wet/monsoon season (October–March). During the monsoon season the main river, inhabited by saltwater crocodiles, and smaller tributaries flooded the surrounding freshwater swamp forest. The combination of physical obstruction by the sheer volume of water and the quantity of sediment within the water probably decreased the conspicuousness of crocodiles. This, in turn, may have reduced site selection and habitat favourability during the wetter months.

Movements and home range

A study on banteng home range was carried out at Huai Kha Khaeng wildlife sanctuary (Thailand) by Prayurasithi (1997). He reported that herds utilized a home range of 20.0–44.0 km^2 in the dry season and 30.0–44.8 km^2 in the wet season. Individuals released from captivity formed a single herd and used a home range larger than 8 km^2.

The pattern of banteng movement was different from one site to another in Java. Alikodra & Sastradipraja (1983) reported that the movement of banteng in Ujung Kulon National Park was not regulated by seasons. Herds of banteng usually moved in the same pattern and used the same area regardless of the season. Regular movement in a relatively small area resulted from a small difference in resource availability in dry and wet seasons. A differential movement pattern was observed in Baluran National Park, Java (Halder 1976). Banteng were strongly but indirectly influenced by season (Halder 1976), and directly by the distribution of water, food and predators, as well as disturbance caused by human activities (Pudyatmoko 2005). Observations in Baluran National Park showed that

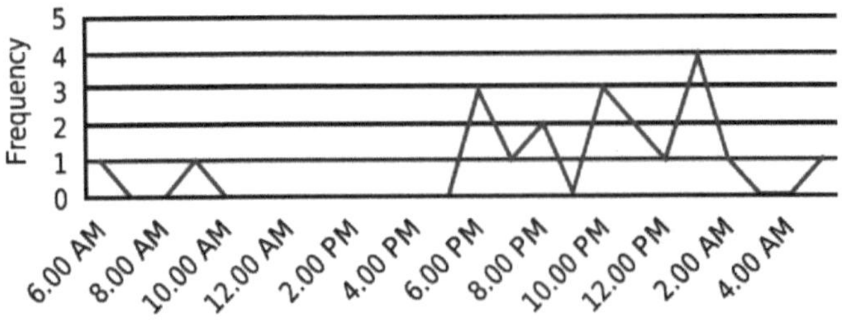

Figure 13.6 Diary of bantengs' activity recorded from camera trapping in Khao Ang Runai Wildlife Sanctuary, Thailand (Bhumpakphan & Yindee, unpublished data).

in the dry season most individuals concentrate in and around Bekol, where food and water are more abundant. In the rainy season, when food and water are available everywhere, the subspecies move throughout the whole area. Banteng require a mosaic habitat within their home range. An ideal habitat should be composed by large secondary forest and grazing area with enough fresh water.

No estimates on home range size or daily travel distances are available for Borneo. Photographic recaptures of recognizable mature bulls in 2009–2013 in Malua Forest Reserve and Tabin Wildlife Reserve indicated that they travelled at least a distance of 10 km and 23 km, respectively (P. Gardner, unpublished). Attempts to capture and collar wild banteng are currently under way in Sabah by Danau Girang Field Centre and the Sabah Wildlife Department; the purpose of this is to document banteng home range size, movement patterns in relation to forest use and habitat disturbance and target areas for conservation management.

Activity patterns

In undisturbed conditions, the banteng's daily activity has a more or less fixed rhythm (Figure 13.6). During the daytime, individuals alternate active and resting periods of 2–3 hours each. Active periods predominantly involve feeding, drinking and social interactions at night. If the pressure by predators and poachers is strong they will shift their activity during the night. Halder (1976) reported that in Ujung Kulon National Park (Java) individuals were active in open areas both at night and during the day. However, he found that banteng in Baluran National Park (Java) were only active in the grassland during the night. This observation was not consistent with the later monitoring conducted by Pudyatmoko (2005). In Baluran National Park, several groups graze in the open area in the early morning until 08:00, and occasionally until 10:00. Banteng then enter the forest to avoid the hot temperatures in the open area. They appear again in the afternoon at about 15:00, and stay there until the night. In Alas Purwo National Park they behave in the same manner as those in Baluran National Park. The most active social activities take place during the morning and afternoon. In Ujung Kulon National Park, during the day individuals spend two hours grazing followed by the same amount of resting period (Halder 1976).

Initial observations of banteng in Borneo by P. Gardner (unpublished) in 2011 and 2012 indicated activity peaked at dawn and during the early morning hours of 05:00–07:00, when banteng emerged from the forest to forage and socialize. No sexual segregation of activity patterns was found. Activity was found to decrease in the later hours of the morning (>09:00) until early afternoon (15:00), when banteng were observed resting and ruminating. In the early evening and at dusk (17:00–18:00) a second peak in banteng activity occurred during which they were observed foraging and socializing. Banteng activity decreased during the night, particularly around 20:00–21:00 and in the early morning at 03:00–04:00; the reasons for this were unknown. No instances of wallowing were observed, despite the monitoring of numerous mud wallows for extended periods of time using camera traps. Furthermore, no evidence of banteng were found at the two mud volcanoes in Tabin Wildlife Reserve.

Feeding ecology

Banteng can be classified as a mix feeder that frequently forages in open grasslands, dry forest and on the edges of evergreen forests. The species sometimes also feed on bamboos, fruits and leaves. In Huai Kha Khaeng (Thailand) they utilize up to 150 different plant species (Prayurasithi 1997; Bhumpakphan & McShea 2011). Banteng are observed at salt licks and waterholes regularly (Tun Yin 1967; Lekagul & McNeely 1988; Prayurasithi 1997).

Faecal analysis and direct observations conducted in Baluran and Alas Purwo National Park (Java) in Ujung Kulon (Hoogerwerf 1970; Halder 1976; Pudyatmoko 2005; Pairah 2007) discovered that the diet consists of grass, tree leaves, herbs, shrubs, twigs and fruit, as well as bark of certain tree species. Table 13.2 lists the main grasses, herb/bush and tree species recorded in Baluran National Park, Java (Halder 1976; Pudyatmoko 2005).

Preliminary observations suggest that *B. j. lowi* in Borneo forage on grasses, herbs, soft vines and wild fruits such as guava (Myrtaceae) (P. Gardner, personal observation).

Within tropical lowland dipterocarp forest in Sabah (Borneo) the preferred foraging areas (characterized by heavy use) were found to be interior openings, because they contained an abundance of suitable grassy forage and were less disturbed by human activity. Foraging behaviour was found to vary with forest structure. Observations of a forest logged in 2009 found banteng repeatedly foraged along internal abandoned logging roads and in interior openings, and banteng were rarely sighted and infrequently detected (by dung deposits) along the main access road (P. Gardner, personal observation). In comparison, observations of an older logged forest which had regenerated over 25 years found that banteng repeatedly foraged along the main access roads (detected by dung and tracks), open river banks and occasional interior openings (P. Gardner, personal observation). It was thought that the early stages of regeneration provided temporary grass-

Table 13.2 Main grass, herb/bush and tree species recorded in Baluran National Park, Java

No.	Grass species	Herb/bush species	Tree species
1	*Brachiaria reptans*	*Aneilema pudiflorum*	*Sesbania sesbans*
2	*Sclerachne punctata*	*Plumbago zeylanica*	*Tamarindus indica*
3	*Themeda arguens*	*Abutilon crispum*	*Acacia nilotica*
4	*Heteropogon contortus*	*Benincasa hispida*	*Schleicera oleosa*
5	*Themeda triandra*	*Portula oleraceae*	*Grewia eriocarpa*
6	*Schlerachne punctata*	*Bidens pilosa*	*Thespesia lampas*
7	*Echlinochloa colonum*	*Aeschynomene indica*	*Phillantus emblica*
8	*Dichantium caricosum*	*Barleria prionitis*	*Corypha utan*
9	*Brachiaria subquadripara*	*Ipomoe obscura*	*Hibiscus tiliaceus*
10	*Eualalia amaura*	*Crotalaria mucronata*	*Azadirachta indica*

Source: Halder 1976; Pudyatmoko 2005.

rich interior foraging sites, which negated the need to forage along the forest edge. Post-logging regeneration provides a temporary abundance of pioneer vegetation, including scrub and grasses which appear to benefit the bantengs' body condition and their ability to breed successfully. The effects of logging activity, however, are likely to be counter-intuitive to the long-term survival of banteng as the act of deforestation, increased human presence and heavy machinery will elevate stress levels that will not only reduce banteng breeding activity, but may even cause mortality. Furthermore, it was not uncommon for those people working at logging camps to hunt banteng for meat.

Reproduction and growth

Banteng cows usually produce a single calf that is weaned at around 6–10 months. Sexual maturity occurs at 2–3 years of age in females and 5–6 years of age in males. Gestation period in *B. j. birmanicus* in Malaya was 285–300 days and oestrus recurs 6–8 weeks after parturition (Medway 1983).

In general, male Javan banteng grow faster and larger than females. The maximum body size of a female will be reached at three or four years, but the bull between five and six years (Choquenot 1993). Sex ratio at birth is 1:1 (National Research Council 1983). However, it might vary from site to site (Hoogerwerf 1970; Choquenot 1993). In some sites, sex ratio of adult animals can change if a specific pressure operates on the population. For example, continuous observation in Baluran National Park (Java) from 1999 to 2005 showed a clear decrease of the female population. Predation by dholes was suspected to be the determining factor (Pudyatmoko 2005). Females were sexually mature at the age of two to four years, and producing a calf at three to five years. Only about 33% of males of 3.5 years had reached maturity, but bulls of more than 4.5 years were all mature. The fecundity of females peaked at 6–7 years then declined among older cows (Choquenot 1993). The proportion of pregnant cows in the Java population was more than 50% (National Research Council 1983; Hedges & Tyson 1996). In the wild, the mortality rate of immature animals was very high. In Cobourg Peninsula, Northern Australia about 26% of calves died in the first six months (Choquenot 1993). However, in captivity bulls can reach 20 years or more (Halder 1976). In the wild, cows can live until 20 years (Hoogerwerf 1970).

In Borneo cows are able to reproduce at three years of age, and young bulls begin to spend periods of time separate from the natal herd at three years of age, when they are also thought to become sexually active (P. Gardner, personal observation). Parturition did not appear to be seasonal in herds located in Tabin Wildlife Reserve and Malua Forest Reserve in Sabah (Borneo); young calves were observed throughout the year (P. Gardner, personal observation).

Behaviour

Social organization (mainland and Java banteng)

Banteng live in groups that can consist of many members, ranging from 2 to 30 individuals (Figures 13.2–13.3). Each herd is usually led by an old cow. Young bulls form a herd with two or three individuals. Adult bulls are solitary but enter a cow herd for mating (Tun Yin 1967; Lekagul & McNeely 1988).

Figure 13.2 shows a mixed herd consisting of adult cows and bulls, subadults and juveniles in Sadengan feeding ground in Alas Purwo National Park (Java). This photo (Figure 13.2) was taken during the early rainy season. Detailed observations of group structure and social behaviour of wild banteng were conducted in Baluran National Park; four types of herd were observed in this park: (1) solitary individuals, (2) small and stable herds, (3) small and temporary herds and (4) large groups which comprised several herds (Pudyatmoko 2005). Solitary animals were mostly old bulls, which were forced by younger bulls to leave the herd. Occasionally in the feeding ground in Baluran National Park an old bull tried to join a herd. It could feed with other members, but when the herd was leaving the feeding area the solitary bull separated from the group and moved in a different direction. A solitary cow was rarely observed. A stable herd was mostly composed of very young animals with their mothers. From regular observations it can be concluded that most herds were not stable; banteng frequently changed their groups. A rather stable herd was a bachelor group that consisted of young bulls. The degree of sexual segregation between male and female groups was low. The reasons sexual segregation was absent was probably due to small population abundance and distribution of resources, the presence of predators and the combination of these factors (Pudyatmoko 2005). Up to 80% of the herds were mixed groups that consisted of one to ten animals. A large group composed of more than 50 individuals was very rare. An example of this kind of group was observed in the feeding ground in Baluran National Park, a large herd which consisted of five herds containing 35, 12, 14, 8 and 3 individuals. This large herd lasted only about an hour (Pudyatmoko 2005). An old cow generally is the leader in the mixed herd (Pudyatmoko 2005). She is the most alert individual while the others are grazing. She alerts the group if a potential danger is detected. In addition, this cow determines the direction of the herd's movement. The bulls have only a minor role as leader. The competition among bulls for cows was difficult to observe in wild banteng. Even in the rut season sparring between bulls was rarely seen. It might be that each individual already understands his position in the social hierarchy of the herd. A dominant bull shows his body and horn size in a lateral display. Actual fights were seen only rarely. When a bull approached a dominant male in the group then they started to push against each other with their heads. The fight lasted a very short time. The loser flees quickly. Usually, the fight is not dangerous for either bull (Pudyatmoko 2005).

Social organization (Bornean banteng)

Social organization is dynamic and herd compositions include solitary individuals, bachelor groups of mixed ages, cow and calf groups and mixed herds. Where mixed herds have been observed, a single bull appeared to maintain dominance by driving off other bulls and juvenile bulls in close proximity to the cows. On some occasions during feeding activity the cows and calves appeared to assert authority over particular mineral licks and the dominant bull displayed considerable tolerance by exhibiting gentle behaviour, particularly towards young calves. Play-fighting was regularly observed between younger bulls of the same age, but also between younger bulls and the dominant bull. At times the cows, calves and young males of the same herd visibly aggravated the dominant bull for no apparent reason.

During breeding and the early stages of a calf's life the dominant (assumed sire) bull maintains close proximity to his progeny and also drives off other bulls within the vicinity. A change in dominance of a small herd containing two calves, two cows, one heifer, one bullock and two bulls was observed in Malua Forest Reserve in 2011 by P. Gardner (personal observation); the mature dominant bull was observed fighting with the younger bull; the end result was the exclusion of the mature dominant bull. For up to five months following his absence from the herd, the mature bull was observed visiting an open grassland site also visited by the herd, but was never again observed accompanying the herd. After this time the mature bull was never observed again and was presumed to have moved out of the area or died of natural causes.

Herd size (Bornean banteng)

Preliminary observations of banteng herd density in different canopy covers were obtained using camera trap photographs in secondary forest in 2011 and 2012. Camera traps established in continuous canopy cover captured photographs of herds with fewer than eight individuals, whereas camera traps in grassland with an open canopy captured herds of fewer than 18 individuals (P. Gardner, personal observation). Variation in herd size was attributed to two factors: (1) detection of all individuals under the canopy was imperfect due to understory vegetation and individuals walking behind the camera trap; and/or (2) herds of large numbers had to split into smaller groups under continuous canopy cover because understory vegetation was dense, but the large herds converged in open areas to socialize and forage (P. Gardner, personal observation).

Parasites and diseases

Banteng have been reported to carry disease and parasites just like other large herbivores, such as tuberculosis, brucellosis, salmonellosis, endo- and ectoparasites. However, banteng populations are exposed to the risk of disease outbreaks from domestic livestock around protected areas. Blackleg (*Clostridium chauvoli*) and mucosal disease have caused heavy losses in captive banteng in European zoos (National Research Council 1983).

The cause of poor physical condition of banteng cannot be explained merely based on malnutrition and unfavourable habitat condition. Veterinary examinations conducted on banteng in very poor condition in Ujung Kulon National Park (Java) found that these animals were infected by the parasites *Fasciola hepatica* and *F. gigantica*. For Bali cattle (the domesticated form) the most dangerous was Jembrana disease, which killed 10–60% of animals in certain localities. This disease was transmitted by cattle tick (National Research Council 1983). To avoid the transmission to wild individuals, direct contact between banteng and domestic cattle should be prevented.

Status in the wild

Worldwide status

The banteng is Endangered (IUCN 2013) in the whole geographical distribution. Small available habitat size, illegal poaching and competition with domestic livestock have negatively affected the banteng population in the last few decades. They have disappeared from many reserves in Thailand and from several countries. Therefore, management plans for both *in situ* and *ex situ* conservation are required for the long-term survival of the species.

The feral population of banteng in the North Territory of Australia consists of about 6000 individuals and seems genetically similar to wild banteng. This population was founded from a small number of Bali cattle (domestic form of wild banteng; Wilson & Mittermeier 2011).

Mainland banteng

In situ and *ex situ* conservation

Meta-populations of banteng are found in many subpopulations throughout its historical range due to habitat fragmentation and habitat loss.

In 2005 the Zoological Organization of Thailand carried out banteng reintroduction with 13 captive-born individuals into Khao Khiew – Khao Chompu Wildlife Sanctuary in southeast Thailand. Over a four-year period the population increased to 34 individuals. They now form a single herd with a home range greater than 8 km^2. Salakphra Wildlife Sanctuary (WS) is now thought to be the main site for future reintroduction. The banteng recovery project is designed under the Management Plan for Salakphra WS in order to reintroduce captive mainland banteng from Khao Nampoo into this former banteng habitat site.

Java banteng

Up until 1970 there were 13 locations known as banteng habitat in Java. Currently, only four locations are inhabited by banteng (Pudyatmoko 2005). The nine extinct populations were previously located in: (1) Cikepuh-Cibanteng; (2) Leuweng Sancang; (3) Pangandaran; (4) Bojonglarang Jayanti; (5) Cimapag; (6) Cikamurang; (7) Kediri; (8) Malang; and (9) Blitar Cikepuh-Cibanteng and Leuweng Sancang Nature Reserves. All of these were previously considered as important habitat for banteng in Java, but the species has been extinct in these locations since 2003. According to historic records, the population size in Cikepuh-Cibanteng was estimated to be about 300 animals in 1970, and the population size in Leuweng Sancang was 200 animals in 1988. Another location with a relatively large population was Pangandaran Nature Reserve, which contained more than 100 banteng up until 1974. However, it was considered that individuals in Pangandaran were hybrids with cattle. Banteng were formerly found in teak plantations on the coast of Blitar, Kediri and Malang, but they are now most probably extinct. The same occurred in Bojonglarang-Jayanti, Cimapag and Cikamurang areas, where the species is locally extinct (Pudyatmoko 2005). The main causes of extinction in some places in Java were habitat degradation and poaching. It seems that there is a strong correlation between habitat size and extinction. All extinct banteng in Java inhabited an area of less than 10 000 ha. This trend should be better investigated for future conservation and management practices.

Currently, Java banteng are living in remnant and isolated habitats in four national parks: Ujung Kulon, Meru Betiri, Baluran and Alas Purwo. The biggest population of wild banteng occurs in Ujung Kulon National Park in a habitat of about 30 000 ha. Although the national park covers about 122 000 ha, not all areas are suitable or occupied by banteng. The highest numbers were 905 individuals recorded in 1997 (Pudyatmoko 2005). Because this park is the only part of the world with Javan rhino (*Rhinoceros sondaicus*), the highest priority of the park is the conservation of this species. Increasing numbers of banteng are considered a threat to Javan rhino, and some measures were planned to reduce the number of banteng in the park, including the capture and relocation of individuals to other areas, but these plans have not yet been carried out.

The second most important habitat for the banteng is Alas Purwo National Park. It covers an area of 43 420 ha, but not all is suitable habitat. In Alas Purwo National Park, a sharp decline in the population took place from 1992 to 1995. During this period the number of banteng decreased from 323 to 119 individuals. The cause of the population decline was unsustainable predation by dhole (Hedges & Tyson 1996). As the population of dhole declined, the population of banteng has been in the process of rapid recovery. This trend was shown by the continuously increasing number of banteng seen in Sadengan feeding ground from 2007 to 2011. Regular monitoring of the populations showed that the maximum number of individuals seen in a single observation was 21 animals in 2007, 56 in 2008, 61 in 2009, 94 in 2010 and 100 in 2011.

Contrary to this, the banteng population in Baluran National Park is decreasing at an alarming rate, indicated by a decreasing encounter rate of individuals in all areas within the national park. Regular observations showed that the number of banteng in Bekol feeding ground decreased from 101 animals in 2002 to 47 in 2003 and to zero in 2006 (Pudyatmoko 2005). However, in the same year regular surveys using concentration counts recorded 15 animals. The result of a survey in 2011 recorded 15 individuals as a minimum number. Currently, it is very difficult to see banteng at Bekol feeding ground or at waterholes. A combination of some factors such as illegal hunting, predation by dhole and degradation of habitat quality were the causes of decline (Pudyatmoko 2005).

The population size of banteng in Meru Betiri National Park was relatively stable. The number fluctuated between 125 to 150 animals. It seems that the future of the subspecies in this park is relatively secure. The management problem of banteng here is the movement of animals crossing the border of the national park and entering cacao and coffee plantations, where forage like young grass and leaves were more abundant. The conflicts between banteng and people arise because this subspecies eat cacao fruits and destroy the bark.

Outside the protected areas, banteng occupy teak, cacao and coffee plantations, especially in Banyuwangi District, east Java. In this area human–wildlife conflict frequently occurs.

People have poisoned banteng because they have raided crops, and there were occasions when some individuals attacked people. Due to the significant economic loss from banteng raiding cacao plantations, a plan was undertaken to capture 19 individuals, with nine being sent to a zoo and ten released in Meru Betiri National Park.

Based on the above findings, it is clear that the population and range of banteng in Java has declined significantly. Because of rapid population decline, an urgent conservation plan should be developed to save the banteng population in Baluran National Park. The decline was predicted with a population simulation using Vortex Program (Pudyatmoko 2005). In the model, the author discovered that the key factor of banteng decline was the very low proportion of young in the population. He inferred that dhole may increase the extinction risk of banteng based on the records of predation. Dhole preferred pregnant cows and juveniles to males or non-pregnant adult cows. As a result, he observed that the proportion of subadults and juveniles in the population were underrepresented, and there was a change of sex ratio towards more males in the population. Currently, the number of dhole in Baluran has decreased. Analysis of dhole scat collected during wet and dry seasons showed that no banteng remains were found. This indicated that predation on banteng has also declined (Pudyatmoko 2012).

Bornean banteng

This subspecies has been hunted to extinction in many areas of Borneo as a result of widespread use of guns since the middle of the last century (Payne *et al.* 1985). Other methods of hunting also included wire snares and homemade bomb devices. Hunting using firearms dramatically increased in the 1940s and 1950s, when net importation of shotgun cartridges to Sarawak quintupled from 444,161 to 2,149,556 (Caldecott 1988). In the following 30 years hunting activity showed no sign of slowing down as importation figures for cartridges remained constantly high at more than 1,300,000 (Caldecott 1988). It is probable that the high level of unregulated hunting activity during this time resulted in the loss of many banteng, particularly as the above figures do not account for the manufacture and re-use of illegal shotgun cartridges, which are favoured by poorer rural communities. Based on these facts, it is probable that hunting caused the extinction of banteng in Brunei; however, no information was found in the literature. Banteng are hunted for meat for subsistence, but more often on a contract-basis for celebrations such as weddings or religious public holidays. The poles of banteng are thought to be of Chinese medicinal value, but there is no information describing the type of condition they treat or the mode/method of treatment. Claims on the use of banteng byproducts are unsubstantiated at best and are held in no regard within the scientific medical community at large. Numerous banteng skulls, horns and other body parts were located in villages in Kalimantan during a survey in 1999 (Hedges & Meijaard 1999), and it is probable that there is some international trade in banteng skulls in Borneo, whereby skulls originating from Kalimantan are brought across the border and sold as trophies in Sabah.

Additional threats include extensive deforestation of banteng habitat for timber production and subsequent conversion to monocrop agricultural plantations, which has caused widespread habitat loss and fragmentation of remnant populations (Caldecott 1988). Banteng that inhabit forest which intersects with plantations are at greater risk of encroachment and disturbance from hunters/opportunistic poachers. In Sabah during 2011–2013, hunters, fishermen and harvesters were observed on many occasions within protected forest reserves. In these areas extensive trail networks had been created and were regularly maintained to facilitate illegal activity (P. Gardner, personal observation).

As a result of habitat fragmentation, the isolation of banteng populations prevents gene flow and renders them highly susceptible to inbreeding, which may result in a lack of genetic diversity and increase their susceptibility to genetic mutation and disease. Banteng that inhabit forest in close proximity to oil palm plantations are at further risk of disease transmission from domestic cattle and water buffalo, which are used to maintain scrubland within the plantation and assist harvesters with the transportation of fruit bunches. These domestic bovids are often grazed along the plantation edge and river banks which are also used by banteng; in Eastern Sabah in 2012 banteng were found to be only 100–200 metres away from domestic cattle (P. Gardner, personal observation). Such close proximity suggests disease transmission and hybridization between domestic and wild cattle to be a real threat.

In Kalimantan, banteng are vulnerable to hunting and are also at risk of disease transmission and hybridization because they inhabit areas that occur outside of conservation areas (Timmins *et al.* 2008; Pudyatmoko *et al.* 2012). These conflict areas are often used for cultivation by local villages; however, the banteng are not thought to cause damage to crops (Pudyatmoko *et al.* 2012).

Status in captivity

By the end of 2010 there were 138 banteng consisting of 43 females and 95 males in captivity centres worldwide (http://app.isis.org/abstracts/abs.asp). According to the Ministry of Forestry Decree of Republic Indonesia Number 58 on Strategy and Conservation Action of Banteng from 2010 to 2020, in Indonesia there were 56 banteng in conservation institutions consisting of 30 females and 26 males. The breeding centre with the largest number of individuals is Ragunan Zoo in Jakarta, with 21 animals. However, the sex ratio was far from ideal at 16 males and five females. A similar situation was found in the third breeding centre in Surabaya Zoo, which houses 12 banteng comprising seven males and five females. The situation was improved in Taman Safari Prigen with 16 individuals and a sex ratio of five males and 11 females.

Inbreeding depression was very pronounced in Ragunan Zoo. For genetic improvement of captive banteng, animal exchange among the breeding centres and the addition of new wild individuals are highly recommended.

In Thailand, mainland banteng are found in four zoological parks with a total of 32 individuals (male:female ratio 15:17). The species is also found in two wildlife breeding stations with 18 individuals, i.e. Khao Kor Wildlife Breeding Station (2:1) and Khao Nampoo Wildlife and Nature Education Station (15) in the Department of National Park and Wildlife and Plant Conservation.

Data from EEP (EAZA), ISB (WAZA) SEAZA breeding programmes are probably an underrepresentation of the total number of banteng in captive collections (see also Chapter 27).

To our knowledge, no known populations of Bornean banteng (*B. j. lowi*) exist in captivity or in semi-free-range conditions.

Acknowledgements

PG, LA and BG thank Houston Zoo (P. Riger), Sime Darby Foundation, the Malaysian Palm Oil Council, the Mohamed bin Zayed Species Conservation Fund and Woodland Park Zoo for supporting their research on the Bornean banteng in Sabah. PG and BG thank the Sabah Biodiversity Centre for allowing them to carry out research in Sabah, and the Sabah Forestry Department, which gave them permission to work in Malua Forest Reserve and Tabin Wildlife Reserve. PC, LA and BG also extend thanks to the following organizations for their support during their work on banteng in Sabah: the NGO HUTAN, Malua BioBank, Sabah Foundation, Leibniz Institute for Zoo and Wildlife Research, Borneo Rhino Alliance, and the following individuals for allowing us to use their camera trap pictures: Andrew Hearn, Joanna Ross, Andreas Wilting, Torsten Bohm and Jedediah Brodie.

SP is indebted to many institutions for their help during his research on banteng. He would like to express his appreciation to the Ministry of Forestry, Indonesia, which permitted him to conduct field research in Baluran National Park. He wishes to thank Nagao Environmental Foundation, Japan, for the financial support for his research. For data collection in the field he would like to thank his students Sugiarto, Mohammad Eko Nurcahyono, Nur Bambang Saputra, Haruna Nur Sasongko, Ali Bosar and Teja Suryo Nugroho. Finally, he would like to thank the Faculty of Forestry, Universitas Gadjah Mada, Indonesia for permission to use the Laboratory of Wildlife Management.

References

Alikodra, H. S. & Sastradipraja, D. (1983). Studi Tentang Beberapa Parameter Faal Pelestarian Satwa Banteng (*Bos javanicus*). [Study on Some Physiological Aspects of Banteng]. Research Report, Faculty of Veterinery Animal, Institute Pertanian Bogor, Indonesia.

Bhumpakphan, N. & McShea, W. J. (2011). Ecology of gaur and banteng in the seasonally dry forests of Thailand. In *The Ecology and Conservation of Seasonally Dry Forests in Asia*. pp. 179–194. Eds. McShea, W. J., Davies, S. J. & Bhumpakphan, N. Washington, DC: Smithsonian Institution Scholarly Press.

Byers, O., Hedges, S. & Seal, U. S. (1995). *Asian Wild Cattle Conservation Assessment and Management Plan Workshop*. Apple Valley, MN: IUCN/SSC Conservation Breeding Specialist Group.

Caldecott, J. O. (1988). *Hunting and Wildlife Management in Sarawak*. Switzerland and Cambridge: IUCN Tropical Forest Programme.

Choquenot, D. (1993). Growth, body condition and demography of wild banteng (*Bos javanicus*) on Couburg Peninsula, Northern Australia. *Journal of Zoology* **231**: 533–542.

Corbet, G. B. & Hill, J. E. (1992). *The Mammals of the Indomalayan Region: A Systematic Review*. Oxford: Oxford University Press.

Davies, G. & Payne, J. (1982). *A Faunal Survey of Sabah*. Malaysia: World Wildlife Fund.

Francis, C. M. (2008). *A Field Guide to the Mammals of Southeast Asia*. London: New Holland Publisher (UK) Ltd.

Grzimek, B. (1987). *Grzimek Enzyklopädie Säugetiere: Register*. Zürich: Kindler Verlag AG.

Halder, U. (1976). *Ökologie und Verhalten des Banteng (*Bos javanicus*) in Java*. Hamburg: Verlag Paul Parey.

Hassanin, A. & Ropiquet, A. (2006). What is the taxonomic status of the Cambodian banteng and does it have close genetic links with the Kouprey? *Journal of Zoology* **271**: 246–252.

Hedges, S. & Meijaard, E. (1999). *Reconnaissance Survey for Banteng (*Bos javanicus*) and Banteng Survey Methods Training Project*. East Kalimantan: Kayan-Mentarang National Park.

Hedges, S. & Tyson, M. (1996). Is predation by ajag a threat to the banteng population in Alas Purwo? Review of the evidence and discussion of management options. Research report. Faculty of Forestry, Universitas Gadjah Mada, Indonesia.

Hoogerwerf, A. (1970). *Udjung Kulon, the Land of the Last Javan Rhinocheros*. Leiden: E. J. Brill.

IUCN (2013). *IUCN Red List of Threatened Species*. Version 2013.2. www.iucnredlist.org, downloaded 22 May 2014.

Lekagul, B. & McNeely, J. A. (1988). *Mammals of Thailand*. 2nd edition. Bangkok: Darnshutha Press.

Lydekker, R. (1913). *Catalogue of the Ungulate Mammals in the British Museum (Natural History)*. Vol. 1: Artiodactyla, Family Bovidae, Subfamilies Bovinae to Ovibovinae (Cattle, Sheep, Goats, Chamois, Serrows, Takin, Musk-Oxen, etc.). London: British Museum (Natural History).

Manatchaiworakul, W., Treeraditthakul, W., Kidsin, K., *et al.* (2011). Genetic variation study of mtDNA of Wild Bovid in Khao Ang Rue Nai Wildlife Sanctuary. Poster presentation, Wildlife Seminar of Thailand.

Medway, L. (1983). *The Wild Mammals of Malaya (Peninsular Malaysia) and Singapore*. Oxford: Oxford University Press.

Medway, L. (1969). *The Wild Mammals of Malaya*. Kuala Lumpur: Oxford University Press.

Nakhasathien, S., Bhumpakphan, N. & Simcharoen, S. (1987). *Ecology of the*

Wildlife in Thung Yai and Huai Kha Khaen Wildlife Sanctuaries. Bangkok: Wildlife Conservation Division, Royal Forest Department.

National Research Council. (1983). *Little-known Asian Animals with Promising Economic future.* Washington, DC: National Academy Press.

National Research Council (2002). *Little-known Asian Animals with a Promising Economic Future.* New York: Books for Business.

Nijman, I. J., Otsen, M., Verkaar, E. L., *et al.* (2003). Hybridization of banteng (*Bos javanicus*) and zebu (*Bos indicus*) revealed by mitochondrial DNA, satellite DNA, AFLP and microsatellites. *Heredity* **90**(1): 10–16. Available at: http://www.ncbi.nlm.nih.gov/pubmed/12522420 (accessed 13 February 2013).

Pairah (2007). Tumpang tindih relung ekologis Banteng (*Bos javanicus*, d'Alton 1823) dan Rusa Timor (*Rusa timorensis*, Blainville 1822) di padang penggembalaan Sadengan, Taman Nasional Alas Purwo, Jawa Timur. [Ecological niche overlap between banteng and rusa timor in grassing area, Baluran National Park, East Java]. Masters thesis. Fakultas Kehutanan Universitas Gadjah Mada. (In Bahasa Indonesia)

Payne, J., Francis, C. M. & Phillips, K. (1985). *A Field Guide to the Mammals of Borneo.* Malaysia: The Sabah Society.

Pedrono, M., Ha Minh, T., Philippe, C. & Frederi, V. (2009). Status and distribution of the endangered banteng *Bos javanicus birmanicus* in Vietnam: a conservation tragedy. *Oryx* **43**(4): 618–625.

Prayurasithi, T. (1987). Ecology of banteng in Huai Kha Khaeng Wildlife Sanctuary. MSc thesis. Kasetsart University.

Prayurasithi, T. (1997). Habitat partitioning between gaur and banteng in Huai Kha Khaeng Wildlife Sanctuary. PhD thesis. Minasota State University.

Pudyatmoko, S. (2005). *Freilanduntersuchungen zur Ökologie und Habitatnutzung Wildlebender Banteng Rinder (*Bos javanicus*) im Baluran National Park in Ost Java, Indonesien.* Goettingen: Cuvillier Verlag.

Pudyatmoko, S. (2012). Influence of dhole (*Cuon alpinus*) on population and distribution of banteng in Baluran National Park. Research Report. Faculty of Forestry UGM-Nagao Environmental Foundation.

Pudyatmoko, S. *et al.* (2012). *National Strategy and Action Plan for Banteng (*Bos javanicus*) Conservation 2010–2020.* Kalimantan: Ministry of Forestry.

Samejima, H., Ong, R., Lagan, P. & Kitayama, K. (2012). Camera-trapping rates of mammals and birds in a Bornean tropical rainforest under sustainable forest management. *Forest Ecology and Management* **270**: 248–256.

Steinmetz, R. (2004). Gaur (*Bos gaurus*) and banteng (*Bos javanicus*) in the lowland forest mosaic of Xe Pian Protected Area in Lao PDR: abundance, habitat use and conservation. *Mammalia* **68**: 141–157.

Timmins, R. J., Duckworth, J. W., Hedges, S., Steinmetz, R. & Pattanavibool, A. (2008). *Bos javanicus.* In *IUCN 2013: IUCN Red List of Threatened Species.* Version 2013.2. www.iucnredlist.org, accessed 4 March 2013.

Tun Yin, U. (1967). *Wild Animals of Burma.* Rangoon: Rangoon Gazette.

van den Bergh, G. D., De Vos, J. & Sondaar, P. Y. (2001). The Late Quaternary palaeogeography of mammal evolution in the Indonesian Archipelago. *Palaeogeography, Palaeoclimatology, Palaeogeology* **171**: 358–408.

Whitemore, T. C. (1990). *An Introduction to Tropical Rainforests.* 2nd edition. Oxford: Oxford University Press.

Whitten, T., Soeriaatmadja, R. E. & Affif, S. A. (1996). *The Ecology of Java and Bali.* Singapore: Periplus Ltd.

Wilson, D. E. & Mittermeier, R. A. (2011). *Handbook of the Mammals of the World.* Vol. **2**. Hoofed Mammals. Barcelona: Lynx Edicions.

Part III Species accounts

Chapter 14

Kouprey *Bos sauveli* A. Urbain, 1937

Mario Melletti, Alexandre Hassanin and Marzia Mirabile

Names

Genus: *Bos* Linnaeus, 1758

Species: Kouprey *Bos sauveli* A. Urbain, 1937

Names in other languages: **French**: Boeuf gris cambodgien; **German**: Kouprey; **Spanish**: Toro Cuprey; **Italian**: Kouprey

Other common names: Grey Ox, Cambodian Grey Ox, Indochinese Forest Ox.

Kouprey in Khmer language means 'forest ox'.

Ecology, Evolution and Behaviour of Wild Cattle: Implications for Conservation, ed. M. Melletti and J. Burton. Published by Cambridge University Press.

Taxonomy

The kouprey (*Bos sauveli*) is a little-known wild cattle species discovered in the nineteenth century in Northern Cambodia. Probably the first to mention this species was Campbell (1860), who described three species of wild cattle in Cambodia: the gaur, the banteng and a 'black or blackish grey' wild ox that frequently occurs 'on the plains in herds of from 50 to 300 at a time'. Later, Dufossé (1918) gave additional information on the species, mentioned that the kouprey is endangered and suggested prohibiting its hunting in order to preserve the remaining populations. The species, *Bos sauveli*, was described by Urbain (1937) based on a young male caught in Preah Vihear Province of Cambodia and maintained in captivity at the Vincennes Zoo (France) until the 1940s. Urbain (1937) placed the kouprey with banteng and gaur into the subgenus *Bibos*. An adult male shot in Cambodia in 1939 was further described by Coolidge (1940), who considered the species sufficiently distinct to put it in the new genus *Novibos*. Subsequently, the validity of *Novibos* was questioned and the kouprey was returned to the genus *Bos* (Bohlken 1961a, 1961b; Pfeffer & Kim-San 1967; Groves 1981) or alternatively placed in the genus *Bibos* (e.g. Geraads 1992).

The holotype of *Bos sauveli* was sequenced by Hassanin & Ropiquet (2004) for two mitochondrial genes (cytochrome b and subunit II of the cytochrome c oxidase) and the promoter of the nuclear lactoferrin gene. The analyses indicated that the kouprey is related to both gaur and banteng, confirming previous hypotheses based on morphology (e.g. Pfeffer & Kim-San 1967; Geraads 1992). Galbreath *et al.* (2006) sequenced the mitochondrial cytochrome b gene of Cambodian banteng and found a close relation with the kouprey. They concluded that the kouprey is not a natural species but a feral hybrid, resulting from the banteng (*Bos javanicus*) with input of zebu cattle (*Bos indicus*). Hassanin & Ropiquet (2007a) suggested, however, that Galbreath *et al.* (2006) misinterpreted the mitochondrial data. They showed that mitochondrial sequences of Cambodian banteng are divergent from those of Javan banteng (>4%), but similar to those of kouprey (1.25%). To explain mtDNA sequence similarity between Cambodian banteng and kouprey, they proposed that the mitochondrial genome of kouprey was transferred into the ancestor of Cambodian banteng by natural hybridization during the Pleistocene period (see Chapter 1 for more details). This hypothesis was confirmed by the analysis of additional nuclear markers on the male holotype of *Bos sauveli* (Hassanin & Ropiquet 2007b), with Y-chromosomal and autosomal sequences different from those of other *Bos* species. Molecular data show therefore that *Bos sauveli*, as represented by its holotype, is a natural species, confirming previous morphological observations (Urbain 1937; Coolidge 1940; Wharton 1957; Hedges *et al.* 2007; Galbreath *et al.* 2007; Timmins *et al.* 2008), and the discovery of Pleistocene fossils in Thailand (Figure 14.1; Vithayanon & Bhumpakphan 2004).

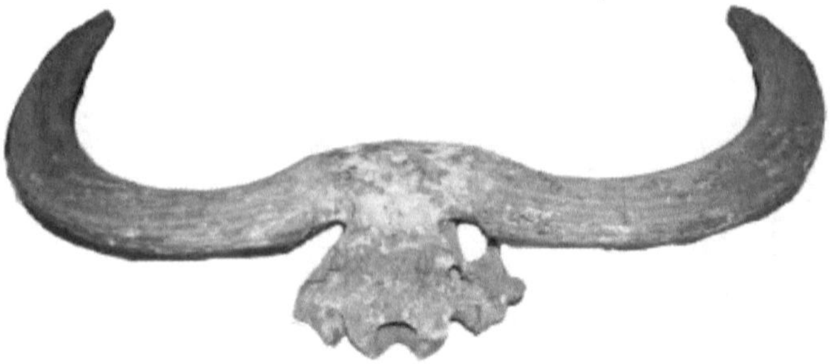

Figure 14.1 Pleistocene horn fossil found in Thailand (photo by C. Vithayanon, Collection of Zoology).

This complex taxonomic situation is exacerbated by the presence in Southeast Asia of two other wild cattle species closely related to the kouprey (gaur and banteng), and by interbreeding between wild, feral and domestic cattle, as highlighted by Bohlken (1961a 1961b), Grigson (1988) and Handschuh & Hassanin (2013). Finally, it is possible that the kouprey was domesticated in some regions temporarily during the Khmer culture (Wharton 1957; Vietmeyer 1983). This assumption was based on phylogenetic analyses of an animal mounted in 1871 at the Muséum national d'Histoire naturelle in Paris (but now conserved at the Muséum d'Histoire naturelle of Bourges), showing phenotypic traits of both kouprey and domestic ox (Hassanin *et al.* 2006). A recent analysis of several nuclear genes has, however, revealed that the specimen in Bourges is a hybrid between a female *Bos sauveli* and a male *Bos taurus* (Handschuh & Hassanin 2013).

Subspecies and distribution

The main stronghold of the kouprey distribution was in Cambodia, with a possible presence in Southern Lao PDR, Southeast Thailand and Western Vietnam (Sauvel 1949; Wharton 1957; Pfeffer 1969; Hoffman 1986; MacKinnon & Stuart 1989; Grubb 2005).

The distribution range based on IUCN (2013) data (illustrated in Plate 6) is very similar to that of the 1950s and 1970s, based on records of local people considered reliable (Figure 14.2).

The kouprey is highly sensitive to human pressure, in particular hunting, and this seems related to its rapid decline in Cambodia (Timmins 2011).

There are no subspecies.

Descriptive notes

Body measurements: **Shoulder height**: 170–190 cm; **Head and body length**: 210–220 cm; **Tail length**: 100–110 cm; **Body mass**: 600–900 kg. Dental formula: I 0/3, C 0/1, P 3/3, M 3/3 (×2)=32 (Wilson & Mittermeier 2011).

There are very few descriptions of this species, among which two dated back to 1929 (two individuals shot in South

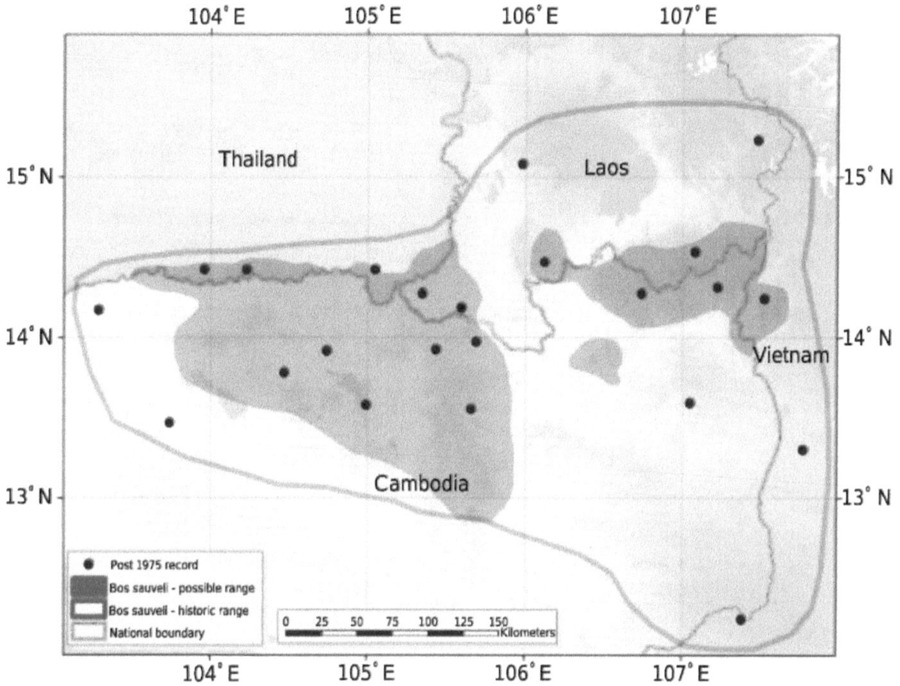

Figure 14.2 A zoomed-in map showing the possible range of the kouprey, along with post-1975 records from villagers and other sources (adapted from MacKinnon & Stuart 1989). A black and white version of this figure will appear in some formats. For the colour version, please refer to the plate section.

Vietnam; Coolidge 1940; Wharton 1957) and another of an individual held in Vincennes Zoo near Paris in 1937 (Urbain 1937).

The colour coat in kouprey varies with age and sex. Males can be grey with black parts or dark brown. Females are grey-brown. The muzzle in males is black with a narrow chestnut area around it. The second half of the lower legs is whitish, but the anterior surface of the fore foot is striped by dark hair. Like other wild cattle, calves are brown and become grey with age. Little horns in calves appear by six months of age. The tail is long. Bulls have a pronounced dewlap much longer than females.

Horns in females can reach 40 cm and the shape is similar to a lyre. Males have spread horns reaching up to 84 cm that form an arch forward and upward and grow up to 81 cm (Dioli 1995; Wilson & Mittermeier 2011). Male horns splinter at the top at around seven years. This characteristic may be due to the habit of rubbing against vegetation and the ground (Wharton 1957).

Habitat

According to Wharton (1957), one of the few studies on kouprey habitat, this species occurs in a mix of dense and open canopy forest with grassy glades, pools and salt licks. Similar habitat requirements have been identified during a recent survey report (Timmins 2011; Table 14.1). On the basis of this study, the extension of kouprey habitat might still sustain a few individuals despite large areas of lowland forest being lost (Timmins 2011). However, kouprey might also be associated to dense dipterocarp forest, gallery and monsoon forests (Wharton 1957; Pfeffer 1969; Timmins 2011). Furthermore, Wharton (1968) suggested that kouprey were also associated with burning grassland habitat.

Movements and home range

Very few data exist on kouprey movements. There is some evidence that this species can move for long distances (up to 15 km within a day) and cross dense forest blocks back and forth from feeding areas (Wharton 1957). Seasonal movements could be undertaken at higher altitude during the wet season, but information is very limited (Wharton 1957; MacKinnon & Stuart 1989).

Activity patterns

Very little information is available on activity; what is available mainly comes from the paper of Wharton (1957). Similarly to other wild cattle species, kouprey alternates periods of feeding, ruminating and resting. Probably, the species is active in early morning and late afternoon with a resting period during midday (Wharton 1957; Wilson & Mittermeier 2011).

Feeding ecology

Only a little information is available on the diet of the kouprey. Grasses of different heights, mainly bamboo grasses (*Arundinaria* spp.), ploong grass (*Arundinella setosa*) and koom grass (*Chloris* sp.) are recorded (Wharton 1957). A percentage of browse (leaves and sprouts) and mineral soil seems to be included in its diet (Wharton 1957; Tran Hong Viet *et al.* 1988; Dang Huy Huynh & Pham Trong Anh 1988; MacKinnon & Stuart 1989).

Table 14.1 Regions and protected areas with the highest probability of kouprey persistence in Cambodia and Laos

Location	Habitat suitability	Protected area coverage	Protection level	Survey coverage	Wild cattle status
Eastern plains south of the Srepok river (Cambodia)	Large DDF,[1] small DDF,[2,3] MDF,[3] SEF[5]	MPF,[6] SPFBA,[7] LWS,[8] PPWS[9]	Above average	Extensive surveys from the mid 1990s	Higher density of wild cattle than other areas; unlikely habitat for kouprey presence
Eastern plains north of the Srepok river (Cambodia)	DDF,[1] MDF,[3] limited SEF[5]	MPF,[6] LWS,[8] OYP[10]	Below average in MPF,[6] not in LWS[8] or OYPF[10]	Limited surveys coverage	Uncertain. Wild cattle tracks recorded. Unlikely habitat for kouprey presence
Northern plains of Cambodia/ adjacent plains of Laos	DDF[1]	PVPF[11]	Above average	Average coverage	Low densities of banteng and gaur. Unlikely habitat for kouprey presence
Kulen Promtep (Cambodia)	Little DDF,[1] Large DDF,[2] Mosaic SEF[5], MDF[3], DDF[3]	KPWS[12]	Above average	Low coverage; survey since the 1990s	Low numbers of gaur and banteng. Unlikely habitat for kouprey presence
Sesan District of Stung Treng Province (Cambodia)	Large DDF, [1] blocks of SEF[5]	None	None; remote area	Small survey	Predict small numbers of wild cattle. Highly unlikely habitat for kouprey presence
Eastern/northern Prey Long (Cambodia)	Little DDF,[2] large SEF,[5] patches of DDF[3]/MDF[3]	None	None; remote area	Little survey (2004–2005)	Low numbers of gaur and banteng. Banteng recorded (2004–2005). Highly unlikely habitat for kouprey presence
Nam Gong/Virachey Mountains (Laos)	DDF,[1,2] large SEF[5]	VNP[13]	Under way; inaccessible area	Very small surveys in 1990s and 2010	Low numbers of gaur. Highly unlikely habitat for kouprey presence
Southeastern Bolaven Plateau (Laos)	Grassland with SEF[5]	None	None	Very limited surveys	Uncertain. Very low numbers or extirpated wild cattle. Improbable habitat for kouprey presence
Western Siem Pang district of Stung Treng province (Cambodia)	Large DDF[2] and SEF[5]	Partial coverage in XPNPA[14]	Limited coverage	Very little coverage	Very low numbers of gaur. Banteng close to extirpation. Improbable habitat for kouprey presence
Eastern Phnom Samkos hills and lowlands (Cambodia)	Large DDF,[2,3] mosaic of SEF[5]	PSWS[15]	Limited coverage	Minimal coverage; ongoing surveys	Uncertain. Gaur vagrant. Banteng in low numbers. Improbable habitat for kouprey presence
Eastern Cardamom Mountains (Cambodia)	DDF,[3] large blocks of SEF[5]	PAWS,[16] CCPF[17]	Limited coverage	Very little coverage	Uncertain. Gaur likely in small numbers. Banteng recorded in 2006, few groups. Improbable habitat for kouprey presence
Central Cardamoms (Cambodia)	Marginal hilly grassland and upland forests[4]	CCPF[17]	Minimal to non-existent; remote area	Few to no surveys	Uncertain. Wild cattle probably extirpated. Improbable habitat for kouprey presence
Xe Kong (Laos)	Mainly DDF[2,3]	None	None	None	Uncertain. Possible residual Banteng. Improbable habitat for kouprey presence

Five categories of suitable habitat type were identified:
(1) Potential suitable kouprey habitat: lowland DDF (deciduous dipterocarp forest) with extensive pools and glades.
(2) Marginal habitat: denser DDF with a high percentage of canopy cover and lower pool and glade densities.
(3) Marginal habitat: dense closed-canopy DDF and MDF (mixed deciduous forest mosaic).
(4) Marginal habitat: hill areas with extensive grasslands.
(5) Marginal habitat: SEF (semi-evergreen forest).

Protected areas:
(6) MPF (Mondulkiri Protected Forest)
(7) SPFBA (Seima Protection Forest and Biodiversity Area)
(8) LWS (Lomphat Wildlife Sanctuary)
(9) PPWS (Phnom Prich Wildlife Sanctuary)
(10) OYPF (O Yadav Protected Forest)
(11) PVPF (Preah Vichear Protected Forest)
(12) KPWS (Kulen Promtep Wildlife Sanctuary)
(13) VNP (Virachey National Park)
(14) XPNPA (Xe Pian National Protected Area)
(15) PSWS (Phnom Samkos Wildlife Sanctuary)
(16) PAWS (Phnom Aural Wildlife Sanctuary)
(17) CCPF (Central Cardamoms Protected Forest)
Source: Timmins (2011), GWC, IUCN/SSC, AWCSG, modified.

Figure 14.3 A 'Sunday hunter', Phnom Penh, with killed adult bull kouprey (photo by James Mellon).

Kouprey were observed feeding in association with banteng and water buffalo, sharing the same foraging sites (Wharton 1957).

Reproduction and growth

Very few data exist on reproduction due to the difficulty studying this species. Mating is probably concentrated around April. Births have a peak before the hottest period of the dry season, between December and February, with a gestation of about 8–9 months. Similarly to other wild cattle species, females give birth to one calf and stay apart from the herd until the calf is about four weeks old (Wharton 1957; MacKinnon & Stuart 1989). The life span is about 20 years.

Behaviour

On the basis of the little information available, behavioural traits are similar to those of other wild cattle species. Mixed herds (females and young) may live separately from males. Bulls can form bachelor herds or become solitary. They may join females herd to form groups of up to 25 individuals in the dry season (Wharton 1957). The herd is led by an adult female (Pfeffer & Kim-San 1967) similarly to other wild cattle species (Melletti *et al.* 2007; see also Chapter 20). Fission–fusion patterns have been described in kouprey herds (Wharton 1957). The young:females ratio recorded by Wharton (1957) was about 1:3. Temporary associations with banteng and water buffalo have been recorded after the breeding season (Wharton 1957).

Parasites and diseases

Almost nothing is known except that Thouless (1987) cites a possible resistance to rinderpest. This was hypothesized by the fact that in some Cambodian regions villagers used to leave domestic cows in the forest to be fertilized by kouprey in order to make the offspring possibly more resistant to diseases. However, this assumption has not been proven.

Status in the wild

The kouprey has been considered one of the most endangered mammals in the world and was declared the national animal of Cambodia in 2004. Nowadays it is considered Critically Endangered (possibly Extinct; Timmins *et al.* 2008). Kouprey appears highly susceptible to hunting pressure (Figure 14.3). Horns still represent a status symbol for many Asian people and are sold for thousands of dollars. Moreover, wild cattle remain a target species for the bushmeat trade in the whole region.

In 1967, prior to the war, the kouprey was known to occur in three areas in Cambodia: the Koulen-Promtep Reserve in the north, the Lomphat Reserve and the Phnom Prich Reserve (Pfeffer & Kim-San 1967).

Investigations in the mid 1970s into possible reliable reports from hunters confirmed the existence in southern Laos, northern Cambodia and the Dongrak Mountain Range in Sisaket Province, eastern Thailand of two herds of kouprey (Enderlein 1976; Neese 1976a, 1976b; Lekagul & McNeely 1977; Thornback 1983).

Table 14.2 Estimates of kouprey populations and sightings in Southeast Asia between the 1940s and 1990s

Country (region)	Estimated population (number of individuals + year)	References
Cambodia (west of Mekong river)	500–600	Sauvel 1949
Cambodia (Samrong region)	~200	Sauvel 1949
Cambodia (west of Mekong river)	400–500	Wharton 1957
Cambodia (Samrong region)	50	Wharton 1957
Cambodia (Srepok river)	200–300	Wharton 1957
Cambodia	5 (1964)	Wharton unpublished; www.wildcattleconservation.org
Cambodia	500 (1952) 200 (1964)	Pfeffer & Kim-San 1967
Cambodia	Numbers unknown (1969)	Pfeffer & Kim-San 1967
Laos–Cambodia–Thailand	Two herds (1974/1975/1976)	Enderlein 1976; Neese 1976a, 1976b; Lekagul & McNeely 1977; Thornback 1983
Vietnam	Few individuals (1980)	Westing & Westing 1980; WWF 1982
Thailand (Cambodia border)	3 (1982)	Jackman 1982; McDermid & Sokha 2006
Vietnam and Laos	Maybe few individuals	MacKinnon 1986
Cambodia (Kampong Sralao district)	Numbers unknown (1984)	Thouless 1987
Vietnam Laos Cambodia	27 (1988) 40–100 (1988) <200 (1988)	Chan Sarun 1985; Dang Huy Huynh & Pham Trong Anh 1988; Le Vu Khoi 1988; Tran Hong Viet *et al.* 1988; Suvanakorn 1988
Cambodia	<12* (1994)	McDermid & Sokha 2006
Cambodia (south Srepok river)	4–5 (1996)** 4–6 (1997)** 3 (1998)**	Weiler *et al.* 1998
Cambodia (Phnom Prich Wildlife Sanctuary	No individuals recorded (2008–2009)	Gray & Phan 2011
Cambodia–Laos–Thailand	No individuals recorded (2010)	Timmins 2011

* Unreliable estimate based on little evidence.
** Second-/third-hand information of questionable reliability.

One of the last records of this species in the wild was a group of three individuals spotted at the Thailand border in 1982, coming from Cambodia; a team tried to follow them but stopped after a ranger was severely injured by a landmine (Jackman 1982; McDermid & Sokha 2006). A second record suggests that a few survived in Vietnam in the Truong Son Mountains of Thua Thien Province in the central part of the country in 1980 (Westing & Westing 1980; WWF 1982). Another sighting was in 1984 in Kampong Sralao district, Cambodia (Thouless 1987). However, according to Lic Vuthy (personal observation) the last confirmed sighting of a kouprey was in 1983 and the species seemed to have completely vanished some time during the late 1980s. In 1983, Sayer visited Laos and stated that the Forest Department believed that the species still occurred in the Attopeu/Champassak area; however, records are based only on sightings or tracks, and it is possible there is some confusion with other wild cattle species (Sayer 1983). Further information on kouprey sightings of doubtful reliability were made at the end of the 1990s in Srepok river area, Cambodia (Weiler *et al.* 1998). One of the largest expeditions in Cambodia looking for kouprey (Cambodian Kouprey Research Project; McDermid & Sokha 2006) took place in 1994; it made no sightings, but found some evidence of a possible herd of, optimistically, fewer than a dozen individuals in a small region of Mondulkiri.

Kouprey were always considered rare since the first description of the species (a review of population estimates is shown in Table 14.2). The first estimates on its population were made by Sauvel (1949) at 500–600 individuals on the western side of the Mekong river and about 200 in the Samrong region of Cambodia. Wharton (1957) considered the population on the west side of the Mekong to be about

400–500 individuals, 50 in the Samrong region and about 200–300 in the Srepok river. However, he gave a conservative number of 500, although the estimate was higher. Pfeffer & Kim-San (1967) estimated 500 individuals in 1952 and only 200 in 1964 in Cambodia. Estimates on kouprey population are not easy to obtain due to the difficulties of conducting research in the geographic area of the species because of the second Indochinese war and longlasting unrest. In the 1980s some estimates from Vietnam and Laos suggested a population between 100 and 300 individuals (Dang Huy Huynh & Pham Trong Anh 1988; Le Vu Khoi 1988; Suvanakorn 1988; Tran Hong Viet *et al.* 1988). Chan Sarun (1985) estimated fewer than 200 individuals in Cambodia and MacKinnon (1986) hypothesized that some individuals still survived in Laos and Vietnam. On the basis of these few estimates, MacKinnon & Stuart (1989) concluded that kouprey probably never exceed 2000 individuals in the twentieth century (Table 14.2).

Extensive camera-trapping surveys in one of the key kouprey habitats was conducted in 2008 and 2009 in Phnom Prich Wildlife Sanctuary, Mondulkiri Province, Cambodia. Despite large, threatened-mammal encounters being recorded, no kouprey were found (Gray & Phan 2011).

An extensive survey was conducted in Laos and Cambodia in 2010 (see Table 14.1; Timmins 2011). About 74 000 pictures were examined. Although banteng, wild water buffalo and gaur were all recorded in these regions, there were no images of kouprey. However, not all pictures were seen and a large number of images from eastern Cambodia were not reviewed. Despite scarce survey coverage in potential kouprey habitat, the probability of finding any individuals is extremely low. If the kouprey still persists, it is probably as scattered individuals in large, remote forest areas. Thus, the possibility of the persistence of some individuals cannot be excluded (Timmins 2011).

In the case that kouprey are detected in any location, what is the best strategy to follow? It is crucial that its presence should not be disclosed to the general public because this could threaten the species. Timmins (2011) identifies three main steps: (1) *in situ* conservation, greatly improving protection efforts; (2) *in situ* conservation using a well-protected fenced area within kouprey habitat; and (3) *ex situ* conservation both internationally and locally. Previously in 1964, Wharton attempted to capture live kouprey for an *ex situ* conservation project. He was able to catch five individuals, but unfortunately two died and three escaped (www.wildcattleconservation.org).

On the basis of previous conservation efforts, Timmins (2011) identifies the main factors that threaten kouprey survival:

- lack of effective protection models even when the resources are immediately available;
- scarce donor interest in supporting long-term protection efforts;
- few country resources for wildlife protection within areas where kouprey might persist;
- great threat on account of the demand for kouprey trophies still being high;
- lack of success in previous *ex situ* conservation projects for species with similar status within the kouprey range countries.

These conclusions suggest that maybe the best conservation option for any possible remnant kouprey is that the species remains undetected. On the contrary, if kouprey are found and

Figure 14.4 A young male kouprey photographed at Vincennes Zoo (Paris) in 1937. The only individual held in a zoo. This animal lived for about five years (photo by Georges Broihanne).

the discovery disclosed, it would likely be hunted to extirpation by poachers before any conservation effort could be undertaken (Timmins 2011).

During a workshop held in Vietnam in June 2008, the IUCN-led Regional Conservation Strategy for Wild Cattle and Buffalo Species in Southeast Asia (IUCN-SSC Asian Wild Cattle Specialist Group 2010) recognized four main goals for kouprey:

(1) Find any remaining kouprey.
(2) Create a genetic bank from trophies held in private and public collections.
(3) Create a fenced protected area within the original range for any kouprey detected.
(4) Evaluate the possibility of cloning kouprey if too few individuals are found.

Status in captivity

The only record of captive kouprey kept in a zoo was a young male captured in Preah Vihear Province, Cambodia. This individual was sent alive by René Sauvel to Vincennes Zoo in Paris in 1937, where it lived for about five years (Figure 14.4). The second captive record was a calf held by Cambodia's Prince Sihanouk that died in the 1950s; it is thought to be the last individual held in captivity (Joseph 2004).

Acknowledgements

The authors are grateful to Ana Maria Denman of Global Wildlife Conservation for her kindness and for the permission to use Figure 14.2. Thanks also to Dr C. Vithayanon for Figure 14.1.

References

Bohlken, H. (1961a). Allometrsiche Untersuchungen an den Schädeln asiatischerWildrinder. *Z. Saügtierk.* **26**: 147–154.

Bohlken, H. (1961b). Der Kouprey, Bos (*Bibos*) sauveli. Urbain. *Z. Säugtierk.* **26**: 193–254.

Campbell, J. (1860). Notes on the antiquities, natural history, &c. of Cambodia. Compiled from manuscripts of the late E. G. J. Forrest, Esq., and from information derived from the Rev. Dr. House, & c. J. *Royal Geographical Society London* **1860**: 182–198.

Chan Sarun (1985). Intervention de la delegation du Kampuchea. In *Conserving Asia's Natural Heritage*. Ed. Thorsell, J. W. Gland: IUCN and Natural Resources.

Coolidge, H. J. (1940). The Indo-Chinese forest ox or Kouprey. *Memoir of the Museum of Comparative Zoology Harvard* **54**: 421–531.

Dang Huy Huynh & Pham Trong Anh. (1988). *Kouprey (Bos sauveli, Urbain 1937): Investigations on Vietnamese, Kampuchean and Laotian Territories.* Hanoi: International Workshop on Kouprey Conservation Program.

Dioli, M. (1995). A clarification about the morphology of the horns of the female kouprey: a new unknown bovid species from Cambodia. *Mammalia* **59**(4): 663–667.

Dufossé, M. (1918). *Monographie de la circonscription résidentielle de Kompong-Thom.* Saigon: Union Nguyen-Van-Cua.

Enderlein, P. (1976). *Report from the Second Expedition to the Dongrak Mountain Range in Eastern Thailand in Search for Kouprey (Bos sauveli)*. Bangkok: FAO, Regional Office for Asia and the Far East.

Galbreath, G. J., Mordacq, J. C. & Weiler, F. H. (2006). Genetically solving a zoological mystery: was the kouprey (*Bos sauveli*) a feral hybrid? *Journal of Zoology* **270**: 561–564.

Galbreath, G. J., Mordacq, J. C. & Weiler, F. H. (2007). An evolutionary conundrum involving kouprey and banteng: a response from Galbreath, Mordacq and Weiler. *Journal of Zoology* **271**: 253–254.

Geraads, D. (1992). Phylogenetic analysis of the tribe Bovini (Mammalia, Artiodactyla). *Zoological Journal of the Linnean Society* **104**: 193–207.

Gray, T. N. E. & Phan, C. (2011). Habitat preferences and activity patterns of the larger mammal community in Phnom Prich Wildlife Sanctuary, Cambodia. *The Raffles Bulletin of Zoology* **59**(2): 311–318.

Grigson, C. (1988). Complex cattle. *Letter to New Scientist*, **4**: 93–94.

Groves, C. P. (1981). Systematic relationships in the Bovini (Artiodactyla, Bovidae). *Zeitschrift Fur Zoologische Systematik Und Evolutionsforschung* **19**: 264–278.

Grubb, P. (2005). Order Artiodactyla. In *Mammal Species of the World: A Taxonomic and Geographic Reference.* 3rd edition, Vol. **2**, pp. 637–722. Eds. Wilson, D. E. & Reeder, D. M. Baltimore, MD: Johns Hopkins University Press.

Handschuh, M. & Hassanin, A. (2013). Pure banteng *Bos javanicus* persist in southern Preah Vihear province, central Cambodia, despite apparent hybridisation with domestic cattle. *Natural History Bulletin of the Siam Society* **59**: 57–60.

Hassanin, A. & Ropiquet, A. (2004). Molecular phylogeny of the tribe Bovini (Bovidae, Bovinae) and the taxonomic status of the kouprey, *Bos sauveli* Urbain 1937. *Molecular Phylogenetics and Evolution* **33**: 896–907.

Hassanin, A. & Ropiquet, A. (2007a). What is the taxonomic status of the Cambodian banteng (*Bos javanicus*) and does it have close genetic links with the kouprey (*Bos sauveli*)? *Journal of Zoology* **271**: 246–252.

Hassanin, A. & Ropiquet, A. (2007b). Resolving a zoological mystery: the kouprey is a real species. *Proceedings of the Royal Society B: Biological Sciences* **274**: 2849–2855.

Hassanin, A., Ropiquet, A., Cornette, R., *et al.* (2006). Has the kouprey (*Bos sauveli* Urbain, 1937) been domesticated in Cambodia? *C.R. Biologies* **329**: 124–135.

Hedges, S., Groves, C. P., Duckworth, J. W., Meijaard, E., Timmins, R. J. & Burton, J. A. (2007). Was the kouprey a feral hybrid? A response to Galbreath *et al.* (2006). *Journal of Zoology* **271**: 242–245.

Hoffmann, R. S. (1986). A new locality record for the kouprey from Viet Nam, and an archaeological record from China. *Mammalia* **50**(3): 391–395.

IUCN-SSC Asian Wild Cattle Specialist Group (2010). *Regional Conservation Strategy for Wild Cattle and Buffaloes in South-east Asia.* www.asianwildcattle.org, accessed 8 April 2014.

IUCN (2013). *IUCN Red List of Threatened Species.* Version 2013.2. www.iucnredlist.org, accessed 1 April 2014.

Jackman, B. (1982). Alive! The ox from the past. *Sunday Times* 22 August, pp. 35.

Joseph, S. (2004). *Husbandry Manual for Wild Cattle Species.* Montgomery, MD: Bison, Buffalo and Cattle Taxon Advisory Group – American Zoo and Aquarium Association (AZA).

Lekagul, B. & McNeely, J. A. (1977). *Mammals of Thailand.* Bangkok: Association for the Conservation of Wildlife.

Le Vu Khoi (1988). *Kouprey (Bos sauveli, Urbain 1937), the current status of knowledge.* Hanoi: International Workshop on Kouprey Conservation Program.

MacKinnon, J. R. (1986). *Report of WWF Project 3696: Kouprey Status Survey.* Mimeo. WWF.

MacKinnon, J. R. & Stuart, N. S. (1989). *The Kouprey: An Action Plan for its Conservation.* Gland: IUCN, SSC, WWF,.

McDermid, C. & Sokha, C. (2006). Search for the kouprey: trail runs cold for Cambodia's national animal. *Phnom Penh Post* **15**(8).

Melletti, M., Penteriani, V., Mirabile, M. & Boitani, L. (2007). Some behavioral aspects of forest buffalo (*Syncerus caffer nanus*): from herd to individual. *Journal of Mammalogy* **88**: 1312–1323.

Neese, H. C. (1976a). The Kouprey (*Bos sauveli*): report on survey in southern Laos. Unpublished report.

Neese, H. C. (1976b). Kouprey clues and Rhino news. *Wildlife* **18**(9): 40.

Pfeffer, P. (1969). Considérations sur l'écologie des forêts claires du Cambodge oriental. *Terre et Vie* **1**: 3–24.

Pfeffer, P. & Kim-San, O. (1967). Le kouprey, *Bos (Bibos) sauveli* Urbain, 1937; discussion systématique et statut actuel. Hypothèse sur l'origine du zébu (*Bos indicus*). *Mammalia* **31**: 521–536.

Sauvel, R. (1949). Distribution géographique du Kouprey (*Bibos sauveli*) Urbain. *Mammalia* **13**: 144–148.

Sayer, J. A. (1983). *Notes on the status and distribution of the larger mammals of Laos. Final report, Forest Management Project, Lao People's Democratic Republic/ Food and Agriculture Organization of the United Nations.* Vietnam, Laos: FAO.

Suvanakorn, P. (1988). *Kouprey in Thailand.* Hanoi: International Workshop on Kouprey Conservation Program.

Thornback, J. (1983). *Wild Cattle, Bison and Buffaloes: Their Status and Potential Value.* Cambridge: The IUCN Conservation Monitoring Centre and Natural Resources.

Thouless, C. (1987). Kampuchea wildlife: survival against the odds. *Oryx* **21**: 223–228.

Timmins, R. J. (2011). *Searching for the Last Kouprey.* Austin, TX: Global Wildlife Conservation, IUCN/SSC, AWCSG.

Timmins, R. J., Hedges, S. & Duckworth., J. W. (2008). *Bos sauveli.* In IUCN 2012. IUCN Red List of Threatened Species. Version 2012.2. www.iucnredlist.org, accessed 15 April 2013.

Tran Hong Viet, Do Tuoc & Pham Mong Giao (1988). *Kouprey Habitat in Vietnam.* Hanoi: International Workshop on Kouprey Conservation Program.

Urbain, A. (1937). Le Kou Prey ou boeuf gris cambodgien. *Bulletin de la Société zoologique de France* **62**: 305–307.

Vietmeyer, N. (1983). *Little-known Asian animals with a Promising Economic Future.* Washington, DC: National Academy Press.

Vithayanon, C. & Bhumpakphan, N. (2004). Historical distribution range of the Kouprey from three fossil locality sites in north Northeastern Thailand. *Journal Wildlife of Thailand* **12**: 214–216.

Weiler, H., Kimmchay, H., Kimsan, O., Masphal, K., Polin, S. & Seiha, U. (1998). *Distribution of Tiger, Leopard, Elephant, and Wild Cattle (Gaur, Banteng, Buffalo, Khting vor and Kouprey) in Cambodia.* Pehnom Penh: Cat Action Treasury and Wildlife Protection Office.

Westing, A. H. & Westing, C. E. (1980). Endangered species of Viet Nam. Unpublished report.

Wharton, C. H. (1957). *An Ecological Study of the Kouprey Novibos sauveli (Urbain).* Manila: Bureau of Printing.

Wharton, C. H. (1968). Man, fire and cattle in Southeast Asia. *Proceedings of the Annual Tall Timbers Fire Ecology Conference* **8**: 107–167.

Wilson, D. E. & Mittermeier, R. A. (2011). *Handbook of the Mammals of the World.* Vol. 2. Hoofed Mammals. Barcelona: Lynx Edicions.

WWF (1982). World's oldest wild cattle believed rediscovered in Southeast Asia, landmines hamper search. *World Wildlife Fund News Release* **12/82**.

Chapter 15

Aurochs *Bos primigenius* Bojanus, 1827

T. van Vuure

Names

Genus: *Bos* Linnaeus, 1758

Species: Aurochs *Bos primigenius* Bojanus, 1827

Subspecies: *Bos primigenius primigenius* Bojanus, 1827; *Bos primigenius namadicus* Falconer, 1859; *Bos primigenius mauretanicus* Thomas, 1881

Names in other languages: **French**: Auroch; **German**: Auerochs; **Spanish**: Uro; **Italian**: Uro; **Polish**: Tur.

Ecology, Evolution and Behaviour of Wild Cattle: Implications for Conservation, ed. M. Melletti and J. Burton. Published by Cambridge University Press.

Unlike most of the other members of the wild cattle family, the aurochs is extinct. That means that not all data concerning this animal and its life history can be described, and that some aspects will be examined here in a different way than in other wild cattle species.

Linnaeus (1758) described domestic cattle under the name of *Bos taurus*. He mentioned that at the time its wild ancestor was briefly described by the Roman commander-in-chief Julius Caesar under the name of *urus* (derived from the Germanic word 'ur'). In 1827, Bojanus made the first osteological research on an aurochs skeleton and gave to this species the name *Bos primigenius*. Because he thought it came 'before the Flood', he added the word *antediluvialis*. Though the name given by Bojanus is widely used so far, more correctly the name given by Linnaeus should be employed, because he described the species first.

The Spanish word for this animal, 'uro', comes directly from the Latin word *urus*. In both the English and the French languages the word 'aurochs' comes from the German word 'auerochs'. For centuries, this latter word was mistakenly used for the wisent (*Bison bonasus*). Around the twelfth and thirteenth centuries, during which the original aurochs became extinct in Germany, the word 'ur' gradually changed into 'auer' and 'auerochs' (Szalay 1915). When people were no longer aware of the original aurochs its name passed to a related animal, similarly impressive – the wisent. Such a process has been observed in other species too.

Taxonomy

Within the tribe Bovini, aurochs are grouped in the subtribe Bovina, together with bison, wisent, banteng, gaur, yak and kouprey. To clarify the taxonomic position of *Bos primigenius* within this subtribe, both osteological and genetic methods were and are used. The question of interspecific relationships among species of the subtribe Bovina remains highly debated among molecular biologists (for more details, see Chapter 1).

Subspecies and distribution

Probably between 610 000 and 850 000 years ago, in India, the wild ancestors of taurine and zebu cattle separated, either from a common ancestor or from each other. The estimated time of this event is based on mtDNA research (MacHugh *et al.* 1997).

Since its origin, the subspecies *primigenius* spread to the north and the west. In Europe the aurochs arrived much earlier in the south than in more northern areas. In Southern Europe the aurochs arrived about 700 000 years ago (Estévez & Saña 1999), and in Germany about 275 000 years ago (Lehmann 1949). It is probable that there could have been a southern and a northern route to this continent. In the course of time, it is likely several subspecies differentiated, but it is not easy to distinguish them from each other due to a lack of bone material for comparison. The Pleistocene subspecies *Bos primigenius namadicus* continued to exist into the Holocene, although bone finds of it from the latter period are very scarce. Nevertheless, it is assumed that this subspecies was domesticated in the area of Pakistan during the Neolithic, and gave rise to zebu cattle. At least in the form of its horns and the back of the head, *Bos p. namadicus* differed from other subspecies (Grigson 1985).

In spite of the large distribution area of the aurochs, only two further names for subspecies are in vogue, namely *Bos primigenius mauretanicus* (in North Africa) and *Bos primigenius primigenius* (in Europe). Since North African aurochs looked very much like European aurochs, at least in horn form and position, the names for these subspecies are only based on the geographical distribution. In view of the large distribution area of the aurochs, without doubt there must have been more subspecies, and maybe this will be elucidated by ongoing genetic research. For the outlines of the distribution area of the aurochs during the Pleistocene and Holocene, see Figure 15.1.

Descriptive notes

Dental formula

I 0/3, C 0/1, P 3/3, M 3/3.

Body size

Since Bojanus (1827) described the aurochs skeleton at Jena, and Wrześniowski (1878) confirmed the former Holocene existence of the aurochs in Europe until 1627, there has been speculation about the size of this extinct animal. Julius Caesar's description of the animal as being 'little smaller than elephants in size' seems to be quite exaggerated, but has appealed much to man's imagination (Caesar, 58–52 BC). Maybe Caesar referred to the (relatively small) North African elephant. Initially, especially in the first half of the twentieth century, but even today, several authors considered the aurochs bull to have stood about 2 m at the shoulder, and the aurochs cow about 1.80 m. Later, smaller sizes were mentioned by others.

In an attempt to bring more certainty about the height at the withers in prehistoric cattle, Matolcsi (1970) undertook an experiment in cattle breeds. He measured the limb bones of several cattle breeds after slaughtering them, and tried to find a relation between these measures and the height at the shoulder of the animals measured when still alive. This provided a number of conversion factors. Multiplication of the length of humerus, radius or metacarpus by their specific conversion factor would give the height at the withers of the animal. However, application of this method to the aurochs was not univocal. If one aurochs specimen is considered, the humerus, the radius and the metacarpus give different values of the shoulder height. For instance, the shoulder height of an aurochs bull from Braunschweig (Germany) was determined through the humerus, radius and metacarpus to be 182, 169 and 164 cm, respectively (van Vuure 2005, p. 114). The humerus turned out to be the bone that nearly always gave the greatest shoulder height. In aurochs skeletons, mounted by

Figure 15.1 The former distribution area of the aurochs, as based on data of many authors and the author's own interpretations. The outermost border of the distribution area during the Pleistocene. For Europe, North Africa and West Asia this border roughly holds for the Holocene period as well. ____ Probable outermost east and southeast border during the Holocene.

Degerbøl and Fredskild (1970), the shoulder height calculated through the humerus turned out to be closest to the height calculated in these mounted skeletons. This is why the calculation through the humerus probably gives the most reliable height at the withers. By this method it could be determined that in the more northern parts of Europe aurochs bulls, in general, had a shoulder height of 160–180 cm and aurochs cows of about 150 cm (see Figure 15.2). Besides that, the size of South European aurochs was somewhat smaller than the northern ones. Furthermore, West European aurochs were somewhat smaller than East European ones.

It can be stated that aurochs were nearly squarely built. This means that the length of the trunk was about the height at the withers.

Differences in body size between aurochs bulls and cows was one of the features of sexual dimorphism that existed in aurochs. The body weight of the aurochs cow and bull was probably comparable to that of the wisent and the banteng.

Horns

Form

From Ice Age rock engravings it is not possible to get a good picture of the three-dimensional form of aurochs horns (see Figure 15.3). But on the basis of hundreds of horn cores and a dozen horn sheaths we are able to imagine the exact form of them. Aurochs horns had a very characteristic form. Both bulls and cows had horns, which generally differed in size and thickness, as part of sexual dimorphism. The basic horn form was the same in both sexes. Von Leithner (1927) gave the first exact description of aurochs horns, which was later confirmed by other authors. This horn form is described as follows: starting from the base, where it attached to the head, the growth direction of the horn was first sideways and a bit upward. After about a quarter of its length it would then curve forward and inward, and in the last quarter further inward and upward. The horn looked a bit like a 'stretched spiral' (see Figure 15.4). The variety in the horn form consisted not only in its length and thickness, but also in the way the horns curved more or less inward. In general, bulls' horns were more curved inward than cows' horns. There was also some variety in the horn position on the head.

Position

With the position of the horn, it means that the angle would have been between the forward curving part of the horns and the forehead. This angle varied between 50° and 70°, with an average of about 60° (van Vuure 2005). This position determines the 'forwardness' of the horns.

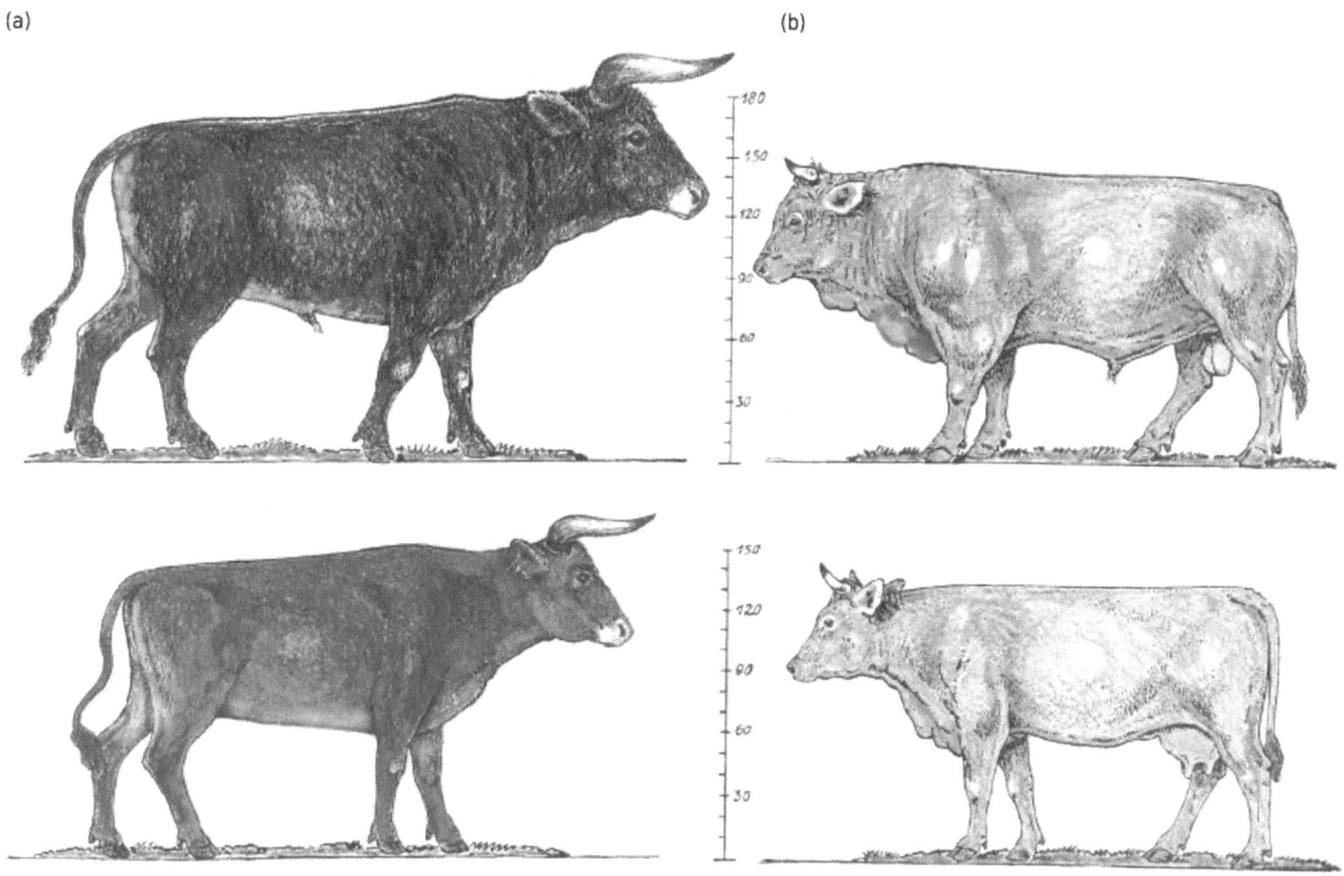

Figure 15.2 Comparison of bulls and cows of the aurochs (a) and modern cattle (b) (after Boessneck 1957, modified and coloured). The differences in body size, udder size, body proportions and horns are clear. A black and white version of this figure will appear in some formats. For the colour version, please refer to the plate section.

Figure 15.3 Rock engraving of an aurochs bull dated 14 000–10 000 BC (Romito Cave, South Italy) (photo by Luca Policastri).

Size

By measuring complete horn cores and horn sheaths, and knowing how long the solid part of the horn sheath was at the end, we are able to determine the original thickness and length of the horns, though not their average (van Vuure 2005). In cows' horns, the diameter (at the base) differed between 7 cm and 10 cm, in bulls' horns between 10 cm and 18 cm. In cows' horns the length differed between 40 cm and 70 cm, in bulls' horns between 60 cm and 107 cm.

Figure 15.4 Aurochs skull with horn cores, from the Baikal area, represented both from the front (top) and the side (bottom) (from Gromova 1931). Such skulls from East Asia bear a striking resemblance to those from Europe.

Colour

From existing horn sheaths we can learn about their original colour. Since many of them were used to hold beer, wine or oil, or were covered with varnish, it is not always easy to determine their original colour. Comparing them with the colour of the horns of aurochs depicted in the Lascaux cave and with that of the horns of the Augsburg aurochs (depicted in Smith 1827), we can determine the colour as having been pale or yellowish white, and black towards the tip.

Coat colour

Since no aurochs coat has been saved, the determination of its colour must be deduced from depictions and descriptions. We have depictions of aurochs in caves like that of Lascaux and furthermore the picture of the Augsburg aurochs. At Lascaux, apart from the non-coloured and spotted aurochs, the artists probably discriminated between reddish-brown cows and black bulls (Grigson 1978). The Augsburg aurochs, a bull, was described by Smith (1827) as being 'sooty black' coloured (see Figure 15.5).

In descriptions, the colour of the aurochs bull was quite clearly rendered, for instance by Von Herberstein (1557) and Schneeberger (in Gesner 1602, pp. 141–142), as 'very dark to black' (German: 'schwarz'; Latin: 'niger'). The nature of the coat colour of the cow is more unclear. In old Polish literature there was the expression 'red like an aurochs' and there was a passage about a 'red aurochs calf'. Apart from this, the best description of the aurochs is from Schneeberger. He mentioned that all aurochs calves were born in a chestnut colour (Latin: 'castanei coloris'). Within half a year the colour of the bull calf turned black, but the cow kept the calf's colour. Very rarely black cows were seen. Taking into account these depictions and descriptions we may conclude that aurochs bulls had a very dark coat colour, blackish-brown or black. Aurochs cows were of a reddish brown colour, maybe comparable with the European red deer (*Cervus elaphus*). According to Schneeberger, the coat colour of the cow was then described with the Polish word 'płowy'. Nowadays in Poland red deer are called 'zwierzyna płowa'. The difference in coat colour between aurochs bulls and cows is a feature of sexual dimorphism.

A striking feature in the coat of the aurochs bull was the dorsal stripe (eelstripe). According to Von Herberstein (1557), who took a close look at a bull's hide, the colour of the stripe was 'grey' ('grahlat'). According to Schneeberger, the dorsal stripe was two fingers wide. The dorsal stripe was not described in aurochs cows. Another feature was the white chin, which was highlighted by Smith (1827) in the Augsburg aurochs bull. This is the only mention of it. In fact, we do not know how widespread this white chin or muzzle was among aurochs, if it only occurred in bulls or in both sexes, or what shape it could take on.

Figure 15.5 The Augsburg aurochs (from Smith 1827). This painting is a copy of the original that was present at a dealer at Augsburg (Germany) in the early nineteenth century. The original probably dates to the beginning of the sixteenth century. It is not known whether the original or the copy still exist.

Other characteristics

Udder

In none of the Pleistocene aurochs cows, portrayed in French and Spanish caves, has the udder been depicted. From this, and from the fact that the other members of the *Bovini* tribe have small udders, we may conclude that aurochs cows had small, hardly visible udders.

Hair on the forehead

Two sources report that the hairs on the forehead between the horns were curled and twisted, which must have made the animal 'terrible to behold' (Schneeberger in Gesner 1602; Święcicki 1634). The hair can be seen in the picture of the Augsburg aurochs. From this part of the skull, hide belts were made, girded by which 'pregnant women could give birth with miraculous ease', according to Schneeberger.

Habitat

Data about the habitat type used by the aurochs stems mainly from Europe. Because of the relatively late occurrence of the aurochs in Europe, the majority of bone finds and records that have been preserved come from European aurochs. Aurochs bones are found mostly in soil layers, and to a lesser degree in caves.

When the overall picture of the aurochs distribution is looked at, we see that it did not occur in steppe areas or savanna landscapes. In steppe regions they probably occurred only in forested areas along rivers. Moreover, they did not occur in high, mountainous areas. In Northern Europe and Asia, aurochs probably did not inhabit the zone of coniferous forests (Gromova 1931). During the Holocene, in North Africa the aurochs only occurred in the northern forested zones. In the area south of it (nowadays Sahara) a savanna landscape existed till 3000 BC (Ritchie & Haynes 1987). In this area wildlife similar to that currently present in savannas of East Africa still occurred. No aurochs bones have been found in these savanna areas of North Africa. The only area in North Africa where aurochs had spread further south was into the valley of the Nile river (Gautier 1988). Looking at other areas of occurrence in the Middle East, a provisional picture of the habitat preference of the aurochs can be formed. They probably used swampy forested habitats with glades that provided water, grasses or sedges and cover.

The vast majority of European aurochs finds stem from the lowlands and low mountain ranges; much fewer were found in high, mountainous areas. Nearly all lowland finds were discovered in wet clay and peat layers, such as river and lake banks, coastal sediments and peat bog layers. Apart from these, bones were found in human settlements. This would seem to warrant the conclusion that the aurochs occurred almost exclusively in wet areas and not outside of them. Nevertheless, the fact that so many findings originate from wet areas does suggest that such habitats at least constituted an important part of the local distribution range of the aurochs. From a study by Hall (2008), comparing bone sites of present mammals with the habitat preference of these animals, strong indications arose that aurochs, just like beavers, were living in low-lying, flat areas, rich in watercourses. In relation to this, we may find a clue in an Anglo-Saxon rune verse from the ninth century AD (Rodrigues 1992):

Ur byþ anmod and oferhyrned;	The ***aurochs*** is fearless and large-horned;
– felafrecne deor – feohteþ mid hornum –	– a fierce animal – it fights with its horns –
mære morstapa; þæt is modig wuht	the ***famous marsh walker;*** it is a brave animal

In this verse, the Germanic word 'mor' means 'fen' or 'marsh'. Apparently it was not unusual to nickname this animal 'marsh walker', presumably because the marshes constituted habitats where these animals were quite regularly seen. Research by Lynch *et al.* (2008), among others on depletion of $\delta^{13}C$ in aurochs and cattle bone collagen, gave as a conclusion that 'a wetland habitat for aurochs is arguably more probable than a forested habitat'. Since the natural Holocene Europe was almost entirely forested (for more details see van Vuure 2005), the only areas where a grazer like the aurochs could sufficiently feed were the grass and sedge-rich marshes along rivers and lakes. Within forests there is little or no grass.

Regarding the forested habitats in two European areas, where aurochs were recorded, we have quite reliable descriptions. These were the Great Wilderness in former East Prussia and the Forest of Jaktorów in Central Poland.

The Great Wilderness was a vast wilderness area that for more than two centuries functioned as a buffer zone between East Prussia and Poland/Lithuania. This area remained nearly untouched by people. It was inhabited by rare species almost extinct elsewhere in Europe. Among them were the wisent (*Bison bonasus*), the wild horse (*Equus caballus ferus*), the elk (*Alces alces*) and the aurochs. We know this from several records concerning trips through this wilderness and from hunting records. The last record of aurochs in this area is from 1521, but long before that these animals must have been already rare. From several authors we can learn that the Great Wilderness was an area that consisted nearly entirely of continuous forests, interspersed with marshes and rivers. Traversing these primeval forests often presented great problems to man because of the many fallen trees and the marshes and rivers (Mortensen & Mortensen 1938; Mager 1960).

The other area was a forest situated near the village of Jaktorów, some 40 km west of Warsaw, Poland (Heymanowski 1972). Here the last wild aurochs population lived in a forest, owned and managed by the Polish kings. Red deer (*Cervus elaphus*), roe deer (*Capreolus capreolus*) and wild boar (*Sus scrofa*) were also present, but probably with small populations. Wolves (*Canis lupus*) occurred there too. Records concerning this forest spoke about 'a forested area' for which the tree species were mentioned. Several brooks and rivulets run through it. Nevertheless, in this forest there must have been parts where grasses or sedges grew, eaten by aurochs. Such parts must have been in the marshy areas along rivulets or in parts where people had cleared the forest to provide grass to the aurochs. Schneeberger (in Gesner 1602) mentioned the presence of exceptionally large oaks (*Quercus* sp.) with a circumference of more than 9 m in this forest. This probably shows that at least parts of the forest were a kind of 'Hudewald', managed as pastures for cattle.

The activities of beavers (*Castor fiber*) probably played a role in benefiting the aurochs. Beavers were able to flood large areas of forest by building dams in rivers. This flooding caused the dying off of these forests and subsequently the origin of fields of grasses and sedges. The same phenomenon could arise when floating tree trunks became stuck and formed dams that blocked a river.

Summarizing, it could be said that in the forested parts of Central and West Europe the aurochs mainly lived in marshy areas along rivers and lakes and/or in salt coastal marshes. Present bovine equivalents with similar habitat preferences as the aurochs are probably the forest buffalo (*Syncerus caffer nanus*) and the wood bison (*Bison bison athabascae*). The former lives in marshes, equatorial savannas and clearings in the rainforests of the Congo Basin. The latter lives in sedge marshes in the otherwise forested boreal landscape of Central Canada.

Movements and home range

Of the last aurochs population living in the wild in the Forest of Jaktorów we know the animal numbers for a certain period. These numbers declined in the course of the second half of the sixteenth century. The highest number of aurochs we know of is from a mention by Anton Schneeberger in 1559. Schneeberger described that in that year the number of aurochs was 'once again over 50'. Since this concerned a managed population, we do not know their natural population density. The Forest of Jaktorów and adjacent forests, where aurochs lived, covered an area of 25 000 ha (250 km^2; Heymanowski 1972). A population of 50 aurochs would mean that their density was one aurochs per 500 ha (0.2 per km^2). We do not know whether all aurochs roamed the whole forest area. In summer they lived singly or in groups; in winter they grouped together ('like some army', as Schneeberger wrote). The latter was probably related to the supplementary feeding of hay in that period. The animals were not allowed to leave the forest. If they did leave the forest area, they were driven back by wardens on horseback, with the help of dogs.

In view of the fact that aurochs ate acorns in autumn and, beside hay, branches of trees and bushes in winter (as Schneeberger wrote), this would mean that during the year they changed habitats. In general, aurochs probably lived more in marshy areas during the summer and more in the forest (edges) in winter. During the winter they probably avoided areas with too much snow. If we look at present cattle types (even the primitive ones), we can observe that cattle cannot deal with snow in too thick layers. Bison and horses handle it by shoving with their heads or pawing, respectively, but cattle do not.

Feeding ecology

Cattle, the domesticated descendants of the aurochs, are considered grazers on the basis of their choice of food and due to specific bodily adaptations (Van Wieren 1996). This is also shown in cattle that live in nature reserves, even if they live there all year round. During the summer their food consists mainly of grasses and graminoids. During the winter grasses are partly replaced by branches of trees and bushes.

What is actually known about the food eaten by aurochs? In a Russian epic poem from the twelfth century aurochs are presented as 'eating grasses in marshes' (Dementiev 1958). Other mentions about aurochs' food are found in the Inspection Reports ('Lustracje') that were made in the sixteenth and seventeenth centuries about the management of the Forest of Jaktorów in Poland (Kędzierska 1959, 1965). In various places in these reports, the relation of the aurochs with the local meadows is mentioned. The farmers of Jaktorów, situated at the edge of the forest, are said to 'mow grass, where the aurochs have their meadows' and 'their horses and cattle graze, where the aurochs stay'. In addition: hunters (game wardens) appointed by the Crown are said to 'have to pasture the aurochs with devotion in the usual places, so that they do no harm to the agricultural fields, also in the meadows of the Crown'. Moreover, the fact that they judged the grazing by cattle and horses as detrimental for the aurochs is mentioned twice. Since the domestic cows happened to come 'where also the aurochs graze', the latter could compete for the feeding places because of the diminished amount of available food resources.

These reports make it reasonable to assume, first of all, that the meadows in question (mainly) contained grass. Second, like the domestic horses and cattle, the aurochs used this grass as their main food source. The grass in the meadows at the village of Jaktorów was used to make hay, which in wintertime was taken to the forest to feed the aurochs. Aurochs also tended to leave the forest to graze in meadows elsewhere. In addition to this, the mown grain that was standing in the fields of the farmers to dry was also taken. This was probably in the month of August, when the grain would ripen and become very attractive to the aurochs. During the autumn, acorns appear to have been a desirable food. Schneeberger wrote that in the forests there were enormously large oaks that produced acorns. During those months the aurochs were 'fatter and more shiny than at other times'.

Regarding the winter food, Schneeberger wrote that 'they eat branches of shrubs and trees'. Besides that, 'they are also supplied with hay that the farmers of the above-mentioned villages have gathered for the aurochs in summer'. Because in some years the hay harvest failed, aurochs were then forced to look for food more in the forest. Because present cattle in nature reserves eat tree bark as an additional food in winter and spring, we may assume that aurochs did this to a certain degree as well.

Reproduction and growth

Around the mating season, a number of changes in the structure of the group take place among wild cattle; the same is likely to have been true in the aurochs. From two sources we know that aurochs bulls were attracted to and mated with domestic cows (Schneeberger in Gesner 1602; Święcicki 1634). Aurochs bulls would go to the herds of domestic cattle to lure away cows in heat to mate with. Aurochs bulls must have gone to the mixed (aurochs) groups to find the cows in heat, in a similar way to current bovine species. From crossings between aurochs and domestic cows, sometimes calves were born. Schneeberger wrote that these cows would 'either miscarry or give birth to non-viable young'. The explanation of this is not clear. The managers of the aurochs population tried to prevent such crossings. Aurochs bulls that covered domestic cows were hunted, according to Schneeberger.

Schneeberger reported about the reproduction of the aurochs:

> In September they mate, while burning with lust, have numerous and very fierce fights and sometimes both fall down dead. But the one animal that they have noticed to be stronger, and constantly provoking others to fights, hunters will hunt under orders of the king. . . . Finally they give birth in May, some in September like domestic cows; this seldom happens, however, and the calves that are born in the autumn are weak and do not stay alive because of the fierce winter cold. When the time to give birth arrives, the cow will retire to the densest areas of the forest and remain there with the calf for about twenty days; as soon as she sees it has become stronger and jumps she will finally take her to the pasture, guarding it closely so that the calf is not caught by hunters or devoured by wolves.

Schloeth (1961) states that in feral Camargue cattle the cow leaves her calf behind in a remote place, joins the herd, and for some days regularly returns to the calf to suckle it.

The months mentioned by Schneeberger here, September for mating and May for birthing, are unlikely to be correct. If conception was in September and birth in May, gestation could only have lasted for eight months. Since gestation in both domestic cows and related bovine species such as gaur and wisent is about nine months, a mating season in the months of August and September, with a birthing season in May and June, is more likely. May and June are the months in which the grass grows most plentiful. In this way the mating and birthing seasons of the aurochs must have coincided with those of the wisent. The Southern European feral Camargue cattle mate in July and August and give birth in the period of March–May (Schloeth 1961). Apparently sometimes some aurochs cows came in heat (again) some three or four months after the regular mating season.

Concerning the number of calves per cow, there is no detailed mention. Schneeberger only wrote that cows had 'a calf'; there was no mention of twins. Schloeth (1961) mentions that in Camargue cattle usually one calf per cow is born; twins are rare. Unlike domestic cattle, aurochs probably were bodily late-maturing, like other members of the Bovini tribe. During the domestication process, cattle became more and more early-maturing. This means that cows were able to calve more often and from an earlier age, and both the bulls and the cows would remain smaller. They would also amass fat at an earlier stage. The cows of early-maturing cattle (e.g. Jersey cattle) often give birth at 24 months or even earlier. Cows of late-maturing breeds generally calve at age three or four. The bulls of some very late-maturing cattle breeds (Maremmana, Pugliese) only reach maturity when they are between the ages of five and seven years. Regarding the age of aurochs, Schneeberger wrote that they were 'not long-living' and 'it is said that only a few have lived beyond the 15th year'.

Behaviour

Social structure

Although there are few data about the social structure and reproduction of the aurochs, they may be just enough to provide us with a global impression of these aspects. The data stem also from the last aurochs population at Jaktorów. From the Inspection Reports and from Schneeberger's record, we may learn about the social structure. The 1564 Inspection Report noted about the animals:

> In the forests at Jaktorów and Wiskitki, where the aurochs hide, during our travels we found a herd of 30 animals, including 22 old cows, three young bulls, five young calves. We were unable to find old bulls, since these had spread out in the forest on account of the thunder, but the hunters had told us that there were eight of these old bulls.

Since 'young calves' were mentioned here this situation must have taken place during the month of June or July. Schneeberger wrote:

> In winter they run in herds and they rest in one place, quite like some army. . . . In summer, however, they roam the wilderness separately.

These 'herds' may indicate the situation, when the animals were supplied with hay during winter. This probably happened at a central place. Whether the word 'separately' ('*singuli*') referred to all the animals or only to the bulls, Schneeberger is not clear about.

Like in other bovine species (see Halder 1976; Sinclair 1977; Prins 1987), probably in aurochs too, there was a sexual segregation between full-grown cows and older bulls. Except for the mating season, in a natural situation there must have been groups of cows with calves and young bulls. The older the bulls became, the more they separated from the cow groups. This is in line with research on the wisent (e.g. Krasińska & Krasiński 2007). Probably the older aurochs bulls roamed through the forest area on their own. In Camargue cattle, Schloeth (1961) researched several aspects of behaviour of these cattle, among other things their social structure. Likely, many of the behavioural features of these cattle (e.g. forming and maintaining group structure and order of rank) were present in aurochs as well.

Sexual behaviour

As mentioned above already, the adult bulls probably joined the mixed cow groups to mate, and during that period the bulls 'mate, while burning with lust, have numerous and very fierce fights and sometimes both fall down dead', according to Schneeberger. This description seems a little exaggerated, but probably contains an element of truth. In Camargue cattle, this fighting until death was exceptional (Schloeth 1961). Another passage in the record by Schneeberger might also refer to the agitated state of mood of the aurochs during mating time. Schneeberger wrote:

> in summer they leave the forests and go into the fields and devour mown grain and when they are satisfied they throw the rest apart with their horns.

This impressive behaviour must have taken place in August, when the grain ripens, right in the mating season of aurochs.

Schloeth (1961) conducted a study on the sexual behaviour of Camargue cattle. Possibly, traits of the sexual behaviour of these cattle (impressing and threatening behaviour, genital inspection, tending bonds, etc.), were also present in aurochs. For wallowing, Schloeth suggested that this behaviour probably occurred in aurochs, like in other wild cattle species (e.g. African buffalo and bison). According to Schloeth, in Camargue cattle typical wallowing no longer occurred. As a relic of former wallowing in aurochs, Camargue cattle only rub their heads and necks against the ground.

Behaviour towards man

Folklore, sayings and records suggest that aurochs, from man's point of view, must have been an impressive and truculent animal, appealing to one's imagination. Nowadays, this impression is often reinforced by experiences with Spanish fighting bulls. Yet this picture has to be put into perspective. From two sources we know that the last aurochs in the Forest of Jaktorów were easily approachable. Schneeberger wrote about this:

> The aurochs is not afraid of humans and will not flee when a human being comes near, it will hardly avoid him when he approaches it slowly. And if someone tries to scare it by screaming or throwing something, this will not scare it in the least, but while it stays in its place it will actually open its mouth, widen it and close it again quickly, as if it is making fun at the human for his attempt [flehming?]. When it is standing in the road or somewhere else, one must go around it, even if one is driving a carriage, since it will not move off the road by itself.

Only if aurochs were provoked and tormented could they become bad-tempered and fierce. In that state they could throw humans, and even a horse and its rider, up into the air with their horns. Illustrative of this is the fact that, in 1472, a Lord Chamberlain was given a sum of money by the Polish king in compensation of a wound that had been inflicted on him by an aurochs during a hunt the previous year (Heymanowski 1972). In the Forest of Jaktorów aurochs were used to people. They were rarely hunted – at the very most by the Polish king, though there was also poaching.

Predation

In its extensive distribution area, the aurochs had to contend with different species of predators. Some records are known from descriptions and pictures, while others derive from their presence within the distribution area of the aurochs and by their choice of prey. Predation by wolves we know from descriptions. Predation by lions (*Panthera leo*) we know from pictures. Judging by their prey, predators which may have preyed on aurochs range in size from lynx (*Lynx* sp.) and wolverine (*Gulo gulo*) to lion, tiger (*Panthera tigris*) and bear (*Ursus* sp.).

During the Holocene, the wolf and to a lesser degree the lion (in the Balkans) were the main predators of aurochs within Europe. Reports about the Forest of Jaktorów mention predation by wolves. Schneeberger wrote about this:

> They [aurochs] do not experience harm from wolves, except when they have just been born, wander around separately, because then they will be torn up by wolves.

Already above, it was written that Schneeberger mentioned that the cows were 'guarding it closely so that the calf is not caught by hunters or devoured by wolves'. Schneeberger's quote, in which he refers to wolves, likely shows that especially calves fell prey to wolves and that adult aurochs apparently suffered little from them. Confrontations between aurochs and wolves are described by Święcicki (1634):

> More than once, a single huge aurochs bull has been seen to have gained the complete victory, at the cost of a number of wolves thrown on the ground and trampled to death.

Here also the description seems a little exaggerated, but probably had an element of truth.

During the day the aurochs were guarded nearly constantly by hunters (game wardens) appointed by the Crown. By night, wolves could get closer to the unguarded aurochs.

In domestic cattle, running wild, cows show a particular behaviour if attacked by wolves. In such situations cows form a circle to defend themselves (Pusch 1837, p. 206) or defended their calves (Frevert 1957, p. 18). This behaviour was likely also present in aurochs. Apparently aurochs did not seek safety in flight, but defended themselves by butting with their horns and kicking with their legs.

Parasites and disease

There is only one mention of disease in aurochs. The 1602 Inspection Report of the Forest of Jaktorów mentioned regarding the remaining four aurochs:

> but many of them have died through other cattle contaminated with the 'infected air'.

Here, 'infected air' is the translation of the Polish term 'powietrze zarażeni', used in the original report. What disease is meant here is not clear, because a further diagnosis is lacking. Maybe this mention of a 'disease' might have been a way of disguising the deteriorating management of aurochs there.

Status in the wild

IUCN Red List Category: Extinct

Last living area

Since aurochs became extinct in the area of East Prussia and adjacent areas (nowadays northeast Poland, west Lithuania and the Russian province of Kaliningrad), its only remaining living area in the wild was the Forest of Jaktorów. Since the fourteenth century this area was the exclusive hunting area of the dukes of Masovia. In 1476 the management of these forests reverted to the Crown. Since old times there was a well-organized game-keeping service. Game wardens had 'to keep an eye on the aurochs, let them graze, collect the hay from Jaktorów from the subjects, to supply the aurochs with this in winter, to know the number of animals and give a report to the district manager or the tenant every three months'. The hay was made by the farmers at Jaktorów, a village established at the time exclusively 'for the aurochs and their welfare'.

The first mention about the number of aurochs was by Schneeberger, who in 1559 mentioned their number as 'once again over 50'. The 1564 Inspection Report mentioned 38 animals (see above). Many Inspection Reports are unfortunately lacking. The 1599 Inspection Report mentioned 24 animals. Then the number of animals diminished quickly. The 1602 Report mentioned only four animals, three bulls and a cow. In 1620 the last aurochs bull died, and one cow was left. This cow eventually died in 1627. The 1630 Report stated:

> In the last Report [from 1620] it was written that there was one aurochs cow, but now the inhabitants of this village [Jaktorów] said that she died three years ago.

Beside this wild population, a number of aurochs lived in captivity for some time in a zoo at the village of Zwierzyniec in southeast Poland. This zoo was owned by the Zamoyski family. We know that both aurochs and wisents were represented there, as a letter is known from 1610 by someone who wanted to visit this zoo and who wrote beforehand (Wrześniowski 1878, p. 518): 'How many aurochs and wisents do you have, will you please let me know, and how many males and females'. It is not known how long aurochs survived there, and whether they outlived the wild population.

Causes of extinction

The aurochs was exterminated by man over a period of thousands of years. Because the final extinction period in Poland is quite well known, the aurochs gained the dubious honour of being the first documented case of extinction of an animal species. This process of extinction consisted of two main elements, hunting and ousting. Both were

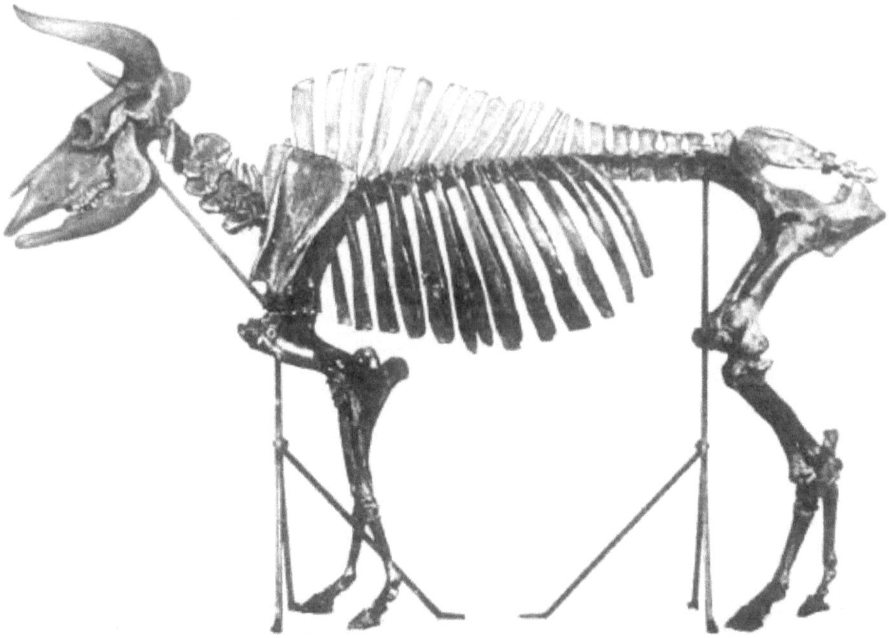

Figure 15.6 Nearly complete skeleton of an aurochs bull from the peat of Hammarlöv, which can now be found in the Zoological Museum of Lund (Sweden) (from Von Leithner 1927).

executed by man and both were strongly related to what is called 'human civilization'.

Probably the first area where an aurochs subspecies became extinct was the living area of the *namadicus* subspecies, possibly somewhere in the area of Pakistan. This must have happened in the Early Holocene, because nothing has been mentioned about this subspecies later in the Holocene. Subspecies that followed this trend were living in the Middle East and North Africa. This happened roughly during the first millennium BC. The decline and disappearance of the aurochs from Europe proceeded from the south and west of Europe to the east and central part. It eventually ended in Poland.

One of the reasons of extinction – hunting – can be observed in the many aurochs bones in former human settlements. These aurochs were killed by hunters. Sometimes aurochs remains have even been found that show stone arrowheads, stuck in the bone and partly overgrown by bone tissue. These aurochs had survived a hunt by man, and sometimes at least twice (the aurochs of Vig; Aaris-Sørensen 1999). The other reason for extinction concerns the ousting of aurochs from their natural feeding areas. Man, and probably also aurochs, preferred to live near rivers and in more or less treeless landscapes that surrounded them. In those landscapes grasses and sedges were growing, and were the main food resources eaten by both aurochs and cattle. The more cattle were kept by man, the more aurochs living area was taken up by cattle and the more the aurochs were ousted.

The game-keeping service that managed the aurochs population at Jaktorów functioned well for probably about two centuries. Only in the course of the second half of the sixteenth century did it tail off (Heymanowski 1972). The Inspection Reports from that period began to show problems concerning management. The forests became increasingly destroyed by excessive cutting, which caused turmoil and disturbance to the aurochs. Even more important for the well-being of the aurochs was the fact that farmers intensified the pasturing of their cattle, horses and pigs (in autumn) in these forests. This probably caused considerable competition with the aurochs. In spite of the feeding with hay, made by the farmers at Jaktorów in summer and brought to the forest in winter, the aurochs faced increasing difficulties. Several times, the authors of the Inspection Reports inquired to the game wardens about the low numbers of aurochs and the challenging situation these animals lived in. The reasons given were various: too severe winters, poaching, disease, food competition with domestic animals, bulls killing each other. Apart from this, in some years the hay harvest failed completely or in part due to bad weather or other reasons (e.g corruption of the area's managers, who were involved in illegal tree felling, pasturing and selling or stealing of hay meant for the aurochs).

The Inspection Reports indicated that negligence and corruption among game wardens, farmers and foresters increased. The supreme authorities, i.e. the Polish kings that ruled in succession, were not free from blame. Especially after 1572, when King Zygmunt August died, a politically very turbulent period began in Poland. This caused the organization of the protection of the aurochs and its living area to be increasingly undermined. It is very likely that this was the final factor that dealt the aurochs their deathblow.

Aurochs remains in museums

In many museums, especially in Europe, aurochs bones can be found (see Figure 15.6). Museums that have a nice collection of such bones are, e.g. the Geologisch-Paläontologisches Museum at Münster (Germany) and the Zoologisk Museum in Copenhagen (Denmark). Both have aurochs skeletons as well. Aurochs skeletons might also be seen at the Phyletisches Museum at Jena (Germany) and at the National Museum in Copenhagen. The former is the one on which Bojanus (1827) based his description of *Bos primigenius*.

One of the horn sheaths of the last aurochs bull, beautifully worked with metal decoration, is preserved at the Livrustkammaren Museum in Stockholm (Sweden). Another aurochs horn sheath, also worked, is preserved at the Salt Mine Museum at Wieliczka (South Poland). Probably, the latter horn sheath originated from the Jaktorów aurochs population as well.

Restitution of the aurochs

Breeding-back experiments by the Heck brothers

During the 1920s, in Germany the brothers Lutz and Heinz Heck started to try to 'breed back' the aurochs. They were driven by curiosity, to find out what the aurochs actually looked like, and by the need to show this species to the public (van Vuure 2005).

For their breeding-back experiments, each of the brothers used a different set of cattle breeds to cross with. Lutz, at the Berlin Zoo, started to breed with Corsican cattle, Camargue cattle and Spanish fighting cattle (L. Heck L 1934). Heinz, at the Munich Zoo, started with the cattle breeds Black Pied Lowland, Angler, Podolic and Hungarian Steppe, Scottish Highland and Corsican cattle (H. Heck 1934). After 12 years of crossing, in 1932 Heinz claimed to have obtained an aurochs-like animal. Apparently he was not content yet, because afterwards he would cross in other cattle breeds, like English Park cattle, Allgäuer, Werdenfelser, Angler and specimens from his brother's experiment. Lutz started later with his crossings and continued these during the 1930s as well. Heinz kept his animals in zoos and wildlife parks. Lutz aimed to release his 'bred-back' aurochs into the wild. This latter was realized in the nature reserves Schorfheide (Germany), Rominten (East Prussia) and the Forest of Białowieża (Poland). All animals of Lutz's population had perished at the end of World War II. Heinz's animals survived the war and are still bred in zoos and nature reserves. In fact, it is fundamentally impossible to get back the aurochs in this way, only because we do not know all the physical characteristics of the original aurochs in detail.

These so-called bred-back aurochs, nowadays called 'Heck cattle', really do not resemble the original aurochs. Heck cattle show various types of horn form (mostly the lyre shape of Hungarian steppe cattle), seldom that of the aurochs type. In body size they are smaller than aurochs were. Cows show different sizes of udder. As regards coat colour, Heck cattle do not show the strict sexual dimorphism that was present in aurochs. In Heck bulls, dorsal stripes may be white, brown or red instead of the aurochs' grey colour. Nowadays two organizations engage in Heck cattle breeding. There are the SIERDA (Syndicat International pour l'Elevage, la Reconnaissance et le Développement de l'Aurochs-reconstitué) and the VFA (Verein zur Förderung des 'Auerochsen'), this latter especially for German breeders. For this breeding, a herdbook is used, in which selected bulls and cows are entered. At the moment a few thousand Heck cattle are kept in zoos, wildlife parks and nature reserves in several European countries.

Ongoing breeding-back experiments

At present, several other breeding-back experiments are still ongoing. The two which got the most publicity are the one in an area along the Lippe River (Germany) and the one in the Province of Brabant (the Netherlands).

In 1991, at the Lippe River, an experiment was started by purchasing some Heck cattle. To upgrade them into more aurochs-like cattle, some years later these were crossed with Chianina cattle from Italy, and Sayaguesa and fighting cattle from Spain. Chianina were used to get larger animals, and the latter two breeds to get the aurochs-like horn form and coat colour. At present, with the offspring from these crossings, and the mentioned breeds (not Heck cattle), it is being bred further under the name of the 'Taurus Project'. Several other breeders in Germany, Denmark and Hungary are working together with the Taurus Project, or carrying out their own crossing experiments with other breeds (M. Bunzel-Drüke, personal communication; see Figure 15.7).

The project in the Netherlands started as the 'TaurOs Project' in 2011. Here, it was not started from Heck cattle, but from Scottish Highland cattle that are being crossed with a number of Italian, Spanish and Portuguese cattle breeds. The breeds from the last three countries are meant to bring in aurochs-like features, like body size, horn form and coat colour. These breeds are: Maremmana (primitive; see Figure 15.8), Limia, Sayaguesa, Pajuna, Maronesa, Podolica and Tudanca. This project is being carried out by the Foundation 'Taurus', and supported and monitored through DNA research by Wageningen University (Wageningen, the Netherlands).

DNA research

As in other zoological fields of study, research on both mitochondrial and nuclear DNA is also going on in current cattle breeds and aurochs remains. DNA research on aurochs remains focused, among other things, on phylogeny, the relationship between aurochs subspecies and present cattle breeds, and research on restitution of the aurochs.

DNA research on bovine phylogeny was done by, e.g. Buntjer (1997) and Buntjer *et al.* (2002). Furthermore, research is being done on discriminating between the different mtDNA haplogroups in aurochs and current cattle breeds, and the place and time of origin of these haplogroups (Bollongino *et al.* 2008; Bonfiglio *et al.* 2010; Mona *et al.* 2010; Gravlund *et al.* 2012). Related to this issue is the question of whether cattle, imported from the Middle East into Europe, have ever been crossed with local aurochs. Moreover, whether there is a difference between animals in Northern Europe and Southern

Figure 15.7 One of the crossings of the back-breeding experiment in the Lippe area (Germany). This bull is half-Sayaguesa and half-Heck (photo by M. Bunzel-Drüke). A black and white version of this figure will appear in some formats. For the colour version, please refer to the plate section.

Figure 15.8 Maremmana bull. One of the cattle breeds used for the back-breeding experiment in the 'TaurOs Project' (photo by Marco Di Luca). A black and white version of this figure will appear in some formats. For the colour version, please refer to the plate section.

Europe. On the basis of the occurrence of the different mtDNA haplogroups and Y-chromosome types, these topics are still being discussed.

During recent years, interest originated in, and research is being done on, the possibility to restitute the aurochs by means of DNA research methods like sequencing and cloning. While a complete bovine genome, of a Hereford cow, was sequenced some years ago, worked is now being carried out on the genome of the aurochs (Słomski *et al.* 2010). Edwards *et al.* (2010) and Lari *et al.* (2011) already sequenced the whole mitochondrial genome of an aurochs from England and Italy, respectively. Whether this research will eventually lead to (a population of) restituted aurochs, we will find out in the future.

Acknowledgements

I would like to heartily thank Margret Bunzel-Drüke for her kindness in providing a photo of one of her back-breeding bulls; Marco Di Luca for the photo of the Maremmana bull; Luca Policastri for the photo of the rock engraving of an aurochs and Sara Marino for information on Romito Cave.

References

Aaris-Sørensen, K. (1999). The Holocene history of the Scandinavian aurochs (*Bos primigenius* Bojanus, 1827). In *Archäologie und Biologie des Auerochsen*, pp. 49–57. Ed. Weniger, G. C. Mettmann: Neanderthal Museum.

Boessneck, J. (1957) Funde des Ures, Bos primigenius Boj. 1827, aus alluvialen Schichten Bayerns. *Säugetierkundliche Mitteilungen* **5**: 55–69.

Bojanus, L. H. (1827). De uro nostrate eiusque sceleto commentatio. *Verhandlungen der Kaiserlichen Leopoldinisch-Carolinischen Akademie der Naturforscher, Bonn* **13**(2): 411–478.

Bollongino, R., Elsner, J., Vigne, J. D. & Burger, J. (2008). Y-SNPs do not indicate hybridisation between European aurochs and domestic cattle. *PLoS One*, **3**(10): 1–5.

Bonfiglio, S., Achilli, A., Olivieri, A., *et al.* (2010). The enigmatic origin of bovine mtDNA haplogroup R: sporadic interbreeding or an independent event of *Bos primigenius* domestication in Italy? *PLoS One*, **5**(12): 1–10.

Buntjer, J. B. (1997). DNA repeats in the vertebrate genome as probes in phylogeny and species identification. Thesis Utrecht.

Buntjer, J. B., Otsen, M., Nijman, I. J., Kuiper, M. T. R. & Lenstra, J. A. (2002). Phylogeny of bovine species based on AFLP fingerprinting. *Heredity* **88**: 46–51.

Caesar, G. I. (58–52 BC). *Commentarii de Bello Gallico, Libri I – VII.*

Degerbøl, M. & Fredskild, B. (1970). The urus (*Bos primigenius* Bojanus) and neolithic domesticated cattle (*Bos taurus domesticus* Linné) in Denmark: with a revision of *Bos* remains from the kitchen middens. *Det Kongelige Danske Videnskabernes Selskab, Biologiske Skrifter* **17**(1): 1–234.

Dementiev, G. P. (1958). Quelques notes sur l'aurochs. *Mammalia*, **22**: 161–167.

Edwards, C. J., Magee, D. A., Park, S. D. E., *et al.* (2010). A complete mitochondrial genome sequence from a mesolithic wild aurochs (*Bos primigenius*). *PLoS One* **5** (2): e9255. doi: 10.1371/journal.pone.0009255.

Estévez, J. & Saña, M. (1999). Auerochsenfunde auf der Iberischen Halbinsel. In *Archäologie und Biologie des Auerochsen*, pp. 119–131. Ed. Weniger, G. C. Mettmann: Neanderthal Museum.

Frevert, W. (1957). *Rominten*. München, Bern, Wien: BLV-Verlag.

Gautier, A. (1988). The final demise of *Bos ibericus*? *Sahara* **1**: 37–48.

Gesner, C. (1602). *Historiae Animalium: Liber I. De quadrupedibus viviparis.* Francofurti.

Gravlund, P., Aaris-Sørensen, K., Hofreiter, M., *et al.* (2012). Ancient DNA extracted from Danish aurochs (*Bos primigenius*): genetic diversity and preservation. *Annals of Anatomy* **194**(1): 103–111.

Grigson, C. (1978). The craniology and relationships of four species of *Bos*: 4. The relationships between *Bos primigenius* Boj. and *Bos taurus* L. and its implications for the phylogeny of the domestic breeds. *Journal of Archaeological Science* **5**(2): 123–152.

Grigson, C. (1985). *Bos indicus* and *Bos namadicus* and the problem of autochthonous domestication in India. In *Recent Advances in Indo-Pacific Prehistory*, pp. 425–428. Eds. Misra, V. N. & Bellwood, P. Oxford, New Delhi, Bombay, Calcutta: IBH Publishing Co.

Gromova, V. I. (1931). Contribution à la connaissance de l'ure (*Bos primigenius* Boj.) de l'Europe orientale et de l'Asie septentrionale. *Annuaire du Musée Zoologique de l'Académie des Sciences de l'URSS, Leningrad* **32**(3): 293–364.

Halder, U. (1976). *Ökologie und Verhalten des Banteng (*Bos javanicus*) in Java – eine Feldstudie.* Hamburg und Berlin: Verlag Paul Parey.

Hall, S. J. G. (2008). A comparative analysis of the habitat of the extinct aurochs and other prehistoric mammals in Britain. *Ecography* **31**(2): 187–190.

Heck, H. (1934). *Der Ur. Das Tier und Wir, Monatszeitschrift für alle Tierfreunde.* März: Sonder-Nummer.

Heck, L. (1934). Über die Neuzüchtung der Ur oder Auerochs. *Berichte der Internationalen Gesellschaft zur Erhaltung des Wisents* **3**(4): 225–294.

Heymanowski, K. (1972). The last mainstay of aurochs and the organization of their protection in the light of contemporary documents. *Sylwan* **116**(9): 9–28.

Kędzierska, Z. (1959). *Lustracje Województwa Rawskiego 1564 i 1570.* Warszawa: Instytut Historii PAN.

Kędzierska, Z. (1965). *Lustracje Województwa Rawskiego XVII wieku.* Warszawa, Wroclaw: Instytut Historii PAN.

Krasińska, M. & Krasiński, Z. A. (2007). *European Bison: The Nature Monograph.* Białowieża: Mammal Research Institute PAN.

Lari, M., Rizzi, E., Mona, S., *et al.* (2011). The complete mitochondrial genome of an 11,450-year-old aurochsen (*Bos primigenius*) from Central Italy. *BMC Evolutionary Biology*: www.biomedcentral.com/1471-2148/11/32.

Lehmann, U. (1949). Der Ur im Diluvium Deutschlands und seine Verbreitung. *Neues Jahrbuch für Mineralogie, Geologie und Palaeontologie, Abhandlungen Abt. B* **90**: 163–266.

Linnaeus, C. (1758). *Systema Naturae: per Regna Tria Naturae. I. Regnum Animale.* Holmiae: Editio decima.

Lynch, A. H., Hamilton, J. & Hedges, R. E. M. (2008). Where the wild things are: aurochs and cattle in England. *Antiquity* **82**: 1025–1039.

MacHugh, D. E., Shriver, M. D., Loftus, R. T., Cunningham, P. & Bradley, D. G. (1997). Microsatellite DNA variation and the evolution, domestication and phylogeography of taurine and zebu cattle (*Bos taurus* and *Bos indicus*). *Genetics* **146**: 1071–1086.

Mager, F. (1960). *Der Wald in Altpreußen als Wirtschaftsraum.* Köln, Graz. Böhlau: Verlag.

Matolcsi, J. (1970). Historische Erforschung der Körpergröße des Rindes auf Grund von ungarischen Knochenmaterial. *Zeitschrift für Tierzüchtung und Züchtungsbiologie* **87**: 89–137.

Mona, S., Catalano, G., Lari, M., *et al.* (2010). Population dynamic of the extinct European aurochs: genetic evidence of a north–south differentiation pattern and no evidence of post-glacial expansion. *BMC Evolutionary Biology*: www.biomedcentral.com/1471-2148/10/83.

Mortensen, H. & Mortensen, G. (1938). *Die Besiedlung des Nordöstlichen Ostpreußens bis zum Beginn des 17. Jahrhunderts. Teil II: Die Wildnis im östlichen Preußen, ihr Zustand um 1400 und ihre frühere Besiedlung.* Hirzel: Verlag von S. Leipzig.

Prins, H. H. T. (1987). The buffalo of Manyara. PhD thesis, Groningen.

Pusch, G. G. (1837). *Polens Paläontologie.* Stuttgart: E. Schweizerbart's Verlagshandlung.

Ritchie, J. C. & Haynes, C. V. (1987). Holocene vegetation zonation in the eastern Sahara. *Nature* **330**: 645–647.

Rodrigues, L. J. (1992). *Anglo-Saxon Verse Runes.* Felinfach: Llanerch Publishers.

Schloeth, R. (1961). Das Sozialleben des Camargue-Rindes. *Zeitschrift für Tierpsychologie* **18**: 574–627.

Sinclair, A. R. E. (1977). *The African Buffalo: A Study of Resource Limitation of Populations*. Chicago, IL, London: University of Chicago Press.

Słomski, R., Dzieduszycki, A. M., Lipiński, D., *et al.* (2010). Analysis of genomic and mitochondrial DNA of aurochs. In *Restoration of Endangered and Extinct Animals*, pp. 176–195. Ed. Słomski, R. Poznań: University of Life Sciences.

Smith, C. H. (1827). Order Ruminantia: the taurine group. In *A Treatise on the Order Ruminantia of the Baron Cuvier*, vol. 4, pp. 411–428. Published in the English edition of the *Regne Animal* of Cuvier, Ed. Griffith, E. Plymouth.

Święcicki, A. (1634). *Descriptio Topographica Ducatus Masoviae*. Varsaviae.

Szalay, A. B. (1915). Der Wisent in Ortsnamen. *Zoologische Annalen* **7**(1): 1–80.

van Vuure, T. (2005). *Retracing the Aurochs: History, Morphology and Ecology of an Extinct Wild Ox*. Sofia, Moscow: Pensoft.

Van Wieren, S. E. (1996). Digestive strategies in ruminants and nonruminants.. PhD thesis. Landbouw Universiteit.

Von Herberstein, S. (1557). *Moscovia der Hauptstat in Reissen*. Wien: Michael Zimmerman.

Von Leithner, O. (1927). Der Ur. *Berichte der Internationalen Gesellschaft zur Erhaltung des Wisents* **2**: 1–139.

Wrześniowski, A. (1878). Studien zur Geschichte des polnischen Tur. *Zeitschrift für Wissenschaftliche Zoologie* **30** (Supplement 45): 493–555.

Chapter 16

Wild water buffalo *Bubalus arnee* (Kerr, 1792)

Anwaruddin Choudhury

Names

Genus: *Bubalus* C. H. Smith, 1827

Species: Wild water buffalo *Bubalus arnee* (Kerr, 1792); *Bubalus bubalis* (Linnaeus, 1758).

Names in other languages: **French**: Buffle d'eau; **German**: Arni; **Spanish**: Búfalo de agua; **Italian**: Bufalo indiano; **Adimising**: Menjek; **Assamese**: Bonoria mah, Pera; **Bengali**: Buno mush, Jangli mohish; **Burmese/Myanmarese**: Kywai, Taw jwai; **Hindi**: Arna; **Lao**: Khoay Pa.

Other common names: Asiatic wild water buffalo, Asian wild buffalo, Indian buffalo, wild buffalo, arna.

In this work 'wild water buffalo' has been given the status of 'first choice' name but most subsequent uses in the text will normally be only 'wild buffalo' unless otherwise required.

Ecology, Evolution and Behaviour of Wild Cattle: Implications for Conservation, ed. M. Melletti and J. Burton. Published by Cambridge University Press.

Linnaeus (1758), who generally used very wide genera, (because he did not use the family rank), originally named it as *Bos bubalis*. Kerr (1792) named it as *Bos arnee*. Later, both were assigned to genus *Bubalus* H. Smith, 1827 (in *Cuvier's Animal Kingdom*, vol. 5: 371). The name *bubalis* is based on a domestic animal. It is also the senior synonym. However, those workers who do not use specific names based on domestic animals prefer to use *arnee* (Corbet & Hill 1992; Groves 1996; Choudhury 1997a, 2001b, 2010).

The Asiatic wild water buffalo *Bubalus arnee* Kerr (=*Bubalus bubalis* Linn.) is among the largest members of the family Bovidae. It is also the ancestor of all the domestic varieties and breeds of buffaloes anywhere. Although the exact time and place of its domestication is uncertain there is little doubt that it has been present for at least 5000 years in India and China. Tame and possibly domesticated buffaloes appear on seals in the Indus valley (Mohenjo-daro) and in Mesopotamian civilisations from about the middle of the third millennium BC (Mason 1974; Lenstra & Bradley 1999; Massicot 2004). According to McIntosh (2007), 'buffaloes occur as images on Indus seals, where they appear to be wild'.

The wild water buffalo (hereafter referred to as 'wild buffalo') is currently distributed in South and Southeast Asia, from India to Cambodia, occurring in India, Nepal, Bhutan, Thailand and Cambodia (Ellerman & Morrison-Scott 1966; Corbet & Hill 1992; Choudhury 1994a, 2010; Groves 1996).

Perhaps the earliest written documentation on wild buffalo was made by the founder of the Mughal Empire in India, Zahirud-din Muhammad Babur (*c.*1529; 1483–1531) in his *The Babur-Nama* ('Memoirs of Babur'). The next written account on the species is found in *Ain-i-Akbari*, authored by Abu'l Fazl (*c.*1590; 1551–1602) (Ali 1927). Linnaeus described the species *Bos bubalis* from a domestic specimen from Rome, Italy in 1758 and Kerr (1792) described it as *Bos arnee* from a wild specimen from India. Further published information came after a long gap in the form of Kinloch (1885, 1892), Richardson (1889), Inverarity (1895) and Sanderson (1896).

General accounts on wild buffaloes are also found in Blanford (1891), Sterndale's *Mammalia of India* from the 1880s (reprinted by Finn 1929), Lydekker (1900, 1907) and Prater (1948). The Maharajah of Cooch Behar (1908), Stockley (1928), Anon (1920) and Hall (1936) provided hunting accounts. Useful information is also found in Forsyth (1889), Stockley (1913) and in Dunbar Brander (1931). Ward (1910) listed the record horns of the wild buffalo from across its range. Phillips (1929) and Mooney (1930) covered the status of buffaloes in Ceylon (now Sri Lanka) and Orissa, respectively.

In the post-1950 period, the notable publications on wild buffalo or general works with information on it were Noronha (1952, 1954), Gee (1953, 1964), Daniel & Grubh (1966), *Oryx* (1967, 1989), Sheshadri (1969), Divekar (1976), Tikader (1983), Divekar & Bhushan (1988) and Ranjitsinh *et al.* (2000).

Choudhury (1994a) provided the first detailed account on the distribution and status of wild buffaloes with mapping of northeast India. Since the 1990s, status reports on several areas of northeast India have been published (Choudhury 1993a, 1994b, 1998a, 1999, 2001a, 2002). Muley (2001) and Flamand *et al.* (2003) studied the genetics of buffaloes in Assam's Kaziranga National Park and Nepal's Koshi Tappu Reserve, respectively. Some information on its behavioural aspects are found in Ashby & Santiapillai (1987), Pandey (1988), Mathur *et al.* (1995), Kotwal *et al.* (2003) and Kotwal & Mishra (2004). The current status in Bhutan has also been assessed (Choudhury 2008). Santiapillai (1990) and Chaiyarat (2002, 2004) provided information on Thailand, and Heinen (1993) on Nepal. Groves (1996) reviewed the taxonomy and also proposed a new subspecies from Thailand. Choudhury (2010) is a comprehensive monograph on the wild buffalo, with maps.

General information on wild buffalo is found in several other synoptic works, notably those of Spillett (1966), Lekagul & McNeely (1977), Mukherjee (1982), Choudhury (1986, 1988, 1989a, 1993b, 1994c, 1997a, 1998b, 2001a, 2003a, 2003b, 2004, 2006, 2009), Corbet & Hill (1992), Graham & Round (1994), Sah (1997), Ranjitsinh (1997), Menon (2003), Wangchuk *et al.* (2004), Choudhury *et al.* (2008a, 2008b) and Choudhury & Dey (2008). There are also unpublished census reports on Kaziranga National Park.

Taxonomy

Type specimen

The wild buffalo was described from a domestic specimen by Linnaeus (1758) as *Bos bubalis*. No type specimen is preserved in any museum. Kerr (1792) described it as *Bos arnee* from a skull in the Edinburgh College Museum and a frontlet and horns from Weir's Museum. The fate of Weir's Museum is not known, but the Royal Scottish Museum, Edinburgh, acquired all the material of the former Edinburgh College Museum, but the specimens are not labelled (Groves 1996). Groves (1996) examined several skulls of wild buffaloes in the Royal Scottish Museum, and only one of them corresponds to Kerr's (1792) figure. Groves (1996) stated that 'It is not unlikely that this really is Kerr's specimen.'

Type locality

Linnaeus (1758) described the species from a domestic specimen from Rome, Italy; however, he stated 'habitat in Asia, cultus in Italia'. Kerr (1792) described it from a skull in the Edinburgh College Museum and a frontlet and horns from Weir's Museum. The description and measurement of the horns and frontlet indicate that it was a wild specimen from somewhere in India or one of the neighbouring countries. Kerr mentioned the type locality as 'India, north from Bengal', restricted to Cooch Behar (Kuch Bihar; Harper 1940). Groves (1996) also mentioned that Kerr's figure (No. 746) shows a typical wild buffalo skull.

Subspecies and distribution

Subspecies

From its distribution pattern, it seems that there could be three subspecies: one in the Gangetic–Brahmaputra plains; one in the plateau of Peninsular India and the third in Indochina (Choudhury 2010). Lydekker (in Wroughton 1920) recognised three subspecies. These are: *B. b. bubalis* Linn., the nominate subspecies (type specimen of *bubalis*) having crescent-shaped or subcircular horns; *B. b. macroceros* Hodgson, from central Assam, having horns which are directed mainly outwards; and *B. b. fulvus* Blanford, from Mishmi Hills (now in Arunachal Pradesh), having dun-brown colour. These buffaloes were reportedly distributed in Assam (*bubalis*), central Assam (*macroceros*) and Mishmi Hills, now in Arunachal Pradesh (*fulvus*).

Choudhury (2010), however, stated that these races are apparently not valid as these were part of a contiguous population till the early twentieth century. There was no zoogeographic barrier to justify any sort of isolation. The differences noted were individual variations and similar characteristics could be seen even in a single herd (variations such as horn shape and body colour are found even within a herd). 'Fulvous' buffaloes have been recorded in east Assam (Dibru-Saikhowa National Park; see photo on p. 76 in Choudhury 2010) as well as in central Assam, especially in Laokhowa and Burhachapori Sanctuaries (see photos on p. 78 in Choudhury 2010). These latter animals are not isolated from Kaziranga buffaloes and there is habitat contiguity. However, all buffaloes in Dibru-Saikhowa do not show a 'fulvous' front (see photos on pp. 76–77 in Choudhury 2010).

Regarding *fulvus* buffaloes, Blanford (1891) never described the subspecies but only mentioned two 'dun coloured' buffalo heads from upper Assam, one of which was in the collection of Mr Hume and the other in the Indian Museum, Calcutta.

Similarly, *macroceros* was mentioned to be from central Assam, although most of the specimens were from west Assam (Manas belt) and north Bengal. These areas had the nominate subspecies as well. There is no evidence so far to suggest that *fulvus* and *macroceros* were ever recorded as exclusive herds or populations. All the evidence, including recent records, indicate that these were variants or at the most forms, which were in herds that comprised animals having characteristics of one, two or even three forms.

In Dibru-Saikhowa, the *fulvus* buffaloes have the entire front of their heads rufous-brown. In Laokhowa-Burhachapori areas it is confined to the forehead only. The body hair of some of the animals shows a lot of rufous-brown in Burhachapori and adjacent areas (see photo on top right of p. 79 in Choudhury 2010). Most of the *macroceros*-like animals seen in Manas were large adult females.

More recently, Groves (1996) described the Thailand animals as a separate subspecies and named them as *Bubalus arnee theerapati*. Comparatively smaller size and widely spreading horns were considered as diagnostic features.

Distribution

The inferred historical range of wild buffalo extended from northern India to Indochina and perhaps as far as the Euphrates River in Iraq towards the west (Cockrill 1984; Corbet & Hill 1992; Choudhury 2010) (Plate 7 and Figure 16.1). Its exact historical range is uncertain due to at least 5000 years of domestication. The more recent historical range of wild buffalo extended from northern Indian *terai* to South Vietnam, with the River Godavari being the southern limit in Peninsular India (Lydekker 1907; Daniel & Grubh 1966; Choudhury 2010). For the Peninsular Indian population, the Wainganga River and Mandla were regarded as their western limits, which is marked by the 80th meridian of longitude (Forsyth 1889; Blanford 1891), although Forsyth (1889) mentioned specimens, presumably of the early part of the nineteenth century, from the present Chhindwara district around 79° longitude. He also mentioned that although Godavari River is their southern limit, he saw tracks of a herd on the western side, i.e. the right bank of the river, in 1865. An analysis of the available literature (Lydekker 1907; Mooney 1930; Daniel & Grubh 1966; Santiapillai 1990; Choudhury 1994a, 2010) indicates that by the turn of the twentieth century, the animals remained in three large disconnected areas: (1) northern India, including Nepal, Bhutan and Bangladesh; (2) Peninsular India; and (3) scattered areas in Indochina. The current distribution is also mainly confined to these three large areas, but with several isolated populations (Gee 1966; Sheshadri 1969; Divekar & Bhusan 1988; Santiapillai 1990; Heinen 1993; Choudhury 1994a, 2010).

The current distribution is a relict and highly fragmented one and is confined to south Bhutan, southeast Nepal, northeast and central India (Assam, Arunachal Pradesh, Meghalaya, Chhattisgarh and north Bengal), western Thailand and eastern Cambodia, and probably north Myanmar and highly doubtfully southwest Vietnam. The 'wild' buffaloes of Sri Lanka, Borneo and Java are of feral origin. The wild buffalo is extinct in Bangladesh. It is also believed that there were true wild populations in Peninsular Malaysia, and on the islands of Sumatra, Java and Borneo, which have also become extinct (Hedges 1995). In north Australia there are fully feral buffaloes of domestic origin. According to Tulloch (1970), these animals are descendants of individuals introduced for meat, first to Melville I and Coburg Peninsula between 1825 and 1843 from Asia by sailing ships.

Country-by-country past and present distributions and status of the wild buffalo are given below (also see Plate 7 and Figure 16.1).

India: the wild buffalo ranges in two disconnected regions, northern India and Peninsular India (Plate 7 and Figure 16.1). In northern India, the main buffalo range states are in the northeast region, in Assam, Arunachal Pradesh, Meghalaya and north West Bengal.

Assam is now the world stronghold of the wild buffalo. About four-fifths of the known population of the species is

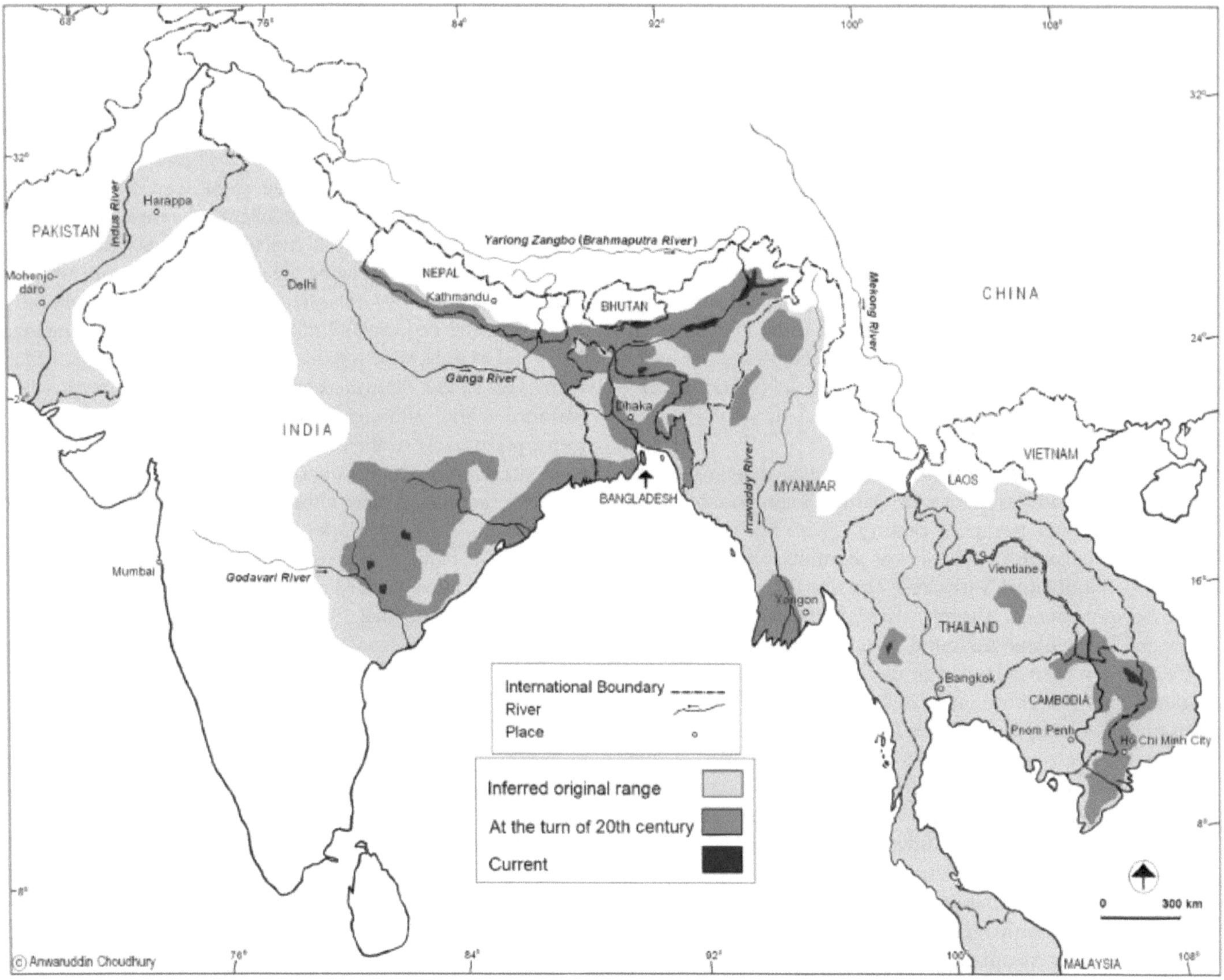

Figure 16.1 Past and present range of the wild buffalo. A black and white version of this figure will appear in some formats. For the colour version, please refer to the plate section.

found in Assam (Choudhury 2010). The historical range of the wild buffalo in Assam covered the entire state. In the nineteenth century, the species occurred in almost all of the plains districts (M'Cosh 1837; Cooper 1873; Maharajah of Cooch Behar 1908; Choudhury 2001b, 2010). In 1872 Kinloch described the wild buffalo in Assam as 'extremely abundant' (Fisher *et al.* 1969). Assam's position as the major buffalo area in India was also mentioned by Hall (1936).

Currently, the wild buffalo is found in Kaziranga, Manas and Dibru-Saikhowa National Parks, Laokhowa and Burhachapori Wildlife Sanctuaries, Dum Duma, Jokai, Chirang and Manas Reserve Forests, Kobo *Chapori* (*Chapori* means riverain islets and tracts) proposed reserve forest, and in a few other scattered locations. The distribution range has five separate areas, some having contiguity with Arunachal Pradesh and Bhutan. The animal has become extinct in Sonai Rupai Wildlife Sanctuary, Pabho Reserve Forest (formerly known as Milroy Buffalo Sanctuary), Kadam Reserve Forest and Barak Valley districts. The protected areas are listed in Table 16.8. The estimated numbers are also mentioned in the Table 16.1.

Kaziranga National Park is the most important habitat of the wild buffalo in the world, containing more than half of the total population. The national park has the largest wet savanna grassland in the entire northeast of India. Kaziranga is among the few areas with a long-term population estimate, in this case for the last four decades or so. Owing to stringent protection given to the park, mainly for the great Indian one-horned rhinoceros, the number of wild buffalo have continued to rise from around 500 in the 1950s to 1963 in 2008. Kaziranga has a long history of conservation beginning in 1902. It was inscribed as a World Natural Heritage Site in 1985.

On Manas National Park, the late E. P. Gee (1964) had stated that 'Wild buffaloes are the animals for which the Manas is justly famous. These animals grow bigger and carry larger

Table 16.1 Estimates of wild buffalo in different areas

Year	Kaziranga	Manas	Laokhowa and Burhachapori	Dibru-Saikhowa	Chhattisgarh	Koshi Tappu	Huai Kha Khaeng
1945						100	
Late 1940s	500	500	50				
1965	700	400	50		250–300	15–20	
1966	471	300–400 (1966–1967)				20–30 to *c.*100	
1968						60	
1969						30–40	
1960s							20–30 to *c.*100
1972	555					40–45	
1975					*c.*50	40–60	
1976						63	
1977						60	
1978	610						
1984	677	1000+ (in 1985–1986)					
1985							35–40
1986						106	
1987						91	
1988			*c.*200 (1989–1990)		*c.* 125	93	
1991	1090	*c.*1200 (pre-1989)					
1993	1034			200 (1993–1994)		158	36–39 (early 1990s)
1995			200+				
1997			150+				
1999	1192						
2000					<75	145	72
2001	1666						
2004		*c.*250				159	
2005		–200					
2008	1963	230 (250+ for entire Tiger Reserve)	75	311 (400+)	30–50		60
2009						219	About 50

Source: Choudhury (2010)

horns than the Wild buffalo of other parts of India. . . . On the wide sandy and shingly beaches of the Manas and Beki Rivers and their many grassfringed channels, you can see plenty of Wild buffalo.' However, in 1989 a slump in the population began following social unrest that resulted in the breakdown of law and order (Choudhury 1989b). Like Kaziranga, the conservation history of Manas dates back to 1902 (Choudhury 2006). It was inscribed as a World Natural Heritage Site in 1985.

The tract located between the Sankosh River and the western boundary of the Manas National Park forms part of the buffer zone of the Manas Tiger Reserve. The local reports and sporadic sightings indicate that the current population in the area is 60–80. A small population (10–15) occurs in Batabari towards the east of Manas National Park (also part of the buffer zone of the tiger reserve).

The Kollolua-Jokai area in Dibrugarh district comprises countryside, small patches of forest and grass along the Burhi-Dihing River (27°17′–23′ N, 94°54′–95°04′ E), and has an estimated 45–50 wild buffaloes (although some villagers think it could be 60–65). During the daytime the main herd mostly remains in a patch of riverain grassland-cum-woodland near Na-Pam Thangaon village (called Elengmari Grazing Reserve; *c.*100 ha), while lone bulls wander over a large area looking for domestic females. Herds also sometimes remain in Jokai Reserve Forest during the day. These buffaloes were first reported by Choudhury (1994a).

Pabho Reserve Forest in Lakhimpur district in east Assam was listed as a sanctuary and was named as 'Milroy Buffalo Sanctuary', covering 51 km^2 in 1941. However, it was not announced as a wildlife sanctuary under the *Wild Life (Protection) Act 1972* until later. Gee (1950) estimated a population of about 100 animals in the late 1940s and also in 1965 (Daniel & Grubh 1966). Due to the high flood of 1972 (when many were drowned and also killed by poachers) and also encroachment, by the early 1980s the buffalo had become extinct in the area.

A small population occurs in Dum Duma Reserve Forest, a rainforest pocket, which is not a normal buffalo habitat. It is believed that these wild buffaloes came to the area from the *chapories* of the Lohit River during high floods in late 1980s. These buffaloes were first reported by Choudhury (1994a). Their current number is greater than 13. In Lonke *tapu* near Burhachapori Wildlife Sanctuary, Sonitpur district, 32 buffaloes were counted during a census in April 2008 (Choudhury *et al.* 2008b). These animals regularly move between Lonke and adjacent *tapus*, as well as Burhachapori and Laokhowa Sanctuaries.

The hilly tracts of Karbi Anglong, more particularly the northern slopes of the eastern Karbi Plateau, are a known stronghold of wild buffaloes, at least seasonally. Every monsoon, a few hundred of these animals from Kaziranga cross the National Highway and take refuge in the highlands of the plateau. They remain there for several days every month from June to September. Once three such buffaloes strayed deep into the plateau and were recorded as far south as in Patradisa Reserve Forest (26°02′ N, 93°17′ E) in 1990–1991 (Choudhury 2010). A few were reported in the Dhansiri Reserve Forest of Karbi Anglong district in 1991 (Sarsing Rongphar, personal communication). In the hilly tracts of the North Cachar Hills (now called Dima Hasao) there were records from the northern areas and Diyung Valley.

No past reference to the wild buffalo in the Barak Plains could be found in the literature before Choudhury (2001a). In the late nineteenth century, a male and a female were shot from the Jabda *Haor*–Andhari *Beel* area (24°55′ N, 92°30′ E) in Cachar district. One was shot dead in 1885–1890 in Rajyeswarpur part VII & VIII (24°35′ N, 92°37′ E). In the early twentieth century, a bull was shot near Itala *Beel* (24°30′ N, 92°37′ E) in Hailakandi district (A. Majid Choudhury, Hasiura, personal communication). Its well-preserved horns were examined by the author. The last animals vanished from Inner Line Reserve Forest in south Hailakandi in about the mid 1970s.

Stray buffaloes are met in Kochmara Reserve Forest, west of Laokhowa and Burhachapori Sanctuaries, in the Dhakuakhana subdivision of Lakhimpur district, Poba Reserve Forest in Dhemaji district, Kobo *Chapori* proposed reserve forest, and in the *chapories* of the Brahmaputra River all over east and central Assam. There are no wild buffaloes in Dhubri district but Munrowd (1945a) mentioned three lone bulls that had killed several domestic bulls near the Dhubri–Garo Hills border. There are six large specimens listed by Ward (1910), which were from near Dhubri (originally written as Dubri), indicating the former importance of the area as a buffalo habitat.

In RG Orang National Park and Pabitora Wildlife Sanctuary there is no recent record of occurrence of wild buffaloes; however, from the 1990s and 1980s, respectively, a few feral buffaloes were seen. The origin of these populations is said to have been nearby buffalo *khuties* and lone wild bulls that strayed from nearby areas. Unlike the feral buffaloes of Australia and elsewhere, which were originated from domestic stock, these Pabitora and Orang animals have their origin in pure wild bulls. Therefore, these animals are very close to real wild buffaloes. They do not mix with the grazing domestic buffaloes and show a great degree of segregation.

The historic range of the wild buffalo in Arunachal Pradesh covered the entire foothills as well as the plains of the Siang, Dibang, Lohit and Noa-Dihing Rivers. In the foothills from the Bhutan border to Siang River, the buffaloes were frequent visitors from nearby Assam areas. The type specimen of one of the forms was obtained from Mishmi Hills (apparently from the plains of the Dibang River, where the animals are still found). Currently, the buffalo is found in D'Ering Memorial Sanctuary and in the *chapories* of Dibang and Lohit Rivers, with a few other scattered locations. It has become extinct in Noa-Dihing Valley (Choudhury 2003a). D'Ering Memorial Sanctuary is a key habitat for the buffalo in the state. This large riverine tract on the Siang River is in fact part of a large river island, a part of which is Kobo *Chapori* proposed reserve forest in Assam. Through Kobo, it is also linked towards the south with Dibru-Saikhowa National Park.

Munrowd (1945b) wrote that in Pakar *Chapori*, downstream of Pasighat, 'A number of buffalo was seen, as many as 16 in one herd . . . and one female had such enormous horns that we debated whether they were not something approaching the record which is nearly 14 ft tip to tip across the forehead'. Gee (1952) saw about 18 from the air after the devastating flood of 1951, which he believed to have killed many buffaloes. According to him, prior to that the area (present D'Ering Memorial Sanctuary) was 'teeming with Wild buffalo'. The population slowly built up and was estimated to be around 100 in the early 1990s (Choudhury 1994a). The devastating flash flood of 2000 killed an unspecified number of wild buffaloes.

Till the early 1970s, wild buffalo was widespread all over the lower areas of the Lower Dibang Valley district (part of the Mishmi Hills area), starting from near Nizamghat to the inter-state border with Assam, mostly in the *chapories* of the Dibang, Deopani and the Sesseri Rivers. The buffalo population declined drastically before being surveyed properly. In March 1993, the author found a few solitary wild bulls near Nizamghat (200 m elevation), where the Dibang River debouches onto the plains (at 28°15′ N it is the northernmost site in the distribution range of the species (Choudhury 2011)). Choudhury (1998a) had estimated 50–70 buffaloes in the mid 1990s.

The *chapories* of the Lohit River (27°48′–54′ N, 96°0′–18′ E) are important for the species. Cooper (1873) had mentioned that 'before leaving Camlang-panee [Kamlang pani] next morning ... wild bull, and it was afterwards killed by some Mishmees with poisoned arrows. The horns, which were very fine, measuring 11ft 2 inches from tip to tip round inside and across forehead ... the dry sandy bed of the *nullah* ... was everywhere cut up with the tracks of buffaloes.' Choudhury (1999) estimated 100–150 wild buffaloes for the Lohit Valley in the mid 1990s. The situation has deteriorated in view of expansion of cultivation, orange orchards, poaching for meat and expansion of townships. The current population is likely to be around 100.

There were stray herds and individuals in some other areas, such as Mebo Reserve Forest in East Siang district, Papum Reserve Forest in East Kameng district (especially near Namorah), Chopai area (part of Doimara Reserve Forest) in West Kameng district; elsewhere, there are past records of occurrence along the Noa-Dihing River in Namsai-Mahadevpur areas until about the 1970s.

From the account of Fazl (1590), who mentioned it in his *Ain-i-Akbari*, it is evident that at that time the entire plain areas of the state of Bihar had wild buffaloes. By the eighteenth century their main concentration remained in the northern areas, especially in the *terai* zone. In the nineteenth and early twentieth centuries their main range was in Purnea, Saharsa and Darbhanga districts in the northeast part of the state (Hall 1936; Daniel & Grubh 1966). Following reorganisation of the districts, undivided Purnea became four districts – Araria, Katihar, Kishanganj and Purnea – and the wild buffalo was present in all, but more numerous in Araria and Kishanganj. In the reorganised Saharsa and Darbhanga districts, the buffalo area of Koshi swamps of the early twentieth century were in Supaul (then part of Saharsa) and Madhubani (then part of Darbhanga). By the second quarter of the twentieth century the only population of Bihar remained in undivided Purnea district (mostly in present-day Araria and Kishanganj districts) (Ara 1949).

Manipur and Mizoram are mainly hilly states and had no previous record of wild buffaloes (Higgins 1934) until Choudhury (2002). The valley of Manipur has supported people for a long time, hence the lack of evidence; although Brown (1874) mentioned an occurrence south of the valley without any details. A large bull was reportedly shot around 1976 in Tipaimukh area inside Manipur, but very close to the inter-state border with Mizoram (approx. 24°12′ N, 93°00′ E). The villagers of the area, in both states, were aware of the presence of a few wild buffaloes in the area in the 1960s and early 1970s. This particular specimen was well known in the Vaitin area of Mizoram and adjacent areas of Manipur till its death for its majestic horns; it was reported to have a massive body. Its horns measured: maximum spread, 152.0 cm; tip-to-tip (span), 105.0 cm; sweep across the forehead, 278.5 cm; individual horn (right), 130.0 cm; individual horn (left), 125.0 cm; girth at base (right), 44.0 cm; girth at base (left), 44.5 cm.

The reporting of the presence of wild buffaloes in the hilly Balpakram National Park of Garo Hills, Meghalaya and adjacent areas (Choudhury 1994a) has been a matter of curiosity and had previously been questioned. Early references to the presence of wild buffaloes in the Garo Hills come from the hunting report of the Maharajah of Cooch Behar (1908). He and his party hunted in Tikri Killa, now in the east Garo Hills district, in 1894 and at Kaee Mari at the Garo foothills (this location is in Goalpara district of Assam, but close to the inter-state border). A bull was shot dead for killing humans near Rewak in the present south Garo Hills district in 1935 (Munrowd 1945a). Interestingly, in the same area another bull was shot for killing several humans in 1975. The head is on display at the Forest Museum, Shillong. Munrowd (1945a) also mentioned another subsequently shot at Rewak by a forest officer; although no measurements were given, it was just short of the 'record' shot in the Assam plains. According to Munrowd, the Rewak locality was 'renowned for buffalo with fine horns'. Later, one more bull was shot by Munrowd himself around Rewak area. Rewak is a hilly forest, but just south of it is a wide river valley which used to be covered with tall grass with marshes – perfect buffalo habitat. Gradually such habitat came under cultivation, affecting grassy and marshy patches, and the remaining animals have moved towards the hills in areas such as Baghmara Reserve Forest, Siju Wildlife Sanctuary and Balpakram National Park. The animals from Balpakram also move to the West Khasi Hills district.

The state of Nagaland is hilly and mountainous, and is not a typical buffalo habitat. Some areas along the foothills and in Dimapur district, however, had historic records (Brown 1874; Allen 1905). The presence of a few individuals along the Dhansiri River, at the northern edge of Intanki National Park, was reported to the author in May 1991. The current status is not known and, owing to disturbance and lack of protection in the area, their extirpation from the area would not be surprising.

There is no recent record of wild buffalo from the state of Tripura; however, in view of past occurrence in the adjacent areas of Bangladesh and Assam, it is apparent that a small population has survived.

In West Bengal, the Maharajah of Cooch Behar (1908) gave a vivid account of wild buffalo hunts in the late nineteenth and early twentieth centuries in Jalpaiguri and Cooch Behar districts of north Bengal. Many of these sites now form part of the Buxa Tiger Reserve and Jaldapara Wildlife Sanctuary. Spillett (1967) stated that wild buffalo was already extirpated from Jaldapara Wildlife Sanctuary. Daniel & Grubh (1966) mentioned occurrence in Dinajpur and Malda areas, also in the first and second quarters of the nineteenth century. However, it seemed that stray individuals still survived near the Assam–Bengal border along the Sankosh River in the 1970s. A small population of 15–20 wild buffaloes now survives in Kumargram area of Buxa Tiger Reserve. These animals' range covers Bhutan forests as well.

The southern areas of Bengal had apparent contiguity with the wild buffalo populations of north Bengal. It perhaps ceased by the early nineteenth century. Daniel & Grubh (1966) mentioned occurrence in the maritime areas of Midnapore and elsewhere (Sunderbans) in the first half of the nineteenth century. It had contiguity with the Orissa population of Balasore district. The Midnapore population may also be grouped with the peninsular populations. This population probably became extinct in the nineteenth century, while the last buffaloes vanished from the Sunderbans at the beginning of the twentieth century (Sarkar 1993). Stockley (1928) was perhaps the last to mention the presence of wild buffaloes in the area.

The first recorded documentation on the occurrence of wild buffalo in Uttar Pradesh was by Fazl (1590) in his *Ain-i-Akbari*. He mentioned that during Akbar's time (1542–1605) the wild buffalo was numerous in Oudh. The Oudh or Avadh now covers a large part of east central Uttar Pradesh, including areas such as present districts of Bahraich, Balrampur, Bara Banki, Gonda, Kheri, Lucknow and Sitapur. Protected areas such as Dudhwa and Katerniaghat are also located in the *terai* zone of erstwhile Oudh. It is not known when the last animals vanished from the state, but it was perhaps in the late nineteenth century as Kinloch (1885) mentioned Pilibhit as the westernmost range. Stebbing (1920) wrote that the buffaloes vanished from the United Provinces 'about the nineties of last century or thereabouts'.

Early information on the buffaloes of Peninsular India is found in Richardson (1889), Forsyth (1889) and Inverarity (1895). The decline was so fast that Dunbar Brander (1931) could not record a single specimen in areas towards the north of Bastar. Gee (1964) stated that there are a few wild buffalo still left in west and south Orissa, as well as in the adjoining areas of Andhra and Madhya Pradesh, near the Godavari River. Relatively lower abundance of wild buffalo in central India than northeast India was aptly summarised by Dunbar Brander (1931), who said 'The great mass of Central Provinces does not suit their requirements.'

In Peninsular India the state of Chhattisgarh (part of Madhya Pradesh until October 2000) was the stronghold of the wild buffalo until the animals were almost extirpated. Bastar district (now trifurcated into Bastar, Dantewara and Kanker districts) (17°48′–20°30′ N, 80°15′–82°15′ E) was the main buffalo habitat in the state. Forsyth (1889), while referring to north Chhattisgarh, stated that 'over all this country roams the Wild buffalo'. The buffaloes were numerous in the tracts lying towards the east and south of Raipur.

Mooney (1930) reported wild buffalo from Bastar state and Raipur district (now trifurcated into Raipur, Mahasamund and Dhamtari districts). Noronha (1952) mentioned a wild male that killed a domestic male and started to accompany a domestic herd in Bastar. Daniel & Grubh (1966) mentioned that in the first and second quarters of the nineteenth century the wild buffalo was abundant in Raipur and Bastar. The main area in west Bastar now forms Indravati National Park, while that in south Bastar is Pamed Wildlife Sanctuary. Daniel & Grubh (1966) found evidence of 86 and 11 wild buffaloes in these two areas, respectively, and estimated that the total number could be 200–250 and fewer than 50, respectively. Udanti Wildlife Sanctuary in Raipur district had a confirmed population. Divekar (1976) found only 23 buffaloes and estimated 'not more than 50' for west Bastar in 1975.

More recently, Ranjitsinh *et al.* (2004) estimated fewer than 75 from Udanti Wildlife Sanctuary and Indravati National Park, while Kotwal & Mishra (2004) found 35–40 in Udanti. They did not mention anything about Bhairamgarh and Pamed. Currently, there are only seven wild buffaloes, all males, in Udanti Wildlife Sanctuary, while there are two to three herds (approximately 15–30 individuals), mostly around Indravati River in Indravati National Park, and one herd living in the forest just outside the Pamed Wildlife Sanctuary. The buffaloes near Indravati River also cross over to Kopela-Kolamarkha proposed wildlife sanctuary in Maharashtra occasionally. Some animals from Udanti Wildlife Sanctuary are also believed to have crossed over to Sunabeda-Khariar Tiger Reserve of Orissa (Kishore Rithe, personal communication, September 2009).

In view of extremist activities by the Naxalites it is virtually impossible to get any current information from the area. A recent report indicated that the extremist activities have indirectly helped the remnant buffalo population as the annual 'parad' festival during which the local tribes engage in mass hunting, has been stopped in several areas. This was due to the poor security situation (the hunters could be caught in the cross-fire) (P. M. Lad, ex-CCF and K. R. Dongaokar, ex-Field Director of Indravati, personal communication, 2009). This indicates that the small surviving population of wild buffaloes has had some reprieve (Table 16.1).

Currently there are no wild buffaloes in Andhra Pradesh. However, their past occurrence in the state has been reported by Forsyth (1889) and Stockley (1928). Gee (1964) stated that there are a few animals still left in Andhra Pradesh adjoining to west and south Orissa, as well as near the Godavari River. The last-mentioned work indicates that the species occurred at least till the 1960s, if not later in the northeast part of the state.

The main buffalo areas of Madhya Pradesh are now in the newly created Chhattisgarh state. However, the range of the animals extended well inside Madhya Pradesh. According to Forsyth (1889), there were wild buffaloes in Denwa Valley in

Mahadeo Hills, Chhindwara district. Forsyth (1889) further reported that in the rains the buffalo used to wander in herds all over the Mandla highlands, where the present well-known Kanha National Park is located. He felt that many of the 'great numbers' of wild buffaloes in the jungles of Kenda and Pedra in Maikala Range were from Mandla uplands, which became unsuitable owing to the presence of large numbers of domestic cattle. This was also corroborated by Dunbar Brander (1931), who mentioned that wild buffaloes were regular visitors to Kanha National Park during the rainy season as late as 1900.

Forsyth (1889) reported wild buffaloes from areas near the confluence of Godavari and Wainganga (and apparently Pranhita also) Rivers, which are located in present Garhchiroli district (erstwhile undivided Chanda district) of Maharashtra. Dunbar Brander (1931) found two small herds of cows and a few solitary bulls in the Allapalli and Ghot (Gatta) forest ranges of south Chanda division (in present Garhchiroli district) in 1904. However, a fairly recent sighting of four in the same general area in Garhchiroli district bordering Chhattisgarh's Indravati National Park (Rithe & Fernandes 2002) indicates that small numbers continue to occur in Maharashtra and there is also movement between the two states. Currently, 2–3 groups (up to 15 individuals) are reportedly living near the Indravati River, mainly in Kopela-Kolamarkha proposed sanctuary (Kishore Rithe, personal communication, September 2009).

The earliest account of the wild buffalo in Orissa was found in Forsyth (1889). He mentioned the shooting of a cow on the road between Raipur (now the capital of Chhattisgarh) and Sambalpur, which is likely to be somewhere on the Orissa–Chhattisgarh border. Forsyth (1889) found 'great herds' in the valley of the Jonk River, a tributary of the Mahanadi River. Mooney (1930) gave a good account and stated that also at that time they were rare and had become extinct in many districts. He found 40–50 animals in two groups on Patna State bordering Raipur, Chhattisgarh State. The area is now part of Balangir and Kalahandi districts. Another small population was found farther east, also in Patna State (now in eastern Balangir). Mooney (1930) also mentioned past instances of wanderers from Raipur to Sambalpur district and occurrence of at least one herd in Gangpur State (now in Sundargarh and Sambalpur districts) near Brahmani River until the early 1890s. This population vanished at the time of construction of the Bengal–Nagpur main railways.

Gee (1964) stated that there were a few wild buffaloes still left in west and south Orissa, as well as in the adjoining areas of Andhra and Madhya Pradesh, near the Godavari River. Daniel & Grubh (1966) reported that they occurred in a very restricted area in the upper reaches of the Sileru River in Koraput district. Daniel & Grubh (1966) quoted a Forest Department official who stated that there were perhaps around 100 buffaloes in the area in the mid 1960s. The buffaloes of Sileru Valley gradually vanished once the Balimela dam and its associated townships came up in the area in the mid 1970s. There is an unregistered specimen from Ganjam (19°28′ N, 85° 05′ E) in the collection of the Natural History Museum, Florence, Italy (Groves 1996). The current status of wild buffaloes in Orissa is not known; owing to extremist activities by the Naxalites, it is nearly impossible to get any report from the area. However, there are reports that some animals from Udanti Wildlife Sanctuary in Chattisgarh are now living in Sunabeda-Khariar Tiger Reserve of Orissa (Kishore Rithe, personal communication, September 2009).

Bhutan: the occurrence of wild buffalo in south Bhutan was mentioned by Blower (1986), Choudhury (1994a) and Wangchuk *et al.* (2004). Bhutan, being largely mountainous, lacks much habitat for the buffalo. There are ten areas where wild buffaloes are still seen or were seen into the recent past. These are: (1) Gabhorukunda (26°46′–48′ N, 90°49–53′ E); (2) Mathanguri (26°47′ N, 90°58 ′E); (3) Nunmati (26°47′ N, 90° 59′ E); (4) Rabang *Nullah* (26°49′ N, 91°04′ E); (5) east of Doimari (26°49′ N, 91°06′ E); (6) Kuklung (26°47′ N, 90°45′ E); (7) Kalamati (26°47′ N, 90°40′ E); (8) Saralbhanga (26°52′ N, 90°15′ E); (9) Jamduar (26°44′ N, 89°52′ E); and (10) west of Kalikhola (26°42′ N, 89°49′ E).

Blower (1986) saw a small herd in Gabhorukunda area in September 1985. During several visits to Gabhorukunda in 2006 and 2007, footprints were seen by this author of lone bulls as well as small herds. A census operation carried out jointly by the Project Tiger authorities and the Rhino Foundation for Nature in Northeast India (an NGO) on 16–17 April 2008 counted 32 animals in three herds in Gabhorukunda area on the Indian side (Choudhury *et al.* 2008a). On the Bhutan side of Gabhorukunda, the animals mostly go for water because in this *bhabar* tract water bodies are limited. There is water in the river only where it debouches onto the flat ground. The reports by forest staff and hunters indicate that during the period of unrest in Manas for a decade from 1989 onwards (Choudhury 1989b, 2006), many buffaloes were killed by poachers near the water body in Gabhorukunda and, hence, the animals still avoid the area during the daytime. This is the only habitat of significance in Bhutan.

Mathanguri has lone bulls as well as one small herd (up to 14 animals) which cross the border to Bhutan only to come back. Since the early 1990s, a buffalo *khuti* has operated on the Bhutan side of Mathanguri, with several domestic female buffaloes. This has prompted a lone wild bull to visit the camp every day. Occasionally another bull also comes. This bull is perhaps the only wild buffalo that actually spends the bulk of its time inside Bhutan territory. The westernmost location is just west of Kalikhola, around Khurul River, where a small population of 15–20 animals now survives in the forests with dry riverbeds and pools. These animals' range covers forests of north Western Bengal as well.

According to Choudhury (2010) Bhutan does not have any resident population of wild buffaloes, mainly because of its limited habitat. However, more than 30 animals could be within Bhutan at any time (e.g. Gabhorukunda). The daily ranges of the herds and lone bulls that occur in Bhutan spread over both sides of the border. Of the ten locations, seven are inside Royal Manas National Park.

Bangladesh: Sanderson (1896) reported a small population of wild buffalo from the northwest of Dhaka as late as 1875, in present Shirajganj district. There are specimens from Faridpur (23°29′ N, 90°0′ E) and Chittagong Hills (22–23°N, 92–92°30′ E) at the Zoological Survey of India, Calcutta and Museum of Natural History, London, respectively (Groves 1996). A specimen was collected from Kukri-Mukri (21°54′ N, 90° 40′ E) in the Bay of Bengal (originally mentioned as 'Kuckri-Muckri, N.W. Vorder-indien'), presently in Bhola district, by the Crown Prince of the German Reich and of Prussia, which was described by Matschie (1912) as a new subspecies *septentrionalis* (Groves 1996).

The wild buffalo occurred in Dinajpur, Dhaka, Faridpur, Noakhali and the maritime areas (Sunderbans) during the first and second quarters of the nineteenth century (Daniel & Grubh 1966). The Sunderbans population probably became extinct at the beginning of the twentieth century (Sarkar 1993), although Jalil (1986) claimed that it was in the 1940s. The records from adjacent areas of northeast India indicate that it was present in the undivided Mymensingh and Sylhet districts even in the early twentieth century. In north Mymensingh, stray animals might have occurred even as recently as the 1970s, because there are wild populations within 5 km of the border and lone bulls reportedly visited cultivations at night. This was also corroborated by Asmat (2001).

Myanmar: according to Stockley (1928), there were a few wild buffaloes in the lower Irrawaddy Valley and delta. There is a specimen (BM 26.9.5.2) from Pegu (17°18′ N, 96°31′ E) in lower Myanmar in the Museum of Natural History, London (Groves 1996). Salter (1983) surmised that the wild buffalo in Myanmar probably became extinct in the mid 1930s. There is a paucity of good documentation on the occurrence of the species in the country, although apparently it occurred in the past. North Myanmar (Kachin State) is among the most poorly surveyed areas in the range of the wild buffalo. There are recent reports of a few wild-living animals independent of human husbandry in the Hukawng Valley of Kachin State (IUCN 2013).

During surveys in Manipur in northeast India, reports of a small 'wild' population were received from Balong River area, not far from the India–Myanmar border in Sagaing State in 2001 (Choudhury 2010)

Nepal: Fazl (1590), in his *Ain-i-Akbari*, mentioned that during Akbar's time (1542–1605) the wild buffalo was numerous in Oudh. The Oudh or Avadh now covers large part of east central Uttar Pradesh, including the Indo-Nepal border areas, with protected areas such as Dudhwa National Park and Katerniaghat Wildlife Sanctuary. This indicates that wild buffaloes occurred in areas such as Bardia and Sukla Phanta grasslands in west Nepal. Heinen (2001) surmised that the animal occurred in Bardia in Karnali River plain till the early twentieth century.

The wild buffalo vanished from Chitwan National Park in the 1960s, probably as a result of disease (Seidensticker 1976). After that the only population, a small one, remained in Nepal, in Koshi Tappu Wildlife Reserve (Gupta & Mishra 1972; Sah 1997) (Figure 16.2). Koshi Tappu was declared as a wildlife reserve in 1976. It is also a Ramsar Site. The reserve spreads over three districts: Sunsari (bulk of the reserve), Saptari and Udaypur (population trend in Table 16.1).

Sri Lanka: The origin of the free-living or 'wild' buffaloes in Sri Lanka is uncertain owing to the fact that there is no evidence of occurrence of wild buffaloes on the mainland south of the Godavari River. Ellerman & Morrison-Scott (1966), Gee (1964) and Maia (1970) considered these buffaloes to be descended from introduced domestic stock. Corbet & Hill (1992) included Sri Lanka within its historical range, but cited no concrete basis for that. The photographs of 'wild' buffaloes of Wilpattu and Yala also indicate that those animals lack the majestic horns and bulk (of body) of the true wild animals. Phillips (1929) stated that these animals are 'most probably the descendent of imported tame buffaloes that have escaped, many generations ago, to the remoter jungles and have established themselves there'. The population of free-living buffaloes in Ruhuna National Park (1500 km^2 including its adjacent reserves) in southeast Sri Lanka comprises about 2000 individuals (de Silva *et al.* 1994).

Thailand: Stockley (1928) stated that there were a few wild buffaloes in west Siam; apparently indicating the general area of the present distribution range. However, there was no report for quite some time. The continued existence of a small population was confirmed in 1964 (Chaiyarat 2004). In 1985 it was found that 35–40 animals were surviving in an area of 70 km^2 in the southern part of Huai Kha Khaeng Wildlife Sanctuary in Uthai Thani province, west Thailand (Nakasatit & Chanard 1985). Currently the number of buffaloes could be 40–50 (sanctuary staff, personal communication, November 2007). The population size is small because the habitat suitable for buffaloes in this large sanctuary (2780 km^2) is very restricted. Lone bulls occasionally stray out towards the south in the northernmost areas of Kanchanaburi Province. Inside Huai Kha Khaeng Wildlife Sanctuary there are two buffalo areas, one near the confluence of Huai Kha Khaeng and Huai Ai Yoh Rivers and the other in the south, near the border (Seidensticker & McNeely 1975).

While it was also surmised that Khao Soi Dao Wildlife Sanctuary may have held buffaloes until recently (Pong Leng-Ee 1978), there has so far been no evidence. Indeed, a very long time ago it is likely that wild buffaloes were widespread throughout the plains of the Chhao Praya River (e.g. the Lower Central Plain around Bangkok would surely have been excellent habitat) but that they disappeared long before the advent of natural history record-keeping. The museum specimens from Mae Wong (15°41′ N, 100°07′ E) in Nakhon Sawan Province in the National Zoological Reference Collection, Bangkok and from Kokan (which is possibly the same as Khon Kaen) (15°41′ N, 100°07′ E) in northeast Thailand in the

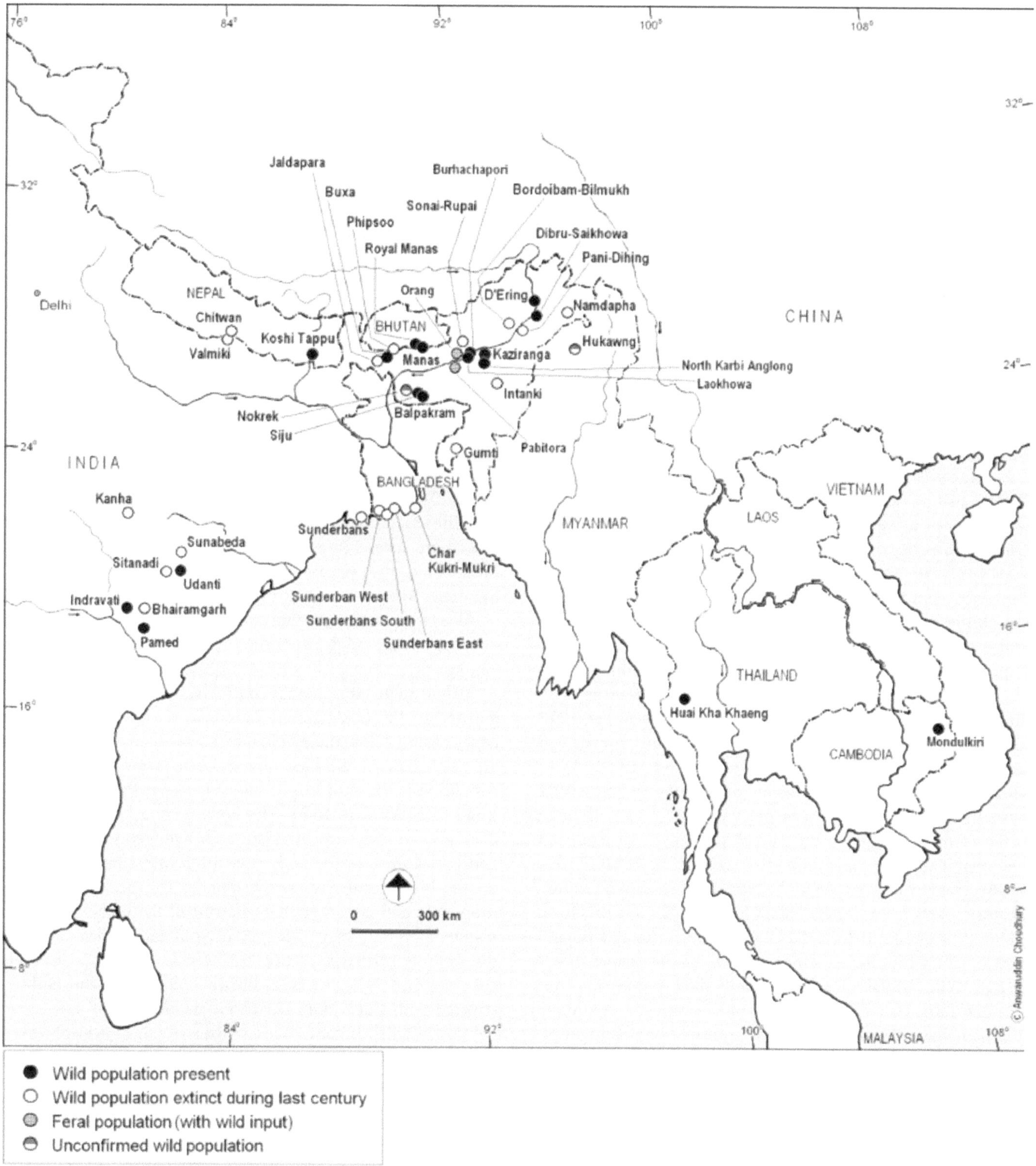

Figure 16.2 Protected areas having wild buffalo present or extinct across their range.

University Zoological Museum, Copenhagen (Groves 1996) also indicate a wider former distribution.

Cambodia: although Corbet & Hill (1992) did not include Cambodia within the range of wild buffalo, this must be seen in the context of the very limited historical zoological exploration of the country (e.g. Walston 2001). Recent camera-traps have confirmed the presence of a small population in that country's Mondulkiri Province. The first single photo was taken in 2001 by a joint survey conducted by the Wildlife Conservation Society and the Cat Action Treasury. In 2005 at least three adults, three subadults and one calf were recorded in a series of photos taken by camera-traps set up by the WWF Srepok Wilderness Area Project (SWAP) (Anon 2005a, 2005b). The past occurrence of some majestic wild buffaloes in the plains of Cambodia could be seen in a footage filmed by the kouprey (*Bos sauveli*) specialist and late US conservationist Dr Charles H. Wharton in 1951. In that year Wharton led a 90-man group (including 60 Royal Government soldiers) on a two-month excursion in the Choam Ksan and Koh Ker areas of Preah Vihear Province. He caught on film kouprey (the only footage of this extinct bovid so far; Anon 2006) as well as wild buffalo. Lekagul (1952) reported wild buffaloes from Tapeng Ra Veng in north Cambodia.

The most complete historical analysis of the species in Cambodia is probably that in Dumas (1944). He cautioned against the difficulties of distinguishing between truly wild animals and long-term feral populations. The population in the country is confined to a small area of easternmost Mondulkiri and possibly Ratanakiri Provinces, and is very small, with perhaps a few dozen individuals remaining (Timmins & Ou Ratanak 2001; Tordoff *et al.* 2005). WWF-Indochina estimated that 30–50 animals are in Srepok area in Mondulkiri Protected Forest (Anon 2005a), which is also the last remaining population in all of Cambodia, Vietnam, Lao PDR and east Thailand. Tom Gray (personal communication, September 2009) felt there could be around 20 individuals only.

Other areas of Southeast Asia: The wild buffalo is probably extirpated in Vietnam and almost certainly in Lao PDR, although it was thought to formerly occur but the date of extirpation is not known (Duckworth *et al.* 1999). Ward (1910) listed two wild bulls from Cochinchina in South Vietnam in his record specimens. In Vietnam, a pair of horns believed to be from a wild animal was found in Yok Don National Park, on the Cambodian border, in 1987 (Groves 1996). There is a specimen at the National Museum of Natural History, Paris from Langna, Song River (11°00′ N, 107°20′ E) in South Vietnam. Before 1999 there was reportedly a population in Chu Mom Ray, Vietnam, which is now extinct (IUCN 2013). Wild buffaloes of the Cambodian population may wander into Dak Lak Province, but no recent records in Vietnam have been confirmed to represent wild buffalo (Le Xuan Canh *et al.* 1997; Eames & Nguyen Duc Tu 2002; Tordoff *et al.* 2005). Brickle *et al.* (1998) mentioned that 'some years ago, a wild buffalo was shot near Cambodian border in Dak Lak Province' while quoting one Nguyan Minh Kha of the Border Guard Station (reported to them on 16 February 1998). Earlier in the twentieth century the Lagna plains were a noted locality among hunters for wild buffalo (e.g. Dumas 1944).

According to Duckworth *et al.* (1999), 'wild' buffaloes were reported during village interviews from Xe Pian NBCA (National Biodiversity Conservation Area) and other areas scattered north almost to Nam Ha NBCA in Lao PDR during 1988–1993, and as present into the 1950s–1970s in some other southern locations. However, no specimens or detailed records of indisputably wild animals are available from the country. Osgood (1932) considered that 'uncertainty apparently exists as to whether or not the Water buffalo of this region are feral', but Delacour (1940), who had (in contrast to Osgood) wide-ranging field experience in Indochina was convinced that buffaloes in the south were truly wild. Deuve (1972) recorded wild buffaloes from Khinak (near Xe Pian NBCA), but there was no concrete evidence.

The large tracts of south Lao that held habitat similar to that in Cambodia and Vietnam inhabited by wild buffaloes suggest very strongly that the species did previously inhabit Lao, too (J. W. Duckworth, personal communication, July 2009).

Descriptive notes

Body measurements: **Shoulder height**: 150–200 cm; **Head and body length**: 240–300 cm; **Tail length**: 60–100 cm; **Body mass**: 700–1200 kg (Prater 1980; Massicot 2004; Wilson & Mittermeier 2011). Dental formula: I 0/3, C 0/1, P 3/3, M 3/3 (×2) = 32. The deciduous dentition is identical, except that true molars are absent. As a bovid, the buffalo does not have upper incisors but three premolars and three molars on either side of the upper jaw.

The cows weigh less than the largest bulls, but some older ones may even weigh more than an average adult bull (up to 800 kg).

Adult male and female buffaloes can be easily distinguished in the field, mainly based on horn shape and body coat colour. Adult males have crescent-shaped horns, which are very thick at the base. These horns are normally highly curved, thus the tip-to-tip distance between them is relatively small (Figures 16.3–16.5). The body coat colour varies from dark grey to slaty and blackish, with light reddish ear pinnae and under parts of the body. The adult males have more hair on the body compared to the female. The neck has slightly more hair and is darker than the rest of the body. Body coat colour is grey to brownish-grey in younger animals. Hair on the body is relatively scanty. In the case of calves, the body coat at birth is a yellowish-brown colour and the body is covered with thick hairs. At about six months of age, calves develop reddish-brown coats. Mathur *et al.* (1995) observed that in males the bones above the eye form pronounced ridges, which have supra-orbital flecks (patch of light-coloured hair). They also have supra-labial and infra-labial flecks, which tend to be obscured with age. The females lack supra-orbital flecks and have supra-labial and infra-labial flecks present.

Mathur *et al.* (1995) mentioned that chevrons are the characteristic features of wild buffalo and helps, not only in age classification, but also in differentiation from domestic buffalo in which chevrons are absent. Chevrons are light and pale cream-coloured markings present on the neck region at two places: below the lower jaw and above the chest. These

Figure 16.3 A wild buffalo of Kollolua-Jokai area, Assam. A black and white version of this figure will appear in some formats. For the colour version, please refer to the plate section.

Figure 16.4 A bull and some cows in Kaziranga National Park, Assam. A black and white version of this figure will appear in some formats. For the colour version, please refer to the plate section.

markings are due to the presence of light-coloured hairs and skin. Chevrons become very distinguished in calves after the age of six months and stay so till the subadult stage. They become obscured and indistinct in older animals. However, in many areas they are of no help in identification of wild animals, because most domestic buffaloes in Assam have such chevrons. J. W. Duckworth (personal communication) reported that in Southeast Asia chevrons are also useless as a feature to discriminate wild animals.

Buffaloes, both wild and domestic, in Assam have white stockings, i.e. their legs are dirty-white up to just above the hocks and the knees, although often not seen due to grass or mud/soil. There are also conspicuous stockings in Thailand and Nepal (Choudhury 2010).

Figure 16.5 A herd of wild buffalo in Manas National Park, Assam. A black and white version of this figure will appear in some formats. For the colour version, please refer to the plate section.

Horn size and life expectancy

The wild buffalo has the maximum horn spread of any living bovid and horns grow throughout the life span (see Appendix 16.1). Horn shape is also an important distinguishing feature of adult male and female buffaloes in the field. Cows have less massive, but often longer, horns. Adult males have crescent-shaped horns, which are very thick at the base. These horns are normally highly curved, thus the tip-to-tip distance between them is relatively small. But there are fine examples of horns which are not highly curved and have relatively more distance between the tips. Adult females have comparatively thin horns. These are often less crescent-shaped, although they can be quite strongly crescentic. Tip-to-tip distance between the horns is normally larger than that of the male (Figure 16.3–16.4).

Although some record horns are listed in Appendix 16.1, perhaps the largest recorded was one bull shot in India in 1955, which had horns measuring 424 cm (13 ft 11 in) from tip to tip along the outside curve across the forehead (Guinness World Records 2006). A cow at Shillong Club, Meghalaya has a tip-to-tip across the forehead measurement of 359.09 cm (11 ft 9.37 in). The record cow (single horn) is 197.49 cm (6 ft 5.75 in). For details of horn records, see Choudhury (2010).

In Sri Lanka, the life expectancy of buffalo in the wild is about nine years for females. A cow may produce up to five calves in her life (de Silva *et al.* 1994). The maximum known life span in the wild is 25 years (Nowak 1999). A maximum age of 29 years has been documented in captivity (Massicot 2004). A wild bull (it was the only bull of its size, and hence was easily identifiable) in Pabitora Sanctuary in Assam was observed for 21 years; however, when it came to the sanctuary it was already an adult, indicating that it survived for 23 or more years.

Habitat

Main habitat characteristics

The habitat of wild buffalo has been described in detail by Choudhury (2010). Their occurrence is strongly linked to grasslands with water bodies, preferably in low-lying areas. The buffaloes inhabit tall grassy areas near swamps, streams and river banks, and open forests and dry thornscrub, preferably in floodplains. Availability of these two basic requirements has also ensured survival of these animals even in the hills and forested tracts. According to Prater (1948), 'tall grass jungles and reed brakes in the neighbourhood of swamps provide the ideal habitat, offering them both food and shelter, pools of water to lie in. In the southern parts of their range they live on drier harder ground, well broken up into *nullahs* and scattered with trees and open expanses of grass.' Lydekker (1907) mentioned that 'it prefers tall grass-jungles or reed brakes ... avoiding hills and rocks, and always seeking the neighbourhood of marshy swamps'. In Indochina the species frequented the lowlands dominated by deciduous forests and with a marked dry season, where it apparently used small pools and marshes, in addition to permanently flowing rivers (Wharton 1957). Sterndale in Finn (1929) said: 'The buffalo never ascends mountains like the bison [gaur], but keeps to low and swampy ground and open grass plains.' According to Forsyth (1889), it is essentially a lover of plains and level plateaux, where the extensive swamps can be found. Inverarity (1895) mentioned that wild buffaloes in the Central Provinces also used to inhabit sal forests. He further said that although they are not partial to hills, they will cross low ranges.

The wild buffalo has generally been described as essentially an animal of the plains; however, it occurs widely in the low hills wherever habitat is available or where it is forced to do so. In Assam's Kaziranga National Park where more than half of

Figure 16.6 A typical Dibru-Saikhowa habitat, Assam. A black and white version of this figure will appear in some formats. For the colour version, please refer to the plate section.

the world population is found, the buffalo's favoured habitat is the short, grassy fields around *beels* (ox-bow lakes and depressions). It also prefers tall elephant grass at different times, but the woodland mostly during travelling and in the monsoon period, when the grasslands are under water. The buffaloes also use the wide, sandy tracts along the Brahmaputra River, where they go for small grassy patches as well as seasonal lagoons. In Manas National Park there are no short, grassy fields around *beels* as, on account of its location in *bhabar* tract, there is no scope for formation of *beels*, in contrast to the ideal conditions in Kaziranga. The tall grassland, short, grassy patches along the streams and the banks, often littered with boulders, of the Beki and Manas Rivers form the habitat of the buffalo. The animals use forests for movement during the day and night. The habitat in Laokhowa and Burhachapori is somewhat like Kaziranga, but owing to human-induced disturbance during the day, the animals use the more open areas around the *beels* at night only. In Dibru-Saikhowa National Park there are no large open expanses of grassland as it was originally a rainforest area – the habitat underwent change after the great Assam earthquake of 1950 (Choudhury 1998b). Over the years deciduous woodland and willow (*Salix*) swamp forest became the dominant vegetation. Buffaloes criss-cross these forests, travelling to grassy patches along the *nullahs* and depressions, and also rest inside woodland (Figure 16.6). The habitat in Dibang and Lohit valleys in Arunachal Pradesh resembles that of Manas National Park to a great extent. In Peninsular India there is no large open expanse of grassland like Kaziranga, but grassy glades and open grasslands amid woodlands form the main habitat. Wild buffaloes in Bastar inhabit tropical dry deciduous forest with grass as the dominant undergrowth. In Kondakamberu Valley, Orissa, the habitat has denser forest with grass as undergrowth (Daniel & Grubh 1966).

Inverarity (1895) reported a solitary bull shot close to him near Udanti River in present-day Chhattisgarh, which inhabited a rocky, stony country, highly atypical habitat. Local people reported to him that the bull had been there for four years and was alone.

The wild buffalo's main habitat remains the grasslands in the plains; however, their presence in the hilly Balpakram National Park of Garo Hills, Meghalaya, especially on the plateau and adjacent areas (Choudhury 1994a) has been a matter of curiosity and has been questioned in the past. But it seems that the small grassy patches along the rivers and streams, in the depressions and on tableland in hilly areas also harbour buffalo populations, apparently in lower density than the plains. Considering their presence in Balpakram, it is plausible that the wild buffalo also occurred in other hilly regions of northeast India and Myanmar in the recent past.

Such preference may be partly due to the earlier conversion of most of the plains. The wild buffalo appears to avoid evergreen rainforest, but there are small populations in pockets such as Dum Duma and Poba, where they graze in small openings and often wallow inside rainforest in jungle pools. These buffaloes move widely through the forests while going to openings with grass and wallows. The feral buffaloes and the free-ranging domestic animals continue to live in similar habitats to their wild cousin (Choudhury 2010).

Wild buffaloes of Nepal, in Koshi Tappu, prefer swampy floodplain with tall elephant grass and scrubby woodland forest in the vicinity (Sah 1997). In Huai Kha Khaeng, Thailand, the animals are mostly confined to the plains of the Huai (river) Kha Khaeng, where they use all vegetation types within a radius of 1 km (Uicharoensak 1992). This river's banks are dominated by various grasses and deciduous forest. The feral buffaloes in Sri Lanka prefer grassland and discontinuous forest areas. In the dry season, they spend more of

Table 16.2 Habitat availability in current wild buffalo range areas

Name of the area	Suitable habitat (km^2)	Remarks
India		
Kaziranga National Park	600	Including adjacent areas
Manas Tiger Reserve	500	About 300 km^2 in the national park area
Dibru-Saikhowa National Park	330	Including adjacent areas
Laokhowa and Burhachapori Sanctuaries	90	
Other areas in Assam	80	
Assam sub-total	**1600**	Additional 90 km^2 of suitable habitat is also available, but the wild buffalo has become extinct in recent decades
D-Ering Memorial Sanctuary	100	
Mebo, Dibang and adjacent areas	150	
Chapories of Lohit River	150+	
Other areas in Arunachal Pradesh	50	
Balpakram–Siju–Baghmara–Rewak–West Khasi Hills complex	70	In Meghalaya
Buxa Tiger Reserve	30	In north West Bengal
Northeast India sub-total	**2150**	Additional 170 km^2 of suitable habitat is also available, but the wild buffalo has become extinct in those areas or there are no recent reports
Indravati Tiger Reserve	150	
Pamed Sanctuary	50+	
Udanti Sanctuary	64	Kotwal & Mishra (2004)
Chhattisgarh sub-total	**260**	Additional 100 km^2 of suitable habitat is also available, but the wild buffalo has become extinct in recent decades
India total	***c.*2500**	Additional 270 km^2 of suitable habitat is also available, but the wild buffalo has become extinct in recent decades
Bhutan		
Royal Manas National Park	10	Choudhury (2008)
Other areas	5	West of Kalikhola
Cambodia		
Srepok Wilderness area	100	In Mondulkiri Protected Forest
Nepal		
Koshi Tappu Wildlife Reserve	100+	H. S. Baral (personal communication 2009), Sah (1997)
Thailand		
Huai Kha Khaeng Sanctuary	70	Nakasatit & Chanard (1985)
Outside India total	**300**	
GRAND TOTAL	**2800 (say *c.*3,000)**	Additional 300 km^2 of suitable habitat is also available, but the wild buffalo has become extinct in recent decades

Favourable habitat outside the known range of wild buffalo has not been included here.
Source: Choudhury (2010).

their time in forested areas (Eisenberg & Lockhart 1972; Ashby & Santiapillai 1987).

Wild buffaloes occurred from sea level (Sunderbans) up to 800 m elevation on the plateau of Balpakram National Park. Wild-living feral buffaloes have been recorded up to at least 1000 m in eastern Java (S. Hedges, personal communication, 2008) and above 1500 m in Western Ghats.

Mathur *et al.* (1995) stated that short grassland was used more intensively, whereas woodland was avoided. *Beels* were also used significantly in proportion to its availability, whereas tall grasslands were used less than their availability. After the monsoon season, when the water starts drying up, short grasslands surrounding the lakes are exposed. These short grasslands around numerous reduced *beels* offer the best grazing grounds for several wild herbivores in Kaziranga National Park, including wild buffaloes.

Habitat availability across the regions is summarised in Table 16.2. It may be noted here that the 'habitat' here means the habitat available in the existing range areas, i.e. the habitat that is occupied. The areas where habitat is available but the animals are hunted out, are not included here.

Table 16.3 Density of wild buffalo in some key habitats

Name of the area	Density per km^2 against the total area	Density per km^2 against potential habitat
Kaziranga National Park	4.51	6.45
Kaziranga National Park and adjacent areas	2.29	3.24
Manas National Park	0.46	0.77
Dibru-Saikhowa National Park and adjacent areas	1.14	1.60
Laokhowa and Burhachapori Wildlife Sanctuaries and adjacent areas	0.62	0.94
D-Ering Memorial Wildlife Sanctuary	0.26	0.5
Koshi Tappu Wildlife Reserve	1.25	2.19
Huai Kha Khaeng Wildlife Sanctuary	0.02	0.71

Source: Choudhury (2010).

Density and fluctuations in population size

Kaziranga National Park, where the largest population survives, has an average density of 6.45 individuals per km^2 in the potential habitat. In the park itself, the density is as high as 12.04 individuals per km^2 in Baguri or West Range area. The density in some of the areas is listed in Table 16.3. Among the feral animals, an average density of 3.8 individuals per km^2 was reported in Sri Lanka, with a concentration of 14.55 per km^2 in very favourable habitats (de Silva *et al.* 1994). Densities of up to 34 feral buffaloes per km^2 occurred in some localities of northern Australia when the population was at it highest (Tulloch 1979). There are about six individuals per km^2 in Pabitora Wildlife Sanctuary.

Variations in habitat use

Seasonal variation in habitat use is not very significant in most of the wild buffalo's range. In central India there are concentrations near water sources during the dry season. However, such a preference has nothing to do with habitat type as it is based upon water source. In areas like Kaziranga, there is seasonal concentration in higher grounds and also in nearby densely forested foothills of Karbi Anglong when the park is covered by flood water. During annual burning of grassland as part of a management tool in Kaziranga and other grass-dominated protected areas, the buffaloes move to unburnt patches and also to edges of open wetlands and sometimes in woodland. However, quantitative data is lacking.

Habitat segregation for age/sex classes

No such segregation is known or observed in the case of this species.

Movements and home range

Movements

Wild buffaloes usually travel in single-file with an adult (normally the lead) female in the lead. Individuals may pass along the same route. Calves normally follow their respective mothers, followed by other females or males (if these latter are with the herd). During movement from one place to the other, individuals were closer to each other than when grazing, probably as an antipredator strategy. In high rainfall areas such as northeast India, they move for short distances, while in drier areas with harsh climate, such as Central India, the herds move seasonally in relation to the depth of water and foraging availability. Hasan's (1980) record of a 'wild' bull in Raipura Forest range of Panna district, which is about 600 km from Bastar, where the nearest wild buffaloes lived, is not being treated as a migration or movement because there is no previous or subsequent record of such a phenomenon. The nearest area with recent historical records (up to around 1900) is Mandla district, about 150 km from Panna. Singh (1980) believes that it was a true wild buffalo, which might have travelled from Torenga and Indagaon blocks of Raipur district (now in Chhattisgarh), about 225 km from Panna. He mentioned two examples of the shooting of wild buffaloes, which were at least 125 km away from their nearest habitat. Prater (1948) mentioned that in the drier areas they are given to wandering, and during the hot weather sometimes cross the low hills in search of grazing.

In Assam there is seasonal migration from Kaziranga National Park to the highlands of Karbi Plateau during the floods every year. Normally the animals travel 2–10 km. A wild bull from Kadam Reserve Forest strayed up to the runway of Lilabari Airport near North Lakhimpur town one day in 1982, travelling at least 10–12 km. During high floods there were records of movement up to Dulung and Kakoi Reserve Forests at the foot of the Himalaya (Choudhury 1994a), about 15 km from Kadam Reserve Forest. As far back as the sixteenth century, Fazl (1590) had observed that 'when the plains are inundated, the animals take to high ground'.

By contrast, similar movement is not noticed in Manas or other areas of Assam, which are not inundated by flood waters. Wild buffaloes of Dum Duma Reserve Forest are believed to have travelled 10–15 km through tea plantations, villages and degraded forest during a devastating flood. Lone bulls occasionally travelling up to 45 km have also been recorded (from east Kaziranga to west Majuli) and once from Laokhowa to near Morigaon. In Pabitora Wildlife Sanctuary a wild bull came during the floods in early 1980s, perhaps having been washed from some way upstream. The nearest habitats are Laokhowa and Burhachapori Sanctuaries, about 60 km upstream along the Brahmaputra River, the possible route of its arrival. Kotwal & Mishra (2004) stated that in Chhattisgarh buffaloes from Udanti Wildlife Sanctuary migrate to Sitanadi Wildlife Sanctuary and surrounding forest in the summer in search of water, fodder and cover (the distance is about 40 km).

It seems that unlike Peninsular India, where a relatively drier climate often forced the buffaloes to migrate longer distances, in northeast India most movements are for short distances and mainly to avoid high floods.

Home range

Long-term research on the home range of wild buffalo is yet to be carried out. Heinen (1993), while studying population viability, has estimated some home ranges in Kosi Tappu Reserve in Nepal. The 'North herd' had a 6 km^2 home range in the rainy season, which increased to 55 km^2 during the dry season in 1976. After a decade, it ranged from 5 km^2 in the rainy season to 7 km^2 in the dry season in 1987–1988. In the case of the 'South herd', it ranged from 8 km^2 in the rainy season to 3 km^2 in the dry season in 1976. After a decade, it ranged from 4 km^2 in the rainy season to 8 km^2 in the dry season in 1987–1988.

From the above, it seems that there is seasonal variation and the range could be as low as 3 km^2 and as high as 55 km^2 (estimated from the maps in Heinen 1993).

No detailed studies to estimate the home range have been done in key habitats such as Kaziranga. A small group of four animals in the Dhakuakhana area was found to use 10.5 km^2 in 1990. The range included farmland interspersed with patches of tall grass and wetlands. A herd in the Kollolua-Jokai area had an estimated home range of 40 km^2 in 2008. In this case, a significant area of farmland is included in their range. Three lone bulls, which also visit domestic buffaloes in the nearby villages, used an area that ranged up to 70 km^2 in 2008.

The feral buffalo groups, called 'clans' in northern Australia, have a home range varying from 1.70 km^2 to 10 km^2 (Jesser *et al.* 2008). In Assam's Pabitora Wildlife Sanctuary, the largest bull (a wild one) in the 1990s had a home range of around 16 km^2. The seasonal (drier winter) home range of a herd near Sohola *beel* in Kaziranga National Park was found to be less than 6 km^2 in 1995–1996. During the monsoon rainy season the range could not be estimated as the animals moved to safer high grounds due to floods.

The home ranges are not defended and there are overlaps; however, the lone bulls, while sharing their range with other lone animals, chase them away when encountered too closely.

Activity patterns

In herbivorous mammals the majority of their time is spent on feeding. Beekman & Prins (1989) found that horses and cattle spend more than 70% and 50–60%, respectively, of their time grazing. Ruminant ungulates must additionally allocate time to ruminating and resting (Ryan & Jordaan 2005). However, detailed studies on the activity budget of the wild buffalo have not been done and quantitative data are lacking. In African buffalo *Syncerus caffer* there are some fine studies such as Stark (1986), Winterbach & Bothma (1998) and Ryan & Jordaan (2005). As observed by Owen-Smith (1988) and Redfern *et al.* (2003) in African buffalo, the wild buffalo is also a large grazer, whose feeding ecology predicts greater selection of high biomass than high-quality grass swards.

General observations of wild buffaloes in Assam indicate that they are both diurnal and nocturnal. In disturbed habitat or where the buffaloes are surrounded by human habitation and farmland, they normally become active after sunset, travelling and feeding, with short bouts of resting, until sunrise. The small group of four animals in the Dhakuakhana area which appeared in patches of tall grass and *beels* spend their entire daytime hours inside a small patch of tall grass. Inside the patch, they were found to be wallowing, simply standing in water and ruminating. After sunset they came out and started grazing and moving, covering more than 3 km (in drier winter months). Similarly, a herd in the Kollolua-Jokai area, another disturbed and unprotected area, spent their entire daytime hours inside a patch of tall grass with small pools and shrubs. Inside the patch, they were found to be wallowing, resting, ruminating and grazing a little. They came out of dense grass around 2–3 hours after sunset and slowly moved towards cultivated fields, mostly of paddy and pulses (*Vigna mungo*), feeding until about an hour before dawn and covering more than 6 km (in drier winter months).

However, in Kaziranga and Manas, after spending the night in woodland or in the open fringe of *beels*, they move out from woodland or spread out from the *beel* edge to graze at dawn. Around mid-morning, they move to the nearby river or *beel* to drink and wallow, returning to graze from mid-afternoon until dusk. Wallow may last from one to more than three hours at a stretch and may be repeated (Figure 16.7). Hence, there are two major daytime grazing bouts. Moderate grazing may also continue during the night. In such well-protected areas, they rest in open areas away from dense vegetation, presumably to avoid tiger attacks (Choudhury 2010). Winterbach *et al.* (1998) have also found two major daytime grazing periods in African buffaloes.

In Peninsular India the buffalo frequents grassy glades and open grasslands amid woodlands in the morning and evening, spending the hot hours of the day inside the forest or in pools. Wallowing in pools, rivers and mud forms an important activity in a buffalo's daily life. It is basically part of the 'rest' period. In the marshy areas, small pools are often created by buffaloes and rhinos by wallowing. The apparent function of wallowing is cooling, as buffaloes normally do not wallow in cool winter mornings or at night. According to some authorities, wallowing has a thermoregulatory function, and mud prolongs the cooling effect by slowing evaporation (Hedges 1995). Studies on feral buffaloes in Australia showed that wallowing may not be essential for the survival of the buffalo as when deprived of wallows the animals behave like cattle and seek shade (Tulloch 1974; Tulloch & Litchfield 1981). An

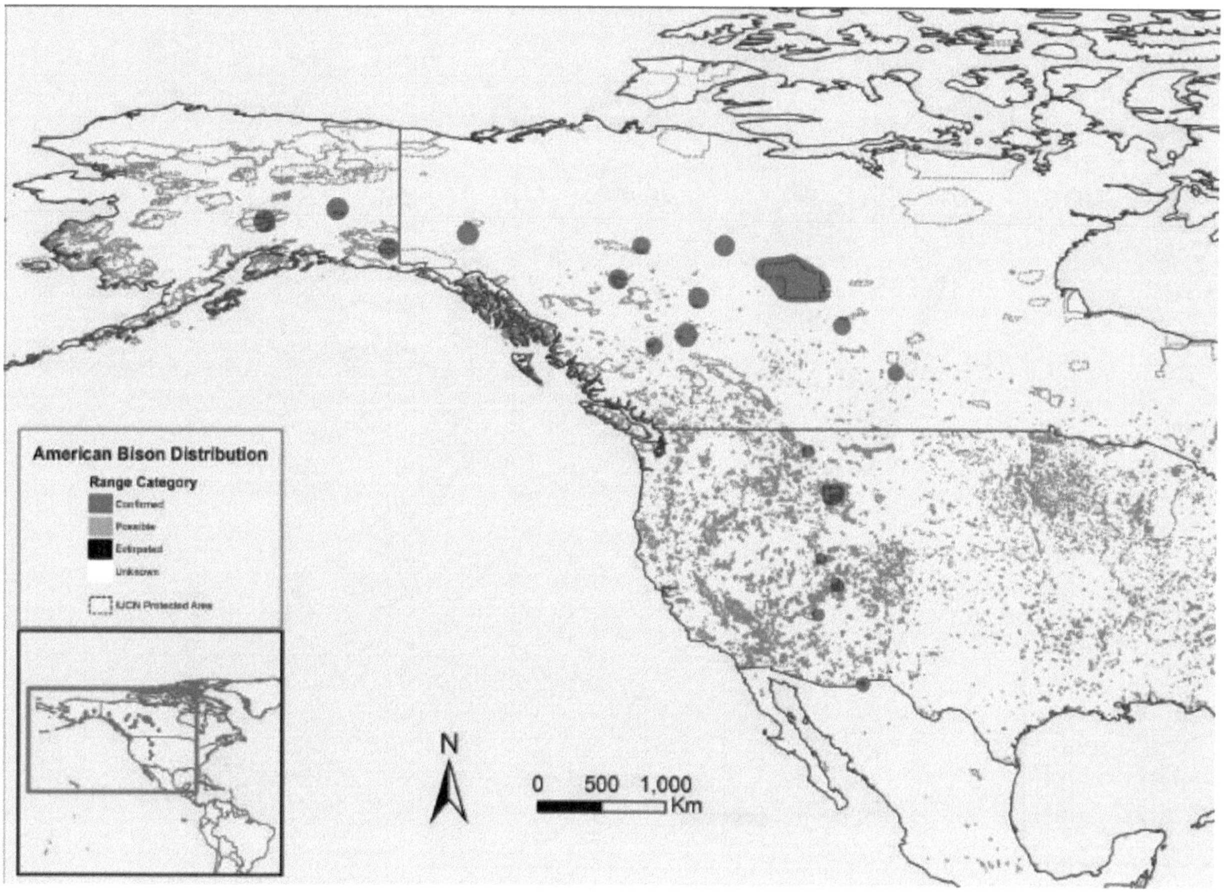

Plate 1 Distribution map of American bison (source: IUCN, 2008, *Red List of Threatened Species*).

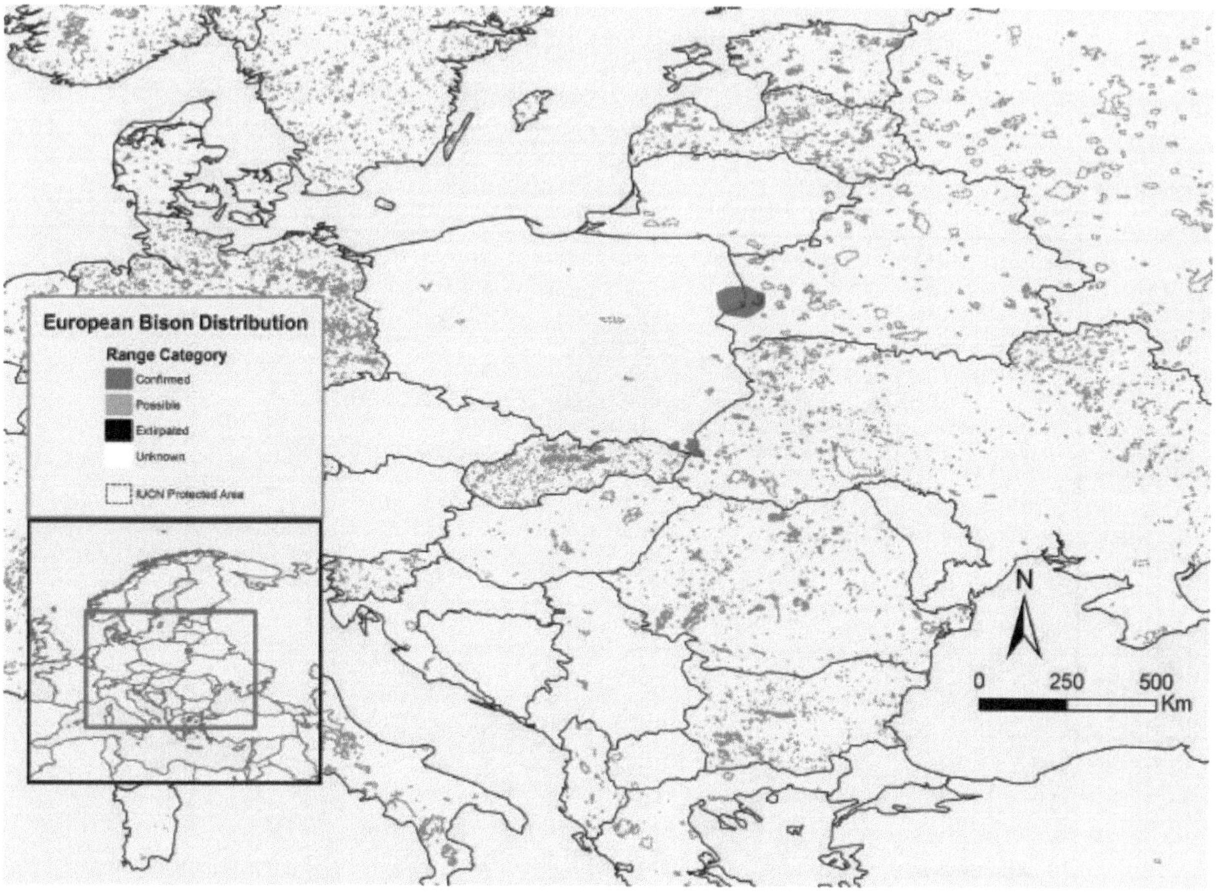

Plate 2 Distribution map of European bison (source: IUCN, 2008, *Red List of Threatened Species*).

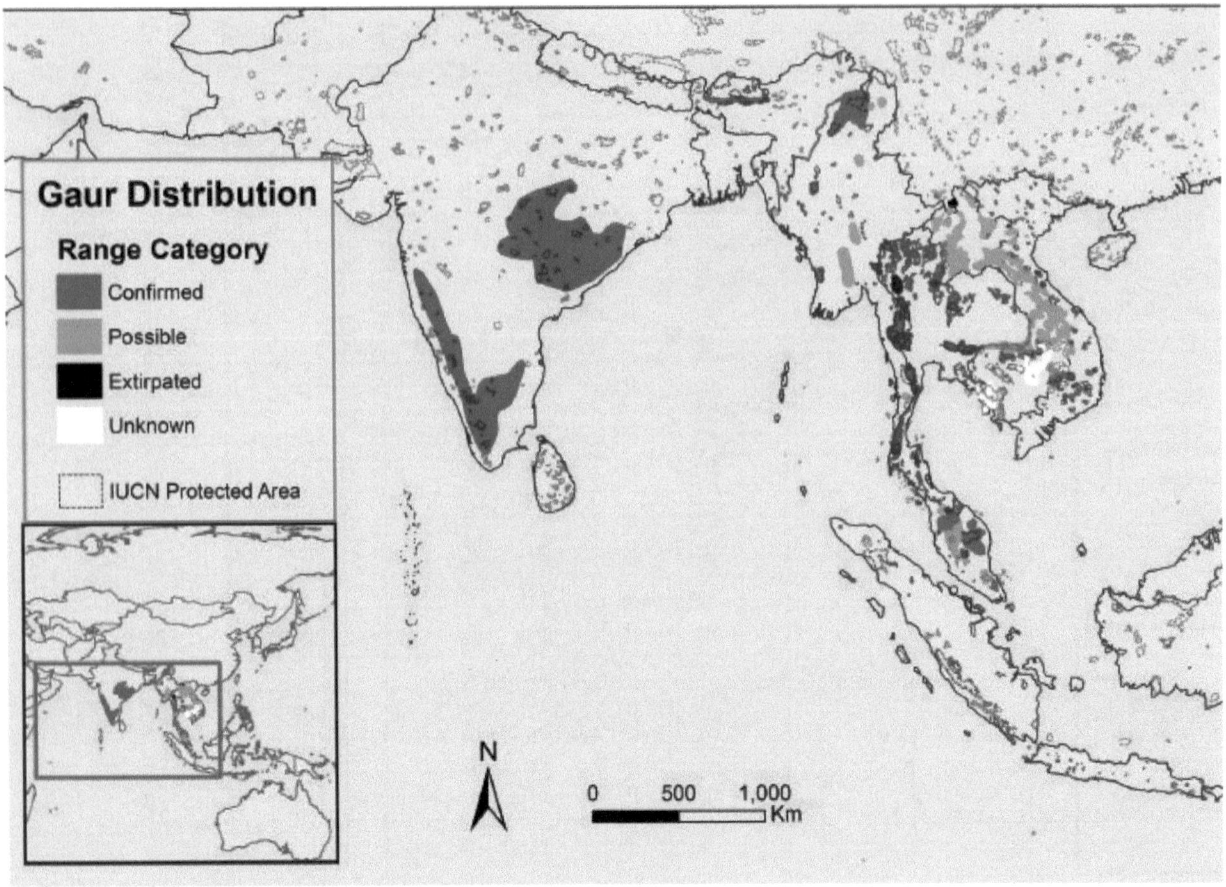

Plate 3 Distribution map of gaur (source: IUCN, 2008, *Red List of Threatened Species*).

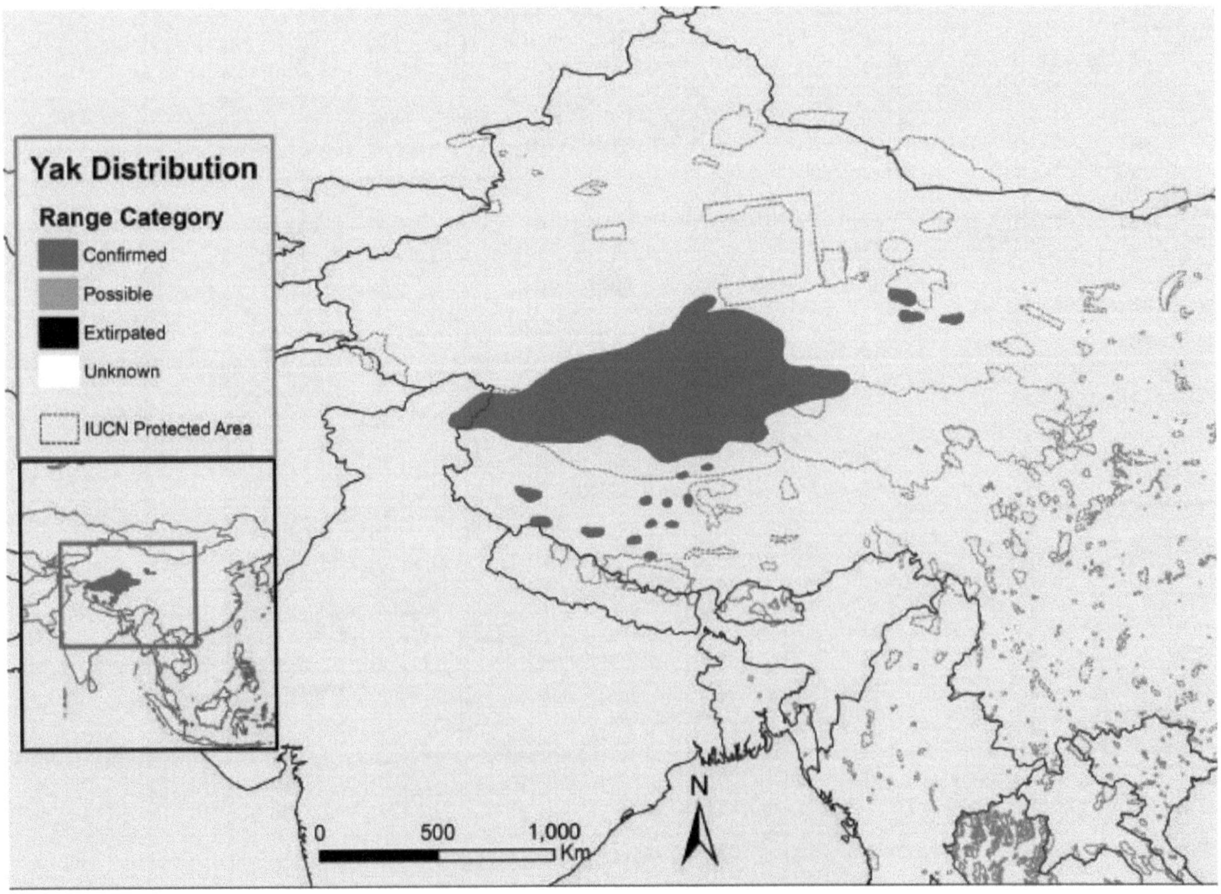

Plate 4 Distribution map of wild yak (source: IUCN, 2008, *Red List of Threatened Species*).

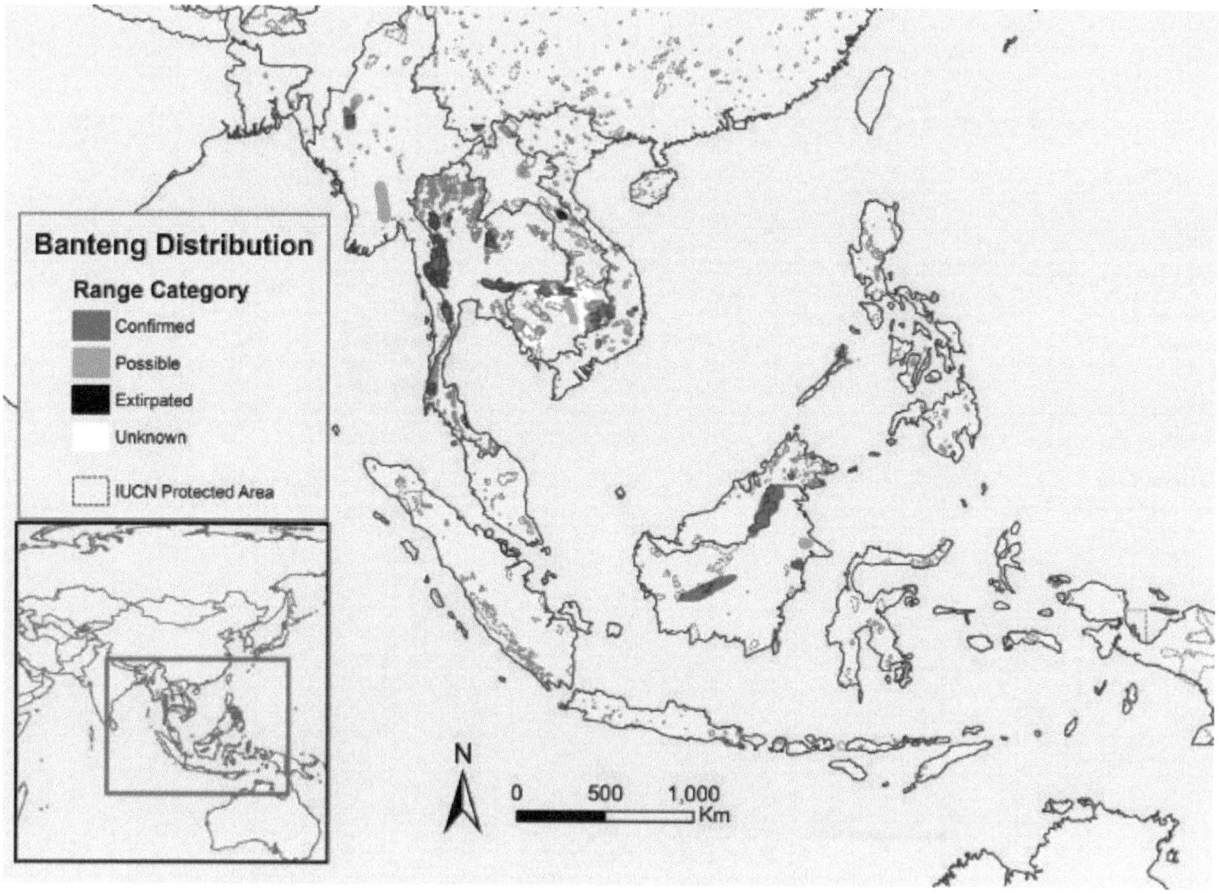

Plate 5 Distribution map of banteng (source: IUCN, 2008, *Red List of Threatened Species*).

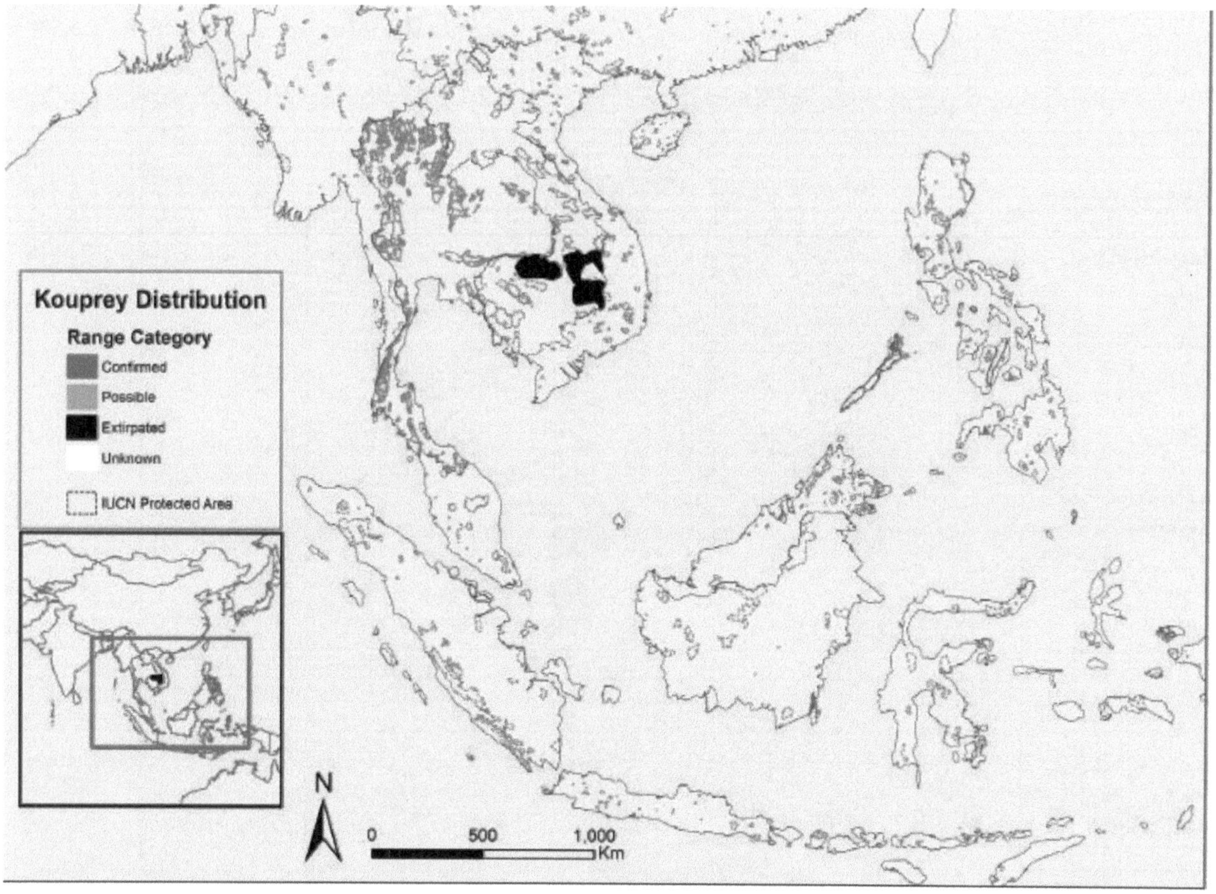

Plate 6 Distribution map of kouprey (source: IUCN, 2008, *Red List of Threatened Species*).

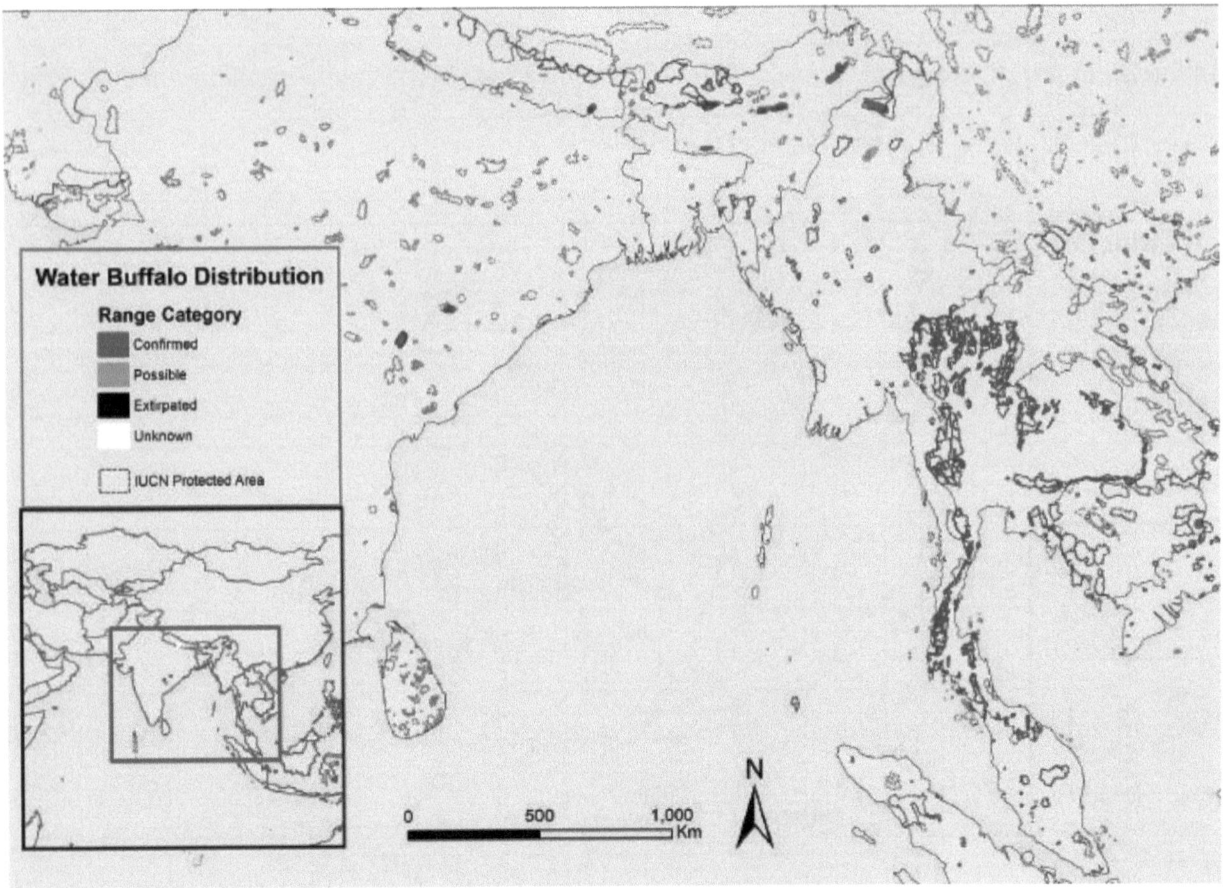

Plate 7 Distribution map of wild water buffalo (source: IUCN, 2008, *Red List of Threatened Species*).

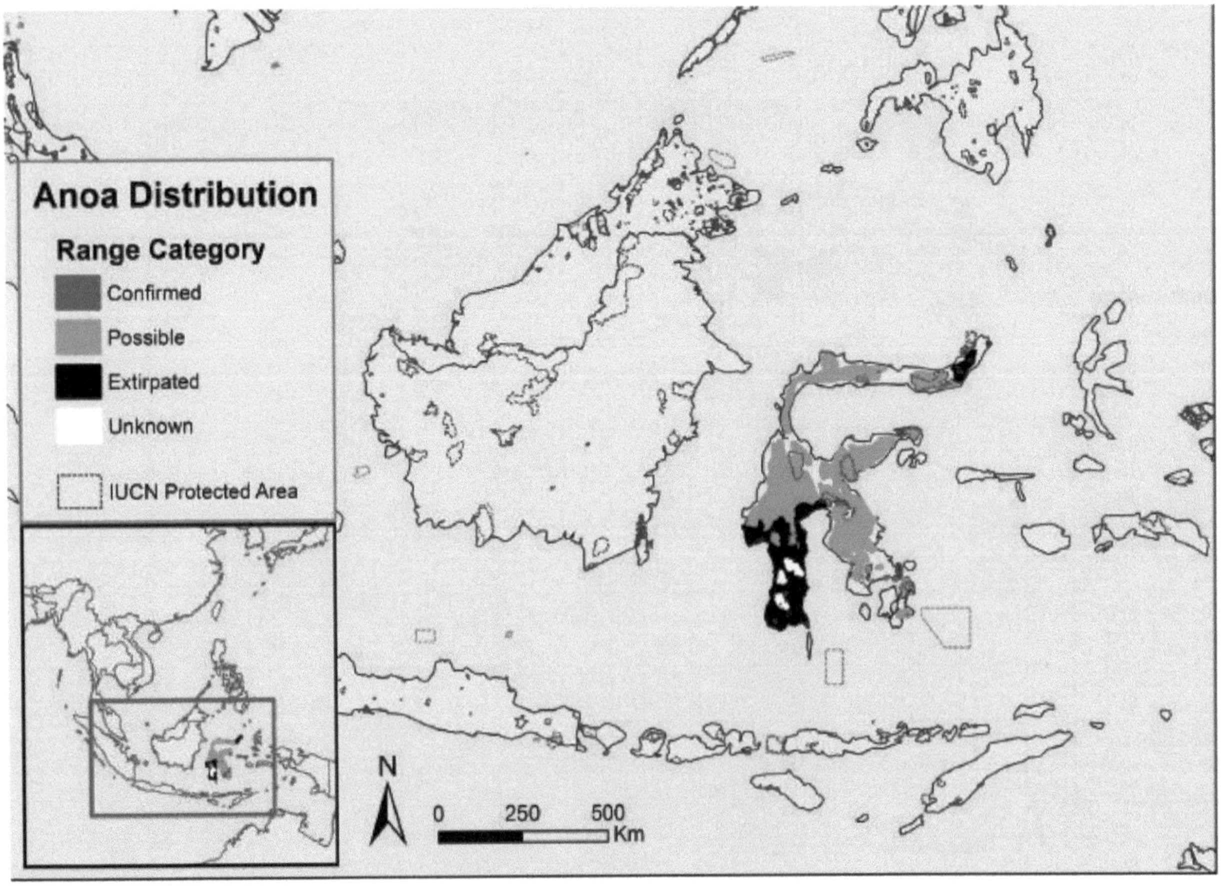

Plate 8 Distribution map of anoa (source: IUCN, 2008, *Red List of Threatened Species*).

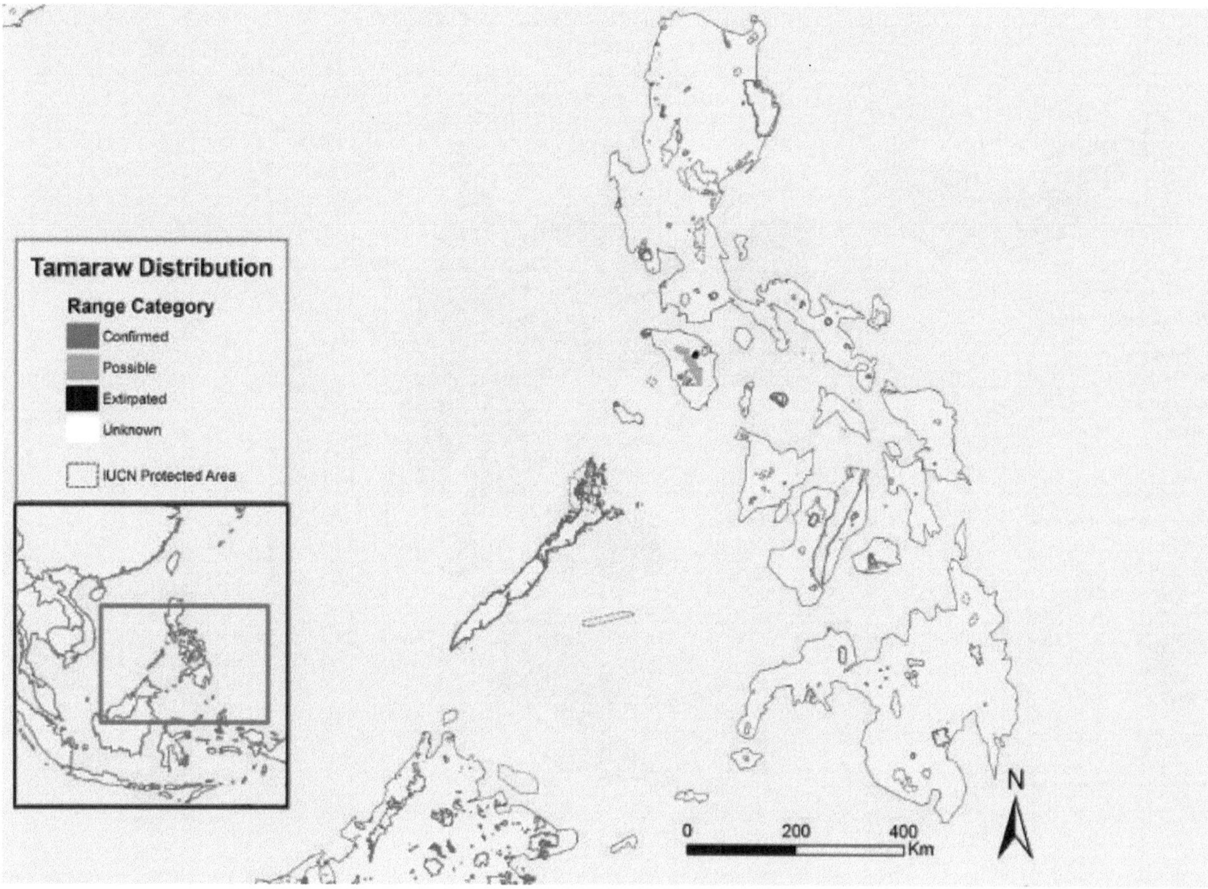

Plate 9 Distribution map of tamaraw (source: IUCN, 2008, *Red List of Threatened Species*). The species occurs only on the island of Mindoro, Philippines. The three areas where it is known to exist are Mts Iglit-Baco, National Park, Sablayah and Mt Calavite.

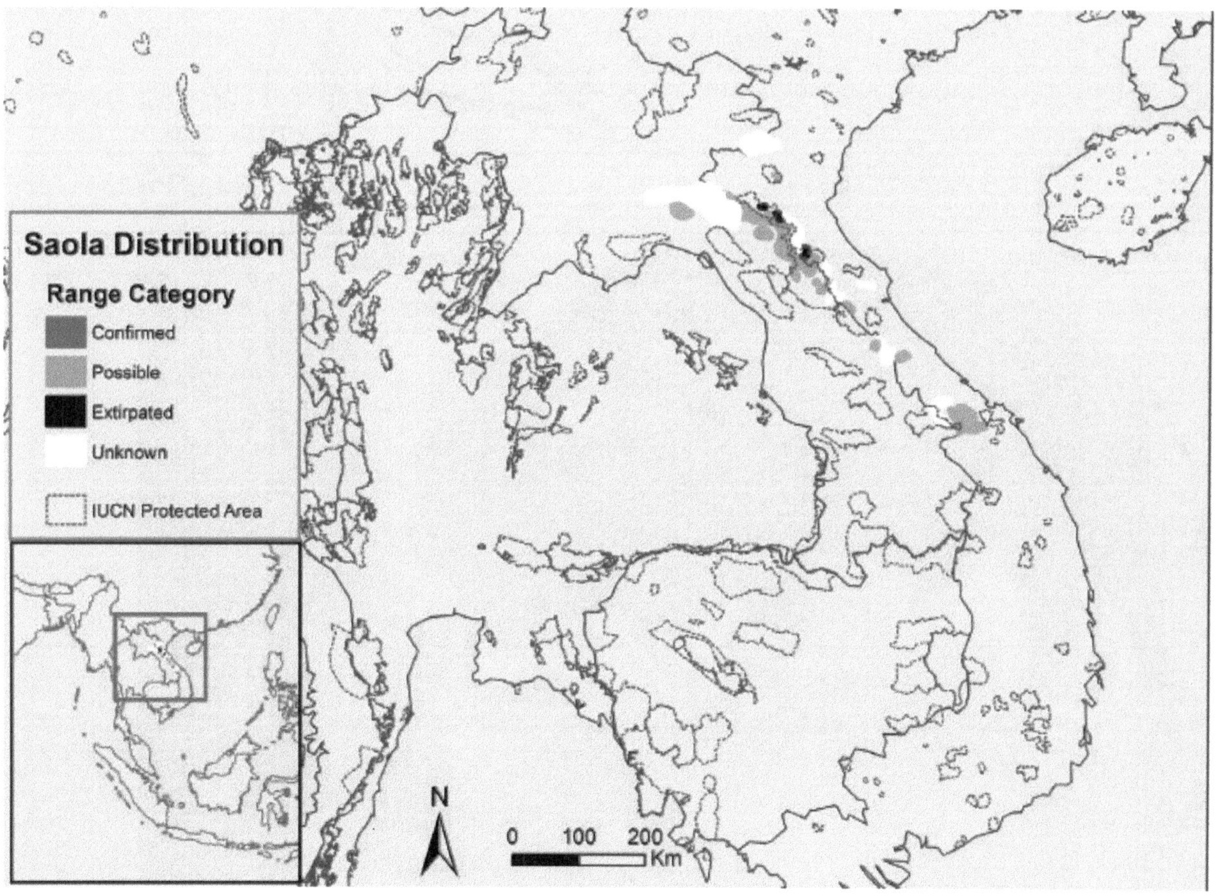

Plate 10 Distribution map of saola (source: IUCN, 2008, *Red List of Threatened Species*).

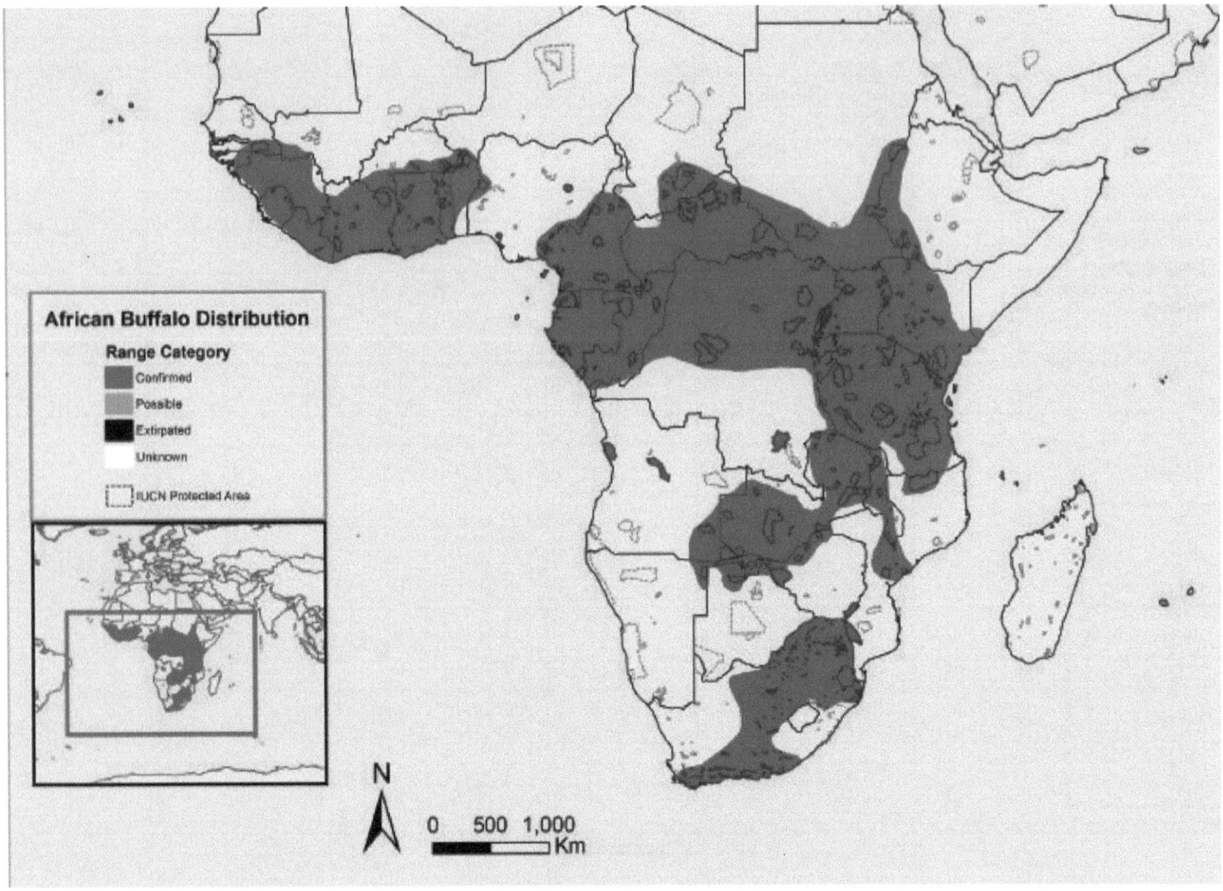

Plate 11 Distribution map of African buffalo (source: IUCN, 2008, *Red List of Threatened Species*).

Plate 12 American bison (illustration by Francesco Rinaldi).

Plate 13 European bison (illustration by Francesco Rinaldi).

Plate 14 Gaur (illustration by Francesco Rinaldi).

Plate 15 Wild yak (illustration by Francesco Rinaldi).

Plate 16 Banteng (illustration by Francesco Rinaldi).

Plate 17 Kouprey (illustration by Francesco Rinaldi).

Plate 18 Aurochs (illustration by Francesco Rinaldi).

Plate 19 Wild water buffalo (illustration by Francesco Rinaldi).

Plate 20 Anoa (illustration by Francesco Rinaldi).

Plate 21 Tamaraw (illustration by Francesco Rinaldi).

Plate 22 Saola (illustration by Francesco Rinaldi).

Plate 23 African Buffalo (illustration by Francesco Rinaldi).

(a) (b)

Plate 24 (Figure 20.3a, b, c) Forest buffalo are forest dwellers, inhabiting different habitats with: (a) grassy glades and watercourse areas in rain forest (photo by M. Melletti); (b) forest muddy areas (photo by A. Turkalo); and (c) coastal and equatorial savannas (photo by Mathiev Bourgarel, Cirad).

(c)

Plate 24 *(cont.)*

Plate 25 (Figure 10.30) Beginning of the contest, bulls with low-bowed heads are pushing at each other (photo by Z. A. Krasiński).

Plate 26 (Figure 10.32) Young European bison released to an acclimatization enclosure at Bieszczady National Park (photo by K. Perzanowski).

Plate 27 (Figure 10.34) Animals in Busolengo Zoo (Italy) (photo by W. Olech).

Plate 28 (Figure 10.20) A young European bison feeds on natural forage after being released to the wild (photo by K. Perzanowski).

Plate 29 (Figure 10.19) Wallowing is the most frequently observed behaviour among adult bulls in the breeding season (photo by M. Hławiczka).

Plate 30 (Figure 17.1) Anoa, female adult, seven years old, from central Sulawesi (photo by A. H. Mustari).

Plate 31 (Figure 17.2) Anoa, male, 19 years old, from central Sulawesi (photo by A. H. Mustari).

Plate 32 (Figure 17.3) Male, one year old, southeast Sulawesi (photo by A. H. Mustari).

Plate 33 (Figure 18.5) An Indigenous Mangyan at Mount Iglit-Baco National Park, Occidental Mindoro, Philippines in 2007 (photo by M. R. Cebrian). For full caption text, see page [313].

Plate 34 (Figure 10.24) A several-days-old calf, rescued after a torrent at Bieszczady Mountains (photo by K. Perzanowski).

Plate 35 (Figure 13.2) A mixed herd of banteng in Alas Purwo National Park (Java) consisting of bulls, cows, subadults and juveniles (photo by S. Pudyatmoko).

Plate 36 (Figure 16.5) A herd of wild water buffalo in Manas National Park, Assam.

Plate 37 (Figure 16.6) A typical Dibru-Saikhowa habitat, Assam.

Plate 38 (Figure 16.9) A wild bull pursuing domestic cows near Dibru-Saikhowa, Assam.

Plate 39 (Figure 16.10) Many domestic water buffaloes in Assam closely resemble their wild cousins owing to the presence of wild buffaloes in different areas.

Plate 40 (Figure 13.3) The banteng form herds with several clans (cows with calfs and juveniles in Thailand); herds range from 2 to 30 individuals. Each herd is usually led by an old cow.

Plate 41 (Figure 13.4) Mature dominant bull in Malua Forest Reserve/Malua BioBank (Borneo), captured on camera trap in 2011 (© Danau Girang Field Centre/Sabah Wildlife Department using Reconyx PC800 HyperFire Professional Semi-Covert IR camera trap).

Plate 42 (Figure 12.1) A group of wild yak around Budongquan station, Kekexili Nature Reserve, western Qinghai, China (photo by D. He). For full caption text, see page [195].

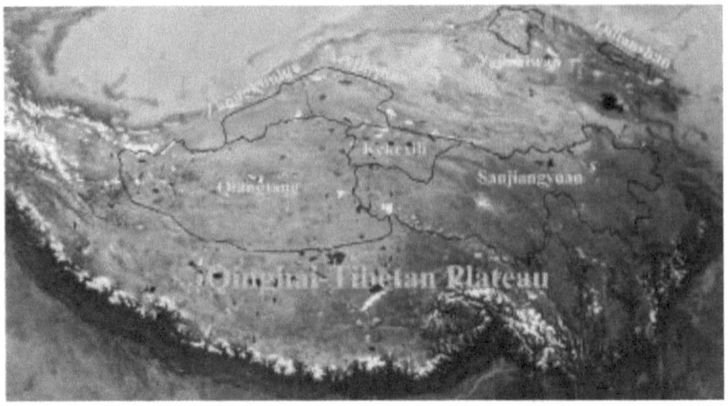

Plate 43 (Figure 12.4) Geographic localities and relative sizes of the six major nature reserves where the majority of wild yak are distributed in China.

Plate 44 (Figure 12.5) A wild yak bull with an atypical light coat in Qiangtang Nature Reserve. The photo was taken in Ritu, Tibet, China in March 2010 (photo by Y. Wang).

Plate 45 (Figure 12.6) A wild yak bull in Kekexili Nature Reserve, western Qinghai, China in late November 2009 (photo by D. He).

Plate 46 (Figure 10.21) A bull mates a cow during the rutting season at the European Bison Breeding Centre in Pszczyna (Poland) (photo by M. Hławiczka).

Plate 47 (Figure 11.3) Adult male gaur with conspicuous dorsal ridge and dewlap – features that distinguish males from females (photo by R. Sreenivasan, location: Valparai, India).

Plate 48 (Figure 11.9) Adult females in a gaur herd (photo by R. Sreenivasan, location: Bandipur, India).

Plate 49 (Figure 11.10) Gaur herd with adult females and calves (photo by M. N. Naveen, location: Bandipur, India).

Plate 50 (Figure 11.4) Montane evergreen grassland-shola habitat (photo by R. Rao, location: Kudremukh National Park, India).

Plate 51 (Figure 10.18) Young bull imported from Sweden and fitted with a radio-collar after its release at Bieszczady Mountains (photo by K. Perzanowski).

Plate 52 (Figure 10.16) Fitting of the radio-collar to a bison cow, Białowieża Forest, winter 2005 (photo by T. Kamiński). For full caption text, see page [140].

Plate 53 (Figure 10.35) Herd in the European Bison Breeding Centre at Pszczyna (Poland) (photo by M. Hławiczka).

Plate 54 (Figure 10.36) Group of European bison foraging at a forest clearing during the rut period; on the left is a radio-collared bull, September 2011 (photo by T. Kamiński).

Plate 55 (Figure 11.1) Adult male gaur (photo by R. Sreenivasan, location: Nagarahole, India).

Plate 56 (Figure 10.15) Adult bulls foraging on the meadow outside of Białowieża Forest, March 2011 (photo by T. Kamiński).

Plate 57 (Figure 11.2) Adult female gaur (photo by R. Sreenivasan, location: Nagarahole, India)

(a)

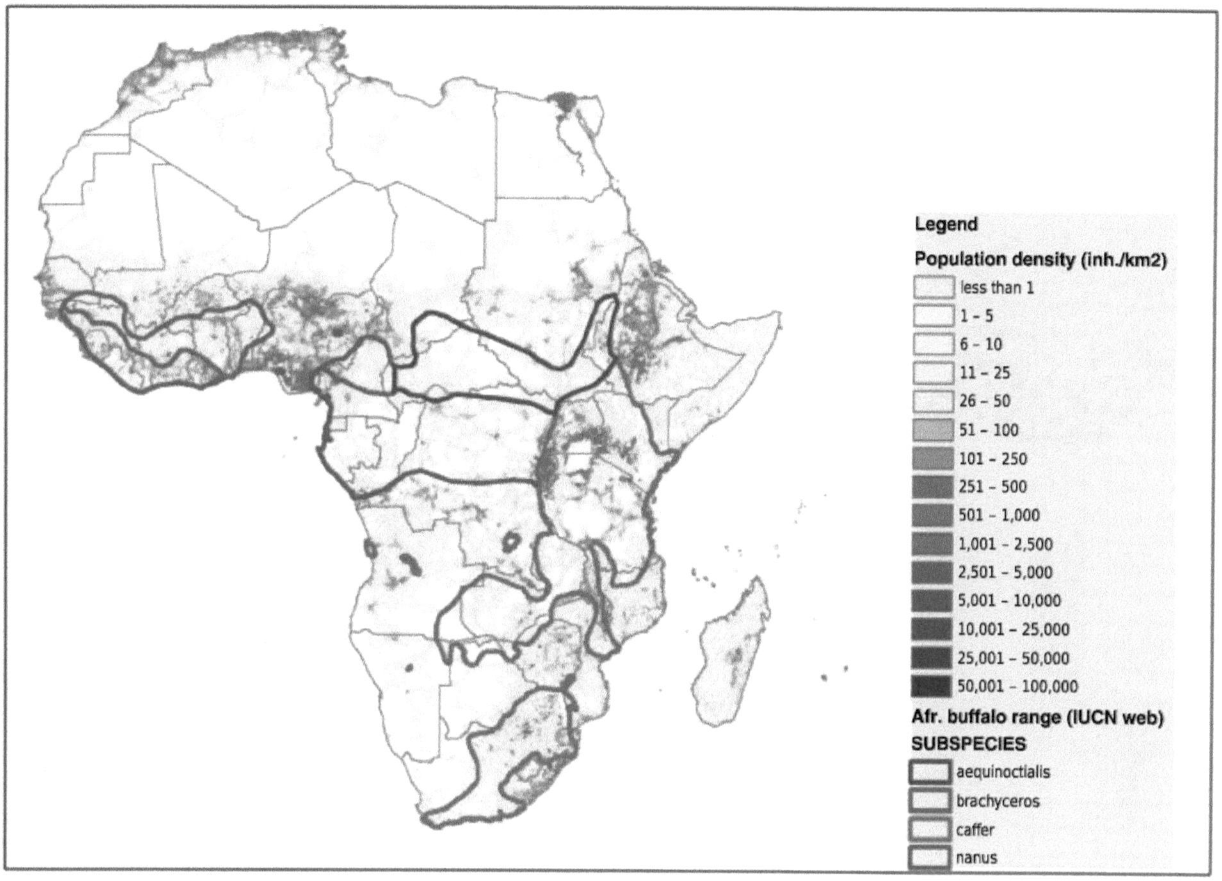

(b)

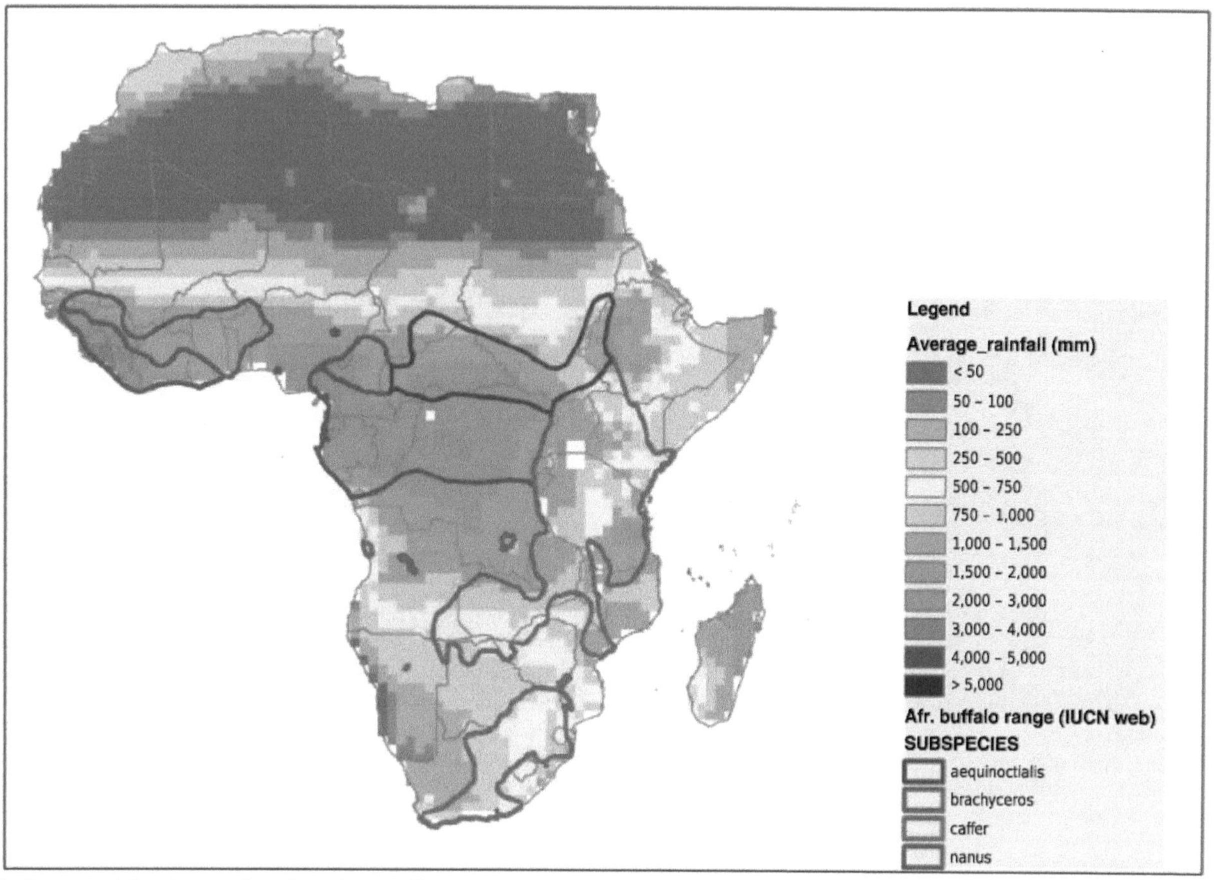

Plate 58 (Figure 20.1) Distribution of the four African buffalo subspecies in relation to: (a) human population density; (b) average rainfall and (c) protected areas (source: IUCN, 2008, *Red List of Threatened Species*).

(c)

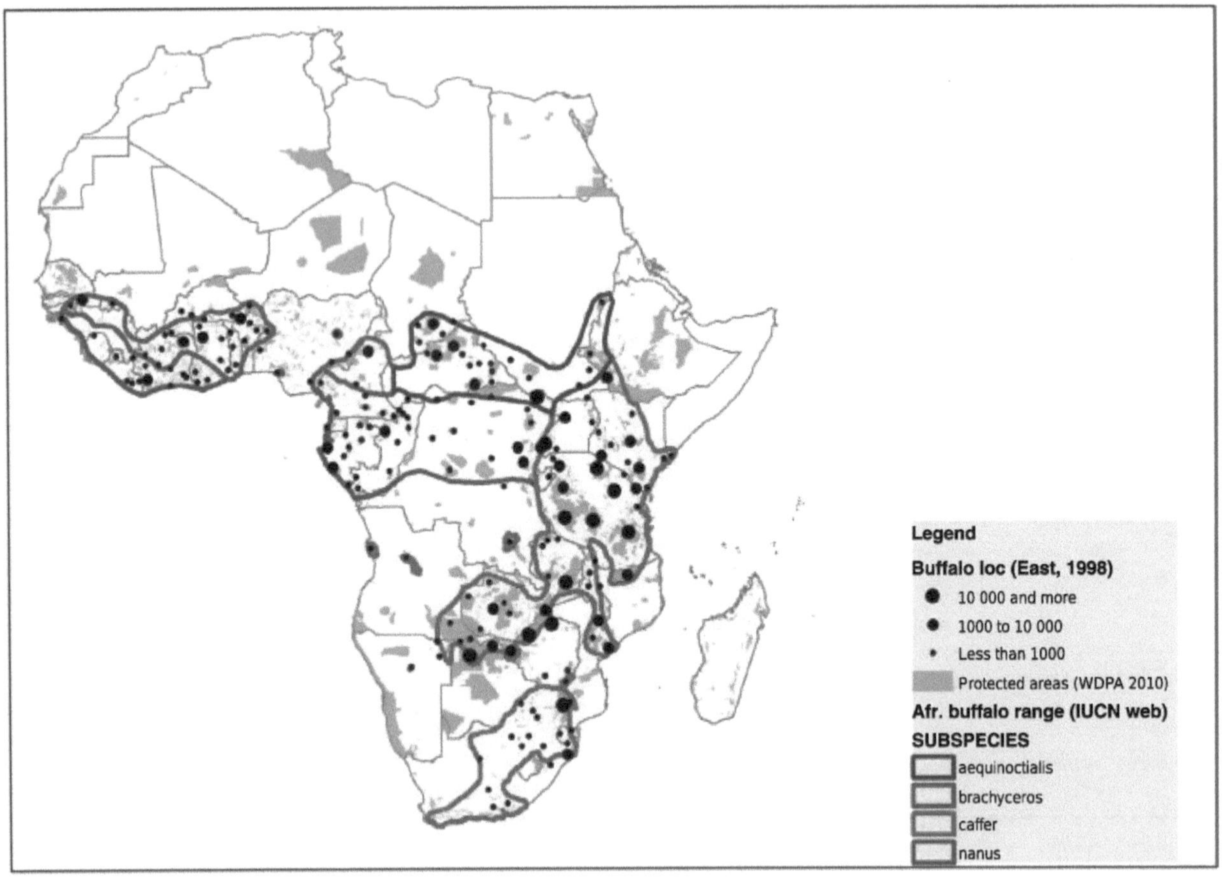

Plate 58 *(cont.)*

(a)

(b)

Plate 59 (Figure 20.2) Differences in body size, horns shape and size of the four African buffalo subspecies; (a) Cape buffalo male; (b) forest buffalo male; (c) West African savanna buffalo male; (d) Central African savanna buffalo male; (e) forest buffalo female; (f) male horns of transitional 'buffalo' from North Angola; (g) horns of Cape and West African savanna buffalo males. (a and g: photo by M. Melletti; b and e: photo by A. Turkalo; c and d: photo by D. Cornèlis; (f) photo by C. Chiarelli).

(c)

(d)

(e)

(f)

(g)

Plate 59 *(cont.)*

Plate 60 (Figure 19.4) Forest stream Au Rong: saola habitat, Quang Nam, Vietnam (source: J. Holden/WWF).

Plate 61 (Figure 19.6) Ka Tu hunter with a saola skull (source: J. holden/WWF).

Plate 62 (Figure 19.5) Tu village in the heart of the saola's range, Quang Nam, Vietnam (source: M. Hoang/WWF).

Plate 63 (Figure 19.7) A captive saola in Vietnam (source: T. Fey/WWF).

Plate 64 (Figure 23.4) Excitement and poor coordination during drug induction (photo by P. Nigam, WII).

Plate 65 (Figure 23.5) Extension of head during drug induction (photo by P. Nigam, WII).

Plate 66 (Figure 23.9) Gaur herd at Bandhavgarh Tiger Reserve after release in the wild (photo by B. Navaneethan, WII).

Plate 67 (Figure 24.1) A river buffalo from Rajasthan (photo by A. Choudhury).

Plate 68 (Figure 24.4) Putative wild water buffalo (*Bubalus arnee*) (courtesy of Port Lympne Wild Animal Park).

Plate 69 (Figure 24.2) Swamp buffalo herd in Sulawesi, Indonesia (photo by J. S. F. Barker).

Plate 70 (Figure 23.1) Gaur herd being moved into open areas employing captive elephants (photo by U.K. Sinha, WII).

Plate 71 (Figure 23.2) Identified gaur darted within the herd (photo by U.K. Sinha, WII).

Plate 72 (Figure 13.5) Tropical lowland dipterocarp forest is the principal habitat type of banteng in Sabah, Borneo (photo by P. C. Gardner).

Plate 73 (Figure 15.7) One of the crossings of the back-breeding experiment in the Lippe area (Germany). This bull is half-Sayaguesa and half-Heck (photo by M. Bunzel-Drüke).

Plate 74 (Figure 16.3) A wild water buffalo of Kollolua-Jokai area, Assam.

Plate 75 (Figure 16.4) A bull and some cows in Kaziranga National Park, Assam.

Plate 76 (Figure 26.1) African Buffalo adult female infected with rinderpest virus in Nairobi National Park 1996 (photo by R. Kock). For full caption text, see page [433].

Plate 77 (Figure 27.2) African buffalo (*Syncerus caffer caffer*) held in Safaripark Beekse Bergen – Hilvarenbeek (photo by R. Veenhuizen, Safaripark Beekse Bergen, Hilvarenbeek).

Plate 78 (Figure 15.8) Maremmana bull. This is one of the cattle breeds used for the back-breeding experiment in the 'TaurOs Project' (photo by M. Di Luca).

Plate 79 (Figure 27.5) Re-created aurochs held at Gaiazoo Kerkrade, the Netherlands (photo by G. Kerkrade).

Plate 80 (Figure 27.3) European bison (*Bison bonasus*) held in Dierenpark Planckendael – Mechelen (photo by H. Schram, Antwerp).

Plate 81 (Figure 12.8) Semi-wild yak bulls in Zhiduo county of western Qinghai, China in August 2012 (photo by D. He).

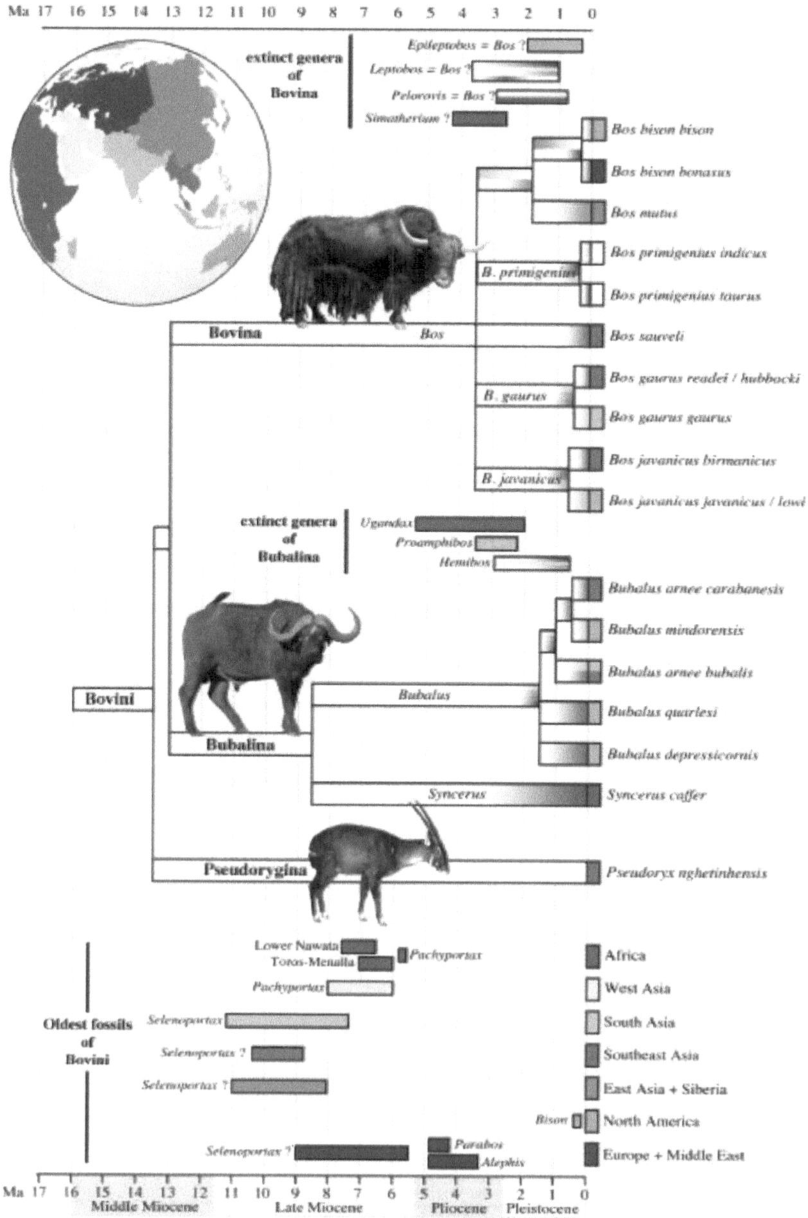

Plate 82 (Figure 1.4) Timetree of Bovini (photo of saola by W. G. Robichaud). For full caption text, see page [16].

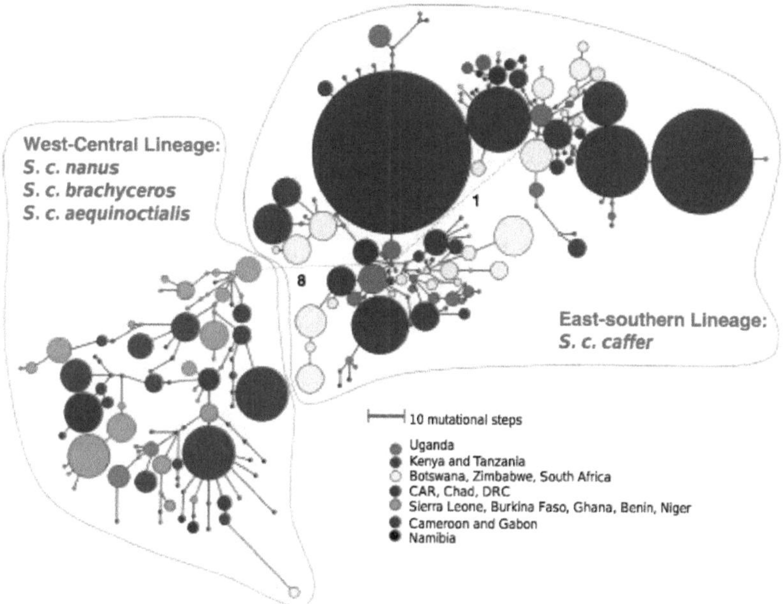

Plate 83 (Figure 25.2) Figure extracted from Smitz *et al.* 2013, based on 787 samples (D-loop), representing an unrooted minimum spanning tree of *Syncerus* sp. haplotypes. For full caption text, see page [415].

Plate 84 (Figure 9.1) Original home range, light brown, and dates of estimated local exterminations of plains bison and wood bison. For full caption text, see page [86].

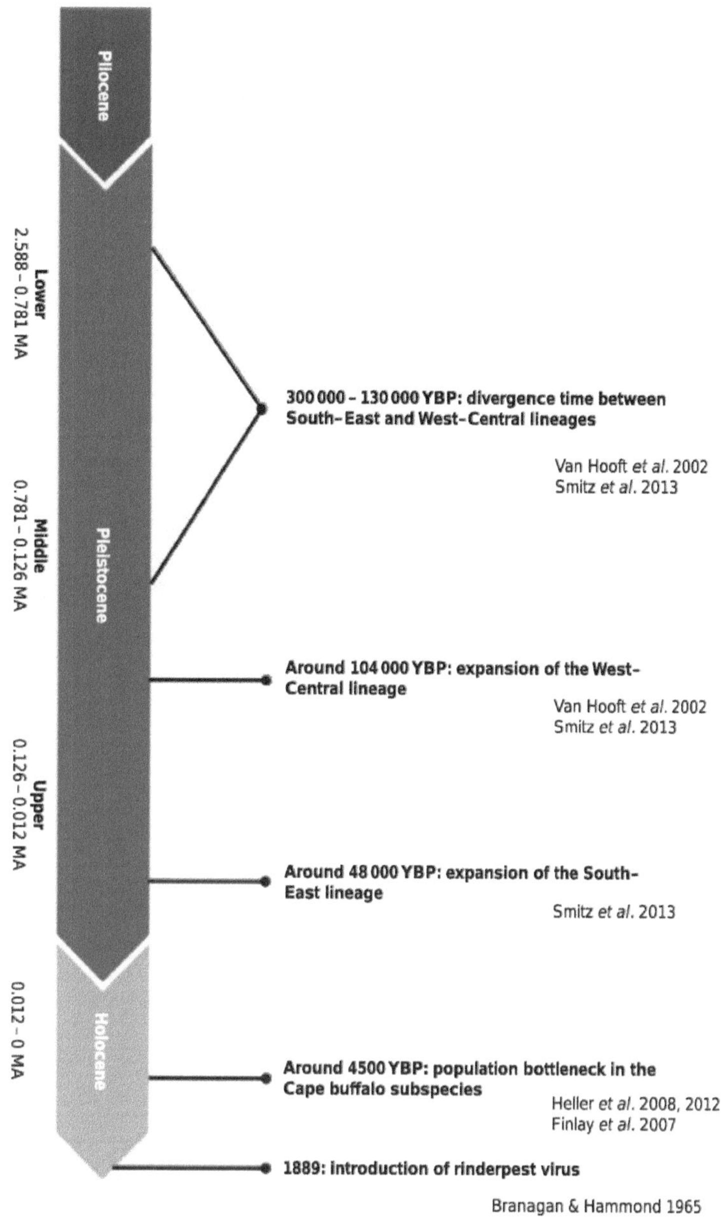

Plate 85 (Figure 25.3) Time scale representing all major evolutionary events in the history of the African buffalo.

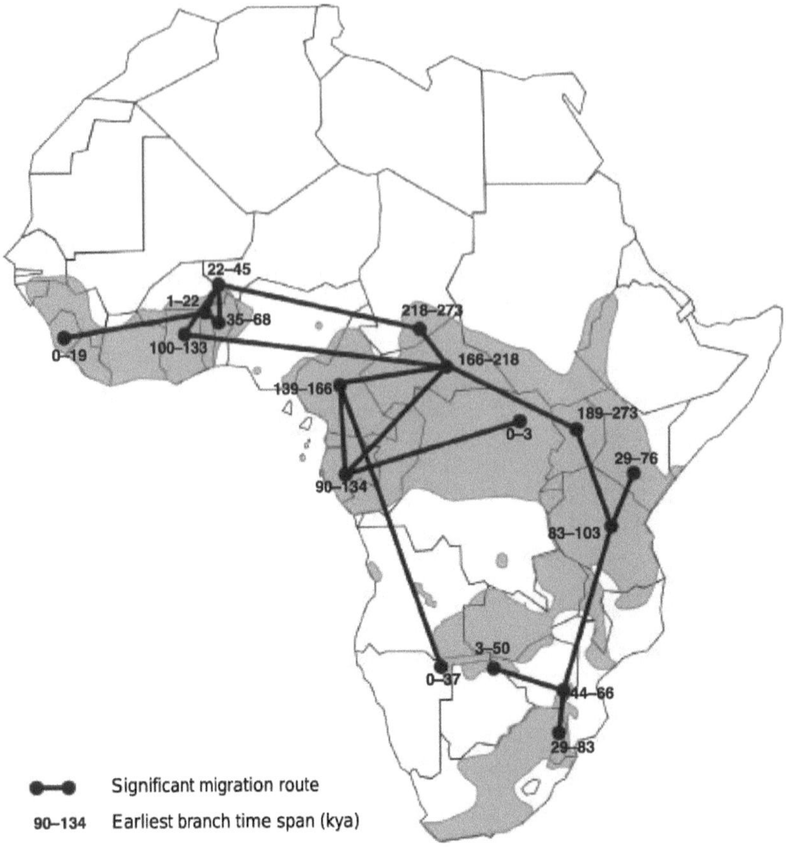

Plate 86 (Figure 25.4) Figure extracted from Smitz *et al.* 2013, based on 787 samples (D-loop), representing a map of the African continent showing the 17 historical migration routes between sampled localities supported by a Bayes factor >3. For full caption text, see page [420].

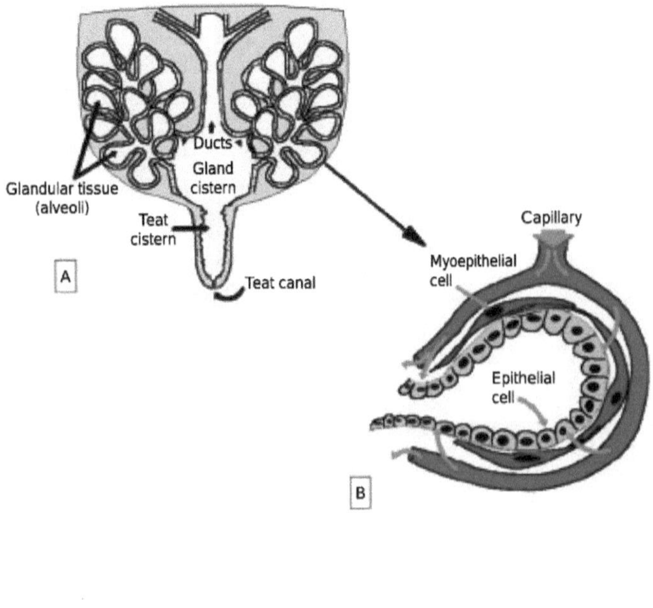

Plate 87 (Figure 5.3) A stylized view (a and b) of the arrangement of lobules of alveoli with major mammary ducts, the gland and teat cistern, as well as the arrangement cells within an alveolus and blood supply (with permission from Nickerson and Akers, 2011).

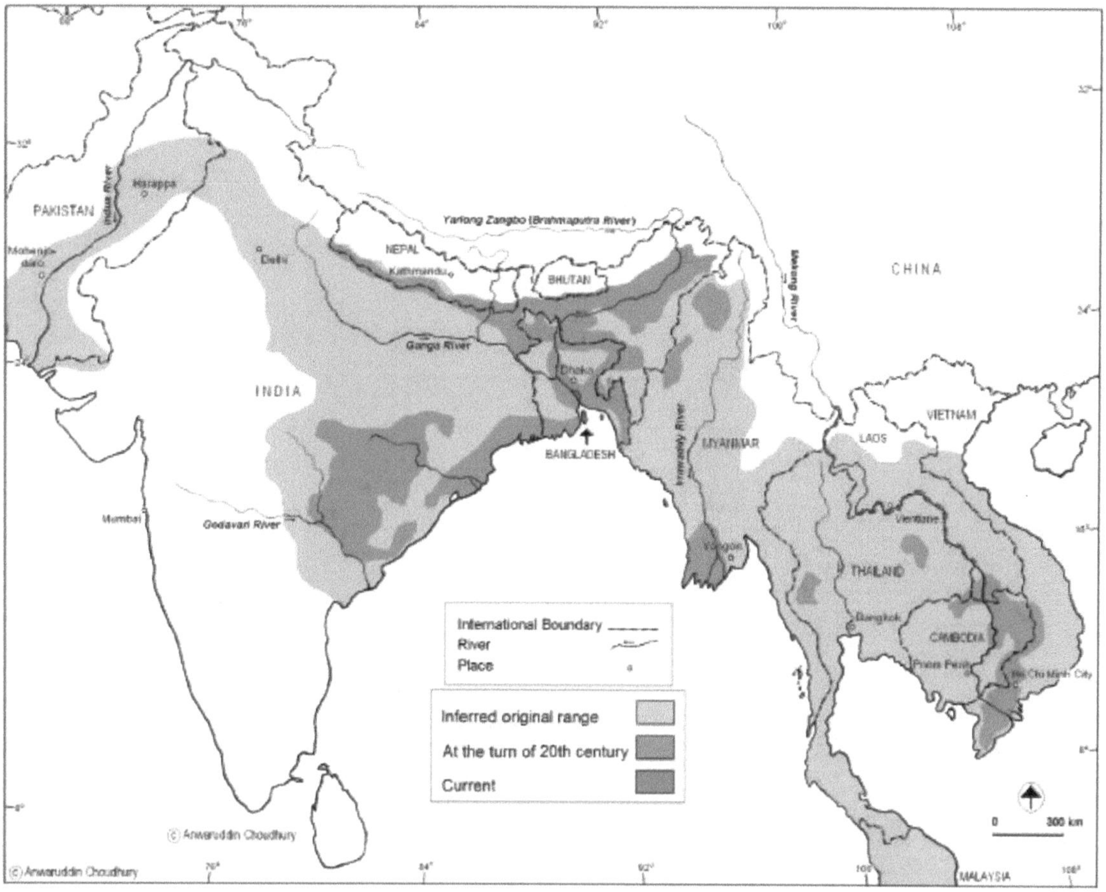

Plate 88 (Figure 16.1)
Past and present range of the wild water buffalo.

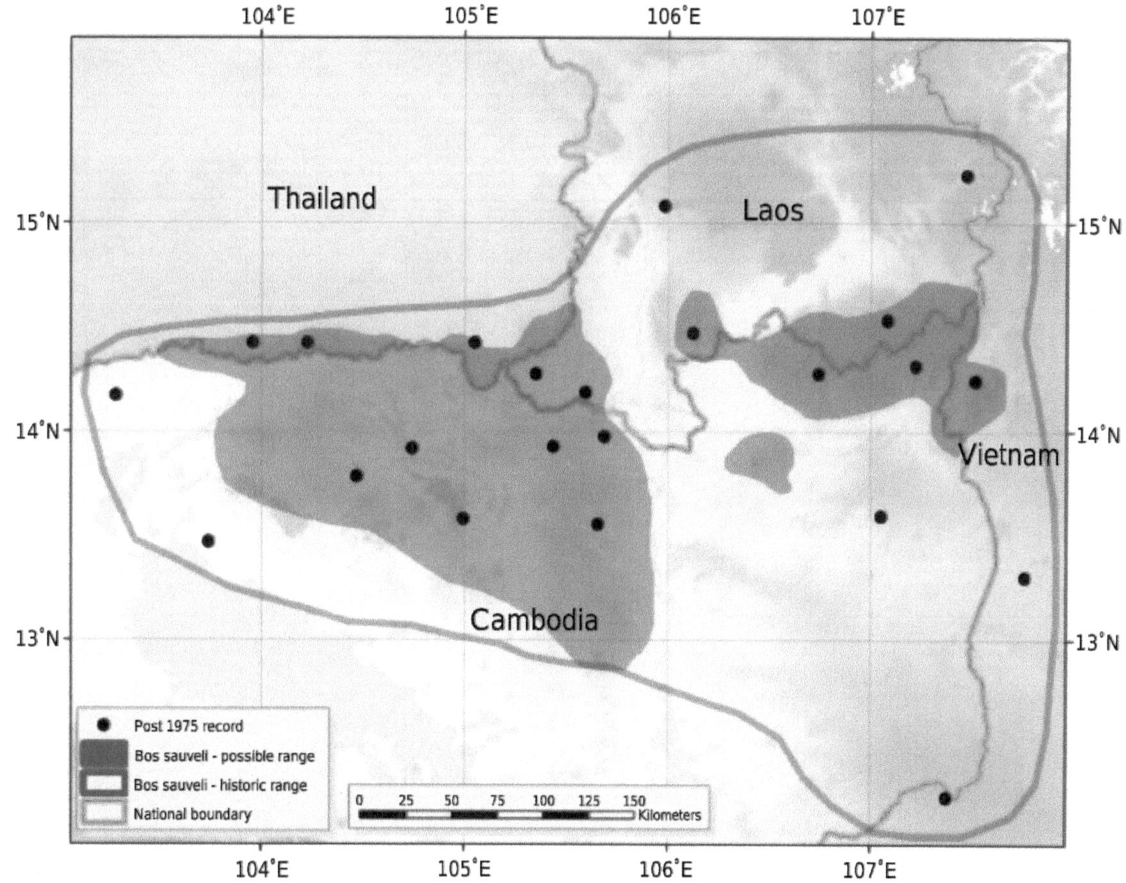

Plate 89 (Figure 14.2)
A zoomed-in map showing the possible range of kouprey, along with post-1975 records from villagers and other sources (adapted from MacKinnon & Stuart, 1989).

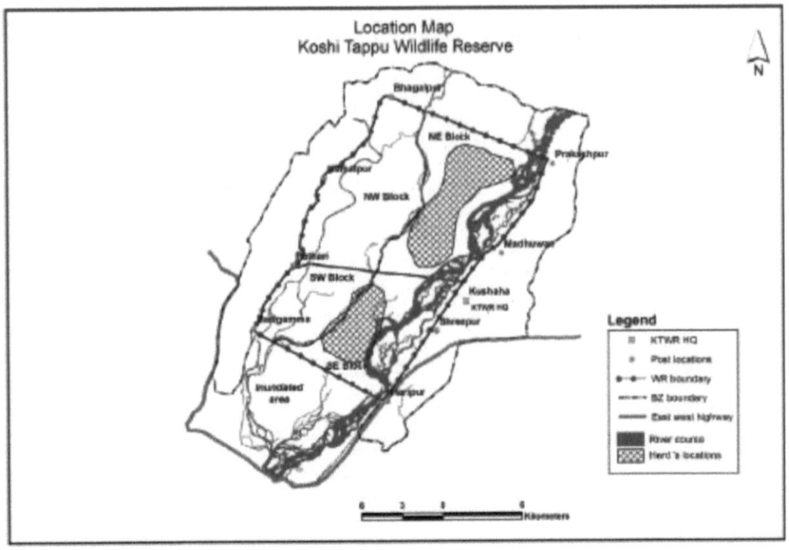

Plate 90 (Figure 24.3) Map of Koshi Tappu Wildlife Reserve Nepal.

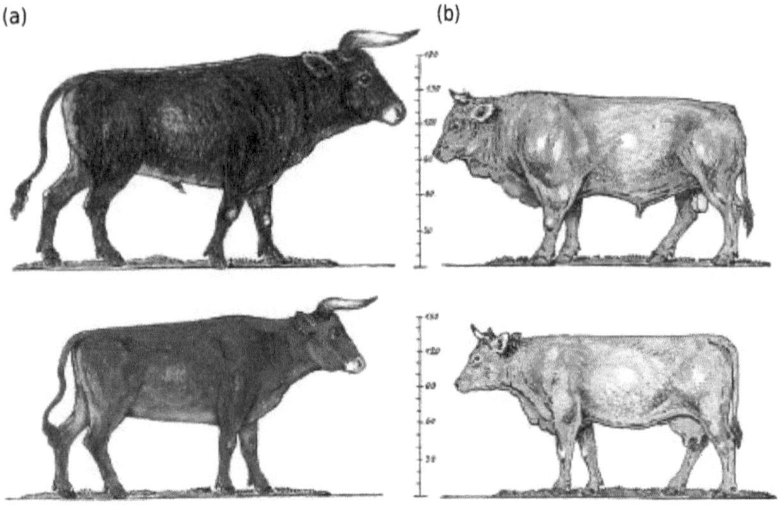

Plate 91 (Figure 15.2) Comparison of bulls and cows of the aurochs (a) and modern cattle (b) (after Boessneck 1957, modified and coloured). For full caption text, see page [243].

Figure 16.7 Such pools inside the forest provide much-needed water for drinking and also a site for wallow (seen here is a pool in Buxa Tiger Reserve, northern West Bengal).

additional and conspicuous advantage of wallowing is the thick coat of mud it leaves on the body, which protects the buffaloes from insect bites for short periods of time. After leaving a wallow, some animals, especially the adults, rub their body along a tree. In drier Peninsular Indian forests, drinking water from village tanks was a common phenomenon and was reported a long time ago by Inverarity (1895).

In the wet (monsoon rains) season (May–September), in floodplains such as Kaziranga, the bulk of the park reels under flood water and the normal activity pattern changes completely and becomes unpredictable.

Ryan & Jordaan (2005) in their analysis of diurnal activity of three herds of African buffalo, found that significantly more time was devoted to feeding than other activities. The herds spent an average of 37.5% or 4.5 hours of the day actively feeding, and another 28.8% (approx. 3.5 hours) standing and ruminating; resting took up another 25.9% (approx. 3.1 hours) of the day. This shows that 75%, or nine hours, of the day is devoted to obtaining and ruminating food.

When they compared diurnal and nocturnal components of the herd activity budget, they found that the buffalo spent a significantly greater proportion of time feeding during the night (44.5% vs 32.0%); and a trend towards a greater proportion of time resting during the day (28.4% vs 16.0%). They also observed that under normal circumstances the difference between day and night feeding and resting activities was not statistically significant.

The feeding time of 9.5 hours (39.41%) out of the 24-hour activity budget in Ryan & Jordaan (2005) also falls within the range for grazing found by both Grimsdell & Field (1976) (9 hours) and Sinclair's (1977) bachelor bull study in the Serengeti (9.7–10.1 hours). This compares with the 40% estimated by Winterbach & Bothma (1998).

Like other ruminants, especially bovids, the wild buffalo ruminates while resting or standing, while also displaying vigilant behaviour. This mainly occurs in bulls if they are with herds, and also adult cows. Such behaviour is exhibited by 'raised head vigilance'. According to Ryan & Jordaan (2005) it is plausible that buffaloes are not only using visual cues, but also olfactory. They further opined that this activity occurs in greater proportion at night, although this is not statistically significant. They also posit that with a reduction in visibility at night more vigilance is necessary. Their study further suggested that vigilant behaviour may comprise a significant proportion of buffalo activity.

Resting includes standing, wallow and lying down on the ground. During the hot hours of the day wild buffaloes tended to rest more, and especially wallow. If rest takes place in woodland, it is typically by standing or lying down on the ground. Except for central India, there is water everywhere in the known buffalo habitats such as Kaziranga; hence, it is not clear how often they drink each day. In such a high density of available water sources, the buffaloes may drink many times in a day.

Feeding ecology

Published information on the diet of wild buffaloes is scarce. Although they are mainly grazers (faecal analysis showed over 90% of their food coming from grass; Chaiyarat 2002), their feeding on herbs and stem as well as browsing trees and shrubs has also been observed. In Assam, the exotic water hyacinth (*Eichhornia crassipes*) has also become part of its diet. Daniel & Grubh (1966) saw buffaloes feeding on the sedge *Cyperus corymbosus*. Kotwal & Mishra (2004) recorded the following food plants in Udanti Wildlife Sanctuary: saplings of *Shorea robusta* (Sal), *Pterocarpus marsupium* and *Bridelia retusa*; cultivated crops such as 'kulthi' (*Dolichhos biflorus*) and mustard (*Brassica campestris*). While raiding cultivations, crops such as rice (*Oryza sativa*), sugar cane (*Saccharum officinarum*), jute (*Corchorus capsularis*), pulses (a favourite is 'Mati-mah' or 'urad', *Vigna mungo*) and vegetables are taken. In Assam, crops of rice, sugar cane and jute, which are all different types of grasses and pulses, are often severely damaged in the fringe of buffalo habitats. After the burning of grassland in Kaziranga and Manas, the buffaloes were observed feeding on tender sprouts of several elephant-grass species. Tender leaves of *Alpinia allughas* (Tora) herb are also taken. Chaiyarat (2002) recorded leaves and stems of several herbs, climbers and shrubs in the diet of wild buffalo in Huai Kha Khaeng Wildlife Sanctuary. Three dominant species of grass – *Arundo donax*, *Saccharum arundinaceum* and *S. spontaneum* – which alone account for 45.4% of buffaloes' diet in Thailand (Chaiyarat 2002), are also the key items in Assam (Appendix 16.2 lists 102 forage species from across its range).

Eisenberg & Lockhart (1972) mentioned that the wild buffalo has a higher tolerance for saline water than many other mammals and will drink from brackish pools. The former occurrence of wild buffalo in the Sunderbans of India and Bangladesh on the Bay of Bengal is consistent with this.

Wild buffaloes also visit salt licks and other mineral licks wherever these are available, especially in Manas National Park in Assam and Royal Manas National Park in Bhutan. In Assam's Kaziranga and Dibru-Saikhowa, wild bulls regularly take salt in the *khuties* (professional buffalo and cattle camps).

Regarding feeding time, see the section on activity patterns. Crop raiding is mostly at night. While grazing, the animals may also split into smaller subgroups. A grazing herd in Kaziranga may be scattered over up to 400 m (as observed), while in Dibru-Saikhowa National Park and Burhachapori Wildlife Sanctuary it is as little as 150 m. In unprotected areas such as Na-pam Thangaon near Jokai Reserve Forest, the animals remain as a close-knit herd unless split up into smaller groups (Choudhury 2010).

Interaction with other species

The wild buffalo generally associates with several other herbivores without any agonistic behaviour. In Assam, grazing and wallowing with great Indian one-horned rhinoceros (*Rhinoceros unicornis*) and Asian elephants (*Elephas maximus*) has been observed in Laokhowa Wildlife Sanctuary and Kaziranga National Park. Peaceful association and grazing with hog deer (*Axis porcinus*) and swamp deer (*Cervus duvaucelii*) in Kaziranga, and with hog deer and gaur (*Bos gaurus*) in Manas, have been observed regularly. Wild pigs (*Sus scrofa*) are also seen near buffaloes without any agonistic behaviour. Waterfowl also freely move around, while several species of birds, mainly cattle egret (*Bubulcus ibis*) and several species of mynas (Sturnidae), ride on the buffaloes. Cattle egrets also follow buffalo herds looking for insects disturbed by them. Forsyth (1889) mentioned a bull nilgai (*Boselaphus tragocamelus*), a type of antelope, along with a large buffalo herd in Orissa, which maintained a vigilant outlook all around. Kotwal & Mishra (2004) also found a nilgai grazing with a herd of buffaloes in Udanti Wildlife Sanctuary. The relation was such that the nilgai waited in the shade of trees when the buffaloes wallowed in a pond. They also observed a herd of chital (*Axis axis*) grazing with a solitary bull buffalo.

Mathur *et al.* (1995) indicated that there is probably no direct competition between the wild buffalo and the four other wild herbivores (elephant, rhino, swamp deer and hog deer) that are common in Kaziranga National Park. While competition with wild herbivores is still debated with no detailed quantitative data (de Silva *et al.* 1994; Mathur *et al.* 1995), there is apparent competition from domestic cattle and buffaloes that graze in the marginal areas of reserves (often going up to 5 km or more inside), such as Kaziranga, Manas and Dibru-Saikhowa, and almost all over in areas such as Koshi Tappu, Laokhowa and Burhachapori. In the drier areas, such as Chhattisgarh and Ruhuna National Park in Sri Lanka, the buffaloes compete with several other wild herbivores for water.

Reproduction and growth

The wild buffalo exhibits a polygynous mating system, with females typically giving birth to single offspring, although twins are possible (at least three cases were recorded by this author in Kaziranga National Park). Tulloch (1979) observed, among the feral animals in N Australia, that if mothers die the calves can be adopted by other females which already have calves. The buffalo is largely a seasonal breeder, typically in October–January, i.e. after the monsoon rains; however, some populations may breed year round (IUCN 2013). The gestation period is 10–11 months, with an interbirth interval of 1–2 years. In cases of premature death of calves due to predation or other reasons, the interval becomes shorter. Weaning occurs after 6–9 months. At about six months of age, calves develop reddish-brown coats and horns start appearing. The males reach sexual maturity at 18 months, while females do so at 2–3 years. The first calving occurs from three to four years of age. However, in the wild, a young male has little or no chance of access to females. Younger males of around 3.5–4 years of age were observed with domestic females in the fringe areas of the buffalo habitat, such as Dibru-Saikhowa National Park and Bolai village. In Nepal also, the peak of mating activity seems to occur in October–November (Anon 2009).

In Kaziranga National Park, several newborn calves were observed in February–March, and in Manas National Park in March–May, but some have also been observed during other months of the year, except during the monsoon (June–September). Kaziranga is normally flooded from May onwards (peak in July–August; occasionally up to October) and no newly born calves were observed in April–May. On the contrary, in Manas National Park, which is located at the edge of the Himalayan foothills, there is no flood like the floodplains in Kaziranga and newborns were seen even in April–May. Inverarity (1895) reported that the wild buffaloes in Bastar produce their calves in March–May, but he believed that at other times of the year there could also be calving. The seasonality seems to be owing to flood risk in Assam, while it is the harshness of the dry season in central India.

De Silva *et al.* (1994) mentioned that in the feral buffaloes of Sri Lanka, a female may produce up to five calves in her life. Calves are born at the end of the rainy season and start of the dry season, when grasses are luxuriant and water is still abundant, to maximise the chances of survival.

Only the dominant bull mates with the female buffaloes of a herd. The same bull may also visit a *khuti* if it is nearby to mate with the females there. Normally one bull can be seen in a wild herd during the peak breeding season (November–January), but a second male may also be seen in a *khuti*, where the number of females is usually large and they are kept in different sheds or open enclosures. These bulls avail themselves of opportunities to breed with as many females as possible during a season.

Behaviour

Social organisation

Wild buffaloes live in herds usually of 5–20 animals; occasionally gatherings of more than 40 have been seen. Some bulls move about alone and are a common feature. A typical herd comprises a lead female, a few adult females, subadult males and females with their calves. Adult males may or may not be seen with the herds. If an adult male remains with a herd during the oestrus period of females, he becomes the dominant animal of the herd. He also ensures that no other male can approach the females. The normal group/herd size varies from 3 to 30 ($n = 65$; also see Table 16.4). The duos are not considered a group here. The mean of 65 groups/herds is 12.9. Of the 65 samples, almost half had 10–19 animals, followed by 24 cases having fewer than nine animals, while only nine had more than 20 animals (Figure 16.6). Congregations (temporary aggregations, usually forced by external circumstances and lacking the stability of 'groups') are rare, but during high floods large numbers (more than 50) could be seen on higher grounds. In 1989 42 buffaloes were seen together in an open river bank in Kadam Reserve Forest, Assam, where the species is now extinct. This has not been treated here as a group/herd but as a congregation (Choudhury 2010). According to Prater (1948), they associate in small herds which may combine to form large assemblages. Herd size in Huai Kha Khaeng Wildlife Sanctuary has been reported to be 2–19 (Chaiyarat 2001), although two cannot be called a herd (because duos are normally temporary associations or part of a herd whose other members were not seen at the time of sighting). Tulloch (1970, 1979) surmised that the females stay within their natal herds for many years. There is no evidence or concrete records of herds or assemblages of 100 animals, although there are local reports of such. Perhaps such gatherings do not happen except in dry areas where water sources are limited (central India) or flood-prone areas with few patches of higher grounds (Brahmaputra floodplains in Assam). The group/herd size in Koshi Tappu Wildlife Reserve typically contained fewer than 40 animals (Dahmer 1978; Heinen 1993).

Table 16.4 Herd size and composition in wild buffaloes (all in Assam)

Date	Place	AM	AF	S/J	C	Total
1987	Kakoi Reserve Forest	1	3	1		5
1988	Borkolia-Dakhingaon, Dhakuakhana	1	2	1	1	5
26 October 1990	Borkolia-Dakhingaon, Dhakuakhana		2	2		4
23 February 1991	Borkolia-Dakhingaon, Dhakuakhana		2	2	1	5
9 May 1991	Patradisa RF	1	1		1	3
10 March 1996	Kapurpura, Manas NP	1	4			5
3 December 2000	Laodubi, Kaziranga NP		8	8	2	18
24 March 2001	Pohufield, Manas NP	1	5	5	4	15
19 March 2008	Sohola, Kaziranga NP		4	3	2	9
27 March 2008	Kathpora, Kaziranga NP	1	11	6	4	22
27 March 2008	Dafflong, Kaziranga NP	1	11	6	4	22
27 March 2008	Block 2-B, Eastern Range, Kaziranga NP		2	2	2	6
27 March 2008	Block 7-A, Eastern Range, Kaziranga NP	1	9	7	6	23
27 March 2008	Block 8-A, Eastern Range, Kaziranga NP	1	8	6	6	21
27 March 2008	Block 9-B, Kaziranga Range, Kaziranga NP	1	4		2	7
27 March 2008	Block 18-B, Kaziranga Range, Kaziranga NP	1	7	5	1	14
27 March 2008	Block 20-A, Kaziranga Range, Kaziranga NP		3		3	6
27 March 2008	Block 26-C, Western Range, Kaziranga NP		5	4	1	10
16 April 2008	Buraburijhar, Manas NP		5	7	3	15
16 April 2008	Chorfuli, Manas NP	2	4	5	1	12
17 April 2008	Latajhar, Manas NP	1	4	8	3	16
17 April 2008	Mathanguri, Manas NP	1	3	7		11
17 April 2008	Dighaljhar, Manas NP	1	4	10		15
17 April 2008	Samajhora, Manas NP	1	4		2	7
17 April 2008	Tangonmara, Manas NP	1	2		2	5
17 April 2008	Sukhanjan-Doimukh, Manas NP	2	3	7		12
17 April 2008	Palangshi Beel, Manas NP	1	5	5	2	13
22 April 2008	Goldubi, Laokhowa Wildlife Sanctuary		3	3		6
22 April 2008	Palashtoli, Laokhowa Wildlife Sanctuary	1	4	3	4	12

Table 16.4 *(cont.)*

Date	Place	AM	AF	S/J	C	Total
22 April 2008	Goldubi, Burhachapori Wildlife Sanctuary		2	1	2	5
22 April 2008	Basabari, Burhachapori Wildlife Sanctuary		10	2	2	14
22 April 2008	Lonke *Tapu*		5	7	4	16
22 April 2008	Lonke *Tapu*		8	4	4	16
24 November 2008	Na-pam Thangaon		10	3	4	17
19 December 2008	Bihguri, Dibru-Saikhowa NP		4	9	3	16
19 December 2008	Tilakbam, Dibru-Saikhowa NP	1	5	7	2	15
19 December 2008	Burhidangori, Dibru-Saikhowa NP	1	9	2	4	16
19 December 2008	Bordubi, Dibru-Saikhowa NP		9	3	7	19
19 December 2008	Bokabil/Kadamoni, Dibru-Saikhowa NP	1	4	3	1	9
20 December 2008	Hollokbari, Dibru-Saikhowa NP		5	3	3	11
20 December 2008	Rongdoi, Dibru-Saikhowa NP		4	2	2	8
20 December 2008	Pulibari, Dibru-Saikhowa NP	1	5	6	2	14
25 January 2009	Sohola, Kaziranga NP	1	7	3	6	17
		27	**219**	**168**	**103**	**517**

AM = adult male; AF = adult female; S/J = subadult/juvenile; C = calf.
Range = 3–23.
NP = National Park; WS = Wildlife Sanctuary; RF = Reserve Forest
Source: Choudhury (2010).

The analysis of census figures obtained for 2008 (Table 16.5) revealed that almost 73.9% (range: 63.2–75.5%) of sexed adults were females, while nearly 17.5% (range: 11.2–21.3%) of the total population was represented by calves. Tulloch (1970) and de Silva *et al.* (1994) observed that the dynamics of wild buffalo populations (they observed feral animals) can severely change during prolonged periods of drought because they are primarily grazers.

As far as the growth rate is concerned, there are regular censuses/estimations from only two sites: Koshi Tappu in Nepal and Kaziranga in Assam. In the former area, the growth rate between 1988 and 2004 was 4.12 buffaloes per year, while in the latter area it was 61.9 between 1993 and 2008. The increase was about 4.43% per year in Koshi Tappu since 1988 and 5.99% per year in Kaziranga since 1993.

Herd and individual behaviour

The herd remains closer when resting and travelling. During grazing they spread out. Mathur *et al.* (1995) recorded an interesting case of social bonds in Kaziranga National Park. When one of the herd members was chemically immobilised for collection of a blood sample, the entire herd remained present at a distance until the immobilised buffalo revived and could join the herd. Male wild buffaloes leave the herds as they mature. Adult and subadult males often form smaller 'bachelor' herds. Older males are often solitary, while such males as well as stronger males also visit domestic females in the fringe villages and *khuties*. It is not always that those males chased away from herds by stronger males visit domestic females. In most cases it is the stronger males which visit both wild and domestic herds, while the weaker ones wait for their chance. In many of the larger *khuties*, two bulls have also been seen, but they maintain a safe distance from one another to avoid direct confrontation.

Solitary bulls have been generally found to be of the following categories: (1) senescent bulls, which are no longer attached to any herd and even cannot visit domestic females; (2) dominant adult bulls, which move alone but attend herds whenever any

Table 16.5 Proportion of various age groups in the population of wild buffalo

	% of adults to total pop.	% of calves to total pop.	% of AM to total pop.	% of AM to total sexed adults	% of AF to total pop.	% of AF to total sexed adults	Sex ratio	% of AF with calves to total F pop.
Kaziranga	64.4	18	15.5	24.5	47.8	75.5	1:3.1	37.6
Manas	45.6	11.2	15.3	35.9	27.4	64.1	1:1.8	40.7
Laokhowa and Burhachapori	46.7	21.3	9.3	20	37.3	80	1:4	57.1
Dibru-Saikhowa	50.8	18	15.4	34	29.9	66	1:1.9	60.2
Other areas of Assam	57.6	15.1	21.2	36.8	36.4	63.2	1:7	41.7
Total	**60.6**	**17.5**	**15.4**	**26.1**	**43.5**	**73.9**	**1:2.8**	**40.2**

Proportion of adult males and females to total population is imprecise as a few adults were listed as 'unsexed', i.e. sex could not be determined during census/observation. AM = adult male; AF = adult female; pop.= population.
Source: Choudhury (2010).

female is in a receptive condition; (3) bulls ousted by another, stronger bull from a herd; and (4) subadult bulls, which have the scope to attend females, especially in *khuties*, and move alone and visit at appropriate times (two such cases were recorded in November–December 2008, one in a *khuti* near Dibru-Saikhowa and another at a village near Jokai Reserve Forest).

Mother–calf behaviour

The females of a herd often keep their young ones in one part of an open area and move away for grazing (as observed in Kaziranga). While grazing, the cows may leave their calves in the care of another adult/adults and move away a few hundred metres.

Agonistic behaviour

The wild buffalo is aggressive in its behaviour; in particular, solitary bulls and cows with calves are dangerous and they very often charge people. This fact was mentioned in the sixteenth century by Babur (1529). Inverarity (1895), by contrast, mentioned that 'A herd of buffalo never charge, sometimes they bolt in the direction you are if they don't know where the shot comes from, and there is a slight chance of being run over. If they see you, they turn off.' A bull with females in oestrous aggressively prevents other bulls from approaching her. In the feral herd of Pabitora Wildlife Sanctuary, the lead bull (of wild origin) was once observed chasing another bull (also of wild origin but slightly smaller) for more than 3 km. Adult bulls also often fight during the mating season for dominance. However, the dominant bull may tolerate the presence of other younger bulls provided they do not approach the females in oestrous.

Sexual behaviour

The sexual behaviour in male wild buffaloes is somewhat similar to domestic buffaloes and also cattle bulls. The normal mating encompasses the following behavioural elements: courtship, erection and protrusion, mounting, intromission, ejaculatory thrust and ejaculation, and dismounting. The alpha bull detects the pro-oestrous cow before oestrus and remains in her general vicinity. He also keeps other low-ranking bulls away from the cow. During courtship, the bull nudges to prompt the female to move forward. It also provides reciprocal stimulation and, once mating has occurred, nudging is stopped. During monitoring by bulls, the females in oestrus are prompted to urinate by licking, nosing of the vulva and perineum. The cow in heat starts to urinate in response to such prompting by the bull. The normal response of the bull to the oestrus scent is through forward extending of neck and muzzle, and curling of the upper lip, exposing the gums and teeth. The bull will rest his chin on her rump and if the cow does not move forward, she is ready for mating and the bull will mount her. Each copulation is short but will take place several times. In a few cases it has been observed that low-ranking males also copulate with breeding cows once the alpha bull loses interest.

Vocal communication

The vocabulary of wild buffalo includes the grunt, with which the members of a herd communicate with each other when moving.

Antipredator behaviour

When a predator, especially a tiger, is nearby, the herd normally forms a line or a circle, with younger animals at the centre. At times they may also bolt, especially in tall grass and woodlands. The adults, especially bulls, raise their heads and look around, adopting an alert posture ('raised head vigilance'). There are instances of tigers being killed by bull buffaloes. Two cases have been recorded in Kaziranga National Park, the most recent being in January 2009.

Sexual segregation

Many consider sexual segregation as the differential use of space by the sexes (Kie & Bowyer 1999; Weckerly *et al.* 2001). There is no apparent habitat segregation among the wild buffaloes on a sexual basis. However, there is some spatial segregation where the subadult and low-ranking males use the peripheral areas. Social segregation is somewhat more conspicuous. The subadults often form bachelor herds. The dominant males generally remain with the herds when there is a female in heat. Low-ranking adult bulls move about alone. In the *khuties*, where a large number of females are kept and where chances of getting cows in heat are high, bulls of all age groups from subadult upwards are seen. However, subadults move away when any adult bull arrives, particularly the alpha bull, who is given a wide berth.

General behaviour

Prater (1948) mentioned that scent is keenly developed in the wild buffalo; they also have good hearing. However, their sight is moderate.

As a rule, the wild buffalo is not shy, and in areas such as Kaziranga National Park, with a long-running history of protection and little poaching, they are easily seen and can be approached up to a reasonable distance (within 5 m on elephant back). However, in areas where they are hunted for meat or where they cannot move freely during the daytime owing to human activities, they are wary and shy. In Manas, including Royal Manas in Bhutan, the buffaloes have become wary and shy after the post-1989 political unrest. In Dibru-Saikhowa, most herds are shy and wary and run for cover at the slightest disturbance. Elsewhere, in Dum Duma Reserve Forest and the Kollolua-Jokai area, even sighting during the daytime is almost impossible, except the lone bulls that come out looking for domestic females. The same was the case with Kadam Reserve Forest and Dhakuakhana, where buffaloes are now extinct. In Nepal also, Heinen (2002) reported that herds of females are extremely wary of any human.

Crop raiding by wild buffaloes is a common phenomenon near their habitat, and lone bulls mate with domestic cows in fringe areas.

Natural predation

For a species as large and strong as the wild buffalo, the only known natural predator of adults is the tiger, which occurs, or used to occur, all over its range. This big cat normally targets female and immature animals, including calves. Adult bulls are rarely attacked and there are instances of tigers being killed by bull buffaloes. There was a recent instance of a calf killed by a leopard (*Panthera pardus*) in Elengmari Grazing Reserve near Jokai Reserve Forest in Dibrugarh district, Assam in October 2008 (Choudhury 2009). In Wilpattu National Park of Sri Lanka, leopard predation on calves is considered a significant cause of mortality (Eisenberg & Lockhart 1972). Dhole (or 'Asian wild dog', *Cuon alpinus*) and python (*Python molurus*) are potential predators and occur or used to occur sympatrically with wild buffaloes. For lone or stray calves, golden jackals (*Canis aureus*) could be a potential predator. The largest inland crocodile of India, the mugger (*Crocodylus palustris*), was also perhaps a likely predator of calves and juveniles in the past, but there seem to be no explicit records of such. In the coastal areas of Sunderbans, estuarine crocodile (*Crocodylus porosus*) was possibly also a predator when the wild buffalo used to occur.

Parasites and diseases

Parasites

No published works on the parasites of wild water buffalo seem to be available. However, there is good information on the domesticated water buffalo. Islam *et al.* (1992) recorded eight trematodes, two cestodes, 14 nematodes, two protozoa and two arthropods. The trematodes are *Fasciola gigantica*, paramphistomes (*Gigantocotyl explanatum, Ceylonocotyl scoliocoelium, Cotylophoron cotylophorum, Gastrothylax crumenifer*), *Schistosoma indicum*, *S. spindale* and *S. nasalis*. The cestodes are hydatid cyst and *Cysticerus tenuicollis*. The nematodes are *Strongyloides papillosus, Capillaria bilobata, C. bovis, Setaria digitata, Onchocerca armillata, Thelazia rhodesii, Gongylonemia pulchrum, Oesophagostomum radiatum*, hookworms (*Agriostomum vryburgi, Bunostomum phlebotomum, Trychostrongylus axei, Mecistocirrus digitatus, Haemonchus contortus* and *Toxocara vitulorum*). The protozoa are *Eimeria zuerni* and *Trypanosoma theileri*. The arthropods are *Haemaphysalis bispinosa* and *Haematopinus tuberculatus*.

Disease

One of the main threats to the wild buffalo population is the outbreak of various contagious diseases spread by domestic livestock, such as rinderpest, foot-and-mouth (FMD) and anthrax. Although FMD is the most frequent, rinderpest has historically taken the heaviest toll. Generally communicated by domestic stock that graze inside forests, such diseases, especially rinderpest, have from time to time also taken a heavy toll. Rinderpest, caused by the rinderpest virus, is also known as cattle plague. Some known cases of rinderpest are listed in Table 16.6. Outside the protected areas, hundreds of wild buffaloes have died, but most of

Table 16.6 Some recorded cases of rinderpest and other diseases in wild buffalo areas

Year	Areas	Remarks
End of the nineteenth century	Sri Lanka	Large numbers of 'wild' [feral] buffaloes died (Phillips, 1929)
1890s	Peninsular India	Inverarity (1895) reported deaths due to cattle diseases
1920s	Central Provinces (Madhya Pradesh and Chhattisgarh)	Wild buffalo population almost wiped out (Stewart, 1927; Stockley 1928)
1948	Balpakram and adjacent areas, Garo Hills, Meghalaya	Many wild buffaloes died due to foot-and-mouth
1963–1964	Golapalli Reserve Forest, Bastar, Chhattisgarh	Several sambar and chital were found dead due to rinderpest. It was suspected that some buffaloes might also be affected
1966	Langting-Mupa Reserve Forest, North Cachar Hills district, Assam	Many gaur died and it is believed that some wild buffaloes also perished*
1967	Kokrajhar and Bongaigaon districts, Assam	Large numbers of gaur died and it is believed that some wild buffaloes also perished in Ripu, Chirang and Manas Reserve Forests*
Late 1960s	Different areas of central India	Many gaur died and it is believed that some wild buffaloes also died†
Mid-1970s	Dhakuakhana area of Lakhimpur district, Assam	More than 100 wild buffaloes died
1972	Orissa	Large numbers of gaur died (Imam, 1985); it is possible that some wild buffaloes also perished
1981	Kaziranga National Park, Assam	At least 150 wild buffaloes died
1984	Kaziranga National Park, Assam	Some rhinoceros and deer died due to unknown disease spread from domestic stock
Mid 1980s	Dhakuakhana area of Lakhimpur district, Assam	Some wild buffaloes died
October 2008	*Khuties* in the *chapories* near Kaziranga National Park, Assam	At least 50 domestic buffaloes died due to unknown disease within a span of just three days

* Obtained locally through discussion with old villagers and confirmed with local veterinary units.
† Forest Departments.
Source: Choudhury (2010).

them went unrecorded. Fortunately, rinderpest has been mostly eradicated from India, with no record of deaths (either domestic or wild animals) during the last decade or more.

Inverarity (1895) reported deaths due to cattle diseases in Peninsular India. There were sporadic cases of anthrax outbreaks in different parts of northeast India in the 1940s and 1970–1971. Daniel & Grubh (1966) stated that such diseases 'have been one of the main reasons for the disappearance of the Wild buffalo in Peninsular India'. Seidensticker (1975) thought that disease was the main cause of extirpation of wild buffalo in Chitwan.

Despite eradication, rinderpest remains a potentially severe risk for smaller populations because of the close overlap of the wild and domestic populations, and the high densities of the latter compared to small and localised wild populations.

Status in the wild

The wild buffalo is listed by the IUCN as Endangered.

The primary factors affecting the range of distribution and population dynamics

With an estimated population of about 3300–3400, the wild buffalo is threatened with extinction across its range. More than half of the total population is inside a single protected area (Kaziranga; Table 16.7), and the wild buffalo part of a continuing process of local extirpation across most of India. Even in some protected areas – for example Sonai-Rupai, Jaldapara and Bhairamgarh – it has become extinct in recent decades. More than 15 separate populations/subpopulations have already been formed from what were once largely contiguous populations. Many of these are non-viable and are susceptible to genetic degradation. At least a few populations such as Pamed, Udanti, Indravati and Poba are heading for certain extirpation. Even the larger populations are separated from each other by many dozens of miles (e.g. Kaziranga and Manas).

This wild buffalo has experienced a severe decline due to habitat destruction and poaching for meat, and perhaps competition and disease transmission from domestic livestock and feral buffaloes. Small-population effects and hybridisation are other potentially major factors threatening the survival of the species in the wild. Some other problems are insurgency, exotic weeds and big dams as causes of habitat loss or hindrances to active protection.

Habitat destruction

The wild buffalo is a dweller of tall grassland with water bodies; hence, its survival depends upon the continued existence of large stretches of wet savanna grassland. The main reasons for grassland destruction are expansion of agriculture, human settlement and developmental activities such as construction of roads, railway and hydroelectric schemes, both in the direct flooding of suitable habitat and in the modification

Table 16.7 Estimates of numbers of wild buffalo

Country	Province/state	Number of wild buffalo*
India	Assam	2800
	Arunachal Pradesh	200
	Meghalaya	50
	Other States	<50
	India total	**3100**
Nepal		200
Thailand		50
Cambodia		<20
Bhutan		<30
	Rest of the world total	**300**
Total		**3400**

* 91.2% in India; 82.3% in Assam; 57.7% in Kaziranga
Source: after Choudhury (2010)

of downstream water dynamics allowing succession from grassland to other habitats. The bulk of the grasslands are suitable for crops, making it impossible to prevent clearance except in the protected areas. The water bodies are also being drained.

A statement made in the nineteenth century by Forsyth (1889) on a fertile valley in central India shows that the habitat of wild buffaloes has been under pressure for a very long time. He said 'it is indeed impossible that such a country as this can long remain a wilderness occupied by herds of buffaloes and wild beasts'. It is the species' misfortune that its habitat use – well-watered, level land, particularly on alluvial soils – corresponds precisely with areas ideal for agriculture. In Nepal and north Bihar the fertile *terai* area has been reclaimed for paddy cultivation since the 1950s, leaving only a few patches of grassland, almost solely inside the protected areas.

Collection of grass, thatch and reeds for building materials and other commercial use is also another threat to the grasslands. Where suitable controls on harvest are instigated, such harvest could be entirely compatible with the survival of even large buffalo populations. The key buffalo habitats in Assam's Dhemaji and Lakhimpur districts have been lost through encroachment by human habitation and cultivation.

Although the wild buffalo is not a forest-dwelling animal, it uses forested areas during travelling and in disturbed areas for shelter during the day. In some areas such as Dum Duma Reserve Forest and Dibru-Saikhowa National Park, both in Assam, it even remains inside the forest for the greater part of the day, only to come out for grazing and wallowing. In Dibru-Saikhowa National Park the 'forest villages' (Laika and Dadhia) have not only encroached upon

Figure 16.8 Poachers handed over arms to the author (centre), who was then District Magistrate, Baksa in Assam, where the bulk of Manas National Park is located.

large areas of the park, but many of the villagers are timber smugglers and poachers.

The buffalo habitat in Cambodia is also facing fragmentation due to human population migration, which is followed by road construction, agricultural expansion and mineral extraction (Tordoff *et al.* 2005). Indeed, such areas have only survived at all because of the withdrawal of people from such areas during the terror regime of the Khmer Rouge. In Myanmar, the potential areas in and south of Hukawng Valley Wildlife Sanctuary, including the vast areas proposed as an extension to form the Hukawng Valley Tiger Reserve, are under very high threat for conversion to agriculture and plantations of cash-crops. There is also likelihood of increase in hunting for wild meat to feed hundreds of thousands of itinerant labourers in this area (J. W. Duckworth, personal communication, 2009).

The ultimate cause of habitat destruction is, however, the very rapid growth of the human population. Every year the population in India increases by more than 15 million. As a very high percentage of the population lives in rural areas (approx. 69% based on the 2011 census), with farming as its main occupation, the large-scale destruction of natural habitat seems inevitable. Between 1950 and now, the wild buffalo has lost at least 90% (approximately) of its habitat outside the protected areas in India. The situation in Nepal, Thailand and Cambodia, and the presumed range states of Lao PDR and Vietnam, is similar.

Hunting and poaching

The wild buffalo is killed in India for its meat. The earliest information on hunting of the species is found in Fazl (1590). According to him, during the reign of Akbar the Great, the commonest method of hunting wild buffalo was to use a tame female on heat as a decoy. When the wild bull comes for mating, it is entangled with nooses slipped round his legs by hunters lying in ambush. The hunting pressure, for hobby, pastime and trophy by the royal dignitaries, is evident from the memoirs of the Maharajah of Cooch Behar (1908). Between 1871 and 1907, his royal hunting parties killed at least 434 wild buffaloes in a relatively small area comprising present north-West Bengal and West Assam. Although no systematic record is available, every year an unspecified number of wild buffaloes is shot outside the protected areas, mostly in central India and in the northeast. Even in some protected areas a few are poached every year. Most such killings go unreported. In the Peninsular India such poaching has nearly extirpated the species from the area.

In Manas National Park, owing to political upheaval, there was an unprecedented increase in the poaching for meat in 1989 (Choudhury 1989b; 2006). The situation worsened when the anti-poaching camps were abandoned by staff due to attacks by underground extremists. An unspecified number – apparently a few hundred – of wild buffaloes were killed until around 2004–2005. The estimated number before 1989 was around 1000 (1200 estimated for the entire tiger reserve area; Deb Roy 1991; Choudhury 1994a), while the number in 2004–2005 was around 200. The meat was sold in the markets, often mixed with deer meat (as latter fetches more money). In Laokhowa Wildlife Sanctuary, most killings take place outside when the animals raid crops. While returning from crop fields, the buffaloes are trapped in hurriedly dug pits. Guns, rifles and spears are used to kill the buffaloes. The local tribes in Lakhimpur and Dhemaji districts in east Assam used spears during high floods when the animals take shelter on higher grounds (Figure 16.8).

Figure 16.9 A wild bull pursuing domestic cows near Dibru-Saikhowa, Assam. A black and white version of this figure will appear in some formats. For the colour version, please refer to the plate section.

The small wild buffalo population of Cambodia is also under severe threat from hunting for trophy horns, both by Cambodians and Vietnamese hunters crossing the border (Timmins & Ou Ratanak 2001; Tordoff *et al.* 2005). The small population in Huai Kha Khaeng Wildlife Sanctuary is also reportedly threatened by poaching (Songsri 1987).

Possible hybridisation

A major potential threat to the surviving wild buffaloes is interbreeding with feral and domestic buffaloes. Genetic studies suggest that despite some levels of cross-breeding, distinct populations of wild buffaloes remain and they are fully worthy of conservation efforts (Muley 2001). Interbreeding is likely to be a greater threat to the smaller populations. There is a general belief that many of the wild buffaloes in Assam and elsewhere are contaminated through hybridisation with domestic animals, but this seems to be incorrect as a rule. This is because domestic males are usually not kept in the fringe villages and *khuties* since they would generally be killed by wild bulls in such circumstances. Even if a domestic bull strays in, it has no chance to compete with wild bulls for access to females. In fact, it is wild bulls which regularly come out of the parks and mate with domesticated females, both in the villages located along the fringe of reserves as well as *khuties*. In these circumstances, gene flow is mostly one-way and is unlikely to affect any substantial wild populations. Generally the hybrid bulls resulting from a backcross with domestic buffaloes cannot dominate wild bulls and hence are not thought to be able to crossbreed with wild females. However, hybrid females would be expected to breed with wild males.

As mentioned earlier, in the sixteenth century Fazl (1590) mentioned that a common method of hunting wild bulls was by using domestic cows as a decoy. This indicates that wild bulls coming out to mate with domestic cows was known even at that time. Cooper (1873) mentioned a large wild bull coming out of the jungle and killing a tame bull to take control of a domestic herd in Arunachal Pradesh. According to Sterndale, 'wild bulls frequently mate with tame cows. This makes for physical improvement, as the wild animal is larger and less scraggy than the tame one' (Finn 1929). These accounts show that wild bulls' associating with domestic cows is an age-old occurrence. However, there is no record of domestic bulls mating with wild cows, which, however, is possible in areas where only some cows remain as the last survivors.

Observations by this author elsewhere (Lakhimpur, Laokhowa Wildlife Sanctuary, etc.) showed that the wild herds reveal a high degree of segregation from domestic animals: they keep to themselves and do not mix with grazing domestic buffaloes. A similar phenomenon has also been observed among Nepal's wild buffaloes in Koshi Tappu (Heinen 2001). The degree of segregation is so high that in 1989–1990, the last survivors, only four animals, in Dhakuakhana confined themselves to a small area of reeds where they remained for the whole day, coming out at night to graze in the nearby fields (Choudhury 1994b). In recent years there have been hardly any domestic buffaloes allowed to go feral by their owners as they are too valuable. An adult domestic buffalo can cost \$150–\$420, which is much more than a whole month's total income for an ordinary village family (Figure 16.9).

The small population of Huai Kha Khaeng Wildlife Sanctuary in Thailand is also interbreeding with domestic water buffalo, but again it is the wild bulls that visit the villages. However, individuals can often be distinguished in the field (Anak Pattanavibool, personal communication in IUCN 2013).

Cross-breeding with wild bulls is encouraged and of economic benefit. In the fringe areas of wild buffalo habitat, the

Figure 16.10 Many domestic water buffaloes in Assam closely resemble their wild cousins, owing to the presence of wild buffaloes in different areas. A black and white version of this figure will appear in some formats. For the colour version, please refer to the plate section.

owners are compelled to sell off their domestic males when these are young in order to avoid them being killed by wild bulls. At the same time, many owners encourage it for economic benefit as hybrid buffaloes are strong and grow bigger, thus fetching more money and making better draught and plough animals (Figure 16.10).

Big dams

A series of large dams (around 170) has been proposed on the Siang (upper reaches of the Brahmaputra), Dibang, Lohit and Subansiri Rivers for generation of hydro-electric power. Works on some of these are ongoing, while the rest are in the pipeline. Key buffalo habitats such as Kaziranga and Dibru-Saikhowa are located downstream and on the rivers and their floodplains. Some of these projects are among the largest in Asia.

According to Odden *et al.* (2005), such developments and resulting changes in water flow and level conditions downstream threaten the ecological maintenance of floodplain areas. The tall grasslands and their associated fauna are mostly confined to the floodplains of the large rivers draining the Himalayan range. Taking the hog deer as an indicator species, they stated that in recent decades cultivation and hydro-electric dam construction have severely reduced their spatial abundance. During 1998–2000 they conducted a detailed study of the habitat use of the deer in the floodplains of the large Karnali River in Bardia National Park, Nepal. Strong preference for tall grass and avoidance of the successional stages following riparian forests throughout the year was revealed by analyses of tracking data from 18 radio-collared deer and from density estimates from block counts ($n = 46$) in different habitat types. Continuously strong selection for tall grass despite the reduced availability of this habitat type during the monsoonal flooding indicated an insignificant effect of density-dependent factors in habitat selection. Reduced availability of cover following traditional cutting and burning of the tall grasslands by local villagers did not produce changes in habitat use or an increase in daily movements, indicating an absence of tradeoffs between habitat preference for food and for predator avoidance. By regulating fluvial action, hydro-electric power development will lead to the loss of tall grasslands due to accelerated secondary succession, which will therefore adversely affect the viability of local populations.

It may be mentioned here that hydro-electric projects have already destroyed part of the buffalo habitat in Peninsular India, such as the Balimela dam in Orissa.

Inadequate attention to this factor could negate all on-site work for these floodplain communities. In protected areas such as Kaziranga National Park, where poaching and habitat loss is under control, hydrodynamics modification may pose the greatest – but potentially most avoidable – threat to the entire park ecosystem, as well as its wildlife, including rhinoceros and buffaloes. Odden *et al.*'s (2005) observation in Nepal clearly shows the fate of the Brahmaputra floodplains once all the hydro-electric projects are completed.

The biggest problems are more likely to come with modification of downstream flows, rather than outright inundation of areas. So far, not a single project has given serious thought to the downstream effects (Vagholikar 2009), and the Environmental Impact Assessment (EIA) reports are also silent about this.

Grazing by domestic stock

Grazing by domestic stock is a major cause of concern as it damages the natural grassland and also reduces food sources for the wild buffaloes. In D'Ering Sanctuary of Arunachal Pradesh, about 200 domestic buffaloes and 1600 cattle grazed in the early 1990s (Choudhury 1994a); now it is more. In Dibru-Saikhowa National Park of Assam, the wild herbivores are competing with about 9000 domestic livestock (mainly buffalo) for grazing (Choudhury 1998b). Even in very well-protected areas such as Kaziranga National Park and Jaldapara Wildlife Sanctuary, grazing by domestic stock could not be fully prevented. Overgrazing has already resulted in severe degradation of the habitat in unprotected grasslands, such as the *chapories* on the Lohit River (Arunachal Pradesh), Sibia *Chapori* (Assam) and in inadequately protected areas such as Laokhowa and Burhachapori Sanctuaries (Assam). Because of the sheer number of domestic animals and professional cattle/buffalo camps, the overgrazed areas have a barren look.

In Nepal's Koshi Tappu Reserve large numbers of domestic buffaloes and cattle graze. In Huai Kha Khaeng Wildlife Sanctuary there is also illegal grazing in its small but important grassland habitat (Humphrey & Bain 1990).

Small-population effects

The presence of small populations is a conservation issue as many of these are non-viable for the long term. For example, there are only three wild buffaloes in the Poba Reserve Forest, Dhemaji district of east Assam. The recently extirpated populations of Dhakuakhana were also very small. The isolated animals in Dum Duma Reserve Forest are estimated to number approximately 13. In the past three decades some excellent buffalo habitats of northeast Assam have been destroyed and encroached upon. Similarly, all the pockets in Peninsular India are non-viable. In Cambodia the available reports indicate a very small population.

Insurgency and political unrest

This is a rather new but serious problem affecting the survival of the wild buffalo and other wildlife in different parts of India, echoing what was probably one of the major causes behind the near-total loss of wild buffaloes from Cambodia, Lao PDR and Vietnam during the middle decades of the twentieth century. From the 1960s until the late 1980s, insurgency by underground guerrillas was a feature of Nagaland and Mizoram. In the late 1980s and 1990s, such problems spread to almost all of northeast India, affecting many well-known protected areas such as Manas, Balpakram, Gumti, Intanki and Sonai-Rupai. The Naxalite (Maoist) movement has severely affected all buffalo habitat in central India. Key reserves such as Indravati in Chhattisgarh, once the home of a large population of water buffalo, are almost out of bounds to staff. While the extremists themselves do not usually harm wildlife, others take advantage of the unpoliced situation for tree-felling and poaching. The Forest Department staff are no match for the extremists and their many modern firearms.

Political unrest has also played a key role in the destruction of grassland habitat and its endangered fauna. In Laokhowa Wildlife Sanctuary, all the rhinos were slaughtered during the early 1980s during the anti-foreigners agitation in Assam. Wild buffaloes in the sanctuary also faced the problem, with a few unreported killings and large-scale disturbance to their habitat. Again in the early 1990s, almost all the rhinos and most of the wild buffaloes (for meat) of Manas National Park fell to poachers during the disturbed atmosphere created by the pushing of the Bodo tribe for a separate state. Unlike Laokhowa, there was involvement of insurgents in Manas who had their camps inside the park, which prevented ordinary Forest Department staff from patrol duty (at least 11 staff members were killed and six seriously injured by extremists between 1989 and 1998, which included a ranger). The situation has improved since then (Choudhury 2010).

Other issues

During high floods in the monsoon period a few wild buffaloes die due to drowning or collisions with vehicles while trying to cross highways. Such cases are mainly experienced in and around Kaziranga National Park. For instance, during the high floods of 1998, as many as 23 drowned between 25 June and 19 September in and around the park. Calves, juveniles and old animals suffer most during such natural calamities. A cow was run over by a night bus near Baghmari Camp of Kaziranga while she was trying to cross the highway on 26 August 2007. In other areas, especially Dibru-Saikhowa, Laokhowa and Burhachapori, the buffaloes also suffer during the floods. However, high floods such as in 1998 do not occur every year, but generally at intervals a few years apart. Many drownings, however, go unreported. In Manas, which is not located in the floodplains and so does not have such high floods, the rivers experience flash-flooding owing to location near foothills. During such flooding, calves suffer and may even die while trying to cross the fast-flowing rivers. Heinen (1993) reported that drowning of calves during floods is a major cause of calf mortality in Kosi Tappu.

Flood is a natural phenomenon in the floodplains such as Kaziranga, Dibru-Saikhowa, Laokhowa and Burhachapori, and is essential for maintaining the diversity of habitats, particularly grasslands and wetlands, by annual inundation and siltation. However, the increasing level of floods owing to perhaps global warming/climate change in recent years is a potentially serious cause of worry. This is particularly so in combination with ongoing fragmentation of the landscape, meaning that during heavy floods movement options for large ungulates to other natural habitats are highly limited. The extirpation of buffaloes from parts of Lakhimpur and Dhemaji was also related to floods, as most of the poachers had killed the animals when they took shelter in small areas of higher ground.

Figure 16.11 Invasion of exotic weeds of *Mimosa* spp. has destroyed a chunk of habitat in areas such as Kaziranga National Park.

Burning of grassland

In the protected areas, especially in Kaziranga and Manas, almost the entire grassland is burnt annually, from December to March, as part of management. This also prevents any growth of scrub and shrub in such areas. Although such burnings are apparently helpful for the buffaloes, late burning (February and March) has been found to be harmful for nesting of some globally threatened birds such as the swamp francolin (*Francolinus gularis*) and the Bengal florican (*Houbaropsis bengalensis*). The long-term effect of such burnings on species composition is not known, which might have some bearing on the favoured grasses of the buffaloes. Conversely, with the taming of natural floodplain activity, such burning might be a positive feature for buffaloes. Formerly, plains grasslands were continually created by flash-floods and ox-bow succession, to replace those colonised naturally by shrubs and eventually regenerating to forest; but with most floodplains having been brought under extensive human management, creation is now rare, while succession is still occurring. One of the fundamental problems facing the plains wet grassland communities of tropical Asia, and why they are so threatened, is that the habitat is fundamentally unstable and, in any given locality, temporary. The modern human practice of designating protected areas, especially rather small ones, is not well suited to the management of such dynamic habitats without heavy management (Choudhury 2010).

Santiapillai (1990) reported that there was similar practice in Thailand's Huai Kha Khaeng also. Currently, however, the management is quite enlightened and successful there (J. W. Duckworth, personal communication, September 2009).

Plantation of trees

Another harmful component of management of grassland in unprotected areas is the plantation of trees by the Forest Department.

Invasion of weeds

The invasive weeds such as *Parthenium hysterophorus*, *Eupatorium odoratum* (*Chromolaena odorata*), *E. adenophorum*, *Mikania micrantha* and *Lantana camara* have been a matter of great concern for many of the national parks and wildlife sanctuaries across India and Nepal. In Manas, *Eupatorium* is a severe invader for grasslands, which are an important habitat for the wild buffalo. However, it is the fairly recent invasion by *Mimosa invisa* (*M. diplotricha*) and *M. inermis* (*M. rubicaulis*) that has spread menacingly, especially in Kaziranga and Orang. In the former park (which has the largest population of wild buffaloes), despite regular uprooting and destruction, 5–10% of the grassland remains heavily affected by these weeds. *Mimosa* spp. were introduced in the tea plantations to enrich the soil in 1960 as these are nitrogen fixers (Figure 16.11).

The wetlands, especially stagnant water bodies such as *beels* and marshes, are affected by water hyacinth, *Eichhornia crassipes*, also an exotic species. Although it is occasionally eaten by the buffaloes, it chokes up wetlands.

Denotification of reserves

Although owing to stringent provisions in the *Wild Life (Protection) Act*, denotification is not easy in India. In Cambodia a part of Mondulkiri Protected Forest has been delisted from conservation status. Such issues pose significant threats to the effective long-term conservation management of the protected areas. There is no certainty that more land would not be removed from the protected areas in the future.

Trade

There were past cases of sale of meat of poached animals in the local markets, especially near Manas National Park in 1989–1995. However, at the Myanmar–Thailand, Cambodia–Laos and Laos–Vietnam borders, buffalo horns are often traded as trophies. It is difficult to quantify or even identify whether these are from true wild or domestic or from feral animals. However, because bigger horns fetch higher prices, and on average wild buffaloes have bigger horns than domestic or feral animals, the incentive to poach wild animals for the horn trade is obvious. Recent (February 2010) seizure of some 400 horns, mostly of domestic origin, in West Bengal, which were from Assam, indicates that there is some trade, not for trophy but for Chinese medicine.

Mining

Mining is a serious issue in central India; however, the wild buffalo has vanished from most parts. There are big coal, iron ore, manganese and other mines in Chhattisgarh, a major mineral-rich area of India. In Assam there are oil wells near buffalo habitats in Dibrugarh district. The Dibru-Saikhowa National Park was covered by a seismic survey for oil exploration and there is a possibility of mining in the near future if any good reserves are found. In Meghalaya, coal mining in the Garo Hills is a potential threat to the small surviving population of wild buffaloes.

Lack of adequate recognition

Although Chhattisgarh State has declared the wild buffalo as the State Animal, the species has never received the recognition and attention it deserves. Even now stray buffaloes are occasionally killed as they are deliberately 'mistaken' for feral animals. The difference is that a wild one has got the highest protection status in India, while a feral animal has none. So it becomes easier for the hunter to kill it without legal hassles, and officials can free themselves from lots of paperwork if they don't need to prosecute the culprits. Wild buffaloes are the origin of all the domestic breeds, which has greatly benefited humankind by its early domestication and subsistence support for livelihood to millions.

Potential misuse of *The Scheduled Tribes and Other Traditional Forest Dwellers (Recognition of Forest Rights) Act, 2006*

There is a threat posed by potential misuse of this Act, which grants ownership of land to tribals or forest dwellers as of 13 December 2005, subject to a maximum of four hectares. Although eligibility to rights under the Act is confined to those who 'primarily reside in forests' and who depend on forests and forest land for a livelihood, it is difficult to distinguish as there is no community in areas such as northeast India who 'primarily reside in forests', as there are in parts of central India and the Andamans. Hence, there is a high likelihood of the misuse of this Act.

Moreover, the allotment of land had already started while the 'Critical Wildlife Habitat', i.e. areas to be set aside for conservation, has yet to be finalised. Hence, there is a likelihood of many key biodiversity areas being encroached upon (e.g. the northern part of Sonitpur district in Assam, including Sonai-Rupai Wildlife Sanctuary).

Human–buffalo conflict

Such conflict is not as serious a conservation problem as is human–elephant conflict. It is mainly confined to some crop depredation. The wild buffalo occasionally strays into human habitations, and occasionally has killed people. However, these phenomena are not new; even in the nineteenth century there were reports of crop raids in central India (Forsyth 1889). This is a result of the aforementioned overlap in habitats ideal for cultivation and those for wild buffaloes. Locally, however, crop depredation can be serious. Another major area of conflict is the attack on domestic male buffaloes. Although domestic males are generally not kept in the fringe areas of buffalo habitats, sometimes farmers are tempted to bring one or two for ploughing as it is more economical than a pair of bullocks.

Minimum viable population (MVP) and minimum area required (MAR)

No such studies on wild buffalo have been carried out so far. However, examples from elsewhere on other mammals, especially ungulates, can give a good insight into the probable MVP of buffalo as well. Estimates of MVP size can be derived by empirical simulation, experiments or long-term monitoring (Traill 2010). An example of a long-term census study is that by Berger (1990), who evaluated the persistence of isolated populations of bighorn sheep (*Ovis canadensis*) over 50 years. Populations with <50 individuals went locally extinct, while those containing >100 individuals generally persisted.

Heinen (1993) made some observations on population viability of wild buffalo in Nepal, but he did not provide any conclusion on MVP. However, he stated that on the basis of the physical nature of the reserve (i.e. Kosi Tappu) and known seasonal movements of buffalo, it suggested that the population (93 animals in 1988, at the end of his study) was not viable for several reasons, such as flood-related mortality, grazing by domestic cattle and buffalo and risk of disease. However, outright classifying an important population as non-viable may not be appropriate, as Dinerstein & McCracken (1990) have shown that genetic diversity can be maintained after recovery from a bottleneck in at least one large mammal population (Indian one-horned rhinoceros in Nepal). Even in

Kosi Tappu, the wild buffalo population has increased significantly to 219 in 2009.

Soulé (1980) has recommended that an N_e (i.e. genetically effective population size) of at least 50, corresponding to a per generation rate of inbreeding of 0.01 (Falconer 1981), is needed to avoid the immediate deleterious effects of inbreeding. Franklin (1980) estimated this to be a genetically effective population size (N_e) of ~500 individuals (50 to avoid inbreeding).

From the above, it appears that some populations which can be termed as non-viable are those of the Chhattisgarh and Meghalaya in India, and Thailand and Cambodia. The remaining populations are either large (Kaziranga, Manas, Dibru-Saikhowa, Kosi Tappu) or have contiguity with a larger population (Laokhowa, Burhachapori, D'Ering) or have occasional inputs in the form of stray bulls (Kollolua-Jokai).

Like MVP, no specific studies on the MAR of wild buffalo have been carried out so far. Density of wild buffalo ranges from 0.5 in low-density habitats such as D'Ering Sanctuary to 6.45/km^2 in high-density habitats such as Kaziranga (Choudhury 2010). If N_e of 50 individuals is accepted as the MVP, the corresponding MAR would be from 7.75 km^2 (in Kaziranga-type ideal habitat complex) to 100 km^2 (in D'Ering- or Huai Kha Khaeng-type habitat complex).

Conservation measures *in situ* and *ex situ*

There are no *ex situ* conservation measures so far; however, *in situ* measures have been undertaken.

Legal protection

The wild buffalo is legally protected in at least Bhutan, India, Nepal, Lao PDR and Thailand. In India it is listed in Schedule I of the *Wild Life (Protection) Act 1972*, which prohibits its killing or capture. This is the highest level of protection accorded to any species in the country. Most of the villagers are not aware of its protected status. In Nepal it is on the fully protected species list in the *National Parks and Wildlife Protection Act 1973*. In Lao PDR it is fully protected, while in Thailand it is on the list of 'reserved' animals under the *Wild Animals Reservation and Protection Act*. Although there are no wild buffaloes in Bangladesh, the *Wildlife (Preservation) Act 1973* provided total protection and any trade is also prohibited. It is listed in Appendix III of CITES, which, however, should have been Appendix I (thus banning all international trade in wild buffalo and wild buffalo products). India was among the earliest parties to CITES (it signed in 1976). It is also listed as Endangered by the IUCN (IUCN 2013).

It has been declared as the State Animal by the Chhattisgarh Government.

Habitat protection

At present at least 15 national parks and sanctuaries protect the wild buffalo, along with other wildlife, in the Indian subcontinent, as does one each in Thailand and Cambodia (Figure 16.2). However, only seven protected areas have viable populations, five in Assam and one each in Nepal and Thailand. The protected areas with wild buffalo are listed in Table 16.8. There are some fairly large contiguous habitats covering more than one protected area that have a viable breeding stock for an indefinite period ('conservation landscapes'). Examples of such areas are listed in Table 16.9. Such habitats often cross international boundaries (for example, the India–Bhutan boundary).

Five notified protected areas in Assam have wild buffalo populations, as do three in Chhattisgarh and one in Arunachal Pradesh. In Meghalaya there are two, while outside India in Bhutan, Cambodia, Nepal and Thailand there is one each. Most, i.e. about 65%, of the known habitat is under the protected area network in Assam. In some sanctuaries, like those in Chhattisgarh, and Laokhowa in Assam, enforcement is also inadequate.

There are two isolated feral populations with wild input in Assam worthy of protection (Orang National Park and Pabitora Wildlife Sanctuary).

Species protection

Owing to stringent protection to prevent poaching of the Indian one-horned rhinoceros, the wild buffaloes in Assam's Kaziranga National Park are perhaps the best protected population, and suffer no systematic poaching. In Manas National Park and the adjoining Royal Manas National Park of Bhutan the situation has improved a lot and there has been no significant poaching of buffalo since 2004. In Dibru-Saikhowa National Park, and in Nepal and Thailand, no recent poaching cases were reported; however, elsewhere in northeast and Peninsular India and in Cambodia the species protection is still inadequate.

Recommended conservation measures

Effective protection and management of the existing protected areas

All such areas should be adequately protected to prevent habitat degradation, encroachment and poaching. To prevent dispersal of buffaloes to cultivations, fencing with metal railings such as those used in expressways is highly recommended in areas where this will not cause risk of flood-associated mortality. Electric fencing is ineffective in high-rainfall areas; hence, for long-term cost-effective benefit, metal railings are the better option. This will also reduce the chance of mixing with domestic stock, thus saving wild populations from potential disease and hybridisation.

Communication systems for protected area staff and their necessary partners in conservation should be improved, with timely replacement and updating of radio networks/mobile phones. To effectively patrol the reserves, well-trained and well-armed guards should be provided in adequate numbers. Presently, many of the reserves are undermanned.

Table 16.8 List of protected areas having wild buffalo populations

Name of reserve	Province/country	Area (km^2)	Population range	Conservation status	Remarks
D'Ering Memorial Wildlife Sanctuary	Arunachal Pradesh, India	190	E	B	
Burhachapori Wildlife Sanctuary	Assam, India	44	E	B	
Dibru-Saikhowa National Park	Assam, India	340	C	B	
Kaziranga National Park	Assam, India	850	A	A	
Laokhowa Wildlife Sanctuary	Assam, India	70	E	B	
Manas National Park	Assam, India	500	C	A	
North Karbi Anglong Wildlife Sanctuary	Assam, India	97	D	D	Some animals migrate from Kaziranga National Park during floods
Balpakram National Park	Meghalaya, India	220	E	B	
Siju Wildlife Sanctuary	Meghalaya, India	5	E	B	Contiguous with Balpakram; occasional visitors
Indravati National Park	Chhattisgarh, India	1258	E	D	
Pamed Wildlife Sanctuary	Chhattisgarh, India	262	E	C	
Udanti Wildlife Sanctuary	Chhattisgarh, India	248	E	C	
Buxa Tiger Reserve	West Bengal, India	759	E	A	In the eastern buffer zone only
Royal Manas National Park	Bhutan	1023	E	B	Contiguous with Manas National Park, Assam, India
Kosi Tappu Wildlife Reserve	Nepal	175	D	A	
Huai Kha Khaeng Wildlife Sanctuary	Thailand	2780	D	A	
Mondulkiri Protected Forest	Mondulkiri, Cambodia	4294	E	B	

Population range: A: >1000; B: 500–1000; C: 250–500; D: 50–250; E: <50.
Conservation status: A: adequately protected; B: inadequately protected; C: very inadequately protected; D: no protection.
Source: Choudhury (2010).

Table 16.9 Contiguous protected areas and adjacent forests forming wild buffalo 'Conservation Landscapes' having viable breeding populations for an indefinite period (also see Figure 16.2).

Landscapes (with detailed list of protected areas and adjacent forests)	Location	Area (km^2)
Kaziranga Landscape: Kaziranga National Park, Laokhowa Wildlife Sanctuary, Burhachapori Wildlife Sanctuary, northern part of North Karbi Anglong Wildlife Sanctuary, Kukurakata Reserve Forest, Bagser Reserve Forest, Panbari Reserve Forest, Kochmora Reserve Forest, Lonke Tapu Proposed Conservation Reserve	Assam	*c.*1100
Manas Landscape: Manas National Park, southern part of Royal Manas National Park, part of Phipsoo Wildlife Sanctuary, Manas Reserve Forest, Daodhara Reserve Forest, Batabari Reserve Forest, Chirang Reserve Forest, Ripu Reserve Forest, eastern areas of Buxa Tiger Reserve	Assam, West Bengal, Bhutan	*c.*3000
Dibru-Saikhowa Landscape: Dibru-Saikhowa National Park, D'Ering Memorial Wildlife Sanctuary, Poba Reserve Forest, Kobo *Chapori* Proposed Reserve Forest, Jamjing Reserve Forest, Sengajan Reserve Forest, Mebo Reserve Forest, part of Turung and Kamlang Reserve Forests, Lohit *Chapories*, Dibang *Chapori* Proposed Reserve Forest, Dibang Reserve Forest, Kerim Reserve Forest, Deopani Reserve Forest, Sirkee Proposed Reserve Forest, Lai Anchal Reserve Forest	Assam, Arunachal Pradesh	*c.*1800

Bhutan should have strong anti-poaching camps at Gabhorukunda and Shipcheni (west of Kalikhola).

Control of poaching

Poaching of wild buffalo for meat should be checked and the known poachers in the area should be apprehended. Special operations against poachers should be organised from time to time. Strict vigilance is needed in Manas, Dibru-Saikhowa and Balpakram in view of the threat from poachers. Whatever number of buffaloes is left in Chhattisgarh should be strictly protected. All unlicensed arms in the fringe areas should be confiscated.

Involvement of local communities

Local communities have already been involved in conservation in Assam's Manas Tiger Reserve with great success. The communities have organised themselves as NGOs and are helping the department to patrol the reserve. They are also engaged in eco-tourism for their sustenance. Interestingly, many of them are ex-poachers and insurgents and have laid down their arms to the author (who was head of administration in the district near Manas). Similar efforts need to be made in other areas.

Detailed surveys in Southeast Asia

It is required in north Myanmar, particularly the upper parts of Chindwin plain, and perhaps the Satay district of Vietnam, southern areas of Lao PDR and much of Cambodia to confirm the occurrence of wild buffaloes. Records from northeast India from a diverse terrain, from floodplains to small grassy areas along rivers in the hills, indicate that there is scope for further surveys in those areas.

Population monitoring

The wild buffalo should be censused all over its range to assess its population status and the trend of population should be monitored at regular intervals (perhaps every four or five years; in smaller pockets with small isolated populations such as Huai Kha Khaeng, Kollolua-Jokai and Dum Duma areas, it should be done every year).

The feral buffaloes and the domestic buffaloes in the fringe villages should also be surveyed and monitored at regular intervals. Since there is little scope for new animals to become feral, the census will help their conservation as well as providing an ecological function in certain areas like Pabitora Wildlife Sanctuary in Assam and in some protected areas of Sri Lanka.

Genetic study

Detailed studies of the genetics of the wild buffalo, by taking samples from all the known populations, are strongly recommended in all the habitats. This should also include the feral, hybrid and domestic buffaloes of the fringe areas of the known habitats of the wild buffalo (see Box 16.1 for more detail).

Removal of exotic weeds

These weeds should be removed regularly by engaging fringe villagers. This will provide part-time employment to these poor people as well. It has been observed that complete eradication of such weeds is almost impossible; hence, regular removal through cutting, uprooting, burning, etc., is the only way until future research finds a form of biological control.

Check on encroachment and eviction of encroachers

No new settlements should be allowed in the buffaloes' ranges, at least in the reserve forests. Fresh encroachments, especially those inside the important buffalo habitats, should be evicted. All the recent encroachments by the 'forest villagers' inside Dibru-Saikhowa and in Laokhowa and Burhachapori should be evicted.

Reintroduction in former habitats

It is time that some wild buffaloes are reintroduced in some of the former habitats where the current protection measures are reasonably good. Some such areas are Chitwan National Park in Nepal, Jaldapara Wildlife Sanctuary in west Bengal, Dudhwa National Park in Uttar Pradesh and Nameri National Park in Assam. In Thailand, small population size is a deterrent factor in any such programme of relocation; however, options should be kept for a second home to avoid any future catastrophe to a single small population.

Of the potential sites shortlisted above for reintroduction, Dudhwa seem to be in the best position. It has a secured, fenced-off area of 29 km^2 (Salukapur) where some great Indian one-horned rhinoceros were successfully reintroduced in 1984 (Singh 1984). Wild buffaloes, which stray out or are problem animals, and those rescued during floods, should be the first animals to be moved to Dudhwa. Dudhwa National Park covers an area of 490 km^2 of grassland, woodland and wetland and is also a tiger reserve.

Big dams

The number of proposed big dams in Arunachal Pradesh should be reduced. A detailed study is required to know the effect on the hydrology of the area and river channel changes, which may have severe impacts on the wildlife (and paddy farmers too!) and their habitat in the floodplains downstream.

Creation of new protected areas

The creation of more protected areas in northeast India is strongly recommended. Since the grasslands are vanishing at an alarming rate, whatever is left should be protected. Parts of Dibang and Kerim Reserve Forests and Sirkee proposed Reserve Forest (totalling 202 km^2) have been recommended for a national park for its importance as the habitat of wild buffalo, as well as other threatened species (Choudhury 1996).

Box 16.1 Genetics of wild water buffalo

J. Stuart F. Barker

The wild water buffalo (Asian buffalo, wild Asian buffalo) – *Bubalus arnee* (Kerr 1792) – is one of a number of *Bubalus* species which were widely distributed in Europe and southern Asia in the Pleistocene, but later was restricted to the Indian subcontinent and Southeast Asia (Mason 1974). Two species, *B. teilhardi* and *B. youngi*, which were morphologically identified from bone remains in the Wei River valley (Shaanxi Province, China) apparently became extinct during the Pleistocene (Yang *et al.* 2008). A third species in China, *B. mephistopheles* (Hopwood, 1925) also found in the Wei River valley and the lower Yangzi River region, survived to the Neolithic and Bronze Age in the Holocene (Liu *et al.* 2004; Yang *et al.* 2008). Extant *Bubalus* species include *B. mindorensis* (Heude 1888) or tamaraw from Mindoro Island, the Philippines, which is critically endangered (Hedges *et al.* 2008a), and two species of anoa from Sulawesi, Indonesia: *B. depressicornis* (C. H. Smith 1827) or lowland anoa and *B. quarlesi* (Ouwens 1910) or mountain anoa. There is still debate about whether these two are distinct species (Burton *et al.* 2005), but both are endangered (Semiadi *et al.* 2008a, 2008b). In historical times, the wild Asian buffalo ranged across South and Southeast Asia from Mesopotamia to Indo-China (Epstein 1971; Mason 1974; Cockrill 1984). It is currently listed as Endangered (Hedges *et al.* 2008b), with a world population of fewer than 4000, potentially fewer than 200, and it is even possible that no purebred wild animals exist (FAO 2000).

Archaeological, anatomical and historical evidence supports the contention that domestic buffalo (*B. bubalis* Linnaeus, 1758) are descended from *B. arnee*. Domestic buffalo have been classified on morphological and behavioural criteria into two types: river and swamp (MacGregor 1941). The two types also differ in chromosome number (Ulbrich & Fischer 1967; Fischer & Ulbrich, 1968) (river – $2n = 50$; swamp – $2n = 48$) and are genetically distinct, based on allozyme and microsatellite genotype frequencies (Barker *et al.* 1997; Zhang *et al.* 2011). Consequently, some workers refer to them as subspecies, naming river buffalo *B. bubalis bubalis* and swamp buffalo *B. bubalis carabanesis*. The reduced diploid chromosome number in the swamp type resulted from a Robertsonian translocation – the fusion of two small chromosomes of the river type (Di Berardino & Iannuzzi 1981). Thus the chromosome number, $2n = 50$, is ancestral and *B. arnee* also would be assumed to be $2n = 50$. One exception to the morphological distinction of the two types is the native buffalo of Sri Lanka. Although phenotypically resembling the swamp type, they have a chromosome number of $2n = 50$ and phylogenetically cluster strongly with river-type populations (Barker *et al.* 1997). They may represent the wild buffalo from which the present breeds of river buffalo were derived, but more likely they are feral buffalo descended from river buffalo introduced to Sri Lanka in historical times.

The time of divergence of the two types (that is, origin of the $2n = 48$ karyotype from the putative $2n = 50$ karyotype of *B. arnee*) has been estimated in various studies as from at least 10 000 years ago to 1.7 Mya, but probably around 128 000–270 000 years ago (Kumar *et al.* 2007). Even given

Box 16.1 *(cont.)*

this uncertainty, divergence occurred well before domestication (Zhang *et al.* 2011).

Morphological data, including horn size and shape, show that *B. arnee* and the swamp type are phenotypically similar. That is, following domestication, the swamp type has not changed phenotypically, except for some reduction in body size, and they are mostly phenotypically homogenous throughout their distribution, while the river type has been subject to human-mediated selection and the development of a number of distinct breeds. The known distribution of *B. arnee*, including areas where it may still exist, overlaps that of the river buffalo distribution in India, Nepal and Bhutan, and overlaps that of the swamp buffalo in Myanmar, Thailand and Cambodia. Thus conservation of *B. arnee* depends on being able to distinguish true wild *B. arnee* from feral river buffalo and hybrids with river buffalo in the former three countries, and from feral swamp buffalo and hybrids with swamp buffalo in the latter three countries. Most wild buffalo are now located in wildlife reserves and national parks. Nevertheless, the major threat to their genetic integrity is crossbreeding with domestic animals, as farmers allow their buffalo females to range into forest areas to mate with wild bulls. F1 progeny that are not found and returned to the domestic herd will become feral and join the wild herd, so that over time the wild herd will contain increasing numbers of animals with varying levels of hybridity.

The difficulty of distinguishing between free-ranging domestic buffalo, feral buffalo, wild × domestic hybrids and truly wild *B. arnee* is well recognised, but few attempts have been made to study the problem – using either phenotypic and/or genetic criteria. For experienced observers, distinguishing wild from domestic, either river or swamp, is not difficult (Muley 2001). But in any field situation, hybrids invariably also are present, and will not be readily distinguishable, particularly backcrosses to *B. arnee*. Muley (2001) used morphometric and genetic data to differentiate wild, hybrid and domestic swamp buffalo in Kaziranga National Park, Assam, India. Animals were first classified as wild, hybrid or domestic on morphological criteria – chevron markings, stockings and tail length. For each animal, morphometric data (linear measurements related to horn shape) were taken from frontal photographs, and subjected to various multivariate analyses. Some analyses showed significant differentiation among the pre-defined groups, but with classification errors of 13% for wild vs domestic and 27.8% for wild vs hybrid. Unfortunately, Muley (2001) does not give results in terms of misclassification of individuals – for example, putative wild classed as hybrid or vice versa. Reanalysis following transfer of misclassified individuals would likely reduce the error rate, and should be investigated. Nevertheless, frontal photographs have the advantage of being non-invasive and safe for the investigator, and the analyses of linear measurements potentially improve discrimination among the three classes, as compared with the initial subjective classification. However, for conservation purposes the critical distinction to be made is between wild and hybrid, and the error rate observed in this study is not acceptable, particularly as it could be higher if the sample size were less than here (about 100 – counted from figure 2.2 in Muley 2001, but not stated in the thesis).

Box 16.1 *(cont.)*

Genetic data provide another possible means to discriminate among wild, hybrid and domestic buffalo. During the domestication process, both river and swamp types were likely subject to a bottleneck, while subsequent selection (particularly for the river type) and genetic drift will have led to increasing differentiation from each other and from *B. arnee*. Muley (2001) assayed 17 microsatellite loci in predefined wild, hybrid and domestic animals (as above for morphometric analyses), and non-descript river buffaloes from a New Delhi slaughterhouse. Principal component analysis of the genotypic data showed the four groups to be different, but this does not mean that each animal was classified correctly. One would expect that domestic animals and F1 hybrids that had been returned to the farmer's herd would be correctly classified. But misclassification is certainly possible, particularly where wild and domestic populations have been in contact for many generations. For example, some of the predefined wild may in fact be hybrids with a high proportion of wild genes, or even predefined domestic animals may be hybrids, due to backcrossing from an F1 individual to domestic animals.

Given such potentially mixed populations, animals can first be classed as wild, hybrid or domestic based on morphological criteria, then genotyped for a set of markers and a genetic admixture and assignment test done to estimate the proportions of wild and domestic ancestry for each individual. This approach has been used by Flamand *et al.* (2003) in trying to identify with high probability those animals most likely to be purebred wild. Hair (tail) samples were collected from 45 water buffalo in or in the vicinity of Kosi Tappu Wildlife Reserve, Nepal. DNA was extracted and genotypes determined for ten microsatellite loci. Applying the assignment method STRUCTURE of Pritchard *et al.* (2000), two of the ten predefined wild and seven of the 28 predefined domestic animals were reclassified as hybrids. A further analysis, assuming that wild and domestic were now correctly defined, checked the inferred ancestry of the hybrid individuals. Subsequently, an additional ten microsatellite loci were assayed and the STRUCTURE analysis repeated, with individuals initially assigned as wild, hybrid or domestic based on the results of the previous ten loci. Five animals were reassigned – one wild to hybrid, one hybrid to wild, two hybrid to domestic and one domestic to hybrid (Zhang *et al.* 2011). As one might expect, additional loci have changed some of the classifications, presumably increasing the accuracy of the assignments.

What is the future? What can we learn from the limited studies that have been done? Clearly phenotypic assessment is not adequate. Rife (1962) provided some information on the inheritance of coat colour, eye colour and horn shape in river and swamp buffalo. More detailed study of the inheritance of coat colour, chevrons and horn traits using crosses between river and swamp (as a proxy for true wild) may improve phenotypic assessment. However, only genotypic data can provide definitive identification of true wild *B. arnee* or hybrid animals with a high proportion of *B. arnee* genes that may be worth including in a conservation programme.

Assignment test results are probabilistic, and conservation management plans will need to consider whether the aim is to attempt a genetically pure wild population, or to accept some degree of introgression from domestic buffalo. In the former case, the costs of obtaining increased accuracy will have to be traded off against other aspects of the programme (including effective size of the founder population), but two approaches are possible: (1) genotyping additional microsatellite loci, and (2) mitochondrial DNA (mtDNA) sequencing. As introgression from the domestic to the wild population will have been primarily female mediated, mtDNA sequencing should be particularly useful (Ward *et al.* 1999). Of 33 mtDNA D-loop haplotypes found in 57 swamp and 23 river buffalo (Lau *et al.* 1998), two of the three most common were present in both swamp and river types, 23 (including one-third of the most common haplotypes) were found in swamp only, and eight in river only. As Lau *et al.* (1998) sequenced only a small portion of the D-loop (158 bp), more extensive analysis of the complete D-loop sequence from a range of river and swamp populations is required to confirm that some haplotypes are specific to river buffalo. If so, presence of a river-only haplotype in a presumed wild animal would identify it as a hybrid, while classification of an animal with any other mtDNA haplotype would have to be based on its microsatellite loci genotypes. Zhang *et al.* (2006) identified a Y-linked SNP that differentiated paternal river and paternal swamp animals. If the paternal swamp nucleotide is present also in *B. arnee*, this marker could be used to identify any male-mediated river buffalo introgression into hybrid populations.

Of course, obtaining samples for genetic analysis is difficult (but see Muley 2001) and the analyses expensive, so priority for such work should be given to those populations (like Koshi Tappu) where the potential for hybridisation is greatest.

Adequate genetic data (microsatellites, SNP or mtDNA) is a prerequisite for any attempt to differentiate wild, hybrid and domestic individuals in a mixed population. In recent years a number of new methods have been developed for genetic mixture and assignment testing (Anderson & Thompson 2002; Price *et al.* 2009; Sohn *et al.* 2012; also see the review by Manel *et al.* 2005), which may provide more accurate discrimination than the method used by Flamand *et al.* (2003). Finally, the accuracy of any assignment method would be dramatically improved if genetic data were available for a sample of true *B. arnee*. Possibly such information may be obtained from bones preserved in archaeological sites, as was done for the extinct *B. mephistopheles* (Yang *et al.* 2008).

References

Anderson, E. C. & Thompson, E. A. (2002). A model-based method for identifying species hybrids using multilocus genetic data. *Genetics* **160**: 1217–1229.

Barker, J. S. F., Moore, S. S., Hetzel, D. J. S., Evans, D., Tan, S. G. & Byrne, K. (1997). Genetic diversity of Asian water buffalo (*Bubalus bubalis*): microsatellite variation and a comparison with protein-coding loci. *Animal Genetics* **28**: 103–115.

Burton, J. A., Hedges, S. & Mustari, A. H. (2005). The taxonomic status, distribution and conservation needs of the lowland anoa *Bubalus depressicornis* and mountain anoa *B. quarlesi*. *Mammal Review* **35**: 25–50.

Box 16.1 (*cont.*)

Cockrill, W. R. (1984). Water buffalo. In *Evolution of Domesticated Animals*, pp. 52–63. Ed. Mason, I. L. London: Longman.

Di Berardino, D. & Iannuzzi, L. (1981). Chromosome banding homologies in swamp and Murrah buffalo. *Journal of Heredity* **72**: 183–188.

Epstein, H. (1971). *Domestic Animals of China*. New York: Africana Publishing Co.

FAO (2000). *World Watch List for Domestic Animal Diversity*. 3rd edition. Ed. Scherf, B. D. Rome: Food and Agriculture Organization of the United Nations.

Fischer, H. & Ulbrich, F. (1968). Chromosomes of the Murrah buffalo and its crossbreds with the Asiatic swamp buffalo (*Bubalus bubalis*). *Zeitschrift für Tierzüchtung und Züchtungsbiologie* **84**: 110–114.

Flamand, J. R. B., Vankan, D., Gairhe, K. P., Duong, H. & Barker, J. S. F. (2003). Genetic identification of wild Asian water buffalo in Nepal. *Animal Conservation* **6**: 265–270.

Hedges, S., Duckworth, J. W., de Leon, J., Custodio, C. & Gonzales, J. C. (2008a). *Bubalus mindorensis*. In *IUCN 2012: IUCN Red List of Threatened Species*. Version 2012.1. www.iucnredlist.org, accessed 7 August 2012.

Hedges, S., Sagar Baral, H., Timmins, R. J. & Duckworth, J. W. (2008b). *Bubalus arnee*. In *IUCN 2012: IUCN Red List of Threatened Species*. Version 2012.1. www.iucnredlist.org, accessed 7 August 2012.

Kumar, S., Nagarajan, M., Sandhu, J. S., Kumar, N., Behl, V. & Nishanth, G. (2007). Mitochondrial DNA analyses of Indian water buffalo support a distinct genetic origin of river and swamp buffalo. *Animal Genetics* **38**: 227–232.

Lau, C. H., Drinkwater, R. D., Yusoff, K., Tan, S. G., Hetzel, D. J. S. & Barker, J. S. F. (1998). Genetic diversity of Asian water buffalo (*Bubalus bubalis*): mitochondrial DNA D-loop and cytochrome b sequence variation. *Animal Genetics* **29**: 253–264.

Liu, L., Chen, X. & Jiang, L. (2004). A study of Neolithic water buffalo remains from Zhejiang, China. *Bulletin of the Indo-Pacific Prehistory Association*, **24**: 113–120.

MacGregor, R. (1941). The domestic buffalo. *Veterinary Record* **53**: 443–450.

Manel, S., Gaggiotti, O. E. & Waples, R. S. (2005) Assignment methods: matching biological questions with appropriate techniques. *Trends in Ecology and Evolution* **20**: 136–142.

Mason, I. L. (1974). Species, types and breeds. In *The Husbandry and Health of the Domestic Buffalo*, pp. 1–47. Ed. Cockrill, W. R. Rome: Food and Agriculture Organization of the United Nations.

Box 16.1 (*cont.*)

Muley, P. D. (2001). Genetic and morphometric studies to differentiate between wild and domestic Asian water buffaloes (*Bubalus bubalis*) and their hybrids in Kaziranga National Park, Assam, India. PhD thesis. University of Wisconsin-Madison.

Price, A. L., Tandon, A., Patterson, N., *et al.* (2009). Sensitive detection of chromosomal segments of distinct ancestry in admixed populations. *PLoS Genetics* **5**(6): e1000519.

Pritchard, J. K., Stephens, M. & Donnelly, P. (2000). Inference of population structure using multilocus genotype data. *Genetics* **155**: 945–959.

Rife, D. C. (1962). Color and horn variations in water buffalo: the inheritance of coat color, eye color and shape of horns. *Journal of Heredity* **53**: 239–246.

Semiadi, G., Mannullang, B., Burton, J., Schreiber, A., Mustari, A. H. & the IUCN SSC Asian Wild Cattle Specialist Group (2008a). *Bubalus depressicornis*. In *IUCN 2012: IUCN Red List of Threatened Species*. Version 2012.1. www.iucnredlist.org, accessed 7 August 2012.

Semiadi, G., Burton, J., Schreiber, A. & Mustari, A. H. (2008b). *Bubalus quarlesi*. In *IUCN 2012: IUCN Red List of Threatened Species*. Version 2012.1. www.iucnredlist.org, accessed 7 August 2012.

Sohn, K.-A., Ghanramani, Z. & Xing, E. P. (2012). Robust estimation of local genetic ancestry in admixed populations using a nonparametric approach. *Genetics* **191**: 1295–1308.

Ulbrich, F. & Fischer, H. (1967). The chromosomes of the Asiatic buffalo (*Bubalus bubalis*) and the African buffalo (*Cyncerus caffer*). *Zeitschrift für Tierzüchtung und Züchtungsbiologie* **83**: 219–223.

Ward, T. J., Bielawski, J. P., Davis, S. K., Templeton, J. W. & Derr, J. N. (1999). Identification of domestic cattle hybrids in wild cattle and bison species: a general approach using mtDNA markers and the parametric bootstrap. *Animal Conservation* **2**: 51–57.

Yang, D. Y., Liu, L., Chen, X. & Speller, C. F. (2008). Wild or domesticated: DNA analysis of ancient water buffalo remains from north China. *Journal of Archaeological Science* **35**: 2778–2785.

Zhang, Y., Sun, D., Yu, Y. & Zhang, Y. (2006). A Y-linked SNP in SRY gene differentiates Chinese indigenous swamp buffalo and introduced river buffalo. *Asian-Australasian Journal of Animal Science* **19**: 1240–1244.

Zhang, Y., Vankan, D., Zhang, Y. & Barker, J. S. F. (2011). Genetic differentiation of water buffalo (*Bubalus bubalis*) populations in China, Nepal and south-east Asia: inferences on the region of domestication of the swamp buffalo. *Animal Genetics* **42**: 366–377.

Jokai and Dum Duma Reserve Forests in Assam should be declared wildlife sanctuaries or conservation reserves. Lonke *Tapu*, a small area near Burhachapori Wildlife Sanctuary which has good habitat and a population of wild buffaloes, should also be brought under a protected area network as a conservation reserve.

Extension of protected areas

In view of large-scale destruction of habitat outside the protected areas, wherever there is a potential buffalo habitat nearby should be included within the park/sanctuary. The Manas National Park in Assam should be extended up to the Kanamakra River towards the west by including most of Manas Reserve Forest. In Meghalaya, the Balpakram National Park–Siju Wildlife Sanctuary needs to be enlarged significantly to cover Baghmara, Rewak and Emangiri Reserve Forests, the corridor between Nokrek National Park and Siju Wildlife Sanctuary, and the unclassed forests of the West Khasi Hills. The total area would be *c.*750 km^2. An additional buffer zone may be worked out in the future.

The grassy parts of Kamlang Reserve Forest, Lai Anchal Reserve Forest and some adjacent areas (totalling about 30 km^2) should be added to Kamlang Wildlife Sanctuary.

The Kumargram and Newlands Blocks should be included within the national park area of Buxa.

Bhutan should declare a small sanctuary around Khurul River (at least 40 km^2), west of Kalikhola.

Rescue of abandoned calves

Every year one or two calves of the Kollolua-Jokai population are lost due to starvation after being abandoned by cows when chased from paddy fields by villagers. Such calves need to be rescued, considering the endangered status and low numbers of the species across its range. These calves may be kept in a new rescue centre that could be set up in Jokai Reserve Forest or shifted to the centre near Kaziranga.

In Kaziranga, Dibru-Saikhowa, Laokhowa and Burhachapori a few calves are washed away or separated from their mothers by flood waters. In Kaziranga such calves are normally rescued, but not elsewhere. Hence, any stranded calf should be rescued and kept in scientifically managed centres for future release into the wild.

Mitigation of human–buffalo conflict

While fencing (as mentioned earlier) could be of great help in reducing such conflicts, the affected families should be paid some compensation for crop loss. In cases of loss of human lives, there is already some mechanism for payment but that needs to be done without delay. Stray animals should not be allowed to be killed under any circumstances.

No tree plantation in the grasslands

No further tree plantation should be done in the grasslands.

Crisis management

In view of past experiences from protected areas such as Manas, Laokhowa, Indravati, Pamed and Udanti, where insurgency and political unrest took a heavy toll on wildlife and their habitat, a crisis management plan should be put in place to tackle such issues. There should be a good network between the management, district/local administration and the security forces fighting the insurgents.

Solution to the problems of forest villages in Dibru-Saikhowa National Park

Dibru-Saikhowa National Park is currently the second most important wild buffalo habitat in its entire range, hence, the problem arising out of the forest villages located inside, namely Laika and Dadhia, should be given more attention. These villagers face severe problems from erosion, flood and lack of communication, while on the other hand, they are involved in illegal felling of trees, encroachment and poaching. Since these were not original human habitations but settled by the Forest Department, the villagers should be provided with a secure package for land and better housing under a relocation programme elsewhere, with their consent.

Experiences elsewhere in India show that often people do not want to live in such marginal areas, but they see themselves as having no credible alternatives, and fear the possibility of ending up in an even worse place if they agree to relocation. Hence, adequate investment in canvassing what the villagers want elsewhere and helping them find it is almost essential for such a relocation programme to succeed.

Close monitoring of potential misuse of Forest Rights Act

Strict monitoring should be there to prevent possible misuse of *The Scheduled Tribes and Other Traditional Forest Dwellers (Recognition of Forest Rights) Act, 2006* during implementation, which has already begun. Determination of bona fide forest dweller status and verification of the date of their occupation must be done thoroughly to check for forgery of documents. All key natural habitats should be brought outside the purview of this Act through suitable amendments.

Other measures

The wild buffalo should be given due recognition and conservation attention across its range by issuing postage stamps, signages and booklets in different local languages. It should also be uplisted to Appendix I of CITES. Conservation education among locals should be implemented, with active involvement of local NGOs. Although rinderpest has been eradicated, cattle populations in the fringes of protected areas should be monitored regularly and inoculated, if necessary, as a preventive measure. Anthrax and FMD should also be monitored and the domestic stock inoculated. The dependency of fringe villagers on grassland and forest should be reduced. One method is large-scale installation of bio-gas, which would greatly reduce pressure on reeds and wood (used as fuel). Stray and problem animals should be relocated after tranquillisation. Further research should be carried out on ecology, behaviour and movement. Use of dangerous pesticides in tea plantations should be banned and monitored. Denotification of any part of any reserve of an endangered species such as wild buffalo should be avoided. Last, but not least, massive human population control measures should be taken in fringe areas.

Status in captivity

The only captive population is in Kathmandu zoo, Nepal where a pair and a grown-up calf are present. There is a female in captivity in Udanti Wildlife Sanctuary, India. In Kaziranga there are also occasional rescued calves. The total captive population in Asia is constituted of two males and two females. In Australia there are 20 undetermined buffaloes. In Europe there are five males, three females and seven undetermined. The worldwide captive population at the time of writing is 39 individuals held in zoological collections in Europe, Asia and Australia (see also Chapter 27).

Appendices

Appendix 16.1 Some measurements of horns of wild buffaloes (measurements in cm/feet and inches)

Locality/area of specimen collection	Max. length of a single horn	Max. spread	Tip to tip (span)	Sweep (across outer side and forehead)	Girth at base (max)	Remarks
1	2	3	4	5	6	7
Male						
India				424 (13′ 11″)		Shot in 1955; perhaps the largest ever recorded (Guinness World Records 2006)
Unknown	195.9 [6′ 5.27″]				44 [1′ 5.32″]	British Museum (Sloane Collection; Ward 1910)
Kamlang pani, Arunachal Pradesh				Approx. 388+ (12′ 9″)		Cooper (1873) had measured tip to tip round inside and across forehead, 340.36 [11′ 2″]
Rangamati, Dhubri, Assam	184.5 [6′ 0.6″]	167 [5′ 5.7″]	93.4 [3′ 0.8″]	381.2 [12′ 6.1″]	51.6	Shot by A. Forbes in 1875 (MNH 1920.5.14.1)
Near Dhubri, Assam	180.34 [5′ 11″]		88.3 [2′ 10.75″]		54.6 [1′ 9.5″]	Measured by A. O.Hume; shot by A. Forbes (Ward 1910)
Near Dhubri, Assam	177.8 [5′ 10″]	182.9 [6′]	152.4 [5′]		47 [1′ 6.5″]	Owned by J. H. Whitehouse (Ward 1910)
Rewak, Meghalaya						It was just short of the 'record' shot in Assam plains (Munrowd 1945a)
Near Dhubri, Assam	171.45 [5′ 7.5″]					Owned by Major H. Gidney (Ward 1910)
Near Dhubri, Assam	167 [5′ 5.75″]				51.44 [1′ 8.25″]	British Museum (Ward 1910), shot in Assam (Lydekker 1907)
Assam, India	160 [5′ 3″]	144 [4′ 8.7″]	237 [7′ 9.3″]	347.7 [11′ 4.9″]	46 [1′ 6.1″]	Beq. Rothschild (MNH 39.4411)
Manas National Park, Assam	160 [5′ 3″]	185 [6′ 0.8″]	148.5 [4′ 10.46″]	332 [10′ 10.7″]	50 [1′ 7.7″]	Displayed at Field Director's Office, Barpeta Road
Near Dhubri, Assam	157.5 [5′ 2″]		256.54 [8′ 5″]		43.2 [1′ 5″]	Owned by Walter Rothschild (Ward 1910)
Assam	156.5 [5′ 1.6″]	156 [5′ 1.4″]	90 [2′ 11.4″]	334.5 [10′ 11.7″]	52 [1′ 8.5″]	Displayed at Tourist Reception Centre, Assam House, Kolkata
Chhattisgarh	152.4 [5′]	182.9 [6′]			55.9 [1′ 10″]	Owned by Major B. Vincent (Ward 1910)
Unknown	152.4 [5′]		101.6 [3′ 4″]		50.8 [1′ 8″]	Owned by Sir Edmund G. Loder (Ward 1910)
West Assam/ north Bengal				327.2 [10′ 8.87″]	45.8 [1′ 6″]	Shot in 1908 (Maharajah of Cooch Behar 1908)
North Bengal				322.6 [10′ 7″]	55.8 [1′ 9.75″]	Shot in 1880 (Maharajah of Cooch Behar 1908)
Unknown	151.1 [4′ 11.5″]	149.86 [4′ 11″]			38.1 [1′ 3″]	Owned by Sir Savile Crossley (Ward 1910)
Parokata, north Bengal				320 [10′ 6″]	48 [1′ 7″]	Shot in 1882 (Maharajah of Cooch Behar 1908)
Cooch Behar, north Bengal				317.5 [10′ 5″]	53.34 [1′ 9″]	Owned by the Maharaja of Cooch Behar (Ward 1910)

Appendix 16.1 *(cont.)*

Locality/area of specimen collection	Max. length of a single horn	Max. spread	Tip to tip (span)	Sweep (across outer side and forehead)	Girth at base (max)	Remarks
1	2	3	4	5	6	7
Jabda, Cachar, Assam	148 [4′ 10.3″]	129.5 [4′ 3″]	61.5 [2′ 0.2″]	317 [10′ 4.8″]	50 [1′ 7.7″]	Sp. 2 in Choudhury (2001a)
Female						
	197.5 [6′ 5.75″]				45.4 [1′ 5.87″]	At NHM (Anon 1940; p. 818)
	196.54 [6′ 5.37″]					Conversion to cm was wrongly shown as 197.6 in Prater (1980)
East Siang, Arunachal Pradesh				*c*.400 [13′+]		Pakar *chapori*, downstream of Pasighat (Munrowd 1945b)
Near Dhubri, Assam	179.1 [5′ 10.5″]	198.1 [6′ 6″]	88.27 [2′ 10.75″]		45.7 [1′ 6″]	Owned by the late Sir A. Campbell Orde (Ward 1910)
Near Dhubri, Assam	163.8 [5′ 4.5″]	243.84 [8′]	107.95 [3′ 6.5″]		45.7 [1′ 6″]	Owned by J. Campbell of Kilberry (Ward 1910)
Manas River, Assam	159.7 [5′ 2.87″]	184.15 [6′ 0.5″]		359.1 [11′ 9.37″]	48.8 [1′ 7.21″]	At Shillong Club (Parsons 1940)
'Garad Haut', north Bengal				351.8 [11′ 6.5″]	38.7 [1′ 3.25″]	Shot in 1886, his finest (Maharajah of Cooch Behar 1908)
Western Assam			250.2 [8′ 2.5″]	333.4 [10′ 11.25″]	42.4 [1′ 4.7″]	A straight-horned cow (Maharajah of Cooch Behar 1908)
Cooch Behar, N Bengal	157.2 [5′ 1.87″]	198.1 [6′ 6″]	56.73 [1′ 10.33″]		40 [1′ 3.75″]	Owned by the Maharaja of Cooch Behar (Ward 1910)
Assam	156.2 [5′ 1.5″		55.9 [1′ 10″]		40.64 [1′ 4″]	Owned by Walter Rothschild (Ward 1910)
North Bengal/ west Assam				331.5 [10′ 10.5″]	38.7 [1′ 3.25″]	Shot in 1907 (Maharajah of Cooch Behar 1908)
Central Provinces	152.4 [5′]				55.9 [1′ 10″]	Editors (1936; p. 818)
Assam	152.4 [5′]	171.8 [5′ 7.62″]	146.05 [4′ 9.5″]		34.3 [1′ 1.5″]	Owned by L. P.Haviland (Ward 1910)
'Bhalka balashi', north Bengal				321.3 [10′ 6.5″]	39.4 [1′ 3.5″]	Shot in 1895 (Maharajah of Cooch Behar 1908)
'Gordongunj', north Bengal				312.4 [10′ 3″]	40.6 [1′ 4″]	His biggest cow at 5′ 11″ (Maharajah of Cooch Behar 1908)
Shirajganj, Bangladesh				276.86 [9′ 1″]		Shot near Kasimpur-Rampur, in 1875 (Sanderson 1896)
Jonk Valley, Orissa				254 [8′ 4″]		Forsyth (1889)
Sex not mentioned						
Assam	177.8 [5′ 10″]					In AMNH, from Assam (Anon 1936).
Assam				356.87 [11′ 8.5″]		Stockley (1928)
Assam				340.36 [11′ 2″]		Stockley (1928)

Source: Choudhury, 2010 (also see this publication for more details and more records).

Appendix 16.2 Food plants and items eaten by wild buffaloes (main food plants in bold)

Species (family)	Parts eaten		Species (family)	Parts eaten	
	Leaf	Stem		Leaf	Stem
Acroceras munroanum (Gram.)	+		*Bambusa arundinacea* (Gram.)		Shoot
Acroceras zizanioides (Gram.)	+		*Brachiaria mutica* (Gram.)	+	
Alloteropsis semialata (Gram.)	+		*Centotheca lappacea* (Gram.)	+	
Arundinella bengalensis (Gram.)	+	+	*Chrysopogon aciculatus* (Gram.)	+	
Arundo donax (Gram.)	+	+	*Coix lacryma-jobi* (Gram.)	+	
Corchorus capsularis (Gram.)	+		*Eupatorium odoratum* (Compositae)	+	+
Cynodon dactylon (Gram.)	+		*Spilanthes acmella* (Compositae)	+	+
Cyrtococcum accrescens (Gram.)	+		*Momordica cochinchinensis* (Cucurbitaceae)	+	+
Cyrtococcum oxyphyllum (Gram.)	+		*Solena heterophylla* (Cucurbitaceae)	+	+
Dactyloctenium aegyptiacum (Gram.)	+		*Cyperus compactus* (Cyperaceae)	+	
Dendrocalamus strictus (Gram.)	+		*Cyperus corymbosus* (Cyperaceae)	+	
Dichanthium glabrum (Gram.)	+		*Cyperus difformis* (Cyperaceae)	+	
Digitaria ciliaris (Gram.)	+		*Cyperus diffusus* (Cyperaceae)	+	
Digitaria setigera (Gram.)	+		*Cyperus rotundus* (Cyperaceae)	+	
Echinochloa colonum (Gram.)	+		*Cyperus tenuiculmis* (Cyperaceae)	+	
Eleusine indica (Gram.)	+		*Fimbristylis mihacea* (Cyperaceae)	+	
Eragrostis atrovirens (Gram.)	+		*Shorea robusta* (Dipterocarpaceae)	+	+
Erianthus ravannae (Gram.)	+		*Cordia dichotoma* (Ehretiaceae)	+	+
Heteropogon contortus (Gram.)	+		*Croton* sp. (Euphorbiaceae)	+	
Hymenachne pseudointerrupta (Gram.)	+		*Euphorbia* sp. (Euphorbiaceae)	+	
Imperata cylindrica (Gram.)	+		*Homonoia riparia* (Euphorbiaceae)	+	+
Leersia hexandra (Gram.)	+	+	*Phyllanthus* sp. (Euphorbiaceae)	+	
Mnesithea leavis (Gram.)	+		*Ricinus communis* (Euphorbiaceae)	+	
Narenga porphyrocoma (Gram.)	+		*Pterocarpus marsupium* (Fabaceae)	+	+
Neyraudia reynaudiana (Gram.)	+	+	*Vigna mungo* (Fabaceae)	+	+
Oryza sativa (Gram.)	+	+	*Dolichos biflorus* (Leguminosae)	+	
Panicum auritum (Gram.)	+		*Lagerstroemia* sp. (Lythraceae)	+	+
Paspalum conjugatum (Gram.)	+		*Sida rhombifolia* (Malvaceae)	+	+
Paspalum paspaloides (Gram.)	+		*Urena lobata* (Malvaceae)	+	+
Paspalum scrobiculatum (Gram.)	+		*Acacia comosa* (Mimosaceae)	+	+
Polytoca digitata (Gram.)	+		*Albizzia procera* (Mimosaceae)	+	
Saccharum arundinaceum (Gram.)	+		*Streblus asper* (Moraceae)	+	+
Saccharum procerum (Gram.)	+	+	*Aeschynomene* sp. (Papilionaceae)	+	
Saccharum nufipilum (Gram.)	+		*Desmodium gangeticum* (Papilionaceae)	+	+
Saccharum officinarum (Gram.)	+	+	*Millettia* sp. (Papilionaceae)	+	+
Saccharum spontaneum (Gram.)	+	+	*Peperomia pellucida* (Peperomiaceae)	+	+
Sclerostachya fusca (Gram.)	+		*Bridelia retusa* (Phyllanthaceae)	+	+
Setaria palmifolia (Gram.)	+		*Monochoria hastata* (Pontederiaceae)	+	

Appendix 16.2 (cont.)

Species (family)	Parts eaten		Species (family)	Parts eaten	
	Leaf	Stem		Leaf	Stem
Sorghum nitidum (Gram.)	+		*Eichhornia crassipes* (Pontederiaceae)	+	+
Themeda arundinaceum (Gram.)	+	+	*Portulaca oleracea* (Portulacaceae)	+	+
Themeda quadrivalvis (Gram.)	+		*Zizyphus cambodiana* (Rhamnaceae)	+	
Vetiveria zizanioides (Gram.)	+		*Zizyphus mauritiana* (Rhamnaceae)	+	
Zea mays (Gram.)	+	+	*Zizyphus oenoplia* (Rhamnaceae)	+	+
Achyranthes aspera (Amaranthaceae)	+	+	*Cephalanthus tetrandra* (Rubiaceae)	+	
Alternanther asessils (Amaranthaceae)	+	+	*Neonauclea* sp. (Rubiaceae)	+	
Amaranthus caudatus (Amaranthaceae)	+	+	*Salix tetrasperma* (Salicaceae)	+	
Amaranthus spinosus (Amaranthaceae)	+	+	*Capsicum frutescens* (Solanaceae)	+	
Ageratum conyzoides (Asteraceae)	+	+	*Colona floribunda* (Tiliaceae)	+	
Brassica campestris (Brassicaceae)	+	+	*Grewia elatostemoides* (Tiliaceae)	+	
Crateva magna (Capparidaceae)	+		*Stachytarpheta indica* (Verbenaceae)	+	
Commelina diffusa (Commelinaceae)	+	+	*Alpinia allughas* (Zingiberaceae)	+	+

+ part eaten. Gram. = Graminae.
Sources: Daniel & Grubh (1966), Choudhury (1997b, 2010), Chaiyarat (2002), Kotwal & Mishra (2004). Identification of grass species of Choudhury (1997b) and unpublished data was from the Botanical Survey of India, Shillong with assistance from Bor (1982) and Shukla (1996).

Acknowledgements

Many thanks go to Joel Heinen for the critical and useful comments on this chapter.

References

Ali, S. (1927). The Moghul emperors of India as naturalists and sportsmen: part II. *Journal of the Bombay Natural History Society* **32**(1): 34–63.

Allen, B. C. (1905 [2002]). *Gazetteer of Naga Hills and Manipur*. New Delhi: Mittal Publications.

Anon. (1920). III: sporting diary. *Journal of the Bombay Natural History Society* **27**(2): 386–391.

Anon (1940). Editor's comments. *Journal of the Bombay Natural History Society* **41**(3): 660.

Anon (2005a). Camera trap catches wild buffalo in Cambodia. www.panda.org/W%20Buffalo/WWF%20-%20Camera%20trap%20catches%20wild%20buffalo%20in%20Cambodia.htm, accessed 3 May 2009.

Anon (2005b). Srepok project: photographing of wild water buffalo. www.panda.org/what_we_do/when_we_work/project/projects_in_depth/srepok_project/project/water_buffalo, accessed 10 July 2009.

Anon. (2006). Hunting the elusive kouprey. *The FCC Cambodia Monthly Newsletter*. September.

Anon. (2009). Wild water buffalo *Bubalus arnee*. www.wildcattleconservation.org, accessed 3 May 2009.

Ara, J. (1949). Wild life reserves in India: Bihar province. *Journal of the Bombay Natural History Society* **48**(2): 283–289.

Ashby, K. R. & Santiapillai, C. (1987). The ecology of free living water buffalo (*Bubalus bubalis*) in Sri Lanka with particular reference to Ruhuna National Park. *Tigerpaper* **14**(3): 20–26.

Asmat, G. S. M. (2001). *Bagladesher bilupto bannyaprani*. Dhaka, Bangladesh: Bangla Academy. (In Bengali)

Babur, Zahiru'd-din Muhammad (1529). *The Babur-nama (Memoirs of Babur)*. Translated from the original Turki text by Annette Susannah Beveridge (1917, 1921). *Vol. II sold by Luzac & Co, 46, Great Russell Street, London, UK. 1922. Section III* Hindustan.

Beekman, J. H. & Prins, H. H. T. (1989). Feeding strategies of sedentary large herbivores in East-Africa, with emphasis on the African buffalo, *Syncerus caffer*. *African Journal of Ecology* **27**: 129–147.

Berger, J. (1990). Persistence of different-sized populations: an empirical assessment of rapid extinctions in bighorn sheep. *Conservation Biology* **4**: 91–98.

Blanford, W. T. (1891). *Fauna of British India including Ceylon and Burma. Mammalia*. London: Taylor & Francis.

Blower, J. H. (1986): *Nature Conservation in Bhutan: Project Findings and Recommendations*. Rome: FAO.

Bor, L. N. (1982). *Flora of Assam*. Vol. V, Graminae (reprint from original of 1940). Delhi: A Von Book Company.

Brickle, N., Cu, N., Quynh, H. Q., Cuong, N. T. T. & San, H. V. (1998). *The Status and Distribution of Green Peafowl* Pavo muticus *in Dak Lak Province, Vietnam.* Hanoi: BirdLife International.

Brown, R. (1874). *Statistical account of Manipur.* Calcutta: Superintendent of Government Printing, India (reprint in 2001 by Mittal Publications, New Delhi).

Chaiyarat, R. (2001). Ecology and habitat utilisation of wild water buffalo (*Bubalus bubalis*) in Huai Kha Khaeng Wildlife Sanctuary, Thailand. Unpublished dissertation. Kasetsart University.

Chaiyarat, R. (2002). Forage species of wild water buffalo (*Bubalus bubalis*) in Huai Kha Khaeng Wildlife Sanctuary, Thailand. *Buffalo Journal* **3**: 289–302.

Chaiyarat, R. (2004). Status of wild water buffalo *Bubalus bubalis* in Thailand. *Tigerpaper* **31**(2): 6–9.

Choudhury, A. U. (1986). Wildlife in north-east India. *North-Eastern Geographer* **18** (1–2): 92–101.

Choudhury, A. U. (1988). Conservation in Manas Tiger Reserve. *Tigerpaper* **15**(2): 23–27.

Choudhury, A. U. (1989a). The Kaziranga Sanctuary. *The India Magazine* **9** (January): 18–28.

Choudhury, A. U. (1989b). S. O. S. Manas! *The Sentinel*, 7 May.

Choudhury, A. U. (1993a). Saving the wild buffalo. *The India Magazine*, September: 75.

Choudhury, A. U. (1993b). *A Naturalist in Karbi Anglong.* Guwahati: Gibbon Books.

Choudhury, A. U. (1994a). The decline of wild water buffalo in Northeastern India. *Oryx* **28**(1): 70–73

Choudhury, A. U. (1994b). Wild buffalo *Bubalus bubalis* in Dhakuakhana, Lakhimpur district of Assam. *Journal of the Bombay Natural History Society* **91**(1): 137–138.

Choudhury, A. U. (1994c). *Checklist of the Mammals of Assam.* Guwahati: Gibbon Books.

Choudhury, A. U. (1996). *Survey of the White-winged Wood Duck and the Bengal Florican in Tinsukia District and Adjacent Areas of Assam and Arunachal Pradesh.* Guwahati: The Rhino Foundation for Nature in North-East India,

Choudhury, A. U. (1997a). *Checklist of the Mammals of Assam.* Revised 2nd edition. Guwahati:: Gibbon Books and Assam Science Technology & Environment Council, India.

Choudhury, A. U. (1997b). Survey of grasslands in some parts of central and southern Assam: to assess their bio-diversity & socio-economic problem. WWF-India North-East Regional Office, Guwahati.

Choudhury, A. U. (1998a). Wild water buffalo *Bubalus bubalis/arnee* in Dibang Valley district of Arunachal Pradesh. *Journal of the Bombay Natural History Society* **95**(1): 110–112.

Choudhury, A. U. (1998b). Mammals, birds and reptiles of Dibru-Saikhowa Sanctuary, Assam, India. *Oryx* **32**(3): 192–200.

Choudhury, A. U. (1999). Status of the wild water buffalo *Bubalus arnee* in Lohit district, Arunachal Pradesh. *Journal of the Bombay Natural History Society* **96**(1): 130–132.

Choudhury, A. U. (2001a). On the occurrence of the wild water buffalo *Bubalus arnee* in the Barak Valley districts of Assam. *Journal of the Bombay Natural History Society* **98**(2): 270–271.

Choudhury, A. U. (2001b). A systematic review of the mammals of North-east India with special reference to the non-human primates. DSc thesis. Gauhati University.

Choudhury, A. U. (2002). Occurrence of the wild water buffalo *Bubalus arnee* in Mizoram and Manipur. *Journal of the Bombay Natural History Society* **99**(3): 512–516.

Choudhury, A. U. (2003a). *The Mammals of Arunachal Pradesh.* New Delhi: Regency Publications.

Choudhury, A. U. (2003b). Meghalaya's vanishing wilderness. *Sanctuary Asia* **23**(5): 30–35.

Choudhury, A. U. (2004). *Kaziranga: Wildlife in Assam.* New Delhi: Rupa & Co.

Choudhury, A. U. (2006). Manas: looking back, looking forward. *Sanctuary Asia* **26**(1): 22–29.

Choudhury, A. U. (2008). Distribution and status of the wild water buffalo *Bubalus arnee* in Bhutan. *Journal of the Bombay Natural History Society* **105**(3): 332–335.

Choudhury, A. U. (2009). *Census of Wild Water Buffalo in Unprotected Areas of Eastern Assam. Final Report.* Guwahati: The Rhino Foundation for Nature in North-East India.

Choudhury, A. U. (2010). *The Vanishing herds: The Wild Water Buffalo.* Guwahati: Gibbon Books & The Rhino Foundation for Nature in North-East India.

Choudhury, A. U. (2011). The northernmost range of gaur *Bos gaurus* and wild water buffalo *Bubalus arnee. Journal of the Bombay Natural History Society* **108**(1): 56–58.

Choudhury, A. U. & Dey, A. (2008). *Census of Wild Water Buffalo in Dibru-Saikhowa National Park. Final Report.* Guwahati: The Environment & Forest Department and The Rhino Foundation for Nature in North-East India.

Choudhury, A. U., Swargiary, A., Bhobora, C. R. & Saikia, B. (2008a). *Census of Wild water Buffalo in Manas National Park. Final Report.* Barpeta Road and Guwahati: The Field Directorate, Manas National Park, and The Rhino Foundation for Nature in North-East India.

Choudhury, A. U., Bonal, B. S. & Muthukumarvel, C. (2008b). *Census of Wild Water Buffalo in Laokhowa and Burhachapori Wildlife Sanctuaries. Final Report.* Guwahati: The Environment & Forest Department and The Rhino Foundation for Nature in North-East India.

Cockrill, W. R. (1984). Water buffalo. In *Evolution of Domesticated Animals*, pp. 52–63. Ed. Mason, I. L. London: Longman.

Cooper, T. T. (1873 [1995]). *The Mishmee Hills.* New Delhi: Mittal Publications.

Corbet, G. B. & Hill, J. E. (1992). *The Mammals of the Indomalayan Region: A Systematic Review.* Oxford: Oxford University Press.

Dahmer, T. D. (1978). Status and ecology of the wild Asian buffalo (*Bubalus bubalis* L.) in Nepal. MS thesis. University of Montana.

Daniel, J. C. & Grubh, B. R. (1966). The Indian wild buffalo, *Bubalus bubalis* (L.) in Peninsular India: a preliminary survey. *Journal of the Bombay Natural History Society* **63**(1): 32–53.

de Silva, M., Dissanayake, S. & Santiapillai, C. (1994). Aspects of the population dynamics of the wild Asiatic water buffalo (*Bubalus bubalis*) in Rahuna National Park, Sri Lanka. *Journal of South Asian Natural History* **1**: 65–76.

Deb Roy, S. (1991). Manas: a monograph. *Tigerpaper* **18**: 6–15.

Delacour, J. (1940). Liste provisoire des mammifères de l'Indochine française. *Mammalia* **4**: 20–29, 46–58.

Deuve, J. (1972). *Les mammifères du Laos.* Vientiane, Lao PDR: Ministère de l'education nationale.

Dinerstein, E. & McCracken, G. F. (1990). Endangered greater one-horned rhinoceros carry high levels of genetic variation. *Conservation Biology* **4**: 417–422.

Divekar, H. K. (1976). A resurvey of the status of wild buffaloes in the West Bastar, Madhya Pradesh. *Journal of the Bombay Natural History Society* **73**(2): 382–385.

Divekar, H. K. & Bhusan, B. (1988). *Status Survey of the Wild Asiatic buffalo Bubalus bubalis in the Raipur and Bastar Districts of Madhya Pradesh.* Bombay: Bombay Natural History Society.

Duckworth, J. W., Salter, R. E. & Khounboline, K. (1999). *Wildlife in Lao PDR: 1999 Status Report.* Vientiane, Lao PDR: IUCN, Wildlife Conservation Society and Centre for Protected Areas and Watershed Management.

Dumas, C. (1944). *La faune sauvage du Cambodge.* Saigon: J. Aspar.

Dunbar Brander, A. A. (1931). *Wild Animals in Central India.* London: Edward Arnold & Co.

Eames, J. C. & Nguyen Duc Tu (2002). An interim biodiversity report for Yok Don National Park, Dak Lak province, Vietnam. Unpublished report to Vietnam PARC Project, Yok Don Component.

Eisenberg, J. F. & Lockhart, M. (1972). An ecological reconnaissance of Wilpattu National Park, Ceylon. *Smithsonian Contributions to Zoology* **101**: 1–118.

Ellerman, J. R. & Morrison-Scott, T. C. S. (1966). *Checklist of Palaearctic and Indian mammals, 1758 to 1946.* 2nd edition. London: British Museum (Natural History).

Falconer, D. S. (1981). *Introduction to Quantitative Genetics.* London: Longman.

Fazl, Abu'l (*c.*1590 [1873]). *Ain-i-Akbari.* Translated by Henry Blochmann (vol. 1) and Henry Sullivan Jarrett (vols 2–3). Calcutta: The Asiatic Society of Bengal.

Finn, F. (1929). *Sterndale's Mammalia of India.* Revised edition. Calcutta and Simla: Thacker, Spink.

Fisher, J., Simon, N. & Vincent, J. (1969). *The Red Book: Wildlife in Danger.* London: Collins.

Flamand, J. R. B., Vankan, D., Gairhe, K. P., Duong, H. & Barker, J. S. F. (2003). Genetic identification of wild Asian water buffalo in Nepal. *Animal Conservation* **6**: 265–270.

Forsyth, J. (1889). *The Highlands of Central India.* London: Chapman & Hall.

Franklin, I. R. (1980). Evolutionary change in small populations. In *Conservation Biology: An Evolutionary-Ecological Perspective*, pp. 135–149. Eds. Soulé, M. E. & Wilcox, B. A. Sunderland MA: Sinauer.

Gee, E. P. (1950). Wildlife reserves in India: Assam. *Journal of the Bombay Natural History Society* **49**(1): 81–89.

Gee, E. P. (1952). The Assam earthquake of 1950. *Journal of the Bombay Natural History Society* **50**(3): 628–635.

Gee, E. P. (1953). Wild buffaloes and tame. *Journal of the Bombay Natural History Society* **51**(3): 727–730.

Gee, E. P. (1964). *The wild life of India.* London: Collins.

Gee, E. P. (1966). A note on the conference on conservation of nature and natural resources in tropical South-east Asia held at Bangkok, Thailand. *Journal of the Bombay Natural History Society* **63**(1): 162–166.

Graham, M. & Round, P. (1994). *Thailand's Vanishing Flora and Fauna.* Bangkok: Finance One Public Co. Ltd.

Grimsdell, J. J. R. & Field, C. R. (1976). Grazing patterns of buffaloes in the Rwenzori National Park, Uganda. *East African Wildlife Journal* **14**: 339–344.

Groves, C. P. (1996). The taxonomy of the Asian wild buffalo from the Asian mainland. *Z. Saugetierkunde (International Journal of Mammalian Biology)* **61**: 327–338.

Guinness World Records (2006). *Longest horns.* http://web.archive.org/web/20051123012556/http://www.guinnessworldrecords.com/content_pages/record.asp?recordid=51205, accessed 7 August 2006.

Gupta, R. B. & Mishra, H. R. (1972). *The Asiatic Wild Buffalo of Koshi Tappu: An Interim Report.* Kathmandu: Department of National Parks and Wildlife Reserve, Nepal.

Hall, J. E. (1936). Shooting wild buffalo in the Kosi river swamps. *Journal of the Bombay Natural History Society* **38**(3): 584–592.

Harper, F. (1940). The nomenclature and type localities of certain old world mammals. *Journal of Mammalogy* **21**: 191–203, 322–332.

Hasan, S. M. (1980). The enigmatic buffalo. *Hornbill* **2**: 26–27.

Hedges, S. (1995). Asian wild cattle and buffaloes: status report and conservation action plan. 2 vols. Working draft for CAMP/AWCSG meeting. WWF Project 3950/Asia, Thailand.

Heinen, J. T. (1993). Population viability and management recommendations for wild water buffalo *Bubalus bubalis* in Kosi Tappu Wildlife Reserve, Nepal. *Biological Conservation* **65**: 29–34.

Heinen, J. T. (2001). A translocation proposal for wild buffalo (*Bubalus bubalis*) in Nepal: recommended management action in the face of uncertainty for a critically endangered species. *Tigerpaper* **28**: 1–8.

Heinen, J. T. (2002). Phenotypic and behavioural characteristics used to identify wild buffalo *Bubalus bubalis* from feral backcrosses in Nepal. *Journal of the Bombay Natural History Society* **99**(2): 173–183.

Higgins, J. C. (1934). The game birds and animals of the Manipur State with notes on their numbers, migration and habits: part VI. *Journal of the Bombay Natural History Society* **37**(2): 298–309.

Humphrey, S. R. & Bain, J. R. (1990). *Endangered Animals of Thailand.* Gainesville, FL: Sandhill Crane Press.

Imam, A. R. H. B. (1985). Seasonal migration of gaur. *Cheetal* **26**: 45–48.

Inverarity, J. D. (1895). The Indian wild buffalo. *Journal of the Bombay Natural History Society* **10**: 41–52.

Islam, F. M. S., Rahman, R. H. & Chowdhury, S. M. Z. H. (1992). Prevalence of parasites of water buffaloes in Bangladesh. *AJAS* **5**(4): 601–604.

IUCN (2013). *IUCN Red List of Threatened Species.* Version 2013.2. www.iucnredlist.org, accessed 29 May 2014.

Jalil, A. F. M. (1986). *Sundarbaner itihash.* Dhaka: Ahmad Publishing House, Bangladesh. (In Bengali)

Jesser, P., Markula, A. & Csurhes, S. (2008). *Pest Animal Risk Assessment Water Buffalo* Bubalus bubalis. Brisbane: Biosecurity Queensland Department of Primary Industries and Fisheries.

Kerr, R. (1792). *The Animal Kingdom of the Celebrated Sir Charles Linnaeus: Class I. Mammalia*. London: J. Murray,

Kie, J. G. & Bowyer, R. T. (1999). Sexual segregation in white-tailed deer: density-dependent changes in use of space, habitat selection, and dietary niche. *Journal of Mammalogy* **80**: 1004–1020.

Kinloch, A. A. (1885). *Large Game Shooting in Thibet, the Himalaya, Northern and Central India*. Calcutta: Thacker, Spink & Co.

Kinloch, A. A. (1892). *Large Game Shooting in Thibet, the Himalaya, Northern and Central INDIA*. 3rd revised and enlarged edition. Calcutta: Thacker, Spink & Co.

Kotwal, P. C. & Mishra, R. P. (2004). Ecobiology of Indian wild buffalo *Bubalus arnee* L. in Udanti Wildlife Sanctuary, Chhattisgarh, India. *Journal of the Bombay Natural History Society* **101**(2): 252–254.

Kotwal, P. C., Naik, M. L. & Mishra, R. P. (2003). Indian wild buffalo (*Bubalus bubalis* L.) an endangered species: habitat and diurnal activity pattern in Udanti Wildlife Sanctuary. *Cheetal* **41**(3–4): 19–28.

Lekagul, B. (1952). On the trail of the kouprey or Indo-Chinese forest ox (*Bibos sauveli*). *Journal of the Bombay Natural History Society* **50**(3): 623–628.

Lekagul, B. & McNeely, J. A. (1977). *Mammals of Thailand*. Bangkok: Association of Conservation of Wildlife.

Lenstra, J. A. & Bradley, D. G. (1999). Systematics and phylogeny of cattle. In *Genetics of Cattle*, pp. 1–14. Eds. Fries, R. & Ruvinsky, A. Wallingford: Commonwealth Agricultural Bureau International.

Le Xuan Canh, Pham Trong Anh, Duckworth, J. W., Vu Ngoc Thanh & Lic Vuthy (1997). *A Survey of Large Mammals in Dak Lak Province, Vietnam*. Hanoi: WWF Indochina Programme and IUCN.

Linnaeus, C. (1758). *Systema naturae per Regna Tria Naturae, secundum Classess, Ordines, Genera, Species, cum Characteribus, Differentiis, Synonymis, Locis. I: Regnum Animale*. 10th edition. Holmia: Laurentius Salvius.

Lydekker, R. (1900). *The Great and Small Game of India, Burma and Tibet*. London: Rowland Ward.

Lydekker, R. (1907). *The Game Animals of India, Burma, Malaya and Tibet*. London: Rowland Ward.

Maharajah of Cooch Behar (1908 [1985]). *Thirty-seven Years of Big Game Shooting in Cooch Behar, the Duars and Assam: A Rough Diary*. Bombay: The Times Press. Reprinted in 1985 as *Big Game Shooting in Eastern and North Eastern India* by Mittal Publications, Delhi, India.

Maia, L. C. M. (1970). The present position of the wild buffalo in Asia. *Loris* **12**: 86–88.

Mason, I. L. (1974). Species, types and breeds. In *The Husbandry and Health of the Domestic Buffalo*, p. 47. Ed. Cockrill, W. R. Rome: FAO Publication.

Massicot, P. (2004). Animal info: wild water buffalo. www.animalinfo.org/species/artiperi/bubaarne.htm.

Mathur, P. K., Malik, P. K. & Muley, P. D. (1995). *Ecology and Population Genetics of Asian Wild buffalo (*Bubalus bubalis*) in India*. Dehradun: Wildlife Institute of India.

Matschie, P. (1912). Beutestücke aus fernen Ländern. *Dt. Jäger-Zeitung* **59**: 97–103.

M'Cosh, J. (1837 [2000]). *Topography of Assam*. Second Indian reprint, 2000. New Delhi: Logos Press.

McIntosh, J. (2007). *The Ancient Indus Valley*. Santa Barbara, CA: ABC-CLIO.

Menon, V. (2003). *A Field Guide to Indian Mammals*. Delhi: Dorling Kindersley (India) Pvt. Ltd.

Mooney, H. F. (1930). The distribution of Wild Buffalo in Orissa. *Journal of the Bombay Natural History Society* **34**: 242–244.

Mukherjee, A. K. (1982). *Endangered Animals of India*. Calcutta: Zoological Survey of India.

Muley, P. D. (2001). Genetic and morphometric studies to differentiate between wild and domestic Asian water buffaloes (*Bubalus bubalis* L.) and their hybrids in Kaziranga National Park, Assam, India. PhD dissertation. University of Wisconsin.

Munrowd H. G. H. (1945a). Some reminiscences of sport in Assam: pt II. *Journal of the Bombay Natural History Society* **45**: 322–332.

Munrowd H. G. H. (1945b). Some reminiscences of sport in Assam: pt III. *Journal of the Bombay Natural History Society* **45**: 485–495.

Nakasatit, N. & Chanard, T. (1985). *Proceedings of the Forestry Seminar*. Bangkok: Kasetsart University.

Noronha, R. P. (1952). The Indian wild buffalo. *Journal of the Bombay Natural History Society* **51**(1): 266–267.

Noronha, R. P. (1954). More notes on the Indian buffalo. *Journal of the Bombay Natural History Society* **52**(1): 202–204.

Nowak, R. (1999). *Walker's Mammals of the World*. 6th edition. Baltimore, MD and London: Johns Hopkins University Press.

Odden, M., Wegge, P. & Storaas, T. (2005). Hog deer *Axis porcinus* need threatened tallgrass floodplains: a study of habitat selection in lowland Nepal. *Animal Conservation* **8**: 99–104.

Oryx (1967). Plight of wild buffalo in India. *Oryx* **9**: 176.

Oryx (1989). Pure buffalo need protection in India. *Oryx* **23**: 166.

Osgood, W. H. (1932). Mammals of the Kelley-Roosevelts and Delacour Asiatic expeditions. *Publications Field Museum of Natural History Zoology* **18**: 193–339.

Owen-Smith, R. N. (1988). *Megaherbivores: The Influence of Very Large Body Size on Ecology*. Cambridge: Cambridge University Press.

Pandey, R. K. (1988). Habitat utilization and diurnal activity pattern of Indian wild buffalo (*Bubalus bubalis* L.) in Indravati National Park, India: a study of habitat/animal interaction. *Journal of Tropical Forestry* **4**(3): 269–280.

Parsons, R. E. (1940). A large head of the Indian buffalo (*Bos bubalis*). *Journal of the Bombay Natural History Society* **41**(3): 659.

Phillips, W. W. A. (1929). The present state of the game and wild life in Ceylon. *Journal of the Bombay Natural History Society* **33**(4): 942–946.

Pong Leng-Ee (1978). The conservation of protected large mammals in Thailand. In *Wildlife Management in Southeast Asia: Proceedings of the BIOTROP symposium on animal populations and wildlife management in Southeast Asia*, pp. 47–60. Bogor: BIOTROP.

Prater, S. H. (1948). *Book of Indian Animals*. 2nd edition. Bombay: Bombay Natural History Society.

Prater, S. H. (1980). *The Book of Indian animals.* 4th edition. Bombay: Bombay Natural History Society.

Ranjitsinh, M. K. (1997). *Beyond the Tiger: Portraits of Asian Wildlife.* New Delhi: Brijbasi Printers Pvt. Ltd.

Ranjitsinh, M. K., Chauhan, S. K. S., Banubakode, S. B., *et al.* (2000). *Status and Conservation of the Wild buffalo (*Bubalus bubalis*) in Peninsular India.* Bombay: Madhya Pradesh Forest Department, WII, BNHS and BSCN.

Ranjitsinh, M. K., Verma, S. C., Akhtar, S. A., Patil, V., Sivakumar, K. & Bhanubhakude, S. (2004). Status and conservation of the wild buffalo *Bubalus bubalis* in Peninsular India. *Journal of the Bombay Natural History Society* **101**(1): 64–70.

Redfern, J. V., Grant, R., Biggs, H. & Getz, W. M. (2003). Surface-water constraints on herbivore foraging in the Kruger National Park, South Africa. *Ecology* **84**: 2092–2107.

Richardson, W. S. J. (1889). Correspondence: wild buffalo. *Journal of the Bombay Natural History Society* **4**(3): 231.

Rithe, K. & Fernandes, A. (2002). On the trail of the wild buffalo. *Sanctuary Asia* **22**(4): 46–51.

Ryan, S. J. & Jordaan, W. (2005). Activity patterns of African buffalo *Syncerus caffer* in the Lower Sabie Region, Kruger National Park, South Africa. *Koedoe* **48** (2): 117–124.

Sah, J. P. (1997). *Koshi Tappu Wetlands: Nepal's Ramsar Site.* Bangkok: IUCN.

Salter, R. E. (1983). *Survey of Currently Available Information on Internationally Threatened Wildlife Species in Burma.* Rangoon: FAO.

Sanderson, G. P. (1896). *Thirteen Years Among the Wild Beasts of India.* London: W. H. Allen & Co. Ltd.,

Santiapillai, C. (1990). Status, distribution and conservation of wild cattle in Thailand. Unpublished report. WWF-Asia Programme, Bogor, Indonesia.

Sarkar, S. U. (1993). *Ecology of Wildlife.* Chittagong: Institute of Forestry and Environmental Sciences.

Seidensticker, J. (1975). Ungulate populations in Chitwan Valley, Nepal. Unpublished report. Office of Zoological Research, National Zoological Park, Smithsonian Institution, Washington, DC.

Seidensticker, J. (1976). Ungulate populations in Chitwan Valley, Nepal. *Biological Conservation* **10**: 183–210.

Seidensticker, J. & McNeely, J. (1975). Observations on the use of natural licks by ungulates in the Huai Kha Khaeng Wildlife Sanctuary, Thailand. *Natural History Bulletin of the Siam Society* **26**: 25–34.

Sheshadri, B. (1969). *The Twilight of India's Wildlife.* London: John Baker Publishers.

Shukla, U. (1996). *Grasses of North-Eastern India.* Jodhpur: Scientific Publishers.

Sinclair, A. R. E. (1977). *The African Buffalo: A Study of Resource Limitation of Populations.* Chicago, IL: University of Chicago.

Singh, T. A. (1980). The enigmatic buffalo: 'feedback'. *Hornbill* **4**: 4.

Singh, S. (1984). A flight of rhinos. *Sanctuary Asia* **4**(3): 250–260.

Smith, C. H. (1827). The Carnassiers. In *Cuvier's Animal Kingdom: Synopsis of the Species of the Class Mammalia.* Vol. **2**. Order VII: Ruminantia (Pecora, Lin.). (Vol. **4**: pp. 1–428; Vol. **5**: pp. 296–376). Eds. Griffith, E. *et al.* London: Whittaker.

Songsri, S. (1987). The water buffalo *Bubalus bubalis* of Thailand. BIOTROP Special Publication No. 30: Conservation and management of endangered plants and animals.

Soulé, M. E. (1980). Thresholds for survival: maintaining fitness and evolutionary potential. In *Conservation Biology*, pp. 151–169. Eds. Soulé, M. E. & Wilcox, B. Sunderland, MA: Sinauer Associates.

Spillett, J. J. (1966). The Kaziranga Wildlife Sanctuary, Assam. *Journal of the Bombay Natural History Society* **63**(3): 494–527.

Spillett, J. J. (1967). A report on wildlife surveys in North India and southern Nepal: the Jaldapara Wild Life Sanctuary, West Bengal. *Journal of the Bombay Natural History Society* **63**: 534–556.

Stark, M. A. (1986). Daily movement, grazing activity and diet of savanna buffalo, *Syncerus caffer brachyceros*, in Benoue National Park, Cameroon. *African Journal of Ecology* **24**: 255–262.

Stebbing, E. P. (1920). *The Diary of a Sportsman Naturalist in India.* London: John Lane.

Stewart, A. E. (1927). *Tiger and Other Game.* London: Longmans, Green Ltd.

Stockley, C. H. (1928). *Big Game Shooting in the Indian Empire.* London: Constable & Co., Ltd.

Stockley, V. M. (1913). *Big Game Shooting in India, Burma and Somaliland.* London: Horace Cox.

Tikader, B. K. (1983). *Threatened Animals of India.* Calcutta: Zoological Survey of India.

Timmins, R. J. & Ou Ratanak (2001). *The Importance of Phnom Prich Wildlife Sanctuary and Adjacent Areas for the Conservation of Tigers and Other Key Species.* Phnom Penh: WWF Cambodia Conservation Program.

Tordoff, A. W., Timmins, R. J., Maxwell, A., Keavuth, H., Vuthy, L. & Hourt, K. E. (2005). *Biological Assessment of the Central Indochina Dry Forest Ecoregion.* Phnom Penh: WWF Cambodia Conservation Program.

Traill, L. W., Brook, B. W., Frankham, R. R. & Bradshaw, C. J. A. (2010). Pragmatic population viability targets in a rapidly changing world. *Biological Conservation* **143**: 28–34.

Tulloch, D. G. (1970). Seasonal movements and distribution of the sexes in the water buffalo, *Bubalus bubalis*, in the northern territory. *Australian Journal of Zoology* **18**: 399–414.

Tulloch, D. G. (1974). The feral swamp buffaloes of Australia's Northern Territory. In *The Husbandry and Health of the Domestic Buffalo*, pp. 493–505. Ed. Cockrill, W. R. Rome: FAO.

Tulloch, D. G. (1979). The water buffalo, *Bubalus bubalis*, in Australia: reproductive and parent–offspring behaviour. *Australian Wildlife Research* **6**: 265–287.

Tulloch, D. G. & Litchfield, R. T. (1981). Wallow for buffalo. *Australian Wildlife Research* **8**: 555–565.

Uicharoensak, V. (1992). *Ecology and Current Distribution of Wild Water Buffalo (*Bubalus bubalis*) in Huai Kha Khaeng Wildlife Sanctuary, Uthai Thani Province.* Bangkok: Kasetsart University. (In Thai)

Vagholikar, N. (2009). 'Dam virus' engulfs Arunachal Pradesh. *Hornbill* April–June: 42–46.

Walston, J. L. (2001). Mammals of Cambodia. In *Biodiversity, the Life of Cambodia: Cambodian bIodiversity Status Report 2001*, pp. 135–152. Ed. Smith, J. D. Phnom Penh: Cambodia Biodiversity

Enabling Activity, Food and Agriculture Organisation.

Wangchuk, T., Thinley, P., Tshering, K., Thsering, C. & Yonten, D. (2004). *A Field Guide to the Mammals of Bhutan*. Thimphu: Department of Forestry, Ministry of Agriculture, Royal Government of Bhutan.

Ward, R. (1910). *Records of Big Game*. 6th edition. London: Rowland Ward Ltd.

Weckerly, F. E., Ricca, M. A. & Meyer, K. P. (2001). Sexual segregation in Roosevelt elk: cropping rates and aggression in mixed-sex groups. *Journal of Mammalogy* **82**(3): 825–835.

Wharton, C. H. (1957). *An Ecological Study of the Kouprey* Novibos sauveli *(Urbain): Monograph 5*. Manila: Monographs of the Institute of Science and Technology.

Wilson, D. E. & Mittermeier, R. A. (2011). *Handbook of the Mammals of the World*. Vol. 2. Hoofed Mammals. Barcelona: Lynx Edicions.

Winterbach, C. H., Winterbach, H. E. K. & Theron, G. (1998). Seasonal variation in home range, movement and habitat selection of the cape buffalo *Syncerus caffer caffer* in the Willem Pretorius Game Reserve, Free State. *Submitted to Koedoe*.

Winterbach, H. E. K. & Bothma, J. Du P. (1998). Activity patterns of cape buffalo *Syncerus caffer caffer* in the Willem Pretorius Game Reserve, Free State. *South African Journal of Wildlife Research* **28**(3): 73–81.

Wroughton, R. C. (1920). Summary of the results from the Indian mammal survey of the Bombay Natural History Society: part VI. *Journal of the Bombay Natural History Society* **27**(1): 57–85.

Chapter 17

Anoas *Bubalus depressicornis* (C. H. Smith, 1827); *Bubalus quarlesi* (Ouwens, 1910)

Philip M. Wheeler, Abdul Haris Mustari and James Burton

Names

Genus: *Bubalus* C. H. Smith, 1827

Species: Lowland anoa *Bubalus depressicornis* (C. H. Smith, 1827)

Names in other languages: **French**: Anoa des plaines; **German**: Anoa; **Spanish**: Anoa de llanura; **Italian**: Anoa di pianura

Species: Mountain anoa *Bubalus quarlesi* (Ouwens, 1910)

Names in other languages: **French**: Anoa de quarle, Anoa des montagnes; **German**: Berganoa; **Spanish**: Anoa de Montaña; **Italian**: Anoa di montagna

Other common names: Quarles anoa

Ecology, Evolution and Behaviour of Wild Cattle: Implications for Conservation, ed. M. Melletti and J. Burton. Published by Cambridge University Press.

Taxonomy

The taxonomy of the anoas is uncertain and has been the cause of much debate (summarised by Burton *et al.* 2005), largely because of the substantial morphological and genetic variation displayed by wild and captive individuals. There is consistent agreement among authors since Groves (1969) that anoas should be regarded as members of the genus *Bubalus*. The subgenus *Anoa* has been proposed to clearly separate the anoas as distinct from other members of *Bubalus*, although it is not considered that there is adequate support for *Anoa* as a genus in its own right (Hollister 1919; Groves 1969). Currently accepted taxonomy recognises two species (Burton *et al.* 2005; Grubb 2005), 'lowland' and 'mountain', but past taxonomies have proposed both a single species and three or more species (e.g. Dolan 1965; Groves 1969). Recent studies have indicated that the genetic basis for the two species split is not strong, and that there may be greater support for distinct races or subspecies separated geographically (Mustari *et al.* 2012). It is relevant to note that the Indonesian National Action Plan for Anoas (Mustari *et al.* in press) was based on the identification of priority regions relating to putative genetically distinct geographically separated populations rather than known ranges of the two currently recognised species. These questions require significant further investigation in order to confirm what constitutes evolutionarily significant units for anoa, but until they are resolved there is no strong basis on which to abandon the two-species classification.

Subspecies and distribution

Anoas are endemic to the island of Sulawesi, Indonesia, where they were formerly widely distributed, and to surrounding islands, where their historical presence is uncertain on all but Buton (Plate 8; Hedges 1996; Burton *et al.* 2005). Anoas' reported good swimming ability (Mustari 2003) and the proximity of some of these islands to the mainland suggest that they must have at some stage supported anoa populations. The current distribution is restricted to mainland Sulawesi apart from the southern 'arm' of the mainland and Buton Island. The English names of the two species indicate altitudinal differentiation in distribution, though there is substantial overlap in the altitudinal as well as geographical range of both species (Burton *et al.* 2005; Semiadi *et al.* 2008a, 2008b).

No subspecies are currently recognised, though three have been proposed by those authors who recognise a single anoa species (*B. depressicornis depressicornis*, *B. d. quarlesi* and *B. d. fergusoni* (Dolan 1965; Groves 1969)). The possibility remains that a future taxonomic revision will identify as yet undescribed races or subspecies. Evidence to date suggests genetically distinct populations in north, central and south Sulawesi and on Buton Island (Mustari *et al.* 2012).

Table 17.1 Main morphological differences between mountain and lowland anoa

	Species	
Characteristic	*B. depressicornis*	*B. quarlesi*
Head/body	170–188 cm	122–153 cm
Horn Length (male)	27–37 cm	15–20 cm
(female)	18–26 cm	15–20 cm
Weight (max.)	300 kg	150 kg
Shoulder height	80–100 cm	75 cm
Life span (captive)	25–30 years	
Dental formula	I 0/3, C 0/1, P 3/3, M 3/3	

Descriptive notes

Body measurements: **Shoulder height**: 60–100 cm; **Head and body length**: 122–188 cm; **Body mass**: <300 kg (Wilson & Mittermeier 2011).

Anoas are the smallest wild cattle species. The *B. depressicornis* type is larger than *B. quarlesi* (Table 17.1). The skin and pelage varies from being dark brown to black and sparsely hairy ('*B. depressicornis* type', Figure 17.1) to lighter brown and densely haired ('*B. quarlesi* type', Figure 17.2). Young animals are densely haired and brown in colour (Figure 17.3) and darken as they mature, the extent being dependent on the species. *B. depressicornis* has distinct but variable white patches on the front of the lower forelimbs.

The horns of *B. depressicornis* type animals have transverse ridges and are triangular in cross-section, while those of *B. quarlesi* type are more round in cross-section and smooth (Groves 1969). There is substantial geographical and altitudinal variation in body size, pelage and horn shape, but it is currently unclear how much of this variation is genetic or environmental in its origin (Schreiber *et al.* 1999; Burton *et al.* 2005).

Habitat

Use of a broad range of habitats has been recorded, including primary and secondary tropical moist forest, swamp forest, grassland and farmland (Figures 17.4–17.5; Mustari 1995; Rahman & Mustari 2002). In north Sulawesi track counts indicated that use of riparian forest was substantially greater than either bamboo or other lowland forest (Rahman & Mustari 2002). On Buton Island, anoas appeared to use forests independent of their primary or secondary nature, the primary determinant of use being human activity in the forest (Dwiyahreni 2006; Wheeler *et al.* 2009). In Lore Lindu National Park, central Sulawesi track counts were highest in higher altitude areas away from the park's fringes (Burton & MacDonald 2009; Okarda 2010).

Figure 17.1 Anoa, female adult, seven years old from central Sulawesi (photo by Abdul Haris Mustari). A black and white version of this figure will appear in some formats. For the colour version, please refer to the plate section.

Figure 17.2 Anoa, male, 19 years old, from central Sulawesi (photo by Abdul Haris Mustari). A black and white version of this figure will appear in some formats. For the colour version, please refer to the plate section.

Data on population density are very limited and restricted to small-scale studies of a limited number of individuals or restricted localities. The most robust data come from southeast Sulawesi, where estimates range from 1.1–1.6 animals per km^2, but with wide confidence limits (Mustari 2003; Burton *et al.* 2005), and Buton where density was estimated at 1.1 animals per km^2 (Wheeler 2004). Further estimates are as low as 0.07 animals per km^2 in north Sulawesi (O'Brien & Kinnaird 1996) and as high as 2.4 animals per km^2 from southeast Sulawesi (Mustari 1995).

In some locations, seasonal shifts in habitat use have been recorded, associated with changes in water availability between dry and wet seasons (e.g. Rahman & Mustari 2002), but such patterns are probably restricted to specific localities where water resources are limiting. There is no evidence that age or sex classes differ in their habitat use.

Figure 17.3 Male, one year old, from southeast Sulawesi (photo by Abdul Haris Mustari). A black and white version of this figure will appear in some formats. For the colour version, please refer to the plate section.

Figure 17.4 Riverine forest, important habitat for anoa in Tanjung Peropa Wildlife Reserve, southeast Sulawesi (photo by Abdul Haris Mustari).

Movements and home range

There are no data on home range or movement of individuals.

Activity patterns

The species has been recorded to be active during both day and night. Direct observations suggest most activity occurs in the early morning: 16 of 18 encounters in the Lambusango wildlife reserve, Buton, were between the hours of 06:00 and 08:00 (P Wheeler, unpublished data), and Mustari (1996) recorded most activity before 10.00 and after 16.00, while individuals were recorded wallowing throughout the day.

Feeding ecology

Feeding preferences appear to be broad and dependent primarily on habitat. No specific studies of food selection in

Figure 17.5 Tanjung peropa Wildlife Reserve, habitat of anoa (photo by Abdul Haris Mustari).

Figure 17.6 *Merremia peltata*, liana, among the food frequently eaten by anoa in Sulawesi (photo by Abdul Haris Mustari).

relation to availability have been carried out. Analyses of diet from eaten plants, direct observation and faeces have demonstrated that anoas eat a wide range of mosses, grasses, forbs, fruit and seeds (Figure 17.6; Syam 1978; Foead 1992; Mustari 1995 2003; Pujaningsih *et al.* 2009). Mustari (2003) identified 146 species of plants consumed by wild anoa in southeast Sulawesi, with remains in faeces composed of between 50% and 80% dicots, the remainder being predominantly monocots and ferns. The digestive anatomy of anoa is somewhat different from other intermediate feeders, indicating that anoas may be more adapted to consuming grasses than other similarly sized species (Clauss *et al.*, 2009); this is supported by feeding studies on captive animals (Flores-Miyamoto *et al.* 2005). As with many herbivores, anoas are known to supplement low (primarily sodium) mineral content in food with feeding and drinking at mineral springs, licks and even drinking salt water

(Mustari 1995; Whitten *et al.* 2002). Daily food intake requirements are known only from captivity: Mustari and Mas'yud (2011) estimated intake at 6–8 kg per day, depending on the composition of food, figures which match the intake of nine animals at Taman Safari Indonesia (Manansang *et al.* 1996b).

Reproduction and growth

There is no evidence for a defined breeding season: breeding in captivity is non-seasonal, with births reported from most months of the year (Dolan 1965). Young animals sighted in the wild have been reported across several months, indicating that the same holds true for wild individuals (Mustari 1995; P. Wheeler, personal observation). The only reliable data on breeding come from captive animals: females can become sexually mature at 2–2.5 years old, males at three, with a median age at first reproduction of four years for both sexes (Dolan 1965). Females are in oestrus for around 24 hours every 22–30 days and gestation is between 275 and 315 days (National Research Council 1983; Jahja 1987). A single calf is usually born, though there are rare reports of twins being born and successfully reared in captivity (Hedges 1996). There are no specific data on calving intervals, beyond a single report of a captive animal producing two calves in 13 months (Feer 1994), but the length of gestation and oestrus period suggests that no more than one calf per year can be produced. In the wild calving intervals are probably significantly longer than this. There are also no data on other important reproductive parameters such as lactation period and length of dependency on the mother.

As individuals develop, the skin colour darkens, especially in *B. depressicornis* type animals, which can become almost black with white patches on the lower forelegs. *B. quarlesi* type animals stay lighter brown (Groves 1969; Burton *et al.* 2005).

Behaviour

Adults are either solitary or associate with a single member of the opposite sex (presumably during breeding periods) or mother with calf (Mustari 2003). There are very occasional reports of groups of more than two adult individuals (Hedges 1996). Mineral licks and wallows are said to attract several anoa at a time and it has been hypothesised that these sites provide an important social function in bringing oestrus females into contact with males (Hedges 1996). It is unclear how long calves stay with mothers, though Mustari (2003) reports sightings of relatively large young anoas with mothers.

There are very few reports of agonistic behaviour in the wild, though Feer (1994) described social interactions between captive animals in some detail. These included posturing, display of horns, charging and attacks with the horns to the flank and stomach of opponents. Such encounters occurred between males, but also between males and females though the latter were less common. Agonistic encounters involving actual contact were relatively rare, though Dolan (1965) reports several instances of captive animals attacking and killing their conspecifics, as well as other species.

Information on sexual display behaviour also comes exclusively from captivity, where animals display patterns of behaviour similar to other bovids (Feer 1994). Similarly, other social behaviours have not been reported from wild animals. Other than a general reputation for aggression, there is no information on interspecific interactions of anoas with other species in the wild, no predators of adult anoas, other than man, are native to Sulawesi and any distinct antipredator behaviour is unlikely.

Parasites and diseases

No major diseases have been reported from wild or captive animals.

Status in the wild

Both species are classified under the IUCN Red List as Endangered and decreasing on the basis of the world population being estimated at fewer than 2500 individuals, ongoing declines of greater than 20% over two generations and no single population being greater than 250 mature individuals (Semiadi *et al.* 2008a, 2008b). The primary cause of decline is hunting for meat (Semiadi *et al.* 2008a, 2008b), though there is some limited evidence of a past trophy or souvenir trade (Melisch 1995). It is likely that deforestation as a factor which increases human accessibility to forests may be a secondary driver of decline. As a result of the almost complete loss of lowland primary rainforest on Sulawesi (Whitten *et al.* 2002), the latest IUCN assessment regards the lowland anoa as probably more threatened than the mountain anoa (Semiadi *et al.* 2008b). In the absence of detailed, reliable data on population structure and behaviour in the wild, assessments of minimum viable populations or minimum areas required for conservation are difficult, and this uncertainty was reflected in the cautious conclusions of the 1996 Population and Habitat Viability Assessment (PHPVA) and its recommendations (Manansang *et al.* 1996a).

Sulawesi and Buton are well served with protected areas (Charles *et al.* 2007), but the effectiveness of these areas in protecting the species they contain is questionable (Burton *et al.* 2005). Protected areas in north Sulawesi have seen extensive declines in anoa density (O'Brien & Kinnaird 1996) while the Lambusango reserve on Buton has seen rapid local declines over a period of five years (Wheeler & Dwiyahreni 2010). Both species have been protected in law since Dutch colonial times (Jahja 1987) and are currently listed on CITES Appendix I (Semiadi *et al.* 2008a, 2008b). Despite this, anoas have been lost from large parts of mainland Sulawesi. It is likely that few or no populations persist in most of South Sulawesi, and that the species has been extirpated from significant parts of the north and central provinces. There are no current conservation programmes actively targeting anoa, though the species have been flagships associated with conservation programmes and are the symbol of the province of southeast Sulawesi. There are no evaluations of the effectiveness of protecting anoa of any conservation programmes.

A series of high-priority sites for *in situ* conservation of anoa have been identified, 11 in mainland Sulawesi and two in Buton (Mustari *et al.* 2012). These sites were selected based on representation of known subpopulations, extent of forest cover and connectivity between forested areas and status of management of area (whether sites were national parks, protected forests, restoration concessions, etc.; Mustari *et al.* 2012); these are the basis for the current national species action plan (Mustari *et al.* in press).

Status in captivity

There is a modest captive population of anoas. In Indonesia 20 individuals are kept in zoos and 15 are known to be kept in captivity by other individuals, including regional forestry offices, universities and members of the public (Mustari *et al.* 2012). Outside Indonesia there are 150 anoas in captivity (71 males, 67 females, 12 young) in 33 zoos, 22 in Europe, ten in North America, one in Asia. These figures are higher than indicated by the data provided from EEP (EAZA), SSP (AZA) and ISB (WAZA) breeding programmes, probably due to the fact that not all institutions are included in these programmes. The ongoing taxonomic uncertainty coupled with unclear origins and provenance of much of the captive stock (Nötzold 1999; Schreiber *et al.* 1999) makes it difficult to assess the true value of the captive population to wider conservation efforts.

References

Burton, J. A. & MacDonald, A. A. (2009). Variation in distribution and abundance of large mammals in Lore Lindu National park, Central Sulawesi, Indonesia. *Malayan Nature Journal* **61**: 295–305.

Burton, J. A., Hedges, S. & Mustari, A. H. (2005). The taxonomic status, distribution and conservation of the lowland anoa *Bubalus depressicornis* and mountain anoa *Bubalus quarlesi*. *Mammal Review* **35**: 25–50.

Charles, H. C., Summers, M., Harting, J. R. & Kessler, P. J. A. (2007). Developing conservation priorities based on forest type, condition, and threats in a poorly known ecoregion: Sulawesi, Indonesia. *Biotropica* **39**: 747–759.

Clauss, M., Reese, S. & Eulenberger, K. (2009). Macroscopic digestive anatomy of a captive lowland anoa (*Bubalus depressicornis*). *Zoo animal nutrition* **4**: 257–265.

Dolan, J. (1965). Breeding of the lowland anoa, *Bubalus* (*Anoa*) *d. depressicornis* (H. Smith, 1927) in the San Diego Zoological Garden. *Zeitschrift für Säugetierkunde* **30**: 241–248.

Dwiyahreni, A. A. (2006). Anoa abundance in the Lambusango forests and the influencing factors. Report to BKSDA Southeast Sulawesi, Kendari.

Feer, F. (1994). Observations éthologiques sur *Bubalus (Anoa) quarlesi* Ouwens, 1910 (Ruminantia, Bovidae) en captivité. *Zeitschrift für Säugetierkunde* **59**: 139–152.

Flores-Miyamoto, K., Clauss, M., Ortmann, S. & Sainsbury, A. W. (2005). Nutrition of captive lowland anoa (*Bubalus depressicornis*): a study on ingesta passage, intake, digestibility, and a diet survey. *Zoo Biology* **24**: 125–134.

Foead, N. (1992). Foead, N. 1992. Studi Habitat Dan Pakan Anoa Gunung [*Bubalus (Anoa) Quarlesi*, Ouwen] Di Taman Nasional Lore Lindu, Sulawesi Tengah. Unpublished MSc thesis. Fakultas Kehutanan, Universitas Gadjah Mada.

Groves, C. P. (1969). Systematics of the anoa (Mammalia, Bovidae). *Beaufortia* **17**: 1–12.

Grubb, P. (2005). Order Artiodactyla. In *Mammal Species of the World: A Taxonomic and Geographic Reference*, pp. 637–722. 3rd edition. Eds. Wilson, D. E. and Reeder, D. M. Baltimore, MD: Johns Hopkins University Press.

Hedges, S. (1996). Anoas Bubalus depressicornis *and* Bubalus quarlesi*: Asian Wild Cattle and Buffaloes. Status Report and Conservation Action Plan.* Gland: IUCN/SSC Asian Wild Cattle Specialist Group.

Hollister, N. (1919). The generic names *Anoa* and *Bubalus. Proceedings of the Biological Society of Washington*, **32**: 45.

Jahja, M. M. (1987). The possibility of breeding anoa in captivity: an alternative for conservation of the species. *Biotrop Special Publication* **30**: 101–108.

Manansang, J., Hedges, S., Dwiatmo, S., Miller, P. & Seal, U. (1996a). Population and Habitat Viability Assessment Workshop for the Anoa (*Bubalus depressicornis* and *Bubalus quarlesi*) report. Population and Habitat Viability Assessment Workshop for the Anoa

Manansang, J., Sumampouw, T., Masnur, I. Y., Agustine, N. & Prastiti, S. (1996b). The anoa at Taman Safari Indonesia (TSI). Anoa Species *Bubalus quarlesi* and *Bubalus depressicornis* population and habitat viability assessment workshop report Taman Safari Indonesia.

Melisch, R. (1995). Anoa threatened by souvenir trade in South Sulawesi, Indonesia. *Oryx* **29**: 224–225.

Mustari, A. H. (1995). Population and behaviour of lowland anoa (*Bubalus depressicornis* Smith) in Tanjung Amolengu Wildlife Reserve South-East Sulawesi, Indonesia. MSc thesis. George-August University.

Mustari, A. H. (1996). Socio-ecological behaviour of lowland anoa (*Bubalus depressicornis* Smith) in Tanjung Amolengu wildlife reserve Southeast Sulawesi. Anoa Species *Bubalus quarlesi* and *Bubalus depressicornis* Population and Habitat Viability Assessment Workshop report, Taman Safari Indonesia.

Mustari, A. H. (2003). Ecology and conservation of lowland anoa, *Bubalus depressicornis*, in Sulawesi, Indonesia. Unpublished PhD thesis. University of New England.

Mustari, A. H. & Masy'ud, B. (2011). The nutritional requirement of anoa. *Media Konservasi*, **7**.

Mustari, A. H., Burton, J., Rejeki, I. S. & Komara, W. Y. (2012). *High Priority Sites for Conservation of Anoa (*Bubalus depressicornis *and* Bubalus quarlesi*) in Sulawesi and Buton: Understanding and Sustaining the Biodiversity and Ecosystems of the Wallacea Region for Human Well-being.* Makassar: Hasanuddin University.

Mustari, A. H., Burton, J., Santoso, H., *et al.* (in press). *National Strategy and Action Plan for Anoa Conservation (*Bubalus depressicornis *and* Bubalus quarlesi*) 2010–2020.* Jakarta: Ministry of Forestry.

National Research Council. (1983). *Little-known Asian Animals with a Promising*

Economic Future. Washington, DC: National Academy Press.

Nötzold, G. (1999). *International Studbook: Anoa, (*Bubalus [Anoa] depressicornis *Smith, 1827)*. Germany: Leipzig Zoo.

O'Brien, T. G. & Kinnaird, M. F. (1996). Changing populations of birds and mammals in North Sulawesi. *Oryx* **30**: 150–156.

Okarda, B. (2010). Potential habitat and spatial distribution of anoa (*Bubalus* spp.) in Lore Lindu National Park. Unpublished MSc thesis. Institut Pertanian Bogor.

Pujaningsih, R. I., Sutrisno, C. I., Supriondho, Y., *et al.* (2009). Diet composition of anoa (*Bubalus* sp.) studied using direct observation and dung analysis method in their habitat. *Journal of the Indonesian Tropical Animal Agriculture* **34**: 223–228.

Rahman, A. M. & Mustari, A. H. (2002). Studi karakteristik habitat anoa dataran rendah (*Bubalus depressicornis*, Smith) di pinogu, Taman Nasional Boganinani Wartabone. Prosiding Seminar Nasional Bioekologi dan Konservasi Ungulata, Institut Pertanian Bogor.

Schreiber, A., Seibold, I., Notzold, G. & Wink, M. (1999). Cytochrome b gene haplotypes characterize chromosomal lineages of anoa, the Sulawesi dwarf buffalo (Bovidae: *Bubalus* sp.). *Journal of Heredity* **90**: 165–176.

Semiadi, G., Burton, J., Schreiber, A. & Mustari, A. H. (2008a). *Bubalus quarlesi*. In *IUCN 2012: IUCN Red List of Threatened Species*. Version 2012.2. www.iucnredlist.org, accessed 1 October 2012.

Semiadi, G., Mannullang, B., Burton, J., Schreiber, A., Mustari, A. H. & The IUCN SSC Asian Wild Cattle Specialist Group. (2008b). *Bubalus depressicornis*. In IUCN 2012: IUCN Red List of Threatened Species. Version 2012.2. www.iucnredlist.org, accessed 1 October 2012.

Syam, A. (1978). *Study of Habitats and Population of Anoa (*Anoa depressicornis*) H. Smith in Mount Tangkoko Batuangus Nature Reserve, North Sulawesi*. Bogor: Lembaga Penelitian Hutan.

Wheeler, P. (2004). Estimating anoa abundance. In *Monitoring Forest Degradation and Animal Populations in the Forests of Central Buton: Preliminary Results from the Pilot Study*, pp. 15–20. Ed. Seymour, A. Old Bolingbroke: Operation Wallacea.

Wheeler, P. M. & Dwiyahreni, A. A. (2010). *Documenting the Decline of the Anoa on Buton: Implications for Conservation*. Bali: *Association for Tropical Biology and Conservation.*

Wheeler, P. M., Purwanto, E., Coles, T., *et al.* (2009). *State of the Lambusango Forests 2008*. Old Bolingbroke: Operation Wallacea.

Whitten, A. J., Mustafa, M. & Henderson, G. S. (2002). *The Ecology of Sulawesi*. Hong Kong: Periplus Editions.

Wilson, D. E. & Mittermeier, R. A. (2011). *Handbook of the Mammals of the World*. Vol. **2**. Hoofed Mammals. Barcelona: Lynx Edicions.

Chapter 18

Tamaraw *Bubalus mindorensis* Heude, 1888

Merben R. Cebrian, Rodel M. Boyles, Josefina L. de Leon and James Burton

Names

Genus: *Bubalus* C. H. Smith, 1827

Species: Tamaraw *Bubalus mindorensis* Heude, 1888

Names in other languages: **French**: Tamarau, Tamarao; **German**: Mindoro-Büffel; **Spanish**: Tamarao, Búfalo de Mindoro; **Italian**: Tamaraw

Other common names: Mindoro dwarf buffalo.

Ecology, Evolution and Behaviour of Wild Cattle: Implications for Conservation, ed. M. Melletti and J. Burton. Published by Cambridge University Press.

Taxonomy

Bubalus mindorensis Heude (1888), type specimen Mindoro Island, Philippines. Despite early debates on its proper taxonomic classification (Everett 1878; Meyer 1878; Bartlett 1878), the tamaraw is classified as a distinct species within the genus *Bubalus* based on morphological characteristics (Groves 1969; Custodio *et al.* 1996; Braun *et al.* 2002; Wilson & Reeder 2005) and genetic analyses (Namikawa *et al.* 1995; Tanaka *et al.* 1996, 2000). Steere (1888a) also independently described the tamaraw, placing it in the genus *Anoa*, but was superseded by Heude (1888). Heude's type specimen was lost and then rediscovered (Braun *et al.* 2002) in Xujiahui, China, in the Shanghai Natural History Museum, which acquired the museum Heude founded (Zikawei Museum of Natural History).

Subspecies and distribution

The tamaraw is endemic to the island of Mindoro, Philippines (Plate 9; Heaney *et al.* 1987; Custodio *et al.* 1996). Thought to be previously widely distributed in Mindoro (Everett 1878; Steere 1888b, 1891), by 1949 there appears to have been fewer than 1000 animals left on the island, with an estimated 244 in Occidental Mindoro (Manuel 1957) in at least three areas (Harrisson 1969a). By 2007 the three areas where tamaraw are now thought to remain are in Mounts Iglit-Baco National Park (MIBNP), the Mount Calavite Wildlife Sanctuary and the Aruyan-Malati area (R. M. Boyles, unpublished; J. de Leon, unpublished).

There is no recognized subspecies. Isolated teeth recovered from Luzon were attributed to *B. mindorensis* (Beyer 1957) but they have not been critically analyzed (Custodio *et al.* 1996). In 2006 the fossilized partial skeleton of a new species of extinct *Bubalus* (*B. cebuensis*; Croft *et al.* 2006) from Cebu, Philippines suggested that the distribution of *Bubalus* in the Philippines may not be restricted to Mindoro Island.

Descriptive notes

Body measurements: **Shoulder height**: 105 cm; **Head and body length**: 220 cm; **Tail length**: 60 cm; **Body mass**: 200–274 kg (Steere 1888b, 1891; Namikawa *et al.* 1995; Wilson & Mittermeier 2011). The dental formula is: I 0/3, C 0/1, P 3/3, M 3/3 (× 2) = 32 (Rabor, 1986 in Custodio *et al.* 1996).

The tamaraw (Figure 18.1–18.3) is the largest extant land mammal endemic to the Philippines. Considered a dwarf buffalo, adult females are estimated to weigh, on average, approximately 244 kg (Talbot & Talbot 1966; Roth & Montemayor-Taca 1971; Namikawa *et al.* 1995). Body color is not sexually dimorphic; adults are grayish-black overall, with a grayish-white stripe parallel to and just above the inner corner of both eyes. Light markings are also seen above the hooves, the inner ears and on the throat. Subadults and juveniles are reddish-brown to gray, with a dark dorsal stripe, while calves are generally reddish in color. In the adults, the V-shaped horns are triangular and stout at the base, with the dorsal side flattened. The horns turn slightly toward each other as they taper to a rounded cross-section at the tips. Field observations indicate that horn configuration is sexually dimorphic (Steere 1888b; Keuhn 1976; Custodio *et al.* 1996; Suchomel 2005). Bulls have thicker horns that are more flattened and are closer together at the base. However, Stefen & Feiler (2004) examined museum specimens and observed that horn configuration is not sexually dimorphic, but rather they change with age. The base of the horns of younger individuals

Figure 18.1 A captive cow in 2007 from Occidental Mindoro (photo by M. R. Cebrian).

Figure 18.2 A wild bull in 2010 (photo by M. R. Cebrian).

Figure 18.3 A family group in 2010 consisting of a cow, a yearling and a juvenile (photo by M.R. Cebrian).

are more rounded with more pronounced sculptures on the dorsal surface, while horns of older individuals are more triangular in cross-section from the base to the tips. Their life span is estimated to be 15–20 years (Stefen & Feiler 2004).

Habitat

Tamaraw inhabit open grasslands and secondary forests along a wide altitudinal gradient, from sea level to approximately 2000 m above sea level on Mindoro Island (Figure 18.4). In the MIBNP, upland areas over 600 m above sea level are dominated by open grasslands interspersed with secondary forests. On moderate to steep slopes, these open grasslands are generally composed of tall grasses dominated by cogon (*Impetra cylindrica*) and talahib (*Saccharum spontaneum*), with *Paspalum*, *Themeda* and *Alloteropsis* genera and other short grasses found on mountaintops and ridges (del Barrio *et al.* unpublished; Talbot & Talbot 1966; Suchomel 2005). Tall bamboo (*Dinochloa* spp.) thickets and secondary dipterocarp forests follow seasonally wet riparian areas and rivers that tamaraw may use as travel corridors. These secondary forests, as well as the open grasslands, are seasonally burned by local farmers who practice 'kaingin' or slash-and-burn-farming on moderate slopes to this day (Talbot & Talbot 1966; Manuel 1957; R. M. Boyles, personal communication). Tamaraw that occur in low-lying areas may intermingle with the domestic water buffalo (*B. bubalis*) near farmlands, although this has historically been discouraged by ranchers shooting the tamaraw. Tamaraw have

Figure 18.4 A typical tamaraw habitat in MIBNP, Occidental Mindoro, Philippines in 2010. The park at elevation is dominated by grasslands interspersed with secondary forests along riparian corridors (photo by M. R. Cebrian).

Figure 18.5 An Indigenous Mangyan at MIBNP, Occidental Mindoro, Philippines in 2007. Tamaraw regularly traverse steep terrain as evidenced by hoof marks and droppings on trails in the mountains (photo by M. R. Cebrian). A black and white version of this figure will appear in some formats. For the colour version, please refer to the plate section.

no known predators. However, intraspecific conflicts and mortalities via illegal hunting are known to occur.

Movements and home range

Tamaraw movement and home range are not well understood in the wild. In the MIBNP trails used by tamaraw are found along river valleys and on the tops of mountains, indicating extensive travel through different habitat types (Figure 18.5). These trails sometimes lead up steep gradients and on mountain ridges, indicating dexterity in traversing steep slopes. Although no studies have yet been made on their home range, it is reasonable to predict that the numerous riparian corridors that cut through the landscape would serve as travel corridors through which tamaraw can disperse, and that steep slopes are not an impediment to their dispersal. Current tamaraw surveys indicate that tamaraw move through different habitat types. Their home range may be limited by human encroachment, as well as human disturbance associated with kaingin farming and other resource-extraction activities on the island.

Activity patterns

Activity patterns are not well documented in the wild. Although time spent sleeping was minimal (1–6 hours per day), Momongan & Walde (1993) observed that captive tamaraw slept more at night than during the day, and longer during the dry season than during the wet season. The following activities were observed for captive tamaraw in a 24-hour period: feeding (23.55%), time spent on rumination (26.4%) and idling (30.62%). The rest of the time was spent wallowing, walking, sleeping, urinating, defecating and drinking. Feeding time was significantly longer during the 6:00–10:00 and 18:00–22:00 periods than any other times of the day. Feeding was followed by rumination, mostly between 14:00–18:00, usually associated with wallowing activity.

Feeding ecology

The feeding ecology of the tamaraw in the wild is not well known. Grasslands comprise the majority of the available habitat in the MIBNP above 600 m in elevation. In these upland areas, *I. cylindrica* and *S. spontaneum* are the most abundant available forage, with an average protein content of 6.15% (del Barrio *et al.*, unpublished). Where kaingin farming gets out of hand and fire escapes into open grasslands, tamaraw may be seen feeding in burned areas where succulent grass shoots are emergent. In captivity, tamaraw have been observed to feed longer in the morning and early evenings than at any other time of the day (Momongan & Walde 1993).

Reproduction and growth

The age of first reproduction of female tamaraw in the wild is unknown. As a point of comparison, the first estrus of *B. bubalis* is typically at two years of age and varies according to nutrition and body condition (Jainudeen 1986; Singh *et al.* 1986). Estrus in *B. bubalis* is difficult to detect (Drost 2007), and the same observation has been noted in the captive tamaraw (Sarabia *et al.* 1998). Sexual maturity for tamaraw bulls is not known, although *B. arnee* bulls reach sexual maturity at three years of age (McCool & Entwistle 1989).

The mating habits and reproductive traits of the tamaraw in the wild are largely unknown. Sanchez (1892), after a one-month expedition to Mindoro in May 1892, reported in the Spanish language newspaper *La Ilustración Filipina* that the tamaraw estrous cycle is 21 days in the wild, which is similar to what has been observed in captivity (Sarabia *et al.* 1998). Sanchez (1892) also reported that the tamaraw estrus lasted 24 hours, while Sarabia *et al.* (1998) noted that estrus lasted 21.8 hours in captivity. The gestation period is 317 days, with a calving interval of 712 days (Sarabia *et al.* 1998). Cows typically bear one calf; twinning has not been observed in the wild (M. R. Cebrian, personal comment). Sanchez (1892) also reported that the mating season for the tamaraw is from April to July. Talbot & Talbot (1966) agreed with this observation, while reporting that local farmers have seen newborn calves from December through January. However, Custodio *et al.* (1996) report that the tamaraw calving season is from June to November, while breeding occurs from December to May. By contrast, field observations in the MIBNP from 2009 to 2012 have recorded calves in late April (M. R. Cebrian, unpublished). For various reasons, reproduction in captivity has been very difficult; only two calves were ever born in captivity at the Gene Pool Farm in Mindoro since the facility's inception in 1979 (Suchomel 2005; R. M. Boyles, personal communication). Considering the uncertainty in the literature, more research is needed in order to understand the reproductive cycle of the tamaraw in the wild. It may be that tamaraw breed year-round if habitat conditions allow.

Behavior

Of 263 tamaraws counted in the MIBNP in April 2006, 29 (11%) were solitary while 234 (79%) were in groups of 2–7 individuals (Ishihara *et al.* 2007). By contrast, Keuhn (1986) observed 172 (82%) lone bulls in 221 days of field observations with no account for double sampling. Of the 11% that were found to be solitary in 2006, about half of them were adult males (15/29).

Females were present in 92% of the groups for which the sexes were identified, and bull–cow pairs were seen in 55% of these groups. Of the groups where cows were present, subadults (juveniles and yearlings) were also present 79% of the time.

Agonistic behavior between bulls was observed in the MIBNP, although no fights were actually seen. Possible explanations include mutual distancing mechanisms between bulls, and the high availability of forage. One observation of a dead bull discovered with a possible goring injury was recorded in 2008 (M. R. Cebrian, personal communication). However, a necropsy was not allowed and the carcass was buried.

The tamaraw is thought to be nocturnal (Manuel 1957; Talbot & Talbot 1966; Alvarez 1970). However, Momongan & Walde (1993) showed in a captive behavioral observation study that the tamaraw is mostly diurnal, moving into open grassland to graze in the early morning. Tamaraw may also be found wallowing in slow-moving creeks and mudholes in mid-morning, and some may return to the riparian zones presumably to ruminate and to seek cover from the heat of the midday sun. Otherwise, they resume grazing in late afternoon and early evening. The same behavior was observed in the wild in the MIBNP in 2006–2010 (M. R. Cebrian, unpublished).

Parasites and diseases

An outbreak of rinderpest, a highly contagious viral disease which can cause spontaneous abortions in cattle, in the 1930s was first mentioned by E. D. Merrill (oral communication,

Table 18.1 The summary of observations from annual tamaraw counts at Mts. Iglit-Baco National Park, Occidental Mindoro, Philippines from 2001 to 2012; these observations are minimum counts of overlapping count areas

Year†	Total	Bull	Cow	Unk	Juveniles	Yrlg/calf*
2001	187	57	53	0	15	62
2003	261	59	94	2	35	71
2004	232	61	74	8	36	53
2005	269	60	97	5	55	52
2006	263	53	97	21	44	48
2007	239	54	78	0	49	58
2008	263	49	98	8	66	42
2009	274	49	99	3	51	63/9
2010	314	60	108	4	74	53/15
2011	274	48	99	17	53	49/8
2012	327	57	103	7	72	55/33
Mean	264	55	91	7	50	55/16
S.E.	11.39	1.46	4.86	2.02	5.25	2.43/5.79

* Calves were recorded separately since 2009.
† No data for 2002
Source: Tamaraw Conservation Project

1938) in Harper (1945); Merrill thought that the disease resulted in the decline of the tamaraw population in Mindoro. Infected cattle were thought to be the vector for the disease's transmission and it is likely that tamaraw occurring in the vicinity of cattle may have contracted the disease. Rinderpest, as a cause of tamaraw decline, was not mentioned again in subsequent literature reports (Manuel 1957; Harrisson 1969a, 1969b; Alvarez 1970; Kuehn 1976) on the status of the tamaraw until Buchholtz (1990).

Anunciado *et al.* (1994) examined tamaraw in captivity and found ticks (*Boophilus microplus, Rhipicephalus sanguineus* and *Amblyomma* spp.), a louse (*Haematopinus tuberculatus*), trematodes (*Calicophoron calicophorum, Fasciola* spp.), a cestode (*Moniezia* spp.) and nematodes (*Mecistocirnus digitatus, Oesophagostomum radatum* and *Tricuris* spp.). Ishihara *et al.* (2007) examined fresh fecal samples and found *Coccidia* and *Fasciola* eggs.

Status in the wild

IUCN status: Critically Endangered.

The species qualifies for inclusion as Critically Endangered under Criteria C1+2a(ii), given the number of mature individuals is estimated to be fewer than 250, with a continuing decline estimated at over 25% over the next three generations (generation length estimated at ten years). In addition, over 90% of individuals are presumed to be in one subpopulation, MIBNP.

The tamaraw is listed in Appendix I of the Convention on International Trade of Endangered Species of Wild Flora and Fauna (Hedges *et al.* 2008) due to a marked apparent decline in population since the early 1900s. Manuel (1957) estimated that there were fewer than 1000 individuals left in Mindoro in 1949. By 1969, with no reliable population estimates having been made, the Tamaraw Conservation Program (TCP) was launched to protect the three known areas where the tamaraw was known to occur – Mount Calavite, Sablayan and the MIBNP. Local knowledge indicates that tamaraw may occur in other areas, but these have yet to be verified.

The MIBNP is the most monitored of the three locations that tamaraw are known to occur in Mindoro. Since 2001 the TCP has counted tamaraw (Table 18.1) in the MIBNP using a simultaneous multi-vantage point method in which 16 vantage points are placed in strategic locations in the park. Tamaraw are counted twice a day for five days in April. At each vantage point, observers note the sex and estimated age of tamaraw seen between 5:30 and 7:00, and again from 17:00 to 18:30. Data are then collated and individuals that were double counted are corrected via a discussion forum wherein observers from adjacent vantage points compare notes. In its latest count in April 2012, 327 animals were observed in the study area. This represents a 75% increase from 2001 (M. R. Cebrian, unpublished; Table 18.1). However, care should be taken in interpreting these figures. Although the count areas are consistent over time, the results can only show a general trend in those areas, and the status of the tamaraw on the whole island has yet to be evaluated. Factors that may affect the results of the annual count include the number of observers at a given location, unequal distribution of binoculars and spotting scopes and recent fires in nearby grasslands. The count does not shed light on species productivity or survival. More studies are needed to fully understand the sustainability of the tamaraw population in Mindoro.

The primary factors affecting the apparent restriction in tamaraw distribution are habitat loss and illegal hunting. For example, the encroachment of farms and cattle ranches competes for space with the tamaraw, while extensive logging and uncontrolled fires associated with kaingin farming clear forested areas that may serve as tamaraw refugia. Illegal hunting still occurs, although the number of tamaraw killed is difficult to quantify. In the 1930s the reported outbreak of rinderpest, combined with the availability of high-powered rifles and automatic weapons, may have pushed the tamaraw population to no more than 500 individuals in the wild. Therefore, it was important for tamaraw conservation that laws and regulations were enacted for its own protection and that of its habitat.

In 1936, *Commonwealth Act No. 73*, which prohibited 'the killing, hunting, wounding, or taking away of *Bubalus mindorensis*, commonly known as tamaraw', was signed into law. Deemed to be full of loopholes, the defense of life and property was used as a flimsy excuse to circumvent the law (Manuel 1957). In a surprising reversal to policy, the *1939 Forestry Administrative Order No. 17* declared that for a nominal fee of 50 pesos to the forestry and game warden, anyone can have a special license for the taking and/or selling of dead or live tamaraw. In 2001 the *Republic Act No. 9147*, known as the *Wildlife Act*, was passed into law. This time, stiff penalties are imposed on the hunting, killing or trading of tamaraw, including the destruction of their habitat. The following year, October was officially declared as the special month for the conservation and protection of the tamaraw.

A secondary threat to the tamaraw is human encroachment due to the continued apparent growth of the Philippine population. With a population of 92,337,852 as of 2010 (Philippine National Statistics Office), overcrowded population centers indirectly drive people into rural areas. Concomitant to human encroachment is the likely exposure of tamaraw to cattle and other domesticated livestock that increase the risk of disease transmission. This scenario has already been implicated in the decline of the tamaraw population in the 1930s.

Recommended steps for tamaraw conservation include those that have already been identified in the Tamaraw Population and Habitat Viability Assessment (PHVA; de Leon 1996). The most needed information regarding the tamaraw remains that of its ecology in the wild, including its habitat use, home range, survival and fecundity. The following measures can help achieve the goal of concerted tamaraw conservation: (1) conduct surveys in areas where the tamaraw is likely to occur; (2) promote innovative studies to understand the tamaraw ecology; (3) create meaningful linkages with Indigenous peoples to promote sustainable management goals in areas tamaraw are known to inhabit; (4) convert the gene pool farm into a wildlife facility for multiple use and visitor information; (5) develop a management plan to include improved wildlife protection measures for the MIBNP and other areas where tamaraw are known to occur: (6) create capacity-building initiatives for communities that surround the MIBNP and other areas where tamaraw are known to occur; and (7) promote eco-tourism industries compatible with the purposes of the MIBNP.

Status in captivity

Only one tamaraw, an adult male born in captivity, remains in the holding facility formerly designated as the tamaraw 'gene pool' in Mindoro. The facility was established in 1980 with a seed population of 19 tamaraw (de Leon *et al.* 1996). However, poor facility maintenance, difficult access, inadequate staffing and funding and inadequate training of staff for the proper handling of the tamaraw were several reasons given for its failure as a breeding program (Cox & Woodford 1990). There are no current plans to revitalize the captive breeding program due to limited success of the facility and to the existing political climate.

References

Alvarez, J. B. Jr. (1970). Philippine tamaraw: here to stay. *International Union for the Conservation of Nature and Natural Resources (New Series)* **18**: 46–51.

Anunciado, R. V. P., Eduardo, S. L., Momongan, V. G. & Escalada, R. F. (1994). Parasites of the captive tamaraw, *Bubalus mindorensis* (Heude, 1888) (Mammalia: Ruminantia). *Philippine Journal of Veterinary Medicine* **31**: 5–10.

Bartlett, A. D. (1878). Letter to the Secretary. *Proceedings of the Zoological Society of London* **1878**: 882–883.

Beyer, H. O. (1957). New finds of fossil mammals from the Pleistocene strata of the Philippines. *University of the Philippines Bulletin* **41**: 1–23.

Braun, A., Groves, C. P. & Grubb, P. (2002). Rediscovery of the type specimen of *Bubalus mindorensis* Heude, 1888. *Mammalian Biology* **67**: 246–249.

Buchholtz, C. (1990). Cattle. In *Grzimek's Encyclopedia of Mammals.* Vol. **5**, pp. 360–417. Ed. P. S. Parker. New York: McGraw-Hill.

Cox, R. & Woodford, M. (1990). A technical evaluation of the Philippine Tamaraw Conservation Program. A report to the Department of Environment and Natural Resources, Republic of the Philippines by IUCN; Zoological Society of London; and Bristol, Clifton, and West England Zoological Society.

Croft, D. A., Heaney, L. R., Flynn, J. J. & Bautista, A. P. (2006). Fossil remains of a new, diminutive *Bubalus* (Artiodactyla: Bovidae: Bovini) from Cebu Island, Philippines. *Journal of Mammalogy* **87**(5): 1037–1051.

Custodio, C. C., Lepiten, M. V. & Heaney, L. R. (1996). *Bubalus mindorensis. Mammalian Species* **520**: 1–5.

de Leon, J., Lawas, N., Excalada, R., *et al.* (Eds.). (1996). *Tamaraw (*Bubalus mindorensis*): Population and Habitat Viability Assessment Report.* Apple Valley, MN: IUCN/SSC Conservation Breeding Specialist Group.

Drost, M. (2007). Bubaline versus bovine reproduction. *Theriogenology* **68**: 447–449.

Everett, A. H. (1878). Mr. A. H. Everett on *Anoa Depressicornis. Proceedings of the Zoological Society of London* **1878**: 792.

Groves, C. P. (1969). Systematic of the anoa (Mammalia, Bovidae). *Beaufortia* **17**: 1–12.

Harper, F. (1945). Extinct and vanishing mammals of the Old World. *American Committee for International Wildlife Protection, Special Publication* **12**: 548–550.

Harrisson, T. (1969a). The tamaraw and its survival. *International Union for the Conservation of Nature and Natural Resources (New Series)* **2**: 85–86.

Harrisson, T. (1969b). The tamaraw and Philippine conservation. *Biological Conservation* **1**: 317–318.

Heaney, L. R., Gonzales, P. D. & Alcala, A. C. (1987). An annotated checklist of the taxonomic and conservation status of land mammals in the Philippines. *Silliman Journal* **34**: 32–66.

Hedges, S., Duckworth, J. W., IUCN SSC Asian Wild Cattle Specialist Group, de Leon, J., Custodio, C. & Gonzales, J. C. (2008). *Bubalus mindorensis*. In *IUCN 2012: IUCN Red List of Threatened Species*. Version 2012.2. www.iucnredlist.org, accessed 27 June 2013.

Heude, P. M. (1888). Note sur le petit buffle sauvage de l'île de Mindoro (Philippines). *Memoires concernant l'histoire naturelle de l'Empire Chinois* **2**: 50–51.

Ishihara, S., Ishida, A., del Barrio, A. N., *et al.* (2007). Field survey on tamaraw (*Bubalus mindorensis*) present population size and herd behavior in wild. *Italian Journal of Animal Science* **6**: 1249–1251.

Jainudeen, M. R. (1986). Reproduction in the water buffalo. In *Current Therapy in Theriogenology*. Vol. **2**, pp. 443–449. Eds Morrow, D. A. Philadelphia, PA: W.B. Saunders.

Keuhn, D. W. (1976). Tamarao: endangered buffalo of the Philippines. *National Parks and Conservation Magazine* **50**: 18–249.

Keuhn, D. W. (1986). Population and social characteristics of the tamarao (*Bubalus mindorensis*). *Biotropica* **18**: 263–266.

Manuel, C. G. (1957). Status of tamaraw, *Anoa Mindorensis* (Heude). *Proceedings of the Eighth Pacific Science Congress* **3**: 1463–1474.

McCool, C. J. & Entwistle, K. W. (1989). The development of puberty and sexual maturity in the Australian swamp buffalo bull. *Therioenology* **32**: 171–184.

Meyer, A. B. (1878). Letter to the secretary. *Proceedings of the Zoological Society of London* **1878**: 881–882.

Momongan, V. G. & Walde, G. I. (1993). Behavior of the endangered tamaraw (*Bubalus mindorensis* Heude) in captivity. *Asia Life Sciences* **2**: 241–250.

Namikawa, T., Masangkay, J. S., Maeda, K., Escalada, R., Hirunagi, K. & Momongan, V. G. (1995). External characters and karyotypes of the captive tamaraws *Bubalus (B.) mindorensis*, at the gene pool in the island of Mindoro, Philippines. *Journal of Animal Genetics* **23**: 19–28.

Roth, H. H. & Montemayor-Taca, B. (1971). Immobilization of the tamaraw (*Anoa mindorensis*). *Philippine Journal of Veterinary Medicine* **10**: 45–48.

Sanchez, D. (1892). El tamarao. *La Ilustracion Filipina* **2**: 271.

Sarabia, A. S., Escalada, R. F. & Momongan, V. G. (1998). Some reproductive parameters in female tamaraw. *Philippine Journal of Veterinary Medicine* **35**: 74–75.

Singh, J., Nanda, A. S. & Adams, G. P. (1986). The reproductive pattern and efficiency of female buffaloes. *Animal Reproduction Science* **60–61**: 593–604.

Steere, J. B. (1888a). The 'Tamaron' of the Philippine Islands. *Nature* **38**: 363–364.

Steere, J. B. (1888b). Letter to the secretary. *Proceedings of the Zoological Society of London* **29**: 413–415.

Steere, J. B. (1891). The island of Mindoro. *The American Naturalist* **25**: 1041–1054.

Stefen, C. & Feiler, A. (2004). Observations on the postnatal changes of skull morphology and tooth wear of the tamara, *Bubalus mindorensis*, (Mammalia: Bovidae). *Zoologische Abhandlungen (Dresden)* **54**: 317–330.

Suchomel, J. (2005). Some notes relating to the observation of tamaraw (*Bubalus mindorensis*) in the Mt. Iglit-Baco National Park, the Philippines, in 2003. *Lynx (Praha), New Series* **36**: 103–110.

Talbot, L. M. & Talbot, M. H. (1966). The tamarau (*Bubalus mindorensis* (Heude). Observations and recommendations. *Mammalia* **30**: 1–12.

Tanaka, K., Solis, C. D., Masangkay, J. S., Maeda, K., Kawamoto, Y. & Namikawa, T. (1996). Phylogenetic relationship among all living species of the genus *Bubalus* based on DNA sequences of the cytochrome b gene. *Biochemical Genetics* **34**: 443–452.

Tanaka, K., Matsuda, Y., Masangkay, J. S., *et al.* (2000). Cytogenetic analysis of the tamaraw (*Bubalus mindorensis*): a comparison of r-banded karyotype and chromosomal distribution of centromeric satellite DNAs, telomeric sequence, and 18s-28s rRNA genes with domestic water buffalos. *The Journal of Heredity* **91**: 117–121.

Wilson, D. E. & Mittermeier, R. A. (2011). *Handbook of the Mammals of the World*. Vol. **2**. Hoofed Mammals. Barcelona: Lynx Edicions.

Wilson, D. E. & Reeder, D. M. (2005). *Mammal Species of the World: A Taxonomic and Geographic Reference*. Baltimore, MD: Johns Hopkins University Press.

Chapter 19

Saola *Pseudoryx nghetinhensis* Dung, Giao, Chinh, Touc, Arctander & Mackinnon, 1993

William Robichaud and Barney Long

Names

Genus: *Pseudoryx* Dung *et al.* 1993

Species: Saola *Pseudoryx nghetinhensis* Dung *et al.* 1993

Names in other languages: **French**: Saola; **German**: Vu Quang-Wildrindes; **Spanish**: Saola; **Italian**: Saola; **Laotian**: Yang

Other common names: Vu Quang Ox, Spindlehorn

Ecology, Evolution and Behaviour of Wild Cattle: Implications for Conservation, ed. M. Melletti and J. Burton. Published by Cambridge University Press.

Taxonomy

The saola became known to science only in 1992, and was one of the most spectacular and surprising zoological discoveries of the twentieth century.

The species is in a highly distinctive monotypic genus, with morphological and genetic characteristics that place it as a primitive member of the family Bovidae (Vu Van Dung *et al.* 1993; Schaller & Rabinowitz 1995; Robichaud 1998).

The original scientific description of the species by Vu Van Dung *et al.* (1993) used analysis of the mitochondrial cytochrome b gene to propose that saola was allied to the subfamily Bovinae (cattle and spiral-horned antelopes), with a provisional placement among the tribe Boselaphini. Thomas (1994) analysed morphological features of the cranium and placed the species as a sister taxon to the subfamily Caprinae (goats, sheep and their allies), but this postulation was not supported by subsequent studies. Schaller & Rabinowitz (1995), for example, presented information on morphological characters that suggested affinity with the tribe Bovini (cattle) of the Bovinae. Robichaud (1998) also supported placement in the Bovinae, based on morphological and behavioural characteristics of a captive animal. Hassanin & Douzery (1999) concluded the species was most probably a sister taxon of the cattle and bison, excluding water buffaloes *Bubalus* and African Buffalo *Syncerus caffer*, based on nuclear and mitochondrial genes from four loci; Gatesy & Arctander (2000) analysed seven gene loci, placing saola in the subfamily Bovinae and probably the tribe Bovini. Consequently, the present consensus is that the saola is a member of the subfamily Bovinae, likely in the tribe Bovini, but possibly meriting its own monogeneric tribe.

Subspecies and distribution

Saola are restricted to the north and central Annamite Mountains of Vietnam and the Lao Peoples' Democratic Republic (Lao PDR; Plate 10; Grubb 2005). It is known from Nghe An, Ha Tinh, Quang Binh, Quang Tri, Thua-Thien Hue and Quang Nam Provinces in Vietnam, and Bolikhamxay, Khammouan, Savannakhet and Xekong Provinces in Lao PDR; it might also occur in Xieng Khouang Province, but suitable areas in the province have never been surveyed. Saola are restricted to the wetter areas of the mountains from about 200 m to 1200 m. The species was possibly historically distributed over a wider area, but its current distribution is highly localized due to poaching pressures and historical deforestation. The existing scant knowledge of its extant distribution is primarily based on interview data from indigenous communities and hunting trophies, yet it appears that no more that 15 forest blocks still hold subpopulations, and this number may be significantly lower.

Descriptive notes

Body measurements: **Shoulder height**: height at top of shoulder 84 cm (one live adult female; Figures 19.1–19.2) to 95.5 cm (one dead adult male); **Head and body length**: 143 cm (one adult male) to 150 cm (one adult female, Figure 19.1); **Tail length**: 16.5 cm (to tip of bone, one adult male) to 23 cm (to end of hair, one adult female); **Body mass**: 70 kg (one dead adult male) to 100 kg (estimated).

Dental formula is: I 0/3, C 0/1, P 3/3, M 3/3 (×2) = 32.

Girth at the broadest point around the belly is 104 cm (one dead adult male) to 108 cm (one weakened, pregnant female),

Figure 19.1 Female saola captured by villagers in Bolikhamxay Province, Lao PDR in August, 2010, and photographed shortly before she died (photo by William Robichaud).

Figure 19.2 Female saola in captivity. Bolikhamxay Province, Lao PDR (photo by William Robichaud).

Figure 19.3 Saola hunting trophy in a Ka Tu house, Quang Nam, Vietnam (source: WWF-Vietnam/Minh Hoang).

ear 10.0 cm (one adult male) to 11.5 cm (one adult female). Horns are up to at least 55 cm long measured along the curve and 51 cm measured across the chord (Figure 19.3). The iris of the eye is dark brown. There are four mammae. The body of this bovid is fairly compact, and the neck relatively long (Figure 19.2). Pelage is somewhat short (1.5–2.5 cm), mostly straight (curlier on the belly and inner forelegs) and soft. Most of the pelage is medium brown in colour, but includes striking white patches on the sides of the face, muzzle and above the eye, white pasterns, cream and black bands on the tail and rump, a black 'chinstrap' over a whitish chin and throat and a black dorsal midline stripe. Horns are blackish, long, straight, sharply pointed and with a long bone core (Figure 19.2). There is apparently no sexual dimorphism in, at least, horn length. Saola possess what may be the largest (in absolute size) preorbital glands of any extant species, measuring 9 cm × 3.5 cm × 1.5 cm deep. Each gland is covered by a muscular flap, which the saola can raise at will. Captive saola have been observed to raise the flaps and expose the glands when scent marking (by rubbing the underside of the flap on edges of rock), or when the animal appears nervous, or while exhibiting defensive behaviour (in the observed case at the approach of a small domestic dog). There are also small nodules, which may be secretory glands (although this function is not confirmed), associated with the white markings on the face. A captive adult female saola left tracks in modelling clay measuring 5.0–6.0 cm long by 5.3–6.4 cm wide for the forehooves and 6.0 cm long by 5.7–6.0 cm wide for the hindhooves. The tracks were widest across the middle.

Habitat

Saola inhabit climatically wet evergreen, broadleaf forest of the Annamite Mountains, along the Vietnam–Lao PDR border. These forests experience unusually long rainy seasons, up to ten months per year, with no month receiving less than 40 mm of rain. Distribution of saola across the Annamite Mountains is largely determined by this rainfall; distribution to the south is bounded by the east–west Bach Ma mountain chain, and western distribution into Lao PDR does not extend far beyond the Annamite Mountains crest, except where the mountain chain dips, allowing penetration of northeast winter monsoon from Vietnam. Saola probably occur up to 1000 m elevation, and perhaps reach 1200 m. The species' lower altitudinal limit in the absence of human disturbance and deforestation could be 200 m or less. However, most reported records today are from above 400 m, probably because very little forest remains in Vietnam below 400 m, and in Lao PDR forests at lower altitudes are not sufficiently wet.

Within this restricted habitat, saola reportedly favour the vicinity of small, deeply shaded streams (Figure 19.4; one of only four photographs of a saola in the wild is of an animal standing in a stream). Due to the paucity of ecological data on the species, the reasons for this preference can only be speculative; however, reports from local human residents indicate that one of the saola's preferred food plants is an Araceae identified as belonging to the genus *Schismatoglottis* (Peter Boyce, Royal Botanic Gardens, Kew, personal communication), which grows primarily along such streams. Saola are also commonly reported to flee to streams when pursued by domestic hunting dogs (and presumably also when pursued by dhole, *Cuon alpinus*, or other wild predators).

Indigenous knowledge across the saola's distribution indicates that saola need access to mineral licks and springs. The first photograph of a saola in the wild was taken by a camera trap set in the vicinity of a presumed mineral spring, in Vietnam's Pu Mat Nature Reserve (Long 2000).

Movements and home range

Seasonal altitudinal movements have been reported by local people, but little more can be said without scientific telemetry studies.

Activity patterns

Limited camera-trapping results and observations of a captive individual indicate that the species is diurnal and/or

Figure 19.4 Forest stream Au Rong: saola habitat, Quang Nam, Vietnam (source: Jeremy Holden/WWF). A black and white version of this figure will appear in some formats. For the colour version, please refer to the plate section.

crepuscular. Local hunters familiar with the animal report that it is most active in the morning, late afternoon and sometimes at night.

Feeding ecology

The saola is probably a browser, of both ground vegetation and the leaves of lower-storey woody plants. Local people commonly report that saola readily feed on the leaves of an aroid in the genus *Schismatoglottis*.

Reproduction and growth

Saola seem to have a fixed breeding season, and, at least in Lao PDR (the west side of the Annamite Mountains), births probably occur in the late dry or early rainy season, between April and June, based on local reports and examination of a pregnant animal that died after her capture in Lao PDR. Intraspecific sparring with the horns has been observed in captive juveniles. Other life history parameters (e.g. age at first breeding) are unknown.

Behaviour

Very little is known about saola behaviour. There has been no sustained scientific study of the species in the wild or in captivity, and in fact no biologist has ever reported seeing a saola in the wild. There are no saola in captivity anywhere in the world, and there is only one published report of observations of a captive animal (Robichaud 1998).

According to local people, saola are usually solitary, except females with calves and possibly loose, extended associations of males and females (two or three animals?) during the breeding season. Home range size is unknown. The function of the large premaxillary gland is not conclusively known, but it may be for territorial scent-marking.

Parasites and diseases

Nothing is known directly about diseases in saola.

Status in the wild

The saola is classified as Critically Endangered on the IUCN Red List A2cd+3cd+4cd;C2a(i) (IUCN 2013) and is listed on Appendix I of CITES. No biologist has ever reported seeing a saola in the wild and there are no saola in captivity. The few animals that have been brought into captivity have not survived, although their level of care was not high.

The global saola population is not known, but at most is probably no more than a few hundred, and could be substantially less – the animal is so difficult to detect (solitary, shy, at low density in thick, steep, remote forest) that narrow population estimates of any confidence are not currently possible. The development of a detection method and standard monitoring protocols are a high priority for the species' conservation. Surviving animals are fragmented into several subpopulations in Lao PDR and Vietnam. Even within sites with presumed extant saola populations, poaching pressure has possibly reduced numbers to such a level that surviving individuals could be isolated in 'sub-micropopulations', which could interfere with breeding. At the time of this writing (June 2012) the most recent confirmed record of the species was an animal captured by villagers in Bolikhamxay Province, Lao PDR in August, 2010, and photographed shortly before it died (Figures 19.1–19.2). This was the first indisputable record of a live saola since 1999, when two camera-trap photographs of saola were taken in Bolikhamxay Province.

Of several saola brought into captivity intermittently since the species' discovery, none have survived very long (at most a few months); however, they did not receive high-quality, expert care.

Hunting and snaring are the main threats to saola. Throughout the Annamite Mountains (Figure 19.5), commercial hunters

Figure 19.5 Tu village in the heart of saola range, Quang Nam, Vietnam (source: Minh Hoang/WWF). A black and white version of this figure will appear in some formats. For the colour version, please refer to the plate section.

Figure 19.6 Ka Tu hunter with saola skull (source: Jeremy Holden/WWF). A black and white version of this figure will appear in some formats. For the colour version, please refer to the plate section.

illegally set wire snares by the thousands. These snares are usually made from the type of wire used for bicycle brake cables. The primary target is animals valued in the local bushmeat trade and as traditional Asian medicines. Historically, offtake of saola by subsistence hunting with spears and crossbows would have driven limited offtake (Figure 19.6), but the introduction of wire snares and the unprecedented increase in demand for wildlife products has put saola under extreme threat. The explosion of the region's wildlife trade and the widespread use of wire snares are relatively recent phenomena in the saola's range; local inhabitants cite the mid-1980s to early 1990s as the start of rapid expansion of the region's wildlife trade.

The wildlife meat trade primarily supplies urban demand, where it is consumed as a delicacy and seen as a status symbol. Paradoxically, although saola is one of the most endangered mammals in Southeast Asia, it is one of few larger animals in the region without a significant value in the wildlife trade. Saola are not sought specifically for their meat, nor is the species known in the traditional pharmacopeia. It is killed largely as by-catch by poachers either targeting other animals, or hunting and snaring generally for wildlife meat (e.g. wild pigs and muntjacs). There have been recent reports of demand for saola horns for their ornamental value, and captive animals, presumably for private menageries, but this does not (yet) appear widespread or constant over time.

According to local people, the species is also susceptible to hunting with dogs, since it commonly responds to the pursuit of dogs by coming to bay and making a stand in a stream, where it is easily killed or caught by pursuing hunters.

In the years since the saola's discovery, hunting and wildlife trade have been exacerbated in the region by human population growth, regional economic growth (which has increased disposable incomes for luxuries such as bushmeat and wildlife medicines) and a flurry of track and road construction in the Annamite Mountains of Vietnam and Lao PDR, opening up previously remote areas to more poaching and making transportation of wildlife to markets more efficient. Some forest areas are now so intensively blanketed with snares that it is unlikely a local saola could long survive.

Killing or capture of saola is prohibited by law in both Vietnam and Lao PDR, and the sale of wildlife in both countries is generally illegal, but there is insufficient enforcement of these laws in both countries.

Other increasing direct or indirect threats to saola are logging, commercial agriculture, mineral mining, hydropower dam construction and human population growth. Long-term climate change could further endanger the saola's cool, wet habitat – either directly, or as a consequence of lowland human communities (especially in Vietnam) moving further into the mountains to follow more favourable conditions for agriculture. Nonetheless, for saola conservation, the undetermined, future effects of climate change are dwarfed in urgency and priority by the present-day certainty and extremity of poaching and wildlife trade.

National action plans for the conservation of saola were written for Lao PDR (Robichaud 1999) and Vietnam (Anon 2006), but were not officially endorsed or implemented by either range governments. The IUCN Wild Cattle Specialist Group conducted a range-wide conservation action in 2008 (IUCN-SSC Asian Wild Cattle Specialist Group 2010). Today, there may be no other large mammal in Asia that shares the saola's combination of phylogenetic distinctiveness, degree of endangerment and paucity of conservation attention. In 2006, in recognition of the perilous status of saola and the need for increased attention to save the species from extinction, the Asian Wild Cattle Specialist Group of the IUCN Species Survival Commission established a Saola Working Group (SWG). The SWG has identified the following priorities for *in situ* conservation of saola (Saola Working Group 2009, 2011), and the SWG is currently working with partners to implement several of them:

(1) Immediately improve anti-poaching measures, especially reduction of snaring in all sites where saola are thought to remain.
(2) Improve protected area management in the eight protected areas where saola are thought extant.
(3) Prevent the illegal sale of wildlife in restaurants across Vietnam and Lao PDR.
(4) Stop the illegal trafficking of wildlife.
(5) Significantly reduce demand for wildlife products in China, Vietnam and Lao PDR.
(6) Reduce illegal or otherwise poorly managed forest clearance.
(7) Research to learn more about saola distribution, ecology and behaviour.
(8) Establish an *ex situ* conservation breeding programme bringing to bear the extensive international expertise on species recovery programmes.
(9) Increase motivation and technical capacity for saola conservation in Vietnam and Lao PDR, and increase support from the international community and donors.

Status in captivity

No saola are currently known in captivity. The species has proven fragile in captivity (Figures 19.1–2, 19.7), but it may be possible to overcome this with the application of international best-practice standards of wild ungulate husbandry; something that has never been attempted with saola. With the

Figure 19.7 Captive saola in Vietnam (Source: Toon Fey/WWF). A black and white version of this figure will appear in some formats. For the colour version, please refer to the plate section.

Box 19.1 Using Community Forest Guards to safeguard the saola's future in Thua-Thien Hue Saola Nature Reserve, Vietnam

Luong Viet Hung, Van Ngoc Thinh, Le Ngoc Tuan and Barney Long

The greatest threat to saola populations is uncontrolled hunting. The newly established Thua-Thien Hue Saola Nature Reserve lacks the manpower, operational budget and equipment to provide the effective protection required for saola to persist within its borders. A new approach for Vietnam was piloted using hired 'Forest Guards' co-managed by WWF-Vietnam and Thua-Thien Hue Saola Nature Reserve. The objective of this model is to provide intensive protection to the saola population of the reserve through the removal of direct threats.

A total of 20 Forest Guards were hired, primarily from communities surrounding the nature reserve, all having at least a secondary-school-level education. They were extensively trained in patrolling, field skills, wildlife conservation, legal issues, navigation, safety procedures and first aid. Four teams were formed, each comprising six members; one team leader, three forest guards, one camp manager and one Saola Nature Reserve ranger. Through a rotation system these teams conduct patrols across the reserve, with each team spending eight hours per day for 16 days per month on patrol with six days working on patrol reporting and training. The teams are monitored using the law enforcement monitoring tool MIST, allowing for adaptive management through monthly assessments and work plans. This is the most intensive patrol effort afforded any protected area in Vietnam. Forest Guards are not government rangers, so do not have power of arrest; however, they do have the right to confiscate equipment and, being accompanied by a ranger from the Nature Reserve, means that one person per team has legal authority to enforce the law.

The Forest Guards started full operations in February 2011 and in their first year of operation they completed 94 patrols totalling 603 days. This led to the removal of 10,929 snares, the destruction of 149 illegal hunting/logging camps, 118 people turned out of the forest and 12 animals released from snares alive. This provides a clear indication of the scale of the threat facing saola and how, with relatively little input, it is possible to start to control these threats. The saola is facing immense pressure in all locations where it is found and without replicating and scaling up initiatives like this the saola is likely to be driven to extinction in the coming years.

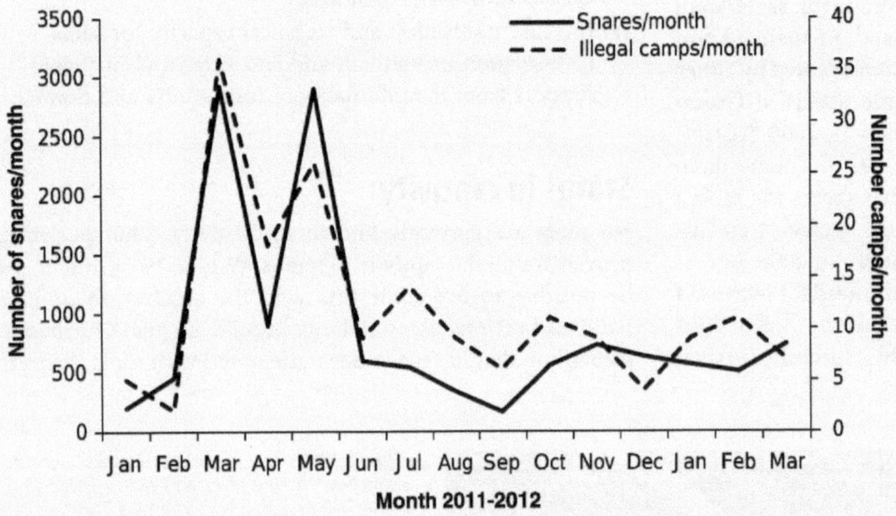

Figure 19.1.1 Results of Forest Guard Patrols over first 15 months of operation in Thua-Thien Hue Saola Nature Reserve in Vietnam.

critical status of the species in the wild and the lack of government commitment to address the threats to the species in the wild, it would be prudent to establish an *ex situ* population, and this is a recommendation of the SWG. However, before *ex situ* management of saola can credibly be attempted, much more basic information needs to be learned about the species, such as its diet and ranging and social behaviour; one reason why field studies on the saola including telemetry studies are so important to undertake as a matter of priority. Any *ex situ* conservation programme should be conducted following international best practice and would require significant and long-term funding before it commences, with funding also being paired with *in situ* sites so that re-introduction to the wild is at the forefront of the programme from the outset.

Acknowledgements

We owe gratitude to Dr Simon Stuart, Chair of the IUCN SSC, Dr James Burton, Chair of the IUCN SSC Asian Wild Cattle Specialist Group (and former Chair, Dr Simon Hedges) and Dr Weston Sechrest of Global Wildlife Conservation, for their consistent advocacy for saola conservation. And thanks to Dr Will Duckworth for his numerous acts of generosity and inspiration over the years.

References

Anon (2006). Vietnam national saola conservation action plan 2005 to 2010.

Gatesy, J. & Arctander, P. (2000). Hidden morphological support for the phylogenetic placement of *Pseudoryx nghetinhensis* with bovine bovids: a combined analysis of gross anatomical evidence and DNA sequences from five genes. *Systematic Biology* **49**: 515–538.

Grubb, P. (2005). Artiodactyla. In *Mammal Species of the World: A Taxonomic and Geographic Reference*, pp. 637–722. 3rd edition. Eds. Wilson, D. E. & Reeder, D. M. Baltimore, MD: Johns Hopkins University Press.

Hassanin, A. & Douzery, E. J. P. (1999). Evolutionary affinities of the enigmatic saola (*Pseudoryx nghetinhensis*) in the context of the molecular phylogeny of Bovidae. *Proceeding of Royal Society of London Series B: Biological Sciences* **266**: 893–900.

IUCN (2013). *IUCN Red List of Threatened Species*. Version 2013.2. www.iucnredlist.org, accessed 10 April 2014.

IUCN-SSC Asian Wild Cattle Specialist Group (2010). *Regional conservation strategy for wild cattle and buffaloes in South-east Asia*. www.asianwildcattle.org/documents/SE.Asia_Cons.Strtgy_AWC.pdf, accessed 10 April 2014.

Long, B. (2000). Large mammals. In *Pu Mat: A Biodiversity Survey of a Vietnamese Protected Area*. Ed. Grieser Johns, A. Vinh, Nghe An Province: SFNC Project.

Robichaud, W. (1998). Physical and behavioral description of a captive saola *Pseudoryx nghetinhensis*. *Journal of Mammalogy* **79**: 394–405.

Robichaud, W. (1999). *Saola Conservation Action Plan for Lao PDR: 1999 Revision*. Vientiane: Wildlife Conservation Society and IUCN.

Saola Working Group (2009). *From Plans to Action: Proceedings of the First Meeting of the Saola Working Group. IUCN Lao PDR Country Programme and the Saola Working Group of the IUCN Species Survival Commission*. Vientiane: Asian Wild Cattle Specialist Group.

Saola Working Group (2011). *Saola Conservation: A Portfolio of Priority Initiatives; v. 2011.1*. Vientiane: Saola Working Group and IUCN Lao Programme.

Schaller, G. B. & Rabinowitz, A. (1995). The saola or spindlehorn bovid *Pseudoryx nghetinhensis* in Laos. *Oryx* **29**: 107–113.

Thomas, H. (1994). Anatomie crânniene et relations phylogénétiques du nouveau bovidé *Pseudoryx nghetinhensis* découvert dans la cordillère annamatique au Vietnam. *Mammalia* **58**: 453–481.

Vu Van Dung, Pham Mong Giao, Nguyen Ngoc Chinh, Do Touc, Arctander, P. & MacKinnon, J. (1993). A new species of living bovid from Vietnam. *Nature* **363**: 443–445.

Chapter 20

African buffalo *Syncerus caffer* (Sparrman, 1779)

Daniel Cornélis, Mario Melletti, Lisa Korte, Sadie J. Ryan, Marzia Mirabile, Thomas Prin and Herbert H. T. Prins

Names

Genus: *Syncerus* Hodgson, 1847

Species: African buffalo *Syncerus caffer* (Sparrman, 1779)

Subspecies: Cape buffalo *Syncerus caffer caffer* (Sparrman, 1779)

Names in other languages: **French:** Buffle du Cap; **German:** Kaffernbüffel; **Spanish:** Búfalo cafre; **Italian:** Bufalo cafro.

Subspecies: West African savanna buffalo *Syncerus caffer brachyceros* (Gray, 1837)

Names in other languages: **French:** buffle de savane d'Afrique de l'Ouest; **German:** Savanne Westafrika Büffel; **Spanish:** Búfalo de Sabana de África Occidental; **Italian:** Bufalo di savana dell'Africa Occidentale

Subspecies: Central African savanna buffalo *Syncerus caffer aequinoctialis* (Blyth, 1866)

Names in other languages: **French:** buffle de savane d'Afrique Centrale; **German:** Savanne Zentralafrika Büffel; **Spanish:** Búfalo de Sabana de África Central; **Italian:** Bufalo di savana dell'Africa Centrale, Bufalo equinoziale.

Other common names: Nile buffalo.

Subspecies: Forest buffalo *Syncerus caffer nanus* (Boddaert, 1785)

Names in other languages: **French:** buffle de forêt; **German:** rotbüffel; **Spanish:** bùfalo de bosque; **Italian:** Bufalo di foresta; **BaAka:** Mboko; **Swahili:** Nyati.

Other common names: dwarf buffalo, red buffalo.

Ecology, Evolution and Behaviour of Wild Cattle: Implications for Conservation, ed. M. Melletti and J. Burton. Published by Cambridge University Press.

Taxonomy

The name *Bos caffer* was attributed by Sparrman in 1779. Since then, 92 species names have been given to the African buffalo. Taxonomists initially thought that each buffalo form represented a distinct species. Brooke (1873, 1875), who established the first classification of the African buffalo, reduced the number to three. Later, Blancou (1935) described up to 12 subspecies of buffalo.

Haltenorth (1963), Ansell (1972) and Grubb (1972) summarized the first classifications of Christy (1929), Schouteden (1945) and Blancou (1935, 1954), concluding that all forms should be considered as monospecific. Although there are considerable morphological variations in body size, fur colour, horn shape and size throughout the range of distribution, the African buffalo is currently considered as a single species by various authorities (IUCN 2013; Prins & Sinclair 2013), with a subdivision into four subspecies: Cape buffalo (*S. c. caffer*), forest buffalo (*S. c. nanus*), West African savanna buffalo (*S. c. brachyceros*), and Central African savanna buffalo (*S. c. aequinoctialis*). Additionally to those four subspecies, a mountain form (*S. c. mathewsi*) was also described in East Africa and may be distinct (Kingdon 1982).

A recent work of Groves & Grubb (2011), based on morphological measures, separates the African buffalo into four species, including the *mathewsi* form. In contrast, recent work of Smitz *et al.* (2013) conducted over the whole geographical range of the African buffalo, analysing mitochondrial D-loop sequences, revealed the existence of two lineages in the species, matching with the *caffer* subspecies of East and South Africa, and the *nanus* subspecies of West and Central Africa. These lineages seem to have diverged between 132and 286 kya (for further insight into the evolutionary history of the African buffalo, see Chapters 1 and 25). On the basis of the evolutionary history of both lineages, the authors suggest that the most prudent treatment of the African buffalo would be a subdivision into two subspecies, or management units, namely *S. c. caffer* (East-Southern Africa's lineage), and *S. c. nanus* (West Central Africa's lineage), which then should be extended to comprise *S. c. brachyceros* and *S. c. aequinoctialis* subspecies following standard nomenclature rules (*S. c. nanus*: Boddaert 1785; *S. c. brachyceros*: Gray 1837; *S. c. aequinoctialis*: Blyth 1866; for more details see Chapter 25). Further investigation using microsatellites may reveal finer-scale population structuring in West-Central and East-Southern Africa, which may be attributed to evolutionary significant units (ESUs) appropriate in terms of conservation management (Moritz 1994).

Hybrids *caffer* x *nanus* have been described in captivity (Cribiu & Popescu 1980). Although the different forms seem to be able to interbreed, the existence of hybridization in wild populations has yet to be conclusively demonstrated by molecular work. The occurrence of red calves in, for instance, the Queen Elizabeth National Park is often quoted as evidence for hybridization, but proof has never been provided. In Lake Manyara National Park, some 800 km further to the east, about 30% of the calves are red when young.

Subspecies and distribution

Historical distribution

Rainfall is the main biophysical factor limiting the distribution and abundance of African buffalo at a large geographical scale. Prior to the nineteenth century, the range of buffalo extended to most sub-Saharan ecosystems with annual rainfall above 250 mm (thus excluding the Horn of Africa and Namib/Kalahari deserts). The African buffalo is essentially a grazer, so its distribution strongly depends on the availability of herbaceous forage. As a consequence, core areas of the closed rainforests in Central Africa (i.e. areas without herbaceous cover) did not historically provide suitable conditions for buffalo (Sinclair 1977; Mloszewski 1983; Prins 1996; Melletti *et al.* 2007a). Buffalo did not colonize islands such as Zanzibar or Mafia, although they colonized Bioko Island (Equatorial Guinea), where they are now extinct (extirpated from Bioko Island sometime between 1860 and 1910 (Butynski *et al.* 1997)).

There is no palaeontological evidence of the presence of the African buffalo in North Africa or in the Nile Valley to the north of Khartoum (Prins & Sinclair 2013). In North Africa, aurochs (*Bos primigenius*; wild ancestor of domestic cattle) occupied a similar niche (Gautier 1988), perhaps preventing the buffalo's spread to the north (for more details on Bovini evolution, see Chapter 1). Buffalo could have expanded their range in eastern and southern Africa during the last Ice Age due to the extinction of possible competitors, such as *Pelorovis antiquus* and *Elephas recki* (for more details on evolution, see Chapter 1; Klein 1988, 1994; Prins 1996). Drawings of buffalo are rare in cave paintings, although representations can be seen in Kondoa in Northern Tanzania (Leakey 1983).

We present below the distribution and abundance of each of the four subspecies. For consistency and comparison purposes, this presentation is based on the geographical contours of the IUCN (Plate 11). Actually these contours are not defined by geographic boundaries, but are gradual and diffuse. For example, the occurrence of two subspecies (*S. c. caffer* and *S. c. aequinoctialis*) in Ethiopia or in the Democratic Republic of Congo is somewhat arbitrary.

Present distribution

Large discontinuities have emerged in the historical distribution range of African buffalo in areas of high human population densities, such as Nigeria or Central Ethiopia (Plate 11) African buffalo populations have undergone a severe reduction in size and geographical distribution since the nineteenth century, as a result of the combined effects of anthropogenic impacts such as land conversion, poaching, disease outbreaks and climatic events such as droughts. At present, around 70% of the savanna buffalo population is confined to a patchwork of protected areas and well-managed surrounding hunting zones, mostly loosely connected to one another (East 1998). Figure 20.1 shows the distribution of the four buffalo

(a)

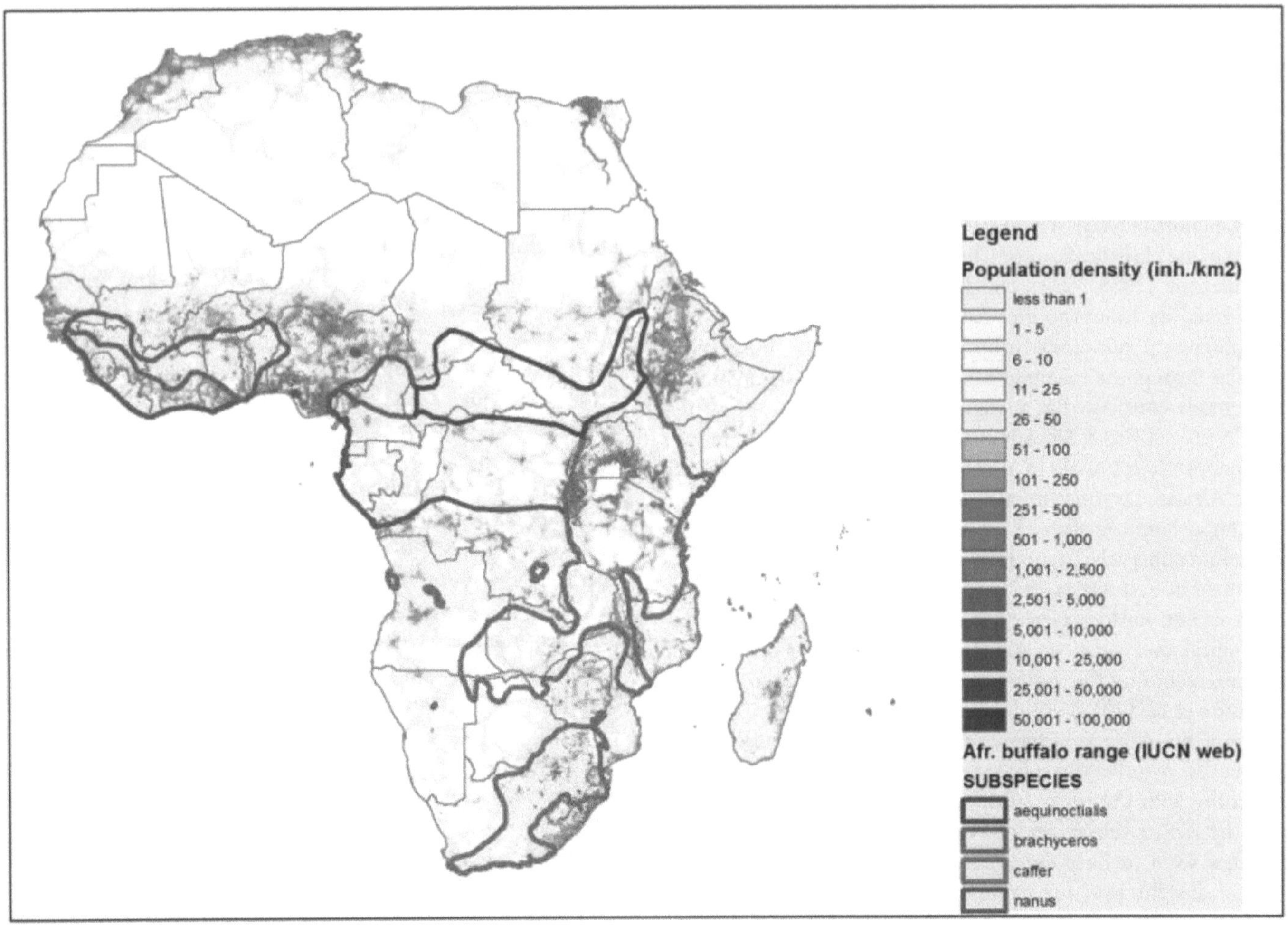

Figure 20.1 Distribution of the four African buffalo subspecies in relation to: (a) human population density; (b) average rainfall and (c) protected areas (source: IUCN SSC 2008). A black and white version of this figure will appear in some formats. For the colour version, please refer to the plate section.

subspecies in relation to human population density, rainfall and protected areas (East 1998; IUCN SSC 2008).

The expansion of livestock production gradually generated direct competition for space and resources and led to large and destructive epidemics in native African buffalo populations. Rinderpest was historically the most devastating disease for buffalo populations throughout Africa, leading to extreme reductions in population densities, and local extinctions. The most severe population collapse occurred in the 1890s, with mortality rates estimated at 90–95% across the continent (Sinclair 1977; Prins & van der Jeugd 1993; Winterbach 1998). This was followed by other episodes throughout the twentieth century. Rinderpest was declared eradicated in Africa by the World Animal Health Organization in 2011. During the twentieth century, the geographic distribution of buffalo has been actively reduced in several countries by large-scale culling operations and veterinary fences, in efforts to limit the transmission of several pathogens, such as foot-and-mouth disease (FMD) and trypanosomiasis, to cattle (Taylor & Martin 1987).

Recent climate fluctuations such as the drought that affected Sahelian and Sudanese regions at the end of the 1960s and Southern Africa in 1992 (Dunham 1994; Mills *et al.* 1995) have also strongly impacted buffalo populations over the past few decades. Last but not least, armed conflicts, the feeding of armies and labourers during peace time, the trafficking of weapons and bushmeat trade have strongly contributed to the reduction of buffalo populations.

Cape buffalo (*S. c. caffer*)

Cape buffalo occurs in southwestern Ethiopia (particularly Omo National Park), southern Somalia and northern Kenya southwards to South Africa. This subspecies is patchily distributed throughout East Africa, southwards to Zambia, Malawi, some parts of Mozambique (small populations south of the Zambezi river with the bulk of it in the north of the country, Niassa Reserve and Cabo Delgado areas; Skinner & Chimimba, 2005), northeastern Namibia (mainly eastern and western Caprivi, Mamili National Park and an introduced population in the Waterberg

(b)

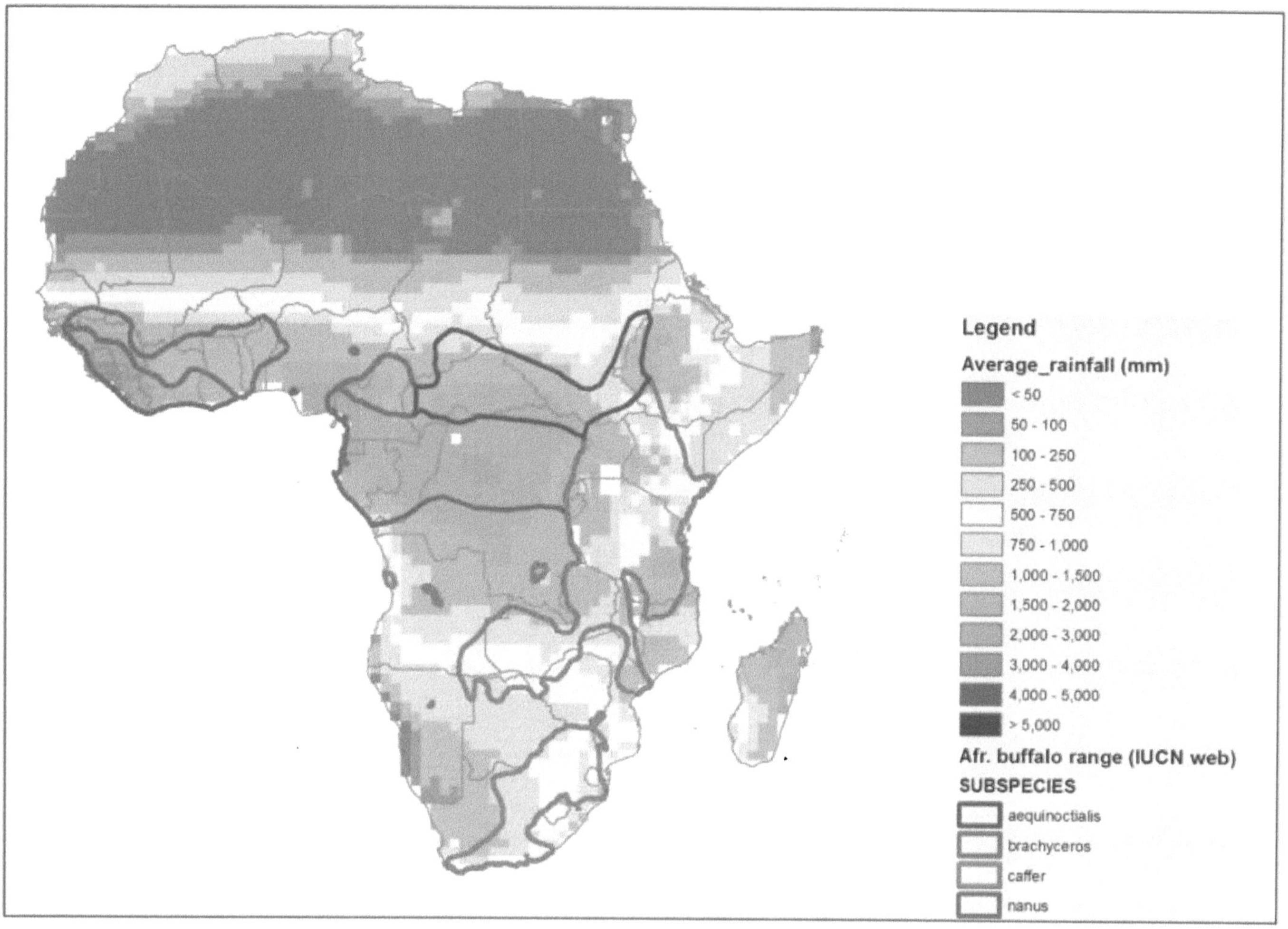

Figure 20.1 *(cont.)*

Plateau Park, central Namibia), Botswana (north of 20° S, Okavango Delta area, Moremi Game Reserve and the Chobe National Park) and Zimbabwe (Hwange National Park; Matabeleland, Zambezi Valley, south Lake Kariba and Gonarezhou National Park; Skinner & Chimimba 2005). Small populations of Cape buffalo still survive in some areas of southeast Angola bordering Zambia and Namibia. In South Africa, Cape buffalo have been reintroduced to some areas from which they were formerly extirpated. They are currently widely distributed in Kruger National Park and smaller populations persist in KwaZulu-Natal (Winterbach 1998) and in many private reserves. The current population in Swaziland was also reintroduced after extirpation.

West African savanna buffalo (*S. c. brachyceros*)

This subspecies still occurs locally in western countries within the Sahelo-Sudanian band (savannas and gallery forests), including southeastern Senegal, northern Ivory Coast, southern Burkina Faso, Ghana, northern Benin, extreme south of Niger, Nigeria (very locally), northern Cameroon and a small part in the Central African Republic (East 1998).

Central African savanna buffalo (*S. c. aequinoctialis*)

This subspecies still locally populates Central African countries within the Sahelo-Sudanian band (savannahs and gallery forests): southeast Chad, northern Central African Republic (East of Chari River), northern Democratic Republic of Congo, southeast Sudan and western Ethiopia (Ansell 1972; East 1998). The subspecies is now extinct in Eritrea.

Forest buffalo (*S. c. nanus*)

This subspecies occurs in two disjoint distribution ranges in West and Central Africa, respectively in the relict coastal rainforest belt and in the large basin of the Congo River. The distribution ranges are mainly located in areas with an annual rainfall of more than 1500 mm, in transition areas between dense tropical forests and savanna ecosystems, for instance in Gabon and Uganda (East 1999). In West Africa, forest buffalo is distributed from Guinea-Bissau to southwestern Ghana. In Central Africa, forest buffalo persist in south Nigeria, south Cameroon, Equatorial Guinea (extinct on Bioko Island), south and southwest Central African Republic, western Uganda,

(c)

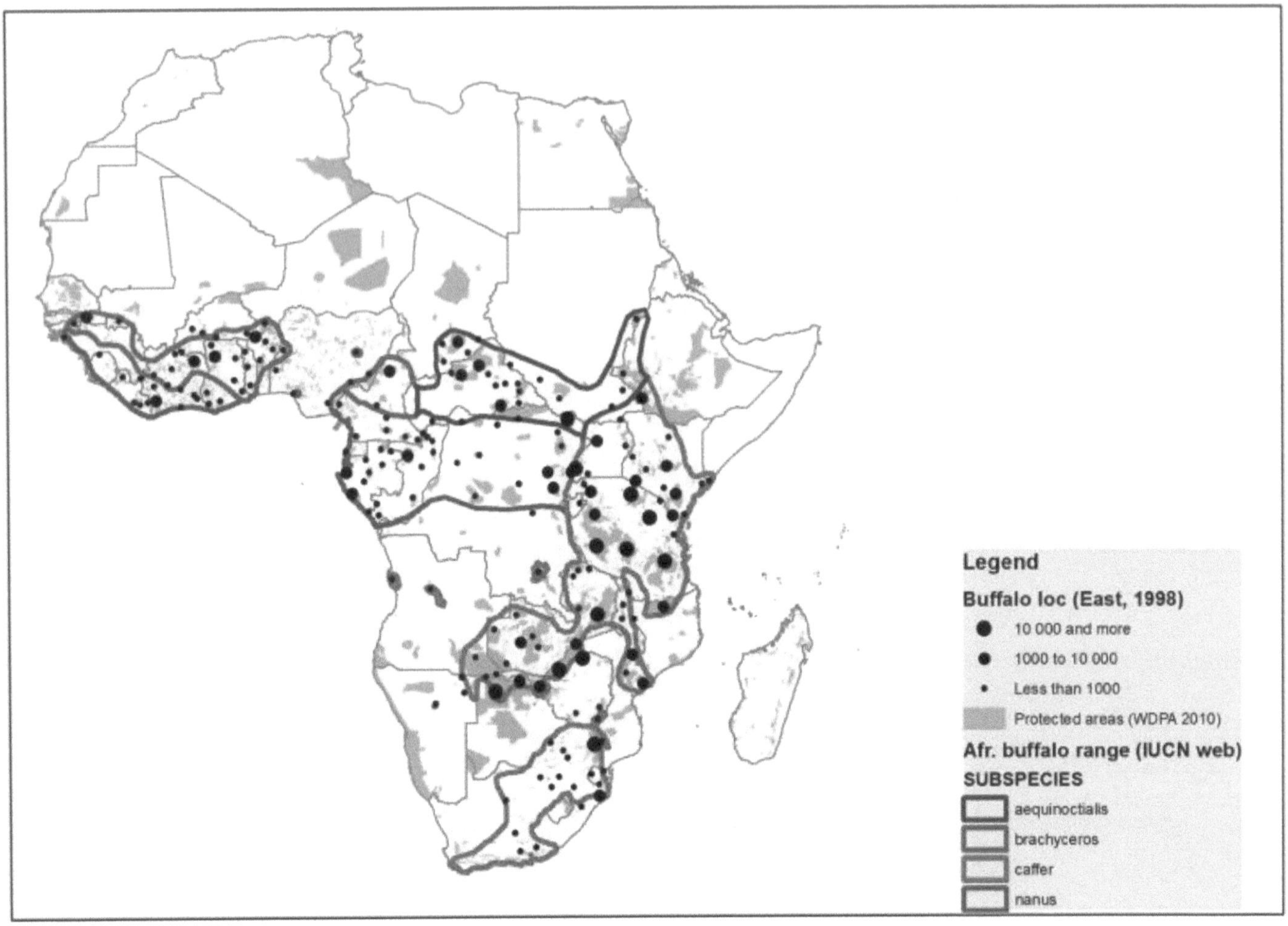

Figure 20.1 *(cont.)*

western Rwanda and possibly northwest Burundi, the Democratic Republic of Congo, the Republic of Congo and Gabon. The greatest part (i.e. 75%) of the population lives in and around protected areas in Cameroon, Central African Republic, Gabon, Democratic Republic of Congo and Republic of Congo (Wilson & Mittermeier 2011). In both distribution ranges, numbers have declined substantially due to poaching and deforestation (East 1999), but their status is generally poorly known (Blake 2002; Melletti *et al.* 2007a, 2007b; Korte 2008a).

Abundance

Over the past 60 years aerial counts have been widely used to estimate population sizes of savanna buffalo. As with any estimation technique, these methods are subject to potentially important biases (Jachmann 2002; Redfern & Viljoen 2002). During aerial counts, significant numbers of animals may be missed due to obstruction by vegetation canopy cover (e.g. high grass, forest galleries), and due to lower probability of detecting small herds. Furthermore, estimates from aerial sampling counts (ASCs) are generally less accurate (especially for small populations and low sampling rates) than aerial total counts (ATCs). Despite these limitations, aerial counts remain the most reliable approach to estimate buffalo populations in savanna ecosystems. We cite the animal numbers from the original reports, in the text, in a rounded-off form, and report to the closest 100 individuals for all figures larger than 200, while we report reserve size to the closest 10 km^2 so as to prevent reporting non-significant digits.

In the following sections and Table 20.1, we present an update of buffalo abundance and distribution per country and protected area, based on the most recent and available census data. The total estimated numbers per country are presented in Table 20.2 and compared with the last update (East 1998) in the section 'Status in the wild'.

Cape buffalo

In **Ethiopia**, the population of Omo National Park and surrounding buffer zone (7850 km^2) was estimated at 700 individuals, based on an ASC carried out in 2007 (Renaud 2007);

Table 20.1 Abundance and distribution of the four African buffalo subspecies per country and protected area based on the most recent and available census data. Original data are presented.

Subspecies	International PA complex	Country	Protected area/ hunting blocks	Date	Method	Sampling effort[a]	Number obs.	Estimated population	Error	Error statistics	Area (km^2)	Density (ind/ km^2)	Source
S. c. aequinoctialis	/	Central African Republic	Bamingui-Bangoran NP, Manovo-Gounda-St. Floris NP and adjacent PAs	2010	ASC	7%	738	4048	± 27%	CV	94 962	0.04	Bouché 2010
	/	Chad	Zakouma NP	2012	ATC	100%	8091	8091	/	/	3026	2.67	Potgieter *et al.* 2012
	/	Democratic Republic of Congo	Garamba NP	2012	ATC	100%	5980	5980	/	/	4920	1.22	Bolanos 2012
	/	Ethiopia	Gambella NP Akobo block	2010	ASC	6%	91	1378	± 2641	SE	20 000	0.07	EWCA 2010
S. c. brachyceros	WAPOK complex (Burkina Faso, Bénin, Niger, Togo)	Benin	W NP	2003	ATC	100%	21	424	/	/	5872	0.07	Bouché *et al.* 2003
		Benin	Pendjari NP	2003	ATC	100%	124	3421	/	/	2827	1.21	
		Benin	Hunting zones (Djona, Pendjari, Konkombri and Mékrou) + Goungoun Classified Forest	2003	ATC	100%	715	715	/	/	5243	0.14	
		Burkina Faso	W NP	2003	ATC	100%	1400	1400	/	/	2412	0.58	
		Burkina Faso	Arli NP	2003	ATC	100%	399	399	/	/	485	0.82	
		Burkina Faso	Hunting zones (Tapoa djerma, Kourtiagou, Koakrana, Pagou-Tandougou, Ougarou, Pama, Singou and Konkombouri)	2003	ATC	100%	3031	3031	/	/	8124	0.37	
		Niger	W NP and Tamou Reserve	2003	ATC	100%	1170	1170	/	/	3083	0.38	
		Togo	Keran NP & Oti-Mandori Reserve	2003	ATC	100%	0	0	/	/	2886	0.00	
		Burkina Faso	Comoé-Léraba complex	2005	ATC	100%	90	90	/	/	1847	0.05	Bouché 2005
		Burkina Faso	Ponasi Complex (Nazinga GR, Kabore-Tambi NP and adjacent hunting zones)	2004	ATC	100%	145	145	/	/	3105	0.05	Bouché *et al.* 2004

Table 20.1 *(cont.)*

Subspecies	International PA complex	Country	Protected area/ hunting blocks	Date	Method	Sampling effort[a]	Number obs.	Estimated population	Error	Error statistics	Area (km^2)	Density (ind/ km^2)	Source
		Cameroon	Bouba Ndjida, Faro, Benoue NPs + hunting blocks	2008	ATC	100%	630	(4000)	/	/	22 137	0.18	Omondi *et al.* 2008
		Ghana	Mole NP	2006	ATC	100%	700	700	/	/	4500	0.16	Bouché 2006
		Ivory Coast	Comoé NP	2010	ASC	6%	56	899	± 905	SD	11 945	0.08	N'Goran *et al.* 2010
		Nigeria	Yankari NP	2011	ATC	100%	174	174	/	/	2244	0.08	Bergl *et al.* 2011
		Senegal	Niokolo Koba NP	2006	ASC	40%	15	457	± 447	SE	9130	0.05	Renaud *et al.* 2006
		Senegal	Niokolo Koba NP	2003	ASC	7%	68	957	± 864	95% IC	9130	0.10	Mauvais & Ndiaye 2004
S. c. caffer		Botswana	North country (Moremi GR, Chobe National NP, Makgadikgadi Nxai Pan NP and surrounding WMAs in the Ngamiland, Chobe and Central districts)	2010	ASC	14% [5–22]	7958	39 579	± 27%	95% CI	73 478	0.54	Chase 2011
		Democratic Republic of Congo	Virunga NP	2010	ASC	?	?	2154	?	?	7823	0.28	Plumptre *et al.* 2010
		Ethiopia	Chebera Churchura NP	2005–2006	Pedestrial (LT)	5%	130	2617	± 426	SE	1215	2.15	Megaze *et al.* 2012
		Ethiopia	Omo NP and buffer zone	2007	ASC	40%	330	675	± 31%	SE (%)	7853	0.09	Renaud 2007
		Kenya	Laikipia/Samburu ecosystem	2002	ATC	100%	2012	2012	/	/	10 000	0.20	Omondi *et al.* 2002
		Kenya	Masai Mara NR/ community areas	2010	ATC	100%	4649	4649	/	/	4716	0.99	Kiambi *et al.* 2010
		Kenya	Meru Conservation Area	2007	ATC	100%	2288	2288	/	/	12 000	0.19	Mwangi *et al.* 2007
		Kenya	Nasolot/south Turkana/Rimoi/ Kamnarok	2010	ATC	100%	0	0	/	/	4551	0.00	Edebe *et al.* 2010
		Mozambique	Chipanje Chetu	2011	ASC	11%	10	/	/	/	6641	/	Craig 2011b
		Mozambique	Niassa Reserve	2011	ASC	10%	644	6214	[3462–8967]	95% CI	42 296	0.15	Craig 2011a

Mozambique	East Musengezi River (south Lake Cabora Bassa)	2010	ASC	25%	41	143	± 177%	95% CI	13 962	0.01	Dunham 2010
Mozambique	West the Musengezi River (south Lake Cabora Bassa)	2010	ASC	12%	587	4483	± 107%	95% CI	2621	1.71	Dunham 2010
Mozambique	Banhine NP	2009	ASC	19%	16	/	/	/	1192	/	Stalmans & Peel 2009
Mozambique	Quirimbas NP	2011	ASC	10%	0	0	/	/	20 834	0.00	Craig 2011c
Mozambique	Marromeu Complex	2010	ATC	100%	10 300	10 300	/	/	11 270	0.91	Beilfuss *et al.* 2010
Mozambique	Zinave NP and surrounding area	2010	ASC	15.2% [6–21]	0	0	/	/	9 231	0.00	Dunham *et al.* 2010
Namibia	Caprivi River Systems	2007	ATC	100%	5971	5971	/	/	1778	3.34	Chase 2007
Rwanda	Akagera National Park	2010	SRF		?	?			1120		Viljoen *et al.* 2010
South Africa	Kruger NP	2011	ATC	100%	40 922	40 922	/	/	20 000	2.05	Ferreira *et al.* 2012
Tanzania	Selous ecosystem	2002	ASC	?	?	280 701	± 61 278	SE	94 009	2.99	Debonnet & Wilson 2008
Tanzania	Selous ecosystem	2006	ASC	?	?	113 463	± 25 613	SE	80 883	1.40	Debonnet & Wilson 2008
Tanzania	Selous-Niassa Wildlife Corridor	2006	ASC	5%	94	1456	± 949	SE	9096	0.16	Tawiri 2006
Tanzania	Selous-Niassa Wildlife Corridor	2009	ASC	10%	418	4095	± 2116	SE	9096	0.45	Tawiri 2009
Tanzania	Serengeti Ecosystem (Serengeti NP, Maswa and Ikorongo-Grumeti GRs, Ngorongoro CA, Liliondo GCA and adjacent areas)	2009	ATC	100%	32 068	32 068	/	/	26 827	1.20	Tawiri 2010
Tanzania	Katavi-Rukwa ecosystem	1988–2002	ATC	?	?	39 557	?	?	8 494	4.65	Caro 2008
Uganda	Murchison Falls PA (Murchison Falls NP, Bugungu & Karuma WRs)	2010	ATC	?	356	9192	± 3789	SE	5044	1.82	Rwetsiba & Nuwamanya 2010
Uganda	Kidepo NP	2008	ATC	100%	3643	3643	/	/	1128	3.23	WCS Flight Programme 2008

Table 20.1 *(cont.)*

Subspecies	International PA complex	Country	Protected area/ hunting blocks	Date	Method	Sampling effort[a]	Number obs.	Estimated population	Error	Error statistics	Area (km^2)	Density (ind/ km^2)	Source
		Uganda	Queen Elizabeth NP	2010	ATC	100%	10 282	10 282	/	/	2113	4.87	Plumptre *et al.* 2010
		Zambia	Kafue Ecosystem	2011	ASC	3–12%	/	4566	± 687	SE	70 000	0.07	Frederick 2011
		Zambia	Luanga Valley ecosystem	2008	ASC	12%	1540	17 155	± 3,673	95% CI	49 700	0.35	WCS Flight Programme 2009
		Zimbabwe	Gonarezhou complex (Gonarhezou NP, Malipati SA and surrounding CLs)	2010	ASC	15.2% [6–21]	558	2742	± 75.8%	95% CI	7112	0.39	Dunham *et al.* 2010
		Zimbabwe	Sebungwe Region (Chizarira and Matusadona NPs, Chiziza/Chete SAs, P&W Estate, Sajarira FA, North Gokwe, Binga Kariba CLs)	2006	ASC	15%[7–23]	1771	10 395	± 37.6%	95% CI	15 622	0.67	Dunham *et al.* 2006
		Zimbabwe	Hwange-Matetsi complex and adjacent areas (Zakuma NP, Deka SA, Ngamo and Sikumi Forest Areas and Tsholotsho and Maitengwe CL)	2007	ASC	8%[3–15]	1461	24 500	±120%	95% CI	24 570	1.00	Dunham *et al.* 2007
	Amboseli-West Kilimanjaro/ Magadi – Natron (Kenya, Tanzania)	Kenya/ Tanzania	Amboseli NP	2010	ATC	100%	235	235	/	/	8797	0.03	WCS Flight Programme 2010
			Namanga-Magadi	2010	ATC	100%	62	62	/	/	5513	0.01	
			Lake Natron	2010	ATC	100%	37	37	/	/	7047	0.01	
			West Kilimanjaro	2010	ATC	100%	0	0	/	/	3014	0.00	

	Tsavo/ Mkomazi ecosystem	Kenya	Tsavo East and West, Chyulu NP, South Kitui NR, Galana, Taita and Rombo blocks	2011	ATC	100%	7281	7281	/	/	45 126	0.16	Ngene *et al.* 2011
		Tanzania	Mkomazi NP				121	121			3193	0.04	
	Zambezi Heartland (Mozambique, Zambia, Zimbabwe)	Mozambique	Lake Cabora Bassa	2003	ASC	15%	265	1561	± 90%	95% CI	6329	0.25	Dunham 2004
		Zambia	Lower Zambezi NP, Chiawa and Rufunsa GMAs/Tonga-Sikongo CL		ASC		844	6658	± 102%	95% CI	11 954	0.56	
		Zimbabwe	Mana Pools NP, Hurungwe, Sapi, Charara, Dande, Chewore SAs and part of Guruve District		ASC		3537	19 527	± 37%	95% CI	13 373	1.46	
S. c. nanus		Cameroon	Campo Ma'an NP	2000–2002	Pedestrial (LT)	/	/	20	/	95% CI	650	0.01	Bekhuis *et al.* 2008
		Central African Republic	Dzanga-Ndoki NP (Dzanga sector)	2002–2004	Pedestrial (RT)	5%	32	40	/	/	495	0.08	Melletti 2005; Melletti *et al.* 2007a
		Gabon	Lope´ NP (northeast sector)	2002–2004	Ground survey	/	/	342	/	/	72	5	Korte 2008b
		Gabon	Réserve de Faune de Petit Loango	2006	Ground survey	/	/	/	/	/	500	1.70	Morgan 2007
		Republic of Congo	Noubale-Ndoki NP	1993–2000	Pedestrial (LT)	/	/	/	/	/	3920	0.01	Blake 2002
		Republic of Congo	Odzala NP	1990s	Ground survey	/	/	500	/	/	13 600	0.04	Chamberlan *et al.* 1995

ASC: aerial sampling counts; **ATC**: aerial total counts; **CA**: conservation area; **CL**: communal land; **FA**: forest area; **GCA**: game controlled area; **GR**: game reserve; **LT**: line transect; **NP**: national park; **NR**: national reserve; **RT**: recce transect; **SA**: safari area; **SRF**: systematic reconnaissance flight; **WMA**: wildlife management area; **WR**: wildlife reserve.

[a] Sampling efforts generally are weighted means of several block counts sampled at different intensities

Table 20.2 The total estimated numbers of three African buffalo savanna subspecies per country are presented and compared with the last update of East (1998)

		East (1998)	Cornélis, Melletti, Korte, *et al.* (this study)
S. c. aequinoctialis		**>59 000**	**>23 000**
	Central African Republic	19 000	4050
	Chad	1020	8090
	Democratic Republic of Congo	39 180	5980
	Eritrea	Ex	Ex
	Ethiopia	x	4380
	Sudan	>100	(8900)
S.c. brachyceros		**>20 000**	**> 17 000**
	Benin	>2000	4560
	Burkina Faso	1620	5070
	Cameroon	3210	4000
	Gambia	Ex	Ex
	Ghana	C	700
	Guinea	V	V
	Guinea Bissau	x	?
	Ivory Coast	8330	900
	Mali	120	Ex?
	Niger	500	1170
	Nigeria	>200	>170
	Senegal	>4000	460
	Togo	U/R	x
S. c. caffer		**> 548 000**	**> 473 000**
	Angola	<500	x
	Botswana	26 890	39 580
	Burundi	500	Uk
	Democratic Republic of Congo	No data	2150
	Ethiopia	2330	3600
	Kenya	>19 560	>16 560
	Malawi	>3150	Uk
	Mozambique	9570	23 310
	Namibia	1000	6000
	Rwanda	1200	R
	Somalia	U	Uk
	South Africa	30 970	>77 800
	Swaziland	U	Uk
	Tanzania	>342 450	>189 230
	Uganda	>20 220	23 120
	Zambia	>40 090	>28 330
	Zimbabwe	50 330	63 000
Total		**>627 000**	**>513 000**

-: absent; **?**: may occur but current presence not confirmed; **Ab**: abundant; **C**: Common; **Ex**: Extinct; **Ex?**: probably extinct; **R**: rare; **U**: uncommon; **Uk**: unknown; **V**: occurs only as a vagrant; **x**: definitely present but abundance unknown

this represents a 64% decrease since an ASC conducted in 1996 (Graham *et al.* 1996). At most 300 individuals are estimated to range within Mago National Park and Tama Wildlife Reserve (respectively 1940 and 1670 km^2, east of Omo National Park; Ludwig Siege, EWCA, personal communication). Further north, in Chebera Churchura National Park (1220 km^2), the buffalo population was estimated (line transect; LT) in 2006 at 2600 individuals (Megaze *et al.* 2012).

In **Uganda**, data from the 2010 ASC of Murchison Falls Protected Area estimated a population of 9200 buffalo within an area of 5040 km^2 (Rwetsiba & Nuwamanya 2010). In Kidepo National Park and surrounding areas (1130 km^2) the last ATC estimated a total population of 3600 individuals, primarily located inside the park (WCS Flight Programme 2008). In Queen Elizabeth National Park (2110 km^2), the most recent ATC reported 10 300 buffalo (Plumptre *et al.* 2010).

In the Akagera National Park (**Rwanda**), the buffalo population was estimated at 7800 individuals in the late 1970s (Monfort 1979). During civil unrest in the 1990s, their population was severely reduced. In 1997 Akagera National Park was reduced in size from more than 2500 km^2 to its current extent of 1120 km^2 (East 1999). An aerial survey (SRF; systematic reconnaissance flight) was carried out in 2010 and small numbers of buffalo were sighted in Akagera National Park (average group size 23 individuals $n = 38$) and Gabiro area (average group size 39 individuals $n = 9$), but no reliable population estimate is available (Viljoen *et al.* 2010).

In Virunga National Park (Democratic Republic of Congo, 7820 km^2), an ASC carried out in 2010 reported 2100 buffalo

(Plumptre *et al.* 2010). In the south of the country (Katanga province), the buffalo population in Upemba National Park (10 000 km^2) and Kundelungu National Park (2200 km^2) is on the brink of extinction with just 15 individuals observed in 2008 surveys (Vanleeuwe *et al.* 2008). In these parks wildlife was abundant throughout the 1980s but then rapidly declined in the 1990s and 2000s due to economic and political instability, and subsequent insecurity (Vanleeuwe *et al.* 2008).

In **Kenya**, the status of buffalo varies with the locality considered. In the northern part of the country (Nasolot, south Turkana, Rimoi and Kamnarok region; 4550 km^2) no buffalo were recorded during an ATC conducted in 2010 (Edebe *et al.* 2010). In Laikipia/Samburu ecosystem (an area covering about 10 000 km^2), 2002 aerial surveys reported 2000 buffalo, with an estimated 30% decrease between 1999 and 2002; the largest reported decline (55%) was in Samburu area (Omondi *et al.* 2002). In Meru Conservation Area (4010 km^2, covering Meru National Park, Kora National Park, Bisinadi, Mwingi and North Kitui National Reserves), there were 1800 individuals (wet season), and 2300 (dry season) (Mwangi *et al.* 2007). In Masai Mara National Reserve and community areas (4720 km^2), the buffalo population was estimated at 4600 individuals in 2010 (Kiambi *et al.* 2010).

In Amboseli-West Kilimanjaro/Magadi-Natron protected areas (Kenya/Tanzania border, 24 370 km^2) an ATC estimated the population at 300 individuals (90% in the Kenya section) (WCS Flight Programme 2010). In Tsavo/Mkomazi ecosystem (Kenya/Tanzania border, 48 320 km^2) the last three ATCs reported estimates of 9400 (2005), 5600 (2008) and 7400 (2011) individuals (Ngene *et al.* 2011). During the last ATC, 98% of the buffalo (7300 individuals) were observed in the Kenyan section of the Tsavo/Mkomazi ecosystem.

Tanzania has the largest African buffalo population, with over 342 000 individuals recorded in the last update (East 1998). Buffalo are common in most of the country's major wildlife areas, e.g. Selous Game Reserve, Serengeti, Tarangire and Katavi National Parks, the Ruaha ecosystem and the Katavi-Rukwa and Kilombero Game Controlled Areas. The Selous ecosystem (over 90 000 km^2, southern Tanzania) certainly has the largest buffalo population in Africa and over half of Tanzania's. Recent ASCs reported a population estimate of 280 700 individuals in 2002 (94 000 km^2) and 113 500 (80 890 km^2) individuals in 2006 (Debonnet & Wilson 2008). In Selous-Niassa Wildlife Corridor (9100 km^2), a region bordering Mozambique, dry season ASCs conducted in 2006 and 2009 reported 1500 and 4100 buffalo, respectively (Tawiri 2006, 2009). In the Serengeti ecosystem (north Tanzania), ATCs conducted in 2008 and 2009 (covering 26 830 km^2) recorded 32 900 and 32 100 buffalo, respectively. The population numbers for 2008 and 2009 surveys were the highest for the Serengeti ecosystem in 14 years (Taiwiri 2010). In Katavi-Rukwa ecosystem (8500 km^2, Tanzania) the buffalo population was estimated at 39 600 individuals (surveys 1988–2002; Caro 2008). A review of aerial census data collected during the late 1980s to early 2000s in eight large census zones of Tanzania suggests that buffalo populations globally remained stable in Tarangire census zone, but declined in Burigi Biharamulo, Greater Ruaha, Katavi and Moyowosi-Kigosi census zones during the time period considered (Stoner *et al.* 2006).

In **Mozambique**, buffalo populations occurred throughout the country until the 1970s, but suffered greatly from the former civil conflict (1977–1992). In Niassa Reserve (42 300 km^2) buffalo were successively estimated at 2300 (2006), 6800 (2009) and 6200 (2011) individuals (Craig 2011a). In Quirimbas National Park and adjacent areas, an ASC undertaken in 2011 over 20 830 km^2 reported no buffalo observations (Craig 2011c). In Chipanje Chetu area, an ASC conducted in 2011 (area surveyed: 6640 km^2) reported the observation of a single herd (ten individuals; Craig 2011b). In southern Cabora Bassa region (northwest Mozambique), an ASC undertaken in 2010 estimated the buffalo population at 4600 individuals (area surveyed: 16 590 km^2) (Dunham 2010). In Gilé National Reserve, no buffalo observation was reported by the ASC conducted in 2007 (Mesochina *et al.* 2008), but a reintroduction programme is currently ongoing (20 buffalo reintroduced from Gorongosa National Park in 2012). Central Mozambique hosts residual populations of buffalo, except the open floodplains of the Marromeu Complex (11 270 km^2), with an estimated population >10 300 individuals (Beilfuss *et al.* 2010). Buffalo were extinct in Gorongosa National Park (4000 km^2) and the population (360 individuals; 2010 estimate) was restocked (2006–2011) with 186 buffalo from Kruger and Limpopo National Parks (Carlos L. Pereira, personal communication). In south Mozambique, 16 individuals were recorded by ASC in Banhine National Park (1190 km^2) in 2009 (Stalmans & Peel 2009). In the Limpopo National Park (11 230 km^2) an ASC conducted in 2007 estimated the buffalo population at around 200 (Swanepoel 2007).

In **Zambia**, aerial surveys were conducted in the Kafue ecosystem (Kafue National Park and surrounding game management areas combined). In 2008, 6300 buffalo were recorded in an area of 48 000 km^2, and in 2011, 4600 individuals were recorded in an area of 70 000 km^2. The 2011 survey used a combination of total and sample counts and was more accurate than the previous 2008 survey (Frederick 2009, 2011). In Luanga Valley ecosystem (49 700 km^2; eastern Zambia), an ASC conducted in 2008 reported an estimation of 17,200 buffalo (WCS Flight Programme 2009).

A transfrontier ASC conducted in 2003 in the Zambezi Heartland between Lakes Kariba and Cabora Bassa (Zimbabwe, Mozambique and Zambia; 31 660 km^2) estimated the buffalo population at 27 700 individuals (Dunham 2004). In Zimbabwe, the buffalo population estimate was 19 500 (13 373 km^2, including Mana Pools National Park, Hurungwe, Sapi, Charara, Dande, Chewore Safari Area and part of Guruve District). In Mozambique the estimate was 1 600 (6330 km^2; communal lands south and north of Lake Cabora Bassa and east of the Luangwa River). No buffalo were observed north of Lake Cabora Bassa. In Zambia the buffalo population estimate was 6700 (11 950 km^2, including Lower Zambezi National Park, Chiawa and Rufunsa Game Management Areas and Tonga-Sikongo communal land).

In **Zimbabwe**, an ASC conducted in 2006 in the Sebungwe region (15 620 km^2, including Chizarira and Matusadona National Parks, Chiziza and Chete Safari Areas, P&W Estate, Sajarira Forest Area and North Gokwe, Binga Kariba communal lands) reported an estimate of 10 400 buffalo (Dunham *et al.* 2006). A 2007 ASC conducted in Hwange-Matetsi complex and adjacent areas (24 570 km^2, including Zakuma National Park, Deka Safari Area, Ngamo and Sikumi Forest Areas and Tsholotsho and Maitengwe communal lands) estimated 24 500 buffalo (Dunham *et al.* 2007). In Gonarezhou complex (7110 km^2, including Gonarhezou National Park, Malipati Safari Area and surrounding communal lands), an ASC conducted in 2010 reported 2700 buffalo (Dunham *et al.* 2010). Other substantial buffalo populations also occur in several conservancies in southeast Lowveld, among which are Save Valley (3340 km^2, 1900 individuals in 2002), Bubye Valley (3430 km^2, about 3000 individuals) and Nuanetsi Conservancies (1490 km^2, about 1000 individuals) (Chris Foggin, personal communication).

In **Angola**, buffalo have been wiped out by uncontrolled hunting and poaching throughout the country, including Kissama National Park, where East (1998) reported a declining population of the transitional 'red' buffalo. Residual populations may still range in the areas north of Luanda, in Cabinda and Cangandala National Park. In the southeast, typical Cape buffalo may still occur in Quando Cabando (Wouter Van Hoven, personal communication), but the numbers are unknown.

In **Botswana**, buffalo are found only north of 20° S in the Okavango-Chobe region. In 2010, an ASC covering northern Botswana, (73 480 km^2, including Moremi Game Reserve, Chobe National Park, Makgadikgadi Nxai Pan National Park and surrounding wildlife management areas) estimated the population at 39 600 individuals (Chase 2011).

In **Namibia**, buffalo were largely eradicated as part of a veterinary campaign. The construction of the veterinary cordon fences along the international border with Botswana is thought to be a factor responsible for buffalo decline (Martin 2002). An ATC conducted in 2007 on the Caprivi River systems (area covered: 1780 km^2) reported 5900 buffalo (Chase 2007). A small population (about 50 buffaloes) was introduced in Waterberg Plateau Park (400 km^2) between 1981 and 1991 and increased to 200 in 2000, and a relict population of 70 animals still occurs in Bushmanland (Tsumkwe area; 24 km^2; Martin 2002).

In **South Africa**, buffalo were extirpated from their former range except for a few areas such as Kruger National Park (20 000 km^2), where the country's major population now occurs (37 500 in 2009 and 40 900 in 2011; Ferreira *et al.* 2010, 2011, 2012). About 4000 buffalo persist in Hluhluwe-Imfolozi Park (900 km^2; Jolles 2007). Buffalo from Hluhluwe-Imfolozi and Kruger National Park are known carriers of diseases transmissible to cattle and thus cannot be translocated from one region of South Africa to another. The buffalo population in Mokala National Park (270 km^2) was estimated at 400 in 2012. In Addo Complex area (1640 km^2) 300 were counted in 2011. In Karoo National Park (770 km^2) only four individuals were observed from 2010 to 2012. In Mountain Zebra National Park (280 km^2) 77 individuals were counted in 2012. Mount Camdeboo Game Reserve (140 km^2) presented stable numbers at 30–36 in 2010–2011 surveys (Ferreira *et al.* 2012). There are approximately 32 000 disease-free buffalo in South Africa. Six thousand of these are in national or provincial parks, while the remaining 26 000 are privately owned and resident on approximately 2700 private game ranches or reserves (South African Private Buffalo Owners Association, personal communication).

West African savanna buffalo

In West Africa, buffalo numbers have declined alarmingly during the last decade. Since East's (1998) assessment, populations larger than 1000 individuals persist in only two of five sites. In West Africa, more than anywhere else in Africa, savanna buffalo populations are restricted within the borders of a limited number of protected areas.

In **Senegal**, the last buffalo population is located in Niokolo Koba National Park (9130 km^2). This population was estimated by ASC at 1000 and at 500 in 2003 and 2006, respectively (Mauvais & Ndiaye 2004; Renaud *et al.* 2006). An estimation dating from the early 1990s by pedestrian sampling count (PSC) was of 11 000 individuals (Galat *et al.* 1992).

This subspecies is now extinct in the **Gambia** (Jallow *et al.* 2004).

In **Mali**, the buffalo is extinct in the Boucle de la Baoulé National Park, but a relict population may still occur along the border with Senegal in the riverine forest of the Faleme Hunting Zone, and maybe in Bafing National Park (Bourama Niagate, personal communication).

In **Guinea Conakry**, Sudanian savannas still locally host small populations in protected areas (Kankan Forest reserve, Bakoun classified forest or Bakoy sector), but no population estimate is available (Brugière 2012).

In **Ivory Coast**, savanna buffalo still range in Comoé National Park (11 950 km^2), where the population was estimated (ASC) at 900 individuals in 2010, representing an 80% decline since the 1970s (N'Goran *et al.* 2010).

The largest buffalo population of West Africa is located in WAP Regional Park (28 350 km^2)in **Benin, Burkina Faso** and **Niger**. This complex comprises three national parks ('W', Arly and Pendjari) and several hunting zones, containing a buffalo population estimated by ATC in 2003 at 10 600 individuals (Bouché *et al.* 2003). Of these, about one-third (3400) were located in Pendjari National Park (2 830 km^2). Counts have not been repeated for the entire area since 2003. However, recent counts conducted at Pendjari National Park suggest a decreasing trend in population (Sinsin *et al.* 2008), while the 'W' National

Park population has remained fairly stable or increased (Bouché 2012).

In **Burkina Faso**, ATCs conducted on Ponasi (3100 km^2) and Comoé-Leraba (1850 km^2) protected area complexes estimated the buffalo population at 145 and 90 individuals, respectively (Bouché *et al.* 2004; Bouché 2005).

In Mole National Park (4500 km^2, **Ghana**), 500 and 700 individuals were counted in 2004 (ASC) and 2006 (ATC), respectively (Mackie 2004; Bouché 2006), whereas the population was estimated (ASC) at 1700 individuals in 1993 (Wilson 1993), suggesting an overall decline in numbers. The presence of buffalo has been reported in 2013 in Kyabobo National Park along the Togo border (Fazao Malfakassa National Park; D. Brugière, personal communication).

In **Togo** no buffalo were observed during the ATC carried out in Keran National Park (1400 km^2) and in Oti-Mandori Faunal Reserve (1800 km^2; Bouché *et al.* 2003).

In **Nigeria**, the West African savanna buffalo formerly occurred widely in the northern savannas, but has been extirpated from most of its former range. To our knowledge, a relict population remains in the National Parks of Yankari (2240 km^2; 170 individuals) (Bergl *et al.* 2011) and Kainji Lake (not estimated; Aremu *et al.* 2007).

In Northern **Cameroon**, an ATC conducted in 2008 in the complex of protected areas of Faro, Benoue and Bouba Ndjidda National Parks and their adjacent hunting blocks (22 140 km^2 in total) found 600 individuals (Omondi *et al.* 2008). It is likely that this figure is a gross underestimate due to inappropriate transect inter-distance (3–5 km). On the basis of empirical correction factors (effective bandwidth of 200–300 m), the size of the population can be estimated at around 4000 individuals.

Central African savanna buffalo

Most Central African savanna buffalo populations declined alarmingly during the last decades, and few protected areas still encompass significant (>1000) numbers. However, in contrast to West Africa, the savannas of Central Africa are less degraded overall, and still host diffuse (low density, small size) populations of buffalo in many areas.

One of the largest populations is located in **Chad** (Zakouma National Park; 3030 km^2), where the ATC carried out in 2012 reported an estimate of 8100 individuals (Potgieter *et al.* 2012). Interestingly, the Zakouma population is still increasing (due to intense conservation efforts), with a mean annual growth rate of 8% during the four last years.

A second important population is located in the northeast of the **Central African Republic**, in a complex of protected areas (94 960 km^2) comprising Bamingui-Bangoran National Park, Manovo-Gounda-St. Floris National Park and protected areas of diverse statuses. This complex includes a population estimate in 2010 (ASC) of 4000 individuals (Bouche 2010). The comparison with prior ASCs emphasized a population decline of 31% between 1985 and 2005 (Douglas Hamilton *et al.* 1985) and a drastic decline (76%) between 2005 and 2010 (Renaud 2005). In 2010 most herds were found in hunting zones, while the national parks and reserves were occupied by transhumant cattle. No information currently exists from Bangassou Forest Reserve (12 080 km^2) where East (1998) reported a population greater than 1000 individuals.

In **South Sudan**, ASCs carried out in 2007 (Fay *et al.* 2007) in an area larger than 150 000 km^2 revealed sizeable buffalo population estimates (8900), mainly located in Boma National Park (22 800 km^2), but these figures may be inferred from very few sightings and thus strongly overestimated. However, comparison with an ATC conducted over the same area (using the same methods) in 1981 indicates that buffalo populations declined by at least 80% in 25 years. No buffalo were observed in 2007 in Southern National Park (23 000 km^2) although 60 000 were counted in 1981. In northern **Democratic Republic of Congo**, ATCs carried out in Garamba National Park (4 920 km^2) in 2007 and 2012 reported estimates of 5200 and 6000 buffalo, respectively (Bolanos 2012). In **Ethiopia**, an ASC conducted in 2010 in Gambella National Park (20 000 km^2) and adjacent areas reported 1400 individuals (EWCA 2010), but this inference was based on very few observations and is to be taken with precaution. Roughly 3000 buffalo are estimated to range in Dati area, north of Gambella (EWCA, unpublished).

Forest buffalo

Forest buffalo population estimates are available for only a few sites. At Lopé National Park (**Gabon**), the population is using a mosaic landscape of forest and equatorial savanna (70 km^2) in the northeast corner of the park, and is estimated to be 300 individuals, organized into 18 herds (Korte 2008b). In Odzala National Park, **Republic of Congo**, the population is estimated to be 500 buffalo subdivided mainly into three large herds (Chamberlan *et al.*, 1995).

Bekhuis *et al.* (2008) estimated a population of 20 buffalo in an area of 650 km^2 in **Cameroon** at Campo-Ma'an National Park. At the Dzanga sector of Dzanga-Ndoki National Park, **Central African Republic**, the population is estimated between 32 and 40 individuals in two herds in 500 km^2 of rainforest (Melletti 2005; Melletti *et al.* 2007a).

Density

Buffalo density is dependent on many factors, including the spatiotemporal distribution of resources (water, forage quality and quantity), and also in conjunction with epidemiological, conservation and management contexts.

Cape buffalo may locally reach very high densities (20 individuals/km^2) in optimal habitats such as Lake Manyara National Park (Tanzania), where both high-quality forage and water are available year-round (Prins 1996). In contrast, in the woodlands of the nearby Serengeti National Park, lower densities were reported (8 individuals/km^2; Sinclair 1977). In the northern section of Akagera National Park (Rwanda), high

Table 20.3 Main differences between the four African buffalo subspecies

African buffalo subspecies	Syncerus caffer caffer	Syncerus caffer brachyceros	Syncerus caffer nanus	Syncerus caffer aequinoctialis
Main habitat	Different types of woodland and open arid savannas, montane grassland	Different types of woodland and open savannas	Rainforest, natural forest clearings, equatorial savanna, forest logging roads, coastal forest	Different types of woodland and open savannas
Home range size (km^2)	Range 0.5–1537 [1]	Range 172–620	Range 2.3–8.0	
Herd size (individuals)	Range 10–1654 2000–4000 [2]	Range 19–150	Range 3–24	Range 2–800 Mean 24.8 [5]; 7.7 [4]
Density (individuals/km^2)	Range 0.7–20	Range 0.8 [4]–1.19 [5]	Range 0.01–7.4	Range 2.55 [5]; 1.75 [4] Mean 2.15
Daily movement (km)	Range 1–15 [1]	Range 5–15 [3]	Range 1–4 [3]	
Calving interval (months)	Range 12–36	~24	~24	~24
Herd structure	Many males in mixed herds; bachelor groups; lone males; fission–fusion patterns	Many males in mixed herds; bachelor groups; lone males; fission–fusion patterns	Only 1–2 males in mixed herds; no bachelor groups; few lone males; fission–fusion patterns	Many males in mixed herds; bachelor groups; lone males; fission–fusion patterns
References	Prins & Sinclair 2013; Bennitt 2012; Naidoo *et al.* 2012; Chase 2011; Ryan *et al.* 2007; Hunter 1996; Prins 1996, 1989b; Funston *et al.* 1994; Prins & Douglas-Hamilton 1990; Conybeare 1980; Sinclair 1977; Cobb 1976	Cornelis *et al.* 2011; Sinsin *et al.* 2008; Stark 1986	Korte 2008a, 2008b; Melletti 2008; Bekhuis *et al.* 2008; Melletti *et al.* 2007a; Morgan 2007; Van Der Hoeven *et al.* 2004; Tutin *et al.* 1997	

[1] Migratory subpopulation combined for different areas
[2] Possible temporary associations of two or more herds
[3] Mixed herds
[4] Dry season (Stark 1986)
[5] Wet season (Stark 1986)

densities were also reported (13 individuals/km^2) until the 1990s. To date, in well-managed protected areas such as Selous reserve (South Tanzania), Kruger National Park (South Africa), or Hwange National Park (Zimbabwe), the average density of Cape buffalo ranges between 1 and 3 individuals/km^2.

West and Central African savanna buffalo range on soils poorer in nutrients than those of Cape buffalo habitats and display lower densities (0.5–1.5 individuals/km^2) in the well-managed protected areas (e.g. within WAPOK complex). Historically, higher densities were observed in several large floodplains (e.g. South Sudan, north Central African Republic), but these areas are now dominated by cattle and crop production, or were heavily poached. At present, high buffalo densities are still observed in the floodplains of Zakouma National Park (Chad; 2.4 individuals/km^2).

In Campo Ma'an National Park (Cameroon) the density of forest buffalo is 0.01–0.04 individuals/km^2 in 650 km^2. Most of this forest has a closed canopy without grass in the understory (Bekhuis *et al.* 2008), leading to low buffalo densities. The same holds for Nouabalé-Ndoki National Park in adjacent Republic of Congo (Blake 2002) and the coastal rainforests of Gabon (Prins & Reitsma 1989). In Dzanga-Ndoki National Park (Dzanga sector, 490 km^2) in Central African Republic the density is between 0.06–0.08 individuals/km^2 (Melletti 2005). In Lopé National Park (Gabon), buffalo density is locally high (5 individuals/km^2; 70 km^2 of the northeast corner of the park) in the forest–savanna mosaic, but lower (0.4 individuals/km^2) in the surrounding forest (White 1992; Korte 2008b). In the Réserve de Faune de Petit Loango (Gabon) the density was locally estimated at 1.7 individuals/km^2 in a coastal habitat (Morgan 2007).

More details on density in different areas are summarized in Tables 20.1 and 20.3.

Descriptive notes

Among African mammals, the buffalo has (with the African elephant) a very marked polymorphism across its range. The size, pelage colour and shape of horns differ greatly by region.

The African buffalo is the largest and most massive bovid of the African continent, with a broad head and short, thick legs and neck. Horns are present in both sexes. The front hooves are larger than the hind to support the heavy weight of the trunk and the head. The muzzle is wide, with a wet nose, large eyes, and drooping fringed ears. Females have a small, rounded udder with two pairs of inguinal mammae. Adult males have a tufted penis. Buffalo have sweat glands but no specialized olfactory skin glands. The tail is tufted with dark hair, and adult males sometimes appear to have a beard. The sense of smell in the buffalo is highly developed. Buffalo also seem to have good eyesight, and can spot predators from a great distance (Prins & Sinclair 2013).

The deciduous dentition is: I 0/3, C 0/1, P 3/3 = 20; by the ninth month of age the temporary teeth have erupted. Crown height and enamel ridge pattern, together with stages of tooth eruption, can be used for ageing buffalo and results are identical for populations of Western, Eastern and Southern Africa (Grimsdell 1969, 1973a; Spinage & Brown 1988; Taylor 1988). The permanent molars erupt at about 3.5 years of age. The lower canines erupt between 4.5 and 5.5 years old. Each year, one cementum line develops, making dental analysis a useful tool for ageing buffalo. Dental formula is: I 0/3, C 0/1, P 3/3, M 3/3 (× 2) = 32. The structure and adaptation of the stomach is described in Chapter 6.

Cape buffalo

Body measurements: **Shoulder height**: 140–160 cm (Ansell 1972); **Head and body length**: 210–300 cm; **Tail length**: 75–120 cm; Body mass: females up to 500 kg, males from 650 kg (Ansell 1972) to 900 kg. Their skin can be thicker than 2–3 cm. The main morphological differences between the four buffalo subspecies are shown in Figure 20.2.

The world trophy record (horn span) comes from Lake Manyara (Tanzania): 165 cm (bull) and 150 cm (cow). The mean live mass of Southern African adult bulls is 753 kg. Muscular tissues represent 40.6% of the live mass and fat 5.6% of carcass mass (Du Toit 2005).

The Cape buffalo is the largest subspecies, heavily built, with the biggest horns, including strong development of bosses in the horns of the males (Prins & Sinclair 2013). Bulls have a hump above the shoulders that supports the heavy head and horns. The pelage colour varies from black to reddish-brown. Horns are large in both sexes. The body is covered by short hair, with some areas where it is absent. Juveniles have a denser coat than adults and may go through colour changes from yellowish-brown to dark brown before they attain the typical adult black. Red calves are also present up to 30% in some Cape buffalo populations (Prins 1996). Colour mutations have been described in two individuals from Luangwa valley (Zambia), with a white band around the body (Skinner & Chimimba 2005), considered partial albinism.

Cape buffalo males have laterally extending curved horns, large bosses, and when adult, little or no hair between the horns. Females have more slender horns, also laterally extending, with hair on the skin between the horns. The tips of the horns often show signs of breakage. Bosses are very variable in size and do not develop until males are 3–5 years old. There is no relation between the age of an adult male (older than five years) and the size of the boss (Prins & Sinclair 2013). Male's horns become conspicuously roughened after some 4–5 years. In particular in the southern savanna males, the basal horn sheaths expand medially and anteposteriorly across heavily roughened frontals that are up-arched to form prominent bosses (Klein 1994). These bosses are not formed in females. When bulls are sparring, they place their bosses against each other and push. Extensive bossing is a recently derived feature that developed only in the Late Quaternary (Gentry 1978, 1990). It does not occur in the *nanus* subspecies, where fights are infrequent (Kingdon 1997; Van Hooft *et al.* 2002). Since extensive bossing only occurs in adult males, it probably evolved through sexual selection; it occurs particularly in Cape buffalo in east and southern Africa, which are characterized by large group sizes. This suggests that bossing has evolved in the context of a complicated male hierarchy (for details see the section on behaviour; Prins & Sinclair 2013). The frontal bones under the bosses have a honeycomb structure. The height of the bony walls of the cells is about 15–20 mm, while cell diameter is about 7 mm. It is thought to be an adaptation to absorb strong hits and for moving the head in any direction to face attack. While the average weight of a dry skull (including horns) is 8 kg in Lake Manyara National Park, the heaviest skulls were 16 kg (Prins & Sinclair 2013). Horn size was studied as an indicator of health status in male and female Cape buffalo (Ezenwa & Jolles 2008). In both sexes, horn size was significantly negatively correlated with the number of different parasites infecting an individual. These findings support the idea that horn condition plays a role as an indicator of health in both sexes. In particular, Ezenwa & Jolles (2008) suggest that this is a possible role for sexual selection in the maintenance of horns in female ungulates; however, this hypothesis needs further research.

West and Central African buffalo

Body measurements: **Shoulder height**: 120–145 cm; **Head and body length**: 200–245 cm; **Tail length**: 55–70 cm; **Body mass**: 300–600 kg. Smaller than Cape buffalo and bigger than forest buffalo, these intermediate subspecies also have widely spread horns, but these are not curved down as are those of Cape buffalo. Horn span measures 56–103 cm and horn length along the curve is 53–94 cm. The skull measures 43–60 cm.

There are also populations that appear to be intermediate between *brachyceros* and *nanus* subspecies in South Nigeria, Central African Republic and North Cameroon (Wilson & Mittermeier 2011). The red forms of savanna buffalo may be found from Guinea to Chad until North Uganda in the Sahel-Soudan savanna (Grubb 1972; Mloszewski 1983; R. Kock, personal communication).

Forest buffalo

Body measurements: **Shoulder height**: 100–130 cm; **Head and body length**: 170–220 cm; **Tail length**: 60–90 cm; **Body mass**: 265–320 kg. Among the four *Syncerus* subspecies, forest buffalo is the smallest, at around half the size of Cape buffalo

(a)

(b)

(c)

(d)

Figure 20.2 Differences in body size, horns shape and size of the four African buffalo subspecies. (a) Cape buffalo male; (b) forest buffalo male; (c) West African savanna buffalo male; (d) Central African savanna buffalo male; (e) forest buffalo female; (f) male horns of transitional 'buffalo' from North Angola compared with (g) horns of West African savanna and Cape buffalo males (above and below, respectively). (*a* and *g*: photo by M. Melletti; *b* and *e*: photo by A. Turkalo; *c* and *d*: photo by D. Cornèlis; *f*: photo by C. Chiarelli). A black and white version of this figure will appear in some formats. For the colour version, please refer to the plate section.

(e)

(f)

(g)

Figure 20.2 *(cont.)*

(Sinclair 1977; Mloszewski 1983; Prins 1996; Wilson & Reeder 2005; Figure 20.2).

Horns are oriented towards the rear in the same plane as the forehead. Horns span ranges of 34–65 cm and horn length along the curve is 41–69 cm and 35–63 cm, in Central and West African populations, respectively (Wilson & Mittermeier 2011). The skull is smaller than that of a savanna buffalo, but still short and wide. The skull length ranges between 39 cm and 49 cm (Wilson & Mittermeier 2011).

Morphological differences within forest buffalo populations are less pronounced than within other subspecies. In addition, the sexual dimorphism (e.g. horn shape and body size) is much less pronounced in the forest buffalo than in the other subspecies. In forest buffalo, the horns are much smaller, back swept and with no boss. These characteristics may reflect the infrequency of fighting, and instead, adaptation to dense rainforest (Kingdon 1997). It has been suggested that horn size tends to increase from West Africa to East and Southern Africa (Grubb 1972); this remains to be tested empirically.

The pelage colour ranges from shades of red and brown to black. Calves and juveniles can have a denser and redder coat than adults, although they may also display dark patterns. Adult bulls show variability in colour patterns even within the same herd. They may be almost all black or completely reddish-brown, with many intermediate shades possible. Females are generally red brown with different shades and show a less pronounced colour gradation than males, but can sometimes be black (L. Korte, personal observation). Another characteristic are the tufted hairs of the ears.

Habitat

African buffalo live in a wide range of habitats, from open grasslands to rainforests, including all intermediate vegetation types: scrublands, woodlands and deciduous forests. They appear not to be bounded by elevation constraints, and their altitudinal distribution ranges from coastal to the boundaries of forests on the highest mountains, except areas with annual rainfall less than 250 mm (e.g. Namib and the Sahara deserts). African buffalo persist in semi-arid environments, as long as surface water is available within 20–40 km, year-round (Naidoo *et al.* 2012; Prins & Sinclair 2013).

The carrying capacity of savanna ecosystems for large grazers such as buffalo is positively correlated with mean annual rainfall and the soil quality (Sinclair 1977; Grange & Duncan 2006; Winnie *et al.* 2008). In similar water regime conditions, the nutrient content of vegetation and primary production are much lower on poor than rich soils (Breman & De Wit 1983; Le Houérou 2008). The low carrying capacity of wild ungulates in savanna ecosystems of West Africa (compared to those of volcanic areas in East Africa) mainly results from poorer soil conditions (Bell 1982; East 1984; Fritz 1997; Hibert 2007).

Most of what we know today about the behavioural ecology of savanna buffalo comes from monographs focused on the Cape buffalo (e.g. Grimsdell 1969; Sinclair 1977; Mloszewski 1983; Prins 1996; Ryan 2006). To date, the ecology of West and Central African savanna buffalo remains poorly investigated and comes from sporadic publications (Boy 1958; Gillet 1969, Stark 1986; Cornelis *et al.* 2011). For this reason, and because several behavioural traits are similar within savanna buffalo subspecies, we present all savanna subspecies in one grouping.

Savanna (Cape, West and Central African) buffalo

Savanna buffalo are mainly found in habitats with a high herbaceous biomass. In the distribution range of Cape buffalo (Eastern and Southern Africa) suitable grasses are found in several types of woodland, such as mopane (*Colophospermum mopane*), miombo (*Brachystegia* spp.), Acacia (*Acacia* spp.) and *Baikiaea* spp. West and Central African savanna buffalo live in a variety of habitats ranging from typical Sahelian shrub savannas (*Combretum* spp., *Terminalia* spp., *Acacia* spp.) to Sudanian woodlands (e.g. *Isoberlinia doka*, *Daniellia oliveri*, *Burkea africana*).

The availability of surface water and cover are commonly cited as the main constraints on habitat use of savanna buffalo. Buffalo must drink at least every two days, taking in about 45 litres daily; they are unable to survive on the moisture content of their food alone (Prins 1996; Prins & Sinclair 2013). They have not evolved any water-saving mechanisms (buffalo sweat profusely, and their faeces comprise approximately 80% water (Taylor 1970; Prins 1996)).

This constraint of access to water is manifest in studies of buffalo habitat use. For example, a GPS-tracking protocol undertaken in West Africa showed that buffalo herds spend 95% of their time within 5.3 km of permanent water areas and around 50% of their time within 1.4 km (Cornelis *et al.* 2011). Many studies showed a preference for riverine habitats, at least during the dry season. These lowland areas (e.g. riparian galleries and surrounding grasslands) provide water, green grass and cover against climatic extremes and predators, simultaneously (Sinclair 1977; Redfern *et al.* 2003; Ryan *et al.* 2006; Cornelis *et al.* 2011). At first rains, the water constraint relaxes and buffalo generally (but not systematically) leave the depleted areas of the dry season and move to habitats located higher in the toposequence (e.g. plateaus), where both water and new green grass are temporary available (Sinclair 1977; Cornelis *et al.* 2011). In the Okavango Delta (Botswana) GPS data showed that buffalo used habitats for from permanent water during the wet season and flooded habitats close to permanent water during the flood season (Bennitt *et al.* 2014).

It is worth noting that buffalo are also good swimmers, and can forage underwater in floodplains (Chobe River, Okavango, Botswana; Prins & Sinclair 2013). Finally, savanna buffalo may also ascend and descend steep slopes, up to about 50° and may be found at high elevations (Prins & Sinclair 2013). They are known to regularly reach the alpine belt on Mount Kenya and are observed on Kilimanjaro, throughout the forest belt on the northern and western slopes, and tracks were observed up to 4200 m (Grimshaw *et al.* 1995).

(a)

(b)

Figure 20.3 Forest buffalo are forest dwellers, inhabiting different habitats with: (a) grassy glades and watercourse areas in rainforest (photo by M. Melletti); (b) forest muddy areas (photo by A. Turkalo); and (c) coastal and equatorial savannas (photo by Mathieu Bourgarel, Cirad). A black and white version of this figure will appear in some formats. For the colour version, please refer to the plate section.

Forest buffalo

Forest buffalo are forest dwellers, inhabiting rainforests with grassy glades, watercourse areas and mosaics of equatorial forests and savannas (Figure 20.3). This subspecies is absent (or present at very low densities) in continuous forests (Prins & Reitsma 1989; Blake 2002; Melletti *et al.* 2007a, 2009; Melletti 2011).

Suitable habitats for forest buffalo are mosaics of forest with equatorial savannas or clearings, which consist of grassy vegetation and shrubs such as *Palisota* spp. (Reitsma 1988; Blake 2002; Tchouto 2004). Blake (2002) at Noubale-Ndoki National Park found high buffalo abundance close to open grassy areas with low abundance in the closed canopy forest, suggesting it is unsuitable for forest buffalo. Melletti *et al.* (2007a, 2007b, 2008, 2009) also found a significant relationship between buffalo and natural forest clearings at Dzanga-Ndoki National Park, where clearings were the centre of buffalo home ranges and no signs of buffalo presence were recorded more than 500 m from clearings (Figure 20.4; Melletti *et al.* 2007a). Buffalo positively selected these clearings, although the surface area of clearings represented only 1% of the total area. In Cameroon, buffalo rarely penetrated into the forest more than

(c)

Figure 20.3 *(cont.)*

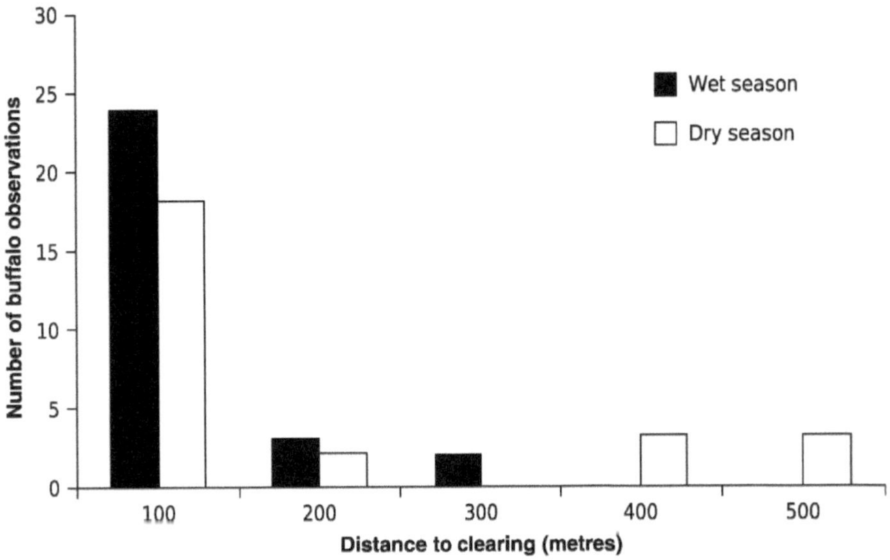

Figure 20.4 Forest buffalo: distribution frequency of the distances to clearings of buffalo observations as revealed by RECCE method in forest habitat during both the dry (from December to February) and wet (from March to November) seasons (Melletti *et al.* 2007a, with kind permission from John Wiley and Sons, *Journal of Zoology*, modified).

300 m from logging roads, which were the main feeding sites for buffalo (Bekhuis *et al.* 2008).

Sites with greater percentages of open areas report high densities of forest buffalo (Morgan 2007; Korte 2008b; Table 20.1). For example, forest buffalo density is 5 individuals/km^2 in the forest–savanna mosaic at Lopé National Park, but only 0.4 individuals/km^2 in the adjacent continuous forest (White 1992). In the coastal area of the Réserve de Faune de Petit Loango (Gabon), Morgan (2007) reports densities of 1.75 individuals per km^2, whereas sites with forest clearing have low densities (Van Der Hoeven *et al.* 2004; Melletti 2005; Bekhuis *et al.* 2008).

At Lopé National Park, Korte (2008a) used radio-tracking data of nine adult females from different herds to quantify habitat preference. Buffalo select proportionally more open habitat (marsh and savanna) and use forest less than would be expected based on available habitat at the landscape level. Habitat use within home ranges varied with season (Figure 20.5). Between March and August >60% of locations of individuals were in forest habitats that represented <50% of home range area. In addition, between the June and August dry season buffalo moved into burned savanna to feed on new grass. These results suggest that forest buffalo use forest fragments, galleries and corridors more often than continuous forest, and habitat preference changes with season based on food availability. Buffalo remained close to the forest clearing areas and did not penetrate deep into the forest, as confirmed

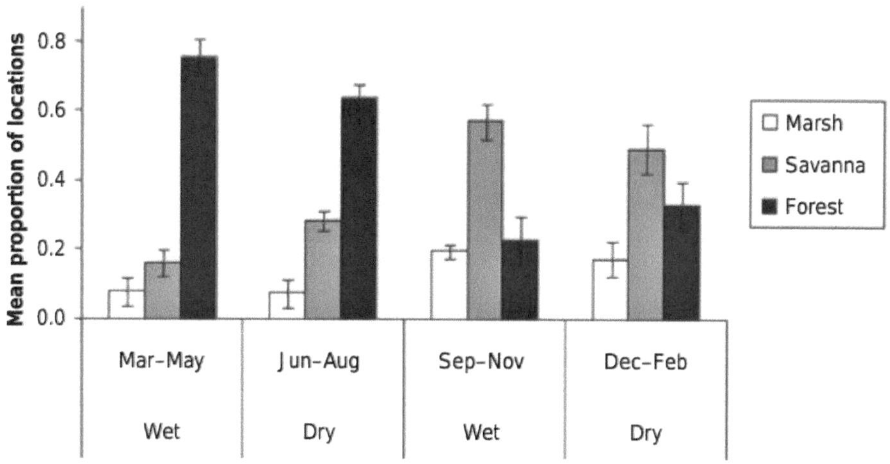

Figure 20.5 Mean proportion of locations (± SE) during daylight hours in each habitat by season for seven radio-collared adult female forest buffalo at Lopé National Park, Gabon, December 2002–December 2004 (Korte, 2008a; with kind permission of the American Society of Mammalogy, Allen Press, Inc).

by studies in other localities (Blake 2002; Melletti *et al.* 2007a; Bekhuis *et al.* 2008; see also Figure 20.4).

Very little information is available on foraging technique. At Dzanga-Ndoki National Park, buffalo use every place within a clearing for feeding, with intermediate areas between the forest edge and the centre of glades used more frequently (Melletti *et al.* 2007b; Melletti 2011). While feeding was mainly within the clearings in Dzanga-Ndoki National Park and in Noubale-Ndoki National Park (Maisels *et al.* 2002; Blake 2002; Melletti *et al.* 2007a; Melletti 2011), sites that lack of clearings in Cameroon and Ivory Coast report signs of feeding on the edges of roads and river banks (Hoppe-Dominik 1992; Bekhuis *et al.* 2008).

Differences in habitat use between age and sex classes are not pronounced. Males are generally permanent members of the herd and share the same habitat and home range with females. Finally, forest buffalo have been found making use of dry caves in the rainforest of Cameroon for resting (C. A. van der Hoeven *et al.*, personal observation).

Movements and home range

The African buffalo is generally considered a sedentary species. In buffalo, both home range (HR) size and movements (either seasonal or daily) are related to habitat and to the spatiotemporal distribution of key resources. As pointed out previously, buffalo habitat use tends to be largely constrained by access to water in most study locations. Small HRs are generally observed in forested or high-rainfall regions, and large ones in open, drier habitats.

Savanna (Cape, West and Central African) buffalo

Large seasonal movements approaching migrations were reported by several authors. Large herds in the Serengeti National Park were shown to undertake seasonal movements (Sinclair 1977). Before the fencing in the 1960s, buffalo in Kruger National Park ranged westward in the dry season to take advantage of wetter areas with persistent forage (Witkowski 1983; Ryan 2006). Using GPS-tracking on Cape buffalo in the Caprivi Strip (northeastern Namibia), Naidoo *et al.* (2012) showed that part of the buffalo population undertook seasonal migrations ('partial migration'). The GPS-tracked animals moved up to 115 km from permanent rivers. This study also emphasized that some individuals behaved as dispersers. Variation in migratory behaviour was explained by a large set of factors: environmental conditions (rainfall, fires, woodland cover, vegetation, biomass), distance to the nearest barriers (rivers, fences, cultivated areas) and social factors (age, herd size). Buffalo in larger herds moved greater distances than those in small ones. A similar study undertaken in the south-eastern part of the Okavango Delta (Botswana) also emphasized contrasting movement strategies (Bennitt 2012). Buffalo formed two subpopulations with divergent migratory strategies (resident in the west of the study area and migrant in the east). According to the author, this contrast was probably driven by differing resource availabilities and levels of anthropogenically induced spatial restriction. Finally, in West Africa (WAP Regional Park), most GPS-tracked herds were shown to perform a large (35 ± 10 km) omnidirectional movement at the onset of the monsoon in response to a large-scale gradient of primary production (Cornelis *et al.* 2011).

At a daily scale, Cape buffalo may transit long distances between different grazing areas and often walk single-file. Buffalo herds can cover distances up to 8 km in a 24-hour period (Sinclair 1977; Stark 1986). Daily average distances of 3 km and 6 km have been reported in Zimbabwe (Conybeare 1980; Taylor 1985), and of 3.35 km in Kruger National Park (Ryan & Jordaan 2005). In contrast, bachelor herds were reported to move 1–2 km/day (Taylor 1985). While moving in the mixed herds, adult bulls are often located in the front and at the rear of the herd, but do not make decisions about where to move (Prins 1996).

Savanna ecosystems generally display strong spatial and temporal variations of resources (Estes 1991). As a result, savanna buffalo herds exhibit variable HR sizes across their continental distribution range. A compilation of the literature reveals that annual HR areas generally range between 50 km^2 and 350 km^2, but sometimes can exceed 1000 km^2 (Table 20.4). Buffalo living in wetter areas appear to have smaller HRs than those in drier habitats (Ryan *et al.* 2006). The larger ranges are generally observed in areas where resources are spatially segregated, which forces buffalo to undertake seasonal movements

(Cornelis *et al.* 2011; Naidoo *et al.* 2012). As in many ungulate species, sexual segregation occurs in herds, and some of the adult and subadult bulls temporarily leave the mixed herds to form bachelor groups (see the Behaviour section). Several studies suggest that bachelor groups form smaller (0.5 km^2 and 4 km^2) HRs (Grimsdell 1969; Sinclair 1977; Taylor 1985; Naidoo *et al.* 2012).

Seasonal differences in home range sizes also varied across studies. For example, in Serengeti National Park (Tanzania), the HR size of mixed herds was smaller in the dry season because space is constrained around water points at this point in the year (Sinclair 1977). In contrast, the opposite was observed in Benoue National Park (Cameroon), with an increase in HR size during the early dry season fires (61 km^2 versus 46 km^2) (Stark 1986).

Long-term longitudinal studies undertaken on buffalo mixed herds have suggested that HRs are stable over time (Sinclair 1977; Prins 1996). Interestingly, studies investigating space sharing between neighbouring mixed herds reported contrasting results. In the Cape buffalo, a tendency to use exclusive HRs was observed by Prins (1996), Halley *et al.* (2002) and Ryan *et al.* (2006), whereas Grimsdell (1969), Conybeare (1980) and Mloszewski (1983) reported overlaps between neighbouring HRs. Cross *et al.* (2005) found considerable exchange of individuals between apparent herds, and a flexible structure over time, suggesting that herd definition itself may vary. In West Africa, (WAP Regional Park), the GPS-tracking of adult females emphasized a marked spatial segregation between neighbouring herds (Cornelis *et al.* 2011). The quasi-absence of direct contacts in this study suggests that HRs were not actively defended. These herds found a way to exploit exclusive HRs without spending substantial energy in their defence. Several ongoing GPS-tracking studies implemented in different places (e.g. Zimbabwe, Mozambique, Botswana, South Africa, Namibia), should soon lead to a better understanding of the strategies of exploitation and sharing of space by this species.

Table 20.4 Home ranges size in different African buffalo populations

Study area	Home range size (km^2), mixed herds	Reference
Okavango Delta (Botswana)	359–1537 [1]; 159–818 [2]	Bennitt 2012
Caprivi strip (Namibia)	50–448 (0.6–125 male herds)	Naidoo *et al.* 2012
Klaserie Private Nature Reserve (South Africa)	170–327	Ryan *et al.* 2006
Botswana	920–1455	Hunter 1996
Lake Manyara National Park, Tanzania	50	Prins 1996
Sabi Sand Game Reserve (South Africa)	40–120	Funston *et al.* 1994
Matusadona National Park, Lake Kariba (Zimbabwe)	60–110 (0.5–3.0, male herds)	Taylor 1989, 1985
Kenya, Zambia (Busanga swamp), Zimbabwe	126–1075	Mloszewski 1983
Sengwa Wildlife Research Area (Zimbabwe)	207–286	Conybeare 1980
Hwange National Park (Zimbabwe)	286	Conybeare 1980
Akagera National Park (Rwanda)	35–60	Monfort 1980
Serengeti (Tanzania)	222	Sinclair 1977
Momella Lakes, Arusha National Park (Tanzania)	10.9	Vesey-Fitzgerald 1974
Rwenzori National Park (Uganda)	9.4–9.6	Eltringham & Woodford 1973
Tsavo National Park (Kenya)	85–350	Leuthold 1972
W Regional Park (Niger, Burkina Faso, Benin)	172–620	Cornélis *et al.* 2011
Benove National Park (Cameroon)	46–61	Stark 1986
Lopé National Park (Gabon)	2–8	Korte 2008a
Dzanga-Ndoki National Park (Central African Republic)	2–8	Melletti *et al.* 2007a

[1] Migratory population
[2] Resident population

Forest buffalo

Few studies investigated HRs and movements of forest buffalo. In Dzanga Ndoki National Park (Central African Republic), a study of a single buffalo herd over two years identified a stable HR size of 8 km^2 (Melletti *et al.* 2007a). In this herd, daily distances travelled by buffalo were generally very short (i.e. 500–1500 m) and were mainly restrained to clearings and surrounding areas (Table 20.3). The maximum distance travelled during a 24-hour tracking period was about 4000 m (Melletti 2008). Forest buffalo males are permanent members of the herd and their HRs thus overlap completely with those of females range: 2–8 km^2; Melletti *et al.* 2007a).

In the mosaics of savanna and forest of Lopé National Park (Gabon), the average HR size was estimated at 5 km^2 (range: 2–8 km^2; $n = 7$ radio-collared adult females) and herds were shown to maintain stable HRs from year to year (Korte 2008a). In this area, the HR overlap was small between radio-collared buffalo belonging to neighbouring herds, suggesting a strong spatial segregation (Figure 20.6).

Long-term observations in the Central African Republic and in the Republic of Congo support the assumption that the same herds occupy the same clearings for long periods (Turkalo & Fay 2001; Blake 2002; Breuer 2008; Geßner 2008).

(a)

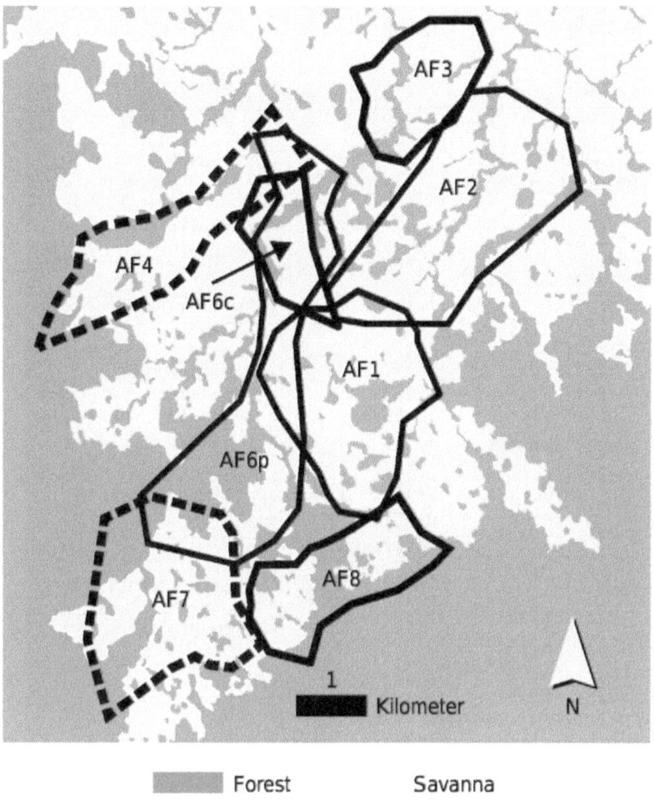

(b)

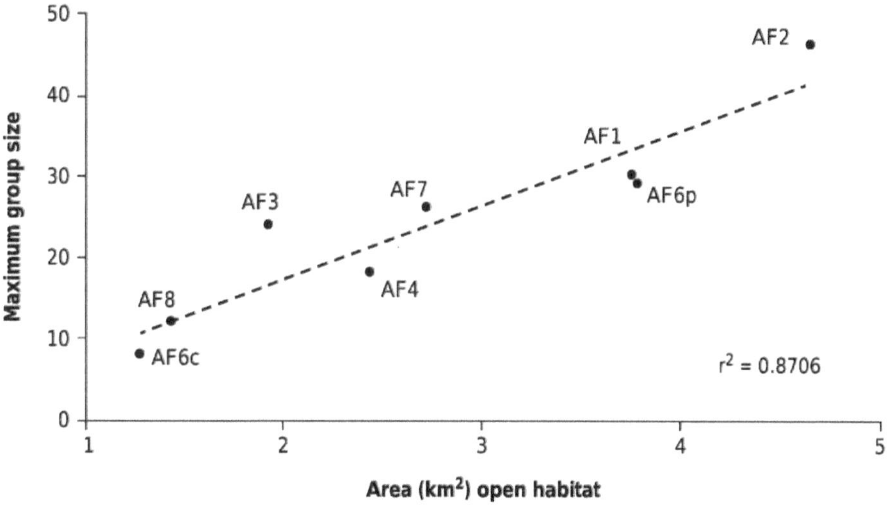

Figure 20.6 (a) Home range of each radio-collared buffalo in the study area at Lopé National Park, Gabon (December 2002 to December 2004); (b) maximum size of group based on observations that included a radio-collared adult female (AF) forest buffalo in relation to the area of open habitat in the home range of those females. Each data point represents maximum group size ($n = 8$) defined by the presence of radio-collared buffalo. Positive correlation between maximum group size and area of open habitat is significant (Korte, 2008b; with kind permission from John Wiley and Sons, *Journal of Zoology*).

Forest buffalo HRs are much smaller than those of the typical savanna subspecies. It is likely that this pattern, common to all studies, primarily results from the spatial arrangement of suitable resources in dense tropical forest, but also to less pronounced seasonality of the environment. Accessible and palatable food for buffalo is scarce in dense tropical forest, except in ecotone areas (e.g. forest clearings, glades, marshes, included savannas, logging roads, etc.: see Bekhuis *et al.* 2008), which often are small in size, patchily distributed and distant from each other (i.e. beyond the locomotion capacity of buffalo). The forest buffalo may thus have adapted several ecological traits to this environment, such as HRs small in size centred on 'islands' of profitable resources, small herds and a lower body mass.

Activity patterns

African buffalo display a large array of activity modes, including feeding, resting/ruminating, relocating between foraging areas (transit), checking for predators (vigilance), wallowing

Table 20.5 Percentages of daytime grazing, resting and ruminating activities for different populations of Cape buffalo

Study area	Grazing	Resting	Ruminating	Reference
Chebera Churchura National Park (Ethiopia)	50%	39%		Megaze *et al.* 2012
Okavango Delta (Botswana)	~40%	~30%		Bennitt 2012
Huhluwe-Umfolozi Game Reserve (South Africa)	21–28% adult females 14–28% adult males	66% adult females 70% adult males		Turner *et al.* 2005
Kruger National Park (South Africa)	37%	26%	29%	Ryan & Jordaan 2005
Willem Pretorius Game Reserve, Free State (South Africa)	40%	34%	34%	Winterbach & Bothma 1998
South Africa	48%			Whyte 1996
Lake Manyara National Park (Tanzania)	22–56%	25% adult females 29% adult males	10–50%	Prins 1996; Prins & Iason 1989
Matusadona National Park (Zimbabwe)	51% bachelor males[1] 47% herds[1]		25% bachelor males[1] 41% herds[2]	Taylor 1989
Western Zambia- Zimbabwe	44–52%[2]	53%[1]	36%	Mloszewski 1983
Serengeti (Tanzania)	38–44%			Sinclair 1977

[1] Mean for dry and wet seasons
[2] Dry season
Males, females and mixed herds are indicated. The range of percentage represents minimum and maximum values.

and drinking. Their respective proportion in both daily and seasonal activity budgets is mainly influenced by spatiotemporal changes in resources quality and availability, interspecific competition, weather conditions and predation pressure (Sinclair 1977; Prins 1996; Ryan & Jordaan 2005; Valeix *et al.* 2009a; Owen-Smith *et al.* 2010). Ruminant ungulates such as buffalo spend large proportions of their time feeding, and must additionally allocate time to ruminating, which results in an overall total time (feeding plus ruminating) of 70–80% (Beekman & Prins 1989; Prins 1996).

Savanna (Cape, West and Central African) buffalo

In Cape buffalo, most studies report feeding time accounting for 35–45% of the 24-hour activity budget (Table 20.5; Grimsdell & Field 1976; Sinclair 1977; Mloszewski 1983; Prins 1996; Winterbach & Bothma 1998; Ryan & Jordaan 2005; Bennitt 2012).

Grazing most often takes place in the early morning and late afternoon, and during the first half of the night, suggesting that buffalo cease feeding during the hottest part of the day and during the coolest part of the night for thermoregulation purposes. Note that an exception to this feeding pattern was observed at Lake Manyara National Park (Tanzania), with the main grazing bout occurring between 10.00 and 14.00 hours (Prins 1996). In most studies, buffalo herds appear to spend an equal or greater proportion of time feeding at night than during the day (Sinclair 1977; Taylor 1985; Prins & Iason 1989; Ryan & Jordaan 2005). At a seasonal scale, most authors reported a trend toward more time spent feeding in the dry season than the wet in response to lower vegetation quality and quantity and decreased intake rates. Other modes of activity, such as resting, rumination and vigilance, are not mutually exclusive, thus making it hard to individualize ratios and to compare across studies. Most studies reported average rumination times of around 30–35% of the 24-hour activity budget (Sinclair 1977; Prins 1996; Winterbach & Bothma 1998; Ryan & Jordaan 2005). Seasonal trends in rumination time are the subject of debate; Sinclair (1977) reported that buffalo spend more time ruminating during the dry season, whereas Beekman and Prins (1989) found the opposite. Buffalo must drink at least every two days, but when they drink appears to vary by study site. Ryan and Jordaan (2005) found two main periods of diurnal drinking (the early morning and mid-morning) and their observations suggest more time devoted to drinking during the day than at night. In contrast, Winterbach and Bothma (1998) found that buffalo drank in the early afternoon, and Grimsdell and Field (1976) found that they drank in the mid-morning. In one study, buffalo were shown to modulate hourly drinking patterns according to the risk of predation by preferentially avoiding waterholes during high-risk hours of the day (dawn and dusk), when lions were likely to be in their vicinity (Valeix *et al.* 2009b). To our knowledge, no seasonal trend in drinking frequency was reported in literature. Mud wallowing is frequent in buffalo males, contrary to females and young. Males generally wallow during the hottest

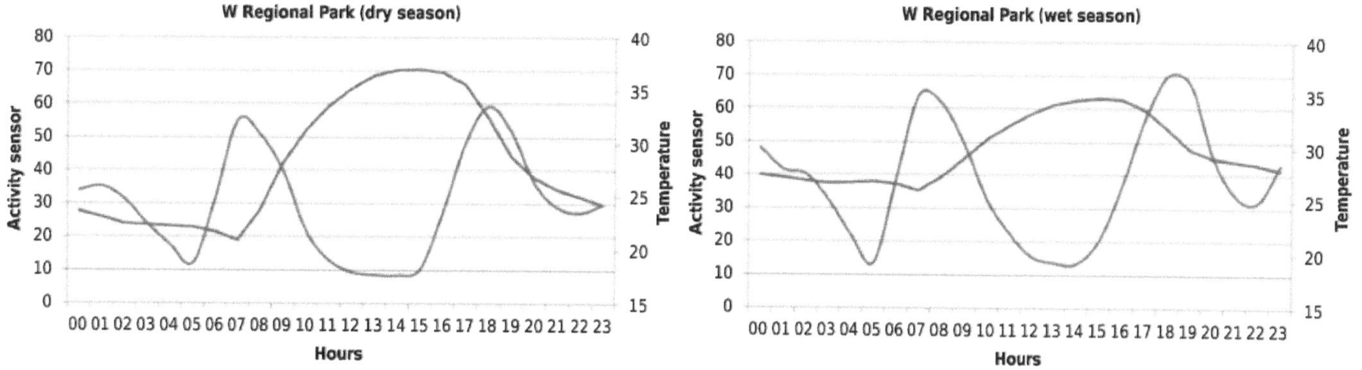

Figure 20.7 Daily bigeminus activity pattern of a buffalo cow tracked in WAP transfrontier conservation area (Burkina Faso, Benin, Niger) from March 2007 to February 2008 (using temperature and activity sensors). The first peak of activity started at dawn and the second finished after dusk, both lasting on average about 3–4 hours. This daily bigeminus pattern was particularly marked during the dry season, with the occurrence of a major resting period between 09:00 and 15:00, corresponding to the hottest hours of the day (Cornelis, 2011).

period of the day for periods lasting up to three hours. This practice is less effective in temperature regulation than the use of shade, which mixed herds generally seek at this time of the day (Sinclair 1977).

Fewer studies investigated the activity patterns of West and Central African buffalo. In Central Africa (Benoue National Park, Cameroon), observations over a 24-hour period conducted during the dry season suggested that buffalo graze approximately 33% of the time, with temporal patterns similar to those of Cape buffalo. In West Africa (WAP Regional Park), the seasonal and daily activity patterns of buffalo herds were explored using biorhythm indices derived from GPS location data and activity sensors (Cornelis 2011). Results suggest that the herds moved faster and were more active when food conditions were most favourable (beginning of the wet season). At a daily scale, herds were equally active during night and day, and were mostly crepuscular, with two main active periods per day (Figure 20.7). However, one herd locally facing a risk factor (daytime human disturbance) clearly switched to a nocturnal activity mode.

Forest buffalo

Observations of forest buffalo activity show similar patterns to those recorded for savanna buffalo: both usually alternate periods of grazing with resting/ruminating throughout the day and the night. At Dzanga-Ndoki, Melletti *et al.* (2007b) recorded higher figures of resting and ruminating time in the dry season than in the wet season and, for this activity, the differences between seasons were statistically significant. At Dzanga-Ndoki, around 50% of the time spent in the clearings by buffalo was dedicated to resting and ruminating activities (both sexes). Observations performed in clearings also suggest than females spent more time feeding (50%) than males (20%). The proportion of grazing individuals increased with increasing group size. Forest buffalo do not show an increase in grazing during the dry season, suggesting that season does not influence grazing activity. A decrease in grazing rate followed by resting in the middle of the day, when temperatures were highest, was observed both in CAR (Melletti *et al.* 2007b) and in Gabon (Molloy 1997).

Korte (2008a) observed similar patterns of feeding activity at Lopé National Park, where buffalo spent >30% of the time feeding. Feeding varied with the habitat and time of the day. Buffalo were most active in savanna habitat during the late morning (09:30–12:30). Periods of inactivity also varied with habitat and period of the day (Korte 2008a). Inactive behaviours were most often observed in marshes in the late afternoon. During daylight hours, the herds spent >38% of their time inactive. Buffalo were relatively inactive during the day between March and August, when >60% of their locations were recorded in forest habitat. Korte (2008a) found that the proportions of time spent in different behaviours did not differ between years or among seasons. However, activity varied significantly among individuals. Forest buffalo wallow in mud and water during the hottest daylight hours mainly in the dry season (December–March; Melletti *et al.* 2007a). Korte (2008a) recorded similar mud and water wallowing within marsh habitat, especially between September and February. Buffalo feed in the savanna areas in the early morning, retire to a wallow and rest in marsh areas until sunset.

Feeding ecology

Buffalo are ruminants, essentially feeding on grass and roughage. This species is capable of subsisting on pastures too coarse and too tall for most other herbivores (Bothma 2002), and as a 'bulk grazer' they ingest around 2.2% of their body mass daily. This represents, on average, 6.5 kg for a forest buffalo of 300 kg, 11 kg for a West or Central African buffalo of 500 kg, and 15 kg for a Cape buffalo of 700 kg.

Buffalo are very efficient grazers (adapted dentition and mobile tongue allowing the ingestion of high quantities of grass in a short time). Optimal feeding conditions for buffalo prevail when the grass forms swards and leaf heights reach and exceed 10 cm, as in flood plains or in forest glades (Prins

1996). However, buffalo cannot cut pastures as short as other species. The African buffalo thus occupies an important niche, opening up habitats that are preferred by short-grass grazers (Estes 1991; Prins 1996). Their primary competitors are cattle, *Bos taurus*, African elephant, *Loxodonta africana*, plains zebra, *Equus quagga*, and wildebeest, *Connochaetes taurinus* (Sinclair 1977; De Boer & Prins 1990; Plumptre 1995; Prins 1996).

Savanna (Cape, West and Central African) buffalo

In savanna buffalo, optimal dietary conditions occur during the rainy season, while the end of the long dry season is a period of food scarcity (both in quantity and quality) (Prins & Sinclair 2013). Detailed diet studies conducted in Eastern and Southern Africa indicate that buffalo are resource-limited and constrained by a minimum dietary protein concentration of 7–8% to make fermentation in the rumen possible (Sinclair 1977; Prins 1996). In such a situation, buffalo thus face a tradeoff of quality and quantity in obtaining an appropriate protein-to-fibre ratio in their diet (Redfern *et al.* 2006; Ryan *et al* 2012). During the dry season, savanna buffalo are generally constrained to become more selective, and to partially switch their diet to browse (see below). Despite this, buffalo were reported to live at or below nitrogen requirements for prolonged periods, accounting for visibly losing body condition (Prins 1989b; Ryan 2006; Ryan *et al.* 2012). In natural ungulate communities, the regulation of populations is driven by two processes: a control by available food resources on the one hand ('bottom-up regulation') and a control by predation ('top-down' regulation) on the other (Hunter & Price 1992). In the case of savanna buffalo, most authors agree that resources play a dominant role in the regulation of populations (Sinclair *et al.* 2003; Hopcraft *et al.* 2010).

There is moderate consensus in feeding studies on Cape buffalo about the preferred grass species (Sinclair 1977). Usually species that are avoided contain low nutritious quality or presence of aromatic oils (Prins 1996). During the dry season, usual food resources are of poor quality, due to lignifications and high standing biomass. During this period, floodplain species (e.g. *Leersia hexandra*) and riverine forest species (e.g. *Setaria sphacelata*) become important. In areas with upwelling groundwater, species such as *Cyperus laevigatus* may also form a mainstay in the dry season (Prins & Sinclair 2013). Buffalo prefer grasses such as *Cynodon dactylon*, but may also eat broadleaved grasses such as *Panicum maximum*. Sinclair (1977), in the Serengeti, reported that buffalo preferred soft, nutritious grass such as *Digitaria macroblephora*. Although there is little information on seasonal changes of diet, Sinclair (1977) analysed stomach contents and showed that Cape buffalo select more grass leaf at the end of the rainy season. Finally, Taylor (1985) found that grasslands dominated by *Panicum repens* were the best habitat on the shores of Lake Kariba (Matusadona National Park, Zimbabwe). In Cape buffalo, the diet of the different age and sex classes does not appear to differ, and dry matter food intake ranges between 1.2% and 3.5% of body mass, similar to other ruminants, but subject to food quality (Sinclair 1977; Prins & Beekman 1989).

Few studies have investigated the feeding ecology of West and Central African buffalo. In West Africa (WAP Regional Park), habitat selection analysis emphasized the importance of perennial grasses (e.g. *Andropogon gayanus*; Cornelis *et al.* 2011). In this study, a gradient in primary production appeared to determine large-scale movements of herds at the onset of the wet season, but its action clearly was modulated by the proportion of perennials available. Buffalo herds were shown to establish early wet season HRs at locations where this proportion reached a 9% threshold. At Benoue National Park (Cameroon), Stark (1986) similarly reported a very high proportion of grasses in the diet, particularly *Andropogon gayanus*, which represented 51% of the volume ingested in the dry season, versus 40% in the wet season. At Zakouma National Park (Chad), Gillet (1969) noted a preference for *Setaria anceps* (particularly new shoots after fire) and *Andropogon gayanus*, but also *Vetiveria nigritana*. In the dry season, the unburned straws of *Echinochloa obtusiflora* appeared to be preferred.

For more details on grass species eaten by savanna buffalo in different locations, see Table 20.6.

The savanna buffalo are primarily grazers, but partially switch to browse when grasses become tall and lignified (Field 1972; Leuthold 1972; Sinclair 1977; Mloszewski 1983; Hashim 1987; Prins 1996). Shrub and tree leaves have been shown to contain higher protein (nitrogen) content than dry grasses during the dry season (Kone *et al.* 1990; Prins 1996). As mentioned above, the contribution of nitrogen from browse facilitates the fermentation of fibrous grass in the rumen, which they eat in high quantities during the dry season. According to Estes (1991), the browse can represent up to 5% of the total diet, but higher figures were reported in several sites (e.g. 26% in the dry season in Cameroon; Stark 1986). A wide range of species of shrubs and trees are consumed throughout the African savanna, including *Grewia* spp., *Heeria* spp., *Combretum* spp., *Capparis* spp., *Piliostigma* spp. (Field 1968; Pienaar 1969; Jarman 1971; Graaff *et al.* 1973; Ryan 2006). In drier habitats in the Eastern Cape, buffalo appear to be adapted to eat woody species because grassy vegetation is scarce. In this particular habitat, during the dry season, up to 33% of their diet comprises species such as *Acacia* sp., *Plumbago* sp. and *Grewia* sp.

In some areas buffalo can maintain or create 'grazing lawns' if the feeding interval is short enough (Prins 1996). Evidence of this 'returning' behaviour has also been described for buffalo in Kruger National Park and Klaserie Private Nature Reserve in South Africa (Bar-David *et al.* 2009). They may also create favourable lawns in conjunction with other large herbivores such as elephant at Lake Rukwa, Tanzania (Vesey-FitzGerald 1960) or hippopotamus in Uganda (Eltringham 1999) and Benoué National Park in central Cameroon (H. H. T. Prins, personal observation).

Geophagy has been reported at several sites, where clay or substrates rich in iron also may explain this preference. On

Table 20.6 Main grass species recorded in the diet of savanna buffalo in different locations

Okavango Delta, Botswana[15]	Serengeti, Tanzania[1]	Lake Manyara National Park, Tanzania[2,5]	Tsavo NP, Kenya[3]	Gonarezhou National Park, Zimbabwe[4]	Mount Meru, Tanzania[1]	Ruwenzori Uganda[6]	Kafue National Park, Zambia[4]	Zambezi Valley, Zimbabwe[4]
Aristida. adscensionis	*Cynodon dactylon*	*Chloris gayana*	*Digitaria macroblephora*	*Cenchrus ciliaris*	*Cynodon dactylon*	*Capparis tomentosa*	*Andropogon gayanus*	*Eragrostis rigidior*
Cenchrus ciliaris	*Panicum coloratum*	*Cynodon dactylon*	*Panicum coloratum*	*Chloridum cameronii*	*Cyperus laevigatus*	*Cynodon dactylon*	*Cenchrus ciliaris*	*Hyparrhenia filipendula*
Cynodon dactylon	*Panicum maximum*	*Cynodon plectostachyus*	*Panicum maximum*	*Colophospermum mopane*	*Themeda trianda*	*Hyparrhenia filipendula*	*Chloridum cameronii*	
Dactyloctenium giganteum	*Panicum infestum*	*Cyperus laevigatus*		*Eragrostis rigidior*		*Sporobolus pyramidalis*	*Cynodon dactylon*	
Eragrostis rigidior	*Setaria chevalieri*	*Leptocarydion vulpiastrum*		*Heteropogon* sp.			*Heteropogon contortus*	
Panicum repens	*Sporobolus spicatus*	*Sporobolus spicatus*		*Hyparrhenia filipendula*			*Hyparrhenia filipendula*	
Panicum maximum	*Sporobolus pyramidalis*	*Sporobolus helvolus*		*Panicum maximum*			*Panicum maximum*	
Stipagrostis uniplumis	*Themeda triandra*	*Sporobolus cordofanus*		*Schizachyrium sanguineum*			*Parinari capensis*[a]	
Sporobolus fimbriatus		*Sporobolus pyramidalis*		*Schmidtia bulbosa*			*Sporobolus pyramidalis*	
Tragus berteronianus		*Urochloa mosambicensis*					*Themeda trianda*	
Urochloa trichopus							*Vossia cuspidata*	

Kruger National Park, South Africa[7,13,14]	Hwange National Park, Zimbabwe[8]	Lake Kariba, Zimbabwe[9]	Sabi Sand Wildtuin, South Africa[10]	Akagera National Park Rwanda[11]	Huhluwe-Umfolozi G.R. South Africa[12]	Zakouma National Park, Chad[16]	Benoue National Park, Cameroon[17]
Andropogon gayanus	*Andropogon gayanus*	*Panicum repens*	*Panicum maximum*	*Chloris gayana*	*Panicum deustem*	*Andropogon gayanus*	*Andropogon gayanus*
Cenchrus ciliaris	*Cynodon dactylon*		*Themeda trianda*	*Cynodon dactylon*	*Panicum maximum*	*Echinochloa obtusiflora*	*Andropogon tectorum*
Digitaria eriantha	*Cyperus laevigatus*			*Hyparrhenia filipendula*	*Themeda trianda*	*Setaria anceps*	*Chloris robusta*
Euclea sp.	*Schizachyrium sanguineum*			*Panicum maximum*		*Vetiveria nigritana*	*Digitaria* spp.
Heteropogon contortus				*Sporobolus pyramidalis*			*Echinochloa* spp.
Panicum coloratum				*Themeda trianda*			*Hyparrhenia* spp.
Panicum maximum							*Loudetia* spp.
Schmidtia bulbosa							*Panicum* spp.
Themeda trianda							
Urochloa mosambicensis							

[1] Sinclair 1977; [2] Prins 1996; [3] Leuthold 1972; [4] Mloszewski 1983; [5] Vesey-Fitzgerald 1969; [6] Field 1968; [7] Pienaar 1969; [8] Wilson 1975; [9] Taylor 1985; [10] Funston *et al.* 1994; [11] Monfort 1979; [12] Perrin & Brereton-Stiles 1999; [13] Macandza *et al.* 2004; [14] Bowers 2006; [15] Bennitt 2012; [16] Gillet 1969; [17] Stark 1986
[a] Fruit also eaten

Mount Kenya, geophagy is reported in the bamboo belt (2100–3000 m), where clay soils are rich in iron and aluminium (Grimshaw *et al.* 1995). This rare activity is carried out mainly by solitary individuals on Mount Kenya (Mahaney 1987). In Lent Valley in Kilimanjaro, buffalo enter into caves rich in sodium bicarbonate and chew off the soda deposits.

Forest buffalo

Few data exist on forest buffalo feeding ecology because direct observations are rarely feasible in rainforest habitats. Blake (2002) and Melletti (2008) observed buffalo feeding mainly on Poaceae and Cyperaceae within clearings, in particular on *Rhyncospora corymbosa*, *Kyllinga* sp. and *Cyperus* sp., in Noubale-Ndoki National Park and in Dzanga-Ndoki National Park. Blake (2002) also recorded several species of Marantaceae, including *Marantochloa purpurea*, *M. cordifolia*, *M. filipes* and *Halopegia azurea*. In addition, species of Commelinaceae family such as *Commelina diffusa*, *Palisota brachythyrsa* and a species of algae (*Spirogyra* sp.) were recorded. Melletti (2008) also found indirect signs of feeding activity on *Commelina* and *Palisota* sp. in the understory of *Gilbertodendrum dewevrei* mono-dominant forest.

Bekhuis *et al.* (2008) used micro-histological faecal analysis in a two-month study to determine the diet of forest buffalo at Campo-Ma'an National Park (southern Cameroon). They found that the most important part of the diet was composed of graminoids (43%, with *Leptochloa caerulescens* representing 15% of the total diet), non-graminoid monocots (21%, mainly Commelinaceae such as *Palisota* spp.), dicotyledonous plants (33%, mainly leaves) and cryptogamous plants (3%). The composition of the diet suggests that buffalo fed mainly along logging roads and river banks (Bekhuis *et al.* 2008). Using a similar method at Lopé National Park (Gabon), Lustenhouwer (2008) and Van der Hoek *et al.* (2012) found that the majority of plants consumed by forest buffalo were monocotyledons, primarily grasses (*Poaceae*) and sedges (*Cyperaceae*), with a low proportion of dicotyledonous plants in the diet. In the same study area, Van der Hoek *et al.* (2012) emphasized the importance of savanna habitat, noting that controlled burning is a key tool for maintaining open areas.

Details of forage species composition and percentage identified in buffalo dung are shown in Table 20.7.

So far, no evidence of the existence of grazing lawns was found in the different studies focused on the feeding ecology of forest buffalo, even in areas of high density, such as Lopé National Park (L. Korte and M. Melletti, personal observations).

Table 20.7 Major diet categories of forest buffalo (Campo Ma'an National Park, Cameroon)[1]

Major categories[2]	Percentage
Graminoids	42.9
Leptochloa	15.1
Other graminoids	27.8
Non-graminoid monocots	21.3
Commelinaceae	18.2
Other non-graminoid monocots	3.1
Dicotyledons	32.1
Dicotyledon leaves	26.5
Dicotyledon stems	6.2
Cryptogamous plants	3.1

[1] Bekhuis *et al.* (2008)
[2] Data refers to April–May 2002
Kind permission John Wiley and Sons, *African Journal of Ecology*, modified.

Reproduction and growth

Savanna (Cape, West and Central African) buffalo

In most of the Cape buffalo range there is a birthing peak during the wet season; however, in the more mesic conditions of Lake Manyara National Park (Tanzania) such a peak is not apparent (Prins 1996). In Matusadona National Park (Zimbabwe), conceptions occurred between December and May, with births between November and April, of which almost two-thirds of them were in January/February (Taylor 1985). These patterns have also been observed in northern Botswana (Carmichael *et al.* 1977), Kruger National Park and neighbouring Klaserie Private Nature Reserve in South Africa (Fairall 1968; Pienaar 1969; Ryan *et al.* 2007). In the Serengeti there is a clear calving peak, with about 50% of all calves born within seven weeks (Sinclair 1977). This peak may be tied to differences in social organization among bulls (see above). In Klaserie Private Nature Reserve, South Africa, births were quite highly synchronized, and the birthing peak was found to be correlated to the highest protein content of grass, which mostly coincided with the month of January (Ryan *et al.* 2007).

Experiments on captive buffalo males in South Africa showed a seasonal cycle in testosterone levels, endocrine function and seminal quality; but this pattern is not clear enough to explain a breeding or calving season (Brown *et al.* 1991). More recently Skinner *et al.* (2006) found a calving period in January for captive buffalo, similar to what was observed in wild populations (Ryan *et al.* 2007). Pro-oestrus may last two or three days and oestrus one day in captive Cape buffalo, and the oestrous cycle from 18 to 22 days (Knechtel 1993). The gestation period is about 340 days (Vidler *et al.* 1963; similar in captivity; Knechtel 1993). Buffalo give birth of a single calf, which weighs 30 kg, and twins are extremely rare. Suckling bouts of calves last about ten minutes when they are four weeks old, decreasing to five minutes when they are six months old (Prins 1996). Calves suckle from the rear; when the female starts walking with the herd the calf can continue receiving milk and the mother does not lose the protection of the herd. The age of weaning depends on many factors, including population density and intraspecific competition. In Lake Manyara National Park, where the buffalo density was high and the population stationary, many calves were still not weaned at

18 months old, while in Serengeti, where the density was much lower and the population increasing, fast weaning took place at the age of ten months. The calving interval in Manyara was therefore about 36 months (Prins 1996) and in Serengeti only 15 months (Sinclair 1977). Considering these patterns, birth rates seem to be density dependent.

Buffalo young are called 'calves' until they reach the inguinal fold at its highest point on the flank of the cow (Pienaar 1969). Juveniles are defined as weaned, but younger than three years (if testicles are not observable it is difficult to sex a juvenile in the field). Subadults begin to show secondary sexual characteristics and horn shape begins to diverge in males and females. For example, in Cape buffalo, the horns of cows start extending laterally whereas those of males curve strongly, upwards, and bosses start to develop. Subadult savanna buffalo bulls still have skin covered with hairs between the bosses, while adult bulls normally do not. Ovulating in females starts after three years of age, and the age at first breeding is five years (Grimsdell 1973b; Sinclair 1977; Taylor 1985). Males reach sexual maturity at 5–6 years of age, but usually do not breed until seven years or older, at which time the horns are fully developed.

Foetal sex ratio is approximately 1:1 in Kruger National Park and Lake Manyara National Park, Tanzania, and appears to be independent of rainfall, maternal lactation status or density (Visscher *et al.* 2004). In Manyara National Park, with increasing age, the sex ratio in herds becomes increasingly slanted in favour of females (Prins 1996). Van Hooft *et al.* (2007, 2010) found that during dry years, calves are sired by bulls with haplotypes that differ from bulls that sire calves during wet years (Van Hooft *et al.* 2007 2010). Bulls spend considerable amounts of time outside herds, which may partly explain the differentiation in fatherhood of calves. Life tables show that mortality patterns across ages in males and females are similar (Grimsdell 1969; Sinclair 1977); the average age of death (excluding calf mortality because remains of calves are too rare to find) is about 12 years. Maximum longevity in the wild is about 20 years and in captivity 29.5 years (Jones 1993).

Forest buffalo

What is known on reproduction and growth in forest buffalo is largely based on incidental observations during field studies of habitat, home range or behavioural activities rather than comprehensive studies focused on reproduction or growth. These initial observations give an indication of trends, but we lack long-term data on forest buffalo reproduction and growth.

It is difficult to ascribe a breeding season to forest buffalo because observations of mating are rare and births do not, so far, seem seasonally limited (Melletti *et al.* 2007a, 2007b; Korte 2008a, 2008b). At Lopé National Park in Gabon, Korte (2008a, 2008b) reported just one observed mating among nine radio-collared buffalo during a two-year study. The mating was observed in August 2003, with a calf born in June 2004. This June birth preceded the September through November wet season, when the lactating female would have access to a seasonal flush of new grass. Other collared females were also observed with new calves during this period, suggesting a possible calving season in the June through August dry season. Two of the buffalo had calves with them when collared in December 2002, two other animals birthed in July 2004, and a fourth female birthed in August 2004. In 1996 at Lopé, new calves were reported in August, September and December, during a six-month study of three herds. However, Korte (2008a, 2008b, 2008c) observed new calves in March 2003 and April 2003, ruling out a definitive calving season.

Sites in the Central African Republic and Cameroon report births during the wet season (Melletti 2008; Geßner 2008). At the Dzanga-Ndoki National Park, Melletti (2008) observed mating three times between June through August in 2002 and 2003. Although the same months as observations at Lopé National Park (Korte 2008a, 2008b), at Dzanga this period is the wet season with a light decreasing of rainfall. Also during the April through August wet season, Melletti (2008) reported new calves; however, no new calves were reported during the December through February dry season. Two births, one on 20 April and a second on 25 June, were reported at Ikwa Bai in Cameroon during a four-month study between 2 April and 6 July 2007 (Geßner 2008). Although calves were born during the wet season when food availability is probably at its peak, small sample size and short study period limit conclusions for calving seasons in Cameroon.

To the best of our knowledge, the literature reports births of only a single calf. During two years of observation of a population of about 342 buffalo at Lopé National Park in Gabon, no multiple births were observed (Korte 2008b). Melletti *et al.* (2007b) report a herd growing from 16 to 24 individuals with the addition of eight calves over a two-year period with no twins. Calving intervals are about two years, with a gestation of at least ten months (Melletti *et al* 2007b; Korte 2008b). Calves remain with cows for at least two years (Korte 2008b). Only adult female buffalo are observed with calves, suggesting that age of first reproduction is at least three years of age, if not older (Melletti *et al.* 2007b; Korte 2008b).

Studies of forest buffalo have followed guidelines developed for Cape buffalo to age individuals (i.e. calves: <12 months; juveniles: 12–36 months; subadults: 3–5 years; adults: >5 years; Pienaar 1969; Sinclair 1977). Little has been done to develop methods for ageing forest buffalo using cementum lines, tooth wear or tooth eruption sequences, to validate or calibrate ageing methods. Tooth wear is not found to be constant within species, differing with factors such as heredity, nutrition, habitat and the severity of the seasons (Hillson 1986). Thus, forest buffalo may have a different tooth wear pattern than Cape buffalo, given their very different habitat, i.e. forest versus savanna. Although horn size can be used as an indication of sex for individuals, visual observation of testicles is the most reliable method to determine a male individual. Adult females can have similar horns and body size to subadult males, so it is best to not rely solely on horn size and/or shape to sex individuals (Molloy 1997). When herds are observed over several months, it is possible to identify distinct individuals using horn morphology, scars and pelage characteristics to consistently record individuals (Prins 1989a; Melletti *et al.* 2007b; Korte 2008a, 2008b; Geßner 2008).

Behaviour

Savanna (Cape, West and Central African) buffalo

African savanna buffalo are gregarious animals living in mixed herds, and core members include adult females, subadults, juveniles and calves. Young females are known to maintain post-weaning bonds with their mother until the birth of their first calf, and certainly longer. In contrast, young males gradually become independent, and are likely to form subadult male groups within the mixed herd. At the age of about 4–5 years (and occasionally earlier), the males temporarily leave the herd to form bachelor groups. The elder males (from about ten years) sometimes permanently leave the herd, but this behaviour is not systematic (Sinclair 1977; Prins 1996).

Herd size in Cape buffalo varies across their distribution, from as few as 20 to as many as 2000 individuals in the floodplains of eastern and southern Africa (Sinclair 1977, Prins 1996). In West Africa (WAP Regional Park), the mean herd size was about 45 individuals when excluding bachelor males, and the largest herds were estimated to contain about 150 individuals (Cornelis *et al.* 2011). Similar figures were reported for Central African savanna buffalo, except in floodplain areas such as Zakouma National Park (Chad), where herds of up to 800 individuals were observed. Within savanna buffalo populations, differences in herd sizes were also observed, and herd sizes tend to be smaller where food resources are poor (Winnie *et al.* 2008).

Mixed herds were shown to be long-lasting social structures displaying a high degree of fidelity to their home range (Sinclair 1977; Prins 1996). The herd as a social unit has a much longer life span than any individuals, and recognizable herds were known to have existed for over 50 years (Prins 1996). Although adult females generally exhibit a high degree of philopatry to their native herd, herd switching was reported by a telemetry study in northern Botswana, where 7 of 45 adult cows switched herds, covering distances up to 133 km (Halley *et al.* 2002). In Cape buffalo, the migration rate per generation between herds was estimated by a genetic approach to be 5–20% for females and close to 100% for males (Van Hooft *et al.* 2003).

Fusion–fission group dynamics operate within mixed herds, meaning that herds may temporarily split into sub-herds and then merge again. Contrasted patterns of splitting and merging were described across studies. In the Serengeti, Sinclair (1977) reported that mixed herds tend to split into sub-herds when resources are fragmented, i.e. during the dry season. In contrast, Bothma (2002) reported that the herds tend to congregate during the dry season in areas where water and forage are abundant, and split into sub-herds during the wet season to exploit the available habitat more efficiently. Halley *et al.* (2002) made the same observation in Chobe National Park (Botswana), where buffalo were shown to range in herds of more than 1400 individuals during the dry season. Such large herds might in fact be temporary associations of two or more mixed herds.

At Lake Manyara National Park (Tanzania), where herds are less constrained by resources, the pattern of fusion–fission appears related to herd size: large herds split more frequently than smaller ones (Prins 1996). In Kruger National Park, South Africa, a longitudinal monitoring of multiple radio-collared individuals showed that herds were far less discrete than in previous studies (Cross *et al.* 2005). New herds may arise from these fusion–fission processes, comprising mixed subsets of previous herds, or single subsets.

As described above, bulls can be found within herds or smaller groups called 'bachelors' (Sinclair 1977; Prins 1989a, 1996). This sexual segregation operated by males is related to the energy cost of reproduction activities, which occur at the expense of the time normally allocated to foraging. Males competing for females are forced to leave the herd once their body condition decreases (Prins 1996; Turner *et al.* 2005). Individuals living in bachelor groups thus can spend more time foraging, but also utilize patches of habitat too small for the herd. The temporal patterns of sexual segregation appear contrasted between study areas. In Serengeti, where buffalo display a clear peak in calving, males join the mixed herds during the wet season (about eight months) and form stable bachelor groups during the dry season (Sinclair 1977). Bachelor herds in Serengeti comprise, on average, 5–10 individuals, the largest ones up to 50 (Sinclair 1977). They have a stable membership and there is evidence of linear dominance hierarchies in both bachelor and herd males (Sinclair 1977). In contrast, in more profitable areas such as Manyara (Tanzania), where buffalo are able to cover maintenance costs year-round, calving (and thus rutting) does not show a clear seasonal pattern, and adult males move in and out of mixed herds at a higher frequency (the so-called 're-entrant consecutive polygyny system'; Prins 1996). In this situation, buffalo males do not form stable associations, do not show fidelity to the same herd and do not have a particular home range. In the re-entrant consecutive polygyny system, the average size of bachelor groups is 3.5 individuals, and the largest groups comprise as many as 50 individuals (Prins 1996). These differences in patterns of sexual segregation depend on the seasonal patterns in resources availability. These two reproductive states may also explain the great differences in testosterone level between bulls during a season (Brown *et al.* 1991).

Cows, juveniles and calves stay together in large to very large herds, and even subadults of both sexes stay within their maternal herd. When the herd moves between feeding areas one or more adult males may precede it. However, adult cows are quite often in the lead as well. Buffalo leaders may stop an entire herd by not moving on or by giving a stopping signal, which involves standing and blocking the direction of movement. These 'individuals' are not to be viewed as 'true leaders' – for example, taking the decision of where to drink. It seems to be a communal decision taken by the adult cows in a procedure that has been called 'voting' (Prins 1996). Communal decision making seems to take place also in other mammals, and may be linked to information exchange. 'Voting' in buffalo takes place after a long resting bout, typically at the end of the afternoon when almost all individuals lie down. In this case females rise and elevate their head as if staring in a particular direction. Prins (1996) describes the voting posture as when the head position is halfway between a resting position and full alert, when the individual has the head elevated at a height as if gazing over its nostrils. The voting posture is maintained for a few minutes, after which the cow beds down again. We can distinguish two patterns of voting postures: (1) when we have a short vector that implies little consensus within the herd of voting females; and (2) where a long vector means a high degree of consensus (i.e. all heads were pointing in the same direction). Little consensus is often followed by a splitting of the herd and the resulting sub-herds then merge again after a few days (Prins 1996).

Social interactions between subadult cows are quite rare and restrained. Subadult bulls show intense sparring behaviour, similarly to adult bulls. During sparring, horns, and especially the bosses, are placed against each other, after which the animals start pushing. The 'winner' often is able to hook the other buffalo. Sparring may function as testing dominance between individuals, although it looks as if both contestants enjoy the game. Most interactions between adult bulls appear to be limited to evaluating the size of each other, often observed by other bulls. A more intense form of 'sizing each other up' is the parallel walk. Fights only happen when both contestants are equally strong and large. They are exceptionally rare and can result in fatal injuries to both bulls. Buffalo calves seldom play.

Females and their offspring have a rather fixed location inside a herd, which may vary with age and sex classes. Such a location may be inherited, because female offspring stay with their mothers up to maturity (Prins 1989b). Aggressive (agonistic) behaviour may also play a role in determining position within the herd. Agonistic behaviour consists of hooking and butting with the horns; among females this is typically aimed at the flanks or the rump of other adult females (Prins & Sinclair 2013; a summary of the main behaviours recorded are shown in Table 20.8).

Among males there is a distinct ritualized threat and submissive behaviour. Dominant males threaten by standing with the head up but muzzle pointing down, thus emphasizing the shoulders, a posture characteristic also of other bovine species, including the gaur (*Bos gaurus*), banteng (*Bos javanicus*) and bison (*Bison bison*, *B. bonasus*). A submissive male, when threatened, approaches the dominant with head held low and parallel to the ground and will often place the muzzle between the hind legs of the dominant one. At the same time the subordinate animal utters a loud and long bellow, sometimes continuing for the entire duration of the submissive display. The behaviour is similar to that of small calves when suckling or searching for their mother. This ritualized submission seems unique to African Buffalo; it is not shown by water buffalo. Buffalo are generally silent. They occasionally grunt, especially when a cow is hooked by another animal. They snort and cough when alarmed and when they start running. Bulls that are injured or about to die can make very loud bellows. Other bulls may or may not react to the sound. In about one-third of the instances that a bull is wounded by lions or hunters another adult bull will try to help him. Buffalo that are badly wounded may also seek the protection of the herd. They do not have an alarm call, but have a very efficient warning system, especially at night, by becoming silent (Prins & Sinclair 2013).

Mating depends on degree of willingness of a cow to be mounted. When the female is receptive, bulls show extensive tending behaviour ('consorting'); only adult bulls exhibit tending. Bulls show flehmen when testing urine of receptive females. When receptiveness increases the bull tries to lay his chin on her tail base. In general she then walks away, after which the bull tries again. When the cow is willing to stay put, the tending bull is often supplanted by a more dominant one who then mates with the cow. Other bulls may attempt intromission during mating, but females start running away, making copulation impossible. Mounting and ejaculation usually take a few seconds only. Females mating with more than one male has not been reported, but given the interrupted tending behaviour described above, this is a possibility (see Van Hooft *et al.* 2010). The cow remains with the herd to calve, although if the herd moves the females and their calves may be left behind temporarily. Calves may associate with age mates, and these associations are especially noticeable in females and may last for many years (Prins 1996), similarly to zebu cattle (Reinhardt 1983). Sometimes these coalitions result in coincident death at a much older age, when both 'friends' perish in so-called 'multiple kills' by lions; in Lake Manyara National Park multiple kills comprised about 6% of kills (H. H. T. Prins, personal observation).

Buffalo exhibit quite aggressive behaviour when attacked by predators, chasing and sometimes killing lions. In particular, they may respond to calf distress calls by chasing off the predators (Sinclair 1977). Buffaloes can reach 48–55 km/hr at top speed, and they can jump about 1.5 m in height. They also defend themselves against lions quite well in dense thickets, standing with their hindquarter to trees and bushes. Buffalo in open habitats adopt the antipredator strategy of forming large herds – predator swamping (Sinclair 1977; Melletti *et al.* 2007b; Prins & Sinclair 2013).

Lions use terrain features as cover to stalk buffalo herds (Prins & Iason 1989; Mugangu *et al.* 1995). Lions may take buffalo from the middle of a herd, sometimes without other

Table 20.8 Main behaviours recorded in the African buffalo

Agonistic behaviour	Vocal communication	Sexual behaviour	Parent/juvenile behaviour	Antipredator behaviour
High horn presentation[1]	Signal to move[9]	Urine testing by bulls	Calf croaking[19]	Alert posture[20]
Submissive bellow[2]	Direction giving signal[10]	Licking vulva by bulls[18]	Pushing off calf	Head tossing[4]
Wallowing[3]	Water call[11]	Cow head on bull rump		Wheeling and flight
Circling[4]	Position signal[12]	Cow pushing belly of bull		Individual/herd attack
Rolling (mud/dust)	Warning call[13]			
Horning (vegetation)[4]	Aggressive call[14]			
Head tossing[4]	Mother to calf call[15]			
Rubbing face and neck in mud[4]	Calf distress call			
Lateral display[5]	Danger call[16]			
Head low posture[6]	Grazing vocalizations[17]			
Flight/alarm posture				
Charging – chin raised[7]				
Front-pressing[7]				
Sparring[8]				

[1] The head is held at shoulder level, associated with head tossing
[2] Mouth is opened and the tongue is curled
[3] Mud and dust baths mainly carried out by dominant males (social significance)
[4] Dominance, threat and aggressive displays
[5] Head is held at or above shoulders, chin pulled, tossing and hooking movements, means of asserting dominance
[6] Defensive and submissive displays performed by a low-ranking male
[7] During fighting
[8] No charge, milder exercise, little horn-tangling and half-hearted pushing, mainly performed by subadults
[9] Call given for herd coordination during movements
[10] Intermittent sound given by leaders at the beginning of herd movement
[11] Call emitted many times during movement to drinking places
[12] Sound emitted by a high-ranking buffalo within a herd that announces its presence and location
[13] Sound given to an encroaching submissive individual
[14] Grunt given by a dominant bull after a stampede
[15] Croaking call emitted by cows seeking their calves
[16] When lions are detected
[17] Variety of sounds during grazing
[18] Bulls can prompt cows to urinate
[19] Call emitted when losing contact with mothers
[20] Advancing to investigate the presence of a possible predators, etc. Alert behaviour mainly performed by adult females in forest buffalo
Sources: Sinclair 1977; Mloszewski 1983; Estes 1991; Prins 1996; Melletti *et al.* 2007b

buffalo paying attention, while in other instances buffalo chase lions and sometimes kill them. It seems that large numbers make buffalo safer, and adult bulls seem invulnerable when in prime condition. However, lions often kill males in bachelor groups (Schaller 1972). In the Serengeti, predation by lions accounted for 44 males, versus six females, six young and six indeterminate sex (Wilson & Mittermeier 2011). Usually, two or more lions attack from the rear, and they have to take the buffalo down quickly, otherwise he can drive them off. Lions unable to take down a standing buffalo have been seen perched on top of the standing buffalo (H. H. T. Prins, personal observation). Sometimes buffalo will try to push lions against *Acacia* thorns. If lions successfully put down the buffalo they try to suffocate it by holding the muzzle because they cannot break the neck.

The main predators of Cape buffalo are lions, but predation by spotted hyena (*Crocuta crocuta*) and leopard (*Panthera pardus*), has been recorded, usually targeting young animals (Kruuk 1972; Schaller 1972; Prins & Iason 1989; Ruggiero 1991; Funston *et al.* 2003). In forest areas such as Mount Meru, Tanzania, predation is insignificant. In the Serengeti National

Park, predation by lions accounts for only 25% of mortality (Sinclair 1977). In Lake Manyara National Park lion predation is very high, up to 85%, in the absence of major diseases. Thus, the population regulatory effect of predation depends strongly on location. Other infrequent (less than 5%) causes of natural mortality include getting stuck in mud, falling down cliffs or fights with other buffalo (Prins & Sinclair 2013). Mass mortality of buffalo has been reported in Zambia when they were stampeded over cliffs by lions on two occasions (63 and 42 buffalo, respectively), and also by mass entanglement in floating grass while trying to cross a lagoon (Prins & Sinclair 2013).

Forest buffalo

Forest buffalo herds are small, stable groups with relatively little switching between herds (Melletti *et al.* 2007b; Geßner 2008; Korte 2008b, 2009). Adult cows and their offspring live in herds that vary between about 3 and 25 individuals in the forest buffalo (Happold 1973; Melletti *et al.* 2007a; Korte 2008a, 2008b). Herds comprise several adult females with their young and with one or two bulls (Dalimier 1955; Blake 2002; Melletti *et al.* 2007a; Korte 2008a). This may be the best reproductive strategy when females occur in small herds that are widely scattered in grassy patches within a rainforest that is otherwise unsuitable to forest buffalo. The ratio of adult males to adult females at Lopé National Park in Gabon is 1:6, with an overall ratio of 1:2 for adults and subadults to juveniles and calves (Korte 2008a). At Bai-Hokou of the Dzanga-Ndoki National Park in the Central African Republic, a herd of 24 individuals included one adult male, nine adult females, six juveniles and eight calves (Melletti *et al.* 2007b). Geßner (2008) reported a stable herd of at least 23 buffalo, including an adult male, at least one subadult male, two subadult females and ten adult females with at least three juveniles and four calves at Ikwa Bai at Nki National Park in Cameroon. Forest buffalo herds split into smaller subgroups and not all individuals are together at all times.

Herd size appears stable within herds, but can vary widely across herds (Korte 2008b). During a two-year study at Lopé National Park in Gabon, the mean group size for 18 herds was 12 (range of means 3–24), considerably smaller than Cape buffalo herds. Although mean group size varied little with time of day, across season or between savanna and marsh habitat, herd size varied widely across herds, from fewer than ten individuals in the smallest herds to more than 20 in the largest herds. Melletti *et al.* (2007b) also report that herd size did not vary with season or food availability. This herd at Dzanga-Ndoki National Park in the Central African Republic increased from 16 to 24 due to the birth of eight calves during the two-year study, with no immigration or emigration. Although the herd remained small, individuals split into smaller subgroups with fission patterns lasting 1–3 days before the herd merged again (Melletti *et al.* 2007b; Melletti 2008). This trend seemed more frequent when the herd grew to 24 individuals at the end of the study period (Melletti 2008). Melletti *et al.* (2008) also report that group size was smaller and individuals were more dispersed within the same forest clearing during the wet season.

Although this subspecies is forest-dwelling, forest buffalo are dependent on open areas adjacent to continuous forest (Melletti *et al.* 2007a, 2007b; Bekhuis *et al.* 2008; Geßner 2008; Korte 2008a, 2008b). At a local level, the quantity and dispersion of food resources appear to influence group size (figure 20.6a,b from Korte 2008b). For eight radio-collared buffalo at Lopé National Park there was a significant positive correlation between maximum group size and the area of open habitat in each collared animal's home range. These open habitats are savannas and marshes, where food resources are abundant. Thus, aggregation patterns appear to be linked to food resources. At other forested sites, buffalo depend on forest clearings or grassy areas along roads, where food resources are present (Hoppe-Dominik 1992; Melletti *et al.* 2007a; Bekhius *et al.* 2008; Geßner 2008; Korte 2008b).

Antipredator strategies in forest buffalo seem to consist of moving and resting together (Melletti *et al.* 2007b). At Dzanga-Ndoki National Park, buffalo were closer to each other when moving compared to distance between individuals when grazing. In addition, an adult female always led the group in flight with the adult male often (i.e. in over 90% of cases) the final individual, possibly an antipredator strategy. Melletti *et al.* (2007b) also report larger group size when buffalo are a greater distance from forest edges, noting that every individual was surrounded by other buffalo, probably indicating an antipredator strategy. In addition, the adult male was commonly closer to the females than to juveniles (Melletti *et al.* 2007b). Melletti *et al.* (2007b, 2008) also report observations of the herd standing in alert postures for a few minutes and then fleeing a few hundred metres within the same clearing (i.e. when the herd smelled the observers or a predator), but on some occasions the buffalo moved to another clearing, with an adult female always leading the group.

Leopard predation on buffalo is rarely observed in the rainforest; however, at Lopé National Park, Gabon, where buffalo density is high in the savanna areas, buffalo make up over 10% of the biomass consumed (Henschel *et al.* 2005).

Geßner (2008) reports that forest buffalo displayed increased vigilance at the start of her three-month study when an observation mirador was constructed at the forest clearing. Vigilance decreased during the study as animals adjusted to the presence of observers; however, levels of vigilance increased when a non-herd subadult female visited the clearing and at the births of two calves. At Dzanga-Ndoki National Park levels of vigilance increased when buffalo were approached by lowland gorillas feeding in the same clearing (Melletti 2008).

Forest buffalo are more often observed as part of a herd than as individuals (Melletti *et al.* 2007b; Korte 2008b; Geßner 2008). For example, solitary males and females are observed at Lopé National Park, but the majority of observations are of two or more buffalo (Korte 2008a, 2008b). Although solitary adult males are observed more often than solitary adult females, adult males were not observed leaving herds to form bachelor groups, and males consistently stayed with the same herds. During two years of field work, Melletti *et al.* (2007b) encountered neither solitary adult males nor bachelor groups. Solitary adult females and juveniles were observed on a few occasions, but they joined the herd after a few days. Geßner (2008) reports a solitary subadult male at the Ikwa Bai in Cameroon, but did not observe bachelor herds. Thus, so far, field studies have produced little evidence for bachelor herds in forest buffalo, unlike Cape buffalo studies.

Forest buffalo herds usually include more than one generation of young. Calves stay with their mothers for at least two years and juveniles will often stay with their mothers after the birth of the next calf (Melletti *et al.* 2007b; Korte 2008a, 2008b). Calves stay close to the females after birth and integrate into the herd with little attention from other herd members (Geßner 2008). Melletti (2008) observed females leaving their herd for a few days at the time of birth, returning to the herd when the new calves are able to walk properly. During these periods, females remained apart from the herd and concealed their calves in tall grass or bushes when grazing. Cape buffalo can display similar behaviour when calves are born (Sinclair 1977; Mloszewski 1983; Prins 1996). Forest buffalo are attentive to calves, but the literature reporting on births of calves does not describe specific care of calves by females (Melletti *et al.* 2007b; Geßner 2008; Korte 2008c).

Few observations of agonistic behaviours are reported for forest buffalo (Molloy 1997; Melletti *et al.* 2007b; Geßner 2008; Korte 2008a, 2008b). Geßner (2008) defined the sequence of a buffalo placing it nose between the hind legs of a second individual and jumping away as a submissive behaviour. This is similar to the appeasement ceremony observed in Cape buffalo (Estes 1991). At Ika Bai in Cameroon, subadult buffalo males displayed submissive behaviour to the adult male of the group; adult females from different subgroups also displayed submissive behaviour when the subgroup joined; and juveniles displayed submissive behaviour to adult buffalo, establishing hierarchy within the group. Geßner (2008) also reports a peak of chasing between the adult male and subadult male in June, when the adult male is establishing his position in the herd. Melletti (2008) also reports submissive behaviour when a subadult or juvenile approached the herd. Defensive displays are rare, with Melletti *et al.* (2007b) reporting the most frequent interaction as adult females displacing juveniles from resting areas at Bai-Hokou (Dzanga-Ndoki National Park) in the Central African Republic. Despite distinct home ranges and the occasional encounters of different herds, territorial defence was not observed at Lopé National Park (Korte 2008a). Field studies report a limited number of behavioural displays among forest buffalo. Melletti (2008) at Bai-Hokou in the Central African Republic observed the only adult male performing urine testing just before copulation on a few occasions. He also reported short vocal signals when the herd was moving from one clearing to another, during grazing, while resting and with all activities associated with the displacement of young by adults at Dzanga-Ndoki National Park.

There are few data on intraspecific and interspecific competition for food resources in forest buffalo. This buffalo lives in smaller herds than savanna buffalo, which could facilitate better inter-individual tolerance and reduce competition for food resources. However, this hypothesis is not supported by the strong spatial segregation of home ranges observed at all study sites (see above). Melletti *et al.* (2007b) observed that most frequent interaction between the members of the herd during feeding and resting was the driving off of other individuals, mainly initiated by adult females and usually directed towards a juvenile. The authors conclude that this kind of interaction could be due to competition for the best feeding and resting spots. Within a clearing these social interactions between buffalo were relatively common. Moreover, several other large mammals use this habitat type and can congregate in the same clearing. During feeding, buffalo, sitatunga, red river hogs, gorillas, bongos and even forest elephants may tolerate each other even within a few metres. Among these, sometime elephants are less tolerant, mainly towards buffalo, driving them off or simply annoying them.

Spatial aggregation patterns in forest buffalo have been studied by Melletti *et al.* (2008, 2010) at 45 resting sites of a single herd in Dzanga-Ndoki National Park, where both habitat structure and season influenced these patterns. In open habitats such as clearings, groups cover a larger area when resting and are more rounded in shape compared to group properties noted in forest during the wet season. Forest buffalo also have a more aggregated spatial distribution when resting in clearings than when in the forest. Individual positions within the herd in the clearing habitat varied with age and sex. In the clearings, the adult male was on most occasions located in the centre of the herd ($n = 20/24$ observations). In contrast, females occupied intermediate ($n = 57/80$), peripheral ($n = 14/80$) and central positions ($n = 9/80$) within the group. Juveniles also occurred in intermediate ($n = 64/77$) and peripheral positions ($n = 13/77$). Although these results are limited by small sample size (one herd), it can give an indication that habitat characteristics and social behaviour can have relevant effects on the spatial distribution of animals within a group.

Parasites and diseases

Savanna (Cape, West and Central African) buffalo

The savanna buffalo is susceptible to several infectious diseases of importance in Africa. Some diseases (hereafter referred to as indigenous) originated in Africa and co-evolved with African buffalo populations while others (hereafter referred to as alien) have been introduced by imported cattle breeds. The savanna buffalo is known to be a reservoir for foot-and-mouth disease (FMD), corridor disease (theileriosis), bovine tuberculosis (BTB) and bovine brucellosis. The role of savanna buffalo in the epidemiology of pathogens such as anthrax, Rift Valley fever and lumpy skin disease is still unclear and requires more investigation (Prins & Weyerhaeuser 1987; de Garine-Wichatitksy *et al.* 2012; Caron *et al.* 2013).

The transmission of infectious diseases between individuals can occur though different routes, including direct inter-individual contact (e.g. sperm, milk, aerosols), indirectly by contaminated material (e.g. faeces, saliva) released on food and in water, or by haematophagous vectors (e.g. ticks, mosquitoes). In buffalo, several ecological traits (e.g. highly gregarious behaviour, fusion–fission group dynamics, dispersal and mating strategies) are conducive to the transmission and the diffusion of pathogens.

Rinderpest, an alien disease imported in the Horn of Africa in the 1890s, had a devastating effect on many African cloven-hoofed ungulate species and is thought to have caused up to 95% mortality of savanna buffalo populations throughout Africa in the early twentieth century. Subsequent rinderpest epizootics caused major die-offs throughout the twentieth century (until the mid 1980s), but the disease was recently declared eradicated (Morens *et al.* 2011). The extreme pathogenicity of rinderpest and high mortality prevented the development of a wildlife reservoir (De Vos & Bengis 1994). In contrast, BTB and bovine brucellosis successfully established themselves in buffalo, which act as a maintenance host. Bovine tuberculosis is a chronic, progressively wasting disease with prevalence rates up to 60% in buffalo herds (Jolles 2004). Although BTB-infected buffalo can suffer declines in body condition (Caron *et al.* 2003), adult survival and fecundity (Jolles *et al.* 2005), longitudinal studies undertaken in a high-prevalence area such as the Kruger National Park (South Africa) suggest little demographic impact (Cross *et al.* 2009). Similarly, current knowledge of the impact of bovine brucellosis (a chronic disease also known as 'contagious abortion') does not indicate an immediate threat to buffalo survival at the population level (Michel & Bengis 2012). However, the impacts of these chronic diseases may be a non-linear function of environmental conditions such that they are only detectable in stressful periods (e.g. prolonged droughts) (Cross *et al.* 2009). For example, anthrax outbreaks were locally shown to cause mass mortalities in buffalo during drought periods (Prins & Weyerhaeuser 1987; De Vos & Bryden 1996).

Indigenous diseases (e.g. FMD, corridor disease and African trypanosomiasis) generally do not pose a threat to the survival of buffalo populations because of the evolutionary development of unique coping mechanisms (Michel & Bengis 2012). These diseases are generally carried asymptomatically by buffalo, which act as biological reservoirs of transmission to cattle, thus posing particularly severe problems at wildlife–domestic interfaces (Miguel *et al.* 2013; see Chapter 26 for more details). In Southern Africa, diseases transmitted from buffalo to cattle have contributed greatly to the eradication of buffalo from cattle production areas and to the confinement of buffalo to well-fenced areas, thus strongly limiting their movements and distribution range (Laubscher & Hoffman 2012). Due to their known or perceived role in the epidemiology of many infectious diseases, buffalo populations have been the target of control measures by veterinary services to protect cattle populations, crucial livelihood assets for most African national economies, but also for poor small-scale farmers living in the periphery of protected areas.

Finally, savanna buffalo are known to be hosts of a long list of endoparasites, including trematodes, cestodes, nematodes and pentastomids (Penzhorn 2000; Belem & Bakoné 2009). These infections do not normally result in pathogenic symptoms or mortality, although co-infection by gastrointestinal worms and bovine tuberculosis was shown to generate synergistic effects (increased mortality) (Jolles *et al.* 2008).

Forest buffalo

The susceptibility of forest buffalo to infectious diseases is expected to be similar to the savanna buffalo due to their phylogenetic relatedness. However, forest buffalo have not been reported to be infected by diseases of importance such as FMD and BTB. Forest buffalo sustain little exposure to alien infectious diseases compared to savanna buffalo because habitat within its distribution range is not suitable for cattle production. Interspecific contacts are thus rare. However, the recent increase in human encroachment in rainforest habitat (e.g. the logging industry) is currently boosting cattle production along the forest block of the Congo Basin, thus potentially leading to higher risks of disease transmission.

Ectoparasites in the forest buffalo have been recorded from an adult male found dead in the Nouabale-Ndoki National Park, Republic of Congo (Deem *et al.* 2005). This individual was affected by ticks of genus *Amblyomma* sp., *Rhipicephalus* sp., trematodes *Carmyerius gregarius* and oocysts of *Eimeria* spp. Intestinal schistosomes can also heavily infest buffalo, especially old bulls that spend considerable time around water. At Lopé National Park in Gabon, Lustenhouwer (2008) reported that forest buffalo used moist areas despite the risk of exposure to endoparasites. A genus of rumen fluke, *Paramphistomum* sp., was the most common parasite based on dung sampling.

Status in the wild

IUCN: Least Concern

African buffalo is considered a species of Least Concern according to the IUCN Red List (IUCN SSC 2008), not qualifying for Critically Endangered, Endangered, Vulnerable or Near Threatened status; buffalo are not a listed CITES species (UNEP-WCMC 2012). Though African buffalo appear to be widespread and abundant, many countries recognize the need to protect forest buffalo at the national level (Djeukam 2007). For example, buffalo are a protected species in Gabon, with no hunting allowed, and in Cameroon buffalo are a partially protected species with regulations on hunting (Kamdem-Toham *et al.* 2003; Djeukam 2007).

Savanna (Cape, West and Central African) buffalo

Savanna buffalo populations are likely to be impacted by a large array of factors, among which are land use changes, competition with livestock, poaching, drought and disease. Buffalo numbers suffered their most severe collapse during the great rinderpest epidemic of the 1890s, and subsequent outbreaks that have spread from cattle to wildlife during the twentieth century (Winterbach 1998). As above mentioned ('Parasites and diseases' section), local eradication programmes, corridor barriers and fencing of protected areas have further fragmented the initial meta-population.

During the last decades, civil wars (e.g. Angola, Sudan, Ivory Coast, Central African Republic, Mozambique, Democratic Republic of Congo, etc.) and national policies allowing armies to exploit bushmeat largely contributed to locally decimate buffalo populations. Opening roads through natural ecosystems facilitates uncontrolled hunting, fragments populations and generates barriers to dispersal (Blom *et al.* 2004). Poaching is not only conducted to support a local demand, but is often carried out by informal but well organized market channels delivering meat to towns (Dublin *et al.* 1990; FitzGibbon *et al.* 1995; Mduma *et al.* 1998). Increasing agricultural activities, high livestock numbers, deforestation (e.g. charcoal production) and increasing human population densities progressively contribute to the degradation of areas suitable for buffalo. In many locales, livestock overgrazing makes the habitat unsuitable for the buffalo and other large mammals, both through straightforward competition and due to impacted soils and erosion. Protected areas thus increasingly become isolated and disconnected, and buffalo subpopulations are left vulnerable to local extinction due to disease outbreaks, hunting or fires. In much of rural Africa, programmes for livestock vaccination have come to a complete standstill. In many regions a combination of disease and lawlessness is likely to present a major threat to the future survival of buffalo populations (Prins & Sinclair 2013).

Buffalo have a high commercial value as trophies throughout Africa, a value enhanced by their reputation of being one of the most dangerous African animals. For example, in 2013 the price of a two-week safari including the opportunity to shoot one buffalo globally amounts to €7000–35 000 (depending on the region, quality of service, trophy size, etc.; for more details on buffalo hunting, see Table 20.9). In Southern Africa, many private ranches and communal 'game conservancies' have been stocked with disease-free buffalo and make a good profit (e.g. Hearne & McKenzie 2000; Du Toit 2005 – see also the section 'Status in captivity'). Most hunting areas, if well managed, play an important role in conserving habitats, species richness, employment and income for local people around protected areas. Buffalo also have an unrealized potential value due to their resistance (both apparent and inherent) against many tropical diseases compared to livestock.

The conservation prospects of the three savanna subspecies of buffalo are quite different. Summation of the country estimates given above (Tables 20.1–20.2) gives a total population of >513 000 for the three subspecies of savanna buffalo, a figure slightly lower than the last global estimate (627 000; East 1998). As mentioned above, these estimates are likely to be conservative, for two main reasons. The first is that aerial surveys tend to underestimate buffalo populations due to detection bias from the air. The second is that population estimates are unavailable for some parts of these subspecies' ranges. Although comparisons between estimates are complex in areas presenting a dissimilar level of information (e.g. Sudan), global trends emerge from this exercise. The Cape buffalo (*S. c. caffer*) population globally appears stable and even increasing in countries such as Mozambique or South Africa. The slight decrease in the overall number of Cape buffalo mainly results from an apparent decrease of the total estimate in Tanzania, the country hosting the largest population. The situation of the West African savanna buffalo (*S. c. brachyceros*) also appears globally stable at the scale of the distribution range, but marked contrasts appear by country. In Burkina Faso, Niger and Benin, sustained efforts brought to several protected areas by the international community (e.g. the European Commission) during the last decade seem to bear fruit. In contrast, the West African savanna buffalo population of Senegal (Niokolo National Park) and Ivory Coast (Comoe National Park) has alarmingly dropped by 90%. Most worrying is the future of the Central African savanna buffalo (*S. c. aequinoctialis*) population, which has faced an estimated decrease of 60% during the last 15 years. Although uncertainties remain about the status of the population in Sudan, the large population previously observed in Democratic Republic of Congo and the Central African Republic fell sharply due to internal problems of governance (civil war, civil unrest).

Forest buffalo

The forest buffalo is declining across its geographic range (IUCN SSC 2008). Based on only a few population estimates, East (1999) estimated a total population of 60 000 forest buffalo with about 75% of the population in nominally protected areas, including Lobeke National Park (Cameroon), Dzanga-Ndoki National Park (CAR), Nouabalé-Ndoki-Kabo (Congo-Brazzaville), Lopé National Park, Wonga-Wongue and Gamba (Gabon), Odzala National Park (Congo-Brazzaville) and Maiko National Park (Congo-Kinshasa).

The future of this subspecies depends on well-managed protected areas and hunting zones with special attention to forest clearings and mosaics of forest and savannas, where critical food resources are abundant (Blake 2002; Melletti 2008; IUCN SSC 2008). In addition to habitat loss, poaching is a major threat to forest buffalo populations (IUCN SSC 2008). Buffalo are hunted for their meat and many rural populations depend on bushmeat for protein (Foerster *et al.* 2012). For example, in the Gamba Protected Areas Complex, Gabon, forest buffalo is a target species in the bushmeat trade, and is the fourth highest species present in the markets (Thibault & Blaney 2003). Thus, appropriate hunting regulations

Table 20.9 The range of costs for a single buffalo hunt in different African countries where the species is legally hunted

	Guinea[1]	Burkina Faso	Benin	Cameroon	Central African Republic	Ethiopia	Uganda	Tanzania	Mozambique	Zambia	Zimbabwe	Botswana	Namibia	South Africa (RSA)
Trophy fees	1120	€430–480	€600	800–1800 €1500–2000[1]	€800–2000	1000–3300	2000–4000[2]	2500–3900	3000–5000	3000–3500	2600–4500	5000	3250–8500	6500–30 000[3]
Daily rates	?	€850	€1000–1300	944–1300 €2230[1]	€1900	1950–3584	1200–1500	2000	1000–1550	1000–1500	850–1250	3900	1650–2000	200–640
Total hunter days	8	7–10	7	7–13	7–10	16	10	10	7–10	7–10	10	10	8	5–10
All-inclusive hunting cost per buffalo	10 000	€6380–8930	€7600–9700	7408–10 900 ~€17,110[1]	~€14 100	~17 950	~14 000	15 000–22 000	10 000–14 500	10 000–18 500	10 000–17 000	~14 000	11 500–14 000	9750–48 000

Buffalo are a key species in the safari hunting industry. The economic activity that can be generated around a buffalo hunt is much higher than only the buffalo trophy fee. This activity creates also jobs, recruiting people from local communities such as drivers, trackers, skinners, cooks, cleaning ladies, etc.
All hunts are considered to be carried out by a single hunter, with one buffalo bull included.
All prices are in US dollars unless otherwise specified.
The trophy fees and daily rates listed come from the websites of the main hunting outfitters. These values may change by a small percentage from year to year.
Daily rates depend on the kind of accommodation, service, quality of hunting areas, logistics and so on and vary from country to country.
A buffalo hunt starts from a minimum of five to a maximum of 16 days.
A percentage of trophy fees and total hunting cost go to local communities and to the government. This percentage varies from country to country.
[1] Trophy and daily fees and total hunting cost are related to the forest buffalo (*Syncerus caffer nanus*).
[2] 75% of trophy fees goes to local communities and 25% to the Ugandan government (source: Uganda Wildlife Authority).
[3] The trophy fees are determined by trophy size (horns size measured in inches) and location. This is mainly applied in private, fenced reserves.
Both the government reserves and private fenced lands are included.

and enforcement of these regulations are also critical for maintaining forest buffalo populations (Wilkie *et al.* 2011).

Status in captivity

Savanna (Cape, West and Central African) buffalo

In South Africa, numerous disease-free buffalo breeding projects have been supported by the government since 1996, with the objective of restocking areas where buffalo populations were not viable anymore, or disappeared due to disease or exclusion by cattle farmers (Laubscher & Hoffman 2012). Disease-free animals were produced using calves taken at birth from captive mothers and fostered by surrogate disease-free dairy cows (mainly Jersey cows due to their good mothering abilities and high milk yield). The initial stocks were built from the Kruger National Park population, offering the best gene pool, but animals (record-class trophies) were also imported from Tanzania. These projects rapidly had a snowball effect through the involvement of the private sector (mainly driven by trophy hunting businesses) and buffalo breeding is now part of a wholly commercial system. Around 26 000 buffalo are privately owned and resident on approximately 2700 private game ranches or reserves in 2013 (South African Private Buffalo Owners Association, personal communication). In recent years, the increasing demand for disease-free buffalo has gradually led to an increase of their price in South Africa. In 2008 disease-free buffalo were being sold for just over ZAR 160 000 (ZAR 1 ≈ $0.11), while in 2010 this price increased to over ZAR 325 000 per buffalo at the same farmers' auctions (Cloete 2011). In 2012, ZAR 20 million was paid for a buffalo cow and her bull calf, as well as a record price of ZAR 26 million for a buffalo bull (Laubscher & Hoffman 2012). The buffalo cow that fetched the 2012 record price had a horn spread of 109 cm, making it the biggest of any cow in captivity. In this buffalo breeding system, bulls are selected on temperament and semen quality as well as horn length, and cows are primarily selected on reproductive performance. Although game farmers argue that selective breeding is restoring what years of culling has eliminated, a selection process based on specific traits (here trophy quality) and implemented at an industrial scale may have unexpected consequences for the genetic integrity of the species.

In 2012 there were 166 Cape buffalo in captivity across five regions of the world. In Africa, 48 (22 males, 24 females, 2 unknown) are kept in different zoos (ISIS: www2.isis.org/Pages/Home.aspx). Furthermore in Africa there are hundreds of Cape buffalo that are sold as game annually. For example, if we only look at the data from www.gamefarmnet.co.za/veiling.htm in 2012, about 350 Cape buffalo were sold.

Europe has 62 (17 males, 45 females), North America 26 (7 males, 19 females), Asia 18 (9 males, 9 females) and Latin and Central America report 12 (4 males, 8 females) (ISIS: www2.isis.org/Pages/Home.aspx; see also Chapter 27 and, in particular, Table 27.1). These numbers reflect individuals managed in cooperation with zoological breeding collections, and therefore are likely an underrepresentation of the total number of buffalo in captive collections.

Forest buffalo

In 2012 there were 135 forest buffalo in captivity across four regions and 26 institutions of ESB (EAZA) (see also Chapter 27). In Africa, six buffalo (two males, four females) are in two zoological gardens. In Australia there are four males and one female forest buffalo in two institutions. Europe has the greatest number of forest buffalo with 32 males, 84 females and 4 undetermined housed across 20 institutions. In the United States there are just four females in two different institutions. San Diego Zoological Society reports high calf mortality and a decision to no longer maintain forest buffalo due to little hope for maintaining high genetic diversity (G. P. Carmi & R. G. Rieches, personal communication).

References

Ansell, W. F. H. (1972). Family Artiodactyla. In *The Mammals of Africa: An Identification Manual*, part 2.15, pp. 1–84. Eds. Meester, J. & Setzer, H. W. Washington, DC: Smithsonian Institution Press.

Aremu, O., Onadeko, S., Ola-Adams, B. & Inah, E. (2007). Population parameters and biomass of African buffalo (*Syncerus caffer*) in Kainji Lake National Park. *Nigeria. Journal of Applied Sciences* 7: 1809–1812.

Bar-David, S., Bar-David, I., Cross, P. C., Ryan, S. J., Knechtel, C. U. & Getz, W. M. (2009). Methods for assessing movement path recursion with application to African buffalo in South Africa. *Ecology* **90**(9): 2467–2479.

Beekman, J. H. & Prins, H. H. T. (1989). Feeding strategies of sedentary large herbivores in East Africa, with emphasis on the African buffalo. *Journal of African Ecology* **27**: 129–147.

Beilfuss, R. D., Bento, C. M., Haldane, M. & Ribaue, M. (2010). Status and distribution of large herbivores in the Marromeu Complex of the Zambezi Delta, Mozambique. World Wide Fund for Nature.

Bekhuis, P. D. B. M., de Jong, C. B. & Prins, H. H. T. (2008). Diet selection and density estimates of forest buffalo in Campo-Ma'an National Park, Cameroon. *African Journal of Ecology* **46**: 668–675.

Belem, A. M. G. & Bakoné, E. U. (2009). Parasites gastro-intestinaux d'antilopes et de buffles (*Syncerus caffer brachyceros*) du ranch de gibier de Nazinga au Burkina Faso.*Biotechnology, Agronomy, Society and Environment* **13**: 493–498.

Bell, RHV. (1982). The effect of soil nutrient availability on community structure in African ecosystems, pp 193–216. In *Ecology of Tropical Savannas*. Eds. Huntley, B. & Walker, B. Berlin: Springer-Verlag.

Bennitt, E. (2012). The ecology of African buffalo (*Syncerus caffer*) in the Okavango Delta, Botswana. PhD thesis. School of Biological Sciences, University of Bristol.

Bennitt, E., Bonyongo, MC. & Harris, S. (2014). Habitat Selection by African Buffalo (*Syncerus Caffer*) in Response to Landscape-Level Fluctuations in Water Availability on Two Temporal Scales. *PLoS* ONE 9(7): e101346. doi: 10.1371/Journal.pone.0101346.

Bergl, R. A., Dunn, A. Haruna, S. & Mshelbwala, J. G. N. (2011). *Aerial Survey of Elephants and Other Large Mammals at Yankari Game Reserve, Bauchi State, Nigeria*. Nigeria: Bauchi State.

Blake, S. (2002). Forest buffalo prefer clearings to closed canopy forest in the primary forest of northern Congo. *Oryx* **36**: 81–86.

Blancou, L. (1935). Buffles de L'Oubangui-Chari-Tschad. *Terre Vie* **5**: 202–221.

Blancou, L. (1954). Buffles d'Afrique. *Zooleo* **27**: 425–434.

Blom, A., van Zalinge, R., Mbea, E., Heitkonig, I. M. A. & Prins, H. H. T. (2004). Human impact on wildlife populations within a protected Central African forest. *African Journal of Ecology* **42**: 23–31.

Bolanos, N. C. (2012). Aerial animal census 2012, Garamba National Park. Democratic Republic of Congo. Institut Congolais pour la Conservation de la Nature, African Parks Network.

Bothma, J. (ed.) (2002). *Game Ranch Management*. Vol. **4**. Pretoria: Van Schaik Publishers.

Bouché, P. (2005). *Inventaire aérien des populations de grande faune dans les sites du PAGEN. Projet de partenariat pour l'amelioration de la gestion des ecosystemes naturels*. Burkina Faso: Ministère de l'Environnement et du Cadre de Vie.

Bouché, P. (2006). Mole wildlife survey. Northern Savannah Biodiversity Conservation Project, IUCN.

Bouche, P. (2010). Inventaire aerien 2010 des grands mammiferes dans le Nord de la Republique Centrafricaine. Programme ECOFAC IV – FINANCEMENT 9th FED. Composante Zones Cynetiques Villageoises Nord.

Bouché, P. (2012). Inventaire aérien de l'écosystème W-Arly-Pendjari (Mai – Juin 2012). Programme CITES MIKE.

Bouché, P., Lungren, C., Hien, B. & Omondi, P. (2003). Aerial total count of the "W"-Arli-Pendjari-Oti-Mandouri-Keran (WAPOK) ecosystem in West Africa. Mike-Paucof Project.

Bouché, P., Lungren, C. & Hien, B. (2004). Recensement aérien total de la faune dans l'Ecosystème naturel Po-Nazinga-Sissili (PONASI), Burkina Faso. Mike-Paucof Project.

Bowers, J. A. (2006). Feeding patch selection of African buffalo (*Syncerus caffer caffer*) in the central region of the Kruger National Park. Master Technology, Tshwane University of Technology.

Boy, A. (1958). Les buffles du pays Gourma. *Rev. Bois et Forêts des Tropiques* **58**: 3–15

Breman, H. & De Wit, C. (1983). Rangeland productivity and exploitation in the Sahel. *Science* **221**: 1341.

Breuer, T. (2008). Mbeli Bai study: annual report. Wildlife Conservation Society, Congo Program.

Brooke, V. (1873). On African buffaloes. *Proceedings of the Zoological Society (London)* **2**: 474–484.

Brooke, V. (1875). Supplementary notes on African buffaloes. *Proceedings of the Zoological Society (London)* **3**: 454–457.

Brown, J. L., Wildt, D. E., Raath, J. R., *et al.* (1991). Impact of season on seminal characteristics and endocrine status of adult free-ranging African buffalo (*Syncerus caffer*). *Journal of Reproduction and Fertility* **92**: 47–57.

Brugière, D. (2012). Identifying priority areas for the conservation of antelopes in the Republic of Guinea, West Africa, using the complementarity approach. *Oryx* **46**: 253–259.

Butynski, T., Schaaf, C. & Hearn, G. (1997). African buffalo *Syncerus caffer* extirpated on Bioko Island, Equatorial Guinea. *Journal of African Zoology* **111**: 57–61.

Carmichael, I. H., Patterson, L., Drager, N. & Breton, D. A. (1977). Studies on reproduction in the African buffalo (*Syncerus caffer*) in Botswana. *South African Journal of Wildlife Research* **7**: 45–52.

Caro, T. (2008). Decline of large mammals in the Katavi-Rukwa ecosystem of western Tanzania. *African Zoology* **43**: 99–116.

Caron, A., Cross, P. C. & Du Toit, J. T. (2003). Ecological implications of bovine tuberculosis in African buffalo herds. *Ecological Applications* **13**: 1338–1345.

Caron, A., Miguel, E., Gomo, C., *et al.* (2013) Relationship between burden of infection in ungulate populations and wildlife/livestock interfaces. *Epidemiology and Infections*, **141**(7): 1522–1535.

Chamberlan, C., Maurois, C. & Maerechal, C. (1995). *Etude mammologique dans le Parc National d'Odzala. Programme ECOFAC*. Congo: Composante.

Chase, M. (2007). Aerial wildlife census of the Caprivi River systems: a survey of rivers, wetlands and floodplains. Elephants Without Borders.

Chase, M. (2011). *Dry season Fixed-wing Aerial Survey of Elephants and Wildlife in Northern Botswana, September–November 2010*. Kasane: Elephant Without Borders.

Christy, C. (1929). The African buffaloes. *Proceedings of the Zoological Society of London* **30**: 445–462.

Cloete, F. (2011).Tendense van lewendewildverkope. *Wild & Jag* **17**(2): 41–43.

Cobb, S. (1976). The distribution and abundance of the large herbivores of Tsavo National Park, Kenya. PhD thesis. Oxford University.

Conybeare, A. (1980). Buffalo numbers, home range and daily movement in the Sengwa Wildlife Research Area, Zimbabwe. *South African Journal of Wildlife Research* **10**: 89–93.

Cornelis, D. (2011).Ecologie du déplacement du buffle de savane ouest-africain (*Syncerus caffer brachyceros*). PhD thesis. Université Montpellier II.

Cornelis, D., Benhamou, S., Janeau, G., Morellet, N., Ouedraogo, M. & Visscher, M.-N. (2011). Spatiotemporal dynamics of forage and water resources shape space use of West African savanna buffaloes. *Journal of Mammalogy* **92**(6): 1287–1297.

Craig, C. G. (2011a). *Aerial Survey of Wildlife in the Niassa Reserve, Mozambique, October 2011*. Mozambique: Sociedade para a Gestao e Desenvolvimento da Reserva do Niassa.

Craig, G. C. (2011b). Aerial survey of chipanje chetu, Mozambique, October 2011; annex 1 of the aerial survey of wildlife in the Niassa game reserve. Prepared for Sociedade para a gestão e desenvolvimento da reserva do niassa Moçambique.

Craig, G. C. (2011c). Aerial survey of Quirimbas National Park and adjacent areas; annex 2 of the aerial survey of wildlife in the Niassa game reserve. Prepared for WWF Mozambique.

Cribiu, E. P. & Popescu, C. P. (1980). Chromosome constitution of a hybrid between East African buffalo (*Syncerus caffer caffer*) and dwarf forest buffalo (*Syncerus caffer nanus*). *Annales de Genetiqueet de Selection Animale* **12**: 291–293.

Cross, P. C., Lloyd-Smith, J. I. & Getz, W. M. (2005). Disentangling association patterns in fission–fusion societies using

African buffalo as an example. *Animal Behaviour* **69**: 499–506.

Cross, P. C., Heisey, D. M., Bowers, J. A., *et al.* (2009). Disease, predation and demography: assessing the impacts of bovine tuberculosis on African buffalo by monitoring at individual and population levels. *Journal of Applied Ecology* **46**: 467–475.

Dalimier, P. (1955). *Les Buffles du Congo Belge*. Bruxelles: Institut des Parcs Nationaux du Congo Belge.

De Boer, W. F. de & Prins, H. H. T. (1990). Large herbivores that strive mightily but eat and drink as friends. *Oecologia* **82**: 264–274.

de Garine-Wichatitksy, M., Miguel, E., Mukamuri, B., *et al.* (2012). Coexisting with wildlife in transfrontier conservation areas in Zimbabwe: cattle owners' awareness of disease risks and perception of the role played by wildlife. *Comparative Immunology, Microbiology and Infectious Diseases* **36**(3): 321–332.

De Vos, V. & Bengis, R. G. (1994). The disease status of African buffalo in South Africa. In *Wildlife Ranching: A Celebration of Diversity*. Eds. Van Hoven, W., Ebedes, H. & Conroy, A. Pretoria: University of Pretoria.

De Vos, V. & Bryden, H. B. (1996). *Anthrax in the Kruger National Park: temporal and spatial patterns of disease occurrence. Salisbury Medical Bulletin*, special supplement **87**: 26–30.

Debonnet, G. & Wilson, E. (2008). Reactive monitoring mission. Selous Game Reserve (United Republic of Tanzania) 23–30 November 2008. UNESCO World Heritage Centre, IUCN.

Deem, S., Maisels, F. & Robbins, R. G. (2005). Necropsy and parasitic findings from an adult forest buffalo (*Syncerus caffer nanus*) found dead in the Republic of Congo. *European Journal of Wildlife Research* **51**: 60–62.

Djeukam, R. (2007). Le cadre legislatif et reglementaire de l'utilisation des produits forestiers non ligneux au Cameroun. Ministère Fédéral d'Allemagne pour l'Alimentation, l'Agriculture et la Protection des Consommateurs.

Douglas Hamilton, I., Froment, J.-M., Doungoube, G. & Root, J. (1985). Recensement aérien de la faune dans la zone Nord de la République Centrafricaine.In: Aménagement de la faune. Unpublished report. République Centrafricaine.

Du Toit, J. G. (2005). *Intensive Buffalo Farming*. http://bigfive.jl.co.za/buffalo%20farming.htm, accessed 28 January 2013.

Dublin, H. T., Sinclair, A. R. E., Boutin, S., Anderson, E., Jago, M. & Arcese, P. (1990). Does competition regulate ungulate populations? Further evidence from Serengeti, Tanzania. *Oecologia* **82**: 238–288.

Dunham, K. M. (1994). The effect of drought on the large mammal populations of Zambezi riverine woodlands. *Journal of Zoology* **234**: 489–526.

Dunham, K. M. (2004). Aerial survey of elephants and other large herbivores in the Zambezi Heartland (Zimbabwe, Mozambique and Zambia), 2003. African Wildlife Foundation.

Dunham, K. M. (2010). *Aerial Survey of Elephants and Other Large Herbivores South of Lake Cabora Bassa, Mozambique: 2010*. Brussels: AGRECO G. E. I. E.

Dunham, K. M., Mackie, C. S., Musemburi, O. C., *et al.* (2006). Aerial survey of elephants and other large herbivores in the Sebungwe Region, Zimbabwe, 2006. WWF-SARPO Occasional Paper 12, Parks & Wild Life Management Authority and the WWF-SARPO.

Dunham, K. M., Mackie, C. S., Musemburi, O. C., *et al.* (2007). *Aerial Survey of Elephants and Other Large Herbivores in North-west Matabeleland, Zimbabwe: 2007*. Harare: WWF-SARPO.

Dunham, K. M., van der Westhuizen, E., van der Westhuizen, H. F. & Gandiwa, E. (2010). Aerial survey of elephants and other large herbivores in Gonarezhou National Park (Zimbabwe), Zinave National Park (Mozambique) and surrounds: 2009. Gonarezhou Conservation Project.

East, R. (1984). Rainfall, soil nutrient status and biomass of large African savanna mammals. *African Journal of Ecology* **22**: 245–270.

East, R. (1998). *African Antelope Database 1999*. Gland and Cambridge: IUCN.

East, R. (1999). Species accounts: family Bovidae, tribe Bovini, African buffalo *Syncerus caffer* (Sparrman 1779). In *African Antelope Database 1998*, pp. 106–115. Eds. East, R. Gland: IUCN.

Edebe, J., Lala, F., Bett, A.& Makau, D. D. K. (2010). *Total Aerial Count of Elephants and Other Large Mammals in Nasolot/ South Turkana/Rimoi/Kamnarok area, Kenya*. Nairobi: Kenya Wildlife Service.

Eltringham, S. K. (1999). *The Hippos: Natural History and Conservation*. London: Academic Press.

Eltringham, S. K. & Woodford, M. H. (1973). The numbers and distribution of buffalo in the Ruwenzori National Park, Uganda. *African Journal of Ecology* **11**: 151–164.

Estes, R. D. (1991). *Behavior Guide to African Mammals: Including Hoofed Mammals, Carnivores, Primates*. Berkeley, CA: University of California Press.

EWCA (2010). Aerial survey report: Gambella reconnaissance 2009 & census 2010. Trans Frontier Conservation Initiative (TFCI) Task Force Aerial Survey Report.

Ezenwa, V. O. & Jolles, A. E. (2008). Horns honestly advertise parasite infection in male and female African buffalo. *Animal Behaviour* **75**: 2013–2021.

Fairall, N. (1968). The reproductive seasons of some mammals in the Kruger National Park. *African Zoology* **3**: 189–210.

Fay, M., Elkan, P., Marjan, M. & Grossman, F. (2007). Aerial surveys of wildlife, livestock, and human activity in and around existing and proposed protected areas of Southern Sudan, dry season 2007. WCS Southern Sudan Technical Report.

Ferreira, S., Gaylard, A., Greaver, C. & Herbst, M. (2010). *Animal Abundances in Parks: 2009/2010*. South Africa: Scientific Services, SANParks.

Ferreira, S., Gaylard, A., Greaver, C., *et al.* (2011). *Animal Abundances in Parks: 2010/2011*. South Africa: Scientific Services, SANParks.

Ferreira, S., Gaylard, A., Greaver, C., Herbst, M., Alard, G. & Moolman, L. (2012). *Animal Abundances in Parks: 2011/2012*. South Africa: Scientific Services, SANParks.

Field, C. R. (1968). A comparative study of the food habits of some wild ungulates in the Queen Elizabeth Park, Uganda: preliminary report. In *Comparative Nutrition of Wild Large Mammals*, pp. 135–51. Ed. Crawford, M. A. London: Academic Press.

Field, C. R. (1972). The food habits of wild ungulates in Uganda by analyses of stomach contents. *East African Wildlife Journal* **10**: 17–42.

FitzGibbon, C. D., Mogaka, H. & Fanshaw, J. H. (1995). Subsistence hunting in Arabuko-Sokoke Forest, Kenya, and its effects on mammal populations. *Conservation Biology* **9**: 1116–1126.

Foerster, S., Wilkie, D. S., Morelli, G. A., *et al.* (2012). Correlates of bushmeat hunting among remote rural households in Gabon, Central Africa. *Conservation Biology* **26**: 355–344.

Frederick, H. (2009). *Aerial Survey Report: Kafue Ecosystem 2008.* Chilanga, Zambia: Zambia Wildlife Authority.

Frederick, H. (2011). *Aerial Survey: Kafue Ecosystem 2011.* Chilanga, Zambia: Zambia Wildlife Authority.

Fritz, H. (1997). Low ungulate biomass in west African savannas: primary production or missing megaherbivores or large predator species? *Ecography* **20**: 417–421.

Funston, P. J., Skinner, J. D. & Dott, H. M. (1994). Seasonal variation in movement patterns, home range and habitat selection of buffaloes in a semi-arid habitat. *African Journal of Ecology* **32**: 100–114.

Funston, P. J., Mills, M. G. L., Richardson, P. R. K. & van Jaarsveld, A. S. (2003). Reduced dispersal and opportunistic territory acquisition in male lions (*Pantheraleo*). *Journal of Zoology* **259**: 131–142.

Galat, G., Benoit, M., Chevillotte, H., *et al.* (1992). *Dénombrement de la grande faune du Parc National du Niokolo-Koba, Sénégal, 1990–1991.* Dakar, Sénégal: Orstom & Institut Français de Recherche Scientifique pour le Développement en Coopération (ORSTOM).

Gautier, A. (1988). The final demise of *Bos ibericus*? *Sahara* **1**: 37–48.

Geßner, J. (2008). Survey of mammals at Ikwabai-Nki National Park, south-east Cameroon. Diploma thesis. University of Oldenburg.

Gentry, A. W. (1978). Bovidae. In *Evolution of African Mammals*, pp. 540–572. Eds. Maglio,V. J. & Cook, H. B. S. Cambridge, MA: Harvard University Press.

Gentry, A. W. (1990). Evolution and dispersal of African Bovidae. In *Horns, Pronghorns, and Antlers: Evolution, Morphology, Physiology, and Social Significance*, pp. 195–227. Eds. Bubenik, G. A. & Bubenik, A. B. New York: Springer.

Gillet, H. (1969). La végétation du parc national de Zakouma (Tchad) et ses rapports avec les grands mammifères. *La Terre et la Vie* **4**: 373–485.

Graaff, G. de, Schultz, K. C. A., & Walt, P. T. van der (1973). Notes on rumen contents of Cape buffalo, *Syncerus caffer* in the Addo Elephant National Park. *Koedoe* **16**: 45–58.

Graham, A. D., Enawgaw, C. & Netserab, B. (1996). Trends in large animals of Omo & Mago National Parks. Technical Report No.2, Southern Nations Nationalities and Peoples, Regional Government, Bureau of Agriculture, Awasa.

Grange, S. & Duncan, P. (2006). Bottom up and top down processes in African ungulate communities: resources and predation acting on the relative abundance of zebra and grazing bovids. *Ecography* **29**: 899–907.

Grimsdell, J. J. R. (1969). Ecology of the buffalo, *Syncerus caffer*, in Western Uganda. PhD thesis. Cambridge University.

Grimsdell, J. J. R. (1973a). Age determination of the African buffalo, *Syncerus caffer* Sparrman. *East African Wildlife Journal* **11**: 31–53.

Grimsdell, J. J. R. (1973b). Reproduction in the African buffalo, *Syncerus caffer*, in Western Uganda. *Journal of Reproduction and Fertility* **19**, supplement: 301–316.

Grimsdell, J. J. R., & Field, C. R. (1976). Grazing patterns of buffaloes in the Rwenzori National Park, Uganda. *East African Wildlife Journal* **14**: 339–344.

Grimshaw, J. M., Cordeiro, N. J. & Foley, C. A. H. (1995). The mammals of Kilimanjaro. *Journal of East African Natural History* **84**: 105–139.

Groves, C. & Grubb, P. (2011). *Ungulate Taxonomy*. Baltimore, MD: Johns Hopkins University Press.

Grubb, P. (1972). Variation and incipient speciation in the African buffalo. *Zeitschriftfür Saugetierkunde* **37**: 121–144.

Halley, D. J., Vandewalle, M. E. J., Mari, M. & Taolo, C. (2002). Herd-switching and long-distance dispersal in female Africa buffalo *Syncerus caffer*. *African Journal of Ecology* **40**: 97–99.

Haltenorth, T. (1963). Klassifikation der Saugetiere: Artiodactyla 1 (18). *Handbuch der Zoologie* **8**(32): 1–167.

Happold, D. C. D. (1973). *Large Mammals of West Africa.* London: Longman.

Hashim, I. M. (1987). Relationship between biomass of forage used and masses of faecal pellets of wild animals in meadows of the Dinder National Park. *African Journal of Ecology* **25**: 217–223.

Hearne, J. & McKenzie, M. (2000). Compelling reasons for game ranching in Maputaland. In *Wildlife Conservation by Sustainable Use*, pp. 417–438. Eds. Prins, H. H. T., Grootenhuis, J. G. & Dolan, T. T. Boston, MA: Kluwer Academic Press.

Henschel, P., Abernethy, K. A. & White, L. J. T. (2005). Leopard food habits in the Lopé National Park, Gabon, Central Africa. *African Journal of Ecology* **43**: 21–28.

Hibert, F. (2007) Déterminants écologiques et anthropiques de la structuration des peuplements d'ongulés sauvages en Afrique de l'Ouest: cas du Parc Régional du W du fleuve Niger. PhD thesis. Ecole Nationale Supérieure d'Agronomie.

Hillson, S. (1986) *Teeth.* Cambridge: Cambridge University Press.

Hopcraft, J. G. C., Olff, H. & Sinclair, A. (2010). Herbivores, resources and risks: alternating regulation along primary environmental gradients in savannas. *Trends in Ecology & Evolution* **25**: 119–128.

Hoppe-Dominik, B. (1992). Habitatnutzung von Waldbuffel, *Syncerus caffer nanus*, Bongo, *Tragelaphuseurycerus*, und Waldelefant, *Loxodontaafricanacyclotis*, auf Forst-FahrwegenimRegenwaldgebiet der Elfenbeinkuste. *SaugetierkundlicheMitteinlungen* **34**: 31–41.

Hunter, C. G. (1996). Land uses on the Botswana/Zimbabwe border and their effects on buffalo. *South African Journal of Wildlife Research* **26**: 136–150.

Hunter, M. D. & Price, P. W. (1992). Playing chutes and ladders: heterogeneity and the relative roles of bottom-up and top-down forces in natural communities. *Ecology* **73**: 723–732.

IUCN (2013). *IUCN Red List of Threatened Species.* Version 2013.2. www.iucnredlist.org, accessed 28 May 2014.

IUCN SSC(2008). *Syncerus caffer*. In *IUCN 2013 IUCN Red List of Threatened Species.*

Version 2013.2. www.iucnredlist.org, accessed 28 May 2014.

Jachmann, H. (2002). Comparison of aerial counts with ground counts for large African herbivores. *Journal of Applied Ecology* **39**: 841–852.

Jallow, A., Touray, O. & Jallow, M. (2004). *An update of the status of antelopes in the Gambia*. In *Antelope Survey Update*, pp. 18–20. Eds. Chardonnet, B. & Chardonnet, P. Paris: IUCN/SSC Antelope Specialist Group Report.

Jarman, P. J. (1971). Diets of large mammals in the woodlands around Lake Kariba, Rhodesia. *Oecologia* **8**: 157–178.

Jolles, A. E. (2004). Disease ecology of tuberculosis in African Buffalo. Thesis. Princeton University.

Jolles, A. E. (2007).Population biology of African buffalo (*Syncerus caffer*) at Hluhluwe-iMfolozi Park, South Africa. *African Journal of Ecology* **45**: 398–406.

Jolles, A. E., Cooper, D. V. & Levin, S. A. (2005). Hidden effects of chronic tuberculosis in African buffalo. *Ecology* **86**: 2358–2364.

Jolles, A. E., Ezenwa, V. O., Etienne, R. S., Turner, W. C. & Olff, H. (2008). Interactions between macroparasites and microparasites drive infection patterns in free-ranging African buffalo. *Ecology* **89**: 2239–2250.

Jones, M. L. (1993). Longevity of ungulates in captivity. *International Zoo Yearbook* **32**: 159–169.

Kamdem-Toham, A., Adeleke, A. W, Burgess, N. D., *et al.* (2003). Forest conservation in the Congo Basin. *Science* **299**: 346.

Kiambi, S., Kuloba, B., Kenana, L., Muteti, D. & Mwenda, E. (2010). Wet season aerial count of large herbivores in Masai Mara National Reserve and the adjacent community areas. Kenya Wildlife Service Mara Research Station.

Kingdon, J. (1982). *East African Mammals: An Atlas of Evolution in Africa*, Vol. **IIIC**. London: Academic Press.

Kingdon, J. (1997). *The Kingdon Field Guide of African Mammals*. London: Academic Press.

Klein, R. G. (1988). The archaeological significance of animal bones from Acheulean sites in southern Africa. *African Archaeological Review* **6**: 3–25.

Klein, R. G. (1994). The long-horned African buffalo (*Pelorovisantiquus*) is an extinct species. *Journal of Archaeological Science* **21**: 725–733.

Knechtel, V. C. (1993). Brunstverhaltenbei Kaffernbüffeln (*Syncerus caffer caffer*) imTierpark Berlin-Friedrichsfelde. *Der Zoologische Garten* **63**: 32–58.

Kone, A., Richard, D. & Guerin, H. (1990). Teneurs en constituants pariétaux et en matières azotées des ligneux fourragers d'Afrique occidentale. In INRA, pp. 947–948. Ed. Seizièmecongrès international des herbages. Versailles: INRA.

Korte, L. (2008a). Habitat selection at two spatial scales and diurnal activity patterns of adult female forest buffalo. *Journal of Mammalogy* **89**: 115–125.

Korte, L. (2008b). Variation of group size among African buffalo herds in a forest-savanna mosaic landscape. *Journal of Zoology* **275**: 229–236.

Korte, L. (2008c). Calving and inter-birth intervals of forest buffalo at Lopé National Park, Gabon. *African Journal of Ecology* **46**: 676–678.

Korte, L. (2009). Herd-switching in adult female African forest buffalo *Syncerus caffer nanus*. *African Journal of Ecology* **47**: 125–127.

Kruuk, H. (1972). *The Spotted Hyena*. Chicago, IL: University of Chicago Press.

Laubscher, L. & Hoffman, L. (2012). An overview of disease-free buffalo breeding projects with reference to the different systems used in South Africa. *Sustainability* **4**: 3124–3140.

Le Houérou, H. N. (2008). *Bioclimatology and Biogeography of Africa*. Berlin: Springer.

Leakey, M. (1983). *Africa's Vanishing Art: The Rock Paintings of Tanzania*. Garden City, NY: Doubleday & Co.

Leuthold, W. (1972). Home range, movements and food of a buffalo herd in Tsavo National Park. *East Africa Wildlife Journal* **10**: 237–243.

Lustenhouwer, I. (2008). Diet composition and parasite occurrence in the forest buffalo (*Syncerus caffer nanus*) in Lopé National Park, Gabon. Thesis. Resource Ecology Group, Wageningen University.

Macandza, V. A., Owen-Smith, N. & Cross, P. C. (2004). Forage selection by African buffalo in the late dry season in two landscapes. *South African Journal of Wildlife Research* **34**: 113–121.

Mackie, C. (2004). Mole aerial survey. Wildlife Division/IUCN, provisional report.

Mahaney, W. C. (1987). Behaviour of the African buffalo on Mount Kenya. *African Journal of Ecology* **25**: 199–202.

Maisels, F., Inkamba-Nkulu, C., Boudjan, P. & Blake, S. (2002). *Research and Monitoring Report: Nouabale´-Ndoki Project, Nouabale´-Ndoki National Park*. Congo: WCS.

Martin, R. (2002). Species report for southern savanna buffalo. Study conducted in support of The Transboundary Mammal Project of the Ministry of Environment and Tourism, Namibia, facilitated by the Namibia Nature Foundation and the World Wildlife Fund LIFE Programme.

Mauvais, G. & Ndiaye, A. (2004). Aerial census of wildlife in Niokolo Koba National Park, Senegal. In *Antelope Survey Update, Number 9: November 2004*, pp 14–17. Eds. Chardonnet, B. & Chardonnet, P. Paris: Fondation Internationale pour la Sauvegarde de la Faune.

Mduma, S., Hilborn, R. & Sinclair, A. R. E. (1998). Limits to exploitation of Serengeti wildebeest and implications for its management. In *Dynamics of Tropical Communities*, Symp. Vol. **37**, pp. 243–265. Eds. Newbery, D. M., Prins, H. H. T. & Brown, N. D. Oxford: Blackwell Scientific.

Megaze, A., Belay, G., & Balakrishnan, M. (2012). Population structure and ecology of the African buffalo (*Syncerus caffer* Sparrman, 1779) in Chebera Churchura National Park, Ethiopia. *African Journal of Ecology*. doi: 10.1111/aje.12049.

Melletti, M. (2005). Habitat preferences and characteristics of resting places of forest buffalo in Central Africa: conservation implications.MSc thesis. University of Rome, 'La Sapienza'.

Melletti, M. (2008). Habitat use and behaviour of forest buffalo (*Syncerus caffer nanus*) in Dzanga-Ndoki National Park, Central African Republic. PhD dissertation. University of Rome, 'La Sapienza'.

Melletti, M. (2011). Ecology and social organization of forest buffalo (*Syncerus caffer nanus*). Buffalo Meeting, FAO, Rome, 7–8 June.

Melletti, M., Penteriani, V. & Boitani, L. (2007a). Habitat preferences of the secretive forest buffalo (*Syncerus caffer* nanus) in Central Africa. *Journal of Zoology* **271**: 178–186.

Melletti, M., Penteriani, V., Mirabile, M. & Boitani, L. (2007b). Some behavioral aspects of forest buffalo (*Syncerus caffer nanus*): from herd to individual. *Journal of Mammalogy* **88**: 1312–1323.

Melletti, M., Penteriani, V., Mirabile, M. & Boitani, L. (2008). Effects of habitat and season on the grouping of forest buffalo resting places. *African Journal of Ecology* **47**: 121–124.

Melletti, M., Mirabile, M., Penteriani, V. & Boitani, L. (2009). Habitat use by forest mammals in Dzanga-Ndoki National Park, Central Africa Republic. *African Journal of Ecology* **47**: 797–800.

Melletti, M., Delgado, M. M., Penteriani, V., Mirabile, M. & Boitani, L. (2010). Spatial properties of a forest buffalo herd, and individual positioning as a response to environmental cues and social behaviour. *Journal of Ethology* **28**: 421–428.

Mesochina, P., Langa, F. & Chardonnet, P. (2008). Preliminary survey of large herbivores in Gilè National Reserve, Zambezia Province, Mozambique. IGF foundation, Direccao Provincial Do Turismo, Zambezia, Mozambique.

Michel, A. L. & Bengis, R. G. (2012). The African buffalo: a villain for inter-species spread of infectious diseases in southern Africa. *Onderstepoort Journal of Veterinary Research* **79**: doi: 10.4102/ojvr.v79i2.453.

Miguel, E., Grosbois, V., Caron, A., *et al.* (2013). Contacts and foot and mouth disease transmission from wild to domestic bovines in Africa. *Ecosphere* **4** (3): doi: dx.doi.org/10.1890/ES1-00239.1

Mills, M., Biggs, H. & Whyte, I. (1995). The relationship between rainfall, lion predation and population trends in African herbivores. *Wildlife Research* **22**: 75–87.

Mloszewski, M. J. (1983). *The Behavior and Ecology of the African Buffalo.* Cambridge: Cambridge University Press.

Molloy, L. M. (1997). Forest buffalo, *Syncerus caffer nanus*, and burning of savannas at Lope´ Reserve, Gabon. MSc thesis. University of Florida.

Monfort, P. N. (1979). Etudes des populations de buffles *Syncerus caffer* (Sparrman) du Parc National de l'Akagera (Rwanda). *ZeitschriftfürSäugetierkunde* **44**: 111–127.

Monfort, P. N. (1980). Etudes des populations de buffles *Syncerus caffer* (Sparrman) du Parc National de l'Akagera (Rwanda). *ZeitschriftfürSäugetierkunde* **45**: 173–188.

Morens, D. M., Holmes, E. C., Davis, A. S. & Taubenberger, J. K.(2011). Global rinderpest eradication: lessons learned and why humans should celebrate too. *Journal of Infectious Diseases* **204**: 502–505.

Morgan, B. J. (2007). Group size, density and biomass of large mammals in the Réserve de Faune du Petit Loango, Gabon. *African Journal of Ecology* **45**: 508–518.

Moritz, C. (1994). Defining evolutionarily significant units for conservation. *Trends in Ecology and Evolution* **9**: 373–374.

Mugangu, T. E., Hunter, M. L. & Gilbert, J. R. (1995). Food, water, and predation: a study of habitat selection by buffalo in Virunga National Park, Zaïre. *Mammalia* **59**: 349–362.

Mwangi, P., Ngene, S. & Esau, K. (2007). *Wet Season Aerial Count of Large Mammals in the Meru Conservation Area (MCA).* Nairobi: Kenya Wildlife Service.

N'Goran, K. P., Maho, N. R., Kouakou, Y. C., Dowd, D. & Herbinger, I. (2010). Etat des ressources naturelles du Parc National de la Comoé et de sa zone périphérique. Rapport de l'inventaire faunique par survol aérien; mars 2010. Wild Chimpanzee Foundation, GTZ & OIPR, Abidjan, Côte d'Ivoire.

Naidoo, R., Du Preez, P., Stuart-Hill, G., Jago, M. & Wegmann, M. (2012). Home on the range: factors explaining partial migration of African buffalo in a tropical environment. *PLoS ONE* 7(5): e36527.doi: 10.1371/journal.pone.0036527.

Ngene, S., Ihwagi, F., Nzisa, M., Mukeka, J., Njumbi, S. & Omondi, P. (2011). *Total Aerial Census of Elephants and Other Large Mammals in the Tsavo-Mkomazi Ecosystem.* Nairobi: Kenya Wildlife Service.

Omondi, P., Bitok, E., Kahindi, O. & Mayienda, R. (2002). Total aerial count of elephants in Samburu-Laikipia EcoSystem: June 2002. Save The Elephants (STE), National Environment Management Authority (NEMA) & Laikipia Wildlife Forum.

Omondi, P., Bitok, E. K., Tchamba, M. R. M. & Lambert, B. B. (2008). *The Total Aerial Count of Elephants and Other Wildlife Species in Faro, Benoue and Bouba Ndjidda National Parks and Adjacent Hunting Blocks in Northern Cameroon.* Yaounde, Cameroon: WWF and Cameroon Ministry of Forestry and Wildlife.

Owen-Smith, N., Fryxell, J. M. & Merrill, E. H. (2010). Foraging theory upscaled: the behavioural ecology of herbivore movement. *Philosophical Transactions of the Royal Society B: Biological Sciences* **365**: 2267–2278.

Penzhorn, B. L. (2000). Coccidian oocyst and nematode egg counts of free-ranging African buffalo (*Syncerus caffer*) in the Kruger National Park, South Africa. *Journal of the South African Veterinary Association – Tydskrif Van Die Suid-AfrikaanseVeterinereVereniging* **71**: 106–108.

Perrin, M. R. & Brereton-Stiles, R. (1999). Habitat use and feeding behaviour of the buffalo and the white rhinoceros in the Hluhluwe-Umfolozi Game Reserve. *South African Journal of Wildlife Research* **29**: 72–80.

Pienaar, U. de V. (1969). Observations on the developmental biology, growth and some aspects of the population ecology of African buffalo in the Kruger National Park. *Koedoe* **12**: 29–52.

Plumptre, A. J. (1995). The chemical composition of montane plants and its influence on the diet of the large mammalian herbivores in the Parc National des Volcans. *Rwanda. Journal of Zoology* **235**: 323–337.

Plumptre, A., Kujirakwinja, D., Moyer, D., Driciru, M. & Rwetsiba, A. (2010). Greater Virunga landscape large mammal surveys, 2010. US Fish and Wildlife Service WCS, UWA,CITES/MIKE.

Potgieter, D., Evans, B., Dogringar, S., Kassar, M. & Labuschagne, J. (2012). *Dry Season Aerial Total Count, Zakouma National Park, Chad, 10–15 April 2012.* New York: Wildlife Conservation Society.

Prins, H. H. T. (1989a). Condition changes and choice of social environment in African buffalo bulls. *Behaviour* **108**: 297–324.

Prins, H. H. T. (1989b). Buffalo herd structure and its repercussions for condition of individual African buffalo cows. *Ethology* **81**: 47–71.

Prins, H. H. T. (1996). *Ecology and Behaviour of the African Buffalo: Social Inequality and Decision Making.* London: Chapman & Hall.

Prins, H. H. T. & Beekman, J. H. (1989). A balanced diet as a goal of grazing: the food of the Manyara buffalo. *Journal of African Ecology* **27**: 241–259.

Prins, H. H. T. & Douglas-Hamilton, I. (1990). Stability in a multi-species assemblage of large herbivores in East Africa. *Oecologia* **83**: 392–400.

Prins, H. H. T. & Iason, G. (1989). Dangerous lions and nonchalant buffalo. *Behaviour* **108**: 262–296.

Prins, H. H. T. & Reitsma, J. M. (1989). Mammalian biomass in an African equatorial rain forest. *Journal of Animal Ecology* **58**: 851–861.

Prins, H. H. T. & Sinclair, A. R. E. (2013). *Syncerus caffer*. In *The Mammals of Africa*. Eds. Kingdon, J. S. & Hoffmann, M. Amsterdam: Academic Press.

Prins, H. H. T. & van der Jeugd, H. P. (1993). Herbivore population crashes and woodland structure in East Africa. *Journal of Ecology* **81**: 305–314.

Prins, H. H. T. & Weyerhaeuser, F. J. (1987). Epidemics in populations of wild ruminants: anthrax and impala, rinderpest and buffalo in Lake Manyara National Park, Tanzania. *Oikos* **49**: 28–38.

Redfern, J. & Viljoen, P. (2002). Biases in estimating population size from an aerial census: a case study in the Kruger National Park, South Africa. *South African Journal of Science* **98**: 455–461.

Redfern, J., Grant, V. R., Biggs, H. & Getz, W. M. (2003). Surface-water constraints on herbivore foraging in the Kruger National Park, South Africa. *Ecology* **84**: 2092–2107.

Redfern, J. V., Ryan, S. J. & Getz, W. M. (2006). Defining herbivore assemblages in the Kruger National Park: a correlative coherence approach. *Oecologia* **146**: 632–640.

Reinhardt, V. (1983). Movement orders and leadership in a semi-wild cattle herd. *Behaviour* **83**: 251–264.

Reitsma, J. M. (1988). Forest vegetation of Gabon. Tropenbos Foundation, Ede, the Netherlands.

Renaud, P. C. (2005). Recensement aérien de la faune dans les préfectures de la région Nord de la République Centrafricaine. ECOFAC III. Commission Européenne.

Renaud, P. C. (2007). Omo National Park report for the wet season aerial survey. African Parks Conservation.

Renaud, P. C., Gueye, M., Hejcmanová, P., Antonínová, M. & Samb, M. (2006). Inventaire aérien et terrestre de la faune et relevé des pressions au Parc National du Niokolo Koba. Plan d'Urgence, Rapport Annexe A. Ministère de l'Environnement et de la Protection de la Nature Protection-African Parks Foundation Dakar.

Ruggiero, R. G. (1991). Prey selection of the lion (*Panthera leo* L.) in the Manovo-Gounda-St. Floris National Park, Central African Republic. *Mammalia* **55**: 23–33.

Rwetsiba, A. & Nuwamanya, E. (2010). Aerial surveys of Murchison Falls Protected Area, Uganda, March 2010. *Pachyderm* **47**: 118–123.

Ryan, S. J. (2006). *Spatial ecology of African buffalo and their resources in a savanna ecosystem*. PhD dissertation. Department of Environmental Science, Policy and Management, UC Berkeley.

Ryan, S. J. & Jordaan, W. (2005). Activity patterns of African buffalo *Syncerus caffer* in the Lower Sabie Region, Kruger National Park, South Africa. *Koedoe* **48** (2): 117–124.

Ryan, S. J., Knechtel, C. U. & Getz, W. M. (2006). Range and habitat selection of African buffalo in South Africa. *Journal of Wildlife Management* **70**(3): 764–776.

Ryan, S. J., Knechtel, C. U. & Getz, W. M. (2007). Ecological cues, gestation length, and birth timing in African buffalo (*Syncerus caffer*). *Behavioral Ecology* **18**: 635–644.

Ryan, S. J., Cross, P. C., Winnie, J. Jr., Hay, C., Bowers, J. & Getz, W. M. (2012). The utility of NDVI for predicting African buffalo forage quality. *Journal of Wildlife Management* **76**(7): 1499–1508.

Schaller, G. B. (1972). *The Serengeti Lion*. Chicago, IL: University of Chicago Press.

Schouteden, H. (1945). De zoogdieren van Belgisch Congo en van Ruanda-Urundi. *Annales du Musée du Congo belge* **3**(2): 169–332.

Sinclair, A. R. E. (1977). *The African buffalo: study in resource limitations of populations*. Chicago, IL: University of Chicago Press.

Sinclair, A., Mduma, S. & Brashares, J. S. (2003). Patterns of predation in a diverse predator–prey system. *Nature* **425**: 288–290.

Sinsin, B., Sogbohossou, E. A., Nobime, G. & Adi, M. (2008). Denombrement aerien de la faune dans la R reserve de biosphere de la Pendjari. GTZ, Pendjari project.

Skinner, J. D. & Chimimba, C. T. (2005). *The Mammals of the Southern African Subregion*. 3rd edition. Cambridge: Cambridge University Press.

Skinner, J. D., Dolt, H. M., Matthee, A. & Hunt, L. (2006). Captive breeding of the white rhinoceros, *Ceratotherium simum*, and the Cape buffalo, *Syncerus caffer*. *Onderstepoort Journal of Veterinary Research* **73**: 237–239.

Smitz, N., Berthouly, C., Cornelis, D., *et al.* (2013). Pan-African genetic structure in the African buffalo (*Syncerus caffer*): investigating intraspecific divergence. *PLoS ONE* **8**(2):e56235. 10.1371/journal.pone.0056235

Spinage, C. A. & Brown, W. A. B. (1988). Age determination of the West African buffalo *Syncerus caffer brachyceros* and the constancy of tooth wear. *African Journal of Ecology* **26**: 221–227.

Stalmans, M. & Peel, M. (2009). Parque Nacional de Banhine, Moçambique, Wildlife survey November 2009. ACTF-MITUR.

Stark, M. A. (1986). Daily movement, grazing activity and diet of savanna buffalo, *Syncerus caffer brachyceros*, in Benoue National Park, Cameroon. *African Journal of Ecology* **24**: 255–262.

Stoner, C., Caro, T., Mduma, S., *et al.* (2006). Changes in large herbivore populations across large areas of Tanzania. *African Journal of Ecology* **45**: 202–215.

Swanepoel, W. T. (2007). An aerial census of the Shingwedzi catchment and Limpopo – Elefantes confluence area of the Parque Nacional do Limpopo in Mozambique 30th October–08th November 2007. South African National Parks.

Tawiri (2006). Aerial census in the Selous-Niassa wildlife corridor, dry season, 2006. Tanzania Wildlife Research Institute Aerial Survey Report, Arusha, Tanzania.

Tawiri (2009). Aerial census in the Selous-Niassa wildlife corridor, dry season, 2009. Tanzania Wildlife Research Institute Aerial Survey Report, Arusha, Tanzania.

Tawiri (2010). Aerial census in the Serengeti ecosystem, wet season 2010. Tanzania Wildlife Research Institute Aerial Survey Report, Arusha, Tanzania.

Taylor, C. R. (1970). Dehydration and heat: effects of temperature regulation of East

African ungulates. *American Journal of Physiology* **219**: 1136–1139.

Taylor, R. D. (1985). The response of buffalo, *Syncerus caffer* (Sparrman), to the Kariba lakeshore grassland (*Panicum repens* L.) in Matusadona National Park. PhD thesis. University of Zimbabwe.

Taylor, R. D. (1988). Age determination of African buffalo, *Syncerus caffer* (Sparrman) in Zimbabwe. *African Journal of Ecology* **26**: 207–220.

Taylor, R. D. (1989). Buffalo and their food resources: the exploitation of Kariba lakeshore pastures. *Symposia of the Zoological Society of London* **61**: 51–71.

Taylor, R. D. & Martin, R. B. (1987). Effects of veterinary fences on wildlife conservation in Zimbabwe. *Environmental Management* **11**: 327–334.

Tchouto, G. P. (2004). Plant diversity in a Central African rain forest: implications for biodiversity conservation in Cameroon. PhD thesis. Wageningen University.

Thibault, M. & Blaney, S. (2003). The oil industry as an underlying factor in the bushmeat crisis in Central Africa. *Conservation Biology* **17**: 1807–1813.

Turkalo, A. & Fay, J. M. (2001). Forest elephant behaviour and ecology: observations from the Dzanga saline. In *Conservation and Ecology of African Rain Forest*, pp. 207–213. Eds. Weber, W., White, L., Vedder, A. & Naughton, L. New Haven, CT: Yale University Press.

Turner, W. C., Jolles, A. E. & Owen-Smith, N. (2005). Alternating sexual segregation during the mating season by male African buffalo (*Syncerus caffer*). *Journal of Zoology (London)* **267**: 291–299.

Tutin, C. E. G., White, L. J. T. & Mackanga-Missandzouu, A. (1997). The use by rain forest mammals of natural forest fragments in an Equatorial African savanna. *Conservation Biology* **11**: 1190–1203.

UNEP-WCMC. (2012). *UNEP-WCMC Species Database: CITES-Listed Species*, 26 June. www.unep-wcmc-apps.org/isdb/CITES/Taxonomy/index.cfm/isdb/CITES/Taxonomy/index.cfm?, accessed 10 April 2014.

Valeix, M., Loveridge, A., Chamaille-Jammes, S., *et al.* (2009a). Behavioral adjustments of African herbivores to predation risk by lions: spatiotemporal variations influence habitat use. *Ecology* **90**: 23–30.

Valeix, M., Fritz, H., Loveridge, A. J., *et al.* (2009b). Does the risk of encountering lions influence African herbivore behaviour at waterholes? *Behavioral Ecology and Sociobiology* **63**: 1483–1494.

Van der Hoek, Y., Lustenhouwer, I., Jeffery, K. J. & van Hooft, P. (2012). Potential effects of prescribed savannah burning on the diet selection of forest buffalo (*Syncerus caffer nanus*) in Lopé National Park, Gabon. *African Journal of Ecology* **51**: 94–101.

Van Der Hoeven, C. A., de Boer, W. F. & Prins, H. H. T. (2004). Pooling local expert opinions for estimating mammal densities in tropical rainforests. *Journal for Nature Conservation* **12**: 193–204.

Van Hooft, W. F., Groen, A. F. & Prins, H. H. T. (2002). Phylogeography of the African buffalo based on mitochondrial and Y-chromosomal loci: pleistocene origin and population expansion of the Cape buffalo subspecies. *Molecular Ecology* **11**: 267–279.

Van Hooft, W. F., Groen, A. F. & Prins, H. H. T. (2003). Genetic structure of African buffalo herds based on variation at the mitochondrial D-loop and autosomal microsatellite loci: evidence for male-biased gene flow. *Conservation Genetics* **4**: 467–477.

Van Hooft, P., Greyling, B. J., Prins, H. H. T., *et al.* (2007). Selection at the Y chromosome of the African buffalo driven by rainfall. *PLoS ONE* **2**(10): e1086.

Van Hooft, P., Prins, H. H. T., Getz, W. M., *et al.* (2010). Rainfall-driven sex-ratio genes in African buffalo suggested by correlations between Y-chromosomal haplotype frequencies and foetal sex ratio. *BMC Evolutionary Biology* **10**: 106.

Vanleeuwe, H., Henschel, P., Pélissier, C. & Gotanègre, A. (2008). Aerial and ground survey in the Upemba-Kundelungu National Park. USFWS-AfECF, DRC.

Vesey-Fitzgerald, D. F. (1960). Grazing succession among East African game animals. *Journal of Mammalogy* **41**: 161–172.

Vesey-Fitzgerald, D. F. (1969). Utilization of habitat by buffalo in Lake Manyara National Park. *East African Wildlife Journal* **7**: 131–145.

Vesey-Fitzgerald, D. F. (1974). Utilization of the grazing resources by buffaloes in the Arusha National Park. *East African Wildlife Journal* **7**: 131–145.

Vidler, B. O., Harthoorn, A. M., Brocklesby, D. W. & Robertshaw, D. (1963). The gestation and parturition of the African buffalo. *East Africa Wildlife Journal* **2**: 122–123.

Viljoen, P., Gruner, J. & Gruner, S. (2010). Akagera National Park aerial survey 21–22 August 2010. Akagera Management Company Ltd, Akagera National Park.

Visscher, D. R., van Aarde, R. J. & Whyte, I. J. (2004). Environmental and maternal correlates of foetal sex ratios in the African buffalo (*Syncerus caffer*) and the savanna elephant (*Loxodonta africana*). *Journal of Zoology* **264**: 111–116.

WCS Flight Programme (2008). *Aerial Surveys of Kidepo National Park, Lipan Controlled Hunting Area & Madi Corridor, March–April 2008.* Uganda: Wildlife Conservation Society Flight Programme.

WCS Flight Programme (2009). *Aerial Survey Report: Luangwa Valley 2009. Wildlife Conservation Society.* New York: Wildlife Conservation Society.

WCS Flight Programme (2010). *Aerial Total Count Amboseli-West Kilimanjaro and Magadi-Natron Cross Border Landscape Wet Season, March 2010.* Nairobi: Kenya Wildlife Service and Tanzania Wildlife Research Institute.

White, L. J. T. (1992). Vegetation history and logging disturbance: effects on rain forest mammals in the Lope Reserve, Gabon (with special emphasis on elephants and apes). PhD thesis. University of Edinburgh.

Whyte, I. J. (1996). The management of large buffalo populations, pp. 21–36. In *Proceeding of a Symposium on the African Buffalo as a Game Ranch Animal.* Republic of South Africa: South African National Parks.

Wilkie, D. S., Bennett, E. L., Peres, C. A., & Cunningham, A. A. (2011). The empty forest revisited. *Annals of the New York Academy of Sciences* **1223**: 120–128.

Wilson, D. E. & Mittermeier, R. A. (2011). *Handbook of the Mammals of the World.* Vol. 2. Hoofed Mammals. Barcelona: Lynx Edicions.

Wilson, D. E. & Reeder, D. M. (2005). *Mammal Species of the World.* Baltimore, MD: Johns Hopkins University Press.

Wilson, V. J. (1975). *Mammals of the Wankie National Park, Rhodesia.*

Salisbury: Trustees of the National Museums and Monuments of Rhodesia.

Wilson, V. J. (1993). *A Zoological Survey of Mole National Park North Western Ghana: Part 1. Large Mammals*. Accra, Ghana: Forest Resource Management Programme Game and Wildlife Dept.

Winnie, J. A., Cross, P. & Getz, W. (2008). Habitat quality and heterogeneity influence distribution and behavior in African buffalo (*Syncerus caffer*). *Ecology* **89**: 1457–1468.

Winterbach, H. E. K. (1998). Research review: the status and distribution of Cape buffalo *Syncerus caffer caffer* in southern Africa. *South African Journal of Wildlife Research* **28**(3): 82–88.

Winterbach, H. E. K. & Bothma, J. D. U. P. (1998). Activity patterns of the Cape buffalo *Syncerus caffer caffer* in the Willem Pretorius Game Reserve, Free State. *South African Journal of Wildlife Research* **28**(3): 73–81.

Witkowski, E. T. F. (1983). Ecology of the Klaserie Private Nature Reserve. PhD thesis. University of Witwatersrand

Chapter 21

What is a wild bison? A case study of plains bison conservation in Canada

C. Cormack Gates

Introduction

Globally, the conservation and restoration of viable populations of wild cattle species faces numerous common challenges. Among them are over-exploitation, habitat loss, population decline, hybridization and domestication. Domestication is prevalent among wild cattle species. As a state, it is the antithesis of wildness. As a process, it is driven by several factors, including selective breeding in captivity to propagate traits desirable for the owner and the market (Diamond 2002), and more subtly under captivity through non-intentional mechanisms (Theodorou & Couvet 2004; O'Regan & Kitchener 2005). Human exploitation of free-ranging animals can also induce changes in heritable traits (Darimont *et al.* 2009). Once lost from the wild or substantially altered by these forces, re-creating a wild species from domestic stock through selective breeding in captivity may involve heroic efforts to produce animals with external morphology similar to an original wild-type, but cannot restore traits previously shaped by natural selection. An example is Hellebrun cattle bred by the Heck brothers in Germany during the 1920s and 1930s in an attempt to 're-create' the extinct aurochs (*Bos primigenus*) by selectively crossing domestic cattle breeds (Heck 1934; cited by van Vuure 2002; see also Chapter 15).

Historically, the American plains bison (*Bison bison bison*) was perhaps the most abundant wild cattle species in the world, numbering in excess of 30 million (Shaw 1995). Numerous authors have chronicled its near extinction following European colonization and market hunting in the second half of the nineteenth century (Roe 1970; Isenberg 2000). Despite conservation efforts, today there are fewer than 20 000 plains bison in herds managed primarily for conservation, and >96% of plains bison (~400 000) exist under private captive ownership managed as livestock, primarily for commercial purposes (Gates *et al.* 2010; see also Chapter 9). Even including commercial herds, today American bison occupy less than 1% of their historical range (Sanderson *et al.* 2008). Despite its numerical abundance, Freese *et al.* (2007) declared that the plains bison is 'for all practical purposes ecologically extinct across its former range'. Interest in restoring bison as wildlife and using them as an agent for restoring ecological services has increased during the last decade (Sanderson *et al.* 2008), yet there has been limited consideration of criteria by which bison can be considered wild in the sense of foundational evolutionary processes.

In this chapter the concept of 'wild by nature' used in the Canadian *Species at Risk Act* (SARA) 2002 is explored as it relates to the status and conservation of plains bison as wildlife in Canada. The numerical recovery of plains bison, largely in captive commercial herds under private ownership, is lauded as a conservation success. The implications of this management system for sustaining wild-type bison warrant critical comment. Finally, the status of 'wild' plains bison in Canada is addressed in response to the questions: what is a wild plains bison, and are there any wild plains bison in Canada?

'Wild by nature': a key concept in conservation praxis

Concerns about the loss of biodiversity through species' extinction have existed for decades. Canada and many other countries acknowledged the importance of biodiversity conservation by ratifying the 1992 United Nations Convention on Biological Diversity (United Nations 1992). A decade later, Canada passed a federal law, the *Species at Risk Act* or SARA in 2002 in which the concept of wildlife species is defined broadly as 'a species, subspecies, variety or geographically or genetically distinct population of animal, plant or other organism, other than a bacterium or virus, that is *wild by nature* [emphasis mine] and (a) is native to Canada; or (b) has extended its range into Canada without human intervention and has been present in Canada for at least 50 years'. Canada's species at risk listing authority is the Committee on Species of Endangered Wildlife in Canada (COSEWIC). Its definition of wildlife species also includes the term 'wild by nature', echoing the SARA. Yet there is limited guidance about what this means. Here concepts fundamental to understanding and defining the term are explored, emphasizing the importance of evolutionary ecology.

Ecology, Evolution and Behaviour of Wild Cattle: Implications for Conservation, ed. M. Melletti and J. Burton. Published by Cambridge University Press.

Understanding how inter- and intraspecific diversity evolves is a central task of evolutionary ecology and is essential to conserving biodiversity. The species is a fundamental unit in biology and in biodiversity conservation (Sites & Crandall 1997; Sites & Marshall 2003). Speciation, the processes by which species arise, is a lengthy phenomenon, not just a point in time represented in extant forms. For mammals speciation may require two million years or more for completion (Avise & Walker 1998, 2000). Species' subpopulations are genetically, demographically, spatially and ecologically heterogeneous in ways that current taxonomies may or may not capture. Dichotomous or trinomial approaches to defining species, such as the practice of employing Latin bi- and trinomial names, whereby populations are classified either as conspecifics or separate species or subspecies, may fail to embrace evolutionarily intermediate states, i.e. the intermediate forms or local variants that arise through complex formational processes.

Biodiversity conservation has as an objective maintaining variation through reproductive separation of genealogical lineages defined by taxonomic classification, but fully construed it also embraces conservation of species-formational processes, including natural selection factors that influence individual fitness and local adaptation. While systematicists will continue to debate and improve species and subspecies classification schemes (Hey *et al.* 2003; Zink 2004), the real work of biodiversity conservation has increasingly focused on maintaining meaningful abundance of indigenous populations and the ecological conditions under which they can continue to evolve. Hence, biodiversity conservation can be defined broadly as 'the protection, maintenance, and rehabilitation of genetic diversity, species, and ecosystems to sustain biodiversity and the continuance of evolution and other natural processes' (adapted from Department of Fisheries and Oceans (DFO) 2009). Moritz (2002) cast the problem of conservation in terms of protecting both 'pattern' and 'process'. Pattern can be equated to local adaptations and geographic variations, and its conservation can be accomplished by the identification and protection of groups of populations, at least over the short term. In contrast, protecting process requires maintaining what Moritz terms the 'context' for natural selection to operate, namely viable populations, habitat integrity and connectedness, infra- and interspecies interactions, and other environmental factors constituting the ecological fabric of species formation. In this context, natural selection allows the constituent diversity of a population to express itself and to be acted upon, with the result that local adaptation is maintained as a process.

Several schemes have been proposed to accommodate the objective of maintaining variation, including ecologically significant units (ESUs) (Ryder 1986), ecologically exchangeable units (Crandall *et al.* 2000) and, in Canada, conservation units (CUs) (DFO 2009) and designatable units (DUs) (Green 2005). The underlying purpose of these schemes is to recognize distinct populations for assessing their conservation status so that appropriate actions may be planned and undertaken (Mace 2004). The ESU has been used as a separate population unit that merits special attention for protection and conservation action. ESUs were first described as population units that 'represent significant adaptive variation' based on 'concordance between sets of data derived by different techniques' (Ryder 1986). Over time, the definition of an ESU gradually changed to encompass two components: reproductive and historical isolation, and adaptive distinctiveness (Crandall *et al.* 2000). Rapid advancements in molecular technology in the past two decades have allowed increasingly finer resolution of distinctions between populations and interpretation of their phylogenies based on neutral genetic variation. These technologies provide more information for diagnosing distinctive population units (patterns). However, as argued by Crandall *et al.* (2000), accounting for molecular phylogeny (patterns) alone is insufficient, and criteria based on both ecological data and genetic variation of adaptive significance are more relevant for conservation. They proposed population categorization based on both ecological and genetic exchangeability. Exchangeability involves the idea that individuals can be moved between populations when populations share life history traits, ecological requirements, morphologies, demographic characteristics, and whether there has been historical gene flow between the populations. Exchangeability is rejected when there is evidence for population differentiation owing to genetic drift or natural selection, and there is evidence of restricted gene flow between populations over meaningful evolutionary time. Geographic variation in life history characteristics associated with individual fitness occurs in many species and is often characterized as 'local adaptation' (Allendorf & Waples 1996). Even small fitness effects can leave a strong pattern on an evolutionary time scale (Nielsen 2005). In reviewing the significance of evolutionary biology to conservation, Hendry *et al.* (2010) stated that 'current biodiversity is the product of past evolution, just as future biodiversity will be a product of contemporary evolution'. The preservation of local adaptations as patterns and adaptation as process are both important objectives of conservation. Arising from the concept of exchangeable units is the principle that conservation action should aim to preserve both adaptive diversity and evolutionary (formational) processes across the geographic range of a species.

Manipulation of animal populations by humans alters formational processes. Hunting, captive breeding and commercial propagation are common practices among members of the Bovini, the wild cattle tribe, that have had significant impacts on conservation of 'wild-type' individuals (Heinen & Srikosamatara 1996; IUCN-SSC Asian Wild Cattle Specialist Group 2010). The concept of wild-type used here refers to the genetic, morphological and behavioural attributes of individuals in a population in which natural selection has and continues to shape adaptation of these attributes. Selection occurs if the fitness of genotypes is unequal (Nielsen 2005). Divergent selection has the effect of driving genetic and phenotypic divergence (Barraclough 2010). Units of diversity are more distinct if divergent selection acts on gene frequencies or traits rather than if divergence occurs solely through drift or uniform selection pressures (Fontaneto *et al.* 2007).

Selection forces under captivity and commercial propagation differ from pressures in nature. In socio-ecologically functioning populations individuals interact behaviourally with each other through processes such as mate competition and intraspecific resource competition, and with other species through processes like predation, nutrient cycling and grazing. Seasonal and spatial nutrient gradients and physical disturbance, and abiotic factors such as snow, drought, changes in climate and their interactions result in ecological feedback on species' evolution, referred to as eco-evolutionary dynamics (Pelletier *et al.* 2009; Post & Palkovacs 2009). Under captivity, selection pressures differ from socio-ecologically functioning populations and can lead to unintentional adaptation, or changes may be induced by intentional selection for desired morphological or behavioural traits. Control of breeding (artificial selection) under domestication is ultimately reflected in the appearance of breeds. Results of a recent genome-wide SNP analysis demonstrated that crossbreeding of cattle lineages by pastoralists occurred early in the history of agriculture and that breed formation involved complex relationships attributed to bottlenecks, evolution in isolation, co-ancestry, migration and selective admixture (Decker *et al.* 2009).

Domestication as a state may be understood as the antithesis of wildness, but domestication as a process can be considered a continuum of human–animal relationships (Jarman & Wilkinson 1972). Artificial selection for traits best suited to captivity begins immediately upon animals being contained in a captive environment (Lynch & O'Hely 2001). The critical elements of domestication as a process are human control of movements and breeding (Russell 2002), which create circumstances for genetic changes underlying domestication as a state. Genetic and linked phenotypic changes (e.g. temperament, morphology, physiology) resulting from captivity occur when a controlled (captive) population is isolated from wild individuals over multiple generations, whether they are managed for conservation or for commercial propagation (O'Regan & Kitchener 2005; McDougall *et al.* 2006). Supplemental feeding to increase fertility and survival moderates the influence of natural selection on a population and on individual fitness. Both environment and gene-by-environment interactions can affect the fitness functions of different temperaments, and hence selection (Archard & Braithwaite 2010). Intentional or inadvertent selection for tameness and reduced flight responses leads to lack of predator recognition or reduced ability to flee (Menzel & Beck 2000). Differences in temperament traits are heritable, and linked to fitness and to several other traits of importance to ecology and evolution (Reale *et al.* 2007). Temperament is expressed in individual variation in behaviours relevant to conservation, such as aggressiveness, avoidance of novelty, willingness to take risks, sociality, and antipredator, foraging and exploratory behaviour.

To be useful for biodiversity conservation a definition of the term 'wild by nature' ought to serve the objectives of maintaining patterns of adaptive variation and formational processes existing in nature. In this context, species, subspecies, varieties or geographically or genetically distinct populations of animals that are 'wild by nature' represent patterns of local adaptation and geographic variation arising from species-formational processes, and occur in locations where ecological and socio-ecological conditions support natural selection and continued evolution of the species. Viable populations, habitat integrity and connectedness, species interactions including predation, and other environmental factors constitute the ecological fabric for forming and maintaining a wild-type mammal.

What is a wild plains bison?

The central questions for assessing if a plains bison population is 'wild by nature' involve the degree to which natural selection has been and continues to be the primary formational process, and by corollary, the extent to which humans control breeding and movements. Natural selection is expressed in terms of fitness, a relative term referring to differences between individuals in their production of offspring that in turn survive to reproduce. While the factors and operating mechanisms involved in fitness, natural selection and species evolution are exceedingly complex, and perhaps unknowable in their full extent, some indicators have been used to evaluate the presence of natural selection in bison populations (Sanderson *et al.* 2008; Gates *et al.* 2010; Gross *et al.* 2010). They include the presence of effective predators (predation), mating competition among males, population size and structure, the absence of supplemental feeding and occurrence of resource limitation, differential fecundity and survival and the ability of bison to move in response to spatio-temporal resource gradients. In addition, genetic introgression from cattle (*Bos taurus*) has been identified in many modern plains bison herds (Halbert & Derr 2007) and may affect phenotypic traits or fitness of bison. Derr *et al.* (2012) found that bison with cattle mtDNA were consistently smaller (height and weight) across all populations sampled, even accounting for different environmental conditions. Hence, genomic integrity is important for conservation of wild-type bison.

Wolves (*Canis lupus*) have been the main predator of bison in North America for at least the past 10 000 years (Lott 1991; Krech 1999). Although other large predators may be sympatric with bison, they rarely prey on them. The presence of wolves in a bison population's range is an important criterion for assessing the functioning of eco-evolutionary processes. Predation is an important selective force and a formational process in the evolution of species (Dawkins & Krebs 1979). In general, within a prey species predators select for substandard individuals, the young, the weak and the sick (Genovart *et al.* 2010). The effects of predation on ungulate prey are not limited to lethal and demographic effects, but also include risk responses that can indirectly affect prey fitness (Preisser & Bolnick 2008; Creel & Christianson 2008). Prey responses can include shifts in foraging strategy, habitat use, heightened vigilance associated with increased foraging costs, changes in grouping behaviour, altering movement patterns and elevated stress levels (Hernandez & Laundré 2005).

Behavioural responses of ungulates to habitat structure are shaped by cost–benefit tradeoffs between foraging efficiency and predation risk (Gude *et al.* 2006; Isvaran 2007). Bison forage mostly in open areas where forage (grasses and sedges) is most abundant, regardless of predation risk (Hernandez & Laundré 2005). In the presence of wolves, frequent movements diminish predation risk by reducing the predictability of prey location (Mitchell & Lima 2002; Gude *et al.* 2006). The consequence of more frequent movements is reduced residency time in primary foraging areas (e.g. open meadows) and reduced foraging time (Fortin *et al.* 2009). Bison compensate by forming larger group sizes in open habitat, reducing individual predation risk and allowing greater residency time (Fortin *et al.* 2009). Conversely, individual bison should perceive risk as relatively higher when in small groups rather than in large groups, leading to a shorter residence time, more time spent in the forest matrix and a weaker selection for meadows by individuals in smaller groups (Fortin *et al.* 2009).

Cursorial hunters like wolves and hyenas rely on evaluating prey flight performance to make selection decisions generally leading to the capture of prey in poorer condition (Estes & Goddard 1967; Kruuk 1972; Mech *et al.* 1998; MacNulthy *et al.* 2007). The bison is the wolf's largest and most aggressive (dangerous) prey (Smith et al. 2000; MacNulthy et al. 2007). In Yellowstone National Park (YNP), Wyoming, wolf attacks on elk groups occasionally led to selecting an individual. In contrast, attacks on bison groups mainly led to watching (surveillance of prey), followed rarely by attacking again (MacNulthy *et al.* 2007). In YNP, elk are three times as vulnerable to predation as bison (Smith *et al.* 2004); wolves attacked elk twice as often as bison (0.69 versus 0.31 attacks/approach; MacNulthy *et al.* 2007).

To be fully functioning as a wild population, predation must be present as a selective pressure. Through purging of deleterious genes, predation and other natural selection factors such as seasonal resource limitation can mitigate the effects of reduced heterozygosis and increased frequency of deleterious alleles associated with small population size and inbreeding (Charlesworth & Charlesworth 1987; O'Grady *et al.* 2006). Inbreeding depression describes the reduced fitness of offspring of closely related individuals compared to the offspring of randomly mated individuals (Hedrick & Kalinowski 2000). The most common effects of reduced heterozygosis are decreased fecundity or increased juvenile mortality (Ralls & Ballou 1986; Mitton 1993; Halbert *et al.* 2004), and increased disease susceptibility. In European bison (*B. bonasus*), a severe population bottleneck after World War I may have led to the loss of allelic variation that has given rise to the male reproductive disease posthitis, which threatens recovery of the species (Radwan *et al.* 2007; Daleszczyk & Bunevich 2009). The current free-living Białowieża population of European bison (910 individuals in 2011, 481 in the Polish sector and 429 in the Belarusian; see also Chapter 10) is characterized by lower genetic diversity than the historic population, impaired reproductive capacity and increased disease susceptibility (Luenser *et al.* 2005). Among plains bison populations in North America, some small, closed captive populations present symptoms caused by inbreeding depression (e.g. Badlands National Park – Berger & Peacock 1988; Texas State herd – Halbert *et al.* 2004).

As a general recommendation, inbreeding depression can be avoided when the effective size of a bison population exceeds 1000 (Hedrick 2009). Using a simulation model, Gross and Wang (2005) estimated that a minimum population of 400 bison is needed to retain 90% of selectively neutral genetic variation over 200 years with a 90% probability. On average, a high proportion of alleles with an initial frequency of less than 0.05 were lost in herds with fewer than 400. Gates *et al.* (2010) considered three size classes of bison herds for conservation: populations of fewer than 400 are small; 400–1000 are medium sized; and populations >1000 are large.

Competition for mating opportunities is an important selective force contributing to fitness. Male bison achieve breeding maturity (adulthood) beginning at 6–7 years of age (Komers *et al.* 1992). Adulthood is reflected in mature body size, risky aggressive behaviours and seasonal segregation, which allow adult males to compete for mating opportunities. Although males can be fertile at 16 months (Helbig *et al.* 2007), participation by juvenile and subadult males in breeding is suppressed by adult bulls (Komers *et al.* 1994a, 1994b). Subadult males may participate in breeding activities if they have an opportunity in the absence of older males (Komers *et al.* 1994b). In wild populations not subject to selective harvesting pressure, the adult bull:cow ratio can exceed 50:100 (Gates *et al.* 1995; Gogan *et al.* 2010).

The pre-Columbian and early historical distributions of bison populations were naturally limited by landforms and ecological features (e.g. mountains, oceans, grasslands and forests) (Reynolds *et al.* 2003; Gates *et al.* 2010), and to some extent by indigenous human populations (e.g. cordillera boundary proposed by White *et al.* 2001). Plains bison populations were partially migratory; some segments undertook large-scale seasonal movements, while others were more sedentary (Epp 1988; Epp & Dyck 2002). By early in the twentieth century the Great Plains was settled for agriculture and today presents highly fragmented landscapes no longer capable of supporting long-distance migration by bison. The current operational scale relevant for restoration of plains bison involves much smaller areas limited by land use, fences and jurisdictional and ownership boundaries (Freese *et al.* 2007). However, even at these smaller scales the capacity of landscapes to support daily and seasonal movements of bison responding to resource gradients is considered an important aspect of the evolutionary ecology of bison (Gross *et al.* 2010).

Domestic bison in Canada (captive commercial herds)

Curiosity about the agricultural potential of bison originated early in the documentary history of North America. Well before bison were significantly impacted by market hunting, Samuel Herne, George Simpson and other early explorer-entrepreneurs (*c.*1790–1820) speculated on prospects for domesticating bison because of its similarity to Old World cattle (Potter *et al.* 2010). Wild plains bison were rapidly depleted between 1840 and 1880 in Canada and the United States, owing primarily to market hunting to serve demands for meat and hides (Hornaday 1889; Isenberg 2000). In the United States, the American military establishment quietly encouraged market hunting with the explicit approval of political leaders anxious to reduce subsistence food supplies for combative, territorial First Nations people for whom the plains bison was a staple resource (McHugh 1972; Dary 1989, Danz 1997; Isenberg 2000). In Canada, extirpation followed a parallel course driven by commercial demand for hides, machine belt material for the industrial heartland of America (Roe 1970), and abetted by ineffective conservation legislation and enforcement (Hewitt 1919). Extirpation in Canada was swift. Foster (1992) described events in March and April 1879 when Metis and First Nations hunters set forth from their winter camps in search of the few bison remaining in southern Alberta. Diminishing numbers were encountered until the end of April that year, when there were none to be found. The search spilled over into Montana where, with the exception of a small number of bison that survived in the interior of YNP, the last wild bison were taken in 1883 (Flores 1996). So ended 10 000 years of bison hunting by peoples of the Great Plains.

With extirpation looming in the early 1870s, Charles Alloway, a former soldier and trading post manager in Winnipeg Manitoba, turned his attention from the hide trade to capturing and propagating live specimens. He participated in three expeditions into Saskatchewan and Alberta between 1873 and 1883. The first two resulted in five calves being successfully translocated to the farm of his partner James McKay (Neufeld 1992). The herd increased to 13 by 1878 when it was sold to Samuel Bedson, who moved them to Stoney Mountain, about 100 km north of Winnipeg. There, the bison increased to over 100. Then, owing to failing health, Bedson decided to disperse the herd. Six went to the New York and London Zoos, 27 were given to Lord Stratchcona (Donald Smith) to repay a debt and the remainder were sold to C. J. 'Buffalo' Jones of Kansas, who was interested in slaughtering some to sell meat and hides, and in propagating bison for commercial hunting on a large private ranch (Neufeld 1992). So ended the first Canadian foray into private commercial bison production. Lord Stratchcona eventually donated most of his herd to the government of Canada, which were moved to a display paddock in Banff National Park in Alberta. Thus began Canadian government efforts to conserve plains bison through captive propagation and establish a central role in bison conservation for the national park system that continues to today.

The next round of commercial interest in bison involved experimental cross-breeding between plains bison bulls and female domestic cattle, carried out between 1894 and 1914 by M. Boyd of Bobcaygeaon, Ontario (Boyd 1914). The government of Canada acquired his stock in 1915 and continued research on hybridization in Wainwright, Alberta. Subsequently, the operation was moved to another federal experimental farm near Manyberries, Alberta, where research continued until 1964. In 1929, in the Red River Settlement in southern Manitoba, captive bison were being crossbred with cattle and a short-lived enterprise called the Buffalo Wool Company was established (Hudson 1999).

Private interest in captive rearing of bison occurred at a low level in Canada until the 1970s, when western Canadian agriculture ministries began to promote alternative livestock production and supporting research (Renecker *et al.* 1989). Beginning in the 1970s in the United States, hundreds of surplus park animals were sold annually to ranchers. In the early 1980s Canada's Elk Island National Park began regular sales of surplus plains bison to agricultural producers. By 1985 there were about 5000 bison on farms in Canada (Figure 21.1). The farmed bison population peaked in 2006 at 195 728, and then declined slightly to 124 848 in 2011 at the time of the last national census. The number of farms reporting that year was 1211. The greatest number of farmed bison was in Alberta (46%), followed by Saskatchewan (32%) then Manitoba (11%). In both the United States and Canada there are about 400 000 bison on ~4000 farms (Freese *et al.* 2007; Gates *et al.* 2010).

As the bison industry prospered, emphasis increased on improving production efficiency through intensive inputs (e.g. feed and mineral supplementation, vaccinations), selection for high fecundity, auctions for breeding stock and technological advancements such as semen testing to support high prices for breeding bulls. The bison is the only indigenous North American ungulate species extensively selected for commercial traits such as carcass composition and docility (Hedrick 2009). Market forces provide incentives for bison producers to emulate beef production, rewarding producers who produce calves off-season, and who finish animals in feedlots rather than on pasture. In the wild, breeding and other physiological cycles are strongly seasonal. Wild bison cows produce two calves on average every three years (Fuller 1962) and there is considerable individual variation in fecundity (Lott & Galland 1985). Calf:cow ratios reported for wild populations vary between 28:100 and 55:100 (Gates *et al.* 1995; Larter *et al.* 2000; DelGiudice *et al.* 2001; Bradley & Wilmshurst 2005; Fuller *et al.* 2007). Under commercial production through selective culling and supplemental feeding of minerals and other nutrients, weaning percentages of >85% are typical. Weaning weights of calves in commercial herds are higher than in the wild. As with domestic cattle production, predation and

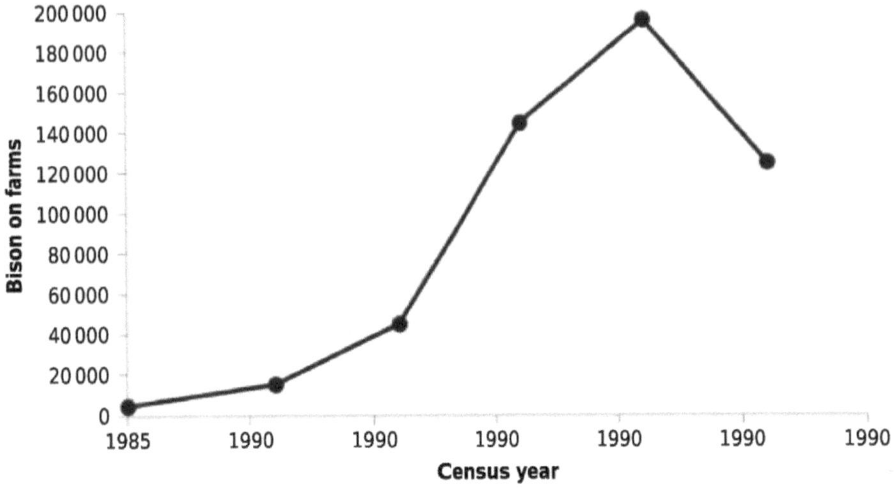

Figure 21.1 The number of bison on farms in Canada (sources: Renecker *et al.* 1989 for 1985; Statistics Canada national farm census data for 1991, 1996, 2001, 2006 and 2011).

the presence of predators – the chief of which are wolves – are poorly tolerated by bison producers.

The age and sex structure of commercial herds are typically manipulated to maximize annual calf production by maintaining a high proportion of mature cows and low but sufficient bull:cow ratios to ensure all cows are bred (typically between 1:10 and 1:20). In production herds, owners generally alter the age structure of the male population by removing males once they reach physical and behavioural maturity. Adult males are dangerous to handle and to contain, and are aggressive to other bison. Young males (2–4 years) fight less than adult bulls and are more gregarious (Komers *et al.* 1994a, 1994b), yet are capable of breeding. Dehorning is becoming a more common practice (Hudson 1999).

Emulating the conventional livestock industry, producers commonly finish bison on grain for ~100 days to achieve greater consistency in carcass quality in accordance with a regulated carcass grading system. Reasons offered by the industry for developing the grading system include providing a consistent product to purchasers and a marketing advantage over competing countries (Clarke 2001). Standards for grading bison carcasses are defined in the Canadian Livestock and Poultry Carcass Grading Regulation. Official grading under this system began in 1995. There are nine grades. Grading evaluates maturity, muscling, meat quality and fat cover. The European Economic Community approved the importation of bison into Europe following implementation of these standards (Clarke 2001).

The effects of the industry on prospects for conserving wild bison are not benign. The historic bison commons of the Great Plains was rapidly replaced by the cattle industry at the turn of the nineteenth and twentieth centuries. Cattle grazing remains the dominant use of public rangelands in western North America. However, with rapid expansion in the number of privately owned bison and bison producers, access to public rangelands for grazing private bison herds is improving. Recently, domestic bison were added to the list of eligible livestock for grazing on public dispositions in Alberta. The grazing of commercial bison on public lands competes for space with the potential for conservation and recovery of bison as a wild species.

During 2009–2011, a group of bison ranchers, industry experts, bison researchers and conservationists concerned with bison conservation developed draft guidelines for managing herds for conservation outcomes (Bison Conservation Working Group 2011). The draft guidelines address maintaining wild characteristics of bison, conserving genetic diversity and integrity and restoring and maintaining biodiversity and ecosystem functions. They have not been subject to scientific peer review. The influence they may have on private herd management for conservation remains to be evaluated. The agricultural paradigm of maximizing production through ecosystem simplification and high-intensity inputs presents a significant challenge for designing a management system(s) that could provide the ecological conditions for evolutionary processes to function while sustaining the economic viability of bison farms.

The numerical abundance and wide distribution of commercially propagated bison in Canada and North America may give a public perception that the taxon is secure (Aune *et al.* 2010). Yet the foundational processes under which wild plains bison evolved are severely diminished or absent in private herds. Hence, the conservation of wild plains bison rests with public herds where eco-evolutionary processes are functioning. Indeed, COSEWIC generally does not consider as part of a wildlife species being assessed any manipulated population established for purposes other than species conservation and for which there is no intention to contribute to the wild population. In 2004 COSEWIC assessed the status of plains bison in Canada and proposed listing it under the formal category 'Threatened'. Strong opposition from the bison industry over concerns that listing under SARA would negatively impact trade led to a decision by the Canadian Minister of Environment to not list the plains bison.

Are there any wild plains bison in Canada?

Freese *et al.* (2007: 178) proposed that the plains bison is 'for all practical purposes ecologically extinct' in the context of large migratory herds freely interacting with biotic and abiotic components of regional-scale ecosystems. The question remains whether some 'wild' populations exist in Canada at smaller scales where plains bison can continue to evolve under ecological and socio-ecological conditions that support natural selection and continued evolution of the species. Five plains bison herds in Canada appear to be functioning as wild populations in the sense of their evolutionary ecology, albeit with some limitations. In aggregate there are about 2200 plains bison in these herds.

Elk Island National Park population

In 1906 the government of Canada purchased the largest herd of bison in the United States. Located in the Flathead area in northwestern Montana, the Pablo herd was founded mostly from the descendants of bison captured in central Saskatchewan and Alberta by Charles Alloway and Samuel Walking Coyote (Neufeld 1992). Coder (1975: 174–251) described the history of the herd's translocation to Canada in detail. It was privately owned by Michel Pablo, who grazed the bison for portions of the year in the Bitterroot Range and on the Flathead Indian Reservation in Montana. The US Congress passed a bill in 1904 that allocated this land for development, thus forcing Pablo to seek additional grazing lands or dispose of the herd. Pablo elected to sell the herd to the government of Canada. Elk Island National Park (EINP) received 410 bison from Pablo in 1907. In 1909, 325 bison were shipped from EINP to the much larger Wainwright Buffalo Park in east central Alberta, where their numbers were augmented with additional shipments from two herds in Montana (Pablo – 306; Conrad – 30) and from Banff National Park (87). Coder (1975: 229) noted that a small, unspecified number of bison bulls were translocated to EINP from Banff at the same time as shipments from Montana. Hence the number of founders for the EINP herd was approximately 90. The EINP herd was the source of plains bison for all other public herds in Canada and for two recent translocations (2011 and 2012) to the American Prairie Reserve in northeastern Montana. The Wainright herd was depopulated during the 1940s.

The EINP plains bison population is confined to the northern portion of the park in a 136 km^2 fenced area of parkland habitat. The herd is subject to periodic roundups to remove surplus animals through translocation. Since the 1970s the herd has been maintained at between 400 and 500. The herd is not supplementally fed except for baiting during roundups. There are no effective predators of bison in the park. Population size is mainly regulated by non-selective removal of calves and yearlings during annual or semi-annual roundups. Older animals are allowed to succumb to old age in the park. Occasionally a bison bull is removed if it is considered dangerous for park visitors. Habitat management is effected with prescribed burning. There are no other manipulations of this herd. It can be considered to be functioning as a wild population with limitations (no predators, relatively small area, medium population size).

Sturgeon River population

In 1969, approximately 50 plains bison were translocated from EINP to the Thunder Hills north of Prince Albert National Park in Saskatchewan. Following release, the bison moved south and settled in several areas in small herds, one of which was in Prince Albert National Park. The number of founders of the Sturgeon River herd is uncertain, but is estimated at between 10 and 22 individuals (Parks Canada: www.pc.gc.ca/pn-np/sk/princealbert/natcul/natcul4/a.aspx; accessed 9 February 2012). The Sturgeon River herd, numbering 200–250 in 2012, is one of two free-ranging plains bison populations within the original range of plains bison in Canada, both of which are in Saskatchewan.

The population ranges in a 700 km^2 area in and around southeastern Prince Albert National Park (R. Tether, Saskatchewan Ministry of Environment, personal communication; S. Cherry, Parks Canada, personal communication). Suitable plains bison habitat is largely restricted to southern portions of the park. Most of the area used by bison is inside the park. However, during the fall a large proportion of the population may concentrate in a small area outside the park on provincial or private land. Animals moving onto private land are hazed back into the park. Expansion of the herd's range to adjacent provincial Crown land or privately owned land may be possible in the future through conservation easements or purchase of land by conservation organizations (S. Cherry, personal communication). Currently conflicts arise when bison leave the national park and damage crops or destroy fencing on private land.

On average, Aboriginal people harvest 10–20 plains bison from the population annually outside the national park, and lone bulls considered a threat by landowners may be removed (S. Cherry, personal communication; R. Tether, personal communication). No other manipulations or management interventions occur. This population is subject to all natural ecological and evolutionary processes. Wolf predation is an important limiting factor (Fortin *et al.* 2009). An outbreak of anthrax occurred in 2008; 28 carcasses were found (Shury *et al.* 2009). A privately owned bison ranch is located on the eastern boundary of the park (Shury *et al.* 2009), which represents a potential source of escapees that could mix with the wild herd.

Parks Canada (Prince Albert National Park) and the provincial government are developing a management plan for this herd. An advisory group has been formed comprising representatives of First Nations, Metis, NGOs, landowners and rural municipalities. This herd can be considered to be functioning as a wild population, with limitations (small population size).

McCusker River population (Primrose Air Weapons Range)

Plains bison were translocated from EINP to the Thunder Hills north of Prince Albert National Park in 1969. The animals scattered and moved south, forming small groups in and around the national park. Seventeen bison settled in the Big River Community Pasture west of the park. The government of Saskatchewan captured this group and moved the animals by truck to Vermette Lake about 50 km north of the Primrose Air Weapons Range (PAWR), near the central western border of Saskatchewan. The bison moved south into the PAWR. The PAWR encompasses an area of about 11 600 km^2 in central Alberta and Saskatchewan. This remote area is leased by the Department of National Defence from the provinces and is primarily used for air weapons testing and military training. The closest communities to the PAWR are Cold Lake, Alberta and Meadow Lake, Saskatchewan.

More than 75% of the area currently used by the McCusker herd is in the PAWR, to which the Department of National Defence restricts public access. This provides some protection from otherwise unregulated hunting. Sightings of bison have increased east of PAWR in Saskatchewan over the last ten years (R. Tether, personal communication). There is limited industrial activity and access to provincial forest lands in this area. However, some development restrictions are being lifted, which will result in increased road development by the forest and natural gas industries. Public access to the area will consequently increase and with it the potential for increased hunting pressure on the bison. Increased sightings on Saskatchewan provincial forest lands suggest the herd may be growing in terms of size and area occupied. Hunting pressure has increased in recent years in Saskatchewan. First Nations people are permitted to access the PAWR on weekends, when they are allowed to hunt there, but road and trail access is limited so hunting pressure is minimal (R. Tether, personal communication). This population of about 150 is free-ranging, and wolves are present (Saskatchewan Environment and Resource Management 2000: 3). Sex and age structure are not manipulated. There are currently no provincial or federal management plans for the population. The McCusker River population is one of two free-ranging wild populations of plains bison near the northern limit of the original range of the subspecies in Canada. It can be considered to be functioning as a wild population with limitations (small population size).

Pink Mountain population

This introduced free-ranging population of about 1000 bison in northeastern British Columbia is extra-limital with respect to the original range of plains bison (Figure 21.2). It is located within the original range of the wood bison. The Pink Mountain population was established in 1971 from 48 plains bison purchased from EINP by a private rancher (COSEWIC 2004). Within a year of their translocation to the ranch the bison escaped captivity to become a free-ranging population. The population is considered wildlife (big game) under the *British Columbia Wildlife Act* and population size and distribution are managed through regulated hunting. The 2011–2012 Limited Entry Hunting Synopsis for British Columbia listed 396 tags available for hunters. The herd ranges in a 2000 km^2 area in the upper Sikanni and Halfway River valleys (Figure 21.2), which provide grass and sedge meadow habitat. Potential exists for intermingling between plains bison from this population and wood bison populations in the region, although provisions in a provincial management plan provide for mitigation of this potential problem.

The Pink Mountain plains bison population is subject to a full range of natural socio-ecological processes, including wolf predation, and its movements are unrestricted by fences (D. Fraser, BC Ministry of Environment, personal communication). Salt blocks are used in winter to encourage bison to stay off the highway (C. Thiessen, BC Ministry of Natural Resource Operations, personal communication). This herd can be considered to be functioning as a wild population with no limitations.

Grasslands National Park population

Grassland National Park (GNP) is situated near the southern border of Saskatchewan in the mixed grasslands natural subregion (Figure 21.2). In December 2005 the park received 71 plains bison from EINP. The animals were released into a 181 km^2 fenced area (west Larson Block) within which they are free-ranging. The population grew to about 350 by the fall of 2012. The target population is 300–350, which may be increased pending the outcome of research on the impacts of grazing at this stocking level (W. Olson, Parks Canada Agency, personal communication). There is potential for establishing a second population in the east block of the park. There have been no removals for population management to date. However, a draft management plan (Olson 2012) calls for nonselective removal of juveniles to control the population at the target level. Hence the stable population structure should be similar to wild populations. Baiting will be used to lure animals into a handling facility to achieve removals. There are no effective predators of bison in this system. This herd is functioning as a wild population with limitations (no predators, limited area, medium population size).

Conclusions

Biodiversity conservation can be defined broadly as the protection, maintenance and rehabilitation of genetic diversity, species and ecosystems to sustain biodiversity and the continuance of evolution and other natural processes. Simply stated, the conservation of a wildlife species requires protecting both 'pattern' and 'process'. The numerical recovery of the plains bison from near extinction at the end of the nineteenth century to in excess of 400 000 has been lauded as a

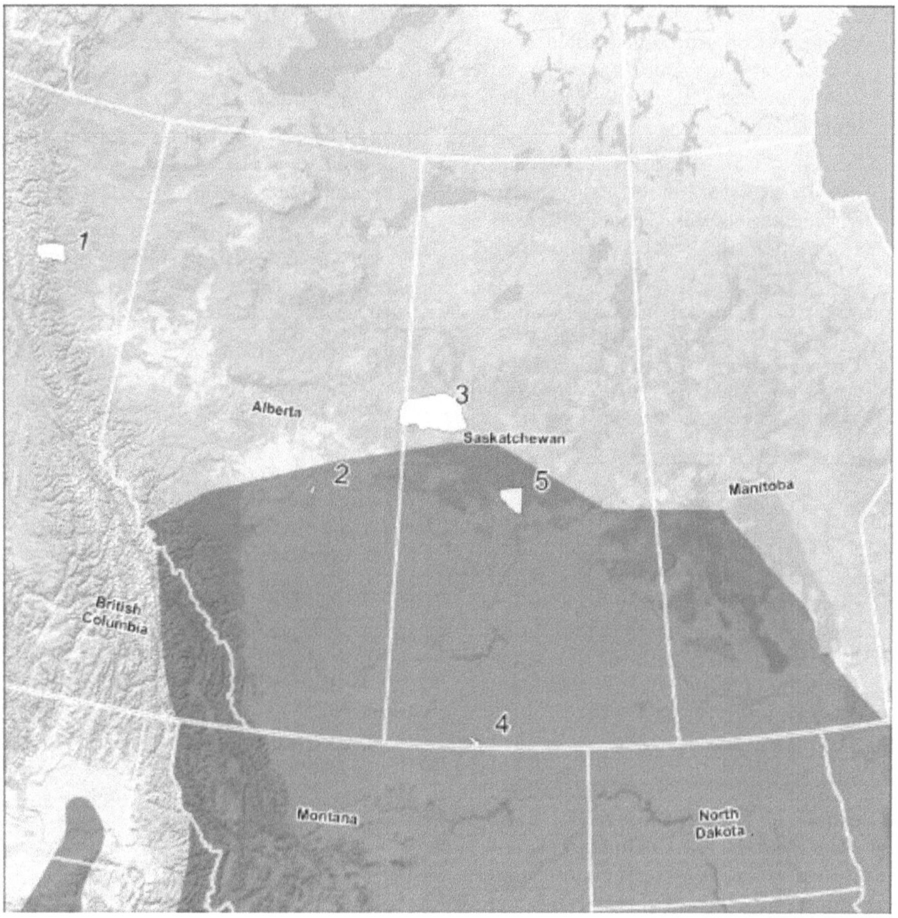

Figure 21.2 Areas occupied by wild plains bison populations in Canada in relation to the original range of the sub-species mapped by Sanderson *et al.* (2008). Populations: (1) Pink Mountain; (2) Elk Island National Park; (3) McCusker River; (4) Sturgeon River; (5) Grasslands National Park. Image created by Greg Chernoff, Senior Research Associate, Miistakis Institute, Calgary, Alberta, Canada.

conservation success. On closer scrutiny, in excess of 96% of American bison are under private ownership, managed primarily for captive commercial propagation where they are subject to artificial selection for heritable market traits and intentional or unintentional selection for traits suited to captivity. Many natural selection factors present in wild populations are absent or diminished in captive herds. Among them are reduced mate competition among adult males, reduced exposure to seasonal resource limitation and natural mortality, absence of predation, artificially skewed sex and age structures, individual variation in fecundity and other traits, and control of movements in response to spatio-temporal resource gradients.

The *Canada Species at Risk Act* (2002) uses the term 'wild by nature' in its definition of a wildlife species. The perspective of evolutionary ecology provides a basis for defining the concept (synonymous in this chapter with 'wild' and 'wild type'). Populations that are 'wild by nature' represent patterns of local adaptation and geographic variation arising from species-formational processes, and occur in locations where ecological and socio-ecological conditions support natural selection and continued evolution of the species. In Canada, five populations of plains bison are functioning as wild populations; all but one with limitations. In aggregate they number approximately 2200. The only population that is fully functioning in terms of evolutionary ecology is the Pink Mountain herd, which is extra-limital with respect to the original range of the subspecies. The other populations occur within the original range of plains bison, but natural selection is limited by small population size or the absence of predation. In conclusion, wild-type plains bison are far from secure in Canada. Further development of conservation and recovery programmes are confounded by widespread private ownership, commercial interests in bison as livestock and weak legislative support for public conservation programmes.

Acknowledgements

The following people contributed information on the status and management of wild plains bison populations in Canada: Seth Cherry (Sturgeon River herd); Wes Olson and Adrian Sturch (Grasslands National Park); Robert Tether (McKusker River herd); Normand Cool and Archie Handel (Elk Island National Park); and Conrad Thiessen (Pink Mountain herd). Herd range map data were provided by Jenny Wu with the Committee on Endangered Species of Wildlife in Canada. Greg Chenoff of the Miistakis Institute for the Rockies prepared the range map.

References

Allendorf, F. W. & Waples, R. S. (1996). Conservation genetics of salmonid fishes. In *Conservation Genetics: Case Histories from Nature*, pp. 238–280. Eds. Avice, J. C. & Hamrick, J. L. New York: Chapman & Hall.

Archard, G. A. & Braithwaite, V. A. (2010). The importance of wild populations in studies of animal temperament. *Journal of Zoology* **281**: 149–160.

Aune, K., Wallen, R., Gates, C. C., Ellison, K., Freese, C. H. & List, R. (2010). Legal status, policy issues and listings. In *American Bison Status Survey and Conservation Guidelines 2010*, pp. 63–84. Eds. Gates, C. C., Freese, C. H., Gogan, J. P. & Kotzman, M. Gland: IUCN.

Avise, J. C. & Walker, D. (1998). Speciation durations and Pleistocene phylogeographic effects on avian populations and the speciation process. *Proceedings of the Royal Society of London Series B: Biological Sciences* **265**: 457–463.

Avise, J. C. & Walker, D. (2000). Abandon all species concepts? A response. *Conservation Genetics* **1**: 77–80.

Barraclough, T. (2010). Evolving entities: towards a unified framework for understanding diversity at the species and higher levels. *Philosophical Transactions of the Royal Society B* **265**: 1801–1813.

Berger, J., & Peacock, M. (1988). Variability in size-weight relationships of *Bison bison*. *Journal of Mammalogy* **69**: 618–624.

Bison Conservation Working Group (2011). Bison conservation management: proposed guidelines for herd managers. www.canadianbison.ca/producer/documents/BisonGuidelinesVersion10ForPublicComment_000.pdf, accessed 15 September 2011).

Boyd, M. M. (1914). Crossing bison and cattle. *Journal of Heredity* **5**: 189–197.

Bradley, M. & Wilmshurst, J. (2005). The fall and rise of bison populations in Wood Buffalo National Park: 1971 to 2003. *Canadian Journal of Zoology* **83**: 1195–1205.

Charlesworth, D. & Charlesworth, B. (1987). Inbreeding depression and its evolutionary consequences. *Annual Review of Ecology & Systematics* **18**: 237–268.

Clarke, D. (2001). The Canadian bison grading system: how it presently works. *Smoke Signals*, February.

Coder, G. D. (1975). The national movement to preserve the American buffalo in the United States and Canada between 1880 and 1920. PhD dissertation. Ohio State University.

COSEWIC (2004). *COSEWIC Assessment and Status Report on the Plains Bison Bison bison bison in Canada*. Ottawa: Committee on the Status of Endangered Wildlife in Canada.

Crandall, K. A., Bininda-Emonds, O. R. P., Mace, G. M. & Wayne, R. K. (2000). Considering evolutionary processes in conservation biology. *Trends in Ecology & Evolution* **15**: 290–295.

Creel, S. & Christianson, D. (2008). Relationships between direct predation and risk effects. *Trends in Ecology & Evolution* **23**: 194–201.

Daleszczyk, K. & Bunevich, A. N. (2009). Population viability analysis of European bison populations in Polish and Belarusian parts of Białowieza Forest with and without gene exchange. *Biological Conservation* **142**: 3068–3075.

Danz, H. P. (1997). *Of Bison and Man*. Niwot, CO: University Press of Colorado.

Darimont, C. T., Carlson, S. M., Kinnison, M. T., Paquet, P. C., Reimchen, T. E. & Wilmers, C. C. (2009). Human predators outpace other agents of trait change in the wild. *Proceedings of the National Academy of Sciences* **106**: 952–954.

Dary, D. A. (1989). *The Buffalo Book: The Full Saga of the American Animal*. Chicago, IL: Swallow Press.

Dawkins, R. & Krebs, J. R. (1979). Arms races between and within species. *Proceedings of the Royal Society of London Series B: Biological Sciences* **205**: 489–511.

Decker, J. E., Pires, J. C., Conant, G. C., *et al.* (2009). Resolving the evolution of extant and extinct ruminants with high-throughput genomics. *Proceedings of the National Academy of Sciences* **106**: 18644–18649.

DelGiudice, G. D., Moen, R. A., Singer, F. J. & Riggs, M. R. (2001). Winter nutritional restriction and simulated body condition of Yellowstone elk and bison before and after the fires of 1988. Bethesda, MD: Wildlife Society.

Department of Fisheries and Oceans (2009). *Framework for Characterizing Conservation Units of Pacific Salmon (*Oncorhynchus *spp.) for Implementing the Wild Salmon Policy*. Ottawa: Fisheries and Oceans Canada.

Derr, J. N., Hedrick, P. W., Halbert, N. D., *et al.* (2012). Phenotypic effects of cattle mitochondrial DNA in American bison. *Conservation Biology* **26**: 1130–1136.

Diamond, J. (2002). Evolution, consequences and future of plant and animal domestication. *Nature* **418**: 700–707.

Epp, H. T. (1988). Ways of the migrant herds: dual dispersion strategy among bison. *The Plains Anthropologist* **33**: 309–320.

Epp, H. T. & Dyck, I. (2002). Early human-bison population interdependence in the Plains ecosystem. *Great Plains Research* **12**: 323–337.

Estes, R. D. & Goddard, J. (1967). Prey selection and hunting behavior of the African wild dog. *Journal of Wildlife Management* **31**: 52–70.

Flores, D. (1996). The great contraction. In *Legacy: New Perspectives on the Battle of the Little Bighorn*, pp. 3–22. Ed. Rankin, C. E. Helena: Montana Historical Society Press.

Fontaneto, D., Herniou, E. A., Boschetti, C., *et al.* (2007). Independently evolving species in asexual bdelloid rotifers. *PloS Biology* **5**: 914– 921.

Fortin, D, Fortin, M.-E., Beyer, H. L., Duchesne, T., Courant, S. & Dancose, K. (2009). Group-size-mediated habitat selection and group fusion-fission dynamics of bison under predation risk. *Ecology* **90**: 2480–2490.

Foster, J. E. (1992). The Metis and the end of the plains buffalo in Alberta. *Alberta: Studies in the Arts and Sciences* **3**: 61–77.

Freese, C., Aune, K. E., Boyd, D. P., *et al.* (2007). Second chance for plains bison. *Biological Conservation* **136**: 175–184.

Fuller, J. A., Garrot, R. A., White, P. J., Aune, K. E., Roffe, T. J. & Rhyan, J. C. (2007). Reproduction and survival of Yellowstone bison. *Journal of Wildlife Management* **71**: 2365–2372.

Fuller, W. A. (1962). *The Biology and Management of the Bison of Wood Buffalo National Park*. Ottawa: Canadian Wildlife Service.

Gates, C. C., Elkin, B. T. & Dragon, D. C. (1995). Investigation, control and epizootiology of anthrax in a geographically isolated, free-roaming bison population in northern Canada. *Canadian Journal of Veterinary Research* **59**: 256–264.

Gates, C. C., Ellison, K., Freese, C. H., Aune, K. & Boyd, D. P. (2010). Numerical and geographic status. In *American Bison Status Survey and Conservation Guidelines 2010*, pp. 55–62. Eds. Gates, C. C., Freese, C. H., Gogan, J. P. & Kotzman, M. Gland: IUCN.

Genovart, M., Negre, N., Tavecchia, G., Bistuer, A., Parpal, L. & Oro, D. (2010). The young, the weak and the sick: evidence of natural selection by predation. *PLoS ONE* 5: 1–5. e9774.

Godin, J. & Dugatkin, L. A. (1996). Female mating preference for bold males in the guppy, *Poecilia reticulata*. *Proceedings of the National Academy of Sciences* **93**: 10262–10267.

Gogan, J. P., Larter, N. C., Shaw, J. H., Gross, J. E., Gates, C. C. & Truett, J. (2010). General biology, ecology and demographics. In *American Bison Status Survey and Conservation Guidelines 2010*, pp. 39–54. Eds. Gates, C. C., Freese, C. H., Gogan, J. P. & Kotzman, M. Gland: IUCN.

Green, D. M. (2005). Designatable units for status assessment of endangered species. *Conservation Biology* **19**: 1813–1820.

Gross, J. E. & Wang, G. (2005). *Effects of Population Control Strategies on Retention of Genetic Diversity in National Park Service Bison (*Bison bison*) Herds*. Bozeman, MT: USGS-Biological Resources Divsion.

Gross, J. E., Halbert, N. D., Derr, J. N., *et al.* (2010). Conservation guidelines for population, genetic, and disease management. In *American Bison Status Survey and Conservation Guidelines 2010*, pp. 85–101. Eds. Gates, C. C., Freese, C. H., Gogan, J. P. & Kotzman, M. Gland: IUCN.

Gude, J. A., Garrott, R. A., Borkowski, J. J. & King, F. (2006). Prey risk allocation in a grazing ecosystem. *Ecological Applications* **16**: 285–298.

Halbert, N. D. & Derr, J. N. (2007). A comprehensive evaluation of cattle introgression into US federal bison herds. *Journal of Heredity* **98**: 1–12.

Halbert, N. D., Raudsepp, T., Chowdhary, B. P. & Derr, J. N. (2004). Conservation genetic analysis of the Texas State Bison Herd. *Journal of Mammalogy* **85**: 924–931.

Heck, L. (1934). Über die NeuzÜchtung des Ur oder Auerochs. *Berichte der Internationalen Gesellschaft zur Erhaltung des Wisents* **3**: 225–294.

Hedrick, P. (2009). Conservation genetics and North American bison (*Bison bison*). *Journal of Heredity* **100**: 411–420.

Hedrick, P. W. & Kalinowski, S. T. (2000). Inbreeding depression in conservation biology. *Annual Review of Ecology and Systematics* **31**: 139–162.

Heinen, J. T. & Srikosamatara, S. (1996). Status and protection of Asian wild cattle and buffalo. *Conservation Biology* **10**: 931–934.

Helbig, L., Woodbury, M. R., Haigh, J. C., Collins, J., & Barth, A. D. (2007). The seasonal fertility of North American bison (*Bison bison*) bulls. *Animal Reproduction Science* **97**: 265–278.

Hendry, A. P., Lohmann, L. G., Conti, E., *et al.* (2010). Evolutionary biology in biodiversity science, conservation, and policy: a call to action. *Evolution* **64**: 1517–1528.

Hernandez, L. & Laundré, J. W. (2005). Foraging in the 'landscape of fear' and its implications for habitat use and diet quality of elk *Cervus elaphus* and bison *Bison bison*. *Wildlife Biology* **11**: 215–220.

Hewitt, C. G. (1919). The coming back of the bison. *Natural History* **19**: 553–565.

Hey, J., Waples, R. S., Arnold, M. L., Butlin, R. K. & Harrison, R. G. (2003). Understanding and confronting species uncertainty in biology and conservation. *Trends in Ecology & Evolution* **18**: 597–603.

Hornaday, W. T. (1889). *The Extermination of the American Bison, with a Sketch of its Discovery and Life History: Annual Report (1887)*. Washington, DC: Smithsonian Institution.

Hudson, R. J. (1999). From prairie to paddock. *Smoke Signals*, October.

Isenberg, A. C. (2000). *The Destruction of the Bison: An Environmental History 1750–1920*. Cambridge: Cambridge University Press.

Isvaran, K. (2007). Intraspecific variation in group size in the blackbuck antelope: the roles of habitat structure and forage at different spatial scales. *Oecologia* **154**: 435–444.

IUCN-SSC Asian Wild Cattle Specialist Group (2010). *Regional Conservation Strategy for Wild Cattle and Buffaloes in South-east Asia*. Gland: IUCN Species Specialist Commission.

Jarman, M. R. & Wilkinson, P. F. (1972). Criteria of animal domestication. In *Papers in Economic Prehistory: Studies by Members and Associates of the British Academy Major Research Project in the Early History of Agriculture*, pp. 83–96. Ed. Higgs, E. S. Cambridge: Cambridge University Press.

Komers, P. E., Messier, F. & Gates, C. (1992). Search or relax: the case of bachelor wood bison. *Behavioural Ecology and Sociobiology* **31**: 195–203.

Komers, P. E., Messier, F. & Gates, C. C. (1994a). Plasticity of reproductive behaviour in wood bison bulls: on risks and opportunities. *Ethology Ecology & Evolution* **6**: 481–495.

Komers, P. E., Messier, F. & Gates, C. C. (1994b). Plasticity of reproductive behaviour in wood bison bulls: when subadults are given a chance. *Ethology Ecology & Evolution* **6**: 313–330.

Krech, S. (1999). *The Ecological Indian: Myth and History*. New York: Norton and Company Ltd.

Kruuk, H. (1972). *The Spotted Hyena*. Chicago, IL: University of Chicago Press.

Larter, N. C., Sinclair, A. R. E., Ellsworth, T., Nishi, J. & Gates, C. C. (2000). Dynamics of reintroduction in an indigenous large ungulate: the wood bison of northern Canada. *Animal Conservation* **4**: 299–309.

Lott, D. E. & Galland, J. C. (1985). Individual variation in fecundity in an American bison population. *Mammalia* **49**: 300–302.

Lott, D. F. (1991). American bison socioecology. *Applied Animal Behaviour Science* **29**: 135–145.

Luenser, K., Fickel, J., Lehnen, A., Speck, S. & Ludwig, A. (2005). Low level of genetic variability in European bison (*Bison bonasus*) from the Bialowieza National Park in Poland. *European Journal of Wildlife Research* **51**: 84–87.

Lynch, M. & O'Hely, M. (2001). Captive breeding and the genetic fitness of natural populations. *Conservation Genetics* **2**: 363–368.

Mace, G. M. (2004). The role of taxonomy in species conservation. *Philosophical Transactions of the Royal Society of London Series B* **359**: 711–719.

Mace, G. M. & Lande, R. (1991). Assessing extinction threats: toward a re-evaluation of IUCN threatened species categories. *Conservation Biology* **5**: 148–157.

MacNulthy, D. R., Mech, D. & Smith, W. (2007). A proposed ethogram of large-carnivore predatory behavior, exemplified by the wolf. *Journal of Mammalogy* **88**: 595–605.

McDougall, P. T., Reale, D. & Reader, S. M. (2006). Wildlife conservation and animal temperament: causes and consequences of evolutionary change for captive, reintroduced, and wild populations. *Animal Conservation* **9**: 39–48.

McHugh, T. (1972). *The Time of the Buffalo.* New York: Alfred A. Knopf, Inc.

Mech, L. D., Adams, L. G., Meier, T. J., Burch, J. W. & Dale, B. W. (1998). *The Wolves of Denali.* Minneapolis, MN: University of Minnesota Press.

Menzel, C. R. & Beck, B. B. (2000). Homing and detour behavior in golden lion tamarin social groups. In *On the Move*, pp. 299–326. Eds. Boinski, S. & Garber, P. A. Chicago, IL: Chicago University Press.

Mitchell, W. A., & Lima, S. L. (2002). Predator–prey shell games: large-scale movement and its implications for decision-making by prey. *Oikos* **99**: 249–259.

Mitton, J. B. (1993). Theory and data pertinent to the relationship between heterozygosity and fitness. In *The Natural History of Inbreeding and Outbreeding*, pp. 17–41. Ed. Thornhill, N. W. Chicago, IL: University of Chicago Press.

Moritz, C. (2002). Strategies to protect biological diversity and the evolutionary processes that sustain it. *Systematic Biology* **5**: 238–254.

Neufeld, P. L. (1992). Bison conservation: the Canadian story. *Manitoba History* **24**. www.mhs.mb.ca/docs/mb_history/24/bisonconservation.shtml, accessed 20 June 2012.

Nielsen, R. (2005). Molecular signatures of natural selection. *Annual Review of Genetics* **39**: 197–218.

O'Grady, J. J., Brook, B. W., Reed, D. H., Ballou, J. D., Tonkyn, D. W. & Frankham, R. (2006). Realistic levels of inbreeding depression strongly affect extinction risk in wild populations. *Biological Conservation* **133**: 42–51.

Olson, W. E. (2012). *Grasslands National Park of Canada, Plains Bison Management Plan: 2011.* Saskatchewan: Parks Canada Agency.

O'Regan, H. & Kitchener, A. (2005). The effects of captivity on the morphology of captive, domesticated and feral mammals. *Mammal Review* **5**: 215–230.

Pelletier, F., Garant, D. & Hendry, A. P. (2009). Eco-evolutionary dynamics. *Philosophical Transactions of the Royal Society of London B: Biological Sciences* **364**: 1483–1489.

Post, D. M. & Palkovacs, E. P. (2009). Eco-evolutionary feedbacks in community and ecosystem ecology: interactions between the ecological theatre and the evolutionary play. *Philosophical Transactions of the Royal Society of London B: Biological Sciences* **364**: 1629– 1640.

Potter, B. A., Gerlach, S. C., Gates, C. C., Boyd, D. P., Oetelaar, G. A. & Shaw, J. H. (2010). History of bison in North America. In *American Bison Status Survey and Conservation Guidelines 2010*, pp. 5–12. Eds. Gates, C. C., Freese, C. H., Gogan, J. P. & Kotzman, M. Gland: IUCN.

Preisser, E. L. & Bolnick, D. I. (2008). When predators don't eat their prey: nonconsumptive predator effects on prey dynamics. *Ecology* **9**: 2414–2415.

Radwan, J., Kawałko, A., Wójcik, J. M. & Babik, W. (2007). MHC-DRB3 variation in a free-living population of European bison *Bison bonasus. Molecular Ecology* **16**: 531–540.

Ralls, K. & Ballou, J. D. (1986). Captive breeding programs for populations with a small number of founders. *Trends in Ecology & Evolution* **1**: 19–22.

Reale, D., Reader, S., Sol, D., McDougall, P. T. & Dingemase, N. J. (2007). Integrating animal temperament within ecology and evolution. *Biological Review* **82**: 291–318.

Renecker, L. A., Blyth, C. B. & Gates, C. C. (1989). Game production in western Canada. In *Wildlife Production Systems: Economic Utilisation of Wild Ungulates*, pp. 248–267. Eds. Hudson, R. J., Drew, K. R. & Baskin, L. M. Cambridge: Cambridge University Press.

Reynolds, H. W., Gates, C. C. & Glaholt, R. D. (2003). Bison (*Bison bison*). In *Wild Mammals of North America: Biology, Management and Conservation*, pp. 1009–1060. Eds. Feldhamer, G. A., Thompson, B. C. & Chapman, J. A. Baltimore, MD: Johns Hopkins University Press.

Roe, F. G. (1970). *The North American Buffalo: A Critical Study of the Species in its Wild State.* 2nd edition. Toronto: University of Toronto Press.

Russell, N. (2002). The wild side of animal domestication. *Society & Animals* **10**: 285–302.

Ryder, O. A. (1986). Species conservation and systematics: the dilemma of subspecies. *Trends in Ecology and Evolution* **1**: 9–10.

Sanderson, E. W., Redford, K. H., Weber, B., *et al.* (2008). The ecological future of the North American bison: conceiving long-term, large-scale conservation of wildlife. *Conservation Biology* **22**: 252–266.

Saskatchewan Environment and Resource Management (2000). Primrose Lake and McCusker River representative areas concept management plan. www.environment.gov.sk.ca/adx/aspx/adxGetMedia.aspx?DocID=735,720,623,247,94,88,Documents&MediaID=309&Filename=primrose+lake+and+McCusker+river.pdf&l=English, accessed 8 April 2014.

Shaw, J. H. (1995). How many bison originally populated western rangelands? *Rangelands* **17**: 148–150.

Shury, T., Fransden, D. & O'Brodovich, L. (2009). Anthrax in free-ranging bison in the Prince Albert National Park area of Saskatchewan in 2008. *Canadian Veterinary Journal* **50**: 152–154.

Sites, J. W. & Crandall, K. A. (1997). Testing species boundaries in biodiversity studies. *Conservation Biology* **11**: 1289–1297.

Sites, J. W. & Marshall, J. C. (2003). Delimiting species: a Renaissance issue in systematic biology. *Trends in Ecology & Evolution* **18**: 462–470.

Smith, D. W., Mech, L. D., Meagher, M. *et al.* (2000). Wolf–bison interactions in Yellowstone National Park. *Journal of Mammalogy* **81**: 1128–1135.

Smith, D. W., Drummer, T. D., Murphy, K. M., Guernsey, D. S. & Evans, S. B. (2004). Winter prey selection and estimation of wolf kill rates in Yellowstone National Park. *Journal of Wildlife Management* **68**: 153–166.

Theodorou, K. & Couvet, D. (2004). Introduction of captive breeders to the wild: harmful or beneficial? *Conservation Genetics* **5**: 1–12.

United Nations (1992). Convention on biological diversity. Secretariat of the Convention on Biological Diversity.

van Vuure, T. (2002). History, morphology and ecology of the aurochs (*Bos tausus primigenius*). *Lutra* **45**: 3–17.

White, C. A., Langemann, E. G., Gates, C. C., Kay, C. E., Shury, T. & Hurd, T. E. (2001). Plains bison restoration in the Canadian rocky mountains? Ecological and management considerations. In *Crossing Boundaries in Park Management. Proceedings of the 11th Conference on Research and Resource Management in Parks and on Public Lands*, Ed. Harmon, D. Hancock, MI: The George Wright Society.

Zink, R. M. (2004). The role of subspecies in obscuring avian biological diversity and misleading conservation policy. *Proceedings of the Royal Society London B: Biological Sciences* **271**: 561–564.

Chapter 22

The case study: the restitution of the wisent (*Bison bonasus*) to the Carpathians

Kajetan Perzanowski and Wanda Olech

Introduction

The history of the native Carpathian (Transylvanian) population of the wisent, regarded as a subspecies – *Bison bonasus hungarorum* Kretzoi, 1946 – ended most probably by the second half of the eighteenth century, when the last remaining animals were captured or killed in the territory ruled then by the Austro-Hungarian Empire (Szalay 1913; Nahlyk 1991, 1992).

The scene for the restoration of this species, which was initiated in the 1920s at Białowieża (Poland), was for about 40 years lowland forests of Poland, Byelorussia, Russia and Ukraine. The first release of wisents to the mountains of Europe occurred in the Carpathian forests of Poland and Ukraine in 1963 and 1965. Behind the reasons for selection of this region for free-ranging wisent herds was undoubtedly a desire to improve the attractiveness of hunting grounds established there for high-ranking officials, but also an aspiration to place descendants of the Caucasian subspecies in appropriate environmental conditions (Krasińska & Krasiński 2007).

The first reintroduction site – the Polish Bieszczady Mountains – was particularly suitable for such an initiative, because after the resettlement of native inhabitants and destruction of local villages in World War II, the density of the human population there dwindled to a mere five people per km^2, and the anthropopression there was reduced to an insignificant level (Augustyn & Kozak 1997). Quite similar conditions occurred within the Ukrainian site at Skolivsky Beskydy, some 30 km to the east of Bieszczady. Both populations grew slowly, and were periodically culled (officially the control of population numbers) until the end of the 1980s, when new economical and political circumstances arose. In the meantime, two other herds were created in the Ukrainian part of the range (Nadvirna at Prut River valley, and Bukovyna near the Romanian border), but in other parts of the Carpathians (Romania and Slovakia) there were only a few breeding enclosures built, with no intention of establishing wild herds. During that time there were no collaboration or coordination efforts among managers of free-ranging animals or wisent breeders in the region.

The first programme aimed at initiation of joint efforts to restore the species within the whole eco-region began in 1999, when thanks to the support of the WWF's Large Herbivore Initiative, the team of the Carpathian Division of the International Centre of Ecology, Polish Academy of Sciences, launched a project on monitoring and population assessment of Carpathian wisents.

Description of the area

The Carpathians are outstanding among other large mountain ranges of Europe. The range extends over almost 210 000 km^2, over the territory of six countries (from the southeast: Romania, Ukraine, Slovakia, Poland, Hungary and Czech Republic). The topography of the range is dominated by gentle slopes, with elevations ranging from ~100 m to 2665 m. A major part of the Carpathians constitute peaks and ridges of medium height (up to 1500 m) covered by forest or vast alpine meadows. Climate is of a temperate-continental type, with strong altitudinal gradients in mean annual temperatures (9 °C in the lower elevation to <0 °C at highest peaks) and precipitation (<500 mm to >2000 mm annually). Climate and topography have shaped four distinct zones of potential natural vegetation: foothills (<600 m) dominated by beech (*Fagus sylvatica*), hornbeam (*Carpinus betulus*) and oak (*Quercus* spp.); montane mixed forests with beech and fir (*Abies alba*) (600–1100 m in the north and 1400 m in the south); sub-alpine coniferous forests (up to 1500 m/1800 m) with Norway spruce (*Picea abies*), stone pine (*Pinus cembra*); and alpine vegetation above the timberline (Webster *et al.* 2001; UNEP 2007).

Its area remains comparatively undeveloped, with mostly traditional, small-scale agriculture and fairly well-maintained, highly productive, native deciduous (beech-dominated) and mixed forests. About 16% of this area is protected as national, nature or landscape parks which, being usually of a considerable size and excluded from economic activities, provide suitable conditions for wildlife refuges. Since accession of the Czech Republic, Hungary, Poland, Slovakia and Romania to the EU, a network of Natura 2000 sites has been implemented there, mostly overlapping with already existing parks, but

Ecology, Evolution and Behaviour of Wild Cattle: Implications for Conservation, ed. M. Melletti and J. Burton. Published by Cambridge University Press.

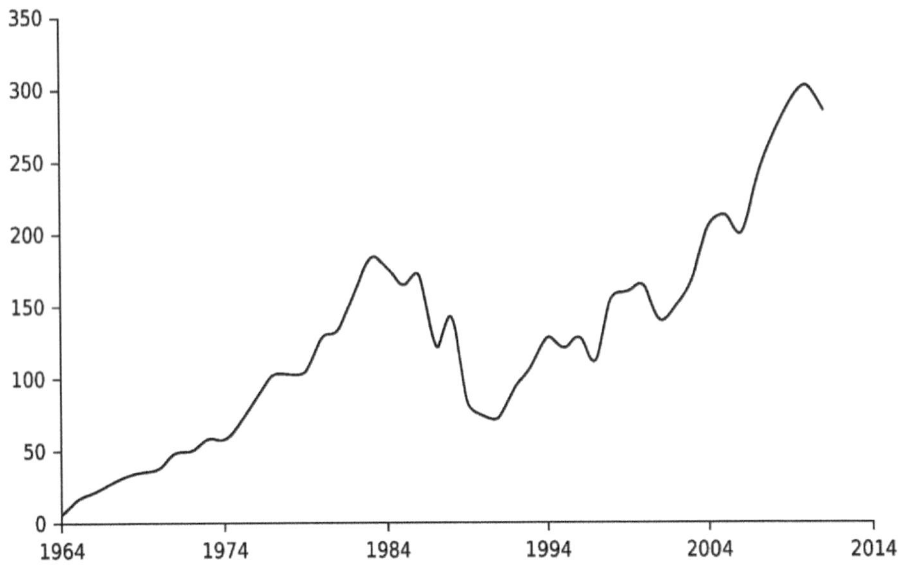

Figure 22.1 Population dynamics of wisents in Polish Bieszczady Mountains.

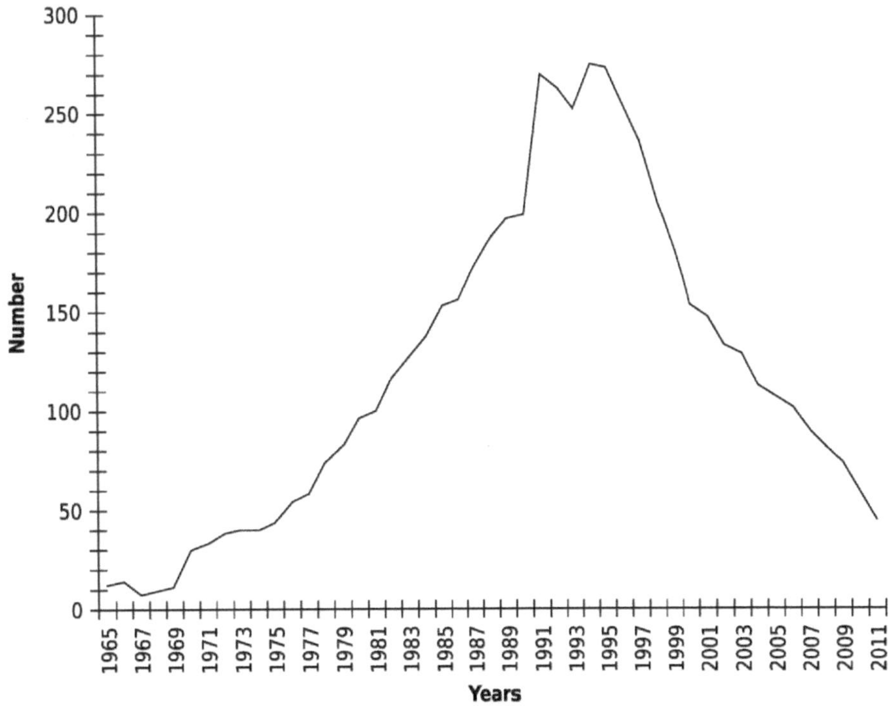

Figure 22.2 Population dynamics of wisents in Ukrainian Carpathians (after Khoyetsky, 2012).

according to the Habitat Directive these emphasize the protection of key habitats for priority species like the wisent. The Carpathian fauna is very rich compared to other European mountain ranges (comprising a complete set of large predators: brown bear, wolf, lynx and wildcat) (Vera 2000; Webster *et al.* 2001).

The status of the species in the Carpathians

By the beginning of the restitution project, there were just two wisent herds at Polish Bieszczady – counting together about 170 animals – and three herds in the Ukrainian part of the range – estimated in total at over 150 animals (Perzanowski & Paszkiewicz 2000; Khoyetsky 2012). Since then, the Polish population has grown to about 300 individuals; two small herds in Ukraine (Nadvirna and Maydan) ceased to exist, and the largest Ukrainian herd at Bukovyna dwindled to about 30 animals (Figures 22.1–22.2).

Just two years ago, as a result of an introduction, the Maydan herd was restored (presently it counts 15 individuals). In 2004 the creation of a new herd started at Slovakia (presently nine wisents in the wild), and in spring 2012 the first group of five animals was released to the wild at Chitele within Vanatori-Neamt Nature Park, Romania. All breeding facilities

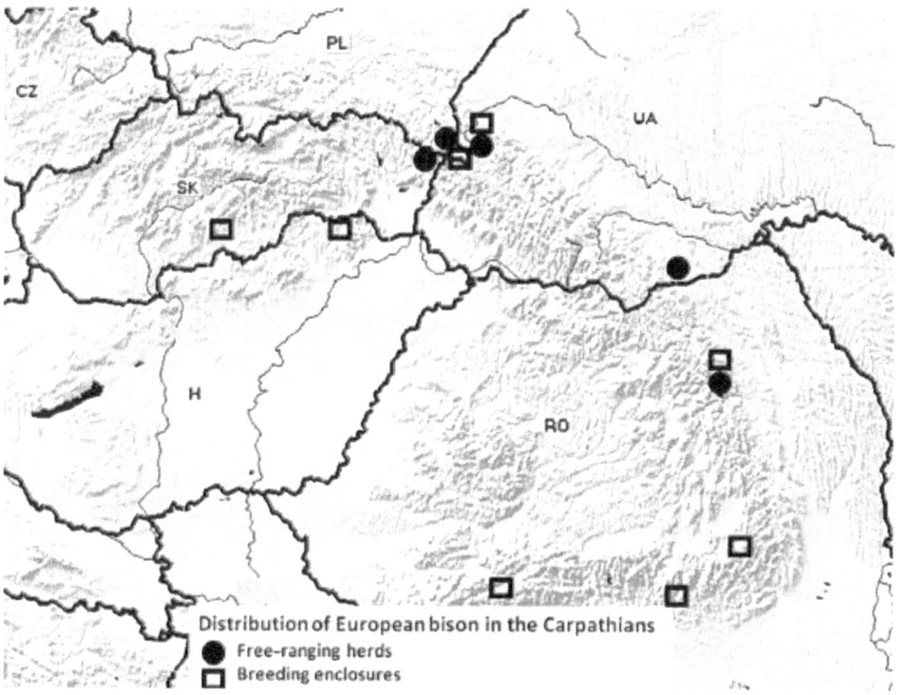

Figure 22.3 Distribution of free-ranging herds and breeding facilities for wisents in the Carpathians.

(three in Romania: Hateg Slivut, Dragoš Voda and Neagra-Bucsani; and one in Slovakia: Topol'cianky) that existed by the end of the 1990s are still functional, and several new enclosures have appeared (in Romania: Vanatori-Neamt, Vama Buzaului; Hungary: Füzérkomlós; and Poland: Muczne) (Figure 22.3). In all countries of the Carpathian eco-region, wisent has the status of a species protected by law.

An important rule which so far has been observed in all countries of the region is that all animals introduced to the Carpathians belong to the mixed Lowland–Caucasian line. Therefore, when the Carpathian metapopulation of wisents becomes functional, i.e. when conditions (habitat connectivity) allow for a natural exchange of individuals, the principle of avoiding mixing the genetic lines of the species will be maintained.

A good sign is that the Carpathian population of wisents is still growing, which reflects the general habitat suitability for this large herbivore. Unfortunately, however, herds' ranges are all spatially isolated, so without ensuring contact or exchange of individuals in the foreseeable future, effective numbers of even the largest Carpathian herds will be too low to ensure their genetic and demographic stability in the long term (Perzanowski & Kozak 2000). The analysis of habitat patches along the Carpathian arch, taking into account not only the type of habitat but also the size of particular patches, revealed 26 patches of a suitable habitat unoccupied so far by wisents (Habitat Suitability Index >0.5), larger than 50 km^2 with a carrying capacity of over 50 animals each. The majority of those areas are situated in the southeastern part of the range – in Ukraine and even more so in Romania (Figure 22.4).

Large parts of the Carpathians, however, are dissected by dispersal barriers (e.g. railroads and other communication routes, linear settlements) preventing in many cases a spontaneous recolonization of the range by wisents (Figure 22.5).

Simulation of various protective scenarios has shown that anti-poaching measures increased a probability of colonization of adjacent patches, but only when large patches were protected. According to the model, the most cost-effective option to increase the metapopulation viability appeared to be reintroductions. Construction of wildlife overpasses, although improving substantially the population viability, was estimated as costly (Kuemmerle *et al.* 2010, 2011).

Goals of the restitution programme

The main aim of the restitution programme of the wisent to the Carpathians is an increase of its numbers to an effective size, ensuring long-term viability of this population. This requires proper design of refuges for the species, and an improvement of the present genetic structure of the population (Pucek *et al.* 2004).

The programme is based on existing home ranges of free-ranging bison herds. The first ('western') group, inhabits the International Biosphere Reserve 'Eastern Carpathians' (consisting of three national parks: Polish 'Bieszczady', Slovak 'Poloniny' and Ukrainian 'Uzhansky' and three landscape parks: two in Poland and one in Ukraine), and situated slightly to the east is Skolivsky Beskydy National Park in Ukraine. The 'eastern' group of herds lives in the 'Zubrovica' Reserve in Ukrainian Bukovyna region and Vanatori Neamt Natural Park in Romania. The meta-population of the western region consists of herds in Polish Bieszczady Mountains, connected with the small herd from Poloniny National Park, and, living some 30 km to the east, the newly created herd at Skolyvski Beskydy

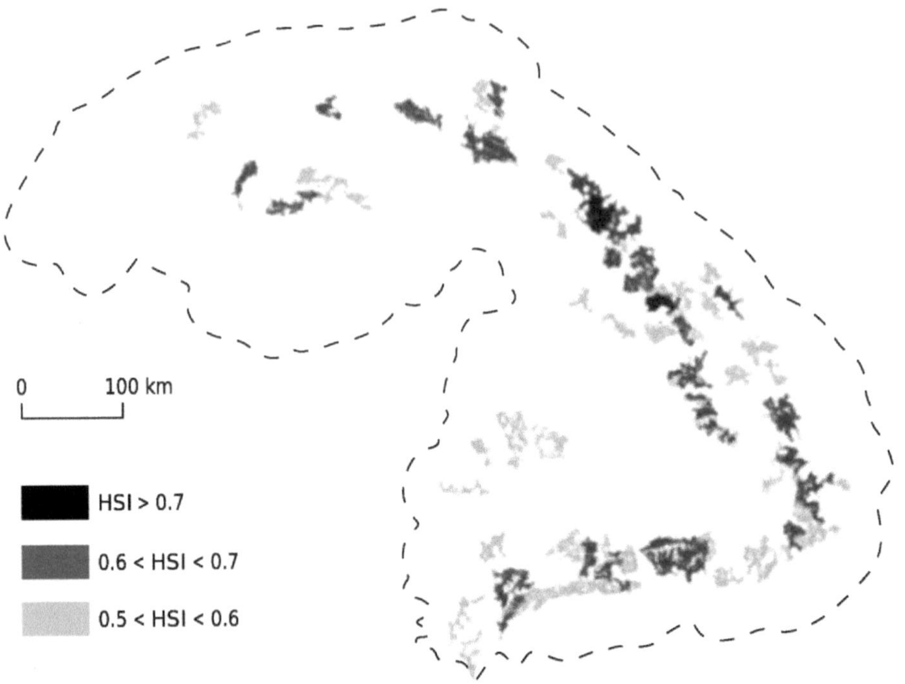

Figure 22.4 Identified patches of a suitable habitat for wisent introduction in the Carpathians (after Kuemmerle *et al.* 2010).

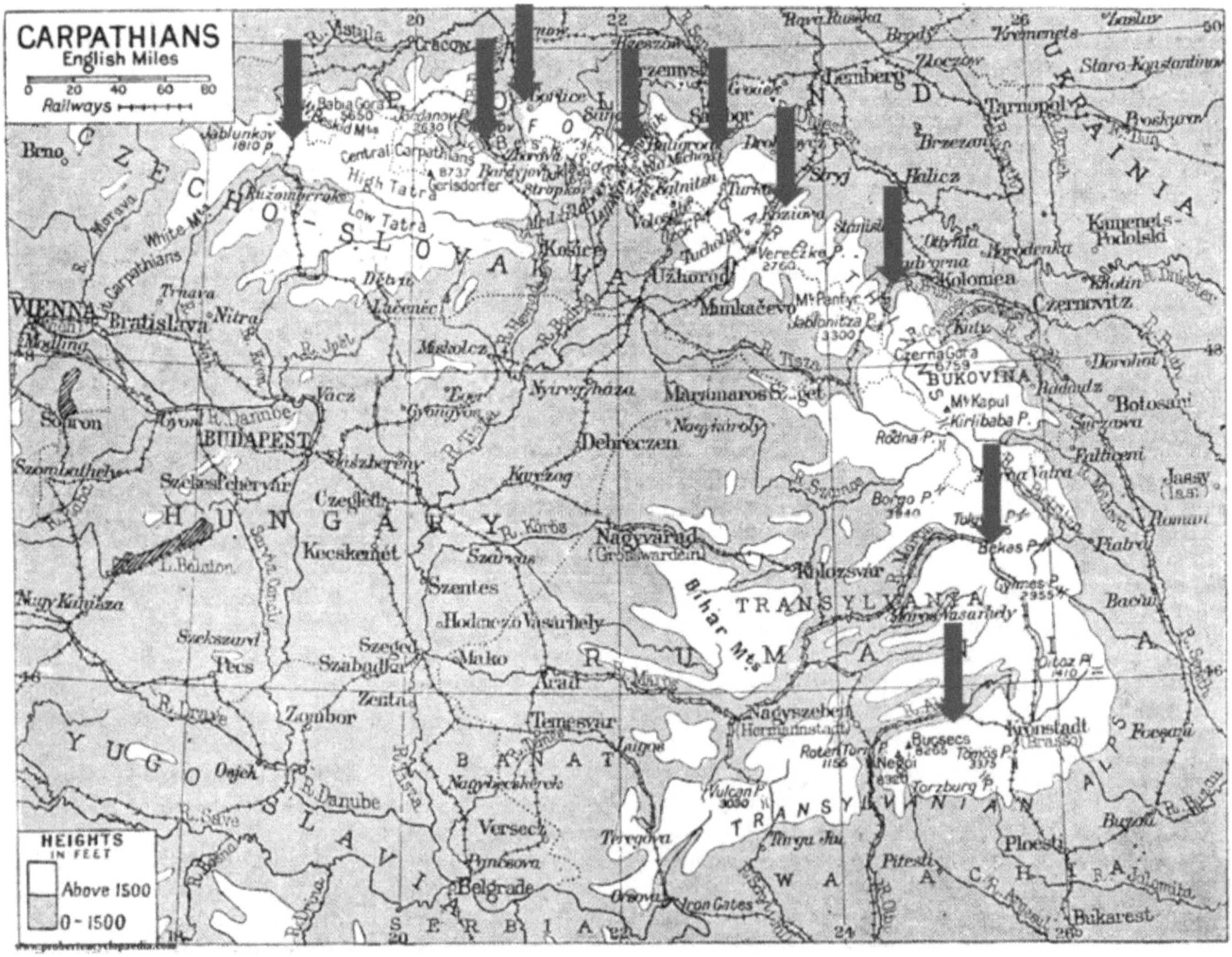

Figure 22.5 Main communication routes fragmenting the Carpathians.

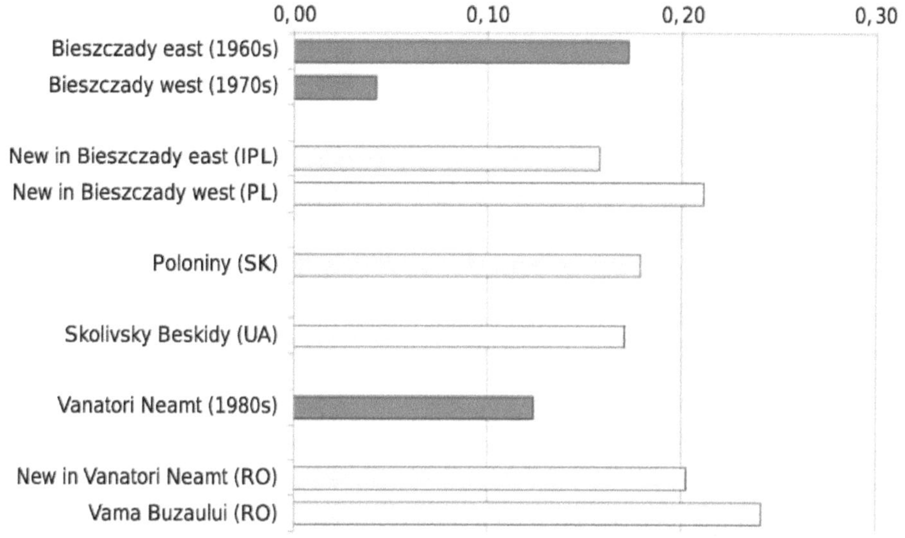

Figure 22.6 The contribution of five founders unique for the Lowland–Caucasian line (No. 100, 96, 95, 35, 46) to the gene pool of wisent herds in Bieszczady PL and Vanatori Neamt RO at the moments of their creation, and in groups of animals transferred to other locations in the Carpathians in 2001–2012. Dark bars represent animals that were introduced between the 1960s and 1980s, and light bars represent wisents introduced after 2000.

National Park. It now accounts for about 300 animals. The 'eastern' group consists of 28–60 (depending on the source) animals in the Ukrainian reserve and five animals recently released to the wild at Vanatori-Neamt National Park. The genetic composition of all those animals is strongly skewed towards over representation of seven founders common to both lines (altogether providing almost 90% of the gene pool) (Olech 1989; Perzanowski & Paszkiewicz 2000; Olech & Perzanowski 2002).

Therefore, the main effort should be oriented towards the supplementation of those herds with animals bearing genes that are lacking or underrepresented in the founders (five founders that are exclusive for the Lowland–Caucasian line). This part of the programme was initiated in 2002, when four young wisents selected from Scandinavian breeding centres were released at Bieszczady. Since then, 30 animals originating from various European countries (Denmark, Sweden, Czech Republic, Ireland and Germany) have joined this population. Subsequently, a free-ranging herd, being genetically a supplement to the neighbouring Polish population of Bieszczady, has been established at Slovak 'Poloniny' National Park with five animals selected from Dutch, Italian and Swiss breeding centres, released in December 2004. In Ukrainian Skolyvski Beskydy, wisents brought in 2010 and 2011 formed a herd that now stands at 15 individuals. In total, 78 animals were brought to the Carpathians from European breeding centres in order to supplement the gene pool of this metapopulation (Table 22.1).

The genetic value of those animals can be measured by the proportion of unique founders' genes within their gene pool. This value can be calculated on the basis of pedigree data and compared with such values for animals released in the 1960s in Bieszczady. It must be explained that the Lowland–Caucasian line was derived from 12 founders, and five of them are unique for this line (the remaining seven are common with the Lowland line). The share of those five founders to the gene pool is shown in Figure 22.6. It is possible to see that animals which were introduced in recent years represent a higher percentage of genes from these rare and important founders (Olech & Perzanowski 2002).

The future of the 'eastern' group is now a little problematic since the numbers in the Ukrainian herd are decreasing every year, and probably further efforts will have to be based mostly on the development of the herd at Vanatori-Neamt National Park. Apart from free-ranging animals, there are currently about 20 wisents in the acclimatization enclosure, which are supposed to breed and become a source of new animals for the supplementation of the wild herd.

A target for Carpathian populations is an effective number of 200–250 individuals in western and eastern subpopulations. Assuming the present sex structure (1:1.2) and a proportion of young and non-breeding animals (about 40%), this means the total population size needs to be larger by 2–2.5 times, i.e. about 1000 bison across all Carpathian herds.

Since in Poland all wisents in captivity belong to the Lowland line, animals suitable for reintroduction are available only in captive breeding centres of Western Europe. The genetic structure of those animals is much less skewed than in Carpathian herds, and the contribution of five founders (exclusive for the Lowland–Caucasian line – No. 100, 35, 46, 95 and 96) is on average twice as high. Estimated median and quartiles for a founder contribution show that a high proportion of captive animals bear even higher percentages of those underrepresented genes and such individuals will be preferred for the programme.

Apart from the genetic aspects of the programme, its second main aim is to ensure a sufficient area and appropriate habitat conditions for the growing population. Except for spatial requirements, wisent home ranges should be protected against an excessive human-related disturbance, so the animals can display their natural behaviour and habitat preferences. The first choice, therefore, will be protected areas, where foraging of wisents upon vegetation is not considered as damage. Such potential exists within the home range of the western population where the International Biosphere Reserve

Table 22.1 Wisents introduced to the Carpathians from European breeding centres in 2001–2012, under the framework of the project on an improvement of the gene pool, coordinated by the European Bison Conservation Centre

Year	Introduction site	M	F	Source
2001	Komańcza PL*	3	1	Denmark, Sweden
2004	Poloniny SK	3	2	Holland, Italy, Switzerland
2005	Poloniny SK	1	1	Czech Republic
	Vanatori Neamt RO	2	6	Germany, Switzerland
	Komańcza PL*	1	1	Czech Republic
2006	Vanatori Neamt RO	1	2	Germany
	Poloniny SK	0	2	Czech Republic, Slovakia
	Komańcza PL*	0	1	Czech Republic
2008	Vama Buzulaui RO	2	3	Austria, Germany, Italy, Switzerland
	Komańcza PL*	4	4	Germany, Ireland, Switzerland
2009	Vama Buzulaui RO	3	3	France
	Skolyvski Beskid UA	3	3	Germany
	Vanatori Neamt RO	0	4	Sweden
2010	Skolyvski Beskid UA	3	2	Austria, Germany
2011	Bukowiec PL*	1	2	Czech Republic
2012	Komańcza PL*	4	1	Germany
	Muczne PL*	3	4	France, Germany, Switzerland
	Muczne PL*	2	0	Germany
	Total	**36**	**42**	

* Bieszczady PL is divided into the western (Komańcza) and eastern part (Bukowiec and Muczne); PL: Poland; SK: Slovakia; RO: Romania; UA: Ukraine.

'Eastern Carpathians' occupies over 200 000 ha in three countries. The situation of the future Ukrainian–Romanian population in the east seems to be more difficult; optimal migration corridors, allowing for contacts between both herds, are still to be established. However, within Romania, where the possibilities for unrestricted migration along forested mountain ridges are considerable, sites suitable for introduction of wisents in the country's national parks have been already identified (Figure 22.7). The development of a Romanian network of Natura 2000 sites, after the accession of the country to the EU, should make this process fairly easy. Undoubtedly, this country has the highest potential for the development of free-ranging wisent populations.

Although wisents (especially bulls) are capable of long-range migrations, natural exchange of animals can be expected at first only between herds for which home ranges are not too distant from one another (around 10 km). To create a link between the eastern and western subpopulations, further introductions will be necessary along the Ukrainian Carpathians (e.g. Skolyvsky Beskydy). This should allow for a metapopulation dispersing along the Carpathian chain, and gradually occupying the most suitable sites. The expected effect would be an increase of the effective size of the population without an excessive congregation of wisents, undesirable for the habitat and possible damages to agriculture, and threatening the population with possible outbreaks of diseases. The ecological benefit would be the influence of a numerous population of this large grazer upon forest habitats, initiating the processes leading to structural changes in present close-canopy forest towards more patchy and diversified habitat (Vera 2000).

The re-establishment of a truly self-sustainable, wild population of wisents in the centre of the continent requires the active protection of this highly threatened species. Taking into account all drawbacks connected with its genetic structure, low resistance to diseases, habitat and spatial requirements and – last but not least – being an easy target for poachers, it would be unreasonable to expect that without human assistance this population may achieve demographically and genetically safe levels by itself. Nevertheless, its restitution may become incorporated into a general plan of nature conservation at the eco-regional scale, with the wisent being a major umbrella species. This can be done through associating home ranges of wisent herds with major protected areas of the region. Numerous wisents could play in the future the role of the 'ecological tool' contributing to the maintenance of diversity of Carpathian habitats. Finally, the restitution of the viable population of this species would be the first step in the reconstruction of something missing in the Carpathians for at least 200 years – the ecological niche of large grazers (wisents, aurochs and wild horses).The restoration of this ecological niche, apart from bringing back these keystone species to the ecosystem, would be an important contribution to an increased level of biodiversity in the region, and may facilitate an introduction of ecologically sound management of local forests.

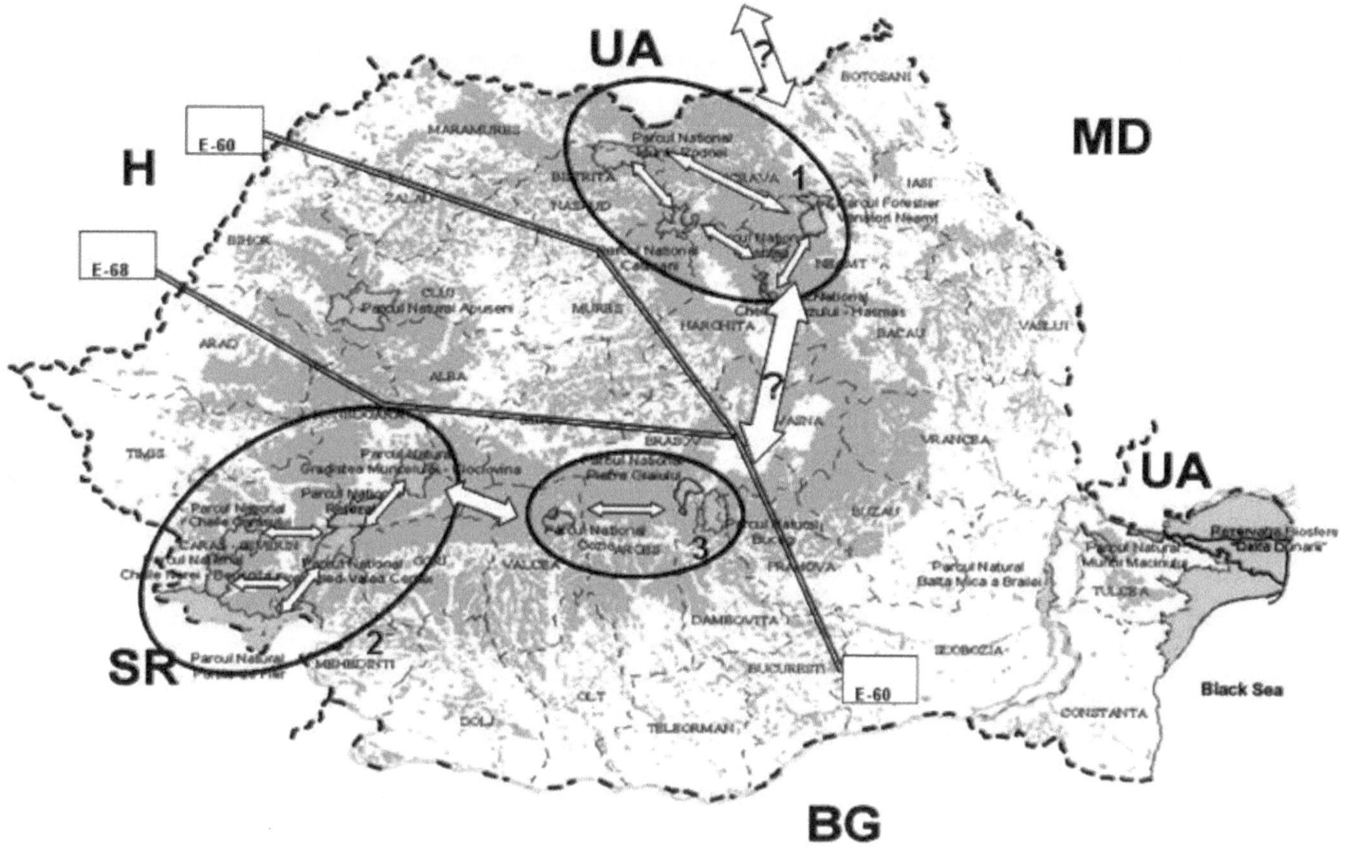

Figure 22.7 Proposed future breeding regions for wisents in Romania, with potential directions of animals' dispersion, and barriers created by two main highways of Romania (after Perzanowski & Deju, 2006, modified).

References

Augustyn, M. & Kozak, I. (1997). The trends of anthropogenic pressure in Polish and Ukrainian Carpathians. In *Proceedings of the Second Annual Meeting of the International Centre of Ecology, Polish Academy of Sciences*, pp. 15–22. Ustrzyki Dolne, Poland: Carpathian Branch ICE PAS.

Khoyetsky, P. (2012) The prospects for *Bison bonasus* (L.) restoration to the Eastern Carpathians. *European Bison Conservation Newsletter* **5**: 73–78.

Krasińska, M. & Krasiński, Z. A. (2007). *European Bison: The Nature Monograph*. Białowieża: Mammal Research Institute PAS.

Kuemmerle, T., Perzanowski, K., Chaskovskyy, O., *et al.* (2010) European bison habitat in the Carpathian Mountains. *Biological Conservation* **143**: 908–916.

Kuemmerle, T., Perzanowski, K., Akçakaya, H. R., *et al.* (2011). Cost-effectiveness of different conservation strategies to establish a European bison metapopulation in the Carpathians. *Journal of Applied Ecology* **48**: 317–329.

Nahlyk, A. (1991). W sprawie transylwańskiego żubra. [About Transylvanian European bison]. *Sylwan*, **85**: 63–68. (In Polish)

Nahlyk, A. (1992). Pro Transilvanskoho zubra. [Transylvanian European bison]. *Lisove hospodarstwo, lisowa, paperova i derevoobrona promyslovist*, **23**: 22–27. (In Ukrainian)

Olech, W. (1989). The participation of ancestral genes in the existing population in European bison. *Acta Theriologica*, **34**: 397–407.

Olech, W. & Perzanowski, K. (2002) A genetic background for reintroduction program of the European bison *Bison bonasus* in the Carpathians. *Biological Conservation* **108**(2): 221–228.

Perzanowski, K. & Deju, R. (2006). Is there a future for European bison in Romania? In *Proceedings of the Workshop 'European Bison and Local Communities in Carpathian Mountains'. Vanatori N. P. 23-26. 08.2006*, pp. 11–16 Piatra-Neamt: Editura Nona.

Perzanowski, K. & Kozak, I. (2000). The Carpathian bison: its past and future perspectives. *Biosphere Conservation* **2**: 75–81.

Perzanowski, K.& Paszkiewicz, R. (2000). Restytucja i współczesny stan populacji żubrów w Bieszczadach. [Restitution and present status of European bison population in Bieszczady]. In Monografie bieszczadzkie: Kręgowce Bieszczadów Zachodnich. Vol. 9: 219–229. Ed. Głowaciński, Z. Ustrzyki Dolne: Bieszczadzki National Park. (In Polish)

Pucek, Z. (ed.), Belousova, I. P., Krasińska, M., Krasiński, Z. A. & Olech, W. (2004). *European Bison: Status Survey and*

Conservation Action Plan. Gland and Cambridge: UK IUCN/SSB Bison Specialist Group IUCN.

Szalay, B. (1913). Bölènyadatok kritikája. *Vadászlap* **14**: 260. (In Hungarian)

UNEP (2007). *Carpathians Environment Outlook*. Geneva: UN Environment Program.

Vera, F. W. M. (2000). *Grazing Ecology and Forest History*. Wallingford: CABI Publishing.

Webster, R., Holt, S. & Avis, C. (2001). The status of the Carpathians: WWF-The Carpathian Ecoregion Initiative Report, November.

Chapter 23

Capture and translocation of gaur (*Bos gaurus*) in India

Parag Nigam, Sankar Kalyansundaram, Dave Cooper, Les Carlisle and Harbhajan Singh Pabla

Introduction

The gaur (*Bos gaurus*), also known as the 'Indian bison', inhabits a wide variety of habitat across South and Southeast Asia, with a presence in three disjoint regions: southwestern India, central India and northeastern India (including Nepal, Bhutan and Bangladesh) (Choudhury 2002a). The habitats broadly include the tropical wet evergreen and semi-evergreen forests in the Western Ghats; tropical moist deciduous and dry deciduous forests in Central India; and tropical evergreen, semi-evergreen, subtropical broadleaf, dry deciduous and temperate broadleaf forests in northeastern India (Hussain & Khader 1969; Johnsingh 1979; Choudhury 2002a). Though gaur were once widely distributed across India, there has been a decline in their number and range, which has been primarily attributed to habitat destruction, poaching, predation – especially at calf and subadult stage – disease, insurgency, straying and various developmental activities (Imam 1985; Choudhury 2002a, 2002b).

The turn of the twentieth century witnessed a remarkable decrease in the population densities, as well as local extinctions of gaur from some protected areas (PAs), namely Bandhavgarh Tiger Reserve (BTR) and Sanjay Gandhi National Park in Madhya Pradesh; Thattakad Wildlife Sanctuary in Kerala and Kanger Valley National Park in Chhattisgarh (Sankar *et al.* 2001). These known extinctions have happened in a relatively short span of time and have necessitated urgent measures to revive the populations. The possible solution besides habitat improvement, habitat connectivity and law enforcement involves active management of the species by reintroduction, or replenishing them in vacant suitable habitats.

Bandhavgarh Tiger Reserve, located in central India, with its tropical dry deciduous forest, provides an excellent habitat for the gaur. Though the reserve supported a small population of gaur (38 individuals) till 1995, the population disappeared between 1995 and 1998 due to unknown reasons. As a part of the conservation initiative, an attempt was made to establish the species in its historical range at BTR, from where it had been extirpated. The IUCN/SSC Guidelines for Reintroduction (IUCN 1998) formed the basis for these translocations. These guidelines propose establishment of 'a viable, free ranging population in the wild of a species, subspecies or race, which has become globally or locally extinct' as the goal for reintroduction programmes.

The objective of the project was to enhance the long-term survival of the gaur; to maintain and/or restore natural biodiversity of BTR, and also to provide long-term economic benefits to the local and/or national economy, to some extent.

In all, 50 gaur – 36 females and 14 males – were captured from Kanha Tiger Reserve (KTR) and translocated to BTR during January 2011 and March 2012. The animals had to be transported approximately 250 km by road and required 8–10 hours of travel time to reach BTR. This was the first successful mass capture and relocation of gaur carried out in India.

Preparatory phase

The initial phase of the project started in 2006–2007 and included carrying out a feasibility study to establish ecological requirements relevant to gaur; assessing the availability of food, cover and water during lean periods at BTR; evaluation of protection strategies against anthropogenic factors (human disturbances, interaction with domestic livestock, poaching) in BTR; identifying source for founder stock, developing an action plan and veterinary protocols for gaur capture and translocation and getting necessary administrative approvals and permissions for capture and translocation.

Analysis and simulation of gaur population viability analysis (PVA) at BTR was assessed using the software VORTEX 9.50 (Lacy *et al.* 2005) under different scenarios (Pabla *et al.* 2011). The simulation of scenarios indicated reintroduction of 50 animals with a sex ratio of 60% females and 40% males at a time, or reintroduction of 20 animals with same sex ratio and subsequent supplementation of another 30 at an interval of ten animals every two years. The overall objective was to maintain a viable population of approximately 50+ animals at BTR.

Ecology, Evolution and Behaviour of Wild Cattle: Implications for Conservation, ed. M. Melletti and J. Burton. Published by Cambridge University Press.

Infrastructure and logistics

The project started with the design of a specialized, compartmentalized transport truck with dimensions suiting to the animals' needs. The major considerations included adequate ventilation, proper flooring and escape-proof doors and locks. Simultaneously, efforts were made to design a stretcher with suitable provision for weighing and carrying immobilized animals from the site of capture to the loading vehicle. A holding boma with a loading ramp was constructed at the site of capture in KTR to hold animals prior to loading on to the transport truck. As the project involved soft release of animals in BTR, a 50 ha enclosure with good vegetation, cover and water was selected and enclosed with a powered fenced.

Capture timings

Winter months (January to March) were selected as the time for the capture of animals as they coincided with the animals coming down from hilly areas to plains, thereby provided optimum temperature conditions for the capture and drier grounds for vehicular access. Capture and translocations were timed to coincide with the cooler hours of the day (below 25 ° C) to avoid heat-related complications (Dinerstein *et al.* 1990; Caulkett 2001). Morning captures were advantageous as the animals have limited rumen contents during early morning hours, thereby potentially decreasing the chances for tympany and regurgitation. It also decreases the chances for passive regurgitation, thereby preventing the potential occurrence of aspiration pneumonia (Curro 2007). Another important consideration during captures was to time the capture operation and movement of the animals in such a manner that the animals could be offloaded without any delay upon arrival at the release site.

Selection of animals

It was envisaged that 20 gaurs were to be captured for initiating a population at BTR. Kanha Tiger Reserve was identified to provide the source population for the reintroduction programme as the area had a well-established gaur population of 1600+ animals in 2006 (Pabla *et al.* 2011). Apparently healthy animals in different age and sex classes from spatially distinct areas were selected. Care was taken to avoid capture of old, pregnant and lactating females with calves, taking due care not to disturb herd dynamics. The animal herd was located during the early hours of the morning and subsequently approached using captive elephants (Figure 23.1). The herd was usually not disturbed by the close approach of captive elephants and the animals could be approached up to a distance of 15–20 m, though with caution. Captive elephants were used to gently push the herd into open areas that had access to the loading vehicle. Captive elephants could efficiently hold the animals within the herd and guided the darted animal away from waters or difficult terrain. This was important as the entire herd became cautious and excited if any member was sneaked out of the herd.

Capture procedure and immobilization

Drug delivery

Remote drug delivery was facilitated using all-plastic darts delivered through an air-pressurized Dan Inject (Model JM) projector. Selection of the dart projector was based on proven

Figure 23.1 Gaur herd being moved into open areas using captive elephants (photo by Ujjwal Kumar Sinha, WII). A black and white version of this figure will appear in some formats. For the colour version, please refer to the plate section.

Figure 23.2 Identified Gaur darted within the herd (photo by Ujjwal Kumar Sinha, WII). A black and white version of this figure will appear in some formats. For the colour version, please refer to the plate section.

accuracy, compactness, easy manoeuvring on elephant back, low weight, minimal traumatization and negligible disturbance of the animal. Though the effective range of the equipment is up to 60 m, the majority of animals were darted from a distance of <15 m. All the animals were darted in the hindquarters as the area provides good muscle mass for efficient drug delivery (Figure 23.2). Plastic darts of 3 ml capacity were used and the needle length was 40 mm.

Criteria for drug dosage calculations

Body health condition, physiological status, age and estimated weight were considered while determining the drug dosage. Correlation between drug dosage and body weight has been proved to be important for several wildlife species (Harthoorn & Luck 1962; Pienaar & Van Niekerk 1963; Blane *et al.* 1967; Wallach *et al.* 1967; Yusof 1990). Weight of the selected animal was estimated based on published information and past experience and ranged between 250 kg and 800 kg for subadults and adults (Curro 2007).

Drugs for immobilization

The detailed published information on immobilization drugs and medicaments for use in gaur are minimal (Armstrong 1989; Wilson *et al.* 1993; Napier *et al.* 2011). Gaur (*Bos gaurus*) immobilizations have primarily been carried out in the captive stock using opioids, α2-adrenoceptor agents and dissociative anaesthetics. Drug combinations comprising Immobilon® (etorphine hydrochloride)-xylazine (Wiesner *et al.* 1982); carfentanil-xylazine (Armstrong 1989; Wilson *et al.* 1993; Napier *et al.* 2011); thiafentanil-medetomidine (Napier *et al.* 2011), tiletamine/zolazepam with or without xylazine (Wilson *et al.* 1993) and xylazine-ketamine (William *et al.* 2011) have been successfully used in captive populations. Free-ranging studies on Malayan gaur (*Bos gaurus laosiensis*) demonstrated effective use of Immobilon® (etorphine + acepromazine) alone and in combination with azaperone and Rompun® (Conry 1986).

The important concerns while selecting immobilizing drugs for use in free-ranging gaur in the operation were its appropriateness and local availability. The majority of drugs (opioids and neuroleptics) for use in gaur are not available in India. Their procurement and use required administrative clearances, approval and licence that required a considerable period of time.

As the immobilizations were to be carried out in a free-ranging situation, low drug volume, rapid induction and minimal knockdown time, availability of reversal drug and rapid recovery time were considered to be important criteria prior to selection. Opioids have been known to possess these qualities and hence were considered as the drug of choice for immobilizing free-ranging gaur (Haigh 1990). The ability to have immediate reversal of the immobilization has also proved to be valuable for their use (Wiesner *et al.* 1982). Etorphine hydrochloride and subsequently thiafentanil oxalate were used as primary drugs for immobilization. The procedures followed in either case are discussed later in the chapter.

Another important concern in capture and translocation of animals was to identify use of short- and long-acting tranquilizers to support capture and transport, and facilitate adaptation of the animal in the new environment. Studies have shown that capture, confinement, transportation or integration into a new herd or group are traumatic events and result in considerable excitement and stress to the animal. Animal may injure themselves, others in the group or may exhibit exhaustion with fatal consequences (Hofmeyr 1981; Ebedes & Raath 1999; Read *et al.* 2000; Read 2002). Long-acting tranquilizers have been effectively used in capture and transport, and produce calming in animals by moderating excitement and reducing aggression and motor activity (Ebedes 1993). Besides these, tranquilizers have aided the animal in adjusting to newer environments by appropriately addressing the issues of exhaustion that may result from constant pacing and loss of condition due to refusal to eat and drink (Read 2002). Phenothiazine derivatives (acetylpromazine, propionylpromazine, perphenazine enanthate, zuclopenthixol acetate), butyrophenones derivatives (azaperone, haloperidol) and benzodiazepines (flumazenil, sarmazenil) have been used in a variety of wild animals as tranquilizers (Ebedes 1993; Ebedes & Raath 1999; Kock & Burroughs 2012). Based on the published information and past experience of their use in South African ungulates, azaperone, haloperidol and perphenazine enanthate were identified to provide tranquilization to support captures and subsequent procedures in gaur.

Azaperone was used as an adjunct to primary anaesthetic to hasten smooth induction and recovery, and provides tranquilization support to the animal during the initial time of capture. The drug has also been reported to counteract some of the respiratory depression caused by opioids and has the advantage of lack of effect upon thermoregulation (Marsboom 1969; Haigh 1990). The sedative effect of azaperone has been reported within ten minutes after intramuscular injection, and the duration of the effect is up to 6–8 hours (Kaandorp 2005). Haloperidol, another short-acting tranquilizer, was preferred as a loading dose to prolong the sedative effect and provide effective sedation during transport that would be maintained by the long-acting tranquilizer. The drug has been used in a variety of ungulates (Blumer 1991; Curro 2007) with onset of action within 5–10 minutes that lasts for 8–18 hours (Curro 2007). Perphenazine, a long-acting tranquilizer given intramuscularly, provided sedative effect to the animals at the release site and during the adaptive phase in the release boma. The drug effect is slow and commences after 10–12 hours, and the duration of tranquilization is 7–10 days (Read *et al.* 2000; Curro 2007).

Capture and handling of gaur

During January 2011, a total of 22 gaurs were immobilized using a drug combination containing etorphine hydrochloride (9.8 mg/ml, Captivion™, Wildlife Pharmaceuticals (Pty) Ltd.) and azaperone (40 mg/ml, Stressnil®, Janssen Pharmaceutica (Pty) Ltd). Of these, 19 gaur were translocated to BTR. The dose rate for etorphine ranged from 0.007 to 0.012 mg/kg for all the animals with mean ± SD of 0.009 ± 0.001 mg/kg. Azaperone up to 100 mg was mixed to make up the 3 ml dart; however, in two subadult animals 60 mg of azaperone was used. Hyaluronidase (1500 IU, Hyalase®, Wockhardt, UK Ltd.) up to 2500 IU was concurrently added to the drug mixture during initial capture to increase the rate of absorption and reduce induction time; however, its use was subsequently discontinued owing to excessive excitement resulting from rapid and fast induction. Subadult animals required slightly higher dose of etorphine compared to adults.

Subsequent gaur captures during March 2012 were carried out using thiafentanil oxalate (10 mg/ml, Thianil™, Wildlife Pharmaceuticals (Pty) Ltd) at a dose rate of 0.007–0.013 mg/kg, with mean ± SD of 0.01 ± 0.001 mg/kg. The drug was mixed with 100 mg of azaperone in the 3 ml dart. Thiafentanil was selected owing to pronounced opiate agonist activity, rapid induction, less cardiopulmonary depressant effect (Napier *et al.* 2011) and quick antagonism by naltrexone (Wolfe *et al.* 2004). A total of 31 gaurs were successfully immobilized and translocated to BTR.

The animal was left undisturbed till proper induction for safe handling was achieved. The dosages of neuroleptic analgesic used were successful in bringing the animal into sternal or lateral recumbency. However, all the animals were assisted during induction by covering their eyes using blindfolds to avoid visual stimulation, plugging ears to avoid auditory stimulation (Figure 23.3), grasping both horns and hindquarters to restrain and rolling the animal back into sternal recumbency. The sequence of response to immobilization included initial excitement, poor coordination, aimless walking, a high-stepping "hackney" gait, extension of the head, dog-sitting position followed by dropping into sternal recumbency (Figure 23.4–23.5). Similar findings have also been reported for other species (Haigh 1990; Zainuddin *et al.* 1990; Napier *et al.* 2011; Kock & Burroughs 2012). Induction time after etorphine and thiafentanil ranged from four to seven minutes and two to five minutes, respectively, for all animals. However, two animals had longer induction following etorphine immobilization owing to animal excitement prior to darting and local disturbance during induction. The field study revealed that variation in induction time was influenced by drug dosage and use of hyaluronidase in the drug mixture. The dosing of animals was based on estimated body weight (Dinerstein *et al.* 1990; Zainuddin *et al.* 1990). The animal was said to be stable and adequately immobilized for safe handling when the animal showed minimal response to tactile stimulation of head and hindquarters and physiological parameters were within acceptable range.

The vital physiological parameters such as rate and depth of respiration, rectal temperature and pulse rate were monitored throughout the operation. During immobilization, the respiration was deep and regular for the majority of animals and ranged from 16–18 per minute following opioid sedation. However, in two cases marked respiratory depression with rates below 12 per minute was recorded. Supplementation of oxygen was helpful in stabilizing the animals. Rectal temperature ranged between 38.3 °C and 38.9 °C (101 °F and 102 °F), whereas the pulse ranged between 62 and 72 per minute in both cases. Hyperthermia did not occur with either immobilization. Besides this, engorgement of the ear veins and pink conjunctival/oral mucous membranes were good external indicators of cardiac and respiratory functions and were continuously monitored till the animals were crated.

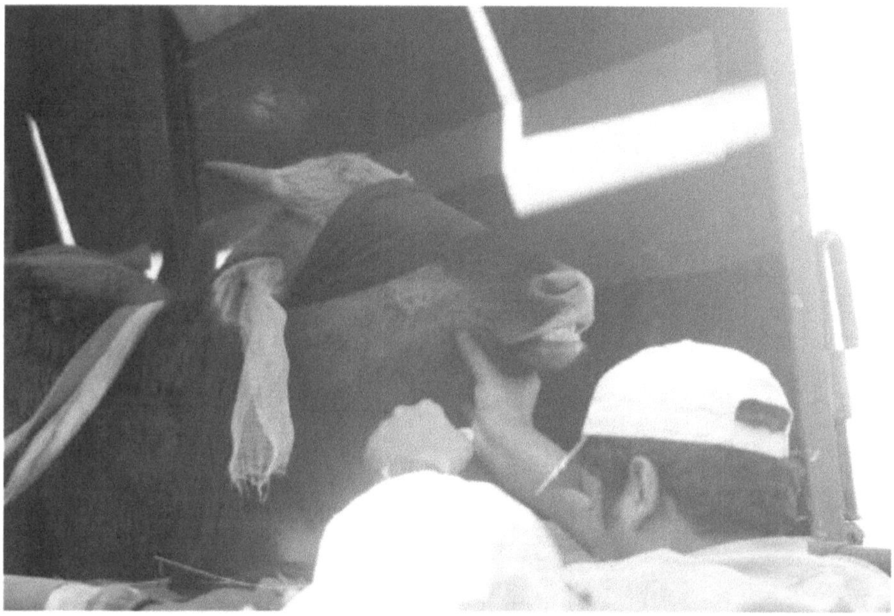

Figure 23.3 Plugging of ears and blindfolding of immobilized animal to avoid visual and auditory stimulation (photo by Aniruddha Majumder, WII).

Figure 23.4 Excitement and poor coordination during drug induction (photo by Parag Nigam, WII). A black and white version of this figure will appear in some formats. For the colour version, please refer to the plate section.

Supplementation of ketamine (100 mg/ml, Ketamil-100®, Troy laboratories, Australia) intravenously at doses of 0.10–1.0 mg/kg was beneficial in animals that were insufficiently immobilized as the drug acted fast and provided sedation levels safe for handling and carrying out further procedures. Ketamine (100–200 mg) given intravenously has been reported to be effective as a 'top-up' drug in animals that are not sufficiently immobilized by the opioid and sedative/tranquilizer mixture (Burroughs *et al.* 2012). The ketamine effect was noticed within 20–30 seconds after intravenous administration and lasted for about 10–15 minutes.

Partial reversal was achieved in an adult female that showed severe cardiac and respiratory depression (muscle tremors, collapsed peripheral veins, extension of head, slight regurgitation of gut contents) by administering 6 mg of butorphanol tartarate (2 mg/ml, Butrum-2®, Aristo pharmaceuticals, Mumbai, India). The drug was administered intravenously and provided immediate relief as manifested by resuscitation of respiration and improvement in vital physiological functions.

Maintaining the animal in sternal recumbency helped in decreasing the severity of hypoxemia and ruminal tympany

Figure 23.5 Extension of head during drug induction (photo by Parag Nigam, WII). A black and white version of this figure will appear in some formats. For the colour version, please refer to the plate section.

(Curro 2007). At all times, efforts were made to keep the head of the animal high so as to avoid regurgitation of gut content into the trachea and to ensure patent airways. In case the animal was immobilized on a slope, the animal was immediately positioned in a manner that the head was positioned in the downward-most direction of the slope to allow the gut contents to flow out of the oral cavity and decrease chance of aspiration. The oral cavity was inspected for any blockages.

Following stabilization, the animals were gently pushed over onto the stretcher and maintained in sternal recumbency. The animals were secured using ropes and the stretcher was moved onto the loading vehicle (Figures 23.6–23.7) for either taking them to boma or directly loaded onto the transport truck (Figure 23.8). The animals were radio-collared for subsequent monitoring at the release site. Subsequently, biological sampling for health assessment, disease screening and genetic studies was done. Intra-dermal tuberculin testing, taking body measurements and filing of horn tips to avoid injury during transport were also part of routine procedures. The dose was based on the body weight and each animal received an additional injection of antibiotic to counteract secondary infections, as well as vitamin and mineral supplements.

The length of immobilization required to carry out these procedures varied from 11 to 26 minutes, except in two cases where the immobilization time was more (34 and 43 minutes). This was attributed to long induction time due to animal excitement, local disturbance and a long distance to the loading truck. The animals were then transferred to the transport truck, positioned in sternal recumbency and administered with the antidote.

Naltrexone (50 mg/ml, Trexonil™, Wildlife Pharmaceuticals (Pty) Ltd) was administered as an antidote to negate the effect of both etorphine and thiafentanil. The drug was given in the ratio of 20:1 and 10:1 of etorphine and thiafentanil, respectively, either intravenously or intramuscularly. Signs of recovery included twitching of ears, holding up of head and movement of limbs, which occurred within 1–5 minutes after intravenous administration. There was no renarcotization noted in any of the animals and all the recoveries were smooth and uneventful.

Once the animals were crated, they were left undisturbed to settle. The animals were subsequently moved during evening hours. The transport truck carried 4–5 animals during each trip. Care was taken to keep the adult bull and cow with calf in separate compartments during the journey to avoid aggression and minimize transport stress to the mother and calf.

Late evening hours were selected as they ensured movement during cool hours, minimized disturbance due to humans and vehicles on the road during the night and also allowed for timed early-morning releases at BTR. Special care was taken to maintain uniform speed during the journey, with speed adjusted to avoid sudden jerking and huddling together of animals. The animals were monitored throughout the journey to ensure they were comfortable and did not injure each other. Most of the animals were calm during the journey and did not display any overt signs of distress. The distance between Kanha and Bandhavgarh was approximately 250 km and it took an average of 8–10 hours to reach the release site.

During the January 2011 captures, one adult female was revived immediately after capture as a young calf was seen following her after darting, whereas another female was immobilized again for loading on to the container as she resisted leaving the boma even after repeated efforts. A subadult female died as a result of aspirating rumen contents. The animal

Figure 23.6 Animal loaded on to the stretcher, being taken to the loading truck (photo by Aniruddha Majumder, WII).

Figure 23.7 Animal loaded on to truck for transport to boma (photo by B. Navaneethan, WII).

aspirated while in lateral recumbency and before the body position could be corrected.

A total of 51 immobilizations were carried out at KTR. Of these, 19 animals – comprising three subadult males, six subadult females, ten adult females and two adult males – were successfully translocated to BTR during January 2011 and seven subadult males, 13 subadult females, nine adult females and two adult males were successfully translocated to BTR during January 2012.

Gaur release at BTR and subsequent monitoring

The gaurs were brought in batches and released in a 50 ha power-fenced enclosure at BTR for acclimatizing in the new area, which provided ample food, water and shelter. The herd was finally released into the wild after one month. To study the home range, habitat use and food habits of reintroduced gaurs, in total 27 gaurs (six males and 21 females) were radio-collared

Figure 23.8 Loading of immobilized gaur directly into the transport container (photo by B. Navaneethan, WII).

Figure 23.9 Gaur herd at Bandhavgarh Tiger Reserve after the release in wild (photo by B. Navaneethan, WII). A black and white version of this figure will appear in some formats. For the colour version, please refer to the plate section.

(GPS/Argos/VHF). The collared animals were tracked intensively through 'homing in' and 'triangulation' techniques (White & Garrot 1990).

Till March 2012 the gaurs reintroduced during January 2011 ranged over an area of 268.7 km^2. The annual home ranges of the males and females were 254.4 km^2 and 200.4 km^2, respectively.

The group size was found to be fluid, ranging from 1 to 19 individuals (Figure. 23.9). The two gaur bulls were found solitary on many occasions, but joined the family groups after 2–3 days. In total, 68 plant species were recorded as food plants of gaur, which included 28 trees, six shrubs, ten herbs, 21 grasses and three climbers. Among the initial reintroduced stock there were three mortalities, two attributed to natural deaths (both adult females) and one subadult female was predated by a tiger. One uncollared female went missing from the park. There were three births, two from the first reintroduced stock and one from the second stock of animals. The present population of gaur in BTR is 50.

Acknowledgments

The gaur reintroduction programme was achieved through the generous cooperation and assistance received from individuals and bodies working together such as Madhya Pradesh Forest Department, &Beyond, Taj Safaris, KZN Wildlife, Wildlife Institute of India (WII), Centre for Wildlife Health and Forensics (CWHF), Jabalpur and Ministry of Environment & Forests, Government of India. Many people assisted in the planning and execution of the gaur reintroduction programme. The authors would like to specially thank Mr Jeff Cooke, Manager Game Capture Unit for Ezemvelo KZN Wildlife, for sharing his expertise during capture and translocation operation. The leadership provided at local levels by Mr Himmat Singh Negi, ex-Field Director, KTR, Mr Chandrakant Patil, Field Director, BTR and Mr Jasbir Singh Chauhan, Field Director, KTR; veterinary support provided by field veterinarians Dr Sandip Aggarwal, Dr Akhilesh Mishra, Dr Atul Gupta and Dr Sanjiv Gupta; laboratory support provided by Dr A. B. Shrivastav, Director, CWHF and assistance provided by Mr Sarath Champati and Mr Kartikeya Chauhan, Naturalists, Taj Safaris are duly acknowledged. We are also thankful to WII researchers Mr B. Navaneethan, Ms Preeti S. Virkar, Mr Manas P. Manjrekar, Dr Aniruddha Majumder, Ms Neha Awasthi and Mr Ujjwal Kumar Sinha for assistance during capture and translocation operations.

References

Armstrong, D. L. (1989). An evaluation of carfentanil as an immobilizing agent for gaur (*Bos gaurus*). *Proceedings of the American Association of Zoo Veterinarians Annual Conference* **8**.

Blane, G. F., Boura, A. L., Fitzgerald, A. E. & Lister, R. E. (1967). A potent morphine-like agent. *British Journal of Pharmacology and Chemotherapy* **30**: 11–22.

Blumer, E. S. (1991). A review of the use of selected neuroleptic drugs in the management of nondomestic hoofstock. *Proceedings of the American Association of Zoo Veterinarians* **1991**: 326–332.

Burroughs, R., Hofmey, M., Morkel, P., Kock, M. D., Kock, R. & Meltzer, D. (2012). Chemical immobilization: individual species requirements. In *Chemical and Physical Restraint of Wild Animals: A Training and Field Manual for African Species*, pp. 143–264. Eds. Kock, M. D. & Burroughs, R. Greyton: International Wildlife Veterinary Service,.

Caulkett, N. (2001). Bison (Artiodactyla: Bovidae). In *Zoological Restraint and Anesthesia*. Ed. Heard, D. Ithaca, NY: International Veterinary Information Service.

Choudhury, A. (2002a). Distribution and conservation of the gaur *Bos gaurus* in the Indian subcontinent. *Mammal Review* **32**: 199–226.

Choudhury, A. (2002b). The status of gaur *Bos gaurus* in north Cachar hills district of Assam. *Journal of Bombay Natural History Society* **99**: 506–509.

Conry, P. J. (1986). Immobilizing gaur with an etorphine and tranquilizer mixture. *Journal of Bombay Natural History Society* **83**: 499–504.

Curro, T. G. (2007). Non-domestic cattle. In *Zoo Animal and Wildlife Immobilization and Anesthesia*, pp. 635–642. Eds. West, G., Heard, J. D. & Caulkett, N. Ames, IA: Blackwell Publishing.

Dinerstein, E., Shrestha, S. & Mishra, H. (1990). Capture, chemical immobilization and radio-collar life in greater one-horned rhinoceros. *Wildlife Society Bulletin* **18**: 36–41.

Ebedes, H. (1993). The use of long-acting tranquilizers in captive wild animals. In *The Capture and Care Manual*, pp. 65–70. Ed. McKenzie, A. A. Pretoria: The South African Veterinary Foundation.

Ebedes, H. & Raath, J. P. (1999). Use of tranquilizers in wild herbivores. In *Zoo and Wild Animal Medicine Current Veterinary Therapy*, pp. 575–584. 4th editon. Eds. Fowler, M. E. & Miller, R. E. Philadelphia, PA: Saunders.

Haigh, H. C. (1990). Opioids in zoological medicine. *Journal of Zoo and Wildlife Medicine* **21**: 391–413.

Harthoorn, A. M. & Luck, C. P. (1962). The handling and marking of the wild East African elephant (*Loxodonnta Africana)* with the drug-immobilizing technique: second preliminary report. *British Veterinary Journal* **118**: 526–530.

Hofmeyr, J. M. (1981). The use of haloperidol as a long-acting neuroleptic in game capture operation. *Journal of South African Veterinary Association* **52**: 273–282.

Hussain, K. M. & Khader, H. M. (1969). The southern Asiatic gaur or Indian bison (*Bos garus garus* H.Smith). *Cheetal* **12**: 29–47.

Imam, B. A. R. H. (1985). Seasonal migration of gaur (*Bos gaurus* H. Smith): a comparative study between observations at Gaurimara Wildlife Sanctuary, West Bengal and Megpal and Bamni in Raiakhol, Orissa. *Cheetal* **16**: 45–48.

IUCN. (1998). *Guidelines for Reintroduction*. Gland and Cambridge: IUCN/SSC Reintroduction Specialist Group.

Johnsingh, A. J. T. (1979). The gaurs of my study area. *Hornbill* **10**: 30–32.

Kaandorp, J. (2005). The use of neuroleptic tranquilizers in zoological medicine. *Verhandlung ber erkrankungen zootiere* **42**: 16–26.

Kock, M. D. & Burroughs, R. (2012). *Chemical and Physical Restraint of Wild Animals, A Training and Field Manual for African Species*. South Africa: South African Veterinary Foundation.

Lacy, R. C., Borbat, M. & Pollak, J. P. (2005). *Vortex: A Stochastic Simulation of the Extinction Process*. Version 9.50. Broodfield, IL: Chicago Zoological Society.

Marsboom, R. (1969). On the pharmacology of azaperone, a neuroleptic used for the restraint of wild animals. *Acta Zoologica Et Pathologica Antverpiensia* **48**: 155–161.

Napier, J. E., Loskutoff, N. M., Simmons, L. G. & Armstrong, D. L. (2011). Comparison of carfentanil-xylazine and thiafentanil-medetomidine in electroejaculation of captive gaur (*Bos gaurus*). *Journal of Zoo and Wildlife Medicine* **42**: 430–436.

Pabla, H. S., Carlisle, L., Cooper, D., *et al.* (2011). Reintroduction of gaur (*Bos gaurus gaurus*) from Kanha Tiger Reserve to Bandhavgarh Tiger Reserve. Technical report.

Pienaar, U. de. V. & Van Niekerk, J. W. (1963). The capture and translocation of three species of wild ungulates in the Eastern Transvaal with special reference to R05-2807/B-5F (Roche) as a tranquillizer in game animals. *Koedoe-African Protected Area Conservation and Science* **6**: 83–90.

Read, M. (2002). Long acting neuroleptic drugs. In *Zoological Restraint and Anesthesia*. Ed. Heard, D. Ithaca, NY: International Veterinary Information Service (IVIS).

Read, M., Caulkett, N. & McCallister, M. (2000). Evaluation of zuclopenthixol acetate to decrease handling stress in wapiti. *Journal of Wildlife Disease* **36**: 450–459.

Sankar, K., Qureshi, Q., Pasha, M. K. S. & Areendran, G. (2001). *Ecology of Gaur (*Bos gaurus*) in Pench Tiger Reserve, Madhya Pradesh*. Dehradun: Wildlife Institute of India.

Wallach, J. D., Frueh, R. & Lentz, M. (1967). The use of M 99 as an immobilizing and analgesic agent in captive wild animals. *Journal of the American Veterinary Medical Association* **151**: 870–876.

White, G. C. & Garrot, R. A. (1990). *Analysis of Radio Tracking Data*. San Diego, CA: Academic Press.

Wiesner, H., Rietschel, W. & Gatesman, T. (1982). Practical experiences with the combination of 'Immobilon' and 'Rompun' in zoo animals. *Zeitschrift Des Kolner Zoo* **25**: 45–62.

William, J., Rao, B. G. D., Simon, M. S., Thilagar, T. S. & Kumar, R. S. (2011). Surgical management of evisceration under xylazine-ketamine anaesthesia in an Indian gaur (*Bos gaurus*). *Tamilnadu Journal Veterinary & Animal Sciences* 7: 213–215.

Wilson, S. C., Armstrong, D. L., Simmons, L. G., Morris, D. J. & Gross, T. S. (1993). A clinical trial using three regimens for immobilizing gaur (*Bos gaurus*). *Journal of Zoo and Wildlife Medicine* **24**: 93–101.

Wolfe, L. L., Lance, W. R. & Miller, M. W. (2004). Immobilization of mule deer with thiafentanil (A-3080) or thiafentanil plus xylazine. *Journal of Wildlife Diseases* **40**: 282–287.

Yusof, E. (1990). Drug dosage regression analysis for seladang. *The Journal of Wildlife and Parks* **9**: 112–127.

Zainuddin, Z. Z., Rahman, M. A. & Daim, M. S. (1990). Field immobilization of serow using etrophine hydrochloride and acepromazine maleate. *The Journal of Wildlife and Parks*, **9**: 65–70.

Chapter 24

Status and management of the endangered wild water buffalo (*Bubalus arnee*) in the Koshi Tappu Wildlife Reserve, Nepal

Ram Chandra Kandel, J. Stuart F. Barker and Mario Melletti

Introduction

Asian wild water buffalo (*Bubalus arnee*) are large ungulates, and the progenitors of all domestic water buffalo (*Bubalus bubalis*). There are two domestic types: the river buffalo of the Indian sub-continent and further west to the Balkans and Italy (Figure 24.1), and the swamp buffalo of Assam in the west, through Southeast Asia to the Yangtze Valley of China (Figure 24.2). All populations of *Bubalus arnee* are considered Endangered (IUCN 2013), but in Nepal this species is protected by the *National Parks and Wildlife Conservation Act* (His Majesty's Government Ministry of Law and Justice 1977).

There is evidence of buffaloes in the Indus Valley at least 5000 years ago (Nowak 1999; Lenstra & Bradley 1999). Although the historic range is uncertain, the species may have occurred from Mesopotamia to Indochina (Sinclair 1977). River and swamp buffalo were domesticated independently from different wild stocks that diverged anywhere from 10 000–15 000 (Barker *et al.* 1997) to over one million years ago (Amano *et al.* 1994), but probably around 128 000–280 000 years ago (Kumar *et al.* 2007a). River buffalo were domesticated around 6300 BP in the western region of the Indian subcontinent (Kumar *et al.* 2007b). Microsatellite and mtDNA diversity analyses (Yindee 2010; Zhang *et al.* 2011) combined with archaeological evidence (Higham 2002) indicate domestication of the swamp buffalo in southern China/northern Indo-China about 2000 BC.

The total number of pure Asian wild water buffalo has been assumed to be around 3300–3400 (Choudhury 1994; see also Chapter 16), although FAO considers there could be fewer than 200 (FAO 2000). The major threat to their genetic integrity is crossbreeding with domestic animals, as farmers allow their domestic buffalo females to range into forest areas to mate with wild bulls (see also Box 16.1). Other threats are competition for food and space with domestic cattle and hybrid feral buffaloes, habitat fragmentation, poaching, poisoning (using pesticides in paddy or sugarcane fields), human disturbances, thatch collection, illegal collection of wood and sand and gravel mining.

Figure 24.1 A River buffalo from Rajasthan (Photo by Anwaruddin Choudhury). A black and white version of this figure will appear in some formats. For the colour version, please refer to the plate section.

Ecology, Evolution and Behaviour of Wild Cattle: Implications for Conservation, ed. M. Melletti and J. Burton. Published by Cambridge University Press.

Figure 24.2 Swamp buffalo herd in Sulawesi, Indonesia (Photo by J. S. F. Barker) A black and white version of this figure will appear in some formats. For the colour version, please refer to the plate section.

Among the current putative wild populations, Koshi Tappu Wildlife Reserve (hereafter KTWR) in Nepal has a good chance of maintaining pure wild stock (Heinen & Singh 2001). But this small, relict population is extremely vulnerable to human pressure and hybridisation with domestic buffalo. Other populations are present in Assam and Madhya Pradesh, India (Choudhury 1994; see also Chapter 16) and in western Thailand, although there is a debate about the genetic integrity of some of these stocks (Diwaker & Bhusan 1988; Hedges 1995; Muley 2001). The main problem from a conservation viewpoint is distinguishing pure wild animals from hybrids with domestic buffalo. Muley (2001) used a combination of morphometric and genetic data to differentiate wild, hybrid and domestic swamp buffalo in Kaziranga National Park, India; and Heinen (2002) describes both behavioural and morphological characteristics used to distinguish them. Experienced observers can distinguish wild from domestic animals, either river or swamp, but Muley (2001) found for the identification of wild and hybrid animals a classification error of 49.1%. Using genetic data, Flamand *et al.* (2003) distinguished three populations in KTWR – wild, domestic and hybrid – but more extensive genetic data are necessary to confirm the distinctions between wild and hybrid, recognising that some hybrids will have a high proportion of wild genes due to extensive backcrossing.

Regarding the status of this species in KTWR, Dahmer (1978) described two herds in 1976, and Heinen (1993a) found these herds in 1987/1988 in highly overlapping home ranges. Heinen & Singh (2001) relocated the herds again in 2000, with little change in the home ranges in spite of the fact that the Koshi River had changed its main course. Heinen & Kandel (2006) censused again in 2004 and KTWR authorities in 2009 (Khatri *et al.* 2012). KTWR authorities carried out massive sweep operations in 2004, 2005 and 2006; and again in 2009 and 2010 inside the protected area, which removed almost 80% of feral cattle, and massive slaughtering operations of feral (hybrid and domestic) buffaloes were carried out in 2003 and 2004.

The southern part of KTWR was flooded during August 2008, resulting in total washout of the southeastern embankment and massive devastation in the southeast sector, threatening the wild buffalo population and its habitat. The Department of National Parks and Wildlife Conservation (DNPWC) authorised new surveys in 2010 and 2011, which were done following the methods applied by Heinen & Kandel (2006) and previous surveys. The results of these surveys are presented here, and compared with earlier censuses: (1) to determine the present status of the wild buffalo and their habitats; (2) to evaluate population threats; and (3) to recommend management prescriptions for the future.

Study area

Koshi Tappu Wildlife Reserve is located in southeastern Nepal, at a latitude of 26° N (Figure 24.3). The climate is subtropical and monsoonal. There are basically three seasons: winter (October–February), spring (March–June) and summer (July–September). Koshi Tappu Wildlife Reserve lies in the floodplain of the Koshi River in the southeastern *Terai* (lowlands) of Nepal, in parts of Saptari,

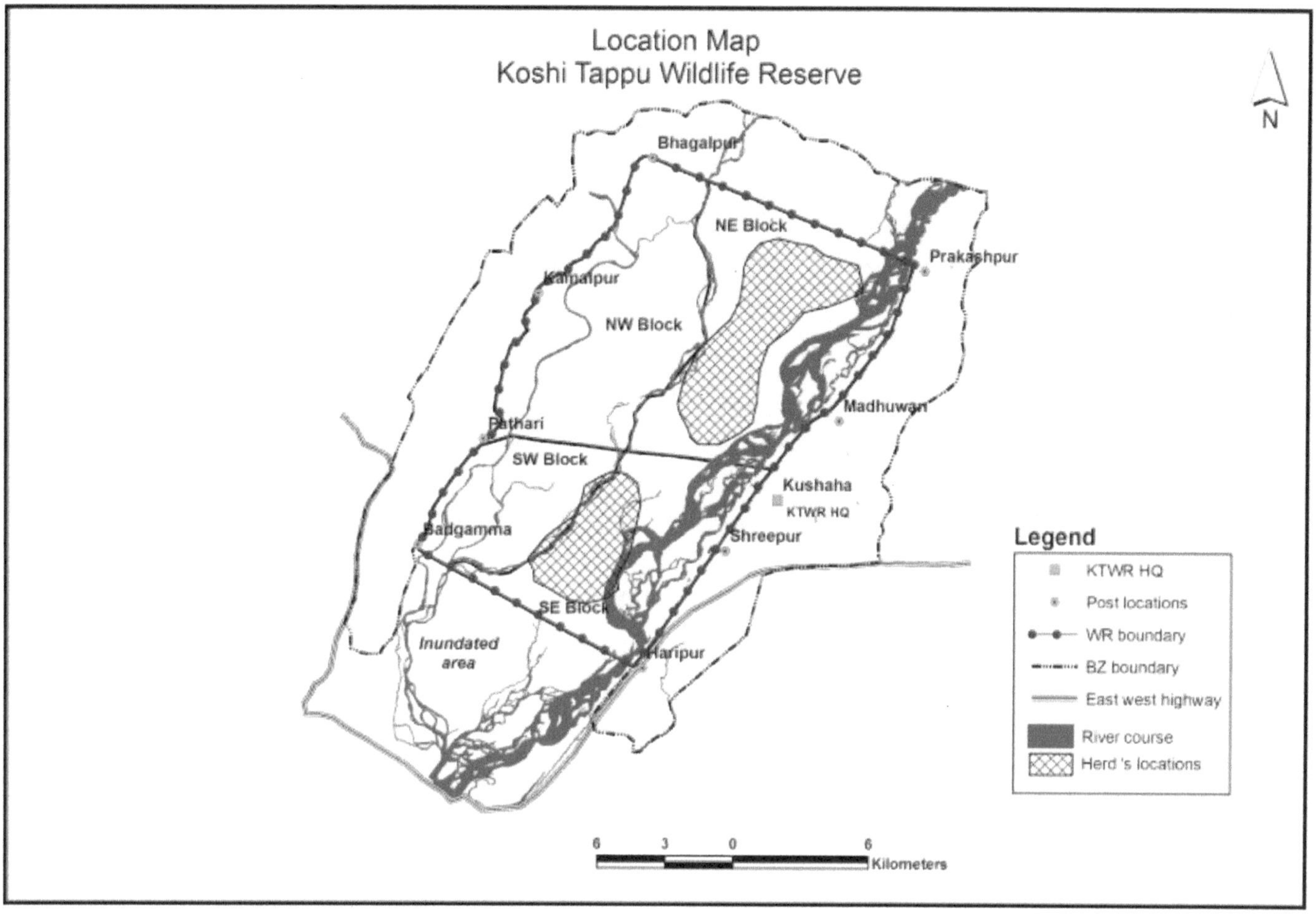

Figure 24.3 Map of Koshi Tappu Wildlife Reserve Nepal. A black and white version of this figure will appear in some formats. For the colour version, please refer to the plate section.

Sunsari and Udayapur Districts. It was established in July 1976 with an area of 176 km^2 under the *National Parks and Wildlife Conservation Act* (His Majesty's Government Ministry of Law and Justice 1977), highlighting the importance of the reserve for the conservation of the last remnant population of endangered wild water buffalo and its habitat. Koshi Tappu Wildlife Reserve is the smallest among Nepal's wildlife reserves and its surrounding wetlands were designated as a Ramsar Site in 1987.

In 2004 the government of Nepal declared a buffer zone with an area of 173.5 km^2 surrounding KTWR. There are 16 village development committees (local political units) with 215 settlements and 108 wards, 14 000 households and 93 300 people in the buffer zone. Although agriculture is the major occupation of local communities, with production occurring in 87.3% of households, only 20% of households are secured for food, and the density of domestic buffalo and cattle is very high (i.e. 1.5 animals per household).

The reserve is subject to extreme flooding in the summer monsoon season. The topography ranges from 80 m above sea level to 100 m. Grassland is the dominant habitat in the reserve, with tall riverine grasses, mainly *Saccharum*, *Phragmites* and *Typha*. Small patches of mixed forests of *Bombax*, *Dalbergia* and *Acacia* occur along the floodplains. Changes in the habitat types are abrupt due to the meandering and braiding of the Koshi River. The flora and fauna are quite rich: 514 species of plants, 21 species of invertebrates, 77 species of butterflies, 200 species of fishes, 45 species of herpeto-fauna, 492 species of birds and 31 species of mammals have been recorded in KTWR.

Methods

Surveys were conducted by covering the whole reserve through the main habitats along the Koshi River and through the open areas including grasslands, to ensure that the vast majority of buffalo were detected. For this purpose, the reserve was divided into two sectors – east and west of the Koshi River – and then each of these divided into north and south blocks as follows: Southeast Block A (Haripur north to Kushaha); Northeast Block B (Prakashpur south to Kushaha); Northwest Block C (Bhagalpur south to Pathari) and Southwest Block D (Badgamma north to Pathari).

Field surveys of wild buffaloes were conducted from March through May, when the visibility remains high after cutting and burning of thatch grasses. During the survey every

Figure 24.4 Putative Wild water buffalo (*Bubalus arnee*) (Courtesy of Port Lympne Wild Animal Park). A black and white version of this figure will appear in some formats. For the colour version, please refer to the plate section.

individual was identified, and age, sex, herd size and other information was recorded when possible. The methodology adopted in previous studies (Dahmer 1978; Heinen 1993a; Heinen & Singh 2001; Heinen & Kandel 2006) was followed to track the herds, and phenotypic characteristics (e.g. shape and size of horns, coat colour and white markings, scars and behavioural characteristics; Heinen 2002) were used to distinguish wild from feral backcrossed and domestic buffaloes (Figures 24.1, 24.2, 24.4).

The timing for the census was between 07:00 and 11:00, and again between 14:00 and 19:00. Ground teams were assigned to the four blocks. In each block, team members were on average at a visible distance of 50–100 m apart and walked each transect in the same direction. The length of walked transects was about 5 km on average.

Wild buffaloes spend long time periods (6–8 hours daily) wallowing, especially during the hot season. Therefore, waterhole observations were used to locate the major herds and to supplement the ground transects. The waterhole observations were carried out from elephant-back (elephants are used in tall grassland to improve visibility in this dense vegetation), rafts, wooden boats, by foot tracking and using four-wheel drive vehicles.

Population growth rate

The population growth rate was calculated per capita per year by the exponential growth rate formula:

$$ri = \frac{\ln(Nj/Nk)}{t}$$

where,

ri = growth rate calculated for that time interval
Nj = population size at the end of the interval
Nk = population size at the beginning of the interval
t = number of years in the interval.

Demography of wild buffalo in Koshi Tappu Wildlife Reserve

All census results are presented in Table 24.1. Calf:cow ratios decreased from 1976 to 2009, and then the ratio gradually increased in 2010 and 2011. Similarly, the numbers of both sexes of adults increased from 1976 to 2011, although at a lower rate for males.

The censuses in 2009 and 2010 were made following the flood of 2008, and relative to the previous survey (Heinen & Kandel 2006), unusually large numbers were recorded in Blocks C and D (Table 24.2). As these 2009 and 2010 results may be less reliable, growth rates are given for intervals excluding 2009 and 2010. Population growth rates for various time periods are given in Table 24.3. The growth rate over the 35 years from 1976 (0.038) is quite low, but it is encouraging that it is at least stable.

All three of the post-flood censuses show an excess of females (ratio about 2:1), with the number of males essentially constant from 2000. There is no obvious reason why death rates of males would have increased during this period, but it is possible that the female counts are biased upwards. It would be expected that hybrid animals with higher proportions of wild genes would increase in frequency over time (due to backcrossing to wild males), and it is likely that distinguishing wild from highly backcrossed hybrids is more difficult in females than in males.

After the 2008 flood, the vegetation status in the eastern sector of KTWR improved and invasion by domesticated animals and other human activities were checked by the ongoing reconstruction of the embankment. Rates of crop raiding by wild buffalo decreased compared with previous years (J. Yadav, personal communication). Further, the authorities involved local people in management of community forests and conservation practices since 2004, so that the incidence of snares, poisoning and poaching was reduced.

Table 24.1 The population status of wild buffalo at KTWR from 1976 to 2011

Year	Adults		Calves		Total	Calf to cow ratio
	Male	Female	Second year	First year		
1976	12	18	22	11	63	0.61
1987	32	29	14	16	91	0.55
1988	37	33	8	15	93	0.45
2000	56	53	17	19	145	0.36
2004	54	63	24	18	159	0.29
2009	55	119	22	23	219	0.19
2010	57	108	24	26	215	0.24
2011	66	117	15	39	237	0.33

Table 24.2 Number of wild individuals in different blocks in the KTWR, 2009–2011 censuses

Blocks	2009	2010	2011
A (southeast)	99	89	137
B (northeast)	87	77	92
C (northwest)	24	13	8
D (southwest)	9	36	1
Total	**219**	**215**	**237**

Table 24.3 Population growth rate of wild buffalo at KTWR (1976–2011)

Time interval	Growth rate (*ri*)
1976–1987	0.033
1987–1988	0.022
1988–2000	0.037
2000–2004	0.023
2004–2011	0.057
1976–2011	0.038

Table 24.4 Numbers of backcrossed feral buffaloes near the borders of KTWR

Years	Backcrossed (feral) buffaloes
2000	131
2004	49
2009	74
2011	18

Table 24.5 Numbers of domestic buffalo and cattle near to the borders of the different blocks in 2011 (KTWR records)

Blocks	Domestic buffaloes	Cattle
A	215	305
B	195	1025
C	0	300
D	0	0

The number of backcrossed feral buffalo near to the borders of KTWR has decreased since 2000 (Table 24.4). However, there were still substantial numbers of domestic buffalo and cattle in 2011 (Table 24.5), particularly near Blocks A and B, close to the main herds of wild buffalo.

Management perspectives

Nepal's *NPWLC Act* (His Majesty's Government Ministry of Law and Justice 1977) rendered the country a leader among developing countries in conservation practices (Lehmkuhl *et al.* 1988; Yonzon & Hunter 1991), improving the management of protected areas and offering wildlife protection (Heinen 1993b).

Regular patrolling along the border of the KTWR focused on minimising human-induced interference and keeping wild buffalo free from any contact with domestic stocks has achieved some success in controlling encroachment. Further, the authorities succeeded in removing 80% of feral cattle. Moreover, management interventions administered in the KTWR and incentives regularly given to the local people under the buffer zone development programme have helped to improve relationships between the KTWR and local communities. Under this scheme, the major approaches are community development activities, conservation programmes, income and employment of local people and conservation education programmes. For example, there is a formal and long-term programme for the buffer zone community forest where local communities can fulfil their daily requirement of forest resources, which they used to acquire illegally. Further, a community-managed livestock health service centre provided by KTWR in 2005 through IUCN Nepal, with the financial support of the World Bank DM Project, is an effective means for providing community incentives, which has had significant impact on reducing human–wildlife conflicts. These are crucial aspects when local people are involved in conservation activities and thus understand the potential benefits of the species and the habitats they use, and for sustainable resource use by their communities. However, the lack of regular options for livelihoods, income generation and self-employment such as local enterprise development are still shortfalls of conservation in KTWR.

Some threats to KTWR are excessive pressure on the wild buffalo population in the southeast and southwest sectors due to extensive farming in the inundated area and also due to excessive grazing by domestic animals. Fuel wood, timber and thatch collection pose further issues, while KTWR has also faced serious threats due to negative attitudes of some local political leaders.

More people in the world depend on the domestic buffalo for all or part of their livelihood than on any other species. If the wild ancestor (*B. arnee*) is to be conserved, it is essential that genetic data be obtained to determine the true status of the presumed wild buffalo in KTWR and elsewhere. As pointed out by Flamand *et al.* (2003), conservation plans need to consider whether the aim is to define a genetically pure wild population, or to accept some degree of introgression from domestic buffalo. In either case, identification of pure (or near pure) wild animals would be greatly facilitated if comparative genetic data (microsatellites, mtDNA and Y-chromosome markers) were obtained from known *B. arnee* in museums and archaeological sites. Comparative analysis of these data with similar data from the putative wild animals in KTWR would provide conclusive evidence on the genetic status of the latter.

Until these genetic data are obtained, and to further safeguard the Nepalese population of putative wild buffalo, we reiterate the recommendation of Heinen & Kandel (2006) that some 30 or more animals should be translocated, preferably to Chitwan National Park, which is a large protected area, has extensive areas of riverine grasslands, as well as abundant upland areas that would offer a flood refuge. Other measures that should be implemented in KTWR to assist conservation efforts include:

(1) Re-establishing security posts in the western part of the reserve which were evacuated during the Maoist insurgency.
(2) Regular monitoring of the buffalo distribution and habitat conditions, at least annually.
(3) Conducting surveys every two years.
(4) Further controlling grazing and limiting contacts between wild and domestic buffalo.
(5) Providing subsidies and grants for local people to improve cattle farming and to adopt stall feeding.
(6) Reducing illegal or otherwise poorly managed forest clearance.
(7) Increasing motivation and technical capacity for buffalo conservation in Nepal, and increasing support from the international community and donors.

Acknowledgements

Many thanks go to Joel Heinen for critical and useful comments on the first draft.

Reference

Amano, W., Miyakoshi, Y., Takada, T., Kikkawa, Y. & Suzuki, H. (1994). Genetic variants of ribosomal DNA and mitochondrial DNA between swamp and river buffaloes. *Animal Genetics* **25** Suppl. (1): 29–36.

Barker, J. S. F., Moore, S. S, Hetzel, D. J. S., Evans D., Tan, S. G. & Byrne, K. (1997). Genetic diversity of Asian water buffalo (*Bubalus bubalis*): microsatellite variation and a comparison with protein-coding loci. *Animal Genetics* **28**: 103–115.

Choudhury, A. (1994). The decline of the wild water buffalo in north-east India. *Oryx* **28**: 70–73.

Dahmer, T. D. (1978). Status and distribution of the wild Asian buffalo (*Bubalus bubalis*) in Nepal. Unpublished MS Thesis. University of Montana.

Diwaker, H. K. & Bhusan, B. (1988). Status of the wild Asiatic buffalo (*Bubalus bubulis*) in Raipur and Bastar districts of Madhya Pradesh. Technical report of Bombay Natural History Society, SANCF Report No. 3/1988, Bombay, India.

FAO (2000). *World Watch List for Domestic Animal Diversity*. 3rd edition. Ed. Scherf, B. FAO: Rome.

Flamand, J. R. B., Vankan, D., Gairhe, K. P., Duong, H. & Barker, J. S. F. (2003). Genetic identification for wild Asian water buffalo in Nepal. *Animal Conservation* **6**: 265–270.

Hedges, S. (1995). *Asian Wild Cattle and Buffaloes: Draft Status Report and Conservation Action Plan*. Vols **1–2**. Gland: IUCN.

Heinen, J. T. (1993a). Population viability and management recommendations for wild water buffalo *Bubalus bubalis* in Koshi Tappu Wildlife Reserve, Nepal. *Biological Conservation* **65**: 29–34.

Heinen, J. T. (1993b). Park–people relations in Koshi Tappu Wildlife Reserve, Nepal: a socio-economic analysis. *Environmental Conservation* **20**: 25–34.

Heinen, J. T. (2002). Phenotypic and behavioural characteristics used to identify wild buffalo *Bubalus bubalis* from feral backcrosses in Nepal. *Journal of the Bombay Natural History Society* **99**: 173–183.

Heinen, J. T. & Kandel, R. C. (2006). Threats to a small population: a census and conservation recommendations for wild buffalo *Bubalus arnee* in Nepal. *Oryx* **40**: 324–330.

Heinen, J. T. & Singh, G. R. (2001). A census and some management implications for wild buffalo (*Bubalus bubalis*) in Nepal. *Biological Conservation* **101**: 391–394.

Higham, C. (2002). *Early Cultures of Mainland Southeast Asia*. Bangkok: River Books.

His Majesty's Government Ministry of Law and Justice (1977). *National Park and Wildlife Conservation Act of 2029 (1972):*

Official Translation. Kathmandu, Nepal: Law Books Management Committee.

IUCN (2013). *IUCN Red List of Threatened Species*. Version 2013.2. www.iucnredlist.org, downloaded on 24 May 2014.

Khatri, T. B., Shah, D. N. & Mishra, N. (2012). Wild water buffalo *Bubalus arnee* in Koshi Tappu Wildlife Reserve, Nepal: status, population and conservation importance. *Journal of Threatened Taxa* **4**: 3294–3301.

Kumar, S, Nagarajan, M., Sandhu, J. S., Kumar, N., Behl, V. & Nishanth, G. (2007a). Mitochondrial DNA analyses of Indian water buffalo support a distinct genetic origin of river and swamp buffalo. *Animal Genetics* **38**: 227–232.

Kumar, S., Nagarajan, M., Sandhu, J. S., Kumar, N. & Behl, V. (2007b). Phylogeography and domestication of Indian river buffalo. *BMC Evolutionary Biology* 7: 186.

Lehmkuhl, J. F., Upreti, R. K. & Sharma, U. R. (1988). National parks and local development: grass and people in Royal Chitwan National Park, Nepal. *Environmental Conservation* **15**: 143–148.

Lenstra, J. A. & Bradley, D. G. (1999). Systematics and phylogeny of cattle. In *The Genetics of Cattle*, pp. 1–14. Eds. Fries, R. & Ruvinsky, A. Oxon and New York: CABI Publishing.

Muley, P. D. (2001). Genetic and morphometric studies to differentiate between wild and domestic Asian water buffalo (*Bubalus bubalis*) and their hybrids in Kaziranga National Park, Assam, India. Unpublished PhD Dissertation. University of Wisconsin.

Nowak, R. (1999). *Walker's Mammals of the World*. 6th edition. Baltimore, MD and London: Johns Hopkins University Press.

Sinclair, A. R. E. (1977). *The African Buffalo: Study of Resource Limitation of Populations*. Chicago, IL: University of Chicago Press.

Yindee, M. (2010). High diversity of the mitochondrial DNA of the Thai swamp buffalo indicates a domestication in Southeast Asia. In *Reproduction and Genetic Diversity of the Swamp Buffalo*, pp. 63–76. Academic thesis. Utrecht University. http://igitur-archive.library.uu.nl/dissertations/2011-0121-200244/yindee.pdf, accessed 15 June 2013.

Yonzon, P. B. & Hunter, M. L. Jr (1991). Cheese, tourists, and red panda in the Nepal Himalayas. *Conservation Biology* **5** (2): 196–202.

Zhang, Y., Vankan, D., Zhang, Y. & Barker, J. S. F. (2011). Genetic differentiation of water buffalo (*Bubalus bubalis*) populations in China, Nepal and south-east Asia: inferences on the region of domestication of the swamp buffalo. *Animal Genetics* **42**: 366–377.

Chapter 25

Genetic structure of the African buffalo (*Syncerus caffer*) at continental and population scales: an evolutionary and conservation approach

Nathalie Smitz, Daniel Cornélis, Philippe Chardonnet, Ettore Randi and Johan Michaux

Inferring the past, recent and future evolutionary history of species

At the end of the last century, the renowned scientist John Avise introduced the phylogeography concept in reference to the study of the historical principles and processes that govern the geographical distribution of genealogical lineages (Avise *et al.* 1987; Avise 2000). Considered as a biogeography subdiscipline, phylogeography emphasizes historical aspects of the contemporary distribution of lineages (Avise & Hamrick 1996; Avise 2000; Hewitt & Butlin 1997). Widely studied, climatic changes have been shown to engender major fluctuations in the demography and distribution of wild species on an evolutionary time scale. Climatic changes (e.g. influence of Quaternary climatic fluctuations) sometimes entail long-lasting isolation of populations, subsequently leading to ecological differentiation or speciation (Lister 2004; Vrba 1995a, 1995b). These phylogeographic patterns shape the evolutionary history of contemporary taxa. Phylogeographical studies have flourished over the last few decades, especially in Europe, but not in Africa, although this continent still hosts large megafaunal communities (Barnosky *et al.* 2004). Understanding how African wildlife responded to climatic changes in the past facilitates the identification of biodiversity hotspots (i.e. glacial refugia), forecasting future population dynamics and developing adapted management strategies.

Recently, in Africa, the combined effects of human pressure (over-exploitation and habitat loss), disease epidemics and climatic fluctuations have strongly impacted wildlife populations and confined many large mammals to restricted protected areas (Hilborn *et al.* 2006; Chantal *et al.* 2007). Substantial climatic variations have been recorded – for example, the cycle of droughts that started in the late 1960s markedly affected Sahelian and Sudanese regions. Africa concomitantly experienced unprecedented population growth, and the resulting extension of agriculture, livestock production and deforestation largely and increasingly contributed to the reduction and fragmentation of natural ecosystems. Moreover, armed conflicts, the trafficking of arms and hunting weapons, as well as the ivory and bushmeat trade have also strongly contributed to the erosion of wildlife populations. It was estimated that the overall number of large mammals in Africa within protected areas fell by about 60% between 1970 and 2005 (Craigie *et al.* 2010). Most wildlife populations are presently confined within a network of protected areas that are loosely connected and benefit from varied conservation conditions (Joppa *et al.* 2008). The resulting disruption/prevention of gene flows between wildlife populations (meta-populations) is likely responsible for genetic erosion (McArthur & Wilson 1967; Young & Clarke 2000; Frankham *et al.* 2002). Isolated populations are likely to have lower reproductive fitness (inbreeding depression) and lose their adaptive genetic variation (Avise & Hamrick 1996), while having a higher risk of extinction (Allendorf 1986; Soulé & Mills 1992; Amos & Balmford 2001; Keller & Waller 2002). In a climate-change setting, the isolation of species within restricted protected areas clearly underlines the question of their future survival. Understanding how the above-mentioned constraints have impacted the genetic structure of species is important to be able to identify conservation priorities and develop adequate management strategies.

The African buffalo (*Syncerus caffer*) is a key species, as it is a major African savanna ungulate biomass constituent and attraction for the photographic and hunting tourism industries (Ernest *et al.* 2012). The African buffalo provides us with a powerful model to enhance our understanding of African biogeography and conservation requirements: it occurs throughout most of sub-Saharan Africa, occupies very different types of habitat, from savanna (Prins 1996) to rainforest (Melletti *et al.* 2007), and exhibits wide morphological variability across its range.

Conservation genetics and molecular tools

Conservation genetics investigations help in identifying and promoting appropriate methods to reduce the risks of species extinction through studies on variations in the spatial

Ecology, Evolution and Behaviour of Wild Cattle: Implications for Conservation, ed. M. Melletti and J. Burton. Published by Cambridge University Press.

distribution of alleles/mutations between and among populations. This research topic is based mainly on phylogeography and population genetics methods. Molecular tools are useful to gain insight into migration patterns, evolution and the phylogenetic relatedness of populations. The African buffalo is the model that was adopted for the present studies. Since different issues were addressed, different molecular marker resolutions were used (details in Tables 25.1–25.2). Each of these molecular markers is briefly described hereafter, with its advantages and limitations.

On a large spatiotemporal scale, phylogeographical studies are principally based on the mitochondrial D-Loop sequences. The control region is the most variable part of the mammalian mtDNA genome (Aquadro & Greenberg 1983), with mutation rates in mammals ranging from 3% to 39% substitutions per million years (Slade *et al.* 1998; Burzynska *et al.* 1999). MtDNA has been extensively used to investigate the structure of animal populations because it evolves rapidly enough to reveal informative variations, while helping to gain insights into the relationship at the species/subspecies level of past historic events occurring during the Quaternary (Avise 1994; Arctander *et al.* 1996a, 1996b; Smith & Wayne 1996). As using mtDNA only infers the female evolutive history, using mtDNA implies that the population structure and the evolutionary history view will be female-biased. Such approaches have been applied to the African buffalo (see Table 25.1 for an overview), with most studies focused on small geographic scales (in East and Southern Africa; *S. c. caffer* subspecies), except those of Van Hooft *et al.* (2002) and Smitz *et al.* (2013).

Highly variable genetic markers, such as microsatellites (i.e. short tandem repeats on nuclear DNA; Bruford & Wayne 1993), are generally required to investigate recent evolutionary processes operating at smaller spatiotemporal scales (e.g. population level). Here, mutations take place through changes in the number of repeats rather than in nucleotide sequences. Microsatellites were shown to be biparentally inherited according to Mendelian segregation patterns (Queller *et al.* 1993; Gotelli *et al.* 1994; Roy *et al.* 1994; Paetkau *et al.* 1995). Their hypervariability in repeat lengths (i.e. alleles) makes it possible to investigate processes at temporal scales ranging from a few thousand to a few hundred years. Microsatellites were shown to be useful markers for fine-scale population structure studies (Bruford & Wayne 1993; Avise & Hamrick 1996; Smith & Wayne 1996; Simonsen *et al.* 1998). The inferred parameters include statistical estimations of population differentiation, effective population sizes (i.e. number of individuals in a population who contribute offspring to the next generation), migration rates and population bottlenecks (i.e. when the size of a population undergoes a drastic reduction). Specific microsatellite markers have not yet been developed for the African buffalo. Van Hooft *et al.* (1999) tested the applicability of previously described bovine autosomal microsatellite markers on the African buffalo (Barendse *et al.* 1997; Kappes *et al.* 1997). Overall, a high level of polymorphism was observed within the African buffalo, with 91 microsatellites being polymorphic within the panel of 168 bovine microsatellites tested. This high success rate of cross-species amplification can be explained by the close phylogenetic relatedness between these two species (divergence time estimated around 12.2 million years, calculated by TimeTree (Pitra *et al.* 2002; Hassanin & Ropiquet 2004; Hedges *et al.* 2006; Kumar & Hedges 2011; Robinson & Ropiquet 2011). A few other Y-chromosomal-specific microsatellites

Table 25.1 Genetic studies performed on the African buffalo, reporting the number of samples involved, the countries sampled and the genetic markers used for the analysis in each of them

Study	No. individuals	Countries sampled	Molecular markers
Grobler & Van der Bank 1996	58	AFS	3 polymorphic loci
O'Ryan *et al.* 1998	105	AFS	7 microsatellites
Simonsen *et al.* 1998	226	UGD, KEN, TAZ, ZIM, BOT	mtDNA fragment 6 microsatellites
Wenink *et al.* 1998	150	AFS, KEN, TAZ	DRB3 gene (MHC)
Van Hooft *et al.* 2000	162	KEN, TAZ, ZIM, AFS, CAM, GAB	14 microsatellites
Van Hooft *et al.* 2002	381	UGD, KEN, TAZ, AFS, BOT, ZIM, CAM, GAB, ANG, NAM	mtDNA fragment Y-microsatellite
Van Hooft *et al.* 2003	89	AFS, TAZ	mtDNA fragment 14 microsatellites
Heller *et al.* 2008	352	KEN, UGD	mtDNA fragment 17 microsatellites
Heller *et al.* 2010	209	KEN, UGD	17 microsatellites
Heller *et al.* 2012	43	ETH, KEN, BOT, ZIM	mtDNA genome
Ernest *et al.* 2012	68	TAZ	mtDNA fragment 15 microsatellites
Smitz *et al.* 2013	787	SIR, GAN, BUF, BEN, NIG, CAM, GAB, RCA, CHD, RDC, UGD, KEN, TAZ, ZIM, BOT, AFS, NAM	mtDNA fragment

AFS: South Africa; BEN: Benin; BOT: Botswana; BUF: Burkina Faso; CAM: Cameroun; CHD: Chad; ETH: Ethiopia; GAB: Gabon; GAN: Ghana; KEN: Kenya; NAM: Namibia; NIG: Niger; RCA: Central African Republic; RDC: Democratic Republic of the Congo; SIR: Sierra Leone; TAZ: Tanzania; UGD: Uganda; ZIM: Zimbabwe

Table 25.2 Precise sampling locality of each genetic study

Country	Locality	Genetic studies
Angola	Unknown	Smitz *et al.* 2013; Van Hooft *et al.* 2002
Benin	Pendjari	Smitz *et al.* 2013
Botswana	Chobe	Smitz *et al.* 2013; Simonsen *et al.* 1998; Van Hooft *et al.* 2002; Heller *et al.* 2012
Burkina Faso	Arly, Pama, Singou	Smitz *et al.* 2013
Cameroon	Benoue	Smitz *et al.* 2013
	Campo ma'an	Smitz *et al.* 2013
	Elephant Camp	Van Hooft *et al.* 2000, 2002
	Zone de Rhinoceros	Van Hooft *et al.* 2000, 2002
Central African Republic	Bamingui-Bangoran	Smitz *et al.* 2013
	Bria	Smitz *et al.* 2013
	Koukourou-Bamingui	Smitz *et al.* 2013
	Mbari	Smitz *et al.* 2013
	Ndji River	Smitz *et al.* 2013
	Ngoto Forest	Smitz *et al.* 2013
	Ouadda	Smitz *et al.* 2013
	Sangha	Smitz *et al.* 2013
	St-Floris	Smitz *et al.* 2013
Chad	Zakouma	Smitz *et al.* 2013
Democratic Republic of the Congo	Garamba	Smitz *et al.* 2013
Ethiopia	Omo valley	Heller *et al.* 2012
Gabon	Gamba	Smitz *et al.* 2013
	Lope	Smitz *et al.* 2013; Van Hooft *et al.* 2000, 2002
Ghana	Kpetsu	Smitz *et al.* 2013
	Mole	Smitz *et al.* 2013
Kenya	Amboseli	Smitz *et al.* 2013; Heller *et al.* 2008, 2010; Simonsen *et al.* 1998; Wenink *et al.* 1998; Van Hooft *et al.* 2000, 2002
	Laikipia	Smitz *et al.* 2013; Heller *et al.* 2008, 2010; Simonsen *et al.* 1998; Van Hooft *et al.* 2002
	Masai Mara	Smitz *et al.* 2013; Heller *et al.* 2008, 2010; Simonsen *et al.* 1998; Van Hooft *et al.* 2000, 2002; Heller *et al.* 2012
	Mount Kenya	Van Hooft *et al.* 2002
	Nakuru	Smitz *et al.* 2013; Heller *et al.* 2008, 2010; Simonsen *et al.* 1998; Wenink *et al.* 1998; Van Hooft *et al.* 2000, 2002
	Nairobi	Smitz *et al.* 2013; Heller *et al.* 2008, 2010; Simonsen *et al.* 1998; Van Hooft *et al.* 2002
	Tsavo	Smitz *et al.* 2013; Van Hooft *et al.* 2000, 2002
Namibia	Okahandja-Windhoek	Smitz *et al.* 2013; Van Hooft *et al.* 2002
Niger	W	Smitz *et al.* 2013
Sierra Leone	Gola Forest	Smitz *et al.* 2013
South Africa	Addo	O'Ryan *et al.* 1998
	Hluhluwe-Imfolozi	Smitz *et al.* 2013; O'Ryan *et al.* 1998
	Kruger	Smitz *et al.* 2013; O'Ryan *et al.* 1998; Wenink *et al.* 1998; Van Hooft *et al.* 2000, 2002, 2003; Grobler & Van der Bank 1996
	St Lucia	O'Ryan *et al.* 1998
	Willem Pretorius	Grobler & Van der Bank 1996
Tanzania	Arusha	Smitz *et al.* 2013; Van Hooft *et al.* 2002, 2003
	Kizigo	Smitz *et al.* 2013; Simonsen *et al.* 1998; Van Hooft *et al.* 2002
	Maswa	Smitz *et al.* 2013; Simonsen *et al.* 1998; Van Hooft *et al.* 2002; Ernest *et al.* 2012
	Selous	Smitz *et al.* 2013; Van Hooft *et al.* 2002
	Ikorongo/Grumeti	Ernest *et al.* 2012
	Loliondo	Ernest *et al.* 2012
	Ngorongoro	Ernest *et al.* 2012
	Serengeti	Smitz *et al.* 2013; Grobler & Van der Bank 1996; Van Hooft et *al.* 2000, 2002; Ernest *et al.* 2012
Uganda	Kidepo Valley	Smitz *et al.* 2013; Heller *et al.* 2008, 2010; Van Hooft *et al.* 2002
	Lake Mburo	Smitz *et al.* 2013; Heller *et al.* 2008, 2010
	Mount Elgon	Smitz *et al.* 2013
	Murchison falls	Smitz *et al.* 2013; Heller *et al.* 2008, 2010
	Queen Elizabeth	Smitz *et al.* 2013; Heller *et al.* 2008, 2010; Simonsen *et al.* 1998; Van Hooft *et al.* 2002
Zimbabwe	Gonarezhou	Smitz *et al.* 2013; Simonsen *et al.* 1998; Van Hooft *et al.* 2002
	Hwange	Smitz *et al.* 2013; Simonsen *et al.* 1998; Van Hooft *et al.* 2002; Heller *et al.* 2012
	Save Valley	Van Hooft *et al.* 2000

The localization of each of these parks/reserves is recapitulated in Figure 25.1.

were also tested, thus shedding light on male buffalo migration patterns (Van Hooft *et al.* 2002, 2007). Microsatellites have been used in numerous studies on the African buffalo with the aim of shedding light on population interrelationships (see Table 25.1 for an overview).

Finally, recent improvements in molecular and statistical techniques have paved the way for investigating individual relationships on a very fine scale. These new markers are characterized by variations in DNA sequences occurring on the scale of a single nucleotide in the genome, i.e. so-called single nucleotide polymorphism (SNP). These markers have not yet been applied to the African buffalo, but two libraries have recently been generated at different scales (Le Roex *et al.* 2012; Smitz *et al.*, unpublished data). These markers offer better accuracy in population screening (Santure *et al.* 2010), with their widespread abundance in coding and non-coding regions (Luikart *et al.* 2003), simple mutation models and low mutation rates (Morin *et al.* 2004). Such markers are expected to enhance the understanding of inter- and intra-herd dynamics in future studies (e.g. within regional conservation area networks).

When molecular markers are finally successfully amplified and genotyped or sequenced, the data set obtained can be analysed using different statistical computation methods. Powerful methods have been developed to analyse these genetic data, using several approaches. A complete review of a large set of available population genetic software packages, describing their principles, statistics computed, assumptions, types of molecular marker handled and some of their limitations, has been published by Excoffier & Heckel (2006). In summary, population genetics analysis can range from the computation of descriptive statistics (i.e. generalist methodologies to uncover the basic indices of genetic diversity: degree of heterozygosity, number of alleles or polymorphic loci, allelic richness, identification of private alleles, tests of Hardy–Weinberg equilibrium at all loci and of linkage disequilibrium between pairs of loci, etc.) to more specific model-based analyses that aim to reconstruct the past demography of a set of populations (i.e. specialized methodologies: population comparisons using various genetic-distance measures, genetic structure, demographic inference and neutrality tests, detection of recent immigrants or of loci under selection, etc.). In the latter case, most of these programs use a Bayesian framework for parameter inference (which requires the incorporation of some prior knowledge on the parameters of interest). Within those coalescent-based analyses, statistical computations are based on the coalescent theory. As defined by Excoffier & Heckel (2006), this theory describes the structure of the genealogy of a sample of genes from the present to the most recent common ancestor. It describes a probability distribution of ancestral genealogies given a population history. Coalescent-based approaches have often been used in different studies performed on the African buffalo to evaluate various parameters, such as population sizes or migration rates between populations (e.g. Finlay *et al.* 2007; Heller *et al.* 2008, 2012, Smitz *et al.* 2013).

Taxonomic controversies

The African buffalo (*Syncerus caffer*) is the only member of the bovini tribe found on the African continent (except for introduced domestic cattle). This species is naturally distributed throughout most of sub-Saharan Africa, as illustrated in Figure 25.1 (according to the IUCN Antelope Specialist Group, 2008; www.iucnredlist.org), and has been locally reintroduced at a number of sites in countries within its range, such as Mozambique, Namibia, South Africa, Swaziland, etc. The African buffalo is found in both savannas and forests, but is also present in other habitats, including mountains, clearings in dense tropical forests and semi-desert areas where water is available (Prins 1996; Melletti *et al.* 2007; Bekhuis *et al.* 2008). Non-colonized habitats include deserts and arid areas with less than 200 mm of annual rainfall. The African buffalo is strongly water dependent and its distribution thus strongly depends on available water supplies (Meester & Setzer 1971; Sinclair 1977; Redfern *et al.* 2003; Cornélis *et al.* 2011).

The African buffalo exhibits extreme morphological variability across its range, greater than most other African mammals, both in body size and weight, but also in coloration and horn size/curvature (Grubb 1971; Buchholtz 1990; Nowak 1991; Alden *et al.* 1995; Prins 1996; Kingdon 1997). Due to the extreme morphological variability, up to 52 different subspecies belonging to two species have been described (Prins 1996). The subspecies number has decreased considerably over time, down to four (Kingdon 1997; East 1998), three (Prins 1996) or two (Wilson & Reeder 2005) subspecies (Table 25.3). Most authors agree on two subspecies with contrasted morphology, namely *S. c. nanus* and *S. c. caffer*. The forest (dwarf) buffalo (*S. c. nanus* – Boddaert 1785) inhabits West and central African rainforests and shows adaptations to forest life, with a smaller body size and unobtrusive swept-back horns (Melletti *et al.* 2007). The Cape buffalo (*S. c. caffer* – Sparrman, 1779) in East and Southern Africa is about twice the size of the forest buffalo, with large downward curved horns and a dark black coat. Differences in karyotypes between *S. c. nanus* and *S. c. caffer* have been reported, with the former having 54–56 chromosomes and the latter 52 (Wurster & Benirsch 1968; Hsu & Benirschke 1970; Cribiu & Popescu 1980). The third and fourth subspecies, which remain controversial, are morphologically midway between the other two (Stark 1986): the West African savanna buffalo (*S. c. brachyceros* – Gray 1837) and the Central African savanna buffalo (*S. c. aequinoctialis* – Blyth 1866). It is worth noting that intermediate ecophenotypes are commonly observed in contact zones between the four subspecies (Prins 1996) and that at least one cross between *S. c. caffer* and *S. c. nanus* has been reported in captivity (Cribiu & Popescu 1980).

Recent genetic studies based on the mtDNA D-loop hypervariable region aimed to clarify this taxonomical controversy.

Table 25.3 Morphological characteristics of the four presently recognized subspecies

Morphological characteristics	*S. c. caffer*	*S. c. nanus*	*S. c. aequinoctialis*	*S. c. brachyceros*
Weight (kg)	700–900 (Nowak 1991; Kingdon 1997)	<320 (Nowak 1991; Kingdon 1997)	Intermediate	Intermediate
Coat	Blackish (Alden *et al.* 1995)	Red to reddish-brown (Buchholtz 1990; Nowak 1991)	Dark brown/black (Alden *et al.* 1995)	Dark brown/black (Alden *et al.* 1995)
Body length (cm)	<340 (Nowak 1991; Kingdon 1997)	170–220 (Nowak 1991; Kingdon 1997)	Intermediate	Intermediate
Width of horn at base (mm)	224 ± 21.5 (Grubb 1971)	128 ± 9.75 (Grubb 1971)	212 ± 17.2 (Grubb 1971)	166 ± 18.3 (Grubb 1971)
Length of skull (mm)	503 ± 17.9 (Grubb 1971)	426 ± 10.7 (Grubb 1971)	500 ± 18.2 (Grubb 1971)	455 ± 22.9 (Grubb 1971)

Source: Smitz *et al.* 2013.

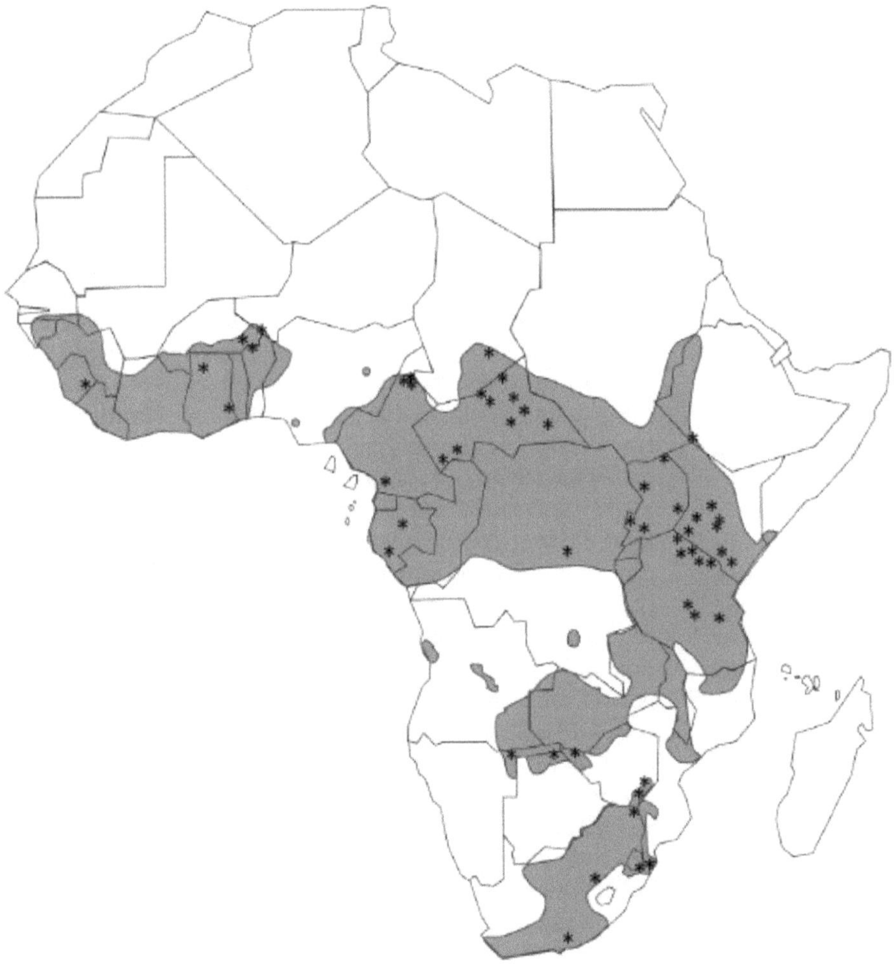

Figure 25.1 Map of Africa with the sampling origins of all studies merged. The distribution range of the African buffalo was redrawn from the distribution map of the IUCN Antelope Specialist Group, 2008.

Both Van Hooft *et al.* (2002) and Smitz *et al.* (2013) pointed out a genetic discontinuity between the West-Central and South-East populations ($F_{CT} = 0.42$ – fixation indices among lineages; Smitz *et al.* 2013). These West-Central and South-East lineages respectively correspond to the geographical distribution of *S. c. caffer* and *S. c. nanus*. The haplotype network shows a low amount of genetic differentiation between *S. c. brachyceros*, *S. c. aequinoctialis* and *S. c. nanus* (see haplotype network (Figure 25.2) and F_{ST} between subspecies (Table 25.4); Smitz *et al.* 2013). F_{ST} and F_{CT} measure the level of genetic differentiation. F_{CT} and F_{ST} refer to the distance among groups, and within subgroups, respectively. This extreme phenotypic variability within the African buffalo West–Central lineage suggests a marked capacity for adaptation to habitat changes on an evolutionary scale. The African savanna buffalo of West–Central Africa (*S. c. brachyceros* and *S. c. aequinoctialis*) appears to be genetically closer to the forest buffalo (*S. c. nanus*) than to the Cape buffalo (*S. c. caffer*) (Table 25.4).

Table 25.4 Pairwise F_{ST} comparison between subspecies (from Smitz *et al.* 2013).

	S. c. nanus	*S. c. brachyceros*	*S. c. aequinoctialis*
S. c. brachyceros	0.12		
S. c. aequinoctialis	0.02	0.08	
S. c. caffer	0.51	0.45	0.44

A coalescent-based analysis performed using an 'isolation with migration' model (Hey & Nielsen 2004; Hey 2005) revealed gene flow between the two main lineages. The gene flow rate was estimated at approximately five mitochondrial genomes per generation since they diverged (Smitz *et al.* 2013). The lineage sorting process does not seem to be complete, thus supporting the subspecies status of the two lineages. This is further substantiated by a low amount of genetic differentiation between the two lineages (Smitz *et al.* 2013), with F_{CT} values comparable to those observed within other large African mammals (Arctander *et al.* 1999; Birungi & Arctander 2000; Nersting & Arctander 2001; Nyakaana *et al.* 2002; Muwanika *et al.* 2003; Alpers *et al.* 2004; Eriksson *et al.* 2004; Okello *et al.* 2005). This assumption is based exclusively on single-locus mtDNA, so no final conclusion may be drawn concerning the taxonomic status of these subspecies without using complementary nuclear markers.

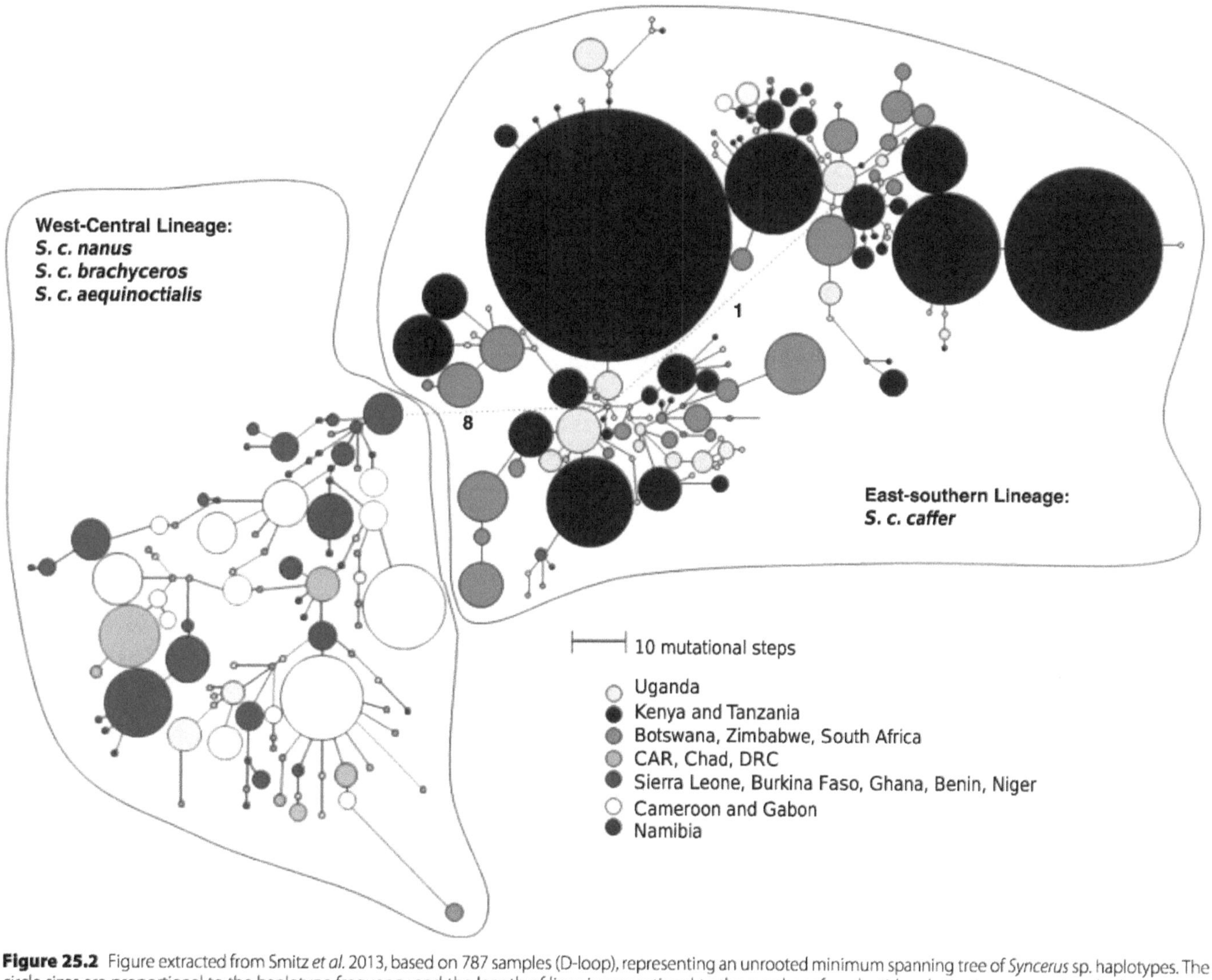

Figure 25.2 Figure extracted from Smitz *et al.* 2013, based on 787 samples (D-loop), representing an unrooted minimum spanning tree of *Syncerus* sp. haplotypes. The circle sizes are proportional to the haplotype frequency and the length of lines is proportional to the number of nucleotide substitutions separating the haplotypes, except for the dotted lines, where the numbers of mutational steps joining the circles are indicated above the connecting branches. CAR: Central African Republic; DRC: Democratic Republic of the Congo. A black and white version of this figure will appear in some formats. For the colour version, please refer to the plate section.

In addition, although based only on a very small sample size, some evidence seems to suggest that Angolan and Namibian buffalo are genetically more closely related to the West–Central African lineage than to the South–East lineage. This is a more surprising result as no major geographical barriers appear in southern Africa, except in the arid Kalahari Desert and the Karoo (Van Hooft *et al.* 2002). Van Hooft *et al.* (2002) suggested the past presence of an arid corridor stretching from Namibia towards Tanzania during glacial periods (van Zinderen Bakker 1978; Kingdon 1989), forming a major barrier to buffalo migration between these two areas. This assumption is supported by a morphological study, which revealed the existence of a dwarf-buffalo-like population in North Angola (Prins 1996; see also Chapter 20). Nevertheless, as this study was only based on two tissue samples (one from Antwerp Zoo with a mother from Okahandja, Namibia), this assertion should be regarded as putative and further investigated (Smitz *et al.* 2013).

Phylogeography of the African buffalo

The African continent hosts large megafaunal communities (Barnosky *et al.* 2004), with the elephant (*Loxodonta africana*) and buffalo (*Syncerus caffer*) being the dominant species in terms of herbivore biomass in savannas and open woodlands (Prins & Douglas-Hamilton 1990). The following section sums up what is known about the evolutionary history and population dynamics of the African buffalo. Investigation of its past demographic fluctuations is crucial for inferring the biogeographical history of regions (Storz *et al.* 2002) and identifying potential management units and barriers to gene flow. To our knowledge, four major specific studies (summarized hereafter in Figure 25.3) have investigated the phylogeography of the African buffalo.

Divergence time and historical refugia

As mentioned above, mtDNA data seem to support the existence of two lineages on a continental scale, namely the *S. c. nanus* subspecies (West–Central Africa) and the *S. c. caffer* subspecies (South–East Africa). A parsimonious explanation for this pattern is that an ancient allopatric separation between a South–East and a West–Central population was the point of divergence between the two lineages (Smitz *et al.* 2013). The divergence time was estimated to occur around 180 000–130 000 BP (Van Hooft *et al.* 2002), and around 150 000–300 000 BP based on a coalescent approach allowing migration between lineages over time (Smitz *et al.* 2013). Fossil data (Kingdon 1982; Klein 1988; Vrba 1995a) supports a recent origin (middle to late Pleistocene) of the Cape buffalo, with the oldest known fossil estimated to date from the same period (Gentry 1978; Kingdon 1982; Klein 1988). Moreover, the close resemblance between the Pleistocene-dated fossils and the present-day West–Central African savanna buffalo suggests that the Cape buffalo might have recently derived from *S. c. brachyceros* stock (Kingdon 1982). Although not significant, a coalescent-based analysis under an isolation with migration model did indicate that the *S. c. caffer* subspecies originated from the isolation of a subpopulation of the West–Central population (Smitz *et al.* 2013), supporting the assumption based on fossil records.

The separation into two lineages, distributed in West–Central and South–East Africa, is a pattern observed in many other African mammal species, such as the kob (*Kobus kob*), African lion (*Panther leo*), roan antelope (*Hippotragus equinus*), bushbuck (*Tragelaphus scriptus*), waterbuck (*Kobus ellipsiprymnus*) and savanna elephant (*Loxodonta africana*) (Birungi & Arctander 2000; Nyakaana *et al.* 2002; Alpers *et al.* 2004; Lorenzen *et al.* 2006, 2007; Moodley & Bruford 2007; Bertola *et al.* 2011). As pointed out earlier in this chapter, the evolutionary history of species can be highly influenced by climate and habitat fluctuations. Nine major and 12 minor glacial periods occurred during the late Quaternary (Sosef 1996), generally fluctuating between cold and dry, and warm and wet periods (van Zinderen Bakker 1978; deMenocal 1995). It was proposed that the current pattern of intraspecific genetic structure in a number of species resulted from successive climate oscillations during the Pleistocene, the alteration of vegetation being the main driver of population expansions and contractions in savanna species during interglacial and glacial phases, respectively (Muwanika *et al.* 2003; Bertola *et al.* 2011). Secondary contact and hybridization between lineages could have occurred during a lineages overlapping phenomenon. Congruent phylogeographical patterns among species assemblies strongly indicated that common historic Pleistocene refugia likely existed where species survived through unfavourable periods – these are proposed to be located in Western, Eastern, Southern and Southwestern Africa (Lorenzen *et al.* 2012). More specifically, species such as the African buffalo presenting a split into two major phylogeographic lineages (e.g. the kob (*Kobus kob*), hartebeest (*Alcelaphus buselaphus*), topi (*Damaliscus lunatus*) and roan antelope (*Hippotragus equinus*) (Arctander *et al.* 1999; Birungi & Arctander 2000; Alpers *et al.* 2004; Lorenzen *et al.* 2007), might have all evolved in West–Central and South–East refugia, likely located in West and East Africa. Population isolation within refugia in West and East Africa can be followed by subsequent dispersal, leading to secondary contact and hybridization between lineages. East Africa seems to be a melting pot of long-diverged lineages across many taxa (Lorenzen *et al.* 2012). The present-day Uganda/Republic of Central Africa buffalo populations displayed the highest genetic diversity, and these areas were proposed to be good candidates for historic refugia within this species (Smitz *et al.* 2013).

Pleistocene population expansion signal

Summary statistics (mismatch distribution and Fu's *Fs* values; Rogers & Harpending 1992; Fu 1997) suggest a population expansion after their Pleistocene divergence, as confirmed in both lineages (Van Hooft *et al.* 2002; Hey & Nielsen 2004;

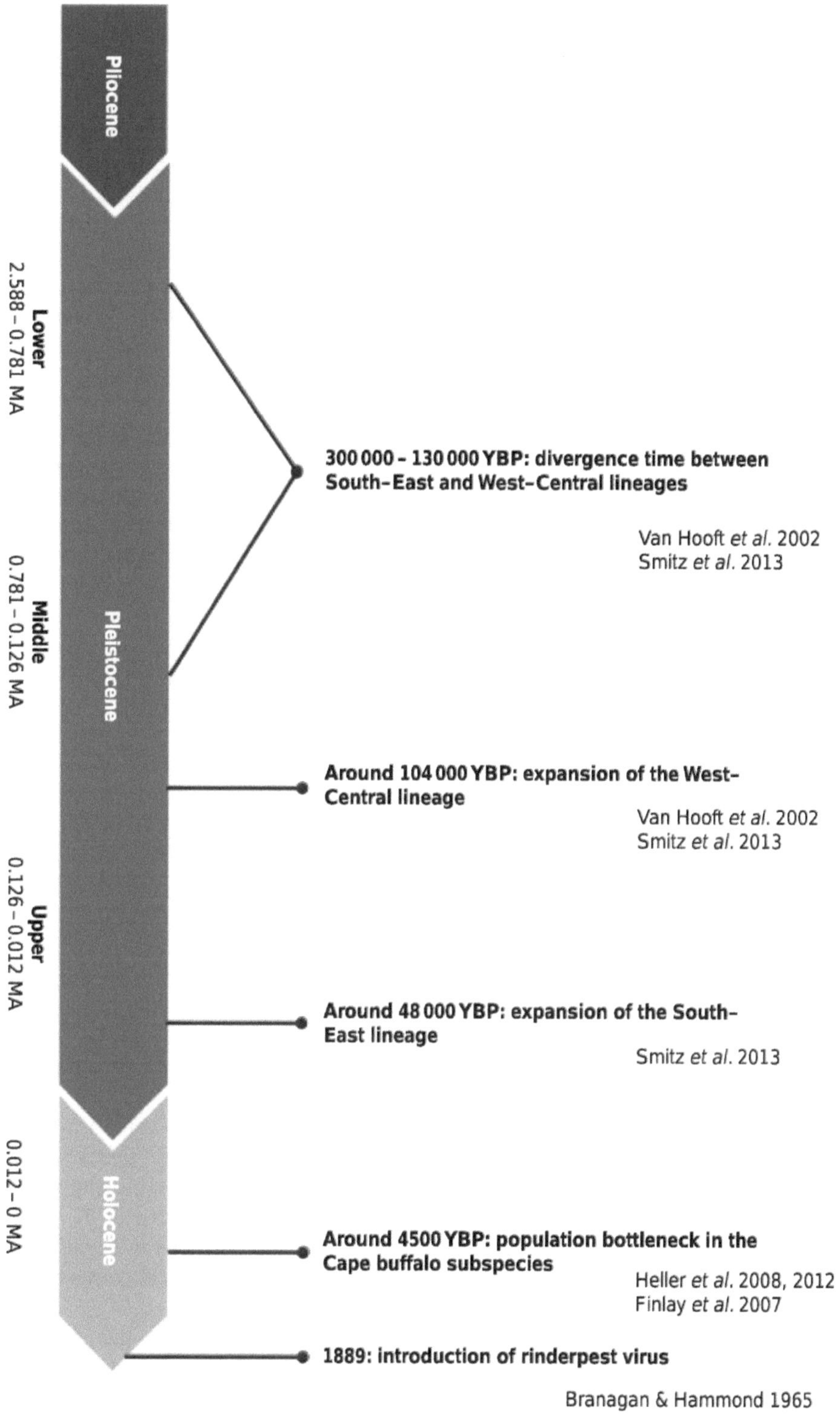

Figure 25.3 Time scale representing all major evolutionary events in the history of the African buffalo. A black and white version of this figure will appear in some formats. For the colour version, please refer to the plate section.

Hey 2005; Smitz *et al.* 2013). This is in agreement with three earlier studies focused on *S. c. caffer* subspecies at a smaller geographical scale, based on an alternative coalescent approach (Finlay *et al.* 2007; Cohen *et al.* 2007; Ernest *et al.* 2012). A tenfold expansion signal was found to have occurred about 1600 generations ago by Ernest *et al.* (2012). Interestingly, Heller *et al.* (2008), based on the shape of a Bayesian skyline plot, suggested that a sudden population expansion in the late Pleistocene is unlikely. On the contrary, it was demonstrated that the expansion during the Pleistocene was progressive (low growth rates) in Kenyan and Ugandan samples (Heller *et al.* 2008).

On the continental scale, this expansion was estimated to have started around 104 000 BP for the West–Central lineage and around 48 000 BP for the South–East lineage (Van Hooft *et al.* 2002; Smitz *et al.* 2013). Both expansion signals could be related to major climatic changes. For the West–Central lineage, the expansion time could be related to major vegetation changes (to more varied and open habitats) recorded after 220 000 BP (Dupont & Agwu 1992; Dupont *et al.* 2000; deMenocal 2001). Before 220 000 BP, the persistence of a rainforest zonal belt during both dry and wet periods was hypothesized, with large westward expansions of *Podocarpus* forest (Dupont & Agwu 1992). Thereafter, the marked reduction in rainforest may have led to the development of savannas in West Africa, thus promoting the expansion of the West African buffalo populations (Smitz *et al.* 2013).

In contrast, the expansion period of the South–East lineage is consistent with the development of open grassland habitats after an extremely arid period (135 000 and 90 000 BP) (Smitz *et al.* 2013). Dry climatic conditions decreased progressively after 95 000 BP, reaching near modern conditions around 60 000 BP (Cohen *et al.* 2007; Scholz *et al.* 2007). The spatial expansion of suitable habitat (open grasslands) may thus have allowed southward colonization of the buffalo populations (Endler 1977; Vrba 1992). The expansion of the South–East lineage is closely in line with the divergence time between the two lineages, which is in agreement with the hypothesis of a rapid evolution of the Cape buffalo based on fossil data, as discussed above. It was also proposed that the development and expansion of the African buffalo populations could have been facilitated by the extinction of the giant long-horned buffalo (*Peloveris antiquus*) (Van Hooft *et al.* 2002), which inhabited African savanna regions until the late Pleistocene (Kingdon 1982; Klein 1988, 1995). Finally, it was proposed that the expansion signal could also have resulted from recovery from a prior bottleneck (Heller *et al.* 2012). This latter proposal is supported by historical records of climatic disturbance, with the hypothesis of a volcanic winter causing significant ecosystem breakdown 74 kya (i.e. the greatest registered eruption during the Pleistocene is Toba, Indonesia; Rampino & Self 1992; Ambrose 1998). The different expansion factors are not mutually exclusive. In any case, it seems that Palaeolithic humans did not have an adverse impact on the buffalo (Heller *et al.* 2012), although they were certainly hunted (Faith 2008; Dusseldorp 2010). Both species expanded simultaneously by about fivefold between 80–10 kya (Heller *et al.* 2012).

Holocene population genetic bottleneck

A congruent genetic bottleneck pattern starting in the mid Holocene (approximately 4500 years ago) was highlighted in the coalescent-based studies of Heller *et al.* (2008) focused on the Kenyan and Ugandan populations, and of Finlay *et al.* (2007) focused on South African populations, indicating that the buffalo experienced similar demographic changes throughout its range, or at least in southeastern Africa. This was a strong signal of population decline, around 75–98% (Heller *et al.* 2008). Moreover, other African species displayed a similar signal during the mid Holocene, particularly the African elephant (*Loxodonta*) (Okello *et al.* 2008) and savanna baboon (*Papio cynocephalus*) (Storz *et al.* 2002). As mentioned above, this common pattern between several species could possibly be explained by common external factors, such as climatic changes or human activities, that could have led to an interspecific population decline (community-wide collapse) (Heller *et al.* 2008).

Heller *et al.* (2008, 2012) proposed that the extent of the mid Holocene decline in African megafaunal biomass could be comparable to the Pleistocene mass extinction that occurred on other continents (Barnosky *et al.* 2004). This Holocene extinction is in agreement with the last great extinction pulse for megafauna on other continents (Barnosky 2008). The decline of the buffalo population may have resulted from a climatic shift or human activities. In the first case, a major climatic shift from moist (African humid period) towards increasingly drier conditions was recorded around 4000–6000 BP (Marchant & Hooghiemstra 2004; Burroughs 2005; Kiage & Liu 2006), thus concomitantly with the buffalo decline (Holocene aridification (Heller *et al.* 2012; Smitz *et al.* 2013)). During this period, the previously savanna-covered Sahara region deteriorated into the present-day desert, and a number of large lakes and waterways in the region dried up completely. It is likely that this considerable and quite rapid climatic shift occurred on a large spatial scale, as declines in the African buffalo effective population size (i.e. size of a stable random mating population that results in the observed degree of drift over a time interval) were recorded in several African regions (Finlay *et al.* 2007; Heller *et al.* 2008). Both a decrease in rainfall and an increase in the duration of dry season affected primary production and ungulate dynamics (Marchant & Hooghiemstra 2004; Ogutu *et al.* 2008). This aridification strongly affected East African wildlife, especially water-dependent species such as the buffalo (Ogutu *et al.* 2008). Climate change in the Holocene was proposed to be definitely severe enough to have had profound ecological consequences (Mayewski *et al.* 2004).

This bottleneck was also concomitant to the Neolithic revolution (i.e. explosive growth in the human population size and cultural transition (Atkinson *et al.* 2008, 2009; Barnosky 2008; Scheinfeldt *et al.* 2010; Gignoux *et al.* 2011; Soares *et al.* 2012), as proposed by Heller *et al.* (2012). The proposed explanation implies Barnosky's theory, stipulating that there is a limited carrying capacity for all megafauna. Domination of megafauna biomass started with the development of humans and their domesticates. The increase in human activities and land use via the adoption of pastoralism and agriculture (Scheinfeldt *et al.* 2010) may have disrupted the ecological balance. Climatic and anthropic factors may also have simultaneously prompted the collapse of the Cape buffalo (Heller *et al.* 2012).

Genetic diversity and population differentiation

Genetic studies have shown that this species harbours high levels of genetic diversity, but low population differentiation throughout its range (Simonsen *et al.* 1998; Van Hooft *et al.* 2000, 2002, 2003). The high nucleotide diversity was accompanied by no excess of heterozygosity (Simonsen *et al.* 1998). In the African buffalo, the overall nucleotide diversity was estimated at around 6% (Smitz *et al.* 2013), which is high compared to other large African mammals (Arctander *et al.* 1996b; Simonsen 1997; Nyakaana & Arctander 1999; Alpers *et al.* 2004). This high genetic diversity suggests a large ancestral population size (Simonsen *et al.* 1998; Van Hooft *et al.* 2000, 2002; Muwanika *et al.* 2003), which was confirmed by applying coalescent-based approaches to samples from Uganda and Kenya, suggesting an effective population size of tens of thousands of individuals during the late Pleistocene (Heller *et al.* 2008).

Furthermore, higher nucleotide diversity was observed within the West–Central lineage as compared to the South–East one. Van Hooft *et al.* (2002) proposed that this may have resulted from fragmentation of the West African habitat, with ancestral isolation of populations in allopatry followed by subsequent contacts (Georgiadis *et al.* 1994). This explanation is not unlikely, as West Africa witnessed a long period of aridity during the last glacial cycle, resulting in an ill-assorted mosaic of wet habitats (Kingdon 1989; Stokes *et al.* 1998; Dupont *et al.* 2000). Moreover, both the lower haplotype diversity and higher between-population genetic distances in forest buffalo (*S. c. nanus*) compared to the other subspecies suggest that fragmentation and genetic drift occurred, indicating that rainforests may act as an important biogeographical barrier to gene flow in dwarf buffalo (Smitz *et al.* 2013). This is further supported by ethological findings concerning the dependency of forest buffalo on food resources only available in small and scattered clearings in the rainforest, thus limiting migration events (Melletti *et al.* 2007; Bekhuis *et al.* 2008; Korte 2008a, 2008b).

Besides the exceptionally high nucleotide diversity, the extent of population differentiation was the lowest within African mammals (Van Hooft *et al.* 2002). In different studies, the congruent pattern of an absence of population structure obtained with microsatellites (nuclear biparental markers) and mtDNA (maternally inherited) indicates that, even though the buffalo was believed to be strongly philopatric (Sinclair 1977; Kingdon 1982; Estes 1991; Prins 1996) with male-biased dispersal (Estes 1991; Prins 1996) and very little interchange between units, both genders have been shown to be dispersive on an evolutionary time scale (Templeton & Georgiadis 1996; Simonsen *et al.* 1998; Van Hooft *et al.* 2002; Smitz *et al.* 2013). Fixation indices among populations (F_{ST}) for mtDNA were low compared to those observed in other African mammals, indicating high levels of female gene flow (Simonsen *et al.* 1998; Matthee & Robinson 1999; Nyakaana & Arctander 1999; Van Hooft *et al.* 2002). Moreover, an in-depth genetic study performed by Van Hooft *et al.* (2003) revealed that when comparing Kruger buffalo herds, the estimated migration rate per generation was 5–20% for females and close to 100% for males, whose migration distance increases with age (Van Hooft *et al.* 2003). This is supported by ethological observations (telemetry) of frequent herd switching among adult females in Botswana, estimated to reach 15% (tracking period: 20 months), with migration distances ranging from 49 km to 133 km (Halley *et al.* 2002), but also with 130 collared animals in South Africa showing frequent herd switching among all sex and age groups (monthly dispersal among females around 2%; Cross & Getz 2006). All available data indicate that the buffalo is likely a vagile species, since it has one of the lowest levels of genetic differentiation of all large African mammals (Funston *et al.* 1994; Prins 1996; Simonsen *et al.* 1998; Van Hooft *et al.* 2000; Halley *et al.* 2002).

Migration and (re-)colonization pattern

Besides indicating an expansion signal (star-like topology), a network reconstruction (Figure 25.2) also highlighted inter-population connectivities. The observation of large fractions of Southern African haplotypes at tips on the network reconstruction suggests a recent origin of the Southern African haplotypes (Van Hooft *et al.* 2002; Smitz *et al.* 2013). This indicates the existence of southward gene flows between East and Southern Africa on an evolutionary time scale. The southward extension of savanna and woodland areas (Vrba 1992) might explain this connectivity and the small genetic distance between East and Southern African populations (Van Hooft *et al.* 2002).

Moreover, coalescent-based analysis made it possible to reconstruct the colonization history of buffalo populations using recent statistical improvements to simultaneously infer the genealogy and ancestral geographic state at each node of the phylogeographical tree (see Smitz *et al.* 2013). This analysis supported the expected role of the present-day Uganda area for the South–East lineage and the present-day Central African Republic area for the West–Central lineage in the colonization history of the African buffalo. Uganda represented the ancestral historical geographical state within the phylogeographical tree reconstruction (Smitz *et al.* 2013). The earliest occurrence of a non-Ugandan branch in the South–East lineage was estimated to be around 83 000–103 000 BP, roughly 170 000 years after the divergence of both lineages. Uganda thus seems to have served as a primary refuge for *S. c. caffer* for a long time, followed by southward expansion and colonization, probably initiated when climatic and habitat conditions were more favourable (Smitz *et al.* 2013). The recent colonization of Southern Africa by *S. c. caffer* is supported by the lack of true Cape buffalo characteristics, i.e. the sweeping horns and

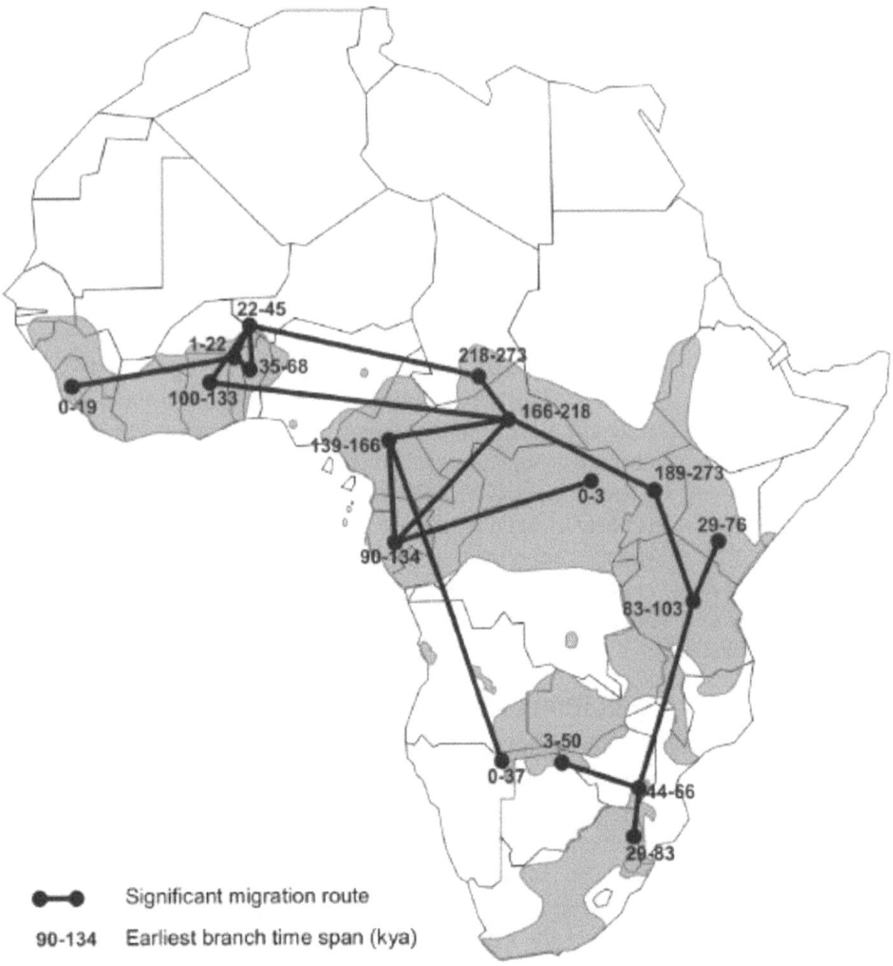

Figure 25.4 Figure extracted from Smitz *et al.* 2013, based on 787 samples (D-loop), representing a map of the African continent showing the 17 historical migration routes between sampled localities supported by a Bayes factor >3. Grey shapes on the map represent the actual distribution of the African buffalo according to the IUCN Antelope Specialist Group, 2008. Numbers on the map indicate the median time endpoints over all BEAST trees of the earliest branch with a given locality state. Hence, it provides an estimate of the earliest migration into each locality. A black and white version of this figure will appear in some formats. For the colour version, please refer to the plate section.

pronounced horn boss, in all buffalo fossils from this region (Kingdon 1982; Moritz 1994). Moreover, the absence of monophyly of the Southern African external branches on the tree and on the network reconstruction suggests that this southward colonization might have occurred several times (Smitz *et al.* 2013). It is worth noting that fossil records also indicate the presence of *Syncerus*-like species in Southern Africa around 542 000 BP (Porat *et al.* 2010). Buffalo seem to have occupied Southern Africa earlier than could be inferred based on mtDNA (Smitz *et al.* 2013), which could indicate multiple expansions towards Southern Africa. This finding also supports the hypothesis that the previous southward colonization signal could indicate a recolonization phenomenon. The hypothesis of major demographic fluctuations in response to climatic change is thus plausible, with refugia located in East and West Africa. The West–Central lineage did not display such a clear geographical pattern (Smitz *et al.* 2013). Significant migration routes and branch time spans are represented in Figure 25.4.

In summary, the following tentative phylogeographic scenario was proposed by the authors (Smitz *et al.* 2013): the ancestral population of all living buffalo occupied Central Africa before being separated (for unknown reasons) around 130–300 kya into two isolated populations. The first one (ancestors of the West–Central lineage) started expanding westward around 100 kya or earlier. The other one (ancestors of the South–East lineage) retained its core population in East Africa, probably due to the arid conditions of Southern Africa until about 50–80 kya, from where it expanded south and east.

Population genetics and conservation implications

Although still considered as 'at least concern' by the IUCN (according to the IUCN Antelope Specialist Group, 2008; www.iucnredlist.org), the African buffalo has suffered major population losses during the last century, impacting all (but unevenly) of the above-mentioned subspecies. Of the more than three million buffaloes that roamed the continent in the nineteenth century, only a few hundred thousand have survived (Lessard *et al.* 1990). For example, recent declines of 15% were observed in Kenya between 1977 and 1994 (Van Hooft *et al.* 2002) and of 50% in Southern Africa between 1991 and

1996 (Winterbach 1998). The species is already extinct in Eritrea and Gambia (East 1998).

Habitat loss and poaching are the main challenges currently threatening the species. Habitat loss can be due to anthropogenic factors (Alroy 2001; Godfrey & Jungers 2003; Surovel *et al.* 2005) or to climatic changes (Rubenstein 1992; Meijard 2003; Barnosky *et al.* 2004; Thirgood *et al.* 2004; Lovett *et al.* 2005; Vanacker *et al.* 2005), such as the increasing drought observed in Africa since the 1990s, which is the most ecologically important climate variable in tropical Africa. The African buffalo, a highly drought-sensitive species (Ogutu *et al.* 2008), has undergone a sharp climate-mediated population decline, as demonstrated by a decrease in the Masai Mara population from 10 000 to 2400 individuals during the severe drought of 1993–1994 (East 1997). Currently around 75% of the global population (estimated around 900 000 animals) is located in protected areas (i.e. National Parks and Game reserves) (East 1999), with many populations completely isolated from each other. Fragmentation of the natural habitat into small patches endangers the populations through increasing genetic drift, resulting in loss of genetic diversity, consequently leading to a reduction in the evolutionary potential of the species (Frankham *et al.* 1999; Hedrick 2005).

Besides generating direct competition for natural resources, the introduction of foreign-born species, such as domestic cattle, is also a serious problem due to the concomitant introduction of pathogens. Indeed, disease transmission is a major problem due to the close phylogenetic relatedness with the African buffalo; for instance, the introduction of the rinderpest morbillivirus in 1889 by an Italian military expedition to Ethiopia (Branagan & Hammond 1965, in Sinclair 1977). No species on the African continent has been more affected than the African buffalo (extreme regional fluctuations in population density, along with many local extinctions; Wenink *et al.* 1998), with the most severe collapse occurring in the 1890s when 90–95% mortality rates were recorded throughout the continent (Mack 1970; Sinclair 1977; Plowright 1982; Prins & Van der Jeugd 1993; Shigesada & Kawasaki 1997; O'Ryan *et al.* 1998; Winterbach 1998). As an illustration, buffalo populations were nearly wiped out in the Serengeti National Park (Sinclair 1977), and only 75 individuals were reported in the Hluhluwe-Umfolozi National Park in 1929 (O'Ryan *et al.* 1998). This high mortality rate may be partly explained by the gregarious lifestyle of the African buffalo, which facilitates lateral transfer of pathogenic agents. The impact of successive rinderpest pandemics (1890, with later outbreaks throughout the twentieth century) on the genetic diversity of buffalo has been largely studied, mainly focusing on the *S. c. caffer* subspecies. The results are discussed below.

Impact of rinderpest epidemics on genetic structure

According to numerous previous studies (Simonsen *et al.* 1998; Wenink *et al.* 1998; Van Hooft *et al.* 2000; Ernest *et al.* 2012; Smitz *et al.* 2013), and despite the extremely high mortality rates reported on the continental scale, it appears that the nineteenth-century rinderpest pandemic had very little impact on the genetic diversity of the African buffalo in terms of allelic or haplotype diversity. Considerable within-population genetic variation was found in East and Southern Africa (Simonsen *et al.* 1998; Van Hooft *et al.* 2002; Heller *et al.* 2008; Ernest *et al.* 2012), with evenly distributed nucleotide diversity ranging from 3.3% to 5.5% (Simonsen *et al.* 1998). Generally, bottlenecks involve a drastic reduction in effective population size, which should be genetically marked by a decrease in allelic diversity and later in heterozygosity (Wright 1931). High within-population nucleotide diversity paired with a heterozygosity level similar to that found in other species indicates that the demographic bottleneck due to the rinderpest outbreak did not result in a genetic bottleneck (Simonsen *et al.* 1998; Van Hooft *et al.* 2000). The absence of genetic signature of population declines could be explained by a very large ancestral population size (Simonsen *et al.* 1998; O'Ryan *et al.* 1998), as also supported by the mtDNA analyses presented above (Van Hooft *et al.* 2002; Smitz *et al.* 2013). This limited genetic signature could also be explained by an overestimation of the population crash (Simonsen *et al.* 1998; Heller *et al.* 2008) and/or by the fact that the crash only affected a few generations (Simonsen *et al.* 1998). Rapid population regrowth together with gene flow between populations may have re-established genetic variability via the introduction of rare alleles, thus distorting the signal of a demographic bottleneck (Heller *et al.* 2008). Calculations have indicated that the buffalo may be so productive that survivors may have recolonized their initial range and recovered their population size within 20–30 years (Sinclair 1977; Estes 1991; Simonsen *et al.* 1998; Van Hooft *et al.* 2000).

The genetic impact of rinderpest epidemics was further investigated by Wenink *et al.* (1998), who studied polymorphism in the highly polymorphic major histocompatibility complex (MHC) *DRB3* gene involved in the immune response. High allelic diversity is observed within most vertebrate populations (Klein 1986; Hughes & Nei 1988) and low MHC molecular variability is generally interpreted as an indication of increased genetic risk (Wenink *et al.* 1998). This study revealed the current high allelic diversity, as well as a high heterozygosity level, in agreement with previous results. Moreover, Wenink *et al.* (1998) assumed that, since only few generations left since the major pandemic, it is unlikely that the buffalo population has offset the effects of strong genetic drift. Thus, strong directional selection by the rinderpest virus for particular *DRB3* alleles seems unlikely – there is no obvious link between MHC heterozygosity and the incidence of rinderpest epidemics (Wenink *et al.* 1998).

The findings of a precise case study of the Kruger National Park buffalo population (South Africa) support all previous assumptions. Kruger was highly affected by the rinderpest epidemics, with especially high mortality, leaving only a small reported number of survivors in 1902 (Stevenson-Hamilton 1911). Despite the fact that high genetic erosion could be expected within this park, genetic studies have highlighted that genetic variation in the Kruger population is currently extensive (Bruford *et al.* 1996; Grobler & Van der Bank 1996; O'Ryan *et al.* 1998; Wenink *et al.* 1998). Wenink *et al.* (1998) proposed that the reported number of survivors could be an underestimation. This is supported by genetic studies, which revealed that at least 15 Y-chromosomal haplotypes (Van Hooft *et al.* 2007) and 34 mtDNA haplotypes (see Greyling 2007) had been found within Kruger. This indicates an effective population size (N_e) during the bottleneck of at least 40 individuals. Considering that the N_e:N_c ratio (i.e. ratio between genetically effective and census population size) is around 10% in most mammal species, the number of survivors in the Kruger region must have been at least in the hundreds, and this was reportedly the most affected population. A similar heterozygosity level was also observed in Lake Nakuru National Park, which had only 30 head of buffalo in the 1970s (Wenink *et al.* 1998). A complementary explanation could be a possible recolonization from a neighbouring population, thus re-establishing high genetic diversity (McCauley 1991). Van Hooft *et al.* (2000) came to the same conclusion, while making a distinction between northern and southern Kruger buffalo populations. A gradient of decreasing heterozygosity along a transect between the Save Valley Conservancy, northern Kruger and southern Kruger was observed (Van Hooft *et al.* 2000). Both an absence of differentiation and intermediate heterozygosity in northern Kruger could be explained by historical data (Van Hooft *et al.* 2000). Mortality was shown to be especially high in northern Kruger, where only a few buffalo were left in the 1930s, while southern Kruger showed quick demographic recovery (Van Hooft *et al.* 2000). Northern Kruger could have undergone recolonization from southern Kruger, but also from Mozambique and Zimbabwe (Van Hooft *et al.* 2000). This is not unlikely as herd settlement has been observed on several occasions, with bachelor bulls readily travelling over 50 km between herds (Prins 1996). In conclusion, the *S. c. caffer* subspecies did not show a genetic bottleneck signal related to rinderpest pandemic/epidemics, but instead displayed a high ability to overcome demographic bottlenecks without undergoing genetic erosion. This could at least partly be explained by a relatively high potential growth rate (Jolles 2007) and good dispersal abilities (Halley *et al.* 2002; Naidoo *et al.* 2012).

Impact of habitat fragmentation

The confinement of buffalo populations within protected areas due to habitat fragmentation and increasing population densities by the early 1900s (Smithers 1983) raises the question concerning the risk of genetic erosion (Meester & Setzer 1971; Sinclair 1977; Haltenorth & Diller 1980; Simonsen *et al.* 1998). The susceptibility of a species to habitat fragmentation depends on different demographic characteristics: their historical population size, dispersal capabilities and phylogeographic history (Galbusera *et al.* 2004; Luoy *et al.* 2007; Heller *et al.* 2012). The buffalo has been shown to be characterized by a large ancestral population size, high levels of genetic diversity, high dispersal capabilities on an evolutionary time scale and low population differentiation throughout its range (Simonsen *et al.* 1998; Van Hooft *et al.* 2000, 2002, 2003), as described earlier in this chapter. It has thus been proposed that the buffalo should be less affected by fragmentation than other savanna ungulates.

Many studies have aimed at gaining greater insight into the impact of habitat fragmentation on the genetic variability of the African buffalo on different geographical scales (Figure 25.5). Two of these focused on a larger geographical scale – i.e. Van Hooft *et al.* (2000) and Simonsen *et al.* (1998); these revealed significant isolation by distance between East and Southern Africa. It seems that, at the continental level, genetic variation reflects an effect of geographical distance between populations (Templeton & Georgiadis 1996), as confirmed by a significant correlation between genetic and geographical distances (Simonsen *et al.* 1998). Nevertheless, along with the plains zebra (Lorenzen *et al.* 2008), the African buffalo displays the lowest genetic differentiation on the continental scale, probably because of its considerable dispersal capabilities and its ability to cope with contrasting environmental conditions by modulating a large array of biological traits (Van Hooft *et al.* 2003; Cornélis *et al.* 2011).

On a regional scale, within either southern or eastern regions, populations showed a low level of differentiation and high heterozygosity values, especially when compared to the findings of other studies of large African savanna ungulates (Simonsen *et al.* 1998; Van Hooft *et al.* 2000; Heller *et al.* 2010). It is commonly proposed that the low genetic differentiation (weak isolation by distance) must have resulted from a high migration rate between populations with interbreeding across considerable distances (Simonsen *et al.* 1998; Van Hooft *et al.* 2000) or result from habitat fragmentation due to human activity of an originally panmictic population displaying homogenizing gene flow between localities (Simonsen *et al.* 1998). Based on these results, it was proposed that, at the regional level, fragmentation of a previous panmictic population is the primary force affecting the genetic structure (Heller *et al.* 2010), whereas on a continental scale, geographical distance was shown to better explain the overall genetic structure (Simonsen *et al.* 1998). Moreover, a few geographical natural barriers to gene flow have been noted. First, the observation of significant differentiation between two adjoining populations in Amboseli and Tsavo National Parks was explained by the dryness of the habitat, which is congruent with what has been observed in other large mammals, such as the Grant's gazelle (Arctander *et al.* 1996a) and elephant (Siegismund & Arctander 1995). Buffalo migration could be seriously limited by habitat dryness. Second, in the same geographical region, Heller *et al.* (2010) demonstrated that the Victoria Nile might also have acted as a historical barrier to dispersal, with significant genetic differentiation on both sides of the river

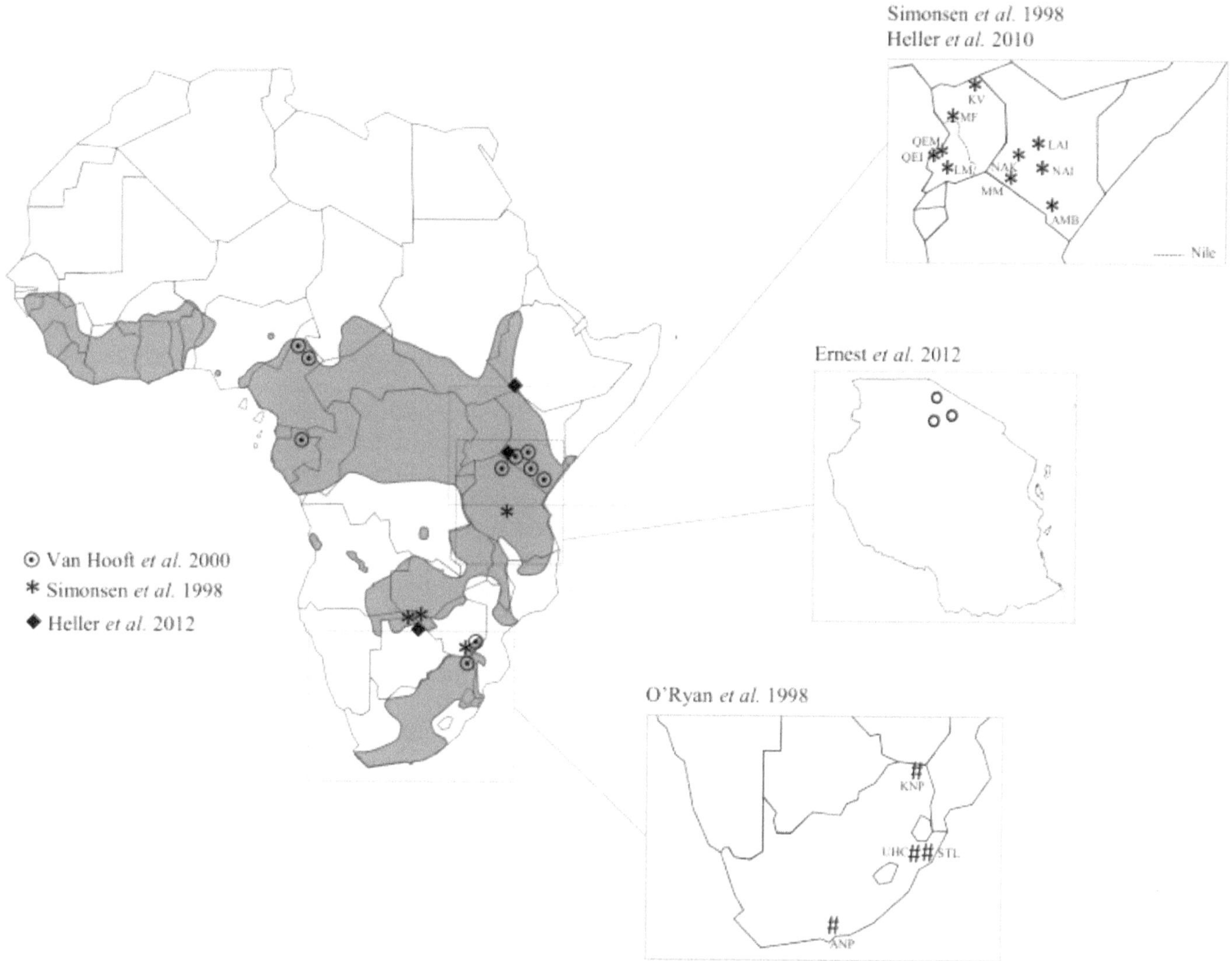

Figure 25.5 Map of sampling localities of three studies aimed at determining the impact of fragmentation on the genetic diversity of buffaloes. QEI: Queen Elizabeth- Ishasha sector; QEM: Queen Elizabeth-Mweya sector; LM: Lake Mburo; MF: Murchison Falls; KV: Kidepo Valley; LAI: Laikipia; NAK: Nakuru; MM: Masai Mara; NAI: Nairobi; AMB: Amboseli; KNP: Kruger; UHC: Hluhluwe- Umfolozi; ANP: Addo Elephant National Park; STL: St Lucia.

(Figure 25.5). Finally, the tropical dense humid forest may also play an important role in the dispersal of savanna species.

Interestingly, a recent study conducted by Heller *et al.* (2010) aimed to evaluate the impact of habitat fragmentation and the confinement of buffalo populations within small protected areas on the distribution of the genetic diversity in Kenya and Uganda. It was assumed that genetic drift would affect populations in smaller conservancies more rapidly than in larger ones. A positive correlation between the size of the protected area and two measures of genetic diversity, as well as the negative correlation with genetic connectivity, indicate that genetic drift had a more pronounced effect on buffalo populations in smaller parks, where they were subjected to more pronounced genetic erosion (Heller *et al.* 2010). Moreover, the author concluded that the suppression or restriction of gene flow by confinement within small areas could also have an ethological impact by disturbing the behaviour of natural dispersion in response to seasonal variations in food availability (Sinclair 1977; Halley *et al.* 2002; Ryan *et al.* 2006; Heller *et al.* 2010). Nevertheless, it was expected that, within confined populations, loss of allelic diversity by genetic drift would occur faster than loss of heterozygosity (Nei *et al.* 1975; Allendorf 1986; Cornuet & Luikart 1996). Heller *et al.* (2010) proposed that the time elapsed since the beginning of conservancy confinement establishment (number of generations estimated to be close to 13 in the African buffalo) was not sufficient to observe such a substantial loss of heterozygosity (Heller *et al.* 2010). Genetic erosion is expected to progressively become significant in the near future (Heller *et al.* 2010). A previous, similar study focusing on South African populations also revealed a relationship between the current population size and allelic diversity – allelic diversity and heterozygosity showed a significant positive relationship with population size (O'Ryan *et al.* 1998). These two studies demonstrated that buffalo populations were strongly affected by fragmentation caused by recent confinement within small conservation areas (Heller *et al.* 2010).

A more recent study (Ernest *et al.* 2012) conducted within the Serengeti ecosystem in Tanzania (SNP: Serengeti National Park; NCA: Ngorongoro Conservation Area; MGR: Maswa Game Reserve) also aimed at testing the influence of habitat fragmentation on the genetic structure of African buffalo populations, highlighting significant genetic differentiation between NCA and the SNP/MGR complex. The differentiation of the NCA buffalo population from the SNP/MGR complex was attributed to isolation by fragmentation, whereas no physical barrier between SNP and MGR seemed to prevent gene flow. This indicates that fragmentation has become serious enough to have a genetic impact (Ernest *et al.* 2012). Two separately or simultaneously acting factors were proposed to have induced this isolation. The first is anthropogenic, with a marked increase in human populations and intensification of land use in response to food scarcity (McCabe *et al.* 1997; Neumann 1998; Boone *et al.* 2006), blocking natural dispersal/gene flow between NCA and SNP/MGR. In fact, fertile lands, which would have provided a good habitat for buffalo, are now cultivated by humans. Second, migration patterns may be governed by the distribution of dry habitats (Van Hooft *et al.* 2000); for drought-intolerant species like the African buffalo, drier environments can form a natural barrier against dispersal.

An interesting relationship between genetic variability and human population density could have been revealed in the buffalo populations of Kenya. The Laikipia National Park buffalo population was shown to be poorly defined, with high admixture of different genetic contributors (Heller *et al.* 2010). This park is situated in northern Kenya, a region where wildlife is not restricted to protected areas, with a low human population density and seasonal migration events (Heller *et al.* 2010). On the contrary, limited gene flow between the Nakuru population and that in southern Kenya, where the human population density is higher, was highlighted (Heller *et al.* 2010). For wild species, which cannot coexist with humans, it seems that conservancies isolated within areas with a high human density are more prone to genetic drift.

Conclusion: species management

The African buffalo has been shown to present a high capacity to disperse on an evolutionary time scale, but also on a scale of a few decades. This characteristic, associated with its high degree of behavioural plasticity (Cornélis *et al.* 2011), could likely explain its capability to recover within only a few generations after a major demographic bottleneck (i.e. disease epidemics and droughts (e.g. Ernest *et al.* 2012)), with immigrants from neighbouring populations restoring part of the lost genetic diversity. The high genetic diversity still observed today indicates a historically large population size, with current population sizes still being over a critical threshold (Heller *et al.* 2010). It has been suggested that the buffalo is capable of maintaining a non-critical population size thanks to its high growth rate, good dispersal capability and a high degree of behavioural plasticity. Nevertheless, estimation of the long-term effective population size (N_e) reached 2600–24 000 individuals (Van Hooft *et al.* 2000), assuming a mean mutation rate of 4.5×10^{-5} to 15×10^{-5} per generation (Crawford & Cuthberston 1996; Schug *et al.* 1997; Lehmann *et al.* 1998). In many protected areas throughout Africa, particularly in West Africa (see Chapter 20), the size of inventoried buffalo populations (N_c) does not reach the minimum threshold of 2600 individuals (Van Hooft *et al.* 2000). This implies that, without migration between confined protected areas, the African buffalo will not be able to maintain historical levels of genetic diversity at microsatellite loci (Van Hooft *et al.* 2000). In agreement with this estimation, O'Ryan *et al.* (1998) demonstrated that the large current size of the buffalo population within Kruger National Park (around 40 000 individuals in 2011; Ferreira *et al.* 2012) may be sufficient to maintain substantial variability at microsatellite loci. However, genetic variability in populations of the three other parks sampled within this study (Hluhluwe-Umfolozi – 4000 individuals; Jolles 2007; St Lucia – 175 individuals; Addo Complex area – 348 individuals in 2011; Ferreira *et al.* 2012) should be maintained by regular introduction of individuals from Kruger National Park.

Even though the effect of habitat fragmentation on the genetic variability of buffalo populations is currently low, the number of generations elapsed since the marked increase in human communities and cultivation is still small and may possibly not yet allow detection of more profound effects (Van Hooft *et al.* 2000). Confinement within small protected areas necessarily obstructs seasonal natural migrations, which will lead to genetic erosion. To maintain historical levels of genetic diversity, dispersal corridors should be developed or programmes should ensure regular translocation from one park to another, while taking special precautions regarding disease transmission risks (Van Hooft *et al.* 2003; Heller *et al.* 2010). Preference may be given to (adult) males since they have been shown to naturally have the highest migration rate (100% per generation; Van Hooft *et al.* 2003). Overall, ongoing habitat fragmentation is likely to affect many other African savanna species, characterized by lower dispersal capacities. There is a need to maintain large conservation areas where human disturbance is kept to a minimum. There is a specific need for the facilitation of gene flow between protected areas to prevent further genetic differentiation – a conservation initiative that may also be essential for all other African savanna ungulates.

Acknowledgements

This study is supported by research grants from the FRS-FNRS of Belgium (Fond National pour la Recherche Scientifique) provided to J. R. Michaux (A5/5-MCF/BIC-11561) and N. M. R. Smitz (F3/5/5-MCF/ROI/BC-20.003) (www1.frs-fnrs.be).

References

Alden, P. C., Estes, R. D., Schlitter, D. & McBride, B. (1995). *National Audubon Society Field Guide to African Wildlife.* New York: Chanticleer Press.

Allendorf, F. W. (1986). Genetic drift and the loss of alleles versus heterozygosity. *Zoo Biology* **5**: 181–190.

Alpers, D. L., Jansen van Vuuren, B., Arctander, P. & Robinson, T. J. (2004). Population genetics of the roan antelope (*Hippotragus equinus*) with suggestions for conservation. *Molecular Ecology* **13**: 1771–1784.

Alroy, J. (2001). A multispecies overkill simulation of the end-Pleistocene megafaunal mass extinction. *Science* **292**: 1893–1896.

Ambrose, S. H. (1998). Late Pleistocene human population bottlenecks, volcanic winter and differentiation of modern humans. *Journal of Human Evolution* **34**: 623–651.

Amos, W. & Balmford, A. (2001). When does conservation genetics matter? *Heredity* **87**: 257–265.

Aquadro, C. F. & Greenberg, B. D. (1983). Human mitochondrial DNA variation and evolution: analysis of nucleotide sequences for seven individuals. *Genetics* **103**: 287–312.

Arctander, P., Kat, P. W., Aman, R. A. & Siegismund, H. (1996a). Extreme genetic variations between populations of Grants gazelle (*Gazella granti*) in Kenya. *Heredity* **76**: 465–475.

Arctander, P., Kat, P. W., Simonsen, B. T. & Siegismund, H. R. (1996b). Population genetics of Kenyan impalas: consequences for conservation. In *Molecular Genetics in Conservation*, pp. 399–412. Eds. Wayne, R. K. & Smith, T. B. Oxford: Oxford University Press.

Arctander, P., Johansen, C. & Coutellec-Vreto, M. A. (1999). Phylogeography of three closely related African bovids (tribe Alcelaphini). *Molecular Biology and Evolution* **16**: 1724–1739.

Atkinson, Q. D., Gray, R. D. & Drummond, A. J. (2008). mtDNA variation predicts population size in humans and reveals a major southern Asian chapter in human prehistory. *Molecular Biology and Evolution* **25**: 468–474.

Atkinson, Q. D., Gray, R. D. & Drummond, A. J. (2009). Bayesian coalescent inference of major human mitochondrial DNA haplogroup expansions in Africa. *Proceedings of the Royal Society of Biology* **276**: 367–373.

Avise, J. C. (1994). *Molecular Markers, Natural History and Evolution.* New York: Chapman & Hall.

Avise, J. C. (2000). *Phylogeography: The History and Formation of Species.* Cambridge, MA: Harvard University Press.

Avise, J. C. & Aquadro, C. F. (1982). A comparative summary of genetic distances in the vertebrates: patterns and correlations. *Evolutionary Biology* **15**: 151–185.

Avise, J. C. & Hamrick, J. L. (1996). *Conservation Genetics: Case Histories from Nature.* London: Chapman & Hall.

Avise, J. C., Arnold, J., Ball, R. M., *et al.* (1987). Intraspecific phylogeography: the population genetics and systematics. *Annual Review of Ecological Systems* **18**: 489–522.

Barendse, W., Vaiman, D. & Kemp, S. J., *et al.* (1997). A medium-density genetic linkage map of the bovine genome. *Mammalian Genome* **8**: 21–28.

Barnosky, A. D. (2008). Megafauna biomass tradeoff as a driver of Quaternary and future extinctions. *Proceedings of the National Academy of Sciences* **105**: 11543–11548.

Barnosky, A. D., Koch, P. L., Feranec, R. S., Wing, S. L. & Shabel, A. B. (2004). Assessing the causes of late Pleistocene extinctions on the continents. *Science* **306**: 70–75.

Bekhuis, P. D. B. M., de Jong, C. B. & Prins, H. H. T. (2008). Diet selection and density estimates of forest buffalo in Campo-Ma'an National Park, Cameroon. *African Journal of Ecology* **46**: 668–675.

Bertola, L. D., Van Hooft, W. F., Vrieling, K., *et al.* (2011). Genetic diversity, evolutionary history and implications for conservation of the lion (*Panthera leo*) in West and Central Africa. *Journal of Biogeography* **38**(7): 1356–1367.

Birungi, J. & Arctander, P. (2000). Large sequence divergence of mitochondrial DNA genotypes of the control region within populations of the African antelope, kob (*Kobus kob*). *Molecular Ecology* **9**: 1997–2008.

Boone, R. B., Galvin, K. A., Thornton, P. K., Swift, D. M. & Coughenour, M. B. (2006). Cultivation and conservation in Ngorongoro conservation area, Tanzania. *Human Ecology* **34**: 809–828.

Branagan, D. & Hammond, J. A. (1965). Rinder pest in Tanganyika: a review. *Bulletin of Epizootic Diseases of Africa* **13**: 225–246.

Bruford, M. W. & Wayne, R. K. (1993). Microsatellites and their application to population genetic studies. *Current Opinion in Genetics and Development* **3**: 939–943.

Bruford, M. W., Cheesman, D. J., Coote, T., *et al.* (1996). Microsatellites and their application to conservation genetics. In *Molecular Genetic Approaches in Conservation*, pp. 278–297. Eds. Smith, T. B. & Wayne, R. K. New York: Oxford University Press.

Buchholtz, C. (1990). Cattle. In *Grzimek's Encyclopedia of Mammals*, pp. 360–417. Ed. Parker, S. P. New York: McGraw-Hill.

Burroughs, W. (2005). *Climate Change in Prehistory: The End of the Reign of Chaos.* Cambridge: Cambridge University Press.

Burzynska, B., Olech, W. & Topczewski, J. (1999). Phylogeny and genetic variation of the European bison *Bison bonasus* based on mitochondrial DNA D-loop sequences. *Acta Theriologica* **44**: 253–262.

Chantal, S., Tim, C., Simon, M., Charles, M., George, S. & Markus, B. (2007). Assessment of effectiveness of protection strategies in Tanzania based on a decade of survey data for large herbivores. *Conservation Biology* **21**: 635–646.

Cohen, A. S., Stone, J. R., Beuning, K. R. M., *et al.* (2007). Ecological consequences of early Late Pleistocene megadroughts in tropical Africa. *Proceedings of the National Academy of Sciences* **104**: 1–6.

Cornélis, D., Benhamou, S., Janeau, G., Morellet, N., Ouedraogo, M. & de Vissher, M.-N. (2011). Spatiotemporal dynamics of forage and water resources shape space use of West African savanna buffaloes. *Journal of Mammalogy* **92**: 1287–1297.

Cornuet, J. M. & Luikart, G. (1996). Description and power analysis of two tests for detecting recent population bottlenecks from allele frequency data. *Genetics* **144**: 2001–2014.

Craigie, I. D., Baillie, J. E. M., Balmford, A., *et al.* (2010). Large mammal population declines in Africa's protected areas. *Biological Conservation* **143**: 2221–2228.

Crawford, A. M. & Cuthbertson, R. P. (1996). Mutations in sheep microsatellites. *Genome Research* **6**: 876–879.

Cribiu, E. P. & Popescu, C. P. (1980). Chromosome constitution of a hybrid between East African buffalo (*Syncerus caffer caffer*) and dwarf forest buffalo

(*Syncerus caffer nanus*). *Annales de Génétique et de Sélection Animale* **12**: 291–293.

Cross, P. C. & Getz, W. M. (2006). Combining data and models of bovine Tb in African buffalo to assess research directions and management options. *Ecological Modelling* **196**: 494–504.

deMenocal, P. B. (1995). Plio-Pleistocene African climate. *Science* **270**: 53–59.

deMenocal, P. B. (2001). Cultural responses to climate change during the late Holocene. *Science* **292**: 667–673.

Dupont, L. M. & Agwu, C. O. C. (1992). Latitudinal shifts of forest and savanna in NW Africa during the Brunhes chron: further marine palynological results from site M 16145 (9°N, 19°W). *Vegetation History and Archaeobotany* **1**: 163–175.

Dupont, L. M., Jahns, S., Marret, F. & Ning, S. (2000). Vegetation change in equatorial West Africa: time-slices for the last 150 ka. *Palaeogeography Palaeoclimatology, Palaeoecology* **155**: 95–122.

Dusseldorp, G. L. (2010). Prey choice during the South African Middle Stone Age: avoiding dangerous prey or maximising returns? *African Archaeological Review* **27**: 107–133.

East, R. (1997). Antelope survey update no. 5. In *Antelope Specialist Group Report.* Gland: IUCN.

East, R. (1998). *African antelope database.* Gland: IUCN/SSC Antelope Specialist Group.

East, R. (1999). *African Antelope Database 1999.* Gland and Cambridge: IUCN.

Endler, J. A. (1977). *Geographic Variation, Speciation, and Clines.* Princeton, NJ: Princeton University Press.

Eriksson, J., Hohmann, G., Boesch, C. & Vigilant, L. (2004). Rivers influence the population genetic structure of bonobos (*Pan paniscus*). *Molecular Ecology* **13**: 3425–3435.

Ernest, E. M., Haanes, H., Bitanyi, S., *et al.* (2012). Influence of habitat fragmentation on the genetic structure of large mammals: evidence for increased structuring of African buffalo (*Syncerus caffer*) within the Serengeti ecosystem. *Conservation Genetic* **13**: 381–391.

Estes, R. D. (1991). *The Behavior Guide to African Mammals.* Berkeley, CA: University of California Press.

Excoffier, L. & Heckel, G. (2006). Computer programs for population genetics data analysis: a survival guide. *Genetics* **7**: 745–758.

Faith, J. T. (2008). Eland, buffalo, and wild pigs: were Middle Stone Age humans ineffective hunters? *Journal of Human Evolution* **55**: 24–36.

Ferreira, S., Gaylard, A., Greaver, C., Herbst, M., Alard, G. & Moolman, L. (2012). *Animal Abundances in Parks: 2011/2012.* Skukuza, South Africa: Scientific Services, SANParks.

Finlay, E. K., Gaillard, C., Vahidi, S. M. F., *et al.* (2007). Bayesian inference of population expansions in domestic bovines. *Biology Letter* **3**: 449–452.

Frankham, R., Lees, K., Montgomery, M. E., *et al.* (1999). Do population size bottlenecks reduce evolutionary potential? *Animal Conservation* **2**: 255–260.

Frankham, R., Ballou, J. D. & Briscoe, D. A. (2002). *Introduction to Conservation Genetics.* Cambridge: Cambridge University Press.

Fu, X. Y. (1997). Statistical tests of neutrality of mutations against population growth, hitchhiking and background selection. *Genetics* **147**: 915–925.

Funston, P. J., Skinner, J. D. & Dott, H. M. (1994). Seasonal variation in movement patterns, home range and habitat selection of buffalos in a semiarid habitat. *African Journal of Ecology* **32**: 100–114.

Galbusera, P., Githiru, M., Lens, L. & Matthysen, E. (2004). Genetic equilibrium despite habitat fragmentation in an Afrotropical bird. *Molecular Ecology* **13**: 1409–1421.

Gentry, A. W. (1978). Bovidae. In *Evolution of African Mammals*, pp. 540–572. Eds. Maglioi, V. J. & Cooke, H. B. S. Cambridge, MA: Harvard University Press.

Georgiadis, N., Bischof, L., Templeton, A., *et al.* (1994). Structure and history of African elephant populations: I. Eastern and Southern Africa. *Journal of Heredity* **85**: 100–104.

Gignoux, C. R., Henn, B. M. & Mountain, J. L. (2011). Rapid, global demographic expansions after the origins of agriculture. *PNAS* **108**: 6044–6049.

Godfrey, L. & Jungers, W. (2003). The extinct sloth lemurs of Madagascar. *Evolutionary Anthropology* **12**: 252–263.

Gotelli, D., Sillero-Zubiri, C., Applebaum, G. D., *et al.* (1994). Molecular genetics of the most endangered canid: the Ethiopian wolf *Canis simensis. Molecular Ecology* **3**: 301–312.

Greyling, B. J. (2007) Genetic variation, structure and dispersal among Cape buffalo populations from the Hluhluwe-Imfolozi and Kruger National Parks of South Africa. PhD thesis. University of Pretoria. http://upetd.up.ac.za/thesis/available/etd-07152008-090238, accessed 6 September 2007.

Grobler, J. P. & Van der Bank, F. H. (1996). Genetic diversity and isolation in African buffalo (*Syncerus caffer*). *Biochemical Systematics and Ecology* **24**: 757–761.

Grubb, P. (1971). Variation and incipient speciation in the African buffalo. *Zeitschrift für Säugetierkunde* **37**: 121–144.

Halley, D. J., Vandewalle, M. E. J., Mari, M. & Taolo, C. (2002). Herd-switching and long-distance dispersal in female African buffalo, *Syncerus caffer. African Journal of Ecology* **40**: 97–99.

Haltenorth, T. & Diller, H. (1980). *Elseviers Gids van de Afrikaanse zoogdieren.* Revised and enlarged by Smeenk, C. Amsterdam: Elsevier.

Hassanin, A. & Ropiquet, A. (2004). Molecular phylogeny of the tribe Bovini (Bovidae, Bovinae) and the taxonomic status of the Kouprey, *Bos sauveli* Urbain 1937. *Molecular Phylogenetics and Evolution* **33**(3): 896–907.

Hedges, S. B., Dudley, J. & Kumar, S. (2006). TimeTree: a public knowledge-base of divergence times among organisms. *Bioinformatics* **22**: 2971–2972.

Hedrick, P. W. (2005). *Genetics of Populations.* Sudbury: Jones and Bartlett Publishers.

Heller, R., Lorenzen, E. D., Okello, J. B. A., Masembe, C. & Siegismund, H. R. (2008). Mid-Holocene decline in African buffalos inferred from Bayesian coalescent-based analyses of microsatellites and mitochondrial DNA. *Molecular Ecology* **17**: 4845–4858.

Heller, R., Okello, J. B. A. & Siegismund, H. R. (2010). Can small wildlife conservancies maintain genetically stable populations of large mammals? Evidence for increased genetic drift in geographically restricted populations of Cape buffalo in East Africa. *Molecular Ecology* **19**: 1324–1334.

Heller, R., Brüniche-Olsen, A. & Siegismund, H. R. (2012). Cape buffalo mitogenomics reveals a Holocene shift in the African human–megafauna dynamics. *Molecular Ecology* **21**: 3947–3959.

Hewitt, G. M. & Butlin, R. K. (1997). Causes and consequences of population structure. In *Behavioural Ecology. An Evolutionary Approach*, pp. 350–372. Eds. Krebs, J. R. & Davies, N. B. Malden, MA: Blackwell Science.

Hey, J. (2005). On the number of New World founders: a population genetic portrait of the peopling of the Americas. *PloS Biology* **3**: 965–975.

Hey, J. & Nielsen, R. (2004). Multilocus methods for estimating population sizes, migration rates and divergence time, with applications to the divergence of *Drosophila pseudoobscura* and *D-persimilis*. *Genetics* **167**: 747–760.

Hilborn, R., Arcese, P., Borner, M., *et al.* (2006). Effective enforcement in a conservation area. *Science* **314**: 1266.

Hsu, T. C. & Benirschke, K. (1970). *An Atlas of Mammalian Chromosomes*. New York: Springer-Verlag.

Hughes, A. & Nei, M. (1988). Pattern of nucleotide substitution at major histocompatibility complex loci reveals overdominant selection. *Nature* **335**: 167–170.

Jolles, A. E. (2007). Population biology of African buffalo (*Syncerus caffer*) at Hluhluwe-iMfolozi Park, South Africa. *African Journal of Ecology* **45**: 398–406.

Joppa, L. N., Loarie, S. R. & Pimm, S. L. (2008). On the protection of 'protected areas'. *Proceedings of the National Academy of Sciences* **105**: 6673–6678.

Kappes, S. M., Keele, J. W. & Stone, R. T., *et al.* (1997). A second-generation linkage map of the bovine genome. *Genome Research* **7**: 235–249.

Keller, L. F. & Waller, D. M. (2002). Inbreeding effects in wild populations. *Trends Ecology and Evolution* **17**: 230–241.

Kiage, L. M. & Liu, K. (2006). Late Quaternary paleoenvironmental changes in East Africa: a review of multiproxy evidence from palynology, lake sediments, and associated records. *Progress in Physical Geography* **30**: 633–658.

Kingdon, J. (1982). *East African Mammals: An Atlas of Evolution in Africa*. Vol. **IIIC**. Bovids. London: Academic Press.

Kingdon, J. (1989). *Island Africa, the Evolution of Africa's Rare Animals and Plants*. Princeton, NJ: Princeton University Press.

Kingdon, J. (1997). *The Kingdon Field Guide to African Mammals*. London and New York: Academic Press.

Klein, J. (1986). *Natural History of the Major Histocompatibility Complex*. New York: Wiley.

Klein, R. G. (1988). The archaeological significance of animal bones from Acheulean sites in Southern Africa. *The African Archaeological Review* **6**: 3–25.

Klein, R. G. (1995). Mammalian extinctions and Stone Age people in Africa. In *Quaternary Extinctions: A Prehistoric Revolution*, pp. 553–573. Eds. Martin, S. & Klein, R. G. Tucson, AZ and London: University of Arizona Press.

Korte, L. M. (2008a). Habitat selection at two spatial scales and diurnal activity patterns of adult female forest buffalo. *Journal of Mammalogy* **89**: 115–125.

Korte, L. M. (2008b). Variation of group size among African buffalo herds in a forest-savanna mosaic landscape. *Journal of Zoology* **275**: 229–236.

Kumar, S. & Hedges, S. B. (2011). TimeTree2: species divergence times on the iPhone. *Bioinformatics* **27**: 2023–2024.

Le Roex, N., Noyes, H., Brass, A., *et al.* (2012). Novel SNP discovery in African buffalo, *Syncerus caffer*, using high-thoughput sequencing. *PLoS ONE* **7**: e48792.

Lehmann, T., Hawley, W. A., Grebert, H. & Collins, F. H. (1998). The effective population size of *Anopheles gambiae* in Kenya: implications for population structure. *Molecular Biology and Evolution* **15**: 264–276.

Lessard, P., L'Eplattenier, R., Norval, R. A. I., *et al.* (1990). Geographical information systems for studying the epidemiology of cattle diseases caused by *Theileria parva*. *Veterinary Record* **126**: 255–262.

Lister, A. M. (2004). The impact of Quaternary Ice Ages on mammalian evolution. *Philosophical Transactions of the Royal Society* **359**: 221–241.

Lorenzen, E. D., Simonsen, B. T., Kat, P. W., Arctander, P. & Siegismund, H. R. (2006). Hybridization between subspecies of waterbuck (*Kobus ellipsiprymnus*) in zones of overlap with limited introgression. *Molecular Ecology* **15**: 3787–3799.

Lorenzen, E. D., De Neergaard, R., Arctander, P. & Siegismund, H. R. (2007). Phylogeography, hybridization and Pleistocene refugia of the kob antelope (*Kobus kob*). *Molecular Ecology* **16**: 3241–3252.

Lorenzen, E. D., Arctander, P. & Siegismund, H. R. (2008). High variation and very low differentiation in wide ranging plains zebra (*Equus quagga*): insights from mtDNA and microsatellites. *Molecular Ecology* 17: 2812–2814.

Lorenzen, E. D., Heller, R. & Siegismund, H. R. (2012). Comparative phylogeography of African savannah ungulates. *Molecular Ecology* **21**: 3656–3670.

Lovett, J. C., Midgley, G. F. & Barnard, P. (2005). Climate change and ecology in Africa. *African Journal of Ecology* **43**: 167–169.

Luikart, G., England, P. R., Tallmon, D., Jordan, S. & Taberlet, P. (2003). The power and promise of population genomics: from genotyping to genome typing. *Nature Reviews Genetics* **4**: 981–994.

Luoy, D., Habel, J. C., Schmitt, T., Assmann, T., Meyer, M. & Muller, P. (2007). Strongly diverging population genetic patterns of three skipper species: the role of habitat fragmentation and dispersal ability. *Conservation Genetics* **8**: 671–681.

Mack, R. (1970). The great African cattle plague epidemic of the 1890s. *Tropical Animal Health and Production* **2**: 210–219.

Marchant, R. & Hooghiemstra, H. (2004). Rapid environmental change in African and South American tropics around 4000 years before present: a review. *Earth Science Reviews* **66**: 217–260.

Matthee, C. A. & Robinson, T. J. (1999). Mitochondrial DNA population structure of roan and sable antelope: implications for the translocation and conservation of the species. *Molecular Ecology* **8**: 227–238.

Mayewski, P. A., Rohling, E., Stager, J. C., *et al.* (2004). Holocene climate variability. *Quaternary Research* **62**: 243–255.

McArthur, R. H. & Wilson, E. O. (1967). *The Theory of Island Biogeography*. Princeton, NJ: Princeton University Press.

McCabe, J. T., Mollel, N. & Tumaini, A. (1997). Food security and the role of cultivation. In Multiple Land-use: The Experience of the Ngorongoro Conservation area, Tanzania, pp. 397–416. Ed. Thompson, D. M. Gland: IUCN.

McCauley, D. E. (1991). Genetic consequenecs of local population extinction and recolonization. *Trends in Ecology and Evolution* **6**: 5–8.

Meester, J. & Setzer, H. W. (1971). *The Mammals of Africa: An Identification Manual*. Washington, DC: Smithsonian Institution Press.

Meijard, E. (2003). Mammals of SE Asian islands and their late Pleistocene environments. *Journal of Biogeography* **30**: 1245–1257.

Melletti, M., Penteriani, V., Mirabile, M. & Boitani, L. (2007). Some behavioral aspects of forest buffalo (*Syncerus caffer nanus*): from herd to individual. *Journal of Mammalogy* **88**: 1312–1318.

Moodley, Y. & Bruford, M. W. (2007). Molecular biogeography: towards an integrated framework for conserving pan-African biodiversity. *PLoS ONE* **5**: e454.

Morin, P. A., Luikart, G., Wayne, R. K. & the SNP Workshop Group (2004). SNPs in ecology, evolution and conservation. *Trends in Ecology and Evolution* **19**(4): 208–216.

Moritz, C. (1994). Defining 'evolutionary significant units' for conservation. *Trends in Ecology and Evolution* **9**: 373–375.

Muwanika, V. B., Nyakaana, S., Siegismund, H. R. & Arctander, P. (2003). Phylogeography and population structure of the common warthog (*Phacochoerus africanus*) inferred from variation in mitochondrial DNA sequences and microsatellite loci. *Heredity* **91**: 361–372.

Naidoo, R., Du Preez, P., Stuart-Hill, G., Jago, M. & Wegmann, M. (2012). Home on the range: factors explaining partial migration of African buffalo in a tropical environment. *PLoS ONE* **7**(5): e36527.

Nei, M., Maruyama, T. & Chakaborty, R. (1975). The bottleneck effect and genetic variability of populations. *Evolution* **29**: 1–10.

Nersting, L. G. & Arctander, P. (2001). Phylogeography and conservation of impala and greater kudu. *Molecular Ecology* **10**: 711–719.

Neumann, R. P. (1998). *Imposing Wilderness: Struggles Over Livelihood and Nature Preservation in Africa*. Berkeley, CA: University of California.

Nowak, R. M. (1991). *Walker's Mammals of the World*. 5th edition. Baltimore, MD: Johns Hopkins University Press.

Nyakaana, S. & Arctander, P. (1999). Population genetic structure of the African elephant in Uganda based on variation at mitochondrial and nuclear loci: evidence for male-biased gene flow. *Molecular Ecology* **8**: 1105–1115.

Nyakaana, S., Arctander, P. & Siegismund, H. R. (2002). Population structure of the African savannah elephant inferred from mitochondrial control region sequences and nuclear microsatellite loci. *Heredity* **89**: 90–98.

Ogutu, J. O., Piepho, H. P., Dublin, H. T., Bhola, N. & Reid, R. S. (2008). Rainfall influences on ungulate population abundance in the Mara-Serengeti ecosystem. *Journal of Animal Ecology* **77**: 814–829.

Okello, J. B. A., Nyakaana, S., Masembe, C., Siegismund, H. R. & Arctander, P. (2005). Mitochondrial DNA variation of the common hippopotamus: evidence for a recent population expansion. *Heredity* **95**: 206–215.

Okello, J. B. A., Wittemeyer, G., Rasmussen, H. B., *et al.* (2008). Effective population size dynamics reveal impacts of historic climate events and recent anthropogenic pressure in African elephants. *Molecular Ecology* **17**: 3788–3799.

O'Ryan, C., Harley, E. H., Bruford, M. W., *et al.* (1998). Microsatellite analysis of genetic diversity in fragmented South African buffalo populations. *Animal Conservation* **1**: 85–94.

Paetkau, D., Calvert, W., Stirling, I. & Strobeck, C. (1995). Microsatellite analysis of population structure in Canadian polar bears. *Molecular Ecology* **4**: 347–354.

Pitra, C., Hansen, A. J., Lieckfeldt, D. & Arctander, P. (2002). An exceptional case of historical outbreeding in African sable antelope populations. *Molecular Ecology* **11**: 1197–1208.

Plowright, W. (1982). The effects of rinderpest and rinderpest control on wildlife in Africa. *Symposia of the Zoological Society of London* **50**: 1–28.

Porat, N., Chazan, M., Grun, R., *et al.* (2010). New radiometric ages for the Fauresmith industry from Kathu Pan, Southern Africa: implications for the Earlier to Middle Stone Age transition. *Journal of Archaeological Science* **37**: 269–283.

Prins, H. H. T. (1996). *Ecology and Behaviour of the African Buffalo: Social Inequality and Decision Making*. London: Chapman & Hall.

Prins, H. H. T. & Douglas-Hamilton, I. (1990). Stability in a multi-species assemblage of large herbivores in East Africa. *Oecologia (Berl.)* **83**: 392–400.

Prins, H. H. T. & Van der Jeugd, H. P. (1993). Herbivore crashes and woodland structure in East Africa. *Journal of Ecology* **81**: 305–314.

Queller, D. C., Strassman, J. E. & Hughes, C. R. (1993). Microsatellites and kinship. *Trends in Ecology and Evolution* **8**: 285–288.

Rampino, M. P. & Self, S. (1992). Volcanic winter and accelerated glaciation following the Toba super-eruption. *Nature* **359**: 50–52.

Redfern, J. V., Grant, R., Biggs, H. & Getz, W. M. (2003). Surface-water contraints on herbivore foraging in the Kruger National Park, South Africa. *Ecology* **84**: 2092–2117.

Robinson, T. J. & Ropiquet, A. (2011). Examination of hemiplasy, homoplasy and phylogenetic discordance in chromosomal evolution of the Bovidae. *Systematic Biology* **60**(4): 439–450.

Rogers, A. R. & Harpending, H. (1992). Population growth makes waves in the distribution of pairwise genetic differences. *Molecular Biology and Evolution* **9**: 552–569.

Roy, M. S., Geffen, E., Smith, D., Ostrander, E. A. & Wayne, R. K. (1994). Patterns of differentiation and hybridization in North American wolflike canids, revealed by analysis of microsatellite loci. *Molecular Biology and Evolution* **11**: 553–570.

Rubenstein, D. (1992). The greenhouse effect and changes in animal behavior: effects on social structure and life-history strategies. In *Global Warming and Biological Diversity*, pp. 180–192. Eds. Peters, R. & Lovejoy, T. New Haven, CT: Yale University Press.

Ryan, S. J., Knechtel, C. U. & Getz, W. M. (2006). Range and habitat selection of African buffalo in South Africa. *The Journal of Wildlife Management* **70**(3): 764–776.

Santure, A. W., Stapley, J., Ball, A. D., Birkhead, T. R., Burke, T. & Slate, J. (2010). On the use of large marker panels to estimate inbreeding and relatedness: empirical and simulation studies of a pedigreed zebra finch population typed at 771 SNPs. *Molecular Ecology* **19**: 1439–1451.

Scheinfeldt, L. B., Soib, S. & Tishkoff, S. A. (2010). Working toward a synthesis of archaeological, linguistic, and genetic data for inferring African population history. *Proceedings of the National Academy of Sciences* **107**: 8931–8938.

Scholz, C. A., Johnson, C., Cohen, A. S., *et al.* (2007). East African megadroughts between 135 and 75 thousand years ago and bearing on early-modern human origins. *Proceedings of the National Academy of Sciences* **104**: 16416–16421.

Schug, M. D., Mackay, T. F. C. & Aquadro, C. F. (1997). Low mutation rates of microsatellite loci in *Drosophila melanogaster*. *Nature Genetics* **15**: 99–102.

Shigesada, N. & Kawasaki, K. (1997). *Biological Invasions: Theory and Practice*. New York: Oxford University Press.

Siegismund, H. R. & Arctander, P. (1995). Structure of African elephant populations. *Journal of Heredity* **86**: 467–468.

Simonsen, B. T. (1997). Population structure and history of African bovids. PhD thesis. University of Copenhagen.

Simonsen, B. T., Siegismund, H. R. & Arctander, P. (1998). Population structure of African buffalo inferred from mtDNA sequences and microsatellite loci: high variation but low differentiation. *Molecular Ecology* **7**: 225–237.

Sinclair, A. R. E. (1977). *The African buffalo: a study of resource limitation of populations*. Chicago, IL: University of Chicago Press.

Slade, W. S., Moritz, C., Hoelzel, A. R. & Burton, H. R. (1998). Molecular population genetics of the southern elephant seal *Mirounga leonina*. *Genetics* **149**: 1945–1957.

Smith, T. B. & Wayne, R. K. (1996). *Molecular Genetic Approaches in Conservation*. New York: Oxford University Press.

Smithers, R. H. E. (1983). *The Mammals of the Southern African Subregion: 663–666.* Pretoria: University of Pretoria.

Smitz, N., Berthouly, C., Cornélis, D., *et al.* (2013). Pan-African genetic structure in the African buffalo (*Syncerus caffer*): investigating subspecies divergence. *PLoS ONE*, **8**(2): e56235, doi: 10.1371/journal.

Soares, P., Alshamali, F., Pereira, J. B., *et al.* (2012). The expansion of mtDNA haplogroup L3 within and out of Africa. *Molecular Biology and Evolution* **29**: 915–927.

Sosef, M. S. M. (1996). Begonias and African rain forest refuges: general aspects and recent progress. In *The Biodiversity of African Plants*, pp. 602– 611. Eds. Van der Maesen, L. J. G., van der Burgh, X. M. & van Mendenbach de Rooy, J. M. Wageningen: Kluwer Academic Publishers.

Soulé, M. E. & Mills, L. S. (1992). Conservation genetics and conservation biology: a troubled marriage. In *Species and Ecosystem Conservation*, pp. 55–65. Eds. Sandlund, O. T., Hindar, K. & Brown, A. H. D. Oslo: Scandinavian University Press.

Stark, M. A. (1986). Daily movement, grazing activity and diet of savanna buffalo, *Syncerus caffer brachyceros*, in Benoue National Park, Cameroon. *African Journal of Ecology* **24**: 255–262.

Stevenson-Hamilton, J. (1911). The relation between game and tsetse flies. *Bulletin of Entomological Research* **2**: 113–118.

Stokes, S., Haynes, G., Thomas, D. S. G., *et al.* (1998). Punctuated aridity in southern Africa during the last glacial cycle: the chronology of linear dune construction in the northeastern Kalahari. *Palaeogeography, Palaeoclimatology, Palaeoecology* **137**: 305–322.

Storz, J. F. & Beaumont, M. A. (2002). Testing for genetic evidence of population expansion and contraction: an empirical analysis of microsatellite DNA variation using a hierarchical Bayesian model. *Evolution* **56**: 154–166.

Storz, J. F., Beaumont, M. A. & Alberts, S. C. (2002). Genetic evidence for long-term population decline in a savannah-dwelling primate: inferences from a hierarchical Bayesian model. *Molecular Biology and Evolution* **19**: 1981–1990.

Surovel, T., Waguespack, N. & Brantingham, P. (2005). Global archaelogical evidence for proboscidean overkill. *Proceedings of the National Academy of Sciences* **102**: 6231–6236.

Templeton, A. R. & Georgiadis, N. J. (1996). A landscape approach to conservation genetics: conserving evolutionary processes in the African bovidae. In *Conservation Genetics: Case Histories from Nature*, pp. 398–430. Eds. Avise, J. C. & Hamrick, J. L. New York: Chapman & Hall.

Thirgood, S., Mosser, A., Tham, S., *et al.* (2004). Can parks protect migratory ungulates? The case of the Serengeti wildebeest. *Animal Conservation* **7**: 113–120.

Van Hooft, W. F., Hanotte, P. W., Groen, A. F., Sugimoto, Y., Prins, H. H. T. & Teale, A. (1999). Applicability of bovine microsatellite markers for population genetic studies on African buffalo (*Syncerus caffer*). *Animal Genetics* **30**: 214–220.

Van Hooft, W. F., Groen, A. F. & Prins, H. H. T. (2000). Microsatellite analysis of genetic diversity in African buffalo (*Syncerus caffer*) populations throughout Africa. *Molecular Ecology* **9**(12): 2017–2025.

Van Hooft, W. F., Groen, A. F. & Prins, H. T. (2002). Phylogeography of the African buffalo based on mitochondrial and Y-chromosomal loci: Pleistocene origin and population expansion of the Cape buffalo subspecies. *Molecular Ecology* **11**: 267–279.

Van Hooft, W. F., Groen, A. F. & Prins, H. H. T. (2003). Genetic structure of African buffalo herds based on variation at the mitochondrial D-loop and autosomal microsatellite loci: evidence for male-biased gene flow. *Conservation Genetics* **4**: 467–477.

Van Hooft, P., Greyling, B. J., Prins, H. H. T., *et al.* (2007). Selection at the Y chromosome of the African buffalo driven by rainfall. *PLoS ONE* **10**: e1086.

Van Zinderen Bakker, E. M. (1978). Quaternary vegetation changes in southern Africa. In *Biogeography and Ecology of Southern Africa 1*, pp. 133–143. Eds. Werger, M. J. A. & Van Bruggen, A. C. The Hague: Dr W. Junk Publishers.

Vanacker, V., Linderman, M., Lupo, F., Flasse, S. & Lambin, E. (2005). Impact of short-term rainfall fluctuation on interannual land cover change in sub-Saharan Africa. *Global Ecology and Biogeography* **14**: 123–135.

Vrba, E. S. (1992). Mammals as a key to evolutionary theory. *Journal of Mammalogy* **73**: 1–28.

Vrba, E. S. (1995a). The fossil records of African antelopes (Mammalia, Bovidae) in relation to human evolution and paleoclimate. In *Paleoclimate and Evolution, with Emphasis on Human Origins*, pp. 385–324. Eds. Vrba, E. S., Denton, G. H., Partridge, T. C. & Burckle, L. H. New Haven, CT: Yale University Press.

Vrba, E. S. (1995b). On the connections between paleoclimate and evolution. In: *Paleoclimate and Evolution, with Emphasis on Human Origins*, pp. 24–48. Eds. Vrba, E. S., Denton, G. H., Partridge, T. C. & Burckle, L. H. New Haven, CT: Yale University Press.

Wenink, P. W., Groen, A. F., Roelke-Parker, M. E. & Prins, H. H. T. (1998). African buffalo maintain high genetic diversity in the major histocompatibility complex in spite of historically known population bottlenecks. *Molecular Ecology* 7: 1315–1322.

Wilson, D. E. & Reeder, D. A. M. (2005). *Mammal Species of the World: A Taxonomic and Geographic Reference.* Baltimore, MD: Johns Hopkins University Press.

Winterbach, H. E. K. (1998). The status and distribution of Cape buffalo *Syncerus caffer caffer* in Southern Africa. *South African Journal of Wildlife Research* **28**: 82–88.

Wright, S. (1931). Evolution in Mendelian populations. *Genetics* **16**(2): 97–159.

Wurster, D. H. & Benirsch, K. (1968). Chromosome studies in superfamily bovidea. *Chromosoma* **25**: 152–171.

Young, A. G. & Clarke, G. M. (2000). *Genetics, Demography and Viability of Fragmented Populations.* Cambridge: Cambridge University Press.

Chapter 26

Livestock and buffalo (*Syncerus caffer*) interfaces in Africa: ecology of disease transmission and implications for conservation and development

Richard Kock, Michael Kock, Michel de Garine-Wichatitsky, Philippe Chardonnet and Alexandre Caron

Introduction

Wildlife–livestock–human interfaces are widespread and diverse in Africa, often leading to human–animal conflicts (Kock 2005a; de Garine-Wichatitsky *et al.* 2013b, 2013c). The origins of land use challenges are complex. They combine the interplay between conservation and development policies, leading to changes that benefit only a few and to the concentration of pastoral or agropastoral communities in the periphery of protected areas or in wild rangeland (Homewood & Thompson 2010), creating animal health challenges (Kock *et al.* 2009). These marginalised communities, often politically disenfranchised, bear the brunt of the problem and few benefits of the policies, despite having socio-economic models far better than modern ranching, or agriculture using similar land (Catley *et al.* 2012).

Whatever the cause, conflict at the interface can threaten the success of conservation and development objectives (Osofsky *et al.* 2005). Old dogma on wildlife as a source of disease remains deeply entrenched in the psyche of many people, but mostly those who are peripheral to the actual keeping of livestock, commercially developed and with large investments at risk from disease events. Recently, in Southern Africa the increased rate of reversion to more integrated landscape approaches, such as transfrontier conservation areas (TFCAs), have facilitated wildlife, livestock and people movement, potentially exacerbating issues at these interfaces (Andersson and Cumming 2013). In most other areas of Africa, open systems have been the normal situation and this has had ecological benefits, but also costs in lost livestock export potential. Understanding the relative costs (lost trade) to benefits (ecosystems services) of more open systems is fundamental to future land use policy in Africa. While Southern Africa begins to tear down fences and open up land, in East Africa policies are tending towards enclosing protected areas or excluding wildlife, threatening ecosystem integrity (Kock 2010). Here we focus on the importance of the sanitary issue at buffalo–cattle interfaces. We describe the main diseases thriving in this context, explore the mechanisms of disease transmission in relation to the type of interface and discuss the implication of this sanitary issue for an integrated conservation and development management of these interfaces across Africa in the face of intense agricultural and human development (Wallace & Kock 2012).

Infections and pathogens at the buffalo–cattle interface

A wide range of viral, bacterial and parasitic infections of free-ranging African buffalo have been described (Bengis *et al.* 2002, 2004; see Table 26.1). Most studies relate to one subspecies, the Cape buffalo (*S. c. caffer* sparrman 1779), while the forest buffalo (*S. c. nanus*), West African and Nile buffalo (*S. c. brachyceros* and *S. c. aequinoctialis*) are rarely studied. African buffalo phylogeny is discussed (Prins 1996; Smitz *et al.* 2013). Historically, diseases shared with cattle and of known importance to buffalo ecology are few; rinderpest is a major exception. In fact, the most important cause of recurrent mortality in buffalo populations has probably been severe drought and starvation (Metzger *et al.* 2010) and the role of diseases in these die-outs is at best sometimes hypothesised. For example, there is evidence that buffalo anthrax infections are not invariably fatal, with relatively high seroprevalence among wild herbivores in the Serengeti (Lembo *et al.* 2011). On the other hand, buffalo are considered an important reservoir host for foot-and-mouth disease (FMD) virus, *Theileria* sp., *Trypanosoma* sp., *Mycobacterium bovis* (bTB) and *Brucella* sp., and provide an amplification host for the Rift Valley Fever virus. For most of these pathogens buffalo are highly tolerant and there are few if any impacts on their population.

Questions remain on the role of buffalo in the epidemiology of most shared diseases with cattle, even those long

Ecology, Evolution and Behaviour of Wild Cattle: Implications for Conservation, ed. M. Melletti and J. Burton. Published by Cambridge University Press.

Table 26.1 Infectious diseases at the buffalo–livestock interface

Disease, *causative agent* and (mode of transmission)	Traditional or endemic host(s)	Key references
Foot-and-mouth disease, *Apthovirus (seven types)* (direct)	Indigenous cattle and buffalo	Thomson *et al.* 2003
Bluetongue, *Orbivirus 24 serotypes* (arthropod)	Wild ruminants and cattle	Hamblin *et al.* 1990
Rift Valley Fever, *Phlebovirus* (arthropod)	Aedine mosquitoes, wildlife?	Olive *et al.* 2012
Bovine Herpes Virus 4 (transmission route unknown) Ovine Herpes Virus 2	Cattle and buffalo Sheep and buffalo	Dewals *et al.* 2006 Pfitzer *et al.* 2013 Hamblin and Hedger 1982
Bovine petecchial fever (arthropod vector implicated)	Cattle, bush buck buffalo possible amplification host	Sumption & Scott 2004
Lumpy skin disease	Cattle, buffalo spill-over host	Hedger & Hamblin 1983
Adenovirus	Cattle	Baber & Condy 1981
PI3	Cattle	Hamblin *et al.* 1990
Bovine viral papillomatosis	Cattle	Williams *et al.* 2011
Heartwater rickettsia, *Cowdria ruminantium* (tick)	Wild ruminants, chelonians and gallinaceous birds, livestock	Wesonga *et al.* 2001
Trypanosomiasis, *Trypanosoma* spp. (tsetse fly)	Cattle, wild bovids, suids, elephant and black rhino	van den Bossche *et al.* 2010
Corridor disease, *Theileria parva lawrenci* (tick)	Buffalo	Pascucci *et al.* 2011
Anthrax, *Bacillus anthracis* (environment, biting flies)	Soil saprophyte	McConnell *et al.* 1972
Bovine tuberculosis, *Mycobacterium bovis* (environment, direct)	Cattle, buffalo, other wild ruminants and suids	De Vos *et al* 2001; Renwick *et al.* 2007
Brucellosis, *Brucella* spp. (environment)	Cattle, sheep, goats, camels, buffalo, humans	Gradwell 1977
Sarcosporidiosis, *Sarcosporidia* (environment)		Perrotin *et al.* 1978
Fascioliasis, *Fasciola gigantica* (environment, snail)	Herbivorous animals, humans	Bindernagel 1972
Parafilariosis, *Parafilaria* spp. (Fly – *Musca*)	Cattle, buffalo	Keet *et al.* 1997
Cryptosporidiosis, *Cryptosporidia* (environment)	>152 mammalian hosts	Mtambo *et al.* 1997; Abu Samra *et al.* 2011
Thelaziosis, *Thelazia rhodesii* (Fly – *Musca*)	Cattle, camels, buffalo	Munang'andu *et al.* 2011
Mange, *Sarcoptes scabei* and *Demodex* (environment – host specificity transmission between species is rare)	Wide host range, several taxa	Wolhuter *et al.* 2009; Munang'anda *et al.* 2010

known to be significant. Hereafter, the historical and current important diseases at the buffalo–cattle interface are presented.

Rinderpest (Figure 26.1) was historically the most devastating disease of buffalo and cattle in Africa, with mortalities confirmed at 60% of the buffalo population in one outbreak (Kock *et al.* 1999). The epidemiology of rinderpest in cattle was rarely, if ever, studied due to its severity and rapid mass vaccination in the face of an outbreak, and there have been few experimental infection studies confirming the potential for cross transmission (Wamwayi 2002). Buffalo are highly susceptible to this introduced cattle disease, even to so-called 'mild' viruses circulating in cattle populations. The evidence for spill-over of infection from cattle to buffalo and to other wildlife is mostly circumstantial and often associated with drought and cattle movements into wildlife areas (Kock 2006). Rinderpest has only been maintained in buffalo populations for three to four years before being 'burned out'. The last cases confirmed globally were in African buffalo in Meru, Kenya in 2001 (Kock *et al.* 2006). Today, this plague is considered eradicated.

Foot-and-mouth disease is an affliction of cattle and of concern to international trade in livestock. FMD control policies in Southern African countries require wildlife reservoir and livestock to be separated and the extensive fencing that has resulted from this has led to serious fragmentation of ecosystems and negative ecological impacts (Ferguson & Hanks 2010). This is due to buffalo being known maintenance hosts of three serotypes (SAT 1, 2 and 3; Thomson *et al.* 2003), with

Figure 26.1 African Buffalo adult female infected with rinderpest virus in Nairobi National Park 1996, resulting in nasal discharge, diarrhoea, dehydration and poor coat condition; 25% of the herd died (photo by R. Kock). A black and white version of this figure will appear in some formats. For the colour version, please refer to the plate section.

calves infected by nine months of age and individuals capable of excreting virus for up to five years. Buffalo are insensitive to FMD unless they are immuno-compromised, but clinical cases have been reported in captivity (Vosloo *et al.* 2007). Disease-free populations of cattle are sometimes victim of outbreaks due to presumed spill-over of SAT viruses from buffalo at shared waterholes or from wildlife breaking through barriers (Hargreaves *et al.* 2004); these data are convincing but in fact are based on circumstantial evidence. Experimental buffalo to cattle infection has been shown in at least seven studies (Dawe *et al.* 1994; Vosloo *et al.* 1996; Tenzin Dekker *et al.* 2006). In mixed or open systems in East and West Africa, where FMD is endemic, there have been few isolates of virus from buffalo. Out of 522 reported isolations of FMD virus from buffalo between 1970 and 2010, only 13 (SAT 1–3) are from East Africa (K. Pirbright, personal communication 2011) and it is believed SAT 1 was introduced to the region from Southern Africa on two separate occasions (Sangula *et al.* 2010), but the pan-African molecular epidemiology remains to be determined. Serological evidence of SAT 1–3 infection in buffalo exists across the whole region in every population sampled (Bronsvoort *et al.* 2008; Ayebazibwe *et al.* 2010), but despite this most outbreaks reported in cattle in East Africa over the last 50 years are of other serotypes (O and A), but with a significant number of SAT 2 (Sangula *et al.* 2010). When looking at SAT 1 types there is considerable variation in the virus across even a single region – e.g. East Africa (Sahle *et al.* 2007) – suggesting both spatially discrete endemic foci, but showing evidence of epidemiological linkages between host populations and reintroductions of the virus periodically. Therefore, even in open ranges it would appear that FMD virus populations are heterogeneous and spill-over between buffalo and cattle is perhaps less common than dogma suggests; however, when it occurs it might significantly impact virus distribution and evolution. Long latent periods, which can occur after transmission (Dawe *et al.* 1994), confuse epidemiology even further, making tracing the virus difficult without using molecular testing. More virus isolates are needed from buffalo (and cattle) in sub-Saharan Africa to confirm the relative status and topotypes in mixed populations and to better understand antigenic diversity in animals and what drives this variation. This has important implications for vaccination and control strategies.

Buffalo are a maintenance host for bTB in South Africa and Uganda; this has been a particular problem in Kruger National Park (KNP), where the disease spread from cattle to buffalo across the southern Crocodile River, probably between 1960 and the 1980s (Vosloo *et al.* 2001), and was discovered in buffalo in the 1990s (Bengis *et al.* 1996). bTB subsequently spread through the KNP buffalo population, from south to north, reaching the extreme north in 2005 (Michel *et al.* 2006) (Figure 26.2). Spread of bTB into the buffalo population of Zimbabwe was seen by 2009 (de Garine-Wichatitsky *et al.* 2010) and it is probably of Kruger origin, based on similarities of the strains isolated. Evidence suggests that ten other small and large mammalian species, including large predators, are spill-over hosts (Keet *et al.* 1996, 2000). This has been particularly significant in the lion population. It appears that with increasing bTB herd prevalence in buffalo there is a decrease in body condition; this apparent correlation is even more marked during the dry season (Beechler *et al.* 2012). Therefore, the buffalo most affected by the disease are easiest for lions to kill (Caron *et al.* 2003) and the frequent exposure of lions to large amounts of infective material leads to spatial spread of bTB within lion prides.

In Uganda relatively high bTB prevalence in buffalo has been known of for many decades (Woodford 1982) but significant declines in numbers of wildlife during politically unstable periods resulted in a period without reported disease. Populations have recovered to some extent in recent decades and the

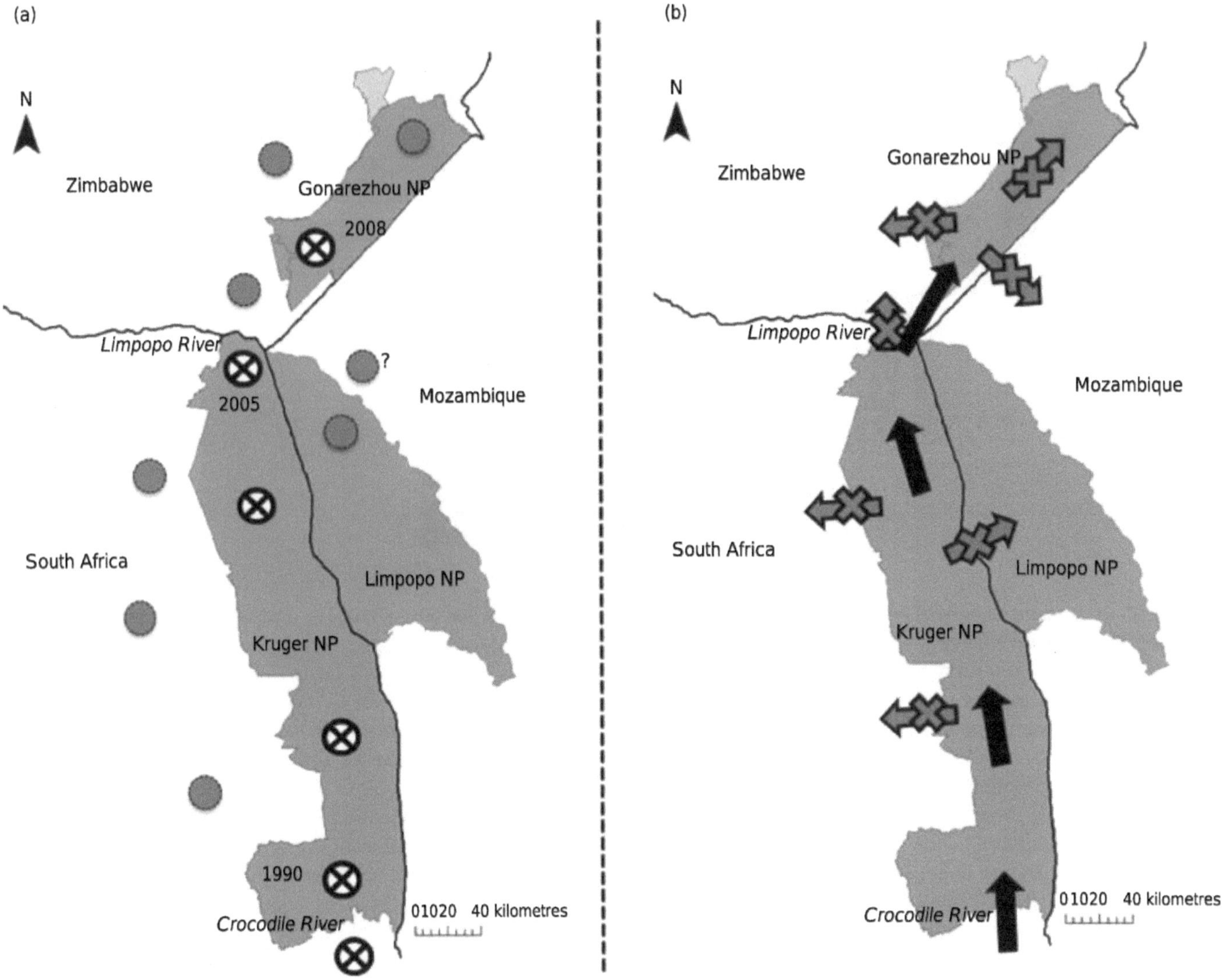

Figure 26.2 Spread and risk of spread of bovine tuberculosis in the Great Limpopo Transfrontier Conservation Area. (a) grey areas represent protected areas and black lines represent international borders; at-risk buffalo populations are represented as circles; crosses denote bTB infected populations. (b) grey arrows with a cross represent interfaces with cattle and buffalo outside the park, but where there is no evidence for bTB transmission to date; black arrows show bTB transmission route over time through buffalo populations. There is a risk of bTB spread: (1) to cattle in the Sengwe communal land from the KNP and GNP buffalo populations; (2) to the northern GNP buffalo population from the southern population and subsequently to northern buffalo and cattle populations; (3) to buffalo and cattle populations in Mozambique. So far, bTB surveys have not confirmed disease in cattle and buffalo in contact with infected buffalo, but recent serology of buffalo and skin testing of cattle in and out of LNP suggest there are positive cases; however, these results are as yet not definitive (Tanner *et al.* 2014). Figure adapted from de Garine-Wichatitsky *et al.* 2013b.

ecosystem has seen bTB re-emerging, notably in Uganda Kob (*Kobus kob thomasi*), and this is coincident with recent declines in herbivore populations (P. Atemnedi, personal communication). Limited data from buffalo surveys show a similar infection status to earlier studies (Kalema-Zikusoka *et al.* 2005), but its presence did not prevent buffalo increasing in numbers after poaching pressure was reduced. In dry land ecosystems and with less compressed (unfenced) populations of animals, the disease incidence seems low and seroprevalence also follows this pattern (Cleaveland *et al.* 2007; Tschopp *et al.* 2010). bTB prevalence in buffalo across Africa appears to vary according to quite specific ecological, epidemiological and environmental conditions, moisture being important to survival of the bacteria (Morris *et al.* 1994). It is hypothesised that current infection in cattle and buffalo in Africa arose from infection acquired from imported cattle during colonisation (Anon 2002), but this remains unconfirmed. With high infection levels, the disease can be easily detected on post mortem, but low incidence can be difficult to detect – the sensitivity of the available blood (antibody) and sputum (antigen) tests are poor, especially for early cases of bTB (Kock *et al.* 2012). Management of diseased buffalo is restricted to culling to reduce densities, or to targeted removal of infective hosts. However, implementation of such control measures in wildlife remains controversial (de Lisle *et al.* 2002; Michel *et al.* 2006).

Theileriosis (East Coast Fever – ECF) is an economically important disease of cattle caused by *Theileria parva*, a protozoan transmitted through the brown ear tick *Rhipecephalus appendiculatus* and *R. zambeziensis*. Epidemiology of ECF in cattle is not dependent on buffalo, and cattle have their own subtype (Pelle *et al.* 2011). Buffalo asymptomatically carry a variant of the cattle parasite, and where cattle are free of *Theileria* infection and come into contact with buffalo and their ticks (*R. appendiculatus* and *R. zambeziensis*), corridor disease can result. There is considerable debate about the relationship of cattle and buffalo parasites with historical confusion in the nomenclature. ECF is a disease mainly of pastoral livestock in East Africa and the role of buffalo is still poorly understood, whereas the role of buffalo as an asymptomatic carrier of *T. p. lawrencei*, and reservoir of corridor disease for sympatric cattle has been clearly demonstrated. The development of better control measures using antibiotic/vaccine combinations, leading to reduced infection in cattle, might increase the relative importance of spill-over infection using buffalo populations as a source.

Trypanosomiasis remains a major constraint to cattle production in sub-Saharan Africa, but equally the disease has provided a natural protection for many wildlife-rich habitats sustaining ecosystem services. Wildlife is tolerant to infection. Variants of the protozoan infect a wide range of species, including man, where it causes sleeping sickness. Trypanosomes require the tsetse vector for transmission and this biting fly requires specific ecological conditions for persistence, basically thick bush vegetation. Buffalo are a preferred species for the fly and they participate in the maintenance and amplification of the parasite. Wide host range means few control measures are possible other than those directed at the fly and domestic host (cattle) specifically. Earlier attempts at control through eradication of wild hosts failed and caused significant ecological damage. Overall the distribution of trypanosomes remains the same now as a century ago, despite disappearing wildlife hosts. Only where there is a dramatic decline in vegetation (associated with severe land degradation) (Ducheyne *et al.* 2009) or highly focused and intensive trapping/chemical targeting or application of the sterile insect technique (SIT), does tsetse and trypanosomiasis disappear (Bourn *et al.* 2005). SIT is based on the knowledge that laboratory-reared male tsetse flies, sterilised by gamma irradiation or by certain chemical compounds and then released, will seek out and mate with wild females, without producing offspring, thereby disrupting the reproductive cycle. It is recommended for use during the final phase of elimination of tsetse from a designated area once natural fly populations have been suppressed by other techniques. The reason is that even in the absence of wild animal hosts, livestock continue to be infected, suffer debilitating disease and provide a reservoir. The number of human African trypanosomiasis (HAT) cases reported has declined by 75% over the last decade with intensive human diagnostic treatment regimes and, in the case of *Trypanosome brucei rhodesiense*, attention to cattle from where the majority of human cases are derived (Welburn *et al.* 2009; Simarro *et al.* 2011) has proved beneficial. An epidemiological role of animal reservoirs in the most common cause of HAT associated with *T. b. gambiense* infection of West and Central Africa is not proven. Buffalo remain a concern with regards to persistence of tsetse and trypanosomiasis affecting livestock, but not for the human disease where only rare isolates of human infective parasites have been made (Anderson *et al.* 2011).

Rift Valley Fever virus can affect buffalo and there is a possibility that buffalo provide an amplification host during inter-epidemic periods for RVF (LaBeaud 2011).

Buffalo can contract brucellosis infection, with up to 30% seroprevalence reported in endemic areas with high livestock numbers, and they show symptoms including hygromas (Herr & Marshall 1981; Waghela and Karstad 1986), but it appears that buffalo sharing habitat with *Brucella*-infected cattle can be free of the infection (Gomo *et al.* 2012).

Other diseases are transmitted directly or otherwise between buffalo and cattle, such as anthrax and other tick-borne diseases, but the economic concern is muted (Table 26.1). If buffalo appears often as the 'black sheep' for disease maintenance, it should be remembered that cattle and their exotic diseases also threaten the health of buffalo, wildlife economies and ecosystems services. The latter includes:

- loss of *provisioning* services: food, genetic resources from associated decline in wildlife from land or forage competition and disease introduction;
- impacts on *regulating* services: disease prevention capacity is reduced since cattle contribute a relatively large environmental load of pathogens, are unregulated by nature and artificially supported;
- effects *cultural* services: loss of natural systems (spiritual) and tourism from loss of wildlife populations.

Ecology of disease transmission between livestock and buffalo

When wild and domestic ungulates interact, through the sharing of grazing or water points, meeting directly or using common pastures at different times, they create an interface hereafter referred as a wildlife–livestock interface. The transmission of microorganisms between host species is dependent on the frequency and intensity of contacts, as well as the taxonomic proximity between the species. Buffalo and cattle are closely related taxonomically, and share many genetic characteristics, including in the immune system (Wenink et al. 1998; Suzuki *et al.* 2012). Overlap in their susceptibility to pathogens is therefore high. Direct or indirect contacts provide transmission opportunities for different pathogens, depending on their transmission mode(s) (e.g. direct contact, environmental, vector-borne, wind-borne), and the understanding of the patterns of contacts between wild and domestic animal hosts are crucial to determine the type of pathogens that can be transmitted between them (Miguel *et al.* 2013). In an endemic wildlife setting infections by many pathogens do not result in severe diseases in wildlife and local heads of livestock, as host–pathogen coevolution tends to select for low pathogenicity (van Baalen 1998) (e.g. trypanotolerant

Ndama cattle in West Africa). When we introduce non-indigenous domesticated species into an ecosystem, their vulnerability to potential pathogens from wildlife can be high and livestock can also introduce novel pathogens into a native wildlife ecosystem with dire consequences.

Buffalo and cattle have similar ecological niches and tend to be attracted by the same type of food resources. Grazing and water are limiting factors in most semi-arid areas of Africa where buffalo/cattle live sympatrically. These resources potentially create opportunities for buffalo–cattle contacts and pathogen transmission (Sinclair 1977), and seasonal profiles of contacts are usually observed, with higher rates of contact during the dry season when resources are depleted (Kock 2005b).

However, buffalo and cattle movements, and the resulting contacts, are not just driven by access to resources. Buffalo–cattle contacts are the result of a complex interplay between biophysical (e.g. animal behaviour, distribution of resources) and human (e.g. wildlife management, livestock husbandry practices) drivers (Miguel *et al.* 2013).

Three main types of buffalo–cattle interfaces are present in Africa, each of them encompassing a range of different contexts. Namely (1) a fenced (ranch) system, where land use is designated for wildlife or livestock or both, (2) an open rangeland with free mixing of wildlife and livestock, (3) protected area boundaries where livestock are excluded or access controlled on one side and wildlife unwanted and partially accepted on the other. In Africa, these different types of interface are not evenly distributed due to different conservation and natural resource policies. Fencing is a widespread management option used in Southern Africa to separate wildlife from livestock (and people) for conservation and/or disease-control purposes (Ferguson & Hanks 2010). Many types of fence have been used (e.g. double fence, electrified), which can efficiently prevent buffalo–cattle contacts, although fences are never 100% ungulate-proof, are costly to put up and require maintenance due to wildlife challenges (in particular elephant) or intentional destruction by local populations (Dube *et al.* 2010). In Southern Africa, extensive fencing has been used to isolate buffalo reservoirs for FMD control. Fencing is also used to protect wildlife from human encroachment, as at KNP in South Africa. In East Africa, fencing has been ad hoc and largely to provide general protection from wildlife incursion and not specifically for disease control.

Open rangeland allows livestock and wildlife coexistence sometimes throughout the year or on a seasonal basis with nomadism. No physical barrier separates buffalo from cattle and contacts between species can occur freely. A wide range of management options can be implemented through adaptive farming practices and use of rangeland: allowing cross infection and encouraging endemic stability, targeted prophylaxis such as vaccination to prevent disease transmission or controlled cattle grazing away from potential contact areas with buffalo. Mixed systems represent, therefore, a large variety of buffalo–cattle interfaces depending on management options selected. This type of buffalo–cattle interface is extensively seen in East Africa. This system can combine cattle production and wildlife use, such as meat production, hunting and photographic tourism.

The last type of interface relates to contexts where there is no physical barrier, but instead a legal boundary with some separation of livestock from wildlife. Often, park boundaries are a mere line or a buffer zone where limited human activity is authorised. A priori, this type of interface should be permissive to buffalo–cattle contacts as in a free-ranging context. However, behavioural adaptations by both cattle and buffalos often modify this situation. As observed for other wildlife species such as elephant, buffalo appear to detect and alter behaviour at park boundaries, even when no physical barrier exists: buffalo home range will often stop at roads, at the periphery of villages or at old fence lines. Cattle's behaviour is influenced at two levels. First, they may detect an increased risk of predation or parasite infestation associated with wild animal presence, which they tend to avoid (Fritz *et al.* 1996; de Garine-Wichatitsky *et al.* 1999; Zanette *et al.* 2011) when crossing park boundaries. Second, cattle are herded and risk perception by cattle owners can influence their decision to send their cattle to areas where risks of predation, wildlife authority sanctions or disease transmission exist. The behaviour of wild and domestic ungulates can have significant impacts on buffalo–cattle interfaces as contacts can be much more limited than dogma suggests. Contact described in the literature is often modelling simulation or just observation of both species in the same area (Dion *et al.* 2011) and in reality there is little or no direct physical contact. This type of interface is widespread in Africa and is dominant in West Africa, where buffer zones such as hunting areas envelop national parks and limit buffalo–cattle contacts. It is expected that each type of interface will result in some contact patterns permissive of some disease transmission and preventing others (Figure 26.3). A double-fenced interface (such as for FMD control) should preclude all direct contacts and associated pathogen transmission between cattle and buffalo, with the exception of vector- and air-borne pathogens transmitted by vectors that can actively move across the double-fence. A single fence cannot fully avoid direct contact between hosts, has little effect on vector-borne disease transmission, but should considerably limit environmental transmission. When there is no fence, the intensity of the interactions is difficult to predict. It will depend on the spatial and temporal distribution of key resources in the landscape (e.g. when a river is the park boundary) and of above-mentioned adapted behaviours of buffalo, cattle and their owners to the different risks perceived. Therefore, the presence of physical boundaries cannot alone be used as reliable indications of the extent of buffalo–cattle contacts.

Caron *et al.* (2013) have recently compared the disease burden of important diseases at different buffalo–cattle interfaces (fenced and unfenced) in Zimbabwe. They conclude that the risk of disease spread at the interface varies among

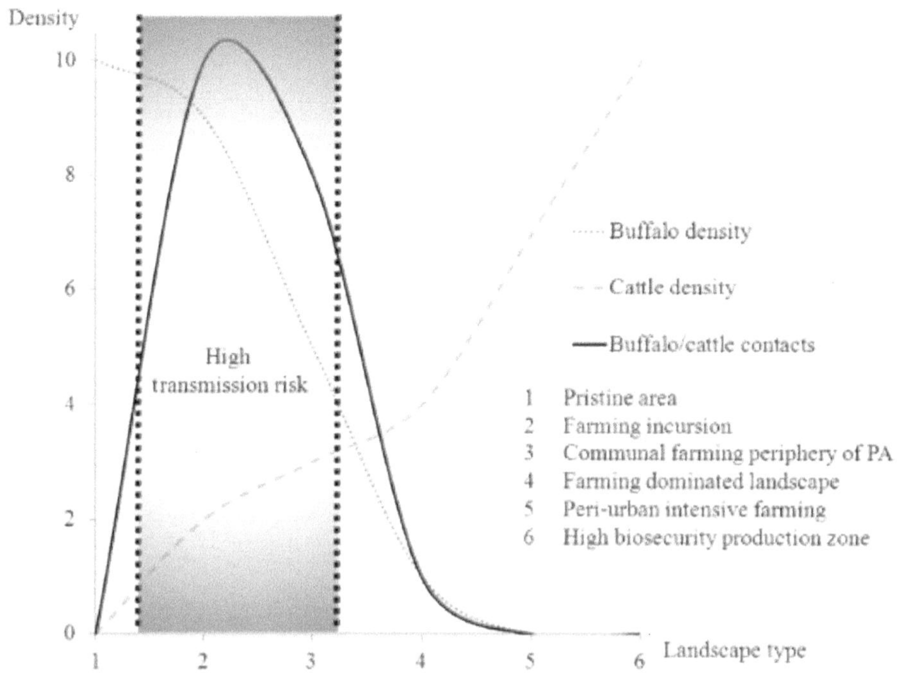

Figure 26.3 Theoretical densities of buffalo and cattle populations, buffalo–cattle contacts in relation to different human–nature interactions in African contexts (PA = protected area).

pathogens: the buffalo population represents a risk for cattle infection with bTB, FMD and theileriosis, but cattle can also represent a risk for buffalo for brucellosis, and maybe RVF in some specific situations (e.g. in South Africa) and lumpy skin disease. The level of risk remains ill-defined and probably varies according to each system and context. These differences are a consequence of the different transmission modes of the pathogen considered and of the different types of interface studied. The effect of fences on disease transmission may also be blurred by cattle-to-cattle disease transmission at the local level. Further protocols in this study site aim at better understanding the ecology of disease transmission at wildlife-livestock interfaces: telemetry on both buffalo and cattle will provide information on species home range, intraspecies fusion–fission patterns and individual interspecies contacts, whether direct or indirect; waterhole and road counts will provide estimates of contact rates in hotspot areas and during high-risk seasons; and social studies on farmers' perceptions of risk associated with the interface should provide insights on cattle herding strategies at the interface (Murwira *et al.* 2013). This approach is process-oriented and takes into account the modes of disease transmission in relation to the type of interface (Caron *et al.* 2012). In particular, it will provide a risk-based approach to assess the probability of bTB spreading to cattle in Zimbabwe after the spread of the disease from KNP to Zimbabwe through buffalos (Figure 26. 2).

Current human and environmental changes, including climate warming and the expected increased scarcity of natural resources (notably surface water), will have consequences for the buffalo–cattle interfaces and the disease transmission processes. The ecology of disease transmission, infection, co-infection (Jolles *et al.* 2008), cofactors and co-evolution of infection in multi-host species systems needs to be better understood in the future to improve the policy, management and control of infectious diseases at wildlife-livestock interface for all species and ecosystems concerned. The risk of disease spread needs to be assessed according to (1) the type of interface determined by human management and perception and buffalo and cattle animal behaviour; and (2) the modes of transmission of pathogens (Table 26.2).

Implications of diseases at the interface: management of cattle–buffalo interfaces in relation to FMD and bTB disease control and biodiversity management

In Africa, specifically in relation to the buffalo–cattle interface, there are a number of key diseases of concern, including bTB, FMD and corridor disease. bTB has zoonotic potential and trade implications, while FMD is the single most important disease at the interface in relation to international trade in animal products. Corridor disease impacts on livestock health but will not be dealt with here in any detail.

With regards to FMD, the ability of a country to export beef to the lucrative European markets and in subsectors such as dairy can be jeopardised by disease impacts on production. In endemic African rangeland settings, FMD is not a constraint on production and barely causes serious loss. As described earlier, buffalo, as carriers of SAT viruses, have been targeted by control programmes due to the potential spread of the virus to cattle, particularly in southern Africa. No more so is this apparent than in Botswana (Kock *et al.* 2002). Cordon fences and disease control areas have been in existence for 50 years (Condy *et al.*

Table 26.2 Heat map of risk of pathogen transmission and impact at buffalo/cattle interface in relation to the type of buffalo–cattle interface and the pathogen transmission modes

	Double-fenced e.g. FMD fence	Single-fenced e.g. game fence	Linear unfenced boundary 1 e.g. park boundary	Linear unfenced boundary 2 e.g. river boundary	Buffer zone e.g. hunting/ safari area at the periphery of PA	Mixed system 1 e.g. year- long sharing of pasture	Mixed system 2 e.g. sequential use of pasture
Expected intensity of contact	light grey	dark grey	dark grey	dark grey	dark grey	black	dark grey
Direct transmission e.g. FMD, bTB	light grey	dark grey	dark grey	dark grey	dark grey	black	dark grey
Environmental transmission e.g. brucellosis, bTB	light grey	light grey	dark grey	dark grey	dark grey	black	black
Water-borne transmission e.g. schistosomiasis	light grey	light grey	black	black	black	black	black
Vector-borne transmission e.g. tick-borne diseases, RVF	dark grey	dark grey	black	black	black	black	black
Risk of disease impact on cattle if transmission occurs e.g. FMD, bTB, brucellosis, tick-borne	black	black	dark grey	dark grey	light grey	light grey	dark grey

Level of risk: black = high; dark grey = medium; and light grey = low.
Risk variability observed is dependent on the spatial distribution of resources at or close to the interface (e.g. in mixed system 2: do buffalo and cattle share a water point?) and human risk management and perception and policies (e.g. are livestock activities allowed in the buffer zone?).

1969), with new fences being built recently (Scott Wilson and the Environment and Development Group 2000; Ferguson & Hanks 2010). These fences have had a significant impact on biodiversity, particularly on migrating wildlife such as blue wildebeest (*Connochaetes taurinus*), red hartebeest (*Alcelaphus caama*) and zebra (*Equus quagga*). As much of Botswana's surface is the Kalahari Desert, water and forage are key drivers in wildlife movements, with fences severely restricting them and causing catastrophic declines in some wild species (Williamson & Williamson 1984; Mbaiwa & Mbaiwa 2006).

As the role of the African buffalo in the epidemiology of FMD has become clearer (Thomson *et al.* 2003), so have control measures to keep buffalo and cattle separate in several southern African countries. Indeed, even when speculation abounded in the 1950s on the possible role of wild animals in FMD transmission to livestock, fences were being built. The Northern Buffalo fence in Botswana was completed in 1995 and prevents buffalo moving out of the Okavango Delta and, therefore, coming into contact with cattle. The role of buffalo as carriers of FMD has been well described earlier, but transmission dynamics between these species remain less certain. The dilemma is balancing developmental needs with environmental and biodiversity concerns, especially in relationship to disease control methods (Ferguson & Hanks 2010; Ferguson *et al.* 2013). In Botswana, tourism currently earns the country more in revenue than livestock farming, and disease control measures are hugely expensive with major perverse subsidies going into the cattle industry (Kolanye & Mullins 2000).

In this regard, on a broader scale and in the Southern and East African contexts, understanding both positive and negative impacts of FMD control methods is essential if one hopes to optimise the potential for the rural poor to benefit from both wildlife conservation as well as trade in products derived from livestock (AHEAD Update, March–May 2012, www.wcs-ahead.org; Ferguson *et al.* 2013). As has been outlined using Botswana as an example, the prevailing approach to managing FMD has been designed on a geographic basis, i.e. the creation of areas (disease-free countries, zones or compartments) with the objective of progressive FMD eradication from domestic livestock. This strategy is supported by the assumption that imports of livestock commodities and products can be safely sourced from such disease-free areas. While this is true in some rural settings, in others, especially where large numbers of free-living cloven-hoofed wildlife (some of which maintain the infection, especially buffalo) are dispersed over vast geographic areas, achieving freedom from FMD is often technically impossible.

Commodity-based trade (CBT) (Thomson *et al.* 2013a, 2013b), involving safe and stringent processing of animal-derived products, increases biological safety and options for beneficiation (value addition). It also alleviates the need for some of the fencing that has been required to separate livestock from wildlife, often with significant negative environmental impacts. It's noteworthy that the OIE's Terrestrial Animal Health Code already provides clear guidance in terms of the acceptability of CBT of beef from FMD-infected countries or zones in Article 8.5.25; this is a recent development (OIE 2013). Thus the historical conflict between buffalo and cattle, and the need to keep them separate, may not be a necessity in order to trade. Therefore, issues at the interface would become less compelling.

It is not just diseases of economic importance that buffalo harbour, but also those that have a potential public and ecosystem health impact. In this regard, the disease of most concern is bTB. The ecological impacts are still being evaluated (Caron *et al.* 2003) but the most politically sensitive aspect is the human and livestock health risk from buffalo bTB (de Garine-Wichatitsky *et al.* 2013a), not only along the western boundary of KNP, but also with regards to the joint development of the Greater Limpopo Trans-frontier Conservation Area (GLTFCA) with Zimbabwe and Mozambique (Bengis 2005; Tanner *et al.* 2013).

The previous section discussed the ill-defined risk of infected buffalo in KNP, providing spill-over infection to neighbouring communal cattle both in South Africa, Zimbabwe and Mozambique, which makes policy on this subject all the more difficult. The management of wildlife and livestock diseases such as bTB within the GLTFCA, including individual parks and the envisaged larger landscape, has remained unresolved and presents a new challenge on approaches to disease control with an impact on existing disease control policies (Michel *et al.* 2006). In this case, fences between KNP and the Limpopo National Park (LNP-Mozambique) have been removed and buffalo have moved eastwards into the LNP. LNP was originally a hunting area, which was populated by villagers during Mozambique's civil war. These communities still exist within the park, with plans for resettlement outside the park within the next 5–10 years. A significant number of herds of cattle associated with these communities live within the park and evidence suggests that there is increasing potential for disease transmission through, most probably, indirect contact between buffalo and cattle.

HIV prevalence in this region is among the highest in Africa and the world. HIV is known to be driving the tuberculosis epidemic in this part of the globe (UNAIDS 2006), although the rate of increase in cases and fatalities is declining (UNAIDS 2011), which gives some hope for this region. Nevertheless several communities drink unpasteurised milk, increasing the possibility of bTB transmission, especially in immuno-compromised people. The movement of bTB-infected buffalo into LNP provides the Mozambique veterinary authorities with a further complication in terms of transboundary animal disease control, although the risk to cattle is unquantified and there is no evidence of zoonotic transmission from buffalo. In terms of health and regulatory issues, the combination of bTB (de Garine-Wichatitsky *et al.* 2013a), wild African buffalo, cattle and their owners provides a heady mix and emphasises the importance of adopting a multi-disciplinary One Health approach in solving issues at the interface (Clifford *et al.* 2011; de Garine-Wichatitsky *et al.* 2013a).

Disease control programmes have sought to keep buffalo and cattle separate, but none of these control programmes seeks to eliminate buffalo – buffalo are recognised as a valuable tourist asset and are highly sought after by sport hunters. Many countries in Africa allow both non-consumptive (eco-tourism) and consumptive (licensed resident and tourist sport hunting, and the cropping and culling of game for meat production) uses of wildlife. Trophy buffalo are the most important species driving the economics of the hunting industry in Southern and East Africa (Booth 2009). A 5–7-day hunt just for buffalo will generate $6000–12 000; a buffalo plus plains game hunt in Tanzania may generate $10 000–20 000. When many of these hunts are linked to a community-based natural resource management programme, benefits to these communities can be significant – even more so when coupled with eco-tourism (Child *et al.* 2012). In Southern Africa, the development of disease-free herds of buffalo (FMD free) has increased the value of this species. In 2011 auctions in South Africa, disease-free buffalo were sold with cow/calf pairs averaging a price of $250 000, cows and bulls ~$151 000 (Anon 2012). African buffalo, therefore, have tremendous potential to contribute to livelihoods and development in Africa, highlighting the need to balance traditional livelihood options such as livestock with conservation.

Unfortunately, due to illegal activities, as a species the African buffalo has declined to low levels in many African countries (see Chapter 20). Both conflict ('weaponisation' of societies) and a growing bushmeat trade have placed this species on the brink of extinction in countries such as the Republic of South Sudan (P. Elkan, personal communication). Buffalo provide less of a threat to either traditional or pastoralist communities than, for example, lion or elephant. Being bulk grazers, they are important components of healthy ecosystems (Hofmann 1973), reducing grass levels to a height preferred by selective grazers. Sustainable off-take of buffalo can also be a valuable source of protein to rural communities, providing an additional buffer in times of food insecurity.

Studies in Africa, especially Southern Africa, have attested to the value of multi-species systems as part of a diverse land use plan as compared to a single-species system (Cumming 1995, 1999). The more diverse a system the more likely it will be able to weather shocks, especially in a time of concern over climate change. The African buffalo has always been a pivotal species in this regard, but the overriding issue has been the potential of disease spread, with FMD especially and other diseases (e.g. ECF). Pastoralist communities, typically from East and Central Africa, have historically lived with wildlife, including buffalo. Pressures are mounting on these traditional extensive lifestyles, especially with regard to marketing of livestock, locally and regionally. In this regard, a CBT approach would be far more appropriate than trying to fulfil stringent standards set by other non-African countries (Thomson *et al.* 2004; Scoones *et al.* 2010). Improving health education within these pastoralist and other African communities will also reduce potential for disease transmission (Clifford *et al.* 2011).

Is there a future for the African buffalo?

Hundreds of thousands of years of co-evolution on the continent have shaped the African buffalo (*Syncerus caffer*) into a continuum of taxa, each of them adapted to its own specific environment including habitats, predators and diseases (see Chapter 20). This highly successful, primarily bulk grazer undoubtedly helped shape the vast savannas and the mammal communities that are still visible in almost their pre-Ice-Age state today and continue to do so. They can still be seen, with many thousands of individual families, coalescing in herds at breeding, in the remaining intact ecosystems of Tanzania, Uganda, Kenya and Southern Africa, but in few other places. Elsewhere in Africa they survive in small bands of nervous animals clinging on to dwindling bushed grasslands and fragmented habitats. In Neolithic times (10 000 BP), comparatively recently in evolutionary history, domestication of progenitors of modern cattle occurred, and on the African continent this may have included *Bos taurus*, independent of the Eurasian, contrary to common belief (Bradley *et al.* 1996). These animals supported new human civilisations, cultures and societies associated with pastoralism in many parts of the continent. This point marked the slow and steady decline of buffalo as humans gradually dominated and exploited the landscape. The species was perceived negatively both as a physical threat (untameable) and a grazing competitor for livestock, but also positively to this day, as a natural resource, highly valued supplementary food in times of climatic stress or loss of domestic stock from disease or drought. At what point people became aware of shared diseases is debatable, but modern pastoralists understand common susceptibilities (de Garine-Wichatitsky *et al.* 2012, 2013), even if these are relatively benign in endemic settings. In more recent history, during colonial and postcolonial times, emerging veterinary science reinforced domestic animal husbandry and more and more land was exploited. As diseases were controlled or prevented and land was cleared of wildlife, buffalo were eradicated, seen as a significant disease threat and constraint to Western-style development.

In the meantime, the buffalo was largely overlooked by scientists, with relatively few studies done. And when the buffalo did attract scientific interest, it was often for negative reasons related to disease risks associated with the interface between cattle and buffalo. Politicians, scientists and agriculturalists joined forces promoting segregation of cattle from buffalo and/or buffalo elimination, with the sole purpose of promoting the monoculture of beef for export to Europe where, ironically, European beef is a surplus production. In a way, the African buffalo has been and still is suffering from a controversial political economy between Europe and Africa. Surprisingly, even conservationists ignored the buffalo, regarded as a sort of second-rate species in conservation programmes because it is not as threatened as other large mammals.

Fortunately, changing attitudes on biodiversity and ecosystem services are gradually recognising the buffalo as a high-value species. Increasingly, buffalo are being seen as one of the most promising African mammals for its potential contribution to sustainable development, understood as the synthesis/reconciliation of conservation and development. Being an indigenous species, the buffalo is well adapted to thrive in the various ecosystems and habitats on the African continent. Ironically, some of these most striking characteristics have been responsible for its persecution by cattle breeders, agriculturalists and other developers. Its resistance or tolerance to indigenous diseases such as trypanosomiasis, tick-borne diseases or foot and mouth disease are considered as a threat to cattle, while it could be regarded as an asset for veterinary science to investigate immunity factors and processes. Also, the buffalo has very much suffered from exotic diseases brought by livestock to Africa, such as rinderpest and tuberculosis. Furthermore, the distribution of the buffalo appears to be one of the broadest of the large mammals in Africa (IUCN SSC 2008). Very few others have succeeded in adapting themselves to such extremely dissimilar ecosystems as equatorial rainforests and tropical dry savannas. Divergent buffalo taxa are evidence of evolutionary adaptability to the point that, for example, the dwarf forest buffalo *Syncerus caffer nanus* is less than half the live weight of the large savanna buffalo, e.g. *Syncerus caffer caffer*, (Nowak 1991) and occupies an entirely different habitat. Such a high intraspecific diversity is a real wealth to be preserved, especially considering that each taxon is able to cope with the particular constraints of its local environment in terms of climate, diet, predation and disease.

While the economic value of cattle mostly relies in the primary sector (beef and milk markets), the value of the wild buffalo is multifaceted, comprising not only the primary sector (venison), but also the tertiary sector (wildlife-viewing tourism and hunting tourism). The latter has benefited from listing as a 'non-CITES species', i.e. not listed in Appendices of the Washington Convention on International Trade in Endangered Species of Wild Fauna and Flora, and with little chance for a change. The prospect of the buffalo remaining as a species suitable for sustainable hunting, and as a vital component of protected areas, is good and this helps its population and long-term resilience. Despite the potential of the species, there is a paradox: (1) despite the outstanding assets of the buffalo resulting from its indigenous nature, the conservation status of the species is steadily degrading overall, with a few notable exceptions; (2) despite all the disadvantages due to its exotic origin (e.g. strict reliability on human care, growing and costly dependence on veterinary assistance), the livestock herd is fast growing in Africa, estimated at 250 million cattle and 500 million sheep and goats (FAO 2007), and showing shifts to small livestock as rangeland has been squeezed by other land uses (FAO 2013). This outcome results from human choice and is understandable, but the high social (husbandry), environmental (degradation) and economic (from disease impacts) costs of cattle when compared to harvesting buffalo on natural ranges challenge its efficiency. If for no other reason, the buffalo should be kept in a viable state as a library of adaptive traits which one day might be needed to resolve genetic weakness in the sympatric domestic bovine, allowing it to cope with ever more challenging environmental degradation and disease challenges.

An even more constructive attitude would be to take advantage of the adaptation of the buffalo for innovating and developing new sustainable development schemes involving the buffalo as a prime production species complementing cattle husbandry. Namibia and South Africa are currently experiencing a steady trend of converting cattle ranches into wildlife ranches, with the buffalo as either a major or potential driver. As an African species, the buffalo could provide African countries and African people with a competitive advantage and a genuine contributor to food security rather than relying solely on domestic species, all of the latter being of exotic origin (with the exception of the guinea fowl and donkey) and/or costly imports. However, there are considerable risks associated with recently practised intentional genetic manipulations of the buffalo. Untimely translocations of buffalo, e.g. introductions of individuals from exotic origin into indigenous populations, are responsible for genetic pollution, homogenisation of the species and loss of intraspecific diversity. Also, intentional selections of peculiar features such as trophy size are likely to affect genetics of the managed population with a risk to reduce species capacity for adaption to environmental changes. Such risks are real when considering the increase of record prices reached for an individual in auction sales, from more than $2 million in 2011 to more than $4 million in 2013 (http://www.bdlive.co.za/business/2013/09/20/buffalo-bull-gets-R40m-as-demand-rockets).

In conclusion, although diseases have been used as a reason to persecute buffalo, its resilience in the face of pathogens indigenous to the African continent should, on the contrary, be considered as a vital and valuable asset for Africa. If the current trend remained unchanged, the well-adapted indigenous buffalo will be replaced by less well adapted exotic cattle. Reversing the trend would require the buffalo to be rediscovered and our attitude to be revisited. By better conserving the buffalo and making it a new and complementary instrument to food security and poverty alleviation, a truly African species could provide Africa with an innovative contribution to sustainable development.

Acknowledgements

RK's work on buffalo was undertaken during employment as a wildlife veterinarian and epidemiologist with the Zoological Society of London, the Kenya Wildlife Services and the African Union InterAfrican Bureau of Animal Resources, with support from the Overseas Development Administration (now DFID) of the British government and the EU PARC and PACE programmes and the support of the World Conservation Union (IUCN) Species Survival Commission (SSC) Antelope (and Buffalo) and Wildlife Health Specialists Groups. AC and MdGW's work was conducted within the framework of the Research Platform 'Production and Conservation in Partnership' (RP-PCP) and the Animal and Human Health for the Environment and Development (AHEAD) programme. The Ministère Français des Affaires Etrangères through the French Embassy in Zimbabwe is thanked for supporting sustainably the RP-PCP. MK acknowledges the opportunities afforded to his work on buffalo while at the Department of National Parks and Wildlife Management in Zimbabwe, with PanLivestock services and the Department of Animal Health and Production in Botswana and with the Wildlife Conservation Society, New York, in various African countries. PC acknowledges the support of the International Foundation for Wildlife Management (France) for his ongoing work on buffalo, the Antelope Specialists group of the IUCN SSC and EU PARC and PACE programmes.

References

Abu Samra, N., Jori, F., Samie, A. & Thompson, P. (2011). The prevalence of *Cryptosporidium* spp. oocysts in wild mammals in the Kruger National Park, South Africa. *Veterinary Parasitology* 175 (1–2): 155–159.

Anderson, N. E., Mubanga, J., Fevre, E. M., *et al.* (2011). Characterisation of the wildlife reservoir community for human and animal trypanosomiasis in the Luangwa Valley, Zambia. *PLoS Neglected Tropical Diseases* 5(6):e1211.

Andersson, J. A. & Cumming, D. H. M. (2013). Defining the edge: boundary formation and TFCAs in southern Africa. In *Transfrontier Conservation Areas: People Living on the Edge*, pp. 25–61. Eds. Andersson, J. A., de Garine-Wichatitsky, M., Cumming, D. H. M., Dzingirai, V. & Giller, K. E. London: Earthscan.

Andersson, J., de Garine-Wichatitsky, M., Dzingirai, V., Giller, K. & Cumming, D. (2012). People at wildlife frontiers in southern Africa. In *Transfrontier Conservation Areas: People Living on the Edge*, pp. 1–11. Eds. Andersson, J., de Garine-Wichatitsky, M., Dzingirai, V. G. K. & Cumming, D. London: Earthscan.

Anon (2002). Disease risk management directive of tuberculosis in buffalo (*Syncerus caffer*) in South Africa. www.nda.agric.za/vetweb/Disease%20Control/Buffalo/20th%20draft%20protocol%20-%20buffalo%20policy.pdf, accessed 10 April 2014.

Anon (2012). Annual average game auction prices 2011. In *Wild & Jag/Game & Hunt* January 45–47. www.wildlifehunt.co.za, accessed 20 July 2012.

Ayebazibwe, C., Mwiine, F. N., Tjørnehøj, K., *et al.* (2010). The role of African buffalos (*Syncerus caffer*) in the maintenance of foot-and-mouth disease in Uganda. *BMC Veterinary Research* **6**: 54.

Baber, D. J. & Condy, J. B. (1981). Isolation and characterisation of bovine adenoviruses types 3, 4 and 8 from free-living African buffaloes (*Syncerus caffer*). *Research in Veterinary Science* **31**(1): 69–75.

Beechler, B. R., Broughton, H., Bell, A., Ezenwa, V. O. & Jolles, A. E. (2012). Innate immunity in free-ranging African buffalo (*Syncerus caffer*): associations with parasite infection and white blood cell counts. *Physiological and Biochemical Zoology* **85**: 255–264.

Bengis, R. (2005) Transfrontier conservation area initiatives in sub-Saharan Africa: some animal health challenges. In *Conservation and Development Interventions at the Wildlife/Livestock Interface Implications for Wildlife, Livestock and Human Health*, pp. 15–19. Eds. Osofsky, S. A., Cleaveland, S., Karesh, W. B., *et al.* Gland: IUCN

Bengis, R. G., Kriek, N. P. J., Keet, D. F, Raath, J. P., De Vos, V. & Huchzermeyer, H. F. A. K. (1996). An outbreak of bovine tuberculosis in a free-living African buffalo (*Syncerus caffer* – Sparrman) population in the Kruger National Park: a preliminary report. *Onderstepoort Journal of Veterinary Research* **63**: 15–18.

Bengis, R. G., Kock, R. A. & Fisher, J. (2002). Infectious animal diseases: the wildlife/livestock interface. *Revue Scientifique et Technique de l'OIE* **21**: 53–65.

Bengis, R. G., Leighton, F. A., Fischer, J. R., Artois, M., Morner, T. & Tate, C. M. (2004). The role of wildlife in emerging and re-emerging zoonoses. *Revue Scientifique et Technique de l'OIE* **23**: 497–511.

Bindernagel, J. A. (1972). Liver fluke *Fasciola gigantica* in African buffalo and antelopes in Uganda, East Africa. *Journal of Wildlife Diseases* **8**(4): 315–317.

Booth, V. R. (2009). Buffalo safaris. In *A Comparison of the Prices of Hunting Tourism in Southern and Eastern Africa*. pp. 14–17. Budapest: FAO and CIC.

Bourn, B. D., Grant, I., Shaw, A. & Torr, S. (2005). Cheap and safe tsetse control for livestock production and mixed farming in Africa. *Aspects of Applied Biology* **75**: 1–12.

Bradley, D. G., Machugh, D. E., Cunningham, P. & Loftus, R. T. (1996). Mitochondrial diversity and the origins of African and European cattle *Proceedings of the National Academy of Sciences* **93**: 5131–5135.

Bronsvoort, B. M. D. C., Parida, S., Handel, I., *et al.* (2008). Serological survey for foot-and-mouth disease virus in wildlife in Eastern Africa and estimation of test parameters of a nonstructural protein enzyme-linked immunosorbent assay for buffalo. *Clinical and Vaccine Immunology* **15(60)**: 1003–1011.

Caron, A., Cross, P. C. & du Toit, J. T. (2003). Ecological implications of bovine tuberculosis in African buffalo herds. *Ecological Applications* **13**: 1338–1345.

Caron, A., Morand, S. & Garine-Wichatitsky, M. (2012). Epidemiological interaction at the wildlife/livestock/human interface: can we anticipate emerging infectious diseases in their hotspots? A framework for understanding emerging diseases processes in their hot spots. In *New Frontiers of Molecular Epidemiology of Infectious Diseases*. pp. 311–332. Eds. Morand, S., Beaudeau, F. & Cabaret, J. Dordrecht: Springer.

Caron, A., Miguel, E., Gomo, C., *et al.* (2013). Relationship between burden of infection in ungulate populations and wildlife/livestock interfaces. *Epidemiology and Infections*, **141**: 1522–1535.

Catley, A., Lind, J. & Scoones, I. (2012). Development at the margins: pathways of change in the Horn of Africa. In *Pastoralism and Development in Africa: Dynamic Change at the Margins*. Eds. Catley, A., Lind, J. & Scoones, I. Oxford: Routledge.

Child, B., Musengezi, J., Gregory, D. P. & Graham, F. T. C. (2012). The economics and institutional economics of wildlife on private land in Africa. *Pastoralism: Research, Policy and Practice* **2**: 18

Cleaveland, S., Shaw, D. J., Mfinanga, S. G., *et al.* (2007). *Mycobacterium bovis* in rural Tanzania: risk factors for infection in human and cattle populations. *Tuberculosis (Edinburgh, Scotland)* **87**(1): 30–43.

Clifford, D. L., Wolking, D. J. & Muse, E. A. (2011). *HALI Wildlife Health Handbook: Recognizing, Investigating, and Reporting Diseases of Concern for Wildlife Conservation and Human Health*. UC Davis, Health for Animals and Livelihood Improvement Project. www.vetmed.ucdavis.edu/whc/programs/hali.cfm, accessed 10 April 2014).

Condy, J. B., Herniman, K. A. J. & Hedger, R. S. (1969). Foot-and-mouth disease in wildlife in Rhodesia and other African territories. *Journal of Comparative Pathology* **79**: 27–312.

Cumming, D. H. M. (1995). Are multi species systems a viable land use option for southern African savannas? In *Proceedings of the International Symposium on Wild and Domestic Ruminants in Extensive Land Use Systems, 3–4 October 1994*, pp. 202–234. Eds. Hofmann, R. R. & Schwartz, H. J. Berlin: Humboldts University.

Cumming, D. H. M. (1999). Living off 'biodiversity': whose land, whose resources and where? *Policy Forum: Environment and Development Economics* **4**: 220–223.

Dawe, P. S., Flanagan F. O., Madekurozwa R. L., *et al.* (1994). Natural transmission of foot-and-mouth disease virus from African buffalo (*Syncerus caffer*) to cattle in a wildlife area of Zimbabwe. *Veterinary Record* **134**: 230–232.

de Garine-Wichatitsky, M., De Meeus, T., Guegan, J. F. & Renaud, F. (1999). Spatial and temporal distributions of parasites: can wild and domestic ungulates avoid African tick larvae? *Parasitology* **119**: 455–466.

de Garine-Wichatitsky, M., Caron, A., Gomo, A., *et al.* (2010). Bovine tuberculosis in buffaloes, Southern Africa. *Emerging Infectious Diseases* **16**: 884–885.

de Garine-Wichatitsky, M., Caron, A., Kock, R., *et al.* (2013a). A review of bovine tuberculosis at the wildlife/livestock/human interface in sub-Saharan Africa. *Epidemiology and Infection* **141**(7): 1342–1356.

de Garine-Wichatitksy, M., Fritz, H., Chaminuka, P., *et al.* (2013b). Consequences of animals crossing the edges of Transfrontier Parks. In *Transfrontier Conservation Areas: People Living on the Edge*, pp. 137–162. Eds. Andersson, J., de Garine-Wichatitsky, M., Cumming, D. H. M., Dzingirai, V. & Giller, K. E. London: Earthscan.

de Garine-Wichatitksy, M., Miguel, E., Mukamuri, B., *et al.* (2013c). Coexisting with wildlife in Transfrontier Conservation Areas in Zimbabwe: cattle owners' awareness of disease risks and perception of the role played by wildlife. *Comparative Immunology, Microbiology and Infectious Diseases* **36**: 321–332.

de Lisle, G. W., Bengis, R. G., Schmitt, S. M. & O'Brien, D. J. (2002). Tuberculosis in free-ranging wildlife: detection, diagnosis and management. *Revue Scientifique et Technique OIE*, **21**: 317–334.

De Vos, V., Bengis, R. G., Kriek, N. P., *et al.* (2001). The epidemiology of tuberculosis in free-ranging African buffalo (*Syncerus caffer*) in the Kruger National Park, South Africa. *Onderstepoort Journal of Veterinary Research* **68**(2): 119–130.

Dewals, B., Thirion, M., Markine-Goriaynoff, N., *et al.* (2006). Evolution of bovine herpesvirus 4: recombination and transmission between African buffalo and cattle. *The Journal of General Virology* **87** (6): 1509–1519.

Dion, E., Vanschalkwyk, L. & Lambin, E. F. (2011). The landscape epidemiology of foot-and-mouth disease in South Africa: a spatially explicit multi-agent simulation. *Ecological Modelling* **222**(13): 2059–2072.

Dube, T., Mirwira, A., Caron, A. & Degarine-Wichatitksy, M. (2010). Preliminary results on the permeability of veterinary fences to buffalo (*Syncerus caffer*) and cattle in Gonarezhou National Park, Zimbabwe. In *Fencing Impacts: A Review of the Environmental, Social and Economic Impacts of Game and Veterinary Fencing in Africa with Particular Reference to the Great Limpopo and Kavango-Zambezi Transfrontier Conservation Areas*, pp. 228–231. Eds. Ferguson, K. & Hanks, J. Pretoria: Mammal Research Institute.

Ducheyne, E., Mweempwa, C., De Pus, C., *et al.* (2009). The impact of habitat fragmentation on tsetse abundance on the plateau of eastern Zambia. *Preventive Veterinary Medicine* **91**(1): 11–18.

FAO (2007). *Gridded Livestock of the World.* Rome: FAO.

FAO (2013). *World Livestock 2013: Changing Disease Landscapes.* Rome: FAO.

Ferguson, K. & Hanks, J. (2010). Executive summary. In *Fencing Impacts: A Review of the Environmental, Social and Economic Impacts of Game and Veterinary Fencing in Africa with Particular Reference to the Great Limpopo and Kavango-Zambezi Transfrontier Conservation Areas*, pp. 12–18. Eds. Ferguson, K. & Hanks, J. Pretoria: Pretoria Mammal Research Institute.

Ferguson, K. J., Cleaveland, S., Haydon, D. T., Kock, R. A., Lembo, T. & Hopcraft, J. G. C. (2013). Evaluating the potential for the environmentally sustainable control of foot and mouth disease in sub-Saharan Africa. *Ecohealth* **10**(3): 314–322.

Fritz, H., De Garine-Wichatitsky, M. & Letessier, G. (1996). Habitat use by sympatric wild and domestic herbivores in an African savanna woodland: the influence of cattle spatial behaviour. *Journal of Applied Ecology* **33**: 589–598.

Gomo, C., de Garine-Wichatitsky, M., Caron, A. & Pfukenyi, D. M. (2012). Survey of brucellosis at the wildlife-livestock interface on the Zimbabwean side of the Great Limpopo Transfrontier Conservation Area. *Tropical Animal Health Production* **44**(1): 77–85.

Gradwell, D. V., Schutte, A. P., van Niekerk, C. A. & Roux, D. J. (1977). The isolation of *Brucella abortus* biotype I from African buffalo in the Kruger National Park. *Journal of the South African Veterinary Association* **48**(1): 41–43.

Hamblin, C. & Hedger, R. S. (1982). Prevalence of neutralizing antibodies to bovid herpesvirus 2 in African wildlife. *Journal of Wildlife Diseases* **18**(4): 429–436.

Hamblin, C. E., Anderson, C., Jago, M., Mlengeya, T. & Hirji, K. (1990). Antibodies to some pathogenic agents in free-living wild species in Tanzania. *Epidemiology of Infection* **105**: 585–594.

Hargreaves, S. K., Foggin, C. M., Anderson, E. C., *et al.* (2004). An investigation into the source and spread of foot and mouth disease virus from a wildlife conservancy in Zimbabwe. *Revue Scientifique et Technique de l'OIE* **23**: 783–790.

Hedger, R. S. & Hamblin, C. (1983). Neutralising antibodies to lumpy skin disease virus in African wildlife. *Comparative Immunology & Microbiology of Infectious Diseases* **6**(3): 209–213.

Herr, S. & Marshall, C. (1981). Brucellosis in free-living African buffalo (*Syncerus caffer*): a serological survey. *Onderstepoort Journal Veterinary Research* **48**(3): 133–134.

Hofmann, R. R. (1973). *The Ruminant Stomach.* Vol **2**, pp. 153–164. Nairobi: East African Literature Bureau.

Homewood, K. & Thompson, M. (2010). Social and economic challenges to conservation in East African rangelands: land use, livelihoods and wildlife change in Maasailand. In *Wild Rangelands: Conserving Wildlife While Maintaining Livestock in Semi-arid Ecosystems*, pp. 340–366. Eds. Deutsch, J., Du Toit, J. & Kock, R. Oxford: Wiley-Blackwell.

IUCN SSC Antelope Specialist Group (2008). *Syncerus caffer.* In *IUCN 2012. IUCN Red List of Threatened Species.* Version 2012.1. www.iucnredlist.org, accessed 1 August 2012.

Jolles, A. E., Ezenwa V. O., Etienne R. S., Turner, W. C. & Olff, H. (2008) Interactions between macroparasites and microparasites drive infection patterns in free-ranging African buffalo. *Ecology* **89**: 2239–2250.

Kalema-Zikusoka, G., Bengis, R. G., Michel, A. L. & Woodford, M. H. (2005). A preliminary investigation of tuberculosis and other disease in African buffalo (*Syncerus caffer*) in Queen Elizabeth National Park, Uganda. *Ondestepoort Journal of Veterinary Research* **72**: 145–151.

Keet, D. F., Kriek, N. P. J., Penrith, M.-L., Michel, A. & Huchzermeyer, H. (1996). Tuberculosis in buffaloes (*Syncerus caffer*) in the Kruger National Park: spread of the disease to other species. *Onderstepoort Journal of Veterinary Research* **63**: 239–244.

Keet, D. F., Boomker, J., Kriek, N. P., Zakrisson, G. & Meltzer, D. G. (1997). Parafilariosis in African buffaloes (*Syncerus caffer*). *Onderstepoort Journal of Veterinary Research* **64**(3): 217–225.

Keet, D. F., Kriek, N. P. J., Bengis, R. G., Grobler, D. G. & Michel, A. L. (2000). The rise and fall of tuberculosis in a free-ranging chacma baboon troop in the Kruger National Park. *Onderstepoort Journal of Veterinary Research* **67**: 115–122.

Kock, M. D., Mullins, G. & Perkins, J. (2002). Veterinary disease control in Botswana: impacts on wildlife health, ecosystems, and rural livelihoods. In *Conservation Medicine: Ecological Health in Practice*, pp. 265–275. Eds. Aguirre, A. A., Ostfeld, R. S., House, C. A., Tabor, G. M. & Pearl, M. C. New York: Oxford University Press.

Kock, R. A. (2005a). The wildlife domestic animal disease interface: should Africa adopt a hard or soft edge? *Transactions of the Royal Society of South Africa* **59**(1): 10–15.

Kock, R. (2005b). What is this infamous 'wildlife/livestock interface?' A review of current knowledge. In *Conservation and Development Interventions at the Wildlife/ Livestock Interface: Implications for Wildlife, Livestock and Human Health*, pp. 1–13. Eds. Osofsky, S., Cleaveland, S., Karesh, W. B., *et al.* Cambridge: IUCN

Kock, R. A. (2006). Rinderpest and wildlife. In *Rinderpest and Peste des Petits Ruminants Virus: Plagues of Large and Small Ruminants, Biology of Animal Infections*, pp. 144–162. Eds. Barrett, T. Pastoret, P.-P. & Taylor, W. London: Academic Press.

Kock, R. A. (2010). The newly proposed Laikipia disease control fence in Kenya. In *Fencing Impacts: A Review of the Environmental, Social and Economic Impacts of Game and Veterinary Fencing in Africa with Particular Reference to the Great Limpopo and Kavango-Zambezi Transfrontier Conservation Areas*, pp. 71–75. Eds. Ferguson, K. & Hanks, J. South Africa: Pretoria Mammal Research Institute.

Kock, R. A., Wambua, J. M., Mwanzia, J., *et al.* (1999). Rinderpest epidemic in wild ruminants in Kenya 1993–7. *Veterinary Record* **145**: 275–283.

Kock, R. A., Wamwayi, H. M., Rossiter, P. B., *et al.* (2006). Rinderpest in East Africa: continuing re-infection of wildlife populations on the periphery of the Somali ecosystem. *Preventive Veterinary Medicine* **75**(1–2): 63–80.

Kock, R., Kock, M., Cleaveland, S. & Thomson, G. (2009). Health and disease in wild rangelands. In *Wild Rangelands: Conserving Wildlife While Maintaining Livestock in Semi Arid Ecosystems*, pp. 98–128. Eds. Du Toit, J. T., Kock, R. A. & Deutsch, J. C. Chichester: John Wiley & Sons.

Kock, R. A., Croft, S., Dixon M., *et al.* (2012). Prioritising the need for new diagnostics, medicine, vaccines and management practices of zoonoses which have significant impact in the developing world. In *Report (6)*. pp. 39–40 and Appendix 1. London: Department for International Development.

Kolanye, M. G. & Mullins, G. R. (2000). *The Cost to the Government of the Control of Foot and Mouth Disease in Botswana: A Study Conducted by the Economics Unit of the Veterinary Epidemiology and Economics Section.* Gaborone: Department of Animal Health and Production, Ministry of Agriculture.

LaBeaud, A. D., Cross, P. C., Getz, W. M., Glinka, A. & King, C. H. (2011). Rift Valley Fever virus infection in African buffalo (*Syncerus caffer*): herds in rural South Africa. Evidence of interepidemic transmission. *American Journal of Tropical Medicine & Hygiene* **84**(4): 641–646.

Lembo, T., Hampson, K., Auty, H. *et al.* (2011). Serologic surveillance of anthrax in the Serengeti ecosystem, Tanzania, 1996–2009. *Emerging Infectious Diseases* **17**(3): 387–394.

Mbaiwa, J. S. & Mbaiwa, O. I. (2006). The effects of veterinary fences on wildlife populations in Okavango Delta, Botswana. *International Journal of Wilderness* **12**(3): 17–41.

McConnell, E. E., Tustin, R. C. & de Vos, V. (1972). Anthrax in an African buffalo (*Syncerus caffer*) in the Kruger National Park. *Journal of the South African Veterinary Association* **43**(2): 181–187.

Metzger, K. L. Sinclair, A. R. E., Hilborn, R., Hopcraft, J. G. C. & Mduma, S. A. R (2010). Evaluating the protection of wildlife in parks: the case of African buffalo in Serengeti. *Biodiversity Conservation* **19**: 3431–3444.

Michel, A. L., Bengis, R. G., Keet, D. F., *et al.* (2006). Wildlife tuberculosis in South African conservation areas: Implications and challenges. *Veterinary Microbiology* **112**(2–4): 91–100.

Miguel, E., Grosbois, V., Caron, A., *et al.* (2013). Contacts and foot and mouth disease transmission from wild to domestic bovines in Africa. *Ecosphere*, **4**: article 51.

Morris, R. S., Pfeiffer, D. U. & Jackson, R. (1994). The epidemiology of *Mycobacterium bovis* infections. *Veterinary Microbiology* **40**: 153–177.

Mtambo, M. M., Sebatwale, J. B., Kambarage, D. M., *et al.* (1997). Prevalence of *Cryptosporidium* spp. oocysts in cattle and wildlife in Morogoro region, Tanzania. *Preventive Veterinary Medicine* **31**(3–4): 185–190.

Munang'andu, H. M., Siamudaala, V. M., Matandiko, W., Munyeme, M. Chembensofu, M. & Mwase, E. (2010). Sarcoptes mite epidemiology and treatment in African buffalo (*Syncerus caffer*) calves captured for translocation from the Kafue game management area to game ranches. *BMC Veterinary Research* **6**: 29.

Munang'andu, H. M., Chembensofu, M., Siamudaala, V. M., Munyeme, M. & Matandiko, W. (2011). *Thelazia rhodesii* in the African buffalo, *Syncerus caffer*, in Zambia. *Korean Journal of Parasitology* **49**(1): 91–94.

Murwira, A., de Garine-Wichatitksy, M., Zengeya, F., *et al.* (2013). Resource gradients and movements across the edge of transfrontier parks. In *Transfrontier Conservation Areas: People Living on the Edge*, pp. 123–136. Eds. Andersson, J. A., de Garine-Wichatitsky, M., Cumming, D. H. M., Dzingirai, V. & Giller, K. E. London: Earthscan.

Nowak, R. M. (1991). *Walker's Mammals of the World.* 5th edition. Baltimore, MD: Johns Hopkins University Press.

OIE (2013). Recommendations for importation from FMD infected countries or zones, where an official control programme for FMD, involving compulsory systematic vaccination of cattle, exists. In *Terrestrial Animal Health Code.* Vol. **1**, chapter 8.5, article 8.5.25. Paris: OIE. www.oie.int/fileadmin/Home/eng/Health_standards/tahc/2010/en_chapitre_1.8.5.htm, accessed 10 April 2014

Olive, M.-M., Goodman, S. M. & Reynes, J. (2012). The role of wild mammals in the maintenance of Rift Valley Fever. *Journal of Wildlife Diseases* **48**(2): 241–266.

Osofsky, S. A., Kock, R. A., Kock, M. D., *et al.* (2005). Building support for protected areas using a 'One health' perspective. In *Friends for Life: New Partners in Support of Protected Areas*, pp. 65–81. Ed. McNeely, J. A. Gland and Cambridge: IUCN.

Pascucci, I., Capobianco Dondona, A., Cammà, C., *et al.* (2011). Survey of ixodid ticks and two tick-borne pathogens in African buffaloes, *Syncerus caffer*, from the Caprivi Strip, Namibia. *Journal of Zoo & Wildlife Medicine* **42**(4): 634–640.

Pelle, R., Graham, S. P., Njahira, M. N., *et al.* (2011). Two *Theileria parva* CD8 T cell antigen genes are more variable in buffalo than cattle parasites, but differ in pattern of sequence diversity. *PLoS One* **29**(6): e19015.

Perrotin, C., Graber, M., Thal, J. & Petit, J. P. (1978). Sarcosporidiosis in African buffaloes (*Syncerus caffer*). *Revue Elevage Medicine Veterinaire Pays Tropicaux* **31** (4): 423–426.

Pfitzer, S., Last, R., Espie, I. & Van Vuuren, M. (2013). Malignant catarrhal fever : an emerging disease in the African buffalo (*Syncerus caffer*). *Transboundary and Emerging Diseases*, 1–7. doi:10.1111/tbed.12131

Prins, H. H. T. (1996). *Ecology and Behaviour of the African Buffalo.* London: Chapman & Hall.

Renwick, A. R., White, P. C. & Bengis, R. G. (2007). Bovine tuberculosis in southern African wildlife: a multi-species host–pathogen system. *Epidemiology of Infection* **135**(4): 529–540.

Sahle, M., Dwarka, R. M., Venter, E. H. & Vosloo, W. (2007). Comparison of SAT-1 foot-and-mouth disease virus isolates obtained from East Africa between 1971 and 2000 with viruses from the rest of sub-Saharan Africa. *Archives Virology* **152**: 797–804.

Sangula, A. K., Belsham, G. J., Muwanika, V. B., *et al.* (2010). Evolutionary analysis of foot-and-mouth disease virus serotype SAT 1 isolates from East Africa suggests two independent introductions from Southern Africa. *BMC Evolutionary Biology* **10**: 371.

Scoones, I., Bishi, A., Mapitse, N., *et al.* (2010). Foot and mouth disease and market access: challenges for the beef industry in Southern Africa. *Pastoralism* **1**(2): 135–164.

Scott Wilson and the Environment and Development Group (EDG). (2000). *Environmental Impact Assessment of Veterinary Fences in Ngamiland: Summary Report.* Edinburgh and Oxford: Scott Wilson and EDG.

Simarro, P. P., Cecchi, G., Franco, J. R., *et al.* (2011). Risk for human African trypanosomiasis, Central Africa, 2000–2009. *Emerging Infectious Diseases* **17**(12): 2322–2324.

Sinclair, A. (1977). *The African Buffalo: A Study of Resource Limitation by Populations.* Chicago, IL: University of Chicago Press.

Smitz, N., Berthouly, C., Cornelis, D., *et al.* (2013). Pan-African genetic structure in the African buffalo (*Syncerus caffer*): investigating intraspecific divergence. *PLoS ONE* **8**(2): e56235. doi: 10.1371/journal.pone.0056235.

Sumption, K. J. & Scott, G. R.(2004). Lesser-known rickettsias infecting livestock. In *Infectious Diseases of Livestock.* Vol. **1**, pp. 543– 544. 2nd edition. Eds. Coetzer, J. A. W. & Tustin, R. C. Cape Town: Oxford University Press Southern Africa.

Suzuki, S., Konnai, S., Okagawa, T., *et al.* (2012). Molecular cloning and characterization of Th1 and Th2 cytokines of African buffalo (*Syncerus caffer*). *International Journal of Immunogenetics* **39**(2): 170–182.

Tanner, M., Inlameia, O., Michel, A., *et al.* (2014). Bovine tuberculosis and brucellosis in cattle and African Buffalo in the Limpopo National Park, Mozambique. *Transboundary and Emerging Diseases* **60**: 492–506.

Tenzin Dekker, A., Vernooij, J. C. M., Bouma, A. & Stegeman, J. A. (2006). Foot-and-mouth disease virus transmission by carriers: a metaanalysis of experimental data. In *Proceedings of the 11th International Symposium on Veterinary Epidemiology and Economics.* www.sciquest.org.nz, accessed 10 April 2014.

Thomson, G. R., Vosloo, W. & Bastos, D. S. (2003). Foot and mouth disease in wildlife. *Virus Research* **91**(1): 145–161.

Thomson, G. R., Tambi, E. N., Hargreaves, S. K., *et al.* (2004). International trade in livestock and livestock products: the need for a commodity-based approach. *Veterinary Record* **155**(14): 429–433.

Thomson, G. R., Penrith, M.-L., Atkinson, M. W, *et al.* (2013a). International trade standards for commodities and products derived from animals: the need for a system that integrates food safety and animal disease risk management. *Transboundary and Emerging Diseases* **60**(6): 507–515.

Thomson, G. R., Penrith, M.-L., Atkinson, M. W, *et al.* (2013b). Balancing livestock production and wildlife conservation in and around Southern Africa's transfrontier conservation areas. *Transboundary and Emerging Diseases* **60**(6): 492–506.

Tschopp, R., Aseffa, A., Schelling, E., *et al.* (2010). Bovine tuberculosis at the wildlife–livestock–human interface in Hamer Woreda, South Omo, Southern Ethiopia. *PloS One* **5**(8): e12205.

UNAIDS (2006). *Fact Sheet 06, Sub-Saharan Africa, December 2006.* Geneva: WHO.

UNAIDS (2011). *World AIDS Day Report.* Geneva: WHO.

van Baalen, M. (1998). Coevolution of recovery ability and virulence. *Proceedings of Biological Science* **265**(1393): 317–325.

van den Bossche, P., de la Rocque, S., Hendrickx, G., & Bouyer, J. (2010). A changing environment and the epidemiology of tsetse-transmitted livestock trypanosomiasis. *Trends in Parasitology* **26**(5): 236–243.

Vosloo, W., Bastos, A. D, Kirkbride, E., *et al.* (1996). Persistent infection of African buffalo (*Syncerus caffer*) with SAT-type foot-and-mouth disease viruses: rate of fixation of mutations, antigenic change and interspecies transmission. *Journal of General Virology* **77**(7): 1457–1467.

Vosloo, W., Bastos, A. D. S., Michel, A. & Thomson, G. R. (2001). Tracing movement of African buffalo in southern Africa. *Revue scientifique et technique Office international des Epizooties* **20**(2): 630–639.

Vosloo, W., de Klerk, L. M., Boshoff, C. I., *et al.* (2007). Characterisation of a SAT-1 outbreak of foot-and-mouth disease in captive African buffalo (*Syncerus caffer*): clinical symptoms, genetic characterisation and phylogenetic comparison of outbreak isolates. *Veterinary Microbiology* **120**(3–4): 226–240.

Waghela, S. & Karstad, L. (1986). Antibodies to *Brucella* spp. among blue wildebeest and African buffalo in Kenya. *Journal Wildlife Diseases* **22**(2): 189–192.

Wallace, R. & Kock, R. (2012). Whose food footprint? Capitalism, agriculture and the environment. *Human Geography* **5**(1): 63–83.

Wamwayi, H. M. (2002). *Rinderpest Virus Africa Lineage-2 Transmission Experiments in Cape Buffalo and Cattle. Wildlife and Rinderpest: Field Observations and Experimental Work. East African Workshop on Mild-Rinderpest 17–19th June 2002 AU-IBAR Pan African Programme for the Control of Epizootics.* Nairobi: AUIBAR.

Welburn, S. C., Maudlin, I. & Simarro, P. P. (2009). Controlling sleeping sickness: a review. *Parasitology* **136**: 1943–1949.

Wenink, P. W., Groen, A. F., Roelke-Parker, M. E. & Prins, H. H. T. (1998). African buffalo maintain high genetic diversity in the major histocompatibility complex in spite of historically known population bottlenecks. *Molecular Ecology* **7**(10): 1315–1322.

Wesonga, F. D., Mukolwe, S. W. & Grootenhuis, J. (2001). Transmission of *Cowdria ruminantium* by *Amblyomma gemma* from infected African buffalo (*Syncerus caffer*) and eland (*Taurotragus oryx*) to sheep. *Tropical Animal Health and Production* **33**(5): 379–390.

Williams, J. H., van Dyk, E., Nel, P. J., *et al.* (2011). Pathology and immunohistochemistry of papillomavirus-associated cutaneous lesions in Cape mountain zebra, giraffe, sable antelope and African buffalo in South Africa. *Journal of the South African Veterinary Association* **82**(2): 97–106.

Williamson, D. T. & Williamson, J. E. (1984). An assessment of the impact of fences on large herbivore biomass in the Kalahari. *Botswana Notes Record* **13**: 107–110.

Wolhuter, J., Bengis, R. G., Reilly, B. K. & Cross, P. C. (2009). Clinical demodicosis in African buffalo (*Syncerus caffer*) in the Kruger National Park. *Journal of Wildlife Diseases* **45**(2): 502–504.

Woodford, M. (1982). Tuberculosis in wildlife in the Ruwenzori National Park Uganda (part I). *Tropical Animal Health and Production* **14**(2): 81–88.

Zanette, L. Y., White, A. F., Allen, M. C. & Clinchy, M. (2011). Perceived predation risk reduces the number of offspring songbirds produce per year. *Science* **334**: 1398–1401

Chapter 27

Ex situ conservation of wild cattle: roles, status, management successes and challenges

Daniel C. de Man

Introduction

This chapter will describe the various roles that *ex situ* populations of wild cattle have or could have for conservation of the species in the wild. It will continue with an overview of wild cattle held in zoological gardens and provide background on the status of these populations. Third, the chapter will highlight the successes and future challenges related to the *ex situ* conservation of wild cattle.

Ex situ conservation roles of wild cattle

History of zoos and wild cattle

The first zoological gardens were established in the late eighteenth and early nineteenth century, e.g.: Tiergarten Schönbrunn,Vienna (1755), Menagerie du Jardin des Plantes, Paris (1793), London (1828), Amsterdam (1838), Antwerp (1843) and Berlin (1844) (Baratay & Hardouin-Fugier 2002). Since their establishment zoological gardens have held domestic and wild cattle species in their collections.

London Zoo first imported a banteng (*Bos javanicus*) in 1863 (Flower 1929). The international banteng studbook indicated that a few years later Artis Zoo in Amsterdam imported the first pair of banteng to Europe in 1867. The species has been held and bred successfully in many European collections since. Menagerie du Jardin des Plantes in Paris imported the first anoa (*Bubalus depressicornis*) in 1844. Also according to Flower, the first gaur (*Bos gaurus*) arrived at London Zoo in 1889. The first forest buffalo (*Syncerus caffer nanus*) arrived in 1908 in Tiergarten Hagenbeck in Hamburg and the first record of a Cape buffalo (*Syncerus caffer caffer*) in a zoological collection dates from 1839 and was held in London Zoo, presented by the Earl of Derby. Arriving from Saigon as 'gaur calf' in 1937, Zoo de Vincennes in Paris housed the only kouprey (*Bos sauveli*) ever known to be held outside of the range countries. Other European zoos with a long tradition of housing cattle include Antwerp, Vienna and Berlin Zoos.

As described by Baratay & Hardouin-Fugier (2002), Zimmermann *et al.* (2007) and many others, zoological gardens have evolved tremendously over the centuries, from housing small numbers of many species to displaying fewer species more naturalistically (see, for example, Figure 27.2). The traditional 'cattle galleries' (rows of small paddocks containing various wild and domestic cattle species – see Figure 27.1) have slowly disappeared in recent decades as zoos became increasingly aware of their crucial role in species conservation.

The conservation roles of zoos

The World Zoo Conservation Strategy (IUDZG/CBSG 1993) and its successor, the World Zoo and Aquarium Conservation Strategy (WZACS) (WAZA 2005), stress conservation as the main aim of modern zoos and aquariums. Governmental bodies such as the European Union recognise this and

Figure 27.1 'Rundergebouw' Zoo Antwerpen, presenting an example of a traditional cattle gallery (indoor building) in Antwerp Zoo, Belgium (photo by Beeldbank Zoo Antwerpen).

Ecology, Evolution and Behaviour of Wild Cattle: Implications for Conservation, ed. M. Melletti and J. Burton. Published by Cambridge University Press.

Figure 27.2 African buffalo (*Syncerus caffer caffer*) held in Safaripark Beekse Bergen – Hilvarenbeek (photo by Rolf Veenhuizen, Safaripark Beekse Bergen, Hilvarenbeek). A black and white version of this figure will appear in some formats. For the colour version, please refer to the plate section.

conservation is at the core of European legislation for zoos and aquariums (European Union 1999). According to WZACS, 'the major goal of zoos and aquaria will be to integrate all aspects of their work with conservation activities', including population management, visitor education, internal and external communication, research, field projects, operations, ethics and animal welfare.

EAZA zoos and aquariums are visited by 130–140 million visitors annually. Globally, over 600 million people visit zoos and aquariums annually (WAZA 2005). An economic impact assessment study carried out by EAZA (2010) showed that EAZA members annually:

- collectively contribute €2.5 billion to the European economy and directly employ 32 000 European citizens;
- support hundreds of conservation projects in the wild and contribute more than €100 million to biodiversity conservation;
- teach five million children about biodiversity, animal welfare and conservation;
- conduct and fund biodiversity and welfare research amounting to €30 million invested;
- run conservation campaigns raising awareness and funds for species and habitat conservation globally.

Zoos across the world contribute to field (*in situ*) conservation by means of direct financial support, equipment or sharing expertise. Projects supported by members of EAZA in 1997–2012 are recorded in the EAZA Conservation Database. Of these, 0.84% focus on wild cattle specifically. EAZA zoos' combined support of cattle field projects amounts to €350 000–400 000 as of 2012. The database includes some €14 million contributed to projects in cattle range states (e.g. for the protection of tiger habitat), of which a portion will likely have an indirect positive effect on cattle conservation (EAZA, unpublished data).

Introducing wild cattle *ex situ* conservation roles

The zoo and aquarium community defines conservation as 'the securing of long-term populations of species in natural ecosystems and habitats wherever possible' (WAZA 2005). When describing the conservation role of wild cattle held in zoological gardens (and other *ex situ* facilities), it is therefore relevant to look into the spectrum of activities they undertake to secure long-term populations of wild cattle species in their natural ecosystems and habitats.

European bison (*Bison bonasus*)

One of the most significant contributions of zoological gardens to the conservation of wild cattle was in preventing the extinction of European bison. A society was formed in 1923 by the directors of the zoos holding the species to establish a coordinated breeding programme, following the extinction of the last wild population in Poland in 1919; the species completely disappeared from the wild by 1927 with the extinction of the population in the Caucasus. After World War II there was a mere 54 animals (derived from 12 founders) held in zoological gardens and private collections. Since 1952, European bison recorded in the European Bison Pedigree Book (EBPB), which was first published in 1932 and was the first formal studbook for a non-domestic species – and since the 1990s also from the European Endangered Species Programme (EEP) – have been reintroduced into the Bialowieza Forest in Poland and gradually into other European countries. Due to

Figure 27.3 European bison (*Bison bonasus*) held in Dierenpark Planckendael – Mechelen (photo by H. Schram, Antwerp). A black and white version of this figure will appear in some formats. For the colour version, please refer to the plate section.

the long-term cooperation between Polish and international scientists, conservationists, zoological gardens, governmental and non-governmental organisations, the population as of the end of 2011 totals some 4600 European bison and the immediate danger of extinction has passed. Approximately 66% of the European bison are in 39 free-roaming/semi-free-roaming herds, with the remainder spread among 228 captive herds (breeding centres, zoos and private holders) maintained in 29 European countries (Raczynski 2012). The total number of individuals kept in captivity was 1497 at the end of 2011 (see also Chapter 10). Three of the individuals in the studbook are presented in Figure 27.3. Important remaining threats are fragmentation, decreasing genetic variability, disease and inappropriate management. A major initiative currently under way is the establishment of a meta-population of European bison in the Carpathian region of Romania, Ukraine, Slovakia, Poland, Hungary and the Czech Republic (Olech & Perzanowski 2002; Pucek *et al.* 2004). European bison are also held in semi-free ranging herds in various other countries in Europe, most of which are in Germany, totalling some 500 animals.

American bison (*Bison bison*)

As with European bison, zoological gardens and private holders also played an important role in ensuring the survival of the American bison. The population underwent drastic declines in the late nineteenth century and fewer than 300 wood bison and perhaps only 200 plains bison remained at the turn of the nineteenth century (Freese *et al.* 2007). No more than 25 bison remained in the wild, in the remote Pelican Valley in Yellowstone National Park (Meagher 1973). Together with other wildlife advocates, William Hornaday, then the director of the New York Zoological Society (today the Wildlife Conservation Society – WCS) and a founding father of zoos' involvement in species conservation, formed the American Bison Society in 1905. The society successfully lobbied for the creation of several public conservation herds in the United States. These herds were stocked with bison from various private and public places, including the Bronx Zoo (Isenberg 2000; Freese *et al.* 2007). The protected herds bred successfully under protection and the number of individuals grew steadily, helping to prevent the extinction of the species (Hornaday 1927; Boyd 2003).

As of 2010 approximately 400 000 plains bison exist in North America. However, few are free of cattle gene introgression, 96% are subject to anthropogenic selection for commodity production and only 4% are in herds managed primarily for conservation purposes (Freese *et al.* 2007; Gates *et al.* 2010; see also Chapter 21). To secure the survival of the American bison and ensure its ecological restoration, the WCS re-established the American Bison Society in 2005 (www.americanbisonsocietyonline.org). The American Bison Society was also used as a model for the formation of the comparable organisation established in 1923 to save the European bison.

Asian wild cattle (*Bos gaurus, Bos javanicus, Bos sauveli, Bubalus arnee, Bubalus depressicornis, Bubalus mindorensis, Bubalus quarlesi, Pseudoryx nghetinhensis*)

The *ex situ* conservation role for Asian wild cattle species was first documented in the 'Asian Wild Cattle Conservation Assessment and Management Plan Workshop Report', published by the IUCN Conservation Breeding Specialist Group (CBSG) in 1995. More recently, the IUCN/SCC Asian Wild Cattle Specialist Group (AWCSG) published the 'IUCN Regional Conservation Strategy for Wild Cattle and Buffaloes in Southeast Asia, 2011–2020', refining these *ex situ* conservation roles. Zoological gardens are well placed to assist with many of the strategy's goals, including: increasing species profiles to the public and governmental organisations, capacity building, providing conservation management, veterinary, husbandry and small population management expertise, offering research opportunities and providing material for banking.

The lowland anoa *ex situ* population has several clearly defined roles in terms of species conservation. As a population that is demographically and genetically diverse, and maintains behavioural and health characteristics, it can serve as a source for eventual release in the framework of a global population management plan. Other roles include an ambassadorial role for fundraising and education, as well as a research role where animals held *ex situ* can help progress science and assist in developing tools for potential use in the field.

Relocation and reintroduction of banteng and gaur is part of the conservation strategy for both species. Animals held *ex situ* are a potential source for reintroduction and therefore it is important to evaluate their suitability. Although not defined as clearly as for anoa, the banteng and gaur populations held in zoological collections also serve an ambassadorial role for fundraising, education and research aiding the conservation of the species.

Creating contingency plans for developing captive or semi-wild populations under emergency conditions (e.g. precipitous decline of a key wild population due to factors which would not operate in captivity) is needed for the conservation of anoa, tamaraw and kouprey. Although not yet defined as such, this role might in the future also be relevant for saola (IUCN/SSC/AWCSG Saola Working Group, unpublished report; W. Robichaud, personal communication). This is a role where the *ex situ* conservation community is able to contribute by sharing experiences with keeping and breeding species of other wild cattle and ungulates.

The EAZA Cattle and Camelid Taxon Advisory Group (TAG) coordinates several actions related to these conservation roles. In partnership with the IUCN SSC AWCSG, the banteng EEP is undertaking a molecular genetics study to determine the genetic diversity of the banteng population held in Europe, to evaluate whether the founder stock was hybridised with Bali cattle, and if so, to what extent. As the banteng held in North American collections derive from Europe, the outcome of the study will also apply to this population. The EAZA Cattle and Camelid TAG has offered assistance with preparing contingency plans for saola to develop captive or semi-wild populations under emergency conditions.

The EAZA and IUCN Species Survival Committee (SCC) Southeast Asia campaign (2011–2013) raises the profiles of wild cattle, raises funds and builds capacity for their conservation through participation from more than 140 zoological collections across the European continent. The saola is featured in the campaign's logo and illustrated in Figure 27.4 (www.southeastasiacampaign.org).

Figure 27.4 EAZA/IUCN SCC Southeast Asia campaign logo.

Yak (*Bos mutus*), Asian buffalo (*Bubalus arnee*) and African buffalo (*Syncerus caffer*)

Ex situ conservation roles for yaks, Asian wild water buffalo and African buffalo have not as of yet been as clearly defined, but are in progress through the AWCSG and the African Buffalo Initiative Group (AfBIG) of the IUCN Antelope Specialist Group.

In conclusion, *ex situ* conservation is important for the conservation of wild cattle species through several different roles that can be categorised into:

- research role – contributing to science of the species and/or its conservation;
- education role – educating zoo visitors, governments and the general public;
- field (*in situ*) support role – providing expertise, funds, equipment, etc.;
- population management (*ex situ*) role – including husbandry and management expertise, back up populations, restocking and reintroduction.

The following section will provide an overview of wild cattle populations held in zoological collections.

Status of wild cattle populations in zoological collections

Population management in modern zoological gardens

Small, fragmented populations need careful demographic and genetic management to remain healthy and viable in the future (Ballou *et al.* 2010). Small populations are susceptible to demographic stochastic events such as normal environmental and intrinsic demographic variation in mortality rates, reproductive rates and sex ratios at birth, as well as catastrophes such as environmental (e.g. floods and fires), disease outbreaks, political and/or economic instability in a region, country or institution etc. (Lande 1988; Lacy 2000; Lande *et al.* 2003). Zoological populations therefore need to be large enough to minimise the effect of random demographic variation, and also large enough to retain a large amount of the gene diversity of the wild population. Gene diversity makes natural selection possible and thus represents the evolutionary potential of a population. In addition, genetically diverse populations tend to have higher fitness (Hedrick *et al.* 1986; Allendorf & Leary 1986; Vrijenhoek 1994; Frankham 1995; Saccheri *et al.* 1998; Frankham *et al.* 2002).

The zoo community has integrated small population management science into regional breeding programme structures and has a wealth of expertise in this field. The European, North American and Australasian zoo associations have specific taxonomic-based expert groups, the so-called TAGs. These manage species under their remit for conservation, education, research and/or recreation purposes. Sustaining healthy populations requires close management of genetics and demography, which is only possible if the target species lives and reproduces reliably in zoological collections. The number of species that can be managed sustainably is restricted by enclosure availability (as defined by institutions), population parameters, resources and knowledge. In other words, there is a limit to the number of species that can be managed sustainably. TAGs must therefore prioritise which species to manage regionally and decide the most appropriate management level; a process that is called regional collection planning. The end of this process is the publication of a Regional Collection Plan (RCP). This so-called regional collection planning process is followed by EAZA, the Association of Zoos and Aquaria (AZA) in North America and the Zoo and Aquarium Association (ZAA) of Australasia.

The most intensive population management level are breeding programmes such as EAZA's EEPs, AZA's Species Survival Plans (SSPs) and ZAA's Australian Species Management Plans (ASMPs). Less intensively managed programmes include EAZA's European studbooks (ESBs) and International studbooks (ISBs) overseen by WAZA.

Although not based on a TAG and RCP system, other regions run breeding programmes as well. The Southeast Asian Zoo and Aquarium Association (SEAZA) has set up breeding programmes for banteng and gaur under the auspices of the Zoological Parks Organisation (ZPO) of Thailand. The Central Zoo Authority India (CZAI) has recently established a studbook for gaur. Hence it is to be expected that more extensive and reliable records of banteng and gaur holdings will become available in the near future.

Population sizes for wild cattle species held in zoological collections

Wild cattle have been held and bred in zoological collections for decades. Table 27.1 presents the current holdings around the world, taken from the International Species Information System (ISIS – November 2012 and April 2013 for European bison) and indicates the breeding programmes that exist for wild cattle.

Although most significant zoological collections around the world are members of ISIS, not all are and Table 27.1 thus gives lower numbers than reality. As relatively few Asian zoological collections have subscribed to ISIS yet, their cattle holdings are most likely to be underrepresented, with several of the species naturally occurring in Asia.

There are no *ex situ* breeding programmes run by zoos for tamaraw (*Bubalus mindorensis*), Asian water buffalo, wild yak (*Bos mutus*), kouprey (*Bos sauveli*), saola (*Pseudoryx nghetinhensis*) or American bison (*Bison bison*). Tamaraw are not held outside of Mindoro, where a population of 19 individuals was established in 1980, of which only one male remains at the facility at present (Byers *et al.* 1995; de Leon *et al.* 1996; see also Chapter 18).

Asian water buffalo and yak are held in zoological collections, but the vast majority of these are in fact domestic animals. The 5/2 wild yak included in the ISIS database are almost certainly domestics and entered erroneously as wild specimens. A small number of wild yak are held in Chinese collections (Joseph 2002). Similarly, the majority of wild Asian water buffalo in ISIS are likely to be incorrectly entered into the database and are, in fact, domestics, with the exception of a small number held in Kathmandu Zoo in Nepal (1/1 and their calf) and one female held in Udanti, India that is not on ISIS (see Chapter 16). In the absence of clearly defined *ex situ* conservation roles, both the AZA and EAZA regional collection plans recommend phasing the domestic forms of these species out in favour of species serving a conservation role (Joseph 2002; EAZA, unpublished data).

Other than the individual in Zoo de Vincennes, kouprey have not been held in zoological collections outside the range countries. A kouprey calf held by Cambodia's Prince Sihanouk on his palace grounds at Phnom Penh died in the 1950s and is thought to have been the last individual held in captivity (Joseph 2004).

Saola have never been held in zoological collections and have not been successfully kept under human care in range states. The four individuals known to have been in holding facilities in its range all died within weeks or months of their arrival (Saola Working Group, unpublished data).

American bison are held in relatively large numbers in zoological collections, particularly in North America and Europe. The plains bison is well represented in the private sector and on federal lands. However, as described above, only 4% of the plains bison are held in so-called conservation herds and there seemingly is a clear role for zoological collections,

Table 27.1 Wild cattle populations in global zoological collections

Species		Population (male/female/unknown sex)							
Common name	**Scientific name**	**Europe**	**North America**	**Asia**	**Australia**	**Africa**	**South and Central America**	**Global**	**Breeding programme(s)**
Asian water buffalo (wild)	*Bubalus arnee*	5/3/7	–	2/2/0	0/0/20	–	–	7/5/27	–
Asian water buffalo (domestic)	*Bubalus arnee bubalis/ carabanesis*	33/61/6	22/26/0	2/1/0	2/6/4	1/2/0	6/19/0	66/115/10	–
Mountain anoa	*Bubalus quarlesi*	2/0/0	–	–	–	–	–	2/0/0	EEP (EAZA), ISB (WAZA)
Lowland anoa	*Bubalus depressicornis*	23/23/0	22/26/2	2/1/0	–	–	–	47/50/2	EEP (EAZA), SSP (AZA), ISB (WAZA)
Yak (domestic)	*Bos mutus grunniens*	68/192/8	26/41/2	2/0/0	–	–	–	96/133/10	-
Yak (wild)	*Bos mutus*	5/10/0	–	–	–	–	–	5/10/0	–
Banteng	*Bos javanicus*	31/46/0	14/42/0	18/29/0	0/2/0	–	–	63/119/0	EEP (EAZA), ISB (WAZA) SEAZA
Gaur	*Bos gaurus*	21/20/0	46/58/2	35/20/4	–	–	1/2/0	103/100/6 [2]	EEP (EAZA), ISB (WAZA), SEAZA, CZAI
African buffalo (Cape)	*Syncerus caffer caffer*	17/45/0	7/19/0	9/9/0	–	22/24/2	4/8/0	59/105/2 [3]	ESB (EAZA)
African buffalo (forest)	*Syncerus caffer nanus*	32/84/4	0/4/0	–	4/1/0	2/4/0	–	38/93/4	ESB (EAZA)
European bison	*Bison bonasus*	132/227/2	–	–	–	–	2/5/0	134/232/2 (1497) [1]	EEP (EAZA) pedigree book (EBPB)
American bison	*Bison bison*	79/157/15	101/233/21	3/6/1	15/8/0	-	17/22/0	215/426/37	–
Re-created Aurochs	*Bos tauros primigenius*	8/23/0	–	–	–	–	–	8/23/0	–

Source: ISIS (November 2012, except European bison: April 2013)
[1] EBPB (2011). These data also include EAZA figures (total of European breeding centres and zoos).
[2] These figures are probably an underestimate of all gaur kept globally in captivity
[3] These figures are probably an underrepresentation of the total number of buffalo in captive collections

particularly in range states, in terms of raising awareness, conservation education and fundraising. In North America the American Bison Conservation Society is actively involved in this regard. The EAZA RCP recommends phasing out American bison in favour of increasing enclosure spaces for European bison.

The re-created aurochs was created in the 1920s and 1930s in an attempt to breed-back domestic cattle to their ancestral form, *Bos primigenius primigenius*. Breeding programmes to achieve this were initiated by the German brothers Lutz and Heinz Heck, director of Berlin Zoological Gardens and the Hellabrunn Zoological Gardens in Munich, respectively (Heck 1951; van Vuure 2005). Currently ongoing experiments attempting to breed-back aurochs are the 'TaurOs Project' in the Netherlands, started in 2011, and the Lippe River project in Germany. These projects cross several cattle breeds and try to bring in aurochs-like features like body size, horn form and coat colour (for more details, see Chapter 15). At present a small number of re-created aurochs are held by zoological collections, including at GaiaZoo Kerkrade (see Figure 27.5).

Ex situ breeding programmes for wild cattle species

In this section, reproductive and genetic parameters for wild cattle species derived from the *ex situ* breeding programmes are presented. It should be emphasised that as part of the *ex situ* programmes these populations are managed intensively, which influences the data presented in this section. Mountain

Figure 27.5 Re-created aurochs held at Gaiazoo Kerkrade, the Netherlands (photo by Gaiazoo Kerkrade). A black and white version of this figure will appear in some formats. For the colour version, please refer to the plate section.

anoa (*Bubalus quarlesi*), despite being part of the Anoa EEP and ISB, are excluded given that only two males are held. The European bison is not included here as its conservation programme is described in detail in Chapter 10.

Reproductive parameters

Table 27.2 gives an overview of the following reproductive parameters: median age at first reproduction, median age of reproduction, growth rate (lambda), generation length (*T*) and first-year mortality for each of four wild cattle species (anoa, banteng, gaur and African buffalo). The African buffalo breeding programme manages two populations separately, per subspecies held: (*S. c. caffer* and *S. c. nanus*). The data derive from the SPARKS (Single Population Analysis and Record Keeping System – ISIS) data sets as maintained by EAZA regional or WAZA international studbook keepers, and looks at 20-year intervals from the date of currentness of the data set backwards. It is important to stress that these populations are managed intensively, influencing parameters like growth rates, generation length (*T*) and life expectancy. The sample size (*n*) is presented following the presented parameter when relevant.

The median ages at first reproduction for these species lie between 2 years, 4 months (banteng males) and 5 years, 1 month (forest buffalo females). The median age at first reproduction is lowest for banteng and highest for lowland anoa and African buffalo (females). Males start breeding at younger ages than females. The average age of reproduction ranges from 5 years, 6 months (banteng males) to 11 years, 1 month (Cape buffalo females).

In zoological collections, banteng females and males start reproducing at younger ages, while anoa males and African buffalo (*S. caffer*) females start reproducing at older ages compared to the other species. Gaur and both African buffalo subspecies bulls start reproducing at around the same age in zoos. African buffalo continue breeding into higher age classes compared to the other species. The mean generation time, *T* (averaged for males and females), is presented in the next column. *T* is the average age of reproduction (production of offspring). It is also the average time from reproduction in one generation to reproduction in the next generation. The mean generation times are close to the median ages of first reproduction for all species included here.

The lambda indicates that in the 20-year period up until the currentness date of the dataset, the five populations are stable/slightly increasing. This is an indicator that reproductive management and husbandry routines for wild cattle held in zoological collections are successfully established. Husbandry routines in zoological collections are described in the AZA Husbandry Manual for Wild Cattle (Joseph 2004), which confirms that routines are successfully established with the exception of anoa, for which it suggests that incomplete understanding of husbandry requirements is hampering more breeding success.

First-year mortality is higher for bulls than for cows, with the exception of anoa where first-year mortality is the same for both sexes. There are different factors that influence first-year mortality, including still births, neonatal deaths and potentially also inbreeding. Nevertheless, it is expected that management culling explains the difference in first-year mortality between males and females. Whereas banteng, gaur and African buffalo are kept in herds matching the natural herd structure and social behaviour, anoa are kept mostly in pairs, which would explain why first-year mortality in anoa is similar for males and females.

Table 27.3 gives further details related to survival (*Lx*) of the populations during the exported date window. The first

Table 27.2 Reproductive parameters from wild cattle populations in *ex situ* breeding programmes

Species		Median age at first reproduction		Median age of reproduction		Growth rate (lambda)	Generation length (*T*)	First-year mortality		Source
Common name	Scientific name	Males	Females	Males	Females	Males	Males	Males	Females	
Lowland anoa	*Bubalus depressicornis*	4Y, 0M (*n* = 67)	4Y, 1M (*n* = 81)	7Y, 2M (*n* = 321)	7Y, 2M (*n* = 342)	1.02	9.2 years	24%	24%	International Anoa studbook SPARKS data set (interval 01/09/1992 till 01/09/2012)
Banteng	*Bos javanicus*	2Y, 4M (*n* = 90)	2Y, 8M (*n* = 222)	5Y, 6M (*n* = 852)	5Y, 10M (*n* = 944)	1.04	7.1 years	27%	20%	International Banteng studbook SPARKS data set (interval 01/06/1991 till 01/06/2011)
Gaur	*Bos gaurus*	3Y, 5M (*n* = 98)	3Y, 0M (*n* = 238)	6Y, 6M (*n* = 795)	6Y, 0M (*n* = 967)	1.01	7.8 years	44%	30%	International Gaur studbook SPARKS data set (interval 20/01/1990 till 20/01/2010)
Cape buffalo	*Syncerus caffer caffer*	3Y, 10M (*n* = 26)	4Y, 11M (*n* = 64)	9Y, 7M (*n* = 403)	11Y, 1M (*n* = 384)	1.08	9.3 years	20%	17%	European African buffalo studbook SPARKS data set (interval 01/01/1992 till 01/01/2012)
Forest buffalo	*Syncerus caffer nanus*	3Y, 3M (*n* = 38)	5Y, 1M (*n* = 85)	9Y, 1M (*n* = 404)	8Y, 8M (*n* = 396)	1.05	9.7 years	30%	16%	European African buffalo studbook SPARKS data set (interval 01/01/1992 till 01/01/2012)

column presents average life expectancy at birth. The next column presents the age to which 25% of the populations survive. The third column presents the oldest recorded animals for each species for both males and females.

The life expectancy for cows is higher than for bulls in all species, although for anoa this is comparable for both sexes and adverse-looking at *Lx* 25%. All species can live until their mid to late twenties, with the oldest anoa female recorded living well into her 33rd year.

Information on birth seasonality of the four species is presented in Table 27.4. A Chi-square statistical test on each distribution is conducted to determine the probability that the observed data would occur by chance. With the exception of the forest buffalo, the Chi-square test indicated a non-random pattern of event distribution across months. Yet the differences between the number of births over the given month are relatively small and strong birth peaks are absent for all species.

Genetic parameters

Table 27.5 gives an overview of the following genetic parameters: percentage of pedigree known, number of founders, genetic diversity maintained (*GD*), potential *GD*, and mean inbreeding (*F*) for each of four wild cattle species: anoa, banteng, gaur and African buffalo (both subspecies *S. c. caffer* and *S. c. nanus*). Per parameter, three values are given. The first (1) presents the known data in the EAZA breeding programme SPARKS data set. The second (2) uses the first set of data but adds educated assumptions to complete existing holes in the pedigrees. The third (3) presents data from the WAZA international studbook (where existing) SPARKS data set. In the absence of complete pedigrees, some data are not reliable (n.r.) and hence excluded from the table. The international data set for gaur and the European data set for forest buffalo also have rather incomplete pedigrees and should thus be interpreted with much care.

From the table it can be concluded that for all but the gaur EEP and lowland anoa ISB, there is a lack of pedigree data. Educated assumptions can solve this to some extent – the data sets may become useful for genetic management, with the exception of the forest buffalo ESB (*S. c. nanus*). The banteng ISB and gaur ISB data sets are not reliable for genetic management either. The lowland anoa assumptions used for the EEP

Table 27.3 Life expectancy (survival – *Lx*) parameters of wild cattle in *ex situ* breeding programmes

Species		Average life expectancy (years)		25% live to (years)		Oldest recorded (years)		Source
Common name	Scientific name	Males	Females	Males	Females	Males	Females	
Lowland anoa	*Bubalus depressicornis*	16.3	16.4	24.1	23.3	29.5	33.5	International Anoa studbook SPARKS data set (interval 01/09/1992 till 01/09/2012)
Banteng	*Bos javanicus*	8.2	12.3	12.5	18.1	23.4	26.6	International Banteng studbook SPARKS data set (interval 01/06/1991 till 01/06/2011)
Gaur	*Bos gaurus*	7.4	10.4	10.3	16.4	24.6	24.6	International Gaur studbook SPARKS data set (interval 20/01/1990 till 20/01/2010)
Cape buffalo	*Syncerus caffer caffer*	7.9	14.6	13.5	21.7	27.7	29.8	European African buffalo studbook SPARKS data set (interval 01/01/1992 till 01/01/2012)
Forest buffalo	*Syncerus caffer nanus*	7.6	13.8	10.2	21.5	28.2	29.7	European African buffalo studbook SPARKS data set (interval 01/01/1992 till 01/01/2012)

data set were transferred to the anoa ISB data set and increase the percentage known from 76% to 96%.

Second, the number of founders for all programmes varies from relatively low to very low. Sampling and genetic theory indicates that 20 unrelated wild individuals are sufficient to capture 97.5% of the gene diversity of the wild population within the founder population (Crow & Kimura 1970; de Boer 1989; Lacy 1994; Frankham *et al.* 2002). In practice, contributions of founders to the living descendant population are uneven and hence more than 20 founders are likely necessary to maintain a high level of genetic diversity (usually 90% over 100 years) and a secure level of population sustainability. This is reflected in the *GD* and potential *GD* columns. The banteng ISB, gaur ISB and Cape buffalo (*S. caffer caffer*) ESB are the only populations with more than 20 founders at present, although the lowland anoa ISB has a high potential *GD* as well.

Mean inbreeding (*F*) levels vary among the species from medium (Cape buffalo, banteng, anoa) to high (gaur and forest buffalo).

Conservation successes and challenges for the future

Wild cattle and buffalo species play major roles in maintaining functioning ecosystems, both as important prey species and by maintaining habitat diversity through grazing and browsing, and such behaviours as creating wallows. These species are also important as major reservoirs of genetic material that could help safeguard and improve domestic cattle breeds throughout the world (IUCN 2010).

The *ex situ* conservation of wild cattle plays an important role in the overall conservation of these species. Notable successes include the reintroduction of European and American bison that contributed to saving these species from the very brink of extinction. Successful breeding programmes have been established for several species of wild cattle and are managed professionally under the auspices of the regional and global zoo associations. Husbandry routines for keeping wild cattle species held in zoological collections are well established and most species show stable population trends over the past 20 years. Zoological collections contribute to raising the profile of wild cattle and building capacity for their conservation, demonstrated by the EAZA/IUCN SSC Southeast Asia campaign.

A challenge that both the *in situ* and *ex situ* cattle conservation communities are taking by the horns at present is the integration of conservation activities following the scheme referred to as the One Plan approach (CBSG 2012; Lacy 2012; Pritchard *et al.* 2012). The conservation strategy for wild cattle and buffalo from Southeast Asia looks at species conservation holistically, clearly defining roles and actions for various areas, levels and stakeholders, including for the *ex situ* community, and as such sets an example in this respect.

For the *ex situ* conservation of Southeast Asian wild cattle species in particular, one of the main priorities is to increase the number of holdings in range countries, for example breeding centres, setting up and further professionalising management programme structures. Second, these breeding programmes should communicate with programmes in other regions, sharing data, knowledge, expertise and resources to build healthy global *ex situ* wild cattle populations. The recently initiated WAZA Global Species Management Plan (GSMP) structure can serve as a model for integrated *ex situ* conservation management plans. Third, the *ex situ* programmes should be linked with *in situ* populations to establish a form of meta-population management where relevant. The use of tools and techniques such as cryopreservation and molecular genetics should be integrated into management and meta-population management plans. General challenges and research opportunities in this respect are described by Leus *et al.* (2011).

Table 27.4 Birth seasonality of wild cattle in *ex situ* breeding programmes

Species		January	February	March	April	May	June	July	August	September	October	November	December	*N*	Source
Common name	**Scientific name**														
Lowland anoa	*Bubalus depressicornis*	6%	6%	11%	8%	8%	8%	12%	9%	9%	7%	8%	6%	325	International Anoa studbook SPARKS data set (interval 01/09/1992 till 01/09/2012)
Banteng	*Bos javanicus*	10%	12%	13%	10%	9%	9%	9%	8%	6%	4%	5%	6%	1058	International Banteng studbook SPARKS data set (interval 01/06/1991 till 01/06/2011)
Gaur	*Bos gaurus*	7%	7%	9%	9%	12%	11%	8%	10%	9%	7%	6%	6%	992	International Gaur studbook SPARKS data set (interval 20/01/1990 till 20/01/2010)
African buffalo	*Syncerus caffer caffer*	8%	8%	11%	12%	11%	5%	10%	9%	5%	7%	7%	7%	323	European African buffalo (Cape) studbook SPARKS data set (interval 01/01/1992 till 01/01/2012)
	Syncerus caffer nanus	7%	5%	8%	10%	8%	10%	11%	9%	10%	8%	6%	10%	400	European African buffalo (forest) studbook SPARKS data set (interval 01/01/1992 till 01/01/2012)

Table 27.5 Genetic parameters from wild cattle populations in *ex situ* breeding programmes

Species		% Pedigree known			# founders			*GD*			Potential *GD*			Mean *F*			Source
Common name	**Scientific name**	**1**	**2**	**3**	**1**	**2**	**3**	**1**	**2**	**3**	**1**	**2**	**3**	**1**	**2**	**3**	
Lowland anoa	*Bubalus depressicornis*	58%	100%	96%	(5)	6	(14, 12 add.)	n.r.	0.84	0.878	n.r.	0.89	0.979	n.r.	0.09	0.1236	1 + 2: International Anoa studbook – EAZA. Current up to 01/09/2012 (only *depressicornis*). 3: International Anoa studbook current up to 01/09/2011 (+ assumptions from EAZA added)
Banteng	*Bos javanicus*	31%	97%	47%	(2)	16	(30, 3 add.)	n.r.	0.84	n.r.	n.r.	0.92	n.r.	n.r.	0.21	n.r.	1 + 2: European studbook, current up to 2012. 3: International Banteng studbook, current up to 1/06/2011
Gaur	*Bos gaurus*	99%	–	78%	8	–	(22, 6 add.)	0.57	–	0.92^4	0.815	–	0.98^4	0.47	–	0.23^4	1 + 2: European Gaur studbook, current to 20/01/2010. 3: International studbook, current up to 31/12/2010
African buffalo	*Syncerus caffer caffer*	50%	98%	–	(9)	22	–	n.r.	0.9	–	n.r.	0.96	–	n.r.	0.12	–	European studbook African buffalo, November 2012 (Man 2012)
	Syncerus caffer nanus	31%	79%	–	(5)	7	–	n.r.	0.86^4	–	n.r.	0.92^4	–	n.r.	0.21^4	–	European studbook African buffalo, November 2012 (Man 2012)

1 = EAZA real data
2 = EAZA educated assumptions
3 = International
4 = Take with caution due to missing pedigree information
n.r. = not reliable due to lack of pedigree information

Exchange of genetic material within and between countries is an important criterion for establishing long-term management plans, but is increasingly complicated and often legally impossible due to biosecurity legislation. Existing breeding programmes need new founder material sooner rather than later. Biosecurity legislation is, as such, hampering species conservation. Government authorities should take responsibility in evaluating the risks and benefits of importation of wild cattle into zoological collections, adding species conservation benefits into the equation as well.

Maintaining and increasing holding space and resources for wild cattle species in zoological collections is an important challenge for the future. More breeding facilities are needed and space should be better prioritised to benefit species conservation (Conway 2007, 2011; Vince 2008). Models, like the Conservation Centres for Species Survival (C2S2 – www.conservationcenters.org) recently set up in North America, where a select number of institutions dedicate space for larger numbers of individuals and resources, needs to be investigated for wild cattle species. A significant number of spaces in zoological collections are filled by species with no clearly defined conservation role, such as domestic yak, domestic Asian water buffalo and American bison, indicating a potentially larger capacity for species such as banteng, gaur and European bison.

Gippoliti (2012) describes how, gradually, fewer species are being maintained in zoos and aquaria, resulting in a competition for the remaining spaces. The trend favours charismatic species, leaving few spaces available for many species threatened with extinction (Lernould *et al.* 2003; Baker 2007). Additionally, zoos are increasingly dependent on revenue from visitors for their self-maintenance, resulting in species selection being constrained more and more by public perception rather than objective conservation criteria (Ratajszczak 2008). Mammals such as apes, elephants and large carnivores, or animals that trigger fear such as spiders and snakes, consistently win out as the most charismatic species drawing visitor attention (Small 2011, 2012; Snijder & Henckens 2012). Wild cattle, as charismatic as they are, unfortunately never score high. It is important that governments such as the European Union actively promote and support *ex situ* conservation activities, particularly for species perceived as less charismatic by the public (Gippoliti 2012). As wild cattle species are a major reservoir of genetic material to safeguard and improve domestic cattle breeds throughout the world, the livestock industry also has a responsibility to contribute in this regard (IUCN 2010). With less than 1% of total field conservation expenditures of the (European) zoo community supporting wild cattle, there is room for stepping up to the mark in this respect as well, not least because *ex situ* conservation roles are so clearly defined in wild cattle conservation strategies.

Acknowledgements

I would like to thank the editors for creating the opportunity to develop this chapter and their important advice and assistance towards completing it. Elmar Fienieg MSc (EAZA volunteer) assisted greatly with collating the reproductive and genetic parameters, for which I am grateful. EAZA Cattle and Camelid TAG chairs Terry Hornsey (Africa Alive) and Gary Batters (Banham Zoo) are acknowledged for their support and advice, as are all EEP coordinators and ISB keepers that maintain the data sets for the breeding programmes that were used in this chapter. Valuable comments that helped improve the chapter were received from Mario Melletti, Laurie Bingaman Lackey, Douglas Richardson and Kristine Schad. Last but not least I would like to thank all colleagues and zoos that submitted pictures for use in this chapter.

References

Allendorf, F. W. & Leary, R. F. (1986). Heterozygosity and fitness in natural populations of animals. In *Conservation Biology: The Science of Scarcity and Diversity*, pp. 57–76. Ed. Soulé, M. E. Sunderland, MA: Sinauer Associates Inc.

Baker, A. (2007). Animal ambassadors: an analysis of the effectiveness and conservation impact of ex situ breeding efforts. In *Zoos in the 21st Century: Catalyst for Conservation?*, pp. 139–153. Eds. Zimmermann, A,. Hatchwell, M., Dickie, L. & West, C. Cambridge: Cambridge University Press.

Ballou, J. D., Lees, C., Faust, L. J., *et al.* (2010). Demographic and Genetic Management of Captive Populations. In *Wild Mammals in Captivity: Principles and Techniques for Zoo Management*, pp 219–252. 2nd edition. Eds. Kleiman, D. G., Thompson, K. V. & Bear, C. K. Chicago, IL and London: University of Chicago Press.

Baratay, E. & Hardouin-Fugier, E. (2002). *Zoo: A History of Zoological Gardens in the West.* London: Reaktion books Ltd.

de Boer, L. E. M. (1989). *Genetics and Breeding Programmes: Genetic Guidelines and Their Background for EEP Coordinators.* Amsterdam: National Foundation for Research in Zoological Gardens.

Boyd, D. P. (2003). Conservation of North American bison: status and recommendations. Masters thesis. University of Calgary.

Byers, O., Hedges, S. & Seal, U. S. (eds) (1995). *Asian Wild Cattle Conservation Assessment and Management Plan Workshop Report.* Apple Valley, MN: IUCN SSC Conservation Breeding Specialist Group (CBSG).

CBSG (2012). One plan approach to conservation planning: ex situ programmes as part of an overall species conservation strategy. *CBSG News* **23**.

Conway, W. G. (2007). Entering the 21st century. In *Zoos in the 21st Century: Catalyst for Conservation?* pp. 12–21. Eds. Zimmermann. A., Hatchwell, M., Dickie, L. & West, C. Cambridge: Cambridge University Press.

Conway, W. G. (2011). Buying time for wild animals with zoos. *Zoo Biology* **30**: 1–8.

Crow, J. F. & Kimura, M. (1970). *An Introduction to Population Genetics Theory*. New York: Harper and Row.

EAZA (2010). 21st century zoos and aquariums; good for animals, good for people – the socio-economic impact of EAZA zoos and aquariums for the European Union. www.eaza.net, accessed December 2012.

European Bison Pedigree Book (2011). Ed. Raczyński, J. Białowieża: Białowieża National Park.

European Union (1999). Council Directive 1999/22/EC of 29 March 1999 relating to the keeping of wild animals in zoos. *Official Journal of the European Communities L* 94/24 (9.4.1999).

Flower, S. (1929). *List of the Vertebrated Animals Exhibited in the Gardens of the Zoological Society of London, 1828–1927*. London: Zoological Society of London.

Frankham, R. (1995). Conservation genetics. *Annual Review of Genetics* **29**: 305–327.

Frankham, R., Ballou, J. D. & Briscoe, D. (2002). *Introduction to Conservation Genetics*. Cambridge: Cambridge University Press.

Freese, C. H., Aune, K. E., Boyd, D. P., *et al.* (2007). Second chance for the plains bison. *Biological Conservation* **136**(2): 175–184.

Gates, C. C., Freese, C. H., Gogan, P. J. P. & Kotzman, M. (eds) (2010). *American Bison: Status Survey and Conservation Guidelines 2010*. Gland: IUCN.

Gippoliti, S. (2012). Ex situ conservation programmes in European zoological gardens: can we afford to lose them? *Biodiversity and Conservation* **20**(1): 1359–1364.

Heck, H. (1951). The breeding-back of the aurochs. *Oryx* **1**: 117–122.

Hedrick, P. W., Brussard, P. F., Allendorf, F. W., Beardmore, J. A. & Orzack, S. (1986). Protein variation, fitness and captive propagation. *Zoo Biology*. **5**: 91– 99.

Hornaday, W.T (1927). *Hornaday's American Natural History*. New York: Charles Scribner's Sons.

Isenberg, A. C. (2000). *The Destruction of the Bison: An Environmental History 1750–1920*. Cambridge: Cambridge University Press.

IUCN (2010). Regional conservation strategy for wild cattle and buffaloes in South-east Asia. IUCN-SSC Asian Wild Cattle Specialist Group.

IUDZG/CBSG. (1993).The world zoo conservation strategy – the role of the zoos and aquaria of the world in global conservation.

Joseph, S. (2002). Bison, Buffalo and Cattle Taxon Advisory Group Regional Collection Plan. American Zoo and Aquarium Association (AZA).

Joseph, S. (2004). *Husbandry Manual for Wild Cattle Species*. Silver Springs, MD: Advisory and Steering Committees, AZA Bison, Buffalo and Cattle Advisory Group.

Lacy, R. C. (1994). Managing genetic diversity in captive populations of animals. In *Restoration of Endangered Species*, pp. 63–89. Eds. Bowles, M. L. & Whelan, C. J. Cambridge: Cambridge University Press.

Lacy, R. C. (2000). Considering threats to the viability of small populations using individual-based models. *Ecological Bulletins* **48**: 39– 51.

Lacy, R. C. (2012). Achieving true sustainability of zoo populations. *Zoo Biology* **32**: 1–13.

Lande, R. (1988). Genetics and demography in biological conservation. *Science* **241**: 1455–1460.

Lande, R., Engen, S. & Saether, B. (2003). *Stochastic Population Dynamics in Ecology and Conservation*. Oxford: Oxford University Press.

de Leon, J., Lawas, N., Escalada, R., *et al.* (1996). *Tamaraw (*Bubalus mindorensis*): Population and Habitat Viability Assessment*. Apple Valley, MN: IUCN/SSC Conservation Breeding Specialist Group.

Lernould, J.-M., Heckel, J.-O. and Wirth, R. (2003). More participants needed for Philippine spotted deer conservation programme. *EAZA News* **43**: 12–13.

Leus, K., Traylor-Holzer, K. & Lacy R. C. (2011). Genetic and demographic population management in zoos and aquariums: recent developments, future challenges and opportunities for scientific research. *International Zoo Yearbook* **45**: 1–13.

Man, D. C. de. (2012). *Dwarf Forest (Congo) Buffalo (*Syncerus caffer nanus*) and Southern Savannah (Cape) Buffalo (*Syncerus caffer caffer*) European Studbook*. 9th edition. Amsterdam: European Association of Zoos and Aquaria (EAZA).

Meagher, M. (1973). *The Bison of Yellowstone National Park*. Washington, DC: National Park Service.

Olech, W. & Perzanowski, K. (2002). A genetic background for reintroduction program of the European bison (*Bison bonasus*) in the Carpathians. *Biological Conservation* **108**: 221–228.

Pritchard, D. J., Fa, J. E., Oldfield, S. & Harrop, S. R. (2012). Bring the captive closer to the wild: redefining the role of ex situ conservation. *Oryx* **46**: 18–23.

Pucek, Z., Belousova, I. P., Krasinski, Z. A. & Olech, W. (eds) (2004). *European Bison: Status Survey and Conservation Action Plan*.Gland and Cambridge: IUCN/SSC Bison Specialist Group.

Raczynski, J. (ed.) (2012). *European Bison Pedigree Book*. Narodowy: Bialowieski Park.

Ratajszczak, R. (2008). Disappearing animals: what's next? *EAZA News* **64**: 31–32.

Saccheri, I., Kuussaari, M., Kankare, M., Vikman, P., Fortelius, W., and Hanski, I. (1998). Inbreeding and extinction in a butterfly metapopulation. *Nature* **392**: 491–494.

Small, E. (2011). The new Noah's Ark: beautiful and useful species only. Part 1: biodiversity conservation issues and priorities. *Biodiversity* **12**: 232–247.

Small, E. (2012). The new Noah's Ark: beautiful and useful species only. Part 2: the chosen species. *Biodiversity* **13**: 37–53.

Snijder, R. & Henckens, E. (2012). The attraction value of zoo animals. Final thesis. Hogeschool van Hall-Larenstein.

Vince, M. (2008). Making the case for off-exhibit bird facilities. *EAZA News* **64**: 40–41.

Vrijenhoek, R. C. (1994). Genetic diversity and fitness in small populations. In *Conservation Genetics*, pp. 37–53 Eds. Loeschcke, V., Tomiuk, J. & Jain, S. K. Basel: Birkhäuser Verlag.

Vuure, T. van (2005). *Retracing the Aurochs: History, Morphology and Ecology of an Extinct Wild Ox*. Sofia: Pensoft.

WAZA (2005). *Building a Future for Wildlife: The World Zoo and Aquarium Conservation Strategy*. Bern: WAZA.

Zimmermann, A, Hatchwell, M., Dickie, L. & West, C. (eds) (2007). *Zoos in the 21st Century: Catalyst for Conservation?* Cambridge: Cambridge University Press.

Index

activity patterns 89, 142, 181, 224, 233, 272, 305, 314, 320, 349
Africa 7, 14
African buffalo 1, 3–4, 7, 9, 21, 25, 40, 79, 84, 181, 248, 272–273, 326–328, 337, 340, 344, 347, 349, 352, 360–361, 410–411, 413, 416, 418–422, 424, 431–432, 438–440, 449, 452–453; cape buffalo, 326–328, 330, 341, ***see also*** ***S. c. caffer***; central African savanna buffalo 326–327, 329, 339, ***see also*** ***S. c. aequinoctialis***; forest buffalo 326–327, 329, 339, 341, 345, 348, 351, 354–355, 359, 361–362, 446, ***see also*** ***S. c. nanus***; west African savanna buffalo 326–327, 329, 338, ***see also*** ***S. c. brachyceros***
aggregation 93, 134, 139, 153, 359–360
alert posture 277, ***see also*** antipredator behaviour
American bison 1, 4, 7, 9, 21–23, 25, 83–85, 88, 90, 98–99, 103, 105, 116, 120–121, 450, ***see also*** ***Bison bison***; plains bison 83, 85, 98–99, 373, ***see also*** ***B. b. bison***; wood bison 59, 85, 93, 98–99, 105, ***see also*** ***B. b. athabascae***
Angola 329, 338
anoa 1, 9, 59, 303, 452; lowland anoa 7, 9, 14, 23, 302, 307, 449, 453, ***see*** ***B. depressicornis***; mountain anoa 302, 451 ***see also*** ***B. quarlesi***
Asia 1, 7, 14
Asian Wild Cattle Specialist Group 3, 448, ***see also*** AWCSG
aurochs 1, 4, 240, 242, 244–245, 247, 250, 451, ***see also*** ***B. primigenius***
Australia 8, 23, 36, 105, 227, 257
AWCSG 3–4, 9, 234, 448–449, ***see also*** Asian Wild Cattle Specialist Group

bachelor groups 92, 152–153, 205, 226, 348, 356, 358, 360
Bangladesh 175, 189, 217, 257, 264, 393
banteng 4, 7, 11, 22–23, 35, 216, 232, 446, ***see also*** ***Bos javanicus***; Bornean banteng 216–217, 226, ***see also*** ***B. j. lowi***; Java banteng 11, 216, 218, 225, 227, ***see also*** ***B. j. javanicus***; mainland banteng 216–217, 222, 227, 229 ***see also*** ***B. j. birmanicus***
behaviour 5, 93, 151, 153, 184, 225, 235, 248, 275, 307, 314, 321, 356; agonistic behaviour 94, 154, 274, 277, 307, 357, 360, ***see also*** sparring and lateral display; antipredator behaviour 188, 307, ***see also*** tossing and alert posture; sexual behaviour 5, 153, 183, 188, 206, 248, 277 ***see also*** flehmen
Benin 329, 338
Bhutan 175, 258, 263
birth 92, 127, 148, 183–184, 235, 247, 274, 354, 360; rates 355
Bison, 4, 11; ***B. bison*** 83, ***see also*** American bison; ***B. bonasus*** 115, ***see also*** European bison; ***B. b. athabascae*** 83, ***see also*** wood bison; ***B. b. bison*** 83, ***see also*** plains bison
body condition 314, 356, 361
Borneo 217–218, 223
Bos 9, 11, 15; ***B. gaurus*** 174, ***see also*** gaur; ***B. g. frontalis*** 35, ***see also*** mithun; ***B. m. grunniens*** 9, 36; ***B. grunniens*** 194–195; ***B. javanicus*** 216–217, ***see also*** banteng; ***B. j. birmanicus*** 216–217, ***see also*** mainland banteng; ***B. j. javanicus*** 216–217, ***see also*** Java banteng; ***B. j. lowi*** 216–217 ***see also*** Bornean banteng; ***B. p. mauretanicus*** 240; ***B. mutus*** 194, ***see also*** wild yak; ***B. p. namadicus***, 35, 240; ***B. primigenius*** 13, 15, 240, ***see also*** aurochs; ***B. p. indicus*** 9, 35 ***B. indicus*** 21, 197, ***see also*** zebu; ***B. sauveli*** 1, 11, 30, 231–232, 266, 446, ***see also*** kouprey; ***B. p. taurus*** 9, 31; ***B. taurus*** 21, 84, 197, 241
boss 341, 344, size of ***see also*** horns
Botswana 338
bovids 15, 64, 67, 72, 88; browsing 70; large-bodied 78; open-country 40
Bovini 1, 7, 9, 14, 39–40; tribe 40; mtDNA sequences 198
bTB (bovine tuberculosis) 433, 439
Bubalus 9, 13–15, 40, 45–46, 255; ***B. arnee*** 15, 59, 255–256, 289, 403, ***see also*** wild water buffalo; ***B. a. bubalis***, ***see*** river buffalo; ***B. a. carabanesis*** 9, 36, 289, ***see also*** swamp buffalo; ***B. depressicornis*** 9, 15, 45, 69, 302–303, ***see also*** lowland anoa; ***B. mindorensis*** 13, 15, 310, ***see also*** tamaraw; ***B. quarlesi*** 9, 15, 45, 302, ***see also*** mountain anoa
bull 154, 225, 268; aggressiveness 93; aurochs 241; bachelor 273, ***see also*** bachelor groups; covered by a 155; dominant 226; groups 94, 151; killed by 278; solitary 136; wounded by lions 357
Burkina Faso 339
Buton Island 303, 307–308, ***see also*** Sulawesi
calves 93, 148, 184, 268, 377 adoption by other females 274; colour 341; leopard predation on 278; as prey to leopards, wild dog and tigers 188; suckling bouts of 354, ***see also*** suckling; survival rate for 97; wolf predation 93
calving 91, 93, 146–147; intervals 184, 307; peak 147, 354; periods 183; season 147, 314, 355
Cambodia 7, 22, 175, 189, 217, 232, 235, 257, 265
Cameroon 339
Central African Republic 339
Chad 339
climatic changes 410
conservation: action 374; activities 3; anoa 308; ***B. arnee*** 289; biodiversity 373; challenge 22; ex situ 449; free-ranging bison 97; genetics 410; herds 99; in situ 161; in situ and ex situ 227; saola 323; securing of long-term populations 447; status 1; tamaraw 316; tragedy 22; units 374; wild bison 103
Convention on International Trade in Endangered Species of Wild Fauna and Flora (CITES) 98

Democratic Republic of Congo 336
disease 1, 5, 35, 96, 431–432, 436–437, 439–440; Southeast Asian wild cattle species 454
DNA 9, 11, 30–31, 84–85, 120; mtDNA 11, 14, 30, 34–36, 198–199, 232, 403, 411, 416, 420
domestication 1, 4, 30, 36, 39, 51, 198, 247, 257, 289–290, 375, 403, 440
driving off 360
dry season 23, 181, 223, 351–352; hottest period 235; summer months 223

East Coast Fever (ECF) 434
Ethiopia 330
Europe 7
European Association of Zoos and Aquaria (EAZA) 447, 449
European bison 2, 4, 7, 9, 11, 13, 23, 25, 59, 84, 105, 115–118, 120–121, 123–124, 126–127, 130, 132–134, 136, 139, 142–143, 146–147, 150–151, 156, 160–162, ***see also Bison bonasus***
European Bison Pedigree Book (EBPB) 117, 120, 447
evolution: Bovidae 39; Bovini 7; Bubalina-Bovina 14; cape buffalo 418; development of bovids horns 72; grasslands 60; infection 437; morphological characters 14; postcranial skeleton 40; sauropsids 51; specialised dentition 70; species 375; the cattle digestive 60
extinction 1, 4, 98, 121, 249, 373; across its range 279; bison 121 brink of 337; driven to 324; giant long-horned buffalo 418; Holocene 418; last population in Poland in 1919, 447; species 217

feeding ecology 91, 143, 181, 224, 233, 246, 273, 305, 314, 321, 351
fight 154; during the mating season 277; to the death 188; site of 155
flehmen 154, 189; ***see also*** sexual behaviour; posture 155; when testing urine 357
FMD (foot and mouth disease) 207, 278, 360–361, 431, 433, 436–437
fusion–fission 356, 361

Gabon 339
gaur 9, 174, 393, 453, ***see also*** *Bos, gaurus*
gayal 35
Ghana 339
grazing 1, 40, 63, 90, 205, 224, 272, 351, 378; areas 347; bouts 272; niche 70; habit 70; lawns 352
Guinea 338

habitat 88, 130, 177, 222, 233, 245, 268, 303, 312, 320, 344
herd 93, 138, 225, 248, 275, 347, 355, 359, 387, ***see also*** herd size; Białowieza 160; bison 379; bTB 433; density 226; Dzanga-Ndoki N.P. 360; EINP 379; free-ranging 124; leaving the 188, 356 livestock 441; maternal 356; member of a 277; Mexico 99; movement 226; protection of 357; as a social unit 356; splitting of the 357; switching 356, 419; Yellowstone 93
herd size 162, 186, 226, 275, 356, 359, ***see also*** herd
Holstein Friesian 34
horns 8, 72–73, 76, 243, 261, 341; anatomy of the 72; in both sex 72; length 34; shape 149; short 34; swept-back 413; ***see also*** boss
hunting 1, 98, 121, 228, 232, 280, 319, 373; blocks 339; business 364; dogs 320; industry 439; illegal 313; with spears 322; tourism 440; trophy 22; zones 327

inbreeding depression 410
India 175, 190, 256, 393
IUCN/Species Survival Commission (SSC) 3
IUCN Red List of Threatened Species, 4
Ivory Coast 338

Java 217, 223, 227, 257

Kenya 337
kouprey, 1, 231, 450, ***see also Bos sauveli,***
Kruger National Park 329, 424, 433, ***see also*** South Africa

Lao PDR 1, 177, 266, 319, ***see also*** Vietnam
lateral display 186, ***see also*** agonistic behaviour
Leptobos 15
lion 249; predation 359; African 416; kills by 433

Mali 338
mating 95, 146, 183, 206, 225, 355, 357; habits 314; opportunities 376; season 92, 247, 314; system 76, 274
microsatellites 411
Miocene 11, 14, 40, 84; fossils of 15,
mithun 35, ***see also Bos, gaurus frontalis***
movements and home range 88, 136, 180, 223, 233, 246, 271, 313, 320, 347
Mozambique 337

Namibia 338
Nigeria 339
North America 7, 17, 84

parasites and diseases 96, 156, 189, 226, 235, 249, 278, 314, 321, 360
pasture 160, 208
Pelorovis 15
Philippines 289, 311
philogeography 410
Pleistocene 11, 14–16, 36, 39–40, 84, 116; dated fossils 416; Europe 15; Indian subcontinent 14; mass extinction 418; refugia 416
Poëphagus 195
Pseudoryx nghetinhensis 70, 318

rainy season, ***see*** wet season
reproduction and growth 92, 146, 183, 225, 235, 247, 274, 307, 314, 321, 354
Republic of Congo 339
resting 89, 95, 142, 233, 273, 351 areas 360; bout 357; in clearings 360; sites 360
rinderpest 278, 328, 361, 432 epidemic 362
ruminating 143, 233, 273, 349 activities 351
Rwanda 336

saola, 1–2, 9, 318, ***see also Pseudoryx nghetinhensis***
savanna, buffalo 330, 341 burned 346; equatorial 339; grassland 258; ecosystem 329, 344; mosaic 346; Sahel-Soudan 341
Senegal 338
social behaviour 225, 360
social organization 185
sounds 155; calling 189; growling 188; whistling 189
South Africa 338, ***see also*** Kruger National Park
South Sudan 339
sparring 92, 186, 341, ***see also*** agonistic behaviour; behaviour 357; between bulls 226; with horns 321
status in captivity 105, 162, 190, 228, 238, 292, 308, 316, 323, 364
status in the wild 98, 157, 189, 226, 235, 249, 279, 307, 315, 321, 361
subspecies and distribution 85, 120, 175, 200, 217, 232, 241, 257, 303, 311, 319, 327
suckling 51, 93, 148, 153, 189, ***see also*** calves

Sulawesi 1, 7, 289, 303, 307–308, ***see also*** Buton Island
Swiss Brown 34
Syncerus 13–15, 39–40, 45–46, 59, 326; ***S. c. aequinoctialis*** 326–327, 329, ***see also*** central African savanna buffalo; ***S. c. brachyceros*** 326–327, 329, ***see also*** west African savanna buffalo; ***S. c. caffer*** 326, 328, 410, 413, 416, 440, 449, ***see also*** Cape buffalo; ***S. c. nanus*** 326–327, 329, ***see also*** forest buffalo

tamaraw 7, 15, 310, 450, ***see also B. mindorensis***
Tanzania 337
taxonomy 4, 84, 116, 175, 195, 217, 232, 241, 256, 303, 311, 319, 327
teeth 8, 63, 67, 116, 150, 341 cementum annuli in the root of 88
Thailand 14, 175, 190, 217, 222, 232, 256, 264, 450
tossing 188; heads 187; ***see also*** antipredator behaviour, agonistic behaviour;
tsetse 435

Uganda 336

Vietnam 1, 190, 217, 232, 236, 257, 319, ***see also*** Lao PDR

wallowing 1, 24, 179, 248, 314, ***see also*** agonistic behaviour; behaviour 25; among bulls 143; mud 350; in pools 25
water 177, 344, 438; availability 90, 304; demanding 25; dependent 418; flowing 25; holes 224; losses 58; points 348; retention 274; saline source 14, 91, 271; via osmosis 51; wallowing in 351
weaning 94–95, 355–356; age of 354
wet season 183, 223, 351, 355, 359; home ranges 352; monsoon 223
wild cattle 1, 4, 7, 40, 59, 76, 105, 238, 241, 303, 373–374, 446–447, 449, 454
wild water buffalo, 4, 13, 23, 36, 255, 289, ***see also Bubalus arnee***; Asian wild water buffalo 403; feral water buffalo 1; swamp buffalo 9, 13, 15, 30, 36, ***see also,***

B. a. carabanesis;
river buffalo 9, 30, 36, 289, 403, ***see also B. a. bubalis***
wild yak 1, 36, 194, 196, 198–199, 450, ***see also Bos mutus***
Wildlife Conservation Society (WCS) 448
World Zoo and Aquarium Conservation Strategy (WZACS) 446, ***see also*** European Association of Zoos and Aquaria (EAZA)
WWF 266, 324, 385

Zambia 337
zebu 35, ***see also Bos primigenius indicus***
Zimbabwe 337